Handbook of Water and Wastewater Microbiology

Edited by

Duncan Mara

and

Nigel Horan

School of Civil Engineering,
University of Leeds, UK

[redacted]

ACADEMIC PRESS

An imprint of Elsevier

Amsterdam · Boston · Heidelberg · London · New York · Oxford · Paris · San Diego
San Francisco · Singapore · Sydney · Tokyo

This book is printed on acid-free paper

Academic Press
An Imprint of Elsevier
84 Theobald's Road, London WC1X 8RR, UK
http://www.academicpress.com

Academic Press
An Imprint of Elsevier
525 B Street, Suite 1900 San Diego, California 92101-4495, USA
http://www.academicpress.com

ISBN 0-12-470100-0

Library of Congress Catalog Number: 2002114097

British Library Cataloguing in publication data

Handbook of water and wastewater microbiology
1. Drinking water – Microbiology 2. Sewage – Microbiology
I. Mara, Duncan II. Horan, N.J.
628.1′6

ISBN 0-12-470100-0

Typeset by Aldens
Printed and bound in Great Britain
03 04 05 06 07 08 BP 9 8 7 6 5 4 3 2 1

Contents

PART 4: DRINKING WATER MICROBIOLOGY

Contributors

G K Anderson University of Newcastle upon Tyne, UK
J P Arbord Hong Kong University of Science and Technology, Kowloon, Hong Kong
B Atkinson Technikon of Natal, Durban, South Africa
J P Barford Hong Kong University of Science and Technology, Kowloon, Hong Kong (SAR) China
G W Barton University of Sydney, Australia
Jamie Bartram World Health Organization, Geneva, Switzerland
John Binkley Product Design and Development, Bolton Institute, UK
Arthur G Boon Halcrow Water, Crawley, UK
K C A Bromley-Challenor Department of Microbiology, University of Leeds, UK
E C S Chan McGill University, Quebec, Canada
T E Cloete University of Pretoria, Transvaal, South Africa
Franscisc Codony Universitat Autònoma de Barcelona, Spain
Tom Curtis University of Newcastle, Newcastle upon Tyne, UK
B S Drasar London School of Hygiene and Tropical Medicine, London, UK
M M Ehlers University of Pretoria, Transvaal, South Africa
G A Ekama University of Cape Town, South Africa
Badri Fattal The Hebrew University of Jerusalem, Israel
R G A Feachem Global Fund to Fight AIDS, Geneva-Cointrin, Switzerland
Caroline S Fitzpatrick University College London, UK
R J Foot Wessex Water, UK
Paul Gale WRC-NSF, Marlow, UK
A Godfree North West Water Ltd, Warrington, UK
John Gregory University College London, UK
A M Grimason Stobhill Hospital, and University of Strathclyde, Glasgow, UK
Pauline S Handley University of Manchester, UK
Oliver J Hao University of Maryland, USA
John Heritage University of Leeds, UK
Nigel Horan University of Leeds, UK
Guy Howard Loughborough University, UK
Debra E Huffman University of South Florida, USA
David W M Johnstone Swindon, UK
Howard Kator College of William and Mary, Virginia, USA
Joan Kelley CABI Bioscience, Surrey, UK
Charmain J Kerr United Utilities, Warrington, UK
Graham Kinsey CABI Bioscience, Surrey, UK
J S Knapp Department of Microbiology, University of Leeds, UK
Yann Le Bihan Université Laval, Québec, Canada
Paul Lessard Université Laval, Québec, Canada
R E Loewenthal University of Cape Town, South Africa
Paolo Madoni Universita di Parma, Italy
D D Mara University of Leeds, UK
J Mas Universitat Autònoma de Barcelona, Spain
Luís F Melo University of Porto, Portugal
J Mir Universitat Politècnica de Catalunya, Barcelona, Spain
J Morató Universitat Politècnica de Catalunya, Barcelona, Spain
R Morris Consultant Environmental Biologist, Leicestershire, UK
Peter Murchie USEPA, Chicago, USA
John Oragui Harefield Hospital, Harefield, Middlesex, UK

Esther Ortenberg Israel Ministry of Health, Tel-Aviv, Israel
Keith S Osborn United Utilities, Warrington, UK
Russell Paterson CABI Bioscience, Surrey, UK
Pierre Payment INRS-Institut Armand–Frappier, Laval, Quebec, Canada
Howard Pearson Federal University of Rio Grande do Norte, Natal, Brazil
John Pinfold Brighton, UK
Walter Quintero-Betancourt Michigan State University, East Lansing, USA
Martha Rhodes College of William and Mary, Virginia, USA
F Ribas AGBAR (Sociedad General de Aguas de Barcelona, S.A.), Barcelona, Spain
Alex H Rickard University of Manchester, Manchester, UK
M S Robinson Wessex Water, UK
Geoff D Robson University of Manchester, Manchester, UK
Joan B Rose Michigan State University, East Lansing, USA
P J Sallis University of Newcastle upon Tyne, UK
Edward D Schroeder University of California, California, USA
Hillel Shuval The Hebrew University of Jerusalem, Jerusalem, Israel
J A Simpson Bolton Institute, Bolton, UK
H V Smith Stobhill Hospital and University of Strathclyde, Glasgow, UK
Rebecca Stott University of Portsmouth, UK
Huw Taylor University of Brighton, Brighton UK
Benjamin Teltsch Mekorot Water Co, Israel
S Uyanik Harran University, Sanlýurfa, Turkey
J van Heerden University of Pretoria, Transvaal, South Africa
Alison J Vincent Hyder Consulting, Cardiff, UK
M C Wentzel University of Cape Town, Cape Town, South Africa
C H Wong CPG Consultants, Novena Square, Singapore
Stefan Wuertz University of California, California, USA

Preface

Some 2300 years ago Hippocrates wrote: 'My other topic is water, and I now wish to give an account both about waters that cause disease and about those that are healthy, and what bad things arise from water and what good things. For water contributes very much to health.' So our appreciation of a relationship between the water we use and our health has been with us for a very long time. Hippocrates was unlikely to have been the first person to realise the existence of this relationship and we have probably known since our species evolved that water of adequate quantity and quality is essential for our survival and our health.

We now know (and have known for just over 100 years) that water quality is governed by (but, of course, not only by) microorganisms – the viruses, bacteria and parasites that can infect us and may (and very often do) make us ill. Microorganisms are also central to wastewater treatment and the reuse of treated wastewaters – we exploit them to treat our wastes biologically (actually, microbiologically), and we must ensure that pathogenic microorganisms are removed in the treatment processes to a level at which they do not cause any excess disease resulting from wastewater use in agriculture or aquaculture.

Water disinfection, usually with chlorine, has been practised in many parts of the world (but regrettably not all) for over 100 years. Water chlorination is a very efficient process: it kills bacteria very quickly (but viruses more slowly, and protozoa such as *Giardia* and *Cryptosporidium* hardly at all). Faecal bacterial numbers, in particular, are reduced to zero, and thus early water engineers judged the quality of chlorinated water supplies quite simply on whether faecal indicator bacteria – principally coliform bacteria – were present in the disinfected water or not. Zero coliforms, and zero faecal coliforms, quickly became *the* microbiological goal of drinking water quality. No-one would really question the general sense of this goal – chlorinate your water and you get zero coliforms per 100 ml, so everything's OK. End of story.

Life is rarely this simple, and water and wastewater microbiology is no exception. Emerging water-borne pathogens (*Cryptosporidium*, for example) require us to have a deeper understanding of water microbiology. Optimizing (really, maximizing) microbiological wastewater treatment also requires a knowledge of microbiology greater than that possessed by many design engineers. Structural engineers have a pretty good understanding of concrete, for example – so why shouldn't those who design activated sludge plants or waste stabilization ponds have an equal appreciation of the microorganisms whose activities are essential to the treatment process they are designing?

The purpose of this Handbook is to provide an introduction to modern water and wastewater microbiology, especially for water and wastewater engineers. The study of water and wastewater microbiology is very rewarding: better water treatment, better wastewater treatment, safer wastewater reuse, and thus healthier people – in all parts of our world.

Duncan Mara
Nigel Horan

Part I Basic Microbiology

1

Microbial nutrition and basic metabolism

E.C.S. Chan

Department of Microbiology and Immunology, McGill University, Montreal, Quebec, Canada

1 INTRODUCTION

In nature, of all living organisms, microorganisms are the most versatile and diversified with regard to their nutritional requirements. For example, microorganisms, such as bacteria, can be found that represent the entire spectrum of nutritional types. Some microbes have an unconditional need for preformed complex organic compounds while others can thrive with just a few inorganic substances as their sole nutritional requirements. Most microorganisms fall between these two extremes.

Although great variation is found in the specific requirements for growth of the diverse species of microorganisms, in general, the nature and functions of growth substances are common for all cells. In part, this is because the chemical composition of microbial cells is more or less similar. For instance, all microbes require a carbon source because carbon is a component of protoplasm; nitrogen is a component of many major macromolecules, such as protein and nucleic acids. Over 95% of a cell's dry weight is made up of a few major elements, such as C, O, H, S, P, K, Ca, Mg, and Fe. All these substances must be put together by biosynthesis to form cellular material. However, not all growth substances are incorporated into cell material. Some are used instead as energy sources. For example, carbon compounds are frequently used as energy sources by many microorganisms; inorganic sulphur compounds are used in energy metabolism by some microbes.

Assimilated nutrient substances need to be metabolized. *Metabolism* refers to the totality of organized biochemical activities carried out by an organism. Such activities, usually catalysed by enzymes, are of two kinds: those involved in generating energy; and those involved in using energy. Some microorganisms use nutrients absorbed by the cell as the source of energy in a process called *catabolism*, which is the reverse of biosynthesis (*anabolism*). In catabolism, large molecules are degraded in sequential stepwise reactions by enzymes and a portion of the energy released is trapped in the form of chemical energy. Other microorganisms derive their energy from the trapping of light and also convert it into chemical energy. This chemical energy is then harnessed to do work for the cell. A microorganism must perform many different types of work, such as synthesis of physical parts of the cell, repair or maintenance of cell components, and growth. Thus metabolism may be viewed as the coupling of energy generation and energy utilization. The kinds of nutrients and how they are assimilated and fed into the various metabolic pathways for energy production and utilization in microorganisms form the subject matter of this chapter.

2 CHEMICAL ELEMENTS AS NUTRIENTS

It may be generalized that there are three essential nutritional needs of a cell: water,

The Handbook of Water and Wastewater Microbiology
ISBN 0-12-470100-0

energy, and chemical compounds, which may be used as building blocks. Most microorganisms, except for the phagocytic protozoa, have an absorptive type of nutrition. Thus chemical energy sources and chemical compounds must be dissolved in water. Water carries the solutes by transport mechanisms into the cell. Within the cell, water is the solvent in which the cell's biochemical reactions occur. It is also the medium for elimination of soluble waste substances from the cell. It is not surprising then that 70–80% of the microbial cell is water and that it constitutes the major portion of a cell's weight.

Required nutritional chemical compounds include chemical elements. These elements are necessary for both the synthesis of cell material and the normal functioning of cellular components such as enzymes. In order to grow, microorganisms must have proper essential chemical elements. The main chemical elements for cell growth include C, N, H, O, S and P – the same elements that form the chemical composition of the cells.

2.1 Macroelements

The major elements (macroelements), like C, H, N, O, S and P, are used in large amounts by microorganisms. The minor elements (microelements), like K, Ca, Mg, Na and Fe, are used in smaller quantities, while the trace elements, such as Mn, Zn, Co, Mo, Ni, Cu, are used in relatively very much smaller amounts. The amount of an element used by a microbe does not correlate with its relative importance; even one used in a trace amount may be essential to the growth or life of a microbial cell.

2.1.1 Carbon

Carbon is one of the most important chemical elements required for microbial growth. Fifty per cent of the dry weight of any cell is carbon; thus all organisms require carbon in some form. Carbon forms the backbone of three major classes of organic nutrients: carbohydrates, lipids and proteins. Such compounds provide energy for cell growth and serve as building blocks of cell material.

Microorganisms that use organic compounds as their major carbon source are called *heterotrophs*. Heterotrophs obtain such organic molecules by absorbing them as solutes from the environment. Some phagotrophic heterotrophs obtain organic molecules by ingestion of other organisms. Microorganisms that use carbon dioxide (the most oxidized form of carbon) as their major or even sole source of carbon are called *autotrophs*. They can live exclusively on relatively simple inorganic molecules and ions absorbed from the aqueous and gaseous environment.

2.1.2 Nitrogen

All organisms require nitrogen in some form. It is an essential part of amino acids that comprise cell proteins. Nitrogen is needed for the synthesis of purine and pyrimidine rings that form nucleic acids, some carbohydrates and lipids, enzyme cofactors, murein and chitin. Many prokaryotes use inorganic nitrogen compounds such as nitrates, nitrites, or ammonium salts. Unlike eukaryotic cells, some bacteria (like the free-living *Azotobacter* and the symbiotic *Rhizobium* of legume plants) and some archaeons (like the methanogens *Methanococcus* and *Methanobacterium*) can use atmospheric or gaseous nitrogen for cell synthesis by a physiological process called *nitrogen fixation*. Some microbes require organic nitrogen compounds such as amino acids or peptides. Some microorganisms use nitrate as an alternative electron acceptor in electron transport.

2.1.3 Hydrogen, oxygen, sulphur and phosphorus

Other elements essential to the nutrition of microorganisms are hydrogen, oxygen, sulphur and phosphorus. Hydrogen and oxygen are components of many organic compounds. Because of this, the requirements for carbon, hydrogen, and oxygen often are satisfied together by the availability of organic compounds. Free oxygen is toxic to most strict anaerobic bacteria and some archaeons, although aerobic microorganisms use oxygen as a terminal electron acceptor in aerobic respiration. Sulphur is needed for the biosynthesis of the amino acids cysteine, cystine, homocysteine, cystathione and methionine, as well as the vitamins biotin and thiamine. Phosphorus is essential for the synthesis of nucleic acids and adenosine triphosphate. It is a component of teichoic acids and teichuronic acids in the cell walls of Gram-positive bacteria as well as a component of various membrane phospholipids. Reduced forms of sulphur may serve as sources of energy for *chemotrophs* or as sources of reducing power for *phototrophs*. Sulphate may serve as a terminal electron acceptor in electron transport.

2.2 Microelements and trace elements

Many other essential elements, the microelements and trace elements, are required in smaller amounts than the macroelements by microorganisms in their nutrition. Some of their functions in supporting the growth of microorganisms are summarized in Table 1.1.

For example, sodium is required by the permease that transports the sugar melibiose into the cells of the colon bacterium *Escherichia coli*. Sodium is required by marine microorganisms for maintaining cell integrity and growth. Some 'salt-loving' prokaryotes, the red extreme halophiles, cannot grow with less than 15% sodium chloride in their environment. Essential elements are often required as cofactors for enzymes. Because iron is a key component of the cytochromes and electron-carrying iron-sulphur proteins, it plays a key role in cellular respiration. However, most inorganic iron salts are highly insoluble in water. Thus, many microbes must produce specific iron-binding agents, called siderophores, in order to utilize this element. Siderophores are chelating agents that solubilize iron salts and transport iron into the cell. Many enzymes, including some involved in protein synthesis, specifically require potassium. Magnesium functions to stabilize ribosomes, cell membranes and nucleic acids and is needed for the activity of many enzymes.

Trace elements, needed in extremely small amounts for nutrition by microorganisms, include manganese, molybdenum, cobalt, zinc, and copper. For instance, molybdenum is required by nitrogenase, the enzyme that converts atmospheric nitrogen to ammonia during nitrogen fixation. Manganese aids many enzymes to catalyse the transfer of phosphate groups. Cobalt is a component of vitamin B_{12} and its coenzyme derivatives.

3 NUTRITIONAL TYPES OF MICROBES

Microbes can be grouped nutritionally on the basis of how they satisfy their requirements for carbon, energy, and electrons or hydrogen. Indeed, the specific nutritional requirements of microorganisms are used to distinguish one microbe from another for taxonomic purposes.

Microorganisms may be grouped on the basis of their energy sources. Two sources of energy are available to microorganisms. Microbes that oxidize chemical compounds (either organic or inorganic) for energy are

TABLE 1.1 Functions of some microelements and trace elements in the nutrition of microorganisms

Element	*Major functions in some microorganisms*
Sodium	Enzyme activator. Transport across membranes. Maintenance of cell integrity. Facilitates growth. Salt form of some required organic acids
Potassium	Cofactor for enzymes. Maintenance of osmotic balance
Iron	Component of cytochromes, haem-containing enzymes, electron transport compounds and proteins. Energy source
Magnesium	Enzyme activator, particularly for kinase reactions. Component of chlorophyll. Stabilizes ribosomes, cell membranes, and nucleic acids
Calcium	Enzyme activator, particularly for protein kinases. Component of dipicolinic acid in bacterial endospores
Cobalt	Component of vitamin B_{12} and its coenzyme derivatives
Manganese	Enzyme activator, particularly for enzymes transferring phosphate groups
Molybdenum	Enzyme activator for nitrogen fixation

called *chemotrophs*; those that use light as their energy sources are called *phototrophs*. A combination of these terms with those employed in describing carbon utilization results in the following nutritional types:

1. *Chemoautotrophs*: microbes that oxidize inorganic chemical substances as sources of energy and carbon dioxide as the main source of carbon.
2. *Chemoheterotrophs*: microbes that use organic chemical substances as sources of energy and organic compounds as the main source of carbon.
3. *Photoautotrophs*: microbes that use light as a source of energy and carbon dioxide as the main source of carbon.
4. *Photoheterotrophs*: microbes that use light as a source of energy and organic compounds as the main source of carbon.

Microorganisms also have only two sources of hydrogen atoms or electrons. Those that use reduced inorganic substances as their electron source are called *lithotrophs*. Those microbes that obtain electrons or hydrogen atoms (each hydrogen atom has one electron) from organic compounds are called *organotrophs*.

A combination of the above terms describes four nutritional types of microorganisms:

1. *Photolithotrophic autotrophy*
2. *Photo-organotrophic heterotrophy*
3. *Chemolithotrophic autotrophy*
4. *Chemo-organotrophic heterotrophy.*

The characteristics of these types with representative microoganisms as well as other organisms are shown in Table 1.2.

Photolithotrophic autotrophs are also called *photoautotrophs*. The cyanobacteria, algae and green plants use light energy and carbon dioxide as their carbon source but they employ water as the electron donor and release oxygen in the process. The purple and green sulphur bacteria use inorganic compounds as electron donors (e.g., H_2S, S^0) and do not produce oxygen in the process. Thus they are described as *anoxygenic*. Chemo-organotrophic heterotrophs are also called *chemoheterotrophs*. They use organic compounds for energy, carbon and electrons/hydrogen. The same organic nutrient compound often satisfies all these requirements. Animals, most bacteria, fungi, and protozoa are chemoheterotrophs. Photo-organotrophic heterotrophs are also called in short *photoheterotrophs*. The purple and green non-sulphur bacteria are photoheterotrophs and use radiant energy and organic compounds as their electron/hydrogen and carbon donors. These common microorganisms, found in polluted lakes and streams, can also grow as photoautotrophs with molecular hydrogen as electron donor. The chemolithotrophic autotrophs are also called *chemoautotrophs* in brief. They include the nitrifying, hydrogen, iron and sulphur bacteria. They oxidize reduced inorganic compounds, such as nitrogen, iron or sulphur molecules, to derive both energy and electrons/hydrogen. They use carbon dioxide as their carbon source. A few of them, however,

TABLE 1.2 Nutritional types of microbes and other organisms

Nutritional type	*Energy source*	*Electron or hydrogen source*	*Carbon source*	*Examples of organisms*
Photolithotrophic autotrophy	Light	Inorganic compounds, water	Carbon dioxide	Purple and green sulphur bacteria; algae; plants; cyanobacteria
Photo-organotrophic heterotrophy	Light	Organic compounds	Organic compounds	Purple and green non-sulphur bacteria
Chemolithotrophic autotrophy	Inorganic compounds	Inorganic compounds	Carbon dioxide	Nitrifying, hydrogen, iron, and sulphur bacteria
Chemo-organotrophic heterotrophy	Organic compounds	Organic compounds	Organic compounds	Most bacteria, fungi, protozoa, and animals

can make use of carbon from organic sources and thus become heterotrophic. Such bacteria that use inorganic energy sources and carbon dioxide, or sometimes organic compounds, as carbon sources can be called *mixotrophic*, because they combine autotrophic and heterotrophic processes. Chemotrophs are important in the transformations of the elements, such as the conversion of ammonia to nitrate and sulphur to sulphate, that continually occur in nature.

Even though a particular species of microorganism usually belongs to only one of the four nutritional types, some show great metabolic flexibility and can alter their nutritional type in response to environmental change. For example, many purple non-sulphur bacteria are photoheterotrophs in the absence of oxygen but become chemoheterotrophs in the presence of oxygen. When oxygen is low, photosynthesis and oxidative metabolism can function simultaneously. This affords a survival advantage to the bacteria when there is a change in environmental conditions.

The specific nutritional requirements of bacteria are used extensively for taxonomic purposes. Specific identification tests have been designed for particular groups of bacteria, such as the Gram-negative intestinal bacilli, to determine the nature of water pollution.

4 GROWTH FACTORS

Some microorganisms have good synthetic capability and thus can grow in a medium containing just a few dissolved salts. The simpler the cultural medium to support growth of a species of microbe, the more complex or advanced is the microbe's nutritional synthetic capability. Thus, the photoautotrophs are the most complex in their nutritional physiology. With the addition of one organic compound, such as the addition of glucose, a glucose-salts medium can support the growth of many chemoheterotrophs; an example is the bacterial indicator of faecal contamination, *Escherichia coli*. However, many microorganisms lack one or more essential enzymes and therefore cannot synthesise all their nutritional requirements. They must obtain these preformed or supplied in the environment or medium.

Organic compounds required in the nutrition of microorganisms, because they cannot be synthesized specifically, are called *growth factors*. The three major classes of growth factors are *amino acids*, *purines* and *pyrimidines*, and *vitamins*. Proteins are composed of about 20 amino acids. Some bacteria and archeons cannot synthesize one or more of these and require them preformed in the medium. For example, *Staphylococcus epidermidis*, the normal resident on the human skin, requires proline, arginine, valine, tryptophan, histidine and leucine in the medium before it can grow. Requirements for purines and pyrimidines, the nucleic acid bases, are common among the lactic acid bacteria. Vitamins are small organic compounds that make up all or part of the enzyme cofactors (non-protein catalytic portion of enzymes). Only very small, or catalytic, amounts suffice to support growth of the cells. Lactic acid bacteria, such as species of *Streptococcus*, *Lactobacillus* and *Leuconostoc*, are noted for their complex requirements of vitamins and hence many of these species are used for microbial assays of food and other substances. Vitamins most commonly required by microorganisms are thiamine (vitamin B_1), biotin, pyridoxine (vitamin B_6) and cyanocobalamin (vitamin B_{12}). The functions of some vitamins for the growth of microorganisms are summarized in Table 1.3.

5 ENERGY TRAPPING IN MICROORGANISMS

Like all living things, microorganisms require energy to live. The ability of a microorganism to maintain its life processes and to reproduce its own kind depends on its ability to trap energy and to use it to drive the endergonic reactions of the cell. Like all living forms, microbes trap or obtain energy in one of two ways: by *high-energy molecules* or by a *proton motive force* (*proton gradient*) across a cell membrane.

TABLE 1.3 Functions of some vitamins in the growth of microorganisms

Vitamin	*Function(s)*
Folic acid	One-carbon transfers; methyl donation
Biotin	Carboxyl transfer reactions; carbon dioxide fixation, β-decarboxylations; fatty acid biosynthesis
Cyanocobalamin (B_{12})	Carries methyl groups; synthesis of deoxyribose; molecular rearrangements
Lipoic acid	Transfer of acyl groups
Nicotinic acid (niacin)	Precursor of NAD^+ and $NADP^+$; electron transfer in oxidation–reduction reactions
Pantothenic acid	Precursor of coenzyme A; carries acyl groups
Riboflavin (B_2)	Precursor of FMN, FAD in flavoproteins involved in electron transport; dehydrogenations
Thiamine (B_1)	Aldehyde group transfer; decarboxylations
Pyridoxal-pyridoxamine group (B_6)	Amino acid metabolism, e.g. transamination and deamination
Vitamin K group; quinones	Electron transport; synthesis of sphingolipids
Hydroxamates	Iron-binding compounds; solubilization of iron and transport into cell
Haem and related tetrapyrroles	Precursors of cytochromes

5.1 High-energy molecules

The energy liberated by an exergonic reaction can be used to drive an endergonic reaction if there is a reactant common to both reactions. This common reactant in the trapping of energy is called an energy-rich or energy-transfer compound. The energy-transfer compounds of greatest use to a cell are able to transfer large amounts of free energy and are called *high energy-transfer compounds* or *high-energy molecules*. Of these, adenosine triphosphate (ATP) is the most important. Such compounds are found in the cytosol or soluble part of the cell.

High-energy molecules are important because they drive biosynthesis in the cytoplasm, including the synthesis of nucleic acids, proteins, lipids and polysaccharides. They also are involved in the active transport of certain solutes into the cell. High-energy molecules, such as ATP, have bonds that have a high free energy of hydrolysis. In the case of ATP, a large amount of energy is needed to link the third phosphate group to adenosine diphosphate (ADP) because of the electrostatic repulsion of the negative charges on the other adjacent phosphate groups. Reactions in which the phosphate is removed from ATP will be favoured since the electrostatic repulsion is decreased as a result of hydrolysis. Thus, this energy is liberated if ATP is hydrolyzed back to ADP.

There is a reason why the phosphoryl group is a common chemical group that is transferred between molecules. The phosphorus in all phosphate groups carries a positive charge. This is because phosphorus forms double bonds (P=O) poorly so that the phosphorus–oxygen bond exists as the semi-polar bond $P^+–O^-$. The electrons in the bond are shifted toward the electron-attracting oxygen. The positively charged phosphorus is attacked by the electronegative oxygen in the hydroxyl of a water molecule in a hydrolytic reaction resulting in the release of inorganic phosphate. One can compare the tendency of different molecules to donate phosphoryl groups by comparing the free energy released when the acceptor is water, i.e. the free energy of hydrolysis. That is, a scale is used, where the standard nucleophile is the hydroxyl group of water, and the phosphoryl donors are all compared with respect to the tendency to donate the phosphoryl group to water. This gives a free energy change, or energy available to perform useful work and designated ΔG, per mole of substrate hydrolysed. The *standard free-energy change* under standard conditions (reactants and products at 1 M concentration and the reactions take place at 25°C at pH 7) is designated $\Delta G^{\circ\prime}$.

The general role of ATP in providing energy to drive an endergonic reaction may be shown below.

Consider the following reaction:

$$\text{Glucose} + H_3PO_4 \rightarrow \text{Glucose-6-phosphate} + H_2O$$

The $\Delta G^{o\prime}$ is +13.8 kJ/mol; it is an endergonic reaction and will not proceed spontaneously.

However, if ATP is provided as a reactant, the reaction becomes exergonic since the $\Delta G^{o\prime} = -16.7$ kJ/mol (−30.5 (from ATP hydrolysis) +13.8 = −16.7 kJ/mol) and proceeds spontaneously:

$$\text{Glucose} + \text{ATP} \rightarrow \text{Glucose-6-phosphate} + \text{ADP}$$

The compound ADP is also sometimes used by cells as a high energy-transfer compound since its hydrolysis also liberates an equally large quantity of energy as ATP ($\Delta G^{o\prime} = -30.5$ kJ/mol). However, adenosine monophosphate (AMP) is a low-energy molecule; its hydrolysis yields only a small amount of energy ($\Delta G^{o\prime} = -8.4$ kJ/mol).

Table 1.4 lists some high energy-transfer compounds with their standard free-energy values upon hydrolysis. Each of them can transfer its energy of hydrolysis directly or indirectly to ATP synthesis, as in the following example:

$$\text{1,3-Diphosphoglyceric acid} + \text{ADP} \rightarrow \text{3-phosphoglyceric acid} + \text{ATP}$$

The synthesis of high-energy-transfer compounds, such as ATP, involves phosphorylation. ATP is formed by phosphorylation of ADP, with energy for the phosphorylation being provided by an exergonic reaction.

TABLE 1.4 Some high energy-transfer compounds with their standard free energy release upon hydrolysis

Compound	*$\Delta G^{o\prime}$, kJ/mol*
Adenosine triphosphate (ATP)	−30.5
Guanosine triphosphate (GTP)	−30.5
Uridine triphosphate (UTP)	−30.5
Cytidine triphosphate (CTP)	−30.5
Acetyl phosphate	−42.3
1,3-Diphosphoglyceric acid	−49.4
Phosphoenolpyruvic acid (PEP)	−61.9

There are two general ways in which this phosphorylation of ADP can occur:

- *Substrate-level phosphorylation*, a reaction in which the phosphate group of a chemical compound is removed and directly added to ADP
- Phosphorylation by a membrane-bound enzyme called a proton-translocating ATPase, which uses the energy of an energy-trapping system called the proton motive force (described later).

In substrate-level phosphorylation, the rearrangement of atoms within chemical compounds derived from nutrients may result in a new compound that contains a high-energy phosphate bond. Such rearrangements can occur when cells dissimilate nutrients. The phosphate group involved in the high-energy phosphate bond then can be transferred directly to ADP, forming ATP, which now contains the high-energy phosphate bond. (Note that bond energy is the energy required to break a bond, e.g. by hydrolysis, and it is equal to the energy released when the bond is formed. It is not the energy released when a bond is broken. For example, the P–O bond energy is about +413 kJ, while the hydrolysis energy is −35 kJ.) One example of substrate level phosphorylation is as follows:

$$\text{Phosphoenolpyruvic acid} + \text{ADP} \rightarrow \text{Pyruvic acid} + \text{ATP}$$

It should be added that any chemical group that is electronegative (such as the hydroxyl groups in sugars) can attack the electropositive phosphorus resulting in phosphoryl group transfer if the proper enzyme is present to catalyse the reaction. In this manner, ATP can phosphorylate many different compounds. Enzymes that catalyse phosphoryl group transfer reactions are called *kinases*.

5.2 The proton motive force

The *chemiosmotic theory* was proposed by the biochemist Peter Mitchell in 1961. He received the Nobel Prize in 1978 for this proposal.

The theory described how organisms can trap energy by a means other than the synthesis of high-energy molecules. In brief, the chemiosmotic theory states that energy transducing membranes, such as bacterial cell membranes, mitochondrial membranes, and chloroplast membranes, pump protons across the membrane resulting in the generation of an electrochemical gradient of protons across the membrane. In the bacteria and the Archaea, the protons are pumped across the cytoplasmic membrane to the outside of the cell. In the non-phototrophic Eukarya, the cytoplasmic membrane is not involved; instead the protons are pumped outward across the inner membrane of a mitochondrion into the space between the inner and outer mitochondrial membrane. In the phototrophic Eukarya, the protons are pumped inward across the thylakoid membrane of the chloroplast. This gradient, also known as the *proton potential* or *proton motive force*, can be harnessed to do useful work when the protons return across the membrane through special proton conductors to the lower potential. Some membrane proton conductors are solute transporters, others synthesise ATP, and others are motors that drive bacterial flagella rotation. The proton potential also provides the energy for other membrane activities, such as gliding motility and reversed electron flow. The chemiosmotic theory brings together the principles of physics and thermodynamics to bear on the biological problem of membrane bioenergetics.

There are two components to the proton motive force:

1. The cell can pump or translocate protons (i.e. hydrogen ions, or H^+) across a membrane without also pumping a compensating anion such as OH^- or Cl^- at the same time. The protons are pumped out of the cell by exergonic driving reactions (energy-producing reactions), which are usually biochemical reactions of respiration, photosynthesis, or ATP hydrolysis. Once protons have been pumped across the membrane, they cannot pass back across to equalize the concentrations, because the membrane is impermeable to protons. Since protons have a positive charge, more positive charges exist at the outer surface of the membrane than at the inner surface, thereby establishing a charge separation that results in an electrical potential, $\Delta\Psi$, across the membrane. Because of this electrical potential, the protons remain very close to the outer surface of the membrane and do not diffuse away.
2. The protons that pass to the outside of the cell cannot pass back in because the membrane is impermeable to protons. The protons accumulate to the point where they decrease the external pH (higher acidity) relative to the pH (lower acidity) inside the cell. The concentration of protons may be 100 times greater on one side of the membrane than on the other (a 2-unit pH difference). This proton gradient, i.e. the greater concentration of protons on one side of the membrane and a lower concentration of protons on the opposite side, is referred to as a ΔpH.

The unequal distribution of protons and electrical charges across the membrane represents a form of potential energy called the proton motive force, or Δp. It is the resultant of both the $\Delta\Psi$ and the ΔpH and can be calculated at 30°C in terms of millivolts by the following relationship:

$$\Delta p = \Delta\Psi - 60\Delta pH$$

For example, in a bacterial cell, if the ΔpH is 2, and the $\Delta\psi$ is -160 mV (negative since the inside of a bacterial cell would be negatively charged with respect to the outside), the Δp will be $-160 + (-60)(2) = -160 - 120 = -280$ mV. Either $\Delta\psi$ or ΔpH alone can constitute the proton motive force; the existence of both factors is not necessary.

When the protons return to the inside of the cell, such as the bacterial cell, moving down the concentration gradient and towards the more negative side of the membrane, the membrane potential is dissipated, i.e. energy is given up and work can be done. Since the membrane is impermeable to protons, there are special conditions under which the protons can pass back across. To do this, the protons must carry out some type of work.

In bacteria, specific channels exist in the cytoplasmic membrane by which protons can pass back across the membrane, and the proton flow through these channels is harnessed by the cell to do various types of work, which includes synthesizing ATP from ADP, powering the rotation of bacterial flagella, active transport of nutrients into the cell, secretion of proteins out of the cell, and reversed electron transport.

Any agent that allows protons to pass freely across the cytoplasmic membrane will destroy the proton gradient and hence the proton motive force. *Uncoupling agents* carry protons freely across the membrane. One such agent is CCCP or carbonylcyanide-chloro-phenylhydrazone. In their presence, there is no accumulation of protons outside the cell, there is no proton motive force, and the bacterial cell dies.

One function of the proton motive force is to power the synthesis of ATP from ADP. This is catalysed by an enzyme complex called a proton-translocating ATPase, or simply ATPase synthase, in the following reaction:

$$ADP + P_i \xrightarrow[\text{Proton-translocating ATPase}]{\text{Proton motive force}} ATP$$

$$+ H_2O \ \Delta G^{o\prime} = +30.5 \text{ kJ}$$

The reaction catalysed by the ATPase is reversible. Not only can an influx of protons drive the synthesis of ATP but, in the absence of a proton motive force, the hydrolysis of ATP can drive the export of protons to create a proton motive force that can be used for energy-requiring activities, such as nutrient accumulation and motility.

The proton motive force is usually generated by one of the two main mechanisms: the *oxidation reactions in electron transport systems* (leading to ATP formation) and the *hydrolysis of ATP*(as just stated above). Other mechanisms exist but they are less common.

6 OXIDATION-REDUCTION SYSTEMS IN ELECTRON TRANSPORT

Generation of the proton motive force by the oxidation reactions in electron transport systems depends on the principle that oxidation reactions liberate energy. *Oxidation* is the loss of electrons from an atom or molecule. It can also be the loss of hydrogen atoms from a molecule, since each hydrogen atom contains an electron in addition to its proton. The opposite of oxidation is *reduction*, i.e. the gain of electrons or hydrogen atoms. Unlike oxidation reactions, reduction reactions do not liberate energy but instead require energy in order to proceed. The reverse of any oxidation is a reduction and, it follows, that the reverse of any reduction is an oxidation. In an oxidation-reduction reaction, a pair of substances is involved: one is the oxidized form and the other the reduced form, e.g. $Fe^{3+}$and Fe^{2+}. Each pair of such substances is called an oxidation-reduction (O/R) system. One O/R system may tend to accept electrons from another system, i.e. the first system will be reduced while the second system will be oxidized. The capacity of an O/R system to be oxidized or reduced depends on the relative oxidizing power of each O/R system.

The oxidative power of any O/R system is called its E_h value. The '$_h$' means that it is measured electrically in relation to the H^+/H_2 system, which is the standard system, and is expressed in volts. The more positive the E_h of an O/R system, the greater the oxidizing ability of the system. Knowledge of the relation of each O/R system to the standard H^+/H_2 system permits the comparison of one system with another. To do this, the concentrations of the oxidized and reduced forms in an O/R system, as well as the pH and temperature, must be taken into account. These variables affect the E_h value of an O/R system. When this is done, a value called $E_{O'}$, the *standard oxidation potential*, is obtained and used. It is the particular E_h value when O and R are at the same concentration ([O] = [R]), the temperature is 25°C, and the pH is 7. Under these particular conditions, any system can oxidize any other system having an $E_{O'}$ that is less positive but not more positive. These relationships are very important in recognizing the sequence in which biological oxidations occur, especially in an electron transport system, where the oxidized and reduced forms of the reactants are in approximately a 1:1 ratio.

When one O/R system oxidizes another, energy is released (the $\Delta G^{o\prime}$ value is negative). The amount of energy released or the standard free energy change, the $\Delta G^{o\prime}$ is directly proportional to the difference in $E_O{}'$ values of the two O/R systems:

$$\Delta G^{o\prime} = -nF\Delta E_{O'}$$

Here, n = the number of electrons transferred per molecule, usually 1 or 2; F is the Faraday (96 500 coulombs). Application of this equation to substances whose standard oxidation potentials are known, allows prediction of the direction of the reaction and the amount of energy released from it. Furthermore, the equation allows determination of whether or not the energy released from a particular oxidation is sufficient to allow formation of a molecule of ATP. The following example is instructive. Consider the oxidation of NADH by FAD. The standard oxidation potential ($E_{O'}$) value for NAD is −0.32, while that for FAD is −0.22. The process involves the transport of two electrons. Substituting these values into the above equation, the amount of energy released per mole of NADH oxidation is:

$$\Delta G^{o\prime} = \frac{-2 \times 96\,500(-0.22 - [-0.32])}{4.18}$$

$$= -4{,}617\,\text{cal} = -4.62\,\text{Kcal}$$

where 4.18 is the conversion factor between coulomb-volts (Joules) and calories. In this example, 4.62 kcal of energy are released per mole of NADH oxidized, an amount insufficient to produce an ATP molecule. The formation of an ATP molecule requires 7.5–8 kcal per mole. It should be remembered, however, that the actual amount of energy released from oxidative processes in cells may be different from that obtained by the kind of calculation as above using values for compounds obtained under standard conditions. Such calculations should be used with caution.

7 ELECTRON TRANSPORT SYSTEMS

An *electron transport system* is a sequence of oxidation reactions in which electrons from a source of reducing power pass from one O/R system to another. The source of reducing power may be as diverse as electrons derived from organic compound oxidation, oxidation of an inorganic ion, or radiant energy. In the sequence, each successive O/R system has a greater ability to gain electrons (i.e. has a greater $E_O{}'$ value) than the one preceding it. As the electrons flow through the series of oxidation reactions, much of the free energy of these reactions is trapped in the form of the proton motive force, which can then power the synthesis of ATP, a process called *oxidative phosphorylation*, as distinguished from substrate level phosphorylation. Electron transport systems are always associated with membranes. In prokaryotes they are in the cytoplasmic membrane; in eukaryotes they are in mitochondrial or chloroplast membranes.

Electron transport through the successive O/R systems is finally terminated in one of the two ways. In cyclic electron transport, present in *anoxygenic* (non-oxygen-forming) photosynthetic prokaryotes, the electron removed from the initial chlorophyll molecule is returned to the chlorophyll molecule from which it arose in a cyclic manner. In chemotrophic electron transport, oxidation of the last electron carrier in the chain is used to reduce a *terminal electron acceptor*. In aerobic microorganisms, the terminal electron acceptor is oxygen. The resulting compound is water, formed by the reduction of a half of a molecule of molecular oxygen with two protons and two electrons derived from the transport process. In some anaerobic microorganisms, organic compounds, like fumarate, may serve functions analogous to molecular oxygen resulting in reduced products, like succinate. In addition, some microorganisms that are facultative use nitrate or sulphate as terminal electron acceptors in the absence of oxygen, leading to the formation of reduced forms of nitrogen and sulphur.

7.1 Carriers in chemotrophic electron transport systems

A generalized scheme of the electron transport system in *chemotrophic* microorganisms is depicted in Fig. 1.1. In this system, the first

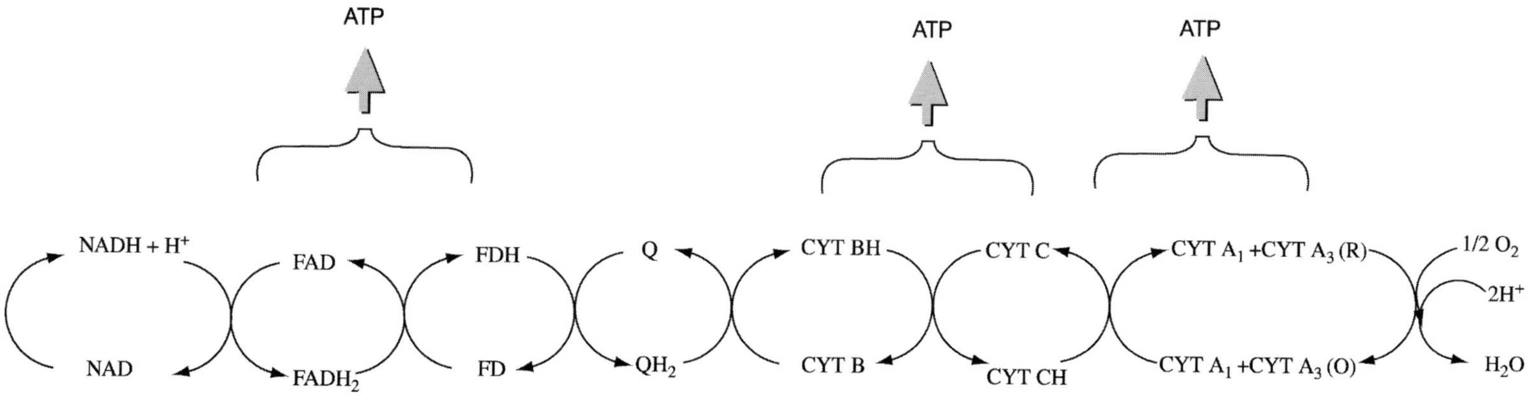

Fig.1.1 Generalized scheme of electron transport in chemotrophic microorganisms.

molecule in the electron chain is typically NAD or NADP. Enzymes, called dehydrogenases, remove electrons and hydrogen ions from substrates and many dehydrogenases use the pyridine nucleotides (NAD) nicotinamide adenine dinucleotide or NADP as their coenzyme. Some dehydrogenases are specific for NAD or NADP, whereas others are non-specific and show activity with either compound. NAD is composed of an ADP, a ribose phosphate and niacinamide. Its structure is shown in Fig. 1.2. NADP is similar to NAD but has an extra phosphate group. NAD can accept two electrons and one proton and thus exists in the reduced form:

$$NAD^+ + 2H \rightarrow NADH + H^+$$

Reduced forms of these pyridine nucleotides are normally reoxidized by transfer of their electrons to a molecule of a flavin-containing compound, typically a flavoprotein.

Flavoproteins contain either flavin adenine dinucleotide (FAD) or flavin mononucleotide (FMN) as prosthetic groups (see Fig. 1.2). The vitamin riboflavin is part of the structure of these prosthetic groups. FAD can accept two hydrogen atoms to exist in a reduced form, $FADH_2$. Similarly FMN can be reduced to $FMNH_2$. The flavoprotein is reoxidized by electron transfer to the quinines, ubiquinone and menaquinone. These are water-soluble compounds, not coenzymes, that move freely within the non-polar matrix of membranes and function as acceptors of two hydrogen atoms, e.g.

$$\text{Ubiquinone} + 2H \rightarrow \text{Ubiquinone}-H_2$$

In some systems, this transfer is mediated by a non-haem-containing iron sulphur protein, most often a ferredoxin or a structurally similar compound. Transfer of electrons from reduced quinone compounds to a series of cytochromes is the next step in electron transport.

A diversity of cytochromes is found in electron transport systems. All of the cytochromes are haem proteins that differ in the nature of both their proteins and the side chains of their haem moieties. Cytochromes are enzymes with a prosthetic group that is a derivative of haem. This prosthetic group has a single iron atom that is responsible for the oxidative or reductive properties of the enzyme (see Fig. 1.2). On the basis of differences in light absorption spectra, cytochromes can be divided into four main categories designated cytochromes a, b, c, and d. There is also cytochrome o, which belongs to the cytochrome b class. Each of these groups has a different function in a respiratory chain and can be further subdivided on the basis of minor differences in light absorption spectra, such as cytochromes c and c_1 or cytochromes a and a_3. Each cytochrome type can exist in either an oxidized or reduced form, depending on the state of the iron atom contained in their structure. Since electrons are usually transferred in pairs, two cytochrome molecules are needed for each electron pair. Certain cytochrome complexes, such as cytochromes a and a_3 (called cytochrome aa_3) and cytochromes b and o (called cytochrome bo) are called cytochrome oxidases, because they can transfer electrons directly to O_2 to form H_2O.

Various electron transport systems may use slightly different chemical substances, but the functions of related substances are physiologically equivalent. It follows that the precise nature of the electron transport systems of prokaryotes differs substantially among microorganisms, but the carrier molecules involved in most electron systems are both structurally and functionally similar.

It will be recalled that energy is released as the electrons move along the carriers in the electron transport system. If the energy release is large, this can be coupled with the movement of H^+ across the membrane, generating a proton motive force resulting in the production of ATP. The number of sites where this can occur depends on the difference between the redox potential of the substrate and the final electron acceptor.

When NADH + H + is the electron donor and oxygen is the electron acceptor, proton movements associated with large free energy changes occur at three sites (see Fig. 1.1). The sites are at NADH-FAD, cytochrome bc_1 complex, and the terminal oxidase. In microorganisms with shorter electron transport systems or where electrons enter at other sites in the system

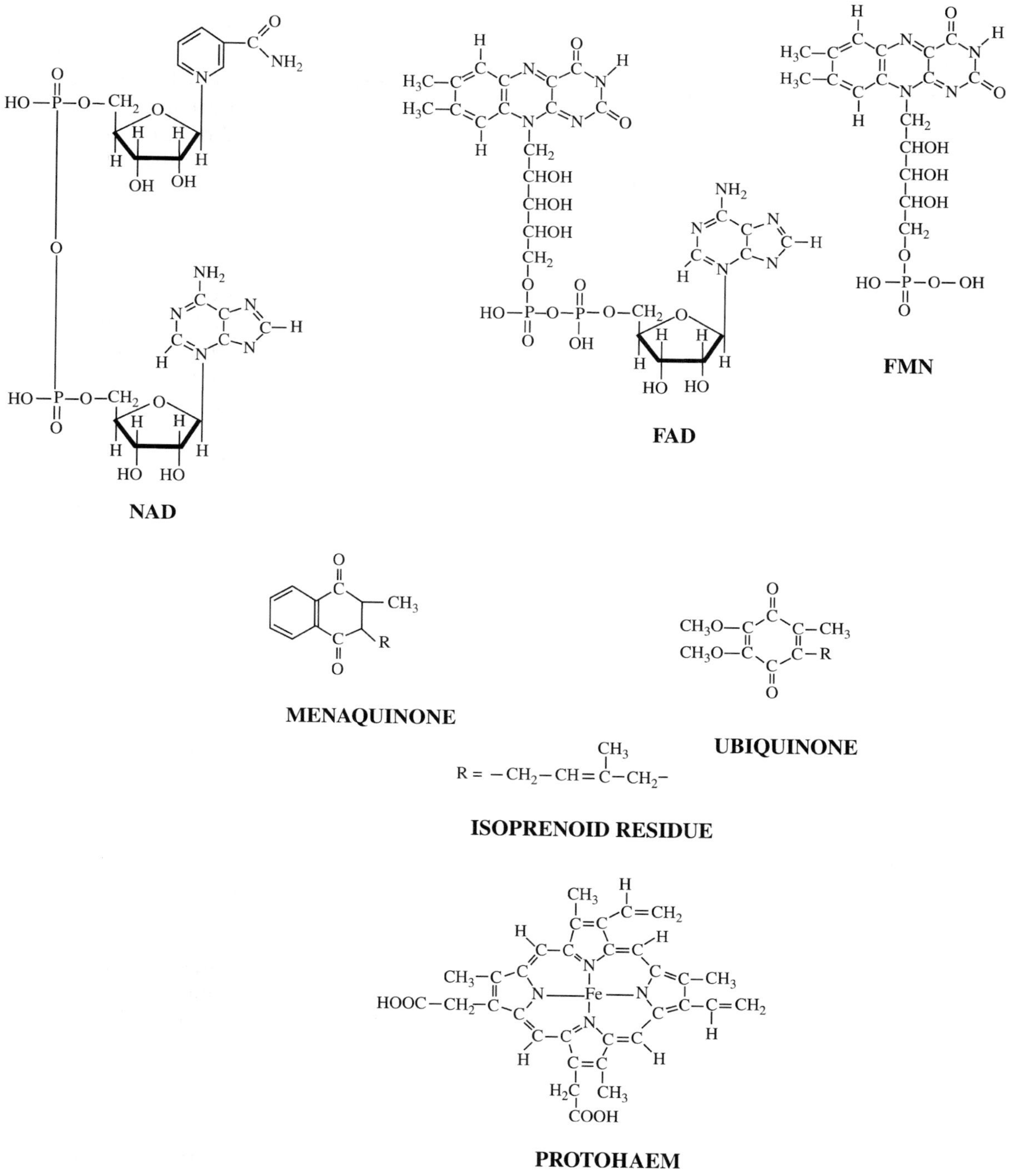

Fig.1.2 Structures of prosthetic groups and compounds in the electron transport system of chemotrophic microorganisms. The various quinones and menaquinones differ from each other by the number of R units attached to the carbonyl-containing rings. Protohaem, or haem B, is a prototype for all of the prosthetic groups of the cytochromes. Haems associated with cytochromes other than cytochromes of the *b*-type have side chains other than those shown for haem B.

or chain (having a more positive redox potential than that of the NADH/NAD + couple), less ATP is produced.

In environments devoid of oxygen, many microorganisms can utilize electron transport mechanisms for ATP synthesis, but they use alternate terminal electron acceptors other than oxygen. This process is called *anaerobic respiration*. The alternate acceptors used include nitrate, sulphate, CO_2, and specific organic compounds. (Facultative microbes will use oxygen in aerobic respiration when it is available.) Microorganisms that use nitrate as an electron acceptor reduce nitrate to nitrite and ammonia by the activity of nitrate and nitrite reductase. Methanogens use CO_2 as the terminal electron acceptor and produce methane in the process. Some sulphur bacteria, like *Desulfovibrio*, reduce sulphate to sulphide when using the former as terminal electron acceptor.

Anaerobic respiration is much less energy-efficient than aerobic respiration. A large amount of substrate is required to produce sufficient ATP for the cell. This is because the pathways for this metabolic process yield less energy for the proton motive force than do pathways with oxygen as the terminal electron acceptor because the E_O' value of the terminal electron acceptor system is more negative than that of the $1/2O_2/H_2O$ system–i.e. it has less oxidizing power.

7.2 Carriers in phototrophic electron transport systems

In *phototrophic* electron transport, the original light-absorbing substance is one or more chlorophyll and bacteriochlorophyll molecules, in tandem with carotenoid pigments and phycobiliproteins. The chlorophylls differ both structurally and in their light-absorbing properties. All of these molecules have a large number of double bonds. This property is essential for their ability to transfer energy from one molecule to another. Briefly, light is absorbed by these pigments, which pass the energy to the photosynthetic reaction centres, which then create the electron transport systems.

Thus, phototrophs have the amazing ability to use light energy to create an electron transport system that provides them with a proton motive force. They do not obtain the terminal electron acceptor for this electron transport system from the environment or the culture medium; instead they use their own oxidized chlorophyll or bacteriochlorophyll molecules.

There are two types of photosynthesis in microorganisms: the *anoxygenic* type does not form oxygen as a product of photosynthesis; the *oxygenic* type forms oxygen. The cyanobacteria carry out oxygenic photosynthesis. Photosynthesis in algae and green plants is similar to that of cyanobacteria. Green and purple bacteria differ from the cyanobacteria and carry out anoxygenic photosynthesis.

7.2.1 Oxygenic photosynthesis

Photosynthetic electron transport in the cyanobacteria is carried out by localized intracytoplasmic membranous organelles called thylakoids. Thylakoids are covered with phycobilisomes containing the light-harvesting pigments. Light-harvesting pigments can be divided into chlorophylls that absorb light in the red (>640 nm) and blue (<440 nm) regions of the light spectrum, and the carotenoids, phycobiliproteins, phycoerythrin and phycocyanin (termed accessory pigments because they absorb light at wavelengths where chlorophylls do not function efficiently (470–630 nm). The pigments are arranged in a highly organized manner to maximize energy transfer from light. Pigments in Photosystem I (PS I), or Reaction Centre I (RC I), absorb light of 700 nm, while pigments in Photosystem II (PS II), or Reaction Centre II (RC II), absorb light at 680 nm.

In oxygenic photosynthesis carried out by cyanobacteria, algae and plants, light energy is used not only to generate a proton motive force but also simultaneously to reduce $NADP^+$ to NADPH. They accomplish this by two kinds of reaction centres, PS I and PS II. The Chl *a* in PS I is energized by light at 700 nm, whereas

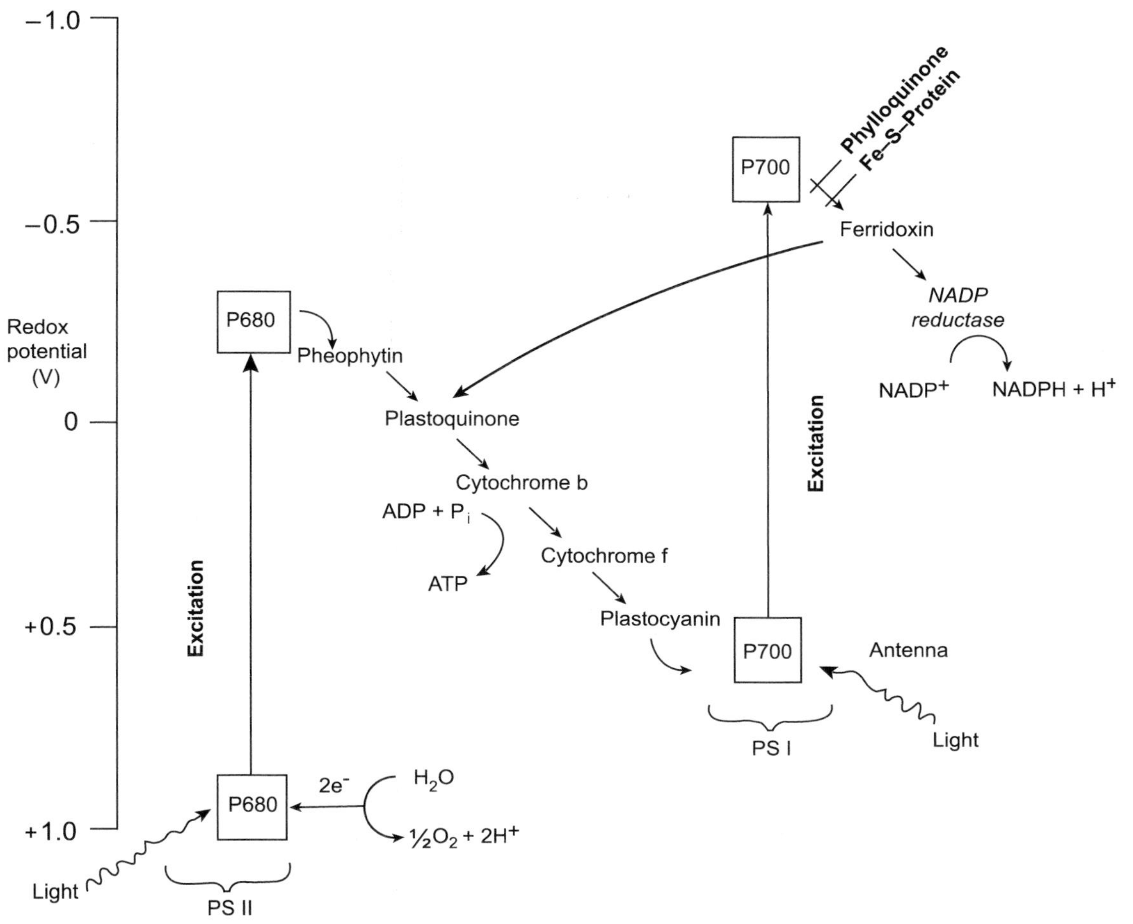

Fig. 1.3 Electron flow in oxygenic photosynthesis carried out by photosynthetic cyanobacteria, algae and green plants.

the Chl *a* in PS II is energized by light at 680 nm. The electrons from the excited reaction centres can then travel through one of two routes as shown in Fig. 1.3.

In one route, light-energized Chl *a* is oxidised by phylloquinone, which then passes the electron to Fe-S proteins and then to ferredoxin. The ferredoxin reduces $NADP^+$ to NADPH. The ferredoxin can also pass the electron to plastoquinone (arrow) instead of $NADP^+$ and generate a proton motive force. In PS II, the light energized Chl *a* is oxidized by pheophytin (a Mg-deficient form of Chl *a*), which passes the electron to plastoquinone and then to a cytochrome bf complex to generate a proton motive force. The electron finally passes to plastocyanin and then to the electron-deficient Chl *a* in PS I. These reactions leave Chl *a* in PS II deficient in electrons. The electrons are replaced in PS II by a process called photolysis of water, in which water serves as an external electron donor. Cyanobacteria, algae and green plants are responsible for producing almost all of the oxygen in the Earth's atmosphere.

7.2.2. *Anoxygenic photosynthesis*

Green and purple sulphur bacteria differ from cyanobacteria because most of them are strict anaerobes and do not use water as an electron source. The green and purple sulphur bacteria

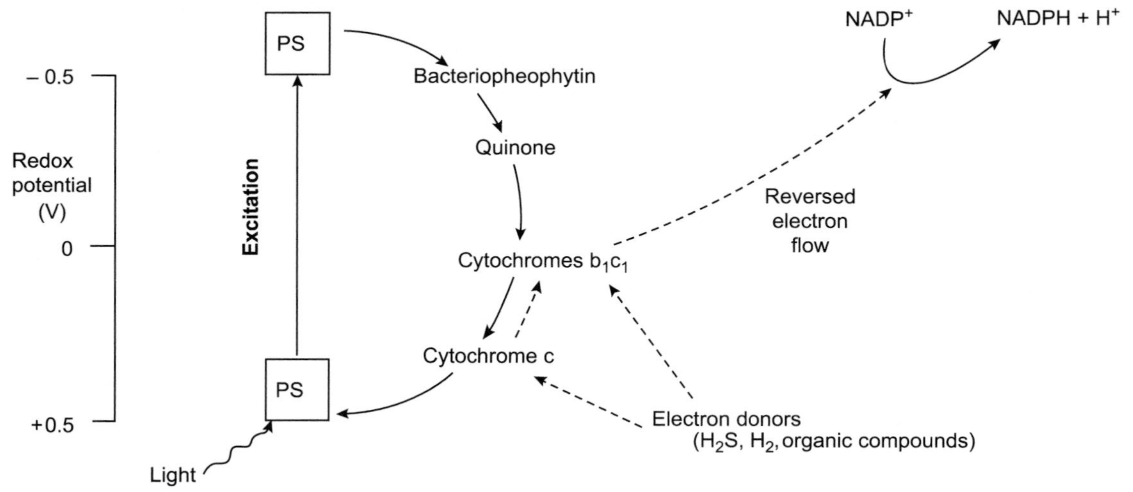

Fig. 1.4 Electron flow in anoxygenic photosynthesis carried out by the purple bacteria.

use H_2, H_2S and elemental sulphur as electron donors and possess different light-harvesting pigments called bacteriochlorophylls. They absorb light at longer wavelengths. Bacteriochlorophyll *a* has an absorption maximum at 775 nm, while bacteriochlorophyll *b* has a maximum at 790 nm. The Photosystem (PS) or Reaction Centre in purple bacteria is called P870, while that in green bacteria is called P840 (denoting the absorption wavelength maxima associated with them). Both bacterial groups exhibit cyclic electron transport which can be used to generate ATP. They are unable to synthesize $NADPH + H^+$ directly by photosynthetic electron movement. The generalized photosynthetic electron flow in anoxygenic photosynthesis is shown in Fig. 1.4. Briefly, an electron is removed from the PS by bacteriopheophytin and passed along to ubiquinone. Electrons pass from ubiquinone through an electron transport system (creating a proton motive force, Δp) and return to the PS restoring it to the reduced state. Purple bacteria generate $NADPH + H^+$ by *reversed electron flow* to drive electrons from organic compounds or inorganic compounds to $NADP^+$ and energized by the proton motive force. Alternatively, in the presence of H_2, $NADPH + H^+$ can be produced directly as H_2 has a reduction potential more negative than NAD^+. In the green sulphur bacteria, the primary acceptor of electrons from the PS is not bacteriopheophytin but an isomer called bacteriochlorophyll 663. The subsequent electron acceptor is not a quinone but ferredoxin. Green sulphur bacteria also exhibit a form of non-cyclic photosynthetic electron flow in order to reduce NAD^+ (not shown in Fig. 1.4). They oxidize sulphide to sulphur, with donation of electrons to the PS, to bacteriochlorophyll 663 and then to an iron sulphur-cytochrome *b* complex with ferredoxin serving as the immediate donor of electrons to NAD^+. The elemental sulphur accumulates as refractile granules outside of the bacterial cells.

8 CATABOLIC PATHWAYS

Besides being the most nutritionally versatile of all living organisms, bacteria are also the most versatile in their ability to extract energy from oxidation of chemical compounds or from phototrophic processes. They exhibit all the metabolic sequences found in other life forms and, as well, possess metabolic sequences that are uniquely bacterial.

Bacteria and other microorganisms use a wide variety of chemical compounds as energy sources. Sometimes these compounds are large molecules like proteins, lipids or polysaccharides that must first be broken down to smaller molecules before they can be dissimilated or used to supply energy. Microorganisms use enzymes to break down proteins to amino acids, fats to glycerol and fatty acids, and polysaccharides to monosaccharides.

Chemotrophic microorganisms derive energy by a series of consecutive enzyme-catalysed chemical reactions called a *dissimilatory pathway*. Such pathways serve not only to liberate energy from nutrients but also to supply precursors or building blocks from which a cell can construct its structure.

Although highly diverse in their metabolism, microorganisms, such as bacteria do not have individually specific pathways for each of the substances they dissimilate. Instead, a relatively few number of *central pathways* of metabolism that are essential to life are found in bacteria and, indeed, in all cellular forms. The ability of bacteria to oxidize a wide spectrum of compounds for energy reflects their ability to channel metabolites from unusual substances into the central metabolic pathways.

The central metabolic pathways are those pathways that provide precursor metabolites, or building blocks, for all of the biosynthetic pathways of the cell. They also generate ATP and reduced coenzymes, such as NADH, that can enter electron transport systems and generate a proton motive force. The most important central metabolic pathways are the Embden-Meyerhof-Parnas pathway (also called the glycolytic pathway), the pentose phosphate pathway, and the Entner-Doudoroff pathway. The last pathway has been found only among the prokaryotes.

8.1 Glycolysis

Glycolysis ('splitting of sugar') is the most common dissimilatory pathway; it occurs widely and is found in animal and plant cells as well as in microorganisms. The majority of microbes utilise the glycolytic pathway for the catabolism of carbohydrates such as glucose and fructose. This series of reactions occurs in the cytosol of microbes and can operate either

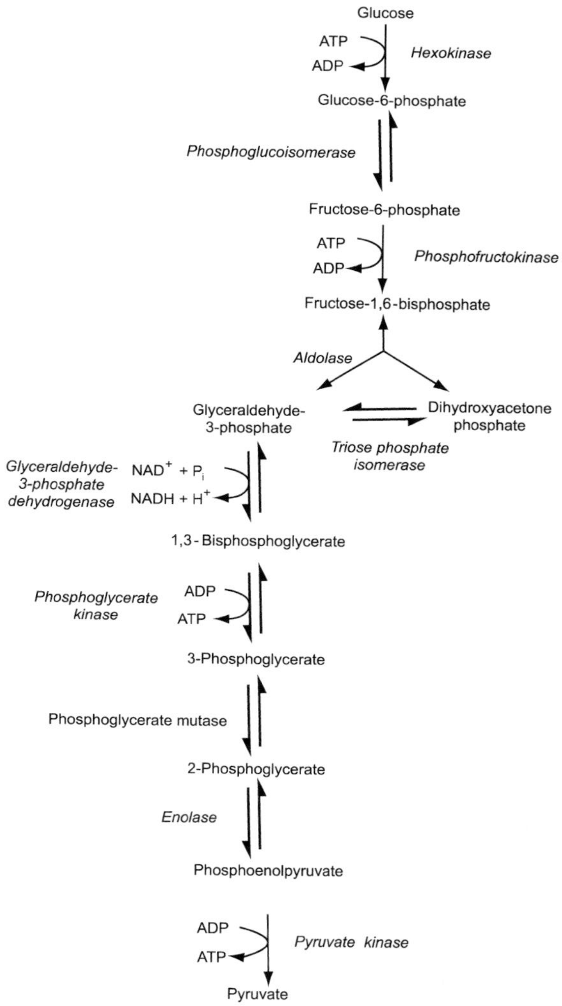

Fig. 1.5 The glycolytic pathway.

aerobically or anaerobically. Fig. 1.5 shows the steps in the glycolytic pathway.

Important features of glycolysis are discussed as follows:

- An enzyme called hexokinase uses the energy of ATP to add a phosphate group to glucose to form glucose-6-phosphate. Similarly, an enzyme, called phosphofructokinase, uses ATP to add a phosphate group to fructose-6-phosphate to form fructose-1,6-diphosphate. This 6-carbon compound is then split by an enzyme, called aldolase, into two 3-carbon moieties, dihydroxyacetone phosphate and glyceraldehyde-3-phosphate. These two compounds are interconvertible by the action of triose isomerase.

- The oxidation of glyceraldehyde-3-phosphate results in the removal of a pair of electrons by NAD^+ and the addition of a phosphate group to form 1,3-diphosphoglyceric acid. This high-energy compound can be used for ATP synthesis by substrate-level phosphorylation. Similarly, the removal of H_2O from 2-phosphoglycerate results in the high-energy compound phosphoenolpyruvate, which also can be used to synthesise ATP by substrate-level phosphorylation.
- For each molecule of glucose metabolized, two molecules of ATP are used up and four molecules of ATP are formed. Therefore, for each molecule of glucose metabolized by glycolysis, there is a net yield of two ATP molecules.

The overall equation for the glycolytic pathway is:

$$\text{Glucose} + 2\text{ADP} + 2\text{NAD}^+ + 2\text{Pi} \rightarrow 2\text{pyruvate} + 2\text{NADH} + 2\text{H}^+ + 2\text{ATP}$$

In the absence of oxygen, the electrons removed from glyceraldehyde-3-phosphate can be used to reduce pyruvic acid to lactic acid or ethanol or other products. In organisms having electron transport systems, the electrons can be used to generate a proton motive force. That is, $NADH + H^+$ can be used to produce energy via oxidative phosphorylation.

8.2 Pentose phosphate pathway

This pathway is also called the oxidative pentose pathway and the hexose monophosphate shunt. It has been called the latter because it involves some reactions of the glycolytic pathway and therefore has been viewed as a shunt of glycolysis. It exists in both prokaryotic and eukaryotic cells. It may operate at the same time as glycolysis or the Entner-Doudoroff pathway. Like glycolysis, it can operate either in the absence or presence of oxygen.

The pentose phosphate pathway is an important source of energy in many microorganisms. Glucose can be oxidized with the liberation of electron pairs, which may enter an electron transport system. However, its major role appears to be for biosynthesis. The pathway does provide reducing power in the form of NADPH, which is required for many biosynthetic reactions instead of NADH, and it provides 4C and 5C sugars for use in nucleotide and aromatic amino acid synthesis. However, the pentose phosphate pathway cannot serve as the sole carbohydrate pathway in an organism since it cannot provide many of the required intermediates for growth. Hence, as mentioned, it usually operates in conjunction with other pathways or sequences. For instance, glyceraldehyde-3-phosphate can be used to generate energy via the glycolytic and the Entner-Doudoroff pathways.

The basic outline of the pathway is depicted in Fig. 1.6.

Glucose is initially phosphorylated to form glucose-6-phosphate. This is oxidized to 6-phosphogluconic acid with the simultaneous production of NADPH. An enzyme, called 6-phosphogluconate dehydrogenase, catalyses the decarboxylation of 6-phosphogluconic acid, yielding NADPH and a 5-carbon monosaccharide, ribulose-5-phosphate. Epimerization reactions yield xylulose-5-phosphate and ribose-5-phosphate (not shown in figure). If not used for RNA and DNA synthesis, these two compounds can be the starting point for a series of transketolase and transaldolase reactions leading subsequently to glyceraldehyde-3-phosphate and fructose-6-phosphate, which can feed into the glycolytic pathway, thereby generating ATP by substrate-level phosphorylation. The overall reaction of the pentose phosphate pathway is as follows:

$$3\text{Glucose} + 3\text{ATP} + 6\text{NADP}^+ \rightarrow 2\text{Fructose-6-phosphate} + 3\text{CO}_2 + \text{Glyceraldehyde-3-phosphate} + 3\text{ADP} + 6\text{NADPH} + 6\text{H}^+$$

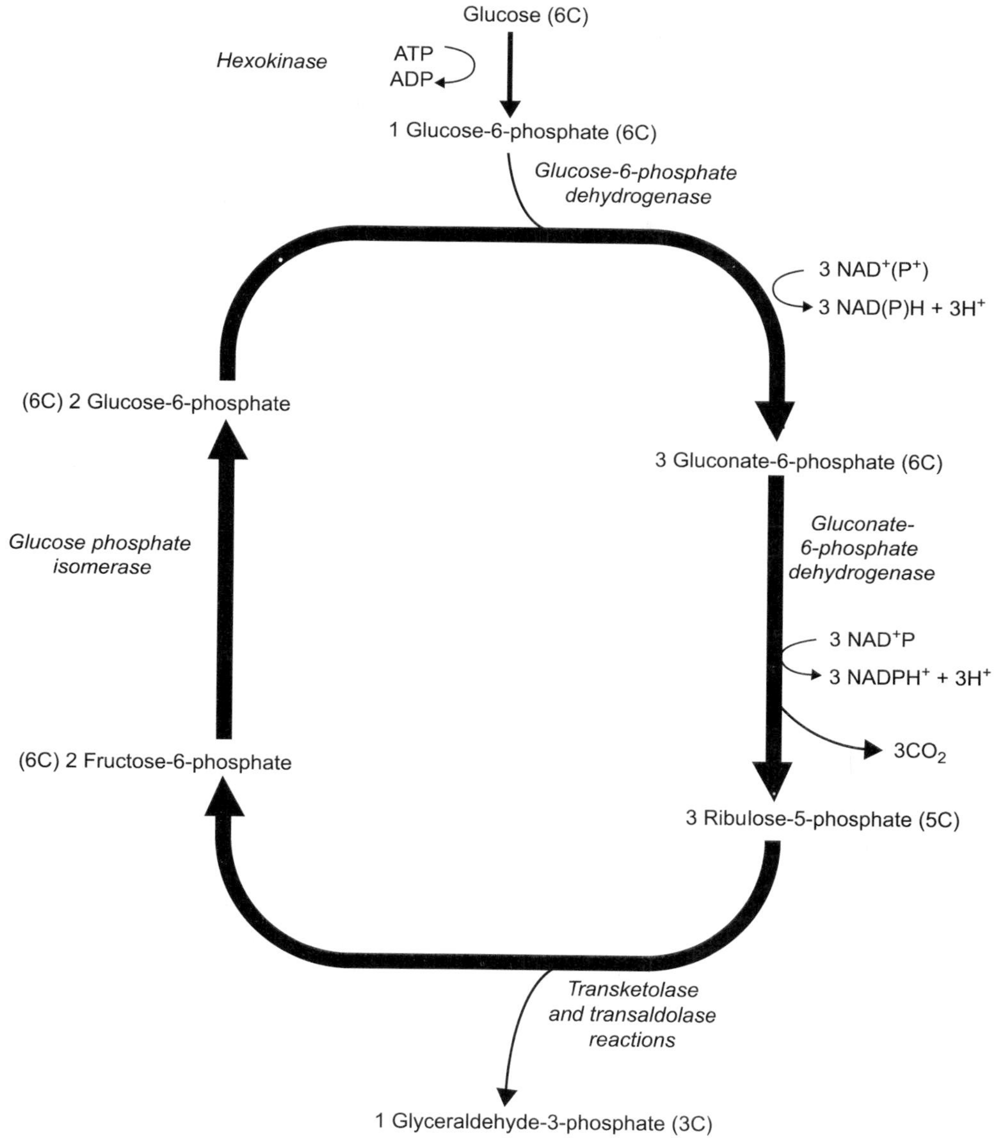

Fig. 1.6 Basic scheme of the pentose phosphate cycle.

8.3 The Entner–Doudoroff Pathway

The *Entner-Doudoroff pathway* occurs in a small number of aerobic Gram-negative bacteria including *Pseudomonas*, *Rhizobium* and *Agrobacter*. It is used mainly by bacteria that are unable to grow on glucose by glycolysis because they lack certain glycolytic enzymes but can grow on gluconic acid. Glucose, however, can be used via the Entner-Doudoroff pathway by being first phosphorylated to glucose-6-phosphate and then oxidized to 6-phosphogluconic acid. The sequence of reactions of the pathway is shown in Fig. 1.7. If gluconic acid is used, it is phosphorylated directly to 6-phosphogluconic acid by an enzyme called glucokinase (not shown in Fig. 1.7). Then 6-phosphogluconate dehydratase catalyses the removal of a molecule of water from 6-phosphogluconic acid to yield 2-keto-3-deoxy-6-phosphogluconic acid (KDPG). KDPG aldolase then catalyses the cleavage of KDPG

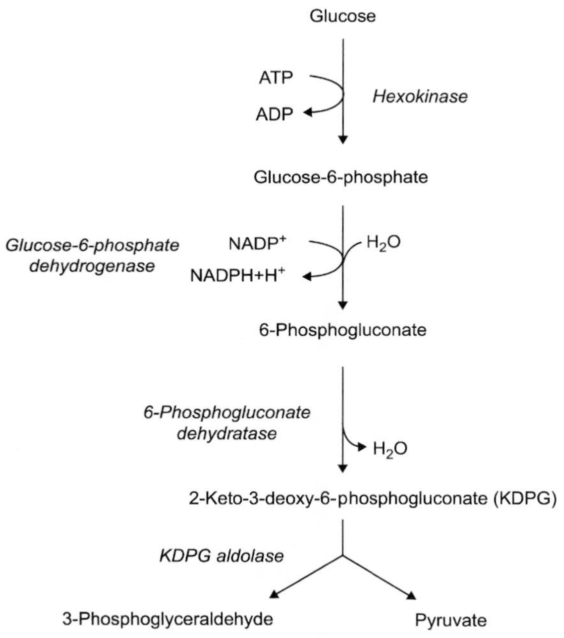

Fig. 1.7 The Entner-Doudoroff pathway.

to pyruvic acid and glyceraldehyde-3-phosphate. The latter is metabolized via glycolysis reactions to produce a second molecule of pyruvate. In many aerobic bacteria the dissimilation is completed via acetyl CoA and the citric acid cycle. The overall reaction of the Entner-Doudoroff pathway is:

$$\text{Glucose} + \text{NADP}^+ + \text{NAD}^+ + \text{ADP} + \text{P}_i \rightarrow 2\ \text{pyruvic acid} + \text{NADPH} + 2\text{H}^+ + \text{NADH} + \text{ATP}$$

8.4 The citric acid cycle

Although the previous pathways discussed provide energy, some of the critical intermediates, and reducing power for biosynthesis, none of them provides all of the critical biosynthetic intermediates. Furthermore, the energy yield obtainable from the dissimilation of pyruvate by one of the previous pathways is limited; a significantly greater yield can be attained in the presence of oxygen from the further oxidation of pyruvate to carbon dioxide via the *citric acid cycle*, also known as the *Krebs cycle* or the *tricarboxylic acid cycle*. This cycle provides the remaining critical intermediates and large amounts of additional energy.

Although the citric acid cycle in its complete form is found in aerobic microorganisms, even those that do not use oxygen in energy metabolism display most, if not all, of the citric acid cycle reactions. For instance, obligate anaerobes typically contain all of the reactions of the citric acid cycle with the exception of the alpha-ketoglutarate dehydrogenase. This is a physiological necessity since the critical intermediates for biosynthesis must be supplied for all organisms, irrespective of the way in which they obtain energy. Such a cycle is termed *amphibolic* because it participates in both dissimilative energy metabolism and in the production of biosynthetic intermediates.

The sequence of reactions of the citric acid cycle is shown in Fig. 1.8. Pyruvate does not enter this pathway directly; it must first undergo conversion into acetyl coenzyme A (acetyl CoA). This reaction is catalysed by a three-enzyme complex called the *pyruvate dehydrogenase complex*. This complex contains three kinds of enzymes: pyruvate dehydrogenase, dihydrolipoic transacetylase and dihydrolipoic dehydrogenase. The overall reaction is as follows:

$$\text{Pyruvate} + \text{NAD}^+ + \text{CoA} \rightarrow \text{Acetyl CoA} + \text{NADH} + \text{H}^+ + \text{CO}_2$$

In the citric acid cycle, initially each acetyl CoA molecule is condensed with oxaloacetic acid to form the 6-C citric acid. For every acetyl CoA entering entering the cycle:

- 3 molecules of NADH are generated
- One molecule of $FADH_2$ is generated in the oxidation of succinic acid to fumaric acid. (FAD is used here because the E_O' of the NAD + /NADH system is not positive enough to allow NAD^+ to accept electrons from succinate)
- One molecule of guanosine triphosphate (GTP) is generated by substrate-level phosphorylation. The GTP is energetically equivalent to an ATP (GTP + ADP → GDP + ATP).

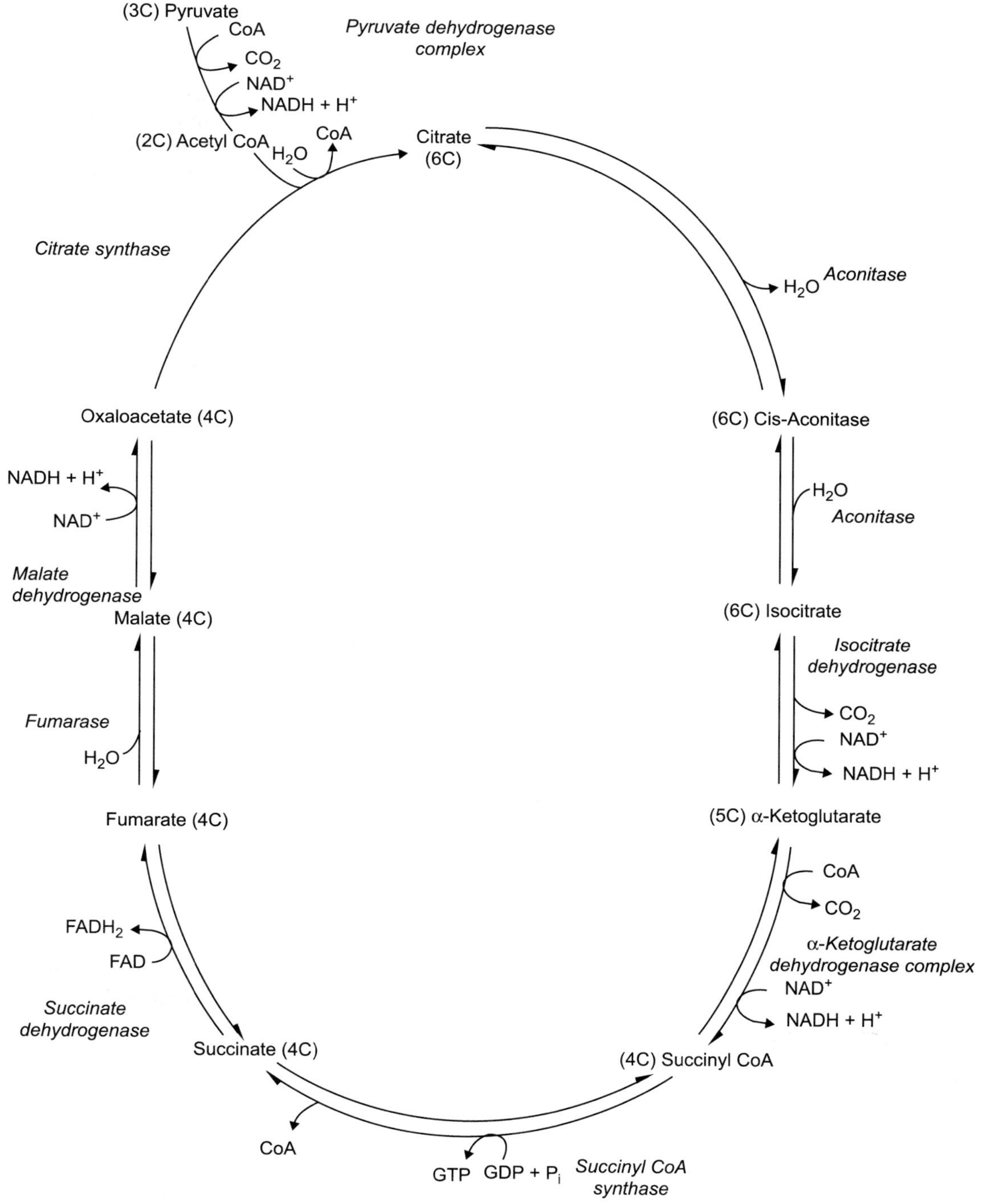

Fig. 1.8 The citric acid cycle (also called the Krebs cycle and the tricarboxylic acid cycle).

The overall reaction of the citric acid cycle is as follows:

$$\text{Acetyl CoA (also written Acetyl-S-CoA)} + 2H_2O + 3NAD^+ + FAD + ADP + P_i \rightarrow 2CO_2 + CoASH + 3NADH + 3H^+ + FADH_2 + ATP$$

8.5 The glyoxylate cycle

The *glyoxylate cycle* or *glyoxylate bypass* is employed by aerobic bacteria to grow on fatty acids and acetate as the sole carbon source. This pathway does not occur in animals because they are never forced to feed on 2-C molecules alone. It occurs also in plants and protozoa.

The glyoxylate cycle resembles the citric acid cycle except that it bypasses the two decarboxylations in the citric acid cycle. For this reason, the acetyl CoA is not oxidized to CO_2. A scheme of sequences of the glyoxylate cycle is shown in Fig. 1.9. The glyoxylate cycle shares with the citric acid cycle the reactions that synthesize isocitrate from acetyl CoA. The two pathways diverge at isocitrate. In the glyoxylate cycle the isocitrate is cleaved to succinate and glyoxylate by the enzyme isocitrate lyase. The glyoxylate condenses with acetyl CoA to form malate, in a reaction catalysed by malate synthase. The malate is then used to replenish the oxaloacetate that was used up at the beginning of the cycle. The overall reaction of the glyoxylate cycle is:

$$2\text{Acetyl CoA} + 2H_2O \rightarrow \text{Malic acid} + 2H + 2CoASH$$

Enzymes such as isocitrate lyase and malate synthase, which carry out replenishment reactions such as this, are known as *anaplerotic enzymes*; they function to maintain a pool of essential intermediates for biosynthesis.

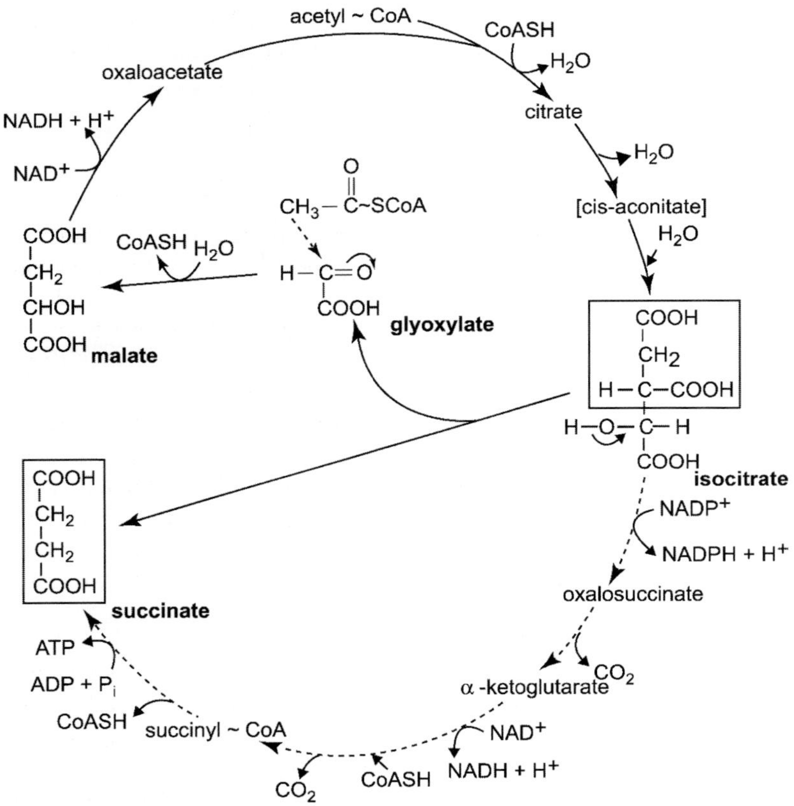

Fig. 1.9 The glyoxylate cycle or bypass. The dotted arrows represent the reactions of the citric acid cycle that are bypassed.

9 FERMENTATION

A major product in the catabolic pathways discussed above is the reduced pyridine nucleotides. A cell contains only a very limited amount of NAD^+ and $NADP^+$. A means for continuously regenerating these from the reduced forms must exist in order for the central metabolic pathways of dissimilation to continue. Microorganisms do this either by *fermentation* or by *respiration*.

Fermentation is a metabolic process in which the reduced pyridine nucleotides produced during glycolysis or other dissimilatory pathways are used to reduce an organic electron acceptor that is synthesized by the cell itself, i.e. an endogenous electron acceptor. Many microbes utilize derivatives of pyruvate as electron and H^+ acceptors and this allows $NAD(P)H + H^+$ to be reoxidized to NAD(P). For example, when yeast cells are grown under anaerobiosis, they carry out an *alcoholic fermentation*. After making pyruvate by glycolysis, they remove a molecule of CO_2 from pyruvate to form acetaldehyde:

$$\text{Pyruvic acid} \rightarrow \text{Acetaldehyde} + CO_2$$

The acetaldehyde is the acceptor for the electrons of the $NADH + H^+$ produced during glycolysis and becomes reduced to ethanol, thus regenerating NAD^+:

$$\text{Acetaldehyde} + \text{NADH} + H^+ \rightarrow \text{Ethanol} + NAD^+$$

Other microorganisms use different fermentations to regenerate NAD^+. Lactic acid fermentation is a common type of fermentation characteristic of lactic acid bacteria and some *Bacillus* species. For instance, *Lactobacillus lactis* carries out a lactic acid fermentation by using pyruvic acid itself as the electron acceptor:

$$\text{Pyruvate} + \text{NADH} + H^+ \rightarrow \text{Lactic acid} + NAD^+$$

Just as the alcoholic fermentation is of great importance to the alcoholic beverage industry, the lactic acid fermentation is important for the dairy industry. The many other types of fermentations carried out by bacteria lead to various end products such as propionic acid, butyric acid, butylene glycol, isopropanol, and acetone.

Knowledge of the kinds and amounts of products made by particular microorganisms is often helpful in identifying that organism. For example, formic acid fermentation is associated with the bacterial family *Enterobacteriaceae*. Formic acid fermentation can be divided into mixed acid fermentation, which results in the production of ethanol, and a mixture of acids such as acetic, lactic and succinic. This is a characteristic of the human intestinal bacteria *Escherichia coli* and *Salmonella* spp. found in polluted water. The second type is butanediol fermentation where 2,3-butanediol, ethanol, and smaller amounts of organic acids are formed. This is characteristic of *Serratia* spp. A different type of bacterial fermentation is that of amino acids by the anaerobic clostridia. These reactions can produce ATP by oxidizing one amino acid and using another as an electron acceptor in a process called the *Stickland reaction*.

Fermentation is an inefficient process for extracting energy by the cell because the end products of fermentation still contain a great deal of chemical energy. For example, the high energy content of the ethanol produced by yeasts is indicated by the fact that ethanol is an excellent fuel and liberates much heat when burned.

10 RESPIRATION

Respiration is the other process for regenerating $NAD(P)^+$ by using NAD(P)H as the electron donor for an electron transport system (which was described previously). It is much more efficient than fermentation for yielding energy. Not only is $NAD(P)^+$ regenerated but the electron transport system generates a proton motive force that can be used to power the synthesis of additional ATP molecules (see Fig. 1.1). For instance, when yeast cells are grown aerobically with glucose, the NADH molecules produced during glycolysis can donate their electrons to an electron transport system that has oxygen as the terminal electron acceptor (aerobic respiration). This system allows not only regeneration of NAD^+, but also the generation of enough of a proton motive force to drive the synthesis of an additional 6 molecules of ATP. Further breakdown

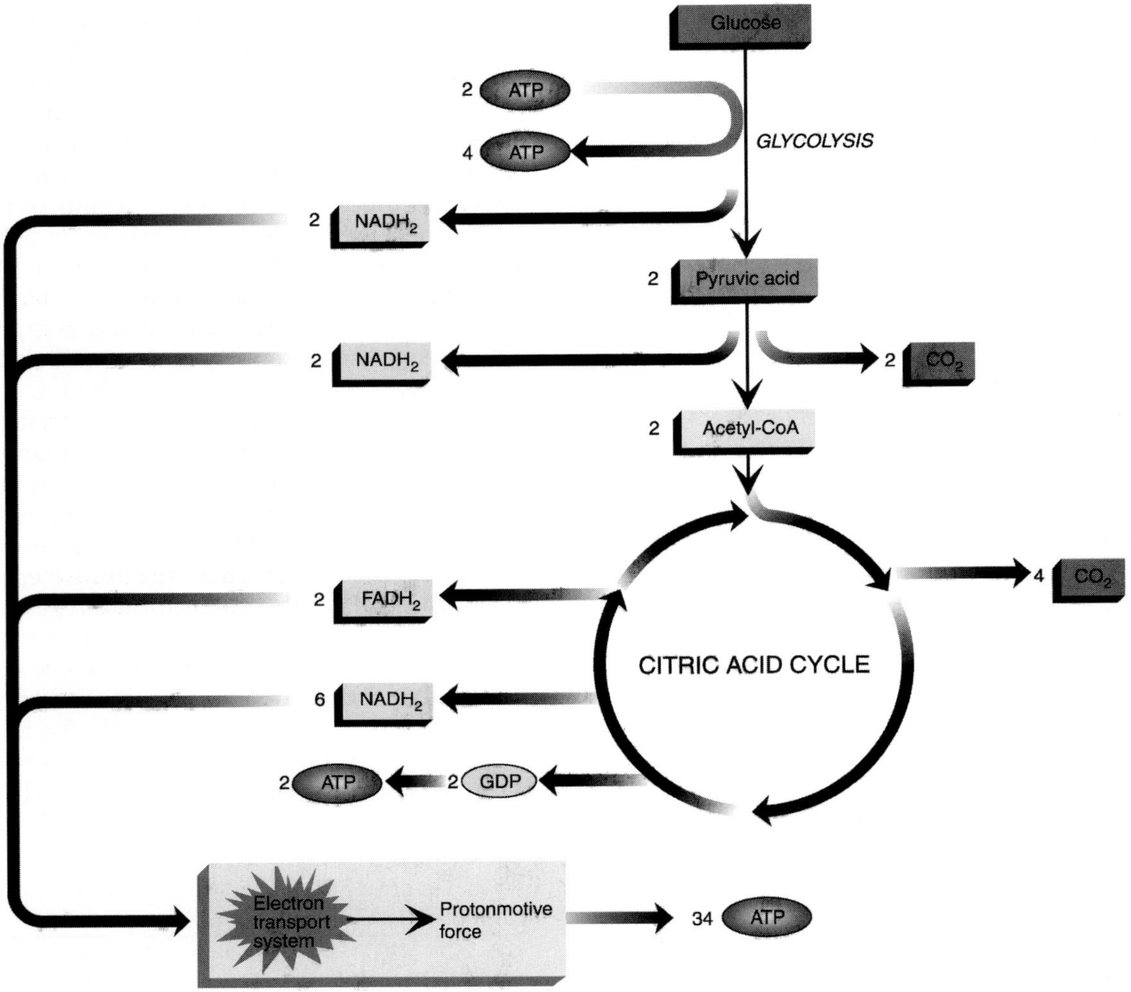

Fig. 1.10 ATP production by aerobic respiration. The complete breakdown of glucose to 6 molecules of CO_2 results in a net yield of 38 ATP molecules.

occurs when pyruvic acid is oxidized to acetyl CoA by pyruvate dehydrogenase. Each of the two molecules of NADH so formed can serve as the electron donor for an electron transport system, creating a proton motive force that can be used for synthesizing 6 molecules of ATP. Oxidation of the acetyl CoA by the citric acid cycle yields 6 more NADH which can be used to make 18 ATP. In addition, two molecules of $FADH_2$ are produced that can provide enough energy for the synthesis of 4 molecules of ATP. Two additional ATPs are made by substrate level phosphorylation. Therefore, the net yield of ATP from complete dissimilation of one glucose molecule is 38 molecules (Fig. 1.10). This is in sharp contrast to the yield of ATP from fermentation when yeast cells are grown anaerobically, where the yield is only 2 molecules of ATP per molecule of glucose.

11 ENERGY-REQUIRING METABOLIC PROCESSES

How microorganisms trap chemical and light energy within high energy transfer compounds and the proton motive force has been discussed up to this point. This chapter concludes with a brief survey of how this energy is used to fuel the various endergonic processes essential to the life of the cell. No attempt is made to provide a comprehensive catalogue of metabolic reactions requiring energy. By intent as well as by the demands of space limitation, only some important energy-requiring processes will be introduced to illustrate how trapped energy is used by microoganisms to maintain themselves and to grow.

Some energy-requiring metabolic processes are biosynthetic ones by which the complex chemical constituents of a cell are constructed. Energy in the form of ATP is needed to power the biosynthesis of many of the various chemical components of the cell, such as DNA, RNA, enzymes, bacterial cell wall peptidoglycan, and cell membrane phospholipids. In addition to being used for biosynthetic processes, energy is also used for several non-biosynthetic functions of the cell, including the inward transport and accumulation of nutrients, the export of certain proteins from the cell, and the powering of the mechanism for flagellar motility.

12 BIOSYNTHETIC PROCESSES

An autotrophic bacterium that can synthesize all its cellular constituents from simple inorganic compounds obviously has great biosynthetic ability. In the same way, a heterotrophic bacterium that can grow in a medium containing only glucose as the carbon and energy source, ammonium sulphate as the nitrogen and sulphur source, and a few additional inorganic compounds, also has great biosynthetic capability. From these nutritional substances, the microbe can synthesize:

1. nitrogenous substances, including proteins (such as enzymes) and nucleic acids
2. carbohydrates, including complex polysaccharides (such as capsules and the peptidoglycan of the cell wall)
3. phospholipids (such as components of the cytoplasmic membrane).

Details of the various specific biosynthetic pathways may be found in biochemistry and microbial physiology textbooks (such as those cited in the references to this chapter). Suffice it here to make some generalizations because all these pathways have fundamental features in common:

- A biosynthetic pathway begins with the synthesis of the biochemical building blocks that are needed to make more complex substances. Such building blocks are small molecules.
- The building blocks are then energized, usually with the energy of ATP molecules. This energy is needed to establish the covalent bonds that subsequently will link the building blocks. In a very real sense, it is these small molecules that constitute the link between energy generation and energy utilisation.
- The energized building blocks are joined together to form complex substances that become structural or functional parts of the cell.

12.1 Biosynthesis of carbohydrates

A heterotrophic bacterium, such as *Escherichia coli*, can convert glucose (supplied as the major carbon source) to various other monosaccharides. For instance, it makes ribose phosphate (needed to synthesise nucleotides) from glucose by the pentose phosphate pathway. However, many bacteria, such as *Campylobacter jejuni*, cannot use glucose or any preformed sugar, probably because sugars are impermeable to the cytoplasmic membrane. Nor do they fix carbon dioxide. Such microorganisms must synthesize all of their sugars from non-carbohydrate precursors by a process called *gluconeogenesis*. The pathway involved is the reverse of glycolysis except for three sites where the glycolytic reaction is irreversible. At these sites the glycolytic enzyme is replaced by another enzyme to bypass the irreversible reaction:

1. Conversion of pyruvate into phosphoenol pyruvate is catalysed by phosphoenol pyruvate carboxykinase (glycolytic enzyme is pyruvate kinase)

2. Conversion of fructose-1,6-bisphosphate into fructose-6-phosphate is catalysed by fructose bisphosphatase (glycolytic enzyme is phosphofructokinase)
3. Conversion of glucose-6-phosphate into glucose is catalysed by glucose-6-phosphatase (glycolytic enzyme is hexokinase).

Fructose can also be synthesized by this route. The synthesis of other sugars can be by simple rearrangements. For example, the enzyme mannose-6-phosphate isomerase catalyses

$$\text{Fructose-6-phosphate} \leftrightarrow \text{Mannose-6-phosphate}$$

In synthesizing polysaccharides from monosaccharides, a cell cannot merely glue sugars together. Energy is required to establish the necessary covalent linkages. This energy is usually in the form of ATP and is expended only inside the cell to energize the monosaccharides. For example, an energized form of glucose is uridine diphosphate glucose (UDPG). Nucleoside diphosphate (ADP-linked) sugars are required for the synthesis of polysaccharides as shown by the following equations:

$$\text{ATP} + \text{Glucose-1-phosphate} \rightarrow \text{ADP-Glucose} + \text{PP}_i$$

$$(\text{Glucose})_n + \text{ADP-Glucose} \rightarrow (\text{Glucose})_{n+1} + \text{ADP}$$

The biosynthesis of bacterial cell wall peptidoglycan is an endergonic process. Although the polymer is located in the cell wall, most of the chemical energy needed for its synthesis is expended inside the cell. The process is highly complex and the resulting macromolecule is a rigid cross-linked material that maintains the shape of the bacterial cell.

12.2 Carbon dioxide fixation by autotrophs

Autotrophic microorganisms are capable of using CO_2, the most oxidized form of carbon, from the atmosphere as their major source of carbon and reducing it to organic compounds. This process is called *carbon dioxide fixation* and is an endergonic process. Autotrophs use ATP as the energy source for CO_2 fixation and NADH or NADPH as the electron source for the reduction.

The main method of CO_2 fixation in autotrophic organisms is the *Calvin cycle*. The reactions of the Calvin cycle are also recognized as the *dark reactions* of photosynthesis because green plants, algae, cyanobacteria, and the purple bacteria use the Calvin cycle for CO_2 fixation. The Calvin cycle is depicted in Fig. 1.11. In the initial reaction of the Calvin cycle, CO_2 is added to the 5-C ribulose bisphosphate resulting in the formation of two molecules of the 3-C phosphoglyceric acid. Then NADPH or NADH, aided by the energy of ATP, provides electrons to reduce the phosphoglycerate to glyceraldehyde-3-phosphate. All of the organic compounds needed by the cell are synthesized subsequently from the phosphoglycerate and the glyceraldehyde-3-phosphate. Each turn of the cycle results in the fixation of one molecule of CO_2. To synthesize one glucose the cycle must operate six times. The overall stoichiometry of the cycle is as follows:

$$6CO_2 + 18ATP + 12NADPH + 12H^+ + 12H_2O \rightarrow C_6H_{12}O_6(\text{glucose}) + 12NADP^+ + 18ADP + 18P_i$$

For a long time, the Calvin cycle was believed to be the sole means of CO_2 fixation in living organisms. Recently, additional modes of CO_2 fixation have been discovered. The *reductive TCA cycle* is known in the green bacterium *Chlorobium limicola* and certain members of the sulphate-reducing bacteria. The *acetyl CoA pathway* is widely distributed among the chemolithotrophic anaerobes including the sulphate-reducing and methanogenic microorganisms.

12.3 Nitrogen fixation and assimilatory nitrate reduction

Seventy-eight per cent of air in the atmosphere is nitrogen, but it cannot be used as a nutrient source of nitrogen by most living organisms. However, some prokaryotes, like the free-living *Azotobacter* and the legume plant symbiont *Rhizobium*, are able to use it by a process

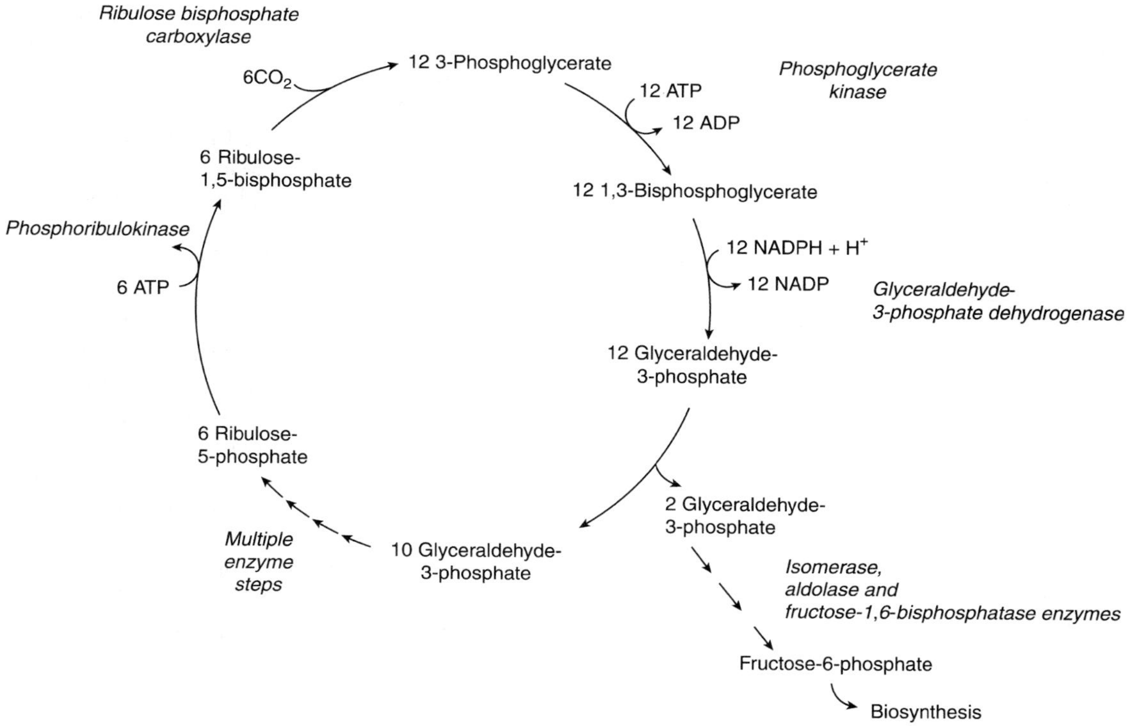

Fig. 1.11 The Calvin cycle.

called *nitrogen fixation*. The process is catalysed by a cytoplasmic nitrogenase complex consisting of two enzymes: one enzyme is dinitrogenase, which contains molybdenum and iron (a MoFe protein); the other enzyme is dinitrogenase reductase, an iron-containing enzyme (an Fe protein). The source of the electrons for N_2 reduction is usually the reduced form of the Fe-S protein ferredoxin (Fd_{red}) which has a very negative E_O' value. Anaerobic or microaerophilic bacteria can provide Fd_{red} from oxidation of pyruvic acid by pyruvate:ferredoxin oxidoreductase. Aerobic bacteria reduce NAD^+ to NADH during pyruvic acid oxidation and thus must use the proton motive force to power reversed electron transport allowing NADH to reduce Fd. The mechanism of nitrogen fixation is shown in Fig. 1.12. The overall reaction for the process is:

$$N_2 + 6H^+ + 6e^- + 12ATP + 12H_2O \rightarrow 2NH_3 + 12ADP + 12P_i$$

Many microorganisms cannot fix nitrogen and assimilate nitrate in a process called

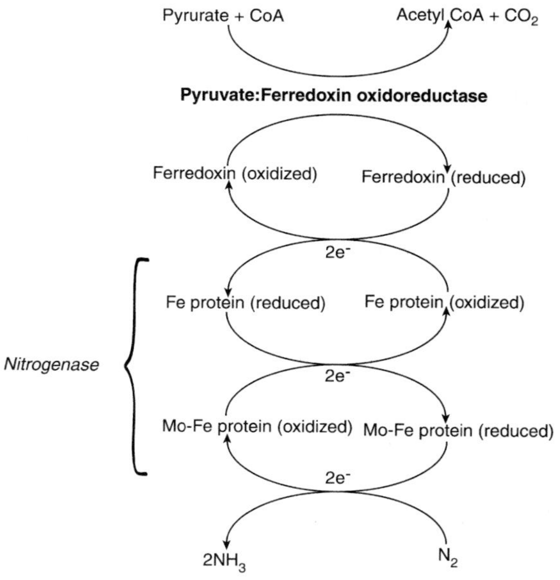

Fig. 1.12 Mechanism of nitrogen fixation.

assimilatory nitrate reduction thereby obtaining ammonia. Assimilatory nitrate reduction uses two cytoplasmic enzymes, nitrate reductase and nitrite reductase. Nitrate reductase catalyzes the reduction of NO_3^- to NO_2^-, and nitrite reductase catalyses a multistep reduction: NO_2^- to NOH (nitroxyl), a postulated intermediate that has never been isolated), NOH to NH_2OH (hydroxylamine), and NH_2OH to NH_3.

Ammonia can be incorporated directly as it is more reduced than other forms. The simplest mechanisms found in all microorganisms involve the formation of alanine and glutamate as shown in the following reactions:

$$\text{Pyruvate} + NH_3 + \text{NAD(P)H} + H^+ \leftrightarrow \text{L-Alanine} + \text{NAD(P)}^+ + H_2O$$

$$\alpha\text{-Ketoglutarate} + NH_3 + \text{NAD(P)H} + H^+ \leftrightarrow \text{Glutamate} + \text{NAD(P)}^+ + H_2O$$

Glutamic acid occupies a key position in amino acid biosynthesis because it is involved in the synthesis of many other amino acids. Microorganisms can use alanine and glutamate to synthesize a number of new amino acids via *transamination reactions* (transfer of α-amino groups). Such amino acids include aspartic acid, serine, valine, isoleucine, leucine, arginine, histidine, phenylalanine and tyrosine. Another way by which glutamic acid can be used to make other amino acids is by the alteration of its molecular structure; proline is made in this manner.

Amino acids need to be energized before they can be linked together to make proteins. Cells energize amino acids by using the energy of ATP:

$$\text{Amino acid} + \text{ATP} \rightarrow \text{Amino acid-AMP} + PP_i$$

A microorganism synthesizes hundreds of different proteins, each protein having its own unique sequence of amino acids.

12.4 Biosynthesis of nucleotides

A nucleotide consists of a nitrogenous base-pentose-phosphate. When the sugar is ribose, a nucleotide is a ribonucleotide and is used for the biosynthesis of RNA. Similarly, when the sugar is deoxyribose, a nucleotide is a deoxyribonucleotide and is used for the biosynthesis of DNA. Ribonucleotides and deoxyribonucleotides that have adenine or guanine as the nitrogenous base are purine nucleotides. Those that have cytosine, thymine or uracil as the nitrogenous base are pyrimidine nucleotides.

The energized form of a nucleotide is its diphosphate derivative. For instance, if the nucleotide is cytidine monophosphate (CMP), the energized form would be cytidine triphosphate (CTP)–the extra phosphate groups supplied by ATP. Prior to energization, the synthesis of pyrimidine nucleotides requires ribose phosphate (from the pentose phosphate cycle), ATP, aspartic acid and glutamine. The synthesis of purine nucleotides requires ribose phosphate, ATP, GTP and the amino acids glycine, aspartic acid and glutamine. DNA and RNA are subsequently assembled from pyrimidine and purine diphosphates.

12.5 Biosynthesis of lipids

The major lipids of microbial cells are the phospholipids that, together with proteins, form the structure of the cytoplasmic membrane. Phospholipids are made from long chain fatty acids. The general way in which microorganisms synthesise phopholipids may be depicted in this manner:

Glucose

↓ glycolysis

Pyruvic acid

↓

Acetyl coenzyme A and

Malonyl coenzyme A

↓

Long chain Fatty Acids

↓ ←Glycerol phospate

Phospholipids

As may be noted, two important building blocks for long chain fatty acids are acetyl coenzyme A (acetyl CoA or acetyl-S-CoA) and malonyl coenzyme A (malonyl CoA or malonyl-S-CoA). Acetyl CoA is made by the oxidation of pyruvic acid. Malonyl CoA is an energized building block made from acetyl CoA. The malonyl group contains three carbon atoms; the extra carbon comes from CO_2 via the bicarbonate ion, HCO_3^-, and energy in the form of ATP is required for the addition.

The synthesis of fatty acids is shown in Fig. 1.13. After malonyl CoA is made, the coenzyme group on each malonyl CoA and acetyl CoA is replaced by a large protein molecule called an acyl carrier protein that serves to anchor the compounds so that appropriate enzymes can act on them. Then a malonyl ACP complex is complexed with an acetyl ACP complex. In this reaction, two of the three carbon atoms of the malonyl group are added to the acetyl group to make a 4-carbon acetoacetyl ACP molecule; the third carbon atom of the malonyl group is liberated as CO_2. The acetoacetyl ACP is eventually converted to butyryl ACP (as shown in Fig. 1.13).

If the butyryl group were to be released from the ACP protein, it would become butyric acid, a 4-C fatty acid. To make a longer fatty acid, the cell must combine the butyryl ACP with another malonyl ACP to add another 2-carbon unit, resulting in a 6-carbon fatty acid ACP. This process of adding two carbons at a time to elongate the fatty acid chain is continued until a fatty acid of the required length is attained (usually 16 to 18 carbon atoms long).

After the fatty acids of the required length are obtained, the cell uses them to synthesize phospholipids. Glycerol phosphate is needed for its synthesis. It is made from dihydroxyace-

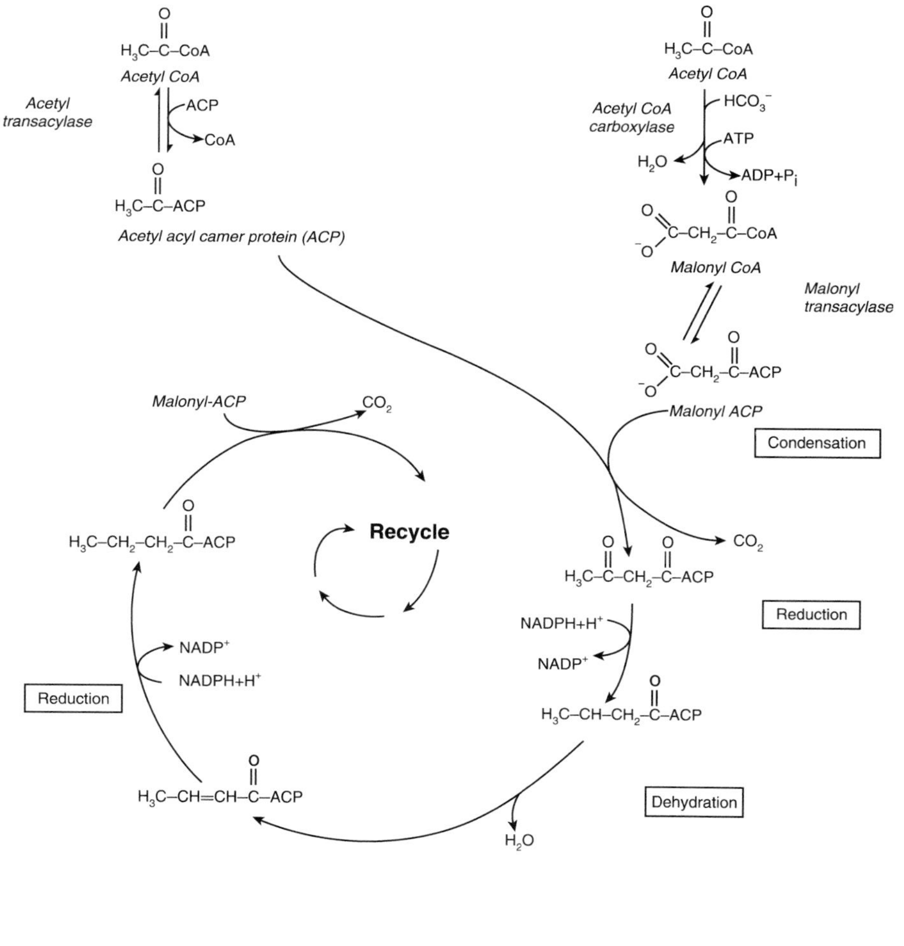

Fig. 1.13 Fatty acid synthesis. The cycle is repeated until the specific chain length is obtained.

tone phosphate, an intermediate in glycolysis:

$$\text{Dihydroxyacetone phosphate} + \text{NADH} + \text{H}^+ \rightarrow \text{Glycerol phosphate} + \text{NAD}^+$$

Two fatty acid molecules are linked to one molecule of glycerol phosphate to form a molecule of phosphatidic acid, a simple phospholipid. The cell can then link other chemical groups to the phosphate group of the phosphatidic acid to make other phospholipids. For example, the amino acid serine can be added to phosphatidic acid to make phosphatidylserine. Energy in the form of cytidine triphosphate (CTP) is required for this reaction.

13 NON-BIOSYNTHETIC PROCESSES

There are non-biosynthetic processes in microbial cells that require energy. These processes include transport of nutrients into cells, protein export and motility in prokaryotic cells.

13.1 Transport of nutrients

Transport of the nutrients into cells essentially means transport of nutrient substances across the cytoplasmic membrane. When substances move across a membrane without interacting specifically with a carrier protein in the membrane, the process is called *passive diffusion*. Gases and water, and some non-ionized organic acids, go in and out of the cell by passive diffusion and are driven only by concentration gradients, i.e. they move down the gradient from high to low concentrations.

The cytoplasmic membrane is an effective barrier to most polar molecules such as the nutrient sugars and amino acids. Transport of these molecules requires the participation of *membrane transport proteins*, or *carrier proteins*, which span the membrane. Three classes of carrier proteins have been identified: *uniporters* move a single type of compound across the membrane; *symporters* move two types of compounds simultaneously in the same direction; *antiporters* move two compounds in opposite directions.

Even with carrier proteins, no metabolic energy is needed for the mere 'downhill' entry into the cell of a solute molecule, i.e. from a greater concentration outside to a lower concentration inside the cell. Such gradient-driven carrier-mediated transport is called *facilitated diffusion*. Carrier proteins exhibit specificity for molecule binding as well as saturation kinetics.

Microbial cells can also concentrate nutrient molecules 'uphill' against a gradient. Consequently, if a bacterium lives in a dilute aqueous environment, with very low levels of nutrients, it can concentrate these nutrients inside the cell to the point where they can be readily used for metabolism. Such internal concentration of nutrients requires metabolic energy and the process is called *active transport*. The energy is used to change the affinity (strength of interaction) of the carrier protein for the particular solute. On the outer surface of the membrane the carrier protein exhibits high affinity for the substrate; during transport, the affinity of the carrier must change to a low one so that the substrate can be released into the cell. The energy for the process to change the conformations of the carrier proteins, and hence their affinities, comes from the proton motive force or ATP hydrolysis.

In the *PEP-phosphotransferase system* of nutrient transport, the solute is chemically altered by acquiring a phosphate group as the result of an enzymatic reaction during its transport across the cytoplasmic membrane. The process is also called *group translocation*. It is the only process in which there is chemical alteration (phosphorylation) of the solute during transport. The carrier has only a low affinity for the phosphorylated form of the solute. The energy for the alteration of the solute molecule comes from the high-energy compound phosphoenolpyruvic acid (PEP), which is an intermediate in glycolysis.

13.2 Export of proteins

The translocation process of proteins synthesized by cytosolic ribosomes into or through the membranes is referred to as *protein export*. The proteins end up in the cytoplasmic membrane, periplasmic space, outer envelope, cell wall, glycocalyx, cell surface appendages and extracellularly into the surrounding environment. Examples of cytoplasmic membrane proteins include cytochromes and other respiratory enzymes as well as carrier proteins. Periplasmic proteins include binding proteins, lipoproteins and various degradative enzymes such as ribonuclease and alkaline phosphatase. Porins are found in the outer membrane. Examples of secreted proteins include exotoxins and enzymes (e.g. pencillinase, amylases, proteinases).

As to be expected, most research work on protein export has been performed with *Escherichia coli*, primarily because of the ease of genetic manipulation. This has led to the formulation of a model for the export of protein; it is called the Sec system.

Proteins destined for export are distinguished by having an extra 20 or so amino acid sequences, called the *signal sequence*, at their NH_2-terminal end when they are being synthesized within the cytoplasm by ribosomes. Signal sequences vary with the particular export protein. They are removed during transport across the cytoplasmic membrane. The protein chain to be exported must not be allowed to fold into its mature form before export. Upon leaving the ribosome, the nascent pre-protein binds to a *chaperone protein* that prevents it from assuming a tightly folded configuration that cannot be translocated. The signal sequence, or peptide, retards the folding of the pre-protein, giving the chaperone protein time to bind. The chaperone protein then delivers the exportable protein to a *docking protein*. Energy is expended when ATP *initiates* the transfer of the exportable protein across the cytoplasmic membrane and the proton motive force *drives* translocation of the rest of the protein chain.

13.3 Motility in prokaryotic cells

A microbial cell is of a very small dimension. As a consequence, water is a very viscous medium to such a cell. Microscopic fins, flippers, or fishtails would be useless for propelling such a cell in such a liquid. The analogy would be that of a human trying to swim or row a boat on a body of thick molasses. However, by rotating its flagellum, a prokaryotic cell can move readily through water in much the same way a corkscrew can easily penetrate a piece of cork. The discs in the basal body of the flagellum are responsible for rotating the appendage. The motor apparatus that causes a flagellum to rotate is associated with the bottom-most disc in the basal body of the flagellum. The motor is a proton motor, driven by a flow of protons from the proton motive force.

14 EPILOGUE

This chapter has provided a comprehensive overview of current information on how microorganisms fill their nutritional needs and extract energy to carry out their life metabolic processes. To this end, the author has drawn freely from his experience in writing textbooks with his co-authors, Dr Michael J. Pelczar, Jr (Professor Emeritus, University of Maryland) and Dr Noel R. Krieg (Professor, University of Virginia and Technical Institute). Their latest combined contribution is *Microbiology: concepts and applications*. This chapter contribution is a grateful acknowledgement of their collaborative efforts through the years, but the author is solely responsible for the accuracy and acceptance of the assembled information in this chapter.

REFERENCES

Caldwell, D.R. (2000). *Microbial Physiology and Metabolism*, 2nd edn. Star Publishing Co., Belmont.

Nicklin, J., Graeme-Cook, K., Paget, T. and Killington, R.A. (1999). *Instant Notes in Microbiology*, Bios Scientific, Oxford.

Pelczar, M.J. Jr, Chan, E.C.S. and Krieg, N.R. (1993). *Microbiology: Concepts and Applications*, McGraw-Hill, New York.

Schlegel, H.G. and Bowien, B., eds. (1989). *Autotrophic Bacteria*, Science Tech, Madison.

White, D. (2000). *The Physiology and Biochemistry of Prokaryotes*, Oxford University Press, New York.

Introduction to Microbes of Sanitary Importance

2

Viruses

John Heritage

Department of Microbiology, University of Leeds, Leeds, LS2 9JT, UK

1 INTRODUCTION

The bacteriological examination of water is routinely used to detect the presence of faecal indicators such as *Escherichia coli, Clostridium perfringens* and enterococci, particularly detection of *Escherichia coli*. These bacteria are easy to manipulate in the laboratory and the bacteriological examination of water is easily performed. In contrast, the virological examination of water involves more specialized techniques and relatively few laboratories undertake these duties. The viruses that can be found in water cause infections that range from those that do not cause noticeable symptoms to those that cause life-threatening infections. It is essential, therefore, that testing water for the presence of viruses is only carried out by personnel who have been properly trained in virology and, where appropriate, have proper vaccination cover.

Because some of the waterborne viruses that will be cultured during the examination of water samples are capable of causing serious disease and also because they may be present in culture extracts in large numbers, proper containment facilities must be used. The pathogens that may be cultivated require Level 2 containment facilities for safe handling. Many laboratories use simian cells in which to cultivate viruses. These must also be handled with caution. All materials used for the virological examination of water must be sterilized before and after the testing procedure to minimize the contamination of samples during testing and to prevent release of infectious viruses from the materials after use. The need for safe handling of viruses and of materials that are potentially contaminated with viruses cannot be overemphasized.

As with the examination of water for faecal bacteria, the virological examination of water depends upon the detection of *indicator viruses*. These viruses are typically found in waters contaminated with human faeces. Not all waterborne viruses are capable of growth in tissue culture and hence the need for exploiting indicators. The most common viruses used as indicators of faecal contamination of water are the enteroviruses, particularly the poliomyelitis virus and the Coxsackie viruses. These will grow relatively easily in tissue culture. Other enteroviruses grow less well in the laboratory. These viruses can cause life-threatening infections. The reference laboratories devoted to water virology help to control such infections and literally play a vital role in the monitoring and maintenance of public health.

2 THE NATURE OF VIRUSES

Viruses are relatively simple structures. They are made up of a nucleic acid core surrounded by a protective protein coat, or capsid, made up of subunits called capsomeres. The nucleic acid of a virus may be either RNA or DNA but virus particles do not carry both nucleic acids. The RNA or DNA carries the virus genome: all the genetic information necessary to complete the virus life cycle. Retroviruses are unusual. One is HIV, the human immunodeficiency

The Handbook of Water and Wastewater Microbiology
ISBN 0-12-470100-0

virus. Infection with this virus is the ultimate cause of AIDS. The retrovirus particle contains an RNA genome but when the virus infects a host cell, a DNA copy is made of the virus genome. This subsequently becomes integrated into the genome of its host cell. This DNA copy of the virus RNA is made by an enzyme known as reverse transcriptase used in many molecular biological techniques.

The virus genome encodes the structural proteins that make up the capsid and all the enzyme functions necessary for virus replication and cell infection. Some viruses have a lipid envelope but this is not the case with all viruses. The envelope material may be closely associated with the capsid, or it may be loose fitting. Although very simple in their construction, viruses can display an astonishing variety of shapes. Amongst the most complex of virus particles is the T-even bacteriophage that infects *Escherichia coli*. These have been said to resemble a lunar landing module (Fig. 2.1).

Viruses are obligate intracellular parasites. This means that to complete their life cycles, viruses depend absolutely on functions provided by cells of a host organism. Not all viruses can be grown in artificial culture. If this can be achieved, the growth of viruses in laboratories requires exploitation of tissue culture technology. When outside a host cell, the only thing a virus can do is to infect a new host. Once cellular infection has been accomplished, however, viruses become highly active and subvert the normal metabolic activity of the cell to produce new virus particles.

Following attachment to and penetration of a new host cell, all viruses show the same general life cycle (Fig. 2.2). Viruses induce the biosynthesis of virus proteins and replication of the virus genome. Typically, the enzymes that are required for virus replication are made

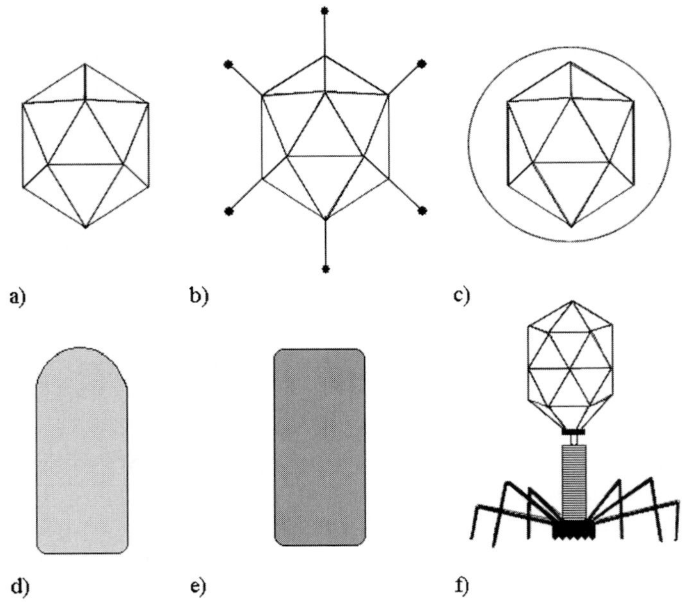

Fig. 2.1 Representative virus particles. Human viruses (a–e) are relatively simple when compared with the T-even bacteriophage that infects *Escherichia coli* (f). Many virus types are icosahedral (a). Icosahedral viruses include enteroviruses and rotaviruses, used in the virological examination of water. Adenoviruses (b) are icosahedral viruses that carry spikes on their surfaces. Some viruses, such as the herpes virus (c) have an envelope around an icosahedral core. Rabies viruses are bullet-shaped (d) and pox viruses are brick-shaped (e). These pictures are not drawn to scale; the smallest human viruses are less than 20 nm in size while the largest are over 400 nm.

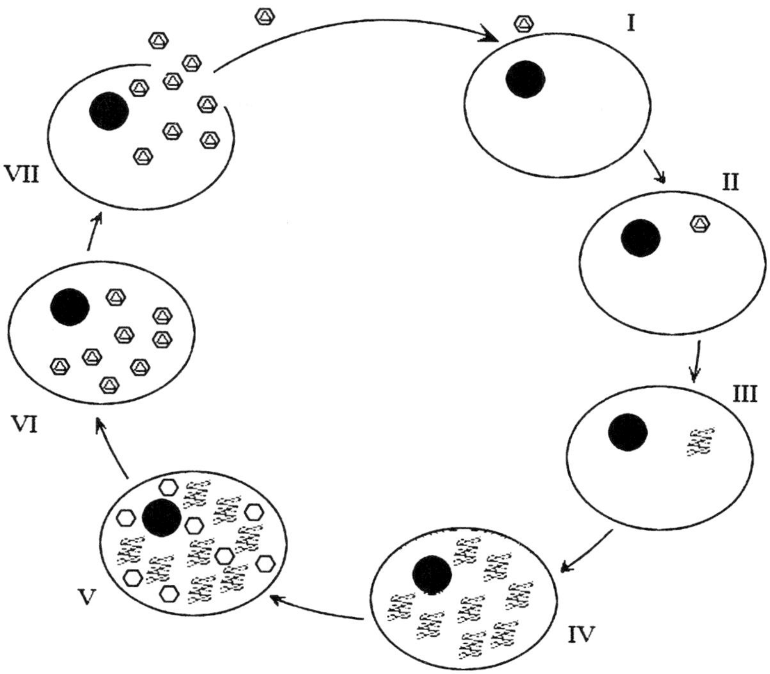

Fig. 2.2 Representative virus life cycle. The enterovirus virion approaches the surface of the cell to be infected (I). It penetrates the cell to infect it (II), and virus RNA passes into the cytoplasm as the virus is uncoated (III). Non-structural proteins are formed during early translation. The virus genome is replicated (IV) and new virus structural proteins are formed (V). New virus particles are assembled (VI) and released by lysis of the infected cell (VII). Reproduced with kind permission from Heritage, J., Evans, E.G.V. and Killington, R.A. (1996). *Introductory Microbiology*, Cambridge University Press.

soon after the virus infects a new host cell and these are referred to as 'early proteins'. The structural proteins that make up the virus capsid are produced later after infection and are thus referred to as 'late proteins'. The assembly of new virus particles and the subsequent release of progeny follows the biosynthesis of virus components. This may result from the complete destruction or lysis of the host cell. Alternatively, new virus particles may 'bud out' from their infected host cell. Viruses that have an envelope sometimes use this strategy, the envelope material being acquired as the newly emerging particles pass through the cell membrane. Not all virus envelopes are derived from the cell membrane. If virus replication occurs within the nucleus of the cell, the envelope material may be derived from the nuclear membrane within the infected cell.

Although the overall pattern of attachment, penetration, replication and release is constant for all viruses, a wide array of strategies has evolved so that viruses can exploit their hosts. For example, enteroviruses can cause the synthesis of messenger RNA from RNA rather than DNA genome templates. The human immunodeficiency virus can make a DNA copy of its RNA genome. The vaccinia virus used for smallpox vaccination can cause the replication of DNA in the cytoplasm of its host cell rather than in its nucleus. There are many other variations of the virus life cycle theme.

All classes of organism have been subject to virus infection. The viruses that infect bacteria are known as bacteriophage and these have recently been exploited as biological markers, for example to trace the source of polluted water. Plant viruses may cause effects that horticulturists find appealing. For example, the striped appearance of tulip flowers prized by many gardeners results from a virus infection.

Animal viruses cause a spectrum of diseases. These range from sub-clinical, where the host is unaware of the infection, through to fatal conditions.

3 VIRUS INFECTIONS THAT SPREAD VIA THE FAECAL–ORAL ROUTE

Viruses that cause waterborne infections are of devastating importance. Viruses are the commonest cause of gastrointestinal infection world-wide and rotaviruses are the most important cause of life-threatening diarrhoea in children under 2 years old. About one-third of all children have experienced an episode of rotavirus diarrhoea by the time they are 2 years old and nearly all children are infected by this age, even though a significant minority show no clinical symptoms. Rotaviruses alone account for 20% of infectious diarrhoea cases in developing countries. In the developed world, rotaviruses are a significant problem for hospitalized patients of all ages. Over half of children suffering from dehydration who are admitted to hospital in the UK are suffering from virus gastroenteritis. Outbreaks of rotavirus diarrhoea are not uncommon in hospitals in the UK. There have been notorious cases associated with cruise liners where a very large number of passengers have developed diarrhoea and vomiting during their holiday. These viruses are particularly infectious. Volunteer studies have shown that fewer than 10 virus particles may cause clinical disease. This makes elimination of the infectious agent very difficult.

Although rotaviruses are a significant cause of gastrointestinal symptoms, there are a number of other viruses that are associated with diarrhoea and vomiting. These include caliciviruses, astroviruses and certain adenoviruses. These viruses are more often associated with infections in older children and adults than in infants. Infections caused by these viruses may show a remarkable seasonal variation in temperate climes. This is reflected in the folk name given to the virus-mediated 'winter vomiting disease'. The most notorious of the caliciviruses causing gastrointestinal disease is the Norwalk agent, named after the Ohio town from which it was first recognized. It causes muscle pain, headache and nausea as well as profuse vomiting and diarrhoea, but the symptoms generally resolve very quickly.

Most of the viruses that cause diarrhoea cannot be grown in tissue culture. They may be visualized in infected faeces by electron microscopy but this is an expensive and specialized technique that is not routinely performed. Latex agglutination techniques can also be used to detect viruses that cause diarrhoea. Latex beads are coated in antibodies specific to the virus to be identified. These are then mixed with a sample of faeces to be tested. If virus particles are present, these will react with the antibodies, causing the latex particles to clump. More frequently, the diagnosis of virus gastroenteritis is made by excluding other causes for the symptoms.

Not all waterborne viruses cause symptoms of gastroenteritis. There are two other important groups of viruses that cause waterborne infections: hepatitis viruses and enteroviruses. A number of viruses are known to cause hepatitis. Clinically, the disease processes seem indistinguishable but the viruses that cause hepatitis have different structures, life-cycles and modes of transmission. Two hepatitis viruses are particularly associated with waterborne infection – hepatitis A and hepatitis E virus. Hepatitis A virus, a small RNA virus, causes acute infective hepatitis. Symptoms of the disease are of rapid onset but the condition generally resolves within a relatively short time. Cases often occur in large clusters. Patients become overwhelmingly tired in the early stages of the illness. They develop a fever and may suffer diarrhoea. Appetite is suppressed and patients complain of feeling awful. About one-third of adults infected with the hepatitis virus experience no further symptoms. The remaining two-thirds go on to develop jaundice, lasting from 1 to 3 weeks. Once the first sign of jaundice appears patients start to feel much better. Progression of the illness is much less common in children with only about one in twelve developing jaundice. Faeces from a patient with acute infective hepatitis caused by hepatitis A are highly infectious. Hepatitis E virus, another RNA virus, is a very rare cause of hepatitis. Like hepatitis A, infection with hepatitis E is associated with exposure to contaminated water and it is most common in the Far East. It, too, causes a relatively mild self-limiting illness, except in pregnant women. The hepatitis E virus can cause a life-threatening infection in such cases.

The hepatitis A virus was at one time classified together with the enteroviruses. It is now considered to be sufficiently different from the enteroviruses to be classified separately. The enteroviruses are a large family of viruses that are spread by water and that cause illness with extra-intestinal symptoms. In tropical countries enterovirus infections are seen throughout the year but in countries with a temperate climate the incidence of enterovirus infection shows a marked seasonal variation with most cases occurring in summer and autumn; at a time when exposure to contaminated water is most likely. The most important of the enteroviruses is the poliomyelitis virus. As well as causing waterborne infections this virus may be transmitted by the respiratory route. Infected faeces are, however, a very significant source of poliomyelitis virus. In most people who are infected with this virus, it causes a mild sore throat and, perhaps, a headache. In about 5% of patients, these symptoms progress to meningitis. The symptoms of severe headache and intolerance of bright light can last from 2 to 10 days but the meningitis then spontaneously resolves. In a very few patients, probably about 0.1%, the infection causes paralytic poliomyelitis. This is an infection that affects the primary motor neurons, causing paralysis and muscle wasting. This damage is not a result of infection of the muscle cells but is secondary to the effects of virus infection of the motor neurones. If the respiratory muscles are infected then the victim is condemned to a life in an iron lung. An effective vaccination programme has almost eradicated poliomyelitis from the globe.

Two other important groups of enterovirus exist. The ECHO viruses cause meningitis. This condition is not uncommon but is much less severe than bacterial meningitis and patients typically make a full recovery from enterovirus meningitis. The Coxsackie viruses infect muscles. Coxsackie B viruses have a predisposition for infecting heart muscle, causing progressive damage. Many heart transplants have been made necessary as a consequence of Coxsackie B infection. These viruses have been associated with a number of other extra-intestinal infections, including respiratory infections and conjunctivitis, although these are generally self-limiting.

Sewage is the main source of virus contamination of water. Raw sewage contains a very large number of viruses and even treated effluent may carry some viruses. Groundwater has only occasionally been found to be contaminated with human viruses. Treatment of potable water removes infectious viruses and the presence of viruses in tap water is an indication of the failure of water treatment processes. Lake and pond water may contain viruses. This depends, at least in part, on exposure to sewage. The occurrence of human viruses in marine and estuarine waters reflects the load of the rivers that feed these bodies of water. Viruses may be found in seawater in the absence of bacterial faecal indicators. If rivers have a large effluent input then the waters they feed will have a correspondingly high virus load.

Shellfish are an important potential source of human virus infection. Many shellfish that we consume are filter feeders. They extract their microscopic food from water by pushing vast quantities of water across their filter gills. There the food particles become trapped in mucus and the beating of cilia on the surface of gills propels food into the gut proper. If filter feeding shellfish live in water that is contaminated with human sewage, then they can very easily concentrate virus particles that are present in the water. In 1998, in a single incident, 61 people in Florida were infected with the hepatitis A virus after eating shellfish that had been exposed to human sewage at very low levels. Although the bacterial examination of shellfish is routine, there is currently no reliable and reproducible method for assaying shellfish for the presence of human viruses.

4 VIRUSES USED AS FAECAL INDICATORS

As with faecal indicator bacteria, viruses chosen to demonstrate faecal contamination of water are chosen because of their resilient properties and their association with human infections. Faecal indicator viruses can survive for long periods in contaminated water. Two groups of viruses are routinely sought as faecal indicators, providing evidence that water is contaminated with human faeces and is thus unsafe. These are the **enteroviruses** and the **rotaviruses**. Although representatives of these two groups may be responsible for mild or even subclinical infection, the enteroviruses are associated with serious and even life-threatening infections and rotaviruses are the most common agents responsible for life-threatening diarrhoea in children under 2 years old worldwide. These infections can be spread by the faecal–oral route, with contaminated water a frequent vector of infection. Rotaviruses are responsible for infections that display the typical gastrointestinal symptoms of vomiting and diarrhoea. In contrast, enterovirus infections are often characterized by extra-intestinal symptoms. The enteroviruses and rotaviruses thus represent useful markers of human faecal contamination of water. Enteroviruses are, in general, much easier to culture than are rotaviruses. There are currently four groups of rotaviruses recognized by their serological reactions. There are over seventy serotypes of enterovirus.

If both enteroviruses and rotaviruses are found in a water sample, this is taken as evidence of gross faecal contamination. When natural waters are examined it is more likely that enteroviruses rather than rotaviruses will be isolated. This is, in part, a reflection of the relative ease with which each of the virus families can be recovered. It also indicates the relative prevalence of these viruses in human faeces and also their differential survival properties.

5 TISSUE CULTURE TECHNIQUES USED IN THE EXAMINATION OF WATERBORNE VIRUSES

Since viruses are obligate intracellular parasites, they must be provided with living cells if they are to replicate. In the laboratory, viruses are often cultivated in tissue cultures. Cells can be removed from animal tissues and can be grown in one of a number of growth media at a suitable temperature and under appropriate atmospheric conditions. Cells in tissue cultures attach to the glass or plastic surface of their culture vessels and spread to form a cell monolayer. In many tissue cultures, cell contact within the monolayer causes an inhibition of further growth. Cells harvested from embryonic tissues tend to grow better than do cells from adult tissues. Once a monolayer is formed, its cells may be harvested and diluted to seed subsequent monolayers. This process is known as passaging. Tissue cultures can typically be maintained for a limited number of passages. Embryonic cells may be passaged up to 100 times. At high passage numbers, such cell lines gradually lose the ability to become infected with viruses. Cancer cells are particularly well suited to tissue culture. If the cells of tumours are maintained in tissue culture, they can be grown indefinitely. They are also not subject to the contact inhibition that causes the formation of monolayers. Once a single layer of tumour cells has developed, the culture will continue to accumulate in layers as cells pile up on top of one another.

Animal cells have exacting requirements for their growth media. In essence, however, the purpose of the growth medium is to provide an environment in which the tissue culture cells can flourish and divide. To do this they must be in an isotonic state where the osmotic pressure of the medium matches that within the cells. The pH must be maintained by a buffer system so that the acid balance suits the cells in culture. Tissue culture media include neutral red as a pH indicator. This is why healthy tissue cultures have a pink colour, whereas infected tissue cultures turn bright yellow. It is important to maintain tissue cultures in the dark since the neutral red indicator becomes toxic to cells if exposed to the light.

It is necessary to provide cells in culture with all their nutritional requirements. The osmotic balance of culture media is provided by a solution of different inorganic salts, referred to as the basal salts solution. If tissue culture cells are to be rested, then they may be maintained in a basal salts medium, provided that it is appropriately buffered. Growth media must contain all the nutrients necessary for cell development. These include vitamins as well as glucose and amino acids. Growth media for tissue cultures include a serum component if cells are to be maintained through a number of passages. Sera may be derived from a number of sources. These include human serum, rabbit serum, calf serum and fetal calf serum. The serum component provides cells in tissue culture with numerous growth factors necessary for the long-term maintenance of the culture. Tissue cultures are typically incubated at 37°C and under an atmosphere where the concentration of carbon dioxide is elevated to between 5 and 10%. This is necessary to provide an environment that models that inside the mammalian body.

Tissue culture media meet the very exacting growth demands of animal cells. Microbes may be much less demanding in their growth requirements. Hence microbial contamination of tissue cultures is a significant problem. Consequently, great attention must be applied to aseptic technique when setting up and maintaining tissue cultures. The risk of microbial contamination may be further reduced by the addition of antibiotics to the tissue culture medium. The antibacterial antibiotics penicillin and streptomycin are routinely added to media used for tissue culture. Together, these two antibiotics have a very broad antibacterial spectrum. They also have a synergistic effect. This means that their combined activity is greater than the additive action of each drug. Care must be employed when using antimicrobial supplements in tissue culture media. They may have an adverse effect on the growth of the cells in culture or they may be detrimental to the yield of virus from the supplemented culture. Although these antibiotics are antibacterial, it is necessary to use them at concentrations that are not toxic to the animal cells in culture. Typically 100 Units of penicillin G and 0.1 mg of streptomycin per millilitre of medium are used. Antifungal agents, such as nystatin at a concentration of 25–50 μg/ml, are also commonly employed in tissue cultures.

Viruses can be grown in a wide array of vessels, depending upon the requirements of particular protocols. If small volumes of cells are required for experiments that need many replicates, then microtitre trays are frequently the container of choice. These are plastic trays measuring approximately 10 cm by 12 cm and containing eight rows of twelve wells: 96 wells in total. Each well will hold slightly more than 300 μl. Alternatively, if large numbers of cells are required, for example to propagate a stock culture of virus, then a Winchester bottle can be used. These are large round bottles that have a 2.5 litre capacity. When such vessels are used, a relatively small volume of culture medium is placed in the seeded bottle. During incubation the bottle is slowly rotated so that its walls will be regularly bathed in the medium. Cells grow as a monolayer on the glass of such bottles.

Viruses can only infect a limited range of cells. This phenomenon is referred to as tropism. It is a consequence of the nature of the infection process. For a virus to infect and penetrate a cell it must first attach to the surface of its target host. To do this, specific receptors on the cell surface must interact with anti-receptors on the virus surface. Infection of the host will only occur if an appropriate interaction between receptor and anti-receptor has occurred, leading to a specific attachment and subsequent penetration of the target host cell. Tropism is a phenomenon that is associated with the attachment and penetration of viruses into host cells. If this step is experimentally by-passed, then cells that cannot normally act as hosts to particular viruses have been shown to support virus replication. For example, poliomyelitis virus is only capable of infecting primate cells under normal circumstances. If, however, virus particles are introduced into mouse cells using microinjection techniques, then the virus will replicate as if it were in a primate cell.

In tissue culture, enteroviruses undergo a complete infectious cycle. Viruses infect cells in the culture, they replicate to form new virus particles and then the infected cell bursts to release new virus particles. The enteroviruses can thus easily be detected using a simple plaque assay. In contrast, the rotaviruses are not capable of undergoing a complete cycle of infection in tissue cultures. Rather, these fastidious viruses infect cells to produce viruses that lack an outer capsid. For this reason, cells in tissue culture that have been infected with rotaviruses are detected using an immunological assay to detect the presence of virus antigens associated with the VP6 antigen of the inner capsid.

Liquid culture can be used for the detection of enteroviruses. Cells in suspension are exposed to samples that may contain virus particles. Enteroviruses will infect cells to cause characteristic changes in the cell morphology. If a number of cultures are infected with aliquots of a sample then reference to published tables can be used to determine the most probable number of virus particles present in the medium. More commonly, simple plaque assays are used to enumerate the viruses in water. This permits slower growing viruses to be detected as well as those that can grow easily and rapidly in tissue culture.

6 PREPARATION OF WATER SAMPLES FOR VIROLOGICAL EXAMINATION

The procedures described in the rest of this chapter relate to the examination of water samples as carried out in the UK (Fig. 2.3). Other methodologies are used elsewhere but no single method has been shown to be superior to other methods. The methods described below are technically demanding and are relatively expensive. Unlike bacteriological testing of water, the examination of water samples for viruses is not undertaken as a routine procedure on a large number of samples. Care must therefore be taken when interpreting the results of such assays.

Water samples should ideally be processed as soon as possible after sampling. It is recommended that samples should be processed within 24 hours of collection and after 36

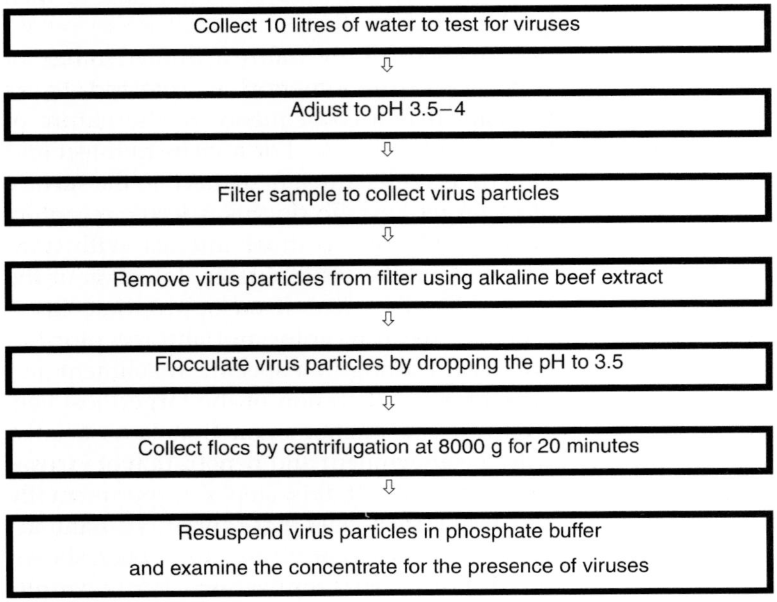

Fig. 2.3 Flow chart for the concentration of virus particles in water samples.

hours if a sample has not been processed it must be discarded as worthless. If they cannot be processed immediately, water samples should be kept cool and in the dark. Preferably they should be maintained at temperatures below 10°C. They should not, however, be frozen.

Samples of water for virological examination may contain very small numbers of virus particles. It is thus necessary to treat the samples to concentrate any viruses that may be present so that they may be detected more easily. If they cannot be tested immediately, sample concentrates may be stored frozen. Concentrates may be frozen at −20°C but if lower refrigeration temperatures are available so much the better. All samples must be properly labelled throughout the examination procedure. If the water to be tested has been previously chlorinated – for example mains water in the UK – then the residual chlorine must be inactivated by the addition of sodium thiosulphate to a final concentration of 18 mg/l.

Typically, 10 litre samples are collected and the pH of the sample is adjusted to lie between pH 3.5 and 4. This is achieved by adding 2 M hydrochloric acid to the sample and monitoring the pH with an accurate pH meter. The suggested method is to take a small subsample of known volume and calculate the volume of hydrochloric acid needed to lower its pH to 3.5. The volume of hydrochloric acid required to treat the whole sample may thus be calculated. At a pH value of 3.5, protein material, including virus particles, will adhere to nitrocellulose. The entire sample is then passed through a nitrocellulose filter with a 45 micron pore size to trap any virus particles that may be present in the sample.

To remove the trapped virus particles, the test filter is treated with 3% (w/v) alkaline beef extract solution (pH 9.5). Not all beef extract can form flocs in acid so it is essential to test each batch that may be used in this procedure for its ability to form flocs. A satisfactory batch will produce a visible precipitate when the pH is lowered to pH 3.5. The beef extract is sterilized by autoclaving prior to use and the pH is raised to pH 9.5 only after autoclaving.

For each 10 litre water sample, 300 ml of alkaline beef extract is reacted with the filter for 2 minutes. After agitation the virus particles become resuspended in the alkaline beef extract but they require further concentration before they may be easily detected. Skimmed milk made up as a 0.5% (w/v) solution in sterile water may be used in place of beef extract. The pH of the beef extract virus mixture is reduced to pH 3.5 by the stepwise addition of 2 M hydrochloric acid. Under these conditions, virus particles within the sample will flocculate and the flocs can be collected by centrifugation at 8000 g for 20 minutes. The supernatant beef extract is decanted from the floccular material containing the virus particles. These are resuspended in 10 ml of a 0.15 molar sodium phosphate buffer at pH 8. This procedure concentrates virus particles 1000-fold and the concentrate is then ready for use in subsequent virus assays. This may be used immediately. Alternatively the concentrate may be frozen and stored for examination in the future. Approximately one-third of the concentrate may be used to test for the presence of rotaviruses: the remainder is available to assay for enteroviruses.

Concentration of water samples using the method described above inevitably leads to the loss of virus particles from the sample. Loss of virus particles may occur at each step of the concentration process. When examining water for its virus content, each experimental run includes a control in which a sample of sterile water is seeded with a known number of virus particles. For the examination of enteroviruses, control samples are typically seeded with 10^6 particles of an attenuated strain of poliomyelitis virus. Attenuated viruses are capable of normal growth in tissue culture but they are incapable of causing clinical disease in humans. The Sabin poliomyelitis vaccine, for example, exploits a live, attenuated virus to confer immunity. The control sample is put through the identical concentration procedure to the test samples. The number of particles recovered from the control sample is then compared with the number that was used in the seeding. From this comparison the percentage recovery of virus particles can be calculated:

$$\frac{\text{Virus particles recovered}}{\text{Virus particles input}} \times 100\%$$

When using the concentration procedure described above the recovery of virus particles is typically between 60 and 70% (Fig. 2.4). When reporting data from the virological examination of water it is often the practice to report the raw data together with the results from the control experiment. This is so that the person commissioning the test can calculate the actual number of viruses in the sample. Alternatively, the tester may calculate the projected number of particles in the original sample. Whichever approach is used it is essential to make clear in the final report whether the count is adjusted or not.

Water samples typically require concentration if viruses are to be recovered, since virus particles are not usually present in water in large numbers. For sludge and samples that contain solid particulate material it is important to dilute samples so that there is no more than 4% (w/v) solid material in the sample. Aluminium chloride, added to a final concentration of 0.5 mM, may be added to aid the flocculation of virus particles in the sample.

Enteroviruses and rotaviruses are almost universally present in sewage. These are both very common virus families and their numbers in raw sewage can be very high. The virological

A 10 litre sample of sterile water is seeded with 10^6 particles of poliomyelitis virus. The control is then subjected to the same concentration protocol as test water from a reservoir suspected of being contaminated with human faeces. A serial 100-fold dilution of the resultant concentrate was made and these dilutions were used in a plaque assay for enteroviruses (poliomyelitis virus is an enterovirus). The following numbers of plaques were observed:

Dilution	Plaque forming units (pfu) per plate
100	Too many to count
10^{-2}	Too many to count
10^{-4}	69

The number of particles recovered is thus the number of plaques seen multiplied by the dilution factor:

$$69 \times 10^4$$
$$=6.9 \times 10^5$$

As 10^6 particles were used to seed the control the percentage recovery in this experiment is:

$$\frac{6.9\times10^5}{10^6}\times100\%$$

= 69%

Fig. 2.4 Sample calculation for the recovery of virus particles from a control sample of water.

examination of sewage is not often undertaken and is probably only important when studying the progress of vaccination programmes, e.g. to monitor the eradication of poliomyelitis virus or if studying the efficacy of novel treatment programmes. Once a sewage sample has been homogenized it is not necessary to subject it to a concentration procedure to isolate viruses of faecal origin. Rather, it is more common practice to use a dilution procedure to detect faecal viruses in samples of sewage. Bacteria and other microbes in the sewage will flourish when inoculated onto a tissue culture. These will outgrow the animal cells and will thus destroy the tissue culture. If microbial growth is not contained, viruses will be unrecoverable from sewage. Dilutions of sewage material are typically made in phosphate buffered saline and the diluted samples are treated with antimicrobial agents. These generally include a β-lactam antibiotic, such as one of the penicillins, together with an aminoglycoside, such as streptomycin. These will kill bacteria in the samples. An antifungal agent is also used to treat the diluted sewage sample before it is used to inoculate a suitable tissue culture.

7 EXAMINATION OF WATER FOR THE PRESENCE OF ENTEROVIRUSES

Enteroviruses are detected in water samples using a tissue culture plaque assay. Primate cells are used as hosts for these viruses. BGM cells or human diploid fibroblasts are the cell lines most commonly used in the laboratory culture of enteroviruses. Other cell lines may be used at the preference of the local virologist. These include PMK, Vero, FL HeLa and Hep-2 cells. Using more than one cell line can improve the detection of human viruses in water concentrates. Tissue cultures are initiated in 75 cm^2 tissue culture flasks and cells are passaged every 5 days. After five passages, the tissue cultures are transferred from stationary flasks to roller bottle cultures. In the examination of water samples for enteroviruses BGM cells are only used between passages 75–120. During this time they have maximal susceptibility to enterovirus infection. With all virus assays, it is important to include both positive and negative controls in the experimental protocol.

Cells from a roller bottle culture are harvested by replacing the normal culture medium with a medium containing 0.05 mg/ml trypsin and incubating for one hour. This loosens the cells from the glass and makes them more susceptible to virus infection. Once the cells have been harvested, the culture yield is determined by counting the cells. This permits a calculation of the volume of cells necessary to initiate a plaque assay. The cell count is undertaken using a counting chamber such as the Improved Neubauer counting chamber shown in Fig. 2.5.

Plaque assays are conventionally carried out in Petri dishes. For enterovirus assays of water samples, 90 mm Petri dishes are used and each dish is seeded with approximately 3×10^7 cells mixed with 15 ml of growth medium. A sample calculation to determine the number of cells in a tissue culture and to determine the volume of cells is shown in Fig. 2.6. This calculation assumes that an Improved Neubauer Counting Chamber is used to determine the cell number. This counting chamber was originally developed for haematology laboratories. Other counting chambers have been developed for different purposes. They all operate on the same principle: determination of the number of particles in a fixed volume. The calculations will, however, require substitution of different factors, depending upon the fixed volume in which particles are counted.

In a plaque assay, a cell monolayer is infected with virus and the culture is overlaid with a semisolid medium to prevent the lateral spread of virus particles. The virus particles within the sample to be tested infect cells within a cell monolayer. Infected cells then produce new virus particles and lyse to release their progeny. These newly released virus particles then infect adjacent cells that subsequently lyse to release even more virus particles. These cannot spread through the tissue culture since they are restrained by a semisolid overlay or by agar within the growth medium, depending upon the restraining method chosen. Consequently, discrete holes appear within the cell

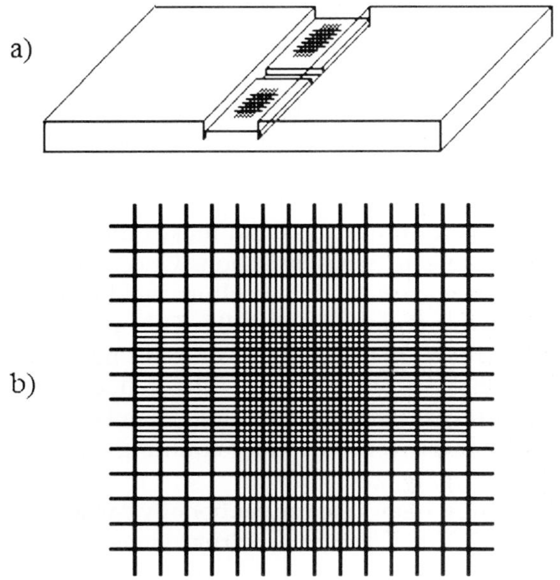

Fig. 2.5 The Improved Neubauer Counting Chamber. This counting chamber (a) is slightly larger than a conventional microscope slide and has indentations for two specimens. In the centre of each chamber is an accurately etched grid (b). The total area of the 25 larger squares in the centre of the grid is 1 mm^2. The sample to be counted is introduced into the chamber and a coverslip is placed over the sample. When placed correctly, the contact between the clean glass of the chamber and the coverslip causes the formation of 'Newton's rings': a rainbow effect between the coverslip and the glass on ether side of the counting chamber. If Newton's rings are seen, the height of the coverslip from the grid is 0.1 mm. The volume of fluid trapped over the central squares of the grid is thus 0.1 mm^3. This is equivalent to 10^{-4} ml. By counting the number of objects within the 25 central squares and multiplying by 10^4, the number of objects per ml can be obtained after taking into account any dilution factor.

monolayer, each originating from one infectious virus unit. These holes are known as plaques and the infectious virus units are referred to as plaque-forming units.

When assessing the number of enterovirus particles in a concentrated water sample, in each assay dish 3×10^7 cells are mixed with 15 ml of growth medium containing 2.4% agar that has been previously cooled to 45°C. The mixture is then used to seed a sterile 90 mm tissue culture quality triple-vented Petri dish. The purpose of the agar is to prevent the spread of newly released virus particles. This causes plaques to develop as discrete entities. A total of 6.5 ml of the concentrated water sample is split between three assay plates. The contents of each of the Petri dishes are then carefully mixed and the tissue cultures are incubated at 37°C in an atmosphere of 5% carbon dioxide in air for 48 hours. The counts resulting from each of the three assays should be of the same order. If one of the three tests differs significantly from the others, then the validity of that assay must be questioned and the sampling must be repeated. This acts as an internal control for the assay procedure. By counting the number of plaques present, the number of virus particles present in the original sample can be calculated and the results presented. A sample calculation and reports are given in Fig. 2.7.

It is necessary to confirm that the plaques that are seen are due to virus activity rather than to a toxic reaction within the tissue culture. Plaques are subcultured to ensure the presence of live virus. Normally, no attempt is made to differentiate viruses isolated from plaque assays. No rapid, inexpensive method is available to identify enteroviruses to serovar level and the presence of any enterovirus is evidence of the faecal contamination of the original sample. It may, however, be desirable to determine which viruses are present in a given sample. For example, if a particular body of water is epidemiologically linked with a particular virus infection, then it may be beneficial to identify the virus isolated from the sample to subfamily level. There are a number of techniques available to differentiate enteroviruses.

A variety of serological methods of identifying enteroviruses may be used but, as there are over 70 distinct serovars of enterovirus, this method requires over 70 antisera to identify fully the isolate in question. In practice, isolates can be identified to group level with a much smaller panel of antisera. Serological techniques used in enterovirus identification include indirect immunofluorescence; enzyme linked immunosorbent assays (ELISA) and neutralization tests. ELISA tests work on a similar

The picture below is a representation of the central squares of an Improved Neubauer Counting chamber with cells from a tissue culture to be used to seed an enterovirus plaque assay. This was derived from a 1:50 dilution of the original cell harvest:

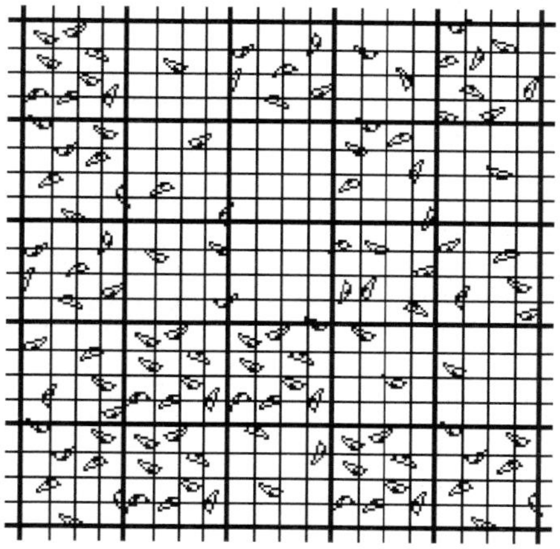

The cell counts in each square may be tabulated thus:

8	1	6	2	8
7	3	0	6	1
6	3	0	6	5
5	8	9	2	1
7	8	3	8	7

There are a total of 120 tissue culture cells within the grid. In the Improved Neubauer Counting Chamber, this grid entraps a volume of 10^{-4} ml. Hence the sample as counted represents a cell count of 120×10^4 cells/ml. The sample counted, however, was a 1:50 dilution of the original cell suspension. Hence, the number of cells in the original suspension would have been $50 \times 120 \times 10^4$ cells/ml. This is equivalent to a count of 6000×10^4 cells/ml. This is more properly written as 6×10^7 cells/ml.

As 3×10^7 cells are required to seed the enterovirus plaque assay, each 90 mm Petri dish requires 0.5 ml of the original cell suspension.

$$\frac{6\times10^7}{3\times10^7}=\frac{1}{2}$$

Fig. 2.6 Sample calculation to decide how many cells to seed for an enterovirus plaque assay.

A party had enjoyed an outward-bound weekend at the Chaldicotes Centre on the River Ouse. This involved participation in water sports, including the practise of a capsize drill. Inevitably, most of the group ingested some of the river water and subsequently a number of the party became unwell. The main symptoms of the illness were a fever and gripping chest pains. One of the party was a microbiologist who suspected that they were suffering from a Coxsackie B virus infection. She considered faecal contamination of the River water to be the common epidemiological link. A virological examination was commissioned and enteroviruses were, indeed, recovered from the water sample examined.

The sample of river water was concentrated and 6.5 ml of concentrate was used in a plaque assay for enteroviruses. A control sample was examined in parallel, as described. This control established that 69% of enterovirus particles were recovered after the concentration procedure. The 6.5 ml of concentrated water sample taken from the River Ouse was divided between three plaque assay plates. The first plate yielded 27 plaques, the second 35 and the third plate 13 plaques. The total count was thus 75 plaque forming units.

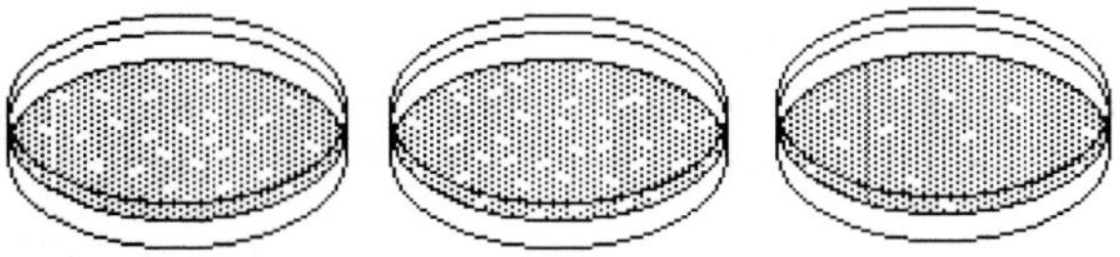

As 6.5 ml from a total of 10 ml of concentrate was used, there would be

$$\frac{75}{6.5}\times 10$$

= 115 virus particles in the whole of the concentrated sample. Since there was a 69% recovery of the control virus, the report may be presented as follows:

There was 115 plaque forming units of enterovirus per 10 litres isolated from the water sample at Chaldicotes on the River Ouse.

Recovery of enteroviruses in a control experiment was 69%.

This is indicative of a total enterovirus load of 167 pfu/ 10 litre, allowing for loss of virus particles during processing.

Fig. 2.7 Sample calculation and specimen report having determined the number of enteroviruses in a water sample.

principle to immunofluorescence but use antibodies tagged with an enzyme rather than a fluorescent dye. The enzyme catalyses a reaction that yields a coloured product. The intensity of the colour produced indicates the strength of the ELISA reaction. In neutralization assays, virus particles are reacted with an array of antisera before being exposed to tissue culture. The antiserum that prevents plaque formation is specific for the virus being investigated. There are currently no standard typing systems based upon the analysis of the virus genome. Enteroviruses readily undergo mutation and the instability of the genome makes nucleic acid-based typing too unreliable for the unambiguous identification of an isolate.

8 EXAMINATION OF WATER FOR THE PRESENCE OF ROTAVIRUSES

In contrast to the plaque assay used to enumerate enteroviruses, when rotaviruses are detected, a cell monolayer is infected and virus particles are allowed to grow within the tissue culture and the cells are subsequently fixed and subjected to indirect immunofluorescence. LIC-MK2 or MA-104B cells are most often used for the culture of rotaviruses. Rabbit serum containing antibodies that react specifically with rotaviruses is then added to the tissue culture. If there are virus particles present within the fixed cells, these will react with the antibodies in the rabbit serum and will trap the antibodies within the fixed cells. The culture is thoroughly washed to remove any unbound antibody and is then flooded with fluorescently labelled anti-rabbit antiserum. The tagged antibodies in this antiserum will react with the rabbit antibodies that have attached to the rotavirus particles. When viewed under ultraviolet light, infected cells will be able to fluoresce and appear brightly coloured. This is because the rotavirus particles will trap the rabbit antibodies that in turn will trap the fluorescently labelled anti-rabbit antibodies. Uninfected cells are unable to fluoresce, as they have no means of trapping the rabbit antibodies (Fig. 2.8). This two-step indirect immunofluorescence technique is used rather than using direct fluorescent microscopy where tagged anti-rotavirus antibodies are used because it permits amplification of the signal. With direct fluorescence each virus particle is quickly saturated with antibody. When an indirect immunofluorescence is used, each virus particle will have numerous anti-rotavirus antibodies attached to it. In turn, each of these antibodies will provide binding sites for the anti-rabbit antibodies.

Human rotaviruses are more difficult to culture than are enteroviruses and they do not undergo a full infectious cycle in tissue culture. Trypsin treatment of the cells prior to infection is essential for the efficient propagation of rotaviruses in tissue culture. At one time, rotaviruses were grown in large-scale tissue cultures but now microtitre plates are used and small volumes of material may be examined.

Rotavirus assays are carried out in microtitre trays. These trays have eight rows of twelve wells each: a total of 96 wells per plate. Each well holds just more than 300 μl. A total of 1.5×10^5 cells in 100 μl of growth medium are

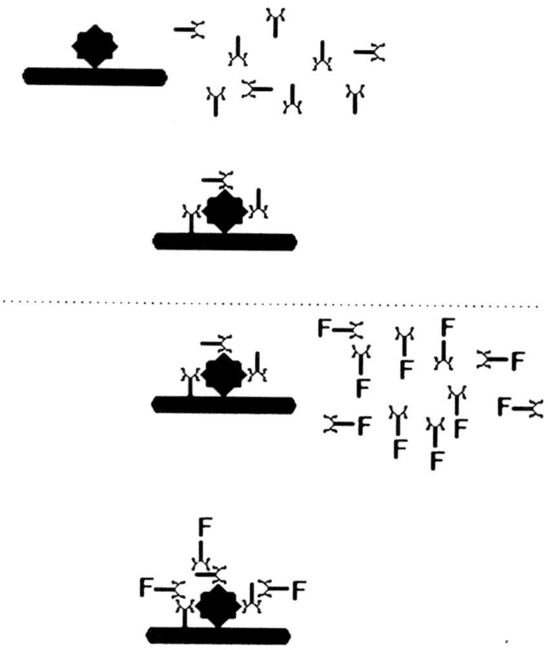

Fig. 2.8 Indirect immunofluorescence used to detect rotavirus particles. The rotaviruses within infected cells provide binding sites for anti-rotavirus antibodies within rabbit antiserum. These are allowed to react and the preparation is thoroughly washed to remove any unbound antibody. The preparation is then reacted with fluorescently tagged anti-rabbit antibodies. These bind to the antibodies that are already attached to the virus particles, thus amplifying the signal and making virus-infected cells easier to see when viewed in a fluorescence microscope.

added to each of the wells of a microtitre tray. The cells are incubated for one hour at 37°C under 5–10% carbon dioxide. This allows the cells to attach to the wells of the microtitre plate. After this, the growth medium is decanted off and is replaced with 100 μl of cell maintenance medium with 0.05 mg/ml of trypsin added. The trypsin is added to make the cells in the culture susceptible to rotavirus infection. After incubation at 37°C in elevated carbon dioxide for a further hour, thirty-two 100 μl aliquots of the concentrated water sample are added to individual wells in the microtitre tray. The water samples and trypsin-containing medium are incubated with the cells for another hour. The liquid in each well is replaced with 100 μl of maintenance medium and the microtitre tray is incubated for another hour to rest the infected cells. Finally, the maintenance medium is replaced with 100 μl of growth medium and the cells are incubated overnight.

On the following day growth medium is removed from the tissue cultures and the cells are fixed. This is achieved by washing the cells thoroughly with three changes of phosphate buffered saline then treating with ice-cold methanol or with absolute ethanol for at least 10 minutes. After ethanol fixation the cells are rehydrated with three washes of 250 μl of phosphate buffered saline. The cells are then ready for virus detection using an indirect immunofluorescence technique.

Fifty microlitres of a 1:500 dilution of anti-rotavirus antiserum raised in rabbits is then added to each well in the microtitre tray. This allows anti-rotavirus antibodies in the anti-serum to attach to virus particles within the fixed tissue culture cells. After incubation at 37°C for one hour the rabbit antiserum is removed from each well and the contents of the wells are washed with three washes of phosphate buffered saline. Fifty microlitres of fluorescently tagged anti-rabbit antiserum is added and the tray is incubated for a further hour. The antibodies in this antiserum will attach to the rabbit antibodies that are, in turn, attached to the virus particles. Using this indirect immunofluorescence technique the signal from each virus particle is amplified and the particles can be detected by fluorescence microscopy. The wells are washed with a further three changes of phosphate buffered saline before microscopic visualization. Plates may be stored at 4°C in the dark for up to 6 months without apparent loss of signal.

Rotaviruses only replicate in the cytoplasm of infected cells. Cells with fluorescent nuclei are thus not counted in this assay. Fluorescent cells with a fluorescent cytoplasm are assumed to be infected with rotavirus. As with the enterovirus assay a control sample is seeded with a known concentration of rotavirus so that the percentage loss during the concentration procedure can be calculated. The virus counts are reported as the number of fluorescent foci per 10 litres of water. It is again important that the report makes clear whether or not allowance has been made for the loss of particles during concentration of the water sample.

9 THE POLYMERASE CHAIN REACTION (PCR) AND DETECTION OF VIRUS PARTICLES IN WATER SAMPLES

There have been many advances in the use of molecular biological techniques in recent years. One of the most significant has been the development of the polymerase chain reaction, known as PCR for short. DNA is the molecule in which almost all genetic information is stored. Four different 'bases' code for genetic information. Groups of three bases on the coding strand of DNA specify which amino acids will become incorporated into proteins. These triplets also provide the punctuation of the code, indicating where protein translation should start and stop. DNA molecules have two strands. This enables it to be accurately replicated as each base can only form a stable pair with one other base: adenine pairs with thymine, guanine with cytosine.

DNA replication is carried out by enzymes known as DNA polymerases and is a chain reaction. Each of the strands of a DNA molecule acts as a template for replication. Thus one molecule of two strands is replicated to produce two molecules with a total of four strands. These are both replicated to give rise to four molecules with a total of eight strands, and so on. DNA replication is initiated when the two DNA strands become separated. A very short sequence of DNA known as an oligonucleotide primer, or just primer for short, then anneals to the strand that is to be replicated. Primers have a sequence that is complementary to their target strand. If the target has adenine, the primer will have thymine, and so on. The primer serves as the starting point for DNA replication and nucleotides are added to the new strand by the action of a DNA polymerase enzyme. DNA replication can take place *in vitro* when single stranded DNA is mixed with appropriate primers in a buffer containing all four nucleotides and magnesium ions together with DNA polymerase.

DNA molecules can easily be separated into their component strands by heating. At temperatures of about 95°C the bonds that maintain the base pairs that hold the two DNA strands together are broken and the strands separate. The strands may re-anneal if the temperature is dropped. The precise temperature of re-annealing depends upon the nucleotides present in the DNA strands. At about 60°C, however, melted strands of DNA can re-anneal to regenerate the familiar double helix structure.

In PCR amplifications, the template DNA is melted in a mixture that contains two primers: one for each DNA strand, and designed to anneal at either end of a target sequence. Primers are generally between 10 and 20 nucleotides long. This is sufficiently long to overcome the problem of primers binding at random to sequences other than the target DNA at the appropriate annealing temperature. This gives PCR a high degree of specificity. Lowering the annealing temperature sufficiently may permit non-specific binding and this may be exploited, for example in PCR typing techniques. This makes PCR a very versatile tool as well as potentially a very specific technique. The temperature of the reaction is lowered to allow the primers to anneal to the target DNA strands. DNA polymerase from *Escherichia coli* was used in the earliest PCR amplifications. DNA replication was thus carried out at 37°C. This was unsatisfactory as for each amplification step new enzyme had to be added to the reaction. This allowed the opportunity for contamination with DNA that could act as a target for amplification, giving rise to false positive results. It was also expensive in its use of DNA polymerase. A typical PCR amplification has between 20 and 30 rounds of replication. The technique was revolutionized when a thermostable DNA polymerase, isolated from a hot water bacterium, *Thermus aquaticus*, was introduced. This enzyme is stable at 95°C and replicated DNA at 72°C. Currently, PCR amplifications involve 20–30 cycles where DNA is melted at 95°C, primers are annealed at an appropriate temperature, typically about 60°C, and DNA is replicated by the *Taq* polymerase at 72°C. This leads to the production of a DNA amplimer of a predicted size, dictated by the positions of the primer pair. The first three cycles of the reaction are illustrated in Fig. 2.9.

PCR amplimers are DNA molecules and can be subjected to the same technologies as DNA isolated from organisms. They are most often detected by agarose gel electrophoresis. DNA molecules are charged and so will migrate when they are placed in an electric field. DNA samples can be loaded into wells in an agarose gel across which an electrical potential difference is applied. The DNA molecules will move out of the well and into the gel. The smaller the DNA molecule, the easier it will pass through the gel. In this way DNA molecules are separated according to size with the smallest molecules migrating the furthest. Having carried out an electrophoretic separation, the DNA may be visualized by soaking the gel in a very dilute solution of ethidium bromide. This dye can intercalate between the base pairs of a DNA molecule, causing it to fluoresce bright orange when illuminated by ultraviolet light. DNA amplimers can be identified and the distance they run can be compared to size standards. More sophisticated techniques may occasionally be applied to verify the nature of an amplimer. DNA carries restriction endonuclease recognition sites. These are points where restriction endonucleases cut DNA into fragments of predictable sizes. Restriction digest analysis can be used to confirm the structure of an amplimer if its nucleotide sequence is known. If the sequence of an amplimer is not known then the amplimer itself may be subjected to nucleotide sequence determination.

As well as amplifying DNA, the polymerase chain reaction may be adapted to detect the presence of RNA in a sample. This is very useful when trying to detect viruses with an RNA genome. In such cases, nucleic acid is extracted from the sample being examined and the RNA is acted upon by reverse

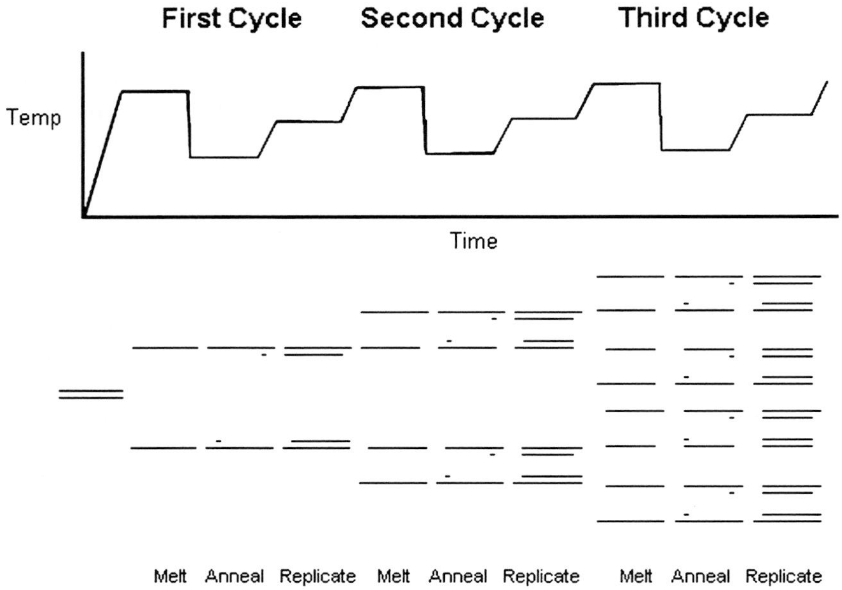

Fig. 2.9 The first three cycles of the polymerase chain reaction. The sample is heated to allow the target DNA to melt into its two separate strands. The reaction temperature is then dropped and primers bind to either end of target sequence, one primer on each strand of DNA. The reaction temperature is raised once more to permit DNA replication mediated by the thermostable *Taq* polymerase. By the end of the third cycle the reaction is dominated by DNA molecules of a size specified by the limits of the primer binding sites. Hence, when the PCR amplification is complete, the product overwhelmingly contains DNA molecules of this defined size.

transcriptase. This enzyme makes a DNA copy of an RNA template. The cDNA produced in this reaction can then subsequently be used as a template in a conventional PCR amplification. The polymerase chain reaction can also be used to amplify DNA from microorganisms that cannot be cultivated. In such cases, providing a unique protein can be characterized and a partial amino acid sequence can be determined, then a set of 'degenerate' primers can be constructed. More than one triplet can code for a single amino acid. Hence the degenerate primers must be designed to anneal with every possible nucleotide sequence that could encode the amino acids in the target protein. Having used degenerate primers to yield an amplimer, the nucleotide sequence of the amplified DNA can be determined in a standard sequence analysis. Specific primers may then be designed to amplify DNA from an organism that cannot be cultivated. The polymerase chain reaction is thus a highly versatile and sensitive technique for the detection of DNA, and, by implication, specific organisms. In theory it is capable of detecting single molecules and in practice it has been found reliable to detect microbes present in very small numbers within a sample. The only problem with using PCR technology to detect viruses is that it detects the total number of virus particles in a sample. It gives no indication of the proportion of virus particles that are potentially infectious. It may thus seriously overestimate the risk of exposure to a water sample.

10 THE USE OF BACTERIOPHAGE IN WATER TESTING

The examination of water for the presence of viruses need not necessarily imply a threat to human health. Humans may harness viruses in the examination of water. Every major group of organisms is susceptible to infection by viruses. This includes bacteria and the viruses that can infect bacteria are known as bacteriophage. These viruses have a very limited host range. Indeed, the susceptibility to infection by a panel of bacteriophage is a very useful method of typing some important bacteria to subspecies level. This specificity of host range can be exploited in the examination of water.

Serratia marcescens is a Gram-negative bacterium, strains of which are susceptible to bacteriophage MS2. If this virus is mixed with a suspension of bacteriophage-sensitive bacteria that are then plated as a lawn onto nutrient agar, the bacteriophage causes the appearance of plaques in the bacterial lawn after overnight incubation of the culture. The bacterial lawn is punctuated with holes, each representing an infectious bacteriophage particle. These plaques are analogous to the plaques seen in tissue culture monoloayers such as are counted in the enterovirus assay described in Section 7. This makes for the easy detection and enumeration of bacteriophage in samples.

Large quantities of bacteriophage can be cultivated by inoculating a broth culture of bacteria that are growing in late exponential phase. Bacteriophage will infect bacterial cells and thus they become replicated. The infected cells lyse to release new bacteriophage particles that can go on to infect other bacteria. The cloudy bacterial culture will become clear as all the bacterial cells become infected and are subsequently lysed to release large numbers of bacteriophage. The culture is centrifuged at low speed to remove cell debris and then at high speed to harvest virus particles.

Bacteriophage MS2 has been used to study the flow of water and other fluids. For example, silage production involves the fermentation of grasses to provide feed for cattle. This fermentation is a bacterial process and it produces acid. This can cause corrosion of the silo and consequently leakage of silage material and its bacteria. This may lead to serious contamination of watercourses. To establish whether leakage of silo contents has occurred and if this has subsequently caused contamination of a particular watercourse, the silo in question can be seeded with MS2 bacteriophage particles. Water samples are then taken at regular intervals and mixed with *Serratia marcescens*–the indicator bacterium. These are then plated onto nutrient agar and incubated so that plaques may subsequently be detected. The appearance of plaques is evidence of bacteriophage in the water sample. Their occurrence after but not before seeding of the silo establishes that the silo is contaminating the watercourse. By taking samples at regular intervals not only can the source of contamination be confirmed, but also the time it takes for the silage material to reach the water may be established. The MS2 bacteriophage marker can also be used, for example, to track the source of water through subterranean courses. It is much more satisfactory than using tracking dyes – the old method for tracking water.

FURTHER READING

Bitton, G. (1980). *Introduction to Environmental Virology*, John Wiley & Sons.

Categorization of pathogens according to hazard and categories of containment. (1990). Advisory Committee on Dangerous Pathogens (HMSO).

Council Directive concerning the quality of bathing waters (1975). European Community 76/160/EEC.

Council Directive relating to the quality of water intended for human consumption. (1980). European community 80/778/EEC.

Guidelines for Microbiological Safety (1986). Portland Press, Colchester.

Heritage, J., Evans, E.G.V. and Killington, R.A. (1996). *Introductory Microbiology*, Cambridge University Press.

Mims, C.A., Playfair, J.H.L., Roitt, I.M. *et al.* (1998). *Medical Microbiology*, Mosby.

Standing Committee of Analysts (1995). *Methods for the isolation and identification of human enteric viruses from waters and associated materials*. HMSO.

3

Bacteria

Ed Schroeder and Stefan Wuertz

Department of Civil & Environmental Engineering, University of California, California 95616, USA

1 INTRODUCTION

Bacteria are indeed the lowest, or at least the simplest, form of life. Their prokaryotic cell structure is significantly different from higher forms such as the single celled algae and protozoa, invertebrates, plants and animals. Only the *Archaea* have a cell structure of similar simplicity. Bacteria are the most numerous of living organisms on earth in terms of number of species, number of organisms, and total mass of organisms. Each cell is small, typically 1–2 μm in diameter and length, and has a total mass of 1–10 pg. One thousand bacteria occupy approximately 10^{-12} ml of volume. Because bacteria are most commonly found in aqueous environments, their size is particularly important. A concentration of 1000 organisms per ml is very significant in terms of protecting health, but is invisible to the naked eye. Bacterial growth or replication is in numbers and concentration rather than mass of a single organism. Under optimal conditions of temperature, pH and nutrient availability, some bacterial species have generation times of less than 30 minutes. Such short generation times account for the rapid progression of infectious diseases.

Bacteria cause many human, animal, and plant diseases. Pathogenic bacteria, those that cause disease, are transmitted through direct contact with an infected host, by ingestion of contaminated food or water, or by the action of an intermediate host or disease vector. An example of a disease vector is a type of leafhopper, the glassy wing sharpshooter, which carries the bacterium *Xylella fastidiosa*, the cause of Pierce's disease in grapes and other woody perennials. Most pathogenic bacteria are relatively host specific and pathogens of non-human animal and plant pathogens generally do not cause diseases in humans. However, human pathogens are often carried by asymptomatic non-human hosts. Such pathogens are termed *zoonoses* and include the organisms causing the most important waterborne diseases. Bacteria of the genus *Salmonella*, which are normal habitants of birds, reptiles and mammals, but are thought to be uniformly human pathogens, are examples of zoonoses.

In this chapter, the focus is on pathogenic bacteria transmitted through water or associated with water in some manner. All waterborne or water-washed diseases are also associated with food and food preparation. In some cases the pathogenic bacteria are normal members of the intestinal flora of food animals and contamination results during processing. Infection results from improper handling during preparation or from inadequate cooking. In other cases organisms may be transferred from a processor to the food directly or by washing the food with contaminated water. In most cases proper cooking and washing of preparation surfaces with soap and hot water will prevent infection. With the exception of cholera and legionellosis, most of the diseases discussed here are more commonly transmitted through food, as indicated in Table 3.1; note that the period 1976–1990 was a period in which water supplies in Israel were greatly improved with respect to public health protection but that food protection lagged considerably.

The Handbook of Water and Wastewater Microbiology
ISBN 0-12-470100-0

TABLE 3.1 Incidence of waterborne salmonellosis, typhoid fever and shigellosis in Israel from 1976 to 1997 (Source: Tulchisky *et al.*, 2000)

Disease	*76–80*	*81–85*	*86–90*	*91–95*	*96–97*
Salmonellosis					
waterborne	979	157	244	260	0
total	10 101	12 386	17 127	28 986	11 481
Typhoid fever					
waterborne	112	76	0	0	0
total	596	629	216	0	0
Shigellosis					
waterborne	6557	10 180	1524	260	0
total	32 839	44 152	29 070	25 874	7274

2 INFECTIOUS DOSE

Infection normally occurs only when a sufficient number of pathogens has been ingested. The number of bacteria required to produce an infection ranges greatly. Based on infections in volunteers, the required dose for *Shigella* is approximately 10 organisms while *Campylobacter* infections require several hundred organisms. Cholera infection requires approximately 10^6 organisms under normal conditions. However, if antacids are taken to raise the stomach pH the infectious dose drops as low as 10^4 (Madigan *et al.*, 2000). High infectious dose requirements make transmission through water more difficult because pathogen concentrations are generally low in natural waters, even when contaminated with human wastes, and because natural waters are, with few exceptions, not good growth mediums for human pathogens.

3 MORBIDITY AND MORTALITY STATISTICS

Morbidity is the incidence of disease, usually stated as cases per 10 000 population per year. Mortality is the rate of deaths from the disease, also commonly stated in units of 100 000 population per year. For most waterborne diseases the morbidity statistics are probably understated. Many cases of all of the diseases are not reported because the symptoms are relatively minor, because the ill person does not see a physician, or because of poor reporting practices within the medical community. Public health agencies in many poor countries have difficulty in maintaining databases and even more difficulty in carrying out investigations of the causes of infection and death. When epidemics occur, the World Health Organization (WHO) and other agencies provide support that often results in improved investigation and record keeping. Thus data for large epidemics are probably more reliable than data for endemic problems.

4 PATHOGENIC BACTERIA OF CONCERN

Several million bacterial species are estimated to exist, with only a few thousand having been identified and to some extent characterized. At present, bacteria are classified by their 16S ribosomal RNA characteristics into 14 *kingdoms*, each of which is made up of many genera, which in turn include many species (Madigan *et al.*, 2000). Human pathogens are scattered among several of the kingdoms. New pathogens appear from time to time and it is not completely clear whether the organisms are new strains or if their pathogenic characteristics had not been observed previously. However, most of the principal bacterial pathogens have been recognized for over 50 years and these organisms account for nearly all of the serious bacterial infections that have been observed.

The principal waterborne diseases and the associated pathogenic bacteria are listed in Table 3.2. The first five diseases listed are all similar, except for severity, in that diarrhoea is the principal symptom. For all diarrhoeal diseases dehydration is a major complication. Even without treatment, keeping patients hydrated results in a high rate of recovery. The sixth disease, haemolytic uraemic syndrome, is also diarrhoeal but a toxin produced by the pathogen attacks the intestinal lining, destroys red blood cells and often results in kidney failure. Legionnaire's disease (also called legionellosis), tullaraemia, and bacterially caused peptic ulcer are not normally classed as waterborne diseases. However, water

TABLE 3.2 Principal waterborne diseases and the associated pathogenic bacteria

Disease	*Bacteria*	*Infectious dose*	*Characteristics*
Cholera	*Vibrio cholerae*	10^6	Severe diarrhoea, dehydration
Salmonellosis	*Salmonella* spp.		Watery diarrhoea often with abdominal cramping, nausea, vomiting fever and chills
Typhoid fever	*Salmonella typhi*	<1000	Fatigue, headache, abdominal pain, elevated temperature. Approximately 4% death rate
Gastroenteritis or campylobacteriosis	*Campylobacter jejuni*	<500	Watery diarrhoea often with abdominal cramping, nausea, vomiting, fever and chills
Dysentery or shigellosis	*Shigella* spp.	≈ 10	Bloody diarrhoea, abdominal cramps, rectal pain. Most virulent species, *S. dysenteriae*, produces toxin causing haemo uraemic syndrome
Haemorrhagic colitis and haemolytic uraemic syndrome	*Escherichia coli* O157:H7	Unknown[a]	Severe, systemic condition that occurs principally in children under 10 years of age
Leptospirosis	*Leptospira* spp.	Unknown	Fever, kidney infection, may result in kidney failure. In some cases there is internal bleeding, including pulmonary haemorrhage
Legionellosis	*Legionella pneumophilia*	Unknown	Acute pneumonia, high fever, headache, cough, little sputum
Peptic ulcer and gastric cancer	*Helicobacter pylori*	Unknown	Sore or hole in the lining of the stomach or duodenum. Cancer

[a] Probably similar to *Shigella.*

is or may be a mode of transmission in each case and all three diseases are widespread throughout the world. Hence they will be discussed here.

4.1 Cholera

Cholera, caused by *Vibrio cholerae*, a Gram-negative, curved and flagellated rod, is perhaps the clearest example of a waterborne disease. Although the disease can be contracted in other ways, by far the most common mode of transmission is ingestion of contaminated water. The infectious dose is very large, at least 10^6 organisms, under normal circumstances due to the sensitivity of *V. cholerae* to low pH environments. Increasing the stomach pH with antacids lowers the infectious dose to about 10^4 organisms. Ingestion of the organisms with food results in a similar lowering of the infectious dose. *V. cholerae* attach to the intestinal lining and secrete an enterotoxin that causes a high rate of discharge of a watery mucus. In addition to severe diarrhoea, symptoms include abdominal cramps, nausea, vomiting, and shock. Fluid and electrolyte loss result in death rates of up to 60% without treatment. Symptoms may begin within 6 hours of ingestion of the organisms, probably because of the high infectious dose requirements.

V. cholerae was first discovered in 1854 by Filippo Pacini. However, the relationship of cholera to contaminated water was elucidated 30 years earlier in 1855 by John Snow in a study of the sources of a cholera epidemic in London. Two private water companies provided water to a London district on a competing basis. The water from the two companies was distributed somewhat randomly, much as long distance telephone service is in the USA today. Snow found that the incidence of cholera in residents of the company taking water from the Thames river near the city centre was approximately 10 times that of the company taking water from the Thames upstream of London.

Several strains of *V. cholerae* cause the disease. Prior to 1960, the strain isolated by Koch (O1) was dominant. In 1960 the *El Tor* strain became the more commonly reported organism. In 1993, *V. cholerae* serogroup O139 (Bengal) became prominent in Bangladesh and eastern India and has since become endemic.

Cholera occurs world-wide but is far more prevalent in areas with inadequate protection of water supplies. The disease is endemic in over 80 countries and is responsible for approximately 10 000 deaths annually. Virtually absent from the Western Hemisphere for most of the 20th century, a cholera epidemic began in Peru in 1991 that gradually moved northward through Central America and Mexico. Over one million cases and 10 000 deaths were reported by 1995. The South American cholera strain has not crossed into the USA other than with infected travellers and through some imported seafood. A snapshot of the worldwide presence of cholera is provided in Table 3.3, which contains incidence and death data for 1999. The total reported cases of cholera in 1999 were 254 310 with 9175 deaths. Of these, five cases were in Australia and New Zealand and 16 were in Europe. Note that 81% of the cases and 95% of the deaths were in sub-Saharan Africa. A number of Latin American nations had infection rates similar to countries in sub-Saharan Africa. Death rates were generally low in Latin America and Asia for those countries reporting.

Cholera is virtually absent in the USA, Europe, Australia and New Zealand where water supplies have been filtered and disinfected since the early part of the 20th century. Of those cases reported in the USA and Europe, a large fraction is associated with travellers from areas where cholera is endemic.

TABLE 3.3 Reported incidence and death due to cholera in 1999 (Source: World Health Organization, 1999)

Africa	*Cases*	*Deaths*	*America*	*Cases*	*Deaths*
Benin	855	25	Belize	12	0
Burkina Faso	93	6	Brazil	3233	83
Burundi	3440	63	Colombia	42	0
Cameroon	326	35	Ecuador	90	0
Chad	217	18	El Salvador	134	0
Comoros	1180	42	Guatemala	2077	0
Congo	4814	20	Honduras	56	3
Dem. Rep. of Congo	12 711	783	Mexico	9	0
Ghana	9432	260	Nicaragua	545	7
Guinea	546	44	Peru	1546	6
Kenya	11 039	350	USA	6	0
Liberia	215	0	Venezuela	376	4
Madagascar	9745	542	American total	8126	103
			Asia		
Malawi	26 508	648			
Mali	6	3	Afghanistan	24 639	152
Mozambique	44 329	1194	Brunei Darussalam	93	0
Niger	1186	85	Cambodia	1711	130
Nigeria	26 358	2085	China	4570	0
Rwanda	217	49	Hong Kong SAR	18	0
Sierra Leone	834	5	India	3839	6
Somalia	17 757	693	Iran	1369	21
South Africa	68	2	Iraq	1985	30
Swaziland	7	0	Japan	40	0
Tanzania U. R. of	11 855	584	Malaysia	535	0
Togo	667	31	Philippines	330	0
Uganda	5169	241	Singapore	11	0
Zambia	11 535	535	Sri Lanka	108	5
Zimbabwe	5637	385	Vietnam	169	0
African total	206 746	8728	Asian total	39 417	344

Non-virulent *V. cholerae* has been isolated from Chesapeake Bay, particularly Baltimore Harbor, quite routinely (Jiang *et al.*, 2000a,b). Similar results have been reported for near shore waters in Peru (Patz, 2000). Their presence was correlated to temperature and increasing salinity. The fact that a virulence factor was missing, together with the temperature dependence, suggests that the population was resident. Increases in cholera cases were correlated with temperature in Mexico (Patz, 2000) and in Peru (Checkley *et al.*, 2000). Isolation of *V. cholerae* from sewage in Lima was determined to have a threshold temperature of 19.3°C and correlated with hospital admissions (Speelmon *et al.*, 2000). In the 1991 epidemic in Peru, initial cases occurring at virtually the same time were geographically scattered. Considering there were no cases in 1990, the geographical scatter may suggest an environmental factor being involved (Seas *et al.*, 2000).

4.2 Gastroenteritis/campylobacteriosis

Gastroenteritis is a general term to describe watery diarrhoea that may be accompanied by abdominal cramping, nausea, vomiting, fever and chills. Among the waterborne and water-washed diseases the most commonly associated bacterium is *Campylobacter*. In the USA 45% of the people complaining of diarrhoea are found to have *Campylobacter* infections. Usually these symptoms occur within 24 hours of ingestion. Illness is usually self-limited and lasts 3 days. Although campylobacteriosis is not a reportable disease in the USA, the Centers for Disease Control and Prevention receive about 10 000 reports of the disease each year, approximately 1 per 100 000 population. The disease generally occurs sporadically and large outbreaks are unusual. In the USA an estimated 2 million cases of campylobacteriosis occur each year with 500 deaths. However, the disease is generally not considered life threatening unless the organisms get into the bloodstream or the infected person has a weakened or underdeveloped immune system.

There are 16 species of *Campylobacter*, all of which are pathogenic to humans or animals. The species most commonly responsible for human infections is *Campylobacter jejuni*. Food, particularly poultry, is probably the most common mode of transmission, although waterborne infections are also common. *Campylobacter* are Gram-negative, motile spirilla and have been placed in the ε-subdivision of the proteobacteria. They are microaerophilic organisms and do not grow well at oxygen tensions greater than 10% of atmospheric. *Campylobacter* produces an enterotoxin similar to that of *V. cholerae* and the resulting diarrhoea is often bloody (Madigan *et al.*, 2000).

4.3 Salmonellosis

Depending on the source, 1000–2000 strains of the genus *Salmonella* have been identified. The discrepancy results from the change in identification methods from phenotyping to genotyping. Many organisms that were considered separate species through phenotyping have been grouped together as a result of information about their genetic makeup. In other cases, genotyping has resulted in organisms being placed in different genera. The recataloguing of large genera, such as *Salmonella*, is far from complete. Bacteria of the genus *Salmonella* are members of the γ-subgroup of the Proteobacteria. Close genetic relatives include the genera *Escherichia, Shigella, Proteus* and *Enterobacter*, and the group is commonly referred to as enteric bacteria. The World Health Organization considers all members of the genus *Salmonella* to be human pathogens. However, three species, *S. typhi, S. enteritidis and S. typhimurium* are the most important. *S. typhi* is the cause of typhoid fever and will be discussed in the following section. The other *Salmonella* infections are relatively mild, similar in their characteristics and severity to campylobacteriosis, and like campylobacteriosis, are more often associated with food than water. The ratio of waterborne to total infections in Israel dropped from nearly 10% in 1976 to less than 1% in 1995 as the result of improvements in water supply systems (Tulchisky *et al.*, 2000). Approximately 30% of the people with diarrhoea who see a physician in the USA are found to have salmonellosis. Whether this statistic is valid in other countries is questionable because of differences in foods, food handling and water supply.

Symptoms of salmonellosis begin from 12 to 72 hours following ingestion. Duration of symptoms is usually less than one week and most people do not require treatment. As with campylobacteriosis, hospitalization is sometimes required. Unlike campylobacteriosis, the disease is notifiable in the USA. Thus statistics are maintained by the US Public Health Service on cases in which the organism is identified (Centers for Disease Control and Prevention, 1994). However, it is likely that only a small minority of infections are treated by a physician and within that group only severe cases would receive a full laboratory work-up. Thus the statistic of 40 000 cases of salmonellosis per year in the USA is probably very much lower than the actual number. Approximately 1000 people die each year in the USA due to salmonellosis. This is a large number for a disease generally considered a mild infection and would lead to speculation that the situation in developing nations is much worse.

Salmonella are zoonoses and are particularly associated with poultry, wild birds and reptiles. An outbreak estimated at over 200 000 infections in the USA in 1994 resulted from using non-pasteurized eggs in ice cream (World Health Organization, 1996).

4.4 Typhoid fever

Typhoid fever, caused by *Salmonella typhi*, is similar to cholera in severity but has much greater incidence and causes approximately 60 times as many deaths. Worldwide, the annual number of cases of typhoid fever is over 16 million, many times that of cholera, of whom nearly 4%, or 600 000, die. Death rates are between 12 and 30% when the disease is not treated and 1 to 4% when treatment is given. As in the other diarrhoeal diseases, the incidence and death rate is far greater in developing countries. There are approximately 400 cases in the USA and a little over 100 cases in the UK per year. Typhoid epidemics have not occurred in developed countries since the 1920s. However, epidemics are relatively frequent in the developing world. There were over 7500 cases in Tajikistan in August 1996. Over 900 cases occurred in January 1996 in Ain Taya, an Algerian city of 45 000.

Typhoid fever is different from salmonellosis in that the responsible organism, *S. typhi*, after initially colonizing the intestines enters the bloodstream causing a systemic infection. Symptoms generally do not occur for at least a week and may take up to 3 weeks to develop. High fever, fatigue, abdominal pain and either constipation or diarrhoea are typical symptoms. Large outbreaks are generally associated with water contamination, as in both the Algerian and Tajikistan cases noted above, or street vendors. Typhoid fever is unusual in that up to 5% of those infected continue to shed infective cells for up to a year following recovery from the disease.

Three typhoid fever vaccines are currently licensed. One is a dead cell variety, one is based on polysaccharides from the cell wall, and the third is a live vaccine. The three vaccines have similar effectiveness but somewhat different periods of protection and side effects.

4.5 Shigellosis (dysentery)

Dysentery is most easily defined as bloody diarrhoea. Although bloody diarrhoea can result from several types of bacterial infection, the most common form is shigellosis, caused by bacteria of the genus *Shigella*. Shigellosis, like campylobacteriosis and salmonellosis, is more commonly associated with direct contact or contaminated food than water, as indicated in Table 3.1. Four species of the genus *Shigella*, *S. dysenteriae*, *S. flexneri*, *S. boydii*, and *S. sonnei*, are most commonly isolated from active cases. Endemic shigellosis is associated with *S. flexneri* and epidemic dysentery is associated with *S. dysenteriae*. Twelve serotypes of *S. dysenteriae*, 13 serotypes of *S. flexneri*, 18 serotypes of *S. boydii* and 1 serotype of *S. sonnei* have been identified on the basis of the O-specific polysaccharide of the lipopolysaccharide. Shigellosis is estimated to cause over 600 000 deaths worldwide each year (World Health Organization, 1998). In the USA, the majority of infections are by *S. sonnei* and direct contact, such as occurs in day care centres, is a major mode of transmission (Centers for Disease Control and Prevention, 2001).

The disease results from growth of *Shigella* spp. in the epithelial cells of the colon. Invaded cells die and the infection spreads to adjacent cells. The result is ulceration, inflammation and bleeding. Classic symptoms of shigellosis include severe diarrhoea, a bloody stool, cramping, and tenesmus. Infection by *Shigella dysenteriae* Type 1, the principal cause of epidemic dysentery, may result in haemolytic uraemic syndrome (HUS), seizures, sepsis and toxic megacolon. A factor in the particular severity of *Shigella dysenteriae* Type 1 is the production of an enterotoxin that inhibits protein synthesis. Haemolytic uraemic syndrome will be discussed below in connection with *Eschericia coli* O15:H7. Up to 15% of *Shigella dysenteriae* Type 1 infections are fatal. As might be expected the most severe problems with dysentery, particularly that associated with *Shigella dysenteriae* Type 1, are in areas where public health facilities and sanitation are poor. Shigellosis in developed countries is a serious but less deadly problem, as indicated by the fact that most cases in the USA are caused by *S. sonnei*. Epidemic dysentery is widespread. Between 1968 and 1972, over 500 000 cases and 20 000 deaths occurred in Central America. The disease surfaced in Congo in 1979 and has been found throughout sub-Saharan Africa since then.

Bacteria in the genus *Shigella* are classified as enteric bacteria and are morphologically similar to other enteric bacteria such as *Escherichia*, *Salmonella* and *Klebsiella*. All are Gram-negative, motile rods and are non-sporulating and facultatively anaerobic. Enteric bacteria are oxidase negative and reduce nitrate only to nitrite (Madigan *et al.*, 2000). *Shigella* and *Escherichia* are closely related genetically with DNA homology generally greater than 70% between members of the two genera.

Shigella are commonly present in soil and water but are not always infective. For infection to occur the organisms must contain the invasion plasmid that allows it to bind to the epithelial cells of the intestine and a smooth lipopolysaccharide somatic antigen. *Shigella* found in natural environments are often unable to initiate the disease.

Members of the genus *Shigella* have shown an impressive ability to develop antibiotic resistance. *Shigella dysenteriae* has shown a particular ability in this regard and often develops resistance to a new antibiotic within 2 years. Thus treatment of *Shigella* outbreaks is difficult and problems increase where endemic conditions exist.

4.6 Haemorrhagic colitis and haemolytic uraemic syndrome

Haemorrhagic colitis and haemolytic uraemic syndrome (HUS) are diseases that can be caused by a number of bacteria. However, the principal cause is six strains of the usually non-pathogenic species *Escherichia coli*. The strains are termed enteropathogenic (EPEC), verotoxigenic (VTEC), enterotoxic (ETEC), entero-invasive (EIEC), diffusely adhesive (DAEC), or enterohaemorrhagic (EHEC), depending on the type of toxin produced. The toxin produced by *E. coli*, Type O15:H7, the EHEC type, is indistinguishable from that of *Shigella dysenteriae* Type 1.

Escherichia coli is one of the predominant species present in the gut of warm-blooded animals and as such is used as an indicator of the possible contamination of food, water or surfaces with human or animal faeces. Pathogenic strains of *E. coli* were not reported prior to 1982 and it is unclear if the diseases are new or simply had not been detected previously. The severity of HUS is such that lack of identification of pathogenic *E. coli* as the cause would seem surprising. However, HUS is caused by other bacteria, most notably *Shigella dysenteriae* Type 1, and the ability to identify strains of species has increased rapidly over the past 50 years.

All of the pathogenic *E. coli* strains cause haemorrhagic colitis that results in abdominal pain and loose stools that may progress to bloody diarrhoea. Severity of the disease varies with minor cases lasting 1–2 days and more severe cases lasting up to 2 weeks. The disease is not distinguishable from shigellosis or salmonellosis except by isolation of the organisms.

Haemolytic uraemic syndrome, caused by *E. coli* Type O15:H7, is a systemic disease that occurs principally in children under 10 years old. Approximately 10% of children with *E. coli* Type O15:H7 infections will develop HUS. The first stage of the disease begins with symptoms similar to gastroenteritis and haemorrhagic colitis and may last up to 2 weeks. Type O15:H7 is pathogenic because of the ability to produce the *Shigella* enterotoxin that inhibits protein production and induces fluid loss from epithelial cells. When the toxins enter the bloodstream red blood cells and the platelets that contribute clotting cells are destroyed. Kidney damage is common and generalized fluid accumulation is a result. Treatment consists of maintenance of fluids and salts and, in some cases, blood transfusions. There is no known way to stop or manage the progress of the disease. Approximately 90% of the children receiving supportive treatment survive the disease but up to 30% have permanent kidney damage that leads to kidney failure.

Outbreaks of *E. coli* Type O15:H7 infections have occurred world-wide. The most common source of infections is incompletely cooked meat, particularly minced (ground) beef. However, outbreaks have been traced to contaminated water. The most notable waterborne outbreak to date occurred in May, 2000 in the small community of Walkerton, Ontario where the community water supply system was contaminated. Six deaths and several hundred illnesses are attributed to the outbreak. The infection route is believed to have been poorly sealed wells, some of which were located downslope from cattle pastures.

4.7 Leptospirosis

Leptospirosis, caused by several species of the genus *Leptospira*, is a widespread disease with its principal reservoir in non-human animals. At least 100 000 cases of leptospirosis occur worldwide annually. Most infections are mild and do not require treatment. However, severe cases may result in kidney failure, internal bleeding and pulmonary haemorrhage. Incidence in developed countries is low – there were 389 cases in the USA between 1985 and 1998. In the UK the rate is about 8 cases per year. In the Caribbean Islands the annual infection rate is between 5 and 30 cases per 100 000 population per year and the death rate is nearly 10%.

Leptospira invades the body through mucous membranes or through broken skin. Infections occur in the central nervous system, kidneys and liver but are most persistent in the kidney tubercules. The organism is discharged in the urine and contact with animal urine or with water contaminated with animal urine is the principal mode of transmission. Although pathogenic *Leptospira* species are commonly found in both wild and domestic organisms, the organism is endemic in rat populations and rats are usually implicated in outbreaks. Individual infections occur among swimmers, fishermen, cavers, rice field workers and abattoir workers. Outbreaks of leptospirosis usually occur following storms and it is presumed that the organisms are washed into watercourses where they come into contact with people. Diagnosis is difficult and the disease is often confused with others having similar symptoms. The incidence of leptospirosis is much greater in developing, particularly tropical, countries than in developed nations. Severe outbreaks occurred in the Azores in 2001, Orissa, India in 1999 and Nicaragua in 1995.

4.8 Legionellosis (Legionnaire's disease)

Legionellosis is a form of pneumonia caused by *Legionella pneumophila*. The name is derived from the first observation of the disease at a convention of the American Legion, a military veterans' organization, in Philadelphia, Pennsylvania in 1976 (Fraser *et al.*, 1977). Over 220 people contracted legionellosis, of whom 34 died, in this first epidemic. The source of the *Legionella pneumophila* was found to be condensate in the air-conditioning system.

Legionella pneumophila is a Gram-negative rod typically measuring from 2 to 20 μm in length. Specific growth requirements are L-cysteine and ferric iron. The organism is a normal inhabitant of natural waters and is commonly found in cooling towers, hot tubs (it grows well up to 45°C), hot water tanks, and sprays in vegetable sections of supermarkets (Madigan *et al.*, 2000; Winn, 1988). *Legionella pneumophila* is a facultative intracellular parasite. It does not multiply in sterile tap water but grows well when free-swimming amoeba cultures are added (Breiman *et al.*, 1990). In the lungs, macrophages engulf *Legionella pneumophila*, but the organisms are able to grow by blocking the fusion of lysosomes with the phagosome.

Legionella pneumophila infections are relatively common. Approximately 1000 confirmed cases of legionellosis occur in the USA each year. However, the Centers for Disease Control and Prevention estimate that the actual rate is 10 000–15 000 cases annually. Although the disease is reportable, most hospitals do not normally check for the organism. Most infections result in subclinical symptoms. However, immunocompromised individuals, young children and older adults are considerably more susceptible to development of severe symptoms. Epidemics, such as the one from which the disease derived its name, are unusual. Most cases occur one or two at a time and rarely are reported in the general media. To date the most common sources have been air-conditioning systems and cooling towers. Not surprisingly, legionellosis is largely a disease of the developed world. Very few data exist on the incidence of the disease in developing countries.

4.9 Peptic ulcer, gastritis and stomach cancer

In the late 1970s Marshall and Warren (1984, 1989) observed *Helicobacter pylori*, a motile Gram-negative rod that is oxidase, catalase and urease positive, consistently to overlay inflamed gastric tissue. They suggested that gastritis and peptic ulcers might be a bacterial disease rather than the result of stress and excessive acid production. The suggestion was received with little enthusiasm from the medical community and probably even less from the pharmaceutical industry. However, Marshall fulfilled Koch's postulates by ingesting a culture of *H. pylori*, developing ulcers, and then curing the disease by taking antibiotics and bismuth and their suggestion began to be taken more seriously (Marshall *et al.*, 1988). Epidemiological studies have established a linkage of *H. pylori* infection to non-cardia gastric adenocarcinoma as well as peptic ulcers. However, removal of *H. pylori* appears to increase the risk of gastro-oesophageal reflux disease, Barrett's oesophagus and adenocarcinoma of the oesophagus (Blaser, 1992, 2000).

H. pylori appear to create relatively high pH microregions by secreting a urease with an unusually low Michaelis constant and by suppressing production of ascorbic acid (Moore, 1994). The organism secretes cytotoxins, proteases and phospholipidases that cause inflammation and damage to epithelial cells and decreasing protection from gastric acids. A suggestion has been made that *H. pylori* obtains sustenance from epithelial cell secretions and therefore irritation of the tissue is beneficial to growth (Blaser, 2000).

Human infection by *H. pylori* is very common. About 2% of children under age nine are infected. Infection rates increase with age until about age 50 when over 50% of all individuals world-wide are infected. Infection is significantly greater in lower economic levels and in developing countries where the rate is as high as 80% of all people. In the developing world, infection rates seem to be decreasing. The incidence of disease in individuals infected with *H. pylori* is quite high. In an Italian study of over 1000 adults, 42% were positive for *H. pylori* antibodies. Of the 420 positive individuals 121 submitted to endoscopy. Only 14 of the 121 were disease free (Vaira *et al.*, 1994).

The mode of *H. pylori* infection is unknown but assumed to be through contaminated food and water. Thus there is a possibility that peptic ulcers and gastric cancer are waterborne diseases. Given the prevalence of these diseases, transmission through water should be investigated.

5 INDICATOR ORGANISMS

Waterborne diseases pose very serious threats to society because of their potential to infect large numbers of individuals over a very short time. The common waterborne diseases are all incapacitating, many have high rates of death for untreated victims, and are particularly difficult problems for children, the elderly and those with challenged immune systems. Testing water for each of the possible pathogens, including viruses, protozoans, and worms is impractical. Early in the development of sanitary practice the need for *indicators of human contamination* was perceived. The issue presented was that if recent human contamination were established then there would be a certain probability of the presence of human pathogens. Accuracy of the relationship would be greater in urban than rural areas because there would be a greater probability of sick people in the general population. Thus the question asked was what could be monitored in water to determine if recent human contamination had occurred. In the early 1900s chemical analysis was nowhere near sensitive enough for chemical compounds to be considered as indicators. However, the ability to identify bacteria characteristic of the colon had been developed and methods were developed to estimate the concentration of these *coliform* bacteria. Consequently coliform bacteria of various types have become the standard method of establishing the biological quality of water (see Chapter 7).

5.1 Drinking water

There is little question that application of the coliform test (APHA, 1998) has been a major factor in the improvement and maintenance of drinking water quality. Whether monitoring a broader group of organisms would have had equally good results is unknown. High concentrations of any bacteria in drinking water would generally be discouraged by public health authorities even if there was no evidence of *E. coli* or enterococci. Water filtration and disinfection result in near sterile water and one usually devoid of coliforms. Microbial films grow in water distribution lines but as long as the system is kept isolated there is not a problem with introduction of pathogens or of coliforms (Camper *et al.*, 1991). Thus samples taken at the tap are almost always free of coliforms in well-maintained and operated systems. However, recent experiences have shown that the lack of coliforms does not provide evidence of the absence of human pathogens, most notably protozoans and viruses. Some protozoans and viruses are much more difficult to remove by filtration and are more resistant to disinfection methods now in use than coliforms, or bacteria in general. The most interesting example is cryptosporidiosis, caused by the protozoon *Cryptosporidium parvum*. Cryptosporidiosis, a normally mild, self-limiting diarrhoeal disease was first recognized as a human disease in 1976, although it had undoubtedly been present for some time. The first report of cryptosporidiosis in association with a water supply that met all standards for coliforms was in 1987 in Carrollton, Georgia, where 13 000 people became ill. In 1993 the largest outbreak of waterborne disease recorded in the USA occurred when an estimated 400 000 people became ill with cryptosporidiosis in Milwaukee, Wisconsin. Again the water supply met all standards for coliform bacteria. Since that time cryptosporidiosis has become the most common known source of waterborne disease outbreaks in the USA.

5.2 Recreational and labour water quality

Individual contact with water occurs through recreational activities such as wading, swimming, boating and water skiing in a variety of venues, pools, ponds, lakes and rivers. Workers come in contact with water in agriculture, food processing and activities requiring washing facilities. Swimming pool water is usually filtered and disinfected but the density of use is greater than in other venues. In more natural venues, such as beaches, there is much less ability to control contamination. For example, beaches typically receive storm runoff from adjacent land and often have one or more small streams or storm drains running across them. In many cases the streams disappear in sand prior to reaching the receiving waters during dry periods. Contaminants mobilized by storm runoff are carried into the streams and storm drains, onto beaches and into adjacent water. Thus animal wastes can be carried into natural recreational waters on a continual basis. In areas where human waste disposal is on land or in cesspools or septic systems there may be direct contamination of recreational waters following storms, also.

Workers contact water in fields during irrigation, in food processing where water is used for washing and for carriage of both products and wastes, and in general commercial activities where water is used to clean surfaces and equipment. There is potential for direct ingestion of contaminated water as well as for infection through the skin or through cuts and abrasions. In areas where water supplies are limited, water recycling is often practised. Without appropriate treatment, water recycling increases the potential for infections to occur.

Problems with using coliforms as indicator organisms in judging the quality of recreational and labour waters are quite different than for judging drinking-water quality. Infections of concern in recreational waters include skin, throat, and ear infections and parasitic diseases as well as the gastrointestinal diseases of concern in drinking water. Sources of contamination are much more diverse and far less controllable than is the case for drinking water and the types of contact with recreational and labour water are also quite diverse. Moreover, natural runoff, including irrigation water from both agriculture and parks, is commonly high in coliforms, *E. coli* and *Enteroccus* concentrations even when it is clear that human contamination is unlikely (Solo-Gabrielle *et al.*, 2000; Schroeder *et al.*, 2002). Suspected sources of *E. coli* and *Enterococcus* are generally listed and include domestic, feral and wild animals, although it is apparent that these organisms can survive and grow in soil.

The microbial population in the guts of warm-blooded animals varies considerably, at least in relative numbers of the various species (Madigan *et al.*, 2000). Considerable effort has been invested in developing methods of determining the source of indicator organisms through identification of strains specific to certain animal species, ratios of *E. coli* and *Enterococcus* concentrations, and application of tools such as neural network analysis to microbial populations. To date a widely accepted method of determining the source of indicator organisms in a stream has not been developed. However, because of the number of zoonoses, the need to differentiate between contamination by humans and animals appears questionable except where epidemics are involved and the source is known.

At present biological standards of quality of recreational waters are based on coliforms, *E. coli* and *Enteroccus* concentrations (USEPA, 2000). The near ubiquitous presence of indicator organisms in both storm and dry weather drainage presents severe problems and would suggest that virtually all drainage needs to be disinfected if current standards are to be met.

REFERENCES

APHA (1998). *Standard Methods for the Examination of Water and Wastewater*, 20th edn. American Public Health Association, Washington.

Blaser, M.J. (1992). Hypotheses on the pathogenesis and natural history of *Helicobacter pylori*-induced inflammation. *Gastroenterology* **102**, 720–727.

Blaser, M.J. (2000). The amphibiotic relationship of *Helicobacter pylori* to humans. In: R.H. Hunt and G.N.J. Tytgat (eds) *Helicobacter pylori: Basic Mechanisms to Clinical Cure 2000*, pp. 25–30. Kluwer Academic Publishers, Dordrecht.

Blaser, M.J. and Berg, D.E. (2001). *Helicobacter pylori* genetic diversity and risk of human disease. *Journal of Clinical Investigation* **107**, 767–773.

Breiman, R.F., Fields, B.S., Spika, J.S., *et al.* (1990). Association of shower use with Legionnaires' disease: possible role of amoebae. *Journal of the American Medical Association* **263**, 2924–2926.

Camper, A.K., McFeters, G.A., Characklis, W.G. and Jones, W.L. (1991). Growth kinetics of coliform bacteria under conditions relevant to drinking water distribution systems. *Applied & Environmental Microbiology* **57**, 2233–2239.

Centers for Disease Control and Prevention (1994). Summary of notifiable diseases. United States, 1993. *MMWR*, **43** (53).

Centers for Disease Control and Prevention (2001). Summary of notifiable diseases, United States, 1999. *MMWR*, **48** (53).

Checkley, W., Epstein, L.D., Gilman, R.H. *et al.* (2000). Effects of the El Niño and ambient temperature on hospital admissions for diarrhoeal diseases in Peruvian children. *Lancet* **355**, 442–450.

Edberg, S.C., Allen, M.J. and Smith, D.B. (1988). The National Collaborative Study – national field evaluation of a defined substrate method for the simultaneous enumeration of total coliforms and *Escherichia coli* from drinking water: comparison with the standard multiple tube fermentation method. *Applied & Environmental Microbiology* **54**, 595–1601.

Fraser, D.W., Tsai, T.R., Orenstein, W. *et al.* (1977). Legionnaires' disease: description of an epidemic of pneumonia. *New England Journal of Medicine* **297**, 1189.

Goodman, K.J. and Correa, P. (1995). The transmission of *Helicobacter pylori*: a critical review of the evidence. *International Journal of Epidemiology* **24**, 875–887.

Jiang, S.C., Louis, V., Choopun, N. *et al.* (2000a). Genetic diversity of *Vibrio cholerae* in Chesapeake Bay determined by amplified fragment length polymorphism fingerprinting. *Applied and Environmental Microbiology* **66**, 140–147.

Jiang, S.C., Matte, M., Matte, G.A. *et al.* (2000b). Genetic diversity of clinical and environmental isolates of *Vibrio cholerae* determined by amplified fragment length polymorphism fingerprinting. *Applied and Environmental Microbiology* **66**, 148–153.

Johnson, C.H., Rice, E.W. and Reasoner, D.J. (1997). Inactivation of *Helicobacter pylori* by chlorination. *Applied and Environmental Microbiology* **63**, 4969–4970.

Madigan, M.T., Martinko, J.M. and Parker, J. (2000). *Brock Biology of Microorganisms*, 9th edn. Prentice-Hall Inc., Upper Saddle River, NJ.

Marshall, B.J. and Warren, J.R. (1984). Unidentified curved bacilli on gastric epithelium in active chronic gastritis. *Lancet* **i**, 1311–1315.

Marshall, B.J., Goodwin, C.S. and Warren, J.R. (1988). Prospective double-blind trial of duodenal ulcer relapse after eradication of *Campylobacter pylori*. *Lancet* **ii**, 1437–1442.

Marshall, B.J. (1989). History of the discovery of *C. pylori*. In: M.J. Blaser (ed.). Campylobacter pylori *in gastritis and peptic ulcer disease*, pp. 7–24. Igaku-Shoin, New York.

Moore, R.A. (1994). Helicobacter pylori *and Peptic Ulcer: A systematic review of effectiveness and an overview of the economic benefits of implementing what is known to be effective*. Pain Research. The Churchill, Headington, Oxford.

Patz, J. (2000). *Integrated Assessment of the Public Health Effects of Climate Change for the United States*, Final Report EPA Grant Number R824995. National Center for Environmental Research, US Environmental Protection Agency, Washington, DC.

Sack, D.A. (1998). Cholera and related. In: S.L. Gorbach, J.G. Bartlett and N.R. Blacklow (eds) *Infectious Diseases*, 2nd edn, W.B. Saunders Company, Philadelphia.

Schroeder, E.D., Thompson, D.E., Loge, F.J. and Stallard, W.M. (2002). *Occurence of Pathogenic Organisms in Urban Drainage*. Report to the California Department of Transportation. Department of Civil Engineering, University of California, Davis.

Seas, C., Miranda, J., Gil, A.I. *et al.* (2000). New insights on the emergence of cholera in Latin America during 1991: the Peruvian experience. *American Journal of Tropical Medicine and Hygiene* **62**, 513–517.

Solo-Gabrielle, H.M., Wolfert, M.A., Desmarais, T.R. and Palmer, C.J. (2000). Sources of *Escherichia coli* in a coastal subtropical environment. *Applied and Environmental Microbiology* **66**, 230–237.

Speelmon, E.C., Checkley, W., Gilman, R.H. *et al.* (2000). Cholera incidence and El Nino-related higher ambient temperature. *Journal of the American Medical Association* **283**, 3072–3074.

Tulchisky, T.H., Burla, E., Sadik, C. *et al.* (2000). Safety of community drinking-water and outbreaks of waterborne enteric diseases: Israel 1976–1997. *Bulletin of the World Health Organization* **78**, 1466–1473.

USEPA (2000). *Draft Implementation Guidance for Ambient Water Quality Criteria for Bacteria – 1986*. Report No. EPA 823-D-00-001, Office of Water, United States Environmental Protection Agency, Washington, DC.

Vaira, D., Miglioli, M., Mulè, P. *et al.* (1994). Prevalence of peptic ulcer in *Helicobacter pylori* positive blood donors. *Gut* **35**, 309–312.

Winn, W.C. Jr (1988). *Legionella*: historical perspective. *Clinical Microbiology Review* **1**, 60.

World Health Organization (1996). *Emerging Foodborne Diseases, Fact Sheet No. 124*, WHO, Geneva.

World Health Organization (1998). *Shigella*. Available at www.who.int/vaccines/intermediate/shigella.htm

World Health Organization (1999). *Communicable Disease Surveillance and Response: Global Cholera Update*. Available at www.who.int/emc/diseases/cholera/choltbl1999.html

World Health Organization (2000). *Weekly Epidemiological Record* **75**, 217–224.

4

Protozoa

Nigel Horan

School of Civil Engineering, University of Leeds, LS2 9JT, UK

1 INTRODUCTION

Protozoa are a large collection of organisms with considerable morphological and physiological diversity. They are all eukaryotic and considered to be unicellular. The majority of them employ chemoheterotrophic nutrition, but as certain species possess chloroplasts they can also practise photoautotrophy. Protozoa are found in almost every aquatic environment and are widely distributed. They play an important role in all aspects of sanitary microbiology ranging from their health impacts as human pathogens through to their role in the treatment of potable waters and wastewaters. With around 35 000 species they demonstrate a wide diversity in form and mode of life and occupy a range of ecological niches. The number of species together with the number of individuals of each species provides an indicator value as to the nature of the habitat in which they are observed. They play an important role in many communities where they occupy a range of trophic levels. As predators upon unicellular or filamentous algae, bacteria and microfungi, protozoa can be both herbivores and consumers in the decomposer link of the food chain. They are also a major food source for microinvertebrates, thus they have an important role in the transfer of bacterial and algal production to successive trophic levels.

Protozoa are defined as single-celled eukaryotic organisms that feed heterotrophically and exhibit diverse mechanisms of motility. As they are unicellular they lack tissues and organs, thus each protozoan cell is able to carry out all the processes that are essential for life. The cell achieves this by means of organelles that are able to carry out specialized functions. Thus there are organelles for feeding (cell mouths), excreting (cytopyge) and locomotion (cilia, flagella, pseudopodia).

2 CLASSIFICATION OF PROTOZOA

In order to understand fully the contribution of protozoa to aquatic ecosystems and to exploit these same properties in engineered systems, it is essential to be able to identify and then classify them. Classification is an aid to identifying similar characteristics in behaviour and response, it proves very useful in the field as it removes the necessity of repeating experiments on each and every species.

Protozoa are classified on the basis of their morphology, in particular as regards their mode of locomotion. However, their classification is not simple as they are not a natural group but simply a collection of single-celled eukaryotes gathered together for convenience. In addition, classification is a fluid area as more information becomes available and ideas about evolution are refined. Classification of protozoa, in particular, is subject to regular revision and this can be an emotive and controversial issue. A complete revision of the protozoa took place in 1980 and this placed them in the Kingdom Protista, sub-Kingdom the Protozoa. They are then further divided into seven Phyla: I Sarcomastigophora, II Labyrinthomorpha, III Apicomplexa, IV Microspora, V Ascetospora, VI Myxozoa, VII Ciliophora. Of these, Phyla I and VII form the free-living protozoa whereas the remainder

The Handbook of Water and Wastewater Microbiology
ISBN 0-12-470100-0

are parasitic protozoa which parasitize plants, animals and algae. The free-living protozoa are aquatic species and are ubiquitous and widely distributed world-wide.

A much simpler scheme, proposed by Jahn (1979) and described in Ross (2000) is used in this article, which recognizes four major groups of protozoa and gives them the status of a Class. The first three Classes are the free-swimming protozoa and the last one comprises the parasitic protozoa.

1. *Mastigophora*: or flagellated protozoa (Euglena)
2. *Sarcodina*: or amoeba-like protozoa (Amoeba)
3. *Ciliophora* or ciliated protozoa (Paramecium)
4. *Sporozoa* or parasitic protozoa (Plasmodium – malaria)

Each of the four classes is discussed in turn in the following sections and the full scheme is illustrated in Fig. 4.1.

2.1 Mastigophora – the flagellated protozoa

The mastigophora are characterized by possession of one or more flagella, which are used both for locomotion and feeding. Some flagellates have a characteristic arrangement of their flagella, e.g., *Dinoflagellates*. In common with the amoebae they usually multiply by longitudinal binary fission. They are the most primitive of the protozoa, believed to be the ancestors of the animals. Although some of them are free living, most are parasitic living in or on other organisms. These include *Trypanosoma gambiense* (the agent of sleeping sickness), *Trichonympha* (found in the guts of termites) and *Giardia lamblia* (causing acute diarrhoea).

Many flagellates are able to feed autotrophically as well as heterotrophically and the dinoflagellates are important primary producers (photosynthesizers) in lakes and oceans, yet they can also ingest prey and feed in an animal-like fashion. This makes them difficult to classify and they are often divided into two classes. The phytomastigophorea usually contain chloroplasts and are capable of autotrophic as well as heterotrophic nutrition. Consequently, many organisms classified as phytomastigophorea are also classified as algae, this includes such organisms as *Euglena, Volvox, Oicomonas* and the dinoflagellates. The remaining class, zoomastigophora, are heterotrophic

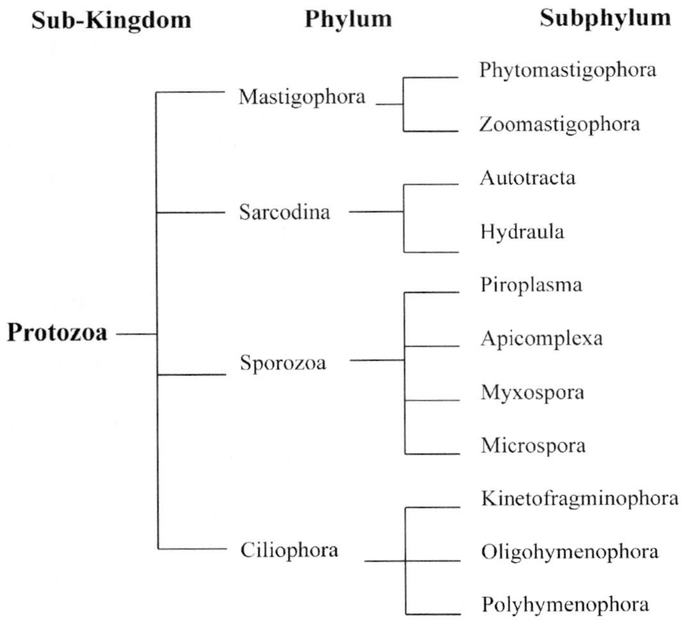

Fig. 4.1 Classification of protozoa which considers them as a sub-kingdom (from Ross, 2000).

and have oval-shaped bodies. They are generally the most numerous protozoa in terms of their total numbers in wastewater treatment plants such as activated sludge and trickling filters (Table 4.1).

2.2 Sarcodina – the amoebae

The Sarcodina possess pseudopodial structures that are used for movement and also for feeding, by means of protoplasmic flow. They demonstrate considerable diversity in that some of them lack any skeletal structure (naked amoeba), whereas others have elaborate shells known as tests (testate amoeba). The shells are composed of proteinaceous, siliceous or carbonaceous material and a wide range of structures have evolved, e.g. *Foraminiferans* contain shells of calcium carbonate which has been extracted from sea water. The shells do not cover the whole body, however, and naked pseudopodia are still used for feeding and locomotion.

Amoebae move by means of flowing cytoplasm, usually with the production of pseudopodia. The pseudopodial morphology has been used in classification of amoebae and the Sarcodina are divided into two major groups or classes. The Rhizopoda have unsupported pseudopodia, whereas the Actinopoda have radiating microtubule-supported axopodia.

TABLE 4.1 Representative protozoa from the three phyla found in activated sludge and trickling filters

Species	Subclass	Phylum
(a) *Bodo caudatus*		
(b) *Oicomonas termo*		Zoomastigophora
(c) *Amoeba proteus*		
(d) *Arcella vulgaris*		Sarcodina
(e) *Chilodenella uncinata*		
(f) *Colpoda cucullus*	Holotrichia	Ciliophora
(g) *Aspidisca costata*		
(h) *Euplotes moebiusi*	Spirotrichia	Ciliophora
(i) *Opercularia microdiscum*		
(j) *Vorticella microstoma*	Peritrichia	Ciliophora
(k) *Popodphyra collini*		
(l) *Acineta tuberosa*	Suctoria	Ciliophora

2.3 Ciliophora – the ciliates

The ciliates form an extremely large group and are the most specialized and complicated of the protozoans. They are characterized by the arrangement of cilia in an ordered fashion over the surface of the cell, and the presence of two kinds of nuclei, a macronucleus and a micronucleus. They also divide by transverse fission. Cilia, which serve to provide locomotion, are usually arranged in rows called kineties and the cilia beat with their effective stroke in the same direction. In addition, cilia are also distributed around what is the protozoal equivalent of a mouth, namely the cytosome. Here they provide an aid to feeding by the production of feeding currents.

The Ciliophora is the largest of the three phyla in terms of the number of species it represents, with over 7000 described in nature. The Ciliophora also provide the greatest species diversity in wastewater treatment plants, although not necessarily the largest number of individuals. Four distinct types of Ciliophora may be identified, based on their locomotion and arrangement of cilia:

1. Holotrichia – free-swimming protozoa which have cilia arranged uniformly over their whole bodies. Typical species are *Tracheophyllum* and *Litonotus.*
2. Spirotrichia – these possess a flattened body with locomotory cilia found mainly on the lower surface. The cilia concerned with feeding are well developed and wind clockwise to the cytosome. They are represented by *Aspidisca, Stentor* and *Euplotes.*
3. Petitrichia – these are immediately recognizable by their inverted, funnel- or bell-shaped bodies which are mounted on a stalk. The other end of this stalk is attached to particulate material such as a sludge floc, and serves to anchor the protozoan. In certain species the stalk is contractile. The wide end of the bell acts as an oral aperture in the Peritrichia and they have cilia arranged around this as an aid in feeding. Typical peritrichs are *Vorticella* and *Opercularia.*
4. Suctoria – the final type of ciliophora is ciliated only in early life, when the cilia

TABLE 4.2 Diseases associated with protozoal infection of humans

Niche	*Protozoa*	*Condition*
Skin	*Leishmania*	Cutaneous leishmaniasis
Eye	*Acanthamoeba*	Corneal ulcers
Gut	*Giardia, Entamoeba, Cryptosporidium*	Giardiasis, Cryptosporidiosis
Bloodstream	*Plasmodium, Trypanosoma*	Malaria, African sleeping sickness
Spleen	*Leishmania*	Visceral leishmaniasis
Liver	*Entamoeba, Leishmania*	Visceral leishmaniasis
Muscle	*Trypanosoma*	Chaga's disease

enable the young suctoria to disperse from their parents. After this they lose their cilia and develop a stalk and feeding tentacles. The stalk is non-contractile and attaches to particulate material, whereas the feeding tentacle is capable of capturing, and then feeding, on other protozoa. This is achieved by piercing them and sucking in organic material from the cytoplasm to form food vacuoles. Suctoria found in wastewaters include *Acineta* and *Podophyra.*

2.4 Sporozoa

All species of the sporozoa are spore producing and have no apparent means of locomotion (in other words they lack cilia or flagella). They are all parasitic protozoa which require the presence of a host organism (such as humans, animals and fish) to complete their life cycle. They generally obtain nutrients by absorbing organic molecules from the host organism.

Sporozoans often have very complicated life cycles and are able to exploit a range of ecological niches in the human body, including the skin, eye, mouth, gut, blood, spleen, liver and muscle, with an associated medical condition (Table 4.2).

3 MOTILITY

Motility plays an important role in protozoal survival. Most protozoa are required to move in order to find food or to avoid unfavourable environmental conditions. The nutrition of protozoa also depends on the use of locomotory appendages to capture or collect food and direct it to the feeding apparatus. As a consequence, all free-living protozoa have some motility and can move freely in the environment at some time during their life cycle.

The different free-living groups have evolved diverse mechanisms for motility and these vary from the pseudopodia, utilized by the Sarcodina, through to the cilia and flagella employed by the ciliated and flagellated protozoa. Flagella and cilia are the most important organelles associated with movement in the protozoa and structurally they are very similar. However, whereas a cell will have only one or two flagella, ciliated cells have large numbers of cilia and this requires a complex system to coordinate their ciliary beat.

The motion of flagella is an undulatory wave beginning at the base of the flagellum and most flagella move only in a planar mode, although for some species, such as *Euglena*, movement is helical.

The pattern of waves generated by cilia is more difficult to discern. The numbers of cilia on each cell are more numerous, they are in close proximity to each other and the nature of the beat they generate is more complex than flagella. A cilium is characterized by a stroke which involves the cilium bending at the base while the rest of the cilium is straight. The cilium is then drawn back to its initial position close to the surface of the cell. This movement of flagella and cilia serves either to propel the protozoa through the water by the viscous forces which act on the cilia, or to generate water currents which permit feeding. The speed at which a protozoan moves is relatively constant and appears to be independent of the cell size. For ciliates this velocity is around 1 mm/s and for flagellates it is about 0.2 mm/s.

Many ciliated species can be observed not swimming, but walking or sliding along a solid surface and this is particularly evident in samples of activated sludge or trickling filter slime. It is thought that van der Waals forces help to hold the protozoa to these surfaces and that the cilia aid in their movement along the surface. Often with these species, the cilia are organized in dense bundles known as cirri.

The Sarcodina have a very different motion involving frontal contraction. The protozoa extend a pseudopodium, which then attaches to the surface. This pseudopodium retracts strongly and pulls the testate form forward. It appears that when two pseudopodia are extended by the same cell, but at 180° to each other, they will compete against each other with the one which is most firmly attached determining the direction of movement.

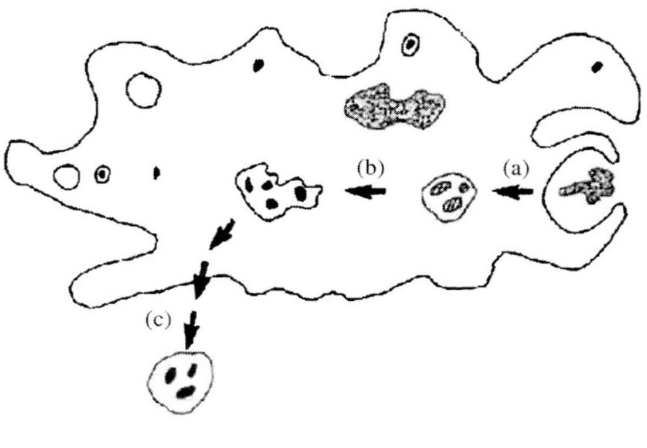

Fig. 4.2 Degradation of a food particle by phagacytosis. (a) Food particle engulfed by pseudopodia and a vacuole formed; (b) enzymic digestion occurs within the vacuole and digestion products released to the cytoplasm; (c) undigested remains ejected from the body.

4 PROTOZOAL NUTRITION

Protozoa demonstrate a wide range of feeding strategies of which four types are represented by the protozoa found in wastewater treatment systems. Certain members of the Phytomastigophorea are primary producers and capable of photoautotrophic nutrition, in addition to the more usual chemoheterotrophic nutrition.

Heterotrophy among the flagellated protozoa contributes to the process of biochemical oxygen demand (BOD) removal, and uptake of soluble organic material occurs either by diffusion or active transport. Protozoa that obtain their organic material in such a way are known as saprozoic, and are forced to compete with the more efficient heterotrophic bacteria for the available BOD. Amoebae and ciliated protozoa are also capable of forming a food vacuole around a solid food particle (which include bacteria) by a process known as phagocytosis. The organic content of the particle may then be utilized after enzymic digestion within the vacuole, a process which takes from 1 to 24 hours. This is known as holozoic or phagotrophic nutrition, and does not involve direct competition with bacteria, which are incapable of particle ingestion.

The final nutritional mode practised by the protozoa is that of predation. These predators are mainly ciliates, some of which are capable of feeding on algae (and are thus herbivores), as well as other ciliate and flagellate protozoal forms.

All protozoa rely on phagocytosis for their energy and carbon for building cellular material (Fig. 4.2). This involves the enclosure of a solid food particle in a vacuole, which is covered with a membrane and in which digestion occurs. Dissolved nutrients are removed from the vacuole leaving the indigestible remains behind. These are removed from the cell by fusion of the vacuole with the cell surface membrane. The typical lifetime of a food vacuole is around 20 minutes, although this time reduces if the cell is not feeding.

In addition to phagocytosis, there are other mechanisms by which a protozoan can obtain energy and cellular building blocks. Some protozoa participate in symbiotic relationships with photosynthetic organisms, whereas others are thought able to take up dissolved nutrients. It is doubtful, however, if this latter mechanism plays any role for the free-living protozoa outside of a laboratory culture.

Although phagocytosis is practised by all the protozoa there are a number of different feeding patterns which are exploited to capture the solids particle and these can be classified into three categories, namely: filter feeders, raptorial feeders and diffusion feeders. Filter feeding involves the creation of a feeding current, which is then passed through a device which acts to filter out the solids particles in the water. In the flagellates this is a collar of straight, rigid tentacles. For the ciliates the water is passed through an arrangement of parallel cilia. The clearance between the tentacles in the collar and the parallel ciliates, dictates the size of particle that is retained. This is typically between 0.3 and 1.5 μm and helps to explain why the presence of a healthy ciliate population in an activated sludge plant generates such a crystal clear effluent with a reduced number of faecal indicator bacteria (Table 4.3).

Raptorial feeding is practised in small flagellates and amoebae, which use it to feed on bacteria. In this mode, water currents are driven against the cell using a hairy anterior flagellum. Particles which make contact with a lip-like structure on the protozoa are phagocytized (Fig. 4.2). As each particle is captured separately (compared to filter feeders which retain all particles of the correct size), this allows the protozoa some discrimination as to what is ingested. This may be based on prey size or type, such as algae or small flagellates. Protozoa are also able to discriminate between different bacterial species with preferred types being selected.

Diffusion feeding is practised by the sarcodines. The suctorians are common protozoa in activated sludge and they feed by diffusion, largely on other ciliates. The suctorians are attached to a floc particle by a stalk and they have bundles of tentacles supported with an internal cylinder of microtubules. Ciliates which touch these tentacles become attached and immobilized. The tentacles then penetrate the attached ciliate and draw the contents through the tentacle into the suctorian.

TABLE 4.3 The effects of ciliated protozoa on the effluent quality from a bench-scale activated sludge plant (from Pike and Curds, 1971)

Parameter	*Without ciliates*	*With ciliates*
BOD (mg/l)	53–70	7–24
COD (mg/l)	198–250	124–142
Organic nitrogen (mg N/l)	14–21	10
Suspended solids (mg/l)	86–118	6–34
OD_{620}	0.95–1.42	0.23–0.34
Viable bacterial count (cfu/ml × 10^6)	106–160	1–9

Within the latter two modes of nutrition, the protozoa display a certain degree of selective feeding. Larger forms of amoebae are carnivorous, eating mainly ciliates and flagellates, whereas the smaller amoebae feed primarily on bacteria. The predatory suctorians are found to feed almost exclusively on holotrichous and spiriotrichous ciliates, with hypotrichs, flagellates and amoebae rarely being captured. Peritrichous ciliates are primarily bacterial feeders, but have a limited number of bacterial species upon which they can feed. Certain bacterial species are capable of supporting growth for long periods, whereas others induce starvation after a short time. In addition, many bacteria, in particular the pigmented types, prove toxic to those ciliates which ingest them.

5 PROTOZOAN REPRODUCTION

For the free-living protozoa, their growth curve involves an increase in cell size followed by some form of asexual reproduction such as binary fission. A single cell divides into two and generates two daughter cells. As the protozoan is not symmetrical the two daughter cells are not identical initially, however, the differences soon disappear. Sexual reproduction is only resorted to in times of stress or adversity, for instance if the food supply diminishes. There are many types of binary fission in which the plane of division varies among the different groups. Division is simplest for the naked amoebae as this has no clear division plane. The cells become rounded and division of cytoplasm into two equal daughter amoebae occurs. It is more complex in the testate amoebae as there is both protoplasm and skeletal structure to reproduce. The flagellated protozoa undergo division along a longitudinal plane and the ciliated protozoa divide along a transverse plane. The Suctoria have a sedentary lifestyle spent attached to particulate material. Thus a method of asexual reproduction involving fission would lead to a rapid increase in their population with increased competition for food and space. Consequently they divide by budding in which the new cell appears as a slight protuberance on the parent cell surface. This is evaginated and liberated, but unlike its parent it is motile in its immature stage, for around 30 minutes, which allows it to migrate and reduce crowding.

The most common form of protozoan reproduction is known as binary fission in which the organism divides into two equal-sized daughter cells. In the ciliated protozoa binary fission is usually transverse with the posterior end of the upper organism forming next to the anterior end of the lower one.

Binary fission generally produces daughter cells with genetic material (DNA) identical to that of the parent. It is an efficient way for protozoa to increase in number during periods when environmental conditions are relatively stable. However, when environmental conditions begin to change, sexual reproduction generally becomes more prevalent. Sexual reproduction allows for the mixing of DNA among the various strains (asexual daughters) of a local protozoan population. Although it has proved possible to maintain populations of protozoa for many years (equivalent to many hundreds of protozoan generations), the introduction of genetic exchange by sexual reproduction produces cells which are genetically different from each other, an important characteristic as genetic make-up determines how cells respond to their environment. A genetically diverse population is more able to adapt to changing conditions. In the competitive world of a wastewater treatment plant the more diverse the gene pool for a species, the greater the likelihood that it will persist over a wide range of changing environmental and operating regimes.

6 IMPORTANCE OF PROTOZOA IN WATER AND WASTEWATER TREATMENT

The importance of protozoa in water treatment systems focuses primarily on their ability to cause disease if ingested in a potable water source. Of particular importance in this respect are the protozoan parasites *Giardia lamblia* and *Cryptosporidium parvum*, which inhabit the gastrointestinal tract and are a major worldwide cause of human morbidity and a significant contribution to mortality. More recent opportunistic pathogens, which infect immunocomprised individuals, include *Cyclospora cayetanensis* and a number of genera from the Phyllum Microspora, such as *Encephalitozoon*, *Entreocytozoon* and *Septata*.

In addition to being parasites, protozoan populations are also subject to parasitism by a large number of fungi, bacteria and other protozoa. The opportunistic human pathogen *Legionella pneumophila* is one such bacterium which is able to parasitize the ciliated protozoan *Tetrahymena pyriformis* and it is thought this mechanism may help its distribution and survival in aquatic environments (Fields *et al.*, 1984).

In wastewater treatment, protozoa are known to play a major role but the full extent of their contribution is not fully quantified. Their major role is probably in maintaining a slime layer in trickling filter systems and in aiding flocculation in activated sludge systems. As shown earlier (Table 4.3), they aid in achieving a good effluent quality by their predatory role in removing bacteria and other small particles. The indicator value of protozoa in treatment systems, together with a fuller account of their role in wastewater treatment, is found in Chapter 22. A simplified key to identifying the protozoa of importance in wastewater treatment systems is found in Patterson (1998), which contains a very useful guide to the influence of organic loading rate and cell residence time on the distribution of protozoal species.

REFERENCES

Betts, W.B., Casemore, D., Fricker, C. *et al.* (1995). *Protozoan Parasites and Water*. Royal Society of Chemistry.

Fenchel, T. (1987). *Ecology of Protozoa: The Biology of Free-living Phagotrophic Protists*. Springer-Verlag.

Fields, B.S., Shotts, E.B., Feeley, J.C. *et al.* (1984). Proliferation of *Legionella pneumophila* as an intracellular parasite of the ciliated protozoan *Tetrahymena pyriformis*. *Applied and Environmental Microbiology* **47**, 467–471.

Jahn, T.L. (1979). *How to Know the Protozoa*. Wm C. Brown Co., Dubuque, IA.

Laybourn-Parry, J. (1984). *A Functional Biology of Free-living Protozoa*. Croom Helm, London and Sydney.

Patterson, D.J. (1998). *Free-living Freshwater Protozoa. A colour guide*. John Wiley & Sons, New York, Toronto.

Pike, E.B. and Curds, C.R. (1971). The microbial ecology of the activated sludge process. In: G. Sykes and F.A. Skinner (eds) *Microbial Aspects of Pollution*, pp. 123–147. Academic Press, London.

Ross, J. (2000). *Protozoans Teachers Handbook*. Insights Visual Productions Inc.

5

Filamentous fungi in water systems

Graham Kinsey, Russell Paterson and Joan Kelley

CABI Bioscience, Egham, Surrey, TW20 9TY, UK

1 INTRODUCTION

Fungi are ubiquitous, occupying every conceivable niche and habitat, displaying ecological strategies that range from micropredators and pathogens to more benign saprobes or symbionts (Dix and Webster, 1995). Such a diversity renders a brief circumscription of filamentous fungi near-impossible, but it includes eukaryotic microorganisms which, for at least some part of their life cycle, display a growth habit comprising a mycelium composed of individual hyphal elements, and which obtain nutrition by the absorption of extracellularly digested material. In many cases the mycelium lacks rigid self-support and is vulnerable to physical damage, but this is met in the majority of fungi by close adherence to surfaces or penetration of the substrate and growth occurring within. A detailed discussion on the biology of the fungal organism is given by Wessels and Meinhardt (1994).

As well as occupying a vast array of habitats, fungi utilize many different media and vectors for dispersal of their spores or other propagules such as sclerotia, sporangia or hyphal fragments. This can involve air (Lacey, 1996), water (Dix and Webster, 1995), plants, including their seeds (Richardson, 1996) animals (Andrews and Harris, 1997), and all kinds of organic and inorganic debris. Despite such a wide range of dispersal mechanisms (Ingold, 1971), some of which are highly-targeted, many fungi appear to rely upon high productivity and subsequent chance encounter with a substrate suitable for growth, and given the preponderance of chance encounter strategies among fungi, it is inevitable that many will find their way into water systems. Moreover, the fungi present are likely to be highly diverse and include more than just those primarily adapted for aquatic habitats.

Water is as essential for the growth of fungi as it is for any other organism and, although fungal dispersal structures are often resistant to desiccation, mycelial hyphae are generally poor at withstanding dry conditions. Most require comparatively high levels of water in the environment to enable vegetative growth (Dix and Webster, 1995). The majority of fungi prefer wet aerobic conditions and most of those that are known in culture can be grown in liquid media, in some cases including unamended water (Kelley *et al.*, 2001). Irrespective of a dependency on high water levels, many fungi excel as oligotrophs, scavenging nutrients from whatever sources that may be available, however dilute or recalcitrant (Parkinson *et al.*, 1989; Wainwright *et al.*, 1992, 1993; Gharieb and Gadd, 1999; Karunasena *et al.*, 2000). It is unsurprising therefore that fungi are frequently found in waters of all sorts and that their activities contribute to many of the microbial processes which are of concern to the water-related industries, including biodeterioration, bioremediation, biofilms, biofouling, as well to health effects and taste and odour issues.

Kelley *et al.* (1997) reviewed the literature on fungi from water distribution systems. They concluded that comparatively few systematic surveys of fungi in water distribution

The Handbook of Water and Wastewater Microbiology
ISBN 0-12-470100-0

systems had been carried out and, in many cases, fungi were detected only incidentally during the course of bacterial isolation programmes. Fungi are to be found in raw, surface and groundwaters and may increase in numbers in storage reservoirs. More fungi are present in surface water than in groundwater. More fungi are present in untreated than in (chlorine) treated water. Fungi are less susceptible to chlorine treatment than are bacteria. Sedimentation and flocculation remove many fungi, but rapid filtration is not an effective treatment. There is evidence to suggest that fungi survive and multiply in distribution systems in surface films and in sediments, particularly where conditions are warmer or flow rates are restricted. Fungi are capable of contributing off-tastes, odours and flavours, but no links to ill health had been reported.

Isolation techniques for fungi from water have also been evaluated (Anon., 1996). This has included an assessment of the diversity of fungi that can be detected and the reproducibility and significance of fungal counts. A study of fungal diversity and significance in the two USA distribution systems has also recently been conducted (Kelley *et al.*, 2001), involving samples taken from several points throughout the systems. This included an examination of the ability of selected fungi to grow in water and to produce metabolites of concern to human health, and also looked at the effectiveness of different control measures. Kinsey *et al.* (1999) have summarized many of the main considerations in sampling fungi from water.

2 STUDY METHODS

The main elements are sampling, detection, identification, quantification and enumeration. Implicit throughout all of these must be an awareness that, due to the exceptional diversity of fungi, there is no single approach that will be effective for all fungi. The study of fungi has relied on the development of technologies and techniques that can be adapted for the purpose (for a historical perspective see Webster, 1996). Increasing sophistication has led to some highly sensitive and specific techniques for particular fungal species (Paterson and Bridge, 1994; Bridge *et al.*, 1998) but, in many cases, these require further development before enabling determinations directly from environmental samples or for resolving mixed populations to species level. Hence, there is still a necessary reliance on more traditional microbiological methods.

2.1 Sampling

Standard procedures for taking water samples for microbiological examination are also adequate for fungi (Anon., 1994). Due to their filamentous nature fungi tend to be less evenly distributed in water than bacteria and so due consideration must be given to representativeness of sample volumes. Ideally, samples should be processed at the place and time of sampling, but, if as is usually the case, storage and transport is unavoidable then samples should be kept cool (2–10 °C and not exceeding the temperature at the site of sampling), and processed as soon as possible, preferably within 6 hours, or exceptionally up to 24 hours and the extent of the delay is recorded. Inadequate storage or delayed processing may mean that fungi present as hyphae are lost, while those present as spores may germinate and proliferate leading to disproportionate recording of the species present. Other factors which may affect fungi present in samples prior to processing include light and vibration. Exposure to light can affect fungal growth and sporulation; hence samples should be stored in the dark. Irregular sudden vibration can cause fragmentation of mycelia and may lead to dispersal of spores; conversely more regular repetitive motion can set up circular rotations which if prolonged can unite the mycelial biomass as a pellet.

There are inevitably interfaces with surfaces of pipework and other objects in any water system, and it is niches such as these where fungi are most likely to be found growing. If access is possible, such surfaces should be sampled by taking swab or sediment samples. Ideally sampling should occur immediately upon exposure to the site, any delay between exposure and sampling should be recorded. Once taken, the swabs or sediment should not be allowed to dry any further; again cool storage and prompt sample handling will help prevent this occurrence.

2.2 Detection

It is not usually possible to distinguish fungi in water samples with the naked eye and some form of concentration is usually required as an initial step. Filtration is often employed, similar to bacteriological methods (Anon., 1994), whereby samples are passed through 0.4 μm pore size filters and the filters retained for subsequent examination. Alternatively, centrifugation at 1500 g for 3 minutes is sufficient to pellet fungal material in 10 ml sample volumes. Although filtration is very convenient, it is possible that some fungi may be 'lost' on the filter, whereas centrifugation has the advantage of resulting in a pelleted mass that can be resuspended for further examination. Swabs or sediment samples can be handled directly. There are other methods that can be used as part of the initial stages of detection, such as enrichment or baiting techniques, and these are designed with specific fungi in mind (see Diversity below).

2.2.1 *Direct observation*

In the majority of cases it is too time-consuming to examine water samples directly. However, for rapid confirmation of the presence of fungi, especially in instances of heavy contamination, it may be worthwhile to examine swabs or scrapings by directly making a slide preparation for microscopy. It is also possible to observe filters directly, either by using light or electron microscopy (Anon., 1996). Similarly, pelleted samples can be directly observed by simply resuspending them in a drop or two of water. However, it is largely impractical to attempt subsequent removal of individual spores or hyphal elements for further study. Hence, direct observation is essentially limited to determination the nature of the fungal material (i.e. hyphal fragment or spore) and describing the morphology of the material (size, shape, colour, etc.). There is an established practice of examining fungal spores from environmental samples, particularly from the air (Lacey, 1996), but it has always been the case that when only spores or small fragments are available, it is rare to be able to identify the fungi with any confidence. The main exceptions are the aquatic and aero-aquatic hyphomycetes (see Diversity below), many of which have complex spores.

2.2.2 *Culture methods*

Many fungi can be cultured from their spores or hyphal fragments and, given suitable growth media (see entry under 'media' in Kirk *et al.*, 2001; Smith and Onions, 1994) and incubation conditions, will germinate, grow and eventually produce spores and spore-bearing structures (Arx, 1981). It is the details of complete reproductive structures that are required for species-level identification. The main disadvantage of culturing fungi is that it requires the intermediate and often time-consuming step of obtaining each organism in isolation on appropriate media. Nor is this guaranteed to result in an identification as not all fungi will grow in laboratory conditions, some never sporulate and, even more frustratingly, some fungi are identifiable only once – details of their primary host or other substrate specificity are known – data which are never likely to be known for cultures derived from spores present by chance in water samples. Nonetheless, culture-based techniques are the main methods for fungal analysis.

Filters through which samples have been passed can be plated directly onto the surface of nutrient agar media and the resultant fungal colonies can be counted and picked off for subculturing. Passing a range of volumes through separate filters avoids the possible extremes of too much or too little material being present as at least some will contain discrete fungal colonies sufficient for enumeration. However, it is preferable to resuspend fungi from filters by placing them in a tube with a known volume of sterile water. By varying the volumes into which the filters are resuspended it is possible to create a dilution series which can then be spread-plated or pour-plated onto the agar media. The resultant fungal colonies can be counted and picked off to prepare pure cultures for identification. The range of dilutions required is a trial and error process depending on the density of fungi present in the particular sample, but a range of 10-fold dilutions from 10^0–10^{-5} is sufficient for most cases. Fungal isolation programmes are prone to suffer from occasional fast-growing colonies which rapidly spread over the entire plate making subsequent counts and subculturing impossible; the only solution is vigilance and prompt picking-off, even to the point of examining plates with a dissecting microscope to detect colonies before they are visible with the naked eye. Colony counting can be continued if required by marking the Petri dish below each point of removal with a permanent pen.

There are many variations of culture methods for fungi, and one of the perennial areas of debate is choice of growth medium. There is a need to standardize media and associated incubation conditions and to some extent this has begun (Anon., 1996), but at the same time an awareness has to be maintained as to the limitations of individual growth media and culture techniques. A similar situation exists for the examination of fungi in soils, where limitations of media and techniques have been known and discussed for some time (Cannon, 1996).

2.2.3 *Biochemical methods*

There are emerging biochemical techniques which are showing considerable promise for use in detecting fungal populations in environmental samples.

Detection of fungi by HPLC analysis of ergosterol is one such technique which has been applied to fungi from a range of different environments (Newell *et al.*, 1987; Johnson and McGill, 1990; Martin *et al.*, 1990; Trevisan, 1996; Dales *et al.*, 1999; Pasanen, et al., 1999; Montgomery *et al.*, 2000; Saad *et al.*, 2001), including fungi from water samples (Gessner and Schwoerbel, 1991; Gessner and Chauvet, 1993; Suberkropp *et al.*, 1993; Newell, 1994; Bermingham *et al.*, 1995; Kelley *et al.*, 2001). Ergosterol is a membrane lipid found in virtually all filamentous fungi (Weete, 1989) but in very few other organisms (Newell *et al.*, 1987). The extraction of ergosterol from a sample is comparatively simple; a microwave-assisted method is among the most convenient (Young, 1995; Montgomery *et al.*, 2000). Using dried filters, this essentially involves the addition of methanol followed by brief (35 s) irradiation in a domestic 750 W microwave oven set at medium power and then extraction by using petroleum ether. Coupled with HPLC this enables rapid and very sensitive detection of any ergosterol present.

Chitin has also been used as an indicator of fungi (Martin *et al.*, 1990), but it is not specific to fungi nor are assays as sensitive or convenient as for ergosterol. There are many other techniques targeted towards gene products and metabolites but, whether employing chromatographic, spectroscopic, substrate-degradation or immunoaffinity methods (Paterson and Bridge, 1994; Kelley *et al.*, 2001), these tend to be specific for individual taxa and, as such, cannot be used for general detection of fungi in a sample. They do, however, have considerable value for analysis where only particular compounds are of concern (see Secondary Metabolites below).

With the continued development of genomic techniques, it is now possible to detect and analyse fungal DNA in environmental samples (Bridge *et al.*, 1998). These techniques are based on the polymerase chain reaction (PCR) (Edel, 1998). A widely used approach is to conduct PCR using oligonucleotide primers specific for the conserved flanking regions of the internally transcriber spacers (ITS) of the fungal rRNA gene (White *et al.*, 1990). The ITS primers are regarded as universal for fungi because rRNA genes appear to be present in most fungi (Gardes and Bruns, 1993). The technique has the potential for very sensitive detection of fungi in environmental samples. However, with environmental samples there is a need for caution as a range of factors can bias the DNA amplification, such as achieving initial DNA extractions for only a subset of the total species present, the presence of compounds inhibitory to amplification, obtaining primer-primer annealment, or the formation of chimeric co-amplified sequences (Wang and Wang, 1997; Hastings, 1999; Head, 1999). As PCR techniques continue to become more widely used, further development will undoubtedly resolve many of these potential problems (Viaud *et al.*, 2000).

It is usual to incorporate further analysis of the PCR product, and this essentially involves either sequence analysis, restriction fragment analysis, or the use of taxon specific probes or primers (Bridge *et al.*, 1998). In many cases the techniques target specific fungal taxa but some, although originally designed for bacteriological studies (examples in Hurst *et al.*, 1997; Edwards, 1999), are targeted much more broadly and are being adapted for analysing community structures in mixed fungal populations.

Suitable techniques include denaturing gradient gel electrophoresis (DGGE) (Hastings, 1999), temperature gradient gel electrophoresis (TGGE) (Hastings, 1999), and single strand conformational polymorphism (SSCP) (Fujita and Silver, 1994). DGGE involves running PCR product (using ITS primers) through a gel containing a gradient of denaturants; the sequence of the bases in the DNA determines at what point on the gel's gradient the strands denature and hinder further migration, so that on visualization the resulting bands reflect the diversity of fungi present in the original sample. DGGE is now beginning to be used for fungal populations (Kowalchuk *et al.*, 1997; Vainio and Hantula, 2000). TGGE is similar except that a fixed concentration of denaturant is used and the temperature is steadily increased during migration to mediate strand separation, which again is sequence-sensitive. Smit *et al.* (1999) applied TGGE to analyse wheat rhizosphere fungi. Both DGGE and TGGE provide profiles of fungal populations, but it is not possible to identify the fungi thus detected without comparison to the behaviour of test strains on the same gel and which therefore requires *a priori* knowledge of the fungi likely to be encountered, or by recourse to sequencing techniques. SCCP uses pre-denatured PCR product added to a non-denaturing gel, and as the strands partially re-anneal, the conformation they adopt is sequence-dependent and alters their mobility through the gel. Walsh *et al.* (1995) have used SSCP to discriminate a range of opportunistic fungal pathogens; identification is effectively by comparison with test strains and, as with DGGE and TGGE, requires *a priori* knowledge of the fungi likely to be encountered.

Another promising technique for population analysis is fluorescent *in situ* hybridization (FISH) (Lindrea *et al.*, 1999). This is somewhat different in that it involves intact cells, permeabilized by means of enzymes or detergents, to which species-specific oligonucleotide probes carrying fluorochromes are added; the probes enter the cells and hybridize with the complementary nucleic acid sequence if present, and the unbound probe is washed away prior to observation. Spear *et al.* (1999) have used FISH to analyse *Aureobasidium* populations on leaves but, while FISH has the advantage over DGGE and TGGE in that it enables immediate identification, it is dependent on developing sufficient probes to detect every possible fungal species that might be present, again requiring *a priori* knowledge of the fungi likely to be present.

Where individual fungal species are of interest it is possible to use much more specific PCR-based techniques, many of which are well-established for use with fungi (Bridge *et al.*, 1998), such as amplified fragment length polymorphism (AFLP), which has been used to study *Aspergillus niger*, *Fusarium graminearum* and *Penicillium chrysogenum* isolates from water systems (Kelley *et al.*, 2001).

2.2.4 *Assessing activity*

In some situations it may be required to detect fungal activity. Direct observation can only provide data on a presence-or-absence basis and does not necessarily even distinguish viable from non-viable. Culture methods only include viable organisms but do not distinguish between resting spores and active hyphae. The molecular methods described above likewise do not give an indication of whole-organism viability or activity. Techniques which can provide an indication of activity include staining of active regions (Cox and Thomas, 1999) or monitoring respiration and ATP (Suberkropp, 1991; Suberkropp *et al.*, 1993); however, with water systems there is the additional challenge, yet to be addressed, of how to sample representative volumes without altering the activity of the fungi in the process of collecting or concentrating the sample.

2.3 Identification

A consequence of the large range of fungal diversity and the need for an expert taxonomic knowledge is that there is a tendency for microbiological surveys to curtail fungal identifications once the level of genus has been reached. However, only in a very few cases does this provide useful information or enable specific cross-referencing to literature data on ecology, metabolites or health implications (for example Table 5.5 shows literature data on fungi producing taste and odour compounds). Identification of fungi is still heavily dependent upon microscopic observation of the physical appearance of both spores and spore-bearing structures, usually achieved by first obtaining the fungi in pure culture. Examination of just the mycelium rarely enables an identification, nor are spores alone sufficient as there are many species producing very similar spores. The UK has an organised network, the UK National Culture Collection (www.ukncc.co.uk), through which it is possible to obtain appropriate taxonomic advice and services on many microorganisms, including fungi.

2.4 Enumeration and quantification

2.4.1 Fungal counts

Enumeration of fungi usually relies on counts of colony-forming units (CFU) obtained by using dilution plates, and attempts critically to assess such methods have been made (Anon., 1996). The data generated by such means are notoriously difficult to analyse as there is very high variability and low reproducibility. The filamentous habit and the production of clustered spore masses by fungi contribute to causing this, i.e. the individual fungus may comprise a discrete sporulating colony but which, during the course of sampling and plating, can become fragmented into many smaller pieces and/or the spores dispersed. Hence, a single individual could give a count greater than one. The degree of fragmentation or spore dispersal is unlikely to be the same for different species, so that some fungi may be over-represented in relation to others. The fungal CFU count from an environmental sample, while it may give an indication of the potential inoculum carried in the sample, is therefore unlikely to reflect accurately the number of individual fungi initially present. Hence a considerable degree of caution is required when interpreting fungal count data. There are methods for statistical analysis, such as most probable number (MPN) estimation, but while this is well established for bacterial counts (Anon., 1994), it has been little used for fungal organisms (Feest and Madelin, 1985; Cooper, 1993). Despite the problems inherent in obtaining fungal counts, they nonetheless provide valuable information, particularly where an assessment is required of the potential for fungal contamination of materials that come into contact with the water.

2.4.2 Quantification

Determining biomass by dry weight is problematic as quantities in water samples are likely to be very low to the point of negligibility for most laboratory balances (i.e. sub-milligram). Metabolite analysis is a possible way forward and HPLC of ergosterol, as well as being a promising technique for fungal detection, has the potential to be used for quantification, albeit without discrimination of individual species (Newell *et al.*, 1987; Nilsson and Bjurman, 1990; Johnson and McGill, 1990; Schnürer, 1993; Pasanen *et al.*, 1999; Montgomery *et al.*, 2000). As a key membrane component, ergosterol is likely to provide a good indication of fungal surface area present in a sample rather than an accurate measure of biomass (Ruzicka *et al.*, 2000). Other measures have been used to estimate biomass including hyphal length (Schnürer, 1993) and ATP (Suberkropp, 1991;

Suberkropp *et al.*, 1993; Ruzicka *et al.*, 2000), but these require the fungi to be in grown in isolation from other organisms.

3 DIVERSITY

3.1 Higher taxa containing filamentous fungi

Fungi and fungi-like organisms are presently classified across three kingdoms, and those which show some form of filamentous growth are found in two of these kingdoms (Table 5.1) Kirk *et al.* (2001) provide background information and further references for all fungal taxa to the genus level.

3.2 Filamentous fungi in water

The term 'aquatic fungi' is somewhat misleading in that many fungi normally considered as terrestrial have been isolated from water samples. In many cases they comprise the majority of fungi detected (see Table 5.2) Similar species have been reported by others with the main genera broadly the same in many other listings of fungi from water samples (references in Kelley, et al., 1997). These 'terrestrial' fungi are predominantly the asexual hyphomycete states (genera such as *Acremonium, Alternaria, Aspergillus, Aureobasidium, Cladosporium, Penicillium* and *Trichoderma*) and coelomycete states (e.g. *Phoma*) of Ascomycota, and also include Zygomycota (*Absidia, Mucor, Mortierella* and *Rhizopus*) and Basidiomycota (see Table 5.3).Conversely, substrates such as soil can yield fungi that are interpreted as aquatic (Hall, 1996). Moreover, for the majority of fungi the ability to survive and grow in water has not been examined.

There is usually an implicit restriction of the term 'aquatic fungi' to mean freshwater fungi, with marine and brackish water fungi receiving separate treatment (see Kohlmeyer and Kohlmeyer, 1979; Moss, 1986; Kohlmeyer and Volkmann-Kohlmeyer, 1991; Hyde, 1997; Kis-Papo, 2001).

There will always be fungi which remain undetected. For example, it would be reasonable to expect common Urediniomycetes and Ustilaginomycetes (rust and smut fungi) to be present by chance in water samples as spores. However, as highly specific biotrophic plant pathogens, there are a few culture-based

TABLE 5.1 Higher taxa containing fungi and fungus-like organisms, taxa containing filamentous fungi are in bold (following Kirk et al., 2001)

Kingdom: PROTOZOA		
Kingdom: **CHROMISTA**		
	Phylum: Hyphochytriomycota	
	Phylum: Labyrinthulomycota	
	Phylum: **Oomycota**	Class: **Oomycetes**
Kingdom: **FUNGI**		
	('**lower fungi**')	
	Phylum: Chytridiomycota	Class: Trichomycetes
	Phylum: **Zygomycota**	Class: **Zygomycetes**
	(EUMYCOTA)	
	Phylum: ('**Symbiomycota**' or **Glomomycota)**	Class: ('**Symbiomycetes**' or **Glomomycetes)**
	Phylum: **Ascomycota**	Class: **Ascomycetes**
		Class: **Neolectomycetes**
		Class: Pneumocystidiomycetes
		Class: **Schizosaccharomycetes**
		Class: **Saccharomycetes**
		Class: **Taphrinomycetes**
	Phylum: **Basidiomycota**	Class: **Basidiomycetes**
		Class: **Urediniomycetes**
		Class: **Ustilaginomycetes**

TABLE 5.2 Numbers of samples in which fungi were detected in UK and US surveys (Anon., 1996; Kelley *et al.*, 2001)

Fungus	*USA (23 sites sampled)*	*UK (20 sites sampled)*	*Where present in treated water samples*	*Classification*
Absidia cylindrospora		3		Zygomycetes
Absidia glauca	2	3		Zygomycetes
Acremoniella atra	1			Ascomycetes
Acremonium cf. *strictum*	2			Ascomycetes
Acremonium simplex		1	UK	Ascomycetes
Acremonium spp	5	5	UK + USA	Ascomycetes
Actinomucor elegans	1			Zygomycetes
Alternaria alternata	12	3	UK + USA	Ascomycetes
Alternaria infectoria		2		Ascomycetes
Alternaria spp	1	1	USA	Ascomycetes
Apiospora montagnei	3	1		Ascomycetes
Arthrinium phaeospermum	1	4		Ascomycetes
Arthrographis cuboidea	1	2		Ascomycetes
Ascochyta spp	2	3	UK	Ascomycetes
Aspergillus aculeatus		2		Ascomycetes
Aspergillus candidus	1			Ascomycetes
Aspergillus clavatus		3		Ascomycetes
Aspergillus flavus		5	UK	Ascomycetes
Aspergillus fumigatus		3	UK	Ascomycetes
Aspergillus japonicus		1		Ascomycetes
Aspergillus niger	20	1	USA	Ascomycetes
Aspergillus parvulus		1		Ascomycetes
Aspergillus spp	1			Ascomycetes
Aspergillus terreus		2	UK	Ascomycetes
Aspergillus versicolor	6	3	UK + USA	Ascomycetes
Asteroma spp		1		Ascomycetes
Asteromella spp		1		Ascomycetes
Aureobasidium pullulans	2	1	UK	Ascomycetes
Basidiomycete	3		USA	Basidiomycetes
Beauvaria bassiana	2	1		Ascomycetes
Beauvaria brongniartii		1		Ascomycetes
Beauvaria spp	1			Ascomycetes
Botrytis cinerea		8	UK	Ascomycetes
Chaetocladium brefeldii		1		Zygomycetes
Chaetomium bostrychodes	1			Ascomycetes
Chaetomium globosum	1	1	UK	Ascomycetes
Chaetomium robustum	1			Ascomycetes
Chaetomium spp	2		USA	Ascomycetes
Chaetomium sphaerale	1			Ascomycetes
Chaetomium sulphureum	1			Ascomycetes
Cladorrhinum spp	2			Ascomycetes
Cladosporium cf. *herbarum*	1			Ascomycetes
Cladosporium cladosporioides	14	11	UK + USA	Ascomycetes
Cladosporium herbarum	20	13	UK + USA	Ascomycetes
Cladosporium oxysporum	16		USA	Ascomycetes
Cladosporium spp	8	4	UK + USA	Ascomycetes
Cladosporium sphaerospermum	8	1	UK + USA	Ascomycetes
Cladosporium tenuissimum	7		USA	Ascomycetes
Cochliobolus pallescens	1			Ascomycetes
Colletotrichum spp	1			Ascomycetes
Cryptococcus albidus	1			Basidiomycetes
Cryptococcus spp	1			Basidiomycetes

TABLE 5.2 (*continued*)

Fungus	*USA (23 sites sampled)*	*UK (20 sites sampled)*	*Where present in treated water samples*	*Classification*
Cyclothyrium spp		1		Ascomycetes
Cylindrocarpon magnusianum	1	1		Ascomycetes
Cytospora spp	1	4		Ascomycetes
Discosporium spp		1		Ascomycetes
Embellisia spp		1		Ascomycetes
Epicoccum nigrum	13	4	USA	Ascomycetes
Eupenicillium spp		1		Ascomycetes
Exophialia jeanselmei		1	UK	Ascomycetes
Fusarium acuminatum	1			Ascomycetes
Fusarium aquaeductuum		1		Ascomycetes
Fusarium avenaceum	1	6		Ascomycetes
Fusarium cf. *oxysporum*		2	UK	Ascomycetes
Fusarium cf. *solani*	1			Ascomycetes
Fusarium culmorum		8		Ascomycetes
Fusarium diamini		1	UK	Ascomycetes
Fusarium equiseti	4	1		Ascomycetes
Fusarium flocciferum		1		Ascomycetes
Fusarium graminearum	7		USA	Ascomycetes
Fusarium moniliforme		1		Ascomycetes
Fusarium oxysporum	2	3	UK + USA	Ascomycetes
Fusarium pallidoroseum	2			Ascomycetes
Fusarium soecheri	2			Ascomycetes
Fusarium solani	1	2	UK	Ascomycetes
Fusarium spp	2	3		Ascomycetes
Fusarium sporotrichioides	2		USA	Ascomycetes
Fusarium torulosum		1		Ascomycetes
Geomyces pannorum	1			Ascomycetes
Geotrichum candidum	3	4		Ascomycetes
Geotrichum spp	3	1		Ascomycetes
Gliocladium roseum	1	3	UK	Ascomycetes
Gliocladium spp		1		Ascomycetes
Gongronella butleri		1		Zygomycetes
Hormonema dematioides	1	1		Ascomycetes
Hormonema spp	1			Ascomycetes
Idriella bolleyi	4			Ascomycetes
Idriella sp.	1			Ascomycetes
Leptodontidium cf. *elatius*	1			Ascomycetes
Leptodontidium elatius	1		USA	Ascomycetes
Leptodontidium spp	3	1		Ascomycetes
Leptodothiorella spp		1		Ascomycetes
Leptosphaeria coniothyrium		2		Ascomycetes
Leptosphaeria fuckelii		2		Ascomycetes
Mauginiella spp		2	UK	Basidiomycetes
Microdochium nivale	1			Ascomycetes
Microdochium spp	1		USA	Ascomycetes
Micromucor isabellina		1		Zygomycetes
Micromucor ramannianus		3		Zygomycetes
Microsphaeropsis olivacea	2	2	USA	Ascomycetes
Microsphaeropsis spp		1		Ascomycetes
Mortierella alpina	1	1	UK	Zygomycetes
Mortierella elongata		1	UK	Zygomycetes
Mortierella zychae		1		Zygomycetes
Mucor circinelloides	5	3	USA	Zygomycetes
Mucor circinelloide sf. *janssenii*		2		Zygomycetes

(*continued on next page*)

TABLE 5.2 (*continued*)

Fungus	*USA (23 sites sampled)*	*UK (20 sites sampled)*	*Where present in treated water samples*	*Classification*
Mucor flavus	1			Zygomycetes
Mucor fuscus		1		Zygomycetes
Mucor genevensis	1			Zygomycetes
Mucor hiemalis	6	11	UK + USA	Zygomycetes
Mucor mucedo	1			Zygomycetes
Mucor plumbeus		1		Zygomycetes
Mucor racemosus f. *racemosus*	5	5		Zygomycetes
Mucor racemosus f. *sphaerosporus*		2		Zygomycetes
Mucor spp	4	3		Zygomycetes
Mucor strictus		1		Zygomycetes
Myrioconium spp	1			Ascomycetes
Nectria ingoluensis	1			Ascomycetes
Neurospora spp		1		Ascomycetes
Nigrospora sphaerica	2			Ascomycetes
Oomycete		1	UK	Oomycetes
Paecilomyces lilacinus	2	1		Ascomycetes
Penicillium aurantiogriseum	7	5	USA	Ascomycetes
Penicillium brevicompactum	8	9	USA	Ascomycetes
Penicillium chrysogenum	17	14	UK + USA	Ascomycetes
Penicillium citrinum	6	12	UK + USA	Ascomycetes
Penicillium corylophilum	1	2	UK	Ascomycetes
Penicillium crustosum		1		Ascomycetes
Penicillium echinulatum	1	1		Ascomycetes
Penicillium expansum	3	5	USA	Ascomycetes
Penicillium funiculosum	1			Ascomycetes
Penicillium glabrum	9	4	USA	Ascomycetes
Penicillium griseofulvum	1	1		Ascomycetes
Penicillium herquei	2			Ascomycetes
Penicillium hirsutum	1	1		Ascomycetes
Penicillium janczewskii		11	UK	Ascomycetes
Penicillium miniolutеum	1	1	UK + USA	Ascomycetes
Penicillium ochrosalmoneum		1		Ascomycetes
Penicillium oxalicum	10		USA	Ascomycetes
Penicillium paxilli	1		USA	Ascomycetes
Penicillium pinophilum	3	1	USA	Ascomycetes
Penicillium purpurogenum		10	UK	Ascomycetes
Penicillium raistrickii		2		Ascomycetes
Penicillium roquefortii	1		USA	Ascomycetes
Penicillium simplicissimum	2	2		Ascomycetes
Penicillium solitum		1		Ascomycetes
Penicillium spp		1		Ascomycetes
Penicillium spinulosum	6	7	UK + USA	Ascomycetes
Penicillium thomii	3		USA	Ascomycetes
Penicillium waksmanii		2		Ascomycetes
Periconia byssoides	1			Ascomycetes
Periconiella spp		1		Ascomycetes
Pestalotiopsis guepinii	1		USA	Ascomycetes
Phialophora cf. *malorum*	1		USA	Ascomycetes
Phialophora fastigiata		5	UK	Ascomycetes
Phialophora malorum	1		USA	Ascomycetes
Phialophora spp	4		USA	Ascomycetes
Phoma eupyrena	1			Ascomycetes
Phoma exigua		3	UK	Ascomycetes
Phoma fimeti	1			Ascomycetes

TABLE 5.2 (*continued*)

Fungus	*USA (23 sites sampled)*	*UK (20 sites sampled)*	*Where present in treated water samples*	*Classification*
Phoma glomerata	5		USA	Ascomycetes
Phoma herbarum	1			Ascomycetes
Phoma jolyana		2	UK	Ascomycetes
Phoma leveillei	1	5	UK	Ascomycetes
Phoma macrostoma	1	2	UK	Ascomycetes
Phoma medicaginis	1	1		Ascomycetes
Phoma nebulosa	1	2		Ascomycetes
Phoma sect. *Phyllostictoides*		1		Ascomycetes
Phoma spp	16	12	UK + USA	Ascomycetes
Phoma tropica	2			Ascomycetes
Phomopsis spp		3	UK	Ascomycetes
Pichia anomala	1		USA	Saccharomycetes
Pichia spp	1			Saccharomycetes
Pilidium concavum		2		Ascomycetes
Pithomyces chartarum	7		USA	Ascomycetes
Pithomyces sacchari	1	1		Ascomycetes
Preussia fleischakii	1			Ascomycetes
Pseudeurotium zonatum		1		Ascomycetes
Pycnidiophora dispersa		1		Ascomycetes
Pyrenochaeta spp		1		Ascomycetes
Rhizopus oryzae	2			Zygomycetes
Rhizopus stolonifer	1	9	UK	Zygomycetes
Robillarda sessilis	1			Ascomycetes
Saprolegnia ferax		1		Oomycetes
Saprolegnia parasitica		3		Oomycetes
Saprolegnia subterranea		1		Oomycetes
Scopulariopsis acremonium		1		Ascomycetes
Scopulariopsis brevicaulis		2		Ascomycetes
Scopulariopsis spp	1		USA	Ascomycetes
Seimatosporium lichenicola		1		Ascomycetes
Sesquicillium spp	1		USA	Ascomycetes
Sordaria fimicola	1			Ascomycetes
Sporothrix spp		2	UK	Ascomycetes
Sporotrichum pruinosum		1		Basidiomycetes
Stereum spp		1	UK	Basidiomycetes
Syncephalastrum racemosum		2		Zygomycetes
Trichoderma harzianum	9	11	UK + USA	Ascomycetes
Trichoderma koningii	3	7		Ascomycetes
Trichoderma piluliferum	3			Ascomycetes
Trichoderma polysporum	1	2		Ascomycetes
Trichoderma pseudokoningii	1	4		Ascomycetes
Trichoderma spp	5	2		Ascomycetes
Trichoderma strigosum	1			Ascomycetes
Trichosporiella sporotrichoides	1			Ascomycetes
Trichosporon spp	1			Ascomycetes
Truncatella angustata		1	UK	Ascomycetes
Ulocladium atrum		1		Ascomycetes
Ulocladium botrytis		1		Ascomycetes
Verticillium lecanii	1			Ascomycetes
Verticillium sp.		1		Ascomycetes
Zygomycete	2			Zygomycetes
Zygorhynchus heterogamus	4			Zygomycetes
Zygorhynchus moelleri	4	3		Zygomycetes

(*continued on next page*)

TABLE 5.2 (*continued*)

Fungus	*USA (23 sites sampled)*	*UK (20 sites sampled)*	*Where present in treated water samples*	*Classification*
Number of species present	UK = 136	USA = 141	Total species = 213	
Number of species in common	64		Species in common/total species = 0.3	

methods likely to detect them, but there is no reason to discount the possibility of their presence.

Fungi that are usually regarded as aquatic are often grouped informally by habitat preference and the nature of the spores.

3.2.1 *Aquatic hyphomycetes*

This is a grouping of asexual states of fungi which typically occur on submerged leaf litter and woody debris in rapidly flowing, well-aerated, non-polluted streams, and produce water-dispersed spores of elaborate appearance. It is thought the spore forms are adaptations to increase the likelihood of attaching to underwater surfaces in flowing waters (Dix and Webster, 1995). Sampling techniques for these fungi include filtration and baiting methods (Shearer and Lane, 1983; Shearer and Webster, 1985). The connected sexual states are ascomycetes and some basidiomycetes; these often occur on exposed plant debris near to streams (Webster, 1992; Shearer, 1993). There has been considerable interest in the ecology of aquatic hyphomycetes and it is known that they have a significant role, interacting with invertebrates, in the degradation of leaf litter and similar debris in streams and other bodies of water (Suberkropp, 1991; Barlocher, 1992; Suberkropp *et al.*, 1993; Dix and Webster, 1995; Bermingham, 1996; Gessner *et al.*, 1997).

3.2.2 *Aero-aquatic hyphomycetes*

These are asexual states of fungi found on plant debris in still and slow-flowing habitats, typically at the edges of ponds and ditches where sporulation occurs on newly-exposed leaf litter which is still moist but exposed to air (Dix and Webster, 1995). The conidia are complex, often helicoid, trapping air as an aid to buoyancy and hence dispersal. As with the aquatic hyphomycetes, the connected sexual states are ascomycetes and basidiomycetes. These fungi grow best under aerobic conditions but can survive prolonged periods of anaerobic conditions and of drought, suggesting that these may be among the most significant decomposer fungi in waters prone to fluctuating levels. Isolation methods include baiting and damp-chamber incubation of litter (Fisher, 1977).

3.2.3 *Zoosporic fungi*

Filamentous fungi producing zoospores belong to the Oomycetes (sometimes referred to as 'water moulds') and include well-known genera such as the plant-pathogenic genera *Pythium* and *Phytophthora*, pathogens of aquatic vertebrates such as species of *Achlya* and *Saprolegnia,* as well as members of the 'sewage fungus' complex such as *Leptomitus lacteus* and allied genera such as the fermentative *Aqualinderella*. Methods for isolation and study have

TABLE 5.3 Classification of fungi detected in UK and US water surveys (Anon., 1996; Kelley *et al.*, 2001)

UK and US survey data combined					
Ascomycetes	*Basidiomycetes*	*Oomycetes*	*Saccharomycetes*	*Zygomycetes*	*Total*
173	6	4	2	28	213
81	3	2	1	13	%

recently been discussed by Hall (1996) and Dix and Webster (1995) and involve specialized baiting and other techniques in order to detect their presence.

There are other fungi-like organisms, such as chytrids, that also produce zoospores but these are non-filamentous and are not considered here (for references see Kirk *et al.*, 2001).

3.2.4 *Other fungi from water*

In some cases fungi are still found which are not easily related to existing descriptions (e.g. Willoughby, 2001), indicating that there is continuing scope for discovery of new fungi in aquatic environments, subject to the development of sampling and culture techniques.

3.3 Structure of populations

The above-mentioned uncertainties surrounding the enumeration of fungi make it difficult to describe mixed fungal populations quantitatively. Often it is only possible to indicate which fungal species are detected; however, as more surveys are completed, which provide detailed species information, it becomes possible to gain an idea of which species and genera are most frequently seen. Tables 5.2 and 5.3 show the fungi detected in two different surveys and, in particular, highlight those fungi present in both the US and UK post-treatment water samples. These include: *Acremonium* spp., *Alternaria alternata*, *Aspergillus versicolor*, *Cladosporium cladosporioides*, *C. herbarum*, *Cladosporium* spp., *C. sphaerospermum*, *Fusarium oxysporum*, *Mucor hiemalis*, *Penicillium chrysogenum*, *P. citrinum*, *P. miniolutеum*, *P. spinulosum*, *Phoma* spp. and *Trichoderma harzianum*. While this list is unlikely to be exhaustive, it is interesting to note, that although these fungi belong to large multispecies genera, only a small subset of the known species for each genus is represented. In many cases therefore, species-level differences may be more important than generic or higher level differences in determining persistent or frequent occurrence in water systems.

TABLE 5.4 Species numbers in samples from UK and US surveys (Anon., 1996; Kelley *et al.*, 2001)

Country	*Mean number of species present per sample ± standard error*	
	Untreated water	*Treated water*
UK	30.7 ± 1.1 (n = 10)	8.5 ± 1.2 (n = 10)
USA	29.6 ± 5.1 (n = 10)	9.9 ± 1.1 (n = 13)

3.3.1 *Comparing samples*

It is often required to analyse the diversity of species present. Samples can be compared in terms of overall numbers of species present, as exemplified in Table 5.4, which shows that samples of treated water in both the US and UK surveys contain approximately one third of the numbers of untreated waters. In order to compare species composition, samples can be analysed in terms of the proportion of the species in common between sites relative to the total number of species detected for both sites; comparing the US and UK species shown in Table 5.2, this ratio is 0.3 (sometimes known as the Jaccard coefficient; see Anon., 1990). Given that the US and UK surveys yielded 141 and 136 species respectively (and the similarity of these numbers of species is also worth noting), this indicates a moderately high overlap with just under 50% of all species present in the US survey also appearing in the UK survey and *vice versa*. There are other numerical treatments, which include weightings for frequency or abundance of individual species, such as calculating species richness estimators (Dighton, 1994; Colwell, 1997), but with fungi the difficulties of enumeration often render this an uncertain exercise.

3.4 Environmental factors

Dix and Webster (1995) and Magan (1997) have reviewed the tolerance of fungi to extreme environmental conditions. Many of the fungi detected in water systems are likely to be allochthonous, with the majority transiently passing through as spores. In which case, the main concern is simply in the ability of the fungi to withstand a range of conditions in order to maintain viability until subsequent deposition. However, there is also an indication that fungi can grow in water systems (Kelley *et al.*, 1997, 2001), and in this case it is important to appreciate the range of conditions that are favourable for fungal growth.

3.4.1 *Habitats and niches within distribution systems*

Fungi have been reported from all types of water systems from raw waters to bottled drinking waters and from heavily polluted waters to distilled or ultra-pure water (Kelley *et al.*, 1997). They may form floating mats of thin finely branched hyphae in static water, or may form pelleted colonies in sustained circular flows. They may also grow attached to a substrate, forming part of microbial biofilms on pipework inner surfaces, debris or sediments, and are particularly likely to become established where there are cracks, pitting or dead ends.

3.4.2 *Temperature*

The majority of fungi are mesophilic, capable of growth between 5 and 35 °C, with the optimum often between 25 and 30 °C (Smith and Onions, 1994; Dix and Webster, 1995; Magan, 1997), which covers the temperatures found in most water systems including the tropics (Yuen *et al.*, 1998). In terms of extremes, there are psychrotolerant fungi, such as some species of *Cryptococcus*, which are capable of growth at or close to 0°C. Psychrophilic fungi, including some species of *Mucor*, are unable to grow above 20°C. However, many fungi are capable of withstanding temperatures down to 0°C or lower, with survival depending on the cell water content and the rate of freezing (Smith and Onions, 1994); generally spores survive better than hyphae. In terms of upper temperatures, the maximum for growth of thermotolerant fungi, such as *Aspergillus fumigatus*, is in the region of 50°C and for some thermophiles, such as *Thermomyces lanuginosus*, approximately 60°C. These thermotolerant and thermophilic species have been reported from heated aquatic habitats such as cooling waters at power stations (Ellis, 1980; Novickaja–Markovskaja, 1995). The majority of fungi are readily killed by temperatures at or above 80 °C, although a few such as *Talaromyces flavus* can withstand this temperature for an hour; survival again depends on the cell water content, with spores surviving better than hyphae (Magan, 1997).

3.4.3 *pH*

Fungi have been detected in extremely acidic and alkaline waters and many fungi when tested can grow over a wide range of pH, often between 2 and 11, although how this correlates to the ability to grow in environments subject to extremes of pH is in many cases still unclear, as other environmental factors are often closely linked (Magan, 1997). An example is the River Tinto in Spain which has been associated with mining for over 5000 years; it presently has a pH of 2–2.5 and contains ferric iron at 2.5 g/l and other heavy metals including Cu, Zn, Co, Mn, Ni as well as sulphate at 10–15 g/l and has high redox potential and conductivity. Nonetheless, a study has yielded 350 fungal isolates belonging to *Bahusakala*, *Heteroconium*, *Mortierella*, *Penicillium*, *Phialophora* (as *Lecythophora*), *Phoma* and *Scytalidium*. Of the isolates, 44% were capable of growth in the conditions present in the river (López–Archilla *et al.*, 1996).

3.4.4 *Aeration*

The majority of fungi are aerobic with a wide range of genera showing the capacity to grow under low oxygen conditions (Dix and Webster, 1995; Magan, 1997). While fungal microaerophilic capabilities have been studied in soils and plant materials it appears that there have been no studies of fungi in water systems in which the oxygen levels have been accurately measured (Kelley *et al.*, 1997). Many fungi, provided appropriate nutrient sources are present (such as fatty acids, sterols and vitamins), are facultative anaerobes. This includes species of *Fusarium*, *Mucor*, *Penicillium* and *Trichoderma* as well as Saccharomycetes. Other facultatively anaerobic fungi found growing in stagnant waters include members of the Oomycete genera *Aqualinderella*, *Mindeniella*, *Rhipidium* and *Sapromyces*. In the case of *Aqualinderella fermentans*, it is incapable of aerobic respiration. Interestingly, the related *Leptomitus lacteus*, although found in heavily polluted waters, is not a facultative anaerobe. The aero-aquatic hyphomycetes mentioned earlier are capable of surviving prolonged anaerobic conditions.

3.4.5 *Nutrients*

Many fungi have the ability to grow in apparently nutrient-free conditions, including unamended water, scavenging nutrients from whatever sources that may be available, however dilute or recalcitrant, even the atmosphere (Parkinson *et al.*, 1989; Wainwright *et al.*, 1992, 1993; Kelley *et al.*, 1997, 2001; Gharieb and Gadd, 1999; Karunasena *et al.*, 2000). The utilization of simple sugars as C-sources occurs in virtually all fungi. Many are also well adapted to make use of more complex C polymers of plant and animal origin, especially cellulose-based materials (see Gessner*et al.*, 1997). The main exceptions are many of the Zygomycetes and saprobic Oomycetes, which appear to be restricted to simple sugars. In terms of N utilization, many fungi can grow over a wide range of concentrations and on a wide range of N-sources. Some show adaptations to high N levels, as in the case of coprophilous fungi, or to very low levels as in wood-decay fungi (Dix and Webster, 1995).

3.4.6 *Flow rates*

For those fungi transiently passing through a system, water flow rate is largely immaterial, although the consequent length of time spent in the system may affect viability. Whether attached to surfaces or otherwise, many other fungi appear to grow equally in either standing or moving water (Kelley *et al.*, 1997, 2001). Extremely high flow rates or turbulence could conceivably cause cellular damage or inhibit colonization, but there is a lack of detailed information for fungi. Aquatic hyphomyetes favour flowing water over standing water and there is an indication that spore shape may in some species be an adaptation to assist anchorage in such conditions (Webster and Davey, 1984).

3.4.7 *Light*

Growth of many species occurs equally in the dark or in daylight. Near-ultraviolet light of wavelengths between 300 and 380 nm is routinely used to induce sporulation in many fungi, while others respond better to darkness or to diurnal cycles (Smith and Onions, 1994). Ultraviolet light has also been used as a treatment against fungi in water (see references in Kelley *et al.*, 1997).

3.4.8 *Pollution and global environmental change*

Wainwright and Gadd (1997) have reviewed the effects of industrial pollutants on fungi, particularly acid rain, heavy metals and radionuclides. Apart from the well known 'sewage fungus' complex (the component organisms include unrelated fungi such as *Leptomitus lacteus* and species of *Fusarium* and *Geotrichum*) found in low pH water heavily contaminated with organic matter, much of the other data concerns terrestrial fungi, especially mycorrhizal species and those which produce large fruitbodies. While there is a strong indication that many fungi are capable of accumulating pollutants, the distinction between survival and tolerance is not always clear, nor yet are the longer-term effects of either toxicity or eutrophication on the composition of fungal communities understood. For example, Bermingham (1996) has observed aquatic hyphomycete communities in a river subject to contamination by heavy metals. An effect of the contamination was a reduction in leaf litter decomposition. Experimental tests showed that sporulation was inhibited by the metals but mycelial growth continued. Since identification of aquatic hyphomycetes is reliant on determination of spore morphologies, it is therefore unclear if the change in community function actually represented a loss of species.

4 CLEANING AND DISINFECTION TREATMENTS

The effectiveness of disinfection treatments in removing fungi from water has been reviewed by Kelley *et al.* (1997). The most commonly used treatment is chlorination, to which fungal spores are more resistant than hyphae. It was suggested that an initial residual level of 1 mg/l chlorine is required for use against fungi. However, they also pointed out examples showing that much higher concentrations can be required for control of certain species such as 129 mg/l for *Botrytis cinerea*. The survey data shown in Table 5.2 indicate that some fungi can pass through treatment systems into distribution systems.

Kelley *et al.* (2001) examined the effectiveness of different treatment strategies to control fungi. This included physical treatments (clarification, rapid gravity filtration) and chemical disinfection (monochloramine, chlorine, chlorine dioxide and ozone), as well as mains cleaning (flushing, air-scouring, swabbing, pigging and flow-jetting). They concluded that physical treatments performed well, with clarification removing over 70% of fungi, while filtration removed over 90% of fungi, irrespective of whether sand or granular activated carbon was used. Efficacy of chemical disinfectants varied from species to species but, overall, chlorine dioxide and ozone were the most powerful of those tested. However, monochloramine appeared more stable in distribution systems and may therefore have a greater long-term efficacy. It was also found that, of the disinfectants, chlorine was the most affected by variation in temperature. Of the mains cleaning treatments, none gave particularly good results. Flushing and air-scouring were least effective, albeit also among the least damaging to the pipe. Swabbing only gave good results on smooth-surfaced plastic pipes but not on rough-surfaced cast iron pipes. Pigging, although very aggressive, also gave poor results due to the limited contact that can be achieved between the bristles and the pipe surface. Flow-jetting was reported as the most effective and was not affected by pipe materials.

5 SIGNIFICANCE OF FUNGAL ACTIVITIES

5.1 Biofilms and biofouling

There have been very few reports of fungi in biofilms (Kelley *et al.*, 1997). In many cases, this is probably a consequence of the use of methods which do not target fungi rather than an absence of fungi. Nagy and Olson (1985) reported fungi belonging to *Acremonium, Alternaria, Epicoccum, Candida, Cryptococcus, Penicillium, Rhodotorula* and other species from pipe surfaces. More recently, Dogget (2000) reported fungi from biofilms in a municipal water distribution system, including members of *Acremonium, Alternaria, Aspergillus, Aureobasidium, Candida, Cladosporium, Cryptococccus, Mucor, Penicillium, Phoma* and *Stachybotrys* as well as other fungal species. In addition, during the surveys reported by Kelley *et al.* (2001), swab samples from pipes at sites of reported taste and odour incidents yielded fungi belonging to *Aspergillus, Cladosporium, Cryptococcus, Pichia, Rhodotorula* and *Yarrowia*. Fungi, such as *Aspergillus, Fusarium, Mucor, Penicillium* and *Trichoderma* species, have been recovered from biofouled membranes of reverse osmosis systems (Dudley and Christopher, 1999). At present there is little information on the role of fungi in causing the formation and stabilization of biofilms in water systems.

5.2 Biodegradation and bioremediation

There are many proposed strategies for biological treatment of wastewater and run-off including constructed wetlands such as reedbeds (Haberl *et al.*, 1997). Fungi are widely recognized as being important organisms for many biodegradative and bioremediative processes (Gadd *et al.*, 1987; Wainwright *et al.*, 1987; Frankland *et al.*, 1996), but there is a surprising lack of information on their activities or potential for use in treatment of water. For example, one of the main macrophytes used in reedbeds is *Phragmites* and about one hundred fungal species have been recorded in, on or around *Phragmites* worldwide (as shown in the databases of the herbaria of CABI Bioscience, incorporating the former International Mycological Institute, and the Royal Botanic Gardens Kew). With such a known list of associated fungal species it is difficult to

TABLE 5.5 Compounds produced by fungi which cause off-tastes or odours (from Kelley *et al.*, 1997a,b)

Fungus	*Compound produced*	*Relative production*	*Reference*
Acremonium spp[1]	oct-1-en-3-ol	major	Kaminski *et al.* (1980)
Alternaria spp	oct-1-en-3-ol	major	Kaminski *et al.* (1980)
Aspergillus spp	2-methylisoborneol	minor	Anderson *et al.* (1995)
Aspergillus niger	oct-1-en-3-ol	major	Kaminski *et al.* (1980)
Aspergillus ocraceus	oct-1-en-3-ol	major	Kaminski *et al.* (1980)
Aspergillus oryzae	oct-1-en-3-ol	major	Kaminski *et al.* (1980)
Aspergillus parasiticus	oct-1-en-3-ol	minor	Kaminski *et al.* (1980)
Basidiobolus ranarum	Geosmin	no data	Lechevalier (1974)
Botrytis cinerea	furfural, benzaldehyde, phenylacetaldehyde, benzyl cyanide	no data	Kikuchi *et al.* (1983)
Chaetomium globosum	geosmin, 2-phenylethanol	no data	Kikuchi *et al.* (1983)
Cladosporium spp	2-methylisoborneol	major	Anderson *et al.* (1995)
Fusarium spp	oct-1-en-3-ol	major	Kaminski *et al.* (1980)
Geotrichum spp	2,4,6-TCA	minor	Anderson *et al.* (1995)
Mucor spp	2,4,6-TCA/2,3,4-TCA	minor	Anderson *et al.* (1995)
Penicillium aethiopicum	Geosmin	minor	Larsen and Frisvad (1995)
Penicillium aurantiogriseum	oct-1,3-diene	minor	Larsen and Frisvad (1995)
Penicillium camembertii	2-methylisoborneol, oct-1-en-3-ol	minor, minor	Larsen and Frisvad (1995)
Penicillium chrysogenum	oct-1,3-diene, oct-1-en-3-ol	minor, minor	Larsen and Frisvad (1995), Kaminski *et al.* (1980)
Penicillium citrinum	oct-1-en-3-ol	major	Kaminski *et al.* (1980)
Penicillium clavigerum	Geosmin	major	Larsen and Frisvad (1995)
Penicillium commune	2-methylisoborneol, oct-1,3-diene, oct-1-en-3-ol	minor, minor, minor	Larsen and Frisvad (1995)
Penicillium coprobium	oct-1,3-diene	minor	Larsen and Frisvad (1995)
Penicillium crustosum	2-methylisoborneol, geosmin, dimethyldisuphide	minor, minor, minor	Larsen and Frisvad (1995)
Penicillium decumbens	oct-1-en-3-ol	not given	Halim *et al.* (1975)
Penicillium discolor	2-methylisoborneol, geosmin	major, major	Larsen and Frisvad (1995)
Penicillium echinulatum	Geosmin	minor	Larsen and Frisvad (1995)
Penicillium expansum	Geosmin	no data	Mattheis and Roberts (1992), Dionigi and Champagne (1995)
Penicillium formosanum	Geosmin	major	Larsen and Frisvad (1995)
Penicillium freii	oct-1,3-diene	minor	Larsen and Frisvad (1995)
Penicillium funiculosum	oct-1-en-3-ol	major	Kaminski *et al.* (1980)
Penicillium glabrum	oct-1,3-diene	minor	Larsen and Frisvad (1995)
Penicillium hirsutum var. *venetum*	Geosmin	minor	Larsen and Frisvad (1995)
Penicillium polonicum	2-methylisoborneol	minor	Larsen and Frisvad (1995)
Penicillium raistricki	oct-1-en-3-ol	major	Kaminski *et al.* (1980)
Penicillium roqueforti var. *carneum*	Geosmin	minor	Larsen and Frisvad (1995)
Penicillium solitum	2-methylisoborneol	minor	Larsen and Frisvad (1995)
Penicillium spp	2,4,6-TCA	no data	Nyström *et al.* (1992)
Penicillium sp. #1	2-methylisoborneol	minor	Anderson *et al.* (1995)
Penicillium sp. #2	2,4,6-TCA/2,3,4-TCA	minor	Anderson *et al.* (1995)
Penicillium tricolor	oct-1,3-diene, oct-1-en-3-ol	minor, major	Larsen and Frisvad (1995)
Penicillium viridicatum	oct-1-en-3-ol*	minor/major	Larsen and Frisvad (1995); Kaminski *et al.* (1980)
Penicillium vulpinum	2-methylisoborneol, oct-1,3-diene	minor, major	Larsen and Frisvad (1995)
Phialophora spp	2,4,6-TCA	no data	Nyström *et al.* (1992)
basidiomycete fungi	oct-1-en-3-ol, oct-1-en-3-one, 1,3-octadiene	high	Jüttner (1990)

[1] as *Cephalosporium* sp.

Note: Oct-1-en-3-ol has also been found in *Agaricus bisporus* and *Boletus edulis* (Kaminski et al., 1980) but it is virtually certain that these fungi would not be found in drinking water.

imagine that fungi do not play a significant role in the functioning of reedbeds, or in other biological treatment systems.

5.3 Secondary metabolites

Fungi produce a wide range of secondary metabolites including toxic compounds (Moss, 1996; Scudamore, 1998), many of which have implications for human health (e.g. Richard *et al.*, 1999). Paterson et al., (1997b) compiled a listing of the mycotoxins known for many of the fungi isolated from water samples. Fungi also produce many compounds liable to cause off-tastes and odours (Table 5.5). There is also concern over oestrogenic compounds and fungi, such as some species of *Fusarium*, are known to produce zearalanone, an oestrogenic compound (Aldridge and Tahourdin, 1998). Kelley *et al.* (2001) have demonstrated production of secondary metabolites by fungi in water including zearalanone by *Fusarium graminearum*, and have detected aflatoxin in a water storage tank contaminated by *Aspergillus flavus* by using immunoaffinity columns (see also Paterson *et al.*, 1997b).

5.4 Water as a source of fungal inoculum

The majority of fungi detected in water samples are plurivorous saprobes of decaying plant materials. Some of these, however, are also secondary/opportunistic pathogenic organisms of plants or animals, including humans (Hoog, 1996; Warnock and Campbell, 1996).

5.4.1 *Pathogenic fungi*

Human pathogenic fungi can be transported in water. Of those listed in Table 5.2, many are also reported as potential secondary pathogens (such as *Aspergillus* species, many of the Zygomycetes, as well as species of *Alternaria*, *Cladosporium*, *Phoma*, and others), for detailed listings and further references see Smith (1989). In most instances these fungi cause no problems for healthy adults, although there may be a potential risk for immunocompromised individuals. In critical environments water can be a source of fungal infection. For example, Warris *et al.* (2001) examined water samples taken from the paediatric bone marrow transplantation (BMT) unit of the National Hospital University of Oslo, Norway; 168 water samples and 20 surface-related samples from taps and showers in the BMT unit and from the main pipe supplying the paediatric department were taken on 16 different days over an 8-month period. Filamentous fungi were recovered from 94% of all the water samples taken inside the hospital. *Aspergillus fumigatus* was recovered from 49% and 5.6% of water samples from the taps and showers respectively. More than one-third of water samples from the main pipe revealed *A. fumigatus* and it was suggested that the source of contamination was located outside the hospital.

Plant pathogenic fungi can also be transported in water, and many of those listed in Table 2.2 are also capable of acting as secondary plant pathogens (such as members of *Alternaria*, *Cladosporium*, *Fusarium*, *Phoma* and others). Untreated water used in irrigation systems may deliver an inoculum of potential plant pathogenic fungi directly to crop plants under ideal conditions for infection (i.e. creating wet conditions). Hydroponic systems are most vulnerable, but any irrigated system, especially in glasshouses, are also liable to be at heightened risk of disease unless appropriate measures are taken (Jarvis, 1992).

6 CONCLUSIONS

Fungi are a significant part of the microbial population in water systems of all types. Their activities influence both water quality and human health. At present there is a need better to understand and quantify the composition and significance of fungal populations.

REFERENCES

Aldridge, D. and Tahourdin, C. (1998). Natural oestrogenic compounds. In: D.H. Watson (ed.) *Natural Toxicants in Food*, pp. 55–83. Sheffield Academic Press, Sheffield, UK.

Anderson, S.D., Hastings, D., Rossmore, K. and Bland, J.L. (1995). Solving the puzzle of an off-spec. 'musty' taste in canned beer. A partnership approach. *MBAA Technical Quarterly* **32**, 95–101.

Andrews, J.H. and Harris, R.F. (1997). Dormancy, germination, growth, sporulation and dispersal. In: D.T. Wicklow and B. Söderström (eds) *The Mycota IV. Environmental and Microbial Relationships*, pp. 3–13. Springer-Verlag, Berlin.

Anon. (1990). *SPSS (Statistical package for social sciences) Reference Guide, Release 3.1.* SPSS, Chicago, USA.

Anon. (1994). *The Microbiology of Water 1994 Part 1 – Drinking Water. Report on Public Health and Medical Subjects No. 71 Methods for the Microbiological Examination of Waters and Associated Materials.* Her Majesty's Stationery Office, London.

Anon. (1996). *Significance of fungi in water distribution systems. (EPG/1/9/69). Final Report to DWI.* Report No. DWI0780. Drinking Water Inspectorate, London.

von Arx, J.A. (1981). *The Genera of Fungi Sporulating in Pure Culture*, 3rd edn. Cramer, Vaduz.

F. Bärlocher (ed.) (1992). *The Ecology of Aquatic Hyphomycetes.* Springer-Verlag, Berlin.

Bermingham, S. (1996). Effects of pollutants on aquatic hyphomycetes colonizing leaf material in freshwaters. In: J.C. Frankland, N. Magan and G.M. Gadd (eds) *Fungi and Environmental Change.* (Symposium of the British Mycological Society held at Cranfield University, March 1994), pp. 201–216. Cambridge University Press, Cambridge.

Bermingham, S., Maltby, L. and Cooke, R.C. (1995). A critical assessment of the validity of ergosterol as an indicator of fungal biomass. *Mycological Research* **99**(4), 479–484.

Bridge, P.D., Arora, D.K., Reddy, C.A. and R.P. Elander, (eds) (1998). *Applications of PCR in Mycology.* CABI Publishing, Wallingford.

Cannon, P.F. (1996). Filamentous fungi. In: G.S. Hall (ed.) *Methods for the Examination of Organismal Diversity in Soils and Sediments*, pp. 125–143. CABI Publishing, Wallingford.

Colwell, R.K. (1997). *Estimates: Statistical estimation of species richness and shared species from samples.* Version 5. User's Guide and application published at: http://viceroy.eeb.uconn.edu/estimates

Cooper, J.A. (1993). Estimation of zoospore density by dilution assay. *The Mycologist* **7**(3), 113–115.

Cox, P.W. and Thomas, C.R. (1999). Assessment of the activity of filamentous fungi using Mag fura. *Mycological Research* **103**(6), 757–763.

Dales, R.E., Millar, D. and White, J. (1999). Testing the association between residential fungus and health using ergosterol measures and cough recordings. *Mycopathologia* **147**, 21–27.

Dighton, J. (1994). Analysis of micromycete communities in soil: a critique of methods. *Mycological Research* **98**(7), 796–798.

Dionigi, C.P. and Champagne, E.T. (1995). Copper-containing aquatic herbicides increase geosmin biosynthesis by *Streptomyces tendae* and *Penicillium expansum. Weed Science* **43**, 196–200.

Dix, N.J. and Webster, J. (1995). *Fungal Ecology.* Chapman and Hall, London.

Doggett, M.S. (2000). Characterization of fungal biofilms within a municipal water distribution system. *Applied and Environmental Microbiology* **66**(3), 1249–1251.

Dudley, L.Y. and Christopher, N.S.J. (1999). Practical experiences of biofouling in reverse osmosis systems. In: C.W. Keevil, A. Godfree, D. Holt and C. Dow (eds) *Biofilms in the Aquatic Environment*, pp. 101–108. Royal Society of Chemistry, Cambridge, UK.

Edel, V. (1998). Polymerase chain reaction in mycology: an overview. In: P.D. Bridge, D.K. Arora, C.A. Reddy and R.P. Elander (eds) *Applications of PCR in Mycology*, pp. 1–20. CABI Publishing, Wallingford.

Edwards, C. (ed.) (1999). *Methods in Biotechnology* 12. Environmental Monitoring of Bacteria. Humana Press, Totowa.

Ellis, D.H. (1980). Thermophilic fungi isolated from a heated aquatic habitat. *Mycologia* **72**, 1030–1033.

Feest, A. and Madelin, M.F. (1985). A method for the enumeration of myxomycetes in soils and its application to a wide range of soils. *FEMS Microbiology Ecology* **31**, 103–109.

Fisher, P.J. (1977). New methods of detecting and studying saprophytic behaviour of aero-aquatic fungi from stagnant water. *Transactions of the British Mycological Society* **68**(3), 407–411.

Frankland, J.C., Magan, N. and Gadd, G.M. (eds) (1996). *Fungi and Environmental Change.* (Symposium of the British Mycological Society held at Cranfield University, March 1994). Cambridge University Press, Cambridge.

Fujita, K. and Silver, J. (1994). Single-strand conformational polymorphism. *PCR Methods and Applications* **4**, S137–S140.

Gadd, G.M., White, C. and Rome, L. de (1987). Heavy metal and radionuclide uptake by fungi and yeasts. In: P.R. Norris and D.P. Kelly (eds) *Biohydrometallurgy, Proceedings of the International Symposium, Warwick*, pp. 421–435. Science and Technology Letters, UK.

Gardes, M. and Bruns, T.D. (1993). ITS primers with enhanced specificity for basidiomycetes – application to the identification of mycorrhizae and rusts. *Molecular Ecology* **2**, 113–118.

Gessner, M.O., Suberkropp, K. and Chauvet, E. (1997). Decomposition of plant litter by fungi in marine and freshwater ecosystems. In: D.T. Wicklow and B. Söderström (eds) *The Mycota IV. Environmental and Microbial Relationships*, pp. 303–322. Springer-Verlag, Berlin.

Gharieb, M.M. and Gadd, G.M. (1999). Influence of nitrogen source on the solubilization of natural gypsum ($CaSO_4{\cdot}2H_2O$) and the formation of calcium oxalate by different oxalic and citric acid-producing fungi. *Mycological Research.* **103**(4), 473–481.

Haberl, R., Perfler, R., Laber, J. and Cooper, P. (eds) (1997). Wetland systems for water pollution control. *Water Science and Technology* **35**(5), 1–347.

Halim, A.F., Narciso, J.A. and Collins, R.P. (1975). Odorous constituents of *Penicillium decumbens*. *Mycologia* **67**, 1158–1165.

Hall, G.S. (ed.) (1996). Zoosporic fungi. *Methods for the Examination of Organismal Diversity in Soils and Sediments*, pp. 109–123. CABI Publishing, Wallingford.

Hastings, R. (1999). Application of denaturing gradient gel electrophoresis to microbial ecology. In: C. Edwards (ed.) *Methods in Biotechnology 12. Environmental Monitoring of Bacteria*, pp. 175–186. Humana Press, Totowa.

Head, I.M. (1999). Recovery and analysis of ribosomal RNA sequences from the environment. In: C. Edwards (ed.) *Methods in Biotechnology 12. Environmental Monitoring of Bacteria*, pp. 139–174. Humana Press, Totowa.

de Hoog, G.S. (1996). Risk assessment of fungi reported from humans and animals. *Mycoses* **39**, 407–417.

Hurst, C.J., Knusden, G.R., McInerney, M.J. *et al.* (eds) (1997). *Manual of Environmental Microbiology.* American Society for Microbiology, Washington.

Hyde, K.D. (ed.) (1997). *Biodiversity of Tropical Microfungi*, Hong Kong University Press, Hong Kong.

Ingold, C.T. (1971). *Fungal Spores their Liberation and Dispersal.* Clarendon Press, Oxford.

Jarvis, W.R. (1992). *Managing Diseases in Greenhouse Crops.* APS Press, St Paul, Minnesota.

Johnson, B.N. and McGill, W.B. (1990). Ontological and environmental influences on ergosterol content and activities of polyamine biosynthesis enzymes in *Hebeloma crustuliniforme* mycelia. *Canadian Journal of Microbiology* **36**, 682–689.

Jüttner, F. (1990). Monoterpenes and microbial metabolites in the soil. *Environmental Pollution* **68**, 377–382.

Kaminski, E., Stawicki, S.T., Wasowicz, E. and Tsubaki, K. (1980). Volatile odour substances produced by microflora. *Die Nahrung* **24**, 103–113.

Karunasena, E., Markham, N., Brasel, T. *et al.* (2000). Evaluation of fungal growth on cellulose-containing and inorganic ceiling tile. *Mycopathologia* **150**, 91–95.

Kelley, J., Hall, G., Paterson, R.R.M. (1997). *The significance of fungi in drinking water distribution systems.* Report DW-01/F for United Kingdom Water Industry Research Limited.

Kelley, J., Kinsey, G.C., Paterson, R.R.M., Pitchers, R. (2001). *Identification and control of fungi in distribution systems.* AWWA Research Foundation: Denver.

Kelley, J., Pitchers, R., Paterson, R. *et al.* (1997). Growth and metabolism of fungi in water distribution systems. In: *Fifth International Symposium on Off-flavours in the Aquatic Environment*. 13–16 October, International Association on Water Quality, Paris, France.

Kikuchi, T., Kadota, S., Suehara, H. *et al.* (1983). Odorous metabolites of fungi *Chaetomium globosum* and *Botrytis cinerea* and a blue-green alga *Phormidium tenue*. *Chemical Pharmacy Bulletin (Tokyo)* **31**, 659–663.

Kinsey, G.C., Paterson, R.R.M. and Kelley, J. (1999). Methods for the determination of filamentous fungi in treated and untreated waters. *Journal of Applied Microbiology Symposium Supplement* **85**, 214S–224S.

Kirk, P.M., Cannon, P.F., David, J.C. and Stalpers, J.A. (2001). *Ainsworth and Bisby's Dictionary of the Fungi.* 9th edn. CABI Publishing, Wallingford.

Kis-Papo, T., Grishkan, I., Oren, A. *et al.* (2001). Spatiotemporal diversity of filamentous fungi in the hypersaline Dead Sea. *Mycological Research* **105**(6), 749–756.

Kohlmeyer, J. and Kohlmeyer, E. (1979). *Marine Mycology, the Higher Fungi.* Academic Press, London.

Kohlmeyer, J. and Volkmann-Kohlmeyer, B. (1991). Illustrated key to the filamentous higher marine fungi. *Botanica Marina* **34**, 1–61.

Kowalchuk, G.A., Gerards, S. and Woldendorp, J.W. (1997). Detection and characterization of fungal infections of *Ammophila arenaria* (Marram grass) roots by denaturing gradient gel electrophoresis. *Applied and Environmental Microbiology* **63**(10), 3858–3865.

Kushner, D.J. (1993). Microbial life in extreme environments. In: T.E. Ford (ed.) *Aquatic Microbiology*, pp. 383–407. Blackwell Scientific, Oxford.

Lacey, J. (1996). Spore dispersal – its role in ecology and disease: the British contribution to fungal aerobiology. *Mycological Research* **100**(6), 641–660.

Larsen, T.O. and Frisvad, J.C. (1995). Characterization of volatile metabolites from 47 *Penicillium* taxa. *Mycological Research* **99**, 1153–1166.

Lechevalier, H.A. (1974). Distribution et rôle des actinomycètes dans les eaux. *Bulletin de l'Institut Pasteur* **72**, 159–175.

Lindrea, K.C., Seviour, E.M., Seviour, R.J. *et al.* (1999). Practical methods for the examination and characterization of activated sludge. In: R.J. Seviour and L.L. Blackall (eds) *The Microbiology of Activated Sludge*, pp. 257–300. Kluwer, Dordrecht.

López-Archilla, A.I., Marín, I., González, A. and Amils, R. (1996). Identification of fungi from an acidophilic river. In: L. Rossen, V. Rubio, M.T. Dawson and J. Frisvad (eds) *Fungal Identification Techniques, Biotechnology (1992–1994)* EUR 16510EN, pp. 202–211. European Commission, Brussels.

Magan, N. (1997). Fungi in extreme environments. In: D.T. Wicklow and B. Söderström (eds) *The Mycota IV. Environmental and Microbial Relationships*, pp. 99–114. Springer-Verlag, Berlin.

Martin, F., Delaruelle, C. and Hilbert, J.-L. (1990). An improved assay to estimate fungal biomass in ectomycorrhizas. *Mycological Research* **94**(8), 1059–1064.

Mattheis, J.P. and Roberts, R.G. (1992). Identification of geosmin as a volatile metabolite of *Penicillium expansum*. *Applied and Environmental Microbiology* **58**, 3170–3172.

Montgomery, H.J., Monreal, C.M., Young, J.C. and Seifert, K.A. (2000). Determination of soil fungal biomass from soil ergosterol analyses. *Soil Biology and Biochemistry* **32**, 1207–1217.

Moore, D. (1998). *Fungal Morphogenesis.* Cambridge University Press, Cambridge.

Moss, M.O. (1996). Mycotoxins. *Mycological Research* **100**(5), 513–523.

Moss, S.T. (ed.) (1986). *The Biology of Marine Fungi.* Cambridge University Press, Cambridge.

Nagy, L.A. and Olson, B.H. (1985). Occurrence and significance of bacteria, fungi and yeasts associated with distribution pipe surfaces. *Proceedings of the American Water Works Association Water Quality Technical Conference*, pp. 213–238. American Water Works Association, Denver.

Newell, S.Y., Miller, J.D. and Fallon, R.D. (1987). Ergosterol content of salt marsh fungi: effect of growth conditions and mycelial age. *Mycologia* **79**(5), 688–695.

Nilsson, K. and Bjurman, J. (1990). Estimation of mycelial biomass by determination of the ergosterol content of wood decayed by Coniophora puteana and Fomes fomentarius. *Material und Organismen* **25**(4), 275–285.

Novickaja-Markovskaja, S. (1995). Aquatic fungi from Lake Druksiai – water cooling reservoir of Ignalina nuclear power station. *Mikologiya i Fitopatologiya* **29**, 22–27.

Nyström, A., Grimvall, A., Krantz-Rülcker, C. *et al.* (1992). Drinking water off-flavour caused by 2,4,6-trichloroanisole. *Water Science and Technology.* **25**, 214–249.

Parkinson, S.M., Wainwright, M. and Killham, K. (1989). Oligotrophic growth of fungi on silica gel. *Mycological Research* **93**, 529–535.

Pasanen, A.-L., Yli-Pietilä, K., Pasanen, P. *et al.* (1999). Ergosterol content in various fungal species and biocontaminated building materials. *Applied and Environmental Microbiology* **65**(1), 138–142.

Paterson, R.R.M. and Bridge, P.D. (1994). *IMI Technical Handbooks No. 1. Biochemical Techniques for Filamentous Fungi.* CABI Publishing, Wallingford.

Paterson, R.R.M., Kelley, J. and Gallagher, M. (1997a). Natural occurrence of aflatoxins and *Aspergillus flavus* (Link) in water. *Letters in Applied Microbiology* **25**, 435–436.

Paterson, R.R.M., Kelley, J. and Kinsey, G.C. (1997b). Secondary metabolites and toxins from fungi in water. In: R. Morris and A. Gammie (eds) *2nd UK Symposium on Health Related Water Microbiology*, pp. 78–94. University of Warwick, 17–19 September 1997.

Paterson, R.R.M., Kinsey and G.C. Kelley, J. (1998). Growth, ergosterol and secondary metabolites (including mycotoxins) from *Fusarium graminearum* in water. *8th International Fusarium Workshop*, CABI Bioscience, UK, 17–20 August 1998.

Richard, J.L., Plattner, R.D., May, J. and Liska, S.L. (1999). The occurrence of ochratoxin A in dust collected from a problem household. *Mycopathologia* **146**, 99–103.

Richardson, M.J. (1996). Seed mycology. *Mycological Research* **100**(4), 385–392.

Ruzicka, S., Edgerton, D., Norman, M. and Hill, T. (2000). The utility of ergosterol as a bioindicator of fungi in temperate soils. *Soil Biology and Biochemistry* **32**, 989–1005.

Saad, D. da, Kinsey, G.C., Paterson, R.R.M. *et al.* (2001). Molecular methods for the analysis of fungal growth on painted surfaces. In: *Proceedings of the 4th Latin American Biodeterioration and Biodegradation Symposium*. Buenos Aires 16–20 April, 2001. (Published on CD ROM).

Schnürer, J. (1993). Comparison of methods for estimating the biomass of three food-borne fungi with different growth patterns. *Applied and Environmental Microbiology* **59**(2), 552–555.

Scudamore, K.A. (1998). Mycotoxins. In: D.H. Watson (ed.) *Natural Toxicants in Food*, pp. 147–181. Sheffield Academic Press, Sheffield.

Shearer, C.A. (1993). The freshwater ascomycetes. *Nova Hedwigia* **55**, 1–33.

Shearer, C.A. and Lane, L.C. (1983). Comparison of three techniques for the study of aquatic hyphomycete communities. *Mycologia* **75**(3), 498–508.

Shearer, C.A. and Webster, J. (1985). Aquatic hyphomycete communities in the river Teign. III. Comparison of sampling techniques. *Transactions of the British Mycological Society* **84**(3), 509–518.

Smit, E., Leeflang, P., Glandorf, B. *et al.* (1999). Analysis of fungal diversity in the wheat rhizosphere by sequencing of cloned PCR-amplified genes encoding 18S rRNA and temperature gradient gel electrophoresis. *Applied and Environmental Microbiology* **65**(6), 2614–2621.

Smith, J.M.B. (1989). *Opportunistic Mycoses of Man and Other Animals.* CABI Publishing, Wallingford.

Smith, D. Onions, A.H.S. (1994). *IMI Technical Handbooks No. 2. The Preservation and Maintenance of Living Fungi*, 2nd edn. CABI Publishing, Wallingford.

Spear, R.N., Li, S., Nordheim, E.V. and Andrews, J.H. (1999). Quantitative imaging and statistical analysis of fluorescence *in situ* hybridization (FISH) of *Aureobasidium pullulans. Journal of Microbiological Methods* **35**, 101–110.

Suberkropp, K. (1991). Relationships between growth and sporulation of aquatic hyphomycetes on decomposing leaf litter. *Mycological Research* **95**(7), 843–850.

Suberkropp, K., Gessner, M.O. and Chauvet, E. (1993). Comparison of ATP and ergosterol as indicators of fungal biomass associated with decomposing leaves in streams. *Applied and Environmental Microbiology* **59**(10), 3367–3372.

Trevisan, M. (1996). Ergosterolo come indicatore di contaminazione fungina. *Informatore Agrario* **52**, 59.

Vainio, E.J. and Hantula, J. (2000). Direct analysis of wood-inhabiting fungi using denaturing gradient gel electrophoresis of amplified ribosomal RNA. *Mycological Research* **104**(8), 927–936.

Viaud, M., Pasquier, A. and Brygoo, Y. (2000). Diversity of soil fungi studied by PCR-RFLP of ITS. *Mycological Research* **104**(9), 1027–1032.

Wainwright, M., Ali, T.A. and Barakah, F. (1992). A review of the role of oligotrophic organisms in biodeterioration. *International Biodeterioration and Biodegradation* **31**, 1–13.

Wainwright, M., Ali, T.A. and Barakah, F. (1993). A review of the role of oligotrophic organisms in industry, medicine and the environment. *Science Progress* **75**, 313–322.

Wainwright, M. and Gadd, G.M. (1997). Fungi and industrial pollutants. In: D.T. Wicklow and B.E. Söderström (eds) *The Mycota IV Environmental and Microbial Relationships*, pp. 85–97. Springer, Berlin.

Wainwright, M., Singleton, I. and Edyvean, R.G.J. (1987). Use of fungal mycelium to absorb particulates from solution. In: P.R. Norris and D.P. Kelly (eds) *Biohydrometallurgy, Proceedings of the International Symposium, Warwick*, pp. 499–502. Science and Technology Letters, UK.

Walsh, T.J., Francesconi, A., Kasai, M. and Chanock, S.J. (1995). PCR and single-strand conformational polymorphism for the recognition of medically important opportunistic fungi. *Journal of Clinical Microbiology* **33**, 3216–3220.

Wang, G.C.-Y. and Wang, Y. (1997). Frequency of formation of chimeric molecules as a consequence of PCR co-amplification of 16S rRNA genes from mixed bacterial genomes. *Applied and Environmental Microbiology* **63**(12), 4645–4650.

Warnock, D.W. and Campbell, C.K. (1996). Medical mycology. *Mycological Research* **100**(10), 1153–1162.

Warris, A., Gaustad, P., Meis, J.F.G.M. *et al.* (2001). Recovery of filamentous fungi from water in a paediatric bone marrow transplantation unit. *Journal of Hospital Infection* **47**(2), 143–148.

Webster, J. (1992). Anamorph–teleomorph relationships. In: F. Bärlocher (ed.) *The Ecology of Aquatic Hyphomycetes*, pp. 99–117. Springer-Verlag, Berlin.

Webster, J. (1996). A century of British mycology. *Mycological Research* **100**(1), 1–15.

Webster, J. and Davey, R.A. (1984). Sigmoid conidial shape in aquatic fungi. *Transactions of the British Mycological Society* **83**(1), 43–52.

Weete, J.D. (1989). Structure and function of sterols in fungi. *Advances in Lipid Research* **23**, 113–167.

Wessels, J.G.H. and Meinhardt, F. (eds) (1994). *The Mycota I. Growth Differentiation and Sexuality.* Springer-Verlag, Berlin.

White, T.J., Bruns, T., Lee, S. and Taylor, J. (1990). Amplification and direct sequencing of fungal ribosomal RNA genes for phylogenetics. In: M.A. Innis, D.H. Gelfand, J.J. Sninsky and T.J. White (eds) *PCR Protocols, a Guide to Methods and Applications*, pp. 315–322. Academic Press, New York.

Wicklow, D.T. Söderström, B. (eds) (1997). *The Mycota IV. Environmental and Microbial Relationships*. Springer-Verlag, Berlin.

Willoughby, L.G. (2001). A new kind of fungus, with giant mycelial cells. *Mycologist*. **15**(2), 52–54.

Young, J.C. (1995). Microwave-assisted extraction of the fungal metabolite ergosterol and total fatty acids. *Journal of Agricultural and Food Chemistry* **43**, 2904–2910.

Yuen, TszKit, Hyde, K.D., Hodgkiss, I.J. and Yuen, T.K. (1998). Physiological growth parameters and enzyme production in tropical freshwater fungi. *Material und Organismen* **32**(1), 1–16.

6

Microbial flora of the gut

Bohumil S. Drasar

London School of Hygiene & Tropical Medicine, London, WC1E 7HT, UK

1 INTRODUCTION

The importance of the intestinal bacteria in the context of water and wastewater derives from the significance of water as a vehicle for the spread of infection. Most important in this context are intestinal infections causing diarrhoea.

Diarrhoea is a major symptom of many intestinal infections. Some gut pathogens may also invade the body from the gut and cause either generalized or localized infections. Such invasion is characteristic of typhoid and some other infections caused by salmonellae. During infections restricted to the gut, bacteria will be excreted in the faeces. When invasion has occurred, bacteria may be passed in the urine as well and can also be found in the bloodstream.

When considering the faeco-oral transmission of an intestinal pathogen by water it is seldom possible to demonstrate directly the pathogen in the water supply. The reasons for this are various but the relatively small number of a pathogen excreted and the limited time in which the excretion occurs are contributory factors. Cholera provides an instructive example of the problem. Patients with severe cholera will be excreting *Vibrio cholerae* in large numbers for a limited time, terminated by death or isolation. Patients with mild disease, or a convalescent carrier, by contrast may look relatively healthy and be mobile while still excreting vibrios in their faeces. Such mobile persons might be thought more likely to be a risk to water supplies, but it must be remembered that the numbers of *V. cholerae* excreted would be small.

A consequence of the vagaries of pathogen detection is the use of normal faecal bacteria as indicators of water pollution. These bacteria are ubiquitous and numerous in the faeces of healthy people. The most widely used indicator has been the faecal coliform *Escherichia coli*, but faecal enterococci are also used as indicators (see Chapter 7). Anaerobic bacteria also, such as *Clostridium*, *Bacteroides* and *Bifidobacterium*, have served as indicators, and each has potential value (Evison and James, 1977).

The bacterial microflora of humans is the most intimate portion of their biological environment and mediates many of their interactions with the chemical environment. This development of our knowledge of the composition of the intestinal microflora can be traced in a series of books (Drasar and Hill, 1974; Clarke and Bauchop, 1977; Hentges, 1983; Hill, 1995). However, an explanation of the effects of the gut bacteria on the body or of the mechanism by which they are controlled remains possible only in outline. Intestinal physiology and host defence mechanisms must play an important role in preventing the microflora from overrunning the host and in determining the distribution of the microflora within the intestine. It seems likely that environmental factors, such as diet, also influence the microflora. The ecological interactions between bacterial species are probably significant in the large intestine. However, their exact involvement is largely obscure.

The Handbook of Water and Wastewater Microbiology
ISBN 0-12-470100-0

2 THE STOMACH

The stomach receives bacteria from the mouth and the external environment. In the context of the faeco-oral transmission of disease it functions as a barrier to infection. The bactericidal action of gastric juice results from secretion of hydrochloric acid. Immediately after a test meal, about 10^5 bacteria per ml of gastric juice can be isolated but, as the pH falls, the bacterial count declines and few viable cells can be recovered after a pH of about 3 has been attained (Drasar *et al.*, 1969). Lactobacilli, which are particularly resistant to acid, persist longer than other bacteria. The achlorhydric stomach, whatever the cause, is usually heavily colonized by bacteria, indicating that gastric acid is an important factor in reducing the numbers of pathogenic bacteria entering the small bowel. Experimental evidence for this is afforded by studies of the infective dose of *Vibrio cholerae* in human volunteers, where neutralization of gastric acid with a solution of sodium bicarbonate enables an infection to be initiated by oral administration of a much smaller challenge dose compared to that needed for untreated subjects (Hornick *et al.*, 1971; Gianelia *et al.*, 1973).

3 THE SMALL INTESTINE

Although the restriction on bacterial proliferation in the small gut stems partly from the efficiency of gastric acid in reducing the bacterial load entering the bowel, small intestinal factors also play a role. Samples of small gut contents obtained from fasting European and North American residents contain few if any cultivable bacteria. The viable count seldom exceeds 10^4 per ml of digesta, but for a short period after a meal, an increased number of bacteria, that are probably transients from the mouth, can be isolated from jejunal samples. Although relatively few bacteria can be cultured, microscopy of jejunal juice reveals large numbers of cells, usually exceeding 10^6 per ml. The importance of these bacteria that can be seen but not grown has not yet been determined.

In healthy persons living in Western Europe and North America, the number of cultivable bacteria is lowest in the upper small bowel and greatest in the terminal ileum. Gram-positive species predominate among those bacteria isolated and those observed microscopically. Streptococci and lactobacilli are most often isolated, though bacteroides and enterobacteria may be found in small numbers. The organisms seen but not identified are Gram-positive rods and cocci.

These results differ markedly from data obtained during studies of the small intestinal flora in residents of developing countries which show that they have a richer and more permanent flora. The relationship of this flora to nutritional status, nutrient utilization and subclinical disease may have important practical consequences.

The flow rate of gut contents contributes to control of bacterial colonization, being greatest at the top of the small intestine where microbial multiplication usually does not exceed the rate at which organisms are removed. Animal experiments have shown that bacteria are cleared from the small intestine wrapped in intestinal mucus. Among the roles that have been suggested for intestinal antibody is the binding of bacteria to this material. The anatomically normal small intestine can only be colonized by adhesion to the epithelium, as occurs with some enteropathogenic serotypes of *E. coli*. In the ileum, water is absorbed, reducing the flow rate, which allows bacterial multiplication. The lower small intestine, the distal and terminal ileum, contain many more cultivable bacteria than the proximal ileum. Although lactobacilli and streptococci are still prominent, bacteroides and enterobacteria occur more constantly. In the terminal ileum viable bacterial counts of 10^5–10^7 per ml are not uncommon, and the flora here qualitatively begins to resemble that of faeces (Gorbach *et al.*, 1967).

4 THE LARGE INTESTINE

In the UK, the adult colon contains approximately 220 g of contents. Bacteria are a major component and about 18 g consist of bacterial dry matter. Bacterial numbers, as estimated by direct microscope counts, increase progressively from the caecum to the rectum. However, most studies on the colonic flora are made on faeces, which can provide some useful information, since the few investigations where viable counts of bacteria were carried out with gut contents have indicated that, with respect to the major taxonomic groups, the gut flora is qualitatively similar to faeces (Bentley *et al.*, 1972). There is some evidence, however, for the existence of specific mucosa-associated populations, whose functional association with the mucus layer and intestinal wall may be of ecological and metabolic significance to the host (Croucher *et al.*, 1983).

In colonic contents and faeces, bacterial counts usually exceed 10^{11} per g dry weight. Several hundred different species are present, but some 30–40 species account for about 99% of the bacterial mass, and most of these are strict anaerobes (Finegold *et al.*, 1983). An important factor that facilitates the establishment of such a luxuriant flora is the increase in transit time of gut contents from about 3–4 hours in the small intestine to about 70 hours in the colon (Cummings, 1978).

Although bacteria are distributed throughout the gastrointestinal tract, the vast majority occurs in the large bowel. The major influences of the gut flora on host metabolism arise from bacterial metabolism of various substances in the colon. Of particular significance in the normal healthy bowel is the fermentation of proteins and carbohydrates. Much recent research has focused on the metabolic significance of the gut bacteria (Hill, 1995; Macfarlane and Cummings, 1999).

5 FAECAL BACTERIA

Many different types of bacteria, representing most bacterial groups, have at some time or other been isolated from faeces. About 400 species are thought to be present usually (Holdeman *et al.*, 1976). Those detected most frequently can be considered as members of the resident flora or as regular contaminants from the environment. The number of bacterial groups that are found is related to the methods used for their isolation and characterization. Few researchers have ever attempted a systematic investigation to determine the composition of the faecal flora, and any list of species present must be incomplete. The scale of the problem, in terms of the numbers of bacteria to be isolated and identified, makes investigations based on classical bacteriological techniques problematic.

A few studies have relied on the identification of large numbers of isolates from non-selective media (Moore and Holdeman, 1974; Holdeman *et al.*, 1976). Other investigators have combined this approach with the use of selective media (Peach *et al.*, 1974; Finegold *et al.*, 1975,1977). More recently the use of molecular techniques has rekindled interest in the problem and provide long-term solutions (Jansen *et al.*, 1999).

Table 6.1 shows the major groups of commonly isolated faecal bacteria. It can be seen that bacterial counts in different individuals range over several orders of magnitude. Most of the bacteria growing in the colon are non-sporing anaerobes and include members of the genera *Bacteroides*, *Bifidobacterium* and *Eubacterium* among many others. Clostridia are also represented, though they are outnumbered by the non-sporing anaerobes, as are facultative anaerobes, such as streptococci and enterobacteria. Quantitatively, the most important genera of intestinal bacteria in

TABLE 6.1 Numbers ($\log_{10}$/g dry wt) of bacteria reported in human faeces (Finegold *et al.*, 1983)

Bacteria	*Description*	*Mean*	*Range*
Bacteroides	Gram-negative rods	11.3	9.2–13.5
Eubacteria	Gram-positive rods	10.7	5.0–13.3
Bifidobacteria	Gram-positive rods	10.2	4.9–13.4
Clostridia	Gram-positive rods	9.8	3.3–13.1
Lactobacilli	Gram-positive rods	9.6	3.6–12
Ruminococci	Gram-positive cocci	10.2	4.6–12.8
Peptostreptococci	Gram-positive cocci	10.1	3.8–12.6
Propionibacteria	Gram-positive rods	9.4	4.3–12.0
Actinomyces	Gram-positive rods	9.2	5.7–11.1
Streptococci	Gram-positive cocci	8.9	3.9–12.9
Fusobacteria	Gram-negative rods	8.4	5.1–11.0
Escherichia	Gram-negative rods	8.6	3.9–12.3

animals and man are the *Bacteroides* and *Bifidobacterium*, which can account for up to 30 and 25% of the total anaerobic counts respectively (Macy and Probst, 1979; Mitsuoka, 1984; Scardovi, 1986). Among the Gram-positive non-sporing rods, several genera are numerically significant; these include *Eubacterium* and *Bifidobacterium*. Species such as *B. bifidum* and *B. infantis* have been isolated from the faeces of breast-fed infants but are probably less common in adults. The genus *Lactobacillus* contains many species that occur in the guts of most warm-blooded animals.

Several types of spore-forming rods and cocci are also normal inhabitants of the gut. The genus *Clostridium* is probably the most ubiquitous: *C. perfringens, C. bifermentans* and *C. tetani* are regularly isolated, albeit in relatively low numbers, from the lower gut of man and animals, and are of significance in human and veterinary medicine. The presence of aerobic members of the genus *Bacillus* is thought to result from contamination from the environment. Facultative and obligately anaerobic Gram-positive cocci are also numerically important. The strict anaerobes include *Peptostreptococcus, Ruminococcus, Megasphaera elsdenii* and *Sarcina ventriculi.* The facultatively anaerobic streptococci are well represented by many species from Lancefield group D, including *Enterococcus faecalis, S. bovis* and *S. equinus* and some from group K such as S. *salivarius*, which is usually associated with the mouth. Gram-negative anaerobic cocci include *Veillonella* and *Acidaminococcus.*

Although they are not numerous, the Gram-negative facultative anaerobic rods include a number of very important pathogens. Members of the Enterobacteriaceae, particularly *Escherichia coli,* are usually thought of as characteristic intestinal bacteria and have historically been the most important indicators of faecal pollution.

6 FAECAL BACTERIA AND INFECTION

Most of the bacteria isolated from faeces do not cause infection and in the present context are important only in so far as they act as surrogate indicators of infective hazard. This is not a straightforward relationship as there are few data on the relationship of the major faecal organisms to the numbers of pathogens or to traditional indicator organisms.

There have been very few systematic attempts to determine the numbers of a pathogen excreted by an infected person, however, in most cases the number of bacteria must be very many fewer than that contributed by the normal flora. A working estimate would be 10^7–10^9 pathogens per gram of faeces during acute infection. These numbers are very similar to those achieved by the usual indicator bacteria, but such organisms can be regarded, for practical purposes, as always present (Tables 6.1 and 6.2). Among the numerically dominant bacterial species (Table 6.3), only *Bifidobacteium adolescentis* has been subject to continuing investigation as an indicator (Resnick and Levin, 1981; Rhodes and Kator, 1999). Studies of other groups such as *Bacteroides* and *Peptostreptococcus* have been disabled by the complexity of the methods needed for their identification.

TABLE 6.2 Rank order of major bacterial indicators (Moore and Holdeman, 1974)

Rank	*Species*	*Approximate number per gram faeces*
37	*Escherichia coli*	10^7
64	*Enterococcus faecalis*	10^7
Unranked	*Clostridium perfringens*	10^5

TABLE 6.3 The 10 most common bacterial species in faeces (Moore and Holdeman, 1974)

1.	*Bacteroides vulgatus*
2.	*Fusobacterium prausnitzii*
3.	*Bifidobacterium adolescentis*
4.	*Eubacterium aerofaciens*
5.	*Peptostreptococcus productus* – 2
6.	*B. thetaiotaomicron*
7.	*E. eligens*
8.	*Peptostreptococcus productus* – 1
9.	*E. biforme*
10.	*Eubacterium aerofaciens* – 3

Thus, in conclusion, it will be clear that the human intestine is the source of a diverse range of bacteria, but that from the practical viewpoint many can be ignored at present in the context of water testing. However, it must be realized that such testing considerably underestimates the actual bacterial load that arises from faecal pollution.

REFERENCES

Bentley, D.W., Nichols, R., Condon, R.E. and Gorbach, S.L. (1972). The microflora of the human ileum and intra-abdominal colon: Results of direct needle aspiration at surgery and evaluation of the technique. *Journal of Laboratory and Clinical Medicine* **79**, 421–429.

Clarke, R.T.J. and Bauchop, T. (1977). *Microbial Ecology of the Gut*. Academic Press, London.

Croucher, S.C., Houston, A.P., Bayliss, C.E. and Turner, R.J. (1983). Bacterial populations associated with different regions of the human colon wall. *Applied and Environmental Microbiology* **45**, 1025–1033.

Cummings, J.H. (1978). Diet and transit through the gut. *Journal of Plant Foods* **3**, 83–95.

Drasar, B.S. and Barrow, P.A. (1985). *Intestinal Microecology*. Van Nostrand Reinhold, Wokingham.

Drasar, B.S. and Hill, M.J. (1974). *Human Intestinal Flora*. Academic Press, London.

Drasar, B.S., Shiner, M. and McLeod, G.M. (1969). Studies on the intestinal flora. 1. The bacterial flora of the gastrointestinal tract in healthy and achlorhydric persons. *Gastroenterology* **56**, 71–79.

Evison, L.M. and James, A. (1977). Microbiological criteria for tropical water quality. In: R.G.A. Feachem, M.G. McGarry and D.D. Mara (eds) *Water, Wastes and Health in Hot Climates*, pp. 30–51. John Wiley, London.

Finegold, S.M., Flora, D.J., Attebery, H.R. and Sutter, L.V. (1975). Faecal bacteriology of colonic polyp patients and control patients. *Cancer Research* **35**, 3407.

Finegold, S.M., Sutter, V.L. and Mathisen, G.E. (1983). Normal indigenous intestinal flora. In: D.J. Hentges (ed.) *Human Intestinal Microflora in Health and Disease*, pp. 3–31. Academic Press, London.

Finegold, S.M., Sutter, V.L., Sugihara, P.T. *et al.* (1977). Faecal flora of Seventh Day Adventist populations and control subjects. *American Journal of Clinical Nutrition* **30**, 1718.

Gianelia, R.A., Broitman, S.A. and Zamcheck, N. (1973). The influence of gastric acidity on bacterial and parasitic infections. *Annals of Internal Medicine* **78**, 271–276.

Gorbach, S.I., Plaut, A.G., Nahas, L. *et al.* (1967). Studies of intestinal microflora. II. Microorganisms of the small intestine and their relations to oral and fecal flora. *Gastroenterology* **53**, 856–867.

Hentges, D.J. (1983). *Human Intestinal Microflora in Health and Disease*. Academic Press, New York.

Hill, M.J. (1995). *The role of gut bacteria in human toxicology and pharmacology*. Taylor & Francis, London.

Holdeman, L.V., Good, I.J. and Moore, W.E.C. (1976). Human faecal flora: variation in bacterial composition within individuals and a possible effect of emotional stress. *Applied Environ.*

Hornick, R.B., Music, S.I. and Wenzel, R. (1971). The Broad Street pump revisited: response of volunteers to ingested cholera vibrios. *Bulletin of the New York Academy of Medicine* **47**, 1181–1191.

Jansen, G.J., Wildeboor-Veloo, A.C.M., Tonk, R.H.J. *et al.* (1999). Development and validation of an automated, microscopy-based method for enumeration of groups of intestinal bacterial. *Journal of Microbial Methods* **37**, 215–221.

Macfarlane, G.T. and Cummings, J.H. (1999). Probotics and prebiotics: can regulating the activities of the intestinal bacteria benefit health? *British Medical Journal* **318**, 999–1003.

Macy, J.M. and Probst, I. (1979). The biology of gastrointestinal bacteroides. *Annual Review of Microbiology* **33**, 561–594.

Mitsuoka, T. (1984). Taxonomy and ecology of bifidobacteria. *Bifid Microf.* **3**, 11–28.

Moore, W.E.C. and Holdeman, L.V. (1974). Human faecal flora: the normal flora of Japanese-Hawaiians. *Applied Microbiology* **27**, 961.

Peach, S., Fernendez, F., Johnson, K. and Drasar, B.S. (1974). The non-sporing anaerobic bacteria in human faeces. *Journal of Medical Microbiology* **7**, 213.

Resnick, I.G. and Levin, M.A. (1981). Quantitative procedure for the enumeration of bifidobacteria. *Applied and Environmental Microbiology* **42**, 427–432.

Rhodes, M.W. and Kator, H. (1999). Sorbitol; fermenting bifidobacteria as indicators of diffuse human faecal pollution in estuarine watersheds. *Journal of Applied Microbiology* **87**, 528–535.

Scardovi, V. (1986). Genus *Bifidobacterium*. In: N.S. Mair (ed.) *Bergey's Manual of Systematic Bacteriology*, Vol. **2**, pp. 1418–1434. Williams & Wilkins, New York.

7

Faecal indicator organisms

Nigel J. Horan

School of Civil Engineering, University of Leeds, LS2 9JT, UK

1 INTRODUCTION

Snow (1855) first demonstrated that gastrointestinal diseases such as cholera were able to be transmitted via faecally-contaminated water. His work, now rightly regarded as a classic in the field of public health medicine in general and water science and engineering in particular, was all the more remarkable as it was done without any knowledge of the causative microorganism, *Vibrio cholerae*. Smith (1895) showed that the presence of the bacterium, then known as *Bacillus coli communis*, could be regarded 'as a valuable indication or symptom of pollution', and he was the first bacteriologist to promote what later became known as the coliform group of bacteria as indicators of faecal pollution. Water bacteriology was then a rapidly advancing field: of the 26 papers presented at the first meeting of the Society for Bacteriology (now the American Society for Microbiology), five were concerned with *B. coli communis* and water and wastewater (Dixon, 1999). The first edition of *Standard Methods* appeared in 1905 and included recommendations for the bacteriological examination of waters (APHA, 1905). In contrast, the first edition of *Report 71*, the UK standard on the bacteriological examination of drinking waters, was not published until 1934 (Ministry of Health, 1934). Emphasis in these early years was on the detection of coliform bacteria using the bile salt–lactose medium developed by MacConkey (1908), although tests for the faecal enterococci and *Clostridium perfringens* were also included.

Faecal indicator organisms remain at the forefront of water and wastewater microbiology. They represent our first and foremost technique for detecting and quantifying aquatic pollution. The American Society for Microbiology has recently declared that 'Microbial pollution of water in the United States is a growing crisis in environmental and public health' (ASM, 1999). The role of faecal indicator organisms is central to the reduction of this crisis which is occurring in all parts of the world.

2 THE IDEAL FAECAL INDICATOR BACTERIUM

Bonde (1962), the World Health Organization (WHO, 1993), Grabow (1996) and Godfree *et al.* (1997) list the following properties of the ideal faecal indicator bacterium:

1. suitable for all categories of water
2. present in wastewaters and polluted waters whenever pathogens are present
3. present in greater numbers than pathogens
4. having similar survival characteristics as pathogens in waters and water and wastewater treatment processes
5. unable to multiply in waters
6. non-pathogenic
7. able to be detected in low numbers reliably, rapidly and at low cost.

Of course, such an ideal organism does not exist. Some, such as faecal coliform bacteria, come close, but even so, difficulties remain with the indication of non-bacterial pathogens (viruses, protozoa and helminths).

The Handbook of Water and Wastewater Microbiology
ISBN 0-12-470100-0

3 COLIFORM BACTERIA

Early water microbiologists defined coliform bacteria as those bacteria able to grow at 37°C in the presence of bile salts (used to inhibit non-intestinal bacteria) and able to produce acid and gas from lactose. Faecal coliforms were considered to be those coliforms which were of exclusively faecal origin and consequently able to grow and ferment lactose at 44°C, and to produce indole from tryptophan. However, the last edition of *Report 71* (Department of the Environment, 1994) defines coliform bacteria as those members of the Enterobacteriaceae possessing the gene coding for the production of β-galactosidase, the enzyme which cleaves lactose into glucose and galactose. This new definition is not dependent on laboratory techniques (acid and gas production from lactose), and recognizes that gene expression may be influenced by environmental (including laboratory) conditions – i.e. under different conditions a coliform bacterium may or may not ferment lactose with the production of acid and gas.

The coliform group, as defined above, includes species of the genera *Citrobacter, Enterobacter, Escherichia, Hafnia, Klebsiella, Serratia* and *Yersinia*. Some members of the genus *Aeromonas* may mimic coliforms in the standard laboratory tests for the latter, but they are oxidase-positive, whereas all coliforms are oxidase-negative.

Faecal coliform bacteria are principally *Escherichia coli*, but some non-faecal coliforms may also grow at 44°C. *E. coli* was found to be a better indicator of the risk of diarrhoeal disease from tropical drinking waters than faecal coliforms (Moe *et al.*, 1991). However, for routine purposes, faecal coliforms, thermotolerant coliforms and *E. coli* may be regarded, if not exactly the same, as generally equivalent indicators of faecal pollution. Total coliforms (i.e. faecal and non-faecal coliforms) are routinely used as indicators of the general bacteriological quality of treated drinking waters, and only waters that have tested positive for this group are examined for the presence of faecal coliforms. Those workers who favour *E. coli* over faecal coliforms (e.g. Gleeson and Grey, 1991) do so on the grounds that some non-faecal coliforms are thermotolerant; however, since humans and many animals excrete around 10^{11} *E. coli* per day and over 95% of excreted thermotolerant coliforms are *E. coli*, in practice this distinction is but a fine one.

Total coliforms may grow in certain non-intestinal environments, e.g. in biofilms in water distribution pipes, but such regrowth of faecal coliforms is rare. Nevertheless, Payment *et al.* (1993) considered regrowth within the distribution system as a major potential cause of gastrointestinal illness attributable to the consumption of treated drinking waters meeting current microbiological water quality standards. The potential for regrowth in distribution systems is best assessed by the numbers of *Pseudomonas* spp.; this group was found to be a better indicator than the usual faecal bacteria or *Aeromonas* spp. (Ribas *et al.*, 2000). Recently *E. coli* has been found to regrow in subtropical coastal environments (Solo-Gabriele *et al.*, 2000).

Despite these certain limitations, faecal coliforms, and particularly *E. coli*, remain the best overall indicators of faecal pollution of waters (Edberg *et al.*, 2000). They are of course especially good indicators of bacterial pathogens, less good indicators of viral pathogens, but very poor indicators of protozoan and helminthic pathogens.

4 FAECAL ENTEROCOCCI

The faecal enterococci comprise species of two genera, *Enterococcus* and *Streptococcus*, which occur in the intestines, and hence faeces, of humans and many animals (Leclerc *et al.*, 1996). Until recently they were all described as faecal streptococci, and the term 'intestinal enterococci' has been proposed (Hernandez *et al.*, 1995); however, the original description (Thiercelin in 1899) was 'enterococcus' and it seems sensible to keep the familiar epithet 'faecal'. The species of *Enterococcus* and *Streptococcus* that are of faecal origin are listed in Table 7.1.

TABLE 7.1 Species of the genera *Enterococcus* and *Streptococcus* of faecal origin

Species	*Intestinal origin*
Enterococcus	
faecium	Man, cattle, pigs, birds
faecalis	Man, cattle, pigs, birds
durans	Man, pigs, birds
hirae	Man, pigs, birds
avium	Man, cattle, pigs, birds
gallinarum	Man, birds
cecorum	Cattle, pigs, birds
columbae	Cattle, pigs, birds
Streptococcus	
bovis	Man, cattle, pigs
equinus	Man, cattle, pigs
alactolyticus	Pigs, birds
hyointestinalis	Pigs
intestinalis	Pigs
suis	Pigs

Source: Godfree *et al.* (1997).

The faecal enterococci fulfill most of the requirements of an ideal indicator – the species listed in Table 7.1 are exclusively associated with the intestinal tract of man or other animals; they do not multiply in the environment; they occur in much higher numbers than bacterial and viral pathogens (around 10^6 per 100 ml of raw domestic wastewater); and they are around twice as resistant to disinfection as faecal coliforms (Godfree *et al.*, 1997).

While the taxonomy of the faecal enterococci has improved recently, as shown in Table 7.1, methods for their isolation from faecal and environmental samples has not – these are essentially tests for the enumeration of the faecal streptococci, and the use of these can lead to erroneous results and misleading conclusions concerning the pollution source (Pinto *et al.*, 1999). Clearly there is a need to evaluate and modify media and methods for faecal streptococci, so that reliable techniques exist for the isolation and enumeration of faecal enterococci (and preferably as a group, i.e. all the species of both genera in Table 7.1, although methods for individual species identification and enumeration would be epidemiologically useful).

5 CLOSTRIDIUM PERFRINGENS

Clostridium perfringens (formerly *Cl. welchii*) is an anaerobic sulphite-reducing spore-forming bacterium. It is present in human and animal faeces, but in lower numbers than faecal coliforms or faecal enterococci. If it can be cultured from a water sample, but faecal coliforms and faecal enterococci cannot be, its presence indicates:

> *either* 'remote' faecal pollution, i.e. pollution that has occurred at sometime past, since when faecal coliforms and faecal enterococci have died off, but not the spores of *Cl. perfringens*,
> *or* faecal pollution in the presence of industrial wastewater toxicants which have killed or inhibited faecal coliforms and faecal enterococci, but not the spores of *Cl. perfringens*.

It is thus generally only used when faecal coliforms and faecal enterococci cannot be detected, or when industrial toxicants are present (or suspected to be present).

6 BACTERIAL INDICATORS OF HUMAN AND ANIMAL FAECAL POLLUTION

Geldrich and Kenner (1969) advocated the use of faecal coliform/faecal streptococci (FC/FS) ratios to differentiate between faecal pollution of human and animal origin. Ratios > 4 were held to be indicative of human pollution and ratios < 0.7 of animal pollution. McFetters *et al.* (1974) questioned the general validity of this interpretation, and Oragui (1978) and Wheater *et al.* (1979) demonstrated unequivocally its invalidity: FC/FS ratios > 4 were found in faecal samples obtained from sheep, pigs, cats, dogs, chickens, ducks, pigeons and seagulls, as well as from humans.

Oragui (1983) found that, in the UK, sorbitol-fermenting strains of *Bifidobacterium breve* and *B. adolescentis* were present only in human faeces, whereas *Streptococcus bovis* and *Rhodococcus coprophilus* were only present in animal faeces (see also Mara and Oragui, 1981, 1983; Oragui and Mara, 1981, 1984). This was confirmed by Mara and Oragui (1985) in Nigeria and Zimbabwe, although very small numbers of *S. bovis* were found in human faeces.

However, these 'new' specific indicators of human and animal faecal pollution have several disadvantages:

1. sorbitol-fermenting bifidobacteria are extremely sensitive in the extra-intestinal environment, disappearing much more rapidly than faecal coliforms (Oragui *et al.*, 1987; Pourcher, 1991), and they are much more sensitive to chlorine (Resnick and Levin, 1981)
2. the incubation period for *R. coprophilus* is very long, 17–18 days at 30°C
3. *S. bovis* does occur in some human faeces, albeit in very small numbers.

Nonetheless, these new indicators have, in practice, proved useful (Jagals and Grabow, 1996; Rhodes and Kator, 1999), and they are included in the current edition of *Report 71* (Environment Agency, 2002).

Recently, Carson *et al.* (2001) reported successful identification of *E.coli* from humans and individual animals (cattle, pigs, horses, chickens, turkeys, dogs and migratory geese) by ribotyping. This extends earlier work on ribotyping and antibiotic resistance profiles to distinguish between human and non-human (i.e. animals collectively) pollution (Parveen *et al.*, 1999). Antibiotic resistance patterns in faecal enterococci were used by Wiggins (1996) to distinguish human and animal faecal pollution in surface waters.

7 VIRUSES AND BACTERIOPHAGES

Viral pollution, especially of ground and coastal waters, is of concern, particularly in the absence of bacterial indicators, due to the high infectivity and virulence of viral pathogens (Friedman-Hoffman and Rose, 1998). Bacteriophages have been used as indicators of enteroviral pollution of waters. The IAWPRC Study Group (1991) recommended the use of male (F^+)-specific bacteriophages, rather than somatic coliphages, as enteroviral indicators (see also Calci *et al.*, 1998). They occur in polluted coastal waters where they are concentrated in shellfish (Chung *et al.*, 1998) and they are found in large numbers in wastewater sludges (Lasobras *et al.*, 1999) and also in aerosols emanating from activated sludge plants (Carducci *et al.*, 1999). Leclerc *et al.* (2000) found that all types of bacteriophage were limited as indicators of enteric viruses in groundwaters due to their low concentration in such waters. They also noted that there are very few field data that permit an understanding of the ecology and survival of phages in the natural environment. Grabow (2001) lists a number of major shortcomings of the use of bacteriophages as indicators of, or surrogates for, enteric viruses in water; thus they cannot be satisfactorily regarded as *absolute* indicators or surrogates, although they can be used in combination with other indicators in the assessment of faecal pollution.

However, Oragui *et al.* (1987) found that faecal coliform numbers in treated tropical wastewaters were good indicators of the level of viral contamination (Table 7.2).

8 FUNGI AND YEASTS

Fungal infections are of increasing concern, particularly in immunocompromised individuals, and an important focus of infection is communal bathing facilities. Research in Portugal (Mendes *et al.*, 1993, 1997, 1998) has led to the fungus *Scopulariopsis* and the yeast *Candida* being proposed as indicators of the microbiological quality of sand on bathing beaches, and to recommendations for maximum levels of these organisms on beaches (10 *Scopulariopsis* and 10 *Candida* per g sand).

9 PROTOZOA AND HELMINTHS

There are no satisfactory indicator organisms for the important protozoa (such as *Giardia lamblia*, *Cryptosporidium parvum* and *Entamoeba*

TABLE 7.2 Geometric mean bacterial and viral numbers[a] and percentage removals in raw wastewater (RW) and the effluents of five waste stabilization ponds in series (P1–P5)[b] in northeast Brazil at a mean mid-depth pond temperature of 26°C

Organism	*RW*	*P1*	*P2*	*P3*	*P4*	*P5*	*Percentage removal*
Faecal coliforms	2×10^7	4×10^6	8×10^5	2×10^5	3×10^4	7×10^3	99.97
Faecal streptococci	3×10^6	9×10^5	1×10^5	1×10^4	2×10^3	300	99.99
Clostridium perfringens	5×10^4	2×10^4	6×10^3	2×10^3	1×10^3	300	99.40
Total bifidobacteria	1×10^7	3×10^6	5×10^4	100	0	0	100.00
Sorbitol-positive bifids	2×10^6	5×10^5	2×10^3	40	0	0	100.00
Campylobacters	70	20	0.2	0	0	0	100.00
Salmonellae	20	8	0.1	0.02	0.01	0	100.00
Enteroviruses	1×10^4	6×10^3	1×10^3	400	50	9	99.91
Rotaviruses	800	200	70	30	10	3	99.63

[a] Bacterial numbers per 100 ml, viral numbers per 10 litres.
[b] P1 was an anaerobic pond with a mean hydraulic retention time of 1 day; P2 and P3–5 were secondary facultative and maturation ponds respectively, each with a retention time of 5 days. Pond depths were 3.4–2.8 m.
Source: Oragui *et al.* (1987).

histolytica). Examination of water and wastewater samples thus has to be for these organisms themselves. However, the cost of routine monitoring outweighs potential benefits (Fairley *et al.*, 1999).

The same is true for helminthic pathogens. When treated wastewaters are used for crop irrigation, the World Health Organization recommends that they contain no more than one human intestinal nematode egg per litre (eggs of *Ascaris lumbricoides*, *Trichuris trichiura*, and the human hookworms) (WHO, 1993), and treated wastewaters are analysed directly for these (Ayres and Mara, 1996).

10 WASTEWATERS

Feachem and Mara (1979) reviewed the usefulness of faecal indicator bacteria in wastewaters, the faecal origin of which is not generally in doubt. Faecal coliform bacteria are regarded as the most useful indicator of the microbiological purifications achieved by wastewater treatment and disinfection. Standards and guideline values for the microbiological quality of recreational waters subject to the discharge of treated wastewaters, and of treated wastewaters used for crop irrigation and fish culture, are commonly expressed in terms of maximum permitted numbers of faecal coliforms (FC). For example, bathing waters in the European Union should contain no more than 2000 FC per 100 ml (Council of the European Communities, 1976), and treated wastewaters used to irrigate vegetables eaten raw no more than 1000 FC per 100 ml (WHO, 1989; see also Mara, 1995; Shuval *et al.*, 1997).

Oragui *et al.* (1987) showed that faecal coliforms were indeed good indicators of bacterial and viral pathogens in wastewater treatment in waste stabilization ponds (WSP; see Chapter 26): when FC numbers were 7000 per 100 ml, salmonellae and campylobacters were absent and faecal viruses present in only very small numbers (see Table 7.2). Oragui *et al.* (1993) found that *Vibrio cholerae* was eliminated in a series of WSP when FC numbers were 60 000 per 100 ml.

11 THE FUTURE?

Haas *et al.* (1999) present order-of-magnitude calculations for waterborne infections in the USA, where there is an estimated number of waterborne illnesses of 1 million. Taking the population of the USA as 280 millions, this is equal to an annual occurrence rate of 0.0036, equivalent to a daily rate of 9.8×10^{-6}. Using dose-response parameters for rotavirus, *Salmonella*, *Giardia* and *Cryptosporidium*, the daily intake of organisms corresponding to this level of daily risk can be determined, as

TABLE 7.3 Daily microbial intakes per person corresponding to an estimated daily rate of waterborne disease in the USA of 9.8×10^{-6} per person

Organism	*Dose-response model and parameter values*	*Daily microbial intake per person corresponding to daily risk of* 9.8×10^{-6} *per person*[a] $(= d)$[b,c]
Rotavirus	Beta-Poisson[b] $N_{50} = 6.17$ $\alpha = 0.263$	1.78×10^{-5}
Salmonella	Beta-Poisson $N_{50} = 23\,600$ $\alpha = 0.313$	0.091
Giardia	Exponential[c] $r = 0.0199$	4.92×10^{-4}
Cryptosporidium	Exponential $r = 0.0042$	2.33×10^{-3}

[a] 1 million waterborne illnesses per year in a population of 280 millions, equivalent to a daily risk of waterborne illness of (1/280 million)/365, i.e. $= 9.8 \times 10^{-6}$ per person.
[b] $P_I(d) = 1 - \{1 + [(d/N_{50})(2^{1/\alpha} - 1)]\}^{-\alpha}$ where $P_I(d) =$ probability (risk) of infection per day (= fraction of persons ingesting d pathogens per day who become infected), $d =$ average daily dose, $N_{50} =$ median infectious dose, and $\alpha =$ a species constant.
[c] $P_I(d) = 1 - \exp(-rd), \approx rd$ for $rd \ll 1$ where $r =$ a species constant.
Source: Haas *et al.* (1999), pp. 383 and 435.

shown in Table 7.3. The computed numbers are small and, given the immense difficulties in measuring low numbers of pathogens in treated drinking waters, it is highly unlikely that these numbers are measurable. For example, in the case of *Cryptosporidium*, assuming all exposure is from drinking water and that water consumption is 2 litres per adult per day, the daily intake of 0.00233 oocyst per person translates into approximately 0.001 oocyst per litre of treated drinking water; this would require the examination of at least one thousand 1-litre samples for a substantial probability of detection (and this assumes that oocyst counting methods are 100% efficient, which they are not; typically they are <50%).

Haas *et al.* (1999) conclude by noting that 'the general inability to measure pathogens in water as [it is] being consumed is not necessarily inconsistent with the estimated prevalence of waterborne infectious diseases.' Given this, it makes little sense to examine treated drinking waters routinely for the presence of pathogens (some of which have long incubation periods – 5–28 days in the case of *Cryptosporidium* and 30–60 days for hepatitis A virus) (see Fairley *et al.*, 1999). This is true also during epidemics of waterborne disease since, as noted by Haas *et al.* (1999), even for the largest known outbreak of waterborne disease (that in Milwaukee in March–April 1993 in which over 400 000 persons became ill with cryptosporidiosis), local public health authorities were unaware of the outbreak during its first 4 days.

The order-of-magnitude calculations of Haas *et al.* (1999) and the above conclusions drawn from them mean, in practice, that *either* microbial indicator organisms must continue to be used for the routine assessment of faecal contamination of treated drinking waters, at least until pathogen detection methods improve to the point that they can reliably detect, e.g. *Cryptospordium* levels of 1 oocyst per 1000 litres (perhaps through the use of gene probes), *or* a train of water treatment processes are specified which can reliably achieve such low pathogen levels. The latter approach is favoured by the USEPA (2002) in its Long Term 1 Enhanced Surface Water Rule, which also includes watershed control and reservoir protection. A useful supplement to this would be Hazard Analysis Critical Control Point (HACCP), which has been widely used in the food industry but only recently proposed for drinking water (Deere and Davison, 1998).

REFERENCES

APHA (1905). *Report of the Committee on Standard Methods of Water Analysis*. American Public Health Association, Chicago.

ASM (1999). *Microbial Pollution in Our Nation's Water: Environmental and Public Health Issues*. American Society for Microbiology, Washington, DC.

Ayres, R.M. and Mara, D.D. (1996). *Analysis of Wastewater for Use in Agriculture: A Laboratory Manual of Parasitological and Bacteriological Techniques*. World Health Organization, Geneva.

Bonde, G.J. (1962). *Bacterial Indicators of Water Pollution*. Teknisk Forlag, Copenhagen.

Calci, K.R., Burkhardt, W., Watkins, W.D. and Rippey, S.R. (1998). Occurrence of male-specific bacteriophage in feral and domestic animal wastes, human feces, and human-associated wastewaters. *Applied and Environmental Microbiology* **64**(10), 5027–5029.

Carducci, A., Gemelli, C., Cantiani, L., Casini, B. and Rovini, E. (1999). Assessment of microbial parameters as indicators of viral contamination of aerosols from urban sewage treatment works. *Letters in Applied Microbiology* **28**(3), 207–210.

Carson, C.A., Shear, B.L., Ellersieck, M.R. and Asfaw, A. (2001). Identification of faecal *Escherichia coli* from humans and animals by ribotyping. *Applied and Environmental Microbiology* **67**(4), 1503–1507.

Chung, H., Jaykus, L.-A., Lovelace, G. and Sibsey, M.D. (1998). Bacteriophages and bacteria as indicators of enteric viruses in oysters and their harvest waters. *Water Science and Technology* **38**(12), 37–44.

Council of the European Communities (1976). Council Directive of 8 December 1975 concerning the quality of bathing water (76/160/EEC). *Official Journal of the European Communities* **L31**/1–7 (5 February).

Department of the Environment (1994). *The Microbiology of Water 1994. Part 1 Drinking Water.* Her Majesty's Stationery Office, London.

Deere, D. and Davison, A. (1998). Safe drinking water: are food guidelines the answer? *Water* Nov/Dec, 21–24.

Dixon, B. (1999). A century of society meetings. *ASM News* **65**(5), 282–287.

Edberg, S.C., Rice, E.W., Karlin, R.J. and Allen, M.J. (2000). *Escherichia coli*: the best drinking water indicator for public health protection. *Journal of Applied Microbiology* **88** (Suppl), 106S–116S.

Environment Agency (2002). The Microbiology of Drinking Water 2002 – Part 1 – Water Quality and Public Health, available at http://www.dwi.gov.uk/regs/pdf/micro.htm

Fairley, C.K., Sinclair, M.I. and Rizak, S. (1999). Monitoring not the answer to cryptosporidium in water. *Lancet* **354**, 967–968 (18 September).

Feachem, R.G. and Mara, D.D. (1979). A reappraisal of the role of faecal indicator organisms in tropical waste treatment processes. *Public Health Engineer* **7**(1), 31–33.

Friedman-Hoffmann, D. and Rose, J. (1998). Emerging waterborne pathogens. *Water Quality International*, (Nov./Dec.), 14–18.

Geldrich, E.E. and Kenner, B.A. (1969). Concepts of fecal streptococci in stream pollution. *Journal of the Water Pollution Control Federation* **41**, R336–R351.

Gleeson, C. and Grey, N. (1991). *The Coliform Index and Waterborne Disease: Problems of Microbial Drinking Water Assessment* E. & F.N. Spon, London.

Godfree, A.F., Kay, D. and Wyer, M.D. (1997). Faecal streptococci as indicators of faecal contamination in water. *Journal of Applied Microbiology (Symposium Supplement)* **83**, 110S–119S.

Grabow, W.O.K. (1996). Waterborne diseases: update on water quality assessment and control. *Water SA* **22**(2), 193–202.

Grabow, W.O.K. (2001). Bacteriophages: update on application as models for viruses in water. *Water SA* **27**(2), 251–268.

Haas, C.N., Rose, J.B. and Gerba, C.P. (1999). *Quantitative Microbial Risk Analysis*. John Wiley & Sons, Inc, New York.

Hernandez, J.F., Delatte, J.M. and Maier, E.A. (1995). *Sea Water Microbiology: Performance of Methods for the Microbiological Examination of Bathing Water (Part 1).* Report No. EUR 16601 (BCR Information Series). European Commission, Directorate General XII, Brussels.

IAWPRC Strategy Group (1991). Bacteriophages as model viruses in water quality control. *Water Research* **25**(5), 529–545.

Jagals, P. and Grabow, W.O.K. (1996). An evaluation of sorbitol-fermenting bifidobacteria as specific indicators of human faecal pollution of environmental water. *Water SA* **22**(3), 235–238.

Lasobras, J., Dellunde, J., Jofre, J. and Lucena, F. (1999). Occurrence and levels of phages proposed as surrogate indicators of enteric viruses in different types of sludges. *Journal of Applied Microbiology* **86**(4), 723–729.

Leclerc, H., Devriese, L.A. and Mossel, D.A.A. (1996). Taxonomic changes in intestinal (faecal) enterococci and streptococci: consequences on their use as indicators of faecal contamination in drinking water. *Journal of Applied Bacteriology* **81**(5), 459–466.

Leclerc, H., Edberg, S., Pierzo, V. and Delatte, J.M. (2000). Bacteriophages as indicators of enteric viruses and public health risk in groundwaters. *Journal of Applied Microbiology* **88**(1), 5–21.

MacConkey, A.T. (1908). Bile salt media and their advantages in some bacteriological examinations. *Journal of Hygiene, Cambridge* **8**, 322–334.

Mara, D.D. (1995). Faecal coliforms–everywhere (but not a cell to drink). *Water Quality International* **3**, 29–30.

Mara, D.D. and Oragui, J.I. (1981). Occurrence of *Rhodococcus coprophilus* and associated actinomycetes in faeces, sewage and fresh water. *Applied and Environmental Microbiology* **42**(6), 1037–1042.

Mara, D.D. and Oragui, J.I. (1983). Sorbitol-fermenting bifidobacteria as specific indicators of human faecal pollution. *Journal of Applied Bacteriology* **55**, 349–357.

Mara, D.D. and Oragui, J.I. (1985). Bacteriological methods for distinguishing between human and animal faecal pollution of water: results of fieldwork in Nigeria and Zimbabwe. *Bulletin of the World Health Organization* **63**(4), 773–783.

McFetters, G.A., Bissonette, G.K. *et al.* (1974). Comparative survival of indicator bacteria and enteric pathogens in well water. *Applied Microbiology* **27**, 823–829.

Mendes, B., Nascimento, M.J. and Oliveira, J.S. (1993). Preliminary characteristics and proposal of microbiological quality standard of sand beaches. *Water Science and Technology* **27**(3/4), 453–456.

Mendes, B., Urbano, P., Alves, C. *et al.* (1997). Sanitary quality of sands from beaches of Azores islands. *Water Science and Technology* **35**(11/12), 147–150.

Mendes, B., Urbano, P., Alves, C., Morais, J., Lapa, N. and Oliveira, J.S. (1998). Fungi as environmental microbiological indicators. *Water Science and Technology* **38**(12), 155–162.

Ministry of Health (1934). *The Bacteriological Examination of Water Supplies*. Report on Public Health and Medical Subjects No. 71. His Majesty's Stationery Office, London.

Moe, C.L., Sobsey, M.D., Samsa, G.P. and Mesolo, V. (1991). Bacterial indicators of risk of diarrhoeal disease from drinking water in the Philippines. *Bulletin of the World Health Organization* **69**(3), 305–317.

Oragui, J.I. (1978). *On Bacterial Indicators of Water and Sewage Pollution*. MSc Thesis. University of Dundee, Dundee.

Oragui, J.I. and Mara, D.D. (1981). A selective medium for the enumeration of *Streptococcus bovis* by membrane filtration. *Journal of Applied Bacteriology* **51**, 85–93.

Oragui, J.I. (1983). *Bacteriological Methods for the Distinction between Human and Animal Faecal Pollution*. PhD Thesis. University of Leeds, Leeds.

Oragui, J.I. and Mara, D.D. (1984). A note on a modified membrane-Bovis agar for the enumeration of *Streptococcus bovis* by membrane filtration. *Journal of Applied Bacteriology* **56**, 179–181.

Oragui, J.I., Curtis, T.P. *et al.* (1987). The removal of excreted bacteria and viruses in deep waste stabilization ponds in northeast Brazil. *Water Science and Technology* **19**(Rio), 569–573.

Oragui, J.I., Arridge, H. *et al.* (1993). *Vibrio cholerae* O1 removal in waste stabilization ponds in northeast Brazil. *Water Research* **27**(4), 727–728.

Parveen, S., Portier, K.M., Robinson, K., Edmiston, L. and Tamplin, M.L. (1999). Discriminant analysis of ribotype profiles of *Escherichia coli* for differentiating human and nonhuman sources of fecal pollution. *Applied and Environmental Microbiology* **65**, 3142–3147.

Payment, P., Franco, E. and Siemiatycki, J. (1993). Absence of relationship between health effects due to tap water consumption and drinking water quality parameters. *Water Science and Technology* **27**(3/4), 137–145.

Pinto, B., Pierotti, R., Canale, G. and Reali, D. (1999). Characterisation of 'faecal streptococci' as indicators of faecal pollution and distribution in the environment. *Letters in Applied Microbiology* **29**(4), 258–263.

Pourcher, A.M. (1991). *Contribution á l'Étude de l'Origine (Humaine ou Animale) de la Contamination Fécale des Eau de Surface*. Doctoral thesis. Université des Sciences et Techniques, Lille.

Resnick, I.G. and Levin, M.A. (1981). Assessment of bifidobacteria as indicators of human faecal pollution. *Applied and Environmental Microbiology* **42**, 433–438.

Rhodes, M.W. and Kator, H. (1999). Sorbitol-fermenting bifidobacteria as indicators of diffuse human faecal pollution in estuarine watersheds. *Journal of Applied Microbiology* **87**(4), 528–535.

Ribas, R., Perramon, J., Terradillos, A. *et al.* (2000). The *Pseudomonas* group as an indicator of potential regrowth in water distribution systems. *Journal of Applied Microbiology* **88**(4), 704–710.

Shuval, H.I., Lampert, Y. and Fattal, B. (1997). Development of a risk assessment approach for evaluating wastewater reuse standards for agriculture. *Water Science and Technology* **35**(11/12), 15–20.

Smith, T. (1895). Notes on *Bacillus coli commune* and related forms, together with some suggestions concerning the bacteriological examination of drinking water. *American Journal of Medical Science* **110**, 283–302.

Snow, J. (1855). *On the Mode of Communication of Cholera*, 2nd edn. John Churchill, London.

Solo-Gabriele, H.M., Wolfert, M.A., Desmarais, T.R. and Palmer, C.J. (2000). Sources of *Escherichia coli* in a coastal subtropical environment. *Applied and Environmental Microbiology* **66**, 230–237.

USEPA (2002). Long Term 1 Enhanced Surface Water Treatment Rule, US Environmental Protection Agency, available at http://www.epa.gov/ogwdw000/mdbp/lt1eswtr.html

Wheater, D.W.F., Mara, D.D. and Oragui, J.I. (1979). Indicator systems to distinguish sewage from stormwater run-off and human from animal faecal pollution. In: A. James and L. Evison (eds) *Biological Indicators of Water Quality*, pp. 1–27. John Wiley & Sons, Chichester.

WHO (1989). *Health Guidelines for the Use of Wastewater in Agriculture and Aquaculture, Technical Report Series No. 778*, World Health Organization, Geneva.

WHO (1993). *Guidelines for Drinking-water Quality, 2nd edn – Volume 1: Recommendations*. World Health Organization, Geneva.

Wiggins, B.A. (1996). Discriminant analysis of antibiotic resistance patterns in fecal streptococci, a method to differentiate human and animal sources of fecal pollution in natural waters. *Applied and Environmental Microbiology* **63**, 3997–4002.

8

Detection, enumeration and identification of environmental microorganisms of public health significance

Howard Kator and Martha Rhodes

School of Marine Science, College of William and Mary, Virginia VA 23062, USA

1 INTRODUCTION

Detection and enumeration of environmental bacteria is a vast and exciting subject area and one that undergoes continual change as new environments are examined and new methods are developed. Recent years have witnessed an ongoing revolution in detection methods based on techniques from molecular biology. These methods hold the potential for reduced analytical time and improved sensitivity and specificity when compared to the traditional cultural methods that now constitute the majority of contemporary detection and enumeration techniques. While molecular methods, based on specific gene sequences, offer the possibility of simultaneous detection of multiple pathogens, thereby addressing a classically vexing problem in the sanitary microbiology of natural waters, their application also raises important questions related to target viability and infectivity. Because organisms must be isolated for study and method development, culture-based techniques will undoubtedly remain important tools in the environmental microbiologist's arsenal of methods. In the near future, Sartory and Watkins (1999) envisioned a melding of conventional culture with biochemical methods for detection of viable cells over shorter time intervals.

This chapter will focus primarily on detection and enumeration topics related to the sanitary microbiology of receiving waters. Problems of surface water sanitation remain significant world-wide and advances in approved method development have been surprisingly limited given the history and continued scope of disease problems encountered. Microbial contamination of groundwater used for drinking purposes is a dominant concern in many countries. The material presented is intended as a general overview, with specific detailed discussion of selected environmental microorganisms of interest used in public health and water quality assessment. It is not intended as a comprehensive literature review.

Microorganisms of sanitary importance present in natural waters are derived from domestic sewage and non-point runoff containing the excreta of humans and animals. Rivers, lakes, and coastal waters serve to receive, dilute and disperse domestic sewage and untreated human and animal waste. The degree of domestic waste treatment varies considerably (Chambers *et al.*, 1997) and its influence on water quality and health remain important problems throughout the world. Improperly operating septic systems, poor well maintenance, surface application of wastewaters and direct injection have led to contamination of

The Handbook of Water and Wastewater Microbiology
ISBN 0-12-470100-0

groundwater. Eutrophication, ecosystem perturbation, toxicities, and disease, in both human and animal communities, are consequences of waste discharge. From a human disease perspective, domestic wastes contribute bacterial (salmonellas, vibrios and aeromonads), viral (Norwalk, polio, hepatitis, enteroviruses), and protozoan (*Cryptosporidium*, *Giardia*) pathogens. Large numbers of indicator organisms (e.g. coliforms, enterococci, streptococci, bacteriophage, *Bacteroides fragilis*, *Bifidobacterium* spp, clostridia) and potential 'marker' compounds (proteins, enzymes, sterols, whitening agents) are also found in these wastes.

2 DETECTION AND ENUMERATION

Microorganisms, especially those normally associated with warm-blooded animals, are likely to become stressed in natural environments from exposure to unfavorable temperatures and abiotic factors such as light, osmotic stress, availability of carbon, and chemical stressors (e.g. disinfection). Importantly, there is a direct link between the efficiency of recovery and degree of stress using traditional media and procedures (Kator and Rhodes, 1991). Many investigators have also observed changes in culturability that can also affect recoverability (Roszak *et al.*, 1984; Rollins and Colwell, 1986; Lopez-Torres *et al.*, 1987; Munro *et al.*, 1987; Roth *et al.*, 1988; Martinez *et al.*, 1989; Desmonts *et al.*, 1990; Garcia-Lara *et al.*, 1991; Cappelier and Federighi 1998; Lleò *et al.*, 1998). Moreover, while most methods listed in compendiums such as Standard Methods (APHA, 1998) can be applied to detection and enumeration of target microorganisms from saline waters there is no reason to assume that detection methods developed and verified for freshwaters will be as effective in saline waters. Various studies, including our own, have proven such an assumption unwise.

2.1 Sampling considerations

Detection and enumeration are final procedures in a chain of events that commence with sample collection and sample storage during transport. Sample collection and sample storage are thus important initial elements of detection/enumeration protocols. Lapses in these activities can affect recovery and enumeration of microorganisms of interest. Improperly prepared or contaminated sampling containers or gear, inadvertent contamination of samples, poor or inappropriate sample storage conditions, excessive transport/storage time, are all factors that can influence ultimate detection and affect quantitative validity. The most careful analyses may be compromised by poor sampling and/or adverse sample storage during transport. Sample volume must be adequate for the organism being sought. True sample replication as opposed to analytical replication might be necessary if the spatial distribution of the microorganism is highly variable.

Sampling locations should be chosen to reflect the desired characteristics of the environment being evaluated. Source evaluation surveys are useful and provide information on which to base a station location process. Consider pollution sources, delivery mechanisms and hydrographic parameters relevant to the aquatic or sediment environment being evaluated. Hydrographic considerations might include factors such as water flow and direction, the existence of vertical or horizontal stratification, local circulation patterns, presence of feeder streams or tributaries, and climatological events such as precipitation and wind speed. Microbiological data should be collected and analyzed with regard to known or potential pollutant sources to confirm the validity of station selection. Advanced modeling techniques, land use data, historical hydrographic data sets, and GIS information systems may be of value in this process.

2.1.1 Sample collection methods

General aseptic principles as outlined in Standard Methods (APHA, 1998) should be followed. Collect water samples for bacteriological examination in clean, sterile containers. Protect the container fully against contamination before, during, and after collection.

Surface water samples can be collected without the aid of special sampling devices. Keep the sample container unopened until immediately before filling. During sampling protect the container stopper or closure from contamination. To collect a sample by hand, hold the bottle near the base and plunge it neck downward below the surface when the water depth permits. Then tilt it with the neck pointing slightly upward and into the direction of flow. During filling, push the container horizontally forward in a direction away from the hand to avoid contamination. If sampling from a vessel, take care to avoid contamination from effluents and sample 'upstream' of the vessel motion. In shallow waters, avoid sampling from waters that are disrupted by sediment suspended by boat activity. An ordinary laboratory clamp affixed to a length of steel or aluminum tubing can be used to hold a bottle for sampling as directed above.

Commercial bottle holding devices are available, but it is very easy to fabricate one's own. Appropriate personal safety precautions should be taken if sampling untreated sewage or similar wastewaters to avoid direct contact. To facilitate sample mixing prior to enumeration, leave a small air space in the bottle before sealing. Precautions against chlorine residuals should be considered when human point source impacted waters or drinking waters are sampled (APHA, 1998).

A variety of sampling devices are available for collection of subsurface samples. Ideally, a subsurface sampler should be of simple (i.e. reliable) mechanical design and able to hold a sterile container that is non-metallic and non-bactericidal, with mechanical devices to open and shut the container at the desired sampling depth. The container should remain sealed until the desired depth is reached. Sampling depth can affect recovery of bacterial populations, especially in dynamic environments where currents vary temporally as a function of tides or winds and bed resuspension may be an important factor. Using *Vibrio* spp. as a target group, Koh *et al.* (1994) found significant variation in cell concentrations occur as functions of both depth and tidal cycle.

A number of subsurface sampling devices are available commercially. Water sampling 'bottles' that remain open prior to triggering and closing at depth are common. However, their use must be considered carefully if contamination during descent will affect results. Bottles should be cleaned between uses with a suitable, non-toxic liquid if contaminant adsorption is a concern. DC-powered peristaltic pumps and sterile tubing can also be used for sampling at depth provided the head capacity of the pump is not exceeded and the tubing is adequately flushed prior to sample collection. Fresh sterile tubing should be used for each new sample. Peristaltic pumps of this type can also be used routinely to sample shallow groundwater wells in the field.

An area that has received little attention is the use of continuous or integrating *in situ* samplers for microbiological sampling. In highly dynamic environments or in those with low densities of bacteria, unattended samplers that collect large multiple sample volumes over selected time intervals 'integrate' water conditions and may provide better means to assess sanitary water quality when compared to traditional grab sampling.

Various studies illustrate that bacteria are concentrated at interfaces such as the surface microlayer (e.g. Hatcher and Parker, 1974; Marshall, 1976; Dutka and Kwan, 1978). The importance of this microhabitat is now recognized as a region regulating air/water gas exchange and the capture of organic compounds of low solubility in water (Sodergen, 1993). There are no standard methods for collection of bacterioneuston, which can be sampled using a variety of techniques ranging from glass plates to mesh screens (van Vleet and Williams, 1980). Care must be taken to use sterile sampling devices to recover microorganisms from microlayers.

Methods for concentration of viruses, chemical indicators and other non-approved indicators may require special techniques. These methods should be evaluated and validated prior to actual field studies. Soil and sediment samples can be collected using sterile spatulas or coring devices. An inexpensive surface corer can be made by cutting off the luer end of a 50-ml autoclavable plastic syringe. Similarly prepared smaller volume syringes can be used for replicate sampling of single core samples. Sampling of groundwater and subsurface soils may require special equipment and great attention must be given to avoid sample contamination.

2.1.2 *Sample storage*

Sample storage is often not considered a vital part of the enumeration/detection process. Simple icing is usually recommended for field samples by many sources. The degree of sample cooling required can vary depending on the activity of a sample. Highly enriched samples with active heterotrophic microbial communities (e.g. domestic sewage, eutrophic lake waters) may change in composition if temperature is not reduced quickly. On the other hand, icing water samples can produce sublethal stress expressed as reduced recovery when target faecal organisms collected from cold waters are placed directly in warm media without an intermediary resuscitation or adaptation step (Rhodes *et al.*, 1983). Samples collected from chlorinated effluents must be treated with sodium thiosulfate to neutralize residual chlorine following recognized protocols (APHA, 1998). In general, it is always best to process samples as soon as possible after collection and never to exceed the specified time limits for a given technique. These and other concerns are described in Standard Methods (APHA, 1998). For non-regulatory methods, sample storage parameters should be evaluated empirically using spiked samples and controls for sample/microorganism interactions.

2.1.3 *Accountability*

A good sampling program must be supported by a consistently applied laboratory and field quality control plan. Thoroughly identify sample containers using an easily understood coding scheme and use field record sheets or water resistant notebooks to provide sample accountability ('chain of custody') from field to laboratory personnel. Use pencils and marking pens that will remain legible under field and laboratory conditions. The chain of custody may be important for quality control or for epidemiological or legal purposes. Insure that all field observations include date, time, and place of collection; identity of person responsible for collection; weather and water conditions at time of collection; and comments on unusual conditions or events observed.

2.2 Relevant physical and chemical parameters

Interpreting field microbiological data requires an understanding of the physical and chemical characteristics of the environment sampled. Routine parameters measured or described during sampling of surface waters can include temperature, conductivity or salinity, pH, and oxygen concentration. Non-routine parameters, such as stage and current of the tide, Secchi depth or water clarity, amount of precipitation, total suspended solids, selected nutrients, chlorophyll a and Eh may be required for specific studies. For sediments, parameters such as sediment dry weight, total organic concentration, or grain size distribution are common variables. Approved methods and instrument specifications for some physical and chemical parameters can be found in Standard Methods (APHA, 1998). References for saline waters (Strickland and Parsons, 1972) and soils (e.g. Weaver, 1994; Parks, 1996) are very useful sources of analytical methods. Many routine parameters can now be measured easily using lightweight portable datasondes, which usually possess comparatively large data storage capability. Salinity can be determined by conductivity or with a hand refractometer if the reduced accuracy of the latter instrument is acceptable. Rainfall often can be site-specific and it is recommended that a simple rain gauge or recording rain gauge be installed and monitored for each study site. Antecedent precipitation should be measured at regular intervals prior to sampling. Remote areas may necessitate installation of a battery-operated recording rain gauge. In all instances, instrumentation should be calibrated using appropriate standards and verified to be within specifications prior to and during sampling, and in some cases, after sampling is completed.

3 METHODS OF ENUMERATION AND DETECTION

The basis of much of environmental microbiology over its brief history has been learned through the removal and isolation of microorganisms under artificial laboratory conditions from natural environments to which they have adapted. While this approach has been functionally necessary, naturally occurring microorganisms belong to natural assemblages where they often exist as synergistic members of mixed populations and their isolation in a defined medium is an obvious oversimplification. Given this background it is not surprising that we can culture but a minor proportion of the total microbial community from a given habitat. Future researchers will no doubt become more adept in developing more sophisticated recovery methods. The recovery of microorganisms from environments to which they are poorly adapted, i.e. pathogens and commensals of warm-blooded animals in surface waters, has been the focus of much enumeration/detection method development for obvious reasons. Even within this group, recognition of the necessity for resuscitation, attention to stress, and even the effects of starvation, has now become part of enumeration/detection methods. The following general discussion contains information related to the application of cultural methods.

3.1 Non-culturability

An important concept influencing the detection of certain enteric pathogens and indicator organisms following environmental exposure was first described by Xu *et al.* (1982), who demonstrated that cells starved in seawater rapidly enter a 'non-recoverable' dormant stage when assayed with standard media and conditions, but remain viable by a direct viable count assay (Kogure *et al.*, 1979). Reduced culturability has been observed by others as well (Roszak *et al.*, 1984; Rollins and Colwell, 1986; Lopez-Torres *et al.*, 1987; Munro *et al.*, 1987; Roth *et al.*, 1988; Martinez *et al.*, 1989; Desmonts *et al.*, 1990; Garcia-Lara *et al.*, 1991; Cappelier and Federighi, 1998; Lleò *et al.*, 1998) and reported to occur over a variety of microbial genera (e.g. Oliver, 1993; Caro *et al.*, 1999). Moreover, the viable-but-non-culturable (VBNC) state for certain species is characterized by a sequence of demonstrable physiological and biochemical changes (Oliver, 1993). If entry into the VBNC state represents a physiologically-programmed survival response to adverse environmental conditions (Morita, 1993), VBNC cells should be capable of resuscitation under appropriate circumstances. Resuscitation attributed to nutrient addition is difficult to demonstrate unequivocally because the complete absence of culturable cells in a population of VBNC cells may be difficult to show. Bogosian *et al.* (1996) discounted entry of *E. coli* into a VBNC state concluding that non-culturable cells were non-viable. Moreover, the extent of entry into a non-culturable state is affected by prior growth conditions (Munro *et al.*, 1995). Contemporary readings on starvation and non-culturability can be found in Colwell and Grimes (2000).

A practical consequence of the VBNC phenomenon is that, although variable numbers of metabolically active cells may remain in a sample, because of non-culturability the 'true' numbers (and consequences) of indicator (e.g. *E. coli*) or pathogenic microorganisms may be underestimated. Thus, the effectiveness of public health standards based on quantitative recovery of microorganisms susceptible to non-culturability could be compromised. *Enterococcus faecalis*, an organism identified as an indicator of risk associated with swimming-associated illnesses has been reported to enter the VBNC state (Lleò *et al.*, 1998). The extent to which the VBNC response affects detection and enumeration of bacteria in natural populations, in wastewater facilities, in low-nutrient environments such as groundwater, and in environmental biofilms remains to be described and evaluated.

3.2 Cultural considerations

3.2.1 *Media components and cultural conditions*

Cultural methods are based upon providing a combination of nutritional and physicochemical conditions that will support the growth of the microorganisms of interest. The range of media developed to culture bacteria reflects their nutritional versatility and range of growth conditions tolerated. The diversity of unique microbial physiologies found in domestic sewage, fecal wastes, and aquatic environments necessitates methods that span a wide range of recovery conditions. Most of the bacteria dealt with in sanitary microbiology are chemoorganotrophs (also called chemoheterotrophs), i.e. bacteria that require organic compounds for energy and as a carbon source. The range of organic compounds an individual species can degrade varies from simple carbohydrates to complex molecules such as polycyclic aromatic hydrocarbons and synthetic halogenated molecules. Bacteria with very specific substrate trace organic and physicochemical chemical requirements are termed 'fastidious'.

Basic components of isolation media include carbon sources to enable growth, an appropriate ionic environment, trace elements, vitamins, a buffer system to stabilize pH as metabolic products accumulate or substrates are removed from the medium, electron acceptors, and nutrients. A general discussion of factors involved in formulation of culture media can be found in Tanner (1996) and Cote and Gherna (1994).

Organic components commonly used in media are classified as either complex or chemically defined. Components such as peptones (products of hydrolytic reactions) or yeast extract are complex organic mixtures whose exact composition will vary from biological source to source. Chemically-defined media contain analytical-grade chemicals of known composition and purity such as glucose, inorganic salts, buffers, metals, dyes or inhibitors.

In addition to a carbon source culture media may contain trace compounds such as vitamins and trace metals required as enzyme cofactors. Salts are commonly added for osmotic balance, buffering of pH, and as sources of nitrogen and phosphorus. Frequently, various compounds are added which are indicators of pH, oxidation/reduction potential or specific metabolism. Biological dyes may be added to enhance contrast between bacteria and the medium. Examples are phenol red as a pH indicator; tetrazolium salts, which change color, are used as redox or electron transport indicators, and a variety of substrates conjugated to chromogens are available to assess species-specific metabolic activities. Examples of the latter include carbohydrate analogs which are conjugated to fluorescent or chromogenic molecules. Enzymatic hydrolysis of specific bonds leads to release of the reporter molecule signaling the presence of the target microorganism. An example of a common method to determine if *Escherichia coli* is associated with positive tubes in the fecal coliform MPN test (APHA, 1998) is based on hydrolysis of a compound called methyl-umbelliferyl-β-glucuronide. Use of this compound is based on the specific presence of the enzyme β-glucuronidase (GUD) in *E. coli* and is the basis for a variety of detection methods (Feng and Hartman, 1982; Watkins *et al.*, 1988; Rice *et al.*, 1990; Sartory and Howard, 1992). A new medium, mTEC (modified mTEC, USEPA, 2000) contains a chromogenic compound (5-bromo-6-chloro-3-indoyl-β-D-glucuronide) which is the basis for enumeration and detection of *E. coli* in one step (USEPA, 2000). The critical assumption in using chromogenic or fluorogenic analogs, or any other substrates to detect a translational product, is that the enzyme is produced in all target microorganisms in the sample. This assumption appears generally correct, but false negative and false positive reactions can occur. For example, Watkins *et al.* (1988) reported that some *Citrobacter* and *Enterobacter* spp. produced GUD.

Some bacteria may also require very low oxygen levels or none at all in order to grow. Bacteria that require oxygen at atmospheric concentrations are called aerobes. Bacteria that require oxygen at very low levels are termed microaerophiles. Bacteria that are intolerant of oxygen are called anaerobes. Coliforms and fecal coliforms are facultative anaerobes, meaning they can grow under aerobic or anaerobic conditions through a metabolic process known as fermentation.

pH is an important variable that must be maintained within a comparatively defined range for most bacteria. Some bacteria require relatively narrow pH ranges. Commercially available media should be checked for appropriate pH values. Media made from individual components for recovery of unique or specialized groups of bacteria must always be checked for pH. Many bacterial genera found in fecal wastes and domestic sewage are facultative anaerobes and require no special considerations for oxygen or Eh. Obligate or aerotolerant anaerobes used as faecal indicators, e.g., *Bacteroides* spp. or *Bifidobacterium* spp., may require complete elimination or reduction of gaseous oxygen, the presence of CO_2 and low Eh values. These incubation conditions can be achieved using water-activated gas generators and portable plate chambers, or more complex and difficult to maintain anaerobe chambers using gas mixtures and catalysts to remove oxygen. Methods for the culture of anaerobes have been described in detail in Holdeman *et al.* (1977).

Salt content and ionic composition is normally not a major issue with recovery of bacteria of sanitary significance and salts are present in media components as contaminants. Environmental exposure to salt has been shown to affect recovery of indicator bacteria from estuarine or marine waters, but resuscitation from salt exposure is not a part of accepted enumeration procedures.

3.2.2 *Media preparation*

Media preparation is an important aspect of the culture-based approaches. The importance of following a reasonable and consistent QA/QC program designed to evaluate reagent water quality, sterilization procedures and effectiveness, pH, temperature, gravimetric measurements, chemical components and other aspects of media preparation cannot be overemphasized. Storage of prepared media should be considered with regard to published shelf life and exposure to light. Bolton *et al.* (1984) demonstrated the production of laboratory light-mediated photochemical products in prepared agar media. Chelala and Margolin (1983) documented that a medium containing riboflavin, aromatic compounds such as tryptophan or phenylalanine, and $MnCl_2$ produced bactericidal products when exposed to visible light. Media should never be stored near windows where exposure to sunlight is possible. Carlsson *et al.* (1978) reported the formation of bactericidal hydrogen peroxide and superoxide radicals in anaerobic broths containing glucose and phosphate when autoclaved in the presence of oxygen. Inhibitors can be avoided using PRAS or prereduced media prepared by removal of oxygen prior to and after autoclaving.

3.2.3 *The recovery process and the design of detection media*

Recovery of bacteria from environmental samples poses several practical problems. First, the medium/incubation conditions should be designed to recover all of the target microorganisms present. Second, these same conditions should not favor the growth of interfering microorganisms (the background microbiota) that are always present. From a practical perspective, for many chemo-organotrophic bacteria, these two objectives are mutually exclusive and ideal recovery is unrealized for many bacterial groups.

Consequently, recovery approaches for environmental bacteria may employ an initial non-selective enrichment phase designed to recover and support the growth of the target organism. This enrichment step may or may not incorporate conditions designed to limit the growth of interfering microorganisms found in environmental samples. If the enrichment medium incorporates compounds inhibitory to such non-target microorganisms, the enrichment is said to be selective. Enrichment and selectivity can be combined in one medium and one step, such as the modified medium for recovery of enterococci by membrane filtration (mEI; USEPA, 2000). The presumptive multiple tube coliform test (APHA, 1998) uses a relatively benign initial enrichment phase (lactose broth) followed by transfer to a harsher, more selective medium and incubation temperatures. Improved recovery can sometimes be obtained from the selective phase because of the primary, benign enrichment step.

Medium components used to provide selectivity by inhibiting growth of unwanted species include organic compounds such as bile salts and sodium deoxycholate, inorganics such as sodium selenite and sodium tetrathionate, and dyes such as brilliant green, malachite green and crystal violet. An example of the influence of inhibitors on recovery of *Salmonella* spp. in various culture media is discussed in Arroyo and Arroyo (1995). A large variety of media has been developed for recovery of fecal streptococci. Audicana *et al.* (1995) evaluated the effect of sodium azide concentration and antibiotic replacement on the selectivity of several agars. Depending upon the organisms being isolated and the environment sampled, inhibitors may be used singly or synergistically.

Inhibitors listed above are commonly employed in media designed to isolate enteric bacteria from environmental water samples. Bile salts, sodium dexoycholate and sodium selenite are used to inhibit Gram-positive organisms in coliform detection media. However, the use of inhibitors is not without consequences. Thus, inhibitors affect the growth of both interfering background and sublethally injured target organisms and can result in reduced recoveries of target bacteria (e.g., Rhodes *et al.,* 1983).

Isolation by streaking or plating samples on solid media differentially supporting the growth of the target organism generally follows and allows for colony isolation and subsequent identification using biochemical, physical characterization, immunological or molecular methods.

Indicator bacteria and some pathogens, especially total and fecal coliforms, can become stressed or altered physiologically in natural waters from exposure to sunlight (Rhodes and Kator, 1990; Sinton *et al.,* 1994; Pommepuy *et al.,* 1996) lack of appropriate substrates (Lopez--Torres *et al.,* 1987), dissolved substances salts or chlorine (McFeters *et al.,* 1986; Munro *et al.,* 1987) and their detection may be compromised. Resuscitation procedures, such as an initial period of incubation at a lower temperature, have been formally incorporated into several methods (e.g. mTEC; USEPA, 2000).

The development and application of microbiological tests for enumerating environmental bacteria should be accompanied by a rigorous evaluation of the method's performance characteristics. Performance characteristics are empirically determined experimental values used to assess the effectiveness of a given enumeration method to recover the target microorganisms from a selected environment. Important performance characteristics include method specificity, selectivity, precision and recovery (ASTM, 1993). In some instances the upper and lower boundaries of counts are also determined. Method specificity is a measure of a procedure's capability to detect all target microorganisms present in a sample with a minimum of false positives. Selectivity refers to a method's ability to primarily support only the growth of the target microorganism and to minimize interfering background growth. Selectivity is conferred through the use of chemical inhibitors, antibiotics or restrictive physical conditions, such as elevated temperature. Method recovery is simply the number of target organisms recovered from a spiked sample compared to a recovery from this same sample using a 'reference' method. Understanding the performance characteristics of a method is important with regard to method application and the quality of quantitative data. For example, using an extremely precise method with low selectivity might be a waste of time when sampling grossly polluted or eutrophic systems because high background counts would defeat the value of high precision.

Detection of target microorganisms obtained through cultural methods can involve a complete variety of techniques available to the modern microbiologist. These include isolation, purification, and biochemical characterization; use of immunologics including fluorescein-labeled antibodies (Wright, 1992; McDermott, 1996), use of enzyme-specific substrates (Toranzos and McFeters, 1996; Sartory and Watkins, 1999), and molecular techniques such as colony lifts followed by hybridization with oligonucleotide probes (Wright *et al.*, 1993).

3.3 Qualitative/quantitative cultural methods

Bacteria are enumerated for a variety of purposes that include detection of pollution, detection of specific phenotypic or phylogenetic groups, and to insure compliance with water quality regulations. Cultural methods provide but a single opportunity to detect and count cells that occur in a freshly collected sample. Beyond maximum holding periods sufficient changes occur and results may be unreliable. The application of appropriate QA/QC programs is essential to enumeration methods. While many methods frequently incorporate verification of isolates randomly selected from samples few, if any, prescribe use of pure cultures to assess responses of recovery media or use of bacterial suspensions of known density to assess recovery efficiency. Donnison *et al.* (1999) consider use of standard bacterial suspensions as a quality control approach to routinely monitor method performance.

3.3.1 Most probable number (MPN)

This is an historically common and important enumeration method used to estimate numbers of viable bacteria in surface waters, soils and sediment, animal scat and shellfish. As a quantitative technique it is extremely versatile and its usage extends well beyond its more common application for estimation of total and fecal coliforms in drinking and surface waters. It has been used to enumerate petroleum-degrading bacteria (Mills *et al.*, 1978), phenol and toluene-degrading bacteria (Fries *et al.*, 1997) and *Staphylococcus aureus* (Bennett and Lancette, 1998). When combined with isotopically-labeled substrates, an MPN method can be used to measure densities of populations with specific degradation potential (Lehmicke *et al.*, 1979). The MPN test is a statistical estimate of culturable units and assumes such units are randomly distributed within a sample (Garthright, 1998). An MPN may or may not represent densities of individuals should the cells occur in groups adsorbed to particles. Functionally, the MPN test is based on dilution of a target microorganism to extinction, i.e. at some dilution, usually by factors of ten, a sample will not contain a single culturable unit. Actual dilutions required are based on experience and MPN tables are commonly based on inoculation of media with 3, 5, 10 or even 12 sample portions at each dilution. The MPN is useful when cells are present at low densities in natural samples. As a statistical estimate based on the binomial distribution, the precision of an MPN value is a function of the number of sample portions assayed for each dilution. In the format generally used in water quality laboratories, the 3 or 5 tube configuration, the 95% confidence intervals of an MPN value are relatively large (APHA, 1998) and significantly better precision can be obtained using a direct count method or increasing the number of portions to 12, 15 or higher values. For improved precision, the MPN method has beeen miniaturized using microplates, which provide for high portion numbers. Hernandez *et al.* (1991, 1993) used 96-well microtitration plates to enumerate *E. coli* and presumptive enterococci. Gamo and Shoji (1999) evaluated the use of multiple BIOLOG substrate plates to obtain MPN estimates of a given microbial community's capability to utilize selected substrates.

Limitations with the MPN method result from departures from assumptions that cells are randomly distributed, interference or inhibition effects, which can reduce the occurrence of positive results, false positives, and the poor precision when small numbers of portions are used. MPN-based methods are much less precise than direct plating or membrane filtration methods. A clear discussion of the MPN technique with useful formulas and tables can be found in Garthright (1998).

3.3.2 Colony or plate counts using solid media

Colony counts on solid media offer significantly better precision when compared to MPN methods and provide for the isolation of individual clones for identification and study. A general reference supplying details concerning the uses of the methods identified below and others can be found in Koch (1994).

Pour plate. This is a comparatively simple enumeration procedure applicable to samples where densities of target organisms are sufficient to yield appropriate numbers of culturable elements for counting purposes. In this procedure, a small sample volume (approximately 0.1–1 ml) containing suspended bacteria or dilutions thereof is mixed with melted or so-called 'tempered' agar (an agar medium is heated or autoclaved to dissolve the agar and then held at a temperature of about 46°C in a water bath to prevent gelling). A known sample volume is placed in a petri dish followed by the tempered agar, carefully mixed to disperse the sample uniformly, and allowed to solidify. Petri dishes should be turned over as soon as feasible to avoid condensation on the lid. Colonies, which will grow both on the agar surface and within the agar, are counted to yield a number of cells per unit sample volume. Limitations of this technique apply to environmentally stressed cells or those sensitive to elevated agar temperatures, which may be killed leading to underestimates of population densities.

Spread plate. A second method to avoid sublethal stress owing to elevated temperatures used in the pour plate is the spread plate. In this technique a small volume of cell suspension, 0.1 ml, is placed on the surface of a slightly 'dried' agar plate and rapidly spread over the surface using a sterile bent rod. If the agar water content is correct, the sample will be rapidly taken up as it is being distributed over the surface. Distribution of the sample can be done by rotating the plate with one hand while continuing to spread the sample using the other. Manual or motorized spread plate 'turntables' are also available for this purpose.

3.3.3 *Membrane filtration*

Membrane filters prepared from a variety of chemical stocks are available for recovery of microorganisms from natural waters where their concentrations are often too low for direct plating and large volume MPN plants would be cumbersome. Membrane filters of various functional pore size distributions can be coupled with appropriate media for the selective recovery of target organisms. The filter pore size commonly used for microbiological methods is 0.45 μm. Depending on the filter, cells are retained owing to mechanical exclusion because of physical pore size or electrochemical forces. Care should be taken to avoid damage to susceptible target microorganisms by unduly high vacuum. Membranes can be placed directly upon solid media or pads saturated with liquid medium. Membranes offer an additional advantage of *in situ* biochemical testing where a membrane can be removed and placed on a second medium or saturated pad containing test reagents to detect specific properties or to calculate a confirmed count. An example of this approach is the urease test used to confirm colonies as *E. coli* on mTEC medium (USEPA, 2000). In this *in situ* test the membrane filter with colonial growth is transferred to a pad saturated with a medium called *urease substrate medium*. After a short incubation period target colonies remain yellow, yellow-green or yellow brown; non-target or urease positive colonies will turn purple or pink. Limitations to the membrane filter method relate to the comparatively narrow range within which method precision applies, e.g. 20–80 colony forming units cfu/plate in the case of USEPA's membrane methods for enumeration of *E. coli* (USEPA, 2000) and interference in colony development and/or counting due to high suspended solids or toxics in a sample.

3.4 Non-cultural methods

3.4.1 *Direct counts*

Direct counts differ from culturable counts in that cells are generally counted by microscopic evaluation and thus do not require the use of complex media to facilitate cell growth to yield densities visible to the unaided eye. Direct counts can be determined on sample suspensions prepared from water, soil or sediment. Cells can be visualized using phase contrast microscopy for unstained cells or by staining cells caught on membrane filters using one of a large variety of fluorescent stains now commercially available. Direct counting methods are eminently amenable to automated counting using image processing hardware and software. The inherent value of direct methods is to reduce the time required for conventional cultural analysis.

In its simplest form quantitative estimates of cell numbers can be determined in a known volume of sample using a special microscope slide/cover slip combination known as a Petroff-Hausser counting chamber. While this technique is simple it does not differentiate live from dead cells, motile cells are difficult to count and should be fixed prior to counting, and sample concentration requires a separate step.

Membrane filter-based assays are now very common and continually expanding in application as new fluorogenic dyes and their conjugates are being developed. Using membrane filters dyed black, common fluorogenic dyes such as acridine orange, 4′,6-diamidino-2-phenylindole (DAPI), Hoechst 33258, or proflavine allow for counting total cells fluorescing against a black background. Kepner and Pratt (1994) provide a comprehensive review of direct enumeration methods, preservation, fluorochromes available, procedures and problems with different kinds of samples.

Specificity can be obtained using fluorescent dyes conjugated to specific target antibodies (Tanaka *et al.*, 2000) or oligonucleotide probes. Langendijk *et al.* (1995) and Franks *et al.* (1998) used 16S rRNA targeted fluorescent oligonucleotide probes to quantify *Bifidobacterium* spp. and major anaerobic groups in suspensions prepared from human fecal samples.

Microscope-based techniques to count rapidly and detect total viable cells, the direct viable count, have been developed that are based on incorporation and chemical modification of dyes or biochemicals that combine the advantages of direct counts with the discrimination of viable from non-viable cells. Autoradiographic techniques have also been used (Meyer-Reil, 1978; Tabor and Neihof, 1982; Roszak and Colwell, 1987) but modern safety concerns and disposal costs associated with use of radioisotopes render their use for routine work prohibitive. Determination of viability can be based on reduction of redox dyes such as tetrazolium salts (Søndergaard and Danielsen, 2001), incorporation of biochemicals that interfere with cell division such as nalidixic acid (Kogure *et al.*, 1979; Roszak *et al.*, 1984; Roszak and Colwell, 1987) and use of conjugated substrates whose hydrolysis by enzymes (e.g. esterase, phosphatase) active in viable cells releases a fluorescent or colored compound that can be detected visually or with photometric techniques (e.g. Gaudet *et al.*, 1996). Nalidixic acid prevents cell division by inhibiting DNA gyrase and, in the presence of appropriate nutrients, cells which can be resuscitated become enlarged and can be separated from 'non-viable' cells using epifluorescenece microscopy. Yokomaku *et al.* (2000) improved the Kogure *et al.* (1979) technique by promoting spheroplast formation selectively in viable cells followed by a lysis step to discriminate viable from non-viable cells. Double staining with a fluorogenic ester and a dye that penetrates inactive membranes can be used to separate active from inactive cell populations (Yamaguchi and Nasu, 1997; Tanaka *et al.*, 2000). The dye, 5-cyano-2,3-ditolyl tetrazolium chloride (CTC), provides advantages of direct counts and discrimination of cells capable of respiring versus non-respiring cells (Søndergaard and Danielsen, 2001).

3.4.2 *Flow cytometry*

Significant progress has been made in adapting flow cytometry to environmental microbiological uses in recent years. Although expensive, flow cytometers offer benefits of individual cell counting, high sample throughput, identification and cell sorting. Advances in direct count techniques now include the use of flow cytometry to quantify total bacteria or when coupled with specific fluorescent dyes can be used to count physiologically active cells (Porter *et al.*, 1995; Yamaguchi and Nasu, 1997; Grégori *et al.*, 2001). CTC has been used with flow cytometry to determine a total viable count (Sieracki *et al.*, 1999). Flow cytometry can be combined with molecular methods through the use of fluorochrome-conjugated oligonucleotide probes (Amann *et al.*, 1990).

3.4.3 *Immunological methods*

Use of fluorescent antibodies for detection and enumeration of specific species or groups of microorganisms has been widely used in environmental microbiology (Wright, 1992). Microorganisms or products thereof have been used to produce polyclonal and more recently, monoclonal antibodies. The latter are considered more specific and potentially more sensitive but require considerably more time, expense and skill to produce. Immunological techniques are useful to confirm the identity of a target microorganism in a purified form or in a mixed cultured population. They can also be used for direct counting when conjugated with fluorochromes, such as fluorescein isothiocyanate (FITC). Direct counting can be accomplished on cells immobilized on glass slides, on membrane filters, or by flow cytometry. Some workers consider routine immunological methods, such as immunofluorescence assays, easier to use and less exacting than molecular methods. Issues of concern with immunological methods commonly encountered when applied to intact environmental samples are those of non-specific staining or background fluorescence. Species in a sample may share reactive sites or epitopes or reactive material derived from the target cells may be present throughout the sample. Immunofluoresence assays have been used to detect VBNC *E. coli* and *Vibrio cholerae* in water (Xu *et al.*, 1982) to study the survival of *Campylobacter* in biofilms (Buswell *et al.* 1998) and to enumerate *V. vulnificus* using a combined MPN-enrichment/selective medium method in samples of estuarine water, sediment and oysters (Tamplin *et al.*, 1991). Fiksdal and Berg (1987) evaluated a fluorescent antibody technique for enumeration of *Bacteroides fragilis* group organisms in water.

3.4.4 *Molecular and biochemical methods*

The evolution and continual development of molecular methods renders a contemporary review of this material a task fundamentally beyond the scope of this chapter. The reader is referred to current literature for specific details. There are many articles covering basic techniques and applications of molecular methods to environmental microbiology (e.g. Hurst *et al.*, 1996; Gerhardt *et al.*, 1994).

Generalized approaches to detection and identification of specific phylogenetic groups using molecular methods can include recovery of target DNA or RNA from environmental or cultured samples, extraction and removal of inhibitors, cloning or amplification of DNA or RNA by PCR, followed by hybridization using specific probes. Direct methods include colony lift (e.g. Wright *et al.*, 1996), fluorescent *in situ* hybridization (FISH), and most recently, flow cytometric methods where DNA probes conjugated to specific dyes (Amann *et al.*, 1990) have been used. Probes have been designed to discriminate specific taxonomic groups using 16S rRNA-targeted and functional genes. Many examples of the former are now found in the literature. Probe design, regardless of the target, requires sequence information and extensive validation. Again, numerous examples of probes, design and verification approaches to avoid hybridization with non-target sequences are found in the literature. Quantitative recovery or enumeration of selected targets remains difficult and is subject to uncertainties related to efficiency of recovery of nucleic acids from environmental samples. Approximate quantitative estimates of organism densities in water samples can be achieved by analysis of diluted samples until no further hybridization is detected.

Fluorescent in situ *hybridization (FISH).* Whole-cell fluorescent *in situ* hybridization or FISH is a non-cultural technique combining attributes of nucleic acid hybridization with various direct detection techniques. For example, fluorochrome-conjugated oligonucleotide probes, designed to target 16S rRNA sequences, can be used to identify individual bacterial cells when visualized by epifluorescence microscopy. Probes can also be designed to detect selected genes and, in the case of ribosomal RNA, to be species genus or group-specific with detection accomplished using epifluorescence microscopy or flow cytometry (Amann *et al.*, 1990). Limitations inherent in direct count microscopy, i.e. the density of the target microorganism must be approximately $\geq 10^4$ cells/ml to see one organism per field, can be overcome using flow cytometry or increasing the volume of sample filtered. Other problems encountered include background autofluorescence, which obscures the target signal, poor signal intensity, non-specific binding of the probe, probe hybridization to unidentified species, or loss of cells during the hybridization process.

To perform FISH by epifluorescence microscopy, a water sample or suspension is filtered through a black polycarbonate membrane, using the same procedure for a membrane filter direct count. The cells on the filter are then hybridized to a selected fluorochrome-conjugated probe following protocols for *in situ* hybridization. Importantly, FISH detects all target cells including those that may be non-culturable or dead. Maruyama and Sunamara (2000) demonstrated that FISH could be combined with direct microscopic counting (called FISH-DV) by coating filters with poly-l-lysine to promote cell retention during the hybridization process. Following hybridization steps and washings to remove non-hybridized probe, salts and reagents, the cells are stained with a non-specific fluorescent dye such as DAPI and washed. Using an epifluorescent microscope total direct counts are obtained using UV illumination and probe-specific counts with excitation wavelengths specific for the fluorochrome-labeled oligonucleotide probe used, e.g. FITC. Multiple probes conjugated to different dyes can be used simultaneously, providing specific multiple and total direct counts. More recently, Maruyama and Sunamara (2000) suggest FISH can be used in a quantitative fashion. DeLong *et al.* (1999) increased the fluorescent intensity of FISH using polyribonucleotide probes instead of singly labeled oligonucleotide probes.

Detection of selected genes. To avoid the use of cultural methods, which are intrinsically variable in recovery and specificity because of shared physiological requirements, investigators have targeted genes to detect selected species, genera or functionally-defined groups. As with direct counting, these methods usually do not provide information on the viability or even expression of the targeted gene in the cells detected. Examples in the recent literature relevant to environmental microbiology include efforts to use the glucuronidase gene (*uidA* gene) to detect and identify *Escherichia coli*. Green *et al.* (1991) used a fragment of the *uidA* gene as probe to detect *E. coli* after a short incubation period on membranes through which water samples had been filtered. Fricker and Fricker (1994) used PCR primers to detect *E. coli* (*uidA*) and coliforms (*lacZ*) isolated from water samples but encountered specificity problems with the latter primer set. All biochemically verified *E. coli* strains tested were positive for the *uidA* gene and a relatively low false positive rate for the *E. coli* primer set occurred. Bej *et al.* (1994) used PCR for improved sensitivity to amplify a segment of the *himA* gene to detect *Salmonella* spp. in oysters. The cytolysin gene of *V. vulnificus* was the target for an enumeration method based on a colony lift procedure (Wright *et al.*, 1993). Following direct inoculation of a sample on a non-selective solid medium, a piece of absorbent paper is placed over the resulting colonial growth, treated to release and retain the DNA from the lifted colonies, and then hybridized with an alkaline phosphatase-labeled oligonucleotide probe. After hybridization and development, areas on the filter showing alkaline phosphatase activity correspond to *V. vulnificus* colonies.

Lipid analysis. Lipid biomarkers have been applied as a direct detection technique to environmental samples from a variety of environments including soil, aquatic sediments, groundwater, and deep subsurface sediment. These methods are based on the assumption that phospholipid fatty acids and fatty acid methyl esters can be used to discriminate microorganisms by group (e.g. Gram-negative versus Gram-positive, eukaryotes versus prokaryotes), based on unique fatty acid structure, chain length, double bond position and number and degree of substitution. Other applications include estimates of biomass, detection of specific physiological groups (Parkes *et al.*, 1993) community structure, and physiological status expressed through phospholipid fatty acid profiles (Nickels *et al.*, 1979). (For details related to the principles, methods, and applications see Frostegaard *et al.* (1991) and White *et al.* (1997).) Commercial systems making specific use of fatty acid methyl esters for identification of microorganisms are available. Gas chromatographic analysis of mycolic acids is a powerful tool for identification of environmental mycobacteria (Butler and Guthertz, 2001).

Enzymatic methods. In addition to targeting specific genes to detect bacterial groups on the basis of characteristic enzymes, the presence of enzymes *in situ* can be assayed by addition of substrate to water samples. In such studies, fluorogenic or chromogenic substrates are added and their hydrolysis products detected, providing the basis for a rapid assay. Berg and Fiksdal (1988) and Fiksdal *et al.* (1994) used several fluorogenic substrates including MUG to detect total and fecal coliform bacteria as an alternative to cultural counting. Van Poucke and Nelis (1977) used chemiluminescent substrates in a presence/absence test to detect β-galactosidase- and β-glucuronidase-containing cells in a variety of water samples over relatively short incubation times. Results suggested significant improvements in sensitivity came at the expense of unfavorable false-positive and false-negative errors. George *et al.* (2000) discuss aspects of the current of the literature concerning enzymatic methods for rapid enumeration and notes such methods detect VBNC as well. Enzymatic methods are uniquely suited to remote sensing applications and may be valuable for this purpose.

4 SELECTED ENVIRONMENTAL MICROORGANISMS OR INDICATORS OF PUBLIC HEALTH SIGNIFICANCE AND EXAMPLES OF METHODS USED FOR THEIR DETECTION/ENUMERATION

4.1 Coliform and enterococcal indicators

There is a large and growing body of literature on coliform and enterococcal indicators and whose size is well beyond the scope of this chapter. Readers are referred to such sources as Standard Methods for the Examination of Water and Wastewater (APHA 1998); Manual of Environmental Microbiology (Hurst *et al.* 1996), the USEPA ICR Microbial Laboratory Manual (USEPA, 1996). The USEPA National Exposure Research Laboratory, Microbiological and Chemical Exposure, maintains a web site (http://www.epa.gov/microbes) which contains the publications of the most recent approved methods for detection and enumeration of microorganisms of public health significance. Toranzos and McFeters (1996) summarize methods for detection and enumeration of indicators in fresh and drinking waters.

4.2 Alternate indicators and selected environmental pathogens

Although, historically, attention has been focused on the indicators within the Enterobacteriaceae and enterococci, recent exploratory research has focused on the potential use candidate indicators such as *Clostridium*, *Bacteroides*, *Bifidobacterium*, and of bacteriophages as model viral indicators. In recent years other environmental pathogens such as mycobacteria, *Campylobacter* and have been the cause for public health concern. The remainder of this chapter provides examples of alternate indicators and pathogens of interest and methods for detection and enumeration.

4.2.1 *Clostridia*

Clostridia are obligately anaerobic bacteria commonly found in soils and the gastrointestinal tracts of warm-blooded animals including man. Clostridia form very resistant structures called endospores, which enable these microorganisms to survive in air, at elevated temperatures and in the presence of chemical disinfectants. Spores of these organisms can persist for extended periods of time in soils, water and aquatic sediments. Although clostridia are fermentative saprophytes, they are commonly associated with food-borne disease in humans owing to their ubiquity, the high temperature tolerance of their spores, and their ability to elaborate extremely potent endotoxins. Typically, spores germinate in food products that are inadequately processed. *Clostridium perfringens* and *C. botulinum* are common causes of food-borne illnesses associated with cooked and stored meats or poultry and seafoods that are inadequately heated before eating. *Clostridium botulinum* toxin is also associated with mass mortalities of waterfowl (Fay *et al.*, 1965; Wobeser, 1997).

The presence of *C. perfringens* in receiving waters has been linked to contamination by faeces and wastewaters (Cabelli, 1977; Bisson and Cabelli, 1980). Bisson and Cabelli (1980) suggested *C. perfringens* spores have value as an indicator of chlorination efficiency and the presence of unchlorinated sources of fecal contamination, and as a 'conservation tracer' delineating the impact and transport of wastewater effluents. Because *Cryptosporidium* oocysts are also resistant to chlorination, it has been suggested that *C. perfringens* be used as an indicator of drinking water treatment (Payment and Franco, 1993). Detection of vegetative cells in environmental waters reflects fresh and untreated fecal matter because the persistence of vegetative cells is very short (Bisson and Cabelli, 1980). Fujioka and Shizumura (1985) and Roll and Fujioka (1997) used *C. perfringens* to detect wastewater discharge into fresh water streams and brackish waters in Hawaii. Hill *et al.* (1993) confirmed the utility of *C. perfringens* to delineate the extent and movement of sewage sludge dumped at the deep ocean '106-Mile Site' off the northeast coast of the USA.

Bisson and Cabelli (1980) developed a membrane filtration method for *C. perfringens* recovery applicable to saline waters based on a highly selective medium (mCP). The composition of this medium was later modified to reduce the concentration of a costly biochemical component without compromising its selectivity (Armon and Payment, 1988). Sartory (1986) compared recovery of *C. perfringens* on mCP and egg yolk-free tryptose-sulphite-cycloserine (TSC) agar as membrane based tests. For a variety of sample types, which included polluted river water and groundwater, egg yolk-free TSC was found as selective as mCP and more efficient. Egg yolk-free TSC was recommended because of its ease of preparation, lower cost, and availability of a simple confirmation scheme. Hill *et al.* (1996) reported a false-positive rate of about 5% using mCP to recover *C. perfringens* from deep ocean dump sites. Compared to TSC, Sartory *et al.* (1998) found mCP to possess a significantly higher false positive rate, to be less efficient, and generally more difficult to use with environmental samples.

4.2.2 *Campylobacter spp*

C. jejuni is the species most frequently associated with campylobacteriosis infections in humans and some consider this organism to be the leading cause of bacterial gastroenteritis in the developed world (Skirrow, 1991; Allos and Blazer, 1995). This microorganism is commonly found in the gastrointestinal tracks of domestic pets (dogs, cats), farm animals (cattle, pigs, poultry), and wild animals such as rodents and birds (Kapperud and Rosef, 1983). Campylobacters can be isolated from sewage and are most likely carried to surface receiving waters in runoff following periods of rainfall. Humans and animals can be asymptomatic carriers. Campylobacters discharged into surface waters used for drinking, contaminated and improperly cooked poultry or meat products, and non-pasteurized milk products, have all been implicated as the cause of campylobacteriosis in humans. Campylobacters have also been found in shellfish beds and market shellfish (Abeyta *et al.*, 1993), but the human health significance of shellfish as a vector remains uncertain.

Campylobacters are microaerophilic organisms with an optimum growth temperature of about 42°C and require complex media supplemented with non-selective constituents, such as whole blood and other compounds (Bolton *et al.*, 1984). Recovery of cells from natural waters is hindered by slow growth rates, competitive background microbiota low numbers, and possible non-culturability (Rollins and Colwell, 1986; Cappelier and Federighi, 1998). Given the complexities for detection of *Campylobacter* spp. from environmental samples, many applications of molecular methods based on PCR exist for its detection in poultry, feces, environmental waters and sewage (e.g. Waage *et al.*, 1999). Other investigators have used immunological-based methods for detection (Buswell *et al.*, 1998). These tools will be valuable in continued efforts to study the persistence of *Campylobacter* cells in natural waters and to understand the health significance of viable and non-culturable states.

4.2.3 Bacteriophages

Human enteroviruses have been implicated as causative agents of waterborne disease transmitted by drinking water and shellfish (IAWPRC, 1991). In the USA, cases of Norwalk virus transmitted through fecally-contaminated shellfish are a recurring concern (Centers for Disease Control, 1995). Routine methods for direct detection of infective enteroviruses, such as Norwalk, from environmental samples remain to be developed and refined. Lacking such methods there has been ongoing interest in the use of bacterial viruses, or bacteriophages, as surrogates for pathogenic enteroviruses, because of the inadequacies of bacteria as indicators of virus. Resistance to chlorination, in particular, is a desirable property of a bacteriophage indicator of treated sewage effluent because hepatitis A and rotavirus are also relatively resistant to chlorination and UV radiation (IAWPRC, 1991). Havelaar *et al.* (1993) reported that concentrations of enteric viruses and F-specific coliphages were highly correlated in contaminated river and lake waters.

Detection of bacteriophages ideally requires a host whose susceptibility range is restricted to a target bacteriophage associated only with the fecal source. In reality this discrimination is difficult to achieve using wild-type hosts because of the diversity of lytic bacteriophages in the environment.

Bacteriophages most recently studied and proposed as viral indicators of fecal contamination are the F-specific or male-specific coliphage and *Bacteroides fragilis* bacteriophage. The former bacteriophages selectively adsorb to the sex pilus which is usually encoded by a plasmid. Superior hosts sensitive to F-specific coliphages and resistant to antibiotics to reduce the background growth of interfering microorganisms have been developed and tested using a variety of environmental waters (Havelaar and Hogeboom, 1984; Debartolomeis and Cabelli, 1991). These hosts differ somewhat in their host range and avoid lysis by somatic phages using different strategies. WG49 (Havelaar and Hogeboom, 1984) is a *Salmonella typhimurium* selected for lack of susceptibility to somatic coliphage and because there are relatively low numbers of salmonella phage in sewage. It was constructed using an F-plasmid which enables the strain to produce sex pili. Debartolomeis and Cabelli (1991) developed an *Escherichia coli* (HS(pFamp)R) host naturally resistant to somatic coliphage. Using relatively uncomplicated procedures, these host strains are capable of routine detection and enumeration of male-specific coliphage from domestic wastewaters and have also been applied to receiving waters (Havelaar and Hogeboom, 1984; Debartolomeis and Cabelli, 1991). Their application to detect male-specific coliphage in environmental waters and other sample types must be carefully monitored to detect lysis by environmental phage (Rhodes and Kator, 1991; Handzel *et al.*, 1993) or FDNA male-specific coliphage. This can be achieved through parallel processing of plates incorporating Rnase. Alterations to the conventional double-agar-overlayer assay have been proposed to allow for sample concentration and promote ease of use (Sobsey *et al.*, 1990; Sinton *et al.*, 1996).

Owing to the labor and time associated with cultivation of *Bacteroides* spp. and its presumed poor survival in the environment, the use of phages to *B. fragilis* as indicators of human fecal pollution has received considerable attention. *Bacteroides* phages occur in human feces and sewage (Booth *et al.*, 1979; Tartera and Jofre, 1987; Tartera *et al.*, 1989; Grabow *et al.*, 1995) and are unable to replicate in natural waters and sediments due to their reliance on a metabolically active host (Jofre *et al.*, 1986; Tartera *et al.*, 1989). *B. fragilis* phages appear to be highly specific to humans and have not been detected in a variety of domestic animals or captive primates (Tartera and Jofre, 1987; Grabow *et al.*, 1995). A strong correlation between *B. fragilis* phages and enteroviruses in sediments and treated wastewater has been reported (Jofre *et al.*, 1989; Gantzer *et al.*, 1998; Pina, 1998; Torroella, 1998).

Resistance of these phages to disinfectants and environmental factors (Jofre *et al.*, 1986; Bosch *et al.*, 1989; Sun *et al.*, 1997) has also promoted their use as indicators. The frequency of recovery of *B. fragilis* phages from chlorinated drinking water suggests these phages are more resistant than somatic and F-specific phages (Jofre *et al.*, 1995; Armon *et al.*, 1997). Although *B. fragilis* phages are more persistent in seawater under dark conditions than F-RNA phages (Chung and Sobsey, 1993) the former are more susceptible to ultraviolet (Bosch *et al.*, 1989) and sunlight inactivation (Sinton *et al.*, 1999). We have observed similar results for *B. fragilis* and F-RNA phages exposed *in situ* in estuarine water but noted prolonged persistence of both phages in estuarine sediments (Kator and Rhodes, 1992). Quantification of phages to enteric bacteria in freshwater environments indicates that *B. fragilis* phages are more resistant to natural inactivation than male-specific or somatic coliphages (Araujo *et al.*, 1997).

Enumeration of *B. fragilis* bacteriophages involves more complex media and methods than used for coliphage assays. *B. fragilis* phages in sewage and fecal samples can be enumerated by conventional plaque assay but less contaminated samples may require concentration or enrichment. The double-agar layer (DAL) assay is more efficient than the MPN technique, but both methods are subject to conditions of the host culture, composition of the host medium, presence of divalent cations and decontamination method used (Cornax *et al.*, 1990). Tartera *et al.* (1992) describe an enumeration procedure based on sample decontamination by filtration through polyvinylidene difluoride membrane filters with subsequent filtrate assay by DAL using a rich medium, *Bacteroides* phage recovery medium (BPRM). Kator and Rhodes (1992) combined the use of antichaotrophic salts to promote phage adsorption to nitrocellulose membrane (Farrah, 1982) with a protracted elution process (Borrego *et al.*, 1991) for application to phage recovery from estuarine waters. Lucena *et al.* (1995) describe concentration or enrichment (presence/absence tests) for use with samples containing low phage densities such as drinking water. Samples are concentrated by filtering a large volume (1 litre) through a beef extract-treated membrane of inorganic material with a honeycomb pore structure and low protein binding activity with subsequent elution of phages and assay of the eluate by DAL or, direct DAL assay of phages retained on the membrane. Enumeration by this concentration technique was significantly affected by sample turbidity. Enrichment-based tests provide good recoveries but require removal of oxygen and are non-quantitative (Armon and Kott, 1993; Lucena *et al.*, 1995; Armon *et al.*, 1997).

A critical shortcoming of using *B. fragilis* phages as indicators is the susceptibility range of the host used for recovery. *B. fragilis* HSP40 (Tartera and Jofre, 1987; Tartera *et al.*, 1992; Cornax *et al.*, 1990; Lucena *et al.*, 1995) used to detect phages in Mediterranean seawater has been ineffective in similar studies conducted in temperate climates (Kator and Rhodes, 1992; Bradley *et al.*, 1999). However, in both of the latter studies, *B. fragilis* hosts recovered from local waters were used successfully to isolate bacteriophage. Also problematic is the high specificity of bacteriophages such that host range among *B. fragilis* strains is restricted (Tartera and Jofre, 1987; Kator and Rhodes, 1992; Bradley *et al.*, 1999).

Infrequent recovery and low densities of *B. fragilis* phages in environmental waters except those proximate to sewage pollution (Cornax *et al.*, 1991; Bradley *et al.*, 1999) may negate the use of this indicator in studies of non-point pollution. However their tendency to adsorb to sediments (Jofre *et al.*, 1986), recovery from sediment samples (Tartera and Jofre, 1987; Jofre *et al.*, 1989; Bradley *et al.*, 1999), and persistence characteristics (Kator and Rhodes, 1992) suggest that sediment analysis might be valuable in determining fecal sources in waters of low to moderate pollution. Lucena *et al.* (1996) demonstrated that persistence of *Bacteroides* phages in sediments contributed to their use as indicators in areas subject to persistent, albeit remote, fecal pollution. These investigators (Lucena *et al.*, 1994) also report that the extended persistence characteristics of *Bacteroides* phages compared with FRNA phages contributes to their usefulness as indicators of human enteric viruses in shellfish.

Some workers suggest that somatic coliphage may be of value (Cornax *et al.*, 1991) as sewage indicators based on observations that somatic coliphages appear to persist longer than FRNA coliphages in natural waters. However, both the early and more recent literature (e.g. Leclerc *et al.*, 2000) suggest there are problems with use of somatic coliphages as fecal or viral indicators. These include significant heterogeneity, broad host range, ability to replicate in the environment, and inconsistent relationships to contaminant sources. In particular, the ability of somatic coliphages to replicate *in situ* or to experience host lysis from indigenous virus in the environment are stumbling blocks to the use of somatic coliphages. Croci *et al.* (2000) compared a variety of indicators, including somatic and FRNA coliphage with enterovirus burden in mussels. They concluded that only direct viral detection is an acceptable method to assess the risk associated with consumption of mussels from the Adriatic Sea.

4.2.4 *Bacteroides spp*

Bacteroides spp. are Gram-negative, anaerobic, pleomorphic rods and represent a major constituent of the colonic flora in normal humans (Jousimies-Somer *et al.*, 1995). In recent years the genus has undergone major taxonomic revisions. The *Bacteroides fragilis* group originally included isolates considered to be subspecies, which have since been assigned species rank based on DNA–DNA homology analyses (Sinton *et al.*, 1998). Consequently early studies referring to the '*B. fragilis* group' as fecal indicators actually may include different species and it is probably more appropriate to use the term *Bacteroides* spp. Interest in *Bacteroides* spp. as fecal indicators is based on their presence at high concentrations in human feces and either absence or occurrence at significantly lower concentrations in most animals (Allsop and Stickler, 1984, 1985; Kreader, 1995). Futhermore as obligate anaerobes, *Bacterioides* spp. are unlikely to multiply in receiving waters and are less persistent in natural waters than *E. coli* (Allsop and Stickler, 1985; Fiksdal *et al.*, 1985). Despite these advantages, complexities associated with cultivation have limited their use as indicators.

The use of prereduced anaerobically sterilized (PRAS) media is superior to prereduced aerobically prepared media for culturing *Bacteroides* spp. (Mangels and Douglas, 1989). A selective and differential medium (WCGP) has been developed that relies on detection of aesculin hydrolysis by *Bacteroides* spp. in the presence of gentamicin and penicillin G, which inhibit other bacteria (Allsop and Stickler, 1984). Recovery of *Bacteroides* spp. by membrane filtration and culturing on WCGP is enhanced by resuscitation initially at a lower temperature. A PRAS agar medium selective for *B. vulgatus* (BVA) has been developed based on this organism's resistance to kanamycin, vancomycin, colistin, ability to grow in the presence of bile, and inability to hydrolyze esculin (Wadford *et al.*, 1995). Isolation of *B. vulgatus* from shellfish meeting the market standard for fecal coliforms suggests it may be a more reliable indicator of fecal contamination

(Wadford *et al.*, 1995) although not necessarily specific for humans since high levels occur in house pets (Kreader, 1995).

Difficulties associated with culturing *Bacteroides* spp. can be overcome by use of alternate detection methods. Culturing has been avoided by enumerating *Bacteroides* spp. with a rapid (2–3 h) fluorescent antibody technique (Fiksdal and Berg, 1987). Species-specific DNA hybridization probes have been used for detection of *Bacteroides* in feces (Kuritza and Salyers, 1985; Kuritza *et al.*, 1986). Kreader (1995) increased the sensitivity of species-specific assays using PCR amplification of 16S rRNA gene sequences followed by detection of specific PCR products through hybridization. Although human feces generally yielded high signals, non-humans generally had lower signals with the exception of house pets. The significance of large fecal outputs from farm animals exhibiting weak signals on distinguishing sewage from farm runoff is not known. Subsequent study showed that *Bacteroides distasonis* DNA, detectable by PCR-hybridization assay, disappeared as a function of temperature and predation with detection possible after 2 weeks exposure of human feces in river water (Kreader, 1998). Analysis of 16S rDNA with PCR amplification has been used to detect *Bacteroides* in coastal lagoons (Benlloch *et al.*, 1995). Bernhard and Field (2000a,b) used host-specific 16S rDNA genetic markers and PCR to discriminate human and ruminant feces and for identification of non-point sources of fecal pollution.

4.2.5 *Fecal streptococci*

The term 'fecal streptococci' has been used to describe a group of taxonomically distinct streptococci that are Gram-positive, catalase negative, non-spore-forming, facultative anaerobes recoverable from the gastrointestinal tracts of humans and animals. This group has been considered as an indicator of fecal contamination in environmental waters. They are present in relatively high densities in human and animal feces, in sewage, and it is generally believed they do not grow and multiply in environmental waters and soils. Recent revisions in the taxonomy of this 'group' have resulted in some members of the group being placed into a new genus called *Enterococcus*. Members of this genus include *E. avium*, *E. casseliflavus*, *E. durans*, *E. faecalis*, *E. faecium*, *E. gallinarum*, *E. hirae*, *E. malodoratus*, and *E. mundtii*. These species usually grow at 45°C in 6.5% NaCl, and at pH 9.6; most grow at 10°C. *S. bovis* and *S. equinus*, fecal streptococci generally associated with animal feces, which are negative for two or more of these characteristics, are retained in the genus *Streptococcus*. Consequently, some species previously considered as fecal streptococci are now enterococci.

Early studies identified the ratio of fecal coliforms to fecal streptococci as useful for differentiation of human from animal fecal pollution sources (Geldreich and Kenner, 1969) but this assumption now appears largely unwarranted. Although it appears that animal gastrointestinal contents and feces have proportionately more fecal streptococci to fecal coliforms than humans, the ratio is not stable and varies with exposure to natural environments. Differential survival of both enterococcal and streptococcal components of fecal streptococci in aquatic environments will occur and further complicates a straightforward interpretation of 'fecal streptococcal' indicator levels. Source specificity is also an issue with this indicator because some fecal streptococcal strains are associated with non-fecal plant, insect, reptile and soil habitats. The presence of these 'atypical' strains in the environment can complicate a straightforward interpretation of the sanitary significance of this group (Mundt *et al.*, 1958; Mundt, 1963; Geldreich *et al.*, 1964; Geldreich and Kenner, 1969). Recent interest in the enterococci (a 'subgroup' of fecal streptococci) as an indicator derives from the USEPA adoption of an epidemiological-based enterococcus criterion for recreational marine and freshwaters waters (USEPA, 1986). Excellent discussions surrounding the use of faecal streptococci as an indicator enumeration and associated problems can be found in Sinton *et al.* (1993a,b; 1994).

Generally, the fecal streptococcal habitat is the intestinal content of both warm- and cold-blooded animals, including insects. None of the enterococci can be considered as absolutely host specific, although some species evidence a degree of host specificity. Although certain genera (e.g. *E. faecalis*, *E. faecium*) have been considered specific to feces from human or other warm-blooded animals, phenotypically similar strains and biotypes can be isolated from other environmental sources (Mundt, 1982; Clausen *et al.*, 1977; Beaudoin and Litsky, 1981). Strains and biotypes of *E. faecalis* and of *E. faecium* can be isolated from a variety of plant materials, reptiles and insects. In comparison, the distribution and viability of *S. bovis* and *S. equinus* in extra-enteral environments appears restricted.

Functionally, 'fecal streptococci' as an indicator are, in effect, defined on the same operational basis as fecal coliforms, i.e. organisms recovered using a given method. The selectivity and specificity of available methodologies both define and complicate the meaning of the term 'faecal streptococci' because not all streptococcal and enterococcal species with a fecal habitat are recovered using any one technique.

Several media have been suggested for the selective isolation and/or enumeration of the fecal streptococci or the enterococci. For the most part, the media and methods that are available presently lack selectivity, differential ability, quantitative recovery, relative ease of use, or a combination of these deficiencies. Some strains of fecal streptococci from anaerobic environments, for example, initiate growth only in the presence of elevated levels of CO_2 until the cultures have been adapted to an aerobic environment. *E. faecalis* and *E. faecium* are the most common enterococci encountered. This undoubtedly reflects the rationale of employing KF streptococcal agar for the estimation of enterococci in foods. m Enterococcus agar is used most often for water.

Validation of the enterococci as an indicator of public health risk in marine waters will require improved recovery methods. The range of selectivity and specificity characteristics of methods for enumeration of enterococci has been mentioned. Current recovery methods require confirmatory testing of presumptive isolates (Ericksen and Dufour, 1986), increasing the time and cost of analysis. Although the mE-based method has been applied to marine and estuarine waters (Cabelli *et al.*, 1983), its utility in non-point source impacted marine and shellfish growing areas is undetermined. The relative occurrence and densities of non-fecal biotypes of *E. faecalis* and *E. faecium* in these waters should be determined. Rapid methods for confirmation of selected enterococcal species, based on serological or biochemical characteristics are needed. Bosley *et al.* (1983) and Facklam and Collins (1989) describe useful and rapid (4 h) identification schemes to separate enterococci from group D non-enterococci based on hydrolysis of L pyrrolidonyl-β-naphthamide (PYR). A colony hybridization method, employing oligonucleotide probes synthesized for specific sequences of 23S rRNA of selected enterococci, was used successfully to detect and identify *E. faecalis*, *E. faecium* and *E. avium* in mixed culture (Betzl *et al.*, 1990).

4.2.6 Non-tuberculosis mycobacteria

In addition to the well-known obligate pathogens, *Mycobacterium tuberculosis* and *M. leprae*, the genus *Mycobacterium* also includes free-living saprophytes that for almost a century received little recognition with regards to public health (Covert *et al.*, 1999). These non-tuberculosis mycobacteria (NTM) are ubiquitous, occurring in a variety of soil and water environments as well as in diseased animals and humans. During the last several decades the clinical significance of these species has become widely acknowledged (Wolinsky, 1992; Falkinham, 1996; Horsburgh, 1996) and water recognized as an important vehicle of transmission (Collins *et al.*, 1984; Covert *et al.*, 1999; Dailloux *et al.*, 1999). Ecological studies of NTM in diverse natural waters show that densities and distribution vary geographically and according to physicochemical characteristics (Kirschner *et al.*, 1992; Iivanainen *et al.*, 1993, 1999; Dailloux *et al.*, 1999). In addition to exposure to mycobacteria indigenous to natural aquatic habitats, individuals are exposed to these organisms in engineered aquatic environments such as treated drinking water systems.

Mycobacteria, relatively resistant to numerous disinfectants including chlorine (von Reyn *et al.*, 1993, 1994; Covert *et al.*, 1999) are particularly persistent in drinking water systems (Dailloux *et al.*, 1999) and are more frequently isolated in water supplies in developed countries (von Reyn *et al.*, 1993).

During recent years, non-tuberculosis mycobacterial infections have increased in actual numbers and in the proportion of total mycobacterial disease, with the highest incidence attributed to the *M. avium* and *M. intracellulare* (MAI) complex (Wolinsky, 1992). Disseminated infections frequently occur in patients with acquired immunodeficiency syndrome and can dramatically affect survival (Horsburgh and Selik, 1989; Horsburgh, 1996). The environment is considered the prime source of these free-living strains and ingestion of contaminated water is a mechanism that could explain severe intestinal infections in AIDS patients (Wallace, 1987; Wolinsky, 1992). One study definitively showed that a human infection by *M. avium* occurred via contaminated potable water (von Reyn *et al.*, 1994). However, overall specific reservoirs and factors affecting human exposure leading to infection have not been defined (Horsburgh, 1996).

The MAI complex can produce a variety of chronic non-tuberculosis mycobacteriosis in both animals and humans, including infections characterized by intestinal colonization. Swine have been implicated as vehicles of infection to humans based on the occurrence of MAI strains of the same serotypes (Ikawa *et al.*, 1989) and similar genotypic characteristics (Bono *et al.*, 1995). *M. avium* and *M. avium* subsp. *paratuberculosis* infections in ruminants and ungulates such as cattle, goats, sheep, deer, and elk cause chronic granulomatous enteritis (Cocito *et al.*, 1994; Essey and Koller, 1994) and result in massive numbers of mycobacteria, e.g. 10^8 organisms per g feces (Whittington *et al.*, 2000). Recently the isolation of *M. avium paratuberculosis* from feces of patients with Crohn's disease and ulcerative colitis has implicated this organism as the etiologic agent (Del Prete *et al.*, 1998). Although members of the MAI complex are also the causative agents of avian tuberculosis in wild birds and domestic poultry, recovered strains differ significantly from those isolated from pigs and humans (Grange *et al.*, 1990; Bono *et al.*, 1995).

Despite the recognition that animals are potential vectors for transmitting mycobacteriosis to humans, concerns have been focused primarily on food-borne or aerosol transmission from animals (Falkinham, 1999; Yoder *et al.*, 1999). Waterborne transmission of mycobacterial infections, primarily *M. avium*, has focused on the aerolization of natural waters and treated water as well as ingestion of the latter (Falkinham, 1999). Public health risks associated with animal slaughterhouse wastewater, stormwater associated with feedlots and intensive animal holding facilities have received little attention.

Recovery of mycobacteria from water samples typically involves three steps: concentration, decontamination and cultivation. Many mycobacteria grow slowly, requiring weeks or months to form visible colonies in culture and their detection can be compromised by background overgrowth. A variety of disinfectants, e.g. 2% HCl, 2% NaOH, and 0.3% Zephiran have been used (Jaramillo and McCarthy, 1986; Schulze-Robbecke *et al.*, 1991; Kamala *et al.*, 1994; Neumann *et al.*, 1997) to eliminate background microbiota, although these chemicals frequently are detrimental to mycobacterial recovery. Our laboratory studies have shown that treatment of aqueous suspensions of pure cultures with widely used decontaminants result in several log reductions after exposure periods routinely used. Observations that mycobacterium species were differentially affected indicates that a particular chemical decontaminant would 'select' for specific species and thus 'bias' the composition of the original sample. A variety of culture media and incubation conditions exist which, likewise, will also affect the types of mycobacteria recovered. It is generally accepted that reported mycobacterial densities are underestimates due to these methodological factors.

Although molecular techniques have been used to detect mycobacteria from diseased fish (Knibb *et al.*, 1993; Talaat *et al.*, 1997) and clinical samples (Crawford, 1999) detection of mycobacteria in natural water samples has been pursued primarily by cultivation (Dailloux *et al.*, 1999). Application of molecular methods developed for detection of bacterial DNA in environmental waters (Belas *et al.*, 1995) to detection of mycobacteria is needed to facilitate ecological and epidemiological studies to increase our understanding of these microorganisms and their transmission via the water route.

4.2.7 *Streptococcus bovis*

Streptococcus bovis is a non-enterococcal, group D streptococcus. High densities occur in feces of ruminants (Wheater *et al.*, 1979) and other domestic animals including dogs, cats, horses and pigs (Clausen *et al.*, 1977; Kenner *et al.*, 1960). Recent investigations demonstrated that feral mammals typically resident in tidal and freshwater marsh habitats (deer, muskrat, raccoon) are sources of both *E. coli* and *S. bovis* (Kator and Rhodes, 1996 unpublished results). Because of its association with animals, *S. bovis* has been proposed as a specific indicator of animal pollution (Cooper and Ramadan, 1955). However, this organism is not unique to animals and is present in the intestines of healthy (11%) and immunocompromised (up to 56%) humans, particularly those with neoplasias of the gastrointestinal tract (Klein *et al.*, 1977; Beebe and Koneman, 1995). *S. bovis* can cause disease, e.g. endocarditis, bacteremia septic arthritis and meningitis in compromised hosts and less frequently in healthy individuals (Klein *et al.*, 1977; Muhlemann *et al.*, 1999; Grant *et al.*, 2000). Bacteremia in patients with colonic neoplasia and bacterial endocarditis has been attributed primarily to *S. bovis* biotype I (Ruoff *et al.*, 1989), although biotype II also is associated with pathological conditions (Cohen *et al.*, 1997; Grant *et al.*, 2000).

Oragui and Mara (1981) developed a selective medium for recovery by membrane filtration, membrane-bovis agar, for the recovery of *S. bovis* based on its ability to utilize NH_4^+ as the sole source of nitrogen and the absence of a requirement for exogenous vitamins when cultured under anaerobic conditions. The original medium formulation was modified (Oragui and Mara, 1984) to contain less sodium azide because of inhibitory effects on *S. bovis* strains from different geographical regions. Although the specificity of the modified medium was 96% when applied to sewage and surface water samples (Oragui and Mara, 1984) lower specificity, 35–47% was observed for samples from freshwater feeder streams and estuarine shellfish growing areas in a different geographic region (Kator and Rhodes, 1991, 1996). The modified medium was not sufficiently selective to inhibit growth of *Enterococcus faecalis* (Oragui and Mara, 1984; Kator and Rhodes, 1991).

Other selective techniques have incorporated fluorogenic substrates to detect *S. bovis*. A medium selective for fecal streptococci provides for the differentiation of *S. bovis* based on its ability to hydrolyze both a fluorogenic galactoside and a chromogenic starch substrate (Little and Hartman, 1983). However strains unable to hydrolyze starch (Oragui and Mara, 1981; Kator and Rhodes, 1991) could lead to an underestimation of *S. bovis* densities using this medium. A novel miniaturized method for the specific detection of *S. bovis* as well as *E. coli* and enterococci is based on the ability of the former to hydrolyze a fluorogenic substrate and its inability to reduce triphenyltetrazolium chloride (TTC) (Pourcher *et al.*, 1991). Enumeration is based on the most-probable-number calculation using microtiter plates (16 wells per dilution).

Distinguishing ruminal from human strains of *S. bovis* has been problematic, relying on biochemical characteristics such as mannitol and melibiose fermentation, growth on starch and amylase activity (Knight and Shales, 1985; Nelms *et al.*, 1995). Recently a PCR technique has been published (Whitehead and Cotta, 2000) to differentiate between human and ruminal strains of *S. bovis* based on 16S rRNA gene sequences (Nelms *et al.*, 1995). Application of molecular techniques to cultural methods could provide for rapid and specific enumeration of *S. bovis* and avoid reliance on time consuming and laborious biochemical verification. For example, related organisms such as *Streptococcus alactolyticus*, *Streptococcus saccharolyticus* and *Enterococcus columbae* would not be identified readily using routine procedures (Knudtson and Hartman, 1992) and require additional biochemical testing, DNA–DNA hybridization studies or 16S rRNA analysis (Farrow *et al.*, 1984; Devriese *et al.*, 1990; Rodrigues and Collins, 1990). Furthermore, reliance on conventional and commercially available miniaturized identification techniques may be misleading. Sodium dodecyl sulfate-polyacrylamide gel electrophoresis analysis of whole-cell proteins of isolates previously identified as *S. bovis* showed that those from human and animal infections were predominantly *Streptococcus gallolyticus* (Devriese *et al.*, 1998), a newly described species originally detected in the feces of koalas (Osawa *et al.*, 1995). Recently ruminal tannin-tolerate streptococci have been isolated from a variety of animals and, on the basis of 16S rRNA, are most closely related to ruminal strains of *S. bovis* and *S. gallolyticus* (Nelson *et al.*, 1998). Such complexities underscore the advantages of using molecular probes to identify and distinguish specific *S. bovis* strains unique to human and animal sources (Whitehead and Cotta, 2000).

Indirect detection of ruminal *S. bovis* by using bacteriophage has been proposed but does not appear promising. Although restriction endonuclease digestion patterns of the PCR amplified 16S rRNA gene of *S. bovis* strains isolated from a variety of ruminants indicated genetic homogeneity isolated lysogenic phages were highly strain specific (Klieve *et al.*, 1999). Previous investigations by others had shown that the host range of lytic *S. bovis* phages was limited (Klieve and Bauchop, 1991; Tarakanov, 1996).

Relative to coliforms and enterococci, *S. bovis* dies off more rapidly in aquatic environments (Geldreich and Kenner, 1969; McFeters *et al.*, 1974; Wheater *et al.*, 1979). *In vitro* experiments conducted at 6°C and 25°C using an isolate of *S. bovis* demonstrated poor survival in filtered fresh and estuarine water, particularly at the higher temperature (Kator and Rhodes, 1991). Persistence experiments (Kator and Rhodes, unpublished data) using non-filtered fresh and estuarine water and conducted *in situ* at 10°C using light transmitting plastic bags, exposed to ambient sunlight immediately below the water surface, showed reduction of four orders of magnitude to undetectable levels within 3 days. Corresponding survival in the dark was enhanced with less than one and two orders of magnitude decrease in fresh and estuarine waters, respectively. Despite poor persistence in environmental waters, we observed prolonged survival of *S. bovis* in deer and muskrat scat exposed in mesh bags at the estuarine intertidal zone with little or no change in concentrations after 1 month of weathering. These observations suggest that animal fecal matter can serve as reservoir of *S. bovis* (and other enteric microorganisms) in watersheds and that, owing, to its poor survival in environmental waters, its detection indicates recent input.

4.2.8 Iron oxidizing bacteria

The *Sphaerotilus-Leptothrix* group of organisms, also known informally and collectively as 'sewage fungus', are sheathed bacteria whose filamentous appearance in contaminated surface waters appears 'fungus-like'. In reality, the filaments are sheaths containing single rod-shaped cells which are Gram-negative, obligate aerobes that divide by binary fission, and do not branch. These bacteria also tend to accumulate poly-hydroxybutyrate which can be seen as inclusions in the cells. The sheaths are composed of protein-polysaccharide-lipid complex (van Veen *et al.*, 1978) and are covered with an exopolysaccharide or slime layer that is associated with deposits of iron or manganese oxides. Whether the oxidized iron associated with *Sphaerotilus natans* represents the outcome of an energy-deriving process remains unconfirmed. This group of microorganisms is important in the cycling of manganese and iron in natural waters.

Waters polluted by sewage contain high concentrations of organic matter and can support indicative populations of filamentous bacteria belonging to the genera *Sphaerotilus* or *Leptothrix*. These bacteria oxidize reduced iron and deposit its oxidized form in conjunction with their mucilaginous filaments. van Veen *et al.*, (1978) recommended isolation by collecting samples of the gray-brown sheaths from a colonized surface and mixing to separate the sheaths so that small pieces of the sheathed cells can be transferred to a dried agar plate containing a low-nutrient medium. Various enrichment media are available for enrichment from water samples (e.g. APHA, 1998). Microscopic examination of water samples, cultures, flocs scraped from rocks or pipes, or material captured on filter surfaces can facilitate identification of these organisms based on their characteristic morphology and encrustations. From a culture-independent perspective, Siering and Ghiorse, (1997) applied 16S rRNA-targeted fluorescein-labeled oligonucleotide probes for detection of *Leptothrix* and *S. natans* by FISH.

4.2.9 *Staphylococci*

Disease symptoms frequently associated with exposure to recreational waters, including swimming pools, are complaints of ear, eye, nose, throat and skin conditions which are of non-enteric etiology. Consequently, indicators such as the staphylococci (Favero, 1985; Borrego *et al.*, 1987; Charoenca and Fujioka, 1993) and *Pseudomonas aeruginosa* (Seyfried and Cook 1984) have been proposed for use as indicators of health risk in recreational waters. Borrego *et al.* (1987) and Charoenca and Fujioka (1993) evaluated a variety of media for detection of *Staphylococcus* spp. Charoenca and Fujioka (1993) subsequently developed a selective medium for use with membrane filtration. The medium was Vogel-Johnson agar and the authors were able to relate numbers of staphylococci to swimmer density in marine waters. Moreover, Charoenca and Fujioka (1993) detected staphylococci in beach waters at night when swimmers were absent implying an ability of these organisms to persist in saline waters. The incidence of staphylococci in beach waters, their association with clinical conditions, and the possibility of persistence, supports the need for continued study of the staphylococci as pathogens in recreational waters.

REFERENCES

Abeyta, C. Jr, Deeter, F.G., Kaysner, C.A. *et al.* (1993). *Campylobacter jejuni* in a Washington state shellfish growing bed associated with illness. *J. Food Prot.* **56**, 323–325.

Allos, B.M. and Blazer, M.J. (1995). *Campylobacter jejuni* and the expending spectrum of related infections. *Clin. Infect. Dis.* **20**, 1092–1101.

Allsop, K. and Stickler, D.J. (1984). The enumeration of *Bacteroides fragilis* group organisms from sewage and natural waters. *J. Appl. Bacteriol.* **56**, 15–24.

Allsop, K. and Stickler, D.J. (1985). An assessment of *Bacteroides fragilis* group organisms as indicators of human faecal pollution. *J. Appl. Bacteriol.* **58**, 95–99.

Amann, R.I., Binder, B.J., Olson, R.J. *et al.* (1990). Combination of 16S rRNA-targeted oligonucleotide probes with flow cytometry for analyzing mixed microbial populations. *Appl. Environ. Microbiol.* **56**, 1919–1925.

Amann, R.I., Ludwig, W. and Schleifer, K.-H. (1995). Phylogenetic identification and in situ detection of individual microbial cells without cultivation. *Microbiol. Rev.* **59**, 143–169.

APHA (1998). *Standard Methods for the Examination of Water and Wastewater*, 20th edn. American Public Health Association, Washington, DC.

Araujo, R.M., Puig, A., Lasobras, J. *et al.* (1997). Phages of enteric bacteria in fresh water with different levels of faecal pollution. *J. Appl. Microbiol.* **82**, 281–286.

Armon, R. and Payment, P. (1988). A modified m-CP medium for enumerating *Clostridium perfringens* from water samples. *Can. J. Microbiol.* **34**, 78–79.

Armon, R. and Kott, Y. (1993). A simple, rapid, and sensitive presence/absence test for bacteriophage detection of aerobic/anaerobic bacteria in drinking water. *J. Appl. Bacteriol.* **74**, 490–496.

Armon, R., Araujo, R., Kott, Y. *et al.* (1997). Bacteriophages of enteric bacteria in drinking water, comparison of their distribution in two countries. *J. Appl. Microbiol.* **83**, 627–633.

Arroyo, G. and Arroyo, J.A. (1995). Selective action of inhibitors used in different culture media on the competitive microflora of *Salmonella*. *J. Appl. Bact.* **78**, 281–289.

ASTM (1993). *Standards on Materials and Environmental Microbiology*. American Society for Testing and Materials, Philadelphia.

Audicana, A., Perales, I. and Borrego, J.J. (1995). Modification of kanamycin-esculin-azide agar to improve selectivity in the enumeration of faecal streptococci from water samples. *Appl. Environ. Microbiol.* **61**, 4178–4183.

Beaudoin, E.C. and Litsky, W. (1981). Faecal streptococci. In: B. J. Dutka (ed.) *Membrane Filtration: Applications, Techniques, and Problems*, pp. 77–118. Marcel Dekker, New York.

Beebe, J.L. and Koneman, E.W. (1995). Recovery of uncommon bacteria from blood: association with neoplastic disease. *Clin. Microbiol. Rev.* **8**, 336–356.

Bej, A.K., Mahbubani, M.H., Boyce, M.J. and Atlas, R.M. (1994). Detection of *Salmonella* spp. in oysters by PCR. *Appl. Environ. Microbiol.* **60**, 368–373.

Belas, R., Faloon, P. and Hannaford, A. (1995). Potential applications of molecular biology to the study of fish mycobacteriosis. *Ann. Rev. Fish Dis.* **5**, 133–173.

Benlloch, S., Rodriguez-Valera, F. and Martinez-Murcia, A.J. (1995). Bacterial diversity in two coastal lagoons deduced from 16S rDNA PCR amplification and partial sequencing. *FEMS Microbiol. Ecol.* **18**, 267–280.

Bennett, R.W. and Lancette, G.A. (1998). *Staphylococcus aureus. Bacteriological Analytical Manual*, 8th edn, Chapter 12. AOAC International, Gaithersburg, Maryland.

Berg, J.D. and Fiksdal, L. (1988). Rapid detection of total and faecal coliforms in water by enzymatic hydrolysis of 4-methylumbelliferone-β-D-galactoside. *Appl. Environ. Microbiol.* **54**, 2118–2122.

Bernhard, A.E. and Field, K.G. (2000a). A PCR assay to discriminate human and ruminant faeces on the basis of host differences in *Bacteroides-Prevotella* genes encoding 16S rRNA. *Appl. Envir. Microbiol.* **66**, 4571–4574.

Bernhard, A.E. and Field, K.G. (2000b). Identification of nonpoint sources of faecal pollution in coastal waters by using host-Sspecific 16S ribosomal DNA genetic markers from faecal anaerobes. *Appl. Envir. Microbiol.* **66**, 1587–1594.

Betzl, D., Ludwig, W. and Schleifer, K.H. (1990). Identification of *Lactococci* and *Enterococci* by colony hydridization with 23S rRNA-targeted oligonucleotide probes. *Appl. Environ. Microbiol.* **56**, 2927–2929.

Bisson, J.W. and Cabelli, V.J. (1980). *Clostridium perfringens* as a water pollution indicator. *J. Wat. Pollut. Control Fed.* **52**, 241–248.

Bogosian, G., Sammons, L.E., Morris, P.J. *et al.* (1996). Death of the *Escherichia coli* K-12 strain W3110 in soil and water. *Appl. Environ. Microbiol.* **62**, 4114–4120.

Bolton, F.J., Coates, D. and Hutchinson, D.N. (1984). The ability of campylobacter media supplements to neutralize photochemically induced toxicity and hydrogen peroxide. *J. Appl. Microbiol.* **56**, 151–157.

Bono, M., Jemmi, T., Bernasconi, C. *et al.* (1995). Genotypic characterization of *Mycobacterium avium* strains recovered from animals and their comparison to human strains. *Appl. Environ. Microbiol.* **61**, 171–173.

Booth, S.J., Van Tassell, R.L., Johnson, J.L. and Wilkins, T.D. (1979). Bacteriophages of *Bacteroides. Rev. Infect. Dis.* **1**, 325–336.

Borrego, J.J., Florido, J.A., Mrocek, P.R. and Romero, P. (1987). Design and performance of a new medium for the quantitative recovery of *Staphylococcus aureus* from recreational waters. *J. Appl. Bact.* **63**, 85–93.

Borrego, J.J., Cornax, R., Preston, D.R. *et al.* (1991). Development and application of new positively charged filters for recovery of bacteriophages from water. *Appl. Environ. Microbiol.* **57**, 1218–1222.

Bosch, A., Tartera, C., Gajardo, R. *et al.* (1989). Comparative resistance of bacteriophages active against *Bacteroides fragilis* to inactivation by chlorination or ultraviolet radiation. *Wat. Sci. Technol.* **2**, 221–222.

Bosley, G.S., Facklam, R.R. and Grossman, D. (1983). Rapid identification of *Enterococci*. *J. Clin. Microbiol.* **18**, 1275–1277.

Bradley, G.J., Carter, D., Gaudie, D. and King, C. (1999). Distribution of the human faecal bacterium *Bacteroides fragilis*, its bacteriophages and their relationship to current sewage pollution indicators in bathing water. *J. Appl. Microbiol.* **85**, 90S–100S.

Buswell, C.M., Herlihy, Y.M., Lawrence, L.M. *et al.* (1998). Extended survival and persistence of *Campylobacter* spp. in water and aquatic biofilms and their detection by immunofluorescent-antibody and -rRNA staining. *Appl. Environ. Microbiol.* **64**, 733–741.

Butler, W.R. and Guthertz, L.S. (2001). Mycolic acid analysis by high-performance liquid chromatography for identification of *Mycobacterium* species. *Clin. Micro. Rev.* **14**, 704–726.

Cabelli, V.J. (1977). *Clostridium perfringens* as a water quality indicator. In: A.W. Hoadley and B.J. Dutka (eds) *Bacterial Indicators/Health Hazards Associated with Water*, pp. 65–79. Special Technical Publication 635, American Society for Testing and Materials, Philadelphia.

Cabelli, V.J., Dufour, P.A., McCabe, L.J. and Levin, M.A. (1983). A marine recreational water quality criterion consistent with indicator concepts and risk analysis. *J. Water Pollut. Control Fed.* **55**, 1306–1314.

Cappelier, J.M. and Federighi, M. (1998). Demonstration of viable but nonculturable state for *Campylobacter jejuni*. *Rev. Med. Vet.* **149**, 319–326.

Carlsson, J., Nyberg, G. and Wrethen, J. (1978). Hydrogen peroxide and superoxide radical formation in anaerobic broth media exposed to atmospheric oxygen. *Appl. Environ. Microbiol.* 36: 223–229.

Caro, A., Got, P., Lesne, J. *et al.* (1999). Viability and virulence of experimentally stressed nonculturable *Salmonella typhimurium*. *Appl. Environ. Microbiol.* **65**, 3229–3232.

Centers for Disease Control (1995). Multistate outbreak of viral gastroenteritis associated with consumption of oysters-Apalachicola Bay, Florida, December 1994–January 1995. Epidemiologic Notes and Reports. *MMWR*, **44**, 37–38.

Chambers, P.A., Allard, M. and Walker, S.L. (1997). Impacts of municipal wastewater effluents on Canadian waters: a review. *Wat. Qual. Res. J. Can.* **32**, 659–713.

Charoenca, N. and Fujioka, R.S. (1993). Assessment of *Staphylococcus* bacteria in Hawaii's marine recreational waters. *Wat. Sci. Tech.* **27**, 283–289.

Chelala, C.A. and Margolin, P. (1983). Bactericidal photoproducts in medium containing riboflavin plus aromatic compounds and $MnCl_2$. *Can. J. Microbiol.* **29**, 670–675.

Chung, H. and Sobsey, M.D. (1993). Comparative survival of indicator viruses and enteric viruses in seawater and sediment. *Water Sci. Technol.* **27**, 425–428.

Clausen, E.M., Green, B.L. and Litsky, W. (1977). Faecal streptococci: indicators of pollution. In: A.W. Hoadley and B.J. Dutka (eds) *Bacterial Indicators/Health Hazards Associated with Water*, pp. 247–264. Special Technical Publication 635. American Society for Testing and Materials, Philadelphia.

Cocito, C.P., Gilot, M. and Coene, M. *et al.* (1994). Paratuberculosis. *Clin. Microbiol. Rev.* **7**, 328–345.

Cohen, L.F., Dunbar, S.A., Sirbasku, D.M. and Clarridge, J.E. III (1997). *Streptococcus bovis* infection of the central nervous system report of two cases and review. *Clin. Infect. Dis.* **25**, 819–823.

Collins, C.H., Grange, J.M. and Yates, M.D. (1984). Mycobacteria in water. *J. Appl. Bacteriol.* **57**, 193–211.

Colwell, R.R. and Grimes, D.J. (2000). *Nonculturable Microorganisms in the Environment*. ASM Press, Washington, DC.

Cooper, K.E. and Ramadan, F.M. (1955). Studies in the differentiation between human and animal pollution by means of faecal streptococci. *J. Gen. Microbiol.* **12**, 180–190.

Cornax, R., Moriñigo, M.A., Paez, I.G. *et al.* (1990). Application of direct plaque assay for detection and enumeration of bacteriophages of *Bacteroides fragilis* from contaminated-water samples. *Appl. Environ. Microbiol.* **56**, 3170–3173.

Cornax, R., Moriñigo, M.A., Balebona, M.C. *et al.* (1991). Significance of several bacteriophage groups as indicators of sewage pollution in marine waters. *Wat. Res.* **25**, 673–678.

Cote, R.J. and Gherna, R.L. (1994). Nutrition and media. In: P. Gerhardt, R.E. Murray, W.A. Wood and N.R. Krieg (eds) *Methods for General and Molecular Biology*, pp. 155–178. American Society for Microbiology, Washington, DC.

Covert, T.C., Rodgers, M.R., Reyes, A.L. and Stelma, G.N. (1999). Occurrence of nontuberculous mycobacteria in environmental samples. *Appl. Environ. Microbiol.* **65**, 2492–2496.

Crawford, J.T. (1999). Molecular approaches to the detection of mycobacteria. In: P.R.J. Gangadharam and P.A. Jenkins (eds) *Mycobacteria I. Basic Aspects*, pp. 131–144. Chapman and Hall, New York.

Croci, L., De Medici, D., Scalfaro, C. *et al.* (2000). Determination of enteroviruses, hepatitis A virus, bacteriophages and *Escherichia coli* in Adriatic Sea mussels. *J. Appl. Microbiol.* **88**, 293–298.

Dailloux, M., Laurain, C., Weber, M. and Hartemann, P.H. (1999). Water and nontuberculous mycobacteria. *Wat Res.* **33**, 2219–2228.

Debartolomeis, J. and Cabelli, V.J. (1991). Evaluation of an *Escherichia coli* host strain for enumeration of F male-specific bacteriophages. *Appl. Environ. Microbiol.* **57**, 1301–1305.

DeLong, E.F., Taylor, L.T., Marsh, T.L. and Preston, C.M. (1999). Visualization and enumeration of marine planktonic Archaea and bacteria by using polyribonucleotide probes and fluorescent in situ hybridization. *Appl. Environ. Microbiol.* **65**, 5554–5563.

Del Prete, R., Quaranta, M., Lippolis, A. *et al.* (1998). Detection of *Mycobacterium paratuberculosis* in stool samples of patients with inflammatory bowel disease by IS900-based PCR and colorimetric detection of amplified DNA. *J. Microbiol. Meth.* **33**, 105–114.

Desmonts, C., Minet, J., Colwell, R. and Cormier, M. (1990). Fluorescent-antibody method useful for detecting viable but nonculturable *Salmonella* spp. in chlorinated wastewater. *Appl. Environ. Microbiol.* **56**, 1448–1452.

Devriese, L.A., Ceyssens, K., Rodrigues, U.M. and Collins, M.D. (1990). *Enterococcus columbae*, a species from pigeon intestines. *FEMs Microbiol. Lett.* **71**, 247–252.

Devriese, L.A., Vandamme, P., Pot, B. *et al.* (1998). Differentiation between *Streptococcus gallolyticus* strains of human clinical and veterinary origins and *Streptococcus bovis* strains from the intestinal tract of ruminants. *J. Clin. Microbiol.* **36**, 3520–3523.

Donnison, A.M., Ross, C.M. and Russell, J.M. (1999). Quality control of bacterial enumeration. *Appl. Environ. Microbiol.* **59**, 922–923.

Dowling, N.J.E., Widdel, F. and White, D.C. (1986). Phospholipid ester-linked fatty acid biomarkers of acetate-oxidzing sulfate reducers and other sulfide-forming bacteria. *J. Gen. Microbiol.* **132**, 1815–1825.

Dutka, B.J. and Kwan, K.K. (1978). Health-indicator bacteria in water-surface microlayers. *Can. J. Microbiol.* **24**, 187–188.

Ericksen, T.H. and Dufour, A.P. (1986). Methods to identify waterborne pathogens and indicator organisms. In: G.F. Craun (ed.) *Waterborne Diseases in the United States*, pp. 195–214. CRC Press, Boca Raton.

Essey, M.A. and Koller, M.A. (1994). Status of bovine tuberculosis in North America. *Vet. Microbiol.* **40**, 15–22.

Facklam, R.R. and Collins, M.D. (1989). Identification of *Enterococcus* species isolated from human infections by a conventional test scheme. *J. Clin. Microbiol.* **27**, 731–734.

Falkinham, J.O. III (1996). Epidemiology of infection by nontuberculous mycobacteria. *Clin. Microbiol. Rev.* **9**, 177–215.

Falkinham, J.O. III (1999). Transmission of Mycobacteria. In: P.R.J. Gangadharam and P.A. Jenkins (eds) *Mycobacteria I Basic Aspects*, pp. 178–209. Chapman and Hall, New York.

Farrah, S.R. (1982). Chemical factors influencing adsorption of bacteriophage MS2 to membrane filters. *Appl. Environ. Microbiol.* **43**, 659–663.

Farrow, J.A.E., Kruze, J., Phillips, B.A., Bramley, A.J. and Collins, M.D. (1984). Taxonomic studies on *Streptococcus bovis* and *Streptococcus equinus*: Description of *Streptococcus alactolyticus* sp. nov. and *Streptococcus saccharolyticus* sp. nov. System. *Appl. Microbiol.* **5**, 467–482.

Favero, M.S. (1985). Microbiologic indicators of health risks associated with swimming. *Am. J. Public Hlth.* **75**, 1051–1053.

Fay, L.D., Kaufmann, O.W. and Ryel, L.A. (1965). *Mass Mortality of Water-birds in Lake Michigan 1963–64*, pp. 36–46. Pub. No. 13, Great Lakes Research Division, The University of Michigan.

Feng, P.C.S. and Hartman, P.A. (1982). Fluorogenic assays for immediate confirmation of *Escherichia coli*. *Appl. Environ. Microbiol.* **43**, 1320–1329.

Fiksdal, L., Maki, J.S., LaCroix, S.J. and Staley, J.S. (1985). Survival and detection of *Bacteroides* spp., prospective indicator bacteria. *Appl. Environ. Microbiol.* **49**, 48–150.

Fiksdal, L. and Berg, J.D. (1987). Evaluation of a fluorescent antibody technique for the rapid enumeration for *Bacteroides fragilis* groups of organisms in water. *J. Appl. Bacteriol.* **62**, 377–383.

Fiksdal, L., Pommepuy, M., Caprais, M.P. and Midttun, I. (1994). Monitoring of faecal pollution in coastal waters by use of rapid enzymatic techniques. *Appl. Environ. Microbiol.* **60**, 1581–1584.

Franks, A.H., Harmsen, H.J.M., Raangs, G.C. *et al.* (1998). Variations of bacterial populations in human faeces measured by fluorescent in situ hybridization with group-specific 16S rRNA-targeted oligonucleotide probes. *Appl. Envir. Microbiol.* **64**, 3336–3345.

Fricker, E.J. and Fricker, C.R. (1994). Application of the polymerase chain reaction to the identification of *Escherichia coli* and coliforms in water. *Lett. Appl. Microbiol.* **19**, 44–46.

Fries, M.R., Forney, L.J. and Tiedje, J.M. (1997). Phenol- and toluene-degrading microbial populations from an aquifer in which successful trichloroethene cometabolism occurred. *Appl. Envir. Microbiol.* **63**, 1523–1530.

Frostegaard, A., Tunlid, A. and Baath, E. (1991). Microbial biomass measured as total lipid phosphate in soils of different organic content. *J. Microbiol. Methods* **14**, 151–163.

Fujioka, R.S. and Shizumura, L.K. (1985). *Clostridium perfringens*: a reliable indicator of stream water quality. *J. Water Pollut. Control Fed.* **57**, 986–992.

Gamo, M. and Shoji, T. (1999). A method of profiling microbial communities based on a most-probable-number assay that uses BIOLOG plates and multiple sole carbon sources. *Appl. Envir. Microbiol.* **65**, 4419–4424.

Gantzer, C., Maul, A., Audic, J.M. and Schwartzbrod, L. (1998). Detection of infectious enteroviruses, enterovirus genomes, somatic colihages, and *Bacterioides fragilis* phages in treated wastewater. *Appl. Environ. Microbiol.* **64**, 4307–4312.

Garcia-Lara, J., Menon, P., Servais, P. and Billen, G. (1991). Mortality of fecal bacteria in seawater. *Appl. Environ. Microbiol.* **57**, 885–888.

Garthright, W.E. (1998). Most probable number from serial dilutions. *Bacteriological Analytical Manual*, 8th edn., Appendix 2. AOAC International, Gaithersburg, Maryland.

Gaudet, I.D., Florence, L.Z. and Coleman, R.N. (1996). Evaluation of test media for routine monitoring of *Escherichia coli* in nonpotable waters. *Appl. Environ. Microbiol.* **62**, 4032–4035.

Geldreich, E.E., Kenner, B.A. and Kabler, P.W. (1964). Occurrence of coliforms, faecal coliforms, and streptococci on vegetation and insects. *Appl. Microbiol.* **12**, 63–69.

Geldreich, E.E. and Kenner, B.A. (1969). Concepts of faecal streptococci in stream pollution. *J. Wat. Pollut. Control Fed.* **41**, R336–R352.

George, I., Petit, M. and Servais, P. (2000). Use of enzymatic methods for rapid enumeration of coliforms in freshwaters. *J. Appl. Microbiol.* **88**, 404–413.

Gerhardt, P., Murray, R.G.E., Wood, W.A. and Krieg, N.R. (eds) (1994). *Methods for General and Molecular Biology*, American Society for Microbiology, Washington, DC.

Grabow, W.O.K., Neubrech, T.E., Holtzhausen, C.S. and Jofre, J. (1995). *Bacteroides fragilis* and *Escherichia coli* bacteriophages: excretion by humans and animals. *Wat. Sci. Tech.* **31**, 223–230.

Grange, J.M., Yates, M.D. and Boughton, E. (1990). The avian tubercle bacillus and its relatives. *J. Appl. Bacteriol.* **68**, 411–431.

Grant, R.J., Whitehead, T.R. and Orr, J.E. (2000). *Streptococcus bovis* meningitis in an infant. *J. Clin. Microbiol.* **38**, 462–463.

Green, D.H., Lewis, G.D., Rodtong, S. and Loutit, M.W. (1991). Detection of faecal pollution in water by an *Escherichia coli* uidA gene probe. *Microbiol. Methods* **13**, 207–214.

Grégori, G., Citterio, S., Ghiani, A. *et al.* (2001). Resolution of viable and membrane-compromised bacteria in freshwater and marine waters based on analytical flow cytometry and nucleic acid double staining. *Appl. Environ. Microbiol.* **67**, 4662–4670.

Handzel, T.R., Green, M., Sanchez, C. *et al.* (1993). Improved specificity in detecting F-specific coliphages in environmental samples by suppression of somatic phages. *Wat. Sci. and Technol.* **27**, 123–131.

Hatcher, R.F. and Parker, B.C. (1974). Microbiological and chemical enrichment of freshwater-surface microlayers relative to the bulk-subsurface water. *Can. J. Microbiol.* **20**, 1051–1057.

Havelaar, A.H. and Hogeboom, W.M. (1984). A method for the enumeration of male-specific bacteriophages in sewage. *J. Appl. Bacteriol.* **56**, 439–447.

Havelaar, A.H., van Olphen, M. and Drost, Y.C. (1993). F-specific RNA bacteriophages are adequate model organisms for enteric viruses in fresh water. *Appl. Environ. Microbiol.* **59**, 2956–2962.

Hernandez, J.F., Guibert, J.M., Delattre, J.M. *et al.* (1991). Evaluation of a miniaturized procedure for enumeration of *Escherichia coli* in seawater based upon hydrolysis of 4-methylumbelliferyl-β-D-glucuronide. *Wat. Res.* **25**, 1073–1078.

Hernandez, J.F., Pourcher, A.M., Delattre, J.M. *et al.* (1993). MPN miniaturized procedure for the enumeration of faecal enterococci in fresh and marine waters. The MUST procedure. Wat. Res. 27, 597–606.

Hill, R.T., Knight, I.T., Anikis, M.S. and Colwell, R.R. (1993). Benthic distribution of sewage sludge indicated by *Clostridium perfringens* at a deep-water ocean dump site. *Appl. Environ. Microbiol.* **59**, 47–51.

Hill, R.T., Straube, W.L., Palmisano, A.C. *et al.* (1996). Distribution of sewage indicated by *Clostridium perfringens* at a deep-water disposal site after cessation of sewage disposal. *Appl. Environ. Microbiol.* **62**, 1741–1746.

Holdeman, L.V., Cato, E.P. and Moore, W.E.C. (1977). *Anaerobe Laboratory Manual*. Anaerobe Laboratory, Virginia Polytechnic Institute and State University, Blacksburg, Virginia.

Horsburgh, C.R. (1996). Epidemiology of disease caused by nontuberculous mycobacteria. *Seminars Respirat. Infect.* **11**, 244–251.

Horsburgh, C.R. and Selik, R.M. (1989). The epidemiology of disseminated nontuberculous mycobacterial infection in the acquired immunodeficiency syndrome (AIDS). *Am. Rev. Respir. Dis.* **139**, 4–7.

Hurst C.J., Knudsen G.R. and McInerney M.J. (eds) (1996). *Manual of Environmental Microbiology*. American Society for Microbiology, Washington, DC.

IAWPRC (1991). Bacteriophages as model viruses in water quality control. IAWPRC Study Group on Health Related Water Microbiology. *Wat. Res.* **25** 529–545.

Ikawa, H., Oka, S., Murakami, H. *et al.* (1989). Rapid identification of serotypes of *Mycobacterium avium–M. intracellulare* complex by using infected swine sera and reference antigenic glycolipids. *J. Clin. Microbiol.* **27**, 2552–2558.

Iivanainen, E.K., Martikainen, P.J., Vaanaen, P.K. and Katila, M.L. (1993). Environmental factors affecting the occurrence of mycobacteria in brook waters. *Appl. Environ. Microbiol.* **59**, 398–404.

Iivanainen, E., Martikainen, P.J., Vaananen, P. and Katila, M.L. (1999). Environmental factors affecting the occurrence of mycobacteria in brook sediments. *J. Appl. Microbiol.* **86**, 673–681.

Jaramillo, V.L. and McCarthy, C.M. (1986). Recovery of *Mycobacterium avium* after treatment with chemical decontaminants. *Can. J. Microbiol.* **32**, 728–732.

Jofre, J., Bosch, A., Lucena, F. *et al.* (1986). Evaluation of *Bacteroides fragilis* bacteriophages as indicators of the virological quality of water. *Wat. Sci. Technol.* **18**, 167–173.

Jofre, J.A., Blasi, M., Bosch, A. and Lucena, F. (1989). Occurrence of bacteriophages infecting *Bacteroides fragilis* and other viruses in polluted marine sediments. *Water. Sci. Technol.* **21**, 15–19.

Jofre, J., Ollé, E., Ribas, F. *et al.* (1995). Potential usefulness of bacteriophages that infect *Bacteroides fragilis* as model organisms for monitoring virus removal in drinking water treatment plants. *Appl. Envrion. Microbiol.* **61**, 3227–3231.

Jousimies-Somer, H.R., Summanen, P.H. and Finegold, S.M. (1995). *Bacteroides, Porphyromonas, Prevotella, Fusobacterium*, and other anaerobic gram-negative bacteria. In: P.R. Murray, E.J. Baron, M.A. Pfaller, F.C. Tenover and R.H. Yolken (eds) *Manual of Clinical Microbiology*, 6th edn, pp. 603–620. ASM Press, Washington, DC.

Kamala, T., Paramasivan, C.N., Herbert, D. *et al.* (1994). Evaluation of procedures for isolation of nontuberculous mycobacteria from soil and water. *Appl. Environ. Microbiol.* **60**, 1021–1024.

Kapperud, G. and Rosef, O. (1983). Avian wildlife reservoir of *Campylobacter feris* subsp. *jejuni*, *Yersinia* spp. and *Salmonella* spp. in Norway. *Appl. Environ. Microbiol.* **45**, 375–380.

Kator, H. and Rhodes, M.W. (1991). Indicators and alternate indicators of growing water quality. In: D.R. Ward and C.R. Hackney (eds) *Microbiology of Marine Food Products*, pp. 135–196. Van Nostrand Reinhold, New York.

Kator, H. and Rhodes, M.W. (1992). *Evaluation of* Bacteroides fragilis *Bacteriophage, a Candidate Human-specific Indicator of Faecal Contamination for Shellfish-growing Waters. Final report (NA90AA-H-FD234) to NOAA*, Southeast Fisheries Science Center, Charleston, SC.

Kator, H. and Rhodes, M.W. (1996). *Identification of Pollutant Sources Contributing to Degraded Sanitary Water Quality in Taskinas Creek National Estuarine Research Reserve, Virginia.* Final report submitted to NOAA/ Sanctuaries and Reserves Division, Washington DC Virginia Institute of Marine Science, Special Report in Applied Marine Science and Ocean Engineering No. 336.

Kenner, B.A., Clark, H.F. and Kabler, P. (1960). Faecal streptococci II. Quantification of streptococci in faeces. *Am. J. Public Health* **50**, 1553–1559.

Kepner, R.L. Jr and Pratt, J.B. (1994). Use of fluorochromes for direct enumeration of total bacteria in environmental samples: past and present. *Microbiol. Rev.* **58**, 603–615.

Kirschner, R.A. Jr, Parker, B.C. and Falkinham, J.O. III (1992). Epidemiology of infection by nontuberculous mycobacteria. *Mycobacterium avium*, *Mycobacterium intracellulare* and *Mycobacterium scrofulaceum* in acid, brown-water swamps of the southeastern United States and their association with environmental variables. *Am. Rev. Resp. Dis.* **145**, 271–275.

Klein, R.S., Recco, R.A., Catalano, M.T. *et al.* (1977). Association of *Streptococcus bovis* with carcinoma of the colon. *N. Engl. J. Med.* **297**, 800–802.

Klieve, A.V. and Bauchop, T. (1991). Phage resistance and altered growth habit in a strain of *Streptococcus bovis*. *FEMS Microbiol. Lett.* **80**, 155–160.

Klieve, A.V., Heck, G.L., Prance, M.A. and Shu, Q. (1999). Genetic homogeneity and phage susceptibility of ruminal strains of *Streptococcus bovis* isolated in Australia. *Lett. Appl. Microbiol.* **29**, 108–112.

Knibb, W., Colorni, A., Ankaoua, M. *et al.* (1993). Detection and identification of a pathogenic marine mycobacterium from the European seabass *Dicentrarchus labrax* using polymerase chain reaction and direct sequencing of 16S rDNA sequences. *Mol. Mar. Biol. Biotech.* **2**, 225–232.

Knight, R.G. and Shales, D.M. (1985). Physiological characteristics and deoxyribonucleic acid relatedness of human isolates of *Streptococcus bovis* and *Streptococcus bovis* (var.). *Int. J. Syst. Bacteriol.* **35**, 357–361.

Knudtson, L.M. and Hartman, P.A. (1992). Routine procedures for isolation and identification of enterococci and faecal streptococci. *Appl. Environ. Microbiol.* **58**, 3027–3031.

Koch, A.L. (1994). Growth measurement. In: P. Gerhardt, R.G.E. Murray, W.A. Wood and N.R. Krieg (eds) *Methods for General and Molecular Biology*, pp. 248–277. American Society for Microbiology, Washington, DC.

Kogure, K., Simidu, U. and Taga, N. (1979). A tentative direct microscopic method for counting living bacteria. *Can. J. Microbiol.* **25**, 415–420.

Koh, E.G.L., Huyn, J.-H. and LaRock, P.L. (1994). Pertinence of indicator organisms and sampling variables to vibrio concentrations. *Appl. Environ. Microbiol.* **60**, 3897–3900.

Kreader, C.A. (1995). Design and evaluation of *Bacteroides* DNA probes for the specific detection of human faecal pollution. *Appl. Environ. Microbiol.* **61**, 1171–1179.

Kreader, C.A. (1998). Persistence of PCR-detectable *Bacteroides distasonis* from human faeces in river water. *Appl. Environ. Microbiol.* **64**, 4103–4105.

Kuritza, A.P. and Salyers, A.A. (1985). Use of a species-specific DNA hybridization probe for enumerating *Bacteroides vulgatus* in human faeces. *Appl. Environ. Microbiol.* **50**, 958–964.

Kuritza, A.P., Shaughnessy, P. and Salyers, A.A. (1986). Enumeration of polysaccharide-degrading *Bacteroides* species-specific DNA probes. *Appl. Environ. Microbiol.* **51**, 385–390.

Langendijk, P.S., Schut, F., Jansen, G.J. *et al.* (1995). Quantitative fluorescence in situ hybridization of *Bifidobacterium* spp. with genus-specific 16S rRNA-targeted probes and its application in faecal samples. *Appl. Envir. Microbiol.* **61**, 3069–3075.

Leclerc, H., Edberg, S., Pierzo, V. and Delattre, J.M. (2000). Bacteriophages as indicators of enteric viruses and public health risk in groundwaters. *J. Appl. Microbiol.* **88**, 5–21.

Lehmicke, L.G., Williams, R.T. and Crawford, R.L. (1979). ^{14}C-most-probable-number for enumeration of active heterotrophic microorganisms in natural waters. *Appl. Environ. Microbiol.* **38**, 644–649.

Little, K.J. and Hartman, P.A. (1983). Fluorogenic selective and differential medium for isolation of faecal streptococci. *Appl. Environ. Microbiol.* **45**, 622–627.

Lleò, M.M., Tafi, M.C. and Canepari, P. (1998). Nonculturable *Enterococcus faecalis* cells are metabolically active and capable of resuming active growth. *Syst. Appl. Microbiol.* **21**, 333–339.

Lopez-Torres, A.J., Hazen, T.C. and Toranzos, G.A. (1987). Distribution and in situ survival and activity of *Klebsiella pneumoniae* and *Escherichia coli* in a tropical rain forest watershed. *Curr. Microbiol.* **15**, 213–218.

Lucena, F., Lasobras, J., McIntosh, D. *et al.* (1994). Effect of distance from the polluting focus on relative concentrations of *Bacteroides fragilis* phages and coliphages in mussels. *Appl. Environ. Microbiol.* **60**, 2272–2277.

Lucena, F., Muniesa, M., Puig, A. *et al.* (1995). Simple concentration method for bacteriophages of *Bacteroides fragilis* in drinking water. *J. Virol. Methods* **54**, 121–130.

Lucena, F., Araujo, R. and Jofre, J. (1996). Usefulness of bacteriophages infecting *Bacteroides fragilis* as index microorganisms of remote faecal pollution. *Wat. Res.* **30**, 2812–2816.

Mangels, J.I. and Douglas, B.P. (1989). Comparison of four commercial brucella agar media for growth of anaerobic organisms. *J. Clin. Microbiol.* **27**, 2268–2271.

Marshall, K.C. (1976). *Interfaces in Microbial Ecology.* Harvard University Press, Cambridge.

Martinez, J., Garcia-Lara, J. and Vives-Rego, J. (1989). Estimation of *Escherichia coli* mortality in seawater by the decrease in ^{3}H-label and electron transport system activity. *Microb. Ecol.* **17**, 219–225.

Maruyama, A. and Sunamura, M. (2000). Simultaneous direct counting of total and specific microbial cells in seawater, using a deep-sea microbe as target. *Appl. Environ. Microbiol.* **66**, 2211–2215.

McDermott, T.R. (1996). Use of fluoresent antibodies for studying the ecology of soil- and plant-associated microbes. In: C.J. Hurst, G.R. Knudsen, M.J. McInerney, L.D. Stetzenbach and M.V. Walter (eds) *Manual of Environmental Microbiology,* pp. 473–481. American Society for Microbiology, Washington, DC.

McFeters, G.A., Bissonnette, G.K., Jezeski, J.J. *et al.* (1974). Comparative survival of indicator bacteria and enteric pathogens in well water. *Appl. Microbiol.* **27**, 823–829.

McFeters, G.A., Kippin, J.S. and LeChevallier, M.W. (1986). Injured coliforms in drinking water. *Appl. Environ. Microbiol.* **51**, 1–5.

Meyer-Reil, L.A. (1978). Autoradiography and epifluorescence microscopy combined for the determination of number and spectrum of actively metabolizing bacteria in natural waters. *Appl. Environ. Microbiol.* **36**, 506–512.

Mills, A.C., Breuil, C. and Colwell, R.R. (1978). Enumeration of petroleum-degrading marine and estuarine microorganisms by the most-probable-number method. *Can. J. Microbiol.* **24**, 552–557.

Morita, R.Y. (1993). Bioavailability of energy and the starvation state. In: S. Kjelleberg (ed.) *Starvation in Bacteria* Plenum Press, New York.

Muhlemann, K., Graf, S. and Tauber, M.G. (1999). *Streptococcus bovis* clone causing two episodes of endocarditis 8 years apart. *J. Clin Microbiol.* **37**, 862–863.

Mundt, J.O., Johnson, A.H. and Khatchikian, R. (1958). Incidence and nature of enterococci on plant materials. *Food Res.* **23**, 186–193.

Mundt, J.O. (1963). Occurrence of enterococci on plants in a wild environment. *Appl. Microbiol.* **11**, 141–144.

Mundt, J.O. (1982). The ecology of the streptococci. *Microb. Ecol.* **8**, 355–369.

Munro, P.M., Gauthier, M.J. and Laumond, F.M. (1987). Changes in *Escherichia coli* cells starved in seawater or grown in seawater-wastewater mixtures. *Appl. Environ. Microbiol.* **53**, 1476–1481.

Munro, P.M., Flatau, G.N., Clement, R.L. and Gauthier, M.J. (1995). Influence of the RpoS (KatF) sigma factor on maintenance of viability and culturability of *Escherichia coli* and *Salmonella typhimurium* in seawater. *Appl. Environ. Microbiol.* **61**, 1853–1858.

Nelms, L.F., Odelson, D.A., Whitehead, T.R. and Hespell, R.B. (1995). Differentiation of ruminal and human *Streptococcus bovis* strains by DNA homology and 16S rDNA rRNA probes. *Curr. Microbiol.* **31**, 294–300.

Nelson, K.E., Thonney, M.L., Woolston, T.K. *et al.* (1998). Phenotypic and phylogenetic characterization of ruminal tannin-tolerant bacteria. *Appl. Environ. Microbiol.* **64**, 3824–3830.

Neumann, M., Schulze-Robbecke, R., Hagenau, C. and Behringer, K. (1997). Comparison of methods for isolation of mycobacteria from water. *Appl. Environ. Microbiol.* **63**, 547–552.

Nickels, J.S., King, J.D. and White, D.C. (1979). Poly-beta-hydroxytbutyrate accumulation as a measure of unbalanced growth of estuarine detrital microbiota. *Appl. Environ. Microbiol.* **37**, 459–465.

Oliver, J.D. (1993). Formation of viable but nonculturable cells. In: S. Kjelleberg (ed.) *Starvation in bacteria,* pp. 239–272. Plenum Press, New York.

Oragui, J.I. and Mara, D.D. (1981). A selective medium for the enumeration of *Streptococcus bovis* by membrane filtration. *J. Appl. Bacteriol.* **51**, 85–93.

Oragui, J.I. and Mara, D.D. (1984). A note on a modified membrane-Bovis agar for the enumeration of

Streptococcus bovis by membrane filtration. *J. Appl. Bacteriol.* **56**, 179–181.

Osawa, R., Fujisawa, T. and Sly, L.I. (1995). *Streptococcus gallolyticus* sp. nov.; gallate degrading organisms formerly assigned to *Streptococcus bovis*. *Syst. Appl. Microbiol.* **18**, 74–78.

Parkes, R.J., Dowling, N.J.E., White, D.C. *et al.* (1993). Characterization of sulphate-reducing bacterial populations within marine and estuarine sediments with different rates of sulfphate reduction. *FEMS Microbiol. Ecol.* **102**, 235–250.

Parks, D.L. (1996) *Methods of Soil Analysis*. Part 3, Chemical methods/editorial committee, D.L. Sparks: Soil Science Society of America Book Series; No. 5. American Society of Agronomy, Madison, WI.

Payment, P. and Franco, E. (1993). *Clostridium perfringens* and somatic coliphages as indicators of the efficiency of water treatment for viruses and protozoan cysts. *Appl. Environ. Microbiol.* **59**, 2418–2424.

Pina, S., Puig, M., Lucena, F. *et al.* (1998). Viral pollution in the environment and in shellfish: human adenovirus detection by PCR as an index of human viruses. *Appl. Environ. Microbiol.* **64**, 3376–3382.

Pommepuy, M., Butin, M., Derrien, A. *et al.* (1996). Retention of enteropathogenicity by viable but nonculturable *Escherichia coli* exposed to seawater and sunlight. *Appl. Environ. Microbiol.* **62**, 4621–4626.

Porter, J., Edwards, C. and Pickup, R.W. (1995). Rapid assessment of physiological status in *Escherichia coli* using fluorescent probes. *J. Appl. Microbiol.* **79**, 399–408.

Pourcher, A.M., Devriese, L.A., Hernandez, J.F. and Delattre, J.M. (1991). Enumeration by a miniaturized method of *Escherichia coli*, *Streptococcus bovis* and enterococci as indicators of the origin of faecal pollution of waters. *J. Appl. Bacteriol.* **70**, 525–530.

Rhodes, M.W., Anderson, I.C. and Kator, H.I. (1983). In situ development of sublethal stress in *Escherichia coli*: effects on enumeration. *Appl. Environ. Microbiol.* **45**, 1870–1876.

Rhodes, M.W. and Kator, H.I. (1990). Effects of sunlight and autochthonous microbiota on *Escherichia coli* survival in an estuarine environment. *Curr. Microbiol.* **21**, 65–73.

Rhodes, M.W. and Kator, H.I. (1991). Use of *Salmonella typhimurium* WG49 to enumerate male-specific coliphages in an estuary and watershed subject to nonpoint pollution. *Water Res.* **25**, 1315–1323.

Rice, E.W., Allen, M.J. and Edberg, S.C. (1990). Efficacy of β-glucuronidase assay for identification of *Escherichia coli* by the defined-substrate technology. *Appl. Environ. Microbiol.* **56**, 1203–1205.

Rodrigues, U. and Collins, M.D. (1990). Phylogenetic analysis of *Streptococcus saccharolyticus* based on 16S rRNA sequencing. *FEMS Microbiol. Lett.* **71**, 231–234.

Roll, B.M. and Fujioka, R.S. (1997). Sources of faecal indicator bacteria in a brackish, tropical stream and their impact on recreational water quality. *Wat. Sci. Tech.* **35**, 179–186.

Rollins, D.M. and Colwell, R.R. (1986). Viable but nonculturable stage of *Campylobacter jejuni* and its role in survival in natural, aquatic environment. *Appl. Environ. Microbiol.* **58**, 531–538.

Roszak, D.B., Grimes, D.J. and Colwell, R.R. (1984). Viable but non-recoverable stage of *Salmonella enteritidis* in aquatic systems. *Can. J. Microbiol.* **30**, 334–338.

Roszak, D.B. and Colwell, R.R. (1987). Survival strategies of bacteria in the natural environment. *Microbiol. Rev.* **51**, 365–379.

Roth, W.G., Leckie, M.P. and Dietzler, D.N. (1988). Restoration of colony-forming activity in osmotically stressed *Escherichia coli* by betaine. *Appl. Environ. Microbiol.* **54**, 3142–3146.

Ruoff, K.L., Miller, S.I., Garner, C.V. *et al.* (1989). Bacteremia with *Streptococcus salivarius*: clinical correlates of more accurate identification of isolates. *J. Clin. Microbiol.* **27**, 305–308.

Sartory, D.P. (1986). Membrane filtration enumeration of faecal clostridia and *Clostridium perfringens* in water. *Water Res.* **20**, 255–1260.

Sartory, D.P. and Howard, L. (1992). A medium detecting β-glucuronidase for the simultaneous filtration enumeration of *Escherichia coli* and coliforms from drinking water. *Lett. Appl. Microbiol.* **15**, 273–276.

Sartory, D.P., Field, M., Curbishley, S.M. and Prtichard, A.M. (1998). Evaluation of two media for the membrane filtration enumeration of *Clostridium perfringens* from water. *Let. Appl. Microbiol.* **27**, 323–327.

Sartory, D.P. and Watkins, J. (1999). Conventional culture for water quality assessment: is there a future? *J. Appl. Microbiol. Sympos. Suppl.* **85**, 25S–233S.

Schulze-Robbecke, R., Weber, A. and Fischeder, R. (1991). Comparison of decontamination methods for the isolation of mycobacteria from drinking water samples. *J. Microbiol. Meth.* **14**, 177–183.

Seyfried, P.L. and Cook, R.J. (1984). Otitis externa infections related to *Pseudomonas aeruginosa* levels in five Ontario lakes. *Can. J. Public Health* **75**, 83–91.

Sieracki, M.E., Cucci, T.L. and Nicinski, J. (1999). Flow cytometric analysis of 5-cyano-2,3-ditolyl tetrazolium chloride activity of marine bacterioplankton in dilution cultures. *Appl. Environ. Microbiol.* **65**, 2409–2417.

Siering, P.L. and Ghiorse, W.C. (1997). Development and application of 16S rRNA-targeted probes for detection of iron- and manganese-oxidizing sheathed bacteria in environmental samples. *Appl. Environ. Microbiol.* **63**, 644–651.

Sinton, L.W., Donnison, A.M. and Hastie, C.M. (1993a). Faecal streptococci as faecal pollution indicators: a review. Part I: taxonomy and enumeration. *New Zealand J. Mar. Freshwater Res.* **27**, 101–115.

Sinton, L.W., Donnison, A.M. and Hastie, C.M. (1993b). Faecal streptococci as faecal pollution indicators: a review. Part II: sanitary significance, survival and use. *New Zealand J. Mar. Freshwater Res.* **27**, 117–137.

Sinton, L.W. and Donnison, A.M. (1994). Characterisation of faecal streptococci from some New Zealand effluents and receiving waters. *New Zealand J. Mar. Freshwater Res.* **28**, 145–158.

Sinton, L.W., Finlay, R.K. and Reid, A.J. (1996). A simple membrane filtration-elution method for the enumeration of F-RNA, F-DNA and somatic coliphages in 100-ml water samples. *J. Microbiol. Methods* **25**, 257–269.

Sinton, L.S., Finlay, R.K. and Hannah, D.J. (1998). Distinguishing human from animal faecal contamination in water: a review. *New Zealand J. Mar. Freshwater Res.* **32**, 323–348.

Sinton, L.S., Finlay, R.K. and Lynch, P.A. (1999). Sunlight inactivation of faecal bacteriophages and bacteria in sewage-polluted seawater. *Appl. Environ. Microbiol.* **65**, 3605–3613.

Skirrow, M.B. (1991). Epidemiology of *Campylobacter* enteritis. *Int. J. Food Microbiol.* **12**, 9–16.

Sobsey, M.D., Schwab, K.J. and Handzel, T.R. (1990). A simple membrane filter method to concentrate and enumerate male-specific RNA coliphages. *J. Am. Water Works Ass.* **82**, 52–59.

Sodergen, A. (1993). Role of aquatic surface microlayer in the dynamics of nutrients and organic compounds in lakes, with implications for their ecotones. *Hydrobiologia* **251**, 217–225.

Søndergaard, M. and Danielsen, M. (2001). Active bacteria (CTC +) in temperate lakes: temporal and cross-system variations. *J. Plankton Res.* **23**, 1195–1206.

Strickland, J.D.H. and Parsons, T.R. (1972). A practical handbook of seawater analysis. Bulletin 167, Fisheries Research Board of Canada, Ottawa. 310 pp.

Sun, Z.P., Levi, Y., Kiene, L. *et al.* (1997). Quantification of bacteriophages of *Bacteroides fragilis* in environmental water samples of Seine River. *Wat. Soil Pollut.* **96**, 175–183.

Tabor, P.S. and Neihof, R.A. (1982). Improved microautoradiographic method to determine individual microorganisms active in substrate uptake in natural waters. *Appl. Environ. Microbiol.* **44**, 945–953.

Talaat, A.M., Reimschuessel, R. and Trucksis, M. (1997). Identification of mycobacteria infecting fish to the species level using polymerase chain reaction and restriction enzyme analysis. *Vet. Microbiol.* **58**, 229–237.

Tamplin, M.L., Martin, A.L., Ruple, A.D. *et al.* (1991). Enzyme immunoassay for identification of *Vibrio vulnificus* in seawater, sediment, and oysters. *Appl. Environ. Microbiol.* **57**, 1235–1240.

Tanaka, Y., Yamaguchi, N. and Nasu, M. (2000). Viability of *Escherichia coli* 0157:H7 in natural river water determined by the use of flow cytometry. *J. Appl. Microbiol.* **88**, 228–236.

Tanner, R.S. (1996). Cultivation of bacteria and fungi. In: C.J. Hurst, G.R. Knudsen, M.J. McInerney, L.D. Stetzenbach and M.V. Walter (eds) *Manual of environmental microbiology*, pp. 52–60. American Society for Microbiology, Washington, DC.

Tarakanov, B.V. (1996). Biology of lysogenic strains of *Streptococcus bovis* and virulent mutants of their temperate phages. *Mikrobiologiya* **65**, 575–580.

Tartera, C. and Jofre, J. (1987). Bacteriophages active against *Bacteroides fragilis* in sewage-polluted waters. *Appl. Environ. Microbiol.* **53**, 1632–1637.

Tartera, C., Lucena, F. and Jofre, J. (1989). Human origin of *Bacteroides fragilis* bacteriophages present in the environment. *Appl. Environ. Microbiol.* **55**, 2696–2701.

Tartera, C., Araujo, R., Michel, T. and Jofre, J. (1992). Culture and decontamination methods affecting enumeration of phages infecting *Bacteroides fragilis* in sewage. *Appl. Envrion. Microbiol.* **58**, 2670–2673.

Toranzos, G.A. and McFeters, G.A. (1996). Detection of indicator microorganisms in environmental freshwaters and drinking waters. In: C.J. Hurst, G.R. Knudsen, M.J. McInerney, L.D. Stetzenbach and M.V. Walter (eds) *Manual of Environmental Microbiology*, pp. 184–194. American Society for Microbiology, Washington, DC.

Torroella, J.J. (1998). Bacteriophages of *Bacteroides* as indicators of pathogenic human viruses in coastal seawaters. Development and testing of sampling and analytical techniques for monitoring of marine pollutants (activity A). Final reports on selected microbiological projects. Map Tech. Rep. Ser. No. 54. UNEP, Athens, Greece.

USEPA (1986). *Ambient water quality criteria for bacteria – 1986*. USEPA Report EPA440/5-84-002. United States Environmental Protection Agency, Washington, DC.

USEPA (1996). *ICR Microbial laboratory manual*. United States Environmental Protection Agency, EPA/600/R-95/178.

USEPA (2000). *Improved Enumeration Methods for the Recreational Water Quality Indicators*: Enterococci *and* Escherichia coli. Report No. EPA/821/R-97/004. Office of Science and Technology, Washington, DC.

Van Poucke, S.O. and Nelis, H.J. (1997). Limitations of highly sensitive enzymatic presence-absence tests for detection of waterborne coliforms and *Escherichia coli*. *Appl. Environ. Microbiol.* **63**, 771–774.

van Veen, W.L., Mulder, E.G. and Deinema, M.H. (1978). The *Sphaerotilus-Leptothrix* group of bacteria. *Microbiol. Rev.* **42**, 329–356.

Van Vleet, E.S. and Williams, P.M. (1980). Sampling sea surface films: a laboratory evaluation of techniques and collecting materials. *Limnol. Oceanogr.* **25**, 764–770.

von Reyn, C.F., Waddell, R.D., Eaton, T. *et al.* (1993). Isolation of *Mycobacterium avium* complex from water in the United States, Finland, Zaire, and Kenya. *J. Clin. Microbiol.* **31**, 3227–3230.

von Reyn, C.F., Maslow, J.N., Barber, T.W. *et al.* (1994). Persistent colonisation of potable water as a source of *Mycobacterium avium* infection in AIDS. *Lancet* **343**, 1137–1141.

Waage, A.S., Vardund, T., Lund, V. and Kapperud, G. (1999). Detection of small numbers of *Campylobacter jejuni* and *Campylobacter coli* cells in environmental water, sewage, and food samples by a seminested PCR assay. *Appl. Environ. Microbiol.* **65**, 1636–1643.

Wadford, D.A., Dixon, B.A. and Cox, M.E. (1995). Techniques for the recovery of *Bacteroides vulgatus* from shellfish. *J. Shell. Res.* **14**, 533–535.

Wallace, R.J. (1987). Nontuberculous mycobacteria and water: a love affair with increasing clinical importance. *Infect. Dis. Clin. North Amer.* **1**, 677–686.

Watkins, W.D., Rippey, S.R., Clavet, C.R. *et al.* (1988). Novel compound for identifying *Escherichia coli*. *Appl. Environ. Microbiol.* **54**, 1874–1875.

Weaver, R.W. (ed.) (1994) *Methods of Soil Analysis*. Part 2, Microbiological and biochemical properties/editorial committee, Soil Science Society of America Book series;

No. 5, Soil Science Society of America, Madison, WI, USA.

Wheater, D.W.F., Mara, D.D. and Oragui, J. (1979). Indicator systems to distinguish sewage from stormwater run-off and human from animal faecal material. In: A. James and L. Evison (eds) *Biological Indicators of Water Quality*, pp. 21–25. Wiley, Chichester.

White, D.C., Pinkhart, H.C. and Ringleberg, D.B. (1997). Biomass measurements: biochemical approaches. In: C.J. Hurst, G.R. Knudsen, M.J. McInerney, L.D. Stetzenbach and M.V. Walter (eds) *Manual of Environmental Microbiology*, pp. 91–101. American Society of Microbiology, Washington, DC.

Whitehead, T.R. and Cotta, M.A. (2000). Development of molecular methods for identification of *Streptococcus bovis* from human and ruminal origins. *FEMS Microbiol. Lett.* **182**, 237–240.

Whittington, R.J., Reddacliff, L.A., Marsh, I. *et al.* (2000). Temporal patterns and quantification of excretion of *Mycobacterium avium* subsp. paratuberculosis in sheep with Johne's disease. Aust. Vet. J. 78, 34–37.

Wobeser, G. (1997). *Diseases of Wild Waterfowl*, 2nd edn. Plenum Publishing Corp., New York.

Wolinsky, E. (1992). Mycobacterial diseases other than tuberculosis. *Clin. Infect. Dis.* **15**, 1–12.

Wright, A.C., Miceli, G.A., Landry, W.L. *et al.* (1993). Rapid identification of *Vibrio vulnificus* on nonselective media with an alkaline phosphatase-labeled oligonucleotide probe. *Appl. Environ. Microbiol.* **59**, 541–546.

Wright, A.C., Hill, R.T., Johnson, J.A. *et al.* (1996). Distribution of *Vibrio vulnificus* in the Chesapeake Bay. *Appl. Environ. Microbiol.* **62**, 717–724.

Wright, S.F. (1992). Immunological techniques for detection, identification, and enumeration of microorganisms in the environment. In: M.A. Levin, R.J. Seidler and M. Rogul (eds) *Microbial Ecology: Principles, Methods, and Applications*. McGraw-Hill, New York.

Xu, H.-S., Roberts, N., Singleton, F.L. *et al.* (1982). Survival and viability of nonculturable *Escherichia coli* and *Vibrio cholerae* in the estuarine and marine environment. *Microbiol. Ecol.* **8**, 313–323.

Yamaguchi, N. and Nasu, M. (1997). Flow cytometric analysis of bacterial respiratory and enzymatic activity in the natural aquatic environment. *J. Appl. Microbiol.* **83**, 43–52.

Yoder, S., Argueta, C., Holtzman, A. *et al.* (1999). PCR comparison of *Mycobacterium avium* isolates obtained from patients and foods. *Appl. Environ. Microbiol.* **65**, 2650–2653.

Yokomaku, D., Yamaguchi, N. and Nasu, M. (2000). Improved direct viable count procedure for quantitative estimation of bacterial viability in freshwater environments. *Appl. Environ. Microbiol.* **66**, 5544–5548.

9

Fundamentals of biological behaviour and wastewater strength tests

M.C. Wentzel, George A. Ekama and R.E. Loewenthal

Water Research Group, University of Cape Town, Rondesbosch 7700, Cape Town, South Africa

1 INTRODUCTION

1.1 Definition of pollution

One definition for pollution is, 'an undesirable in change in the physical, chemical or biological characteristics of our air, land and water that may or will harmfully affect human life or that of desirable species, our industrial processes, living conditions, and cultural assets; or that may waste or deteriorate our raw material resources' (Odum, 1971). However, whatever definition of *pollution* one may choose, all imply that pollution is, (1) implicitly connected with life, and (2) disrupts an existing life order. Nature knows no pollution. Rather, it has a built-in propensity to respond to any stimulation and dispassionately absorbs it. From our human perspective, the response may be beneficial or adverse: if adverse we call the stimulation pollution. Although it is hard in this first decade of the 21st century to conceive a stimulation that is not somehow connected to human activity, the stimulation need not arise from mankind's activities but may be completely natural.

1.2 The role of energy and matter

From a temporal point of view, life requires a continuous throughput of *energy* and *matter* to maintain itself. For any system (e.g. living organism or ecosystem), the energy is received at a high level and leaves at a lower lever, the difference being the energy utilized. Any inefficiency in energy utilization leads to energy being lost as heat. The required continual input of energy arises from the laws of the created order: during any natural process, in the absence of life an isolated system will move naturally towards a state of disorder, where the molecules, atoms and elementary particles are arranged in the most random manner. This is called an increase in *entropy*. However, the living cell (or any other living system, e.g. organism, ecosystem) is anything but random, being a highly ordered structure.

To counter the natural tendency towards disorder, the living cell must continually take in energy from the surroundings. If no energy is available the cell will deteriorate and die. Thus, the cell is constantly tending towards degradation and energy is constantly needed to restore structures, as they are degraded; cells need to maintain structure in order to function. Furthermore, energy is also required for cell reproduction and the various cell functions (e.g. motility). Thus to maintain and reproduce life, a continuous energy flow is required. Matter is inextricably bound up with this energy flow. The matter passing through living systems is principally the macronutrients

The Handbook of Water and Wastewater Microbiology
ISBN 0-12-470100-0

carbon (C), hydrogen (H), oxygen (O), nitrogen (N) and phosphorus (P), but a variety of organic and inorganic micronutrients also take part. If the sources of energy, or matter, diminish, the quantity of life in the sense of mass of living organisms that can be supported diminishes correspondingly as happens in famine and drought, and vice versa.

1.3 Categorization of organisms

Energy is principally derived from three sources and carbon from two sources, and these form convenient criteria to categorize organisms, in particular the microorganisms involved in water quality control and wastewater treatment:

1.3.1 Energy

1. Sunlight radiation – *phototrophs*
2. Organic compounds – *heterotrophs*
3. Inorganic compounds – *lithotrophs*

1.3.2 Carbon

1. Organic compounds – *heterotrophs*
2. Inorganic compounds – *autotrophs*

In the biosphere, the basic source of energy is solar radiation. The photosynthetic autotrophs, e.g. algae, are able to fix some of this solar energy into complex organic compounds (cell mass). These complex organic compounds then form the energy source for other organisms – the heterotrophs. The general process whereby the sunlight energy is trapped, and then flows through the ecosystem, i.e. the process whereby life forms grow, is a sequence of reduction–oxidation ('redox,' reactions). In chemistry the acceptance of electrons by a compound (organic or inorganic) is known as reduction and the compound is said to be reduced, and the donation of electrons by a compound is known as oxidation and the compound is said to be oxidized. Because the electrons cannot remain as entities on their own, the electrons are always transferred directly from one compound to another so that the reduction–oxidation reactions, or electron donation–electron acceptance reactions, always operate as couples. These coupled reactions, known as redox reactions are the principal energy transfer reactions that sustain temporal life. The link between redox reactions and energy transfer can be best illustrated by following the flow of energy and matter through an ecosystem.

1.4 Redox reactions

1.4.1 Phototrophs

The photosynthetic autotrophic organisms, e.g. algae, grow and thereby fix some of the solar energy into organism mass comprising complex organic compounds; oxygen is a byproduct of this process. Of the mass requirements for the photosynthetic process, the hydrogen and oxygen are obtained from water, the carbon from carbon dioxide and phosphorus and nitrogen from dissolved salts of these elements. While carbon dioxide and water are readily available in the biosphere, the remaining two macronutrients, nitrogen and phosphorus, may be limited. Nitrogen (N) is useful to organisms in the nitrate or ammonia forms that may be limited and phosphorus (P) does not readily occur in its useful soluble phosphate form. The restricted availability of the N and P compounds usually limits the mass of photosynthetic autotrophic life a body of water can sustain. For this reason N and P are called eutrophic[1] or life nourishing substances and the excessive growth of algae or water plants in response to the N and P nutrients is called eutrophication. Whereas phosphorus is usually limited by the mass available that enters an ecosystem, certain microorganisms can generate ammonia from dissolved nitrogen gas (called N fixation). For this reason availability of phosphorus in a body of water is usually more crucial in eutrophication than

[1] From the Greek 'eu' meaning life and 'trophikos' meaning nourishment. Trophic and tropic should not be confused: they are distinctly different: tropic (from Greek 'tropikos' meaning turning towards) means attracted to, which is the reason why domestic flies are called phototropic, i.e. attracted to light, an important feature in the design of VP pit latrines: phototrophic means nourished by light.

the availability of nitrogen. Limitations in the availability of phosphorus will limit the amount of life (algae, plants, etc.) the ecosystem can support through photosynthesis. Accordingly, strategies to control eutrophication are usually focused more on limiting the discharge of P than N to the aqueous environment. Thus in photosynthesis, the simple low energy building blocks carbon dioxide and water are transformed to high energy complex organic compounds (algae, plants), the energy for this provided by sunlight radiation. Although there are some photosynthetic organisms, known as photosynthetic heterotrophs, that can use simple organic molecules as carbon source instead of carbon dioxide this is not common. The photosynthetic process is accomplished by redox reactions in which water is oxidized, producing oxygen as a byproduct and the electrons (and protons) are used to reduce carbon dioxide to form complex organic compounds.

1.4.2 *Heterotrophs*

The complex organic compounds formed through photosynthesis (i.e. plant matter) are the source of matter and energy for the heterotrophs. This is a diverse group of organisms, which, given sufficient time and appropriate environmental conditions, will utilize every type of organic material. The group is ubiquitous and in any given situation those members of the group that derive maximum benefit from the specific organic matter and environmental conditions will develop. As the organic source or environmental conditions change, so associated changes in the heterotrophic organism species take place.

To obtain energy from complex organic compounds, the heterotrophs degrade these in a sequence of biochemical pathways. In the redox reactions accompanying this degradation, electron transfer proceeds in the opposite direction to that for photosynthesis. The electrons originally captured in photosynthesis are now donated by complex organic compounds and are passed to an electron acceptor. For example, if the reaction is aerobic the electrons are passed from the complex organic compound to oxygen, forming water. In the process, the original sunlight energy captured in the complex organic molecules by the phototrophs is released and becomes available for use by the particular organism executing the degradation reaction. A number of different types of degradation reactions occur.

1.5 Electron donors and acceptors

In an aquatic environment, if dissolved oxygen is available in the water, the heterotrophic organisms will be aerobic and degrade some of the organic compounds to carbon dioxide and water and consume some of the available oxygen. In this process, more heterotrophic organisms are formed so that some of the energy (and matter) contained in the high-energy organic compounds (algae) is transformed to heterotrophic organism mass and retained as different organic compounds in the ecosystem. The remainder of the energy (about one-third) is lost as heat. In these processes, the organic compound supplies the electrons and is oxidized and the function of the oxygen is to accept the electrons released and thus become reduced. For this reason the organic compound is called the electron donor and the oxygen is called the terminal electron acceptor. Should the dissolved oxygen become depleted and anoxic conditions develop (no dissolved oxygen present), then some of the heterotrophic organisms can utilize nitrate instead of oxygen as the terminal electron acceptor, reducing the nitrate to nitrogen gas. In this way these organisms continue their redox reactions and grow under anoxic conditions. Should both dissolved oxygen and nitrate become depleted and sulphate is present, then certain microorganisms, the sulphate reducers, utilize sulphate as the terminal electron acceptor. The sulphate is reduced to sulphide and hydrogen sulphide gas, which not only has a foul smell but also is toxic to many aquatic life forms. Fortunately, sulphides are relatively insoluble ($pK_a \sim 7.1$) and form a black precipitate with many metals. However, the odours and black appearance in a natural water body are aesthetically unpleasant. The organisms that are unable to reduce sulphate have no external electron acceptors available.

If high-energy organic compounds are still available in the water, the organisms have to generate their own electron acceptor internally in order to gain energy through redox reactions.

1.6 Defining aerobic and anaerobic processes

As a general definition, with organic compounds as the electron donor and when the terminal electron acceptor is available externally, the redox reactions are called respiration. The nature of the terminal electron acceptor defines the kind of respiration; if this is oxygen, it is called aerobic respiration, if not oxygen it is called anaerobic respiration[2]. When the terminal electron acceptor needs to be generated internally to the organisms, the redox reaction is called fermentation. In fermentation both the electron donors and acceptors are organic compounds; e.g. in fermentation of glucose to lactate, the glucose is first oxidized to pyruvate, the electrons then being used to reduce the pyruvate to lactate.

1.7 Predation

The mass of heterotrophic organisms generated via the different degradation reactions detailed above, in turn form the matter and energy sources (prey) for other organisms that live on them (predators), and these in turn become prey to yet other predators. Each predator–prey transformation is accompanied by a substantial loss of energy, lost as heat, due to inefficiencies in energy transfers. In this fashion, through the sequential chain of life there is a continuous reduction of the total organic energy originally fixed by the photosynthetic autotrophs, which in turn is dependent on the quantity of the available macronutrients nitrogen (N) and phosphorus (P). When the organic energy reduces to zero, heterotrophic life ceases. If the quantity of organic energy is small, it can be reduced to zero while the conditions remain aerobic; however, if the quantity is large, the conditions in the water body can become anaerobic.

1.7.1 Lithoautotrophs

The last of the three groups of organisms categorized by energy source and termed the lithotrophs are, like the heterotrophs, a diverse group. Of importance in wastewater treatment are the lithoautotrophs, more commonly called autotrophs. Interesting species in this group are the sulphur, iron, nitrifying and hydrogenotrophic methanogenic organisms. The sulphur and iron organisms are important in sewer and pipe corrosion. They are all obligate (strict) aerobes, e.g. *Thiobacillus (T. thiooxidans* and *T. ferrooxidans)* and have found wide biotechnological application in the bioleaching of ores such as iron from pyrites (FeS). In bioloeaching, these organisms utilize the S from iron sulphide (FeS) as electron donor and oxygen as electron acceptor, with the S being oxidized to sulphate (SO_4^{2-}). The same reaction occurs in the condensate in the top gas space of sewers except in this case the electron donor is hydrogen sulphide (H_2S) gas which is oxidized to sulphate. The sulphates so produced generally form sulphuric acid (H_2SO_4) resulting in a very low pH between 1 and 2 which corrodes the top of the sewers. Surprisingly, the low pH values are not detrimental to the *Thiobacillus* organisms provided the pH value >1; indeed it is their preferred pH. The H_2S gas is generated from sulphate originally present in the wastewater. The sulphate is used as electron acceptor by anaerobic heterotrophic sulphate reducers (e.g. by *Desulfovibrio desulphuricans)* with organics in the wastewater

---

[2] These definitions are those generally used in bacteriology and microbiology. The definitions conventionally used in sanitary engineering are somewhat different in that respiration in an absence of dissolved oxygen but a presence of oxidized nitrogen, nitrite or nitrate, is called anoxic respiration and that in an absence of oxygen and oxidized nitrogen but presence of sulphate is called anaerobic respiration. Furthermore, in biological nutrient removal (BNR) activated sludge systems, the term anaerobic has a different meaning from that used in microbiology. In BNR it means an absence of *both* dissolved oxygen and oxidized nitrogen and not just absence of oxygen. Actually, in BNR, the definitions of anoxic and anaerobic go further in that they include zero input of dissolved oxygen (anoxic) and zero input of dissolved oxygen and oxidized nitrogen (anaerobic).

serving as electron donors; the sulphate is reduced to hydrogen sulphide in the liquid wastewater phase, some of which comes off as a gas and redissolves in the condensate of the top gas space.

Interestingly, both lithoautotrophic *Thiobacillus* species are also capable of utilizing iron as electron donor, oxidizing ferrous (Fe^{2+}) to ferric (Fe^{3+}) while reducing oxygen to water, but the reaction requires the presence of sulphate. This feature of these organisms is therefore an added danger for metal pipes transporting water and wastewaters. Chlorination of municipal water supplies assists in providing protection against these bioreactions. A third *Thiobacillus* species, *T. denitrificans*, can utilize reduced sulphur, such as S or sulphides (H_2S, S^-), as electron donor and nitrate or nitrite as electron acceptor, with the reduced sulphur being oxidized to sulphate and the oxidized nitrogen being reduced to nitrogen gas – a lithoautotrophic denitrification.

While the *Thiobacillus* autotrophs are important in water and wastewater transport systems, only the nitrifying and hydrogenotrophic methanogen lithoautotrophs are of particular importance in municipal wastewater treatment systems. The nitrifiers are strict (obligate) aerobes and the organisms of importance in wastewater treatment systems are generically called Ammonia Oxidizers (AO) and Nitrite Oxidizers (NO). Being autotrophs, the carbon they require to form cellular material is obtained from carbon dioxide like the algae, but their energy requirements are obtained from oxidizing ammonia to nitrite and nitrite to nitrate, respectively – reactions collectively called nitrification. The ammonia oxidizers, e.g. *Nitrosomonas* and *Nitrosococcus* spp., ammonia (NH_4) as an electron donor, and oxygen as an electron acceptor. The ammonia is oxidized to nitrite and the oxygen is reduced to water. The nitrite oxidizers, e.g. *Nitrobacter* and *Nitrospira* spp., which are usually encountered together with the ammonia oxidizers, use nitrite as electron donor, oxidizing it to nitrate with oxygen serving as the terminal electron acceptor. These nitrifying organisms can only execute these redox reactions while dissolved oxygen is present and as a consequence they are obligate aerobes. In contrast, the hydrogenic methanogens are obligate anaerobic organisms (oxygen is toxic to these organisms). To obtain energy they use hydrogen as electron donor and carbon dioxide as an electron acceptor to form methane. Being autotrophs, carbon requirements to form cell material are obtained from carbon dioxide. The organisms are of importance in anaerobic digestion systems, e.g. in digestion of primary and waste activated sludges generated in the treatment of municipal wastewater and in upflow anaerobic sludge bed (UASB) systems.

1.8 Anabolism, catabolism and yield

From the above discussion, a fundamental difference between the heterotrophic organisms is that the former uses organic compounds as electron donor, whereas the latter uses inorganic compounds like sulphides, ammonia or nitrite. Since for the lithotrophs the electron donor does not include carbon, they have to obtain the carbon for building the organic compounds of cell material from molecules separate from the electron donor. Of importance are the lithoautotrophs that obtain carbon from carbon dioxide dissolved in the water. As a consequence, for the lithoautotrophs the energy requirements are derived from reducing the electron donor and different inorganic compounds supply the material requirements. The redox reactions that supply the energy requirements are called *catabolism* and those that supply the material requirements are called *anabolism*. Catabolism and anabolism together are called *metabolism*. In contrast to the lithoautotrophs, the heterotrophs obtain the energy requirements (catabolism) and material requirement (anabolism) from the same organic compounds irrespective of the type of external terminal electron acceptor. This difference in the metabolism of the autotrophic and heterotrophic organisms is the principal reason why the cell yield, (i.e. the organism mass formed per electron donor mass utilized) is different. It is low for autotrophs (e.g. 0.10 mg VSS/mgNH_4 – N nitrified for the nitrifiers) and high for heterotrophs (e.g. 0.45 mg VSS/mg COD utilized). Not only is the energy requirement in anabolism to convert carbon dioxide to cell mass far greater than that required to convert organic compounds to cell mass, but also the energy released (catabolism) in oxidizing inorganic compounds is less than that in oxidizing organic compounds.

2 OBJECTIVES OF WASTEWATER TREATMENT

From the foregoing description of biological behaviour, the consequences of discharging municipal wastewater to a water body can be appreciated. From a chemical point of view, municipal wastewater or sewage contains (1) organic compounds, such as carbohydrates, proteins and fats, (2) nitrogen, principally in the form of ammonia, and (3) phosphorus, which is principally in the form of phosphate from human waste and detergents. Municipal wastewater of course, has many other constituents of a particulate and dissolved nature, such as pathogens, plastics, sand, grit, live organisms, metals, elements, anions and cations. All these constituents have to be dealt with and considered at wastewater treatment plants, but not all of these are important for wastewater treatment plant modelling and design. Rather, attention is focused on the carbonaceous (C), nitrogenous (N) and phosphorus (P) constituents because these are the main ones that influence biological activity and eutrophication in the receiving water.

When municipal wastewater is discharged to a water body, the organic compounds stimulate the growth of the heterotrophic organisms causing a reduction in the dissolved oxygen. While oxygen is present, the ammonia, which is toxic to many higher life forms such as fish and insects, will be converted to nitrate by the nitrifying organisms causing a further demand for oxygen. Depending on the volume of wastewater discharged and the amount of oxygen available, the water body can become anoxic. If the water does become anoxic, nitrification of ammonia to nitrate by the autotrophs will cease. However, some of the heterotrophs now will use nitrate instead of oxygen as terminal electron acceptor and continue their metabolic reactions. Depending upon the relative amount of organics and nitrate, the nitrate may become depleted. In this case the water will become anaerobic with associated sulphate reduction and fermentation. Irrespective of whether or not all of these steps happen, when the organic energy of the wastewater has been depleted, the water body in time recovers, clarifies and again becomes aerobic. But most of the nutrients, nitrogen (N) and phosphorus (P), remain, and these with sunlight penetration into the water body, stimulate aquatic plants like algae to grow in much the same way as farmers fertilize their lands with nitrogen and phosphates to stimulate crop growth. The algae reintroduce organic material into the water body and the pollution cycle starts over again. Only when the nutrients N and P are depleted and the organic energy sufficiently reduced, can the water body be said to be eutrophically stable again.

From these considerations, the objectives of wastewater treatment can be seen to be threefold:

1. reduce organic bound energy to a level where it will no longer sustain heterotrophic growth and thereby avoid deoxygenation
2. oxidize ammonia to nitrate to reduce its toxicity and deoxygenation effects
3. reduce eutrophic substances, ammonia, nitrate and particularly phosphates, to levels where photosynthetic microorganisms are limited in their growth and therefore are limited in their capacity to fix solar energy as organic bound energy.

3 BASIC BIOLOGICAL BEHAVIOUR

3.1 Organism types in wastewater treatment

It was shown above that the nature of the electron donors and electron acceptors, which the various organism types utilize in their biological activities, form a useful means of categorizing organisms. In municipal wastewater treatment incorporating biological nutrient removal (BNR), two basic categories of organisms are of specific interest: the heterotrophic organisms and the lithoautotrophic nitrifying organisms including the ammonia and nitrite oxidizers. The former group utilizes the organic compounds of the wastewater as electron donor and either oxygen or nitrate as terminal electron acceptor depending on whether or not the species is obligate aerobic or facultative and the conditions in the plant are aerobic or anoxic respectively. Irrespective of whether obligate aerobic or facultative the heterotrophic organisms obtain their catabolic (energy) and anabolic (material) requirements from the same organic compounds. In contrast the latter group, being lithoautotrophs, obtain their catabolic and anabolic requirements from *different inorganic* compounds, the catabolic (energy) from oxidizing free and saline ammonia in the wastewater to nitrite and nitrate and the anabolic (material) from dissolved carbon dioxide in the water. Being obligate aerobic organisms, only oxygen can be used as electron acceptor and therefore the nitrifying organisms require aerobic conditions in the plant.

In activated sludge plants irrespective of whether or not biological N and P removal is incorporated, the heterotrophic organisms dominate the biocenosis (mixed culture of organisms) and make up more than 98% of the active organism mass in the system. If the organism retention time (sludge age) is long enough, the nitrifiers may also be sustained but, because of their low specific yield coefficient compared to the heterotrophs (see Section 1.8 above) and relatively small amount of ammonia nitrified compared to organic material degraded, they make up a very small part of the active organism mass (<2%). Therefore, insofar as sludge production and oxygen or nitrate utilization is concerned, the heterotrophs have a dominating influence on the activated sludge system. Three different groups of heterotrophic organisms grow in the nutrient removal activated sludge system: (i) ordinary aerobic, (ii) facultative and (iii) polyphosphate accumulating heterotrophs, designated OHO, FHO and PAO respectively. While these three subgroups have some distinct behavioural characteristics, which result in the particular biological reactions required to achieve the wastewater treatment objectives, their overall behaviour is similar and governed by their basic heterotrophic nature. An understanding of basic heterotrophic organism behaviour from a bioenergetic point of view, will therefore form a sound foundation on which many important principles in biological wastewater treatment are based, such as (1) growth yield coefficient and its association with oxygen or nitrate utilization, and (2) energy balances and its association with wastewater strength measurement.

3.2 Bioenergetics of heterotrophic organism behaviour

Bioenergetics is the study of material pathways occurring within organisms (and biosystems) as a result of an energy flow through a system. In *aerobic* systems the two principal pathways by means of which organic material is processed (metabolized) by organisms are catabolism and anabolism.

3.2.1 *Catabolism and anabolism*

1. *Catabolic pathway*: this is the pathway by which organisms obtain energy in a useful form for carrying out biological work, *inter alia* the work involved in protoplasm (organism mass) synthesis (termed anabolism see below). A fraction of the organic molecules taken up by an organism is enzymically oxidized to CO_2 and water, and a large amount of energy is released. A part of this energy is captured by the organism and is available for performing useful biological work; the part the organism is unable to 'capture' is lost as heat to the surroundings.
2. *Anabolic pathway*: this is the pathway by which organisms synthesize protoplasm (construct new cell mass). A fraction of the organic molecules taken up by the organism is enzymically modified to form part of the biological protoplasm. This synthesis process not only requires an input of organic molecules, but also inorganic molecules (e.g. ammonia (NH_4), phosphorus

(P), and micronutrients), energy, protons and electrons.

In biological aerobic treatment of organic compounds in wastewaters, the two material cycles set out above together constitute the *metabolism* of the organism and are of paramount importance. This is because one of the principal objectives of wastewater treatment is removal of energy (contained in the organic compounds) from the wastewater. Viewed over the biological treatment system as a whole, this objective is attained biologically in two ways:

1. through energy losses (as heat) to the surroundings resulting from the partial capture of energy in catabolism and the partial loss in the use of this energy for anabolism (both due to inefficiencies in energy transfers)[3]
2. through transformations of the organic compounds (dissolved or suspended) in the wastewater to organic molecules incorporated in biological protoplasm (anabolism).

The biomass so formed is settleable and is removed from the treated wastewater by physical separation of the biomass solids and wastewater liquid phases, leaving a clarified effluent virtually free of the original organic compounds. In most wastewater treatment plants, the biomass that is collected from the solid/liquid separation stage (usually a secondary settling tank) is returned to the biological reactor to inoculate the reactor with biomass for treating the continuously incoming wastewater (Fig. 9.1). Should the wastewater contain particulate organic material that is

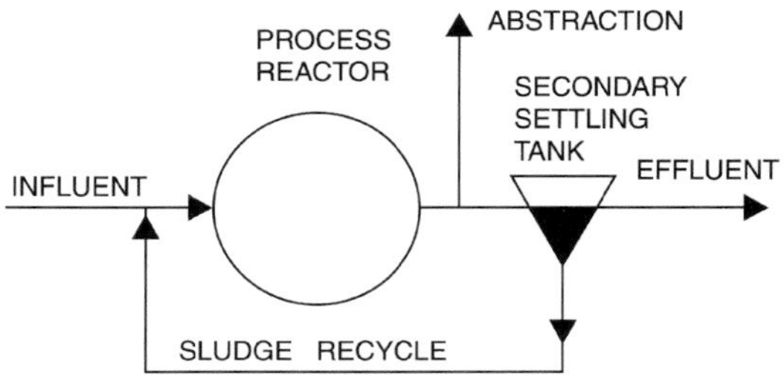

Fig. 9.1 Schematic layout of the activated sludge system in its simplest form showing biological reactor, secondary settling tank, sludge return and waste sludge abstraction from the biological reactor for hydraulic control of sludge age.

degraded slowly, this material will become enmeshed with the biomass, separated from the clarified effluent and returned to the reactor for further treatment. Slowly biodegradable particulate material therefore does not influence the effluent organic concentration. If there is slowly biodegradable organic material in a dissolved form it will escape with the effluent. However, in municipal wastewater treatment, most dissolved organics are readily biodegradable so low effluent organic material concentrations are usually easily achieved.

By means of a simplified bioenergetic model based on thermodynamic principles, an approximate estimate can be obtained of the proportion of the energy in the original wastewater organic compounds that is lost as heat or is transformed into biomass. This proportion is known as the active heterotrophic mass yield coefficient (Y_h, mass of organism formed per mass of organic material degraded). Apart from producing theoretical yield coefficients that can be usefully compared with experimentally measured values, the bioenergetic model also provides insight for many of the fundamental principles accepted as 'laws of nature' in activated sludge design models and procedures.

3.2.2 *Organism electron and energy transport mechanisms*

Energy becomes available to organisms principally through internally mediated reduction–oxidation (redox) reactions involving electron

[3] In municipal wastewater treatment, the volumes of liquid undergoing biological treatment are too great and the heat losses from the biological reactions too small to make a significant difference to the temperature of the activated sludge mixed liquor. However, in autothermal sludge treatment where the organics are much more concentrated (by >100 times) and special precautions taken to minimize heat losses, the bio-heat released in the metabolism of the organisms is sufficient to raise the sludge liquor to greater than 60°C to affect sludge pasteurizaton. This is done in forced air composting, thermophilic aerobic digestion and dual digestion of sewage sludge.

(and proton) transfer under controlled conditions. For heterotrophic organisms, the electrons are transferred from the organic compound, the electron donor (substrate or 'food' for the organisms), via a number of intermediate compounds, to some final or terminal electron (and proton) acceptor. In the course of the electrons moving from the electron donor to the electron acceptor via the various intermediate steps, energy is released progressively, some of which is captured by the organism. Fundamental to the means by which organisms harness the energy released in the redox reactions, are two types of compounds which couple into the redox reactions. The first is a compound called nicotinamide adenine dinucleotide (NAD). The oxidized (NAD) and reduced ($NADH_2$) forms act as *intermediate* electron acceptor and donor respectively. The second is a compound that in the form adenosine diphosphate (ADP) acts as an energy acceptor and in the form adenosine triphosphate (ATP) acts as an energy donor.

3.2.3 Electron transport molecule, $NADH_2$

The organism electron acceptor molecule, NAD, is able to accept two electrons (e^-) and two protons (H^+) thereby becoming reduced, and in the reduced state is depicted as $NADH_2$, (equation 1):

$$NAD + 2e^- + 2H^+ \rightarrow NADH_2 \qquad (1)$$

For the above half reaction to occur the organic compounds provide the electrons and protons in a concomitant second half reaction. Overall the organic compound is oxidized enzymatically to CO_2, protons (H^+) and electrons (e^-). Taking oxidation of glucose as an example, this reaction takes place in two stages. In the first stage the glucose is processed in a biochemical pathway called the Embden–Meyerhof pathway to pyruvate (equation 2a):

$$C_6H_{12}O_6 \rightarrow 2C_3H_4O_3 + 4H^+ + 4e^- \qquad (2a)$$

The pyruvate is then converted to acetyl CoA which enters the main oxidation pathway, called the Krebs or tricarboxylic acid (TCA) cycle, and is completely oxidized as follows (equation 2b):

$$2C_3H_4O_3 + 6H_2O \rightarrow 6CO_2 + 20H^+ + 20e^- \qquad (2b)$$

The above two reactions combined give the net reaction for complete oxidation of the glucose (equation 2c):

$$C_6H_{12}O_6 + 6H_2O \rightarrow 6CO_2 + 24H^+ + 24e^- \qquad (2c)$$

If the organic compound is acetate, this enters the Krebs or TCA cycle directly to give (equation 2d):

$$CH_3COOH + 2H_2O \rightarrow 2CO_2 + 8H^+ + 8e^- \qquad (2d)$$

The electrons and protons from the organic material are captured by the organism in the $NAD/NADH_2$ reaction (equation 1). Balancing the electrons, the sum of the two half reactions in equations 1 and 2c, 2d is (equations 2e and 2f):

Glucose:

$$C_6H_2O_6 + 6H_2O + 12NAD \rightarrow 6CO_2 + 12NADH_2 \qquad (2e)$$

Acetate:

$$CH_3COOH + 2H_2O + 4NAD \rightarrow 2CO_2 + 4NADH_2 \qquad (2f)$$

Once the organism has captured the electrons (and protons) in the form of $NADH_2$, the $NADH_2$ is available for generating new cell mass (anabolism) or energy (catabolism). In the latter case the electrons are donated to terminal electron acceptors either oxygen or nitrate, which are themselves reduced to water and nitrogen gas respectively (equations 3a and 3b).

Oxygen:

$$O_2 + 2NADH_2 \rightarrow 2H_2O + 2NAD \qquad (3a)$$

Nitrate:

$$2NO_3^- + 5NADH_2 + 2H^+ \rightarrow N_2 + 6H_2O + 5NAD \qquad (3b)$$

The energy released in these half reactions is partially captured by the organism via the formation of ATP, its energy transport molecule (see 3.2.4). Thus, the molecule $NADH_2$ provides the link between the substrate oxidation pathway, where it captures the electrons (and protons) released in oxidation of the organic substrate and the energy generation reactions, where it donates the electrons to the terminal electron acceptor. In this process some of the energy released is captured by the organism.

3.2.4 *Energy transport molecule, ATP*

The formation of ATP from ADP ~ and phosphate (PO_4^{3-}) requires an energy input of about 42 to 50 kJ/mol (10 to 12 kcal/mole) ATP formed under physiological conditions. The energy is conserved in the high-energy phosphoryl bond (equation 4a):

$$\text{Adenosine.}PO_4^{3-} \sim PO_4^{3-} + \text{energy} + PO_4^{3-} \rightleftharpoons \text{Adenosine.}PO_4^{3-} \sim PO_4^{3-} \sim PO_4^{3-} \quad (4a)$$

$$\text{ADP} + 42 \text{ to } 50\text{ kJ} + PO_4^{3-} \rightleftharpoons \text{ATP} \quad (4b)$$

Two types of processes are used by organisms to form ATP from ADP and so capture the energy released in reactions, substrate and oxidative phosphorylation. In substrate phosphorylation, the energy capture couples directly into the transformation of organic compounds and ATP is formed without transfer of electrons to the terminal electron acceptor. This energy release/capture is associated with the direct transfer of a high energy phosphoryl group from the substrate to ADP to form ATP. This occurs during the partial breakdown of the organic compound to simpler organics like pyruvate that can enter the TCA (Krebs) cycle (equation 2a). Substrate phosphorylation does not involve the net release of electrons and therefore does not require an electron acceptor (Fig. 9.2). For example, in one step of the Embden-Meyerhof pathway, equation 5:

$$\text{Phosphoenolpyruvate} + \text{ADP} \rightarrow \text{Pyruvate} + \text{ATP} \quad (5)$$

In oxidative phosphorylation, the energy capture reaction couples into the transfer of electrons to the terminal electron acceptor in the electron transport chain (ETC). In the ETC, $NADH_2$ is oxidized and the electrons are released and passed via a series of enzyme complexes to the terminal electron acceptor (equation 3); in this process energy is released and some of it is captured by the organism in ATP. Noting that for aerobic conditions, passing 2 electrons from $NADH_2$ to the terminal electron acceptor oxygen, ideally gives rise to the formation of 3 ATPs (a feature of the ETC) as follows (equation 6):

$$NADH_2 + \tfrac{1}{2}O_2 + 3\text{ADP} + 3PO_4^{3-} \rightarrow \text{NAD} + H_2O + 3\text{ATP} \quad (6)$$

From equation 6 it can be concluded that:

1. Each molecule of $NADH_2$ releases a pair of electrons and protons which are captured by the electron acceptor oxygen.
2. Oxidation of 1 molecule of $NADH_2$ with oxygen ideally yields 3 ATP, but this ATP yield will vary depending on the electron acceptor.
3. Thus, each half molecule of oxygen reduced to water indicates that two electrons and protons have been transferred to it and 3 ATP formed.

3.2.5 *Efficiency of energy capture in catabolism*

Taking the oxidation of glucose with oxygen as an example (equations 7a, b and c):

$$C_6H_{12}O_6 + 6H_2O \rightarrow 6CO_2 + 24e^- + 24H^+ \quad (7a)$$

$$6O_2 + 24e^- + 24H^+ \rightarrow 12H_2O \quad (7b)$$

$$C_6H_{12}O_6 + 6O_2 \rightarrow 6CO_2 + 6H_2O \quad (7c)$$

By considering the free energies of formation (ΔG_f^0) of the reactants ($C_6H_{12}O_6$ and O_2) and products (CO_2 and H_2O), the free energy of the reaction (ΔG_R^0) is $-$ 2893 kJ/mol ($-$689 kcal/mole) glucose (where $-$ve indicates a release of energy) (equation 8):

$$\begin{aligned}\Delta G_R^0 &= \Delta G_f^0(\text{products}) - \Delta G_f^0(\text{reactants})\\ &= 6\Delta G_f^0(CO_2) + 6\Delta G_f^0(H_2O) - \Delta G_f^0(C_6H_{12}O_6) - 6\Delta G_f^0(O_2)\\ &= 6(-395) + 6(-239) - (-611) - 6(0)\\ &= -2893\text{ kJ/mol glucose or}\\ &= -689\text{ kcal/mol glucose}\end{aligned} \quad (8)$$

In aerobic breakdown of 1 mole of glucose by heterotrophic organisms, 12 moles of $NADH_2$

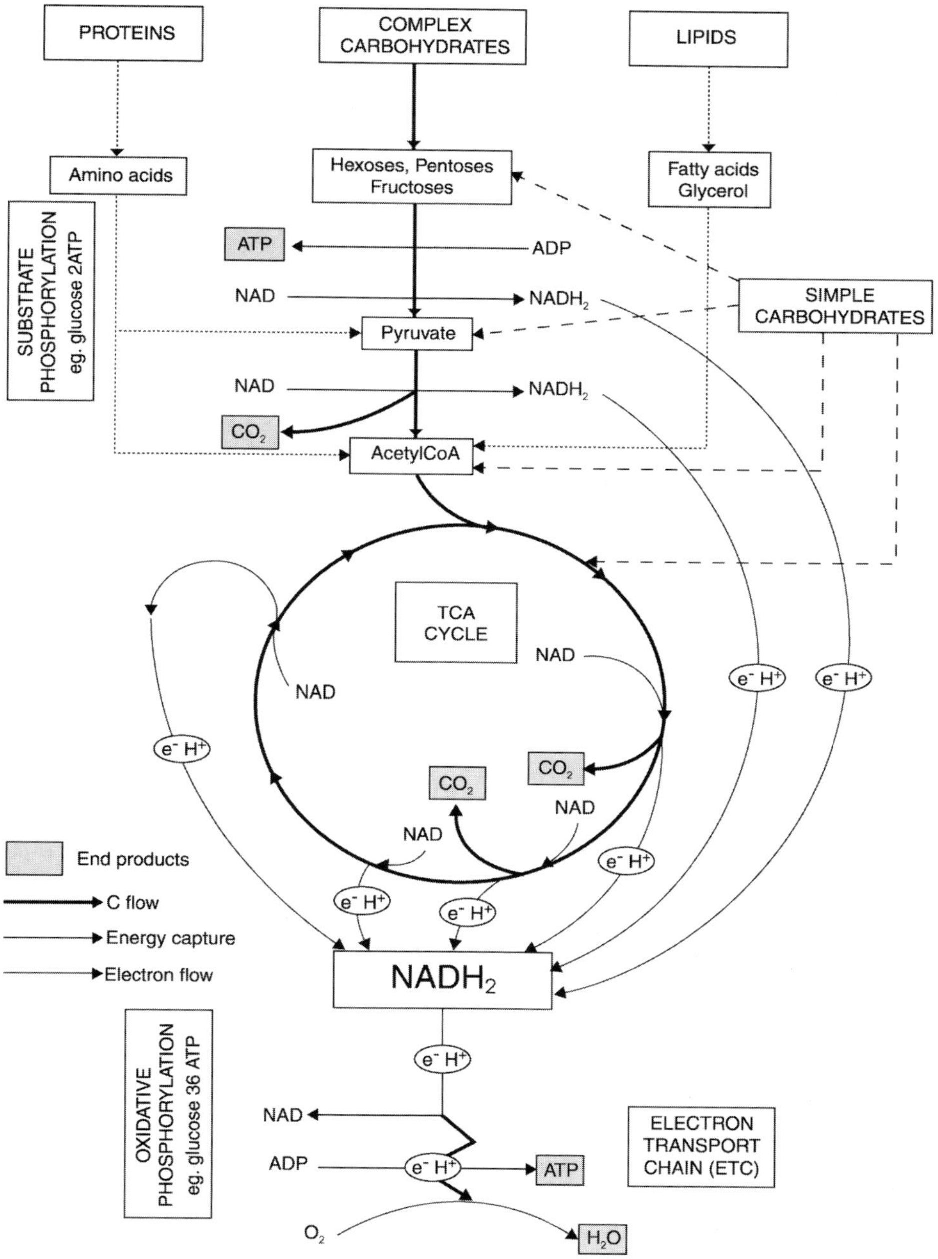

Fig. 9.2 Simplified schematic representation of heterotrophic organism catabolism of organics showing carbon flow from organics (substrate) to carbon dioxide, electron flow from donor (organics) to acceptor (oxygen) and energy capture as ATP in substrate phosphorylation (transformation – no electron release) and oxidative phosphorylation (electron transfer to terminal electron acceptor).

are produced (equation 2e). From equation 5, each mole of NADH, ideally produces 3 moles ATP by oxidative phosphorylation. This gives $12 \times 3 = 36$ moles of ATP produced by oxidative phosphorylation in oxidation of 1 mole of glucose. In addition, 2 moles ATP are produced by substrate phosphorylation in the Embden-Meyerhof pathway (equation 2a).

This gives a total of 38 moles ATP produced for each mole glucose oxidized. Accepting that with each mol ATP formed, 42 kJ (10 kcal) is captured, then the energy capture is 1596 kJ/mol (380 kcal/mole) glucose. From the above, in complete oxidation of one mole of glucose, the free energy released is 2893 kJ/mol glucose. Thus, the energy captured in ATP on oxidation of glucose represents a fraction of 1596/2893 = 0.55 of the free energy released. The organism is therefore able to capture about 55% of the free energy released and this energy is obtained via substrate phosphorylation (2 out of 38 ATP) in the Embden-Meyerhof pathway and via oxidative phosphorylation (36 out of 38 ATP) in passing the 24 electrons donated by the glucose to the electron acceptor oxygen in the transport chain (ETC) where 3 ATP are formed per electron pair transferred.

3.2.6 Organic material oxidation and free energy release

Taking the oxidation of glucose with oxygen as an example, from the section above, the free energy of the reaction (ΔG_R^0) is -2893 kJ/mol glucose where $-$ve indicates a release of energy. Because $24e^-$ equivalents (e^-eq, or moles of e^-) are released in the reaction the free energy released per electron equivalent (e^-eq) is $-2893/24 = 120.5$ kJ/e^-eq (-28.7 kcal/e^-eq). Also, because (equation 9):

$$O_2 + 4H^+ + 4e^- \rightarrow 2H_2O \quad (9)$$

so that $1e^-$eq = 8 gO (see equation 3a), the free energy released per gO is $-2893/(24 \times 3) = -15.1$ kJ/gO (-3.59 kcal/gO). For acetate, the free energy released is -853 kJ/mol (-203 kcal/mole). Noting that 1 mole acetate yields 8 e^-eq and each e^-eq is equivalent to 8 gO, yields for acetate a free energy release of $853/(8 \times 3) = 13.3$ kJ/gO (-3.17 kcal/gO). With 4.2 J/cal, this is equal to -13.3 kJ/gO.

The free energy per gO is termed the free energy of combustion and represents the amount of available energy (energy that can be used to do work) released in the complete oxidation of the organic. It is important to note that the free energy of combustion for many different organic compounds when oxidized to CO_2 and H_2O, is closely constant and ranges from 13.3 kJ/gO for the short-chain fatty acids and alcohols to 15.1 kJ/gO for the sugars and carbohydrates. Even the gaseous non-physiological fuels like methane and the liquid petroleum gases (LPG) give free energies of combustion in this range. For example for methane (CH_4) it is 13.86 kJ/gO and octane (C_8H_{18}) it is 13.65 kJ/gO. Clearly the free energy of combustion (useful energy released) is closely linked and almost proportional to the mass of oxygen consumed by the compound, or equivalently the number of electrons it can donate in its oxidation to CO_2, and water. This is irrespective of whether it is done rapidly such as in combustion or slowly such as in organism respiration.

Experimental support for the proportionality between the free energy released and the electron donating capacity (EDC, i.e. kJ/e^-eq), or equivalently through equation (9), the theoretical chemical oxygen demand (Th.COD, i.e. kJ/gO), was presented by Servizi and Bogan (1963, 1964) and Burkhead and McKinney (1969). Servizi and Bogan (1963) identified two groups of organics, a high energy group comprising carbohydrates, glycolysis products and TCA cycle intermediates and a low energy group comprising alcohols, aromatics, hydrocarbons (benzene ring compounds) and aliphatic acids (short or long chain saturated or unsaturated fatty acids). The difference in the free energy released per Th.COD between the high and low energy groups of organics was only 5% of the two groups mean value. McCarty (1972) found that the range of free energy released per Th.COD for naturally occurring organic compounds was bounded by methane, which yields the lowest, and formate, which yields the highest. The difference between formate and methane is only 10% of their mean value. Clearly there is substantive evidence that the free energy release in the oxidation (respiration to end products carbon dioxide and water) of organic compounds is proportional to the EDC or equivalently the Th.COD, and remains relatively constant for the spectrum of organic compounds.

The conclusion that the free energy released is proportional to the EDC for most organics is fundamental not only to bioenergetics and tracing energy pathways in organisms, but also for measuring the energy strength of organic wastewaters. The application of this principle to both these aspects is discussed below.

3.2.7 Free energy release and energy capture by the organism

It has been concluded above that in complete oxidation of organics the free energy released is proportional to the EDC. In biological oxidation of the organics the proportion of this free energy captured by the organism is approximately constant provided the terminal electron acceptor (e.g. oxygen) is the same. This implies that the energy that becomes available to the organism in biological oxidation of organics is proportional to the EDC of the organic.

In considering biological oxidation, two separate phases can be identified. In the first phase, irrespective of which particular biochemical pathway or combination of biochemical pathways is used, the organic is oxidized to CO_2, electrons and protons. The electrons and protons are captured by the organism in $NADH_2$. In the second phase the electrons and protons from NAH_2, are donated to the terminal electron acceptor (via the ETC) and some of the energy released is captured by the organism in the formation of ATP (oxidative phosphorylation). *Since for a specific terminal electron acceptor the number of ATP molecules formed per pair of electrons passing to the terminal electron acceptor is constant, the energy captured by the organism in oxidative phosphorylation is proportional to the EDC of the organic.*

In Section 3.2.6 above, it was noted that there were two groups of organics; one comprising complex organics, such as carbohydrates and sugars, with about 14% higher heats of combustion than the other group comprising simple organics, such as acetate and pyruvate. There are also some minor variations in the total number of ATPs formed per e^- donated by different organics. These minor differences in the energy captured by the organism are due to the relative contribution to ATP production by substrate and oxidative phosphorylation. In oxidative phosphorylation with the same terminal electron acceptor, the same amount of ATP is generated by the organism per electron donated by the organic material. However, the amount of ATP generated by substrate phosphorylation will vary depending on the nature of the organic compound and the biochemical pathways used in its transformation prior to entering the TCA (Krebs) cycle (see Fig. 9.2). For example, glucose will give 2 ATP via substrate phosphorylation and acetate none. However, substrate phosphorylation plays a relatively minor role in ATP generation – for glucose only 2 ATP out of a total of 38 ATP are generated by substrate phosphorylation and the balance by oxidative phosphorylation. The dominating effect of oxidative over substrate phosphorylation implies that the EDC or Th.COD will provide a good estimate of the energy that will become available to the organism. This dominating effect of oxidative phosphorylation is also the reason why the difference in the heats of combustion between the different organic groups is so small. It is also the reason why energy changes, both free energy released and that captured by the organism in the form of ATP, can be monitored via the changes in EDC or Th.COD.

3.3 Energy utilization in heterotrophic organisms – organism yield and oxygen utilization

In utilization of an organic compound by the organism, both the energy and the electrons must be conserved, i.e. we must be able to trace the fate of both the energy and electrons. The energy is conserved in that, of the free energy available in the organic compound, some is retained as organic compounds of a different form, i.e. the organism mass and the rest is 'lost' as heat. The electrons are conserved in that, of those available in the organic compound, some are retained in the organic compounds of the organism mass formed and the rest are passed on to the terminal electron acceptor (Fig. 9.3). The proportion of the electrons (or energy) retained by the organisms defines the specific organism mass yield coefficient Y_h, i.e. mass of organism mass formed per mass of organic compound

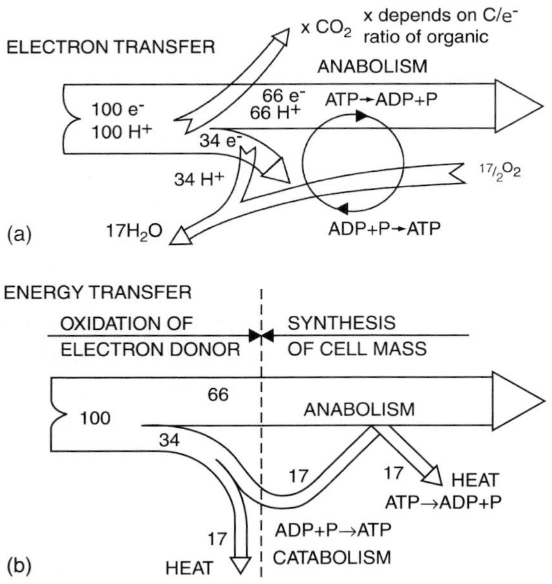

Fig. 9.3 Electron and energy transfer in anabolism and catabolism, showing electron and energy capture in organism mass synthesized (anabolism) and electron 'loss' by transfer to the electron acceptor (a, top) and energy 'loss' as heat to drive the anabolic processes (catabolism) (b, bottom).

oxidized, all in terms of electron equivalents (equation 10):

$$Y_h = \frac{e^{-}\text{(or energy) conserved in organism mass}}{e^{-}\text{(or energy) in substrate utilized}} \tag{10}$$

Loewenthal (1986) developed a procedure from bioenergetic and thermodynamic principles with which it can be shown that the specific yield coefficient (Y_h) for heterotrophic respiration reactions involving oxygen as the terminal electron acceptor (i.e. in which 3 ATPs are formed per pair of electrons transferred) is 0.42 g cells/gTh.COD. This procedure is briefly outlined below in which the anabolic and catabolic pathways are considered separately and then combined to determine the yield.

3.3.1 *Anabolic pathways*

The organic components of organism mass are highly complex and very numerous, but principally are proteins, fats (lipids) and carbohydrates. The formation of these compounds (anabolism) follows a wide variety of biochemical pathways too complex to trace. Consequently, the procedures to estimate heterotrophic yield are developed from global considerations only.

The first requirement for a global approach is a stoichiometric formula representative of organism cell mass. A number of these have been proposed (Table 9.1) each comprising different quantities of C, H, O and N. Of these formulae, $C_5H_7O_2N$ seems the most widely accepted and consequently will be adopted for this discussion. An important requirement of the stoichiometric formula for organisms is its free energy of formation (ΔG_f^0). This value is not reported in the literature, but has been deduced by Loewenthal (1986) to be +40 kcal/mole protoplasm as defined by $C_5H_7O_2N$. With this value it is possible to calculate from thermodynamic principles the fraction of energy or EDC transformed to cell mass and the fraction lost as heat. For such calculations the ΔG_f^0 protoplasm can be accepted to remain constant for different electron donors and acceptors provided the redox reactions are those of respiration.

Recognizing that the anabolic synthesis reactions follow a myriad of pathways too complex to model, the reactions are simplified by assuming they take place in two steps. The first of these is the formation of protoplasm ($C_5H_7O_2N$) from CO_2, NH_3 and electrons (and protons). The second step is the oxidation of substrate to supply the CO_2 and electrons

TABLE 9.1 Theoretical COD values for various empirical formulations for microbial sludge (after McCarty, 1965)

Bacterial formula	*Mol. mass*	*Th. COD (g)*			*Ref*
		Per mol	*Per g vss*	*Per g C*	
$C_5H_7O_2N$	113	160	1.42	2.67	1
$C_5H_9O_3N$	131	160	1.22	2.67	2
$C_7H_{10}O_5N$	156	232	1.48	2.76	3
$C_5H_8O_2N$	114	168	1.47	2.80	4

1. Hoover and Porges (1952) 2. Speece and McCarty (1964) 3. Sawyer (1956) 4. Symons and McKinney (1958).

(and protons) for the first step. In terms of these two steps, the function of the anabolism reactions is to transform electrons and protons in the substrate to electrons and protons in the organism mass.

Formation of protoplasm from CO_2, NH_3 and electrons. To form 1 mole of protoplasm (which has a mass of 113 g – Table 9.1) from CO_2 and NH_3 requires 20 electrons (and proton) (equation 11):

$$5CO_2 + NH_3 + 20H^+ + 20e^- \rightarrow C_5H_7O_2N + 8H_2O \qquad (11)$$

As noted earlier, organisms transfer electrons (and protons) via the $NADH_2/NAD$ molecules. Coupling the $NADH_2 \rightarrow NAD$ half reaction into equation 11 yields equation 12:

$$5CO_2 + NH_3 + 10NADH_2 \rightarrow C_5H_7O_2N + 8H_2O + 10NAD \qquad (12)$$

Equation 12 is *not* intended to show that the organisms are chemical autotrophs obtaining their cell material from carbon dioxide and ammonia. The organisms considered are heterotrophs and so the cell material is obtained from the organic compound. It is only for the purposes of establishing a datum for the free energy calculations that the organic compound is assumed to be broken down to CO_2 and H_2O and the protoplasm built up from these basic products.

Oxidation of substrate to supply electrons and protons. To supply the electrons and protons necessary for the first step above, the organism must oxidize substrate (note that the balance is made on supply of electrons, not carbon since the ratio of moles of CO_2 produced per mole of substrate oxidized varies between substrates). From equation 11 for protoplasm formation, the organism needs to obtain 20 e^-eq (10 moles $NADH_2$). If the organic substrate is glucose or acetic acid, which yield 24 or 8 electrons per mole of substrate respectively (equations 2c and d), then $20/24 = 0.83$ moles of glucose or $20/8 = 2.50$ moles of acetate are required for anabolism. The anabolism reaction (equation 12) is an uphill one and energy must be supplied to drive it. The anabolic energy requirement is calculated from the free energy of reaction for the formation of protoplasm (equation 12). Subtracting the standard free energies of formation (ΔG_f^0) of the reactants (left hand side of equation 12) from those of the products (right hand side of equation 12), yields the standard free energy of reaction (ΔG_f^0) for the formation of protoplasm. In other words, + 465 kJ/mol (+110.6 kcal/mole) protoplasm with the +ve sign indicating that it is an energy requiring reaction. The energy transfers by the organism are not 100% efficient and it can be assumed that the efficiency of the energy transfer in the anabolic reaction is the same as calculated earlier for ATP formation at only about 55%. Hence $465/0.55 = 844$ kJ/mol (201 kcal/mole) protoplasm are required. This energy is supplied by ATP generated in the catabolic pathways.

3.3.2 *Catabolic pathways*

The catabolic pathways supply the energy (ATP) requirements for the anabolic reactions. As mentioned earlier each ATP has an available free energy of 42 kJ/mol (10 kcal/mole) ATP. Thus to supply the energy requirements for anabolism (944 kJ/mol protoplasm), $844/42 = 20.1$ mol ATP must be generated in catabolism. Remembering that ideally 3 ATP are formed per pair of electrons transferred to the terminal electron acceptor oxygen, the catabolic electron equivalent requirement is $20.1/3 = 13.4$ e^-eq. To supply these electron equivalents, organic molecules are oxidized by the organism. If the organic is acetic acid or glucose then $13.4/8 = 1.67$ or $134/24 = 0.56$ moles of organic substrate/mol protoplasm respectively are required.

3.3.3 *Yield*

The combined anabolic and catabolic electron equivalent (e^-eq) requirements to synthesize 1 mole of protoplasm are therefore $20 + 13.4 = 33.4$ e^-eq. At 8 and 24 e^-eq/mole for acetate and glucose respectively, this gives $2.5 + 1.67 = 4.17$ moles acetic acid and $0.83 + 0.56 = 1.39$ moles glucose respectively. Now 1 mole of protoplasm is 113 g mass and because its constituents $C_5H_7O_2N$ are all organic, its volatile solids mass, VSS, is also 113 g (see Table 9.1). Therefore the theoretical specific organism yield is 113/4.17 = 27.1 gVSS/mole acetic acid or 113/1.67 = 67.7 gVSS/mole glucose. Because the organisms respond to the free energy or equivalently the e^-eq, it is better to specify the yield coefficient in terms of e^-eq, i.e. 113/33.4 = 3.38 gVSS/e^-eq. To express e^-eq in terms of oxygen, 1 e^-eq is equal to 8 gO (equation 9), so that the theoretical specific organism yield in terms of the Th.COD is 3.38/8 = 0.42 gVSS/gTh.COD. Using the COD test (Standard Methods, 1985) as an estimate of the Th.COD, Marais and Ekama (1976) measured the yield coefficient for activated sludge grown on municipal wastewater as 0.43 mgVSS/mg COD, very close to the theoretically estimated value. In activated sludge models, a Y_h value for heterotrophs of 0.45 mgVSS/mgCOD is used.

From Table 9.1, it can be seen that the T.hCOD of the $C_5H_7O_2N$ protoplasm formed is 1.42 gO/gVSS. This was obtained by dividing the Th.COD of the protoplasm in its oxidation to CO_2 and water (i.e. 20 e^-eq or 5 moles of O_2 which is equal to 5.32 = 160 gO) by the mass of protoplasm i.e. 160/113 = 1.42 mgTh.COD/mgVSS. With a yield coefficient of 0.42 mgVSS/mgTh.COD, the Th.COD of the protoplasm formed is 0.42 × 1.42 = 0.60 mgTh.COD/mgTh.COD. This means that of the original organic material Th.COD, 60% is transformed to new organism mass. The balance of 40% is electrons passed on to the terminal electron acceptor oxygen. The same proportion (60/40) is obtained by considering the electron equivalents directly, i.e. of the 33.4 e^-eq required to form 1 mole (113 g) of protoplasm, 20 e^-eq (60%) were conserved in protoplasm and 13.4 e^-eq (40%) were passed on to oxygen. The e^-eq passed on to oxygen manifests as oxygen utilization by the organism.

Experimental support for Loewenthal's (1986) approach to estimate theoretically the yield coefficient, Y_h from bioenergetics and thermodynamic data is provided by Bauchop and Elsden (1960). They found that organism yield is linearly related to the number of ATP moles produced in the complete oxidation of organics, i.e. $Y_h \propto K_1 N_{ATP}$, where N_{ATP} is the number of ATP moles generated per mole of organic and Y_h the gVSS generated per mole of organic. The constant K_1 was found to be 2.12 gVSS/mole ATP. Furthermore, for the known aerobic substrate pathways, it can be shown that the number of ATP moles produced per kJ free energy released is constant at about 0.013 moles ATP/kJ, i.e. $N_{ATP} = K_2 \Delta G_f^0 = 0.013\, \Delta G_f^0$. For example, in the breakdown of glucose, 38 moles of ATP are generated and − 2894 kcal of free energy released making the ratio K_2 = 38/2894 = 0.013 mole ATP/kJ. Therefore, through K_1 and K_2 the yield coefficient for heterotrophs, Y_h, is proportional to the free energy released by the organic compound, i.e. $Y_h = K_1 K_2 \Delta G_f^0$. Defining Y_h and ΔG_f^0 in terms of the Th.COD, Y_h for glucose is 0.013 × 2.12 × (2894/192) = 0.41 gVSS/gTh.COD, which is very close to Loewenthal's theoretical value of 0.42 gVSS/gTh.COD. Experimental data from the aerobic metabolism of a wide variety of organic substrates presented by Eckenfelder (1965) give an average coefficient of 0.39 gVSS/gCOD, which is also close to the theoretically estimated value (Fig. 9.4).

3.3.4 *Nitrate as terminal electron acceptor*

When nitrate serves as electron acceptor, approximately the same quantity of free energy

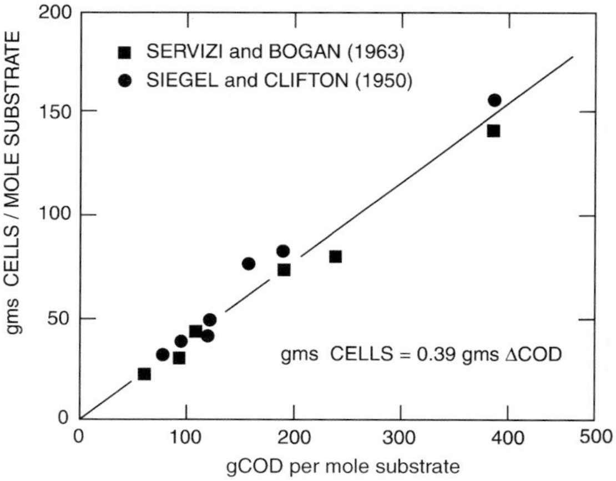

Fig. 9.4 The direct relationship between COD and Yield (Y_h) for a wide variety of substrates (Eckenfelder, 1965).

TABLE 9.2 Comparison of measured free energy transferred by different organics with oxygen and nitrate as terminal electron acceptor (McCarty, 1972)

Energy source electron donor	*Energy transfer (kcal/e⁻)*		
	Electron acceptor		*Ratio*
	Oxygen	*Nitrate*	
1. Methanol	27.6	26.1	0.94
2. Ethanol	26.3	24.7	0.94
3. Domestic WWV	26.3	24.7	0.94

is available (94%) to the organisms as when oxygen serves as electron acceptor (Table 9.2). However, the number of ATP moles formed per electron transferred is reduced. In the calculations above with oxygen as terminal electron acceptor, 3 ATP are formed per pair of electrons transferred. If nitrate is used instead of oxygen, only 2 ATP are formed per pair of electrons transferred to the nitrate (Payne, 1981). Accordingly with nitrate, the catabolic electron equivalent requirement is $20.1 \times 2/2 = 20.1$ e^-eq. To supply these electron equivalents, organic molecules are oxidized by the organism. If the organic is acetic acid or glucose then $20.1/8 = 2.51$ or $20.1/24 = 0.84$ moles of acetic acid or glucose respectively are required.

The combined anabolic and catabolic e^-eq requirements to synthesize 1 mole of protoplasm with nitrate as electron acceptor is therefore $20 + 20.1 = 40.1$ e^-eq, which for 8 and 24 e^-eq for acetate and glucose respectively, gives $2.5 + 2.51 = 5.01$ moles acetic acid and $0.83 + 0.84 = 1.67$ moles glucose respectively. Converting the units as described above, this gives the theoretical specific organism yield in terms of the Th.COD as 0.35 gVSS/gTh.COD. Thus the yield with nitrate as terminal electron acceptor (0.35 gVSS/gTh.COD) is lower than that with oxygen (0.42 gVSS/gTh.COD). This theoretical reduction in organism yield has not been incorporated in steady state activated sludge models and design procedures nor is it recognized in the earlier general kinetic activated sludge simulation models. These include UCTOLD (Dold *et al.*, 1991), UCTPHO (Wentzel *et al.*, 1992) and ASM Nos 1 and 2 of IAWQ (Henze *et al.*, 1987, 1994). In these models, it is accepted that the bioenergetics of aerobic and anoxic respiration are the same in every respect, the only difference is that the electrons are passed to oxygen for aerobic conditions and nitrate for anoxic conditions. More recent simulation models like ASM2d and 3 include a reduction in yield coefficient under anoxic conditions but the differences are relatively small.

3.4 Conclusions from bioenergetics

Two important principles flow from the above discussion of bioenergetics.

3.4.1 The biological oxygen utilization is directly linked to substrate utilization for synthesis

The oxygen equivalent (Th.COD) of the organic compound is not the same as the oxygen demand for synthesis of protoplasm. The former is a measure of the free energy released in the complete oxidation of all the organic compound to CO_2 and H_2O. This is equivalent to the heat of combustion where all the free energy released is lost as heat. The oxygen utilization for synthesis forms part of the heterotrophic biological oxygen utilization (the other part is endogenous respiration required for cell maintenance) and represents only a part (40%) of the original Th.COD of the organic compounds. The balance of the Th.COD is transformed into the new cell mass synthesized. Thus, in a hypothetical 1 m^3 bioreactor wherein only synthesis of heterotrophic biomass is taking place on glucose, in the degradation of 1 mole of glucose (180 g), which has a Th.COD of 192 gO, $0.42 \times 192 = 80.6$ gVSS organism mass ($C_5H_7O_2N$) is formed, comprising 80.6×1.42 gTh.COD/gVSS $= 114.5$ gTh.COD. The biological oxygen utilization (OU) is the difference between the Th.COD of the glucose and the organism VSS mass formed, i.e. OU $= 192 - 114.5 = 77.5$ gO, which is 40% of the original Th.COD of the glucose.

Thus, in general heterotrophic organism synthesis, if the Th.COD concentration of the organic compound is denoted S and it is completely degraded, i.e. $\Delta S = S - 0$, then the concentration of organic material degraded, the concentration of organisms formed and the biological OU are directly linked and given by equations 13a, 13b and 13c.

$$X_a = \cdot Y_h \cdot \Delta S \qquad (13a)$$

where:

X_a = concentration of organisms formed ($gVSS/m^3$)
Y_h = yield coefficient (gVSS/gTh.COD)
ΔS = substrate utilized ($gTh.COD/m^3$)

$$Z_a = fcv \times Y_h \Delta S \qquad (13b)$$

where:

Z_a = Th.COD of X_a formed ($gCOD/m^3$)
fcv = COD/VSS ratio of organisms = 1.42 organism VSS

$$\begin{aligned} OU &= \Delta S - Z_a = \Delta S - fcv\, Y_h \Delta S \\ &= (1 - fcv\, Y_h)\Delta S \qquad (13c) \end{aligned}$$

where:

OU = Biological oxygen utilized for synthesis of the organism concentration (gO/m^3)

3.4.2 Energy balances can be made over biological treatment systems in terms of electron equivalents

By measuring the OU, which represents the electrons passing to oxygen in organism mass synthesis, an energy balance over the biological system is possible in terms of e^-eq or O. If the organic compounds are measured as Th.COD, then for electrons to be conserved, the mass of Th.COD entering the reactor must be equal to the mass of Th.COD leaving the reactor. The former is the mass of Th.COD entering the reactor via the influent flow and the latter is given by the sum of the Th.COD of the undegraded organics in the effluent flow, the OU, and the Th.COD of the organisms removed from the system, all over a fixed period of time. The ability to build e^-eq or Th.COD mass balances is a powerful feature of the Th.COD approach and forms the basis not only of the steady-state activated sludge model (Marais and Ekama, 1976; WRC, 1984), but also of the activated simulation models such as UCTOLD (Dold *et al.*,1991), UCTPHO (Wentzel *et al.*, 1992) and IWA ASMs Nos 1 and 2 (Henze *et al.*, 1987, 1994).

To build from the above two principles in activated sludge kinetic modelling requires a means whereby the EDC or theoretical COD of the organics in wastewater can be measured in a practical way. The issues associated with this and energy measurement in general are discussed in the next section.

4 ENERGY MEASUREMENT

If the organic compounds are defined in terms of both type and concentration, then there is no difficulty in determining the free energy release, EDC or Th.COD. These can be calculated directly from thermodynamic and chemical data as illustrated for glucose and acetate (see Section 3.4). The EDC or Th.COD can even be calculated for a mixture of organics, if each is defined in type and concentration. However, when the mixture of organics is not defined in type and concentration, as is the usual case in wastewater treatment, then the free energy release, EDC or Th.COD cannot be calculated from thermodynamic and chemical data. It is here that the significance of the above discussion comes to the fore because (1) the free energy release in the oxidation of most organics is related to the EDC or Th.COD of the organics, and (2) organism metabolic behaviour is deeply rooted in electron transfer redox reactions. If a method of experimentally determining the EDC or Th.COD of a mixture of organics could be established, then a method of tracing the free energy changes in biological treatment systems would be available.

Currently there are four tests in wide use to estimate the energy content of the organics in wastewaters. Three of these are oxygen based tests and one is a carbon based test, namely: biochemical oxygen demand (BOD), chemical oxygen demand (COD), total oxygen demand (TOD) and total organic carbon (TOC). Each is briefly reviewed to evaluate their advantages and disadvantages for measuring the energy content of organics in wastewaters.

4.1 Biochemical oxygen demand (BOD) test

This test provides a measure of the oxygen consumed in the biological oxidation of organic (carbonaceous) material in a sample at 20°C over a prescribed period of time, usually 5 days. According to Gaudy (1972), the origins of this test can be traced back about a hundred years to Frankland who appears to have been among the first to recognize that the observed oxygen depletion in a stored sample is due to the activity of microorganisms. The Royal Commission on Sewage Disposal appeared to have developed the concept of using the oxygen depletion as a measure of the strength of pollution. Over the past century extensive research towards elucidating the meaning of the test, quantifying it and integrating it in theoretical descriptions of wastewater treatment processes, has been undertaken. To date the biochemical oxygen demand (BOD) test is still the most popular parameter for assessing the organic material strength of influents and effluents and for describing the behaviour of treatment systems. Its continued popularity seems to stem mainly from the body of experience that has built up in its use. While there is some justification for retaining it for setting effluent quality standards for regulatory purposes, in so far as present day modelling, design and operation of biological treatment processes is concerned, few cogent arguments for its continued use over that of the COD test can be advanced.

The BOD test is an empirical one performed under strictly specified conditions and procedures and any deviation may give rise to very uncertain results. In the 5-day BOD test (BOD_5), which is the one commonly employed, the sample of wastewater is suitably diluted with well-oxygenated water and an inoculum of microorganisms, adapted to the wastewater if necessary, is introduced. The initial oxygen concentration is measured and the sample stored in darkness at 20°C for 5 days. Over the 5-day period, the oxygen concentration is measured (which is done automatically with modern BOD apparatus) at regular intervals and at the end of the 5-day period. The difference in oxygen concentration between the beginning and the end of the 5-day test period, taking due account of dilution, gives the BOD_5. For the ultimate BOD test (BOD_u) the sample is stored for about 20 days at 20°C to provide sufficient time for the complete utilization of the biodegradable organic energy (Standard Methods, 1985). Of the two, the BOD_u test probably gives the most stable relative measure of the biodegradable organic energy, but the time factor of 20 days makes the test impractical for routine use. If the BOD_u/BOD_5 ratio remains constant for different wastewaters and different stages of primary (sedimentation) and secondary (biological) treatment, then the BOD_5 test would serve as a satisfactory relative measure. However, this is different between wastewaters and also at different stages of treatment for the same wastewater.

The BOD tests are also very sensitive to test procedure and inoculum preparation. Low BOD values are obtained if the inoculum is not suitably adapted to the wastewater. Some wastewater constituents like heavy metals and toxic organics inhibit organism activity. Also nitrifying organisms may develop in the test sample unless specifically suppressed by chemical additives (usually allyl thiourea, ATU). In particular, nitrification may occur in effluent samples from nitrifying biological treatment plants, or in influent samples where some treatment plant operational practice introduces nitrifying secondary sludge to the influent stream, or in tests where the inoculum is obtained from a nitrifying system. Nitrifying organisms in the wastewater or effluent sample convert ammonium (NH_4^+) to nitrite (NO_2^-) and nitrate (NO_3^-) and utilize oxygen in the process (about 4.6 $mgO_2/mgNH_4 - N$ nitrified to nitrate, $NO_3 - N$) to give an inflated value for the organic 'energy' content. Test procedure, inoculum adaptation, inhibition and nitrification always need to be carefully considered in evaluating BOD test results.

Despite the deficiencies listed above, if the BOD_5 value constituted a consistent estimate it could still have provided a relative measure of practical usefulness, but even this cannot be guaranteed. Gaudy (1972), in a thought provoking review on the BOD, gave a summary of

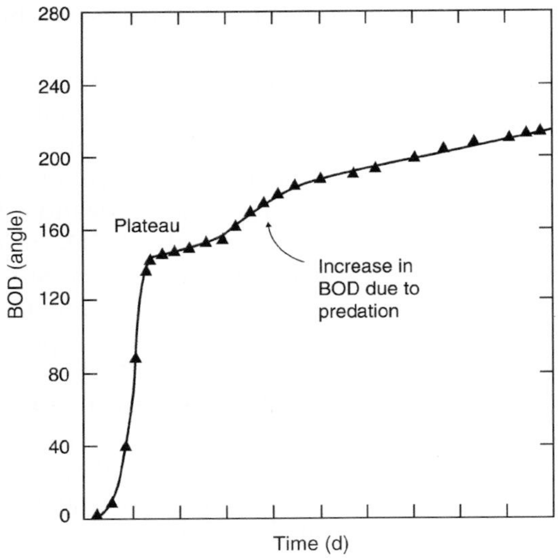

Fig. 9.5 Typical experimental BOD time curve on wastewater with heterogeneous seed clearly showing 'plateau' behaviour after about 1.5 days, followed by an increase in oxygen utilized, considered to be due to predation (Copcutt, 1993).

the work on the BOD test (in which he took a leading part). He concluded that the standard approach to formulating the BOD time curve as a first order reaction is quite invalid. The actual curve arises from two main effects: (1) bacterial synthesis from the biodegradable input – a reaction usually complete in 1–2 days, and (2) predator growth using the synthesized bacteria as substrate. The second phase, being dependent on the first, usually lags the first so that a distinct 'pause' or 'plateau' in the time curve is exhibited (Fig. 9.5). The occurrence and duration of this pause is dependent on the bacteria–predator relationship initially present in the sample, the 'strength' of the organic material, presence of inhibitory substances and so on (Hoover *et al.*, 1953). These compound effects certainly are not reflected by a first order formulation. Furthermore, should nitrification occur during the 5 days, a second plateau in the time curve might arise. This may not be apparent unless the test results are compared with a test in which procedures to suppress nitrification are imposed. Evidently the BOD_5 value is subject to a number of influences, which can be present in varying degrees, resulting in unknown effects on the observed value, which indicates the importance of measuring the complete BOD-time curve in the test to enable a meaningful interpretation. For routine use this is onerous and time consuming.

The main deficiency of the BOD tests is that they do not measure the electron donating capacity (EDC) of the organics in the wastewater. From the discussion of biological behaviour above, the oxygen utilized is only the catabolic energy requirement of the organisms; the greater part (~60%) of the energy or electrons are transformed to new organic compounds in the form of organism mass, the anabolic requirement of the organisms. Even though these compounds become organics for other organisms to utilize in repeated predator–prey cycles with a catabolic oxygen requirement in each cycle, some of the organism mass formed is not degraded within the time period of the test and also some of the organism mass is unbiodegradable and accumulates as unbiodegradable organism (endogenous) (Fig. 9.6). Therefore in the BOD_5 test organics not only is the utilization of the wastewater organics incomplete to some unknown degree, but also some of the electrons (energy) of the wastewater organics become bound up with unbiodegradable organism residue. Even in the BOD_u where the utilization of the wastewater organics may be complete, an unknown fraction of the biodegradable energy originally present in the wastewater does not manifest as oxygen utilization and remains as unbiodegradable residue. Consequently, it is not possible to perform an electron (or energy) balance on the biodegradable organics of the wastewater with the BOD tests. This is a severe deficiency and places the BOD outside the major developments that have taken place in biological treatment process modelling, design and simulation, all of which are based on electron (COD) mass balance principles (Marais and Ekama, 1976; Dold *et al.*, 1980; Van Haandel *et al.*, 1981; WRC, 1984; Henze *et al.*, 1987, 1994; Dold *et al.*, 1991; Wentzel *et al.*, 1992).

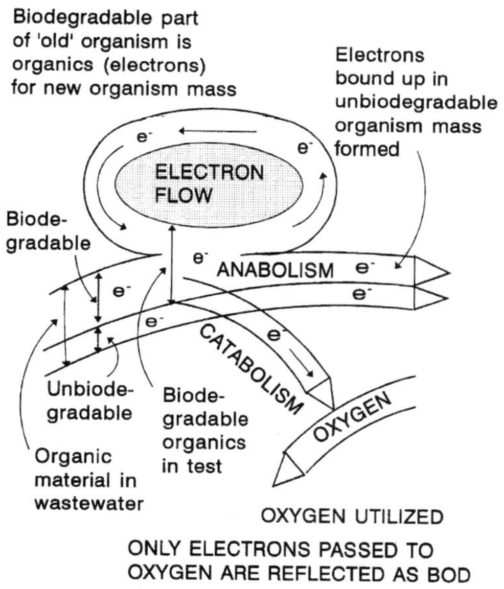

Fig. 9.6 Electron flow in the BOD test showing (1) oxygen utilized (BOD) as a consequence of organism catabolic action only, (2) organic material recycling through predation, and (3) electrons bound in unbiodegradable material from the influent or formed biologically in the test do not manifest as oxygen utilized (BOD).

Many organizations, in particular the regulatory agencies, regard the fact that the (nitrification inhibited) BOD tests measure only the strength of the biodegradable organics in the sample an advantage. For the wastewater treatment plant effluent that may be so, but for the influent this is a considerable shortcoming. Most wastewaters include biodegradable and unbiodegradable organics of a soluble and particulate nature. The soluble unbiodegradable organics (in terms of the time spent in the wastewater treatment system) escape with the effluent, but the particulate fraction becomes enmeshed in the biomass in the biological reactor and manifests as a part of the volatile solids concentration of the biomass. The BOD tests therefore do not incorporate a procedure related to the test whereby an assessment of this accumulation process can be obtained. In BOD_5-based design procedures, this is taken account of by measuring the suspended solids concentrations with 0.45 μm membrane filters and assuming a certain fraction of these solids are not biodegradable. In the COD test (discussed below), because the total electron donating capacity of the organics is measured, whether it is biodegradable or unbiodegradable, it is possible to estimate the soluble and particulate unbiodegradable organic fractions from the COD test results and the kinetic behaviour (sludge production and oxygen utilization) of the activated sludge system.

4.2 Chemical oxygen demand (COD) test

This test gives the electron donating capacity (in terms of oxygen, see equation 7) of practically all the organic compounds in the sample, biodegradable or unbiodegradable and soluble or particulate. A strong oxidizing agent, a hot mixture of dichromate and concentrated sulphuric acid making chromic acid, oxidizes the organic compounds to CO_2, the same end product as biological activity. In the test, a known mass of potassium dichromate (volume times concentration) and measured volumes of wastewater and sulphuric acid are added to the test and refluxed (boiled) for 2 hours. After the sample has cooled, it is titrated with ferrous ammonium sulphate (FAS) and ferrous indicator (Ferroin). The dichromate is the electron acceptor and the organic compounds and FAS the electron donors. During refluxing, the organic compounds are oxidized to CO_2, and their electrons pass to the dichromate, which is reduced. After all the organic compounds are oxidized, some dichromate electron acceptor remains and it is this remaining amount that is measured in the titration with the FAS. The ferrous (Fe^{2+}) in FAS donates electrons to the remaining dichromate and is oxidized to ferric (Fe^{3+}) while the dichromate is reduced. When the dichromate has all been reduced, it can accept no more electrons and the ferrous of the FAS remains as Fe^{2+}. At this point the Ferroin indicator detects the presence of the Fe^{2+} ions and changes colour (yellow to red), which defines the end point of the titration. The difference between the initial electron accepting capacity of the dichromate (measured by titrating a blank in which distilled water instead of sample is refluxed) and number of electrons donated by the FAS titration, gives the number of electrons donated in the oxidation of the organic compounds. This electron donating capacity expressed as O_2 is the COD. The COD test is thus rooted in redox chemistry and therefore is directly related to the Th.COD discussed earlier. But how close is the COD test result to the Th.COD?

An important and time saving feature of the COD test is that ammonia is not oxidized. For most municipal wastewaters the COD result is therefore virtually exclusively associated with the organic material. Nitrite is oxidized to nitrate in the test, a reaction which exerts a COD. However, this COD can be suppressed in the test by adding sulphamic acid to the dichromate, but usually nitrite concentrations are so low in comparison with the organic contents and the oxygen requirement so low (1.2 mgO per $mgNO_2^- - N$ oxidized to $mgNO_3^- - N$) that this influence is negligible. With industrial wastewaters, which have significant concentrations of reduced inorganic compounds such as sulphides, the electrons donated in the oxidation of these inorganics can contribute significantly to the COD result, and therefore caution should be exercised with such wastewaters.

In practical application, some organic material is not oxidized by the COD test under any circumstances but microorganisms do utilize it. This organic material includes aromatic hydrocarbons (e.g. benzene, phenols) and pyridines (water soluble benzene rings with an N in place of one of the carbon atoms). They are mainly encountered in petrochemical and plastic industry wastewaters and are not likely to be encountered in municipal wastewater in significant concentration. A poor estimate for COD can also be obtained with lower fatty acids (the C_2 to C_5 acids) such as acetate. These compounds are in the unionized form at the low pH at which the test is performed – the unionized molecules can be extremely difficult to oxidize. The addition of the catalyst silver sulphate overcomes this problem to a large degree to give virtually 100% oxidation of unionized short-chain fatty acids. However, silver sulphate reacts with the halides (chloride, bromide and iodide) to produce precipitates which do not act as catalysts. This difficulty can be largely overcome by adding mercuric sulphate to complex the halides, to an upper limit of 2000 mg Cl/l.

In performing a COD test it should be remembered that the test involves a time dependent reaction and the specified refluxing period of 2 hours must be adhered to to ensure complete or near complete oxidation. Furthermore, the degree of oxidation is subject to mass action effects – equivalent masses of the same organic compound refluxed at different final test volumes yield different results for the COD. The temperature at which the refluxing step takes place also affects the oxidation rate. The refluxing temperature in turn is dependent on the concentration of sulphuric acid in the test. These factors have all been considered in the development of the COD test procedure. Hence, the COD test will give reliable results only if the test is done in strict accordance with the set procedures, e.g. those in Standard Methods (1985) or later editions.

If the specified procedures for the COD test are strictly followed, experience has shown that it accurately reflects the electron donating capacity (EDC) of a mixture of organic compounds. Over the past 15 years application of COD test results to organic material balances on fully aerobic and anoxic/aerobic activated sludge plants indicate that, with careful experimental work, electron balances (in terms of oxygen) between 95 and 102% are obtainable. In such balances, taking due account of nitrification in the oxygen utilization, the daily influent COD mass can be accounted for by the sum of the daily COD masses in the waste sludge and effluent, the carbonaceous oxygen demand and, if anoxic conditions exist, the COD utilized in the denitrification of nitrate to nitrogen gas. Experience has indicated that where poor COD mass balances are obtained, invariably the causes can be traced to errors in operation of the experimental activated sludge system. When the COD test is conducted in accordance with Standard Methods (1985), very seldom has the error been in the COD test as a measure of the EDC.

4.3 Total oxygen demand (TOD) test

In the TOD test, the sample is oxidized at high temperature in a combustion oven housed in a sophisticated electronic instrument called a TOD meter. In the TOD meter, a very small (10–20 μl) known volume of sample is injected by means of a Hamilton syringe into a gas stream consisting of nitrogen to which oxygen is dosed at a constant rate. The nitrogen-oxygen-vaporized liquid sample mixture is passed through a high temperature (900°C) combustion column with a platinum catalyst. The oxidizable matter is oxidized, consuming oxygen from the gas stream. The reduction in the oxygen content of the gas is detected electronically and is measured, also electronically, as the mass of oxygen required to restore the oxygen content of the gas mixture to its original value. In older instruments the test result is obtained in about 5 minutes but in more modem instruments, automatic injection and faster sample throughput is possible.

The TOD is the mass of oxygen required for the oxidation of practically all oxidizable material contained in a unit volume of wastewater sample. Both carbonaceous and nitrogenous compounds are oxidized. A problem with the test is that the amount of oxygen available in the combustion oven influences the degree of oxidation of ammonia and organic nitrogen. When oxygen is in excess there is a conversion of nitrogenous material to nitric oxide, whereas when oxygen is not in excess, oxidation of the nitrogenous material may be only partial. Furthermore, nitrate, nitrite and dissolved oxygen in the sample influence the TOD test result. Because the test determines both the carbonaceous and nitrogenous oxidation potential, an additional nitrogen test is required to isolate the carbonaceous fraction of the TOD. With the large proportion of particulate matter in municipal wastewater, difficulty is found in obtaining representative and reproducible readings; samples must be thoroughly homogenized to obtain representative reading because the sample volume is extremely small (10–20 μl). However, sample homogenization increases the dissolved oxygen contamination. The difficulties of applying the TOD test to wastewater can be overcome, but the resulting test procedure and the expensive sophisticated instrument required reduces its attraction for general routine and field use.

4.4 Total organic carbon (TOC) test

The TOC test gives a specific and absolute measure of the carbon in the sample. It operates on the same basic principle as the TOD meter except that carbon dioxide, the product of combustion (oxidation), is detected. In the TOC analyser, a small volume (10–20 μl) is injected with a syringe into a catalysed combustion furnace at 900 °C. The organic compounds are oxidized to carbon dioxide and water and the total carbon dioxide concentration is measured. The dissolved carbon dioxide (inorganic) in the sample is therefore included in the result. For high organic strength wastewaters this normally is not a problem because of the low dissolved CO_2 concentrations. If the organic carbon content only is required, then it is necessary to have a pretreatment stage in sample preparation to remove inorganic carbon, by acidification and CO_2 purging by sparging, or, to do an additional test in which the inorganic carbon only is measured. The older instruments can only measure the carbon content of soluble organic material but newer models can deal with particulate organic material.

As a procedure to estimate the electron donating capacity (or free energy release) the TOC test is not useful. This is because the organic carbon content of an organic compound is not proportional to the number of electrons it can donate. This can be demonstrated by comparing glucose and glycerol. Glucose ($C_6H_{12}O_6$) with a carbon content of $6 \times 12 = 72$ gC/mol can donate $24e^-$/mol which is equivalent to a COD of $24/4 \times 32 =$ 192 gCOD/mol, which gives a COD/TOC ratio of $192/72 = 2.67$ gCOD/gTOC. Glycerol ($C_3H_8O_3$) with a carbon content of $3 \times 12 =$ 36 gC/mol, can donate 14 e^-/mol which is equivalent to a COD of $14/4 \times 32 =$

112 gCOD/mol and yields a COD/TOC ratio of 112/36 = 3.11 gCOD/gTOC which is 17% higher than glucose. Therefore 17% more electrons (and therefore energy) are released per gram of C in the utilization of glycerol than in the breakdown of glucose. The organism yield coefficient in terms of TOC (i.e. gVSS synthesized per gTOC oxidized) can therefore be expected to be higher for glycerol than for glucose. However, in terms of COD, the yield coefficient can be expected to be the same for glucose as glycerol because the free energy release per electron transfer is the same. For the short-chain fatty acids $[CH_3(CH_2)_nCOOH]$ the COD/TOC ratio increases as the chain gets longer (n increases from 0 to 9) starting at 2.67 for acetic ($n = 0$) and increasing to 3.1 for propionic ($n = 1$), 3.33 for butyric ($n = 2$), 3.47 for valeric ($n = 3$) up to 3.73 for capric acid ($n = 8$).

Servizi and Bogan (1963) contend that the TOC has similar advantages as the COD with regard to its application to biological growth kinetics. However, according to McCarty (1963), this is only true if the COD/TOC ratio of the organic compounds oxidized is equal to the COD/TOC ratio of the organism mass formed; generally this is not the case, so that the TOC is not a good substitute parameter for COD.

The above discussion demonstrates that, fundamentally, it is inadvisable to use the TOC test to describe the energy based behaviour of biological wastewater systems. While carbon balances over activated sludge systems are possible, provided the CO_2 gas production by the system can be measured, these require more complex measuring and analysis techniques than oxygen (or electron) based mass balances. However, the test is inferior to the COD principally because TOC is not proportional to the free energy changes that take place in biological redox reactions. This does not mean that the test does not have other uses; it can serve a valuable function as a control parameter or as an analytical method in tertiary water treatment and water reclamation where the carbon content in the water is of major significance to assess water quality.

4.5 Advantages of the COD test

From the above discussion, it clear that the COD is the most appropriate test for assessing organic wastewater strength from a theoretical and practical point of view, in that:

- it measures the electron donating capacity (EDC) which is directly related to the free energy changes in the oxidation of organics
- it does not require sophisticated analytical equipment – it is a simple wet chemical titration procedure and, as such, is appropriate for practice in the field
- it gives a result in a short time
- it allows electron balances to be made over the biological system
- it does not oxidize ammonia so it gives only the electron donating capacity of the organic compounds
- it includes the unbiodegradable organics.

Many opponents of the COD test, particularly the proponents of the BOD test, regard the last point as a disadvantage, not an advantage. For surface water quality this may be true, but for wastewaters to be biologically treated it is important to have a measure that embraces both the biodegradable and unbiodegradable organics. This is because even though the unbiodegradable material is inert in the biological reactor, the particulate unbiodegradable material accumulates in the biological reactor and adds to the sludge production of the plant and the soluble unbiodegradable will appear in the effluent. One cannot of course *ab initio* determine what the unbiodegradable components of the COD are, but with simple kinetic models based on the COD and materials balances (Marais and Ekama, 1976) and the response of the activated sludge system to the particular wastewater, it is possible to define the unbiodegradable fractions of organics in the wastewater.

4.6 Relationships between wastewater strength tests

Over the past 10 years a large number of activated sludge kinetic theories and models have been developed. These are practically all based on the COD as the organic energy measure and the electron mass balance feature of the COD has allowed these models to be formulated into computer simulation programs. These programs are capable of simulating the activated sludge system with remarkable accuracy over a wide variety of configurations and applications, including biological N removal. These kinetic models have also led to the development of simpler steady-state design procedures which assist with the determination of the important system parameters required for the simulation models (e.g. WRC, 1984) is one of the steady-state design procedures based on COD for activated sludge in general and biological N and P removal in particular.

The BOD_5 test has been in use for more than a century and consequently remains deeply entrenched in the practice and experience of biological wastewater treatment. However, as a test it is clearly lacking in many respects compared to the COD test. These shortcomings have caused it to fall outside the major developments that have taken place over the past two decades in biological wastewater treatment design procedures and modelling. Indeed, for biological wastewater treatment (as distinct from surface water quality standards) there remain very few cogent reasons for continuing to use the BOD_5. Nevertheless, in order to make the theories and developments in activated sludge biological wastewater treatment based on COD accessible to those continuing to use the BOD, or TOC, some information is given below allowing conversion of wastewater strength to COD.

4.6.1 *BOD_5 and COD*

From data in the literature (Table 9.3) there is an approximate ratio between the COD and BOD, for influent municipal wastewater (equation 14):

$$COD \cong 2BOD_5 \tag{14}$$

The data in Table 9.3 are for settled wastewater which has had a significant proportion of the settlable solids removed. The settlable solids contain a significant proportion of unbiodegradable particulate organics so the COD/BOD_5 ratio can be expected to be lower for settled wastewater than for raw (unsettled) wastewater. With biological growth and endogenous respiration kinetics based on COD and the first order BOD curve, it is possible to devise theoretically a COD/BOD_5 ratio (Hoover *et al.*, 1953; Ekama and Marais, 1978). For settled wastewater (with a low unbiodegradable particulate COD fraction), the theoretical ratio is between 1.8 and 1.9. For raw wastewater (high unbiodegradable particulate COD fraction) the ratio is 2.0 to 2.1. These approximate ratios between COD and BOD_5 are very useful, but caution should be exercised in using them as they can vary between wastewaters depending on the wastewater characteristics, i.e. type and quantity of industrial wastewater contribution, permissibility of garbage grinding and other community practices, etc. The ratios given are strictly for influent municipal wastewater principally of domestic origin. The ratio COD/BOD_5 changes as the wastewater passes through the biological treatment plant, usually increasing. The ratio increases because, as the biodegradable organics are degraded, so the oxygen utilization for this reduces, while the unbiodegradable organics, which exert a COD but not a BOD_5, remain unchanged in their passage through the plant. The COD:BOD_5 ratio for biological wastewater treatment plant effluents ranges between 4 and 12 depending on the unbiodegradable soluble COD fraction, effluent suspended solids concentration and nitrification in the BOD test.

For the effluent of a biological wastewater treatment plant, the BOD_5 gives a relative but

TABLE 9.3 COD:BOD_5 ratio of settled sewage

Reference	*Mean*	*Range*	*No. of data*
Smith and Eilers (1969)	2.06	1.90–2.27	12
Boon and Burgess (1972)	1.88	1.73–2.23	17
Dixon *et al.* (1972)	2.04*	1.95–2.16	

* Average of four years operation, 1967–1970.

uncertain measure of the energy still available for rapid (5 day) degradation in the receiving stream. Unless the BOD_u/BOD_5 ratio for the effluent is known, the 5-day test value is no more than a rough estimate of the pollution potential (i.e. oxygen demand) of the effluent discharge. The BOD gives no estimate of the 'unbiodegradable' (in terms of the time under treatment in the plant) organics. The COD test includes the unbiodegradable organics but the biodegradable and unbiodegradable cannot be determined separately in the test. However, with the aid of kinetic models of the biological treatment process, it is possible to obtain estimates of the unbiodegradable and biodegradable effluent COD fractions (Marais and Ekama, 1976; WRC, 1984). For long sludge age activated sludge systems (>5 days) treating normal municipal wastewater, the effluent biodegradable COD fraction is very small, less than 2% of the influent COD, and usually can be ignored.

4.6.2 *TOC and COD*

The TOC has found very little application in biological wastewater treatment, although there are some instances of its use in research. A ratio TOC to COD is therefore unlikely to be required. In the discussion of energy measurement above the ratio electron donating capacity (EDC) per carbon was mentioned. This is the same as the COD/TOC ratio and therefore for the various organics like glucose, short-chain fatty acids, glycerol and others, ratios between 2.7 and 3.3 were derived. For influent municipal wastewater, which comprises mostly relatively simple organics, a ratio of 3.0 is usually employed. For municipal wastewater effluents, the ratio can change substantially because the unbiodegradable organics, which generally are more complex in structure, form the major constituents.

4.6.3 *COD/VSS ratio*

The great advantage of the COD, which became apparent in the discussion of bioenergetics above, is that it gives a direct measure of the electrons (or energy) built into the organism mass formed. On the basis of the stoichiometric formula assumed for organism mass (C_5H7O_2N) its COD was calculated above to be 1.42 gCOD/g organism mass (see Table 9.1). For a pure culture system growing on a completely biodegradable soluble substrate in a high dilution rate chemostat (i.e. short sludge age), the volatile suspended solids (VSS) formed can be accepted to be virtually entirely organism mass. Therefore the COD/VSS ratio for such pure systems can be measured by measuring the COD and the VSS concentrations of the suspended solids material that accumulate in the reactor. Measured values on such systems give values close to 1.42 mgCOD/mgVSS (Eckenfelder and Weston, 1956).

For municipal wastewater the situation is more complex. The biomass or VSS that accumulates in the activated sludge reactor (aerobic or anoxic-aerobic) comprises three components:

1. the active heterotrophic organisms (nitrifier active mass is too low (<2%) to contribute significantly to the VSS)
2. the unbiodegradable particulate material originating from the organisms themselves called endogenous residue and
3. the unbiodegradable particulate material from the influent wastewater called inert solids.

The proportion of the VSS that these component fractions attain depends on the wastewater characteristics (fraction of unbiodegradable particulate organics in the wastewater) and the sludge age of the system. Generally, the active organism fraction decreases as the sludge age increases, while the two unbiodegradable particulate VSS components increase a different rates with sludge age. Consequently, if each component part of the VSS had a different COD/VSS ratio then it would be expected that the COD/VSS ratio of the mixed liquor would change with sludge age. This was found not to be the case (Fig. 9.7 and Table 9.4). Consequently, Marais and Ekama (1976) concluded that the COD/VSS ratio of the different components of the VSS can be accep-

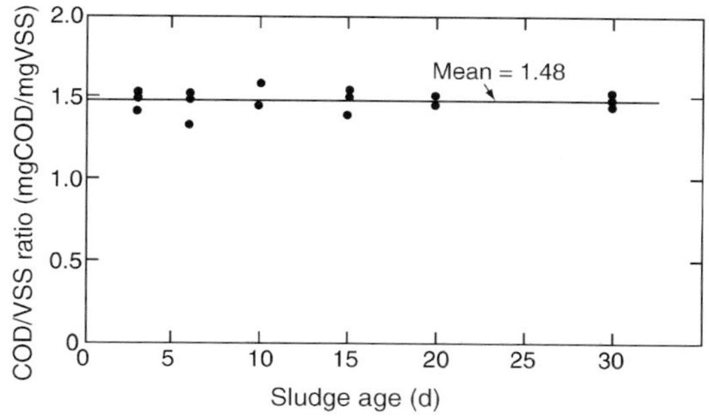

Fig. 9.7 Relationship of COD/VSS ratio of the mixed liquor with sludge age from laboratory scale aerobic completely mixed activated sludge systems. The constancy of the ratio with sludge age indicates that the ratio of the constituent fractions of the mixed liquor are the same and equal to 1.48 mgCOD/mgVSS.

ted for practical purposes to be the same, allowing a global measured value of 1.48 mgCOD/mgVSS to be used. This value is acceptably close to the theoretical 'organism VSS' value of 1.42 mgCOD/mgVSS. The global COD/VSS ratio of 1.48 mgCOD/mgVSS has been found to apply to all activated sludge systems in which heterotrophic organisms dominate the biocenosis, i.e. fully aerobic, anoxic/aerobic and also biological excess P removal systems. Even though the biochemical behaviour of the polyphosphate accumulating organisms is different to that of the aerobic and facultative ones, a very similar COD/VSS ratio (f_{cv}) and yield coefficient (Y_g) have been found applicable to them (Wentzel *et al.*, 1989).

TABLE 9.4 Experimental activated sludge COD:VSS ratios (Marais and Ekama, 1976)

Sludge source	*COD/VSS ratio*	*96% Confidence interval*	*No. of tests*
Bellvile full scale plant	1.43	±0.01	230
Laboratory full aerobic units 2.5–30 days (see Fig. 9.1)	1.48	±0.08	120
Aerobic digested sludge from laboratory system above	1.54	±0.17	80
Raw sewage	1.45	±0.04	180
Wine distillery raw waste	1.43	±0.01	13

5 CLOSURE

By considering the bioenergetics of heterotrophic organisms, the most important group in biological wastewater treatment, it is demonstrated that the chemical oxygen demand (COD) test is the most appropriate and convenient for measuring wastewater organic content (strength). The COD measures the electron donating capacity of the organics and this is not only closely linked to the energy content of the wastewater, but also is intimately associated with the reduction–oxidation (redox) reactions whereby organisms degrade organics and grow. This association establishes two important principles in biological wastewater treatment system modelling:

1. the organism growth yield coefficient and oxygen utilization are directly linked and quantified in terms of biological COD (organic) utilization, and
2. electron, or COD, balances can be conducted over the biological treatment system in which the influent COD mass is reconciled with the effluent COD mass, oxygen mass utilized and COD mass of sludge produced over a defined unit of time.

Combining these principles with the practical simplicity of the test and the rapidity with which it gives a result, makes the COD test an almost ideal test for monitoring the progress of biological reactors in wastewater treatment research and practice. The above two principles form the basis of the steady state and general kinetic activated sludge models of Marais and Ekama (1976), Dold *et al.*, (1980), Van Haandel *et al.* (1981) and WRC (1984). They have paved the way to transform activated sludge design procedures and models from an empirical approach, using a disconnected but associated assemblage of formulae, to an integrated process modelling approach based on mass balance and continuity principles. Modern design procedures are based on these principles and are essentially simplifications for steady-state conditions of the more complex formulations in the general kinetic simulation models such as those of the UCT (UCTOLD – Dold *et al.*, 1991; UCTPHO – Wentzel *et al.*, 1992) and IAW (ASM No. 1 – Henze *et al.*, 1987; ASM No. 2 – Henze *et al.*, 1994).

REFERENCES

Bauchop, T. and Elsden, S.R. (1960). The growth of microorganisms in relation to their energy supply. *J Gen. Microbiol.* **23**(3), 457–469.

Boon, A.G. and Burgess, C.R. (1972). Effects of diurnal variations of flow of settled sewage on the performance of high rate activated sludge plants. *Water Pollution Control* **71**(5), 493–522.

Burkhead, C.E. and McKinney, R.E. (1969). Energy concepts of aerobic microbial metabolism. *J. Sanit. Eng. Div. ASCE* **95**(3A2), 253–268.

Copcutt, P.A. (1983). *The BOD time curve.* BSc Thesis, Dept of Civil Eng., Univ. of Cape Town, Rondebosch, 7701, Cape, RSA

Dixon, H., Bell, B. and Axtell, R.J. (1972). Design and operation of the works of the Basingstoke Department of Water Pollution Control. *Water Pollut. Control* **71**(2), 167–175.

Dold, P.L., Ekama, G.A. and Marais, GvR. (1980). A general model for the activated sludge process. *Prog. Wat. Tech.* **12**, 47–77.

Dold, P.L. Wentzel, M.C. Billing, A.E. *et al.* (1991). *Activated sludge system simulation programs (Version 1.0).* Water Research Commission, Private Bag X03, Gezina 0031, South Africa.

Eckenfelder, W.W. and Weston, R.F. (1956). Kinetics of biological oxidation. In: B.J. Mccabe and W.W. Eckenfelder, Jr (eds) *Biological Treatment of Sewage and Industrial Wastes – I: Aerobic oxidation,* pp. 18–34. Reinhold Publ, Corp., New York.

Eckenfelder, W.W. Jr (1965). Discussion on: Thermodynamics of biological synthesis and growth. In: *Advances in Water Pollution Research, (Proceedings 2nd International Conference on Water Pollution Research, Tokyo, 1964),* **2**, 195–197. Pergamon Press, Oxford.

Ekama, G.A. and Marais, GvR. (1978). The dynamic behaviour of the activated sludge process. *Res. Report. W27,* Dept. of Civil Eng., Univ. of Cape Town, Rondebosch, 7701, Cape, South Africa.

Gaudy, A.F. (1972). Biochemical oxygen demand. In: R. Mitchell (ed.) *Water Pollution Microbiology,* pp. 305–332. Wiley Interscience, New York.

Henze, M., Grady, C.P.L. Jr, Gujer, W. *et al.* (1987). *Activated Sludge Model No.1. IAWQ Scientific and Technical Report No.1,* IAWQ, London.

Henze, M., Gujer, W., Mino, T. *et al.* (1994). *Activated sludge model No.2: Biological phosphorus removal. IAWQ Scientific and Technical Report No.3,* IAWQ, London.

Hoover, S.R. and Porges, N. (1952). Assimilation of dairy wastes by activated sludge. II. The equation of synthesis and rate of oxygen utilization. *Sew. Indus. Wastes* **24**(3), 306–312.

Hoover, S.R., Jasewicz, L. and Porges, N. (1953). An interpretation of the BOD test in terms of endogenous respiration of bacteria. *Sew Indus. Wastes* **25**(10), 1163–1173.

Loewenthal, R.E. (1986). *Bioenergetics.* Lecture notes for 3rd year Course CIV336S Water Engineering III, Dept. of Civ, Eng, Univ. of Cape Town, Rondebosch, 7701, Cape, South Africa.

Marais, GvR. and Ekama, G.A. (1976). The activated sludge process – Part 1: Steady state behaviour. *Water SA* **2**(4), 163–200.

McCarty, P.L. (1963). Discussion on: Free energy as a parameter in biological treatment. *J. Sanit. Eng. Div. ASCE* **89**(SA6), 65–68.

McCarty, P.L. (1965). Thermodynamics of biological synthesis and growth. In: *Advances in Water Pollution Research, (Proceedings 2nd International Conference on Water Pollution Research, Tokyo, 1964),* Vol. 2, 169–199. Pergamon Press, Oxford.

McCarty, P.L. (1972). Energetics of organic matter degradation. In: R. Mitchell (ed.) *Water Pollution Microbiology,* pp. 91–118. Wiley-Interscience, New York.

Odum, E.P. (1971). *Fundamentals of Ecology,* 3rd edn. WB Saunders Co., London.

Payne, W.J. (1981). *Denitritication.* John Wiley & Sons, New York.

Sawyer, C.N. (1956). Bacterial nutrition and synthesis. In: B.J. Mccabe and W.W. Eckenfelder, Jr (eds) *Biological Treatment of Sewage and Industrial Wastes, 1: Aerobic oxidation,* pp. 3–17. Reinhold Publ. Corp., New York.

Servizi, U.A. and Bogan, R.H. (1963). Free energy as a parameter in biological treatment. *J. San. Eng. fliv. ASCE* **89**(SA3), 17–40.

Servizi, U.A. and Bogan, R.H. (1964). Thermodynamic aspects of biological oxidation and synthesis. *J. WPCF* **38**(5), 607–618.

Smith, R. and Eilers, R.G. (1969). *A generalized computer model for steady state performance of the activated sludge process.* Report submitted to WPCF Admin., Advanced Water Treatment Branch, Div. of Research, Cincinnati, Ohio.

Speece, R.E. and McCarty, P.L. (1964). Nutrient requirements and biological solids accumulation in anaerobic digestion. In: *Advances in Water Pollution Research (Proceedings 1st International Conference on Water Pollution Research, London, 1962),* **2**, 305–322. Pergamon Press, Oxford.

Standard Methods (1985). *Standard methods for the examination of water and wastewater,* 16th edn. Prepared and published jointly by American Public Health

Association, American Water Works Association and Water Pollution Control Federation, Washington DC.

Symons, J.M. and McKinney, R.E. (1958). Biochemistry of nitrogen in the synthesis of activated sludge. *Sew. Indust. Wastes* **30**(7), 874–890.

Van Haandel, A.C., Ekama, G.A. and Marais, GvR. (1981). The activated sludge process – Part 3: Single sludge denitrification. *Water Res.* **15**(10), 1135–1152.

Wentzel, M.C., Dold, P.L., Ekama, G.A. and Marais, GvR. (1989). Enhanced polyphosphate organism cultures in activated sludge systems – Part III Kinetic model. *Water SA* **15**(2), 89–102.

Wentzel, M.C., Ekama, G.A. and Marais, GvR. (1992). Processes and modelling of nitrification denitrification biological excess phosphorus removal systems – A review. *Wat. Sci Technol.* **25**(6), 59–82.

WRC (1984). *Theory, design and operation of nutrient removal activated sludge processes*. Water Research Commission, Private Bag X03, Gezina 0031, South Africa.

Part II Water and Excreta Related Diseases

10

Microorganisms and disease

R. Morris

Consultant Environmental Biologist, Barwell, Leicestershire LE9 8DN, UK

1 INTRODUCTION

Many aspects of daily life are dependent upon the activities of the smallest of organisms (microorganisms): from the preparation of food (e.g. yoghurt, bread, cheese) and drink (e.g. wine and beer), the production of antibiotics (e.g. penicillin), the natural breakdown of decaying material (composting), sewage treatment (activated sludge, filtration beds) – the list is endless. Equally, they can also be a nuisance to mankind causing, for instance, the corrosion of concrete pipes, spoilage of food and, of course, disease. Indeed, the history of mankind is inextricably dominated by a close interaction with the microbial world. However, the most effective manifestation of the activities of microorganisms is the range of diseases which they may give rise to in susceptible hosts.

2 HISTORY OF DISEASE

It is impossible to pinpoint exactly when man became aware of the significant effect that microorganisms had on his activities, but it was certainly before the establishment of the science of microbiology. Milk, butter and cheese were used by the Egyptians as early as 3000 BC, while between 3000 and 1200 BC the Jews used salt from the Dead Sea to prevent the deterioration of foods (Jay, 1992). Wines are known to have been prepared by the Assyrians by 3500 BC and fermented sausages were prepared and consumed by the Babylonians and Chinese as far back as 1500 BC.

The recognition of disease goes back at least 5000 years, although the concept of a microorganism was unknown at that time. Smallpox was familiar to the Chinese in 1000 BC and the mummy of Rameses V (who died in 1157 BC) showed clear signs of the disease (Morris and Waite, 1981) before it became a scourge of Western Europe in the 18th century. Yellow fever (a viral disease) has been known ever since European sailors first visited West and Central Africa. Influenza has been recognized in Europe since the 16th century and has shown a remarkable tendency to cause periodic pandemics interspersed with long periods of relatively little infection.

Over 4000 years ago, the Egyptians realized that contamination of drinking water supplies by dead dogs, rats etc. was a cause of illness among the populace and took measures to protect their drinking water supply. Such action was based entirely on the necessity rather than any recognition of a microbial cause. Similarly the Romans recognized the risks associated with drinking polluted water and developed the habit of throwing silver coins into the water which, although they did not understand why, rendered many of the harmful bacteria innocuous.

Girolamo Fracastoro, a 16th century Italian scholar, proposed a disease theory long before the existence of microorganisms was demonstrated. He believed that disease occurs when 'seeds too small to be seen pass from one thing to another' (Atlas, 1995). However, it was not until the late 19th century that Robert Koch showed that specific microorganisms could be related to specific diseases. By the development

The Handbook of Water and Wastewater Microbiology
ISBN 0-12-470100-0

TABLE 10.1 Koch's Postulates

1	The microorganism must be present in all hosts suffering from the disease and absent from those without disease
2	The microorganism must be capable of being grown outside the diseased host as a pure culture
3	When such a pure culture is administered to a healthy host then the symptoms of the disease must develop
4	The microorganism must be capable of re-isolation from the experimentally infected host and shown to be identical to the original microorganism

of his *Postulates*, Koch argued that a microorganism isolated from a disease case, when administered to a healthy host, would initiate the same disease in that host. Koch also demonstrated the possibility of growing disease-causing organisms outside the host which, when introduced into a healthy host, could result in disease. Koch's Postulates are summarized in Table 10.1.

Disease is particularly obvious when populations come under stress, e.g. warfare and natural disasters. Such events are usually correlated with the breakdown of social structure and the failure of hygiene measures. Wars, for instance, often lead to a concentration of people, the intermixing of populations and the diversion of resources. These lead to a decrease in hygiene and medical care, with an increase in malnutrition and even famine. All lead to social disintegration and the stressed individual is particularly susceptible to disease. As an example: before the First World War, malaria and typhus (both with insects as vectors) occurred in Europe, albeit at low levels. However, the conditions endured during the war caused these two diseases to erupt into epidemics (Zinsser, 1934). Similarly, two million Greeks contracted malaria during the Second World War after German occupation of the country despite malaria-free conditions beforehand (Peterson, 1995).

3 WHAT ARE MICROORGANISMS?

The term 'microbe' or 'microorganism' is generally taken to refer to those small biological entities that can only be seen with the aid of magnification – although this is not always the case. They include the viruses, bacteria, fungi, algae and protozoa. (For the purpose of this review, the larger parasitic helminths etc. have not been included.) They can vary in size from less than 20 nm (some viruses) to several microns (protozoal parasites). Nearly all microorganisms show complex organizational structure, possessing the means to reproduce (DNA and RNA) and carry out complex biochemical processes. The exceptions are the viruses, which possess either DNA or RNA (never both) and do not have the means to carry out biochemical processes, i.e. they are obligate intracellular parasites. The bacteria, in contrast, are capable of independent growth and multiplication given a suitable source of nutrients.

4 INFECTION VERSUS DISEASE

The entry of a microorganism into a host body in itself does not necessarily lead to infection, let alone disease. Many inhabit the gut, skin and mucosal surfaces without any apparent invasion of the host tissue. These are essentially commensal organisms, which utilize the available nutrients in their immediate surrounding environment to sustain their viability. When this benign relationship is disrupted it is possible that such organisms may invade the host tissue – however, organisms such as *Escherichia coli* (used as an indicator of faecal pollution) may remain in the host gut without any apparent ill effects on the host.

Some microorganisms actively invade host tissue, i.e. they are pathogenic – this is the process of infection. In such cases the organism multiplies and is released into the environment to infect other susceptible hosts. When the activity of such an infection is such that it causes a reaction in the host, this is termed disease. In most cases, infection at best leads to a mild disease often not recognized. However, in extreme cases the disease that develops may vary from mild, through a recognized sequence of symptoms and may even result in host death.

Manifestation of an infection as a disease may result in stomach upsets, skin rashes, headache, pulmonary problems, muscular paralysis, blindness, encephalitis, arthritis and many other clinical conditions.

The relationships between microorganisms, hosts and disease are dynamic. Usually the host is healthy and environmental conditions, e.g. social factors, genetic, lifestyle etc., do not allow pathogens to become a problem. In theory, virtually any microorganism can lead to infection (and possibly disease) under suitable conditions. The infectious microorganism (because of inherent properties or failure of the host defences) is able to establish itself within the host. Provided sufficient numbers of microorganisms can penetrate into the host and reach the target tissues, then the process of pathogenesis may begin.

Thus, infection indicates the penetration into and the multiplication of an agent within the host and is determined largely by factors governing exposure to the agent and the susceptibility of the host. Disease represents the host's response to infection when it is severe enough to evoke a recognizable pattern of clinical symptoms. The factors influencing the occurrence and the severity of this response vary with the particular pathogen involved and their portal of entry, but the most important determinants for many common infections lie within the host itself. Of these, the age of the host at the time of infection is most crucial, with the young and old being particularly susceptible. Those who have diminished host defences, i.e. are immunocompromised (e.g. transplant patients and AIDS sufferers) are also at great risk of acquiring infections.

5 INFECTIVE DOSE

The number of pathogens that are necessary to enter the host to establish an infection is the *minimum infective dose* (MID). The level necessary to initiate infection varies widely, being dependent upon the pathogen itself, the susceptibility (i.e. the immune status) of the host, route of infection, environmental factors and even the way in which the infective dose was experimentally determined.

It has already been pointed out that in cases of stress (such as war and famine), hosts are particularly vulnerable to infection, often initiated by low levels of microorganisms and depressed body defences. In other cases, ingestion of high levels of organisms may be necessary to initiate infection because of the effect of acid in the stomach – protection from such physiological conditions, e.g. by encasement in food particles, may allow much lower levels of pathogens to initiate infection. Such is the case with salmonellosis, where it is generally accepted that 10^4 microorganisms may need to be ingested. However, *Salmonella typhimurium* was capable of causing disease in 361 people after consumption of contaminated chocolate which had bacterial levels as low as 10/100 g (Kapperud *et al.*, 1990).

That erroneous infective dose values can easily be arrived at by the use of an insensitive assay procedure is illustrated by a study of the MID for Echovirus 12 (Schiff *et al.*, 1984). The report found an apparent low MID but a reassessment of the virus preparation used in the volunteer studies using a different assay technique showed that the administered dose was at least a 100-fold higher! (Ward and Akin, 1984).

Thus, the infective dose is the minimum intake of a pathogen that will elicit a response from the host (this may be a serological conversion with no evidence of disease or may lead to a whole variety of medical conditions). However, infective doses must be viewed with some caution as:

- they are often extrapolated from epidemiological investigations
- they are obtained by feeding of the pathogen to healthy young adult volunteers
- they are only best estimates based on a limited database from outbreaks
- they are worse case estimates
- because of variables of the microorganism and the host, they cannot be directly used to assess risk.

TABLE 10.2 Factors affecting derivation of infective dose

Pathogen variables	Variability of gene expression of multiple pathogenic mechanisms
	Potential for damage or stress of the pathogen
	Interaction of pathogen with food menstruum and environment
	Susceptibility to pH
	Immunological 'uniqueness'
	Interraction with other organisms
Host variables	Age, general health, pregnancy, occupation
	Malignancy
	Metabolic disorders, alcoholism, cirrhosis, gastric acid variation (antacids, natural variation, dilution)
	Nutritional status, amount of food eaten
	Immune competence
	Surgical history
	Genetic disturbances

Some factors affecting the derivation of MIDs are shown in Table 10.2.

The understanding of the lack of certainty of the infective dose is essential if the microbiologist is to carry out meaningful epidemiological investigations of disease outbreaks and assessment of risk of populations to disease.

6 IMMUNITY

The continual exposure to microorganisms through contact in the environment, consumption of contaminated foods and water, inhalation of aerosols etc. does not result in continuous disease in the individual. This is because of a complex network of overlapping defence systems that protect the host from potentially pathogenic microorganisms. This immunity determines whether the host is resistant to pathogens or susceptible to infection and disease. Disease may, therefore, be regarded as a failure of this defence system.

Defences against pathogens are non-specific and immunospecific responses. Non-specific responses are especially important in preventing infections. Physical barriers include the skin, mucosal membranes and fluid flow (e.g. tears from the eyes), which block the entry of the pathogen into the host. Probably the most important of the chemical defences is acidity, as many microbes are sensitive to low pH conditions – thus the stomach is a particularly important barrier to prevent invasion of the small intestine. Other chemicals in the host also play an important role in preventing infection, including enzymes, interferons, iron-binding proteins and complement (a blood component).

The presence of phagocytic cells, macrophages, which occur throughout much of the host body and are capable of engulfing, killing and digesting microorganisms, is an additional non-specific defence to infection by pathogens. Phagocytosis is stimulated by an abnormal rise in body temperature (fever) which increases the rate of enzymatic reactions that lead to degradation of microorganisms and repair of damaged tissue, intensifies the action of interferons and causes a reduction in blood iron concentration (needed by many microorganisms for replication). In some situations, an inflammatory response may result, e.g. in skin abrasions. Inflammation is a generalized response to infection or tissue damage, aiming at localization of the invading pathogen and arresting the spread of infection. Inflammation is usually characterized by reddening of the localized area by blood vessel dilation, swelling, pain and elevated temperature.

The non-specific defences are augmented by a second mechanism – the specific immune response. This is a learned response that recognizes specific substances which trigger the formation of antibodies. This physiological response is especially important against infection and for protection against disease. The system is so essential that failure, such as with infection by organisms affecting the immune system directly (e.g. the human immunodeficiency virus, HIV), can rapidly result in severe disease and mortality caused by secondary infections. The adaptive immune response is characterized by specificity, memory and the acquired learning ability to detect foreign substances. For example, exposure to measles and diphtheria results in life-long immunity. In other cases, e.g. viral gastroenteritis, immunity may be relatively short-lived, perhaps only a few months, allowing many episodes of the disease during a lifetime.

The immune system responds to antigens, which are specific components of each type or group of microorganisms. These activate the immune response ('learning') followed by a rapid and specific secondary response. Thus, the host only acquires immunity after exposure to the pathogen and full immunity is only gained after the secondary immunological response. Acquiring immunity protects the host against future infections by pathogens that the body has learned to recognize. This acquired immunity results in exposure to many pathogens resulting in disease only once in a lifetime – this response has been the basis for the development of vacccines against the more serious disease-causing pathogens. (The body's defence mechanisms against pathogens are described in Atlas, 1995).

7 EPIDEMIOLOGY

The idea that environmental factors can influence the occurrence of disease is not a new one. Over 2000 years ago, Hippocrates and others postulated that this was the case, inadvertently setting the scene for the science of *epidemiology* (Beaglehole *et al.*, 1993). However, it was not until the 19th century that the relationship between disease, human population and the environment in which they lived was studied.

Perhaps the most famous founding father of epidemiology was, appropriately, John Snow, a physician in London who was trying to deal with an outbreak of cholera in 1854. His observation that the distribution of cholera cases correlated with the use of water from the Broad Street Pump, together with determination that the water supply was indeed contaminated with faecal material, led to a most dramatic containment of an outbreak by the simple expediency of removing the pump handle! Cholera cases declined markedly (but not altogether) as a consequence. Snow's studies (Snow, 1855) were only one aspect of a wide-ranging series of investigations being undertaken at the time which involved a consideration of physicochemical, biological, sociological and political factors (Cameron and Jones, 1983).

Snow was able to arrive at his conclusions by painstakingly documenting cholera cases and correlation of the comparative incidence of cholera among subscribers to two of the city's water supply companies. He found that cholera occurred more frequently in customers of one water company, the Southwark & Vauxhall (Table 10.3). This company, he observed, drew its water from the lower reaches of the River Thames where it was contaminated with London's sewage. In contrast, the Lambeth Company's supply came from the relatively unpolluted River Thames further upstream.

The modern day concept of epidemiology has developed considerably since Snow's monumental conclusions. The epidemiology of infectious disease is now concerned with the circumstances under which both infection and disease occur in a population and the factors that influence their frequency, spread and distribution. Such approaches are now regularly applied in the investigation of outbreaks of disease under all situations, e.g. the waterborne outbreak of cryptosporidiosis at Milwaukee in 1993 (MacKenzie *et al.*, 1996) and the *E. coli* O157 outbreak associated with meat products in Scotland in 1994 (Davis and Brogan, 1995).

The excellent book on basic epidemiology published by the World Health Organization is recommended for further reading (Beaglehole *et al.*, 1993). How society has developed the means to monitor, assess and control infectious

TABLE 10.3 Cholera in London 8 July to 26 August 1854 – relationship to water supply (Snow, 1855)

Water supply company	*Number of deaths*	*Death rate/ 100 000*
Lambeth	18	94
Southwark & Vauxhall	844	503

diseases is described by Williams (1985) who details the development of the UK's Public Health Laboratory Service.

8 TRANSMISSION ROUTES

An infectious disease is an illness caused by transmission of the pathogen from an infected host to a susceptible host, either directly or indirectly. Communicable disease is the most important of health problems in all countries. In most developing countries, communicable diseases are the greatest cause of morbidity and mortality in all age groups, whereas in developed countries, diseases associated with the respiratory tract, in particular, are the commonest cause of mortality in the young and old. The emergence of previously unrecognized diseases has greatly stimulated the need to gain epidemiological data, which can lead to risk assessment and to appropriate control programmes.

One of the most important epidemiological factors that needs to be understood before control measures can be undertaken to prevent or control disease outbreaks, is the means by which the pathogens move from infected to susceptible hosts. Some routes of transmission are summarized in Table 10.4.

All routes can be interrupted, so offering a means of disease control and prevention. For example, water- or food-borne disease can effectively be prevented by sound hygiene established through educational programmes. Transfer of diseases, such as AIDS and hepatitis, by the use of contaminated needles by drug addicts can be reduced by using use-once needles and the avoidance of needle sharing.

To be successfully transmitted, the pathogen needs to be able to survive outside the host body without loss of viability until a susceptible host is infected. For some routes of transmission, e.g. genital, the transfer time between hosts can be very short, whereas with others, e.g. those pathogens excreted in the faeces, there is a need to be able to survive for considerable lengths of time. In the latter case, the pathogen has to be able to survive a host of environmental factors, some of which may be designed to prevent their transmission. The level of intervention (whether it be water or food production, personal hygiene, hospital practices etc.) will depend upon the resources available, what is being controlled and the level of protection desired.

9 THE ECONOMIC AND SOCIOLOGICAL COST OF DISEASE

Diseases cause suffering among all populations on the planet. It also carries with it a substantial cost in economic, as well as sociological, terms. During 1995 in Sweden, 25% of a 1400 population reported gastroenteritis caused by *Campylobacter* contamination of the drinking water supply (Andersson *et al.*, 1997). An assessment of the cost of the outbreak was attempted (taking into account

TABLE 10.4 Examples of routes for disease transmission in humans (Beaglehole *et al.*, 1993)

Exit route	*Transmission route*	*Entry route*	*Example*
Respiratory	Bite	Skin	Rabies
	Aerosol	Respiratory	Legionella
	Mouth to hand	Oropharyngeal	Herpes simplex
Gastrointestinal tract	Faeces to hand	Mouth	Salmonella
	Faeces to water	Mouth	Cryptosporidium
Skin	Air	Skin	Warts
Blood	Insects	Skin	Typhus
	Blood transfusion	Skin	Hepatitis B
	Needles	Skin	HIV
Genital	Intercourse	Genital	Syphilis
Placental	Vertical to embryo	Blood	Rubella

the cost of rectifying water treatment failure, medical costs, time off work costs, use of bottled water etc.), with the finding that a conservative final economic burden was in the order of $1 million. The cost of the much larger Milwaukee outbreak of cryptosporidiosis remains to be finalized, but is likely to run into several hundreds of millions of dollars.

Outbreaks in developing countries have little value in an economic sense. Here the cost is predominantly one of human misery with death being all too common.

10 CONCLUSIONS

Infectious diseases continue to be an important public health problem world-wide. Access to sufficient water and food of suitable quality is a prime requirement for the improvement and maintenance of public health. Education of susceptible populations in good hygienic practices (especially personal) is particularly important. Avoidance of situations which would lead to the ideal conditions for disease outbreak, e.g. warfare, famine, floods, social deprivation, would also contribute greatly to safeguarding public health. However, all these require financial and physical resources, as well as a willingness to address public health matters on a personal and a political level.

The benefit of sound public health policies is self-evident as the pioneering work of John Snow adequately demonstrated. An openness of mind to recognize that 'emerging pathogens' may result in a serious threat to public health (e.g. *Cryptosporidium, E. coli* O157, Ebola, antibiotic-resistant tuberculosis – causing *Mycobacterium*) is equally essential. Complacency about the importance of disease to Man's welfare is reprehensible.

REFERENCES

Andersson, Y., de Jong, B. and Studahl, A. (1997). Waterborne *Campylobacter* in Sweden: the cost of an outbreak. *Wat. Sci. Tech* **35**(11–12), 11–14.

Atlas, R.M. (1995). *Microorganisms in Our World*. Mosby, St Louis.

Beaglehole, R., Bonita, R. and Kjellström, T. (1993). *Basic Epidemiology*. World Health Organisation, Geneva.

Cameron, N.D. and Jones, I.G. (1983). John Snow, The Broad Street pump and modern epidemiology. *Int. J. Epidemiol.* **12**, 393–396.

Davis, B.S. and Brogan, R.T. (1995). A widespread community outbreak of *E. coli* O157 infection in Scotland. *Public Health* **109**, 381–388.

Jay, J.M. (1992). *Modern Food Microbiology*, 4th edn. Chapman and Hall, London.

Kapperud, G. *et al.* (1990). Outbreak of Salmonella typhimurium infection traced to contaminated chocolate and caused by a strain lacking the 60-megadalton virulence plasmid. *J. Clin. Microbiol.* **28**, 2597–2601.

MacKenzie, W.R. *et al.* (1996). A massive outbreak in Milwaukee of Cryptosporidium infection transmitted through the public water supply. *N. Eng. J. Med.* **331**, 161–167.

Morris, R. and Waite, W.M. (1981). Environmental virology and its problems. *J. Inst. Wat. Eng. Sci.* **35**, 232–244.

Peterson, R.K.D. (1995). Insects, diseases and military history: the Napoleonic campaigns and historical perception. *Am. Entomol.* **41**, 147–160.

Schiff, G.M. *et al.* (1984). Minimum human infectious dose of enterovirus (echovirus 12) in drinking water. *Monogr. Virol.* **15**, 222–228.

Snow, J. (1855). *On the Mode of Communication of Cholera*. Churchill, London. (cited by Beaglehole *et al.*, 1993).

Ward, R.L. and Akin, E.W. (1984). Minimum infective dose of animal viruses. *Curr. Rev. Environ. Cont.* **14**, 297–310.

Williams, R.E.O. (1985). *Microbiology for the Public Health: the Evolution of the Public Health Laboratory Service 1939–1980*. Public Health Laboratory Service, London.

Zinsser, H. (1934). *Rats, Lice and History*. Routledge, London.

11

Unitary environmental classification of water- and excreta-related communicable diseases

D.D. Mara[*] and R.G.A. Feachem[†]

[*]*School of Civil Engineering, University of Leeds, Leeds LS2 9JT;* [†]*Global Fund to Fight AIDS, Geneva-Cointrin, Switzerland*

1 INTRODUCTION

In the 30 years since Bradley developed his environmental classification of water-related diseases (White *et al.*, 1972), and the 19 years since Feachem *et al.* (1983) presented their environmental classification of excreta-related diseases, several new water-related and excreta-related pathogens have been identified (Table 11.1). Given the recent and continuing concern with the risks to global public health posed by these and other newly emerging (and re-emerging) pathogens (see, for example Ewald, 1994; Wilson *et al.*, 1994; Scheld *et al.*, 1998), it is appropriate to reconsider these earlier classifications of water- and excreta-related diseases and, due to the fact that many (but not all) water-related communicable diseases are also excreta-related, combine them into a unitary classification which is both comprehensive and up-to-date, while remaining useful to engineers in the design of water supply and sanitation systems, especially those for low-income communities in developing countries.

2 ENVIRONMENTAL CLASSIFICATION OF WATER- AND EXCRETA-RELATED COMMUNICABLE DISEASES

An environmental classification of disease groups, such as water-related and excreta-related diseases, is more useful to environmental engineers than one based on biological type (virus, bacterium, protozoon or helminth) as it groups the diseases into categories of common environmental transmission routes. Thus an environmental engineering intervention designed to reduce the transmission of pathogens in a particular category is likely to be effective against all pathogens in that category, irrespective of their biological type.

We present our unitary classification of water- and excreta-related communicable diseases in Table 11.2. Category I comprises the faeco-oral diseases that are transmitted from person to person either by faecally-contaminated water (the classical waterborne transmission route) or through the unavailability (or lack of use) of a sufficient quantity of water for personal and domestic hygiene (Bradley's water-washed route). Category II contains the skin and eye infections which are not faeco-oral but transmitted by the water-washed route.

Bradley's concept of water-washed disease transmission (White *et al.*, 1972) represents the single most important development in our understanding of water-related disease since Snow's (1849, 1855) demonstration of the waterborne transmission of cholera (see Longmate, 1966). In low-income communities in developing countries with inadequate water supplies, the direct person-to-person water-washed

The Handbook of Water and Wastewater Microbiology
ISBN 0-12-470100-0

TABLE 11.1 Major water- and excreta-related pathogens recognized since 1973[a]

Year	Pathogen	Type	Disease
1973	Rotavirus	Virus	Diarrhoea
1976	*Cryptosporidium parvum*	Protozoon	Acute enterocolitis
1977	*Legionella pneumophila*	Bacterium	Legionnaires' disease
1977	*Campylobacter jejuni*	Bacterium	Diarrhoea
1982	*Escherichia coli* O157:H7	Bacterium	Haemorrhagic colitis, haemolytic uraemic syndrome
1983	*Helicobacter pylori*	Bacterium	Gastric ulcers, stomach cancer
1985	*Enterocytozoon bienusi*	Protozoon	Diarrhoea
1986	*Cyclospora cayetanensis*	Protozoon	Diarrhoea
1988	Hepatitis E	Virus	Enteric hepatitis
1991	*Encephalitozoon hellem*	Protozoon	Conjunctivitis
1992	*Vibrio cholerae* O13	Bacterium	Cholera
1992	Hepatitis F	Virus	Enteric hepatitis

[a] Adapted from Satcher (1995), Grabow (1997) and Lederberg (1997).

transmission of faeco-oral diseases is responsible for a much greater proportion of these diseases then their more indirect waterborne transmission. Thus engineering interventions in water supply for these communities should concentrate first on supplying them with adequate volumes (around 30 litres per person per day) of water to minimize water-washed transmission, rather than with high quality treated water to reduce waterborne transmission (Feachem, 1975) – in an ideal world both would be done, but the expense of the latter generally precludes its consideration, especially in the rural tropics.

The water-washed transmission route is also likely to be important even in areas with an adequate water supply but where personal and/or domestic (including food) hygiene is poor – *Operation Clean Hands*, launched by the American Society for Microbiology (Cassell and Osterholm, 1996), is an example of the recognition of the need to reduce water-washed disease transmission in a highly developed society.

As noted by Feachem (1975), 'the term "waterborne disease" has been, and still is, greatly abused by public health and water engineers who have applied it indiscriminately so that it has almost become, synonymous with "water-related disease".' Regrettably this is true today – for example, the WHO Regional Office for Europe's recent report entitled *Eradication of water-related diseases* (WHO (1997)) refers exclusively to waterborne diseases, and so ignores water-washed diseases in particular, even though these are a significant health problem in Europe (Letrilliart *et al.*, 1997). Water engineers need to be aware that water-related diseases are not just simply waterborne diseases, and that other categories of water-related disease, especially water-washed diseases, are generally of greater epidemiological significance, particularly in developing countries. This is not to deny the importance of truly waterborne disease (for example, Payment *et al.* (1997) found that between 14 and 40% of gastrointestinal illnesses were attributable to the consumption of tap water meeting current North American microbiological quality requirements), but to emphasize the greater importance of water-washed diseases.

Category I includes infection with *Helicobacter pylori*, which is transmitted faeco-orally (and also oro-orally) and which causes gastritis and stomach cancer (Blaser, 1998); its prevalence is high (30–90% in industrialized countries and 80–90% in developing countries), and it is the only bacterium designated as a known human carcinogen (IARC, 1994; see also Mara and Clapham, 1997). Category I also includes several other 'new' pathogens (see Table 11.1): rotavirus and *Campylobacter jejuni*

TABLE 11.2 Unitary environmental classification of water- and excreta-related diseases

Category	*Environmental transmission features*	*Examples*	*Control strategies*
I. Faeco-oral waterborne and water-washed diseases	Non-latent (except *Ascaris*) No intermediate host Infectivity: medium to low (bacteria), high (others) Persistence: medium to high (bacteria), low to medium (others, except *Ascaris*: very high) Able (bacteria) and unable (others) to multiply outside host	*Viral:* Hepatitis A, E and F Poliomyelitis Rotaviral diarrhoea Adenoviral diarrhoea *Bacterial:* Campylobacteriosis Cholera *Helicobacter pylori* infection Pathogenic *Escherichia coli* infection Salmonellosis Typhoid and paratyphoid Yersiniosis *Protozoan:* Amoebiasis Cryptosporidiasis *Cyclospora cayetanensis* diarrhoea *Enterocytozoon bienusi* diarrhoea Giardiasis *Isospora belli* diarrhoea *Helminthic:* Ascariasis Enterobiasis Hymenolepiasis	Improve water quantity, availability and reliability (water-washed disease control) Improve water quality (waterborne disease control) Hygiene education
II. Non-faeco-oral water-washed diseases	Non-latent No intermediate host High infectivity Medium to high persistence Unable to multiply	Skin infections (scabies, leprosy, yaws) Eye infections (trachoma, conjunctivitis, including that caused by *Encephalitozoon hellem*) Louse-borne fevers	Improve water quantity, availability and reliability. Hygiene education
III. Geohelminthiases	Latent Very persistent Unable to multiply No intermediate host Very high infectivity	Ascariasis; trichuriasis; hookworm infection	Sanitation. Effective treatment of excreta or wastewater prior to reuse. Hygiene education.
IV. Taeniases	Latent Persistent Able to multiply Very high infectivity Cow or pig intermediate host	Beef and pork tapeworm infections	As III above, plus proper cooking of meat and improved meat inspection

(*continued on next page*)

Table 11.2 (*continued*)

Category	*Environmental transmission features*	*Examples*	*Control strategies*
V. Water-based diseases	Latent Persistent Able to multiply High infectivity Intermediate aquatic host(s)	*Bacterial:* Leptospirosis Tularaemia Legionellosis	Decrease contact with contaminated water. Improve domestic plumbing. Public education
		Helminthic: Schistosomiasis Clonorchiasis Fasciolopsiasis Guinea worm infection	Decrease contact with contaminated waters. Sanitation. Treatment of excreta or wastewater prior to reuse. Public education
		Fungal: Pulmonary haemorrhage due to *Stachybotrys atra* infection	Drying of flood-damaged homes. Public education
VI. Insect-vector diseases		*Water-related:* Malaria Dengue Rift Valley fever Japanese encephalitis Yellow fever African sleeping sickness Onchocerciasis Bancroftian filariasis	Decrease passage through breeding sites. Destroy breeding sites. Larvicide application. Biological control. Use of mosquito netting and impregnated bed nets
		Excreta-related: Fly-borne and cockroach-borne excreted infections[a] Bancroftian filariasis	Improve stormwater drainage Public education
VII. Rodent-vector diseases		Rodent-borne excreted infections[a] Leptospirosis Tularaemia	Rodent control. Hygiene education Decreased contact with contaminated water Public education

[a] The excreted infections comprise all those diseases in Categories I, II and IV and the helminthic diseases in Category V.

(globally the most common viral and bacterial causes of diarrhoea, respectively); new strains or types of known pathogens, such as hepatitis E and F viruses, *Escherichia coli* O157 and *Vibrio cholerae* O139; and two new protozoa – *Cryptosporidium parvum* (responsible for the largest recorded outbreak of waterborne disease in the USA, that in Milwaukee in 1993 when over 400 000 people were infected (Mackenzie *et al.*, 1994), and recently reported to be also disseminated by waterfowl (Graczyk *et al.*, 1998); and *Cyclospora cayetanensis*, which may be waterborne (Benenson, 1995) or foodborne (Majkowski, 1997) and is unusual in that it is not immediately infective upon excretion but is latent for a few days to a few weeks. Diarrhoea due to the protozoon *Isospora belli* is also included, an example of a re-emerging pathogen, especially (but not only) in immunocompromised persons (Marcial-Seoane and Serrano-Olmo, 1995; Goodgame, 1996; Marshall *et al.*, 1997). Two other new protozoa, both mainly affecting the immunocompromised, are also included for completeness: *Enterocytozoon bienusi*, which causes diarrhoea (Collins, 1997), in Category I; and *Encephalitozoon hellem*, which causes conjunctivitis (Didier *et al.*, 1991), in Category II.

Categories III, IV and V are the geohelminthiases, taeniases and water-based diseases, respectively. These are all helminthic infections, although Bradley's original category of water-based disease is broadened to include water-based bacterial diseases such as legionellosis (which can be transmitted in a sauna; Den Boer *et al.*, 1998), leptospirosis (see Mara and Alabaster, 1995), and Buruli ulcer due to *Mycobacterium ulcerans* in swamp waters (Wright, 1998). Category V also includes pulmonary haemorrhage/haemosiderois due to the toxigenic fungus *Stachbotrys atra*, which occurs in infants living in homes subjected to flood-induced water damage which promotes the growth of this fungus (Dearborn, 1997).

Categories VI and VII comprise the water- and excreta-related insect vector and excreta-related rodent vector diseases, respectively. Category VI includes important mosquito-borne diseases such as malaria and Bancroftian filariasis, and also dengue fever and dengue haemorrhagic fever which are now the most important human insect-vector viral diseases; the dengue pandemic has intensified over the last two decades, and currently there are 50–100 million cases of dengue fever and several hundred thousand cases of dengue haemorrhagic fever each year (Gubler and Kuno, 1997). In category VI Bancroftian filariasis appears under both the water-related and the excreta-related subcategories as in some parts of the world its mosquito vector is *Aedes aegypti* which breeds preferentially in 'clean' water, whereas in others it is *Culex quinquifasciatus* which prefers to breed in 'dirty', i.e. excreta- or wastewater-contaminated) water, although this distinction between clean water and dirty water breeding mosquitoes is now less clearcut than it used to be.

3 GLOBAL BURDEN OF WATER- AND EXCRETA-RELATED DISEASES

Murray and Lopez (1996*b*) present data on the incidence and prevalence of several major water- and excreta-related diseases in 1990. These are listed in Table 11.3. While these data are awesome in their sheer magnitude, they tell us little about disease *burden*. WHO (1980) claimed that as much as 80% of all morbidity in developing countries was due to water- and excreta-related disease, but the epidemiological evidence for this figure was not given. More recently, Murray and Lopez (1996a) give data on the global burden of disease in 1990, including that fraction attributable to poor water supply, sanitation and personal and domestic hygiene (Table 11.4). Globally, 5.3% of all mortality in 1990 was due to poor water supply, sanitation and hygiene. In the developing world as a whole the corresponding percentage was 6.7, but much higher in the Middle Eastern Crescent (8.3), India (9.0) and sub-Saharan Africa (10.7), and in developing countries this risk factor was second in importance only to malnutrition (which was responsible for 14.9% of all deaths).

When the burden of disease is expressed in disability-adjusted life years (DALYs), which equal the sum of discounted years of life lost due to premature death and discounted years lived with disability (see Murray and Lopez, 1996a), a similar picture emerges (Table 11.4) since, with water- and excreta-related diseases, the principal component (>90%) of DALYs is years of life lost, rather than years lived with disability. The percentage of total DALYs lost (i.e. DALYs lost due to all causes) attributable to poor water supply, sanitation and hygiene is 6.8 globally and 7.6 in developing countries as a whole, and again higher in the Middle Eastern Crescent (8.8), India (9.5) and sub-Saharan Africa (10.1).

Of the total number of deaths attributable to poor water supply, sanitation and hygiene, i.e.

TABLE 11.3 Global water- and excreta-related disease statistics for 1990[a]

Disease	*Number*	*Remarks*
Diarrhoea	4 073 920 110 episodes	56% in children aged 0–4 94% in developing countries
Malaria	213 743 000 episodes	All in developing countries 87% in sub-Saharan Africa
African sleeping sickness	267 000 persons infected	All in sub-Saharan Africa
Schistosomiasis	208 276 000 persons infected	87% in sub-Saharan Africa
Onchocerciasis	5 802 000 persons infected 478 000 persons with impaired vision 356 000 blind persons	>99% in sub-Saharan Africa >99% in sub-Saharan Africa >99% in sub-Saharan Africa
Leprosy	2 434 000 persons infected	70% in Asia
Dengue	415 000 episodes	92% in Asia
Japanese encephalitis	58 000 000 episodes	All in Asia
Trachoma	292 000 persons with impaired vision 192 000 blind persons	34% in sub-Saharan Africa; 34% in China; 24% in MEC[b] 34% in sub-Saharan Africa; 34% in China; 24% in MEC[b]
Ascariasis	61 847 000 persons with high-intensity infection	73% in children aged 5–14 All in developing countries
Trichuriasis	45 421 000 persons with high-intensity infection	79% in children aged 5–14 All in developing countries
Human hookworm infection	152 492 000 persons with high-intensity infection 36 014 000 persons with anaemia	84% in adults aged 15–59 All in developing countries 72% in adults aged 15–44 All in developing countries

[a] The world population in 1990 was 5.3 billions, of which 3.9 billions (74%) were in developing countries.
[b] Middle East Crescent, covering North Africa, the Middle East, Pakistan and the Central Asian republics of the former Soviet Union.
Source: Murray and Lopez (1996b).

attributable to water- and excreta-related diseases, 99.9% occurs in developing countries. For total DALYs lost, the corresponding percentage is 99.8.

4 APPROPRIATE CONTROL STRATEGIES

Table 11.2 lists appropriate control strategies to reduce the incidence of water- and excreta-related communicable diseases. Generally, and especially in developing countries, these need only to be relatively simple, low-cost interventions that can, in many cases, be done by the communities involved or at least with their participation, and there is much information available on these, e.g. water supplies (Feachem, 1975; Hofkes, 1983; Morgan, 1990), and sanitation (Morgan, 1990; Mara, 1996). Vector control is discussed by Curtis (1990, 1991).

TABLE 11.4 Burden of disease attributable to poor water supply, sanitation and personal and domestic hygiene in 1990

Region[a]	*Deaths*	*Percentage of all deaths*	*DALYs*[b]	*Percentage of all DALYs*
EME	1100	<0.1	101 000	0.1
FSE	2400	0.1	128 000	0.2
IND	839 900	9.0	27 463 000	9.5
CHN	81 400	0.9	4 231 000	2.0
OAI	354 300	6.4	13 192 000	7.4
SSA	875 600	10.7	28 870 000	10.1
LAC	135 300	4.5	5 183 000	5.3
MEC	378 200	8.5	13 224 000	8.8
World	2 668 200	5.3	93 392 000	6.8
Developed regions	3500	<0.1	229 000	0.1
Developing regions	2 664 700	6.7	93 163 000	7.6

[a] EME, established market economies; FSE, formerly socialist economies of Europe; IND, India; CHN, China; OAI, other Asia and islands; SSA, sub-Saharan Africa; LAC, Latin America and the Caribbean; MEC, Middle Eastern Crescent.
[b] DALYs, disability-adjusted life years.
Source: Murray and Lopez (1996a).

Hygiene education is of paramount importance, and this is discussed by Nyamwaga and Akuma (1986), Boot (1990, 1991), Boot and Cairncross (1993) and Hubley (1993).

5 CONCLUDING REMARKS

The unitary environmental classification of water- and excreta-related diseases is a comprehensive categorization of these diseases into seven categories of common environmental transmission patterns. This unitary classification is likely to be useful to tropical public health engineers and other professionals as it highlights the facts that many of these diseases are both water and excreta related, and that these diseases are best controlled in the long term by sustainable (and indeed sustained, i.e. well operated and maintained) improvements in both water supply and sanitation, which are supplemented by effective programmes of hygiene education. Although convincing arguments can be made for broadening the disease concept from water- and excreta-related diseases to housing-related diseases (Mara and Alabaster, 1995) or, more generally, to cover urban or peri-urban health (Harpham and Tanner, 1995), the fact remains that water- and excreta-related diseases are the most important of these diseases. It is in this context that the unitary environmental classification of these diseases is most useful.

REFERENCES

Benenson, A.S. (1995). *Control of Communicable Diseases Manual*, 16th edn. American Public Health Association, Washington, DC.

Blaser, M.J. (1998). *Helicobacter pylori* and gastric diseases. *British Medical Journal* **316**, 1507–1510.

Boot, M. (1990). *Making the Links: Guidelines for Hygiene Education in Community Water Supply and Sanitation.* Occasional Paper No. 5. IRC International Water and Sanitation Centre, The Hague.

Boot, M.J. (1991). *Just Stir Gently: The Way to Mix Hygiene Education with Water and Sanitation.* Technical Paper No. 29. IRC International Water and Sanitation Centre, The Hague.

Boot, M. and Cairncross, S. (1993). *Actions Speak: the Study of Hygiene Behaviour in Water and Sanitation Projects.* IRC International Water and Sanitation Centre, The Hague.

Cassell, G. and Osterholm, M. (1996). A simple solution to a complex problem. *ASM News* **62**(9), 516–518.

Collins, R. (1997). Protozoan parasites of the intestinal tract: a review of Coccidia and Microsporidia. *Journal of the American Osteopathic Association* **97**(10), 593–598.

Curtis, C.F. (1990). *Appropriate Technology in Vector Control.* CRC Press Inc., Boca Raton.

Curtis, C.F. (1991). *Control of Vector Diseases in the Community.* Wolfe Publishing, London.

Dearborn, D.G. (1997). Update: pulmonary haemorrhage/hemosiderosis among infants – Cleveland, Ohio,

1993–1996. *Morbidity and Mortality Weekly Report* **46**(2), 33–35.

Den Boer, J.W., Yzerman, E., Van Belkum, A., Vlaspolder, F. and Van Breukelen, F.J.M. (1998). Legionnaire's disease and saunas. *The Lancet* **351**, 114.

Didier, E.S., Didier, P.J., Friedberg, D.N. *et al.* (1991). Isolation and characterization of a new human microsporidian. *Encephalitozoon hellem* (n. sp.), from three AIDS patients with keratoconjunctivitis. *Journal of Infectious Diseases* **163**(3), 617–621.

Ewald, P.W. (1994). *Evolution of Infectious Disease.* Oxford University Press, New York.

Feachem, R.G. (1975). Water supplies for low-income communities in developing countries. *Journal of the Environmental Engineering Division, American Society of Civil Engineers* **101**(EE4), 687–703.

Feachem, R.G., Bradley, D.J., Garelick, H. and Mara, D.D. (1983). *Sanitation and Disease: Health Aspects of Excreta and Wastewater Management.* John Wiley & Sons, Chichester.

Goodgame, R.W. (1996). Understanding intestinal spore-forming protozoa: cryptosporidia, microsporidia, isospora and cyclospora. *Annals of Internal Medicine* **124**(4), 429–441.

Grabow, G.O.K. (1997). Hepatitis viruses in water: update on risk and control. *Water SA* **23**(4), 379–386.

Graczyk, T.K., Fayer, R. *et al.* (1998). *Giardia* sp. cysts and infectious *Cryptosporidium parvum* oocysts in the feces of migratory Canada geese (*Branta canadensis*). *Applied and Environmental Microbiology* **64**(7), 2736–2738.

Gubler, D.J. and Kuno, G. (1997). *Dengue and Dengue Hemorrhagic Fever.* CAB International, Wallingford.

Harpham, T. and Tanner, M. (1995). *Urban Health in Developing Countries: Progress and Prospects.* Earthscan Publications Ltd., London.

Hofkes, E.H. (1983). *Small Community Water Supplies.* John Wiley & Sons, Chichester.

Hubley, J. (1993). *Communicating Health: An Action Guide to Health Education and Health Promotion.* Macmillan, London.

IARC (1994). *IARC Monographs on the Evaluation of Carcinogenic Risks to Humans, Volume 61: Schistosomes, Liver Flukes and* Helicobacter pylori. International Agency for Research on Cancer. Lyon.

Lederberg, J. (1997). Infectious disease as an evolutionary paradigm. *Emerging Infectious Diseases* **3**(4), 417–423.

Letrilliart, L., Desenclos, J.-C. and Flahault, A. (1997). Risk factors for winter outbreak of acute diarrhoea in France: case-control study. *British Medical Journal* **315**, 1645–1649.

Longmate, N. (1966). *King Cholera: the Biography of a Disease.* Hamish Hamilton, London.

Mackenzie, W.R., Hoxie, N.J., Procter, M.E. *et al.* (1994). A massive outbreak in Milwaukee of *Cryptosporidium* infection transmitted through the public water supply. *New England Journal of Medicine* **331**(3), 161–167.

Majkowski, J. (1997). Strategies for rapid response to emerging foodborne microbial hazards. *Emerging Infectious Diseases* **3**(4), 551–554.

Mara, D.D. (1996). *Low-cost Urban Sanitation.* John Wiley & Sons, Chichester.

Mara, D.D. and Alabaster, G.P. (1995). An environmental classification of housing-related diseases in developing countries. *Journal of Tropical Medicine and Hygiene* **98**(1), 41–51.

Mara, D.D. and Clapham, D. (1997). Water-related carcinomas: environmental classification. *Journal of Environmental Engineering, American Society of Civil Engineers* **123**(5), 416–422.

Marcial-Seoane, M.A. and Serrano-Olmo, J. (1995). Intestinal infection with *Isospora belli. Puerto Rico Health Sciences Journal* **14**(2), 137–140.

Marshall, M.M., Naumovitz, D., Ortega, Y. and Sterling, C.R. (1997). Waterborne protozoan parasites. *Clinical Microbiology Reviews* **10**(1), 67–85.

Morgan, P. (1990). *Rural Water Supplies and Sanitation.* Macmillan, London.

Murray, C.J.L. and Lopez, A.D. (1996a). *The Global Burden of Disease. Global Burden of Disease and Injury Series*, Vol. 1, Harvard University Press, Cambridge, MA.

Murray, C.J.L. and Lopez, A.D. (1996). *Global Health Statistics. Global Burden of Disease and Injury Series*, Vol. 2, Harvard University Press, Cambridge, MA.

Nyamwaga, D. and Akuma, P. (1986). *A Guide to Health Education in Water and Sanitation Programmes.* African Medical and Research Foundation, Nairobi.

Payment, P., Siemiatycki, J., Richardson, L., Renaud, G., Franco, E. and Prévost, M. (1997). A prospective epidemiological study of the gastrointestinal health effects due to the consumption of drinking water. *International Journal of Environmental Health Research* **7**(1), 5–31.

Satcher, D. (1995). Emerging infections: getting ahead of the curve. *Emerging Infectious Diseases* **1**(1), 1–6.

Scheld, W.M., Armstrong, D. and Hughes, J.M. (1998). *Emerging Infections 1.* ASM Press, Washington, DC.

Snow, J. (1849). *On the Mode of Communication of Cholera.* John Churchill, London.

Snow, J. (1855). *On the Mode of Communication of Cholera*, 2nd edn. John Churchill, London.

White, G.F., Bradley, D.J. and White, A.U. (1972). *Drawers of Water: Domestic Water Use in East Africa.* Chicago University Press, Chicago.

WHO (1997). *Eradication of Water-related Disease.* Report No. EUR/ICP/EHSA 02 02 03. WHO Regional Office for Europe, Copenhagen.

WHO (1980). *Drinking Water and Sanitation, 1981–1990: A Way to Health.* World Health Organization, Geneva.

Wilson, M.E., Levins, R. and Spielman, A. (1994). Disease in Evolution: Global Changes and Emergence of Infectious Disease, *Annals of the New York Academy of Sciences.* **740** Academy of Sciences, New York.

Wright, D. (1998). Australia focuses on *Mycobacterium ulcerans. The Lancet* **351**, 184.

12

Emerging waterborne pathogens

Debra E. Huffman*, Walter Quinter-Betancourt† and Joan Rose†

*College of Marine Science, University of South Florida, St Petersburg FL 33701; †Department of Fisheries and Wildlife, Michigan State University, East Lansing MI 48824, USA

1 INTRODUCTION

The English dictionary defines the word 'emerging' to mean, 'to rise out of a state of depression or obscurity; to come to notice; to reappear after being eclipsed'. Thus is the state of infectious diseases throughout the world in the twenty-first century. Not only are new microorganisms being identified every year but some of the old established diseases are again gaining attention. More specifically, emerging diseases have been defined as those 'whose incidence in humans has increased in the past two decades or threatens to increase in the near future' (NIM, 1998).

The reason for this 'emergence' of infectious diseases has been attributed to many factors:

- Sensitive or susceptible populations: there is an increasing number of elderly and immunocompromised individuals (transplant patients, AIDs patients) in our communities, in addition to diabetics, infants, and pregnant women, all of whom may be more susceptible to severe outcomes (Gerba *et al.*, 1996a).
- Global transportation: the food supply comes from all over the world, thus leading to a distribution of pathogens (i.e. *Cyclospora*) and people are much more mobile and can transverse the globe in less than 24 hours, bringing infections with them.
- Antibiotic resistance: due to the widespread application of antibiotics in medicine and in agriculture, antibiotic resistance is spreading. The World Health Organization (WHO) has recently reported that increasingly drug-resistant infections throughout the world are threatening to make once-treatable diseases incurable (http://www.who.int/emc/amr.html).
- Zoonotic transmission: more animals and changes in agricultural practices may lead to greater chances for microbial transmission and spread from animals to humans.
- Evolution of pathogens: RNA viruses, e.g. during replication lack repair mechanisms and can evolve quickly.
- Improved diagnostic tools: diseases previously not recognized or microorganisms that are not culturable can now be studied through the use of powerful molecular tools.

The factors describe above also apply to the growing list of waterborne pathogens. Table 12.1 describes some of the microorganisms that can be spread through contaminated water. In the last three decades, 1970 to 2000, numerous species of bacteria, parasites and viruses have been described that have caused important waterborne outbreaks throughout the world. This chapter will briefly discuss these pathogens. Because most of the microorganisms that are waterborne are spread by the faeco-oral route, their occurrence and control in wastewaters, animal wastes and drinking waters are important. Key in the future to understanding and controlling emerging pathogens in drinking water and wastewater is the application of new tools for monitoring and studying these microorganisms.

In the past, the direct detection of pathogenic microorganisms in the water environment

The Handbook of Water and Wastewater Microbiology
ISBN 0-12-470100-0

TABLE 12.1 Examples of 'emerging' waterborne pathogens of concern

Pathogen	*First described: as human pathogen/ waterborne transmission*	*Methods of detection*
Bacteria		
Escherichia coli 0157:H7 Associated with bloody diarrhea (hemorrhagic colitis) and Hemolytic uremic syndrome	1982*, Riley *et al.*, 1983; Mead and Griffin, 1998. 1991†, Centers for Disease Control, 1991	Cultural methods Molecular methods (AWWA Research Division Microbiological Contaminants Research Committee, 1999)
Helicobacter pylori Diarrhea, peptic and duodenal ulcer disease, gastric carcinoma	1982*, Warren and Marshall, 1983. 1991†, Klein *et al.*, 1991.	Cultural methods (microbiological methods) Molecular methods Autoradiography ATP bioluminescence (Velázquez and Feirtag, 1999)
Mycobacterium avium complex Diarrhea and respiratory disease	1982*, Greene *et al.*, 1982 1994†, Singh and Lu, 1994	Cultural methods Molecular methods Chemical methods (AWWA Research Division Microbiological Contaminants Research Committee, 1999)
Vibrio cholerae Profuse watery diarrhea 'rice water stools', and vomiting.	1600*†, Pollitzer, 1959	Cultural methods Molecular methods (Baumann *et al.*, 1984; Sharma *et al.*, 1997; Basu *et al.*, 2000)
Parasites		
Cryptosporidium Profuse watery diarrhea (cholera-like), fluid loss, fever and abdominal pain	1976*, Meisel *et al.*, 1976. 1984†, D'Antonio *et al.*, 1985.	Immunofluorescence assay Tissue culture methods Molecular methods (Gasser and O'Donoghue, 1999)
Cyclospora Explosive, watery diarrhea, fatigue, anorexia, weight loss, nausea	1979*, Ashford, 1979, 1994†, Rabold *et al.*, 1994	Epifluorescence (UV illumination), acid-fast staining, molecular techniques. (Long *et al.*, 1991; Yoder *et al.*, 1996)

TABLE 12.1 (*continued*)

Pathogen	*First described: as human pathogen/ waterborne transmission*	*Methods of detection*
Microsporidia Gastrointestinal, pulmonary, nasal, ocular, muscular, cerebral and systemic infections	1959*, Matsubayashi *et al.*, 1959 1997†, Sparfel *et al.*, 1997	Chromotrope staining, chemofluorescent agents, Giemsa stain, immunodetection, molecular methods, tissue culture methods (Weber *et al.*, 1999; Dowd *et al.*, 1998a, 1999)
Toxoplasma (Flu-like symptoms). Painful swollen lymph glands in the cervical, supraclavicular and inguinal regions. Fever, headache, muscle pain, anemia, lung complications. Fetus at greatest risk	1979†, Benenson *et al.*, 1982	Mouse bioassay (Aramini *et al.*, 1998) ICR US. EPA cartridge filtration (Isaac-Renton *et al.*, 1998).
Enteric viruses		
Coxsackie virus Aseptic meningitis, herpangina, paralysis, exanthema, hand, foot and mouth disease, common cold, hepatitis, infantile diarrhea, acute hemorrhagic conjunctivitis	1947*, Dalldorf and Sickles, 1948	Tissue culture methods Molecular methods (Hurst *et al.*, 1997; Hurst, 1999; Murrin and Slade, 1997)
Hepatitis viruses Fever, nausea, abdominal pain, anorexia and malaise, associated with mild diarrhea, arthralgias, scleral icterus. Cytologic damage, necrosis and inflammation of the liver (HAV)	1950†, Bradley, 1992	Tissue culture methods Molecular methods Immunofluorescence (Smith, 2001, Sobsey, 1999, Jothikumar *et al.*, 1993, Pina *et al.*, 1998)
Norwalk-like viruses Diarrhea, vomiting, abdominal pain, cramping, low fever, headache, nausea, tiredness (malaise), and muscle pain (myalgia)	1968*, 1968†, Gerba, 1996c	Transmission electron microscopy; enzyme-linked immunosorbent assay Molecular methods (Hurst, 1999).
Rotavirus Vomiting, abdominal distress, diarrhea, dehydration, fever,	1973*, Bern *et al.*, 1992	Tissue culture methods Molecular methods (Gratacap-Cavallier, 2000)

* Pathogen.
† Waterborne transmission.

had been tedious and often required many biochemical analyses to identify the genus and/or species present. In fact, no method for cultivating the microorganism in question existed at all. The use of polymerase chain reaction (PCR, the genetic copying method) now allows for a more rapid and specific detection of microorganisms and their associated public health risks. The advantages of using this method are tremendous and the field is moving forward very quickly. Almost all the emerging pathogens discussed in this chapter have been studied in the water environment using this powerful technique.

2 BACTERIAL PATHOGENS

Early in the 1800s, when it became apparent that disinfection of drinking water could prevent waterborne disease epidemics caused by bacterial pathogens such as *Salmonella typhi* (typhoid fever), most people thought that the days of waterborne bacterial epidemics were gone forever. Unfortunately, this was not to be the case. During the recent decade, there has been a number of waterborne disease outbreaks caused by bacteria such as *Escherichia coli* and *Vibrio cholerae*. Advanced techniques that combine standard cultural methods with enhanced PCR detection methods have revealed the presence of bacteria previously not associated with waterborne disease transmission such as *Mycobacterium avium* and *Helicobacter pylori*.

2.1 *Escherichia coli* O157:H7

Escherichia coli O157:H7 is an enteropathogenic strain of *E. coli* that was first identified as a human pathogen in 1982 (Mead and Griffin, 1998). Infection with the organism can cause severe bloody diarrhea with abdominal cramping. In small children and the elderly, fluid replacement is of the utmost importance for a full recovery. A common, more serious, complication of infection with *E. coli* O157:H7 is hemolytic uremic syndrome (HUS) where there is loss of red blood cells and kidney failure. In severe cases, HUS can cause permanent kidney damage or death.

Escherichia coli O157:H7 has been shown to survive similarly to typical *E. coli* strains under routine drinking water conditions. There have been numerous documented outbreaks of waterborne disease caused by *E. coli* O157:H7. The most recent drinking-water outbreak of *E. coli* O157:H7 took place in Walkerton, Canada (90 miles west of Toronto) in May 2000. It is estimated that there were more than 1000 cases of illness and nine deaths associated with the outbreak. It is believed that the organisms entered the town's drinking water supply when animal waste from nearby farms washed into wells during a flood earlier in the month. The town had been using chlorine to disinfect the drinking water, however, during the time of the outbreak the chlorine disinfection system was malfunctioning.

In the USA there have been several documented outbreaks of waterborne *E. coli*. In 1990, a drinking-water outbreak of pathogenic *E. coli* took place in Cabool, Missouri, with 243 cases, 32 hospitalizations and four deaths (Geldreich *et al.*, 1992; Swerdlow *et al.*, 1993). The source of the water for Cabool was deep groundwater wells with no disinfection in the distribution system. Disturbances in the distribution system included the replacement of 43 water meters and two line breaks. Disturbances to distribution systems can provide a pathway for infiltration from stormwater run-off or sewage contamination. In August 1999, unchlorinated well water, used to prepare beverages and ice at the Washington County Fair near Albany, New York, was contaminated with *E. coli* and caused more than 900 cases of illness with 65 hospitalizations and two deaths. Also in 1999, in Petersburg, Illinois more than 200 people became ill after eating and/or drinking contaminated food and water at a festival that was being held in a cow pasture. In 1996, in Atlanta, Georgia, 26 children playing at a water park became ill after ingesting pool water that had been contaminated with fecal material containing the bacteria (Barwick *et al.*, 2000). Another documented waterborne outbreak of *E. coli* O157:H7 took place in Scotland with 496 cases (272 laboratory confirmed cases) and 19 deaths (Dev *et al.* 1994).

The largest outbreak of *E. coli* O157:H7 occurred in 1996, in Sakai City, Japan. The outbreak was due to the consumption of radish sprouts that had been washed with contaminated water. Approximately 6000 people were infected and three children died as a result of the infection (Michino *et al.*, 1999).

2.2 *Helicobacter pylori*

Helicobacter pylori has been cited as a major etiologic agent for gastritis and has been implicated in the pathogenesis of peptic and duodenal ulcer disease (Taylor and Blaser, 1991). It has also been associated with the development of gastric carcinoma (Eurogast Study Group 1993). The mode of transmission of *H. pylori* is not well characterized. Recent studies suggest that gastrointestinal dissemination may be due to vomiting in childhood (Axon, 1995). Persons living in low socioeconomic conditions have consistently been shown to have a high prevalence of *H. pylori* and the organism has also been found routinely in the feces of children living in endemic areas (Thomas *et al.*, 1992). Klein *et al.* (1991) reported that, in Peru, the water source might be a more important risk factor than socioeconomic status in acquiring *H. pylori* infection. Children whose homes had external water sources were three times more likely to be infected than those whose homes had internal water sources. Among families with internal water sources, there was no difference in *H. pylori* infection associated with income. Children from high-income families whose homes were supplied with municipal water were 12 times more likely to be infected than were those from high-income families whose water came from community wells. These findings show that substandard municipal water supplies may be important sources of *H. pylori* infection.

2.3 *Mycobacterium avium* complex (MAC)

MAC has been isolated from natural water and drinking water distribution systems in the USA and is likely present throughout the world. Members of the MAC are considered opportunistic human pathogens. There has been a dramatic increase in the number of AIDS patients infected with MAC, with 25–50% of late stage AIDS patients now infected. Evidence for environmental transmission of MAC includes: increased frequency of MAC in gastrointestinal tract of advanced AIDS patients, higher frequency of isolation of MAC from the gastrointestinal tract than from the respiratory tract, and gastrointestinal symptoms (i.e. nausea, vomiting and diarrhea) common with MAC infections (Singh and Lu, 1994). *M. avium* strains from infected AIDS patients in Los Angeles have been shown to be genetically related to isolates recovered from water to which the patients were exposed through drinking or bathing (von Reyn *et al.*, 1994; Glover *et al.*, 1994). Control of MAC can be difficult as they are extremely resistant to conventional disinfection methods and have been shown to survive 10 mg/l of free chlorine (duMoulin and Stottmeier, 1983).

2.4 *Vibrio cholerae*

Epidemics of cholera have devastated Europe and North America since the early 1800s. A lack of sanitation and an increasing population, often with limited access to clean water, has brought about numerous disease outbreaks. A total of 1,041,422 cases and 9642 deaths due to cholera in the Americas were reported by 1995 (Morbidity and Mortality Weekly Report, 1995). In 1994 non-O1 cholera was detected for the first time from the Bug River (freshwater) in Poland and Hong Kong has recently reported two outbreaks of cholera (Lee *et al.*, 1996). Although the cause of the Hong Kong outbreaks was not clearly identified, increasing pollution of coastal waters has been implicated. Further concern over the cholera epidemic stems from the discovery of a new strain of *V. cholerae*, O139, which has shown an increased mortality rate (Lee *et al.*, 1996). Inadequate disinfection or the lack of disinfection has contributed significantly to the spread of cholera throughout Africa and Latin America.

3 PROTOZOAN PARASITES

The emerging protozoa associated with waterborne disease all produce an oocyst or spore-like structure, which are the environmental as well as the infectious stage. These life stages are very resistant to routine disinfection in drinking water and little is known about wastewater treatment and how it may affect their survival and removal. Most are associated with animals (with the exception of *Cyclospora*) and therefore the control of animal wastes is as equally important as the control of human wastes. Microscopy had been the most common method for detection of protozoa in clinical specimens. However, in the water environment, while microscopy is often used, future methods are focusing on PCR and cell culture.

3.1 *Cryptosporidium*

Cryptosporidium was first diagnosed in humans in 1976. Since that time, it has been well recognized as a cause of severe watery diarrhea, lasting several days to a week (Dubey *et al.*, 1990). Reported incidence of *Cryptosporidium* infections in the population range from 0.6 to 20% depending on the geographic locale. While there is a greater prevalence in the populations of Asia, Australia, Africa and South America, in the USA and the UK, *Cryptosporidium* has been described as the most significant cause of waterborne disease today. Detection of this organism has been reported in waters throughout the world.

Oocysts have been detected in 4 to 100% of surface waters tested at concentrations ranging from 0.1 to 10 000/100 l depending on the impact from sewage and animals (Lisle and Rose, 1995). Groundwater, once thought to be a more protected source, has shown between 9.5 and 22% of samples positive for *Cryptosporidium* (Hancock *et al.*, 1998). The largest documented drinking water outbreak of *Cryptosporidium* in the USA occurred in Milwaukee, Wisconsin in 1993 where 400 000 people were ill and 100 died due to contamination of the water supply (MacKenzie *et al.*, 1994).

New methods are available for the detection and characterization of *Cryptosporidium*. In particular, immunomagnetic separation techniques for improved recovery (Bifulco and Schaeffer, 1993; Deng *et al.*, 1997; Bukhari *et al.*, 1998); cell culture techniques for evaluating the infectivity of the oocysts (Slifko *et al.*, 1997, 1999; DiGiovanni *et al.*, 1999) and finally, PCR methods for distinguishing the human–human genotype from the animal–human genotype (Wiedenmann *et al.*, 1998). PCR is an attractive diagnostic procedure as it is rapid, sensitive, and pathogen specific (Johnson *et al.*, 1995; Rochelle *et al.*, 1997; Kostrzynska *et al.*, 1999; DiGiovanni *et al.*, 1999). Xiao *et al.*, (1999) evaluated several PCR techniques to determine whether they detected *C. parvum* in environmental samples. The authors concluded that two nested-PCR-restriction fragment length polymorphism (RFLP) based on the small-subunit rRNA (Xiao *et al.*, 1999) and dihydrofolate reductase genes (Gibbons *et al.*, 1998) were more sensitive than single-round PCR or PCR-RFLP protocols for detection and *Cryptosporidium* species differentiation (Xiao *et al.*, 1999).

3.2 *Cyclospora cayetanensis*

Cyclospora (previously termed cyanobacterium-like body) is a single cell coccidian protozoan that has been implicated as an etiologic agent of prolonged watery diarrhea in humans (Ortega *et al.*, 1993). The organism was first described as early as 1977 (Ashford, 1979) and has been reported with increased frequency since the mid-1980s. *Cyclospora* has been described in patients from North, Central and South America, Europe, Asia and North Africa, however, the true prevalence of this parasite in any population is unknown (Soave and Johnson, 1995).

Cyclospora is now known to be an obligate parasite of immunodeficient and immunocompetent humans (Ortega *et al.*, 1993). In an immunocompromised person the parasite can cause profuse, watery diarrhea lasting several months. The infection is much less severe in immunocompetent patients. Symptoms may range from no symptoms to abdominal cramps, nausea, vomiting and fever lasting from 3 to 25 days (Goodgame, 1996).

While *Cryptosporidium* appears to be predominantly waterborne, *Cyclospora* has been related more often to transmission through contaminated produce from a world market and the possible contamination of irrigation waters with human wastes. The differences between the protozoa and their transmission may be due to their biology and structure, size of the oocysts, need for sporulation and presence of animal reservoirs (Table 12.2). The treatment of human wastes in North America compared to lack of treatment in other parts of the developing world may also contribute to the transmission of the organism.

Two waterborne outbreaks have been reported. In June 1994, several cases of diarrhea were detected among British soldiers and dependants stationed in a small military detachment in Pokhara, Nepal (Rabold *et al.*, 1994). The drinking water for the camp was a mixture of river and municipal water that was treated by chlorination. A candle filtration system was also used to remove particulates, but was not guaranteed to filter *Cyclospora*-sized particles (8–10 μm). *Cyclospora* was detected in 75% of the diarrhea samples examined, and a water sample processed by membrane filtration taken from the camp also revealed the presence of *Cyclospora* oocysts. Twenty-one cases of prolonged diarrhea in employees and staff physicians were noted on July 9, 1990 in a Chicago hospital (Huang *et al.*, 1995). Upon investigation, *Cyclospora* oocysts were identified in the stools and epidemiological investigations implicated the tap water in the physicians' dormitory. It was speculated that the storage tank had become contaminated. Although this outbreak has been identified as a waterborne outbreak, a plausible scenario for the contamination of the water has not been developed.

In 1996, a total of 1465 cases of cyclosporiasis occurred in 20 states in the USA and Canada. Some 55 events were identified as a part of this nationwide outbreak. The suspected vehicles of transmission for these outbreaks were Guatemalan raspberries. *Cyclospora* reappeared in 1997 (Rose and Slifko, 1999). Twenty-five clusters of outbreaks have occurred resulting in 1450 cases in nine states. The foods associated with these outbreaks included raspberries, basil, and lettuce. Some of these foods were traced to Central Guatemala. In 1998, few cases were identified in the USA; however, an outbreak did occur in Canada.

The predominant method for detecting and identifying *Cyclospora* has been by microscopy. There are currently no antibodies available for specifically staining the oocyst. However, PCR will likely turn out to be the better method for screening for the presence of this pathogen in wastewaters and potentially surface waters used for irrigation that are receiving human sewage. The amplification reaction for *Cyclospora cayetanensis* has

TABLE 12.2 Comparison of *Cryptosporidium* and *Cyclospora*

Attribute	*Cryptosporidium*	*Cyclospora*
Taxonomy	Intestinal coccidian	Intestinal coccidian
Infective unit	Oocysts 4–5 μm Immediately infectious upon excretion	Oocysts 8–10 μm Requires sporulation* in the environment, not immediately infectious upon excretion
Animal reservoir	*C. parvum* found in most mammals, can cross species barriers	*C. cayetanensis* documented only in humans
Waterborne disease	12 waterborne outbreaks in North America since 1985	1 outbreak in Chicago, 1 in Nepal

* Sporulation is a process by which the oocysts undergo maturation in the environment before becoming infectious.

been based in a nested PCR reaction that included forward and reverse oligonucleotide primers that hybridize to sequences encoding 18S rRNA (Yoder *et al.*, 1996). An oligo-ligation assay (OLA) for the differentiation between *Cyclospora* and *Eimeria* spp. (an animal enteric protozoan which is related to *Cyclospora*) has also been included in the procedure (Jinneman *et al.*, 1999).

3.3 Microsporidia

The microsporidia are obligate intracellular spore-forming protozoa that are capable of infecting both vertebrate and invertebrate hosts. Their role as an emerging pathogen in immunosuppressed hosts is being increasingly recognized. The prevalence of microsporidiosis in studies of patients with chronic diarrhea ranges from 7 to 50% world-wide (Bryan, 1995). It is unclear whether this broad range represents geographic variation, differences in diagnostic capabilities or differences in risk factors for exposure to microsporidia. Microsporidia are single-celled, spore-forming, obligate intracellular parasites that belong to the phylum Microspora. The microsporidia are considered emerging pathogens because new species have been identified as causes of disease in humans during the last 20 years. Because species of microsporidia that were recognized causes of disease in animals are now causing infections in humans, microsporidia are also considered re-emerging pathogens (Didier and Bessinger, 1999).

In the summer of 1995, in France, a waterborne outbreak of microsporidia occurred with approximately 200 cases of disease. The species identified was *E. bieneusi* (Sparfel *et al.*, 1997). While fecal contamination of the drinking water was never detected, contamination from a nearby lake was suspected. Typical symptoms of infection include chronic diarrhea, dehydration and significant weight loss (>10% body weight). Other symptoms include keratitis, conjunctivitis, hepatitis, peritonitis, myositis, central nervous system infection and renal disease. Treatments are available for certain species of microsporidia, however, some species remain resistant to therapy (Cotte *et al.*, 1999).

Microsporidia spores have been shown to be stable in the environment and remain infective for days to weeks outside their hosts (Shadduck and Polley 1978; Waller 1979; Shadduck, 1989). Because of their small size (1–5 μm) they may be difficult to remove using conventional filtration techniques and there is a concern that these organisms may have an increased resistance to chlorine disinfection similar to *Cryptosporidium*. Initial studies using cell culture suggest that the spores may be more susceptible to disinfection (Wolk *et al.*, 2000).

In the USA there are minimal data on the occurrence of human strains of microsporidia in surface waters. There are more than 1000 species of microsporidia, of which 13 are presently known to infect humans (e.g. *Encephalitozoon cuniculi*, *Enterocytozoon bieneusi*, *Encephalitozoon intestinalis* and *Encephalitozoon hellem*). There are two species of microsporidia associated with gastrointestinal disease in humans: *E. bieneusi* and *E. intestinalis* (Dowd *et al.*, 1998a). Dowd *et al.* (1998b) described a PCR method for detection and identification of the microsporidia (amplifying the small subunit ribosomal DNA of microsporidia). They found isolates in sewage, surface waters and ground waters. The strain that was most often detected was *Enterocytozoon bieneusi*, which is a cause of diarrhea and excreted from infected individuals into wastewater.

3.4 *Toxoplasma gondii*

Toxoplasma is considered a tissue protozoan of cats and other felines who become infected mainly from eating infected rodents or birds, or from feces of infected cats. The symptoms of infection include flu-like symptoms, with swollen glands in the neck, armpits or groin area and most people recover without treatment. The cat is known as the definitive host, which means it is the only mammal that is known to excrete oocysts.

Humans and other mammals acquire the disease through ingestion of the oocysts (or ingestion of contaminated meat which has been undercooked and contains the tissue stage). In a healthy adult the illness is a febrile illness of varying severity. However, in the immunocompromised the infection may cause severe disease. The greatest risk is to the fetus from infection of the woman during pregnancy, which can lead to fetal infection, chronic chorioretinitis or death.

Two outbreaks of toxoplasmosis have been associated with contaminated surface water. In 1979, 600 USA soldiers attended a 3-week training course in a jungle in Panama. Within 2 weeks of their return to the USA, 39 out of 98 soldiers in one company came down with a febrile illness. Serological testing revealed 31 confirmed cases of acute toxoplasmosis. The outbreak was attributed to the ingestion of contaminated water while on maneuvers in the jungle (Benenson *et al.*, 1982).

In March 1995, the Capital Regional District of Victoria, British Columbia identified 110 cases of toxoplasmosis (Bowie *et al.*, 1997). The outbreak was attributed to a single drinking water source for the area. The number of newly identified cases of toxoplasmosis declined sharply after the drinking water reservoir suspected of contamination was shut down. An estimated 3000 people (1% of the population) may have been infected by municipal drinking water contaminated with toxoplasmosis. Detection methods for oocysts in water have not been developed, however, the presence of infectious *T. gondii* oocysts in cougars as well as domestic cats living in the affected area was confirmed using mouse bioassays and serological surveys (Aramini *et al.*, 1998; 1999, Isaac-Renton *et al.*, 1998).

4 ENTERIC VIRUSES

There are several hundred enteric viruses that are potentially important agents of waterborne disease. In the past, it was thought that viruses were very species specific for the most part. Thus human sewage and wastes were considered the only source of viruses in water. However, there is new concern about the ability of RNA viruses, in particular, to evolve quickly and potentially jump the species barrier. Such is the case with hepatitis E virus found in pigs. There is limited information regarding the incidence of virus infections in the USA population as well as throughout the world. Bennett *et al.* (1987) have reported 20 million cases of enteric viral infections and 2010 deaths per year.

Contamination of groundwater with viruses is of great concern due to the resistant nature of the viral structure and the colloidal size (20 nm), which makes this group of microorganisms easily transported through soil systems. Viruses also survive up to months in groundwater and are more resistant to water disinfection than are the coliforms (Yates and Yates, 1988; Gerba and Rose, 1989). Studies in the USA have found viruses in 20–30% of the groundwater where coliforms were not predictive of viral contamination (Abbaszadegan *et al.*, 1993). New techniques using PCR (Table 12.3) have shown that there is much more contamination than previously recognized.

The most frequently identified viruses in water or associated with waterborne outbreaks are hepatitis, Norwalk, coxsackie, rotavirus and the echoviruses. The cultivatable enteroviruses, which include the poliovirus, echovirus and coxsackie virus, make up only a small percentage of what is in wastewater and the tried and true cell culture methods used, which elicit a cytopathic effect when the virus is present, have been shown to be insensitive to many viruses, which grow but do not destroy the cells. Thus, PCR is now being utilized in cell culture to detect the non-cytopathic

TABLE 12.3 Virus detection in groundwater in the USA by various methods (Abbaszadegan *et al.*, 1993)

Virus	*Method*	*Percentage of samples positive*
Culturable enteric viruses	Cell culture	6.8 (12/176)
Enteroviruses	RT-PCR	30
Hepatitis A virus	RT-PCR	7
Rotavirus	RT-PCR	13
Total viruses	RT-PCR	39.3 (53/135)

RT-PCR (reverse transcription, polymerase chain reaction) is nucleic acid amplification for detection of the internal components of the virus.

viruses. In addition, new viruses are constantly being identified, which are non-culturable causes of gastroenteritis. These human caliciviruses (of which Norwalk is one) will need to be investigated in water using RT-PCR (reverse-transcription, polymerase chain reaction) methods.

4.1 Coxsackie viruses

Diarrhea has been one of the risks associated with many of the enteric viruses, but the recognition is emerging that more serious chronic diseases are now being associated with viral infections and these risks need to be better defined. Studies have reported that coxsackie virus B is associated with myocarditis (Klingel *et al.*, 1992). This could be extremely significant given that 41% of all deaths in the elderly are associated with diseases of the heart. In recent studies, enteroviral RNA was detected in endomyocardial biopsies in 32% of the patients with dilated cardiomyopathy and 33% of patients with clinical myocarditis (Kiode *et al.*, 1992). In addition, there is emerging evidence that coxsackie virus B is also associated with insulin dependent diabetes (IDD) and this infection may contribute to an increase of 0.0079% of IDD (Wagenknecht *et al.*, 1991).

Coxsackie virus B can be cultivated in cell culture and do cause a destruction of the cells, the cytopathic effect (Payment, 1997). They are commonly isolated from sewage, surface waters and groundwaters. Dahling *et al.* (1989) reported enterovirus concentrations in the discharges from several poorly operated activated sludge and trickling filter plants that were higher than 100,000 plaque-forming units per liter. At one of the plants, the discharge outlet was within one km of a drinking water treatment plant. On average, 95% of the enteroviruses detected were coxsackie virus type B5. Other viruses, like reoviruses and polioviruses, were detected but were found at lower levels. Krikelis *et al.* (1985, 1986) reported that almost 30% of the isolates from raw sewage from central Athens were coxsackie virus types B2, B4 and B5 with an estimated coxsackie virus concentration of 35.8 to 172.8 cytopathic units (CPU) per liter. Approximately 23% of the enteroviruses detected were of the echovirus group.

Between 1970 and1979, four laboratories in different areas of the German Democratic Republic analyzed 1908 surface water samples from 30 sites for the presence of enteric viruses. Coxsackie virus (particularly types B3 and B4) was isolated every year during the sampling period (Walter *et al.*, 1982). Treated sewage discharges into the river were not disinfected. Lucena *et al.* (1985) collected samples from the Liobregat River and the Besos River and found an average coxsackie virus type B concentration of 0.107 and 0.60 most probable number cytopathic units, respectively.

Payment *et al.* (1985) sampled seven drinking water treatment plants twice a month for 12 months and detected coxsackie viruses in 7% (11 of 155) of the finished water samples. Coxsackie virus types B3 and B5 were also detected in well water during an outbreak in Texas in 1980 with an average concentration of 0.31 PFU/liter detected (Hejkal *et al.*, 1982). Compared to other viruses, including those in the enterovirus group, coxsackie viruses are commonly isolated from sewage and surface waters. Their role as a transmitter of waterborne disease is significant.

The cell culture methods for isolation and identification are cumbersome and lengthy, taking several weeks. New methodologies using cell-culture integrated with RT-PCR (reverse transcription, polymerase chain reaction) will provide faster more efficient and specific identification (Reynolds *et al.*, 1996; Grabow *et al.*, 1999; Lewis *et al.*, 2000).

4.2 Hepatitis viruses

Hepatitis A virus (HAV) is the second most commonly reported infection in the USA and hepatitis A is the most commonly reported type of viral hepatitis (Melnick, 1995). The hepatitis A virus is considered to be endemic in most Latin American and Caribbean countries (Craun, 1996). The symptoms of hepatitis A include fever, nausea, anorexia and malaise, often with mild diarrhea. The liver cells are ultimately infected causing cytologic damage, necrosis and inflammation of the liver. Illness usually lasts from 1 to 2 weeks but may last several months. While HAV is an old disease, more recently outbreaks have occurred as a result of contaminated produce (Hutin *et al.*, 1999). Thus the control of HAV in human wastewaters and their elimination from waters used for irrigation of crops is a worldwide concern.

Hepatitis E virus (HEV) is an enteric RNA virus and causes jaundice and clinical symptoms very much like HAV. HEV has caused devastating waterborne disease outbreaks, particularly in tropical and subtropical countries with inadequate sanitation (Aggarwal and Naik, 1997; Balayan, 1997). In Kanpur, India in 1991, there were 79 000 cases of HEV due to sewage contamination of the drinking water. The earliest confirmed outbreaks occurred in the 1950s in India (Bradley, 1992).

Children are often asymptomatic and the mortality rate is between 0.1 and 4% (Grabow *et al.*, 1994). In pregnant women in their third trimester, the mortality rate can exceed 20% (Hurst *et al.*, 1996). There has been speculation that HEV is endemic in various parts of the world and subclinical cases may be contributing to the spread of the disease.

HEV is found in wild and domestic animals. Studies using genetic sequencing have found that human and pig HEV are very similar genetically in the USA as well as in Nepal (Meng *et al.*, 1997; Tsarev *et al.*, 1999). Thus zoonotic transmission seems very possible (Mushahwar *et al.*, 1998; Smith, 2001).

Although the cultivation of HEV has been reported (Smith, 2001), RT-PCR is the method of choice in water and has been used to detect HEV RNA in sewage (Jothikumar *et al.*, 1993; Pina *et al.*, 1998).

4.3 Norwalk-like caliciviruses viruses

The Norwalk-like caliciviruses (NLV) are enteric viruses (also known as small round structured viruses) and have been the cause of numerous food and waterborne outbreaks, associated with fecally-contaminated shellfish (Lipp and Rose, 1997), drinking water (Kaplan *et al.*, 1982; Lawson *et al.*, 1991; LeGuyader *et al.*, 1994) and recreational water (Kappus *et al.*, 1982). A diverse group of RNA viruses, they are a common cause of gastroenteritis, with diarrhea and/or vomiting, fever and respiratory symptoms lasting approximately 2 days. They were first identified with use of the electron microscope and include Norwalk virus, Snow Mountain agent, astroviruses and caliciviruses. There are extensive data showing these viruses are a major cause of shellfish-associated disease and may be the most significant group of viruses causing adult gastroenteritis (Lipp and Rose, 1997). Symptoms include vomiting, diarrhea, fever and, in some cases, respiratory illness. The viruses are heat stable and more resistant to chlorine, which is used to disinfect wastewaters, than the bacteria.

It has been suggested that these viruses are much more prevalent then previously thought. Reported in sewage at 10^7 RNA-containing particles per liter (Lodder *et al.*, 1999), a variety of genotypes could be detected, including those specific to outbreaks in the Netherlands.

4.4 Rotavirus

Rotavirus was first identified in 1973 and is believed to be responsible for the deaths of 4–5 million persons annually world-wide (Bern *et al.*, 1992). It is the major cause of viral gastroenteritis throughout the world and several waterborne outbreaks have been documented. In the USA, it is estimated that over one million cases of severe diarrhea in the 1–4 years age group are caused by rotaviruses annually, with up to 150 deaths (Ho *et al.*, 1988). In Third World countries it has been estimated that over 125 million cases of gastroenteritis due to rotavirus occur in children every year, with 18 million of these cases classified as moderately severe to severe. The very young, elderly, transplant patients and immunocompromised persons appear to be at greater risk of severe disease and mortality from rotavirus. The failure of a vaccine trial for young children continues to highlight the need to understand and prevent rotavirus transmission.

All of the rotavirus outbreaks have been associated with direct fecal contamination of a water supply or suboptimal drinking water treatment. Rotaviruses have been detected in surface waters world-wide with average concentrations ranging from 0.66 to 29 per liter (Gerba *et al.*, 1996b). The highest concentrations have been reported in surface waters receiving untreated sewage discharges. Rotavirus has also been detected in drinking water wells in Mexico and in Colombia (Deetz *et al.*, 1984; Toranzos *et al.*, 1988). The potential source of contamination was septic tank discharge into the groundwater.

While cell culture methods are available for detection of rotaviruses, they are cumbersome. These methods do not necessarily distinguish between human and animal rotaviruses. Throughout the animal kingdom there are many animal rotaviruses and natural reassortants arising from different species may appear including in humans (Vonsover *et al.*, 1993). Thus, RT-PCR is being used to detect and differentiate animal from human rotaviruses in water (Gratacap-Cavallier *et al.*, 2000).

5 FINAL COMMENTS

No doubt this chapter on emerging waterborne pathogens will need to be rewritten in another few years as new microorganisms are discovered or rediscovered and added to the list of concern. The detection of the bacterium *Tropheryma whippelii*, for example in sewage, using PCR (Maiwald *et al.*, 1998) makes one wonder if the faeco-oral and waterborne transmission of Whipple's disease (an intestinal lipodsystrophy disorder) is likely; this needs to be fully investigated. The recent terrorist act associated with the Klingerman virus, which causes dysentery with a high mortality, means that genetically engineered enteric organisms may now enter the wastestream and the water cycle. Only through continued monitoring studies and the application of new technologies can one address the emerging microorganisms that we are likely to see in this next century in the water environment.

REFERENCES

Abbaszadegan, M., Huber, M.S., Gerba, C.P. and Pepper, I. (1993). Detection of enteroviruses in groundwater using PCR. *Appl. Environ. Microbiol.* **59**, 1318–1324.

Aggarwal, R. and Naik, S.R. (1997). Epidemiology of hepatitis E: past, present and future. *Trop. Gastroenterol.* **18**, 49–56.

American Water Works Association Research Division Microbiological Contaminants Research Committee (1999). Committee Report: Emerging pathogens – viruses, protozoa, and algal toxins. *J. Am. Water Works Ass.* **91**, 110–121.

Aramini, J.J., Stephen, C. and Dubey, J.P. (1998). *Toxoplasma gondii* in Vancouver Island cougars (*Felis concolor vancouverensis*): Serology and oocyst shedding. *J. Parasit.* **84**, 438–440.

Aramini, J.J., Stephen, C., Dubey, J.P., *et al.* (1999). Potential contamination of drinking water with *Toxoplasma gondii* oocysts. *Epid. Infec.* **122**, 305–315.

Ashford, R.W. (1979). Occurrence of an undescribed coccidian in man in Papua New Guinea. *Ann. Trop. Med. Parasitol.* **73**, 497–500.

Axon, A.T.R. (1995). Review article: is *Helicobacter pylori* transmitted by the gastro-oral route? *Ailment. Pharmacol. Ther.* **9**, 585–588.

Balayan, M.S. (1997). Epidemiology of hepatitis E virus infection. *J. Viral Hepatitis* **4**, 155–165.

Barwick, R.S., Levy, D.A., Craun, G.F. *et al.* (2000). Surveillance for waterborne disease outbreaks – United States, 1997–1998. *MMWR* **49**, 1–35.

Baumann, P., Furniss, A.L. and Lee, J.V. (1984). Genus 1, *Vibrio*. In: N.R. Krieg and J.G. Holt (eds) *Bergey's Manual of Systematic Bacteriology*, pp. 518–538. Williams and Wilkins, Baltimore.

Basu, A., Garg, P., Datta, S. *et al.* (2000). *Vibrio cholerae* O139 in Calcutta 1992–1998: incidence, antibiograms, and genotypes. *Emerging Infectious Diseases* **6**, 139–147.

Benenson, M.W., Takafuji, E.T., Lemon, S.M. *et al.* (1982). Oocyst-transmitted toxoplasmosis associated with ingestion of contaminated water. *N. Engl. J. Med.* **307**, 666–669.

Bennett, J.V., Homberg, S.D., Rogers, M.F. and Soloman, S.L. (1987). Infectious and parasitic diseases. *Am. J. Preventive Med.* **55**, 102–114.

Bern, C., Martinew, J., deZoysa, I. and Glass, R.I. (1992). The magnitude of the global problem of diarrhoeal disease: a ten year update. *Bull. Wld. Hlth. Org.* **70**, 705–714.

Bilfulco, J.M. and Schaeffer, F.W. (1993). Antibody-magnetite method for selective concentration of *Giardia lamblia* cysts from water samples. *Appl. Environ. Microbiol.* **59**, 772–779.

Bowie, W.R., King, A.S., Werker, D.H. *et al.* (1997). Outbreak of toxoplasmosis associated with municipal drinking water. *Lancet* **350**, 173–177.

Bradley, D.W. (1992). Hepatitis E: epidemiology, aetiology and molecular biology. *Rev. Med. Virol.* **2**, 19–28.

Bryan, R.T. (1995). Microsporidiosis as an AIDS-related opportunistic infection. *Clin.Infect. Dis.* **21**, 62–65.

Bukhari, Z., McCuin, R.M., Fricker, C.R. and Clancy, J.L. (1998). Immunomagnetic separation of *Cryptosporidium parvum* from source water samples of various turbidities. *Appl. Environ. Microbiol.* **64**, 4495–4496.

Centers for Disease Control (1991). Waterborne-disease outbreaks, 1989–1990. *Mortality Weekly Report* **40**, 1–21.

Cotte, L., Rabodonirina, M., Chapuis, F. *et al.* (1999). Waterborne outbreak of intestinal microsporidiosis in persons with and without human immunodeficiency virus infection. *J. Infect. Dis.* **180**, 2003–2008.

Craun, G.F. (1996). Waterborne disease in the United States. In: G.F. Craun (ed.) *Balancing the microbial and chemical risks in drinking water disinfection*, pp. 55–77. ILSI Press, Washington, DC.

Dahling, D.R., Safferma, R.S. and Wright, B.A. (1989). Isolation of enterovirus and reovirus from sewage and treated effluents in selected Puerto Rican communities. *Appl. Environ. Microbiol.* **55**, 503–506.

D'Antonio, R.G., Winn, R.E. and Taylor, J.P. *et al.* (1985). A waterborne outbreak of cryptosporidiosis in normal hosts. *Ann. Intern. Med.* **103**, 886–888.

Dalldorf, G. and Sickles, G.M. (1948). An unidentified, filterable agent isolated from the feces of children with paralysis. *Science* **108**, 61–62.

Deetz, T.R., Smith, E.M., Goyal, S.M. *et al.* (1984). Occurrence of rota- and enteroviruses in drinking water and environmental water in a developing nation. *Wat. Res.* **18**, 567–571.

Deng, M.Q., Cliver, D.O. and Mariam, T.W. (1997). Immunomagnetic capture PCR to detect viable *Cryptosporidium parvum* oocysts from environmental samples. *Appl. Environ. Microbiol.* **63**, 3134–3138.

Dev, V.J., Main, M. and Gould, I. (1994). Waterborne outbreak of *Escherichia coli* O157:H7. *Lancet* **337**, 1412.

Didier, E.S. and Bessinger, G.T. (1999). Host-parasite relationships in microsporidiosis: animal models and immunology. In: M. Wittner and L.M. Weiss (eds) *The Microsporidia and Microsporidiosis*, pp. 225–257. ASM Press, Washington, DC.

DiGiovanni, G.D., Hashemi, F.H., Shaw, N.J. *et al.* (1999). Detection of infectious *Cryptosporidium parvum* oocysts in surface and filter backwash water samples by immunomagnetic separation and integrated cell culture-PCR. *Appl. Environ. Microbiol.* **65**, 3427–3432.

Dowd, S., Gerba, C., Enriquez, F. and Pepper, I. (1998a). PCR amplification and species determination of microsporidia in formalin-fixed feces after immunomagnetic separation. *Appl. Environ Microbiol.* **64**, 333–336.

Dowd, S., Gerba, C. and Pepper, I. (1998b). Confirmation of the human-pathogenic microsporidia *Enterocytozoon bieneusi, Encephalitozoon intestinalis and Vittaforma corneae* in water. *Appl. Environ. Microbiol.* **64**, 3332–3335.

Dowd, S.E., Gerba, C.P., Kamper, M. and Pepper, I. (1999). Evaluation of methodologies including immunofluorescent assay (IFA) and the polymerase chain reaction (PCR) for detection of human pathogenic microsporidia in water. *Appl. Environ. Microbiol.* **35**, 43–52.

Dubey, J.P., Speer, C.A. and Fayer, R. (1990). General biology of *Cryptosporidium*. In: J.P. Dubey, C.A. Speer and R. Fayer (eds) *Cryptosporidiosis of man and animals*, pp. 1–29. CRC Press, Boca Raton.

duMoulin, G.C. and Stottmeier, K.D. (1983). Waterborne mycobacteria: an increasing threat to health. *Am. Soc. Microbiol. News* **52**, 525–529.

Eurogast Study Group (1993). An international association between *Helicobacter pylori* infection and gastric cancer. *Lancet* **341**, 1359–1362.

Gasser, R.B. and O'Donoghue, P. (1999). Isolation, propagation and characterization of *Cryptosporidium. Int J. Parasitol.* **29**, 1379–1413.

Geldreich, E.E., Fox, K.R., Goodrich, J.A. *et al.* (1992). Searching for a water supply connection in the Cabool Missouri disease outbreak of *Escherichia coli* O157:H7. *Wat. Res.* **26**, 1127–1137.

Gerba, C.P. (1996). Pathogens in the environment. In: I.L. Pepper, C.P. Gerba and M.L. Brusseau (eds) *Pollution Science*, pp. 279–300. Academic Press, New York.

Gerba, C.P. and Rose, J.B. (1989). Viruses in source and drinking water. In: G.A. McFeters (ed.) *Drinking Water Microbiology*, pp. 308–396. Springer-Verlag, New York.

Gerba, C.P., Rose, J.B. and Haas, C.N. (1996a). Sensitive populations: who is at the greatest risk? *Int. Jour. Food Microbiol.* **30**, 113–123.

Gerba, C.P., Rose, J.B., Haas, C.N. and Crabtree, K.D. (1996b). Waterborne rotavirus: a risk assessment. *Wat. Res.* **12**, 2929–2940.

Gibbons, C.L., Gazzard, B.G., Ibrahim, M. *et al.* (1998). Correlation between markers of strain variation in *Cryptosporidium parvum*: evidence of clonality. *Parasitol. Int.* **47**, 139–147.

Glover, N., Holtzman, A., Aronson, T. *et al.* (1994). The isolation and identification of *Mycobacterium avium* complex (MAC) recovered from Los Angeles potable water, a possible source of infection in AIDS patients. *Int. J. Environ. Health Res.* **4**, 63–72.

Goodgame, R.W. (1996). Understanding intestinal spore-forming protozoa: *Cryptosporidia, Microsporidia, Isospora* and *Cyclospora. Ann. Intern. Med.* **124**, 429–441.

Grabow, W.O.K., Favorov, M.O., Khudyakova, N.S. *et al.* (1994). Hepatitis E seroprevalence in selected individuals in South Africa. *J. Med Virol.* **44**, 384–388.

Grabow, W.O.K., Botma, K.D., de Villiers, J.C. *et al.* (1999). Assessment of cell culture and polymerase chain reaction procedures for the detection of polioviruses in wastewater. *Bull. Wld. Hlth. Org.* **77**, 973–980.

Gratacap-Cavallier, B., Genoulaz, O., Brengel-Pesce, K. *et al.* (2000). Detection of human and animal rotavirus sequences in drinking water. *Appl. Environ. Microbiol.* **66**, 2690–2692.

Greene, M.B., Gurdip, S.S., Sidhu, S. *et al.* (1982). *Mycobacterium avium-intracellulare*: A cause of disseminated life-threatening infections in homosexuals and drug abusers. *Ann. Intern. Med.* **97**, 539–546.

Hancock, C.M., Rose, J.B. and Callahan, M. (1998). *Crypto* and *Giardia* in US groundwater. *J. Am. Wat. Works Assoc.* **90**, 58–61.

Hejkal, T.W., Keswick, B., LaBelle, R.L. *et al.* (1982). Viruses in a community water supply associated with an outbreak of gastroenteritis and infectious hepatitis. *J. Am. Water Works Assoc.* **74**, 318–321.

Ho, M.S., Glass, R.I., Pinshy, P.F. and Anderson, L.L. (1988). Rotavirus as a cause of diarrheal morbidity and mortality in the United States. *J. Infect. Dis.* **158**, 1112–1116.

Huang, P., Weber, J.T., Sosin, D.M. *et al.* (1995). The first reported outbreak of diarrheal illness associated with *Cyclospora* in the United States. *Ann. Int. Med.* **123**, 409–414.

Hurst, C.J., Clark, R.M. and Regli, S.E. (1996). Estimating the risk of acquiring infectious disease from ingestion of water. In: C.J. Hurst (ed.) *Modeling Disease Transmission and Its Prevention by Disinfection*, pp. 99–139. Cambridge University Press, Cambridge.

Hurst, C.J. (1999). Caliciviruses. In: D. Talley, M. Malgrande and S. Nakauchi-Hawn (eds) *Manual of Water Supply Practices: Waterborne Pathogens*, pp. 231–234. American Water Works Association, Denver.

Hutin, Y.J.F., Pool, V., Cramer, E.H. *et al.* (1999). A multistate, foodborne outbreak of hepatitis *A. N. Eng. J. Med.* **340**, 595–602.

Isaac-Renton, J., Bowie, W.R., King, A. *et al.* (1998). Detection of *Toxoplasma gondii* oocysts in drinking water. *Appl. Environ. Microbiol.* **64**, 2278–2280.

Jakubowski, W., Boutros, S., Faber, W. *et al.* (1996). Environmental methods for *Cryptosporidium*. *J. Am. Wat. Works Assoc.* **88**, 107–121.

Jinneman, K.C., Wetherington, H., Omiescinski, W.E. *et al.* (1999). An oligonucleotide-ligation assay for the differentiation between *Cyclospora* and *Eimeria* spp. polymerase chain reaction amplification products. *J. Food Protection* **62**, 682–685.

Johnson, D.W., Pieniazek, N.J., Griffin, D.W. *et al.* (1995). Development of a PCR protocol for sensitive detection of *Cryptosporidium* oocysts in water samples. *Appl. Environ. Microbiol.* **61**, 3849–3855.

Jothikumar, N., Aparna, K., Kamatcdhiammal, S. *et al.* (1993). Detection of hepatitis E virus in raw and treated wastewater with the polymerase chain reaction. *Appl. Environ. Microbiol.* **59**, 2558–2562.

Kaplan, J.E., Goodman, R.A., Schonberger, L.B. *et al.* (1982). Gastroenteritis due to Norwalk virus: an outbreak associated with municipal water system. *J. Infect. Dis.* **146**, 190–197.

Kappus, K.D., Marks, J.S., Holman, R.C. *et al.* (1982). An outbreak of Norwalk gastroenteritis associated with swimming in a pool and secondary person-to-person transmission. *Am. J. Epidemiol.* **116**, 834–839.

Kiode, H., Kitaura, Y., Deguchi, H. *et al.* (1992). Genomic, detection of enteroviruses in the myocardium studies on animal hearts with coxsackievirus B3 myocarditis and endomyocardial biopsies from patients with myocarditis and dilated cardiomyopathy. *Jap. Circulation J.* **56**, 1081–1093.

Klein, P.D., Graham, D.Y., Gaillour, A. *et al.* (1991). Water source as risk factor for *Helicobacter pylori* infection in Peruvian children. *Lancet* **337**, 1503–1506.

Klingel, K., Hohenadl, C., Canu, A. *et al.* (1992). Ongoing enterovirus-induced myocarditis is associated with persistent heart muscle infection: quantitative analysis of virus replication, tissue damage and inflammation. *Proc. Natl. Acad. Sci.* **89**, 314–318.

Krikelis, V., Spyrou, N., Markoulatos, P. and Serie, C. (1985). Seasonal distribution of enteroviruses and adenoviruses in domestic sewage. *Can. J. Microbiol.* **31**, 24–25.

Krikelis, V., Markoulatos, P. and Spyrou, N. (1986). Viral pollution of coastal waters resulting from the disposal of untreated sewage effluents. *Wat. Sci. Tech.* **18**, 43–48.

Kostrzynska, M., Sankey, M., Haack, E. *et al.* (1999). Three sample preparation protocols for polymerase chain reaction based detection of *Cryptosporidium parvum* in environmental samples. *J. Microbiol. Methods* **35**, 65–71.

Lawson, H.W., Brayn, M.M., Glass, R.I.M. *et al.* (1991). Waterborne outbreak of Norwalk virus gastroenteritis at a Southwest U.S. resort; role of geological formations in contamination of well water. *Lancet* **337**, 1200–1204.

Lee, S.H., Lai, S.T., Lai, J.Y. and Leung, N.K. (1996). Resurgence of *cholera* in Hong Kong. *Epidemiol. Infect.* **117**, 43–49.

LeGuyader, F., Dubois, E., Menard, D. and Pommepuy, M. (1994). Detection of hepatitis A virus, rotavirus and enterovirus in naturally contaminated shellfish and sediment by reverse transcription-seminested PCR. *Appl. Environ. Microbiol.* **60**, 3665–3671.

Lewis, G.D., Molloy, S.L., Greening, G.E. and Dawson, J. (2000). Influence of environmental factors on virus detection by RT-PCR and cell culture. *J. Appl. Microbiol.* **88**, 633–640.

Lipp, E.K. and Rose, J.B. (1997). The role of seafood in foodborne diseases in the United States of America. *Rev. Sci. Tech. Off. Int. Epid.* **16**, 620–640.

Lisle, J.T. and Rose, J.B. (1995). *Cryptosporidium* contamination of water in the USA and UK: a mini review. *J. Wat. Supply Res. Technol.* **44**, 103.

Lodder, W.J., Vinje, J., van de Heide, R. *et al.* (1999). Molecular detection of norwalk-like caliciviruses in sewage. *Appl. Environ. Microbiol.* **65**, 5624–5627.

Long, E.G., White, H., Carmichael, W.C. *et al.* (1991). Morphologic and staining characteristics of a cyanobacterium-like organism associated with diarrhea. *J. Infect. Dis.* **164**, 199–202.

Lucena, F., Bosch, A., Jofre, J. and Schwartzbrod, L. (1985). Identification of viruses isolated from sewage, riverwater and coastal seawater in Barcelona. *Wat. Res.* **19**, 1237–1239.

MacKenzie, W.R., Hoxie, N.J., Proctor, M.E. *et al.* (1994). A massive outbreak in Milwaukee of *Cryptosporidium* infection transmitted through the public water supply. *N. Engl. J. Med.* **331**, 161–167.

Maiwald, M., Schuhmacher, F., Hans-Jurgen, D. and von Herbay, A. (1998). Environmental occurrence of the Whipple's disease bacterium. *Applied Environ. Microbiol.* **64**, 760–762.

Matsubayashi, H., Koike, T., Mikata, I. *et al.* (1959). A case of encephalitozoon-like body in man. *Arch. Pathol.* **67**, 181–187.

Melnick, J.L. (1995). History and epidemiology of hepatitis A virus. *J. Infect. Dis.* **171**, S2–S8.

Mead, P.S. and Griffin, P.M. (1998). *Escherichia coli* 0157:H7. *Lancet* **352**, 1207–1212.

Meisel, J.L., Perera, D.R., Meligro, C. and Rubin, C.E. (1976). Overwhelming watery diarrhea associated with a *Cryptosporidium* in an immunosuppressed patient. *Gastroenterology*, **70**: 1156.

Meng, J., Dobreil, P. and Pillot, J. (1997). A new PCR based seroneutralization assay in cell culture for diagnosis of hepatitis *E. J. Clin. Microbiol.* **35**, 1373–1377.

Michino, I.K., Araki, H., Minami, S. *et al.* (1999). Massive outbreak of *Escherichia coli* 0157:H7 infection in schoolchildren in Sakai City, Japan, associated with consumption of white radish sprouts. *Am. J. Epidemiol.* **150**, 787–796.

Morbidity and Mortality Weekly Report (1995). Update, *Vibrio cholerae* O1 Western hemisphere, 1991–1994, and *V. cholerae* O139-Asia, 1994. *MMWR* **44**, 215–219.

Murrin, K. and Slade, J. (1997). Rapid detection of viable enteroviruses in water by tissue culture and semi-nested polymerase chain reaction. *Wat. Sci. Tech.* **35**, 429–432.

Mushahwar, R., Purcell, M. and Emerson, S.U. (1998). Genetic and experimental evidence for cross-species infection by swine hepatitis E virus. *J. Virol.* **72**, 9714–9721.

National Institute of Medicine (NIM) (1998). *Recommendations for changes in the organization of federal food safety responsibilities, 1949–1997.* National Academy Press, Washington, DC.

Ortega, Y.R., Sterling, C.R., Gilman, R.H. *et al.* (1993). *Cyclospora species* – a new protozoan pathogen of humans. *N. Engl. J. Med.* **328**, 1308–1312.

Payment, P. (1997). Cultivation and assay of viruses. In: C.J. Hurst, G.R. Knudsen, M.J. McInerney, L.D. Stetzenbach and M.V. Walter (eds) *Manual of Environmental Microbiology*, pp. 72–78. ASM Press, Washington, DC.

Payment, P., Trudel, M. and Plante, R. (1985). Elimination of viruses and indicator bacteria at each step of treatment during preparation of drinking water at seven water treatment plants. *Appl. Environ. Microbiol.* **49**, 1418–1428.

Pina, S., Jofre, J., Emerson, S.U. *et al.* (1998). Characterization of a strain of infectious hepatitis E virus isolated from sewage in an area where hepatitis E is not endemic. *Appl. Environ. Microbiol.* **64**, 4485–4488.

Pollitzer, P. (1959). *Cholera*. Monograph Series No. 43. World Health Organization, Geneva.

Quintero-Betancourt, W., Peele, E.R. and Rose, J.B. (2002). *Cryptosporidium parvum* and *Cyclospora cayetanensis*: a review of laboratory methods for detection of these waterborne parasites. *J. Microbiol. Methods* **49**, 209–224.

Rabold, J.G., Hoge, C.W., Shlim, D.R. *et al.* (1994). *Cyclospora* outbreak associated with chlorinated drinking water. *Lancet* **344**, 1360–1361.

Reynolds, K.A., Gerba, C.P. and Pepper, I. (1996). Detection of infectious enteroviruses by an integrated cell culture-PCR procedure. *Appl. Environ. Microbiol.* **62**, 1424–1427.

Riley, L.W., Remis, R.S., Helgerson, S.D. *et al.* (1983). Hemorragic colitis associated with a rare *E.coli* serotype. *N. Engl. J. Med.* **308**, 681–685.

Rochelle, P.A., De Leon, R., Stewart, M.H. and Wolfe, R. (1997). Comparison of primers and optimization of PCR conditions for detection of *C. parvum* and *G. lamblia* in water. *Appl. Environ. Microbiol.* **63**, 106–114.

Rose, J.B. and Slifko, T.R. (1999). *Giardia*, *Cryptosporidium*, and *Cyclospora* and their impact on foods: A review. *J. Food Protect.* **62**, 1059–1070.

Shadduck, J.A. and Polley, M.B. (1978). Some factors influencing the *in vitro* infectivity and replication of *Encephalitozoon cuniculi*. *J. Protozool.* **25**, 491–496.

Shadduck, J.A. (1989). Human microsporidiosis and AIDS. *Rev. Infect. Dis.* **11**, 203–207.

Sharma, C., Nair, G.B., Mukhopadhyay, A.K. *et al.* (1997). Molecular characterization of *V. cholerae* O1 biotype El Tor strains isolated between 1992 and 1995 in Calcutta India: evidence for the emergence of a new clone of the El Tor biotype. *J. Infect Dis.* **175**, 1134–1341.

Singh, N. and Lu, V. (1994). Potable water and *Mycobacterium avium* complex in HIV patients: is prevention possible? *Lancet* **343**, 1110–1111.

Slifko, T.R., Friedman, D.E., Rose, J.B. *et al.* (1997). An *in vitro* method for detection of infectious *Cryptosporidium* oocysts using cell culture. *Appl. Environ. Microbiol.* **63**, 3669–3675.

Slifko, T.R., Huffman, D.E. and Rose, J.B. (1999). A most probable assay for enumeration of infectious *Cryptosporidium parvum* oocysts. *Appl. Environ. Microbiol.* **65**, 3936–3941.

Smith, J.E. (2001). A Review of Hepatitis E virus. *J. Food Prot.* **64**, 572–586.

Soave, R. and Johnson, W.D. Jr (1995). *Cyclospora*: conquest of an emerging pathogen. *Lancet* **345**, 667–668.

Sobsey, M. (1999). Hepatitis A. In: D.V. Talley, M. Malgrande and S. Nakauchi-Hawn (eds) *Waterborne Pathogens: Manual of Water Supply Practices*, pp. 241–246. American Water Works Association, Denver.

Sparfel, J.M., Sarfati, C., Ligoury, O. *et al.* (1997). Detection of microsporidia and identification of *Enterocytozoon bieneusi* in surface water by filtration followed by specific PCR. *J. Eukaryot. Microbiol.* **44**, 78.

Swerdlow, D.L., Woodruff, B.A., Brady, R.C. *et al.* (1993). A waterborne outbreak in Missouri of *Escherichia coli* O157:H7 associated with bloody diarrhea and death. *Ann. Int. Med.* **117**, 812–819.

Taylor, D.N. and Blaser, M.J. (1991). The epidemiology of *Helicobacter pylori* infection. *Epidemiol. Rev.* **13**, 42–58.

Thomas, J.E., Gibson, G.R., Darboe, M.K. *et al.* (1992). Isolation of *Helicobacter pylori* from human faeces. *Lancet* **340**, 1194–1195.

Toranzos, G., Gerba, C.P. and Hansen, H. (1988). Enteric viruses and coliphages in Latin America. *Toxicity Assessment: Int. J.* **3**, 491–510.

Tsarev, S.A., Binn, L.N., Gomatos, P.J. *et al.* (1999). Phylogenetic analysis of hepatitis E virus isolates from Egypt. *J. Med. Virol.* **57**, 68–74.

Velázquez, M. and Feirtag, J.M. (1999). *Helicobacter pylori*: characteristics, pathogenicity, detection methods and mode of transmission implicating foods and water. *Int. J. of Food Microbiol.* **53L**, 95–104.

Vonsover, A., Shif, I., Silberstein, I. *et al.* (1993). Identification of feline and canine-like rotaviruses isolated from humans by restriction fragment length polymorphism assay. *J. Clin. Microbiol.* **31**, 1783–1787.

von Reyn, C.F., Maslow, J.N., Barber, T.W. *et al.* (1994). Persistent colonization of potable water as a source of *Mycobacterium avium* infection in AIDS. *Lancet* **343**, 1137–1141.

Wagenknecht, L.E., Roseman, J.M. and Herman, W.H. (1991). Increased incidence of insulin-dependent diabetes mellitus following an epidemic of coxsackievirus B5. *Am. J. Epidemiol.* **133**, 1024–1031.

Waller, T. (1979). Sensitivity of *Encephalitozoon cuniculi* to various temperatures, disinfectants and drugs. *Lab Anim.* **13**, 227–230.

Walter, R., Dobberkau, H.J., Bartlelt, W. *et al.* (1982). Long-term study of virus contamination of surface water in the German Democratic Republic. *Bull. Wld. Hlth. Org.* **60**, 789–795.

Warren, J.R. and Marshall, B. (1983). Unidentified curved bacilli on gastric epithelium in active chronic gastritis. *Lancet*, 1273–1275.

Weber, R., Schwartz, D.A. and Deplazes, P. (1999). Culture and propagation of microsporidia. In: M. Wittner and L.M. Weiss (eds) *The Microsporidia and Microsporidiosis*, pp. 363–392. ASM Press, Washington, DC.

Wiedenmann, A., Krüger, P. and Botzenhart, K. (1998). PCR detection of *Cryptosporidium parvum* in environmental samples – a review of published protocols and current developments. *J. Industrial Microbiol. Biotechnol.* **21**, 150–166.

Wolk, D.M., Johnson, C.H., Rice, E.W. *et al.* (2000). A spore counting method and cell culture model for chlorine disinfection studies of *Encephalitozoon* syn. *Septata intestinalis*. *Appl. Environ. Microbiol.* **66**, 1266–1273.

Xiao, L., Escalante, L., Yan, Ch. *et al.* (1999). Phylogenetic analysis of *Cryptosporidium* parasites based on the small-subunit rRNA gene locus. *Appl. Environ. Microbiol.* **65**, 1578–1583.

Yates, M.V. and Yates, S.R. (1988). Modeling microbial fate in the subsurface environment. *Crit. Rev. Environ.Control.* **17**, 307–344.

Yoder, K.E., Sethabutr, O. and Relman, D.A. (1996). PCR-based detection of the intestinal pathogen *Cyclospora*. In: D.H. Persing (ed.) *PCR Protocols for Emerging Infectious Diseases, a Supplement to Diagnostic Molecular Microbiology: Principles and Applications*, pp. 169–176. ASM Press, Washington, DC.

13

Health effects of water consumption and water quality

Pierre Payment

Centre de Recherche en virologie, INRS-Institut Armand-Frappier, Laval, Quebec H7V 1B7, Canada

The subject of microbiological safety of water will very soon become an international priority as travel across national boundaries and the sheer numbers of human citizens increase in the decade ahead, placing Promethean demands on water resources.

(Ford and Colwell, 1996)

1 INTRODUCTION

It was only at the beginning of the 19th century with the advent of water filtration, wastewater disposal, disinfection of drinking water with chlorine, pasteurization of milk and food, and by a general enhancement of hygiene, that waterborne pathogenic microorganisms and their diseases were finally controlled to an acceptable level in most countries. Waterborne diseases should not be seen as an independent part of the infectious disease cycle, but as a vehicle for their transmission. While we assume that a significant proportion of these gastrointestinal illnesses may be waterborne, we have no data to estimate the proportion of the overall burden of disease they represent (Figure 13.1). In a holistic approach, a reduction of waterborne disease should also result in a measurable reduction of the overall rate of gastrointestinal illnesses (Mara and Cairncross, 1989). This is the result of limiting person-to-person transmission and also the result of reducing the risk of the contamination of foods by contaminated water or individuals that have been infected through drinking water (Figure 13.2).

At the onset of the third millennium, what do we know of the health effects of drinking water and their impact on our so-called modern societies? There is a pleiad of reports on the impact of waterborne diseases in countries world-wide, revealing thousands of outbreaks due to bacterial, viral, and parasitic microorganisms associated with the consumption of untreated or improperly treated drinking water (WHO, 1993, 1996; Ford and Colwell, 1996; Fewtrell and Bartram, 2001). These reports emphasize the fact that access to water and water quality are in direct relationship to life expectancy and child mortality, as countries with poor access to drinking water have the highest morbidity rates in children under 5 years of age.

2 WATERBORNE OUTBREAKS

Numerous enteric pathogens have been involved in waterborne outbreaks and recognized as such. Others are just emerging as being implicated in diseases transmitted by the water route (LeChevallier *et al.*, 1999a,b). Urbanization, ageing of water treatment plants, increasing numbers of immunocompromised individuals and ageing populations are potential causes for an increased risk of waterborne infectious diseases (Ford and Colwell, 1996; Fewtrell and Bartram, 2001). As we will see later, the endemic level of disease due to

The Handbook of Water and Wastewater Microbiology
ISBN 0-12-470100-0

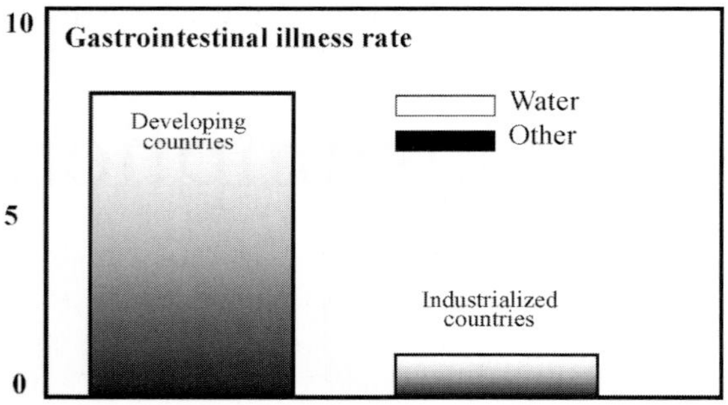

Fig. 13.1 The rate of gastrointestinal illnesses in developing countries is much less than in developing countries, but we still know very little about the proportion that can be assigned to water compared to other routes of transmission.

drinking-water consumption has also been shown to be quite significant and it contributes to maintaining these pathogens in the affected populations. The dilemma of balancing microbial and cancer health risks is also a difficult one to resolve, but it should not result in a reduction of treatment efficiency. The low risk level for cancer is insignificant when compared to the risk of waterborne infectious disease in absence of adequate water treatment (Payment and Hartemann, 1998; Fewtrell and Bartram, 2001).

The microorganisms implicated are described elsewhere in this book as well as in many recent publications on clinical and environmental microbiology (Murray *et al.*, 1995; Hurst *et al.*, 2001). Waterborne diseases are usually described in terms of outbreak reporting in the various countries and the USA has produced most of the available data through the decades-long effort of Gunther Craun (Craun, 1984, 1986, 1990) and a continuous effort to collect data (Herwaldt *et al.*, 1991, 1992). In other countries, data gathering is often performed very poorly because of lack of resources to identify the water-related events as well as the lack of centralized data gathering official authorities (Andersson and Bohan, 2001). Several methods for the detection and investigation of waterborne outbreaks have been described, but are still not widely used (Craun, 1990; Andersson and Bohan, 2001) as resources and funds are critically lacking, even in industrialized countries (Fewtrell and Bartram, 2001). An enormous effort is needed to educate all on the importance of water in the dissemination of disease. All levels of society, from consumer to politicians, must be educated to the necessity of improving water quality (Ford and Colwell, 1996) as a major step in improving the quality of life and health.

3 THE LESSONS OF THE RECENT DECADES

In all countries, a steady decline in waterborne disease was evidenced by the virtual elimination of cholera and the reduction of waterborne outbreaks to very low levels in most countries.

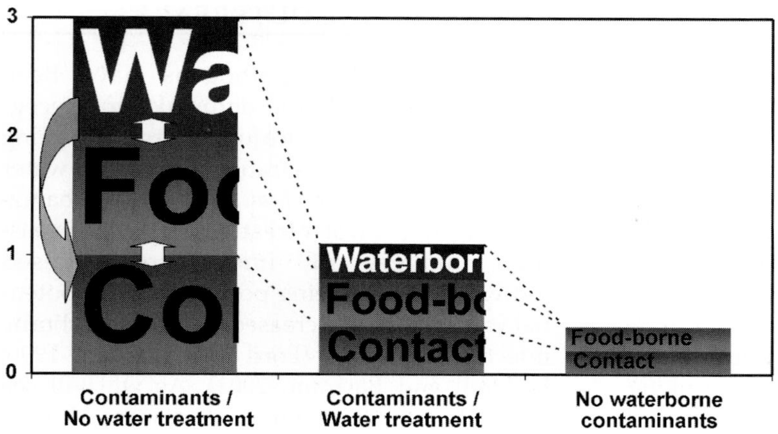

Fig. 13.2 Interrelations between waterborne disease and other routes of infection: reducing the role of water as a vehicle of transmission will significantly reduce the level of infectious disease in the community.

Most bacterial waterborne pathogens have been eliminated by the simple use of chlorine disinfection. There are still reports of bacterial disease outbreaks, mainly attributable to untreated water (non-disinfected groundwater), failure of disinfection when it is used as the sole barrier against microorganisms and finally re-contamination of the water in the distribution system. Major outbreaks such as the Walkerton outbreak of *Escherichia coli* in Canada (Anonymous (2000)) act as a reminder that the barrier must be maintained at all times.

However, we are finding strains of *Vibrio cholerae* that are more resistant to disinfection, *Legionella* has been found in water heaters and the *Mycobacterium avium* complex is now on the list of potential pathogens (LeChevallier *et al.*, 1999b; AWWA, 1999).

Since the 1950s, with the development of methods to detect and identify viruses, many outbreaks of waterborne gastrointestinal illness that would otherwise have been classified as of non-bacterial origin were attributed to enteric viruses. Hepatitis A and E, Norwalk, small round structured viruses, astroviruses, caliciviruses and many others are now well-known names in the water industry (LeChevallier *et al.*, 1999a; AWWA, 1999).

Parasites are also being identified as pathogens of importance, even in industrialized countries (Ortega, 1993; Chiodini, 1994; LeChevallier *et al.*, 1999a; AWWA 1999). The recognition that giardiasis was waterborne, with numerous outbreaks in the USA (Craun, 1986) and several outbreaks of cryptosporidiosis in England (Badenoch, 1990a,b) and in many other countries (Payment and Hunter, 2001) have now shifted the focus to these parasites, which are extremely resistant to the water disinfection process. Dozens of outbreaks of cryptosporidiosis have now been reported world-wide, but most are small compared to the explosive outbreak experienced in Milwaukee (USA) in the spring of 1993 (Edwards, 1993; Mackenzie *et al.*, 1994). Following what appears to be a failure in treatment, water that met the United States Environmental Protection Agency water quality guidelines may have caused gastrointestinal illnesses in over 400 000 people or one-third of the population of this city over a period of one month. Most of these illnesses were cryptosporidiosis, but many were probably of viral origin and the number of illnesses reported may have been overestimated by 100-fold (Hunter and Syed, 2001). The most surprising aspect of this event is that cryptosporidiosis was probably occurring even before it was detected following a report from an inquisitive pharmacist (Morris *et al.*, 1998). This suggests that unless an effort is made to identify waterborne diseases they will remain undetected, buried in the endemic level of illness in the population (Payment and Hunter, 2001).

4 WATERBORNE DISEASE IS NOT ONLY DIARRHOEA

While the most often reported disease associated with drinking water remains gastroenteritis, this is probably due to the very apparent nature of the symptoms and the fact that the attack rates for these infections can often reach over 50% of the exposed population. Even infectious disease specialists often forget that enteric microorganisms are associated with a wide range of symptoms and diseases (Table 13.1). Protozoan parasites, such as amoebae, can cause severe liver or brain infections and contact lens wearers are warned of the dangers of eye infections. Bacteria can cause pneumonia (*Legionella*) and some are suggesting the possibility that *Helicobacter pylori*, which has been associated with gastric ulcers, could be transmitted by the water route. Among the 72 known enteroviruses, many can induce illnesses other than gastrointestinal, such as hepatitis (viral type A or E), poliomyelitis, viral meningitis, enteroviral carditis, epidemic myalgia, diabetes, ocular diseases. Some viruses can also induce abortions, stillbirth and fetal abnormalities (Payment, 1993a,b). Who can say that a death due to myocardial failure is not the result of a waterborne coxsackie virus infection months or years earlier?

TABLE 13.1 Enteric viruses and associated diseases or symptoms

Symptoms/diseases	*Viruses*
Minor malaise, asymptomatic diseases	All
Paralysis, poliomyelitis and polio-like	Poliovirus, coxsackie virus, enteroviruses 70–71
Aseptic meningitis and mild paresis	Poliovirus, coxsackie virus (B1–6, A7, A9)
Eye diseases, conjunctivitis	Echoviruses, coxsackie virus (B2, B4, A24), enterovirus 70,
Exanthems (Hand-foot-and-mouth)	Enterovirus 71, coxsackie viruses A, echoviruses
Cardiac diseases (myocarditis, chronic)	Coxsackie virus A and B, echoviruses
Pleurodynia and epidemic myalgia	Coxsackie virus A and B
Meningoencephalitis	Enterovirus 70–71
Respiratory illnesses	Coxsackie virus, echoviruses, rotaviruses
Viral gastroenteritis	Enteroviruses, calicivirus, astrovirus, rotavirus, coronavirus, adenovirus 40–41, etc.
Neonatal diseases	Coxsackie viruses A and B
Viral hepatitis	Hepatitis A and E viruses, coxsackie viruses A and B
Diabetes	Coxsackie virus B4
Post-viral fatigue syndrome	Enteroviruses, coxsackie viruses B

5 ENDEMIC WATERBORNE DISEASE IN INDUSTRIALIZED COUNTRIES: EARLY EPIDEMIOLOGICAL STUDIES

The fact that waterborne disease is still a major problem in developing countries does not need to be reaffirmed. In this chapter, we will thus emphasize the current state of knowledge of waterborne diseases in countries that have historically been assuming that their populations were not exposed to risks from their drinking water, as this was assumed to be adequately treated.

In the absence of evident acute health effects (i.e. epidemics or outbreaks), epidemiological studies have been centred on long-term effects of potentially carcinogenic chemicals (Crump and Guess, 1982). Birth defects or abortions have also been associated with the consumption of tap water (Swan *et al.*, 1992; Wyndham *et al.*, 1992), even though much of this effect might have been due to recall bias during retrospective and matched-control studies (Hertz-Picciotto *et al.*, 1992). Such epidemiological studies are extremely difficult to assess and are often never confirmed because there are too many confounding variables in studies that attempt to analyse lifetime exposure to various factors.

Batik *et al.* (1979), using hepatitis A virus cases as an indicator, could not establish a correlation with water quality and could not find a correlation between current indicators and the risk of waterborne outbreaks (Batik *et al.*, 1983).

In France, the group of Collin *et al.* (1981) prospectively studied the gastrointestinal illnesses associated with the consumption of tap water using reports from physicians, pharmacists and teachers. Their results were based on more than 200 distribution systems of treated or untreated water and they reported five epidemics (more than 1000 cases) associated with poor quality water. This study is typical of most studies which relied on the detection of epidemics to assess the level of water quality: they do not address the endemic level of gastrointestinal illnesses which may be due to low level contamination of the water. The same group, in a prospective follow-up study on 48 villages for 64 weeks, has evaluated the gastrointestinal illneses associated with untreated ground water and found a relationship between faecal streptococci and acute gastrointestinal disease (Ferley *et al.*, 1986; Zmirou *et al.*, 1987). Faecal coliforms did not appear to be related to acute disease. Total coliforms and total bacteria had no independent contributions to disease but, even in the absence of all measured indicator germs, one-fourth of the cases were still observed.

In Israel, Fattal *et al.* (1988) addressed the health effects of both drinking water and aerosols. Their studies on kibbutz water quality and morbidity were performed in an area with relatively high endemicity of gastrointestinal disease and did not show a relationship between health effects and total or faecal coliforms. This study was, however, based only on morbidity reported to physicians, data that are considered to represent only 1% of the actual cases in a population.

In Windhoek (Namibia), Isaäckson and Sayed (1988) conducted a similar study over several years on thousands of individuals served by recycled wastewater as well as normal drinking water. They did not observe an increased risk of reported acute gastrointestinal illness associated with the consumption of recycled waters. The populations compared had higher incidence rates than the one observed in North America and they were subjected to a high endemicity level due to other causes, thus masking low levels of illnesses.

6 THE CANADIAN EPIDEMIOLOGICAL STUDIES

Two major epidemiological studies were conducted in Canada in 1988–89 and in 1993–94. The results of these studies suggest that a very high proportion of gastrointestinal illnesses could still be attributable to tap water consumption even when water met the current water quality guidelines (Payment *et al.*, 1991a,b, 1997).

6.1 First study

The first study was carried out from September 1988 to June 1989. It was a randomized intervention trial carried out on 299 randomly selected eligible households which were supplied with domestic water filters (reverse osmosis(RO)) that eliminated microbial and most chemical contaminants from their tap water, and on 307 randomly selected households which were left with their usual tap water without treatment. The gastrointestinal symptomatology was evaluated by means of a family health diary maintained prospectively by all study families. The estimated annual incidence of gastrointestinal illness was 0.76 among tap-water drinkers as compared with 0.50 among RO-filtered-water drinkers ($P <$ 0.01). Because participants in the RO-filter group were still exposed to tap water (i.e. about 40% of their water intake was tap water), it was estimated that about 50% of the illnesses were probably tap water-related and preventable. The remaining illnesses were probably attributable to the other possible causes such as endemic infectious illnesses, food-related infections, allergies, etc.

The fact that the participants in the study, while randomized into two groups, were not blinded to the type of water they consumed was considered a confounding factor that could have affected the results. However, the rate of illness increased with the amount of water consumed (i.e. a dose–response effect was demonstrable), suggesting that the effect observed was probably not due to bias.

Attempts were made to correlate microbiological data obtained on water samples from the water distribution system. There was no correlation between the number of episodes and total or faecal coliforms, chlorine or heterotrophic bacteria in the tap water. There was an association between the duration of illness and heterotrophic plate count (HPC) bacteria at 20°C (Payment *et al.*, 1993). However, due to the large number of statistical analyses performed this could have been a spurious result due to chance only.

In the RO-filtered water study group, there was a significant correlation between the rate of illness and the HPC bacteria growing at 35°C on R2A medium, but no relationship with the amount of water consumed (Payment *et al.*, 1991b). The rate of water-related illness increased with distance from the plant and bacterial regrowth was suggested as an explanation. Studies of the virulence of bacteria isolated from the tap water revealed that a small fraction of these bacteria could be considered as potential pathogens (Payment *et al.*, 1994a) and it was suggested that their multiplication to high numbers could be a health risk (Payment, 1995).

Attempts were also made to determine the aetiology of the observed illnesses. Sera had been collected on four occasions from volunteers and they were tested for antibodies to various pathogens. There was no indication by serology for water-related infections by enteroviruses, hepatitis A virus and rotavirus (Payment, 1991) or Norwalk virus infections (Payment *et al.*, 1994).

6.2 Second study

The second study (Payment, 1997) was more complex because it attempted not only to re-evaluate the level of waterborne illness, but also to identify the source of the pathogens responsible for them. It was conducted from September 1993 to December 1994 and it compared the levels of gastrointestinal illness in four randomly selected groups of 250 families, which were served water from one of the following sources:

- Tap water (normal tap water)
- Tap water with a valve on the cold water line (to examine the effect of home plumbing)
- Plant effluent water as it left the plant and bottled (i.e. not influenced by the distribution system)
- Plant effluent water further treated and bottled (to remove any contaminants).

The site was selected for the high level of microbial contamination of the river water that it treats and for the quality of operation of the water treatment plant. Raw water entering the plant was contaminated with parasites, viruses and bacteria at levels found throughout the world in faecally-contaminated waters. The water treatment plant produced water that met or exceeded current Canadian and US regulations for drinking-water quality. The distribution system was in compliance for coliforms but residual chlorine was not detectable at all times in all parts of the distribution system.

The rates of highly credible gastrointestinal illnesses (HCGI) were within the expected range for this population at 0.66 episodes/person-year for all subjects and 0.84 for children 2 to 12 years old. The rate of illness was highest in autumn–winter and lowest in summer. Overall, there were more illnesses among tap-water consumers than among subjects in the purified (bottled) water group, suggesting a potential adverse effect originating from the plant or the distribution system. Children were consistently more affected than adults and up to 40% of their gastrointestinal illnesses were attributable to water. The rates of gastrointestinal illness among consumers of water obtained directly at the treatment plant were similar to the rate of illness among consumers of purified water. Two periods of increased tap-water-attributable illnesses were observed respectively in November 1993 and in March 1994.

Subjects in the two bottled water groups (i.e. purified water and water obtained directly at the water treatment plant) still consumed about one-third of their drinking water as tap water. They were thus exposed to some tap water and its contaminants; as a result the risks due to tap water are underestimated.

Consumers of water from a continuously running tap had a higher rate of illness than any other group during most of the observation period. This was completely unexpected, since the continuously running tap was thought to have been able to minimize the effects of regrowth in home plumbing. Although there are several unsubstantiated theories as to the cause of this effect, it remains unexplained.

The data collected during these two epidemiological studies suggest that there are measurable gastrointestinal health effects associated with tap water meeting current standards and that contaminants originating from the water treatment plant or the distribution system could be the source of these illnesses. Short-term turbidity breakthrough from individual filters at the water treatment plant might explain the observed health effects. Potential follow-up research should further examine the relationship of turbidity breakthrough and should investigate the role of the continuously running tap in the occurrence of gastrointestinal illness.

7 OTHER EPIDEMIOLOGICAL STUDIES

Following the example of these studies in Canada, the Australians (Hellard *et al.*, 2001) and the Americans (Hayward, 2000; Colford *et al.*, 2002) have initiated similar prospective intervention studies.

The Australian study reported no health effects due to unfiltered clean surface waters. The design was modelled on the Canadian studies but participants were blinded to the type of device they had and one group received a sham device. The results suggested that when source water quality is excellent, drinking water that is disinfected does not constitute a significant health risk (Hellard *et al.*, 2001).

Initial results of the first of several ongoing American studies appear to confirm the Canadian studies. This American study also models the Canadian studies and also made efforts to insure that participants were effectively blinded to the type of device installed in their house. This study is closer to the conditions described in the Canadian studies: the treatment plant is a well-operated complete conventional treatment plant that treats water from a river that is significantly polluted (Colford *et al.*, 2002). Preliminary results are very similar to the results of the Canadian studies.

8 ENDEMIC WATERBORNE DISEASES: POSSIBLE CAUSES

In the Canadian studies, it has not been possible to assign the observed effects to a single cause. While in the first study the source of illnesses was a complete mystery, the authors suggested three explanations: low level or sporadic breakthrough of pathogens at the water treatment plant, intrusions in the distribution system (repairs, breaks, etc.) and finally bacterial regrowth in the mains or in the household plumbing. While no single explanation has been given, many answers have been provided in recent years.

8.1 Health significance of bacterial regrowth

Bacterial regrowth is common in water and has been observed even in distilled water. In water distribution systems, the heterotrophic plate count can occasionally be elevated and there have been concerns that this flora could still contain opportunistic pathogens. Data from epidemiological studies involving reverse-osmosis units suggested that there was a correlation between gastrointestinal illnesses and heterotrophic plate counts at 35°C (Payment *et al.*, 1991b). However, a few outliers in the dataset drove the correlation and the study would have to be repeated in order to confirm the relationship. Furthermore, this observation could be limited to such water purification devices in which a rubber bladder is used to accumulate the purified water. Bacteria or fungi growing on rubber could have been selected for virulence factors increasing their potential for initiating disease.

Heterotrophic bacteria isolates from tap water were subjected to a number of tests to determine the presence of virulence factors (Payment *et al.*, 1994). Several isolates contained numerous virulence factors, but these bacteria were found in numbers generally too low to significantly affect human health. For most bacterial pathogens, even frank pathogens, thousands to millions of bacteria are needed for an infection to occur. This is a rare occurrence in drinking water. It was concluded that these bacteria did not present a significant health risk (Payment, 1995; Edberg *et al.*, 1996; WHO, 2002).

The health significance of pigmented bacteria found in drinking water was reviewed in a recent paper (Rusin *et al.*, 1997). The species investigated included the opportunistic pathogens *Flavobacterium*, *Pseudomonas*, *Corynebacterium*, *Nocardia*, *Mycobacterium*, *Erwinia*, *Enterobacter*, *Serratia* and *Micrococcus*. The authors also concluded that, although a number of these bacteria are opportunistic human pathogens, the available data show them to be infrequent causes of disease.

8.2 Health significance of turbidity

Subsequent to the investigation of the main Milwaukee outbreak mentioned earlier, Morris *et al.* (1996) carried out an analysis of hospital records and water turbidity readings over a period of 16 months before the recognized outbreak. They found that attendance of children with gastrointestinal illness at hospital emergency departments showed a strong correlation with rises and falls in turbidity, but did not describe any specific time lag relationships.

Schwartz *et al.* (1997) presented evidence that rises in rates of gastroenteritis in children were preceded by rises in turbidity of the treated drinking water supply. The authors looked for correlations between the number of children attending the emergency department at the Children's Hospital of Philadelphia (CHOP) for gastroenteritis with the turbidity of treated water in the preceding 14 days. The 1.2 million people of Philadelphia receive their drinking water through three water treatment plants supplied by rivers considered to be 'highly contaminated'. The finished waters tested in all these studies met the commonly accepted definition of safe drinking water in terms of turbidity and chlorine treatment and this raised serious questions about their wholesomeness (Franco, 1997).

These findings complement each other. The Canadian studies concluded that a fraction of gastrointestinal illness attributable to drinking water arises from microbiological events in the distribution system but did not discount the treatment plant as a source of pathogens. Schwartz *et al.* (1997) and Morris *et al.* (1996) suggested that variations in rates of illness were due to changes in the numbers of pathogens (carried in or on small suspended particles) coming through the distribution system from the treatment plant. A water treatment plant is a continuously working system and it changes minute by minute according to the demands. Rapid sand filters are not homogeneous and their overall performance varies considerably. The distribution system is subjected to numerous challenges and the intrusion of pathogens is now a highly debated subject.

9 COSTS TO SOCIETY

The societal cost of the so-called 'mild gastrointestinal illnesses' is several orders of magnitude higher than the costs associated with acute hospitalized cases (Payment, 1997). In the USA, it was estimated that the annual cost to society of gastrointestinal infectious illnesses is $9500 million dollars (1985 US dollars) for cases with no consultation by physician, $2750 million dollars for those with consultations, and only $760 million dollars for those requiring hospitalization (Garthright *et al.*, 1988, Roberts and Foegeding, 1991). These estimates do not even address the deaths associated by these illnesses, particularly in children and older adults. Furthermore, this mode of contamination acts as a constant reinfection source for the population, maintaining at an endemic level illnesses that could be further reduced.

From the data collected during the Canadian studies, the economic costs of endemic waterborne diseases were calculated based on reported symptom and behaviour rates between unexposed and tap-water exposed groups (Payment, 1997). The rate of gastrointestinal illness in the unexposed group was taken as the baseline, then the excess illness in the tap water groups was 50% in the first study and 20% in the second study – a difference attributable to tap water. As even the control groups drank some tap water, the difference attributable to tap water is likely to be higher than the observed figures at a range of 25–50%. These estimates were then combined with published figures for the cost of gastrointestinal infectious diseases in the USA (Garthright *et al.*, 1988; Roberts and Foegeding, 1991). Assuming a population of 300 million individuals, the estimate of the cost of waterborne illness ranges from US$269 to $806 million for medical costs and US$40 to $107 million for absences from work. Such figures can only underscore the enormous economic costs of endemic gastrointestinal illnesses, even in societies believed to be relatively free of it.

10 THE FUTURE: RISK ASSESSMENT

Risk assessment in microbiology was almost non-existent until recently (Eisenberg *et al.*, 1996). It has now become an important part of the preparation of guidelines for water quality in all countries (Fewtrell and Bartram, 2001). In their current rule-making process, the USA have incorporated an important place to microbial risk assessment (Regli *et al.*, 1991; Sobsey *et al.*, 1993) and all agencies world-wide are now debating the balance to be maintained between chemical risks and microbial risks (Fewtrell and Bartram, 2001). It is now recognized that the chemical risk reduction objective should never result in a decrease in water disinfection (Guerra de Macedo 1993; WHO 1996; Fewtrell and Bartram, 2001). This could have catastrophic implications as was experienced during the 1991 cholera outbreak in South America. It is important to remember that differentiation between developed and developing countries is an arbitrary process when drinking water safety is considered. Problems and solutions vary from country to country, a fact that must be taken into account when formulating solutions.

The spread of cholera in Peru during the last outbreak may have been attributable to the fact that disinfection may have been abandoned or reduced in response to the report from the US EPA on risk assessment of the carcinogenic potential of disinfection byproducts (Anderson, 1991).

Water is becoming a commodity whose value will grow in the coming years. Wars have been fought for water and will probably continue to occur if the current use of water resources continues. The global decline in water quality due to global climatic changes can easily upset the balance in many parts of the word and the first observable effect is often a microbiological degradation that can affect the health of millions of individuals.

The subject of microbiological safety of water will very soon become an international priority as travel across national boundaries and the sheer numbers of human citizens increase in the decade ahead, placing Promethean demands on water resources. (Ford and Colwell, 1996, preface by Rita R. Colwell).

REFERENCES

Anderson, C. (1991). Cholera epidemic traced to risk miscalculation. *Nature* **354**, 255.

Andersson, Y. and Bohan, P. (2001). Disease surveillance and waterborne outbreaks. In: L. Fewtrell and J. Bartram (eds) *Water Quality: Guidelines, Standards & Health: Risk assessment and management for water related infectious diseases* (Chapter 6). World Health Organization and IWA Publishing, London.

Anonymous (2000). Waterborne outbreak of gastroenteritis associated with a contaminated municipal water supply, Walkerton, Ontario, May–June 2000. *Canada Commun. Dis. Rep.* **26**(20), 170–173.

AWWA (1999) *Waterborne Pathogens.* (AWWA Manual of Water Practices, M48) American Water Works Association, Denver, Colorado.

Badenoch, J. (1990a). Cryptosporidium – a water-borne hazard. *Letters Appl. Microbiol.* **11**, 269–270.

Badenoch, J. (1990b). Cryptosporidium *in Water Supplies.* Dept Environment and Dept of Health, HMSO, London.

Batik, O., Craun, G.F. and Pipes, W.O. (1983). Routine coliform monitoring and water-borne disease outbreaks. *J. Env. Health* **45**, 227–230.

Batik, O., Craun, G.F., Tuthil, R.W. and Kroemer, D.F. (1979). An epidemiologic study of the relationship between hepatitis A and water supply characteristics and treatment. *Amer. J. Publ. Health* **70**, 167–169.

Chiodini, P.L. (1994). A 'new' parasite: human infection with *Cyclospora cayetanensis. Trans. R. Soc. Trop. Med. Hyg.* **88**, 369–371.

Colford, J.M., Rees, J.R., Wade, T.J. *et al.* (2002). Participant blinding and gastrointestinal illness in a randomized, controlled trial of an in-home drinking water intervention. *Emerging Infect. Dis.* **8**, 29–36.

Collin, J.F., Milet, J.J., Morlot, M. and Foliguet, J.M. (1981). Eau d'adduction et gastroentérites en Meurthe-et-Moselle. *J. Franc. Hydrologie* **12**, 155–174.

Craun, G.F. (1984). Health aspects of ground water pollution. In: G. Bitton and C.P. Grerba (eds) *Groundwater Pollution Microbiology*, pp. 135–197. John Wiley & Sons, Inc., New York.

Craun, G.F. (1986). *Waterborne Diseases in the United States.* CRC Press Inc., Boca Raton.

Craun, G.F. (1990) *Methods for the Investigation and Prevention of Waterborne Disease Outbreaks.* EPA/600/1-90/005a. US EPA, Washington DC.

Crump, K.S. and Guess, H.A. (1982). Drinking water and cancer: review of recent epidemiological findings and assessment of risks. *Ann. Rev. Public Health* **3**, 339–357.

Edberg, S.C., Gallo, P. and Kontnick, C. (1996). Analysis of virulence characteristics of bacteria isolated from bottled, water cooler and tap water. *Microb. Ecol. Health Dis.* **9**, 67–77.

Edwards, D.D. (1993). Troubled water in Milwaukee. *ASM News* **59**, 342–345.

Eisenberg, J.N., Seto, E.Y., Olivieri, A.W. and Spear, R.C. (1996). Quantifying water pathogen risk in an epidemiological framework. *Risk Analysis* **16**, 549–563.

Fattal, B., Guttman-Bass, N., Agursky, T. and Shuval, H.I. (1988). Evaluation of health risk associated with drinking water quality in agricultural communities. *Water Sci. Technol.* **20**, 409–415.

Feachem, R.G., Bradley, D.J., Garelick, H. and Mara, D.D. (1983). *Sanitation and Disease: Health Aspects of Excreta and Wastewater Management.* John Wiley & Sons, New York.

Ferley, J.P., Zmirou, D., Collin, J.F. and Charrel, M. (1986). A prospective follow-up study of the risk related to the consumption of bacteriologically substandard water. *Rev. Epidemiol. Sante Publique* **34**, 89–99.

Fewtrell, L. and Bartram, J. (2001). *Water Quality: Guidelines, Standards & Health: Risk Assessment and Management for Water-related Infectious Diseases.* IWA Publishing, London.

Ford, T.E. and Colwell, R.R. (1996). *A global decline in microbiological safety of water: a call for action.* American Academy of Microbiology, Washington DC.

Franco, L.E. (1997). Defining safe drinking water. *Epidemiology* **8**, 607–609.

Gangarosa, R.E., Glass, R.I., Lew, J.F. and Boring, J.R. (1991). Hospitalizations involving gastroenteritis in the United States, 1985: the special burden of the diseases among the elderly. *Amer. J. Epidemiol.* **135**, 281–290.

Garthright, W.E., Archer, D.L. and Kvenberg, J.E. (1988). Estimates of incidence and costs of intestinal infectious diseases. *Public Health Rep.* **103**, 107–116.

Guerra de Macedo, C. (1993). Balancing microbial and chemical risks in disinfection of drinking water: the Pan American perspective. *Bull. Pan. Am. Health Org.* **27**, 197–200.

Hayward, K. (2000). Science supports a National estimate. *Water* **21**(12), 12–14.

Hellard, M.E., Sinclair, M.I., Forbes, A.B. and Fairley, C.K. (2001). A randomized blinded controlled trial investigating the gastrointestinal health affects of drinking water quality. *Environ. Health Perspect.* **109**, 773–778.

Hertz-Picciotto, I., Swan, S. and Neutra, R.R. (1992). Reporting bias and mode of interview in a study of adverse pregnancy outcomes and water consumption. *Epidemiology* **3**, 104–112.

Herwaldt, B.L., Craun, G.F., Stokes, S.L. and Juranek, D.D. (1991). Waterborne diseases outbreaks, 1989–1990. *MMWR* **40**(SS-3), 1–22.

Herwaldt, B.L., Craun, G.F., Stokes, S.L. and Juranek, D.D. (1992). Outbreaks of waterborne diseases in the United States: 1989–1990. *J. Amer. Water Work Ass.* **83**, 129.

Hunter, P.R. and Syed, Q. (2001). Community surveys of self-reported diarrhoea can dramatically overestimate the size of outbreaks of waterborne cryptosporidiosis. *Water Science Tech.* **43**(12), 27–30.

Hurst, J.H., Knudsen, G.R., Melnerney, M.J. *et al.* (2001). *Manual of environmental microbiology,* 2nd edn. American Society for Microbiology, Washington DC.

Isaäcson, M. and Sayed, A.R. (1988). Health aspects of the use of recycled water in Windhoek, SWA/Namibia, 1974–1983: diarrhoeal diseases and the consumption of reclaimed water. *South Afric. Med. J.* **73**, 596–599.

LeChevallier, M.W., Abbaszadegan, M., Camper, A.K. *et al.* (1999a). Committee report: Emerging pathogens – viruses, protozoa, and algal toxins. *J. Amer. Water Work Ass.* **91**(9), 110–121.

LeChevallier, M.W., Abbaszadegan, M., Camper, A.K. *et al.* (1999b). Committee report: Emerging pathogens – bacteria. *J. Amer. Water Work Ass.* **91**(9), 101–109.

Mackenzie, W.R., Hoxie, N.J., Proctor, M.E. *et al.* (1994). A massive outbreak in Milwaukee of *Cryptosporidium* infection transmitted through the public water supply. *N Engl. J. Med.* **331**, 161–167.

Mara, D. and Cairncross, S. (1989). *Guidelines for the Safe Use of Wastewater and Excreta in Agriculture and Aquaculture.* World Health Organization, Geneva.

Morris, J.G. Jr, Sztein, M.B., Rice, E.W. *et al.* (1996). *Vibrio cholerae* O1 can assume a chlorine-resistant rugose survival form that is virulent for humans. *J. Infect. Dis.* **174**, 1364–1368.

Morris, R.D., Naumova, E.N. and Griffiths, J.K. (1998). Did Milwaukee experience waterborne cryptosporidiosis before the large documented outbreak of 1993? *Epidemiology* **9**, 264–270.

Morris, R.D., Naumova, E.N., Levin, R. and Munasinghe, R.L. (1996). Temporal variation in drinking water turbidity and diagnosed gastroenteritis in Milwaukee. *Am. J. Public Health* **86**, 237–239.

Murray, P.R., Baron, E.J. and Pfaller, M.A. *et al.* (1995). *Manual of Clinical Microbiology,* 6th edn. American Society for Microbiology, Washington DC.

Ortega, Y.R. (1993). *Cyclospora* species – a new protozoan pathogen of humans. *N. Engl. J. Med.* **328**, 1308–1312.

Payment, P. (1991). Antibody levels to selected enteric viruses in a normal randomly selected Canadian population. *Immunol. Infect. Dis.* **1**, 317–322.

Payment, P. (1993a). Viruses: prevalence of disease levels and sources. In: G. Craun (ed.) *Safety of Water Disinfection: Balancing Chemical and microbial Risks.* ILSI Press, Washington DC.

Payment, P. (1993b). Viruses in water: an underestimated health risk for a variety of diseases. In: W. Robertson, R. Tobin and K. Kjartanson (eds) *Disinfection Dilemma: Microbiological Control versus By-products.* American Water Works Association, Denver, Colorado.

Payment, P. (1995). Health significance of bacterial regrowth in drinking water. *J. Water Sci.* **8**, 301–305.

Payment, P. (1997). Epidemiology of endemic gastrointestinal and respiratory diseases – incidence, fraction attributable to tap water and costs to society. *Water Sci. Technol.* **35**, 7–10.

Payment, P. and Hartemann, P. (1998). Les contaminants de l'eau et leurs effets sur la santé. *Rev. Sci. l'Eau* **11**, 199–210.

Payment, P. and Hunter, P. (2001). Endemic and epidemic infectious intestinal disease and its relation to drinking water. In: L. Fewtrell and J. Bartram (eds) *Water Quality:*

Guidelines, Standards & Health: Risk Assessment and Management for Water-related Infectious Diseases. IWA Publishing, London.

Payment, P., Franco, E. and Fout, G.S. (1994). Incidence of Norwalk virus infections during a prospective epidemiological study of drinking water related gastrointestinal illness. *Can. J. Microbiol.* **40**, 805–809.

Payment, P., Coffin, E. and Paquette, G. (1994). Blood agar to detect virulence factors in tap water heterotrophic bacteria. *Appl. Env. Microbiol.* **60**, 1179–1183.

Payment, P., Richardson, L., Siemiatycki, J. *et al.* (1991a). A randomized trial to evaluate the risk of gastrointestinal disease due to the consumption of drinking water meeting currently accepted microbiological standards. *Am. J. Public Health* **81**, 703–708.

Payment, P., Franco, E., Richardson, L. and Siemiatycki, J. (1991b). Gastrointestinal health effects associated with the consumption of drinking water produced by point-of-use domestic reverse-osmosis filtration units. *Appl. Env. Microbiol.* **57**, 945–948.

Payment, P., Franco, E. and Siemiatycki, J. (1993). Absence of relationship between health effects due to tap water consumption and drinking water quality parameters. *Water Sci. Technol.* **27**, 137–143.

Regli, S., Rose, J.B., Haas, C.N. and Gerba, C.P. (1991). Modeling the risk from Giardia and viruses in drinking water. *J. AWWA* **83**, 76–84.

Roberts, T. and Foegeding, P.M. (1991). Risk assessment for estimating the economic costs of foodborne diseases caused by microorganisms. In: J.A. Caswell (ed.) *Economics of Food Safety*. Elsevier, New York.

Rusin, P.A., Rose, J.B. and Gerba, C.P. (1997). Health significance of pigmented bacteria in drinking water. *Water Sci. Technol.* **35**, 21–27.

Schwartz, J., Levin, R. and Hodge, K. (1997). Drinking water turbidity and pediatric hospital use for gastrointestinal illness in Philadelphia. *Epidemiology* **8**, 615–620.

Sobsey, M.D., Dufour, A.P., Gerba, C.P. *et al.* (1993). Using a conceptual framework for assessing risks to health from microbes in drinking water. *J. Amer. Wat. Works Ass.* **83**, 44–48.

Swan, S.H., *et al.* (1992). Is drinking water related to spontaneous abortion? Reviewing the evidence from the California Department of Health Services studies. *Epidemiology* **3**, 83–93.

World Health Organization (1993) *Guidelines for Drinking-water Quality – Volume 1: Recommendations*, 2nd edn. World Health Organization, Geneva.

World Health Organization (1996) *Guidelines for Drinking-water Quality – Volume 2: Health Criteria and Other Supporting Information*, 2nd edn. World Health Organization, Geneva.

World Health Organization (2002) *Heterotrophic Plate Count Measurement in Drinking Water Safety Management*. Report of an Expert Meeting, Geneva, 24–25 April 2002. Water, Sanitation and Health, Department of Protection of the Human Environment, World Health Organization Geneva [WHO/SDE/WSH/02.10].

Wyndham, G.C., Swan, S.H., Fenster, L. and Neutra, R.R. (1992). Tap or bottled water consumption and spontaneous abortion: a 1986 case-control study in California. *Epidemiology* **3**, 113–119.

Zmirou, D., Ferley, J.P., Collin, J.F., Charrel, M. and Berlin, J. (1987). A follow-up study of gastrointestinal diseases related to the consumption of bacteriologically substandard drinking water. *Amer. J. Public Health* **77**, 582–584.

14

Drinking-water standards for the developing world

Jamie Bartram* and Guy Howard†

*Department of Protection for the Human Environment, World Health Origanization, CH-1211 Geneva27, Switzerland; †Loughborough University, Leicestershire LE11 3TU, UK

1 INTRODUCTION

In 2000 it was estimated that in excess of 1 billion persons still lacked access to improved drinking water sources (WHO and UNICEF, 2000). Most of these are the rural and urban poor living in developing countries who remain disadvantaged in terms of access to basic services and whose health and well-being is put at risk from poor and contaminated environments. Drinking-water supply and quality both have important impacts upon health and socio-economic development and therefore remain important components of the poverty cycle.

Approaches to the management of drinking water and of drinking-water quality vary widely between countries in response to factors such as environmental and climatic conditions, technical capacity, level of economic development and cultural and societal norms and practices.

In countries where access to improved water sources or supply is low, the role of legislation, regulation and standards is likely to be best oriented towards encouraging extension of access to higher levels of service, ensuring effective use of existing infrastructure and ensuring that the minimum safeguards are in place to prevent outbreaks of waterborne disease. Improving access to improved drinking-water supply that present limited risks to health and fulfil basic rights to a 'safe' and 'adequate' water supply implies making best use of limited resources towards these priorities. Legislation, regulation and standards may promote this directly and may themselves be more or less cost-effective in terms of the resources demanded for their implementation and the impact of their application.

In terms of drinking-water quality, most regulations and standards are dominated by health concerns. The World Health Organization publishes *Guidelines for Drinking-water Quality* to assist countries at all levels of economic development in establishing national approaches to drinking water that are effective in the protection of public health (WHO, 2003a).

In December 2000 formal recognition was given to the human right to water through the publication of a 'General Comment' on this theme. The General Comment creates an obligation upon governments to support *progressive realisation* of universal access to sufficient, safe, acceptable, physically accessible and affordable water and to *respect*, *protect* and *fulfil* the right to water (WHO, 2003b).

1.1 Water supply and quality – differences between developed and developing countries

Challenges and problems in drinking water supply are generally more complex and daunting in poorer countries which are at the same time typically less well equipped (both financially and with technical and policy expertise) to deal with the complexity.

The opinions expressed in this article are those of the author and do not necessarily represent the policies or views of WHO or University of Loughborough.

The Handbook of Water and Wastewater Microbiology
ISBN 0-12-470100-0

Industrially developed nations typically have established drinking-water supply infrastructure for all except some rural areas and small community supplies. The principal technical challenges therefore relate to efficient management of that infrastructure, including rehabilitation and replacement strategies that require significant recurrent investment and to the pursuit of improved water quality. The problems of fecal contamination are largely under control, although outbreaks of disease may occasionally be recognized and it is likely that other outbreaks go unrecognized. Increasing attention will be dedicated to chemical hazards in drinking water, some of which, while attracting significant public concern, may be of limited health significance.

Less industrially developed nations have incomplete water supply infrastructures such that even middle-income countries may have significant populations with limited access to piped drinking water supply at higher levels of service. As a result, significant proportions of their populations utilize water supplies of poor quality and reliability (Table 14.1). Institutions responsible for both supply and oversight of water supply may be less 'mature' than elsewhere and supply of trained personnel of all types may be inadequate. As a result the efficiency of the water supply sector may be low.

In less developed countries a complex juggling act is therefore required to make the best use of limited resources for public health benefit. This requires trade-offs between extension of supply, basic water quality (largely management of fecal contamination) and other water quality issues.

In both more and less industrially developed nations ongoing changes, especially in public and private sector roles, are creating new challenges to established structures and procedures, with the expectation of improved future efficiency.

In rural and poor urban areas of developing countries direct community management of water supply is common and indeed is seen as desirable by many in the sector. Development of effective legislative and regulatory approaches to support effective community management is especially complex.

TABLE 14.1 Examples of countries with significant populations without access to improved sources of drinking water in 2000

Country	*Percentage without safe drinking water*		
	Urban	*Rural*	*Total*
Afghanistan	81	89	87
Central African Republic	20	54	40
Chad	69	74	73
Papua New Guinea	12	68	28
Madagascar	15	69	53
Angola	66	60	62
Mozambique	14	57	40
Sierra Leone	77	69	72
Uganda	28	54	50
Vietnam	19	50	44
Mali	26	39	35
Myanmar	12	40	32
Lao PDR	41	–	10
Nigeria	19	61	43
Iraq	4	52	15
Nepal	15	20	19
Zambia	12	52	36
Malawi	5	56	43
Sri Lanka	9	20	17
Benin	26	45	37
Sudan	14	31	25

Source: based on data from WHO/UNICEF/WSSCC, 1996.

In relation to water quality, microbiological hazards figure highly in rational analysis of health priorities in both developed and developing countries. This is a result of their contribution to both outbreaks of disease, which may be large in scale, and of their contribution to background rates of disease. Diarrheal diseases related to sanitation, water supply and personal hygiene account for 1,73 million deaths annually and are ranked third in importance in disability adjustment life years in the global burden of disease (World Health Report, 2002). Other diseases are related to poor water supply, sanitation and hygiene such as trachoma, schistosomiasis, ascariasis, trichutiasis, hookworm disease, guinea worm disease, malaria and Japanese encephalitis and contribute to an additional burden of disease. In some circumstances chemical hazards such as arsenic, fluoride, lead and nitrate in drinking water may also be of significance.

1.2 Regulating water supply and quality in developing countries

Overall, the management of drinking water supply and quality for public health benefit is more complex in developing countries than within industrially developed nations.

In many industrially developed nations, water quality standards may be developed and, indeed, implemented in isolation from considerations related to the need to extend supplies to unserved populations. This greatly simplifies the development of legislation and of regulations. Legislation in the established market economies, oriented principally toward the regulation of established water supplies administered by recognizable entities, may therefore have little relevance in less industrially developed nations.

While a significant proportion of the world's population does not have access to piped water supply, and many more do not have access to reliable, safe, piped drinking-water supply, most legislation and regulation is oriented towards this means of provision.

Ensuring an adequate piped water supply is dependent upon proper activity throughout a chain that begins with securing adequate volumes of water and promoting maintenance of high quality water resources from which drinking-water can be abstracted. It also depends on management of drinking-water sources, on effective treatment and distribution and on proper household care. The latter means that basic elements such as household water-handling and domestic plumbing may have a significant impact upon effective water supply. The overall approach to regulation may be similar in more industrially developed nations than less-developed nations where water supply coverage is incomplete and more flexible legislative approaches are required to enable trade-offs to be made.

Legislation and regulation supportive of improved non-piped water sources is relatively scarce, although many principles are similar. The tendency to regulate piped but not other forms of water supply may be one of the factors that draws additional resources to the further improvement of already improved supplies rather than towards the extension of improvements to the unserved.

Overall, for many less developed countries, the role of regulation might be argued to be to support the best use of available resources in the provision of incrementally improved service standards to the population as a whole. Available experience and guidance derived from regulatory bodies in the more developed nations may be inappropriate for direct application in developing nations. As a result, countries may be reluctant to adopt foreign practice in legislation because of its limited relevance to their national situation (Caponera, 1992). This is especially valid in the case of drinking water supply.

In all cases, regulation tends to be more effective when it is driven by processes of negotiation and the inclusion of civil society within the debate about both standards and the structure of the water supply services. The use of a range of regulatory tools and instruments will also be commonly required in order for balanced decisions to be made based on priorities within the sector.

1.3 Urban areas of developing countries

While there is a common perception that urban populations receive high-quality water supplies at higher service levels, in reality there is great inequity in access to drinking water supplied on-plot (i.e. at least a yard level of service). In most developing countries a minority of wealthier households enjoy water supplies of high quality and service level, while the majority of the poor rely on communal piped and point water sources (Howard, 2002). Where piped water supplies are used, it is not uncommon to find that the urban poor experience greater unreliability in supply.

Not only are the inequities immense in terms of service level, the consequences for the urban poor of utilizing point sources of water is increased exposure to pathogens and therefore increased risk to health. In addition, there is an increased risk of re-contamination of the water during transport and storage. Furthermore, in most developing countries the urban poor pay proportionately more for poorer water services than those in wealthier areas and are often more likely to be disconnected from utility supplies.

The improvement of water supplies in urban areas, while in some regards simpler than for rural areas, still retains fundamental problems. Sustained access and operation and maintenance is reliant on payment of utility water tariffs and therefore the costs of water become a significant factor in determining access and, in particular, whether charges meet the ability and willingness to pay by poor households and communities. Furthermore, the tariff structure and payment methods, with an emphasis on payment on bills presented periodically (typically monthly or quarterly), are often at odds with the income patterns of the urban poor where incomes may be highly unstable and abilities to save are limited (Howard *et al.*, 1999).

The regulation of the quality, reliability and cost of utility supplies that form a major component of the urban water sector can be effectively achieved utilizing simple and transparent approaches, however, a careful balance must be maintained between competing demands of affordability and access and concerns over quality (Ince and Howard, 1999). More significant problems are found in regulating other aspects of water supply, such as social provision and leakage control. In the case of social provision, there may remain significant conflicts between the demands for financial sustainability of the utility. Where leakage-control programs are implemented, sustaining rehabilitation and replacement strategies (which require significant expense and planning) may be difficult. Consultation with both users and water suppliers may be essential in establishing targets for water quality and other aspects of water supply, in order to ensure that required improvements are made.

1.4 Rural areas of developing countries

There remain particular problems in the development of water supplies in rural areas where 80% of the presently unserved population live. Overall, coverage of the rural population with functioning water supplies remains low in many countries. The rural water supply sector outside the established market economies typically comprises diverse means of supply and with diverse types of infrastructure, many of which fail either permanently or temporarily soon after construction and a high percentage of the accumulated infrastructure out of commission at any given time.

In addition to the frequent functional failure of many small water supplies, water quality has often proved difficult to sustain at acceptable levels as sanitary protection measures deteriorate through a lack of maintenance. Such problems are not unique to developing countries. There is substantial evidence that small, effectively community-managed, supplies in Western Europe and North America fail to meet national and regional water-supply regulations. For instance, in the UK, 47% of small supplies gave at least one unsatisfactory result in a recent assessment (Fewtrell *et al.*, 1998).

Approaches to improving drinking water supply and quality have increasingly tended to emphasize the role of the community as water managers. However, as early as 1980, community participation was referred to as the 'mythology' of the International Water Supply and Sanitation Decade (the 1980s) (Feachem, 1980). In a review of the results of the massive investments in the Decade, it was noted that while community management of water supplies can be effective, it is unlikely to offer more than a medium-term solution to overcome current inadequacies in government and the private sector in meeting rural water demands (Carter *et al.*, 1993).

The special problems of the rural sector are recognized by those specializing in water law (Caponera, 1992), both because of the specific relationship of rural populations with water resources and in terms of water supply and sanitation and their relationship to health and disease transmission. In a book otherwise couched in relatively specific legal terms, Caponara (op. cit.) indicates that:

In such areas legislation by itself is not sufficient to abate the occurrence. It is necessary to educate the users of water points in proper operation and maintenance practices. This often requires a minimum training background in public hygiene and health, in operation and maintenance of the facilities, and in other basic aspects relating to water management. Water legislation should contain the necessary provisions, as well as provisions tending to promote the organization of water users' associations.

For rural water supply it is especially important that ownership and responsibilities related to operation and maintenance of systems constructed by one agency and operated by another (often local) one (such as a water users' association) are clear and unambiguous. Thus, Appleton (1995) refers to the need for legislative changes to provide the authority and autonomy for decentralized operations to function successfully. Attempts have been made to promote and support community operation and maintenance through monitoring, and evidence is accumulating that this can be effective. These approaches suggest that far more resources and time need to be spent in developing capacity with communities to be able to recognize the need for preventative maintenance and identify early signs of major faults and thus be able to rectify these.

1.5 Water supply and quality in international trade

The issues arising from the process of 'globalization' are complex. Of particular relevance to drinking-water supply is international trade in drinking-water supply services. Within the past two decades, the role of international companies in managing water supply to 'foreign' cities has emerged as a significant issue. Unlike other areas of trade, such as food, internationally recognized points of reference for specification or performance evaluation are often lacking. Exceptions include the WHO *Guidelines for Drinking-water Quality*.

There remain unresolved key issues about international trade in water services. While the increased use of such approaches is advocated by many donor agencies and governments, public perception may be far less favorable. Important considerations concern quality standards to be met, their independent surveillance and the pursuit of public health priorities in a privately managed system. The latter may imply significant relaxation in quality standards to enable effective extension of supply into under- or unserved areas. The acceptability of such relaxations may be low despite their evident logical need.

2 DRINKING-WATER LEGISLATION, REGULATION AND STANDARDS

Regulation in general is primarily designed to protect the public interest against undue exploitation or the provision of substandard goods and services that either infringe 'rights' or raise the risks to health. However, an important subsidiary objective is to ensure that there is a transparent system of management where responsibilities and liability are clearly defined, an approach that could be said to be providing a 'free and fair playing field'. Thus effective regulation is often as much in the best interest of water suppliers (or at least those of a reputable nature) as it is for the public at large.

The principal concerns of drinking-water legislation and regulation are which functions must be exercised and what administrative arrangements are required in order that they are exercised effectively and efficiently. While the functions vary relatively little between countries, the exact means of their implementation and, in particular, administrative arrangements, vary widely and these require adaptation to take account of wide differences in basic circumstances.

2.1 Functions

The principal functions that must be exercised include:

- those relating to the management of drinking-water sources and the resources from which drinking water is abstracted – these

typically include volume management (conservation of resources), licensing of abstractions and pollution control (effectively water resource quality management)
- measures relating to the provision of supply services to the population – these often vary in response to administrative arrangements. Essential are those measures that define minimum service standards (continuity, quality of water, tariff controls, leakage targets) and which seek to pursue the objective of increasing population access and service level to a reasonable minimum target such as a single tap within the house or yard
- drinking-water quality standards and their implementation through actions by both the supply agency and others, including information sharing and public reporting
- the regulation of ancillary services, which impact directly upon supply or quality (including for example, those which may impact upon both water security through leakage and safety through proper practice against contamination)
- public health functions (including ongoing public health oversight and response to occurrences endangering public health)
- licensing of materials and chemicals used in the production and distribution of drinking water (such as pipes and treatment chemicals) and of household devices such as taps.

These different functions can be translated into a set of regulations and standards that would typically be incorporated into a regulatory framework. The areas of regulation and their principal features and means of monitoring are outlined in Table 14.2.

2.2 Basic law

It is generally preferred that all provisions concerning water are contained in a single law and so the basic law will typically be far wider than drinking-water supply alone. The basic law that relates to water may take diverse forms (constitution, act, code etc.) and would normally include a statement of purpose (policy) which would refer to the intention to protect human health and welfare and define priorities among different uses. Aspects such as public health protection and consumer rights may be dealt with in other parts of the basic law but be applicable to water supply.

The basic law generally refers to the jurisdictions, responsibilities and authority of specified competent agencies in relation to water and to their relationships to one another and, in particular, to the development of subordinate legislation. This should include water users (consumers), those constructing and administering water supply systems, those regulating different aspects of water and having responsibilities in related domains and agencies involved in data and information management and sector planning. Kandiah (1995) refers to the legal and policy frameworks that assign responsibility to agencies to implement and coordinate their programs as one of the key factors affecting the suboptimal performance of all types of water schemes.

In general, while the principles of drinking-water regulation are contained within the basic law, the actual tools of regulation are more effective when established as statutory instruments. Provisions for the development of such instruments would typically be included within the basic law, which would outline the aspects of drinking water where such statutory instruments can be set up, who the responsible minister will be, the frequency of revision and what procedures and expert advice is required for such revisions. In the case of drinking-water quality for example, standards may be set by the Minister of Health, with:

1. A review of standards and new information undertaken on an annual basis by an expert review body appointed by the minister composed of relevant sector stakeholders (including water suppliers).
2. The review to take into account new evidence of health impact.
3. The review to take into account impacts of new standards on investment and other requirements.

In Ghana, a review of standards undertaken in light of a private sector lease operation recommended that the regulatory instruments

TABLE 14.2 Regulations and standards for drinking-water supply and quality

Area of regulation	*Typical major features of regulations and standards*	*Means of verification/monitoring*
Water sources and resources	• Abstraction licensing • Source protection norms and practice • Polluter-pays principle • Monitoring of land-use, control of encroachment	• Permits and bulk flow meters; annual visits and periodic assessment • Sanitary inspection • Pollution monitoring and enforcement
Continuity of supply	• Set and enforce targets for continuity of supply • Investment requirements for improvement and maintenance of reliability	• Sanitary inspection/community interview • Investment plans and expenditure; evidence of rehabilitation strategies
Drinking-water quality	• Place obligation continuously to supply safe water upon the supply agency • Require supply agency to establish a water safety plan and exercise due care • Define what is 'safe' in terms of acute or long-term exposure • Establishment of limit values, objectives and guidelines • Development of statutory instruments, negotiation and enforcement procedures • Independent verification	• Routine surveillance and monitoring of water quality • Audit of monitoring programs • Best practice approaches and international evidence • Standards set through consultative process • Instruments developed and linked to primary law • Surveillance/audit
Monitoring requirements	• Divide monitoring obligations among a number of agencies or sectors (public sector oversight of water resource management; public health authorities; drinking water suppliers; including local government) • Require information-sharing among these agencies • Define the extent of monitoring required of each agency and define the conditions under which monitoring should be undertaken • Define the technical requirements against which monitoring results would be compared	• Allocation of responsibility in primary law and through statutory instruments • Legal instruments • Limits of responsibility and definition of duty of care • Standards for water quality, process control, source protection and distribution management

(*continued on next page*)

TABLE 14.2 *(continued)*

Area of regulation	*Typical major features of regulations and standards*	*Means of verification/monitoring*
Treatment requirements	• Require supply agency to treat water to achieve the required standards (may be implicit in the requirement of the basic law to supply safe water) • Require that newly constructed or rehabilitated supply systems reach minimum treatment capabilities (these may be formulated in terms of performance criteria or processes required, generally in relation to the characteristics both of the source water and of the supplied population) • Require the upgrading of existing systems to reach the standards required of newly constructed or rehabilitated systems • Define technical details of construction or operating practice	• Minimum treatment requirements for source types and source quality; process control standards; drinking-water quality standards. Verification through surveillance • Minimum treatment requirements for source types and source quality; process control standards; and drinking-water standards. Verification though surveillance • Rehabilitation and upgrading strategies in place with firm investment plans • Standard operating procedures, verification through audit
Contingency plans	• Oblige government agencies to establish contingency plans • Define the conditions under which contingency planning is required, and the types of contingency to be planned for • Describe the basic components of an 'adequate' plan	• Monitoring through regional organization, government watchdog bodies and non-government organizations to ensure plans developed that meet agreed criteria of the basic plan for supply quality under extreme conditions

be set and related to the new water act. The responsibility for developing and applying the regulatory instruments would fall under a public utilities regulatory commission and where consultation in the derivation of standards would include water suppliers, the health sector and local government.

The use of statutory instruments is important for a number of reasons. The setting of water quality standards is a dynamic process (Ince and Howard, 1999). However, basic or primary law tends to be relatively inflexible since its change is generally a long process, usually involving an Act of Parliament. Provisions in primary law cannot therefore reflect the rapidity of change in the available knowledge about, for instance, the health impacts of drinking water, changing socio-economic conditions and society's demand for protection. For similar reasons, penalties to be applied in enforcement actions should not be incorporated into basic law. In many developing countries where such penalties were traditionally incorporated into the basic law, changing economic conditions have reduced fines (originally used as a deterrent) to insignificant amounts which water suppliers themselves may be willing to pay, rather than invest in the upgrading of water supplies.

2.3 Subsidiary legislation

Among the details that form part of the subsidiary legislation on water are standards of various kinds. For example, there may be standards on minimum treatment requirements linked to source type and quality, and for drinking-water quality. However, standard-setting should not be restricted to drinking-water quality, but should also extend to water resource protection and water supply service more generally. Indeed, it should extend beyond the simple assessment of quality and include the adequacy of structures and systems, such as the definition of safe facilities and practices, minimum standards specification, and minimum standard operating systems.

There may also be schedules or annexes to statutory instruments, consisting of even more detailed subject matter. In some countries, drinking water service standards are contained in a sanitary code or code of good practice. It is worth noting that such standards are most likely to receive support in their implementation if consultation with the affected entities (e.g. supply agencies and professional bodies) has occurred.

One of the most important criticisms of current approaches to standards and norms relates to microbiological quality of water and, in particular, the over-reliance on a set of end-product indicator standards that have well-recognized limitations. Such criticisms apply equally to utility and alternative supplies. This has led to a change in approach within the third edition of the WHO *Guidelines for Drinking-Water Quality* to a quality assurance approach based on Water Safety Plans (Davison *et al.*, 2002; WHO, 2003). The Water Safety Plan approach emphasizes control of water safety from the catchment to the consumer through the use of the multiple barrier principle. Water Safety Plans focus on effective process control linked to simple monitoring activities and periodic internal audit. These are accompanied by health-based water quality targets set at national level and periodic independent surveillance/audits.

In the case of community-managed supplies, enforcement of regulations that would typically govern aspects related to ongoing operation and maintenance, such as water quality and continuity, is unlikely to be a workable mechanism to achieve improvements in water supply. Given inherent problems in applying regulatory procedures in circumstances where there is no differentiation between consumers and suppliers, different approaches are required to support water supply improvement. One approach is to develop water supply surveillance programs to act as a supportive role in promoting improvements. Critical to this process is the routine use of sanitary inspection as a measure of operation and maintenance and the support to communities to address such weaknesses. There remains some potential for regulation during the construction phase of such supplies, provided an organization or contractor external to the community is undertaking the work through a variety of mechanisms.

2.4 Institutional aspects

There are diverse approaches to drinking water management – varying between the extremes of centralized state ownership and decentralized private ownership and management. The ongoing trends towards private sector participation and towards community management are attempts to move towards a more demand-driven approach, which may offer better possibilities for regulation and greater public demand for protection.

Other significant trends have been towards *decentralization* (the principle of devolving decision-making to the lowest appropriate level); and the recognition at governmental level that effective policy development and management requires *multisectoral cooperation*. There is general acceptance of the principle that central government should retain responsibility for public health and maintaining quality standards (see for example, Appleton, 1995; IRC, 1995).

As the use of water for drinking is effectively universal, it is perceived as a high priority for effective independent oversight. The oversight functions will vary according to legal, administrative and technical arrangements. There is extreme diversity in approaches to this independent oversight between exercise of direct control by a governmental surveillance agency and auditing approaches requiring relatively limited public sector intervention. However, most 'models' for organization of monitoring, including that proposed by WHO (1976, 1985), envisage two complementary roles: service quality control by the supply agency and an independent surveillance function.

WHO has indicated that 'the water supplier and surveillance agency should be separate and independently controlled' (WHO, 1976; WHO, 1997). Whilst such a model is both reasonable and often viable in urban areas worldwide, a rigid division between surveillance and quality control is rarely applicable for small community supplies in developing countries (Helmer, 1987), except where privatization has occurred to a definable responsible agency or where a small community is supplied by a formal utility as sometimes occurs near major cities or where several rural communities are supplied through a single centralized system.

Within the urban sector, the viability of strict division of roles depends, in part, on the water supply management arrangements. Where water supplies are provided by a utility, surveillance by local public health departments under the overall guidance of a national health body, would generally be the preferred model (Howard, 2002). Where local government provides water, this may become more problematic, although experience in Uganda indicates that provided separate departments within local government undertake the two functions, effective separation of roles can still be achieved (Howard *et al.*, 1999).

In both urban and rural settings, the use of surveillance to promote better hygiene during the collection, transport and storage of water is important. This is particularly true in any setting where the water supply is primarily through communal water supplies.

In all cases duplication of functions may be wasteful of limited resources and procedures for information sharing. Appropriate auditing and quality control are increasingly accepted to enable the independent surveillance agency to focus energies on high-risk areas and population groups.

It is not inevitable that the surveillance function should fall under the direct supervision of the health sector (WHO, 1985, 1996). For example, in the United Kingdom regulation of water quality in utility supplies falls under the remit of the Department of Environment. In many cases, actual implementation of monitoring and surveillance functions is more effective when de-centralized to local levels. This approach is reflected in UK practice in relation to private water supplies and similar models are found in Yemen.

Given the broad range of issues in the water sector that require some form of regulatory practice, there are a number of institutions with an interest in ensuring that this is implemented (see Table 14.3).

TABLE 14.3 Functions and examples of potentially interested agencies concerned with drinking-water supply

Abstraction licensing	Water resources management agency, agricultural authorities, abstraction licensing body, river basin authorities, navigation boards, hydroelectricity generators, water supply agencies
Water resource quality	Water resources management agency, pollution control boards, river basin authority, health, water supply agencies, local governments
Drinking water supply service quality	Water supply oversight body, consumer protection bodies, water supply agencies, health, local government
Ancillary services (e.g. plumbing)	Professional associations, trade associations, water supply agencies, local government (through environmental health)
Public health protection	Health, local government, water supply agencies, consumer protection bodies
Licensing of materials and chemicals	National licensing authority/certification agency; health, water supply agencies

NB: representatives of civil societies including NGOs and professional associations operate at each level

Aspects of the administrative arrangements that may be put in place to ensure that the functions outlined above are exercised effectively may relate, for example, to:

- Arrangements for multisectoral participation and integrated management
- Provisions for participation in management and decision making (including community participation and evolving public–private sector roles)
- Provisions related to the sharing of information
- Approaches to and responsibility for monitoring and assessment in general and independent oversight in particular.

3 STANDARD SETTING AND IMPLEMENTATION

Conflicts may exist between standards and ensuring optimal health effects (Holland, 1991). Resources are finite and, in practice, a supply agency may have to choose, for example, between investing in fulfilling water quality standards despite relatively good quality, and extension of coverage. The choice between these two may have a profound influence on the resulting impact on the health of the population (Esrey *et al.*, 1985, 1991). There has been a change in approach in recent decades among public health professionals to water supply provision from 'all or nothing' to 'something for all' (IRC, 1995).

While regulation often focuses upon direct drinking-water *quality* standards, its scope includes regulation concerning:

- construction standards
- plumbing standards
- service quality standards (cost, continuity)
- water quality standards
- coverage (e.g. when an authority has responsibility for a defined geographical area or population)
- licensing of materials and chemicals used in drinking water supply.

Standards need to be developed to take account of national circumstances, such as health burden, current water qualities (source and distributed), availability of resources (infrastructure, finance, willingness to pay and expertise) to meet standards and implications for upgrading and analytical capacity. Key to the setting of standards is therefore the evaluation of health priorities related to water, of the current ability to meet existing or anticipated standards (reflecting technical capacity, water source qualities, abilities to pay) and the available resources to implement improvement required.

Standards for different components are highly inter-related and the establishment of a standard in one aspect, for instance water quality, has direct impacts on the achievability of other standards. The tariff set, for instance, may determine the ability to meet water quality standards and may directly influence the ability of the supplier to extend services or to be able to reduce unreliability. In the same way, the establishment of a water quality standard will typically have direct influences on the cost of producing water and therefore the tariff that will be applied. Experience suggests that integrated approaches to standard derivations are required.

3.1 Scientific basis for microbiological standards

The scientific basis for establishing standards for microbial safety are the same in all settings, although the final targets established may vary. Overall approaches to establishing effective standards for the microbial safety of water, taking into account the range of exposure routes (drinking water, recreational water and re-use of wastewater), has been recently reviewed: a 'harmonized' framework for future application to developing standards has been proposed (Fewtrell and Bartram, 2001).

Typically, microbiological hazards are of greatest concern and constitute the causative agents of disease (pathogens) that may be transmitted through consumption of contaminated water. They may be protozoa, viruses or bacteria. There are a wide variety of microbiological agents that may be transmitted through drinking water (Table 14.4).

For all of the agents listed in Table 14.4, a single exposure may be significant for public health and, for example, sufficient microbes to cause disease may be consumed in a single glass of apparently innocuous water. Furthermore, this 'infectious dose' may be extremely small – potentially as little as a single viable cyst or virion for some protozoa and viruses.

Analytical methods are available for some of these, but unavailable for many others. Where analytical methods exist they may or may not be quantitative and problems may be encountered in their application to water as an analytical medium.

The principle that no pathogens should be present in drinking water has therefore become widely accepted. This principle, alongside the lack of analytical methods and the fact that almost all of the pathogens of interest are primarily derived from human excreta (feces), led to the development of the concept of 'fecal indicators'. The value of quantitative estimates of fecal indicator bacteria in water was recognized early in the history of sanitary microbiology. The definition of microbiological quality now used in most regulatory and non-regulatory monitoring world-wide is therefore based on the premise that fecal contamination of drinking water is unsafe and that assessing the presence of indicators of fecal contamination provides an indication of the safety of drinking water.

The use of fecal indicators has made a significant contribution to the protection of human health over a sustained period and it continues to be valuable and popular. It is, nevertheless, imperfect both in conception and in application. Its principal limitations relate to well-recognized shortcomings of the principal available indicators and their ability to meet the basic criteria presented in Box 1. The majority of currently used indicators are bacteria and this has important implications regarding their use and the information they provide in relation to non-bacterial pathogens.

BOX 1: ***Characteristics of an ideal microbial indicator***

- Should be present in wastewater and contaminated water when pathogens are present
- Should be present when there is a risk of contamination by pathogens
- Should be present in greater numbers than pathogens
- Should not multiply in environmental conditions under which pathogens cannot multiply
- Should correlate with the degree of fecal contamination
- Survival time in unfavorable conditions should exceed that of pathogens
- Should be more resistant to disinfection and other stresses than pathogens
- Should present no health risk
- Should be easy to enumerate and identify by simple methods
- Should have stable characteristics and give consistent reactions in analysis

(Source: WHO Teaching Pack on Drinking-Water Quality.

TABLE 14.4 Pathogenic microorganisms that may be found in water

Pathogen	*Health significance*	*Persistence in water supply*	*Resistance to chlorine*	*Relative infective dose*	*Important animal reservoir*
Bacteria					
Camplyobacter jejuni, C. coli	High	Moderate	Low	Moderate	Yes
Pathogenic *E. coli*	High	Moderate	Low	High	Yes
Salmonella typhi	High	Moderate	Low	High	No
Other salmonellae	High	Long	Low	High	Yes
Shigella spp.	High	Short	Low	Moderate	No
Vibrio cholerae	High	Short	Low	Moderate	No
Yersina enterocolitica	High	Long	Low	High (?)	No
Pseudomonas aeruginosa	Moderate	May multiply	Moderate	High (?)	No
		May multiply			
Aeromonas spp.	Moderate		Low	High (?)	No
Viruses					
Adenoviruses	High	?	Moderate	Low	No
Enteroviruses	High	Long	Moderate	Low	No
Hepatitis A	High	?	Moderate	Low	No
Enterically transmitted non-A, non-B hepatitis viruses, hepatitis E	High	?	?	Low	No
Norwalk virus					
Rotavirus	High	?	?	Low	No
Small round viruses	High	?	?	Moderate	No (?)
	Moderate	?	?	Low (?)	No
Protozoa					
Entamoeba histolytica	High	Moderate	High	Low	No
Giardia intestinalis	High	Moderate	High	Low	Yes
Cryptosporidium parvum	High	Long	High	Low	Yes
Helminths					
Dracunculus medinensis	High	Moderate	Moderate	Low	Yes

Source: WHO (1993) Guidelines for Drinking-water Quality Volume 1.

There has been increasing evidence of presence of pathogens in water meeting guidelines and standards for the principal fecal indicator bacteria, *E. coli*. In some cases this has led to outbreaks of infectious disease related to the consumption of contaminated water. As a result, greater attention has been placed on alternative methods of defining microbiological quality, including testing for pathogens, identification of alternative indicators for non-bacterial pathogens (e.g. phage as an indicator of potential viral contamination) and other risk assessment approaches, such as sanitary inspection. For instance, in the case of *Cryptosporidium*, monitoring efforts may be better focused on ensuring adequate sanitary completion of groundwater sources and control of turbidity during treatment.

It is increasingly recognized that zero risk is unachievable and its pursuit inhibits the application of risk–benefit approaches. As noted above, it is unlikely that any approach will provide the degree of certainty required to define a 'safe' water supply in all circumstances, while current indicators can be taken as indicative of gross fecal pollution. This then leads to change in the way we view microbiological contamination, which is that no water is 'safe', but rather may be low, intermediate or high risk – a 'tolerable disease burden'.

The idea of tolerable disease burden (TDB) is especially important in drinking-water microbiological safety because of the variable health outcomes from different exposures. Most studies have addressed diarrheal diseases which, although they account for a significant global burden of disease, are often self-limiting and, in some circumstances, of limited concern to the general public. They may be contrasted with more severe health outcomes. Different pathogens may produce diseases of varying public health importance, e.g. enteric hepatitis viruses (hepatitis), *Vibrio cholerae* (cholera), *Cryptosporidium* spp. (cryptosporidiosis), or *Salmonella typhi* (typhoid). Some workers have suggested specific figures for a TDB for diarrhea, but this has not as yet appeared in a form translatable to other disease outcomes (such as disability-adjusted life years or DALYs) and proposals for TDBs for more severe health outcomes have not yet been made.

Increasing attention is therefore being paid to the characteristics of individual pathogens and the concept of TDB as the future basis for definition of microbiological safety (quality) for drinking water. Substantial ongoing work will see this theme develop rapidly in future years. However, translating TDB into practical descriptions of quality is complex. The distribution of microbes in water (individual or clustered on particulates) may have an effect on the probability of infection or of developing disease. The relationship between exposure, infection and disease, especially at low dose exposures, remains poorly understood for most pathogens and inter-relates with external factors such as immunity and the form of exposure (e.g. from food, aerosols etc.) (Fewtrell and Bartram, 2001; Prüss *et al.*, 2001).

Approaches to ensuring microbial safety are increasingly based on a preventive management approach, taking account of factors from source through to consumer. The quality of drinking water may be controlled through a combination of protection of water sources, control of treatment processes and management of the distribution and handling of water.

The WHO Guidelines for drinking-water quality outlines a framework for safe drinking water that comprises five key components (Davison *et al.*, 2002; WHO, 2003):

1. Water quality targets based on critical evaluation of health concerns
2. System assessment to determine whether the water supply chain (from source through treatment to the point of consumption) as a whole can deliver water of a quality that meets the above targets
3. Monitoring of the control measures in the supply chain which are of particular importance in securing drinking-water safety
4. Management plans documenting the system assessment and monitoring; and describing actions to be taken for normal and incident conditions; including upgrade and improvement and documentation and communication

5. A system of independent surveillance that verifies that the above are operating properly.

Component 1 would normally be undertaken by the national authority responsible for public health. Components 2–4 collectively comprise a water safety plan and are the responsibility of the water supplier, whether large or small. Component 5 may be the responsibility of several independent agencies with local and national capacity to undertake auditing and/or compliance testing.

3.2 Scientific basis for chemical standards

There are a few chemicals in drinking water that have been clearly associated with large-scale health effects. These include arsenic and fluoride. In order to direct resources to best effect, identifying those chemicals that may be of public health concern nationally or locally, is important. Guidance on practical approaches employing diverse types of information in addition to water quality analysis is available in documents accompanying the WHO *Guidelines for Drinking-Water Quality.*

The principal focus for controlling chemical safety of drinking water is through establishing numerical standards for substances against which the results of water quality analysis can be compared. An exception to this is where chemicals are either added during treatment or may leach from pipe and pipe joining materials. These would typically be regulated through product specifications.

In order to establish standards, it is necessary to identify the concentration of a substance in drinking water that will represent 'safety'. Two approaches are adopted depending on whether the chemical has a 'threshold effect' or not. A threshold effect is where a concentration can be defined below which no adverse effect on health is seen over a lifetime consumption. Non-threshold chemicals are largely genotoxic carcinogens where any level of exposure represents a theoretical risk.

For threshold chemicals, the concentration at which no adverse effect is detected forms the basis of the standards. Corrections are made for exposure from other sources, such as food. In practice, data from real human populations are rarely available and reliance must be placed on the studies undertaken over short periods on animals. An uncertainty factor is therefore often incorporated into the Guideline Value.

To establish a numerical standard for non-threshold chemicals, a level of 'negligible' risk is defined (often 10^{-5} or 10^{-6} lifetime risk of cancer). In order to accommodate variation in the human population, the concentration derive is in fact an upper bound estimate.

A number of practical considerations are also taken into account when establishing guidelines for standards for chemical substances in water. These include preventing standards that:

1. are below the limit of detection
2. are not technically or economically achievable; and/or
3. would counteract other aspects of water safety (such as discouraging disinfection).

A comprehensive explanation of the derivation of Guidelines Values is presented in the WHO *Guidelines for Drinking-Water Quality* (2003).

3.3 Social/cultural, economic and environmental factors

While risk management is often seen as a highly rational process, it should be recognized that the scientific basis for many of its elements is actually often weak. Furthermore, scientific assessment will not support effective risk management if it fails to address the perceptions and priorities of the society concerned. One schematic explaining the process of risk management is presented in Fig. 14.1. This describes the process as circular showing the feedback between policy evaluation into revised (improved) policy. It places communication at the centre of the decision-making process that is designed to both ensure that information from each stage is communicated

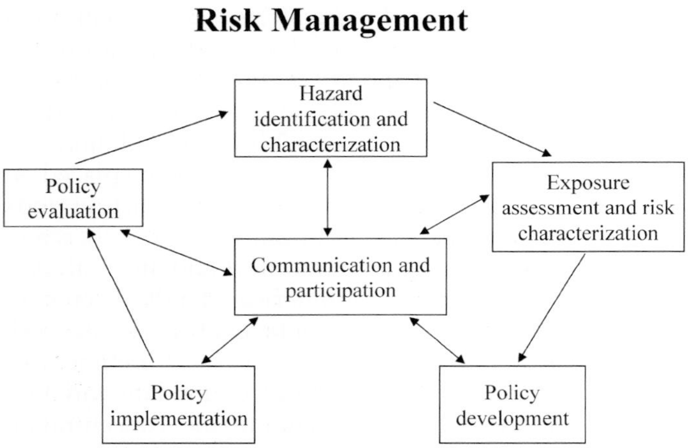

Fig. 14.1 The Risk Management Cycle (adapted from Chorus and Bartram, 1999 and Sobsey *et al.* 1993)

with stakeholders, and that the needs for communication are taken into account when undertaking each stage of the risk assessment and management process.

As well as the scientific aspects of standard development, adaptation is therefore required to take account of social, cultural and economic factors. In some circumstances, societal concerns will lead to the adoption of standards that are stricter than can be justified by simple health risk assessment alone.

A critical component in the development of standards and the balancing of numerous demands for improved water supply, is the systematic assessment of costs of meeting objectives and the implications that this will have for tariffs and access. This is further complicated by the very different forms of water supply commonly found in many developing countries. Unlike their wealthier counterparts in the developed world, a large proportion of supplies will not be expected to meet stringent water quality standards, primarily in the rural sector, while the more sophisticated urban piped water supplies should be able to achieve a quality of water similar to those in developed countries. This, therefore, raises difficult ethical and political problems regarding the application of standards and the dangers of the already significant inequity in the water sector being exacerbated. When standards are set, the process includes negotiation on the quality to be achieved. Such negotiation already often occurs between regulators and suppliers. However, in many countries, this needs to be expanded to include more public consultation to take into account social demands for protection and public perception of risks.

3.4 Progressive implementation

In many developing countries, the need to develop and apply drinking-water standards has increased in importance as the nature of government roles in the sector has changed. The result, in part in response to public pressure effectively to 'control' private sector supply agencies, may be 'strict' standards. However, where water supply legislation is 'utopian' it will tend to promote the perfection of existing water supplies at perhaps great cost despite minimal health risk and at the expense of providing an improved, but perhaps imperfect, supply to (perhaps many) more individuals. In particular, the emphasis on meeting advanced water quality standards may be counter-productive in situations where costs of investment in new infrastructure to meet particular standards may result in reduced access by the poor to services that are yet to be developed or become too expensive.

The use of regulatory instruments, in particular the application of relaxations and exemptions, the development of interim standards and definition of priorities for control, are therefore important if regulation is to be a positive force for change. Essential to this is the identification of priorities with regard to water supply and an emphasis placed on meeting key standards before attention is paid to less significant problems. Often this will lead to an emphasis on microbiological before chemical quality.

A relatively frequent need is to adopt less stringent 'interim standards', either to encourage progressive improvement or to ensure that limited financial and technical resources are not diverted from issues of high to low relative priority for public health.

This is common where the resolution of water quality problems may require significant upgrading of water infrastructure that will take an extended period of time or large-scale investment or enforcement of other legislation (For instance, prevention of encroachment). In such cases, an interim standard for a substance may be set at a higher level to allow improvements to be made progressively. Examples might include requiring immediate implementation of chlorination and thereafter progressively improved treatment where surface waters are supplied without treatment.

In addition to the delay in implementing standards fully or developing interim standards is the use of two specific tools – exemption and relaxation.

In many cases, a failure in water supply may not be easy to resolve in a short period of time without large-scale investment that will eventually be passed onto the consumer. In some cases, this may be supply and substance-specific, in which case an exemption can be applied. This would typically be time-constrained, linked to an agreed and defined program of action and an allowable deviation from the standard set. Such approaches would usually require a case to be made by the water supplier for an allowed deviation.

Relaxations would typically be much broader and cover all supplies, or particular types of water supply. They may in some cases also apply to particular parameters of more limited health concern. Relaxations would usually be phrased in terms of an acceptable percentage of samples where analysis shows that a standard has been exceeded, often with an upper limit of allowed deviation also specified.

3.5 Enforcement

Standards are intended to ensure that the consumer enjoys a reliable supply of safe, potable water; not to shut down deficient systems (WHO, 1984; Wheeler, 1990; Bartram, 1996). However, the most frequently articulated reason for routine monitoring of water supplies is to detect failures in compliance with water supply standards and regulations to be employed to enforce improvement (daSilva and O'Kane, 1989; Ward, 1995). Thus Le Moigne (1996, p. 20) suggests that 'the best water management strategy would mean little without laws and regulations that can be enforced'; while Lloyd and Helmer (1991, p. 58) have gone so far as to claim that in rural areas 'it is only worth considering planning the development of surveillance programs if the political will to implement water legislation exists'.

Many water supply regulations are therefore designed to assess measures of water supply service quality (and especially water quality), to compare results with a standard and to take some form of action upon detection of failure. However, because 'reality' is so far removed from the ideal situation proposed in many water supply standards and legislation (Okun and Lauria, 1992), standards and legislation have lost credibility in many countries (Jensen, 1967). In order to optimize the benefits of their use it is therefore appropriate that explicit schemes for enforcement implementation are provided for and pursued (Jensen, 1967).

3.6 Source protection and minimum treatment requirements

The establishment of drinking-water quality standards alone is not sufficient to provide confidence in the water supply in achieving an adequate level of water quality throughout the system. In addition, best practice norms need to be established for source protection and treatment requirements. Within the emerging paradigm for water safety promoted within the third edition of the WHO *Guidelines for Drinking-Water Quality* ensuring adequate source protection and treatment are given higher profile and are linked to specific control measures.

As noted in Section 2 above, source protection norms would normally be included within the statutory instruments linked to the primary law that governs water. Typically, such norms will make provisions for the requirements at several levels. These would usually include the basic sanitary completion measures designed to prevent direct contamination of the source, a broader area to protect the source from more widespread microbiological contamination, as well as broader measures protection for sources against chemical contamination and broader watershed management. In general, source protection norms have been better developed and enforced for groundwater than surface water, largely because the natural microbiological quality of groundwater is much higher, control of pollution is often simpler and an acceptance that all surface water should be treated prior to consumption. Such a simplistic approach is being increasingly challenged in recognition of the chemical hazards found in groundwater sources of drinking water that pose a significant risk to public health. The need for protection of surface water sources should also not be minimized, as ineffective source protection leads to increasing loss of resources available for drinking-water supply within reasonable levels of treatment costs.

Minimum treatment requirements are often specified in relation to specific sources. However, the basis of most treatment requirements is that the processes should be proven to be able effectively to remove contaminants and that a multiple barrier principle should be applied. This principle is designed to ensure that water will pass through several stages of treatment, thus reducing the risks of supply of untreated water. In some cases, failures in critical processes may lead to automated shutdown. The different treatment processes are not reviewed here, but it should be noted that increasing attention is being paid to 'older' processes that are primarily biological and physical as these may represent more cost-effective approaches than chemical processes.

The final component in most treatment processes is disinfection and this may also be applied to groundwater. Disinfection serves two purposes: it provides the final stage of treatment and is designed to remove any remaining microorganisms that have survived the preceding processes; it may also provide partial protection against re-contamination within distribution. The latter is an important function as water quality failure post-production remains common within many water supplies, particularly those that are older or prone to discontinuity or high leakage. However, it is increasingly recognized that there are some pathogens for which commonly used disinfectants are not effective. Furthermore, relying on disinfectant residuals to control the presence of pathogens in drinking water during distribution is ineffective.

The disinfectant of choice worldwide remains chlorine, although alternatives such as ozone are attracting increasing attention. Disinfection with chlorine is commonly applied as a final treatment step for surface water. In this application a minimum 'contact time' is normally engineered into the treatment plant configuration. In the case of groundwater, the practice has, in general, been only to provide in-line terminal disinfection that has limited contact time and therefore may not inactivate microorganisms entering the distribution system. There is increasing evidence of pathogen survival greatly in excess of those previously thought which indicates that this approach may need to be re-considered.

4 CONCLUSION

The development of standards for drinking-water quality is an important component in water quality control. However, this needs to be considered in light of other needs and, in particular, the extension of supply to those not served. The establishment of standards needs to consider not just health evidence and technical ability, but also impacts on costs and tariffs and social considerations. Routine monitoring and enforcement are essential if standards are to have meaning, In addition to standards governing water quality, complementary standards should address source protection and minimum treatment requirements in order to ensure adequate water quality.

REFERENCES

Appleton, B. (1995). The Influence of Technology on Operation and Maintenance of Rural Water Projects. In: *Integrated Rural Management Volume II: Technical Documents Prepared for the Second Technical Consultation*, pp. 19–25. World Health Organization, Geneva.

Bartram, J. (1996). Drinking Water and Health in the Wider Europe. In: S. Aubry (ed.) *New World Water 1996*, pp. 13–16. Sterling Publications Ltd, London.

Chorus, I. and Bartram, J. (1999). *Toxic cyanobacteria in water. A guide to their public health consequences, monitoring and management*. WHO, E & FN Spon, London.

Caponera, D.A. (1992). *Principles of Water Law and Administration, National and International*. AA Balkema Publishers, Rotterdam.

Carter, R., Tyrrel, S.F. and Howsam, P. (1993). Lessons Learned from the UN Water Decade. *Journal of the Institution of Water and Environmental Management* **7**(6), 646–650.

DaSilva, M.C. and O'Kane, P. (1989). The Design of a Water Quality Monitoring Network. In UNESCO/UNDP/WHO *Proceedings of the International Seminar on Water Quality – Assessment and Management, Lisbon, 17–19 May, 1989*. PGIRHT–DGRN–MARN, Lisbon, Portugal.

Davison, A., Howard, G., Stevens, M. *et al*. (2002). *Water Safety Plans (draft for consultation)* World Health Organization (WHO), Geneva, Switzerland.

Esrey, S., Feachem, R. and Hughes, J. (1985). Interventions for the Control of Diarrhoeal Diseases Among Young Children: Improving Water Supplies and Excreta Disposal Facilities. *Bulletin of the World Health Organization* **63**(4), 757–772.

Esrey, S.A., Potash, J.B., Roberts, L. and Schiff, C. (1991). Effects of Improved Water Supply and Sanitation on Ascariasis, Diarrhoea, Dracunculiasis, Hookworm Infection, Schistosomiasis and Trachoma. *Bulletin of the World Health Organization* **69**(5), 609–621.

Feachem, R. (1980). Rural Water and Sanitation. *Proceedings of the Royal Society, London B* **209**, 15–29.

Fewtrell, L., Kay, D. and Godfree, A. (1998). The microbiological quality of private water supplies. *Journal of the Chartered Institution of Water and Environmental Management* **12**, 45–47.

Fewtrell, L. and Bartram, J. (2001). *Water Quality: Guidelines, Standards and Health*. World Health Organization, IWA Publishing, London.

Holland, R. (1991). Cited in D.P. Maguire (ed.) *Appropriate Development for Basic Needs, Proceedings of the Conference on Appropriate Development for Survival – the Contribution of Technology*, p. 119. Thomas Telford, London.

Howard, G. (2001). Challenges in increasing access to safe water in urban Uganda: economic, social and technical issues. In: F. Craun (ed.) *Safety of water disinfection*, 2nd edn, pp. 492–503. Institute of Life Sciences, Washington DC.

Howard, G. (2002). *Water Supply Surveillance: A Reference Manual*. WEDC, Loughborough University, UK.

Howard, G., Bartram, J.K. and Luyima, P.G. (1999). Small water supplies in Urban Areas of Developing Countries. In: J.A. Cotruvo, G.F. Craun and N. Hearne (eds) *Providing safe drinking water in small system: technology, operations and economics*, pp. 83–93. Lewis Publishers, Washington, DC.

Ince, M. and Howard, G. (1999). Developing realistic drinking water standards. In: J. Pickford (ed.) *Integrated Development for Water Supply and Sanitation, Proceedings of the 25th WEDC Conference, Addis Ababa*, 1999 pp. 294–297.

IRC (1995) *Water and Sanitation for All: a World Priority Vol. 2: Achievements and Challenges*. Ministry of Housing, Spatial Planning and the Environment, The Hague, The Netherlands.

Jensen, P. (1967). Examination of Water Supply and Drinking Water. *Danish Medical Bulletin* **14**(1), 273–280.

Kandiah, A. (1995). Review of Progress on the Implementation of Recommendations of the First Technical Consultation. In: *Integrated Rural Management Volume II: Technical Documents Prepared for the Second Technical Consultation*, pp. 5–14. World Health Organization, Geneva.

Le Moigne, G. (1996). Change of Emphasis in World Bank Lending. In: S. Aubry (ed.) *New World Water 1996*, pp. 17–21. Sterling Publications Ltd., London.

Lloyd, B. and Helmer, R. (1991). *Surveillance of Drinking Water Quality in Rural Areas*, p. 58. Longman, London.

Okun, D.A. and Lauria, D.T. (1992). Capacity Building for Water Resources Management: An International Initiative for the 1990s. In: M. Munro (ed.) *Water Technology International 1992*, pp. 17–24. Century Press, London.

Soby, B.A., Simpson, A.C.D. and Ives, D.P. (1993) *Integrating Public and Scientific Judgements into a Tool Kit for Managing Food-Related Risks, Stage 1: Literature Review and Feasibility Study.* A Report to the UK Ministry of Agriculture, Fisheries and Food, ERAU Research Report No. 16, University of East Anglia, Norwich.

Ward, R. (1995). Water Quality Monitoring as an Information System. In: Adriaanse, J. van der Kraats, P. Stoks and R. Ward (eds) *Monitoring Tailor-made*. RIZA, The Netherlands.

Wheeler, D. (1990). Risk Assessment and Public Perception of Water Quality. In: *Institution of Water and Environmental Management Engineering for Health, Technical Papers of Annual Symposium, 27–28 March 1990, UMIST, Manchester (paper 2)*. Institution of Water and Environmental Management, London.

WHO (1976). *Surveillance of Drinking-water Quality*, WHO Monograph Series No 63. World Health Organization, Geneva.

WHO (1984). *Guidelines for Drinking Water Quality Volume 1: Recommendations*. World Health Organization, Geneva.

WHO (1985). *Guidelines for Drinking Water Quality Volume 3: Drinking-water Quality Control in Small-community Supplies*. World Health Organization, Geneva.

WHO (1993). *Guidelines for Drinking-water Quality Volume 1: Recommendations*, 2nd edn. World Health Organization, Geneva.

WHO (1996). *Guidelines for Drinking Water Quality Volume 2: Health Criteria and Other Supporting Information*, 2nd edn. World Health Organization, Geneva.

WHO/UNICEF/WSSCC (1996). *Water Supply and Sanitation Sector Monitoring Report 1996, Sector Status of as 31 December 1994*. World Health Organization, Geneva.

WHO (1997). *Guidelines for Drinking Water Quality Volume 3: Drinking Water Quality Control in Small Community Supplies*, 2nd edn. World Health Organization, Geneva.

WHO (1999). *Food and Personal Hygiene Account for over 2 Million Deaths Annually*. World Health Organization, Geneva.

WHO (2000). *Guidelines for Drinking Water Quality Training Pack*. World Health Organization, Geneva.

WHO (2003a). *Guidelines for Drinking Water Quality Training Pack*. Third edition, World Health Organization, Geneva. Web site: http://www.who.int/water_sanitation_health/Training_mat/GDWAtrtoc.htm

WHO (2003b). The Right to Water. World Health Organization, Geneva. http://www.who.int/water_sanitation_health/

15

Control of pathogenic microorganisms in wastewater recycling and reuse in agriculture

Hillel Shuval* and Badri Fattal‡

Division of Environmental Science, The Hebrew University of Jerusalem, Jerusalem 91904, Israel

1 INTRODUCTION

As presented in detail, in Chapter 11, raw domestic wastewaters normally carry the full spectrum of pathogenic microorganisms – the causative agents of bacterial, virus and protozoan diseases endemic in the community and excreted by diseased and infected individuals. While recycling and reuse of wastewater for agriculture, industry and non-potable urban purposes can be a highly effective strategy for developing a sustainable water resource in water short areas, nutrient conservation and environmental protection, it is essential to understand the health risks involved and to develop appropriate strategies for the control of those risks. This chapter will concentrate on the control of pathogenic microorganisms from wastewater in agricultural reuse since this is the most widely practised form of reuse on a global basis. However, more and more water specialists, natural resource planners and economists see water as an economic good and, as time goes on, there will be an increased motivation to divert recycled wastewater from low income agriculture to areas where the added value of water is greater, such as industrial and non-potable urban uses including public parks, green belts, golf courses, football fields. As time goes on and water shortages in arid areas increase, there will undoubtedly be an expansion of the reuse of purified wastewater for industrial and a wide variety of urban/non-potable purposes. There are some specific heath guidelines and standards for such uses, but they will not be reviewed in this chapter.

The control measures for agricultural reuse include establishing and enforcing microbial guidelines for effluent quality, regulation of the types of crops to be irrigated, minimizing the potential for crop contamination by various irrigation techniques and the treatment of the wastewater to an *appropriate degree* so as to control potential health risks, both to the farmers and the consumers of crops, from pathogenic microorganisms in the wastewater stream. This chapter will deal with the above questions.

2 PATHOGENIC MICROORGANISMS CAN BE TRANSMITTED BY WASTEWATER IRRIGATION

Pathogenic microorganisms in the wastewater stream can be transmitted to healthy individuals and cause disease if improper regulation and control methods in wastewater irrigation are practised. In order for disease causing microorganisms or pathogens in the wastewater stream flowing from a community to infect a susceptible individual they must be

Lunenfeld-Kunen *Professor of Environmental Science
‡Professor of Environmental Health

The Handbook of Water and Wastewater Microbiology
ISBN 0-12-470100-0

able to survive in the environment (i.e. in water, soil, or food) for a period of time and they must be ingested in a sufficiently high number. Factors that affect the survival of pathogens in soil include antagonism from soil bacteria, moisture content, organic matter, pH, sunlight and temperature. Excreted enteric pathogens such as bacteria, viruses, protozoa, and helminth eggs do not usually penetrate undamaged vegetables but can survive for long periods in the root zone, in protected leafy folds, in deep stem depressions, and in cracks or flaws in the skin.

Data from numerous field and laboratory studies have made it possible to estimate the persistence of certain enteric pathogens in water, wastewater, soil, and on crops. These survival periods in the environment are presented in summary graphic form in Fig. 15.1. For example, it appears that *Campylobacter* may survive in soil or on crops for only a few days, whereas most bacterial and viral pathogens can survive from weeks to months. The highly resistant eggs of helminths, such as *Trichuris*, *Taenia* and *Ascaris*, can survive for 9–12 months, but their numbers are greatly reduced during exposure to the environment.

Field studies in Israel have demonstrated that enteric bacteria and viruses can be dispersed for up to 730 m in aerosolized droplets generated by spray (sprinkler) irrigation, but their concentration is greatly reduced by detrimental environmental factors such as sunlight and drying (Teltsch *et al.*, 1980; Applebaum

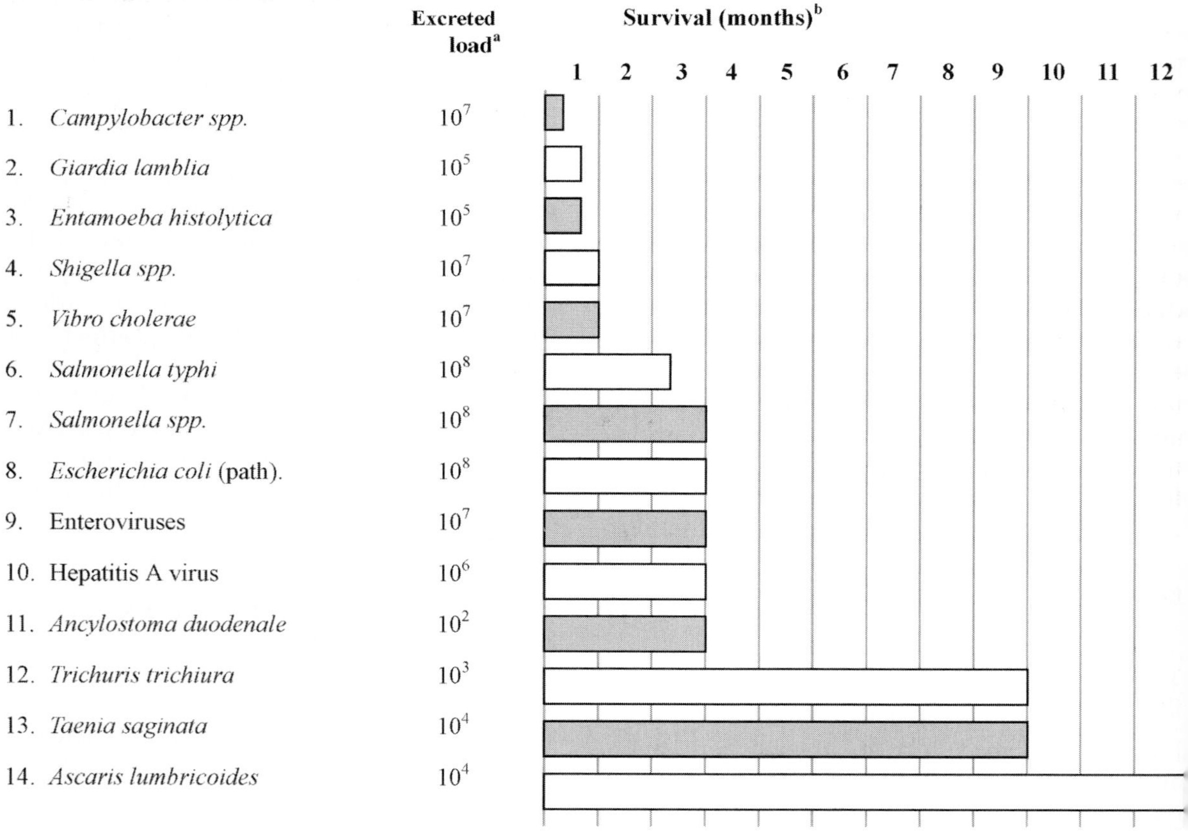

[a] Typical average number of organism/g feces
[b] Estimated average life of infective stage at 20–30°C.

Fig. 15.1 Survival times of enteric pathogens in water, wastewater, soil and on crops (from Shuval *et al.*, 1986 - based partially on data from Feachem *et al.*, 1983).

et al., 1984; Shuval *et al.*, 1988, 1989a). Thus, most excreted pathogens can survive in the environment long enough to be transported by the wastewater to the fields and to the irrigated crops. The contaminated crops eventually reach the consumer, although by then the concentration of pathogens is greatly reduced. The rapid natural die-away of pathogens in the environment is discussed in a later section as it is an important factor in reducing the health risks associated with wastewater reuse.

Theoretical analysis suggests that a number of epidemiological factors determine whether various groups of pathogens will cause infections in humans through wastewater irrigation. We have developed a model to evaluate the empirical epidemiological data and to formulate control strategies (Shuval *et al.*, 1986).

Table 15.1 summarizes the epidemiological characteristics of the main groups of enteric pathogens as they relate to the five factors that influence the transmission and degree of infections and disease resulting from wastewater irrigation. This summary provides a simplified theoretical basis for ranking the groups of pathogens according to their potential for transmitting disease through wastewater irrigation. On this basis, it appears that the helminthic diseases are the ones most effectively transmitted by irrigation with raw wastewater because they persist in the environment for relatively long periods; their minimum infective dose is small; there is little or no immunity against them; concurrent infection in the home is often limited; and latency is long and a soil development stage is required for transmission.

In contrast, the enteric viral diseases should be least effectively transmitted by irrigation with raw wastewater in developing countries with low levels of sanitation in the home, despite their small minimum infective doses and ability to survive for long periods in the environment. Due to poor hygiene in the home, and the prevalence of concurrent routes of infection in some areas, most of the population has been exposed to, and acquired immunity to, most of the enteric viral diseases as infants. Most enteric viral diseases impart immunity for life, or at least for very long periods, so that they are not likely to re-infect individuals exposed to them again, e.g. through wastewater irrigation. The transmission of bacterial and protozoan diseases through wastewater irrigation lies between these two extremes. In developed countries with higher levels of home sanitation and little concurrent disease transmission due to poor hygiene practice, there will be lower levels of immunity to diseases which could be transmitted by wastewater irrigation or vegetables eaten uncooked.

3 REVIEW OF RESEARCH FINDINGS ON DISEASE TRANSMISSION BY WASTEWATER IRRIGATION

This section will provide an extensive review and evaluation of the research findings on disease transmission by wastewater irrigation based on available scientific papers published in recognized journals and in numerous unpublished government reports, university theses, and private papers obtained during an intensive world-wide search carried out with the help of international and national

TABLE 15.1 Epidemiological characteristics of enteric pathogens *vis-à-vis* their effectiveness in causing disease through wastewater irrigation

Pathogen	*Persistence in environment*	*Minimum infective dose*	*Immunity*	*Concurrent routes of infection*	*Latency/soil development stage*
Viruses	Medium	Low	Long	Mainly home contact, food and water	No
Bacteria	Short/medium	Medium/high	Short/medium	Mainly home contact, food and water	No
Protozoa	Short	Low/medium	None/little	Mainly home contact, food and water	No
Helminths	Long	Low	None/little	Mainly soil contact outside home and food	Yes

agencies and individuals. Over 1000 documents, some more than 100 years old, were examined in the course of this study, but few offered concrete or reliable epidemiological evidence of health effects. Most of them based their conclusions on inference and extrapolation. Nonetheless, about 50 of these reports provided enough credible evidence based on sound epidemiological procedures to make a detailed analysis useful. Those studies are reviewed in detail in the UNDP-World Bank report on which this chapter is partially based (Shuval *et al.*, 1986). Our general conclusions of some of the more pertinent studies are presented below.

One of the goals of our studies for the UNDP/World Bank study (Shuval *et al.*, 1986; Shuval, 1990), as described in this chapter, was to re-evaluate all the credible, scientifically valid and quantifiable epidemiological evidence of the real human health effects associated with wastewater irrigation. Such evidence is needed to determine the validity of current regulations and to develop appropriate technical solutions for existing problems.

3.1 Illness associated with wastewater irrigation of crops eaten raw

In areas of the world where the helminthic diseases caused by *Ascaris* and *Trichuris* are endemic in the population, and where raw, untreated wastewater is used to irrigate salad crops and/or other vegetables generally eaten uncooked, the consumption of such wastewater-irrigated salad and vegetable crops may lead to significant levels of infection. Khalil (1931) demonstrated the importance of this route of transmission in his pioneering studies in Egypt. Similarly, a study in Jerusalem (Shuval *et al.*, 1984) provided strong evidence that massive infections of both *Ascaris* and *Trichuris* may occur when salad and vegetable crops are irrigated with raw wastewater. These diseases almost totally disappeared from the community when raw wastewater irrigation was stopped (Fig. 15.2). Two studies from Darmstadt, Germany (Krey, 1949; Baumhogger, 1949) provided additional support for this conclusion.

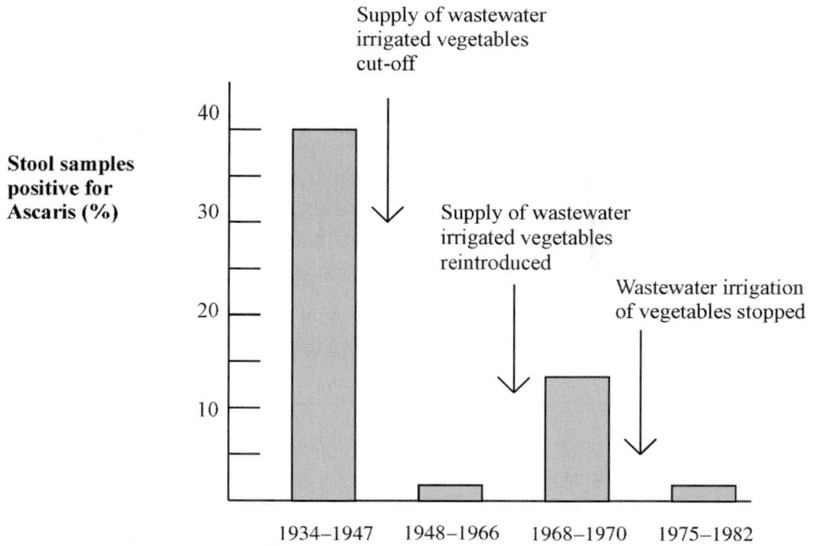

Fig. 15.2 Relationship between Ascaris-positive stool samples in population of western Jerusalem and supply of vegetables and salad crops irrigated with raw wastewater in Jerusalem, 1935–1982 (from Shuval *et al.*, 1986 – based on partially on data from Ben-Ari, 1962; Jjumba-Mukabu and Gunders, 1971; Shuval *et al.*, 1984).

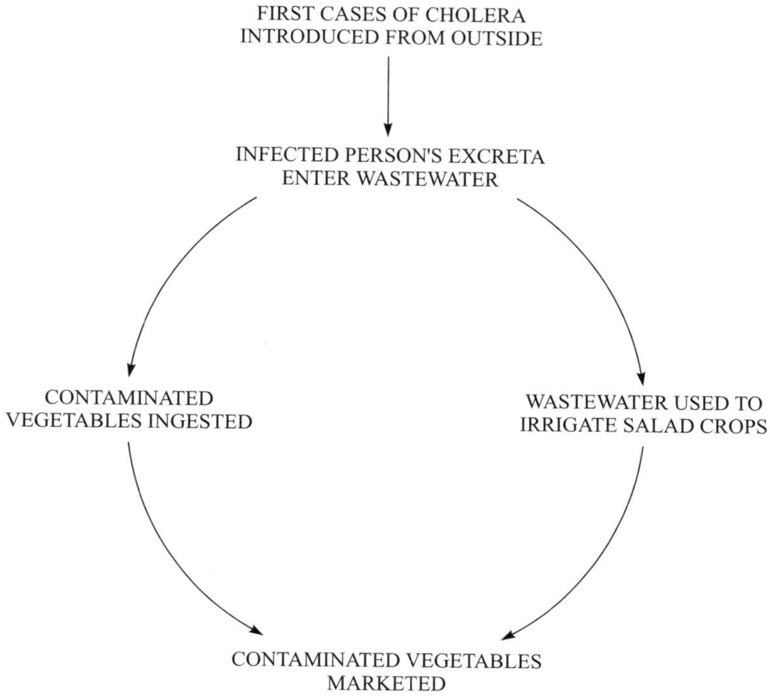

Fig. 15.3 Hypothesized cycle of transmission of Vibrio cholerae from first cholera carriers introduced from outside the city, through wastewater-irrigated vegetables, back to residents in the city (from Fattal *et al.*, 1986b).

These studies also indicate that regardless of the level of municipal sanitation and personal hygiene, irrigation of vegetables and salad crops with raw wastewater can serve as a major pathway for continuing and long-term exposure to *Ascaris* and *Trichuris* infections. Both of these infections are of a cumulative and chronic nature, so that repeated long-term re-infection may result in a higher worm load and increased negative health effects, particularly among children.

Cholera can also be disseminated by vegetable and salad crops irrigated with raw wastewater if it is carrying cholera vibrios. This possibility is of particular concern in non-endemic areas where sanitation levels are relatively high, and the common routes of cholera transmission, such as contaminated drinking water and poor personal hygiene, are closed. Under such conditions, the introduction of a few cholera carriers (or subclinical cases) into a community could lead to massive infection of the wastewater stream and subsequent transmission of the disease to the consumers of the vegetable crops irrigated with the raw wastewater, as occurred in Jerusalem in 1970. The hypothesized cycle of transmission from the first imported case from outside the city through wastewater irrigated vegetables back to the residents is shown in Fig. 15.3 (Fattal *et al.*, 1986b).

Similarly, our study from Santiago, Chile (Shuval, 1984), strongly suggests that typhoid fever can be transmitted by fresh salad crops irrigated with raw wastewater. The number of typhoid fever cases in Santiago rose rapidly *annually* at the beginning of the irrigation season, after 16 000 ha of vegetables and salad crops (usually eaten uncooked) had been irrigated with raw wastewater. The relatively high socioeconomic level, good water supply, and good general sanitation in the city supports the hypothesis that wastewater irrigation can become a major route for the transmission of such bacterial disease.

3.2 Cattle grazing on wastewater irrigated pastures

Wastewater is often used to irrigate pasture for cattle and sheep. What are the health risks associated with this practice? There is only limited epidemiological evidence to indicate that beef tapeworms (*Taenia saginata*) have been transmitted to populations consuming the meat of cattle grazing on wastewater-irrigated fields or fed crops from such fields. However, there is strong evidence from Melbourne, Australia (Penfold and Phillips, 1937), and from Denmark (Jepson and Roth, 1949) that cattle grazing on fields freshly irrigated with raw wastewater or drinking from raw wastewater canals or ponds can become heavily infected with the disease. This condition can become serious enough to require veterinary attention and may lead to economic loss. Irrigation of pastures with raw wastewater from communities infected with tapeworm disease may provide a major pathway for the continuing cycle of transmission of the disease to animals and humans.

3.3 Exposure of wastewater farmers to disease

Obviously the individuals most intensely exposed to the wastewater stream at the farms where wastewater irrigation is practised, are the farmers themselves. Some of the studies with the clearest epidemiological evidence relates to the health of such farm workers. Sewage farm workers exposed to raw wastewater in areas of India, where *Ancylostoma* (hookworm) and *Ascaris* infections are endemic, have much higher levels of infection than other agricultural workers (Krishnamoorthi *et al.*, 1973). The risk of hookworm infection is particularly great in areas where farmers customarily work barefoot, because the broken skin of their feet is readily penetrated by the motile hookworm larva. Sewage farm workers in this study also suffered more from anaemia (a symptom of severe hookworm infestation) than the controls. Thus, there is evidence that continuing occupational exposure to irrigation with raw wastewater can have a direct effect on human productivity and, thus, on the economy.

Sewage farm workers are also liable to become infected with cholera if the raw wastewater being used for irrigation is from an urban area experiencing a cholera epidemic. This situation is particularly likely to arise in an area where cholera is not normally endemic and where the level of immunity among the sewage farm workers is low or non-existent. This proved to be the case in the 1970 cholera outbreak in Jerusalem (Fattal *et al.*, 1986b). In a related study (Fattal *et al.*, 1985), we have shown that irrigation workers exposed to aerosols from spray irrigation of both fresh water and wastewater had significantly higher rates of serum positivity to antibodies of *Legionella pneumophila*, the causative agent of Legionnaires' disease, as compared to the non-exposed control group of farmers and their families. This study does show that virulent pathogens can be transmitted by aerosols from spray irrigation and can infect highly exposed workers. However, it does not suggest that Legionnaires' disease is transmitted by wastewater irrigation any more than by fresh water, where the organisms are very often found.

Studies from industrialized countries have thus far produced only limited, and often conflicting, evidence of the incidence of bacterial and viral diseases among wastewater irrigation workers exposed to partly or fully treated effluent, or among workers in wastewater treatment plants exposed directly to wastewater or wastewater aerosols. Most morbidity and serological studies have been unable to give a clear indication of the prevalence of viral diseases among such occupational exposed groups.

It is hypothesized that many sewage farmers or treatment plant workers have acquired relatively high levels of permanent immunity to most of the common enteric viruses endemic in their communities at a much younger age. Thus, by the time they are exposed occupationally, the number of susceptible workers is small and not statistically significant. Presumably this is also the case among infants and children in developing countries, because they are exposed to most endemic enteric viral diseases by the time they reach working age. Although this is not the case for some bacterial and protozoan pathogens, multiple routes of concurrent infection with these diseases may well mask any excess infection among wastewater irrigation workers in developing countries.

3.4 Exposure of residents in the vicinity of wastewater farms

A number of studies have evaluated the potential negative health effects that might result from living in the vicinity of farms where wastewater irrigation, particularly sprinkler irrigation is practised. There is little evidence linking disease and/or infection among population groups living near wastewater treatment plants or wastewater irrigation sites with pathogens contained in aerosolized wastewater. Most studies have shown no demonstrable disease resulting from such aerosolized wastewater, which is caused by sprinkler irrigation and aeration processes. Researchers agree, however, that most of the earlier studies have been inadequate.

Recent studies in Israel suggest that aerosols from sprinkler irrigation with poor microbial quality wastewater can, under certain circumstances, cause limited infections among infants living near wastewater-irrigated fields. The studies, however, also concluded that these were negligible and could be controlled by better treatment of the wastewater (Fattal *et al.*, 1986a, 1987; Shuval *et al.*, 1988, 1989b).

These findings support the conclusion that, in general, relatively high levels of immunity against most viruses endemic in the community block additional environmental transmission by wastewater irrigation. Therefore, the additional health burden is not measurable. The primary route of transmission of such enteroviruses, even under good hygienic conditions, is through contact infection in the home at a relatively young age. As already mentioned, such contact infection is even more common in developing countries, so that a town's wastewater would not normally be expected to transmit a viral disease to rural areas using it for irrigation.

3.5 Epidemiological evidence of beneficial effects from wastewater treatment

When raw wastewater is used for irrigation there is no doubt that the wastewater stream carries very high concentrations of pathogens. Conventional wastewater treatment plants were not normally designed to reduce the concentration of pathogenic microorganisms, however, such treatment can nonetheless provide a degree of removal up to about 85–95% reduction in coliform bacteria and pathogens. Some epidemiological studies have provided evidence that negative health effects can be reduced when wastewater is treated for the removal of pathogens. For example, Baumhogger (1949) reported that, in 1944, residents of Darmstadt who consumed salad crops and vegetables irrigated with raw wastewater experienced a massive infection of *Ascaris*; but the residents of Berlin, where biological treatment and sedimentation were applied to the wastewater prior to the irrigation of similar crops, did not.

Another study on intestinal parasites was conducted on school children near Mexico City (Sanchez Levya, 1976). The prevalence of intestinal parasites in children from villages that used wastewater irrigation did not differ significantly from that in children from the control villages, which did not irrigate with wastewater. The lack of significant difference between the two groups may have resulted from long-term storage of the wastewater in a large reservoir for weeks or months prior to its use for irrigation. It is assumed that sedimentation and pathogen die-away during long-term storage were effective in removing the large, easily settleable protozoa and helminths, which were the pathogens of interest in this study. This study provides the first strong epidemiological evidence of the health protection provided by microbial reductions achieved in wastewater storage reservoirs.

Furthermore, the absence of negative health effects in Lubbock, Texas (Camann *et al.*, 1983) and in Muskegon, Michigan (Clark *et al.*, 1981), appears to be associated with the fact that well-treated effluents from areas of low endemicity were used for irrigation.

Data from these field studies strongly suggest that pathogen reduction by wastewater treatment, including long-term storage in wastewater reservoirs, can have a positive effect on human health. In all the above studies, this positive effect was achieved despite the use of effluent which had not been disinfected and which contained a few thousand of fecal coliform bacteria per 100 ml. These data agree with water quality data on pathogen removal and suggest that appropriate wastewater treatment resulting in effective reduction of coliforms to the level of a few thousand/100 ml, but not total removal, can provide a high level of health protection.

3.6 Conclusions from the analysis of the epidemiological studies

It is possible to draw certain conclusions from the series of epidemiological studies on the health effects of wastewater reuse in agriculture. The studies from both developed and developing countries indicate that the following diseases are occasionally transmitted via raw or *very poorly* treated wastewater:

1. The general public may develop ascariasis, trichuriasis, typhoid fever or cholera by consuming salad or vegetable crops irrigated with *raw wastewater*, and probably tapeworm by eating the meat of cattle grazed on wastewater-irrigated pasture. There may also be limited transmission of other enteric bacteria, viruses and protozoa.
2. Wastewater irrigation workers may develop ancylostomiasis (hookworm), ascariasis, possibly cholera and, to a much lesser extent, infection caused by other enteric bacteria and viruses, if exposed to raw wastewater.
3. Although there is no demonstrated risk to the general public residing in areas where wastewater is used in sprinkler irrigation, there may be minor transmission of enteric viruses to infants and children living in these areas, especially when the viruses are not endemic to the area and raw wastewater or very poor quality effluent is used.

Thus, the empirical evidence on disease transmission associated with *raw wastewater* irrigation in developing countries strongly suggests that helminths are the principal problem, with some limited transmission of bacterial and viral disease. The above ranking, based on empirical data, agrees with that predicted in our model.

In interpreting the above conclusions, one must remember that the vast majority of developing countries are in areas where helminthic and protozoan diseases such as hookworm, ascariasis, trichuriasis, and tapeworm are endemic. In some of these areas, cholera is endemic as well. It can be assumed that in most developing countries, in populations with low levels of personal and domestic hygiene, the children will become immune to the endemic enteric viral diseases when very young through contact infection in the home.

In conclusion, epidemiological evidence of disease transmission associated with the use of *raw wastewater* in agriculture in developing countries indicates that the pathogenic agents may be ranked in the following order of declining importance:

1. *High risk*: helminths (*Ancylostoma*, *Ascaris*, *Trichuris* and *Taenia*)
2. *Lower risk*: enteric bacteria (cholera, typhoid, shigellosis and possibly others); protozoa (amebiasis and giardiasis)
3. *Least risk*: enteric viruses (viral gastroenteritis and infectious hepatitis).

As pointed out earlier, these negative health effects were all detected in association with the use of *raw or primarily treated* wastewater. Therefore, wastewater treatment processes that effectively remove all, or most, of these pathogens, according to their rank in the above list, could reduce the negative health effects caused by the utilization of raw wastewater. While helminths are very stable in the environment, bacteria and viruses rapidly decrease in numbers in the soil and on crops.

Thus, the ideal treatment process prior to wastewater recycling and reuse, should be particularly effective in removing helminths, even if it is somewhat less efficient in removing bacteria and viruses. Wastewater treatment technologies that can be used to achieve this goal are discussed later in this chapter. In general, the above ranking of pathogens will not apply to the more developed countries or other areas in which helminth diseases are not endemic. In those areas the negative health effects, if any, resulting from irrigation with raw or partly treated wastewater will probably be associated mainly with bacterial and protozoan diseases and, in a few cases, with viral diseases. Whatever the country or the conditions, however, the basic strategies for control are the same – the pathogen concentration in the wastewater stream must be reduced and/or the type of crops irrigated must be restricted.

Overall, our studies have demonstrated that the extent to which disease is transmitted by wastewater irrigation is much less than was widely believed to be the case by public health officials in the past. Moreover, this study does not provide epidemiological support for the use of the much-copied California standard requiring a coliform count of 2/100 ml for effluent to be used in the irrigation of edible crops and even less support for the more recent USEPA/USAID recommended guideline of zero fecal coliforms. No detrimental health effects were detected or reported when well-treated wastewater with much higher coliform counts was used.

4 THE DEVELOPMENT OF HEALTH STANDARDS AND GUIDELINES FOR WASTEWATER REUSE

4.1 The importance of health guidelines and standards for reuse

One of the most important and widely practised administrative methods for protecting the public health from the risks of uncontrolled wastewater irrigation, particularly of vegetables and salad crop consumed uncooked, is the establishment of guidelines or legally binding standards for the microbial quality of wastewater used for irrigation. This section will review the scientific basis and historical and social forces that influenced the evolution of microbial standards and guidelines for wastewater reuse for agricultural purposes. This analysis will draw extensively on World Bank and World Health Organization studies and reports whose goal was a cautious re-evaluation of the credible scientific evidence which could provide a sound basis for establishing safe and feasible health guidelines for wastewater reuse (Engelberg Report, 1985; Shuval *et al.*, 1986; WHO, 1989; Shuval, 1990).

The strict health regulations governing wastewater reuse that have been developed in the industrial countries over the past 60 years, such as those of the Department of Health of the State of California, which requires an effluent standard of 2 coliforms/100 ml for irrigation of crops eaten uncooked, and even the more recent USEPA/USAID (1992) recommended guidelines for unrestricted effluent use in agriculture of zero fecal coliforms, have been based to a great extent on early scientific data indicating that most enteric pathogens can be detected in wastewater and that they can survive for extended periods in wastewater-irrigated soil and crops (see Fig. 15.1). Many health authorities have erroneously concluded that, because pathogens can survive long enough to contaminate crops, even if their numbers are very low and below the minimum ineffective dose level, they still pose a serious risk to public health. However, these regulations were formulated at a time when sound epidemiological evidence was rather scanty. As a result, policy makers used the cautious 'zero risk' approach and introduced very strict regulations that they hoped would protect the public against the potential risks thought to be associated with wastewater reuse. Most industrial countries were not concerned that these regulations were overly restrictive because the economic and social benefits of wastewater reuse were of only marginal interest.

Many of the current standards restrict the types of crops to be irrigated with conventional wastewater effluent to those not eaten raw. Regulations like those in California, requiring the effluent used for the irrigation of edible crops to have a bacterial standard approaching that of drinking water (2 coliforms/100 ml), are usually not technically feasible or sustainable without very highly skilled operators and a high-tech service infrastructure. This is particularly true for developing countries, but even applies to many developed countries. In reality, a standard of 2 coliforms/100 ml for irrigation is superior to the quality of drinking water for the majority of urban and rural poor in developing countries (where fecal coliforms are generally in excess of 10/100 ml of drinking water).

In developed countries, where these crop restrictions can normally be enforced, vegetable and salad crops are not usually irrigated with wastewater. In the developing countries, many of which have adopted the same strict regulations, public health officials do not approve of the use of wastewater for irrigation of vegetable and salad crops eaten raw. However, when water is in short supply such crops are widely irrigated illegally with raw or poorly treated wastewater. This usually occurs in the vicinity of major cities, particularly in semi-arid regions.

Since the official effluent standards for vegetable irrigation are not within the obtainable range of common engineering practice and for economic considerations, new projects to improve the quality of effluent are not usually approved. With the authorities insisting on unattainable, expensive and unjustifiable standards, farmers are practising widespread uncontrolled and unsafe irrigation of salad crops with raw wastewater. The highly contaminated vegetables are supplied directly to the nearby urban markets, where such horticultural products can command high prices. This is a classic case in which official insistence on the 'best' prevents cities and farmers from achieving the 'good'.

Some inconsistency exists between the strict California standards, which require edible crops to be irrigated with wastewater of drinking water quality, and the actual agricultural irrigation with normal surface water as practised in the USA and other industrialized countries with high levels of hygiene and public health. There are few, if any, microbiological limits on irrigation with surface water from rivers or lakes, which may be polluted with raw or treated wastewater. For example, the US Environmental Protection Agency's water quality criteria for unrestricted irrigation with surface water is 1000 fecal coliforms/100 ml (USEPA, 1972). A WHO world survey of river water quality has indicated that most rivers in Europe have mean fecal coliform counts of 1000–10 000/100 ml. And yet none of these industrialized countries has restrictions on the use of such river water for irrigation.

A number of microbial guidelines have been developed for recreational waters considered acceptable for human contact and swimming. In the USA, for example, microbial guidelines for recreational water have, in the past, ranged from 200 to 1000 fecal coliforms/100 ml, although they currently are at about 10 fecal coliforms/100 ml. In Europe, guidelines vary from 100 coliforms/100 ml in Italy to 20 000 coliforms/100 ml in Yugoslavia. The European Community has recommended a guideline of 2000 fecal coliforms/100 ml for recreational waters (Shuval *et al.*, 1986).

It is difficult to explain the logic of a 2 coliforms/100 ml standard for effluent irrigation when farmers all over the USA and Europe can legally irrigate any crops they choose with surface water from free-flowing rivers and lakes, which often have fecal coliform levels of over 1000/100 ml. It is even more difficult to explain the epidemiological rationale of the 2 coliforms/100 ml standard for effluent irrigation, while in Europe recreational water for bathing is considered acceptable at 2000 fecal coliforms/100 ml.

4.2 The World Bank/World Health Organization initiative to re-evaluate wastewater reuse guidelines

Because of the questions raised in Section 4.1 above, and the fact that the strict coliform standards adopted by many countries were rarely enforced and often found to be unfeasible for economic reasons and due to the lack of adequate technical infrastructure, in 1981, the World Bank and the World Health Organization initiated an extensive multidisciplinary study on the health effects of wastewater irrigation. The primary goal was to obtain an up-to-date scientific evaluation of the public health justification and validity of existing standards and guidelines and to develop alternatives if this was deemed to be justified. These studies were carried out by teams of epidemiologists, engineers, agronomists and environmental specialists simultaneously and independently at three different environmental sciences and public health research centers – the London School of Hygiene and Tropical Medicine, the International Reference Centre on Wastes Disposal, Zurich and the Division of Environmental Sciences of the Hebrew University of Jerusalem, Israel. These groups, working independently, prepared reports summarizing their findings, analysis and recommendations (Feachem *et al.*, 1983; Shuval *et al.*, 1986; Blum and Feachem, 1985).

In July 1985, a group of environmental experts, including engineers and epidemiologists, meeting at Engelberg, Switzerland, under the auspices of UNDP, World Bank, WHO, UNEP and IRCWD, reviewed the preparatory scientific studies which provided new epidemiological data and insights and from them formulated new proposed microbiological guidelines for treated wastewater reuse in agricultural irrigation (Engelberg Report 1985). The group accepted the main findings and recommendations of the UNDP-World Bank study (Shuval *et al.*, 1986) and concluded that 'current guidelines and standards for human waste use are overly conservative and unduly restrict project development, thereby encouraging unregulated human waste use'. The new guidelines recommended in the Engelberg Report were later accepted and approved by a WHO Meeting of Experts (WHO, 1989). However, before the WHO Executive approved those recommendations for publication as an official WHO document, the report was sent out for review and approval by a panel of some 100 epidemiologists, public health officials and environmental engineers. Thus, these new recommended guidelines for the microbial quality of effluent used for wastewater irrigation of edible crops carry the stamp of approval of the highest international authority on public health and environmental matters. Other important international technical assistance agencies joined the WHO in supporting the new recommended guidelines for wastewater irrigation including the World Bank, The Food and Agricultural Organization (FAO), The United Nations Environment Program (UNEP) and the United Nations Development Program (UNDP). Meanwhile, a number of governments in developing and developed countries have adopted the new WHO recommended guidelines, including France.

There were a number of innovations in the new recommended guidelines. Since the possibility of transmitting helminth disease by wastewater irrigation of even non-edible crops was identified as the principal health problem, a new, stricter approach to the use of raw wastewater was developed. The new WHO guidelines recommend effective water treatment in all cases to remove helminths to a level of *one or fewer helminth eggs per liter.* The main innovation of the WHO guidelines is: for crops eaten uncooked, an effluent must contain one or fewer helminth eggs per liter, with a *geometric mean of fecal coliforms not exceeding 1000/100 ml.* This is a much more liberal coliform standard than the early California requirement of 2 total coliforms/100 ml.

An attractive feature of the new WHO (1989) effluent guidelines is that they can be readily achieved with low cost, robust waste stabilization pond systems and wastewater storage and treatment reservoirs that are particularly suited to developing countries. The high levels of pathogen removal that can be achieved by such low-cost stabilization pond systems are shown in Fig. 15.4. In conjunction with alternating wastewater storage reservoirs even higher degrees of treatment with an added safety factor can be achieved. Most studies indicate that the critical design parameter to achieve the WHO microbial guidelines in the effluent is a long detention period of up to 25–30 days in maturation ponds or long

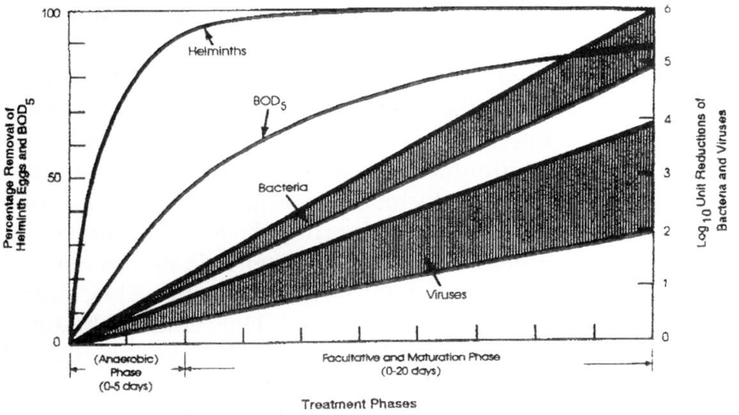

Fig. 15.4 Generalized removal curves for BOD, helminth eggs, excreted bacteria and viruses in wastewater stabilization ponds at temperatures above 20°C (from Shuval *et al.*, 1986).

detention times of several months in batch-operated storage reservoirs prior to the irrigation season (see Chapter 26).

4.3 The USEPA/USAID initiative for wastewater reuse guidelines

The US Environmental Protection Agency (USEPA), with the support of the US Agency for International Development (USAID), established their own rigorous recommended guidelines in 1992 of zero fecal coliforms/100 ml, a BOD of 10 mg/l, a turbidity of 2 NTU and a free chlorine residual of 1 mg/l. This quality of wastewater effluent can only be achieved in very costly high-tech wastewater treatment plants that require a high level of technological infrastructure for operation and maintenance, so that they can continuously meet such very rigorous standards. These guidelines were drafted by one of the leading American consulting engineering firms under contract to USAID. Such consulting engineering firms often tend to favor such high-tech treatment processes. Again, these new American guidelines are essentially as strict as those required for drinking water and reaffirm the 'no risk' or 'fail safe' approach that has been taken by the Americans in setting wastewater reuse guidelines and standards. The fact that little if any natural river water or water at approved bathing beaches in the USA or elsewhere could meet these recommended irrigation guidelines did not seem to bother those who drafted and approved the new American guidelines. No one has suggested that such river water should not be allowed for irrigation purposes, nor has any health risk from such irrigation been reported.

5 THE DEVELOPMENT OF A RISK ASSESSMENT/COST-EFFECTIVENESS METHOD FOR EVALUATION OF GUIDELINES

The WHO Guidelines are supported by numerous international technical assistance agencies and have been adopted by France and a number of other developed and developing countries. However, some groups have favored the stricter 'no risk' USEPA/USAID recommended guidelines and have questioned whether the WHO guidelines provide a sufficiently high level of safety and health protection. The debate over the appropriateness of the various guidelines has so far been on a qualitative level.

We have carried out a study aimed at developing a quantitative risk assessment and cost-effectiveness approach based on a mathematical model and experimental data, to arrive at a comparative risk analysis of the various recommended microbiological guidelines for unrestricted irrigation of vegetables normally eaten uncooked (Shuval and Fattal, 1996; Shuval *et al.*, 1997). The guidelines that were compared are those of the World Health Organization (<1000 fecal coliforms/100 ml) and those recommended by the USEPA/USAID (zero fecal coliforms/100 ml).

5.1 Risk assessment model

For the purposes of this study, the risk assessment model, estimating the risk of infection and disease from ingesting microorganisms in drinking water, developed by Haas *et al.* (1993) was selected. However, certain modifications were required since we are estimating the risk of infection associated with eating vegetables irrigated with wastewater of various microbial qualities. The basic model of Haas *et al.* (1993) for the probability of infection (P_I) from ingesting pathogenic microorganisms in water is:

$$P_I = 1 - [1 + N/N_{50}(2^{1/\alpha} - 1)]^{-\alpha} \quad (1)$$

where

P_I = risk of infection by ingesting pathogens in drinking water
N = number of pathogens ingested
N_{50} = number of pathogens that will infect 50% of the exposed population
α = the ratio N/N_{50} and P_I

Various studies have shown a wide variation in the probability of infected persons becoming ill, with morbidity rates varying between 1 and 97%, depending on the virulence of the pathogen, and the age, nutritional and general health status of the subjects. Since not every person infected by the ingestion of pathogens becomes ill, an independent estimate is made of P_D, the probability of contracting a disease:

$$P_D = P_{D:I} \times P_I \quad (2)$$

P_D = the risk of an infected person becoming diseased or ill
$P_{D:I}$ = the probability of an infected person developing clinical disease.

5.2 Determining the number of pathogens ingested

In order to make an estimate of the number of pathogens ingested from eating wastewater-irrigated vegetables, we first had to determine, through laboratory experiments, the amount of liquid that might cling to vegetables irrigated with wastewater. It was then possible to estimate the concentration of indicator organisms and pathogens that might remain on such wastewater irrigated vegetables and be ingested by subjects eating such vegetables. In doing this we assumed that any microorganisms contained in the residual wastewater remaining on the irrigated vegetables would cling to the vegetables even after the wastewater itself evaporated.

Based on laboratory determinations, we have estimated that the amount of wastewater of varying microbial qualities that would cling to the outside of wastewater irrigated cucumbers would be 0.36 ml/100 g (or one large cucumber) and 10.8 ml/100 g on long leaf lettuce (about three lettuce leaves).

Based on these measurements, we estimated the amount of indicator organisms that might remain on the vegetables if irrigated with raw wastewater and with wastewater meeting the WHO guidelines. In the case of irrigation with raw wastewater, we estimated that the fecal coliform (FC) concentration was 10^7/100 ml. In the case of irrigation with wastewater meeting the WHO guidelines the FC concentration would be 10^3/100 ml. We then estimated that the enteric virus:FC ratio in wastewater, based on various studies (Schwartzbrod, 1995), is $1:10^5$. For the purposes of this preliminary risk estimate we have assumed that all of the enteric viruses are a single pathogen, such as the virus of infectious hepatitis, so that it will be possible to make certain assumptions as to median infectious dose and infection to morbidity ratios. This errs on the conservative side.

We also assumed that under actual field conditions there would be a certain degree of indicator and pathogen die-away and/or removal from the wastewater source and irrigated vegetables until the final ingestion by the subject in the home. These factors include settling, adsorption, dessication, biological competition, UV irradiation from sunlight, and a degree of removal and/or inactivation by washing of the vegetables in the home. A number of studies have indicated that there is rapid die-off or removal of bacterial indicator organisms as well as pathogenic bacteria and viruses in wastewater irrigated soil and on crops by as much as 5 logs or 99.999% in 2 days under field conditions (Rudolfs *et al.*, 1951; Bergner-Rabinowitz, 1956; Sadovski *et al.*, 1978;

Armon *et al*., 1995). Armon *et al*. also suggest that there is the possibility of the regrowth of bacteria on vegetables contaminated with wastewater, but presented no data to support this hypothesis. In any event, *human enteric viruses cannot ever multiply under environmental conditions*. Asano and Sakaji (1990) have determined virus die-off in the environment under field conditions of wastewater reuse, which indicate that in 2 weeks the total virus inactivation reaches some 99.99%, while in 3 days there is a 90% reduction of virus concentration. Even superficial washing of vegetables in the home can remove an additional 99–99.9% of the virus contamination. Schwartzbrod (1995) has estimated that there would be as much as a 6 log reduction (99.9999%) of virus concentration between irrigation with wastewater and consumption of the crops if the total time elapsed reached 3 weeks. To be on the conservative side, we have estimated that the total virus inactivation and/or removal from the wastewater source until ingestion results in a reduction in virus concentration by 3 logs, or 99.9%, although a 99.99% loss is not unreasonable and might occur in most cases.

5.3 Estimates of risk of infection and disease

Based on the above tests results and assumptions we now can estimate the number of pathogens ingested by a subject who eats a 100 g of cucumber or 100 g (three leaves) of long lettuces irrigated with wastewater of various qualities. In this preliminary risk estimate we have selected the enteric virus infectious hepatitis, which can result in serious disease sequelae and which has had a clear epidemiological record indicating the possibility of it being environmentally transmitted and waterborne (Schwartzbrod, 1995). We have assumed that the median infectious dose for 50% of the exposed subjects to become infected (N_{50}) could range between 30 and 1000 PFU. We have also assumed that while the ratio of infections to clinical disease is often as low a 100:1, we shall estimate, as a worst case, that 50% of those infected will succumb to clinical disease ($P_{D:I} = 5$). We also assumed, based on vegetable consumption patterns in Israel, that on an annual basis a subject would consume 100 g of lettuce or cucumbers/day for 150 days($P_{D:I} = 25$). We have assumed that $\alpha = 0.5$, however, even assuming $\alpha = 0.2$ does not lead to a significantly increased risk.

First, as a positive control test of the model, we have estimated the risk of infection and disease from consuming vegetables irrigated with raw wastewater with an estimated initial fecal coliform level of 10^7. Based on the above assumptions, including a 3 log die-away, we have estimated that under such conditions a 100 g cucumber or 100 g (three leaves) of lettuce irrigated with raw wastewater will have a final level of contamination of FC of $10–10^2$. Based on that level of FC contamination and a virus:FC ratio of $1:10^5$, it can be estimated that there is a probability, in the case of irrigation with raw wastewater, that one cucumber in 10 000 will carry a single enteric virus and that one leaf of lettuce in 1000 will carry a single enteric virus. Based on these estimates of ingesting enteric viruses, more specifically the infectious hepatitis virus, we have estimated the risk of infection and disease that might result. The rate of disease, from eating 100 g lettuce, among a population eating such vegetables irrigated with *raw wastewater* is between 10^{-3} to 10^{-4} or about one case per 1000 to one case per 10 000 individuals exposed to risk.

This rate of infection has been found to correlate well with the disease rates detected in our field studies of the cholera outbreak in Jerusalem (Fattal *et al*., 1986b) and typhoid fever in Santiago, Chile (Shuval, 1984), which were clearly associated with the irrigation of vegetables normally eaten uncooked with raw wastewater. Thus our estimates based on the laboratory data, our assumptions and using the Haas mathematical model provided a reasonably close approximation of the degree of risk of disease that occurs in real world situations. This is a vital step in the validation of the reliability and usefulness of such a predictive simulation model and helped assure us that our predictions of risk under various conditions have a reasonably sound basis.

We then proceeded to evaluate the risk of disease if crops were irrigated with wastewater that meets the WHO guidelines. If the effluent is treated to meet the WHO guidelines for irrigation of vegetables to be eaten uncooked of 1000 FC/100 ml, the risk of hepatitis infection and disease estimates for lettuce is reduced to about 10^{-7}–10^{-8}, which means that the chances of becoming infected and diseased from such a low level of exposure is somewhere around one person per million people exposed per year or less. We have also calculated the annual risk of disease from the more infectious, but less severe, rotavirus as 10^{-5}–10^{-6}/year.

Are these a high level of risk or a low level of risk? In order to shed some light on what are considered reasonable levels of risk for communicable disease transmission from environmental exposure it should be noted that the USEPA has determined that microbiological quality guidelines for drinking water microbial standards should be designed to ensure that human populations are not subjected to *risks of infection by enteric disease greater than* 10^{-4} *(or 1 case per 10 000 persons per year)* (Regli *et al.*, 1991). Thus, compared with the USEPA estimates of reasonable acceptable risks for waterborne disease acquired from drinking treated drinking water, the WHO wastewater reuse guidelines appear to be safer by some one to two orders of magnitude.

5.4 Cost/effectiveness analysis

At this stage of our study we have made only some very preliminary estimates of the cost/-effectiveness associated with meeting the various wastewater effluent guidelines. As an example we shall present the hypothetical case of a Third World city of 1 million population about to build a wastewater treatment plant to assure safe utilization of the effluent for agricultural irrigation of vegetable crops, including those eaten uncooked, which would serve the population of the city. It is assumed that they have opted for a waste stabilization pond system that will meet the WHO guidelines. They want to compare the cost and risks at that level of treatment to the cost and risks if they had adopted the USEPA/USAID guidelines for treatment for vegetables eaten uncooked. We have assumed, for the purposes of this illustration only, that the unit cost of wastewater treatment to meet the various guidelines can roughly be estimated as follows:

WHO guidelines – 1000 FC/100 ml
(in stabilization ponds) \$0.125/M^3
or the cost/person/year (assuming
100 M^3/person/year) \$12.5/p/year
US-EPA/US-AID guidelines – 0 FC/100 ml
\$0.40/M^3
or the cost/person/year (assuming 100 M^3/
person/year) \$40.00/p/year.

The estimate of treatment costs for meeting WHO guidelines does not necessarily apply to all situations but is generally illustrative of a situation that may apply in hot sunny climates in developing countries where low cost land is available for effective stabilization pond treatment. According to this estimate, the additional cost for that city to meet the US-EPA/US-AID guidelines would be \$25 500 000/year.

We have assumed that half the population consumes wastewater irrigated vegetables on a regular basis and that the degree of annual risk of contracting a case of infectious hepatitis associated with the use of irrigated vegetables eaten uncooked with wastewater meeting the WHO guidelines is in the worst case some 2×10^{-6} (or about 1 case per year per 500 000 exposed persons) as estimated in this study. If we assume that the USEPA/ USAID guidelines, which require no detectable FC/100 ml, will achieve an essentially zero risk of disease, than we can estimate that the one case of infectious hepatitis per year would have been prevented. Thus the additional cost of treatment would result in a cost of about \$25 000 000 for each case of disease prevented. In the case of rotavirus disease, the cost would be some \$2 500 000 per case prevented. If, however, the true level of risk associated with the WHO guidelines is closer to the 10^{-7} level, estimated by the less conservative interpretation of the results of this study, then no detectable reduction of risk would be gained by the additional investment of $25 000 000 required to meet the strict and expensive USEPA/USAID guidelines.

It is questionable whether this level of additional treatment associated with major additional expenditures is justified to reduce further, the already negligibly low levels of risk of infection and disease that our estimates indicate are associated with the new WHO Guidelines.

Our conclusion is that the new WHO guidelines, which are based on extensive epidemiological evidence and can be achieved with low-cost wastewater treatment technology, provide a high degree of public health protection at a reasonable cost. There is little or no justification, in our opinion, for the unreasonably restrictive 'zero risk' USEPA/USAID guidelines, which are exceedingly expensive to achieve and require costly high-tech wastewater treatment technology, which provides little if any measurable increase in health protection.

Thus, after well over a century, health guidelines for wastewater reuse have gone through a complete cycle from no regulation or control in the 19th century to unreasonably strict standards in the earlier part of this century to what now appears to be a scientifically sound and rational basis with a less restrictive approach as recommended by the WHO guidelines. It is hoped that this new approach will encourage the development of controlled wastewater reuse for the benefit of mankind, while providing an appropriate level of health protection.

6 CONTROL OF CROPS AND IRRIGATION METHODS TO REDUCE HEALTH RISKS

Early in the development of wastewater irrigation for agriculture, methods of reducing health risks by controlling the type of crops grown or the methods of irrigation have been proposed and in some cases used effectively. The risk of transmission of communicable disease to the general public by irrigation with raw or settled wastewater can be reduced by a number of agronomic techniques. Some of these restrict the types of crops grown, and others, through modification and/or control of irrigation techniques, prevent or limit the exposure of health-related crops to pathogens in the wastewater.

6.1 Regulating the type of crops

One of the earliest and still most widely practised remedial measures is to restrict the type of crops irrigated with raw wastewater or with the effluent of primary sedimentation. Since there is ample evidence that salad crops and other vegetables normally eaten uncooked are the primary vehicles for the transmission of disease associated with raw wastewater irrigation, forbidding the use of raw effluent to irrigate such crops can be an effective remedial public health measure. Although such regulations have been effective in countries with a tradition of civic discipline and an effective means for inspection and enforcement of pollution control laws, they will likely be of less value in situations where those preconditions are absent.

In many arid and semi-arid areas near major urban centers, where subsistence farmers irrigate with raw wastewater, the market demand for salad crops and fresh vegetables is very high. Thus, governmental regulations forbidding farmers to grow such crops would be little more than a symbolic gesture. Even under the best of circumstances, it is difficult to enforce regulations that work counter to market pressures; to enforce regulations that prevent farmers from obtaining the maximum benefit from their efforts under conditions of limited land and water resources would be impossible.

6.2 Controlling irrigation methods

Basin irrigation of salad and vegetable crops usually results in direct contact of the crops with wastewater, thus introducing a high level of contamination. Sprinkler irrigation of salad crops also results in the deposit of wastewater spray on the crops and their contamination. The level of contamination may be somewhat less than basin irrigation. Many vegetables that grow on vines (i.e. tomatoes, cucumbers, squash, and the like) can be partially protected from wastewater contact if properly staked and/or grown hanging from wires that keep them off the ground, although some of these vegetables will inevitably touch the ground.

Well-controlled ridge-and-furrow irrigation reduces the amount of direct contact and contamination. These methods cannot completely eliminate direct contact of the wastewater with leafy salad crops and root crops.

Drip irrigation causes much less contamination of the crops than any other irrigation method. In fact, our studies indicate the use of drip irrigation tubes under polyethelene plastic surface sheeting used as a mulch can vastly reduce or totally eliminate crop contamination. In our studies, we determined that the level of enteroviruses and bacterial indicator organisms on cucumbers grown under such protective drip irrigation systems was negligible during the first 24 hours after the introduction of a massive seeding of microorganisms into the wastewater effluent stream used for irrigation. After 24 hours no contamination of irrigated crops was detectable (Sadovski *et al.*, 1976, 1978). Drip irrigation is the most costly form of irrigation, but its hygienic advantages make it attractive as a safe method of wastewater irrigation of sensitive vegetable and salad crops, even when the microbial quality of the effluent is not up to the strictest standards.

Fruit orchards do well with basin or ridge-and-furrow irrigation, but normal overhead sprinkler irrigation leads to direct contamination of the fruit. With low-level low-pressure sprinkler irrigation, however, the main spray is below the level of the branches, and the fruit is less likely to be contaminated. In all cases, windfalls picked from the ground will have been in contact with wastewater-contaminated soil. Another possible control measure is to discontinue irrigation with wastewater at a specified period, such as 2 weeks, before harvesting the crop. This option is feasible for some crops, but the timing of a vegetable harvest is difficult to control. In addition, some types of vegetables are harvested over long periods of time from the same plot.

Some of the above irrigation control techniques can help reduce the danger of crop contamination, but they are feasible only in fairly advanced and organized agricultural economies. Health regulations dependent upon any of the above procedures to protect certain high-risk crops from contamination must be enforced by legal sanctions and frequent inspections. If well-organized inspection and law enforcement systems are not present, as in some developing countries, the value of these options as a major remedial strategy may be limited. However, in the case of large centrally operated sewage farms, managed by the government or large well-organized companies, such procedures can be of value.

7 WASTEWATER TREATMENT

7.1 What are the goals of wastewater treatment for recycling and reuse?

Obviously, the degree of wastewater treatment, particularly as it relates to the effective removal and inactivation of pathogenic microorganisms, will have a critical effect in controlling any possible health risks associated with wastewater irrigation. The microbiology of wastewater treatment is reviewed in depth in Chapters 19–28. Thus it is beyond the scope of this chapter to review all possible wastewater treatment technologies suitable for wastewater recycling and reuse. However, we shall attempt to review some general principles and to emphasize low cost treatment technologies particularly suited to warm climates in developing countries.

In areas with plentiful rainfall, wastewater has traditionally been disposed of or diluted in large bodies of water, such as rivers and lakes. High priority has been given to maintaining the oxygen balance of these bodies of water to prevent serious detrimental effects, such as anaerobic conditions and odors from wastewater pollution. Most of the conventional processes used to treat wastewater in industrial countries have been designed primarily to remove the suspended and dissolved organic fractions, which decompose rapidly in natural bodies of water. The organic matter in wastewater, usually measured as biochemical oxygen demand (BOD), provides rich nutrients to the natural microorganisms of the stream, which multiply rapidly and consume the limited reserves of dissolved oxygen (DO) in the streams. If oxygen levels drop too far, anaerobic conditions may develop, serious odors may evolve and fish may die. However, for recycling and reuse of wastewater for agricultural irrigation, a high degree of BOD removal in the effluent is not directly relevant since, with land disposal, the soil is not harmed, but benefits from high level of organic matter and plant nutrients deposited on it. However, removal of organic matter, as represented by BOD, may be necessary when chemical disinfection processes are required, after the biological treatment stage, to achieve the microbial effluent standards.

A secondary goal of conventional wastewater treatment has been to reduce pathogenic microorganisms in order to protect the quality of the sources of drinking water used by downstream communities. However, conventional biological wastewater treatment systems, such as biological filtration and activated sludge, are not particularly efficient in removing pathogens. Thus, communities that draw their drinking water from surface sources cannot depend upon upstream wastewater treatment plant systems to reduce pathogens to a safe level. Therefore, they must remove the pathogens with their own drinking water treatment plants using a series of highly efficient, technical, and costly processes (i.e. coagulation, sedimentation, filtration and chemical disinfection). The most effective conventional wastewater treatment system is activated sludge, which removes 90–99% of the viruses, protozoa and helminths and 90–99.9% of the bacteria. Conventional processes cannot achieve higher levels of pathogen removal without additional expense for chemical disinfection, such as chlorination, and additional sand filtration. Further research and development are needed to improve the removal of helminths by conventional methods. However, the new micro/nanofiltration processes hold much promise in the effective removal of helminths and all other pathogenic microorganisms from wastewater.

7.2 Low cost wastewater stabilization ponds provide high quality effluent

In contrast to conventional treatment systems, studies have shown that well-designed multi-cell stabilization ponds allowing 20–30 days of retention can remove almost 100% of the helminth eggs (Yanez *et al.*, 1980; Feachem *et al.*, 1983; Mara and Silva, 1986; Fattal *et al.*, 1998). Bacteria, viruses and protozoa are often attached to larger fecal particles that settle out in pond systems. At best, however, only 90% can be removed by sedimentation.

The most effective process for removing bacteria and viruses in stabilization ponds is natural die-off, which increases with time, pH and temperature. Many developing countries have hot climates in which stabilization ponds are exposed to the direct rays of the sun and may reach temperatures up to 40°C. The pH at midday is commonly 9 or higher due to the photosynthetic activity of the algae. Predatory or competing microorganisms may also affect die-off by attacking or damaging pathogens directly or indirectly. Exposure to the ultraviolet rays of the sun may also play a role in killing pathogens in ponds. Long retention times, however, appear to be the most important factor in reducing bacterial concentrations in pond systems.

In warm climates with temperatures in excess of 20°C, a pond system with 4–5 cells and a 20- to 30-day retention time usually reduces the fecal coliform concentration by 4–6 log orders of magnitude – i.e. by 99.99–99.9999%. Thus, if the initial concentration of fecal coliform bacteria in the raw effluent is approximately 10^7/100 ml, the effluent will contain 10^3 or 1000/100 ml. The same pond system will reduce enteric viruses by 2–4 log orders of magnitude (i.e. from an initial concentration of about 1000/100 ml to 10 or fewer/100 ml). Helminth eggs will be removed almost completely, while the BOD will be reduced by about 80%. Fig. 15.4 shows the generalized removal curves for BOD, heiminth eggs, bacteria, and viruses in a multicell stabilization pond system in a warm climate.

Our cooperative studies with Egyptian colleagues in the city of Suez on waste stabilization pond treatment of wastewater for irrigation and aquaculture have demonstrated that with well-designed and well-operated multicell ponds with 30 days of detention, an effluent of 10–100 fecal coliforms/100 ml was consistently achieved (Mancy, 1996). In our studies at a waste stabilization pond pilot plant system, treating primary effluent from the Jerusalem Municipal treatment plant, an effluent of 1000 fecal coliforms/100 ml was achieved with 25–30 days of detention (Fattal *et al.*, 1997).

Waste stabilization ponds are therefore highly suitable for treating wastewater for irrigation. They are more efficient in removing pathogens, particularly helminths, than are conventional wastewater treatment systems. In addition, they produce a biologically stable, odorless, nuisance-free effluent without removing too many of the nutrients. Thus, ponds should be the system of choice for wastewater irrigation in warm climates, especially if land is available at a reasonable price. Ponds are particularly attractive for developing countries because they cost little to build and maintain and are robust and fail-safe. They should never be considered a cheap substitute. In reality they are superior to conventional methods of treatment in almost all respects. Although ponds require relatively large land areas, land costs are, in many cases, not a serious obstacle.

When wastewater will be used to irrigate crops for human consumption, the goals of treatment are the reverse of the goals of conventional treatment. The primary goal for treatment of wastewater to be used for irrigation must be removal of pathogenic microorganisms in order to protect the health of the farmers and consumers. Removal of the organic material, however, which contains valuable agricultural nutrients is neither necessary nor desirable, although aerobic conditions should be maintained because a black, highly odorous, anaerobic wastewater effluent would probably be an environmental nuisance to farmers and nearby residents.

7.3 Public health advantages of wastewater treatment and storage reservoirs

Since wastewater is generated by the community 365 days a year and the irrigation season in most areas is limited to a number of months per year, a means must be found to handle wastewater flows during non-irrigation periods. If allowed to flow unrestricted, the effluent will contaminate the region's natural bodies of water.

A suitable solution is the storage and treatment reservoirs pioneered in Israel (Juanico and Shelef, 1994; Pearson *et al.*, 1996). Such reservoirs are designed to upgrade the quality of the effluent during the long residence time in the reservoirs and to store up to 8–10 months of wastewater flow in the rainy winter months to be used for irrigation during the dry summer season. They are often preceded by settling ponds and/or by conventional stabilization ponds, and may also be designed to catch surface runoff. The operational regimes of such reservoirs vary. Some are operated to change between non-steady-state flow and batch while others are operated sequentially, receiving and storing influent for extended periods, after which the inflow is stopped for a period of stabilization and bacterial decay prior to the discharge for irrigation. Log mean concentration of microorganisms in often partially treated typical wastewater influent at the entrance of the reservoir per 100 ml are: hetrotrophic bacteria 10^7–10^8, fecal coliforms 10^6–10^7, enteroccoci 10^4–10^6 and F^+ bacteriophages 10^5–10^6.

The results at the outlet in such reservoirs show that, in general, the quality of the effluents is much improved and is best at the beginning of the irrigation season when the reservoir is full of old effluent (long detention time), but sharply deteriorates when the water level drops and new wastewater continues to be pumped into the reservoir

(Juanico and Shelef, 1994; Pearson *et al.*, 1996). The die-off rate of microorganisms in the summer is higher than in the winter. We have shown that there is a significant correlation between the mean hours of sunshine in the month and the rate of bacterial die-off. Thus with greater sunshine duration in the summer months the bacterial die-off is highest (Fattal *et al.*, 1996). However, in single reservoirs during the summer the relative volume of the influent, raw or partially treated is higher, leading to poorer bacterial quality of the effluent since the pond levels are at their lowest during the intensive irrigation season. Coliform removal is high in the epilimnion, where high pH values occur due to algal activity, and low in the hypolimnion where pH values are low (Liran *et al.*, 1994). The logs mean reduction of hetrotrophic bacteria, *E. coli*, enterococci and F^+ bacteriophages at the outlet of the reservoir are: 1, 3, 2.5 and almost 4 logs, respectively (Fattal *et al.*, 1993). The mean reduction of BOD is 76%, COD 72% and TSS 45% (Fattal *et al.*, 1993).

In order to improve the quality of the effluent, innovative changes were made in Israel in the design of such stabilization/storage reservoirs (Liran *et al.*, 1994; Juanico and Shelef, 1994; Juanico, 1996; Friedler and Juanico, 1996). The storage of the effluent was changed from seasonal to multiseasonal and/or from single reservoir to two or more reservoirs used sequentially as batch reservoirs, supplying effluents for irrigation only from reservoirs which no longer receive fresh influent. By doing so, the input of the effluent from treatment units into the reservoir is stopped before reservoir effluent is released for irrigation. Another improvement was made by implementing better treatment of the wastewater before entering into the reservoir. The performance of the improved batch stabilization reservoirs, when properly designed and operated, showed that they are able to remove fecal coliform bacteria by up to five orders of magnitude (i.e. 99.999%). Such improved alternating reservoirs systems can normally produce an effluent which can easily meet the WHO guidelines for unrestricted irrigation of 1000 fecal coliforms/100 ml or less.

7.4 Remedial environmental methods assures greater health protection

The history of public health progress has proven that the broadest and most effective public health benefits are obtained from remedial or preventive measures taken by a central authority and involving environmental interventions that lower the level of exposure of large populations to environmentally transmitted disease. Such measures as central plants for the purification of drinking water supplies, pasteurization of milk, and area-wide campaigns for reducing the breeding sites of malaria-carrying mosquitoes are well-known examples of success using this strategy. Any remedial action based on changing personal behavior and lifestyle through education, law enforcement, or both, is a much slower process and, in general, has succeeded only in areas with relatively high educational levels and living standards.

The wastewater storage and treatment/storage reservoir option reviewed above offers this type of centrally managed and engineered form of remedial environmental intervention. It is the only remedial measure that will simultaneously reduce the negative health effects for sewage farm workers and for the public that consumes wastewater-irrigated vegetables. It is also the only measure that can bring about health benefits in a short time without massive changes in personal behavior or restrictive regulations that depend on complex inspection and law enforcement procedures. However, it does require central organizational and management capacity, availability of major financial resources and availability of land.

Although it may be appropriate in some situations to restrict the type of crops grown or to control wastewater irrigation practices, such regulations are difficult to enforce where there is great demand for salad crops and garden vegetables. In arid and semi-arid zones (as well as some humid areas), where irrigation is highly desirable, many economists and agricultural authorities consider it economically prudent to allow unrestricted wastewater irrigation of cash crops in high demand. That goal can only be achieved with an effective high level of wastewater treatment as suggested in this chapter.

REFERENCES

Armon, R., Dosoretz, C.G., Azov, Y. and Shelef, G. (1995). Residual contamination of crops irrigated with effluent of different qualities: a field study. *Water Science and Technology* **30**(9), 239–248.

Applebaum, J., Guttman-Bass, N. and Lugten, M. (1984). Dispersion of aerosolized enteric viruses and bacteria by sprinkler irrigation with wastewater. In: J.L. Melnik (ed.) *Enteric Viruses in Water, Monographs in Virology* **15**, pp. 193–201. Karger, Basle.

Asano, T. and Sakaji, R.H. (1990). Virus risk analysis in wastewater reclamation and reuse. In: H.H. Hahn and R. Klute (eds) *Chemical Water and Wastewater Microbiology* Springer Verlag, Berlin.

Baumhogger, W. (1949). Ascariasis in Darmstadt and Hessen as seen by a wastewater engineer. *Zeitschrift fur Hygiene und Infektionskrankheiten* **129**, 488–506.

Ben-Ari, J. (1962). The incidence of *Ascaris lumbriocoides* and *Trichuris trichuria* in Jerusalem during the period of 1934–1960. *American Journal of Tropical Medicine and Hygiene* **11**, 336–368.

Bergner-Rabinowitz, S. (1956). The survival of coliforms, *S. faecalis* and *S. tennesee* in the soil and climate of Israel. *Applied Microbiology* **4**, 101–106.

Blum, D. and Feachem, R.G. (1985). *Health Aspects of Night soil and Sludge Use in Agriculture and Aquaculture: An Epidemiological Perspective*. International Reference Centre for Wastes Disposal, Dubendorf, Switzerland.

Camann, D.E., Northrop, R.L., Graham, P.J. *et al.* (1983). *An Evaluation of Potential Infectious Health Effects from Sprinkler Application of Wastewater to Land. Lubbock, Texas*. Environmental Protection Agency, Cincinnati.

Clark, C.S., Bjornson, H.S., Holland, J.W. *et al.* (1981). *Evaluation of the Health Risks Associated with the Treatment and Disposal of Municipal Wastewater and Sludge. Report No. EPA-600/S-1 81-030*. Environmental Protection Agency, Cincinnati.

Engelberg Report (1985). *Health Aspects of Wastewater and Excreta Use in Agriculture and Aquaculture*. IRC News (No. 5). International Reference Centre for Wastes Disposal, Dubendorf, Switzerland.

Fattal, B., Bercovier, H., Derai-Cochin, M. and Shuval, H.I. (1985). Wastewater reuse and exposure to Legionella organisms. *Water Research* **1**, 693–696.

Fattal, B., Wax, Y., Davies, M. and Shuval, H.I. (1986a). Health risks associated with wastewater irrigation: An epidemiological study. *American Journal of Public Health* **76**, 977–979.

Fattal, B., Yekutiel, P. and Shuval, H.I. (1986b). Cholera outbreak in Jerusalem 1970 revisited: The case for transmission by wastewater irrigated vegetables. In: J.R. Goldsmith (ed.) *Environmental Epidemiology: Epidemiological Investigation of Community Environmental Disease*, pp. 49–59. CRC Press, Boca Raton.

Fattal, B., Margalith, M., Shuval, H.I. *et al.* (1987). Viral antibodies in agricultural populations exposed to aerosols from wastewater irrigation during a viral disease outbreak. *American Journal of Epidemology* **125**, 899–906.

Fattal, B., Puyesky, Y., Eitan, G. and Dor, I. (1993). Removal of indicator microorganisms in wastewater reservoir in relation to physico-chemical variables. *Water Science and Technology* **27**, 321–329.

Fattal, B., Goldberg, T. and Dor, I. (1996). Model for measuring the effluent quality and the microbial die-off rate in Naan wastewater reservoir. *J Handasat Maim* **26**, 26–30 (in Hebrew).

Fattal, B., Berkovitz, A. and Shuval, H.I. (1998). The efficiency of pilot plant stabilization pond of wastewater. *International Journal of Environmental Health Research* **8**, 153–156.

Feachem, R.G., Bradley, D.J., Garelick, H. and Mara, D.D. (1983). *Sanitation and Disease: Health Aspects of Excreta and Wastewater Management*. John Wiley & Sons, Chichester.

Friedler, E. and Juanico, M. (1996). Treatment and storage of wastewater for agricultural irrigation. *International Water Irrigation Review* **16**, 26–30.

Haas, C.N., Rose, J.B., Gerba, C.P. and Regli, S. (1993). Risk assessment of viruses in drinking water. *Risk Analysis* **13**, 545–552.

Jepson, A. and Roth, H. (1949). Epizootiology of cyiticercus bovis-resistance of the eggs of *Taenia saginata*. In: *Report of the 14th International Veterinary Congress*, Vol. 2. His Majesty's Stationery Office, London.

Jjumba-Mukabu, O.R. and Gunders, E. (1971). Changing patterns of intestinal helminth infections in Jerusalem. *American Journal of Tropical Medicine and Hygiene* **20**, 109–116.

Juanico, M. (1996). The performance of batch stabilization reservoirs for wastewater treatment, storage and reuse in Israel. *Water Science and Technology* **33**, 149–159.

Juanico, M. and Shelef, G. (1994). Design operation and performance of stabilization reservoirs for wastewater irrigation in Israel. *Water Research* **28**, 175–186.

Khalil, M. (1931). The pail closet as an efficient means of controlling human helminth infections as observed in Tura prison, Egypt, with a discussion on the source of ascaris infection. *Annals of Tropical Medicine and Parasitology* **25**, 35–62.

Krey, W. (1949). The Darmstadt ascariasis epidemic and its control. *Zeitschrift fur Hygiene und Infektions krankheiten* **129**, 507–518 (in German).

Krishnamoorthi, R.P., Abdulappa, M.R. and Anwikar, A.R. (1973). Intestinal parasitic infections associated with sewage farm workers with special reference to helminths and protozoa. In: *Proceedings of Symposium on Environmental Pollution*. Central Public Health Engineering Research Institute, Nagpur, India.

Liran, A., Juanico, M. and Shelef, G. (1994). Coliform removal in a stabilization reservoir for wastewater irrigation in Israel. *Water Research* **28**, 1305–1314.

Mancy, H.K. (1996). Comparative research on environmental health in pursuit of peace in the Middle East.

In: Y. Steinberger (ed.) *The Wake of Change*, pp. 736–741. ISEEQS, Jerusalem.

Mara, D.D. and Silva, S.A. (1986). Removal of intestinal nematode eggs in tropical waste stabilization ponds. *Journal of Tropical Medicine and Hygiene* **99**, 71–74.

Pearson, H.W., Mara, D.D., Cawley, L.R. *et al.* (1996). Pathogen removal in experimental deep effluent storage reservoirs. *Water Science and Technology* **33**, 251–260.

Penfold, W.J. and Phillips, M. (1937). *Taenia saginata* and *Cysticercosis bovis*. *Journal of Helminthology* **15**.

Regli, S., Rose, J.B., Haas, C.N. and Gerba, C.P. (1991). Modeling the risk from *Giardia* and viruses in drinking water. *Journal of the American Water Works Association* **83**, 76–84.

Rudolfs, W., Falk, L.L. and Ragotzkei, R.A. (1951). Contamination of vegetables grown in polluted soil – bacterial contamination of vegetables. *Sewage and Industrial Wastes* **23**, 253–263.

Sadovski, A.Y., Fattal, B., Katnelson, E. *et al.* (1976). A study of the distribution and survival of enteric viruses and bacteria during a simulated epidemic in soil and crops irrigated with wastewater by the drip irrigation method. *Proceedings of the seventh Scientific Conference of the Israel Ecological Society*, pp. 221–235. IES, Tel Aviv.

Sadovski, A.Y., Fattal, B. and Goldberg, D. (1978). Microbial contamination of vegetables irrigated with sewage effluent by the drip method. *Journal of Food Protection* **41**, 336–340.

Sanchez Levya, R. (1976). *Use of Wastewater for Irrigation in Districts 03 and 88 and its Impacts on Human Health.* Master's thesis. School of Public Health, Mexico City (in Spanish).

Schwartzbrod, L. (1995). *Effect of Human Viruses on Public Health Associated with the Use of Wastewater and Sewage Sludge in agriculture and Aquaculture. Report No.-WHO/EOS/95.19.* World Health Organization, Geneva.

Shuval, H.I. (1984). *Health Aspects of Irrigation with Sewage and Justification of Sewerage Construction plan for Santiago, Chile.* Unpublished Internal Report. The World Bank, Washington, DC.

Shuval, H.I. (1990). *Wastewater Irrigation in Developing Countries – Health Effects and Technical Solutions – Summary of World Bank Technical Paper no. 51*, Wastewater and Sanitation Discussion Paper No. 2. The World Bank, Washington, DC.

Shuval, H.I. and Fattal, B. (1996). Which health guidelines are appropriate for wastewater recycling to agriculture? A risk-assessment/cost-effectiveness approach. In *Proceedings – Potable Water and Wastewater – AQUATEC Cairo*, Egypt.

Shuval, H.I., Yekutiel, P. and Fattal, B. (1984). Epidemiological evidence for helminth and cholera transmission by vegetables irrigated with wastewater: Jerusalem a case study. *Water Science and Technology* **17**, 433–442.

Shuval, H.I., Adin, A., Fattal, B. *et al.* (1986). *Wastewater Irrigation in Developing Countries: Health Effects and Technical Solutions.* Technical Paper Number 51. The World Bank, Washington, DC.

Shuval, H.I., Wax, Y., Yekutiel, P. and Fattal, B. (1988). Prospective epidemiological study of enteric disease transmission associated with sprinkler irrigation with wastewater: An Overview. In: *Implementing Water Reuse: Water Reuse IV*, pp. 765–781. American Water Works Association, Denver.

Shuval, H.I., Guttman-Bass, N., Applebaum, J. and Fattal, B. (1989a). Aerosolized enteric bacteria and viruses generated by spray irrigation of wastewater. *Water Science and Technology* **21**, 131–135.

Shuval, H.I., Wax, Y., Yekutiel, P. and Fattal, B. (1989b). Transmission of disease associated with wastewater irrigation: A prospective epidemiological study. *American Journal of Public Health* **79**, 850–852.

Shuval, H.I., Lampert, Y. and Fattal, B. (1997). Development of a risk assessment approach for evaluating wastewater recycling and reuse standards for agriculture. *Water Science and Technology* **35**(11–12), 15–20.

Teltsch, B., Shuval, H.I. and Tadmor, J. (1980). Die-away kinetics of aerosolized bacteria from sprinkler irrigation of wastewater. *Applied and Environmental Microbiology* **39**, 1191–1197.

USEPA (1972). *Water Quality Criteria – 1972.* Environmental Protection Agency, Washington, DC.

USEPA/USAID (1992). *Guidelines for Water Reuse.* Environmental Protection Agency, Washington, DC.

World Health Organization (1989). *Health Guidelines for the Use of Wastewater in Agriculture and Aquaculture. Technical Report Series No. 778.* WHO, Geneva.

Yanez, F., Rojas, R., Castro, M.I. and Mayo, C. (1980). *Evaluation of the San Juan Stabilization Ponds: Final Research Report of the First Phase.* Pan American Center for Sanitary Engineering and Environmental Sciences, Lima, Peru.

16

Developing risk assessments of waterborne microbial contaminations

Paul Gale

WRC-NSF, Medmenham, Marlow, Bucks SL7 2HD, UK

1 INTRODUCTION

Pathogens can and do gain entry into drinking water supplies even in industrialized countries. Break-through during treatment and ingress through cracked pipes are well-documented causes of waterborne outbreaks of cryptosporidiosis (Craun *et al.*, 1998) and *Escherichia coli* O157 (Swerdlow *et al.*, 1992), respectively. Microbiological risk assessment (MRA) is the emerging method to predict the risks to public health from these pathogens. Models have been developed for a range of waterborne pathogens in drinking water, including *Cryptosporidium parvum* (Teunis *et al.*, 1997), *Giardia* (Regli *et al.*, 1991) and enteric viruses (Haas *et al.*, 1993). Most of the models represent non-outbreak conditions and model the endemic levels of infection through drinking water. Recently, a model for *C. parvum* has been developed for conditions representative of a waterborne outbreak of cryptosporidiosis (Gale and Stanfield, 2000).

The risk assessment approach can be used to answer questions ('what if?' scenarios), such as how many more people will be infected if part of the drinking water treatment plant fails and by how much will public health be jeopardized if disinfection is eliminated. One application of MRA is to provide a defensive position for new and emerging agents in the absence of epidemiological evidence. For this reason, a risk assessment for the transmission of bovine spongiform encephalopathy (BSE) to humans through groundwater supplies was developed (Gale *et al.*, 1998). In the absence of epidemiological data, the primary objectives of the MRA were therefore to determine whether the BSE agent in the environment could conceivably present a risk to those drinking water consumers supplied by water sources near to rendering plants and to determine whether drinking water supplies were adequately protected. The objective was to develop order of magnitude estimates of risks of infection.

Over the last few years the application of MRA has expanded greatly to cover other environmental routes, including pathogens in:

- sewage sludges applied to agricultural land; e.g. BSE (Gale and Stanfield, 2001), salmonellae (Gale, 2001b) and *E. coli* O157
- manures and animal slurries applied to agricultural land
- composted domestic wastes applied to land
- rendering plant condensates discharged to surface waters.

2 THE SOURCE–PATHWAY–RECEPTOR APPROACH

Environmental risk assessments are based on the source–pathway–receptor approach. The 'source' term defines the amount of infectivity in the environment, and the 'receptor' term

The Handbook of Water and Wastewater Microbiology
ISBN 0-12-470100-0

comprises the various human and animal groups exposed to infectivity. The 'pathway' term describes the routes by which receptors may be exposed. These are best modelled by event trees. Central to environmental MRAs are pathway barriers and biomedical barriers (Gale, 2001b). Pathway barriers include drinking-water treatment processes (chemical coagulation, sand filtration), the hydrogeological substrata above aquifers, sewage sludge treatment processes, decay of the agent in the environment, and dilution. The pathway barriers serve to reduce exposure. The biomedical barriers include the species barrier in the case of BSE (Gale *et al.*, 1998), acquired protective immunity in *C. parvum* infection (Chappell *et al.*, 1999) and natural gut microbiota in salmonella and *Cryptosporidium* infection (Meynell, 1963; Harp *et al.*, 1992). The biomedical barriers serve to protect the receptors from infection after exposure. In general, there is greater uncertainty in the biomedical barriers than the pathway barriers.

2.1 The use of the arithmetic mean exposure

Gale and Stanfield (2000) used a Monte Carlo approach to simulate the variation in doses of *Cryptosporidium* oocysts ingested by individual drinking water consumers under conditions representative of a waterborne outbreak (Table 16.1). Log-normal distributions for the oocyst densities in 100-l volumes of raw and treated waters are shown in Fig. 16.1 The simulation is based on the Poisson-log-normal distribution and takes into account:

- the variation in oocyst densities in the raw water, in this case a river with agricultural drainage (Hutton *et al.*, 1995)
- the variation in removal of oocysts at the drinking-water treatment works based on particle removal data for 67 works (Fig. 16.2)
- the variation in the volumes of tap water ingested daily by drinking water consumers (0.1–3 l/person/day; Roseberry and Burmaster, 1992).

TABLE 16.1 Daily exposures to *Cryptosporidium* oocysts simulated for individual consumers under conditions consistent with an outbreak (Gale and Stanfield, 2000). Doses ingested are compared accommodating the variation predicted using the Poisson-log-normal distribution and simply using the arithmetic mean exposure in the Poisson distribution

Dose (oocysts/person/day)	*Percentage of population exposed to dose*	
	Poisson-log-normal distribution (Monte Carlo simulation)*	*Arithmetic mean exposure (0.37 oocysts/person/day) in Poisson distribution*
0	88.43	68.9
1	6.76	25.7
2–5	3.57	5.4
6–10	0.61	0.00027
11–50	0.57	$<10^{10}$
51–100	0.05	0
>100	0.01	0

* The arithmetic mean of the simulated exposures = 0.37 oocysts/person/day.

According to the Poisson-log-normal distribution of exposures simulated in Table 16.1, the majority of consumers do not ingest any oocysts each day, even during an outbreak. This is consistent with the finding that 66% of 1000 litre volumes did not contain oocysts during the Farmoor (UK) outbreak of cryptosporidiosis (Richardson *et al.*, 1991) and that oocysts are often not detected in outbreaks (Joseph *et al.*, 1991; Craun *et al.*, 1998). The simulation predicts that about half of those consumers who were exposed ingested just a single oocyst each day. However, a small proportion of consumers (1.2%) were exposed to between 6 and 50 oocysts/day. Furthermore, 1 in 10 000 consumers was exposed to very high daily doses, which exceeded 100 oocysts. These high count exposures reflect a combination of temporally high raw water loadings (e.g. after heavy rain or agricultural discharge) and poor oocyst removal efficiencies by treatment, e.g. because of failure or bypass (Craun *et al.*, 1998).

The arithmetic mean for the simulated oocyst exposures was 0.37oocysts/person/day.

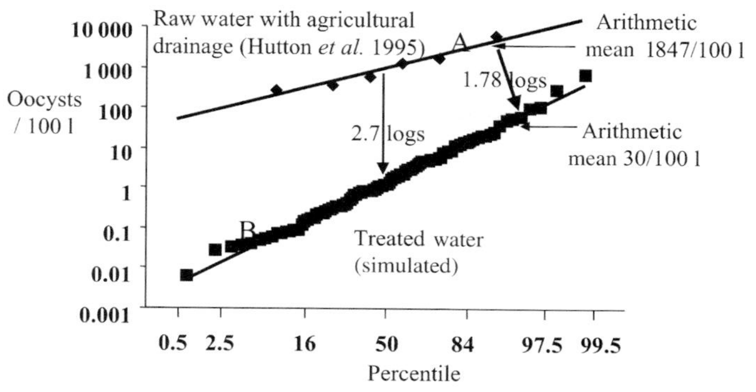

Fig. 16.1 Simulated log-normal distribution of densities of *Cryptosporidium* oocysts in treated water during a waterborne outbreak (Gale and Stanfield, 2000).

Exposures calculated using this single point average and assuming a Poisson distribution are also presented in Table 16.1. MRA models, which use this single point average with the Poisson distribution, and ignore the large variation in the simulated Poisson-log-Normal exposures, would predict that a much higher proportion of the population is actually exposed daily to oocysts (31% compared to

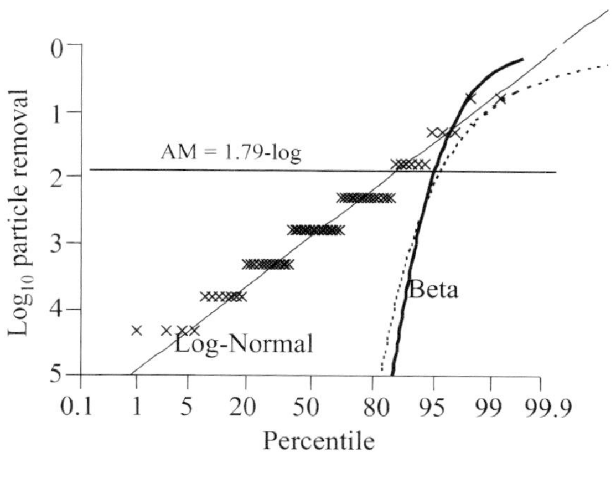

Fig. 16.2 Distribution of particle (>3 μm) removal ratios by drinking water treatment. Data for 67 drinking water treatment plants from LeChevallier and Norton (1995). Log-normal distribution ($\mu = -2.70$; $\sigma = 0.89$) as fitted by Gale and Stanfield (2000) and beta distribution (solid line) fitted to data ($\alpha = 0.0162$; $\beta = 0.986$). Arithmetic mean (AM) removal is same for both models. In addition the beta distribution ($\alpha = 0.025$; $\beta = 4.10$) for removal of spores of sulohite-reducing bacteria as presented by Teunis *et al.* (1997) is shown as dashed line.

11%), but that those who are exposed never ingest more than about 5 oocysts each day. By ignoring the temporal/spatial heterogeneity in exposures, MRA would never predict that a consumer would be exposed to high pathogen doses in a single exposure.

The expected numbers of *Cryptosporidium* infections across the population for the two exposure scenarios in Table 16.1 are very similar. Indeed the arithmetic mean exposure of 0.37 oocysts/person/day may be used directly in the dose–response curve, even though it represents a fraction of a pathogen. Thus, Haas (1996) demonstrated mathematically that the appropriate statistic for risk assessment is the arithmetic mean exposure, even for log-normal distributions of exposures, which show great variation in magnitude across the population (Table 16.1). This 'rule' only holds if the pathogens act independently and breaks down for the more highly infectious agents such as rotavirus.

2.2 Presentation of risks

A Monte Carlo simulation will give a range of predicted risks for individual consumers within the population. Thus, in the outbreak simulation, consumers are exposed to between 0 and >100 oocysts/person/day (see Table 16.1). This raises the question of what value should be quoted for the risk. Typically 5, 50 and 95 percentiles are quoted. However, the actual risk values for these percentiles will depend critically on the distribution of pathogens within the medium. This is shown with an example in

TABLE 16.2 Three exposure scenarios for 50 *C. parvum* oocysts distributed within 100 litre of drinking water and 50 consumers each ingesting 2 litres

Exposure model	*Percentile exposure (oocysts/person)*			*Number of infections* per 50 persons*
	5	50	95	
Arithmetic mean: All 50 persons ingest 1 oocyst each	1	1	1	0.209
Poisson distribution of exposures: 18 persons ingest 0 oocysts 19 persons ingest 1 oocyst 9 persons ingest 2 oocysts 3 persons ingest 3 oocysts 1 person ingests 4 oocysts	0	1	3	0.208
Extreme clustering: 49 persons ingest 0 oocysts 1 person ingests all 50 oocysts	0	0	0	0.189

* Calculated using dose-response relationship in Equation (1) where $r = 0.00419$ (Haas *et al.*, 1996).

Table 16.2 which considers 50 *C. parvum* oocysts in 100 litre of drinking water, and 50 consumers each drinking 2 litre of that water. Thus, in the case of the extreme clustering distribution scenario, the 5, 50 and 95 percentile risks are all zero. This questions the appropriateness of quoting these percentile values and, in particular, the median (or 50% percentile) risk. Indeed, quoting a 5, 50 or 95 percentile risk merely gives information on the least exposed portion of the population and raises the question of what risks are the remaining proportion exposed to. The 'bottom line' is how many infections are there across the whole population. For all three scenarios the group risk across the 50 individuals is similar (i.e. between 0.19 and 0.21 infections per 50 persons). In this respect, the appropriate value is the arithmetic mean risk. This, in effect, represents the number of infections across the population as a whole. The median risk is typically in the region of an order of magnitude less than the arithmetic mean risk for a log-normal model, and may thus give too optimistic a representation (Fig. 16.1).

2.2.1 The arithmetic mean is appropriate for each of the steps in the source–pathway–receptor approach

The source, pathway and receptor terms used by Gale and Stanfield (2000) to simulate *Cryptosporidium* exposures during an outbreak are presented in Table 16.3. Gale and Stanfield (2000) demonstrated by Monte Carlo simulation that the arithmetic means are all that is required for each of the source, pathway and receptor terms. Indeed, providing the dose–response relationship is linear, then the variation in the source, pathway and receptor terms (Table 16.3) may be accommodated simply by using the arithmetic mean. This provides a 'quick and simple' alternative to Monte Carlo simulations. Thus, the arithmetic mean *Cryptosporidium* oocyst exposure in Table 16.3 is calculated as:

$$18.47 \div 61.3 \times 1.26 = 0.38 \text{ oocysts/person/day}$$

which is very similar to the value of 0.37 oocysts/person/day calculated from the simulated Poisson-log-normal distribution of exposures (Table 16.1). This exposure may then be used directly in the dose–response curve (see the nature of the dose-response relationship, p. 267) to calculate the risk of infection (see Integrating pathogen exposures, p. 271). The main sources of uncertainty are underestimating the arithmetic mean oocyst loading in the raw water (or treated water by monitoring programmes) and overestimating the net oocyst removal by drinking water treatment. Both contribute to underestimating the arithmetic mean exposure and hence the risk of infection.

TABLE 16.3 Source, pathway and receptor terms used to simulate exposures to *Cryptosporidium* oocysts through drinking water (Gale and Stanfield, 2000)

Term	*Data*	*Arithmetic mean**	*Variation*	*Uncertainty*
Source	Oocyst concentrations in raw water	18.47 oocysts/l	Natural fluctuations in oocyst concentrations, e.g. after heavy rainfall	Monitoring protocol missing high count spikes and recovery efficiency of analytical method contribute to underestimating net oocyst loading†
Pathway	Removal of oocysts from raw water by drinking water treatment	61.3-fold (arithmetic mean removal ratio)	Fluctuations in removal efficiency between and within individual treatment works	Monitoring protocol may cause overestimation of net removal efficiency‡
Receptor	Volume of tap water ingested daily by individual consumers	1.26 l/person/day	Individual habits, age groups, seasonal effects, amount imbibed after boiling	Method of data collection, e.g. self-assessment
Exposure	Simulated	0.37 oocysts/person/day		

* Calculated from 10 000 samples 'drawn' from log-normal distributions by Monte Carlo simulation.
† Uncertainty–Monitoring may underestimate the arithmetic mean pathogen exposure and hence risk (p. 273).
‡ Uncertainty–Estimating the removal of pathogens by treatment processes (p. 274).

2.3 The nature of the dose–response relationship

Dose–response data are critical for quantitative MRA. Dose–response data from human studies have been obtained for several waterborne pathogens, including rotavirus (Ward *et al.*, 1986) and *C. parvum* (DuPont *et al.*, 1995; Chappell *et al.*, 1999). Dose–response parameters have also been estimated for BSE in humans (Gale, 2001a).

2.3.1 *Cryptosporidium parvum*

In feeding trials, 29 healthy immunocompetent adult volunteers were each given a single dose of between 30 and 10^6 oocysts of a calf strain of *C. parvum* (DuPont *et al.*, 1995). The proportion infected at each dose is plotted in Fig. 16.3. Exposure simulations presented in Table 16.1 suggest that, even under outbreak conditions, most drinking-water consumers ingest fewer than 30 oocysts daily and half of those consumers exposed to oocysts ingest only a single oocyst. Therefore, the doses administered by DuPont *et al.* (1995) are generally too high for the application of modelling risks through water. To estimate the risk from ingestion of just a single *C. parvum* oocyst, a process of low dose extrapolation of fitted mathematical dose response curves is used.

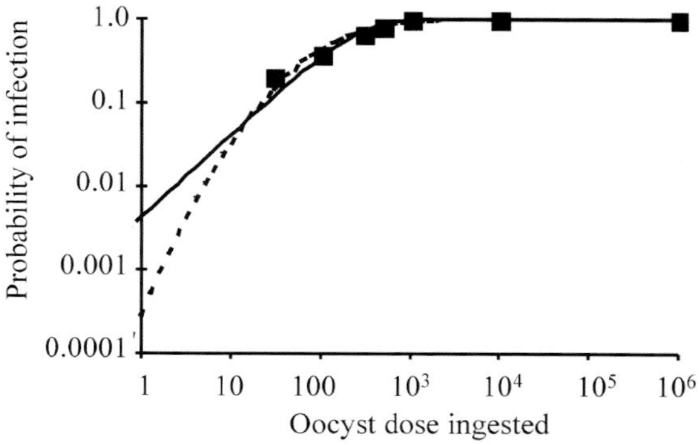

Fig. 16.3 Independent or co-operative action. Negative exponential dose response curve (Equation 1; r = 0.00419) fitted by Haas *et al.* (1996) to *C. parvum* human infectivity data (*N*) obtained for volunteers selected on the basis of having no serological evidence of the past infection with *C. parvum* (DuPont *et al.*, 1995).The log-probit curve (dashed) ($\mu = 2.119$, $\sigma = 0.614$; $\log_{10}$) is shown for comparison.

Haas *et al.* (1996) fitted a negative exponential dose–response curve to the *C. parvum* human infectivity data (Fig. 16.3). According to this model, the probability, P_i, of being infected by ingesting a dose of N pathogens is expressed mathematically as:

$$P_i = 1 - e^{-rN} \quad (1)$$

where r is a parameter specific for the pathogen and the host population and represents the fraction of the ingested pathogens that survive to initiate infections (Regli *et al.*, 1991). This dose–response relationship assumes that the oocysts act completely independently during infection and do not act cooperatively, for example, in overcoming the host protective barriers. Indeed at low doses, Equation (1) transforms to:

$$P_i = rN \quad (2)$$

indicating a direct linear relationship between the ingested dose, N, and risk of infection. The risk of infection from ten oocysts, for example, is tenfold that from the ingestion of just a single oocyst. If pathogens acted cooperatively during infection, then at low doses, the probability of infection would be non-linearly related to the dose. One mathematical model that describes a non-linear dose–response relationship is the log-probit curve (Haas, 1983). This model is defined by two parameters and is also shown in Fig. 16.3. It is apparent that the negative exponential and the log-probit curves do not differ markedly over the range of doses administered in the trial (30 to 10^6 oocysts). However, more considerable differences are observed on extrapolation to the low doses of oocysts to which drinking water consumers are likely to be exposed (see Table 16.1). The risk of infection predicted by the log-probit model from ingestion of a single oocyst is about 15-fold smaller than that predicted by the negative exponential model (Fig. 16.3). This reflects the fact that, if oocysts acted cooperatively, then low doses (e.g. one oocyst) would present greatly diminished risks. According to the log-probit model, the risk of infection from a dose of 10 oocysts is 120-fold higher than from ingestion of only a single oocyst.

2.3.2 The term 'infective dose' is meaningless

The extreme scenario for a non-linear dose–response relationship is the threshold or minimum infective dose. The concept of a minimum infective dose or threshold is misleading in MRA because it suggests zero risk of infection at very low pathogen doses. This does not appear to be the case for *C. parvum*. Teunis (1997) has demonstrated that fitted models assuming threshold doses of 2, 3, 4, or 5 oocysts do not fit the infectivity data in Fig. 16.3 and lead to increasingly steep dose–response curves. There is no experimental evidence for a minimum infective dose and no reason why a *single* oocyst or other waterborne pathogen should not be able to initiate infection, albeit with very low probability in some cases. Indeed, individual bacterial cells and virions contain all the macromolecular machinery and genes for infection. The diverse pathogenic strategies of enteric bacteria are based on a limited array of macromolecular systems (Donnenberg, 2000). Blewett *et al.* (1993) concluded that the minimum infective dose for gnotobiotic lambs is just one oocyst of *C. parvum*. According to Blewett *et al.* (1993), *C. parvum* is very infectious to gnotobiotic lambs and the ID_{50} is less than 10 oocysts. The ID_{50} for *C. parvum* in healthy human adults (with no acquired protective immunity) is 165 oocysts (Fig. 16.4), and single oocysts appear to present much lower risks to healthy human adults than to gnotobiotic lambs. The reason for this may lie in the nature of the host protective barriers and, in particular, the indigenous microbiota. Harp *et al.* (1992) presented data supporting the hypothesis that resistance of adult mice to *C. parvum* infection is mediated by non-specific mechanisms associated with the presence of intestinal microbiota.

In terms of dose–response curves, there is no such quantity as an 'infective dose'. Indeed, each dose has a specific probability of initiating infection according to the dose–response curve (see Fig. 16.3). The 'minimum infective dose' is one pathogen.

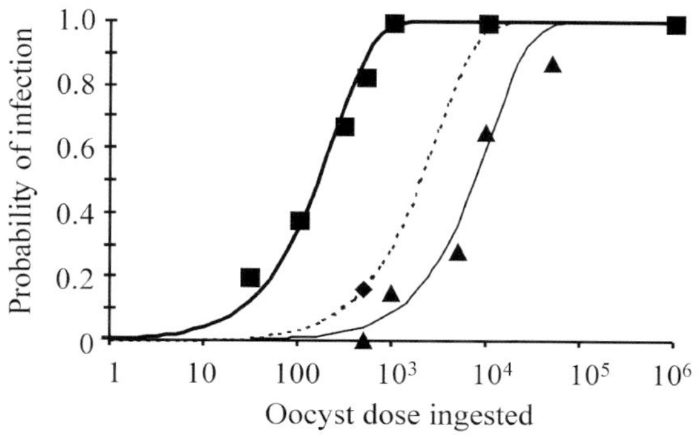

Fig. 16.4 Acquired protective immunity for *C. parvum* (Iowa isolate) in humans. Negative exponential dose response curve (r = 0.00419) fitted by Haas *et al.* (1996) to human infectivity data (*N*) obtained for volunteers selected on the basis of having no serological evidence of past infection with *C. parvum* (DuPont *et al.*, 1995). On rechallenge one year later (Okhuysen *et al.*, 1998), 16% of those volunteers were infected with a dose of 500 oocysts (*U*). The negative exponential curve (r = 0.00035) through this point is presented as the dashed line. Negative exponential curve (r = 0.000091) for confirmed infection (Chappell *et al.*, 1999) in human volunteers selected with pre-existing anti-*C. parvum* IgG (*S*).

2.3.3 *Assessing the evidence that pathogens act independently and do not cooperate during infection*

Meynell (1957) demonstrated that fatal infections from *Salmonella typhimurium* given to mice by mouth arose from a very small number of organisms, or possibly just a single organism. He concluded that inoculated bacteria act independently, not cooperatively. However, it could be argued that, if infection through the oral route were a two-stage mechanism, then his conclusions are only applicable to the second stage and that the initial stage *could* involve events more consistent with a cooperative model. The oral route presents many barriers to infection by salmonellae, including pH and the combined inhibitory effects of a low oxidation-reduction potential and short-chain fatty acids produced by the normal gut microbiota (Meynell, 1963). Indeed Meynell (1955) reported that when the gut microbiota were killed by streptomycin, the oral LD_{50} for a streptomycin-resistant strain of *S. typhimurium* was reduced from greater than one million organisms to less than five organisms.

Bacteria contain a variety of mechanisms for overcoming host barriers, including endotoxins to damage the mucosa, neutralization of antibacterial agents, IgA proteases and acid suppression. In addition, microorganisms produce many substances which give them competitive advantage, or a means of defence, in establishing themselves in natural environments. For example, the bacteriocins are polypeptide antibiotics produced by one species of bacteria, which cause damage to the cellular membranes of other bacteria. It could be speculated that in an oral challenge comprising thousands of salmonellae, some of those salmonellae could use these mechanisms to lower the host barriers, so increasing the chances of a subsequent single bacterium successfully establishing infection. Essentially this would involve a two-stage process for microbial infection. Meynell and Maw (1968) presented results with salmonellae in mice suggesting that infection passed through an initial stage lasting a few hours, in which a varying proportion of the inoculum is killed, followed by a second stage. The initial stage is 'decisive'. Could it be that bacteria act cooperatively during that initial stage, so helping subsequent bacteria to survive and progress to the second stage in which a single bacterium multiplies and causes infection? If so, then exposure to low pathogen doses would present lower risks than those predicted by the negative exponential or beta-Poisson dose–response curves and a non-linear model perhaps resembling the log-probit curve would be more appropriate (see Fig. 16.3, for example). There are currently no experimental data on whether *C. parvum* oocysts act independently or cooperatively during infection. Indeed, oocysts may not even have the biochemical mechanisms necessary. It is generally assumed for MRA that oocysts act independently and that the negative exponential dose–response curve is appropriate (Haas *et al.*, 1996). For the purpose of public health protection, this assumption is acceptable because it gives an added margin of safety for exposures to low doses. However, if there were some cooperative mechanism for oocysts then MRA could overestimate the risks by as much as 15-fold (see Fig. 16.3), which would be unacceptable for water company operators.

The recent discovery of small molecules for pathogen communication demonstrates a potential mechanism for pathogen cooperation, although whether this could occur in the human gut is not known. Thus, cell-to-cell communication is used by bacterial cells to sense their population density (Dong *et al.*, 2001). This is known as quorum-sensing and controls expression of certain genes, which control biological functions such as virulence. The best characterized example is autoinduction of luminescence in the marine bacterium *Vibrio fischeri* (Fuqua *et al.*, 1996). Quorum-sensing signals, such as acyl-homoserine lactones (AHLs), are central to the communication system. The quorum-sensing regulation of virulence is a common strategy that many bacterial pathogens have adopted during evolution to ensure their survival in host–pathogen interactions. The expression of virulence genes in the Gram-negative bacterium, *Pseudomonas aeruginosa*, for example, requires cell-to-cell communication (Passador *et al.*, 1993). Among the extracellular virulence factors produced by *P. aeruginosa* is a protease, elastase, which cleaves elastin, human immunoglobulins, collagens and components of the complement system. These are all components of the host defence system. Passador *et al.*, (1993) speculate that when carried out in response to cell density or other environmental and nutritional stimuli, cell-to-cell communication could result in a concentrated attack on the host. Cellular communication in pathogenic bacteria is an area where discoveries in molecular biology may input to the future development of quantitative MRA. Whether quorum-sensing is important in botulism toxin production by *Clostridrium botulinum* spores, or toxin production by *Bacillus anthracis* spores would be an important factor in MRA, and would necessitate the development of models to take into account the spatial heterogeneity of the spores (see Within-batch or spatial variation, p. 275) in making exposure assessments. There is evidence that campylobacters at high concentrations protect each other. The same would be true for an aggregate of virus particles. Furthermore, there could be cooperative effects of viruses during infection of target cells. Thus binding of one virus to receptors on the cell membrane could facilitate binding of further virus particles, such as in the formation of clathrin-coated pits.

2.3.4 *Acquired protective immunity for* C. parvum

The 29 adults used in the study of DuPont *et al.*, (1995) were selected on the basis of having no serological evidence of past infection with *C. parvum*. Information on the extent of seroprevalence in various drinking-water communities and the quantitative protective effect is necessary for development and validation of MRA models for *C. parvum* in drinking water. Anti-*C. parvum* IgG is typically found in 25–33% of persons in the USA (Chappell *et al.*, 1999). Okhuysen *et al.*, (1998) investigated if infection of humans with *C. parvum* is protective 1 year after exposure. The subjects were 19 of those 29 adults used in the study of DuPont *et al.*, (1995). Rechallenge with a dose of 500 oocysts resulted in infection of three of the 19 adults (16%). Constructing a negative exponential dose–response curve through this one result (the dashed curve in Fig. 16.4) suggests that primary exposure increased the ID_{50} from 165 oocysts to about 1900 oocysts and that the risk of infection (r in Equation (1)) from a single oocyst is decreased from 0.0042 to 0.00035. In further human infectivity studies, groups of volunteers with anti-*C. parvum*-specific serum antibody were challenged with doses between 500 and 50 000 oocysts (Chappell *et al.*, 1999). These points are also shown in Fig. 16.4. The negative exponential dose–response curve ($r = 0.000091$) is shifted to the right, and the ID_{50} is 7600, i.e. 46-fold higher than for volunteers with no serum antibody.

2.3.5 *Human rotavirus*

The beta-Poisson dose–response curve assumes that r in Equation (1) is actually not a constant but is itself described by a beta probability distribution. The probability of infection is written as:

$$P_i = 1 - [1 + \frac{N}{N_{50}}(2^{1/\alpha} - 1)]^{-\alpha} \quad (3)$$

where α and N_{50} characterize the dose–response (Regli *et al.*, 1991). As α increases so the beta-Poisson model becomes closer to the negative exponential model (Equation (1)). N_{50} represents the ID_{50}.

Ward *et al.*, (1986) obtained infectivity data for human rotavirus in 62 healthy adult volunteers. Subjects ingested doses ranging from 0.009 to 90 000 focus-forming units (ffu). The proportions of adults infected at each dose are plotted in Fig. 16.5. The beta-Poisson dose–response curve fitted by Haas *et al.* (1993) is plotted, together with a two-component negative exponential curve, which is written as:

$$P_i = f_1\left(1 - e^{-r_1 N}\right) + f_2\left(1 - e^{-r_2 N}\right) \quad (4)$$

where f_i represents the fraction of the population represented by the negative exponential model defined by r_i. The sum of f_1 and f_2 is always equal to 1.0 in a two-component model. The two-component model assumes there are two human populations which differ fundamentally in their susceptibility to rotavirus infection. This could reflect molecular differences, e.g. in the rotavirus receptor on the human host cells, although there is no evidence currently for this. The sequencing of the human genome and the science of bioinformatics may shed light on this in the future (International Human Genome Sequencing Consortium, 2001). In the case of BSE infection in humans, epidemiological data for vCJD and genetic differences (polymorphisms) in the human prion protein gene at codon 129 provide a theoretical basis for a two-component dose–response curve (Gale, 2001a). The two-component negative exponential model fits the rotavirus data better (Fig. 16.5) than the beta-Poisson model, particularly at the lower doses. The beta-Poisson curve predicts risks 3.5-fold higher for the lower doses.

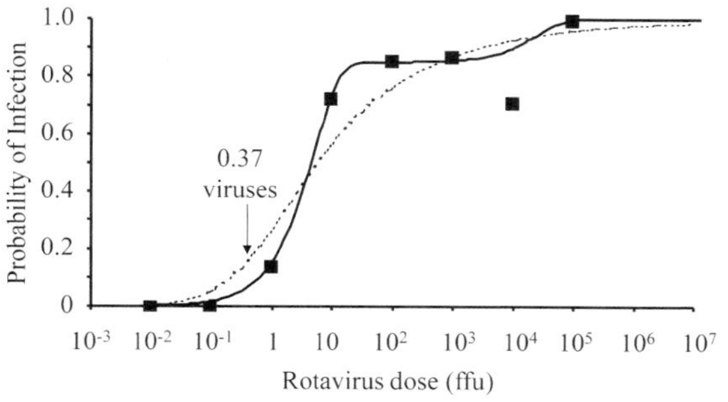

Fig. 16.5 Beta-Poisson dose-response curve ($\alpha = 0.265$; $N_{50} = 5.6$) (dashed line) fitted by Haas *et al.* (1993) to rotavirus infectivity data (*N*) of Ward *et al.* (1986). Two component negative exponential model (solid line) fitted assuming $r_1 = 0.20$ for 85% (f_1) of the population and $r_2 = 0.00004$ for the remaining 15% (f_2) in Equation 4.

3 INTEGRATING PATHOGEN EXPOSURES AND DOSE–RESPONSE CURVES

The effect of the variation in the magnitude of the pathogen doses ingested by individual drinking-water consumers (see Table 16.1) on risk prediction is illustrated using the dose–response curves for *C. parvum* (Fig. 16.4) and rotavirus (Fig. 16.5) to predict the number of infections across the population (Table 16.4).

3.1 Risk prediction for *C. parvum*

Ignoring the log-normal variation in oocyst exposures and just using the Poisson distribution of exposures with the log-probit dose–response curve underestimates the risk of infection by a factor of about ninefold. This is because the Poisson distribution of pathogen exposures does not predict exposures to high pathogen doses approaching the ID_{50} (see Table 16.1). It only predicts exposures to low doses which have a greatly diminished risk according to the log-probit curve (see Fig. 16.3). It is concluded that modelling the log-normal variation in pathogen exposures would be critically important for MRA if pathogens were to act cooperatively during infection. Indeed, if pathogens acted cooperatively then dispersion within the drinking-water supply (i.e. going from Poisson-log-normal to Poisson)

TABLE 16.4 Predicted number of infections (per 10 000 persons/day) using simulated pathogen exposures in Table 16.1 with *C. parvum* (Iowa isolate) and rotavirus dose–response curves

Pathogen	*Dose–response curve*	*Exposure model*		
		Poisson-log-normal ($\mu = -1.66$; $\sigma = 1.04\ \log_{10}$ *oocysts/person/day)*	*Poisson distribution (mean = 0.37 oocysts/person/day)*	*Arithmetic mean (0.37 oocysts/person/day)*
C. parvum (Fig. 16.3 and Fig. 16.4)	Negative exponential (no immunity)	14.97*	15.44*	15.47
	Log-Probit (no immunity)	9.65*	1.70*	0.17
	Negative exponential (volunteers with anti-*C. parvum* IgG in serum)	0.336*	0.336*	0.336
Rotavirus (Fig. 16.5)	Beta-Poisson	410*	886*	1488
	Two component negative exponential model	331*	551*	605

* The risk calculation is described fully in Gale and Stanfield (2000).

would reduce the risk by a factor of fivefold according to Table 16.4. However, it is never appropriate to use the arithmetic mean in the log-probit model. Furthermore, if there were morbidity and mortality thresholds, as suggested by Williams and Meynell (1967), then MRA models need to accommodate the log-normal variation in exposures to cover the possibility of some consumers ingesting high doses which exceed those thresholds.

In contrast, the risk predicted using the negative exponential dose–response curve is little affected by whether the log-normal variation in exposures is accommodated or just the single point arithmetic mean exposure is used in the Poisson distribution (see Table 16.4). For both the Poisson-log-normal and Poisson distributions of exposure, a risk of 15 *Cryptosporidium* infections/10 000 persons/day is predicted. Indeed, the same risk is predicted simply by using the arithmetic mean of 0.37 oocysts/person/day directly in Equation (1). This is because the dose–response relationship is essentially linear (Equation (2)). In effect, the risk of infection is directly related to the total number of oocysts in supply and is not influenced by their spatial distribution (Gale, 2001a).

It should be noted that acquired protective immunity reduces the risk of infection by a factor of 46-fold (see Table 16.4).

3.2 Risk prediction for rotavirus

Using the arithmetic mean exposure of 0.37 virus/person/day directly in the beta-Poisson dose–response curve (arrow in Fig. 16.5) predicts 1488 infections/10 000 persons/day (see Table 16.4). In contrast, accommodating the log-normal variation in exposures in Table 16.1 predicts 410 viral infections/10 000 persons/day. Thus, ignoring the log-normal variation in virus exposures through drinking water would, in the case of the virus risk assessment, over-predict the risk by 3.6-fold. This is partly due to the fact that ignoring the log-normal variation predicts that more consumers would be exposed to highly infectious virus each day (see Table 16.1). In effect, dispersion of the virus across the population increases the number of infected persons. The effect is less with the two-component dose–response curve.

3.3 Dispersion of pathogens serves to increase the risk

In the event of pathogens cooperating during initiation of infection, dispersion of the pathogens within the medium would naturally lower the risk (see Table 16.4). However, assuming pathogens act independently, then dispersion serves to increase the number of infections in situations where the exposure/ID_{50} ratio is high. This is apparent for rotavirus in Table 16.4 (and also to a lesser degree for *C. parvum* in Table 16.2). Thus, dispersing the virus particles (i.e. going from Poisson-log-normal to Poisson) doubles the number of infections in the population predicted by the beta-Poisson dose–response curve. Using the arithmetic mean directly increases the predicted number of infections still further. In this respect, using the arithmetic mean gives the worst case. In general, pathogen exposures to individuals through drinking water are low relative to the ID_{50} (see Table 16.1) and it is acceptable to use the arithmetic mean. Only in the case of highly infectious agents such as rotavirus does the arithmetic mean over-predict the risk. However, consider a 100 g porcine kidney infected with 100 oral ID_{50} of foot and mouth disease virus and discarded into pig swill. Dispersion of the kidney tissue in the swill would increase the number of infected pigs considerably.

4 UNCERTAINTY – MONITORING MAY UNDERESTIMATE THE ARITHMETIC MEAN PATHOGEN EXPOSURE AND HENCE THE RISK

The counts of *Cryptosporidium* oocysts in 100-litre volume 'spot' samples of treated water in an outbreak simulation are presented in Table 16.5. Taking single 100-litre volume spot samples from the drinking-water supply would, in general, underestimate the arithmetic mean oocyst concentration and hence the risk of *Cryptosporidium* infection. Indeed, almost nine out of every ten (88%) 100-1itre samples contain fewer oocysts than the arithmetic mean of 29.75 oocysts/100 litre. Furthermore, 54.75% of the samples contained 0, 1, or 2 oocysts, i.e. >10-fold less than the arithmetic mean (Table 16.5). Further simulations showed that even taking the arithmetic mean of the counts for ten 100-litre random spot samples underestimated the 'true' arithmetic mean (29.75 oocysts/100 litre) by more than fivefold some 25% of the time. Thus a quarter of all sets of ten samples (on taking the average) will underestimate the risk by more than fivefold. In the case of sets of just five samples, the means of 37% of the sets underestimated the true arithmetic mean by more than fivefold. This is why the *Cryptosporidium* Regulations in the UK require continuous monitoring of treated drinking water as it leaves the works (Gale and Stanfield, 2000).

TABLE 16.5 Simulated counts of *Cryptosporidium* oocysts (Poisson-log-normal distribution) in 100-litre volumes under conditions consistent with an outbreak (Gale and Stanfield 2000)

Oocyst counts (100 litre)	*Percentage of samples*
0	32.91
1	13.82
2	8.04
3–5	12.62
6–9	7.87
10–29	12.40
30*–50	4.14
51–100	3.47
101–200	2.17
201–1000	2.05
1001–5000	0.47
5001–16 458	0.02

* Arithmetic mean = 29.75 oocysts/100 litre.

4.1 Risk assessment complements epidemiological observations

Craun *et al.* (1998) reviewed oocyst count data collected during 12 waterborne outbreaks and concluded that there was no clear association between the risk of illness and the *Cryptosporidium* oocyst concentrations measured in the drinking-water supply. This would appear to challenge the very foundations of MRA for *Cryptosporidium* in drinking water. However, it can be explained by the fact that any association will tend to be masked by the large temporal-spatial heterogeneity in oocyst concentrations in drinking water because data from spot sampling in general underestimate the risk of cryptosporidiosis.

4.2 A zero oocyst reading provides little reassurance even during an outbreak

Haas and Rose (1995) proposed an action level of 10–30 oocysts/100 litre in finished water within and above which there is the possibility of a waterborne outbreak of cryptosporidiosis. Craun *et al.* (1998) dismissed this approach on the basis that many samples collected during outbreaks contained fewer counts than the action level. Indeed, in four of the 12 outbreaks studied, no oocysts were ever detected by sampling. According to the model for an outbreak in Table 16.5, a third of 100-litre samples contained zero oocysts, while a small proportion of 100-litre samples contained hundreds and even thousands of occysts. Thus, a zero reading from a Poisson-log-normal distribution gives very little reassurance. Indeed, the heterogeneity explains why oocysts are often not detected during waterborne outbreaks of cryptosporidiosis. If, on the other hand, all the oocysts were evenly distributed in the treated water during an outbreak, then the oocyst concentration would equal the arithmetic mean (29.75 oocysts/100 litre) at all points in space and time. The probability of any 100-litre volume spot sample containing zero oocysts would be remote at 10^{-13} (assuming the Poisson distribution), and most would contain at least 19 oocysts, which is well within the proposed action level for an outbreak of cryptosporidiosis (Haas and Rose, 1995). A single spot sample would therefore be adequate to give a representative picture of the level of contamination and the risk of illness across the supply. A reading of 0 oocysts/100 litre would be very reassuring and representative of very low numbers of oocysts across the whole supply.

5 UNCERTAINTY – ESTIMATING THE REMOVAL OF PATHOGENS BY TREATMENT PROCESSES

Treatment removal efficiencies may vary greatly for a particular process due to 'between-batch' and 'within-batch' variation. As shown in the section on the arithmetic mean exposures, p. 266, the arithmetic mean pathogen removal by treatment may be used directly in the risk assessment. Thus increasing the net pathogen removal by 10-fold will decrease the risk of infection by 10-fold. The problem is how to determine experimentally whether a treatment process has been improved by 10-fold.

5.1 Experimental designs may overestimate the net pathogen removal by treatment

Determining the arithmetic mean pathogen removal is complicated by the large variation in the treatment removal efficiencies, which gives rise to large variations in pathogen densities in the treated water (see Table 16.5). A key assumption of the *Cryptosporidium* outbreak model in Fig. 16.1 is that drinking water treatment not only removes oocysts but also increases the variation in the concentrations of those oocysts remaining in the treated water relative to the raw water. Therefore, monitoring programmes based on spot sampling will tend to underestimate the net pathogen loading in the treated water to a greater degree than in the raw water (Gale and Stanfield, 2000). The result of this is that, without intensive monitoring of the treated water samples, experiments to measure the net pathogen removal by treatment will tend to overestimate the arithmetic mean removal and hence underestimate the risk.

5.2 Modelling 'between-batch' or temporal variation

The variation in particle removal efficiencies by drinking water treated as used for the model in Fig. 16.1 are described by a log-normal distribution (see Fig. 16.2). Although particle removal ratios varied between 1 and 5 logs and the median (50 percentile) removal was 500-fold (2.7 logs), the arithmetic mean or net removal was 61-fold (1.79 logs). It is apparent that 50% of removals are more than eightfold more efficient than the arithmetic mean. Therefore half of the single 'one-off' experiments undertaken to measure the removal efficiency will overestimate the arithmetic mean efficiency by more than eightfold. As an example, consider an experiment in which a single raw water sample is taken at Point A and a single treated water sample at Point B in Fig. 16.1. The result would suggest a >4-log removal, when in fact the net removal is only 1.79-logs. This demonstrates the importance of understanding the variation in removal efficiencies when designing experiments to estimate the net removal by a treatment process.

Beta distributions are typically used to model the variation in the proportions of pathogens breaking through a treatment process (Teunis *et al.*, 1999). The beta distribution does not fit the data in Fig. 16.2 too well. Indeed, data for plate counts in drinking water support the application of the log-Normal distribution (Gale, 1996). Teunis *et al.* (1997) fitted a beta distribution to model the variation in removal efficiencies for spores of sulphite-reducing bacteria by drinking-water treatment. This is also plotted in Fig. 16.2. It is interesting to note how the log-normal and beta models differ at the higher removal efficiencies. Thus, according to the beta distribution, 83% of removals are greater >5 log, while for the log-normal distribution, only 0.5% are >5 log. Indeed, the beta distribution predicts 50% of removals are >18.5 logs. These high removal efficiencies have little effect on the magnitude of the arithmetic mean removal. Indeed, the log-normal and beta distributions fitted to the data in Fig. 16.2 have the same arithmetic mean of 1.79 logs. This is because, irrespective of whether a high removal value is 5 log (log-normal distribution) or 18.5 log (beta distribution), the proportion breaking through, i.e. 10^{-5} or $10^{-18.5}$ respectively, is so small as to tend to zero and has little impact on the arithmetic mean. However, the poor removal efficiencies (i.e. between 0 log and 1 log) have a great influence on the arithmetic mean removal (Gale and Stanfield, 2000; Gale, 2001b). Indeed, the beta distribution of Teunis *et al.* (1997) plotted in Fig. 16.2 has fewer poor removal efficiences in the 0 to 1 log range, and consequently has a better net removal efficiency, the arithmetic mean being 2.2 logs.

5.3 'Within-batch' or spatial variation

The effect of drinking-water treatment on the statistical distribution of spores of aerobic spore-forming bacteria within 100-litre volume samples has been studied (Gale *et al.*, 1997). This can be considered as 'within-batch' or spatial variation. The overall conclusion was that operational drinking water treatment not only removed 94–98% of the spores, but also promoted the spatial association of the remaining spores. Further data from studies undertaken at pilot scale suggested a more complicated picture (Gale *et al.*, 2002). Thus, spore counts were not over-dispersed in all treated water volumes investigated and coliform counts were, in general, Poisson-distributed in the treated water volumes. Such findings contribute to explaining why no 'ideal' surrogate has been identified for treatment plant performance (Nieminski *et al.*, 2000; Gale *et al.*, 2002)

6 THE DEVELOPMENT AND APPLICATION OF EVENT TREES

Event trees simplify the process of modelling the various pathways and visualizing how infectivity in the source term partitions through the various routes to the receptors.

6.1 *Escherichia coli* O157 in sewage sludge

6.1.1 Source term

Gale (2001b) presented an event tree to model partitioning of salmonellae into sewage sludge at a sewage treatment works based on the plentiful data available for salmonella concentrations in raw sewage. There is currently little information on *E. coli* O157 concentrations in raw sewage or in sewage sludge. An alternative approach is to estimate the arithmetic mean pathogen loading in the sewage sludge, as developed by Gale and Stanfield (2001) for BSE in sewage sludge. The main source of *E. coli* O157 in raw sewage is from slaughter of cattle and sheep at abattoirs. The model uses UK Government data that 3.13 million cattle and 15.86 million sheep are slaughtered annually at abattoirs in England and Wales and assumes, on the basis of expert advice, that 5% of faecal material in animals slaughtered at abattoirs enters the sewage treatment works, either directly into sewers or by tankering. Shere *et al.* (1998) report counts of *E. coli* O157:H7 in infected cattle ranging from 200 to 87 000/g. Since the arithmetic mean is nearer to the maximum in magnitude for a log-normal distribution (see Fig. 16.1), this value of 87 000 cfu/g is used as an estimate of the arithmetic mean for the MRA. Assuming 15.7% of cattle are infected (Chapman and Siddons, 1997) and that each bovine contains 10 kg of faeces at time of slaughter, then 2.1×10^{13} *E. coli* O157/year enter sewage works across England and Wales from the slaughter of cattle. The arithmetic mean *E. coli* O157 concentration in faeces from sheep and lambs during the New Deer (Scotland) outbreak may be estimated from data of Strachan *et al.*, (2001) as 365 500/g. (This assumes that the count of $>10^6$/g recorded in one lamb was 10^7/g.) Assuming 2.2% of sheep flocks are infected (Chapman et al., 1997) and each sheep has 1 kg of faeces at point of slaughter, then the arithmetic mean *E. coli* O157 loading across sewage works in England and Wales from slaughter of sheep in abattoirs is 6.3×10^{12} cfu/year. In England and Wales, 967 000 tonnes dry solids (tds) of sewage sludge are produced annually (Gale and Stanfield, 2001). On the basis of the salmonella model (Gale, 2001b), 82.9% of *E. coli* O157 in raw sewage would partition into the raw sludge at the works. Thus, the arithmetic mean loading from cattle and sheep combined in raw sewage sludge in England and Wales is 2.4×10^7 *E. coli* O157/t (dry solids). This is three orders of magnitude lower than the arithmetic mean of 2.0×10^{10} salmonellae/t (dry solids) of raw sewage sludge predicted by the salmonella model (Gale 2001b) based on monitoring data for raw sewage.

6.1.2 *Pathway term*

The pathway for transmission to root crops of *E. coli* O157 in sewage sludge applied to agricultural land is shown in Fig. 16.6. The model assumes that conventional sludge treatment (e.g. anaerobic digestion) destroys 2 logs of *E. coli* O157 according to the UK Code of Practice for Agricultural Use of Sewage Sludge (Department of the Environment, 1996). Strachan *et al.*, (2001) allow for a T_{90} (time for a 1-log decay) of 16 days in a model for an environmental outbreak of *E. coli* O157, and Bolton *et al.* (1999) demonstrated a linear 5-log decay over 50 days on grazing land during the winter months. The model (Fig. 16.6) allows a 5-log decay, which is conservative for the 12-month time interval specified by the Safe Sludge Matrix (www.adas.co.uk/matrix/) between application of treated sludge and harvesting of root crops. Dilution in soil is modelled as the probability of a root crop colliding with a sludge particles compared to the much greater probability of colliding with a soil particle (Gale and Stanfield, 2001). At point of harvest, 0.02 tonnes of soil/sludge is transferred to each tonne of root crops. This is the origin of the 0.02 in Fig.16.6. Following the event tree through, the arithmetic mean incremental exposure to root crops at point of harvest is calculated as:

$$2.4 \times 10^7 \times 0.01 \times 10^{-5} \times 0.0018 \times 0.02 = 8.3 \times 10^{-5} \text{ } E. coli \text{ O157/root crops.}$$

The salmonella model predicted 70 salmonellae/root crops at point of harvest from application of conventionally-treated sewage sludge, although this assumed only a 2-log decay in the soil (Gale, 2001b).

6.2 *Escherichia coli* O157 risk assessment for a rendering plant discharging treated effluent to a surface water

Gale (2001b) estimated the risks of *E. coli* O157 from a rendering plant discharging treated effluent to a river, which is used for recreational water sport activities (including water skiing) and also for municipal drinking water abstraction. The source term is based on the processing of 619 cattle carcasses/day with the same *E. coli* O157 loadings assumed above. The pathway from carcass to raw effluent is set out as an event tree in Fig. 16.7.

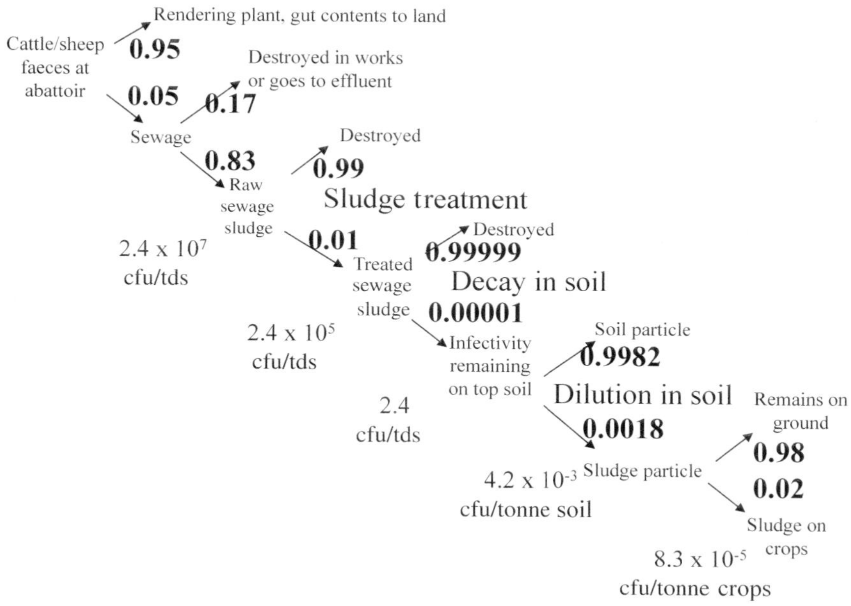

Fig. 16.6 Event tree for transmission of *E. coli* O157 in faeces at abattoirs to root crops via application of treated sewage sludge to agriculture land.

6.2.1 Uncertainty in the fate of the lorry washings has little impact

The model assumes that only 1% of faeces leaking out in the lorry goes to the raw effluent in the form of lorry/yard washings. This may be rather optimistic. A 'What if?' scenario is modelled in Fig. 16.7 to assess the impact of 10% of the faeces leaking out in the lorry going to the raw effluent as washings. The total loading of *E. coli* O157 in the raw effluent increases from 6.5×10^8 to 11.7×10^8/day. Therefore the risk is not even doubled.

6.2.2 By-pass of treatment has a huge impact

The raw effluent is treated on site prior to discharge to the river. The net removal of *E. coli* O157 by treatment (which includes DAF, activated sludge treatment and UV disinfection) is 4.6 logs, i.e. only 0.0025% of the *E. coli* in the raw effluent breakthough treatment. However, if there is a 1% by-pass of treatment such that for 3.6 days of the year the effluent receives no treatment, then the net removal is reduced to just 1.999-logs, i.e. 1.002% of *E. coli*

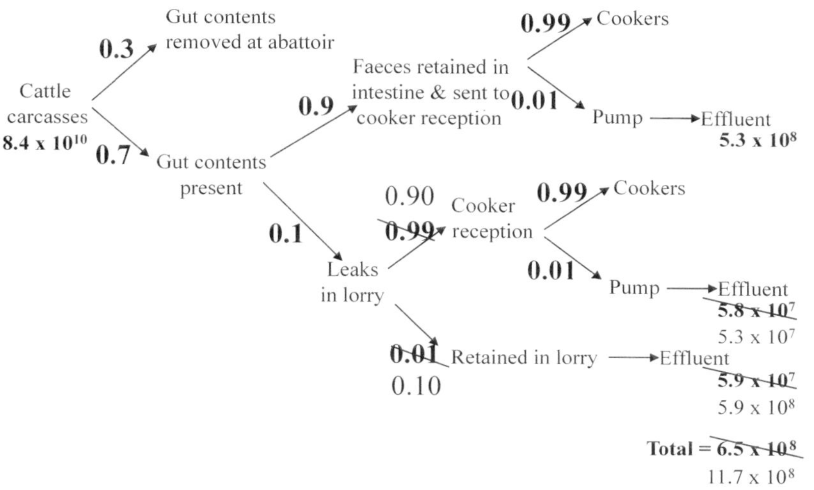

Fig. 16.7 Event tree for modelling partition of *E. coli* O157 in cattle carcasses into the effluent each day at a rendering plant. The model demonstrates a 'What If?' analysis.

break-through. This increases the *E. coli* O157 loading in the treated effluent, and hence the risk, by 400-fold.

7 BSE IN RENDERING PLANT EFFLUENTS

Gale *et al.*, (1998) estimated the risks of BSE transmission to consumers' drinking water from a chalk aquifer above which a rendering plant discharges treated effluent. The rendering plant processes 286 cattle carcasses/day. Assuming the human oral ID_{50} is 1 g of BSE-infected bovine brain and that 0.54% of the cattle are in the final stages of infection, it may be calculated that 1544 human oral ID_{50} enter the plant, on average, daily. An event tree to model breakthrough of BSE infectivity into the aquifer is presented in Fig. 16.8. Rendering destroys 98% of the BSE infectivity (Taylor *et al.* 1995) and 0.6% of the meat and bone meal (MBM) product breaks through into the condensate as suspended solid. The effluent treatment processes include DAF and activated sludge, which together remove 99.7% of particulate material, and hence any BSE agent. The model allows for a 93% decay of BSE agent in the soil (over a period of 2 years) and a 90% attenuation in the chalk substrata. The rendering plant produces 120 000 litre of effluent/day, which is diluted into 10 Ml of groundwater in the aquifer. Dilution is modelled in the event tree as the probability of a consumer ingesting effluent compared to the much greater probability of ingesting (uncontaminated) ground water. The concentration of BSE in the drinking water is calculated from Fig. 16.8 as 4.7×10^{-8} ID_{50} in 120 000 litre. A person ingesting 2 litre of tap water/day for a year would therefore ingest 2.8×10^{-10} ID_{50}. Assuming a negative exponential dose–response curve (Gale, 1998), this translates into a risk of 2.0×10^{-10}/person/year.

7.1 'What if' 1% of the raw material by-passes the rendering process?

According to the model in Fig. 16.8, the rendering process not only destroys 98% of the BSE agent but also partitions 99.4% of the remaining infectivity away from the condensate and into the MBM product. The net breakthrough of BSE agent into the condensate from the rendering process is therefore calculated as $0.02 \times 0.006 = 0.00012$. According to the *E. coli* O157 model in Fig. 16.7, some 1% of the raw material in the cooker reception by-passes

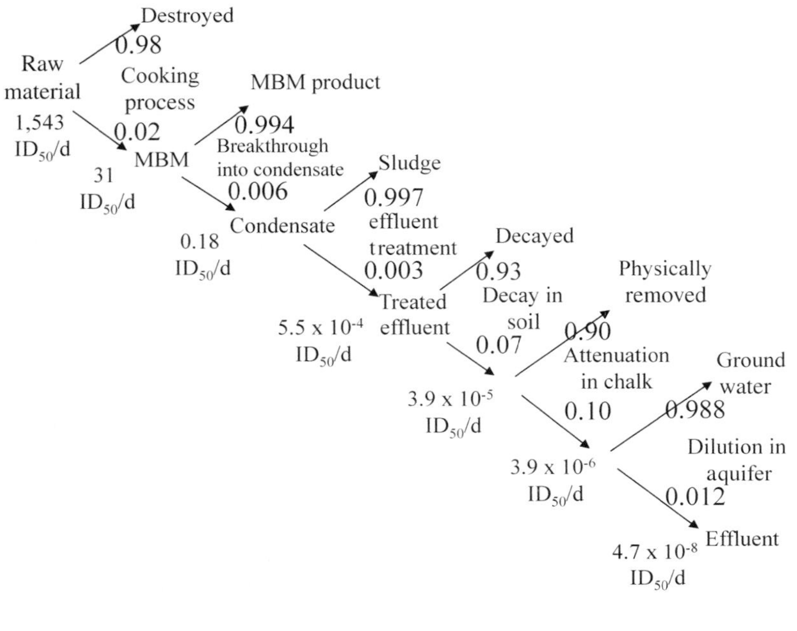

Fig. 16.8 Event tree for modelling the partitioning of BSE infectivity from cattle at a rendering plant into an aquifer (based on Gale *et al.*, 1998).

the cookers and is pumped to the raw effluent. It is extremely unlikely that as much as 1% of the brain and spinal cord from each bovine by-passes the cookers because the head is removed intact and the carcass is no longer butchered up the spinal column at abattoirs (see Gale and Stanfield 2001). However, skulls may be cracked by the Archimedes screws which lift the material from the sump to the cookers. By-pass of 1% of the raw brain/spinal cord from each bovine would mean that the net breakthrough of rendering would be $0.01 \times 1.0 + 0.99 \times 0.00012 = 0.01012$. Thus, the net breakthrough of rendering would be increased 84-fold from 0.00012 to 0.01012. The risks to drinking-water consumers would also be increased 84-fold, to 1.6×10^{-8}/person/year.

REFERENCES

Blewett, D.A., Wright, S.E., Casemore, D.P. *et al.* (1993). Infective dose size studies on *Cryptosporidium parvum* using gnotobiotic lambs. *Water Science and Technology* **27**(3), 61–64.

Bolton, D.J., Byrne, C.M., Sheridan, J.J. *et al.* (1999). The survival characteristics of a non-toxigenic strain of *Escherichia coli* O157:H7. *Journal of Applied Microbiology* **86**, 407–411.

Chapman, P.A., Siddons, C.A., Cerdan Malo, A.T. and Harkin, M.A. (1997). A 1-year study of *Escherichia coli* O157 in cattle, sheep, pigs and poultry. *Epidemiology and Infection* **119**, 245–250.

Chappell, C.L., Okhuysen, P.C., Sterling, C.R. *et al.* (1999). Infectivity of *Cryptosporidium parvum* in healthy adults with pre-existing anti-*C. parvum* serum immunoglobulin G. *American Journal of Tropical Medicine and Hygiene* **60**(1), 157–164.

Craun, G.F., Hubbs, S.A., Frost, F. *et al.* (1998). Waterborne outbreaks of cryptosporidiosis. *Journal of American Water Works Association* **90**(9), 81–91.

Department of the Environment (1996). *Code of Practice for Agricultural Use of Sewage Sludge.* DETR Publications, Ruislip.

Dong, Y.-H., Wang, L.-H., Xu, J.-L. *et al.* (2001). Quenching quorum-sensing dependent bacterial infection by an N-acyl homoserine lactonase. *Nature* **411**, 813–816.

Donnenberg, M.S. (2000). Pathogenic strategies of enteric bacteria. *Nature* **406**, 768–774.

DuPont, H.L., Chappell, C.L., Sterling, C.R. *et al.* (1995). Infectivity of *Cryptosporidium parvum* in healthy volunteers. *New England Journal of Medicine* **332**, 855–859.

Fuqua, C., Winans, S.C. and Greenberg, E.P. (1996). Census and consensus in bacterial ecosytems: the LuxR-LuxI family of quorum-sensing transcriptional regulators. *Annual Review of Microbiology* **50**, 727–751.

Gale, P. (1996). Coliforms in the drinking-water supply: what information do the 0/100-ml samples provide? *Journal of Water Supply Research and Technology – Aqua* **45**(4), 155–161.

Gale, P. (1998). Quantitative BSE risk assessment: relating exposures to risk. *Letters in Applied Microbiology* **27**, 239–242.

Gale, P. (2001a). A review: Developments in microbiological risk assessment. *Journal of Applied Microbiology* **91**(2), 191–205.

Gale, P. (2001b). Microbiological Risk Assessment. In: S. Pollard and J. Guy (eds) *Risk Assessment for Environmental Professionals*, pp. 8–90. Lavenham Press.

Gale, P. and Stanfield, G. (2000). *Cryptosporidium* during a simulated outbreak. *Journal of the American Water Works Association* **92**(9), 105–116.

Gale, P. and Stanfield, G. (2001). Towards a quantitative risk assessment for BSE in sewage sludge. *Journal of Applied Microbiology* **91**, 563–569.

Gale, P., van Dijk, P.A.H. and Stanfield, G. (1997). Drinking water treatment increases microorganism clustering; the implications for microbiological risk assessment. *Journal of Water Supply Research and Technology – Aqua* **46**, 117–126.

Gale, P., Young, C., Stanfield, G. and Oakes, D. (1998). Development of a risk assessment for BSE in the aquatic environment. *Journal of Applied Microbiology* **84**, 467–477.

Gale, P., Pitchers, R. and Gray, P.E. (2002). The effect of drinking water treatment on the spatial heterogeneity of micro-organisms: implications for assessment of treatment efficiency and health risk. *Water Research* **36**(6), 1640–1648.

Haas, C.N. (1983). Estimation of risk due to low doses of microorganisms: A comparison of alternative methodologies. *American Journal of Epidemiology* **118**, 573–582.

Haas, C.N. (1996). How to average microbial densities to characterize risk. *Water Research* **30**(4), 1036–1038.

Haas, C.N. and Rose, J.B. (1995). Developing an action level for *Cryptosporidium*. *Journal of the American Water Works Association* **87**(9), 81–84.

Haas, C.N., Rose, J.B., Gerba, C. and Regli, S. (1993). Risk assessment of virus in drinking water. *Risk Analysis* **13**, 545–552.

Haas, C.N., Crockett, C.S., Rose, J.B. *et al.* (1996). Assessing the risk posed by oocysts in drinking water. *Journal of the American Water Works Association* **88**(9), 131–136.

Harp, J.A., Chen, W. and Harmsen, A.G. (1992). Resistance of severe combined immunodeficient mice to infection with *Cryptospordium parvum*: the importance of intestinal microflora. *Infection and Immunity* **60**(9), 3509–3512.

Hutton, P., Ashbolt, N., Vesey, G. *et al.* (1995). *Cryptospordium* and *Giardia* in the aquatic environment of Sydney, Australia. In: W.B. Betts, D. Casemore, C. Fricker, H. Smith, J. Watkins *et al.* (eds.) *Protozoa Parasites and Water*, pp. 71–75. The Royal Society of Chemistry, London.

International Human Genome Sequencing Consortium (2001). Initial sequencing and analysis of the human genome. *Nature* **409**, 860–921.

Joseph, C., Hamilton, G., O'Connor, M. *et al.* (1991). Cryptosporidiosis in the Isle of Thanet; an outbreak associated with local drinking water. *Epidemiology and Infection* **107**, 509–519.

LeChevallier, M.W. and Norton, W.D. (1995). Plant optimisation using particle counting for treatment of *Giardia* and *Cryptosporidium*. In: W.B. Betts, D. Casemore, C. Fricker, H. Smith and J. Watkins (eds) *Protozoa Parasites and Water*, pp. 180–187. The Royal Society of Chemistry, London.

Meynell, G.G. (1955). Some factors affecting the resistance of mice to oral infection by *Salm. typhi-murium*. *Proceedings of the Royal Society of Medicine* **48**, 916–918.

Meynell, G.G. (1957). The applicability of the hypothesis of independent action to fatal infections in mice given *Salmonella typhimurium* by mouth. *Journal of General Microbiology* **16**, 396–404.

Meynell, G.G. (1963). Antibacterial mechanisms of the mouse gut II: The role of EH and volatile fatty acids in the normal gut. *British Journal of Experimental Pathology* **44**, 209–219.

Meynell, G.G. and Maw, J. (1968). Evidence for a two-stage model of microbial infection. *Journal of Hygiene* **66**, 273–280.

Nieminski, E.C., Bellamy, W.D. and Moss, L.R. (2000). Using surrogates to improve plant performance. *Journal of the American Waterworks Association* **92**(3), 67–78.

Okhuysen, P.C., Chappell, C.L., Sterling, C.R. *et al.* (1998). Susceptibility and serologic response of healthy adults to reinfection with *Cryptosporidium parvum*. *Infection and Immunity* **66**(2), 441–443.

Passador, L., Cook, J.M., Gambello, M.J. *et al.* (1993). Expression of *Pseudomonas aeruginosa* virulence gene requires cell-to-cell communication. *Science* **260**, 1127–1130.

Regli, S., Rose, J.B., Haas, C.N. and Gerba, C.P. (1991). Modelling the risk from *Giardia* and viruses in drinking water. *Journal of the American Waterworks Association* **83**(11), 76–84.

Richardson, A.J., Frankenberg, R.A., Buck, A.C. *et al.* (1991). An outbreak of waterborne cryptosporidiosis in Swindon and Oxfordshire. *Epidemiology and Infection* **107**, 485–495.

Roseberry, A. and Burmaster, D.E. (1992). Log-normal distributions for water intake by children and adults. *Risk Analysis* **12**, 99–104.

Shere, J.A., Bartlett, K.J. and Kaspar, C.W. (1998). Longitudinal study of *Escherichia coli* O157:H7 dissemination on four dairy farms in Wisconsin. *Applied and Environmental Microbiology* **64**(4), 1390–1399.

Strachan, N.J.C., Fenlon, D.R. and Ogden, I.D. (2001). Modelling the vector pathway and infection of humans in an environmental outbreak of *Escherichia coli* O157. *FEMS Microbiology Letters* **203**, 69–73.

Swerdlow, D.L., Woodruff, B.A., Brady, R.C. *et al.* (1992). A waterborne outbreak in Missouri of *Escherichia coli* O157:H7 associated with bloody diarrhea and death. *Annals of Internal Medicine* **117**, 812–819.

Taylor, D.M., Woodgate, S.L. and Atkinson, M.J. (1995). Inactivation of the bovine spongiform encephalopathy agent by rendering procedures. *Veterinary Record* **137**, 605–610.

Teunis, P.F.M. (1997). *Infectious Gastro-enteritis – Opportunities for Dose Response Modelling. Report no. 284 550 003.* National Institute of Public Health and the Environment (RIVM). Bilthoven, The Netherlands.

Teunis, P.F.M., Medema, G.J., Kruidenier, L. and Havelaar, A.H. (1997). Assessment of the risk of infection of *Cryptosporidium* or *Giardia* in drinking water from a surface water source. *Water Research* **31**(6), 1333–1346.

Teunis, P.F.M., Evers, E.G. and Slob, W. (1999). Analysis of variable fractions resulting from microbial counts. *Quantitative Microbiology* **1**(1), 63–88.

Ward, R.L., Bernstein, D.I., Young, E.C. *et al.* (1986). Human rotavirus studies in volunteers: Determination of infectious dose and serological response to infection. *Journal of Infectious Diseases* **154**, 871–880.

Williams, T. and Meynell, G.G. (1967). Time-dependence and count-dependence in microbial infection. *Nature* **214**, 473–475.

17

Health constraints on the agricultural recycling of wastewater sludges

Alan Godfree

North West Water Limited, Warrington WA5 3LP, UK

1 INTRODUCTION

The objective of sewage treatment is to remove solids and to reduce its biochemical oxygen demand (BOD) before returning the treated wastewater to the environment. Sewage sludge, increasingly referred to as biosolids, is an inevitable product of wastewater treatment. Conventional wastewater treatment processes comprise separate process streams for the liquid and solid fractions (sludge). The overall aim of treatment (in terms of solids) is: (i) to reduce to the minimum the amount of solids in the treated effluent in order to achieve discharge standards; and (ii) maximize the level of solids in the sludge in order to minimize the volume requiring further treatment and disposal.

Sludge is produced at various stages within the wastewater treatment process (Fig. 17.1). Usually, these solids are combined and treated as a whole. Dedicated sludge treatment may not be available at all works, particularly smaller plants. In these circumstances it is normal practice to transport the sludge to a larger works for subsequent treatment.

2 BASIC PRINCIPLES OF WASTEWATER TREATMENT

2.1 Preliminary treatment

Preliminary treatment consists of screening through bar screens to remove coarse solids and buoyant materials, such as plastics or rags, which may become trapped in pumps or other mechanical plant. The screenings are usually removed from the process stream and disposed of separately by landfilling or incineration. Occasionally, screenings may be shredded (comminuted) to reduce their size and returned to the process stream. This may cause problems with downstream processes and, as a consequence, is rarely practised at works which incorporate secondary (biological) treatment. The other component of preliminary treatment is grit removal, which is accomplished in chambers (or channels) or by centrifugation, taking advantage of the greater settling velocities of these solids. The material is largely inorganic in nature and is usually disposed of to landfill.

2.2 Primary treatment

Primary treatment is designed to reduce the load on subsequent biological (secondary) treatment processes. Although the design of primary sedimentation tanks differs, they achieve the removal of settleable solids, oils, fats and other floating material, and a proportion of the organic load. Efficiently designed and operated primary treatment processes should remove 50–70% of suspended solids and 25–40% of the BOD (organic

The Handbook of Water and Wastewater Microbiology
ISBN 0-12-470100-0

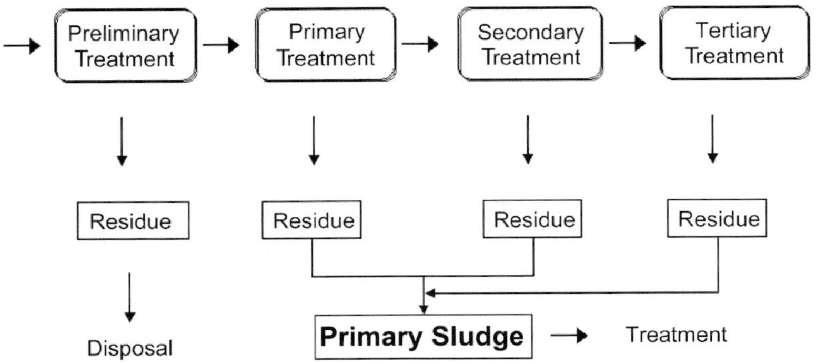

Fig. 17.1 Schematic representation of wastewater treatment showing sludge production.

load). The separated solids have a high organic content and are usually treated in order to stabilize the material prior to disposal.

2.3 Secondary treatment

Biological processes are used to convert dissolved biodegradable organic substances and colloidal material into inorganics and biological solids (biomass). There are several secondary treatment processes, but these may be divided into fixed film (e.g. trickling filters, rotating biological contactors) and suspended processes (e.g. activated sludge). Solids separation is the final stage of many of these treatment systems, which produces a sludge, the nature of which will depend on the upstream treatment process. Sludges arising from secondary treatment are usually combined with primary sludge and treated as a whole.

2.4 Tertiary treatment

Tertiary treatment will only be required for treatment works subject to specific discharge conditions. Many of these processes will not produce sludge and those that do are likely to generate only small amounts requiring dedicated treatment. Solids removed by granular media filtration will be passed to the primary sludge treatment process.

3 SLUDGE TREATMENT

The sludge obtained from the various stages is usually in the form of a liquid containing between 0.5 and 6% dry solids. The typical composition of raw (untreated) and anaerobically digested sludge is shown in Table 17.1. The nature and extent of any treatment depends on the means of final disposal or beneficial use. The aim of treatment is to reduce the water and organic content of the sludge and render it suitable for disposal or reuse. There are several commonly used methods of sludge treatment, including long-term storage (in lagoons), lime stabilization, digestion (aerobic or anaerobic), air drying, thermal drying, and incineration or gasification (for energy recovery). Detailed descriptions of these processes are outside the scope of this handbook. However, the effect of the various treatments on pathogens is described below.

Sewage sludge contains valuable amounts of plant nutrients (nitrogen and phosphorus) and trace elements (Table 17.2). For this reason sludge has historically been applied to agricultural land as part of an integrated farm management plan. Other options for disposal include energy recovery and land reclamation activities. In Europe, North America and elsewhere, the disposal of sewage sludge is subject to strict controls designed to protect soil quality while encouraging the use of sludge in agriculture. Codes of Practice, such as those published by the UK Department of the Environment (DoE, 1996) and the UK Ministry of Agriculture, Fisheries and Food (MAFF, 1998a,b), provide advice on practical aspects of utilizing sewage sludge in agriculture.

Strict limits are set on the amounts of potentially toxic elements permitted in sludge which may be used in agriculture. Application

TABLE 17.1 Chemical composition and properties of untreated and digested sludge

Constituent	*Untreated primary sludge*		*Digested primary sludge*	
	Range	*Typical*	*Range*	*Typical*
Total dry solids (TS) %	2.0–8.0	5.0	6.0–12.0	10.0
Volatile solids (% of TS)	60–80	65	30–60	40
Grease and fats (% of TS)				
Ether soluble	6–30		5–20	18
Ether extract	7–35			
Protein (% of TS)	20–30	25	15–20	18
Nitrogen (N, % of TS)	1.5–4	2.5	1.6–6.0	3.0
Phosphorus (P_2O_5, % of TS)	0.8–2.8	1.6	1.5–4.0	2.5
Potassium (K_2O, % of TS)	0.0–0.1	0.4	0.0–3.0	1.0
Cellulose (% of TS)	8.0–15.0	10.0	8.8–15.0	10.0
Iron (not as sulphide)	2.0–4.0	2.5	3.0–8.0	4.0
Silica (SiO_2, % of TS)	15.0–20.0		10.0–20.0	
pH	5.0–8.0	6.0	6.5–7.5	7.0
Alkalinity (mg/l as $CaCO_3$)	500–1500	600	2500–3500	3000
Organic acids (mg/l as HAc)	200–2000	500	100–600	200
Energy content (Kj/kg)	23 000–29 000	25 500	9300–14 000	11 500

Source: Metcalf and Eddy (1991).

rates are controlled to minimize the accumulation in the soil of toxic metals. Due to the low levels of metals in sludge, application rates are governed in practice by maximum nitrogen application rates (250 kg/ha y^{-1} or 500 kg/ha y^{-1}) and to balance phosphorus addition with crop off-take.

Information on the amounts of sewage sludge produced and its disposal is collected by a number of countries, principally the USA and the member states of the European Union. Annual sludge production in the USA is in the region of 6.8 million tonnes dry solids (M tds) of which 54% is applied to land (Bastian, 1997). The figures for the EU are 5.1 M tds and 48% respectively (CEC, 1999). Within the EU amounts of sludge produced vary considerably, with Germany producing the largest amount of treated sludge followed by the UK and France (Fig. 17.2). The proportion of treated sludge used in agriculture varies across the European Union, with just over 10% of sludge production in Ireland being applied to land compared with 66% in France (Fig. 17.3). Factors affecting the amount of sludge applied to agricultural land include topography, land use, climatic conditions, and the availability of alternative means of disposal. In the UK, sludge production is increasing, principally as a result of the EU Directive on the treatment of urban wastewater (CEC, 1991). The cessation of sea disposal has resulted in a greater proportion of sludge being used in agriculture (Table 17.3), a trend which is projected to continue in the medium term (Fig. 17.4).

TABLE 17.2 Nutrient content of sewage sludge (% dry weight)

Constituent	*Range*	*Typical*
Nitrogen	<0.1–17.6	3.0
Phosphorus	<0.1–14.3	1.5
Sulphur	0.6–1.5	1.0
Potassium	0.02–2.6	0.3

4 REGULATIONS GOVERNING THE USE OF SLUDGE IN AGRICULTURE

4.1 USA

The treatment and ultimate disposal of sewage sludge, including domestic septage, derived from the treatment of domestic sewage, is governed by 40 CFR Part 503 rule (US EPA, 1993). The regulations were developed over

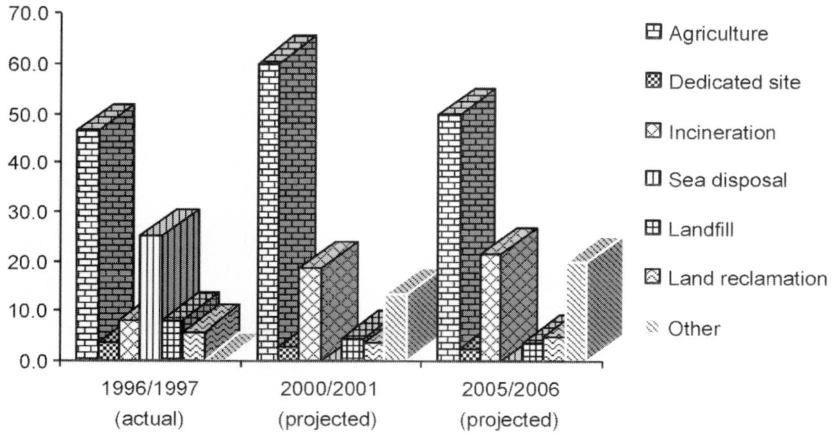

Fig. 17.2 Sludge disposal routes in the UK (Environment Agency, 1999).

a number of years and are designed for uniform application across the country over a wide range of geographic and climatic conditions, from Alaska to Hawaii. For this reason the regulations are extensive and complex. Only treated sludge is permitted to be applied to land of any type. Sludges are categorized as Class A or Class B depending on the level of treatment intended to reduce pathogens.

In order to attain Class A status sludges must have been treated by 'a process to further reduce pathogens' (PFRP). Such a process is considered capable of reducing the number of pathogens to those normally present in the soil. Provided that the treated sludge complies with end product microbiological standards (Table 17.4), it may be applied without restriction to a wide range of land types, including that intended for agricultural or horticultural

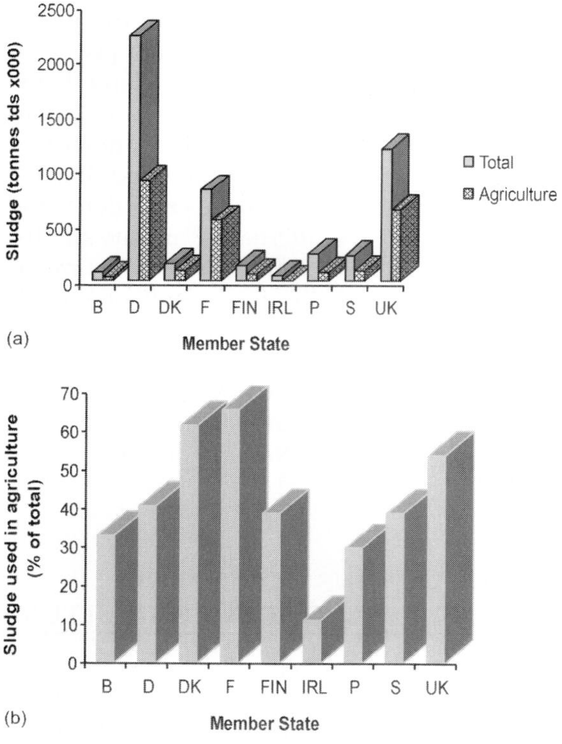

Fig. 17.3 (a) Sludge production within the European Union and amounts recycled to agricultural land; (b) Proportion of EU sludge recycled to agricultural land.

TABLE 17.3 Sludge disposal outlets in the UK

Outlet	*Quantity (%) (tds/y $^{-1}$ × 10^3)*	
	1990/91	*1996/97*
Agriculture	465 (42)	520 (47)
Dedicated site	25 (2)	39 (3)
Sea disposal	334 (30)	280 (25)
Incineration	77 (7)	91 (8)
Landfill	88 (8)	91 (8)
Land reclamation		64 (6)
Forestry		1 (<1)
Horticultural compost		13 (1)
Storage (on site)	50 (5)	15 (1)
Other	68 (6)*	1 (<1)
Total	1107 (100)	1115 (100)

* More general category of 'Beneficial' used which included activities classified separately in 1996/7 survey.
Source: WRc (1998).

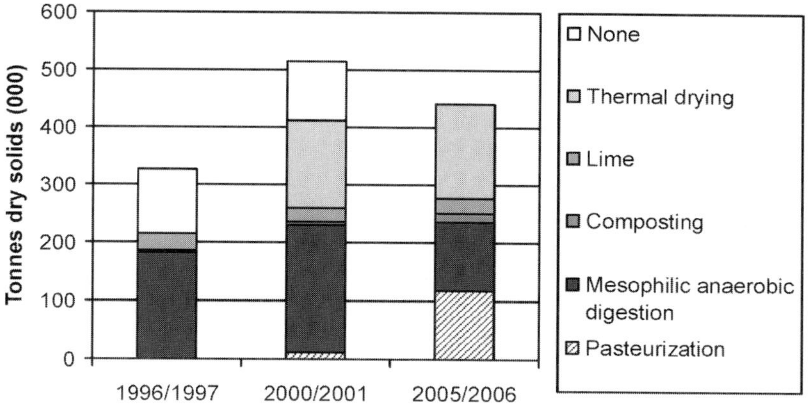

Fig. 17.4 Sludge treatment processes used in the UK.

use. Class B sludges are required to have been treated by 'a process to significantly reduce pathogens' (PSRP). With restrictions, Class B sludge may be applied to agricultural land. The sludge is required to meet end product microbiological standards (Table 17.4). The restrictions on application are:

- No grazing or harvesting of fodder crops with 30 days (of application)
- No harvesting of crops grown above ground within 14 months
- No harvesting of crops grown below ground for 20 months if the sludge remains on the soil for 4 months or longer; 38 months if the sludge remains on the soil for less than 4 months
- No harvesting of turf within 12 months
- No public access within 12 months (parks, playing fields etc.)

The implicit goal of the requirements for Class A biosolids is to reduce the number of pathogens in sewage sludge to below the level of detection (<3 MPN salmonella, <1 PFU enteric viruses, and <1 viable helminth ova – all per 4-g dry weight). The goal for the production of Class B biosolids is the reduction in the number of pathogens to levels that are unlikely to pose a public health risk (US EPA, 1999).

TABLE 17.4 End product microbiological standards for Class A and Class B sludges (US EPA, 1993)

Standard	*Class A*	*Class B*
Faecal coliforms/g ds	Less than 1000	Less than 2 000 000*
Salmonellae 4/g ds	Less than 3	
Enteroviruses pfu 4/g ds	Less than 1	
Parasite ova 4/g ds	Less than 1	

ds Dry solids.
* Geometric mean of seven samples.

4.2 European Union

The controls on the application of sewage sludge to agricultural land within member states derive from Council Directive 86/278/EEC published in 1986 for implementation within 3 years (CEC, 1986). The principal rationale of the Directive was to minimize the accumulation in the soil of heavy metals or other potential toxic elements (PTE) with the objective of protecting soil fertility and public health. However, the Directive included measures for controlling transmissible disease by introducing constraints on the use of sludge. Article 7 of the Directive requires Member States to prohibit the use of sludge or the supply of sludge for use on:

- grassland or forage crops if the grassland is to be grazed or the forage crops to be harvested before a certain period has elapsed. This period, which shall be set by the Member States, taking particular account of their geographical and climatic situation, shall under no circumstances be less than 3 weeks:

TABLE 17.5 Examples of effective sludge treatment processes as defined in the UK Code of Practice

Process	*Conditions*
Pasteurization	Minimum 30 min at 70°C; or Minimum 4 h at 55°C Followed in all cases by mesophilic anaerobic digestion
Mesophilic anaerobic digestion	Mean retention of at least 12 days at 35°C ± 3°C; or Mean retention of at least 20 days at 25°C ± 3°C. Followed in each case by secondary digestion with a mean retention time period of at least 14 days
Thermophilic aerobic digestion	Mean retention of at least 7 days. All sludge to be subjected to a minimum of 55°C for at least 4 h
Composting (windrows or aerated piles)	Compost must be retained at 40°C for at least 5 days including a period of 4 h at a minimum of 55°C. Followed by a period of maturation
Alkaline stabilization (with lime)	pH to be 12 or greater for a period of at least 2 h
Liquid storage	Storage for at least 3 months.
Dewatering and storage	Dewatering and storage for at least 3 months. Storage at least 14 days if sludge previously subjected to mesophilic anaerobic digestion

Source: DoE (1966).

- soil in which fruit and vegetables are growing, with the exception of fruit trees
- ground intended for the cultivation of fruit and vegetable crops which are normally in direct contact with the soil and normally eaten raw, for a period of 10 months preceding the harvest of crops and during the harvest itself
- sludge shall be treated before being used in agriculture[1]. Member States may nevertheless authorize, under conditions laid down by them, the use of untreated sludge if it is injected or worked into the soil.

In the UK, The Sludge (Use in Agriculture) Regulations 1989 directly implement the provisions of the Directive (Anon, 1989). This was accompanied by a Code of Practice (DoE, 1996) which provided practical guidance on how the requirements of the Directive could be met. It recognizes that pathogens may be present in untreated sludges and that their numbers can be reduced significantly by appropriate treatment. Examples of effective treatment processes are given in the Code (Table 17.5).

At the time that the Code was prepared the pathogens of concern were considered to be salmonellae, *Taenia saginata* (human beef tapeworm), potato cyst nematodes (*Globodera pallida* and *Globodera rostochiensis*) and viruses.

The guidance was based on the concept of multiple barriers to the prevention of transmission of pathogens when sludge was applied to agricultural land. The barriers are:

- Sludge treatment, which will reduce pathogen content
- Restrictions on which crops may be grown on land to which sludge has been applied
- Minimum intervals before grazing or harvesting.

The scientific and public health principles which underpin this concept are valid. They recognize that for certain crops the risk of disease transmission is unacceptable, i.e. salad items which have a short growing period and which are to be consumed raw. For other crops the combination of treatment and a suitable period of no harvesting will result in the numbers of pathogenic microorganisms being reduced below a minimum infective dose (MID). The concept of MID is important – it relates to the number of organisms which must be ingested to cause

[1] Treated sludge is defined in Article 2(b) of the Directive as *'sludge which has undergone biological, chemical or heat treatment, long-term storage or any other appropriate process so as significantly to reduce its fermentability and other health hazards resulting from its use.'*

TABLE 17.6 Minimum infective dose (MID) for a range of gastrointestinal pathogens

Organism	*Minimum infective dose*
Salmonella spp.	10^4–10^7
Salmonella typhi	10
Escherichia coli O157:H7	10–10^2
Vibrio cholerae	10^3
Giardia intestinalis	10–10^2
Cryptosporidium parvum	10–10^2
Entamoeba histolytica	10–10^2
Hepatitis A virus	1–10 PFU

PFU, plaque forming unit

disease. It varies widely depending not only on the particular pathogen but also on the susceptibility of the host (Table 17.6). For example in the young, elderly, pregnant or those whose immunity is reduced the minimum number of organisms required to initiate disease is much smaller.

Despite the current concerns surrounding the risks to food safety, it is important to recognize that there have been no instances documented in which disease transmission to man or animals has occurred where the provisions of the relevant UK Regulations and Codes of Practice were followed.

5 PATHOGENS

Pathogens are microorganisms that are capable of causing disease in the host species (man, animals or plants). All the major groups of microorganisms contain species which are pathogenic including viruses (e.g. hepatitis virus), bacteria (e.g. salmonellae), fungi (e.g. *Aspergillus*), protozoa (e.g. *Cryptosporidium*) and helminths (e.g. *Taenia*). Although there are several plant pathogens which may potentially be present in sewage sludge (e.g. brown rot, potato root eelworm and beet rhizomania), this review will deal with pathogens affecting man and animals. Many of these are described as zoonotic, i.e. directly transmissible to man from animals. Examples of zoonotic infections include salmonellosis and cryptosporidiosis. This is a particularly important factor when considering the risks to human health arising from the use of sludge in agriculture.

The type and number of pathogens that are likely to be present in untreated sewage will depend on the inputs to the sewerage system. The spectrum of human pathogens will mirror the incidence of infection in the community. People suffering from diseases of the gastrointestinal tract will excrete large numbers of the pathogen in their faeces. Industrial sources of pathogens include meat processing plants, abattoirs and livestock facilities. The World Health Organization in its review of health risks arising from sewage sludge applied to land described a wide range of pathogens that could be present in sludge (WHO, 1981). This was subsequently updated and expanded by Strauch (1991) and the United States EPA who collated the data shown in Table 17.7 (USEPA, 1989).

The list is extremely comprehensive and, in reality, the risk from many of these microorganisms is very small. The organisms shown in bold are those identified by the US EPA as posing a significant risk to human health and which were taken into account in the development of the current Part 503 Regulations (US EPA, 1992). It is interesting to note that at that time they did not consider *Escherichia* as posing a significant risk to health. It is now known that certain shiga toxin-producing strains, such as *E. coli* O157[2], are capable of being transmitted by contaminated foodstuffs (Armstrong *et al.*, 1996; Tauxe, 1997; Mead and Griffin, 1998; Parry and Palmer, 2000).

In practice, the list of microorganisms that we need to be concerned with is relatively small. The pathogens of concern will vary from region to region depending on the nature and prevalence of endemic infectious intestinal disease within the indigenous population. For example, data for England and Wales collated by the PHLS Communicable Disease Surveillance Centre reveal that over half of all notified infections are due to *Campylobacter* (Fig. 17.5).

In contrast, intestinal parasites, particularly *Ascaris*, are a major disease burden in the developing world. These parasites form cysts or ova which are especially robust and resistant to environmental conditions, attributes which contribute to high levels of re-infection.

[2] Previously referred to as verotoxigenic *E. coli* or VTEC.

TABLE 17.7 Pathogenic microorganisms which may be present in sewage sludge

Bacteria	*Viruses*	*Protozoa*	*Yeasts*
Salmonella	**Hepatitis A**	**Cryptosporidium**	Candida
Shigella	**Enteroviruses**	**Entamoeba**	Cryptococcus
Yersinia	**Poliovirus**	**Giardia**	Trichosporon
Escherichia	**Coxsackie viruses**	**Balantidium**	
Pseudomonas	**Echoviruses**	Toxoplasma	**Fungi**
Clostridia	**Rotavirus**	Sarcocystis	Aspergillus
Bacillus	Adenovirus		Phialophora
Listeria	Reovirus	**Cestodes**	Geotrichum
Vibrio	Astrovirus	**Taenia**	Trichophyton
Mycobacterium	Calicivirus	Diphyllobothrium	Epidermophyton
Leptospira	Coronavirus	Echinococcus	
Campylobacter	Norwalk and Norwalk-like viruses		
Staphylococcus		**Nematodes**	
Streptococcus		**Ascaris**	
		Toxocara	
		Trichuris	
		Ancylostoma	
		Necator	
		hymenolepsis	

Source: US EPA (1999).

The literature contains numerous reports of surveys and investigations into the occurrence of pathogens and indicator bacteria in wastewater and sludge treatment plants. Because of methodological and geographical differences and, in some instances, lack of details regarding treatment processes, many of these data are not directly comparable. However, the data shown in Table 17.8 provide an indication of the likely numbers of indicators and pathogens in domestic wastewater and sludges.

6 EFFECTS OF SLUDGE TREATMENT ON PATHOGENS

It must be recognized at the outset that treatment is designed to stabilize sewage sludge and reduce its putrescence. Pathogens may be inactivated as a consequence of the particular treatment applied. It has not been normal practice to optimize sludge treatment processes for pathogen reduction. Indeed to do so may reduce the effectiveness of the stabilization process.

A review of the literature by Ward and colleagues (1984) showed that the range of pathogen inactivation reported was large, depending on the extent of the treatment process and variation between operating conditions, even for the same generic treatment process (Table 17.9).

There are limited data on the effect of some of these processes on certain pathogens. In practice, research in this area has been restricted to

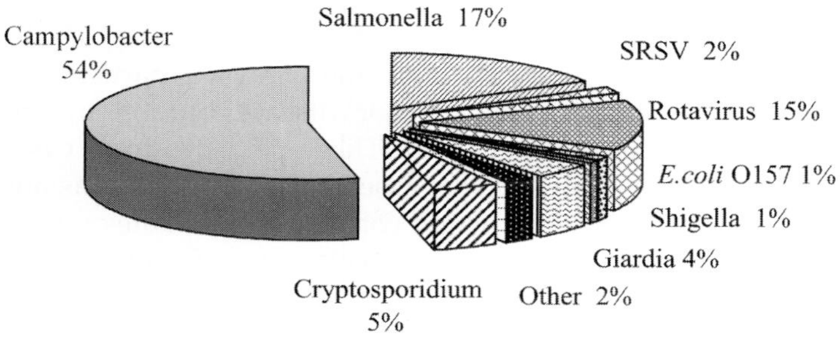

Fig. 17.5 Infectious intestinal disease in England and Wales by aetiological agent.

TABLE 17.8 Typical numbers of microorganisms found in various stages of wastewater and sludge treatment

Microorganism	*Number per 100* ml				*Number per gram*	
	Crude sewage	*Primary treatment*	*Secondary treatment*	*Tertiary treatment*[a]	*Raw*	*Treated*[b]
Faecal coliforms	10^8	10^7	10^6	<2	10^7	10^6
Salmonellae	10^3	10^2	10	<2	10^3	10^2
Shigella	10^3	10^2	1	<2	10^2	3
Listeria	10^4					10^3
Campylobacter	10^5				10^4	10
Enteric virus	5×10^4	1×10^4	10^3	0.002	10^3	10^2
Helminth ova	8×10^2	10	0.08	<0.08	10	10
Giardia cysts	10^4	5×10^3	2.5×10^3	3	10^2	10

[a] Including coagulation, sedimentation, filtration, disinfection.
[b] Mesophilic anaerobic digestion.
Source: Jones *et al.* (1990); Metcalf and Eddy (1991); National Research Council (1996); De Luca *et al.* (1998); Watkins and Sleath (1981).

those pathogens with a high prevalence and likely to cause disease (e.g. salmonellae) and those that are more likely to exhibit resistance to the sludge treatment process (e.g. *Ascaris*). In comparison little is known about the pathogens which have only recently emerged as public health issues, most notably *E. coli* O157. The recent emergence of this pathogen means that there is little information available on the fate of *E. coli* O157 (and other shiga toxin-producing *E. coli* (STEC)) during the treatment of wastewater and sewage sludge. Despite the highly infectious nature of STEC and the presence of multiple virulence factors there is evidence that it is no more resistant to inactivation during sludge treatment than the indigenous populations of *E. coli* in sewage and sludge (Horan, personal communication).

TABLE 17.9 Summary of pathogen reduction during sludge treatment

Treatment	*Log reduction*		
	Bacteria	*Viruses*	*Parasites*
Mesophilic anaerobic digestion	0.5–4	0.5–2	0
Aerobic digestion	0.5–4	0.5–2	0
Composting	2–>4	2–>4	2–>4
Air drying	0.5–4	0.5–>4	0.5–>4
Lime stabilization	2–>4	>4	0

Source: Ward *et al.* (1984)

The inactivation of indigenous *E. coli* in full-scale sludge treatment processes was investigated during a 3-month study, which looked at nine different sludge treatment processes at 35 sites in the UK (UKWIR, 1999). All of the processes surveyed reduced the numbers of *E. coli*. So-called 'enhanced' treatment processes, such as composting, lime addition and thermal drying, were capable of reducing numbers of *E. coli* to the detection limit of the analytical method. For all of these methods, over 90% of results showed bacterial reductions of 6 log or greater. Lagooning of sludge was capable of significantly reducing numbers of *E. coli* and, depending on the method of operation, reductions in the order of 5 log were observed. Mesophilic anaerobic digestion (MAD), the process carried out at the majority of sites surveyed, reduced numbers of *E. coli* by, on average, between 1.4 and 2.3 log depending on the solids content of the product. For sites producing a liquid product (2–4% ds) 78% of all reductions for were in the range 1 to 2 log. Where digested sludge was subsequently dewatered to produce a cake, 89% of results showed reductions in the range 2 to 4 log. The one vermiculture site in the survey showed results intermediate between MAD and the 'enhanced' treatment processes (Table 17.10).

Cryptosporidium parvum is another pathogen of increasing importance. Wastewater discharges and run off from agricultural land are

TABLE 17.10 Effect of sludge treatment processes on numbers of *E. coli*

Treatment	*n*	*Log reduction in E. coli*		*Log reduction in E. coli treated sludge (100/g dry wt)*	
		Mean	95%ile	Mean	95%ile
Lagooning	36	2.65	6.00	5.93	8.32
MAD, liquid	208	1.39	2.36	7.41	8.27
MAD, cake	93	2.29	3.64	6.65	7.46
Vermiculture	14	5.12	6.54	4.50	5.07
Composting	31	6.71	9.10	2.43	4.70
Lime addition	32	7.10	9.05	1.45	3.00
Thermal drying	70	7.14	8.90	1.67	3.56

Source: UKWIR (1999).

an important source of *Cryptosporidium* oocysts found in watersheds. The transmission of cryptosporidiosis is zoonotic and the possibility exists of foodborne infection arising from the use of sewage sludge in agriculture. Stadterman *et al.* (1995) found that a laboratory activated sludge plant removed 98.6% of seeded *Cryptosporidium parvum* oocysts. In a comparison of different treatment regimes, activated sludge and anaerobic digestion were found to be the most effective means of removing oocysts, the latter destroying 99.9% in 24 hours.

Studies of anaerobic mesophilic digestion under laboratory conditions showed that oocysts added to the contents of a digester operating at 35°C rapidly lost viability (as measured by excystation), decreasing to 17% after 3 days from an initial 81% viability (Whitmore and Robertson, 1995). Losses of viability in distilled water and anaerobic sludge at 35°C were similar, amounting to 90% after 18 days, indicating that the principal effect on viability was temperature. Oocysts exposed to mesophilic anaerobic digestion for 3 days and then stored for a further 14 days were completely inactivated. Aerobic digestion or pasteurization, both at 55°C, caused 92% loss of viability in 5 minutes. Thermophilic anaerobic digestion at 50°C resulted in complete inactivation within the first 24 hours (Whitmore and Robertson, 1995).

7 ROUTES OF TRANSMISSION

Recently expressed concerns over the use of sewage sludge in agriculture have focused on the risks to human health arising from the production of foods on land to which sludge has been applied (Anon, 1998a). A number of exposure pathways whereby foodstuffs become contaminated with pathogens can be envisaged. The exposure pathways relevant to sewage sludge in agricultural production are:

1. Sludge → soil → plant → human
2. Sludge → soil → animal → human
3. Sludge → soil → plant → animal → human
4. Sludge → soil → drinking water → human
5. Sludge → soil → irrigation water → plant → human.

These may be developed into a conceptual model which describes the framework for a microbiological risk assessment (Fig. 17.6). Other routes of exposure, which do not involve the food chain, can be identified. These include direct contact with sludge-treated soil or indirect contact via companion animals. The risk of exposure by this route (direct contact) is probably greatest among children.

It can be seen that there are several pathways that are unrelated to the use of sewage sludge, probably the most important of which is the application to land of organic wastes such as animal slurries and manures. The use of such materials in agriculture is less regulated than for sewage sludge and accounts for the majority of organic waste spread to land (Table 17.11). Despite this, the focus of attention has been on the human health risks via the food chain from the application of sewage sludge to agricultural land (RCEP, 1996; Anon, 1998a).

As previously mentioned, at the time that existing controls on the agricultural use of sewage sludge were being formulated, the pathogens of concern were salmonellae and *T. saginata*. In the intervening period, pathogens such as shiga toxin-producing *Escherichia coli* and *Cryptosporidium* have been recognized as important causes of intestinal infectious disease in humans. Any assessment of health

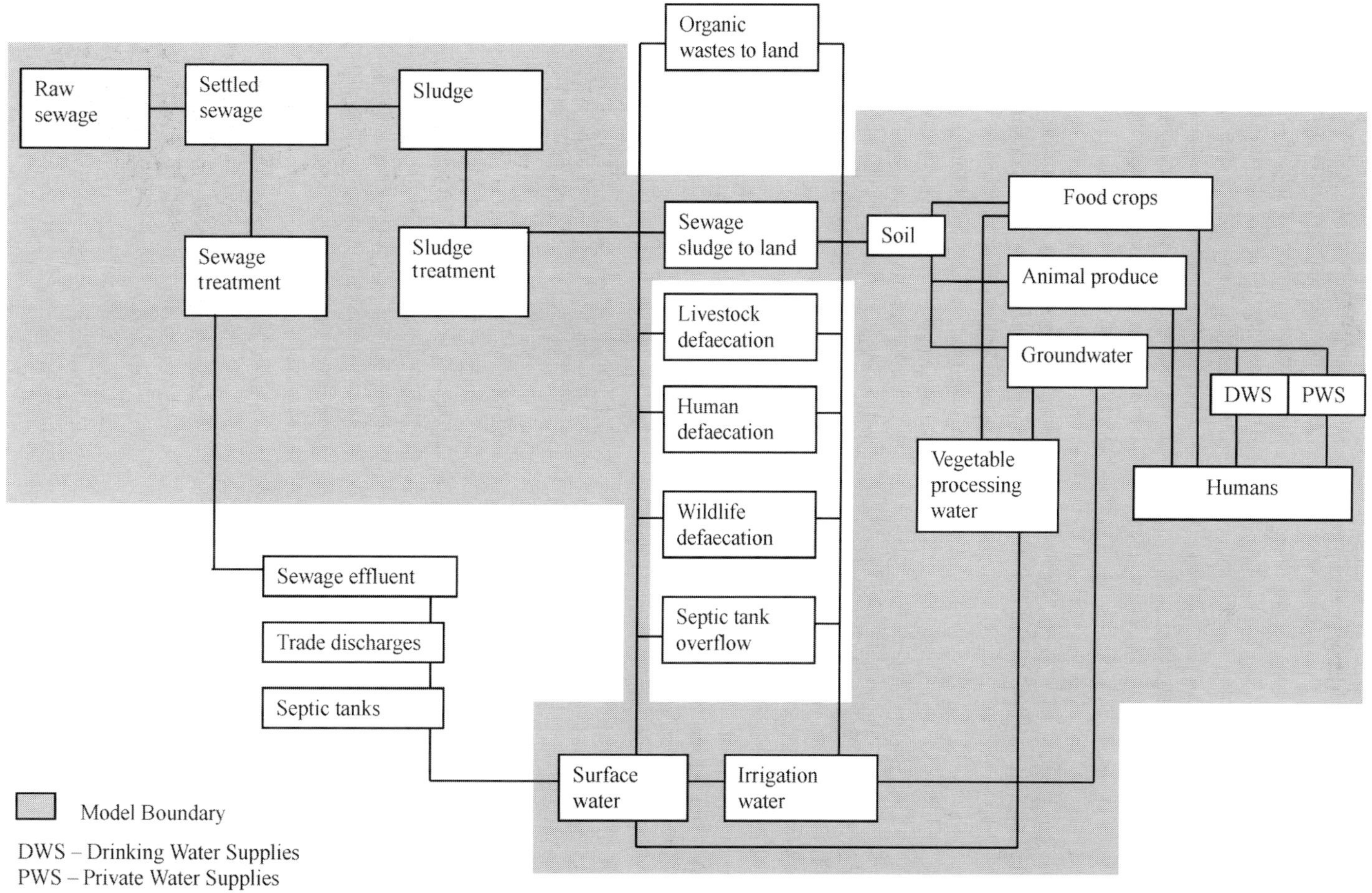

Fig. 17.6 Conceptual model for a microbiological risk assessment into the application of sludge to agricultural land (Pollard, personal communication).

TABLE 17.11 Estimates of the quantities of organic materials applied to land in the UK

Origin	*Quantity (tonne × 10^3 dw)*
Farm animal	21 000
Sewage sludge	430
Paper industry	520
Food industry	600
Sugar industry	200
Others	150

dw: Dry weight.
Source: WRc (1998).

risks associated with the beneficial use of sewage sludge in agriculture must consider these two pathogens.

7.1 Shiga toxin-producing *E. coli*

Shiga toxin-producing *E. coli* (STEC) are now recognized as an important group of enteric pathogens. Although there are many serotypes capable of producing shiga toxins, *E. coli* O157:H7 is the most widely known. This organism was first described in 1982 following an outbreak of haemorrhagic colitis in the USA (Riley *et al.*, 1983). Outbreaks have been associated with the consumption of foodstuffs, drinking water, and swimming in natural surface waters. Zoonotic infections have also been reported. The majority of cases are believed to be foodborne, with an estimated 85% of cases (n = 110 220) in the USA suspected to be food-related (Mead *et al.*, 1999).

There is very little information concerning the presence of STEC in sewage and sewage sludge. It is reasonable to assume that domestic sewage will contain STEC, if not continuously then intermittently, reflecting the incidence of infection in the community. The likelihood of STEC being present will be greater for those wastewater treatment works receiving wastes from animal-handling facilities, such as markets, abattoirs and meat processing plants. Surveillance of *E. coli* O157 in animals presented for slaughter carried out in northern England revealed that 15.7% of cattle, 2.2% of sheep and 0.4% of pigs were positive for the organism (Chapman, 2000).

Survival of *E. coli* O157 in the environment has been investigated by a number of workers. Maule (1997, 2000) showed that survival of the organism was found to be greatest in soil cores containing rooted grass. Under these conditions viable numbers were shown to decline from approximately 10^8/g soil to between 10^6 and 10^7/g soil after 130 days. When the organism was inoculated into cattle faeces it remained detectable at high levels for more than 50 days. In contrast, the organism survived much less readily in cattle slurry and river water where it fell in numbers from more than 10^6/ml to undetectable levels in 10 and 27 days, respectively. Survival of *E. coli* O157:H7 in bulk manures may be prolonged, with the organism being detected for more than one year in static piles of ovine manure (Kudva *et al.*, 1998). Survival was reduced if the manure piles were aerated.

The fate of *E. coli* O157 present in animal slurry applied to pasture was investigated by Fenlon and colleagues (2000). Following application, numbers of both *E. coli* and *E. coli* O157 declined steadily with greater than 2 log reduction within 29 days. Relatively few cells (2% of total) were transported away from the soil surface and into the deeper layers of the soil. Run-off following heavy rainfall resulted in a loss from the soil of 7% of the *E. coli* applied in the slurry. A recent ecological study on predation of *E. coli* O157 by *Acanthamoeba polyphaga* has shown that the bacterial cells are capable of surviving and even replicating within this common environmental protozoan (Barker *et al.*, 1999). This may be important in the dissemination and survival of STEC within the environment.

7.2 Cryptosporidium

Human infection with *Cryptosporidium* was first reported in 1976 (Fayer *et al.*, 2000). It is now apparent that *Cryptosporidium* is a significant cause of infectious intestinal disease (IID), accounting for about 5% of IID in which the causative organism is identified. In England and Wales, during 1999, there were nearly 5000 laboratory confirmed cases of cryptosporidiosis (n = 4759) (Anon, 2000a); in the USA the estimated number of cases is 300 000 of which 10% are believed to be foodborne (Mead *et al.*, 1999).

Oocysts of *C. parvum* are environmentally hardy and, under certain conditions, may remain viable for many months. Survival studies in microcosms containing untreated river water showed that oocysts are extremely persistent, the times for 10 log reductions in viability being 160 days at 15°C and 100 days at 5°C (Medema *et al.*, 1997). Investigations into the survival in soils treated with sewage sludge showed that viability declined by 20–40% at 20°C over 44 days. Temperature was the principal factor affecting oocyst survival (Whitmore and Robertson, 1995).

Little is known about the movement of oocysts through the soil. The most relevant data were reported by Mawdsley and colleagues (1996a) who studied the transport of *Cryptosporidium* oocysts through soil following the application of slurry to a poorly draining silt clay loam soil. Bovine slurry seeded with 5×10^9 *C. parvum* oocysts was applied to the surface of soil blocks (80 cm × 56 cm × 20 cm) removed from a perennial ryegrass ley at an application rate equivalent to 50 m^3/ha. The blocks were irrigated 24 h following slurry application and periodically thereafter. Samples of run-off (at 4 cm depth) and leachate (at 20 cm depth) were collected and the number of oocysts enumerated. After 70 days the blocks were destructively sampled and examined for the presence of oocysts. Experiments were carried out in triplicate. Numbers of oocysts leaching from the blocks declined from 8.4×10^6 on day 1 to 2.3×10^4 at day 70. Oocysts levels were consistently lower in run-off compared with the leachate from the base of the soil blocks. Numbers fell below the limit of detection after 21 days and 28 days in two blocks, but were detectable in the third block for the duration of the experiment (70 days). These results suggest that oocysts tend not to become associated with soil particles, being either transported away in run-off or moving vertically downwards through the soil column. The majority of oocysts were retained in the top 2 cm of soil (Mawdsley *et al.*, 1996b).

8 ASSESSMENT OF HUMAN HEALTH RISKS

The process of microbiological risk assessment is now considered to comprise three phases: problem formulation, analysis, and risk characterization (ILSI, 2000). The analysis phase consists of two elements: characterization of exposure and characterization of human health effects (Fig. 17.7).

Characterization of exposure requires an evaluation of the interaction between the pathogen, the environment and the human population. Factors that need to be considered include the virulence of the pathogen, survival in the environment, route of infection, numbers of pathogen present, effectiveness of control/treatment processes, infectious dose, severity of illness and size of exposed population.

Unlike the field of chemical toxicology, microbiological risk assessment is in its infancy. There exist major gaps in our knowledge about the organisms of concern. This is particularly the case for the emerging pathogens such as STEC and *Cryptosporidium*. There have been attempts to assess the risks posed by *Cryptosporidium* in drinking water supplies (Gale, 1996, 1999; Haas, 2000). However, the foodborne route has not been modelled, probably because of the received view that cryptosporidiosis is primarily waterborne, despite evidence demonstrating the potential for foodborne transmission. *Cryptosporidium* oocysts have been found on the surface of fresh, raw vegetables obtained from retail markets (Ortega *et al.*, 1991; Monge *et al.*, 1996). In the UK, an outbreak of cryptosporidiosis affecting 50 school children was linked to the consumption of improperly pasteurized milk (Gelletli *et al.*, 1997). In the USA, outbreaks have been associated with the drinking fresh-pressed apple juice (non-alcoholic cider) (Millard *et al.*, 1994). Outbreaks of cryptosporidiosis associated with infected food-handlers demonstrate clearly the potential for significant foodborne transmission of *Cryptosporidium* (Besser-Wiek *et al.*, 1996; Quiroz *et al.*, 2000).

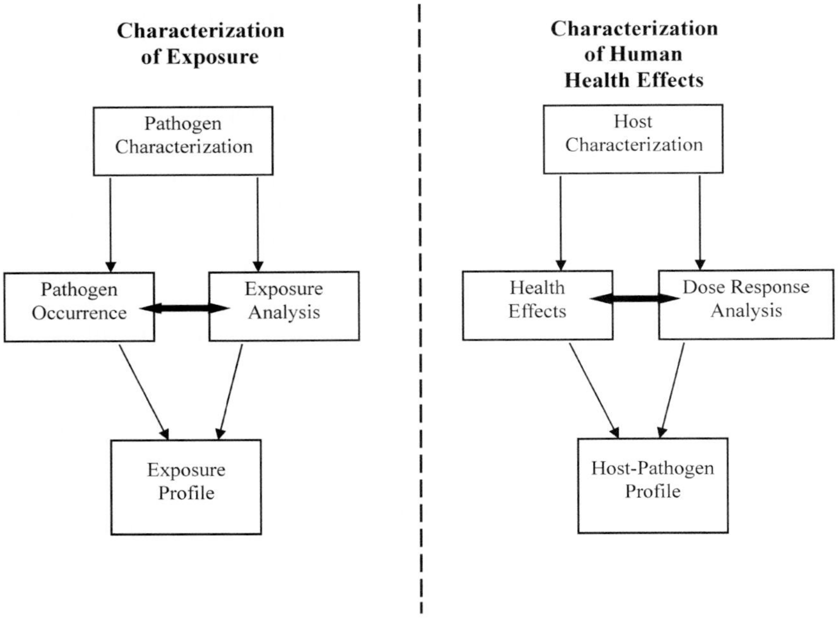

Fig. 17.7 Analysis phase of a microbiological risk assessment for foodborne pathogens (ILSI, 2000).

On the other hand, it is clear that STEC infection is primarily foodborne (Mead *et al.*, 1999). Data are required on the levels of STEC in treated sludge, their survival following land application and the potential for transfer to food crops before a microbiological risk assessment can be performed. Research is currently being undertaken to address these issues.

9 PRECAUTIONARY PRINCIPLE

Against a background of concern over methods of food production in the UK, the water industry, under the auspices of Water UK, and representatives of the food suppliers agreed a set of guidelines matching the level of sewage treatment with the crop under cultivation (Anon, 1998b).

The Safe Sludge Matrix (Anon, 2000b, Tables 17.12 and 17.13) forms the basis of the agreement and consists of a table of crop types, together with clear guidance on the minimum acceptable level of treatment for any sewage sludge (biosolids) based product, which may be applied to that crop or rotation. The agreement was driven by the desire to ensure the highest possible standards of food safety and to provide a framework that gives the retailers and food industry confidence that sludge reuse on agricultural land is safe. The matrix enables farmers and

TABLE 17.12 The safe sludge matrix

Crop group	*Untreated sludges*	*Treated sludges*	*Enhanced treated sludges*
Fruit	✗	✗	✓‡
Salads	✗	✗ (30 month harvest interval applies)	✓‡
Vegetables	✗	✗ (12 month harvest interval applies)	✓‡
Horticulture	✗	✗	✓‡
Combinable and animal feed crops	✗	✓	✓
Grass – grazing	✗	✗† (Deep injected or ploughed down only)	✓†
Grass – silage	✗	✓†	✓†
Maize – silage	✗	✓†	✓†

† 3 week no grazing and harvest interval applies.
‡ 10 months harvest interval applies.

TABLE 17.13 Cropping categories within the Safe Sludge Matrix

Fruit	*Salad (e.g. ready to eat crops)*	*Vegetables*	*Horticulture*	*Combinable and animal feed crops*	*Grassland and maize*	
					Silage	*Grazing*
Top fruit (apples, pears, etc.)	Lettuce	Potatoes	Soil based glasshouse and polythene tunnel production (including tomatoes, cucumbers, peppers, etc.)	Wheat	Cut grass	Grass
	Radish	Leeks	Mushrooms	Barley	Cut maize	Forage Swedes/turnips
Stone fruit (plums, cherries, etc.)	Onions	Sweetcorn	Nursery stock and bulbs for export	Oats	Herbage	Fodder mangolds/beet/kale
	Beans (including runner, broad and dwarf French)	Brussels sprouts	Basic nursery stock	Rye	Seeds	Forage rye and triticale
Soft fruit (currants and berries)	Vining peas	Parsnips		Triticale		Turf production
	Mange tout	Swedes/turnips	Seed potatoes for export	Field peas		
Vines	Cabbage	Marrows	Basic seed potatoes	Field beans		
Hops	Cauliflower	Pumpkins		Linseed/flax		
	Calabrese/broccoli	Squashes	Basic seed production	Oilseed rape		
Nuts	Courgettes	Rhubarb		Hemp		
	Celery	Artichokes		Sunflower		
	Red beet			Borage		
	Carrots			Sugar beet		
	Herbs					
	Asparagus					
	Garlic					
	Shallot					
	Spinach					
	Chicory					
	Celeriac					

growers to continue to utilize the beneficial properties in sewage sludge as a valuable and cost effective source of nutrients and organic matter.

The main impact was the cessation of raw or untreated sewage sludge being used on agricultural land. As from the end of 1999, all untreated sludges have been banned from application to agricultural land used to grow food crops. Treated sludge[3] can only be applied to grazed grassland where it is deep injected into the soil. The regulations require that there will be no grazing or harvesting within 3 weeks of application. Where grassland is reseeded, sludge must be ploughed down or deep injected into the soil.

More stringent requirements apply where sludge is applied to land growing vegetable crops and, in particular, those crops that may be eaten raw (e.g. salad crops). Treated sludge can be applied to agricultural land which is used to grow vegetables provided that at least 12 months has elapsed between application of the sludge and harvest of the vegetable crop. Where the crop is a salad, which might be eaten raw, the harvest interval must be at least 30 months. Where enhanced treated sludges[4] are used, a 10-month harvest interval applies.

10 CONCLUSIONS

From an environmental perspective there is a persuasive argument that, of the disposal options available, recycling nutrients by means of applying sewage sludge to land, with appropriate safeguards, is the Best Practicable Environmental Option (BPEO) (CEC, 1986; RCEP, 1996). The risks to human, animal and plant health were taken into account when developing the current regulations and codes of practice. The fundamental principle (for reducing disease transmission risk) implicit to these controls on the use of sludge in agriculture is the concept of imposing multiple barriers to the recycling of pathogens from sludge to their hosts. The effectiveness of this approach is borne out in practice as noted by the Royal Commission on Environmental Pollution (RCEP, 1996) which concluded that, 'There are no instances in the UK in which a link has been established between the controlled application of sewage sludge and occurrence of disease in the general population through water or food contamination'. However, it is the case that the current controls predate the emergence of pathogens such as *Cryptosporidium* and STEC and may not sufficiently reduce the risk associated with these microorganisms.

More data are required on the numbers and fate of these emergent pathogens before a meaningful microbiological risk assessment can be carried out. Research is being undertaken in the UK, USA and elsewhere to generate these data. The likelihood is that the controls on the use of sewage sludge in agriculture will be strengthened as results of this research and in the light of public perception and expectations about food safety and environmental risks.

[3] There is a range of different treatment processes used to reduce the fermentability and possible health hazards associated with sewage sludge. These rely on biological, chemical or heat treatment. The most common form of treatment is anaerobic digestion.

[4] Enhanced treatment, originally referred to as 'Advanced Treatment', is a term used to describe treatment processes which are capable of virtually eliminating any pathogens which may be present in the original sludge.

REFERENCES

Anon (1989). The Sludge (Use in Agriculture) Regulations 1989. SI No. 1263 as amended by SI No. 880 (1990).

Anon (1998a). House of Commons Environment, Transport and Regional Affairs Committee. Sewage Treatment and Disposal, Second Report. Stationery Office, London.

Anon (1998b). The 'ADAS Matrix': the food and water industry in agreement. *Wastes Management*, December 1998, 28–29.

Anon (2000a). Data on infectious disease in England and Wales. Available on line at http://www.phls.co.uk/

Anon (2000b). The Safe Sludge Matrix. Available on line at http://www.adas.co.uk/matrix/SSM.pdf

Armstrong, G.L., Hollingsworth, J. and Morris, J.G. (1996). Emerging foodborne pathogens: *Escherichia coli* O157: H7 as a model of entry of a new pathogen into the food supply of the developed world. *Epidemiologic Reviews* **18**(1), 29–51.

Barker, J., Humphrey, T.J. and Brown, M.W. (1999). Survival of *Escherichia coli* O157 in a soil protozoan:

implications for disease. *FEMS Microbiology Letters* **173**(2), 291–295.

Bastian, R.K. (1997). The biosolids (sludge) treatment, beneficial use, and disposal situation in the USA. *European Water Pollution Control* **7**(2), 62–79.

Besser-Wiek, J.W., Forfang, J., Hedberg, C.W. *et al.* (1996). Foodborne outbreak of diarrheal illness associated with *Cryptosporidium parvum* – Minnesota, 1995. *Morbidity and Mortality Weekly Report* **45**(36), 783–784.

CEC (1991). Council Directive 91/271/EEC of 21 May 1991 concerning urban waste water treatment. *Official Journal of the European Communities* **L135**/40 (30 May).

CEC (1986). Council Directive 86/278/EEC of 12 June 1986 on the protection of the environment, and in particular of the soil, when sewage sludge is used in agriculture. *Official Journal of the European Communities* **L181**/6 (4 July).

CEC (1999). Report from the Commission to the Council and the European Parliament on the implementation of community waste legislation for the period 1995–1997. Commission of the European Communities, Brussels, COM 752 (final).

Chapman, P.A. (2000). Sources of *Escherichia coli* O157 and experiences over the past 15 years in Sheffield, UK. *Journal of Applied Microbiology,* **88** Supplement, 51S–60S.

De Luca, G., Zanetti, F., Fateeh-Moghadm, P. and Stampi, S. (1998). *Zentrablatt für Hygiene und Umweltmedizin* **201**(3), 269–277.

DoE (1996). *Code of Practice for Agricultural Use of Sewage Sludge,* revised edition. HMSO, London.

Environment Agency (1999). UK Sewage Sludge Survey. *R&D Technical Report P165.* Environment Agency, Bristol.

Fayer, R., Morgan, U. and Upton, S.J. (2000). Epidemiology of *Cryptosporidium*: transmission, detection and identification. *International Journal of Parasitology* **30**(12–13), 1305–1322.

Fenlon, D.R., Ogden, I.D., Vinten, A. and Svoboda, I. (2000). The fate of *Escherichia coli* and *E. coli* O157 in cattle slurry after application to land. *Journal of Applied Microbiology* **88** Supplement, 149S–156S.

Gale, P. (1996). Developments in microbiological risk assessment models for drinking water – a short review. *Journal of Applied Bacteriology* **81**, 403–410.

Gale, P. (1999). Assessing the risk of cryptosporidiosis. *Journal of American Water Works Association* **91**(3), 4.

Gelletli, R., Stuart, J., Soltano, N. *et al.* (1997). Cryptosporidiosis associated with school milk. *Lancet* **350**, 1005–1006.

Haas, C.N. (2000). Epidemiology, microbiology, and risk assessment of waterborne pathogens including *Cryptosporidium. Journal of Food Protection* **63**(6), 827–831.

ILSI (2000). *Revised Framework for Microbial Risk Assessment.* International Life Sciences Institute, Washington DC.

Jones, K., Betaieb, M. and Telford, D.R. (1990). Correlation between environmental monitoring of thermophilic campylobacters in sewage effluent and the incidence of Campylobacter infection in the community. *Journal of Applied Bacteriology* **69**(2), 235–240.

Kudva, I.T., Blanch, K. and Hovde, C.J. (1998). Analysis of *Escherichia coli* O157:H7 survival in ovine or bovine manure and manure slurry. *Appl. Environ. Microbiol.* **64**, 3166–3174.

MAFF (1998a). *Code of good agricultural practice for the protection of water.* HMSO, London.

MAFF (1998b). *Code of good agricultural practice for the protection of soil.* MAFF Publications, London.

Maule, A. (1997). Survival of the verotoxigenic strain *E. coli* O157:H7 in laboratory-scale microcosms. In: *Coliforms and E. coli: Problem or solution?* D. Kay and C.R. Fricker (eds) Royal Society of Chemistry, London.

Maule, A. (2000). Survival of verocytotoxigenic *Escherichia coli* O157 in soil, water and on surfaces. *Journal of Applied Microbiology* **88** Supplement, 71S–78S.

Mawdsley, J.L., Brooks, A.E., Merry, R.J. and Pain, B.F. (1996a). Use of a novel soil tilting table apparatus to demonstrate the horizontal and vertical movement of the protozoan pathogen *Cryptosporidium parvum* in soil. *Biology and Fertility of Soils* **23**(2), 215–220.

Mawdsley, J.L., Brooks, A.E. and Merry, R.J. (1996b). Movement of the protozoan pathogen *Cryptosporidium parvum* through three contrasting soil types. *Biology and Fertility of Soils* **21**, 30–36.

Mead, P.S. and Griffin, P.M. (1998). *Escherichia coli* O157:H7. *Lancet* **352**(9135), 1207–1212.

Mead, P.S., Slutsker, L., Dietz, V. *et al.* (1999). Food-related illness and death in the United States. *Emerging Infectious Diseases* **5**(5), 607–625.

Medema, G.J., Bahar, M. and Schets, F.M. (1997). Survival of *Cryptosporidium parvum, Escherichia coli,* faecal enterococci and *Clostridium perfringens* in river water: influence of temperature and autochthonous microorganisms. *Water Science and Technology* **35**(11–12), 249–252.

Metcalf and Eddy Inc. (1991). *Wastewater Engineering: Treatment, Disposal and Reuse.* McGraw-Hill, New York.

Millard, P.S., Gensheimer, K.F., Addiss, D.G. *et al.* (1994). An outbreak of cryptosporidiosis from fresh-pressed apple cider. *Journal of the American Medical Association* **272**(20), 1592–1596.

Monge, R. and Chinchilla, M. (1996). Presence of *Cryptosporidium* oocysts in fresh vegetables. *Journal of Food Protection* **59**, 1866–1870.

National Research Council (1996). *Use of Reclaimed Water and Sludge in Food Crop Production.* National Academy Press, Washington DC.

Ortega, Y.R., Sheehy, R.R., Cama, V.A. *et al.* (1991). Restriction fragment length polymorphism analysis of *Cryptosporidium parvum* isolates of bovine and human origin. *Journal of Protozoology* **38**(6), 40S–41S.

Parry, S.M. and Palmer, S.R. (2000). The public health significance of VTEC O157. *Journal of Applied Microbiology,* **88** Supplement, 1S–9S.

Quiroz, E.S., Bern, C., MacArthur, J.R. *et al.* (2000). An outbreak of cryptosporidiosis linked to a foodhandler. *Journal of Infectious Diseases* **181**, 695–700.

RCEP (1996). Sustainable use of Soil, Royal Commission on Environmental Pollution Nineteenth Report, Cmnd 3165. HMSO, London.

Riley, L.W., Remis, R.S., Helgerson, S.D. *et al.* (1983). Hemorrhagic colitis associated with a rare *Escherichia coli* serotype. *New England Journal of Medicine* **308**(12), 681–685.

Stadterman, K.L., Sninsky, A.M., Sykora, J.L. and Jakubowski, W. (1995). Removal and inactivation of *Cryptosporidium* oocysts by activated sludge treatment and anaerobic digestion. *Water Science and Technology* **31**(5–6), 97–104.

Strauch, D. (1991). Microbiological treatment of municipal sewage waste and refuse as a means of disinfection prior to recycling in agriculture. *Studies in Environmental Science* **42**, 121–136.

Tauxe, R.V. (1997). Emerging foodborne diseases: an evolving public health challenge. *Emerging Infectious Disease* **3**(4), 425–434.

UKWIR (1999). *A Survey of E. coli in UK Sludges*, Report No. 99/SL/06/3. UK Water Industry Research, London.

US EPA (1989). *Technical Support Document for Pathogen Reduction in Sewage Sludge*. NTIS No. PB89-136618. National Technical Information Service, Springfield VA.

US EPA (1992). *Technical Support Document for Reduction of Pathogens and Vector Attraction in Sewage Sludge. Report No. 822/R-93-004*. US EPA, Cincinatti.

US EPA (1993). Part 503 – Standards for the use and disposal of sewage sludge. *Federal Register* **58**(32), 9387–9401.

US EPA (1999). *Control of Pathogens and Vector Attraction in Sewage Sludge, Report No. EPA/625/R-92/013*. US EPA, Cincinatti.

Ward, R., McFeters, G. and Yeager, J. (1984). *Pathogens in Sewage Sludge: Occurrence, Inactivation and Potential Regrowth*. Sandia Report; DAND83-0557, TTC-0428, UC-71, Sandia National Laboratories, Alberquerque.

Watkins, J. and Sleath, K.P. (1980). Isolation and enumeration of *Listeria monocytogenes* from sewage, sewage sludge and river water. *Journal of Applied Bacteriology* **50**, 1–9.

Whitmore, T.N. and Robertson, L.J. (1995). The effect of sewage sludge treatment processes on oocysts of *Cryptosporidium parvum*. *Journal of Applied Bacteriology* **78**(1), 34–38.

WHO (1981). *The Risk to Health of Microbes in Sewage Sludge Applied to Land. EURO Reports and Studies No 54*. World Health Organisation, Copenhagen.

WRc (1998). *Review of the Scientific Evidence relating to the Controls on the Agricultural Use of Sewage Sludge*. WRc, Medmenham.

18

Effluent discharge standards

David W.M. Johnstone

Swindon SN3 ILG, UK

1 INTRODUCTION

From the Middle Ages until the early part of the nineteenth century the streets in many cities of Europe were fouled with excreta and garbage to the extent that people often held clove-studded oranges to their nostrils in order to tolerate the atmosphere. The introduction of water-carriage systems of sewage disposal in the early nineteenth century merely transferred the filth from the streets to the rivers. The problem was intensified by the Industrial Revolution with the establishment of factories on the banks of the rivers where water was freely available for power, process manufacturing and effluent disposal. As a consequence, the quality of most rivers deteriorated to the extent that they were unable to support aquatic life and some were little more than open sewers. In much of today's developed world there followed a period of slow recovery when many rivers were restored to balanced ecosystems with good fish stocks. However, in most of the developing world there has been no such recovery; urban watercourses still substantially act as open sewers and the position is often one of continuing deterioration, rather than marked improvement.

In countries where there has been recovery it has not been easy nor has it been cheap. It has been based on the application of good engineering supported by the passing and enforcement of necessary legislation and the development of suitable institutional capacity to finance, design, construct, maintain and operate the required sewerage and sewage treatment systems. Such institutional and technical systems not only include provisions for the disposal of domestic sewage, but also for the treatment and disposal of industrial wastewaters and for the integrated management of river systems. A key element in all successful systems has been the establishment and enforcement of appropriate and affordable effluent discharge standards.

In the developing world, improvements to the current situation are often hampered by poor legislation, lack of appropriate institutional arrangements for regulation, lack of resources, insufficient finances and, frequently, lack of political will to enforce control measures. The position is complicated by the fact that many developing countries are at a loss on how to set standards and resort to importation of those from the developed world without either the means of achievement or the faintest idea of the costs of regulation and control. While no doubt well intentioned, the consequent problems exemplify the 'law of unintended consequences' (Cairncross, 1991) and result in slow or non-existent progress.

The objective of this chapter is to describe the general basis of setting standards and to discuss institutional arrangements for their regulation and enforcement. It is aimed primarily at discharges from domestic sewage treatment plants, including those accepting industrial effluents. It is not aimed at setting standards for the direct discharge of industrial effluents to rivers or to other watercourses, but it does give a brief account of the control of such effluents when discharged to public sewers and hence conveyed to a sewage treatment plant. A short historical review is followed by an overview of standard setting and a discussion on the problems

The Handbook of Water and Wastewater Microbiology
ISBN 0-12-470100-0

of implementation. There is also a discussion on the problems of regulating and controlling the private sector, especially where penalties are imposed for failure to comply with contractual standards, and where many of the problems are due to a lack of understanding of sampling procedures and the statistical nature of sewage effluents.

2 HISTORICAL REVIEW

Every country, whether designated as developed or developing, has its own historical perspective on the evolution of environmental control measures. Parallel progress in a number of developed world countries has led to different ways of describing environmental quality objectives and the standards deemed necessary to achieve them. It is impossible, but fortunately unnecessary, to describe each of these different ways: it is sufficient to outline some key developments, to discuss trends that have influenced standards over the years, and to describe significant differences in approach.

Much of the pioneering work was carried out in the UK (see Johnstone and Horan, 1996) where many watercourses had been devastated by the effects of the industrial revolution and had become a serious menace to public health, the extent of which was demonstrated in two great epidemics of cholera in 1866 and 1872. Following a number of studies and Commissions, the government of the day passed the Public Health Act 1875, which is regarded as one of the foremost sanitary measures of its time. Its prime importance was that it recognized, for the first time, that care of public health was a national responsibility. This Act was immediately followed by the Rivers Pollution Prevention Act 1876, which formed the basis of all legal action with regard to river pollution until 1951. A significant element of the 1876 Act was that it prohibited the discharge of solid or liquid sewage matter into a river and determined that *it was no defence to argue that the river had already been polluted upstream*. The significance of this simple statement is profound and today forms a cornerstone of much of the world's environmental legislation.

Although the 1876 Act was comprehensive, it was in many ways ahead of its time. It specified duties with respect to sewage treatment at a time when such treatment was still in the development stage and many technical and scientific issues had not been resolved. In order to resolve these issues and determine a way forward, the government of the day set up a Royal Commission on Sewage Disposal. The Commission commenced its work in 1898 and completed the last of its nine voluminous reports 17 years later in 1915.

By far the most famous, and arguably the most important findings, were presented in the eighth report published in 1912. It dealt with the question of standards and tests to be applied to sewage and sewage effluents, and it introduced the BOD (biochemical oxygen demand) test. However, the first attempt at establishing effluent standards came in the fifth report published in 1908. Preliminary, empirical standards were proposed, based on an analysis of data from existing plants. Thus it was expected that 'a well-operated treatment works effluent':

1. should have no more than 30 mg/l suspended solids and
2. should, after filtering, absorb no more than:

   5 mg/l dissolved oxygen in 24 hours,
   10 mg/l dissolved oxygen in 48 hours, and
   15 mg/l dissolved oxygen in 5 days.

The eighth report was produced *inter alia* to ensure that money was not spent on sewage treatment when the circumstances of the area did not warrant it. This involved gathering data on local stream conditions (volume, depth, velocity and degree of aeration), together with chemical analyses of water quality. The report recommended that the old permanganate value be replaced by the BOD test (or dissolved oxygen absorption test, as it was then known). It recommended that the test be carried out at 65°F (18°C), which was the maximum river temperature during the hottest month in the UK. It also recommended that a 5-day incubation period be used as this was found to give a smaller experimental error than tests carried out at shorter incubation periods. Contrary to popular belief, the 5-day period had nothing to do with the retention period of rivers in the UK.

In the course of the investigations, a number of rivers, and reaches of rivers, were classified as follows:

Very clean	1 mg/l BOD
Clean	2 mg/l BOD
Fairly clean	3 mg/l BOD
Doubtful	5 mg/l BOD
Bad	10 mg/l BOD

Although the Commissioners realized that it was logical to provide individual discharge standards for each sewage treatment works, depending on the nature and condition of the receiving water, this course of action was not adopted because of objections that:

- it would be difficult to administer
- it would place unequal burdens of purification between authorities discharging to the same watercourse
- it would allow a relaxation in discharge quality in some cases.

The final outcome was a general standard based on a mass-balance that assumed an eightfold dilution with river water of 2 mg/l BOD. It was felt that, if the river was not to deteriorate with serious consequences, the water after the discharge point should not absorb more than 4 mg/l BOD. On this basis, a mass balance shows that the sewage effluent should not absorb more than 20 mg/l BOD. Superficially, this would appear to have been a relaxation of the 15 mg/l standard provisionally set in the fifth report. However, the standard of 20 mg/l BOD was to be applied to an unfiltered sample as opposed to the filtered sample proposed in the fifth report.

The Commission retained the 30 mg/l value for suspended solids recommended in the fifth report and thus was born the famous (or even infamous) Royal Commission 20/30 Standard still adopted today as a base standard in much of the world. Thus, by the year 1915, the Royal Commission on Sewage Disposal had established the first standards for effluent discharges and had introduced an elementary river classification system. In a separate report it also laid the basis for the control and disposal of industrial wastewaters.

Over the following 80 years there were many developments in procedures for setting effluent standards and determining environmental quality objectives. In the UK there was a series of, often fragmented, approaches to developing river classifications based on mathematical modelling and the use of river water quality objectives. River quality classification systems developed over a number of years and were changed as and when considered necessary. For example, as a result of a River Quality Survey in 1975, it was felt that the classification then in use should be replaced by a more objective one that would take account of potential uses of river water, and during a subsequent Review of Consent Conditions in 1978, there were moves towards linking effluent standards to the concept of water quality objectives. There were also technical changes, such as the introduction of the inhibited BOD test using allyl thiourea and ammoniacal nitrogen became an important standard for some rivers, such as the Thames, where there is widespread use of the river water for potable supply.

Improvements in the knowledge of sewage treatment processes meant that it was practical for standards to include limits on nitrogen and phosphorus and many countries introduced limits on bacteria, usually specified in terms of faecal coliforms. In 1989 the World Health Organization published health guidelines (Mara and Cairncross, 1989) for use of effluents in agriculture and aquaculture that specified limits on faecal coliforms and helminth eggs. More recently, WHO published a comprehensive set of guidelines on water quality, including those for sewage effluents; it deals, *inter alia*, with risk, risk management, economic evaluation and practical aspects of implementation (WHO (2001) Water Quality: Guidelines (2001)).

Over the years there has been much debate over the advantages and disadvantages of establishing a system based on 'fixed emissions', where all discharges have to comply with one set of standards, and a system based on 'water quality objectives', in which an individual discharge has standards based on the quality needs of the receiving watercourse. Some countries have included a long list of parameters based on 'the precautionary principle', while the BATNEEC principle is used elsewhere, especially in conjunction with the philosophy of integrated pollution control (IPC). The BATNEEC principle recognizes the cost of environmental pollution control and attempts to choose mitigation measures using the 'best available technology not entailing excessive costs'.

For much of the twentieth century, discharge standards were based on numerical limits on chemical parameters, sometimes accompanied by some descriptive limits. In recent years there have been considerable advances on the use of toxicity testing and it is highly likely that it will become an increasingly useful tool in controlling discharges, especially from sources such as the chemical industry or from sewage treatment works that include complex industrial wastewaters. Control of individual chemicals by concentration does not consider the possible synergistic effects such that 'cocktails' may have on the aquatic environment, whereas toxicity testing provides an indication of potential damage. At present the application of toxicity techniques is in its infancy and suitable protocols and procedures need to be formulated.

Today it has become a fact of life that ever-increasing environmental standards have proliferated throughout the world (Johnstone and Horan, 1994). Even when the standards are superficially simple, such as in the European Directive on Urban Waste Water Treatment, their application can vary widely, leading to difficult comparisons and misleading interpretation, as illustrated by Jacobsen and Warn (1999). In the developed world, there has been a major increase in the number of determinands included in effluent standards. Garber (1992) states that the wastewater from Los Angeles wastewater treatment plant in the USA has to be analysed for 140 priority pollutants before it is discharged to sea. Many of the limits are based on a precautionary approach and the achievement of 'zero risk'.

Unfortunately, many of the very high standards are very expensive to achieve and some have been imposed without any logical attempt at a cost–benefit analysis. It is even more unfortunate that many such standards, set for the developed world, have been adopted by some developing countries without consideration of the cost of implementation or the costs or the means of enforcement and regulation. Many developing countries now find that they cannot tackle problems of gross pollution in a phased and affordable way, due to the adoption of inappropriate standards, a subject that is discussed in greater detail later.

3 SETTING STANDARDS

Essentially there are two approaches to setting standards, although there are various options within each. One approach uses Fixed Emission Standards (FES) that are applicable to all discharges, while the other sets Water Quality Standards (WQS) to achieve some predetermined Water Quality Objective (WQO). Water quality standards therefore vary from discharge to discharge depending on the particular objective. There are advantages and disadvantages to both.

A number of issues need to be considered when setting standards, irrespective of the general approach used and whose understanding is important when designing treatment facilities. The most important are as follows:

- the chemical or other intensive parameters to be used as control measures
- the average and maximum daily flows
- the instantaneous rate of flow
- the possibility of using mathematical models

- whether to use concentration or minimum percentage removal
- whether the sampling regime is to be based on 'snap' samples or 'composite' samples
- errors in sampling and analysis
- whether standards are to be based on absolute values or on statistical compliance
- the period of judgement of compliance. Whether it is to be, for example, a single day's sample, a monthly average, a yearly average, or on a percentile value measured over a certain period
- the required frequency of sampling and analysis, for example daily, weekly, monthly, etc.
- the contractual nature of the 'permit', the 'licence' or the 'consent' that encompasses the standards
- the legal and regulatory basis of the 'permit', the 'licence' or the 'consent'
- penalties for failure to achieve the standards
- the cost and affordability of achieving and regulating the standards
- the possibility of phasing standards to suit affordability
- the need to control industrial wastewaters that discharge to the public sewer.

3.1 Fixed emission standards (FESs)

Fixed emission standards are relatively simple to apply and to regulate and can provide an effective way of achieving basic environmental objectives. They can be used as a first step in a phased approach to achieving some long-term environmental objective. They are particularly useful in countries where there is virtually no sewage treatment and where it is difficult or impossible to afford strict standards in the short term. Fixed emission standards usually take either the form of a percentage removal of one or a number of contaminants, or limits based on concentration.

There are some dangers in using this approach if the standards are not set appropriately. If they are set too low, there is a danger that they might not achieve noticeable improvement in the short term, which can lead to problems arising from the public's perception of benefit, especially where expectation of improvement is high. On the other hand, if generally applicable fixed emission standards are set too high they could lead to unnecessary over-expenditure or could be deemed unaffordable with the result that nothing is done.

Fixed emission standards need not be the same for all discharges. It is not unusual to set one FES for inland rivers, another for lakes and another for saline waters. The European Directive on Urban Waste Water Treatment (CEC, 1991) is a good example of an effective fixed emission standard approach. For discharges to inland watercourses there are two fixed standards, a basic one that provides a minimum degree of treatment and a more rigorous one for discharges to 'sensitive areas'. The basic standard is given in terms of BOD or COD, and suspended solids, the latter being optional, while the more rigorous one includes limits on nitrogen and phosphorus. Both standards can be achieved in terms of concentration or minimum percentage removal. An attractive feature of the European standards is that they contain only a few determinands and they are simple. The Directive is meant to establish a minimum degree of treatment in all member states, upon which individual countries can build and impose stricter standards if not already in place. This is a sensible approach in Europe where countries vary from ones with virtually no treatment to ones with high standards and high levels of treatment. A danger, however, is that those with high standards could be tempted to reduce them to the basic ones.

While the EU Directive can be considered as a sensible application of FES, there are examples of poor application in other parts of the world. A major criticism is that they often contain long lists of parameters with concentration limits that could not be achieved in an affordable manner. In many developing countries, urban watercourses are no more than open sewers while the standards in place are aimed at maintaining high-class rivers. Brazil is a good example, especially since it covers such a vast area with rich communities towards the south and very much poorer communities towards the northeast.

The National Council for the Environment passed a Resolution (No. 20) in 1986 (CONAMA, 1986) that established FESs for all discharges to all watercourses. The standards, described in Article 21 of the Resolution, are federal ones, but states are free to adopt their own. However, states either tend to adopt the federal ones or use them as a basis for their own. There are logical limits on such measures as pH and temperature but some restrictions, such as the one that restricts flows to 1.5 times the daily average, are without logical foundation. Strangely, the standards do not include either BOD or COD or any other measure of oxygen demand. The main problems arise from the limits set on 31 individual contaminants that range from heavy metals to organophosphates and chlorinated hydrocarbons. A major problem is the limit of 5 mg/l placed on ammoniacal nitrogen. If this were to be strictly applied, it would completely exclude the ability of the country to develop a phased approach to the improvement of environmental conditions and provision of basic sanitation. The only way to achieve the ammonia limit would be to use aerobic treatment. It could not be achieved with simple primary settlement nor with anaerobic treatment, yet these are both very cost-effective methods of providing significant initial improvements.

Fixed emission standards that contain long lists of determinands are often adopted in an attempt to cover all circumstances, but they sometimes turn out to have negative benefits. In many situations they are impossible to regulate effectively; they cause design and contractual problems and they give rise to application of inappropriate or unaffordable technologies and sometimes nothing is built for fear of prosecution and no progress is made.

3.2 Water quality standards (WQSs)

Much of the world sets its discharge standards to meet some water quality objective (WQO) usually based on the use that would be made of the water in a watercourse or in a particular stretch of a watercourse. WQOs are usually determined in the form of a classification system. In the UK there are four classes, with class 1 divided into 1A and 1B. Brazil has eight classes, four for rivers, two for saline waters and two for brackish waters. There are similar classification systems in other countries but it is not within the scope of this chapter to describe or discuss in detail any such classification. It is sufficient to recognize that standards can be set to meet the individual WQOs. This, of course, means that standards could vary from discharge to discharge, depending on individual water classification.

For discharges to inland rivers, the historical method for determining an effluent quality was to use a simple mass balance that determined the discharge load by balancing the sum of the upstream load and the discharge load with the needs of the downstream quality objectives. Although the Royal Commission Standards are often used as fixed emission ones, they are actually based a WQO approach, using a simple mass balance in which the model had an eight-to-one dilution factor and the downstream quality objective was that the river should not deteriorate by more than 2 mg/l BOD.

Over the years, limitations of the simple mass balance approach have been recognized and more sophisticated methods developed to describe WQOs on a statistical basis, where the value of some determinand would not be exceeded for more than a specified percentage of time, usually 95%. The mass balance to determine the discharge standard is then determined statistically using a combined distribution method such as Warn-Brew or Monte Carlo simulations (Warn and Brew, 1980). The result is that an effluent standard is produced on a percentile basis. While the use of such techniques is logical and environmentally sound, it can cause difficulties in enforcement and a full understanding of the significance of the statistical approach is vital in the design of treatment facilities.

Some legal and institutional systems find it difficult to cope with a statistical discharge standard since it, in effect, 'allows' failures to occur. There is sometimes a fear that significant pollution could occur during the 'allowed' failure days, while the discharge limit is still achieved in overall statistical terms. A way to overcome this problem is to introduce, in addition to the percentile standard, an upper 'maximum admissible concentration', never to be exceeded. Another problem with the approach is that, to be effective, many samples have to be taken to give statistical significance, and this is often beyond the capacity of some countries.

3.3 Sample types and sampling frequency

Discharge standards can be set using either 'snap' samples or 'composite' samples. A 'snap' (also referred to as 'grab' or 'spot') sample is simply one taken at random at a single moment in time. A 'composite' sample is one that attempts to measure the effect of an effluent over a period of time, usually a day. Composite samples can be produced manually by mixing individual snap samples in proportion to the volume that passes in the time between consecutive samples. However, it is more common to use an automatic sampler that can produce a composite sample weighted in proportion to the wastewater flow. Most automatic samplers are also capable of taking a number of discrete samples that could be analysed individually or mixed to produce a composite.

'Snap' samples are simple to take, do not involve the use of expensive apparatus and are particularly useful for audit purposes. However, they only reflect the position at a single instant and not at other times. Flow-weighted composite samples, on the other hand, better reflect the total pollution load but tend to mask individual 'spikes' in concentration. They also require expensive apparatus. The most comprehensive way to monitor a discharge would be to use a flow weighted 'composite' sample to measure the daily pollution load and to analyse a number of discrete 'snap' samples over the period of a day. However, this is prohibitively expensive in many cases. A reasonable compromise for large, significant discharges is to use composites along with occasional diurnal surveys based on 12 or 24 individual 'snap' samples.

It can be argued that, for one-dimensional free-flowing rivers, snap samples are better

TABLE 18.1 Log-normal distribution of BOD and suspended solids

	95%		*Absolute*	
	Composite	*Snap*	*Composite*	*Snap*
BOD (mg/l)	25	31	50	65
	20	25	45	60
	15	19	40	55
	12	15	35	50
Suspended solids (mg/l)	50	70	115	160
	45	63	110	140
	40	56	90	125
	30	42	75	105
	25	35	65	90

than composite since the important environmental criterion is concentration and not load. For tidal estuaries or tidal rivers, where the flow direction changes with tide, load may be more appropriate and a composite sample a better measure.

The description of an effluent quality using a series of snap samples is not the same as the description of the same effluent using a series of composite samples. The matter is further complicated when snap and composite samples are used to describe effluents based on 'absolute' values or on 'percentile' values. Ross (1994) used the information given in Table 18.1 to indicate the nature of the relationship based on a log-normal distribution of BOD and suspended solids.

Thus, an effluent with a composite BOD of 25 mg/l is equivalent to an effluent with a snap of 31 mg/l. The same effluent, if it were judged on an absolute (99.9%) basis, would have BOD values of 50 mg/l and 65 mg/l. These figures are not presented to indicate absolute relationships; rather they indicate the nature of the relationships and the difficulties of both specifying and understanding effluent discharge standards. When setting numerical limits, it is important to be aware of the significance of the sampling regime to be used, as well as the approach used to determine the standards. It is a complex subject far beyond the scope of this chapter.

3.4 Industrial wastewaters

Industrial effluent control is of vital importance, particularly in developing countries where control is often very poor or non-existent. There is an old axiom which states that 'good sewage treatment starts with good control of industrial effluents'. Non-existent control could mean an inability to treat domestic sewage effectively.

The prime reasons for the control of industrial wastewaters are:

- to protect workers working within the sewerage system
- to protect the fabric of the sewer and the fabric of any downstream treatment works
- to prevent fires and explosions due to inflammable or explosive chemicals
- to prevent surcharging or blockages within the sewerage system
- to protect the physical operation of downstream treatment works
- to ensure that industrial discharges do not affect the performance of downstream biological treatment processes
- to protect the environment where sewers eventually discharge to a watercourse or to the sea
- to ensure that industrial discharges do not affect disposal of sludges from treatment works.

The organization responsible for the sewerage system and sewage treatment works must set standards for individual sewer discharges, based upon the following factors:

- the type and nature of the industrial wastewater
- the capacity of the sewerage system to accommodate the wastewater
- the potential of the wastewater to cause surcharging or blockages of the sewerage system
- the potential effect on the capacity of the treatment plant
- the potential effect on biological treatment processes.

For any given sewerage system, the total daily flow, suspended solids, BOD or COD, and ammoniacal nitrogen determine the size and capacity of the required sewage treatment works. Pretreatment of industrial wastewaters may be required to maintain flows and loads within treatment capacities, or extension of the treatment works may be necessary. Substances that affect biological processes must be limited to concentrations below those that would cause problems and the limits need to be set taking account of quantities arising from other sources.

If the standard for the sewage treatment discharge has limits on certain substances, the only recourse may be to control them at the industrial source. Standards for substances such as sulphides, cyanides and heavy metals should be set at very low limits, whereas some, such as the 'red list substances', should be banned altogether. Others including many volatile organic compounds could also be considered for prohibition. Toxic metals arising from industry need to be controlled as they become incorporated into the biosolids (sludge) and can seriously affect disposal to agricultural land. In some countries (e.g. USA) pretreatment standards apply to specific industrial categories, so that all are treated with some degree of equity, but this is not the case in other countries.

4 MEETING THE STANDARDS

Significant developments in process technology in recent years mean that there are now many ways of treating sewage to achieve a wide range of discharge standards. However, the availability of a process is not by itself sufficient; it should be cost-effective, appropriate to the local conditions and affordable. In this sense there needs to be a differentiation made between the developed and developing worlds.

Most of the developed world has extensive sewerage networks, a relatively high degree of sewage treatment and some form of industrial wastewater control. Discharge standards are now aimed at improving conditions in which gross pollution is generally absent. Targets differ from place to place, but the general tendency is towards improving the aquatic environment and minimizing risk to humans using that environment. In general, developed world countries can afford to pay for these improvements, although there are some who may be reluctant. By contrast, most of the developing world has inadequate sewerage systems, very little sewage treatment and almost no effective industrial wastewater control. Urban watercourses are no more than open sewers whose condition is further reduced by large amounts of garbage arising from inadequate collection facilities. Many developing countries either do not have discharge standards or use inappropriate ones imported from the developed world.

Although the conditions in the two worlds are different, the problems are really similar but out of time sequence. The conditions in many developing world cities today are the same as those of the developed world cities 50 to 100 years ago. There are, however, important differences. On the one hand, the developing world's financial problems are likely to make it more difficult to bring about improvements at a rapid rate but, that on the other hand, they can benefit from some of the technological advances made during the intervening years and, hopefully, learn from previous mistakes of the developing world.

4.1 Sewerage and sewage treatment

Critical to the improvement of environmental conditions is the need for effective collection and transportation of sewage for treatment. Most of the developed world has elaborate networks based on either separate or combined sewerage systems with sewage transported to treatment plants by a system of interceptor sewers and pumping stations. When properly installed, operated and controlled the separate system is the more cost-effective as it reduces the amount of sewage to be treated and avoids the problems of discharges from combined sewer overflows (CSOs). In fact, the issue of treating the flows from CSOs is one of the principal problems facing the sanitation industry in the developed world today; the solutions are not easy and tend to be very expensive.

Following the lessons of the developed world, the sewerage systems of many developing world cities were designed and built as separate systems. However, unlike the developed world cities, they have often not been well operated since the control of connections is virtually non-existent. As a result, so-called separate systems have many connections illegally made to the surface water sewers and not to the foul or sanitary sewers as intended. There are also many cross connections and, in many cases, separate systems are effectively combined systems. This causes great problems when collecting sewage in interceptor sewers. If only discharges from recognized sanitary sewers are collected, much of the sewage will continue to be discharged through the surface water system, so diminishing the benefit of collection. Separating out the two systems is extremely difficult and prohibitively expensive and could delay effective collection by many years. An option could be to collect all discharges containing sewage, transport everything for treatment and install stormwater facilities at the treatment plant, but this would completely remove the cost effectiveness of a separate system. It could also cause some urban watercourses to dry up.

It is beyond the scope of this chapter to debate the problems of competing sewerage systems, interception and transportation. However, for towns and cities with such conditions, it should be recognized that the introduction of high discharge standards for treatment plant effluents would be of no avail if a large proportion of the sewage were still discharged directly through stormwater sewers. Under these circumstances it would be much more effective to phase discharge standards over a long period to take account not only of affordability, but also of prolonged delays in getting adequate quantities of sewage to the treatment plant.

With regard to the treatment of sewage, there is a wide range of methods and processes available to cover all of the world's needs provided the selection is appropriate to the environmental conditions, that it is affordable and that the community is willing to pay. Within today's treatment technology it is possible to design plants that can do anything from simple removal of gross solids to ones that would produce high quality water for reuse. The standard should therefore reflect the objectives, and the objectives of the developed world are not the same as those of the developing world, at least not in the short term. The use of developed world standards in the developing world is generally a retrograde step.

While a standard should be set to meet an environmental objective, the way that it is written can have a profound effect on the design and cost of treatment; a subject that should be fully understood by both the standard setter and the treatment plant designer. This is particularly true if the standard is written in statistical terms. The difference in design for an absolute standard and a statistical standard (see Section 3.2) can be very significant and the interpretation of the standard is very important. This was illustrated by Johnstone and Norton (2000) who showed the difference in cost of an aeration system to produce a 50 mg/l BOD standard based on four different ways of describing 50 mg/l, – as an average composite, a 95 percentile composite, an absolute composite and an absolute snap. Designing on an absolute composite basis produced the lowest cost. The design based on a 95 percentile composite was 31% more expensive, while that based on an absolute (99.9 percentile) composite was 92% more expensive. The design based on an absolute (99.9 percentile) snap was 230% more expensive. These are very significant cost differences, which become even greater when the numerical limits are reduced (i.e. when the standards become higher), not only in absolute terms but also in relative terms.

An appreciation of the statistical nature of sewage effluents and a full understanding of the nature of the standard is particularly important in the development of BOOT (Build, Own, Operate and Transfer) schemes for the provision of a sewage treatment plant by the private sector, especially where financial penalties are to be imposed for failures (see Section 5.2). Dunn *et al.* (1998) made studies on the statistical nature of sewage effluents that attempt to predict the risk on the design of new treatment facilities.

4.2 The developed world

In much of the developed world, sewage treatment processes now meet ever-improving standards and the tendency is towards the application of improved aerobic secondary treatment or the addition of tertiary or advanced processes. There is greater use of nitrogen and phosphorus removal techniques, of UV disinfection rather than chlorine and, in many places, of processes that can be contained within buildings or can be built underground. Communities demand better odour control from treatment plants and no longer tolerate disposal of non-sterilized sewage sludges to agricultural land. In addition, more attention is being placed on treating combined sewer overflows.

Although some process developments have been driven by the need to reduce costs, many more in the future will be driven by the need to meet ever-improving discharge standards (Johnstone and Norton, 2000). It is likely that more use will be made of advanced treatment techniques, such those based on membrane technology or granulated activated carbon, as attention is further drawn towards removal of specific compounds. Toxicity testing of effluents will lead to greater control of industrial effluents at source. Whether or not many of the standards, especially those based on the precautionary principle and desire for zero risk, can be justified is a matter of debate (Garber, 1992; Johnstone and Horan, 1994). The cost of meeting them will no doubt increase; whether or not the community will continue to pay for them, or not, is another matter for debate.

4.3 The developing world

Meeting discharge standards in the developing world is often by application of processes at the other end of the technological scale to those used in the developed world, although there are exceptions. In most large cities, land availability is limited and often expensive, so that some of the more basic processes are not applicable. There is no hard-and-fast rule to selecting a cost-effective process, but the decision is sometimes complicated in the developing world. On the one hand, inappropriate standards imported from the developed world would drive the selection towards processes that are inappropriate and unaffordable. On the other hand, there are many advocates of so-called 'appropriate technology' (meaning in this case low technology) who demand that basic technologies be applied everywhere in the developing world. There is a tendency in some quarters to promote the use of processes, such as waste stabilization ponds in large cities, where they may not be practicable and more conventional processes may be more appropriate.

While waste stabilization ponds may not always be appropriate for large cities, they can find great use in many other circumstances, although care must be taken in specifying standards based on BOD. Often waste stabilization ponds discharge algae with the final effluent that exerts a BOD but with little environmental damage. This was recognized by the European Union in the Urban Waste Water Directive by basing the BOD standard for effluents from waste stabilization ponds on filtered samples and not unfiltered samples as is required for all other determinands; the Directive also requires a suspended solids standard for pond effluent of only 150 mg/l.

Another problem arises when some standard setters say that BOD, suspended solids and some of the other basic sanitary parameters, are not important in the developing world and that the only parameter of importance is faecal bacteria. While there is little doubt that reduction in faecal bacteria is important, statements like this underline a lack of understanding of the underlying principles of wastewater treatment systems, and indeed often a lack of understanding of BOD. BOD is a measure of sewage strength; it is a measure of readily degradable organic matter and not just a measure of oxygen demand. Outside the field of waste stabilization ponds, treatment plants do not remove substantial quantities of bacteria, but by removing BOD and suspended solids they can make an effluent amenable to disinfection. To purport that effluents can be efficiently disinfected without first adequately reducing the BOD and suspended solids displays ignorance of the principles of process technology.

A sensible approach to improving environmental conditions in the developing world is to use a strategy based on phasing process units, a strategy given appropriate emphasis by von Sperling and Fattal (2001) in the newly published WHO guidelines. In many cases, the provision of simple preliminary screening could do much for the aesthetics of the area, but it is more usual to have a first stage based on primary treatment or an anaerobic process such as an USAB. However, phasing is only suitable provided that the standards in force are flexible enough to allow for this type of approach. Anaerobic processes can be particularly attractive as they use very little power and they produce minimal quantities of sludge. However, they tend not to be capable of dealing with large flow variations and can be adversely affected by high concentrations of heavy metals, which can be a problem in places with poorly controlled industrial effluents.

Advanced primary sedimentation using chemicals could also provide an initial stage, or it could be used as a first upgrade for a basic sedimentation system. There are a number of proprietary advanced primary treatment processes available.

The addition of secondary treatment or the enhancement of first phase treatment can take place at any time, but it should be introduced in the full knowledge of pollution from other sources. For example, if there is still major pollution arising from contaminated surface water sewers, the benefit from secondary treatment may be marginal and funds would be better spent in tackling the sewerage problem. There is no simple answer but, for developing countries, it is important to consider the problem in the long term and phase treatment facilities according to benefit and affordability. To do so also requires a phased approach to the introduction of standards and, if necessary, to have interim standards and long-term targets.

5 REGULATION

There is little doubt that the setting of appropriate environmental standards is important. However, their effectiveness is only as good as the subsequent enforcement and regulation. Enforcement is principally a matter of political will. In some parts of the world, standards abound without the political will to enforce them and this achieves nothing. Sometimes there are ostensibly good reasons, as the expenditure needed to meet environmental standards must take its place beside other priorities.

Regulation is often seen as being too expensive and it can be if not done effectively. For example, consider the monitoring of discharges and the frequency of sampling. It is reasonable to argue that where society correctly requires a high standard, the frequency of sampling should reflect the importance of that high standard and the costs of achieving it. Equally, when a low standard is required, it seems reasonable to allow a relatively low sampling and monitoring regime. Yet there is little evidence of this being common practice and even when it is practised, it is doubtful if the monitoring frequency is high enough to monitor high standards with justifiable statistical significance. Surely if society deems it appropriate to impose a high standard, then the cost of adequate monitoring should be taken into account as part of the cost of achieving it.

The issue of regulation and enforcement highlights another marked distinction between the developed and developing worlds. Countries in North America, Europe and Scandinavia, for example, have mature and sophisticated environmental agencies, most of which are well funded. They generally have political support and society can afford to pay for them. By contrast much of the developing world is characterized by poor or inadequate environment agencies that are badly funded and without real political support.

Traditionally, sewage treatment works have been built and operated by the public sector, with the discharges regulated and controlled by other public sector organizations, such as an environment agency. This would normally be done by the issue of a licence or some other legal instrument and controlled through a regular monitoring programme. The consequence of failing to comply with a standard would range from virtually none, through reprimands to financial fines.

Over recent years there has been a greater use of private sector financing of treatment plants by way of some form of BOOT (Build, Own, Operate and Transfer) scheme, of which there many variations. Regulation and control of discharges is also through a licence that is usually no different from that for a public sector works. However, regulation and monitoring tend to be much stricter and the consequence of failure much more severe. It is not unknown for daily fines of around US$25 000 to be imposed for failure to achieve the discharge standard. Under these circumstances, the nature and wording of a standard and its interpretation can have profound effects on the design and operation of the plant. Indeed, it can have a profound effect on the success of the whole project, a subject discussed below.

5.1 Permit, licence or consent

Legal instruments for setting out discharge standards and conditions differ from place to place, but they generally take the form of a 'permit', 'licence' or 'consent'. It is a contract between the discharger and, usually, a regulator empowered by government to enforce standards. Although there is a wide variation in such documents, they usually comprise:

- the standards to be achieved; these may be descriptive or numeric
- specific conditions relating to the effluent
- general conditions relation to the institutional requirements.

Descriptive standards usually relate to things that are difficult to measure and could include general statements such as:

- 'there shall be no sign of visible signs of solid matter downstream of the discharge point' and/or
- 'there shall be no signs of visible oil downstream of the discharge point'.

Numerical standards specify limits on physical and chemical parameters, such as daily flow rate, maximum flow rate, pH, temperature, BOD, COD, suspended solids and heavy metals, etc. They might also include toxicity limits but, at present, protocols for setting such limits are not fully developed.

Specific conditions may relate to the exact location of the discharge point and might describe the exact nature of a required outlet, which could be a pipe or spillway. Materials of construction may also be specified. It is often a requirement that a flow metre is installed and that flow records be kept and made available for inspection by the regulator. It would be usual to specify that a suitable sampling point be constructed at a designated position.

General conditions may specify regulations in the law under which the legal instrument was issued, and the powers of entry that allow regulators to inspect premises and to take samples. The frequency of audit sampling could also be specified, as could the penalties for infringement and the steps that have to be taken during infringement. For example, it may specify the exact procedures to be followed when taking samples, especially those to be used in evidence of a breach. There could also be general conditions that absolutely exclude the discharge of certain substances, or groups of substances.

It is important that, no matter what the form of the instrument, it should specify the exact requirements and lay out unambiguously the performance, duties and obligations of the discharger.

5.2 The polluter pays principle

In some areas of the world, such as in Scandinavia and the Baltic States, there are systems for charging for pollution, which is an additional means of regulation and control. In one Baltic country, charges are imposed on the loads of BOD, suspended solids, total nitrogen, phosphorus and on 'oil and grease'. There are separate tariffs per tonne of each parameter and the charge is imposed four times per year on the total load for the period. As an encouragement to meet the standard, the tariff is reduced to 50% if the effluent complies with the standard for the whole period. If it does not, the full tariff is imposed for all days during which the effluent remains within the standard; and for those days during which the effluent fails the standard the tariff is increased tenfold. If the tariffs are set at proper levels, this can prove to be a very effective means of regulation as it can force operators to look towards optimizing operational control. However, if set too low, it merely becomes another form of taxation. It can also be an expensive form of control, since samples of effluent have to be taken every day of the year and, eventually, this cost and the cost of the fines has to be borne by the community.

5.3 Standards, regulation and the private sector

There are many ways that the private sector can participate in the treatment of sewage, one of which is the BOOT contract. In this, the private sector, usually a consortium of contractors and operators, raises the finance to fund the construction of a treatment plant. After construction the operating arm of the consortium operates the plant for some 20 to 30 years, during which time they are paid per m^3 of sewage treated. Payment on a volume basis is usually the only source of income and must be sufficient to pay for loan repayments, dividends and all operational and other expenses. In some contracts, failure to comply with the discharge standard could incur substantial financial penalties.

There are many risks in this type of venture and there are two that are relevant here, which if not recognized and handled properly could cause the failure of this type of scheme. The first is concerned with payment on a volumetric basis. Due to the problems with sewerage and bringing sewage to the treatment plant (see Section 4.1), there could be prolonged delays in achieving sufficient volumes to pay for treatment, which, in combination with failures to meet a suitable standard, could cause the project to fail. If not properly addressed, the private sector may not be prepared to bid for such a project in the first place.

The problems with discharge standards can arise from the nature of the standards, the limits set, type of monitoring to be undertaken, errors in sampling and analysis, and generally the question of 'what constitutes a failure?' When the penalties for failure are high, these issues are of great concern, especially since there is often mistrust between public sector regulators and private sector operators. Often problems arise due to unclear or unspecified rules.

If a standard is set as an absolute one (never to be exceeded) there will inevitably be failures and fines, and the higher the standard, the more frequent the fines. However, it is more practical to have the standards imposed as (for example) a monthly average, or a monthly average with an upper limit, or a percentile compliance over a given period. Whatever is chosen, it needs to be specified unambiguously in the contract and understood clearly by both the operator and regulator. At present, there are many cases where it is not and where problems are inevitable.

Another issue is concerned with what actually constitutes a failure and should this incur a fine. If there are, for example five parameters in the standard where four pass and one fails, is this a failure that should incur a fine? Also, what allowance should be made at the margin of analytical and sampling error? For example, if a BOD limit is 20 mg/l and a sample analysis is 21 mg/l, is this a failure bearing in mind that analytical error may be around 20%? It seems inequitable that a private company should suffer a large daily fine if four out of five determinands are within limits and the fifth is outside only by the margin of analytical error. The environmental consequences of this would be zero. It may appear that such ambiguities would not be real but they are, and they arise from inexperienced regulators setting inappropriate rules for regulation and uninformed contractors agreeing to them without understanding the consequences. The problems would not be so great if disputes were left to reasonable people to resolve, but this is rarely the case. The problems become acute when lawyers and prosecutors get involved, as the law tends to see things as black and white, whereas they are usually grey. However, the regulation of the private sector is still in its infancy and no doubt these issues will be resolved with time and experience.

REFERENCES

Cairncross, F. (1991). *Costing the Earth*. The Economist Books Ltd, London.

CEC (1991) Council Directive concerning urban waste water treatment. (91/271/EEC). *Official Journal of the European Communities*, **L135/40**, (30 May).

CONAMA (1986) *Resolution 20, Article 21*: Effluent Standards Conselho Nacional do Meio Ambiente, Brasilia.

Dunn, A.J., Frodsham, D.A. and Kilroy, R.V. (1998). Predicting the risk to permit compliance of new sewage treatment works. *Water Science and Technology* **38**(3), 7–14.

Garber, W.F. (1992). Factors in environmental improvement: developing and industrial nations. *Water Science and Technology* **26**(7–8), 1941–1951.

Jacobsen, B.N. and Warn, T. (1999). Overview and comparison of effluent standards for wastewater treatment plants in European countries. *European Water Management* **2**(6).

Johnstone, D.W.M. and Horan, N.J. (1994). Standards, costs and benefits: an international perspective. *Journal of Institutional Water & Environmental Management* **8**(5), 450–458.

Johnstone, D.W.M. and Horan, N.J. (1996). Institutional developments, standards and river quality: a UK history and some lessons for industrialising countries. *Water Science and Technology* **33**(3), 211–222.

Johnstone, D.W.M. and Norton, M.R. (2000). Development of standards and their economic achievement and regulation into the 21st Century. In: N.J. Horan and M. Haigh (eds) *Wastewater Treatment Standards and Technologies to meet the Challenges of the 21st Century.* Terence Dalton, UK.

Mara, D.D. and Cairncross, S. (1989). *Guidelines for the Safe Use of Wastewater and Excreta in Agriculture and Aquaculture – Measures for Public Health Protection.* WHO/UNEP, Geneva.

Ross, S.L. (1994). Setting effluent standards. *Journal of Institutional Water & Environmental Management* **8**(6), 656.

von Sperling, M. and Fattal, B. (2001). Implementation of guidelines; some practical aspects. In: *Water Quality: Guidelines, Standards and Health*. World Health Organization, Geneva.

Warn, A.E. and Brew, J.S. (1980). Mass balance. *Water Research* **14**(10), 1427–1434.

WHO (2001). In: L. Fewtrell and J. Bartram (eds) *Water Quality: Guidelines, Standards and Health. Assessment of risk and risk management for water-related infectious disease.* IWA Publishing, London.

Part III Microbiology of Wastewater Treatment

Introduction to Microbiological Wastewater Treatment

19

Fixed film processes

Paul Lessard and Yann Le Bihan

Département de Génie Civil, Université Laval, Québec G1K 7P4, Canada

1 INTRODUCTION

Raw wastewaters have long been discharged directly to receiving waters (and still are, even in some industrialized countries), hoping their self-purification capacity would take care of our waste. Unfortunately, most receiving bodies failed to do so as they were overcharged with organic and nitrogen pollution resulting in a chronic state of degradation. In most cases, wastewater treatment is therefore needed before effluent discharge. One of the most popular types of treatment is the biological one which more or less mimics some of the natural processes found in a self-purifying receiving body, mainly organic degradation and nitrogen conversion through bacterial action. These can be done by attached or suspended microorganisms, giving rise to two main families of wastewater treatment processes (Fig. 19.1): (a) the fixed film processes, discussed here and represented by the trickling filter, and (b) the suspended growth processes, such as the activated sludge, discussed in Chapter 21.

However, one has to remember that to engineer these natural processes into a compact biological sewage treatment processes, one must satisfy the following criteria (Mudrack and Kunst, 1986):

- The biomass concentration in the reactor must be increased
- The increased oxygen demand must be catered for
- An optimal contact between the biomass, the sewage constituents and dissolved oxygen must be ensured
- Inhibitory or toxic substances or inhibiting factors must not be allowed to reach critical levels in the reactor.

These points may seem trivial, but they are in fact not that easy to achieve.

2 FIXED FILM PROCESSES IN WASTEWATER TREATMENT

2.1 Biofilm development

Fixed film processes are based on the capacity of different microorganisms to grow on surfaces. They tend to attach to solid surfaces due to various reasons:

1. substrate availability
2. protection from a harmful environment, particularly at high-velocity water currents
3. interaction of physical forces like attraction, adsorption and adhesion (Senthilnathan and Ganczarczyk, 1990).

Fundamental mechanisms of microbial attachment are reviewed in a subsequent section,

The Handbook of Water and Wastewater Microbiology
ISBN 0-12-470100-0

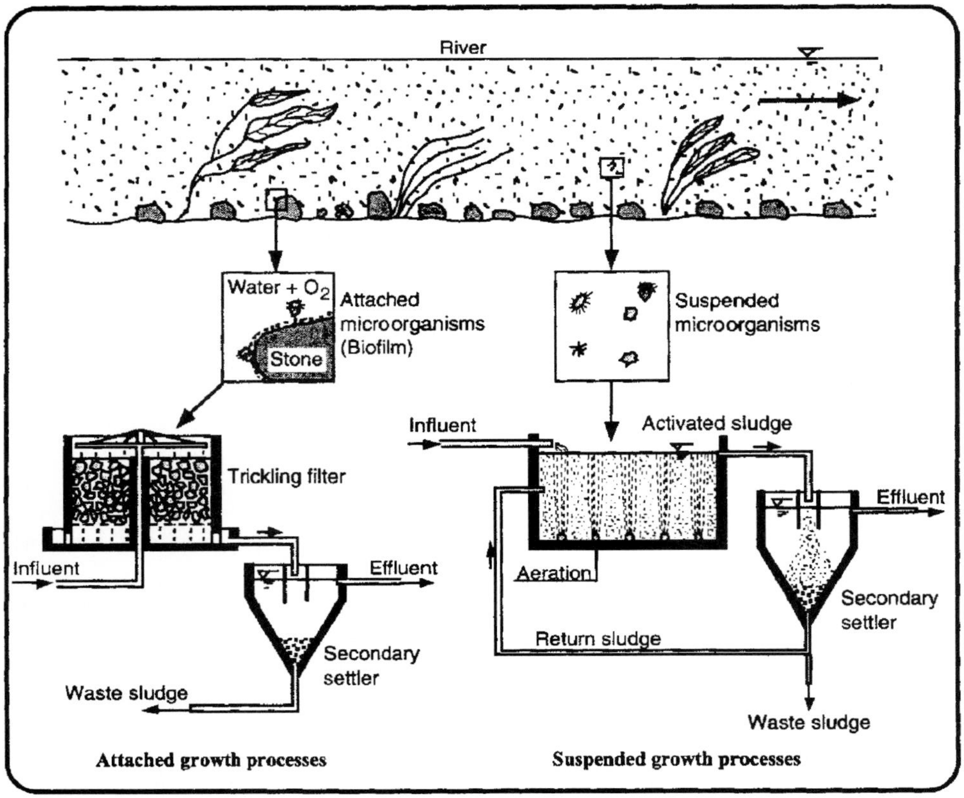

Fig. 19.1 Transfer of self-purification process and organisms from a river to a biological sewage treatment plant (adapted from Mudrack and Kunst, 1986).

but it is necessary here to describe minimally the biofilm formation process before discussing technologies.

Bacterial adhesion is, in general, a three-step process (adsorption, adhesion and adherence) which takes approximately from 20 to 30 minutes under optimal environmental conditions (Brisou, 1995). Fixation mechanisms will depend on the type of bacteria and on the type of surface available for them to colonize.

Adsorbed microorganisms (specifically bacteria) grow, reproduce and produce extracellular polymeric substances, which frequently extend from the cell forming a gelatinous matrix called a 'biofilm' (Bryers and Characklis, 1990). These exopolymers consist mainly of a variety of heterogeneous polysaccharides depending on the type of microorganisms involved (Lazarova and Manem, 1995). A schematic view of a biofilm is shown in Fig. 19.2.

The substrate removal mechanism within a fixed film process is very complex. The phenomena that take place when a biofilm is brought into contact with a wastewater-containing substrate and oxygen are as follows (WPCF, 1988):

- Transport of the substrate and oxygen from the wastewater to the surface of the biofilm
- Internal transport of the substrate and oxygen through the biofilm by diffusional processes
- Oxidation of the substrate within the biofilm
- Diffusion of by-products back to the wastewater.

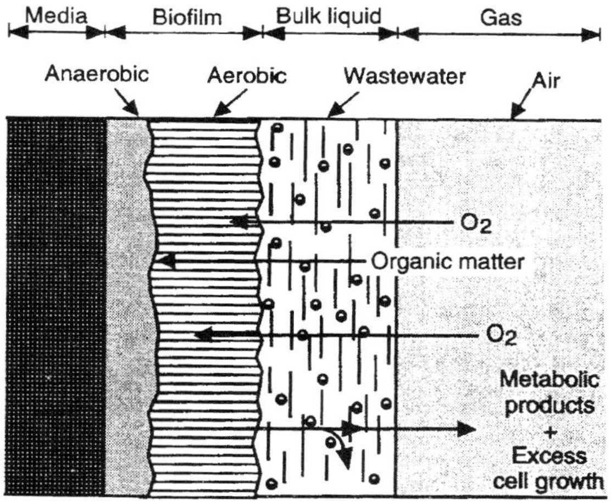

Fig. 19.2 Schematic view of a biofilm.

Typically, after some microorganisms have adhered to a surface medium they start growing, producing the biofilm, which is aerobic. As growth continues, oxygen diffusion across the biofilm may become limiting and an anaerobic layer develops along the surface media. Thus removal of organic matter implies a continuous growth of bacteria and consequently a thicker biofilm. This accumulation process is generally balanced by the biofilm sloughing off. The following conditions are important for the sloughing off to occur and are used technologically, intentionally or unintentionally, to control the biofilm (Henze *et al.*, 1995):

- Hydraulic erosion acts continually on the surface of the biofilm and leads to a steady sloughing off on the outer side
- Degradation of bacteria in endogenous phase at the bottom of the biofilm may cause a weakening of the adhesion
- Gas formation within the biofilm (e.g. methane or nitrogen) may destroy the adhesion.

Development of a bacterial biofilm for substrate degradation is thus a complex phenomenon that is influenced by many factors such as: wastewater characteristics (nature of substrate, nutrients ratio and environmental conditions), operational factors (inoculation, organic loading rate, hydrodynamics) and the support medium.

The development of technologies based on the concept of attached growth therefore necessitates a good knowledge, not only of process behavior but also of microbiology and biofilm formation.

2.2 Usual processes

A possible classification of biological growth systems is shown in Fig. 19.3. Some authors have proposed more specific classifications for fixed film processes (e.g. Crites and Tchobanoglous, 1998; Lazarova and Manem, 1994), but a general one for all types of fixed film processes found in wastewater treatment is difficult to establish as many factors can be considered: type of medium (organic versus inorganic), type of reactors (fixed bed versus moving bed), environmental conditions (aerobic versus anaerobic), wastewater characteristics (municipal versus industrial) and types of pollutants to be removed (C, N, P, xenobiotics). It is difficult to cover in this chapter all possible fixed film technologies for wastewater; only the applications of fixed film processes related to municipal wastewaters are described, which means mainly aerobic processes in packed bed reactor or hybrid systems, and carbon, nitrogen and phosphorus removal.

Fixed-film processes in municipal wastewater treatment are now over a hundred years old, with the introduction in the late nineteenth century of the trickling filter (Peters and Foley, 1983). Since then other processes have appeared, but until the 1980s, trickling filter (TF) and rotating biological contactors (RBC) were almost the only two processes used on a regular basis. Biological aerated filters (BAF) appeared on an industrial scale in the 1980s, while the 1990s saw the arrival of hybrid systems consisting of biomass support systems (BSP) immersed in AS reactors and used mainly for plant upgrading.

2.2.1 *Trickling filters*

The trickling, or percolating, filter, was the first fixed film process that was developed. It is a packed bed reactor (2–3 m deep) filled with a medium consisting of 5–10 cm stones

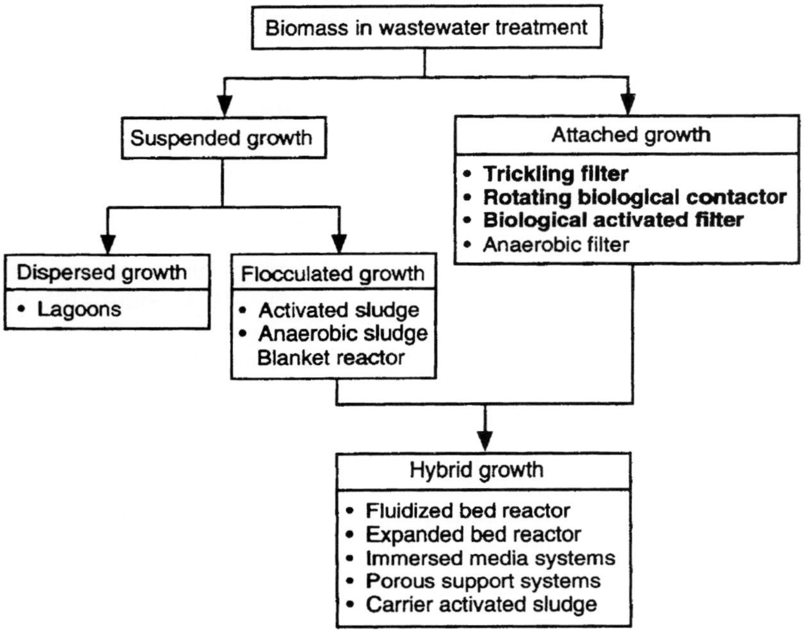

Fig. 19.3 Systematics of biomass forms in wastewater treatment systems (adapted from Senthilnathan and Ganczarczyk, 1990).

with a specific surface area of 40–100 m^2/m^3. The trickling filter is very efficient with respect to adhesion of bacteria, contact between water and biofilm and reaeration of the water (Henze *et al.*, 1995).

After a compulsory primary sedimentation step, the wastewater is distributed over the gravel bed, trickles down to be collected under the filter and flows to a secondary settling tank, also called a humus tank. Aeration is provided through natural drafts resulting from temperature differences between the ambient and the internal air. A typical trickling filter is shown in Fig. 19.4, and its advantages and disadvantages are summarized in Table 19.1.

Trickling filters can be operated in different modes: single pass, alternating double-filtration and recirculation mode (Bitton, 1994). Depending mainly on the organic and hydraulic loading rate, trickling filters will be classified as low, standard (or intermediate) and high rate. Design criteria are given in Table 19.2, and more information can be found in standard textbooks (Metcalf and Eddy, 1991; WEF and ASCE, 1992; Henze *et al.*, 1995; Crites and Tchobanoglous, 1998; Grady *et al.*, 1999). However, according to Parker (1999), engineering practice in trickling filter design has been influenced by the propogation of myths as much as by the analysis of data and factual determinations.

Effluents from well-operated trickling filters range around 20 mg/l for BOD_5 and SS (Crites and Tchobanoglous, 1998). At weak organic and hydraulic loadings (around 0.2 kg BOD/m^3day

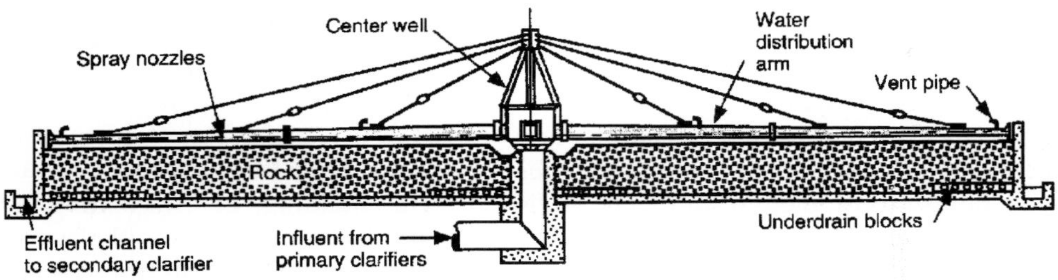

Fig. 19.4 Typical trickling filter.

TABLE 19.1 Some advantages and disadvantages of trickling filters

Advantages	*Disadvantages*
Ease of operation	Filter clogging for high organic loads results of excessive slime bacteria
Low maintenance and energy costs	Potential odor problems
Reliability	Flies
Sloughed biofilm easily removed by sedimentation	No real possibilities for operational control
Able to treat industrial wastewaters	Pretreatment and primary sedimentation necessary
Withstand shock loads	

and 0.66 m^3/m^2 day) trickling filters can nitrify fully (Henze *et al.*, 1995). The process is thus capable of compliance with effluent discharge consents of 2 mg NH_3-N/l, as shown by Pearce and Williams (1999).

The use of plastic media, with a specific surface area around 200 m^2/m^3, has made it possible to create a large volumetric surface in low-weight filters, and these can therefore be built as towers. Higher organic and hydraulic loading rates can be applied: 1–5 kg BOD/m^3 day and 36–72 m^3/m^2 day. Such filters are, however, typically used for the treatment of industrial wastewaters.

2.2.2 *Rotating biological contactors*

Rotating biological contactors (RBCs) consist of a series of circular plastic disks mounted on a horizontal central shaft and distanced at 1.5–2.5 cm. Normally 40% of the disks' surface is submerged and they are rotated in a tank containing the wastewater. Attached microorganisms rotate into the water, where organic matter is adsorbed onto the biofilm, and out of the wastewater, where the oxygen necessary for the conversion of organic matter is obtained by adsorption from the air. Thus, the rotation provides aeration and the shear force that causes sloughing off of the biofilm from the disk surface. The peripheral speed on the disks should not be less than 0.3 m/s (Henze *et al.*, 1995). A typical RBC is shown in Fig. 19.5, and its advantages and disadvantages are given in Table 19.3.

TABLE 19.2 Design criteria for trickling filters

Design criteria	*Intermediate rate**	*Reference*
Hydraulic loading (m^3/m^2 day)	3.52–9.39	Crites and Tchobanoglous (1998)
	14.40–28.80	Henze *et al.* (1995)
	–	Agences de l'eau (1994)
	4.5–25	Edeline (1993)
Organic loading (kg BOD/m^3 day)	0.24–0.48	Crites and Tchobanoglous (1998)
	0.45–0.75	Henze *et al.* (1995)
	–	Agences de l'eau (1994)
	0.40–1.0	Edeline (1993)

* Some call those normal, standard.

Some authors suggest that the suspended biomass found in the reactor may also contribute to some 4 to 10% of pollution removal (Edeline, 1993). However, due to the short retention time, these additional removals can be neglected (Benefield and Randall, 1980).

Design criteria are given in Table 19.4, and more information is given in standard textbooks (Metcalf and Eddy, 1991; WEF and ASCE, 1992; Henze *et al.*, 1995; Crites and Tchobanoglous, 1998; Grady *et al.*, 1999).

Good quality effluents (approximately 15 mg BOD/l) can be obtained with organic loadings in the range given in Table 19.4. However, according to the Bavarian Water Authority, the organic loading should not exceed 10.5 g/m^2 day, if an effluent BOD less than 25 mg/l is desired (Edeline, 1993). Nitrification can be achieved in RBCs, provided a second or third stage of disks is added.

2.2.3 *Biological aerated biofilters*

Over the last 20 years biological aerated filters (BAF) have been developed not only as competitors for other secondary treatment systems, such as tricking filters and activated sludge plants, but also for tertiary treatment (Mann *et al.*, 1999). These biofiltration processes are popular in Canada (Pineau and Lessard, 1994) and France (Canler and Perret, 1994), and they are gaining interest in other countries as well. The possibility of achieving carbon oxidation with nitrification/denitrification in

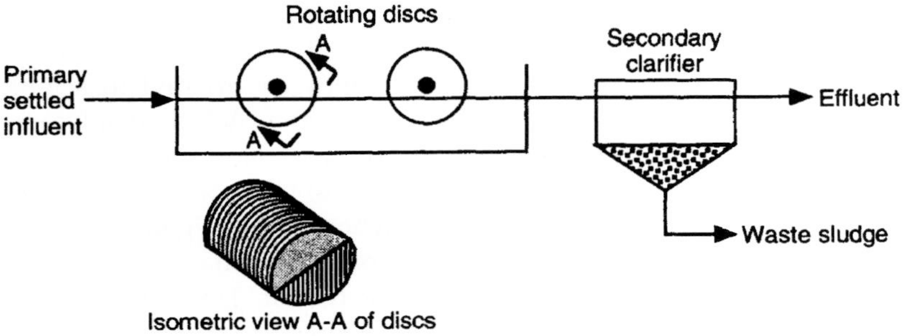

Fig. 19.5 Typical RBC system.

the same reactor is probably a major reason for its increasing usage.

This technology was developed for retention of suspended solids and for oxidation of carbonaceous soluble pollution in the same reactor. The water flows through a packed bed media which is generally inorganic with a particle size around 3–5 mm and a specific surface area of 300–500 m^2/m^3. Aeration for oxidation is done at the base of the biofilter and water flows upwards or downwards, giving two main types of biofilters: upflow or downflow biofilters (Fig. 19.6).

Given that the biofilm grows and suspended solids are retained within the biofilters, clogging occurs and backwashing of the process is necessary. A filtration cycle length is typically between 12 and 48 hours and backwash with air and water takes around one hour (Agences de l'Eau, 1994). The advantages and disadvantages of BAFs are given in Table 19.5.

Many types of BAFs exist and different organic loading rates can be found in the literature (Stephenson, 1997). However, the Agences de l'Eau (1994) suggested the following design criteria for biofiltration processes: for hydraulic loading, 2–5 m^3/m^2 h; and organic loading, 6–8 kg COD/m^3 day. Nitrogen removal (nitrification) is possible with BAFs (Lacamp *et al.*, 1992; Stephenson, 1997; Canler *et al.*, 2002) together with denitrification (Peladan *et al.*, 1997).

The second generation of submerged filters involves a floating support made from a synthetic material with a density lower than that of water. This overcomes some of the disadvantages of the original systems, especially in the filter regeneration phase, for which large amounts of fluid (water and air) and power can be consumed to loosen the bed and wash away the excess sludge (Capdeville and Rols, 1992). Floating media have been found to perform better than sunken media for SS, COD_t and ammonia removal and at higher flow rates under shock loading conditions (Mann *et al.*, 1998, 1999).

The trend in BAFs is now to look at more compact filters able to perform at high water velocities (10–33 m^3/m^2 h) and to achieve nitrification/denitrification within the same biofilter (Peladan *et al.*, 1997; Puznava *et al.*, 2001). Studies are also being undertaken to determine the feasibility of removing phosphorus biologically through sequencing of

TABLE 19.3 Some advantages and disadvantages of RBCs

Advantages	*Disadvantages*
Short residence time	No operational flexibility
Low operation and maintenance costs	Has to be covered for cold climates
Low energy costs	Need of primary sedimentation
Good settling characteristics of sloughed materials	Possible dryness of unsubmerged biofilm portion (warm climates)

TABLE 19.4 Design criteria for RBCs

Design criteria	*Criteria*	*Reference*
Hydraulic loading (m^3/m^2 day)	0.04–0.15	Edeline (1993)
Organic loading (kg BOD/m^2 day)	0.005–0.026	Henze *et al.* (1995)
	0.015–0.060	Edeline (1993)

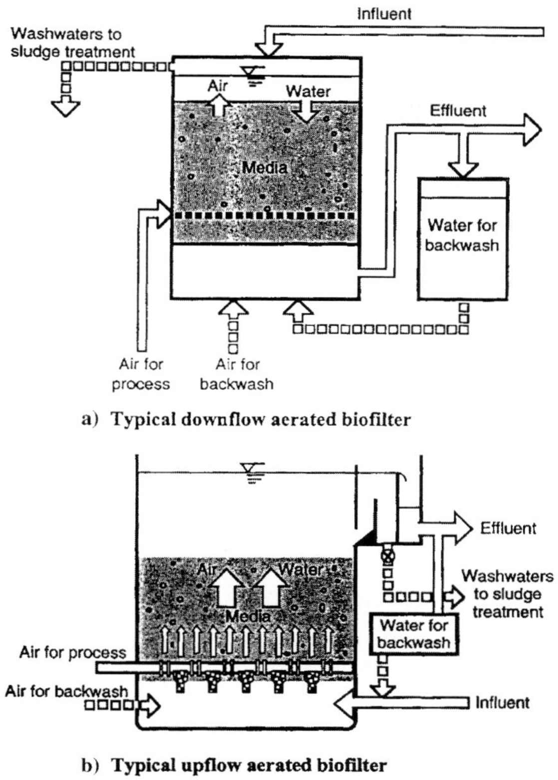

Fig. 19.6 Aerated biofilters. (a) Typical downflow aerated biofilter; (b) typical upflow aerated biofilter.

the filter operation (Morgenroth and Wilderer, 1999).

2.3 Microbiology of standard processes

For all processes, the continuous addition of substrate and microorganisms present in the wastewater encourages the formation of a

TABLE 19.5 Advantages and disadvantages of biological aerated biofilters

Advantages	*Disadvantages*
Good quality of treatment	Transient operation
Compact process, no secondary settlers required	High energy consumption
Fast recovery after problems	Large volumes of washwaters
Start up quick even after several months stopped	

complex biofilm composed of bacteria, fungi, protozoa, macroinvertebrates (larvae, worms) and sometimes algae (Bruce and Hawkes, 1983). Fig. 19.7 shows in which order the colonizing microorganisms appear in a biofilm, which already contains several cellular waste and inert particles from the wastewater being treated.

The microbiology of attached growth processes is somewhat similar from one process to another. Possible differences are attributable to the maturity of the biofilm, which varies between processes, thus influencing microbial colonization.

2.3.1 *Bacteria*

Bacteria are unicellular prokaryotic organisms. Besides being the most abundant organisms in the fixed film treatment processes ($\pm 10^{10}$ CFU/g VSS), they are the first to colonize the biofilm (Mack, 1975). Their small size and large surface with relation to their volume, gives them a clear advantage for the assimilation of the substrate in the liquid phase. Under optimal growth conditions, the doubling time of heterotrophic bacteria is about 20 minutes, which helps them out-compete any other organism in the process.

Generally, for fixed film processes, the dominant bacterial genera are very similar to those found in activated sludge. During the aerobic treatment of domestic wastewater, these genera are often Gram-negative heterotrophic rod-shaped organisms, including *Zooglea*, *Pseudomonas*, *Chromobacter*, *Achromobacter*, *Alcaligenes* and *Flavobacterium*. There are also many coliforms in the treatment processes, but they are not considered indigenous members of the microbial community (Gray, 1989), rather coming in with the *influent* and retained by the process (Tremblay *et al.*, 1996). Filamentous bacteria such as *Beggiatoa*, *Thiotrix* and *Sphaerotilus* genera are also found in the biofilm (Vedry, 1996), as are nitrifying organisms like *Nitrosomonas* and *Nitrobacter* (Bitton, 1994). For a thick biofilm, certain facultative or obligate anaerobic species are susceptible to being buried deep in the anoxic or anaerobic zone of the biofilm. These bacteria include sulfate-reducing bacteria, facultative aerobic bacteria

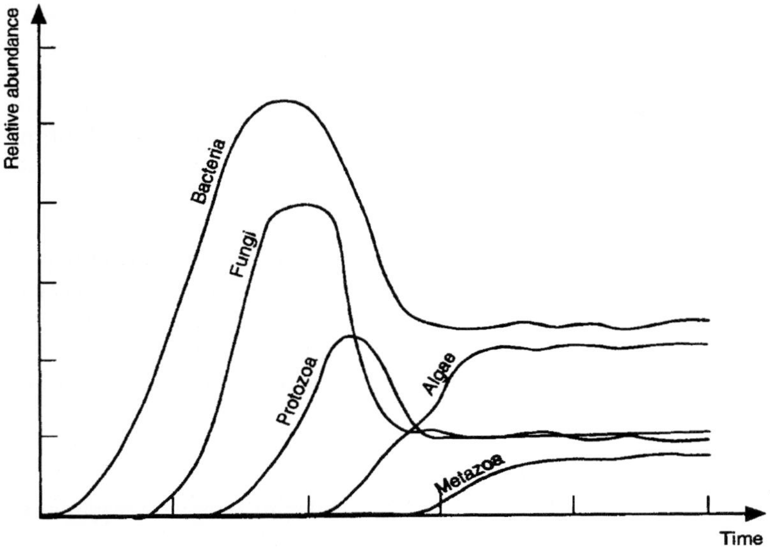

Fig. 19.7 Relative succession of microbes constituting microbial film (adapted from Iwai and Kitao, 1994).

(fermentative) and methanogenic bacteria (Arvin and Harramoës, 1990). However, these anaerobic bacteria are not, of course, dominant in the aerobic processes (Crowther and Harkness, 1975).

Role and nuisance of bacteria. Most bacteria in fixed film processes take part in the removal of the soluble pollutant matter. The colloidal part is adsorbed on the biofilm and hydrolyzed by extracellular enzymes and used later (Bitton, 1994). Certain pH, temperature, aeration and low loading conditions (0.1–0.4 kg BOD_5/m^3/day) can encourage the growth of autotrophic nitrifying organisms. Denitrification can take place in the anoxic zone of the biofilm and methanization, carried out by biofilm organisms, can occur in the anaerobic zone, i.e. close to the support medium.

Generally, bacteria are the dominant organisms in the biofilm, and few species are known to hinder process performance. However, certain species such as *E. coli* or *Nocardia* spp. perform less well than others in removing carbonaceous pollution (James, 1964).

Under certain conditions, filamentous bacteria can develop as a thick biofilm, resisting sloughing or shearing, thus encouraging biofilter clogging (Vedry, 1996). For processes using rotating biological contactors, the presence of organisms, like *Beggiatoa,* increases biofilm accumulation, thus favoring a biomass surcharge at the disk surface, and a decrease in oxygen diffusion into the biofilm (Metcalf and Eddy, 1991); this situation may generate some operational problems (WEF and ASCE, 1992).

2.3.2 Fungi

Fungi are single- or multi-cellular eukaryotic organisms. In fixed film processes, they play a role of purifier, important for the carbonaceous substrate. In lower quantities than bacteria, about 10^5 and 10^7 CFU/g VSS for fungi and yeast respectively (Le Bihan and Lessard, 1998). Their high hydrolytic potential makes them very competitive against bacteria. Their growth is favored over that of bacteria by cold temperatures, an acid pH (<5) and an industrial or toxic *influent*, especially one that is highly concentrated or contains carbohydrates (Tomlinson and Williams, 1975). Except for yeasts, most fungi are obligate aerobes. The average doubling time of fungi varies between 2 and 14 hours.

The most frequent fungi colonizing fixed film processes are: *Sepedonium* spp., *Subarromyces splendens, Ascoidea rubescens, Fusarium aquaeductuum, Geotrichum candidum* and *Trichosporon cutaneum* (Tomlinson and Williams, 1975). Cooke (1963) identified over a hundred species of fungi in wastewater and in the biomass of biological processes.

Role and nuisance of fungi. Fungi take part in the removal of the carbonaceous substrate in a similar manner to heterotrophic bacteria. However, the conversion rate, much higher than for bacteria, encourages a significant production of biomass for a given quantity of substrate. Too great a proportion of fungi in the biomass favors the formation of a highly resistant biofilm that is difficult to slough. As oxygen can diffuse through fungal protoplasm deep into the biofilm, there is an accumulation of fungal biofilm (Gray, 1989). This characteristic helps the fungal biofilm maintain a greater thickness of aerobic conditions, thus limiting the sloughing effect favored by the deep anaerobic zone.

For trickling filter processes, such as biological aerated filters, these growth characteristics help the development of an abundant biofilm, which could accelerate clogging of the process (Tomlinson and Williams, 1975).

It seems fungi have very specific growth cycles based on the season. The maximum number of fungi is reached in winter and in early spring; they become rarer in midsummer (Tomlinson and Williams, 1975). During the warm period (25°C), only *Furasium*, *Geotrichum* and *Trichosporon* spp. have a doubling time short enough (7–8 hours) to compete with zoogleal bacteria with a doubling time of 10–25 hours (Gray, 1989). As their affinity for the substrate is lower than that of the bacteria, their growth is favored when the *influent* is concentrated (>280 mg BOD_5/l) and generally at the surface of the filter, where the substrate is more concentrated (Gray, 1989). Thus, for cold temperatures and a concentrated substrate, the fungal biomass is clearly favored in relation to the bacteria.

2.3.3 *Protozoa*

Protozoa are unicellular eukaryotic organisms widespread in fixed film processes. More than 218 species were identified (Gray, 1989). These organisms, larger than bacteria, are mostly aerobic. They mainly feed on organic particulate matter, bacteria, algae or other *Protozoa*. Certain genera of Mastigophora flagellates are able to use a soluble substrate, but are less competitive than bacteria and fungi (Bruce and Hawkes, 1983).

Protozoan growth is slower than bacterial growth, the latter have an observed average doubling time of 3 to 22 hours (Pike, 1975; Iwai and Kitao, 1994). Their quantity varies with the type of *influent* and aeration conditions but there are generally between 10^6 and 10^8 organisms/g VSS. Protozoa are divided into three branches: Mastigophora (phytomastigina and zoomastigina), Rhizopoda (heliozoa and amoebae), and Ciliophora (ciliated). A few studies have identified ciliated Protozoa as being the most numerous in the biomass (Curds and Cockburn, 1970; Hoag *et al.*, 1983; Hull *et al.*, 1991). The most abundant ciliated protozoa (Ciliophora) in trickling filter processes are fixed organisms of the genera *Vorticella*, *Opercularia*, *Carchesium* and swimmers like *Aspidisca* and *Chilodonella* (Curds and Cockburn, 1970). These authors also showed that *Rhizopoda* (amoebae), which are slightly less abundant than the ciliates, are frequently represented by the *Arcella* (testate amoebae) genus. Naked amoebae are rather rare in fixed film processes. Mastigophora mostly include *Paranema*, *Bodo* and *Trepanomas* spp.

Role and nuisance of protozoa. Protozoa take on the role of predators of bacteria and take part in maintaining bacterial growth by reducing their population density (Gray, 1989). Certain Mastigophora species are able to assimilate particulate and dissolved organic matter. The absence of Mastigophora in a biomass is an indicator of toxic conditions (Vedry, 1996).

The Rhizopoda found in the biomass of fixed film processes are mainly composed of Thecamibae (testate amoeba). These shelled organisms do not swim, but move by using their pseudopod on the biomass. They feed on mainly fixed, agglomerated, filamentous bacteria or unicellular algae. Ciliates feed on suspended particles, loam and free bacteria. All are very sensitive to a lack of oxygen. Some swimming ciliate species also feed on protozoa and flocculated bacteria. Other species (e.g. *Chilodonella cucullulus*) have an oral morphology, which permits them to feed on filamentous bacteria and fungi. Fixed ciliates (peritricha) require a support on which to fix, and they feed mainly on free bacteria.

Only a few studies deal with the nuisance caused by protozoa. However, Vedry (1996) shows the inconveniences of an invasion by fixed ciliates, specifically, colonial peritricha. When they invade a biological aerated filter, they can hardly be removed from the support, even with a wash. This fastening capacity increases their number to a point of equilibrium with the turbulence in the environment. According to this author, accumulation of fixed ciliates increases oxygen consumption and head loss of the process. Other studies also mention that the presence of protozoa and other predators could affect the development of microorganisms with a low growth rate, such as the bacteria responsible for nitrification (Lee and Welander, 1994).

2.3.4 *Algae and cyanobacteria*

Several photosynthetic organisms, algae and cyanobacteria, can grow exposed to light at the surface of a biofilm. The difference between algae and cyanobacteria is that algae are single or multicellular photosynthetic eukaryotes, while cyanobacteria (blue-green) are unicellular photosynthetic prokaryotes. The frequently found genera include *Ulothrix*, *Phormidium*, *Anacystis*, *Euglena*, *Stigeoclonium*, *Chlorella* and *Oscillatoria* (Cooke, 1959). Their doubling time is similar to that of protozoa, varying from 7 to 25 hours, depending on the species (Iwai and Kitao, 1994).

Roles and nuisance of algae and cyanobacteria. Algae and cyanobacteria only play a small role in fixed film processes (Vedry, 1996). Some species, especially *Phormidium cyanobacterium*, develop into multilayers at the surface of the trickling filter and decrease the efficiency of the process, eventually clogging the filter. This layer of photosynthetic organisms also serves as a favorable environment for fly breeding that could cause nuisance during their emergence (Bruce and Hawkes, 1983).

2.3.5 *Metazoa*

Metazoa are multicellular eukaryotic organisms that encompass the phyla of worms (rotifers, nematodes, gastrotricha and oligochaetes) and arthropods (arachnids, tardigrades, crustacea, myriapods and insects). With their relatively high doubling time, between 47 and 238 hours (Iwai and Kitao, 1994), these organisms are only present in an aged biomass. Most of them are aerobic; however, some organisms can survive under microaerophilic conditions (nematodes, oligochaetes).

Roles and nuisance of Metazoa. These predators and detritivorous organisms feed on the biofilm, embrittle it, encourage loosening, decrease biomass volume and accelerate mineralization. Their grazing capacity also helps avoid clogging of trickling filters through biofilm consumption. With their low activity in cold water (winter), trickling filter clogging is often observed during this period, especially in the case of a growing fungal biofilm (Hawkes, 1983). Nematodes are often observed where voluminous, potentially anoxic, bacterial masses with low oxygen permeability are found. Their role in biomass mixing allows the oxygen to reach the anoxic zones inside these bacterial masses.

Metazoa are frequently observed in trickling filter processes, but rarely in aerated biological filter processes. The frequent washings of these filters control the age of the biomass. Thus the presence of oligochaetes, rotifers and nematodes in an aerated biological filter process is an indicator of poor operation of the washing process, which allows the undesirable accumulation of biomass with a high sludge age (Vedry, 1996). Moreover, since they are light, oligochaete dejects could find their way into the effluent of upflow biofilters, thus increasing the concentration of suspended solids.

For arthropods, the *Psychoda* and *Sylvicola* larvae are also excellent biofilm consumers (Bruce and Hawkes, 1983). They are mainly found in non-submerged filters. Their development and emergence into adult insects (filter flies) can cause invasion and nuisance problems for the trickling filter operators. Flow increase or sporadic immersion of the filter helps control and reduce larval development. However, there are very few data regarding the problems caused by arthropods in wastewater treatment processes.

2.3.6 *Environmental effects on biofilm ecology*

Some microorganism species, favoring or not proper process operation, will develop with relation to various operational parameters such as: substrate, nutrients, dissolved oxygen, temperature, pH, and hydraulic loading.

Substrate. Substrate composition imposes a selective strain on biomass species (e.g. heterotroph versus autotroph). Normally, soluble compounds such as sugars, acids and amino acids are much more easily assimilated by a large quantity of heterotrophic organisms than are insoluble or partially soluble compounds such as cellulose, lignin or fatty acids with high molecular weights. These substrates are used as secondary substrates by specialized organisms. Moreover, sugars are consumed by many microorganisms, whereas proteins and fats are restricted to more specialized organisms.

Substrate concentration also has an effect on biofilm nature. Generally, growth of an abundant, easy-to-slough biofilm is obtained using a readily oxidizable substrate (Heukelekian and Crosby, 1956). However, according to these authors, more biofilm is accumulated in the process when fed by a weak influent. Seek Park *et al.* (1998) observed that the biofilm at the surface of a rotating biological contactor was thicker, less dense, filamentous and contained more extracellular polymers (EPS) when the process was fed by a concentrated influent. For a diluted *influent*, the biofilm was thin, but very dense.

From the selection aspect, an easily bioavailable substrate (sugar) favors the growth of filamentous bacteria such as *Sphaerotilus natans* (Gray, 1989). A very concentrated influent favors the growth of fungi.

Nutrients. Generally, a C/N ratio value of 18 or less is required for adequate growth and treatment efficiency. Values above 22 decrease performance and favor the development of filamentous organisms (Hatting, 1963). A fungal biofilm can also develop with an influent with a very high C/N ratio (Tomlinson and Williams, 1975). These conditions are commonly found in the treatment of industrial effluents.

When using a soil to treat wastewater, a high C/N ratio favors the production of extracellular polymers (EPS) and decreases the soil's hydraulic conductivity (Magesan *et al.*, 1999). It seems that similar clogging phenomena could be observed in aerated biological filter processes with a small-sized medium (2–3 mm).

For phosphorus, a C/P ratio lower than 90 to 150 is required to maintain treatment efficiency (Gray, 1989). However, no specific problems related to phosphorus deficiency have been reported in the literature.

Dissolved oxygen. Oxygen is used as a final electron acceptor by aerobic microorganisms and their growth rate increases with increasing dissolved oxygen concentration. In biofilm processes, the anoxic zone normally starts 0.1–0.2 mm below the biofilm surface; in the case of a fungal biofilm, this zone can reach down to 2 mm due to the diffusion of oxygen through the fungal protoplasm (Gray, 1989). In this case, the aerobic layer is thicker and the sloughing caused by the development of the anaerobic zone is delayed and the biofilm thickens. Applegate and Bryers (1991) have shown that a biofilm limited by oxygen sloughs more easily than a carbon-limited biofilm.

Temperature. On average, domestic wastewater temperatures vary between 10 and 20°C and, in general, trickling filter processes are more sensitive to temperature than activated sludge (Bruce and Hawkes, 1983). Influent temperature affects the performance of fixed film processes since this factor has an impact on microbial populations. The first effects are the slowdown of microbial metabolism. Under normal circumstances, the activity doubles with each temperature increase of 10°C between 5 and 30°C. Temperature also has an inhibiting effect on nitrification, which is drastically reduced at temperatures below 10°C (Bruce and Hawkes, 1983).

In addition to decreasing metabolism, cold temperatures favor the accumulation of biofilm in the process (Heukelekian, 1945). Cold winter temperatures decrease the activity of grazing fauna on the biofilm and favor an accumulation of solids in the process (Bruce and Hawkes, 1983), a process critical for biofilm clogging (Hawkes, 1983). Moreover, at cold temperatures, several fungi grow more quickly than bacteria and thus can dominate in the biofilm.

pH. pH effects microbial growth. Fungi develop between a range of pH (2.5–9) wider than that of bacteria, and are favored at pH $<$ 6.5. Bacteria dominate under neutral pH conditions (Gray, 1989). The optimal pH for oxidation of carbonaceous compounds is about 6.5–8.5, which corresponds to the pH of domestic wastewater. For nitrification, an optimal pH of 7.5 and 8.5 has been reported (Bock *et al.*, 1986).

Hydraulic loading. The shear forces generated by the water flow favors sloughing of the biofilm, although this depends on the nature and type of biofilm. A fungal biofilm is more resistant to shear than a bacterial biofilm (Tomlinson and Williams, 1975). Nitrifying biofilms are thinner and denser, but slough more easily than heterotrophic biofilms (Oga *et al.*, 1991).

In rotating biological contactor processes, sloughing is influenced by rotational speed. A rapid flow rate delays initial biofilm formation, but favors biofilm development later on (Heukelekian and Crosby, 1956). In aerated biological filter processes, the flow rate also favors biomass distribution deep into the filter and accelerates clogging (Visvanathan and Nhien, 1995).

2.4 Comparison between attached and suspended growth

It is important at this stage to make a comparison between suspended and attached growth processes before discussing hybrid systems. The main characteristics of both types of processes are given in Table 19.6. To summarize briefly the difference between the two types of processes: fixed film processes (TF and RBC) seem to be less efficient than suspended growth processes (activated sludge), but they are more stable. (The main difficulty with AS is to grow a biomass that will separate well.)

Attached microorganisms possess some advantageous properties compared to suspended microorganisms. Often they exhibit:

1. Increased persistence in the system
2. Faster growth rate
3. Increased metabolic activity
4. Greater resistance to toxicity (Senthilnathan and Ganczarczyk, 1990).

An ecological comparison between the process activated sludge and a trickling filter is made in Fig. 19.8. Biomass diversity is greater for the trickling filter process than for the activated sludge, which means a process that is more stable and less subject to environmental conditions.

Most importantly, the two forms of biomass, fixed and suspended, differ in their main parameter, the solids retention time Θ_x, and this may be very advantageous in cases where fast and slow degradable components have to be combined, e.g. organic removal with nitrification (Wanner *et al.*, 1988).

2.5 Hybrid systems

The optimal growth conditions differ for each type of microorganism involved in biological treament (nitrifiers, heterotrophs, etc.), and several compromises have to be made to enable all the bacteria to achieve as high a degree of activity as possible. The use of hybrid systems can help to achieve that degree of activity. Hybrid systems can be defined by two types of processes, i.e. systems that use: (1) a biofilm reactor, usually a trickling filter (TF), and an activated sludge (AS) process in series, and (2) a biomass support system (fixed or mobile) immersed in an activated sludge reactor.

TABLE 19.6 Biofilm systems vs suspended-growth systems in wastewater treatment (adapted from Shieh (1987) and Bryers (1987))

Biofilm systems	*Suspended-growth systems*
Microorganisms are retained within the biofilm attached to the media.	Microorganisms are suspended in the wastewater by means of mixing and/or aeration.
Fair protection against the adverse effects of toxicants in the influent.	Sensitive to the adverse effects of toxicants in the influent.
Inter- and intra-phase mass transfer is significant because the biofilm structure tends to retard the rate of transport of substrate through it. Liquid-biofilm interface forms another resistance to the transport of substance across it. Therefore, the biofilm systems are heterogeneous systems.	Inter- and intra-phase mass transfer is insignificant. Vigorous mixing and/or aeration reduces the thickness of the liquid film surrounding the flocs and the size of the flocs which, in turn, reduce the effects of mass transfer. Therefore, the suspended-growth systems are treated as homogeneous systems.
The system performance is not affected by the performance of the secondary clarifier. The biomass in the reactor is maintained through the attachment of microorganisms on the media surface and the subsequent growth of biofilm.	The system performance is intimately linked to the performance of the secondary clarifier because the maintenance of the desirable biomass concentration in the reactor depends upon the recirculation of concentrated microbial solids from the secondary clarifier with proper thickening function.
Elimination of wash-out restrictions.	Possible wash-out of biomass.
Secondary clarification in some cases may be eliminated because the suspended solids level in the system effluent is very low. This is possible because most of the biomass is retained within the biofilms rather than suspended in wastewater.	Secondary clarification is required to reduce the effluent suspended solids concentration to the acceptable level.
Unified design approaches are not fully developed yet because the biofilm systems are generally complex.	Unified design approaches, such as F/M approach and SRT approach are now well developed and widely employed for practical applications.

2.5.1 *Systems in series*

Secondary treatment processes combining fixed growth (trickling filter) and suspended growth (activated sludge) systems in series are popular, especially in the USA. These processes offer simplicity of operation and process stability of fixed film processes with the high quality effluent associated with suspended growth processes, and they may also offer economic advantages in comparison to other treatment options (Harrison *et al.*, 1984). Besides this stable performance, sludge with good settling properties is produced with those processes (Harrison *et al.*, 1984), which could be explained as follows (Wanner *et al.* 1988):

- large particles of biofilm, in which filamentous microorganisms are fixed in a matrix of extra cellular polymers, are generated in the preceding biofilm reactor;
- according to a conventional selection theory, preceding biofilm reactor can act as selector.

Various combinations exist, and typical hybrid systems in series are shown in Fig. 19.9. Design information can be found in standard textbooks (Metcalf and Eddy, 1991; WEF and ASCE 1992; Crites and Tchobanoglous, 1998).

These processes have proved to be adaptable to a variety of situations including both large and small plants, cold and warm climates, and a wide variety of effluent requirements (Parker *et al.*, 1993, 1998; Parker and Bratby, 2001). Well-designed and properly operated combined TF and AS plants of all modes are capable of producing secondary effluent quality (30 mg/l BOD and SS) and many can produce even better quality effluents (Harrison *et al.*, 1984). An economic trade-off exists between the more energy-efficient TF/SC and ABF processes and the less capital-intensive RF/AS and BF/AS processes (Grady *et al.*, 1999). The microbiology of such systems is thought to be similar to that of trickling filters and activated sludge processes.

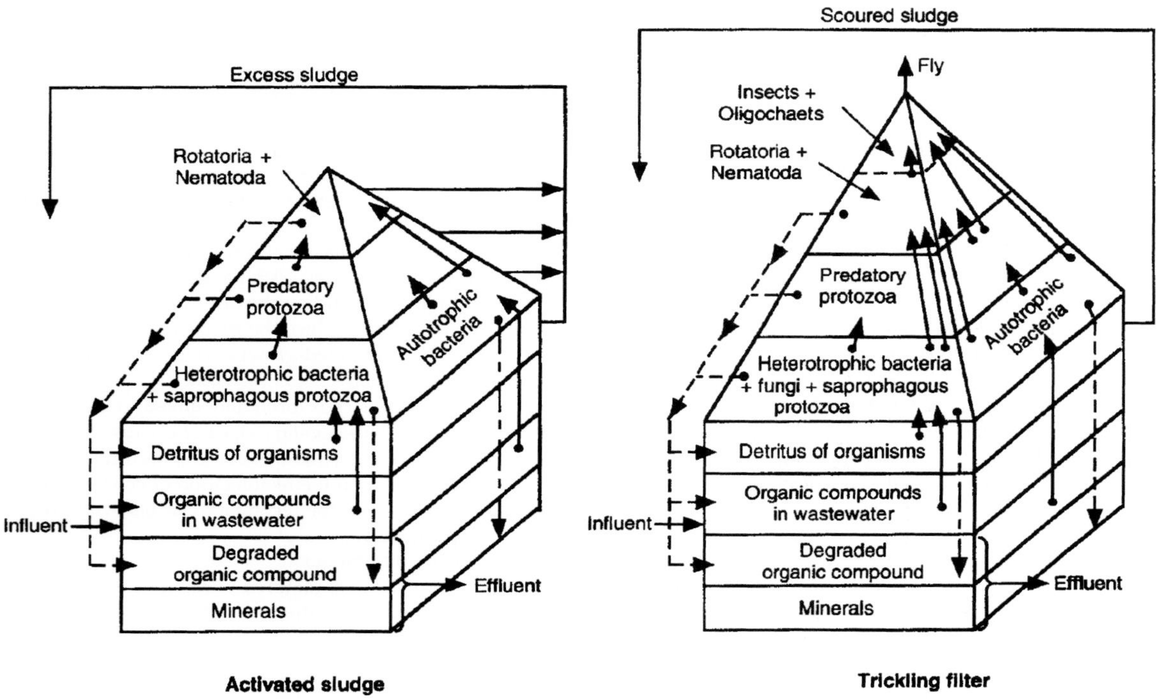

Fig. 19.8 Comparison of biotas between activated sludge and microbial film (adapted from Iwai and Kitao, 1994).

2.5.2 Biomass support systems

Biomass support systems consist of immersing various types of support media in an activated sludge reactor to favor the growth of fixed bacteria. The support can be fixed in the reactor or can consist of mobile media such as foam pads, small carriers, etc. (Fig. 19.10).

These hybrid systems should allow a reduction in the aeration tank volume following the introduction of biomass support to meet a certain objective, and thus an increase in the treatment system stability and performance (Gebara, 1999). The main advantages of these systems are improved nitrification and an increase in sludge settleability (Wanner *et al.*, 1988; Muller, 1998). Nicol *et al.* (1988), based on mathematical simulation, also suggested the following advantages: ability to resist failure from large hydraulic surges; stable nitrification under transient inhibitory conditions resulting from temperature changes, hydraulic surges, and/or toxic chemicals; and the ability to establish stable operating conditions with respect to both carbon oxidation and nitrification at short hydraulic retention times and low sludge ages.

Kinetic analysis of such systems have been made and models proposed (Nicol *et al.*, 1988; Hamoda, 1989), but no clear design guidelines exist for such systems. However, one point that must always be remembered when considering an activated sludge process with biomass supports is that, because of a higher biomass concentration, the volumetric oxygen requirement will be greater in these systems than in conventional activated sludge. Hence, reductions in aeration tank volume proposed for these systems will be limited by the aeration system's ability to handle higher volumetric oxygen demands (Nicol *et al.*, 1988).

These types of hybrid systems are mainly used for the upgradation of activated sludge systems, either to increase the effective biomass (higher treatment capacity) or to implement nitrification. Many examples can be found in the literature (Morper, 1994; Randall and Sen,

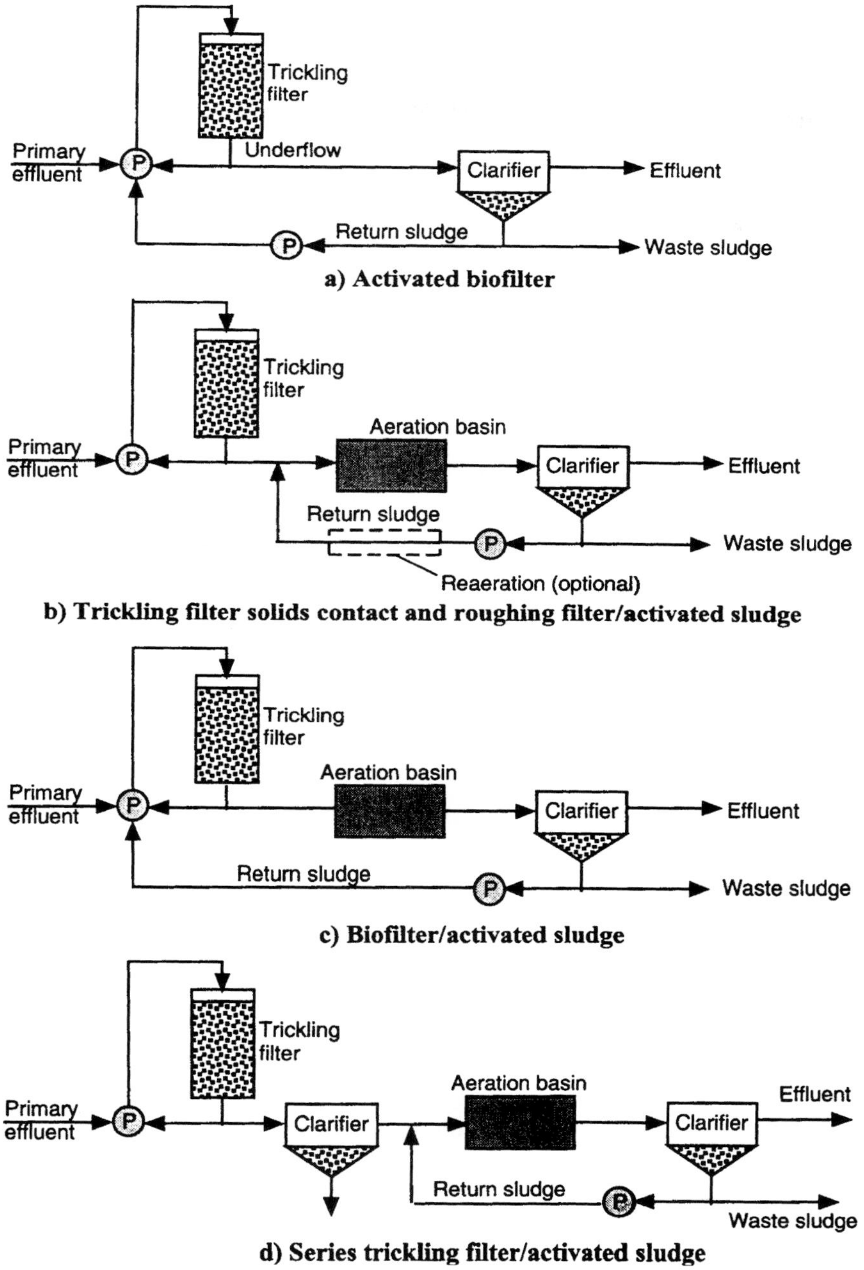

Fig. 19.9 Typical combined attached- and suspended-growth aerobic treatment system flow diagrams. (a) Activated biofilter; (b) trickling filter solids contact and roughing filter/activated sludge; (c) biofilter/activated sludge; (d) series trickling filter/activated sludge (adapted from Crites and Tchobanoglous, 1998).

1996; Jones *et al.*, 1998; Muller, 1998; Rusten and Neu, 1999). Moreover simultaneous denitrification, probably in the deeper layers of the biofilm, has been observed in plants with stable nitrification (Muller, 1998). These types of systems can also be operated in sequencing batch mode to permit phosphorus removal (Garzon-Zuniga and Gonzalez-Martinez, 1996; Wang *et al.*, 1998).

Unfortunately, not much has been written on the microbiology of such systems. Most studies rapidly qualify the biomass. For example,

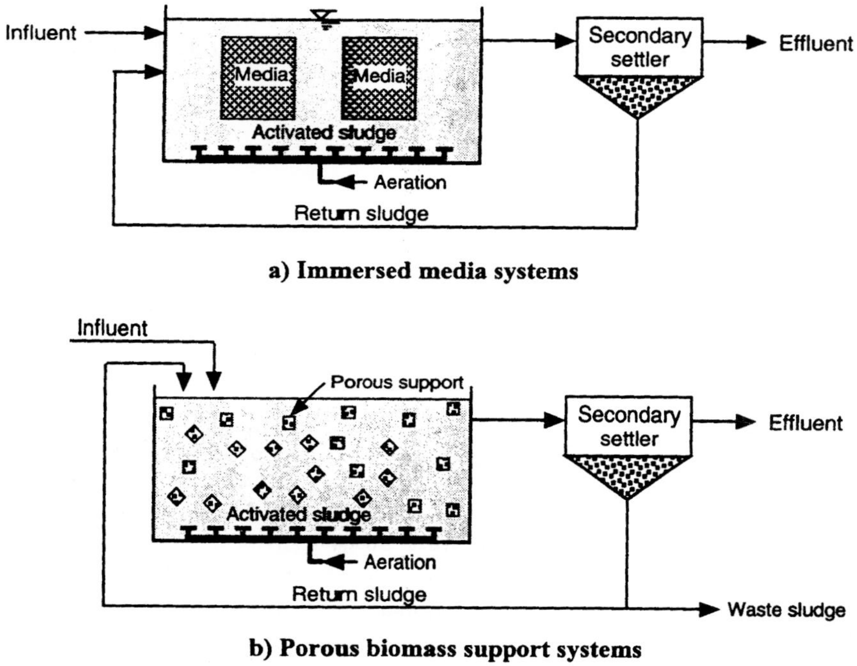

Fig. 19.10 Hybrid biomass support systems.

Gonzalez-Martinez and Duque-Luciano (1992) described the biomass adhering to their support as a mixture of biofilm and filamentous growth, with length between 1–5 cm, attached to the plastic material from one end. However, Reddy *et al.* (1994) drew an interesting comparison between the organisms in the activated sludge liquor and those on the media of their process (Table 19.7). It is to be observed that rotifers and free swimming ciliates are abundant in the biofilm which could explain the low SS and BOD in the process effluent (Muller, 1998).

TABLE 19.7 Organisms found in a hybrid system (from Reddy *et al.*, 1994)

Organisms	*Activated sludge MLSS*		*CAPTOR media MLSS*	
	Abundance	*Rank*	*Abundance*	*Rank*
Rotifers	Few	4	Abundant	1
Free swimming ciliates	Abundant	1	Very common	2
Stalked ciliates	Very common	2	Common	3
Flagellates	Common	3	Rare	5
Nematodes	Rare	5	Rare	5

3 CONCLUSION AND PERSPECTIVES

Growth of biofilms is a natural process that has been engineered in many ways to treat wastewater. Fixed film processes, developed mainly for the treatment of municipal wastewaters, have been presented here along with their microbiology.

Attached growth processes are often compared to, and seen as competitors of, suspended growth processes, but it is interesting to note that currently there is a tendency to couple both processes to make better use of each technology. This in a sense makes the point that there is no such thing as *the technology,* but that each technology has its adavantages and disadvantages and that it is applicable in some cases but not in others.

So far as future trends are concerned, it is always difficult to predict exactly what is going to happen. However, increasing standards for

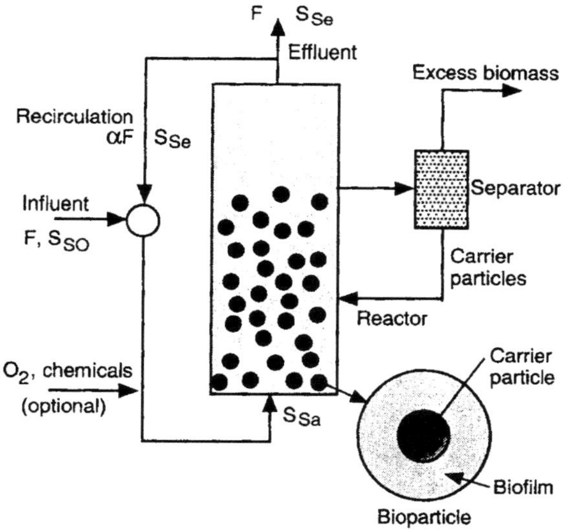

Fig. 19.11 Schematic diagram of a fluidized bed reactor (from Grady *et al.*, 1998).

nutrient and xenobiotic removal, along with more rigorous noise and odor control, have stimulated the development of new intensive advanced technology processes (Lazarova and Manem, 1994).

One of these intensive processes that is attracting increasing attention is the fluidized bed reactor (FBR), which consists of a biofilm growing on small carriers kept in suspension in the fluid by the drag forces associated with the upward flow of water (Fig. 19.11). FBR are a more sophisticated version of the immersed mobile biomass system presented earlier. Their main advantages are:

1. high removal efficiency of carbon and nitrogen through large amounts of fixed biomass with a low hydraulic retention time
2. no clogging
3. better oxygen transfer
4. reduced sludge production (Lazarova and Manem, 1994).

However, the following problems are met at large scale fluidized-bed facilities (Chuboda *et al.*, 1998):

1. a problematic media separation and biofilm control system
2. a lack of suitable bed height control
3. a very high sensitivity of the inlet distribution system to clogging
4. a relatively uncertain control of a stable fluidization.

Despite these problems, many systems have been built, but mainly as two-phase fluidized beds for the treatment of industrial wastewaters (Cooper and Wheeldon, 1980; Sutton and Mishra, 1991, 1994). Time may come when they will be used for municipal wastewater treatment.

Finally, the development of intensive processes may be of interest, but there is a growing need for the development of extensive processes (low cost and low operational requirements) for on-site and small community wastewater treatment. In that perspective, packed-bed reactors using organic media, such as peat or compost, are being studied and used on site. Couillard (1994) carried out a thorough literature review on this subject, and he concluded that the peat bed does have some disadvantages, such as color, high COD and low pH in the effluent, but that, despite these, the effluent quality satisfies criteria for discharge into most receiving waters at a cost outcompeting most processes. However, their low hydraulic loading rate makes them at the moment only applicable for small flows. Laboratory scale and field studies have demonstrated the good performance of organic media (Lens *et al.*, 1994; Talbot *et al.* 1996), and it is likely that they will be used more and more frequently in the future.

REFERENCES

L'assainissement des agglomérations: techniques d'épuration actuelles et évolutions. Agence de l'Eau Artois-Picardie, France.

Applegate, D.H. and Bryers, J.D. (1991). Effects of carbon and oxygen limitations and calcium concentrations on biofilm removal processes. *Biotechnology Bioengineering* **37**, 17–25.

Arvin, E. and Harramoës, P. (1990). Concepts and models for biofilm reactor performance. *Water Science and Technology* **22**(1–2), 171–192.

Benefield, L.D. and Randall, C.W. (1980). *Biological Process Design for Wastewater Treatment*. Prentice-Hall.

Bitton, G. (1994). *Wastewater Microbiology.* Wiley-Liss, New York.

Bock, E., Koops, H.P. and Harms, H. (1986). Cell biology of nitrifying bacteria. In: J.I. Prosser, (ed.) *Nitrification.* SGM IRL Press Oxford, Washington, pp. 17–38.

Brisou, J.F. (1995). *Biofilms – Methods for Enzymatic Release of Microorganisms.* CRC Press.

Bruce, A.M. and Hawkes, H.A. (1983). Biological filters. In: C.R. Curds and H.A. Hawkes (eds) *Ecological Aspect of* Used-Water Treatment, Vol. 3: *The Processes and their Ecology,* pp. 1–113. Academic Press, London.

Bryers, J.D. (1987). Application of captured cell systems in biological treatment. In: D.L. Wise (ed.) *Bioenviromental Systems,* Vol. **IV**, pp. 27–53. CRC Press, USA.

Bryers, J.D. and Characklis, W.G. (1990). Biofilms in water and wastewater treatment. In: W.G. Characklis and K.C. Marshall (eds) *Biofilms,* pp. 671–696. John Wiley and Sons, Inc.

Canler, J.P. and Perret, J.M. (1994). Biological aerated filter: assesment of the process based on 12 sewage treatment plants. *Water Science and Technology* **29**(10/11), 13–22.

Canler, J.P., Perret, J.M., Lengrand, F. and Iwema, I.(2002). La nitrification en biofiltration. Applications à des charges variables et à des basses températures. *Techniques Sciences Méthodes,* no. 10, octobre 2002, 97 eme année.

Capdeville, B. and Rols, J.L. (1992). Introduction to biofilms in water and wastewater treatment. In: L.F. Melo (ed.) *Biofilms – Science and Technology,* pp. 13–20. Kluwer Academic Publishers, Netherlands.

Chuboda, P., Pannier, M., Truc, A. and Pujol, R. (1998). A new fixed-film mobile bed bioreactor for denitrification of wastewaters. *Water Science and Technology* **38**(8/9), 233–240.

Cooke, W.B. (1959). Trickling filter ecology. *Ecology* **40**, 273–291.

Cooke, W.B. (1963). *A Laboratory Guide to Fungi in Polluted Waters Sewage, and Sewage Treatment Systems; Their Identification and Culture.* US Department of Health, Education and Welfare, Cincinnati.

Cooper, P.F. and Wheeldon, D.H. (1980). Fluidized- and expanded-bed reactors for wastewater treatment. *Water Pollution Control* **79**, 286–306.

Couillard, D. (1994). The use of peat in wastewater treatment. *Water Research* **28**, 1261–1274.

Crowther, R.F. and Harkness, N. (1975). Anaerobic bacteria. In: C.R. Curds and H.A. Hawkes (eds) *Ecological Aspect of Used Water Treatment – The Organisms and their Ecology,* Vol. **1**, pp. 65–91. Academic Press, London.

Crites, R. and Tchobanoglous, G. (1998). *Small and Decentralized Wastewater Management Systems.* McGraw-Hill.

Curds, C.R. and Cockburn, A. (1970). Protozoa in biological sewage-treatment processes-i. a survey of the protozoan fauna of british percolating filters and activated-sludge plants. *Water Research* **4**, 225–236.

Edeline, F. (1993). *L'épuration biologique des eaux: Théorie & technologie des réacteurs.* CEBEDOC, Liège.

Garzon-Zuniga, M.A. and Gonzalez-Martinez, S. (1996). Biological phosphate and nitrogen removal in a biofilm sequencing batch reactor. *Water Science and Technology* **34**(1/2), 293–301.

Gebara, F. (1999). Activated sludge biofilm wastewater treatment system. *Water Research* **33**, 230–238.

Gonzalez-Martinez, S. and Duque-Luciano, J. (1992). Aerobic submerged biofilm reactors for wastewater treatment. *Water Research* **26**, 825–833.

Grady, C.P.L., Daigger, G.T. and Lim, H.C. (1999). *Biological Wastewater Treatment,* 2nd edn. M. Dekker, New York.

Gray, N.F. (1989). *Biology of Wastewater Treatment.* Oxford University Press.

Hamoda, M.F. (1989). Kinetic analysis of aerated submerged fixed-film (ASFF) bioreactors. *Water Research* **23**, 1147–1154.

Harrison, J.R., Daigger, G.T. and Filbert, J.W. (1984). A survey of combined trickling filter and activated sludge processes. *Journal of Water Pollution Control Federation* **56**, 1073–1079.

Hatting, W.H.J. (1963). The nitrogen and phosphorus requirements of the microorganisms 1. *Water and Waste Treatment Journal* **9**, 380–386.

Hawkes, H.A. (1983). The applied significance of ecological studies of aerobic processes. In: C.R. Curds and H.A. Hawkes (eds) *Ecological Aspects of Used-Water Treatment – The Processes and their Ecology,* Vol. **3**, pp. 174–329. Academic Press, London.

Henze, M., Harremoes, P., la Cour Jansen, J. and Arvin, E. (1995). *Wastewater Treatment: Biological and Chemical Processes.* Springer-Verlag.

Heukelekian, H. (1945). The relationship between accumulation biochemical and biological characteristics of film, and purification capacity of a biofilter and standard filter. *Sewage Works Journal* **17**(1), 23–38.

Heukelekian, H. and Crosby, E.S. (1956). Stream pollution-slime formation in polluted waters II: factors affecting slime growth. *Sewage and Industrial Wastes* **28**(1), 78–92.

Hoag, G.E., Widmer, W.J. and Hovey, W.H. (1983). Microfauna and RBC performance: laboratory and full-scale systems. In: Y.C. Wu and E.D. Smith (eds) *Fixed-Film Biological Processes for Wastewater Treatment,* pp. 206–226. Noyes Data Corporation.

Hull, M., Klimiuk, E. and Janczukowicz, W. (1991). Comparative studies of activated sludge and rotating biological disc process (II-Microfauna). *45th Purdue Industrial Wastes Conference Proceedings* pp. 339–346. Lewis Publishers Inc.

Iwai, S. and Kitao, T. (1994). *Wastewater Treatment with Microbial Films.* Technomic Publishing Company, Inc.

James, A. (1964). The bacteriology of trickling filters. *Journal of Applied Bacteriology* **27**(2), 197–207.

Jones, R.M., Sen, D. and Lambert, R. (1998). Full scale evaluation of nitrification performance in an integrated fixed film activated sludge process. *Water Science and Technology* **38**(1), 71–78.

Lacamp, B., Hansen, F., Penillard, P. and Rogalla, F. (1992). Wastewater nutrient removal with advanced biofilm reactors. *Proceedings of the 65th Annual Conference, Water Environment Federation,* New Orleans, USA.

Lazarova, V. and Manem, J. (1994). Advances in biofilm aerobic reactors ensuring effective biofilm activity control. *Water Science and Technology* **29**(10/11), 319–327.

Lazarova, V. and Manem, J. (1995). Biofilm characterization and activity analysis in water and wastewater treatment. *Water Research* **29**(10), 2227–2245.

Le Bihan, Y. and Lessard, P. (1998). Étude microbiologique d'un biofiltre à lit ruisselant: représentativité des eaux de lavage et distribution verticale des bactéries hétérotrophes aérobies. *Environmental Technology* **19**, 555–566.

Lee, N.M. and Welander, T. (1994). Influence of predators on nitrification in aerobic biofilm processes. *Water Science and Technology* **29**(7), 355–363.

Lens, P.N., Vochten, P.M., Speleers, L. and Verstraete, W.H. (1994). Direct treatment of domestic wastewater by percolation over peat, bark and woodchips. *Water Research* **28**, 17–26.

Mack, W.N. (1975). Microbial film development in a trickling filter. *Microbial Ecology* **2**, 215–226.

Magesan, G.N., Williamson, J.C., Sparling, G.P. *et al.* (1999). Hydraulic conductivity in soils irrigated with wastewaters of differing strengths: field and laboratory studies. *Australian Journal of Soil Research* **37**(2), 391–402.

Mann, A.T., Mendoza-Espinosa, L. and Stephenson, T. (1999). Performance of floating and sunken media biological aerated filters under unsteady state conditions. *Water Research* **33**, 1108–1113.

Mann, A.T., Mendoza-Espinosa, L. and Stephenson, T. (1998). A comparison of floating and sunken media biological aerated filters for nitrification. *Journal of Chemical Technology and Biotechnology* **72**, 273–279.

Metcalf and Eddy (1991). *Wastewater Engineering: Treatment, Disposal and Reuse*. 3rd edn. McGraw-Hill, New York.

Morgenroth, E. and Wilderer, P.A. (1999). Controlled biomass removal – the key parameter to achieve enhanced biological phosphorus removal in biofilm systems. *Water Science and Technology* **39**(7), 33–40.

Mudrack, K. and Kunst, S. (1986). *Biology of Sewage Treatment and Water Pollution Control*. Ellis Horwood Limited, Chichester.

Morper, M.R. (1994). Upgrading of activated sludge systems for nitrification removal by application of the LINPOR®-CN process. *Water Science and Technology* **29**(12), 167–176.

Muller, N. (1998). Implementing biofilm carriers into activated sludge process – 15 years of experience. *Water Science and Technology* **37**(9), 167–174.

Nicol, J.P., Benefield, L.D., Wetzel, E.D. and Heidman, J.A. (1988). Activated sludge systems with biomass particle support structures. *Biotechnology and Bioengineering* **31**, 682–695.

Oga, T., Suthersan, S. and Ganczarczyk, J.J. (1991). Some properties of aerobic biofilms. *Environmental Technology* **12**(5), 431–440.

Parker, D.S. (1999). Trickling filter mythology. *Journal of Environmental Engineering* **125**, 618–625.

Parker, D.S., Romano, L.S. and Horneck, H.S. (1998). Making a trickling filter/solids contact process work for cold weather nitrification and phosphorus removal. *Water Environment Research* **70**, 181–188.

Parker, D.S., Brischke, K.V. and Matasci, R.N. (1993). Upgrading biological filter effluents using the TF/SC process. *Journal of the Institution of Water Environment Management* **7**, 91–100.

Parker, D.S. and Bratby, J.R. (2001). Review of two decades of experience with TF/Sc process. *Journal of Environmental Engineering*, **127**, 380–387.

Pearce, P. and Williams, S. (1999). A Nitrification Model for Mineral-Media Trickling Filters. *The Journal of the Chartered Institution of Water and Environmental Management* **13**, 84–92.

Peladan, J.G., Lemmel, H., Tarallo, S. *et al.* (1997). A New Generation of biofilters with high water velocities. *Proceedings of International Conference on Advanced Wastewater Treatment Processes*. Leeds, UK, September 8–11, 1997.

Peters, R.W. and Foley, V.L. (1983). Fixed-Film wastewater treatment systems: their history and development as influenced by medical, economic, and engineering factors. In: Y.C. Wu and E.D. Smith (eds) *Fixed-Film Biological Processes for Wastewater Treatment*, pp. 1–44. Noyes Data Corporation.

Pike, E.B. (1975). Aerobic Bacteria. In: C.R. Curds and H.A. Hawkes (eds) *Ecological Aspects of Used-Water Treatment – The Processes and their Ecology*, Vol. **3**, pp. 1–53, Academic Press, London.

Pineau, M. and Lessard, P. (1994). Procédés de biofiltration et applications au Québec. *Sciences Techniques Eau* **27**, 13–17.

Puznava, N., Payraudeau, M. and Thornberg, D. (2001). Simultaneous nitrification and denitrification in biofilters with real time aeration control. *Water Science Technology*, **43**, 269–276.

Randall, C.W. and Sen, D. (1996). Full-scale evaluation of an intergrated fixed-film activated sludge (ifas) process for Enhanced Nitrogen Removal. *Water Science and Technology* **33**(12), 155–162.

Reddy, M.P., Pagilla, K.R., Senthilnathan, P.R. *et al.* (1994). Estimation of biomass concentration and population dynamics in a CAPTOR® activated sludge system. *Water Science and Technology* **29**(7), 149–152.

Rusten, B. and Neu, K.E. (1999). Moving-bed biofilm reactors move into the small-flow treatment area. *Water Environment & Technology* **11**(1), 27–33.

Seek Park, Y., Won Yu, J. and Koo Song, S. (1998). Biofilm properties under different substrate loading rates in a rotating biological contactor. *Biotechnology Techniques* **12**(8), 587–590.

Senthilnathan, P.R. and Ganczarczyk, J.J. (1990). Application of biomass carriers in activated sludge process. In: R.D. Tyagi and K.V. Vembu (eds) *Wastewater Treatment by Immobilized Cells*, pp. 103–141. CRC Press.

Shieh, W.K. (1987). Biofilm kinetics – mass transfer effects and their implication to process design, operation and control. In: D.L. Wise (ed.) *Bioenvironmental Systems*, Vol. IV, pp. 155–179. CRC Press.

Stephenson, T. (1997). High rate aerobic wastewater treatment process – what next?. *Proceedings of the 3rd International Symposium on Environmental Biotechnology*, Ostend, Belgium, **1**, 57–66.

Sutton, P.M. and Mishra, P.N. (1991). Biological fluidized beds for water and wastewater treatment – a state-of-the-art review. *Water Environment & Technology* **3**(8), 52–56.

Sutton, P.M. and Mishra, P.N. (1994). Activated carbon based fluidized beds for contaminated water and wastewater treatment – a state-of-the-art review. *Water Science and Technology* **29**(10/11), 309–317.

Talbot, P., Bélanger, G., Pelletier, M. *et al.* (1996). Development of a biofilter using an organic medium for on-site wastewater treatment. *Water Science and Technology* **34**(3/4), 435–441.

Tomlinson, T.G. and Williams, I.L. (1975). Fungi. In: C.R. Curds and H.A. Hawkes (eds) *Ecological Aspects of Used Water Treatment: The Organisms and their Ecology*, Vol. 1, pp. 93–151. Academic Press, London.

Védry, B. (1996) *Les Biomasses Épuratrices*. Agence de l'Eau Seine-Normandie, Paris, France.

Visvanathan, C. and Nhien, T.T.H. (1995). Study on aerated biofilter process under high temperature conditions. *Environmental Technology* **16**, 301–314.

Wang, B., Li, J., Wang, L. *et al.* (1998). Mechanism of phosphorus removal by sbr submerged biofilm system. *Water Research* **32**, 2633–2638.

Wanner, J., Kucman, K. and Grau, P. (1988). Activated sludge process combined with biofilm cultivation. *Water Research* **22**, 207–215.

WEF and ASCE (1992). *Design of Municipal Wastewater Treatment Plants* (volume 1). Water Environment Federation, Virginia.

WPC (1988). *O & M of Trickling Filters, RBCs, and Related Processes*. Water Pollution Control Federation, Virginia.

20

Biofilm formation and its role in fixed film processes

Luís F. Melo

Department of Chemical Engineering, University of Porto, 4200-465 Porto, Portugal

1 BIOFILMS: CONCEPT AND RELEVANCE

A biofilm is a surface-attached gelatinous matrix composed of microorganisms, the extracellular polymers they excrete (EPS) and foreign substances such as adsorbed molecules and small abiotic particles. Such a matrix is highly hydrated, often containing more than 90% water (mass percentage) and its properties depend on three kinds of factors: biological (microbial species), chemical (fluid composition in contact with the biofilm), and physical (hydrodynamic and thermal conditions under which the biological layer is formed). The thickness of biofilm layers can reach a few millimetres, or even centimetres, but typically ranges from 10 micron to 1 mm. Cell concentrations inside biofilms are commonly in the range of 10^7-10^9 cells/cm^2. EPS often represent more than 50% of the total organic mass of the biological layer. A great diversity of microbial species is found in biofilms in quite different environments, as illustrated in Table 20.1. As can be seen, bacteria predominate in biofilms, but algae and fungi are also present.

Microbial films can have beneficial or detrimental effects, depending on where they build up. Examples of the former can be found in fixed biomass reactors used in wastewater treatment (Harremöes, 1978) or in the production of ethanol (Dempsey, 1990) and citric acid (Briffaud and Engasser, 1979). Unwanted biofilm formation (biofouling) occurs on the surfaces of heat exchangers (Bott, 1995), cooling water towers, valves, tubes and ships' hulls, increasing the resistance to flow and to heat transfer and inducing/enhancing corrosion (Geesey, 1991). Some biofilms can also become a serious problem for human health when they attach on lungs, teeth, urinary catheters, etc. (Costerton *et al.*, 1987). The present chapter will focus essentially on biofilm growth and characteristics in industrial equipment.

2 BIOFILMS VERSUS DISPERSED CELLS

Microorganisms live in dispersed suspensions, in flocs and in attached films. The aggregated forms are clearly predominant in nature (Lappin-Scott and Costerton, 1995) and they probably constitute a more efficient way of surviving in many environments. Two words can be used to explain why microbes may choose to form biofilms or other aggregates, instead of living as dispersed cells: proximity and protection.

The distances between microorganisms in biofilms are considerably smaller than in suspension and this leads to the creation of microenvironments with specific nutrient concentration gradients, pH's, electrical charges distributions, proton concentrations, etc., which affect the metabolism of microorganisms. Natural selection will lead to growth of certain species as a result of the specific local conditions created within the biological matrix. Microbial growth rates may sometimes be lower in biofilms than in suspensions, but

The Handbook of Water and Wastewater Microbiology
ISBN 0-12-470100-0

TABLE 20.1 Examples of microorganisms present in biofilms

Processes	*Microorganisms*
Cooling water systems	*Pseudomonas* spp., *E. coli*, *Staphylococcus*, *Desulfovibrio*, clostridia, green algae
Biological wastewater treatment	*Nitrosomonas*, *Nitrospyra*, *Nitrobacter*, *Pseudomonas*, *Methanotrix*, *Methanosarcina*, *Clostridium*, *Desulfovibrio*, *Staphylococcus*
Food processing	*Legionella*, *Salmonella*, *Listeria*
Chemical and extractive industry	*Acetobacter*, *Thiobacillus ferrooxidans*, *Thiobacillus thiooxidans*
Lungs	*Pseudomonas aeruginosa*
Urinary catheters	*Providencia stuarti*, *E. coli*, *Pseudomonas aeruginosa*, *Klebsiella pneumoniae*
Teeth and mouth	*Prevotella intermedia*, *Actinobacillus*, *Streptococcus sangui*, *Candida albicans*

the production of extracellular polymers is normally higher, resulting in a protective network within which the cells are entrapped.

This leads us to the second specificity of biofilm matrices: protection. The exopolymers, often composed of polysaccharides and glycoproteins, contribute decisively to both the adhesion of microorganisms to the initially 'clean' surface and the internal cohesion of the film. Simultaneously, these biopolymers act as a protective network that makes diffusion of toxic substances more difficult and attenuates the effects of chemical and hydrodynamic aggression from the environment. Additionally, polymers may enhance the adsorption of nutrient molecules and the accumulation of particles in the inner zones of the biofilms. Organic particles, for instance, can be retained long enough inside the matrix in order to be biologically degraded.

It should be stressed that biofilms grown in nature and industrial equipment usually contain a great variety of microbial species and that such variety is stimulated by the existence of zones with distinct microenvironments. One example is the creation of anaerobic zones far away from the water-biofilm interface and the development of aerobic species near this interface. Another example is the simultaneous presence of nitrifying and anoxic denitrifying bacteria in the same biofilm, allowing nitrogen removal (from ammonium to nitrogen gas) in the reactor which otherwise would be quite difficult to achieve. The corrosion of metallic surfaces is favoured by the development of sulphate-reducing bacteria underneath the aerobic layers adjacent to the aqueous environment. These cooperative mechanisms constitute clear advantages over the planktonic growth mode from the point of view of the microbial world.

The protective biopolymer network also acts as a diffusion barrier to the transport of substrates, which explains in part the smaller microbial growth rates sometimes found in biological films.

Thiobacillus ferrooxidans showed growth rates 70 times lower in biofilms on glass particles than in suspended cultures (Karamanev, 1991). However, this is not a general rule, particularly when surfaces are prone to adsorb substrates: Davies and McFeters (1988) proved that *Klebsiella oxytoca* growth rate was much higher (10 times) when attached to activated carbon particles than in liquid suspension when the substrate was glutamate; this did not happen when glucose was used as substrate, because glutamate adsorbs easily to the activated carbon, opposite to glucose.

An important factor that affects the behaviour of cells in microbial films as compared to suspended systems is the attachment process. The presence of a solid surface was found to increase the production of exopolymers (e.g. Vandevivere and Kirchman, 1993). Some authors concluded that adhesion seems to trigger the expression of genes that control the excretion of the biofilm polysaccharides (McCarter *et al.*, 1992; Davies *et al.*, 1993). Adhesion may in fact modify the phenotype expression of a very significant fraction of the cell proteins. An increase in the liquid viscosity was found to stimulate the production of lateral flagella in *Vibrio parahaemolyticus* (McCarter *et al.*, 1992), as a result of special gene expression induced by the presence of a solid surface.

3 BIOFILM PROPERTIES: ARCHITECTURE AND COMPOSITION

The internal structure of biofilms is characterized by a marked heterogeneity: the biological material is organized in clusters containing cells and the excreted polymeric network, while channels and pores filled with the ambient liquid occupy the free spaces between the clusters. This liquid is not necessarily stagnant and, in fact, convective flows were measured within the channels (Stoodley *et al.*, 1994). Each cluster can contain layers with different microbial species, different polymer compositions, different densities of active cells, etc. Often, particularly in aerobic heterotrophic biofilms, filamentous structures ('streamers') protrude out of the film into the external liquid. Stacks of cells have been often found to extend from a thin basal biofilm attached to the solid surface (Keevil and Walker, 1992).

The physical structure of microbial films changes with time and, in 'older' biofilms, biopolymers were observed to form bridges between clusters and increase the density of the fixed biomass (Fig. 20.1).

Nowadays, new analytical tools, based on the use of molecular probes together with confocal laser scanning microscopy, allow detailed studies of microbial population distributions inside biofilm matrices (Wobus *et al.*, 2000; Neu, 2000) and contribute to establish sound relations between the operating conditions of reactors and the biofilm properties.

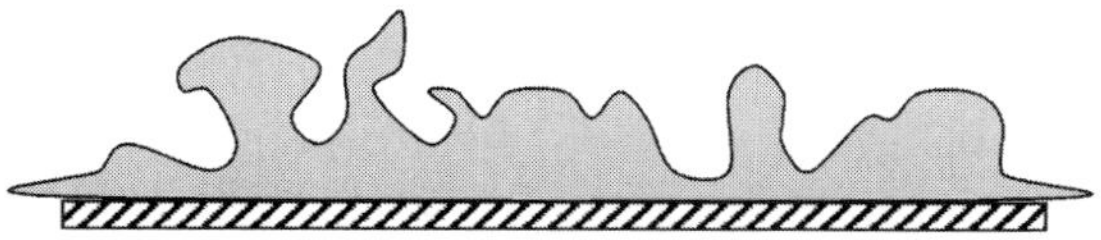

Fig. 20.1 Schematical representation of biofilm structure.

For example, in a combined nitrification and organic carbon removal process, the use of FISH techniques (fluorescence *in situ* hybridization) confirmed that heterotrophic bacteria developed on the top of the nitrifying biofilm under oxygen limiting conditions (Nogueira *et al.*, 2002). In this work, curious interactions between microbial growth and physical phenomena were found: the nitrifying efficiency of the biological reactor decreased significantly when the reactor was operated at high values of the hydraulic residence time, contrary to the low residence time operation. This could be due to the presence of greater amounts of heterotrophs in suspension at higher residence times and therefore to a higher liquid viscosity; ultimately, it led to a decrease in the liquid shear stress and to the formation of a thicker layer of heterotrophs on top of the nitrifying biofilm, inhibiting the activity of the ammonium and nitrite-reducing bacteria located in the inner zones.

Similar techniques were applied to observe how the density of cells varies from the substratum to the biofilm-liquid interface. The results were not always coincident: in some cases, higher cell densities were found near the solid surface (Lawrence *et al.*, 1987; Lawrence and Korber, 1993; Kuehn *et al.*, 1998), whereas in other studies the reverse situation was detected (Neu and Lawrence, 1997).

4 HOW BIOFILMS GROW

The main processes involved in the build-up of biofilms are:

1. formation of the so-called 'conditioning film', a very thin (mono-) layer of organic molecules and ions on the adhesion surface
2. transport of microorganisms (and other particles) to the surface
3. microbial attachment and growth (see also process 5)
4. substrate transport through the external liquid medium and through the biofilm matrix
5. biological reaction at the 'active sites' inside the biofilm, i.e. at the active microbial cells

that consume the substrate and produce not only new cells but also exopolymers
6. mass transfer (from the biofilm back to the liquid) of the products resulting from the biological reactions within the biolayer
7. detachment of parts of the biofilm by erosion and abrasion (continuous processes) and sloughing off (sporadic events).

The last process (detachment) causes loss of fixed biomass while the other six processes contribute directly or indirectly to the growth of the biofilm layer (Fig. 20.2).

The initial conditioning film is established in a few minutes following the transport and adsorption of organic molecules contained in the liquid, some resulting from the decomposition of organisms such algae, bacteria, small plants, etc. These molecules may form polymeric chains that interact with the exopolymers on the surface of microorganisms and form strong bridges that stabilize cell adhesion. The formation of the first microbial layer facilitates the subsequent adhesion of other cells and abiotic material.

The main substrate transport mechanisms from the liquid to the biofilm surface are molecular diffusion (in stagnant or laminar flowing media) and eddy diffusion (when turbulent flow prevails). Within a mature biofilm, molecular diffusion is predominant, but convective flows contribute to enhance mass transport in the biofilm channels.

Once the substrate molecules reach the cells inside the matrix, the production of cellular biomass and of extracellular polymers leads to the growth of the dry mass of the biofilm while its thickness also increases. Usually, the effect of the shear stress (and other detachment mechanisms) tends to level out the growth rate of the fixed biomass and a final 'steady-state' thickness is reached, although subject to periodic fluctuations. Sometimes, the production of gases as a result of microbial metabolism creates pockets of higher pressure inside the matrix and eventually causes the detachment of large portions of the attached biomass. This may happen, for instance, in biofilms containing a consortium of nitrifying and denitrifying bacteria: the latter tend to build up on the top of the biofilm in order to utilize the oxygen dissolved in the water, while the anoxic denitrifiers tend to be located near the initial adhesion surface. In this case, the conversion of ammonium to nitrate by the nitrifiers, followed by the reduction of nitrate to nitrogen gas creates bubbles inside the biofilm that break and disrupt the biomass structure.

Abrasion due to particle collisions is considered a determining mechanism in the detachment of fixed biomass from carrier particles in turbulent air-lift reactors (Gjaltema, 1996).

It should be stressed that the first six processes mentioned above are consecutive (in series), meaning that the slowest step will control the rate of the growth mechanisms. More often than not, one of the following three processes is the controlling one: mass transfer in the liquid, mainly in laminar flow situations;

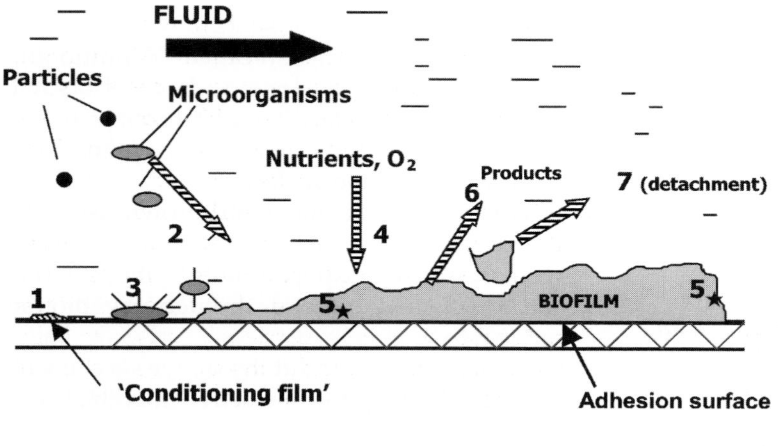

Fig. 20.2 Phenomenology of biofilm formation.

mass transfer through the biofilm; biological kinetics within the biofilm. However, detachment is acting simultaneously, as a parallel process. Therefore, the rate of biofilm formation results from the competition between the overall growth process (comprising the first 6 processes, controlled by one of them) and the detachment process. The faster of these two overall processes will condition the rate at which biofilm builds up. In turbulent flows, for example, the detachment rate is frequently controlling the growth rate of the attached layer.

A typical curve of biofilm growth over time is shown in Fig. 20.3.

The curve in Fig. 20.3 can be modelled by a simple overall phenomenological equation:

$$m_f = m_f^{\infty}[1 - \exp(-b{\cdot}t)] \qquad (1)$$

where:

m_f = mass of attached biofilm per unit surface area at time t

m_f^{∞} = maximum mass of attached film (for $t = \infty$, pseudo-steady state)

b = empirical parameter proportional to the hydrodynamic forces acting upon the biofilm surface and to the amount of attached biomass.

The reciprocal of b ($1/b$) increases with the cohesiveness of the biofilm, which means that $1/b$ is a measure of the biofilm resistance to detachment. The above expression can be derived by considering a few assumptions (Melo and Vieira, 1999; Melo and Oliveira, 2001), one of them being that the fraction of cellular biomass in the biofilm decreases as the biofilm grows. This assumption is justified by the fact that the cells in each layer or cluster use the substrate to produce living material (new cells) and inert material (polymers) and that these exopolymers are not able to create any new mass (Pereira *et al.*, 2002).

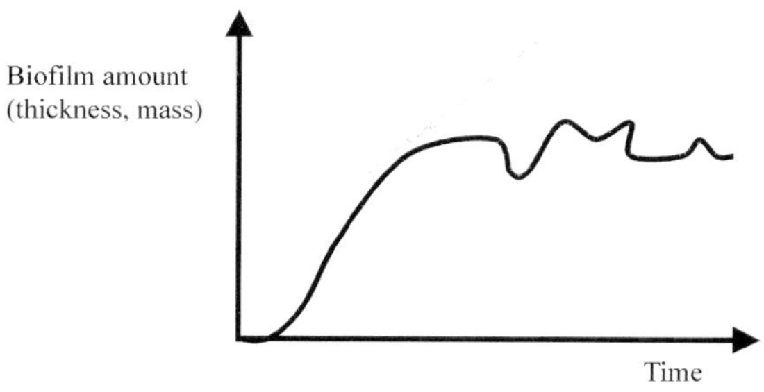

Fig. 20.3 Biofilm growth curve.

This simple overall model may be used if the parameters m_f^{∞} and b are experimentally correlated with the environmental (chemical, biochemical and physical) conditions under which the biofilm is grown. In recent years, structured multidimensional models using powerful computer tools based on 'cellular automata' techniques were developed to describe the growth, morphology and composition of biofilms over time (Picioreanu *et al.*, 1999; Noguera *et al.*, 1999; Hermanowics, 1999). Such models, as many others, including the very simple one presented above, still require basic experimental information on the values of diffusion and kinetic parameters or related variables.

Several factors affect the formation and properties of microbial films. Among others, the following should be emphasized:

1. characteristics of the microbial species and strains
2. composition and roughness of the surface material where the microorganisms attach
3. liquid composition, its pH, temperature, ionic strength
4. hydrodynamic features of the fluid, such as velocity and turbulence.

Microbial species that produce greater amounts of extracellular polymers can more easily attach to solid surfaces. Incorporation of small inorganic particles, such as clays, in the biofilm structure can make them more resistant to external aggressions (Vieira and Melo, 1995). Surface electrical charges of microbes and solid surfaces are affected by pH and, as such, increase or reduce repulsive interactions between them. Naturally, pH has a marked influence on microbial metabolism, which is thus dependent on the specific microenvironments created inside the biofilm matrix. The carbon/nitrogen ratio in the medium affects the ability of microorganisms to attach to solid surfaces, since it conditions the production of extracellular polymers (Veiga *et al.*, 1992).

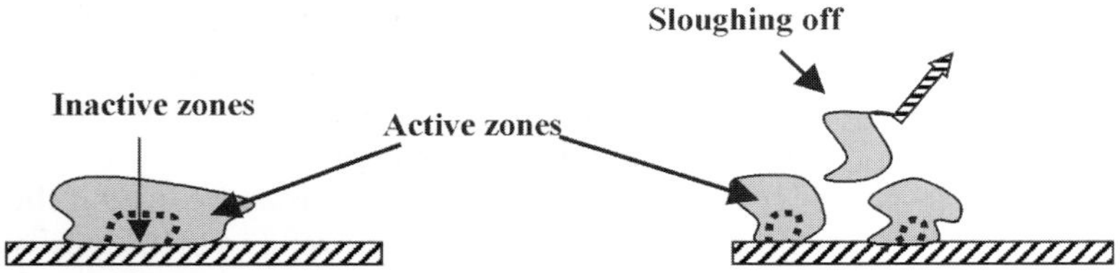

Fig. 20.4 Sloughing off of attached biomass.

Fluid velocity causes a variety of effects:

1. substrate transport rates increase with fluid velocity and turbulence
2. on the other hand, detachment rates are enhanced by higher liquid velocities resulting in thinner biofilms
3. additionally, higher fluid velocities increase the compactness of the biological matrix, which in turn affects internal substrate diffusion (Melo and Vieira, 1999; Vieira and Melo, 1999).

It is not easy to predict which will be the overall effects of the velocity field on the properties and activity of biofilms. Fig. 20.4 is an illustrative example of what may happen when a biofilm is subject to low fluid velocities and high substrate concentrations: it will probably become too thick, and the substrate will not be able to reach its inner zones favouring the appearance of biologically inactive mass. This will enhance the sloughing off of attached biomass, which will produce instability in biofilm performances. In recent years, design of biofilm reactors has favoured turbulent, high shear stress processes that minimize such instabilities (Tijhuis et al., 1995; Lazarova and Manem, 1997).

5 BIOFOULING IN INDUSTRIAL WATER SYSTEMS

Problems caused by biofouling in industry are particularly relevant in water systems, such as cooling water circuits (cooling towers, heat exchangers), paper mill operations, reverse osmosis membranes (Melo and Bott, 1997; Flemming, 1997). Biofilms create additional pressure drops (resistance to flow), increased thermal resistances in heat exchangers and reduced permeability in reverse osmosis membranes. Water used in industry (taken from rivers, lakes, bore holes) contains various macromolecules that result from the breakdown of living material and these substances not only adsorb onto the equipment surfaces, but also act as nutrients to the living cells, thereby starting the biofilm growth process. In cooling water systems, temperatures in heat exchangers are usually quite suitable for microbial growth (from 10–15°C at the inlet to 30–40°C at the outlet).

Biofilm control is an important aspect of industrial equipment where water flows at moderate temperatures. Use of appropriate velocities (above 1 m/s), temperatures and surface conditions can reduce biofouling effects, but often will not be enough, and mechanical and chemical methods will have to be considered to mitigate biofouling effects (see detailed analysis in Bott, 1995). The former include the physical removal of biofilm from the surface, which can be applied on-line by periodically circulating sponge rubber balls (or similar devices) through the tubes during normal operation, or off-line by shutting down the equipment and cleaning it with high pressure water jets or manual procedures when feasible (e.g. in plate heat exchangers). Use of modified solid surfaces (well polished metals or coated with specific polymeric molecules) also reduces the build-up of biofouling layers.

Another way of minimizing biofilm development is to apply disinfectants (biocides) to kill or inactivate the microorganisms. Chlorine is still the preferred biocide, but its use has been under attack because of the carcinogenic by-products of chlorine reactions with organic matter. Hydrogen peroxide and ozone, although more expensive, may prove to be more environmentally acceptable alternatives.

Other techniques based on the use of surfactants and phages (biological predators) are important contributions to biofouling mitigation, but some are still under scientific, technological and economical scrutiny. A methodology based on the use of a mild biocide (carbamate solution), which simultaneously has an aggregative action, was proposed in order to reduce biofouling problems in paper and pulp plants (Pereira *et al.*, 2001). The idea in this specific situation was to aggregate the bacteria before they had time to form slimy EPS layers that eventually incorporate in the cellulose pulp and cause paper machine shutdowns and affect paper quality.

6 BIOFILMS IN WASTEWATER TREATMENT

Biofilm reactors contain particles of the carrier or support material where the film is attached. The particles are sometimes porous and the biofilm also develops within the pores. The wastewater to be treated flows in direct contact with the biofilm allowing nutrients and metabolic products to be exchanged between the film and the liquid. Supports are made of plastic material of a great variety of shapes and dimensions, and also of natural materials such as sand and basalt particles. The specific area of the carriers has increased in the last decades from around 100 m^2/m^3 to more than 500 m^2/m^3.

For the cells to attach to the carrier surfaces in significant amounts, the residence time of the fluid in the reactor should be smaller than the replication time of the cells. In such cases, the cells will tend to adhere to the supports (provided they are good exopolymer producers) to avoid being washed out of the reactor. Most of the biofilm reactors also contain suspended biomass, some of it resulting from the detachment of attached biomass (which has to be purged periodically).

There is a wide range of biofilm reactors available for wastewater treatment (Melo and Oliveira, 2001). They are mostly continuous reactors, the main exception being the sequencing batch biofilm reactor (SBBR) consisting of a tank, which is periodically filled with the feed liquid and discharged (Wilderer, 1995). Continuous reactors include:

1. trickling filters, where the liquid is split into fine streams and percolates downwards through the biofilm while the air flows upwards
2. rotating disc contactors, with the biofilm attached on the surface of vertical discs rotating in the liquid medium
3. submerged beds with biofilm particles immersed in the liquid (upflow or downflow)
4. fluidized beds, where the biofilm layer forms around small solid particles which are kept under fluidization conditions
5. moving beds, that consist of an expanded bed of particles circulating throughout the equipment together with the fluid (often with gas bubbles too), such as the air-lift reactor and the circulating bed reactor
6. membrane reactors, with microbial films attached to microporous membrane surfaces allowing the supply of nutrients and oxygen to both sides of the film.

These reactors are described in a number of publications (Metcalf and Eddy, 1987; Tijhuis *et al.*, 1994; Wilderer, 1995; Lazarova and Manem, 1997; Nogueira *et al.*, 1998). Reactors containing dense microbial granules without support particles, such as UASB (upflow anaerobic sludge blanket) are usually treated as biofilm reactors as well (Letting and Hulshoff-Pol, 1992; Brito and Melo, 1997).

Fig. 20.5 schematically presents examples of some biofilm reactors.

Biofilm reactors used in wastewater treatment are still often designed on the basis of an empirical parameter (the 'eliminated load'), the values of which are known from previous practical experience, without taking into account any phenomenological approaches. Basically, the following mass balance is used:

$$V_R = \frac{Q \cdot (S_1 - S_2)}{B_v} \tag{2}$$

V_R is the reactor volume, Q the volumetric flow rate of wastewater, S_1 and S_2 are the inlet

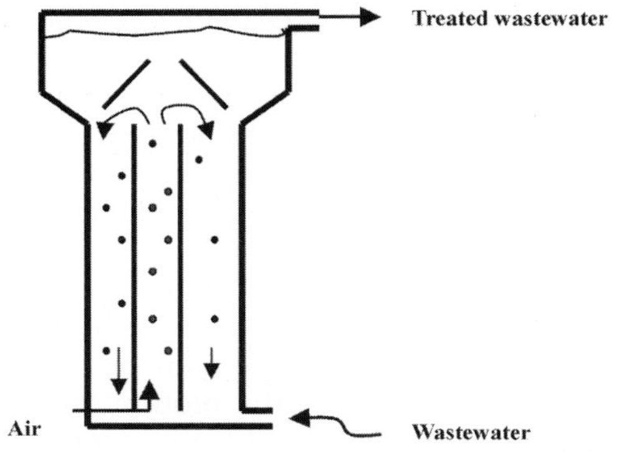

(a)

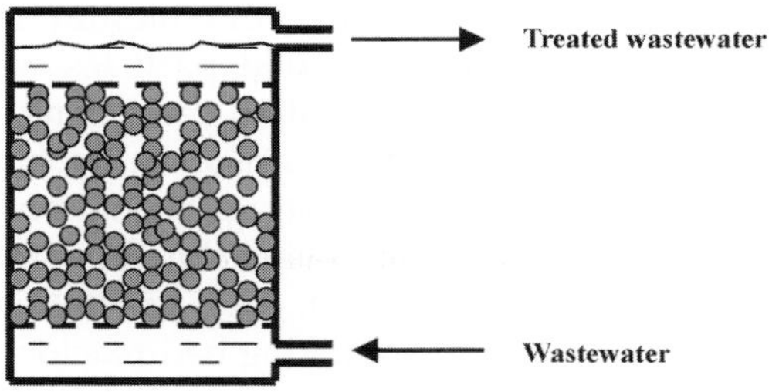

(b)

Fig. 20.5 Examples of two biofilm reactor configurations: (a) air-lift reactor; (b) submerged up-flow fixed bed.

and outlet substrate concentrations and B_v is the mass of substrate consumed per unit time and unit volume of the reactor (eliminated load).

A more rational approach to estimate B_v, or an analogous rate parameter, is to model it in terms of diffusion-reaction concepts similar to the ones developed in heterogeneous catalysis (Harremöes and Henze, 1995). Integrated equations for these models are available for zero order and first order biological kinetics, and are summarized in Table 20.2. They relate the rate of substrate consumption per unit area of biofilm (r_A) to substrate concentration (S), external mass transfer coefficient (k_m), substrate diffusivity within the biofilm (D_f), intrinsic reaction constants (k_{1f}-first-order and k_{0f}-zero order) and biofilm thickness (L_f).

Parameter B_v (eliminated load) is related to the substrate consumption rate (r_A) by:

$$B_v = r_A \frac{A_f}{V_R} \tag{3}$$

where A_f is the surface area of the biofilm, i.e. the interfacial mass transfer area. If the microbial film is attached to a spherical particle (e.g. in a fluidized bed or an air-lift reactor) and is not too thick, it can be considered as a flat biofilm and its surface area will be:

$$A_f = \frac{3V_R(1-\varepsilon)}{L_f + r_p} \tag{4}$$

where L_f is the thickness of the biofilm and ε is the bed porosity (ratio of the liquid volume over the reactor volume). For thin biofilms, $L_f \ll r_p$ and the denominator can be reduced to r_p.

Taking into account Equations 2–4, together with Equations 5–12 in Table 20.2, the reactor volume can be estimated as indicated in Table 20.3.

The external mass transfer coefficient can be calculated for many types of reactors by appropriate correlations (see textbooks on Transport Phenomena and Reaction Engineering). However, there is not yet an established methodology for the prediction of the following variables *as a function of the reactor operating conditions*:

1. reaction constants (the reaction constants depend also on the knowledge of the *mass of active cells* and of the *biomass yield* inside the microbial film)
2. internal diffusivities
3. biofilm thickness.

The joint efforts of experimental researchers and mathematical modellers are required to obtain this type of information in different practical situations.

Harremöes (1978) reported that in many wastewater treatment reactors biofilms were partially penetrated and the apparent reactions were of half order. Values of the 'apparent half-order constant' $(k_{1/2})_{app}$ were presented for several pairs of microbial species-substrates, although the specific hydrodynamic and chemical conditions were not fully known. Table 20.4 shows a list of some of those practical values.

The 'apparent reaction constants' differ from the 'intrinsic' ones because the former

TABLE 20.2 Main equations of the diffusion-reaction model for two limiting cases: zero order and first order intrinsic biological kinetics

First order intrinsic kinetics		*Zero order intrinsic kinetics*	
$r_A = \dfrac{S}{\dfrac{1}{k_m} + \dfrac{1}{\eta_i k_{1f} L_f}}$	(5)	$r_A = \dfrac{k_m S}{2\lambda^2}\left[\sqrt{1+4\lambda^2} - 1\right]$	(8)
$\eta_i = \dfrac{\tanh\phi}{\phi}$	(6)	$\lambda = \dfrac{k_m \sqrt{S}}{\sqrt{2k_{of} D_f}}$	(9)
$\phi = \sqrt{\dfrac{k_{1f} L_f^2}{D_f}}$	(7)	***Special cases*** *When external mass transport is not the limiting step:*	
		(a) Biofilm fully penetrated by the substrate: $r_A = k_{of} L_f = \dfrac{(k_{1/2})^2_{app}}{2D_f}$	(10)
		(b) Biofilm only partially penetrated by the substrate: $r_A = k_{of} L_f \beta = \sqrt{2k_{of} D_f S} =$ *or :* $r_A = (k_{1/2})_{app}\sqrt{S}$	(11)
		$\beta = \dfrac{1}{\phi} = \sqrt{\dfrac{2D_f S_i}{k_{of} L_f^2}}$	(12)

TABLE 20.3 Evaluation of reactor volume based on the expressions of the diffusion-reaction model

Ideal continuous well stirred reactor		*Ideal plug flow reactor*	
$V_R = \dfrac{Q(S_1 - S_2)(L_f + r_p)}{3r_A(1-\varepsilon)}$	(13)	First order reaction:	
(For these reactors: $S_2 = S$)		$V_R = \dfrac{Q(L_f + r_p)}{3(1-\varepsilon)k_{1f} L_f \eta} ln\left(\dfrac{S_1}{S_2}\right)$	(14)
		with: $\dfrac{1}{\eta} = \dfrac{1}{\eta_i} + \dfrac{k_{1f} L_f}{k_m}$	(15)
		Zero order reaction with full substrate penetration and no external mass transfer limitations: $V_R = \dfrac{Q(L_f + r_p)}{3(1-\varepsilon)k_{of} L_f}(S_1 - S_2)$	(16)
		Zero order reaction with partial substrate penetration (apparent half-order) and no external mass transfer limitations: $V_R = \dfrac{Q(L_f + r_p)\sqrt{2S_1}}{3(1-\varepsilon)\sqrt{k_{of} L_f}}\left[1 - \sqrt{\dfrac{S_2}{S_1}}\right]$	(17)

TABLE 20.4 Apparent half-order constants in the diffusion-reaction model

Process, microbial species and limiting biological factor	*Reactor type*	$K_{1/2ap} \cdot 10^5$ ($kg^{1/2}/m^{1/2}/s$)
Aerobic, heterotrophic biomass Oxygen (Grasmick *et al.*, 1982)	Fixed bed	0.12
Aerobic, heterotrophic biomass Toluene (Pederson and Arvin, 1996)	Waste gas trickling filter	0.07–0.11
Nitrification, autotrophic biomass Oxygen (Çeçen and Gönenç 1994)	Submerged filter	0.05–0.10
Nitrification, autotrophic biomass Oxygen (Nogueira *et al.*, 1998)	Circulating bed reactor	0.15
Nitrification, autotrophic biomass Ammonium Gönenç and Harremöes, 1985	Rotating disc	0.06
Anaerobic, methanogenic biomass Acetate (Hamoda and Kennedy, 1987)	Downflow filter	0.12
Anaerobic, methanogenic biomass Molasses (Gönenç *et al.*, 1991)	Upflow filter	0.04–0.48
Anoxic, denitrifying biomass Nitrate (Jansen, 1982)	Rotating drum	0.02–0.14
Anoxic, denitrifying biomass Nitrate (Watanabe and Ishiguro, 1978)	Rotating disc	0.11

encompass both the biological reaction and the internal mass transfer in the biofilm. The definition of the 'half-order apparent reaction constant' is:

$$(k_{1/2})_{app} = \sqrt{2k_{of}D_f} \qquad (18)$$

As regards the prediction of the effective diffusivity in the biofilm matrix (D_f), advances have been achieved in recent years, but their applicability to reactor design is still limited. Published values of the ratio between the effective diffusivity in biofilms and the diffusivity of the same component in water range from 15 to 120% (Melo and Oliveira, 2001). Some authors related biofilm density to diffusivity (Fan *et al.*, 1990; Stewart, 1998). The effect of velocity on biofilm structure and on the effective diffusivity of nutrients was reported by a few authors (Vieira *et al.*, 1993; van Loosdrecht *et al.*, 1995; Brito and Melo, 1999; Casey *et al.*, 2000). The results do not yet allow a unifying theory because, in general, higher velocities produce two opposite effects (thinner but more compact biofilms) on the mass transfer resistance offered by the biological layer. The particular hydrodynamic patterns present around biofilms (laminar versus turbulent flow) seem to have a determinant role not yet duly understood.

Although the above unstructured model and other more sophisticated structured models help in understanding the role of the different variables in the efficiency of substrate removal in biofilm reactors, they are not yet able to provide engineers with values for practical application to reactor design. Alternatively, empirical expressions were developed over the years by the practitioners. For the treatment of domestic wastewaters in trickling filters, the following is commonly used:

$$C_2 = C_1 \exp\left[0.11^{(T-20)} Z.A_v.\frac{A_s}{Q}\right] \qquad (19)$$

with C_1 and C_2 being the inlet and outlet substrate concentrations expressed in mg/l of BOD_5 (biochemical oxygen demand during 5 days), Z the height of the filter, A_v the surface area of the carrier per unit reactor volume (m^2/m^3), A_s the cross-sectional area of the reactor (m^2), Q the volumetric flow rate (m^3/s) and T the wastewater temperature (°C).

For rotating biological contactors:

$$C_2 = C_1 - \frac{A_f}{Q}\frac{p \cdot C_2}{(k + C_2)} \qquad (20)$$

where A_f is the surface area of the discs covered by biofilm and p and k are empirical parameters, which have to be estimated from the operation of similar reactors at pilot-plants or full-size plants.

NOMENCLATURE

A_f	surface area of biofilm (m^2)
A_s	cross-sectional area of the filter (m^2)
A_v	specific area of support per volume of reactor (m^2/m^3)
b	reciprocal of the resistance to detachment (s^{-1})
C_1	substrate concentration at reactor inlet, as soluble BOD_5 (mg/l^{-1})
C_2	substrate concentration at reactor outlet, as soluble BOD_5 (mg/l^{-1})
D_f	effective diffusion coefficient or effective diffusivity (m^2/s^{-1})
k_m	external mass transfer coefficient (m/s^{-1})
k_{0f}	zero order biofilm reaction rate ($kg/m^3/s^{-1}$)
k_{1f}	first order biofilm reaction rate ($/s^{-1}$)
$k_{1/2ap}$	apparent half order constant ($kg^{1/2}/m^{1/2}/s^{-1}$)
L_f	thickness of the microbial layer (m)
m_f	mass of attached biofilm per surface area ($kg/m^2/s^{-1}$)
m_f	maximum mass of biofilm at steady state
Q	volumetric flow rate (m^3/s^{-1})
r_A	reaction rate per unit area of biofilm or surface reaction rate ($kg/m^2/s^{-1}$)
r_p	radius of the bare carrier particles (m)
S	bulk substrate concentration in the solution (kg/m^3)
S_1	substrate concentration at reactor inlet (kg/m^3)
S_2	substrate concentration at reactor outlet (kg/m^3)
t	time (s)
T	temperature (°C)
V_R	reactor volume
Z	depth of the trickling filter (m)

Greek symbols

β	degree of substrate penetration in the biofilm
ε	reactor porosity
ϕ	Thiele modulus
η_i	biofilm internal efficiency
η	biofilm efficiency based on external substrate concentration
λ	external mass transfer rate/internal coupled diffusion-reaction rate

REFERENCES

Bott, T.R. (1995). Biological growth on heat exchanger surfaces. In: *Fouling of Heat Exchangers*, Ch. 12, pp. 223–267. Elsevier, Amsterdam.

Briffaud, J. and Engasser, J.M. (1979). Citric acid production from glucose. II Growth and excretion kinetics in a trickle-flow fermenter. *Biotechnol. d Bioeng.* **XXI**, 2093–2111.

Brito, A.G. and Melo, L.F. (1997). A simplified analysis of reaction and mass transfer in UASB and EGSB reactors. *Environ. Technol.* **18**, 35–44.

Brito, A.G. and Melo, L.F. (1999). Mass transfer coefficients within anaerobic biofilms: effects of external liquid velocity. *Wat. Res.* **33**, 3673–3678.

Casey, E., Glennon, B. and Hamer, G. (2000). Biofilm development in a membrane-aerated biofilm reactor: effect of flow velocity on performance. *Biotechnol. Bioeng.* **67**, 476–486.

Çeçen, F. and Gönenç, I.E. (1994). Nitrogen removal characteristics of nitrification and denitrification filters. *Water Sci. Technol.* **29**, 409–416.

Costerton, J.W., Cheng, K.J., Geesey, G.G. *et al.* (1987). Bacterial biofilms in nature and disease. *Ann. Rev. Microbiol.* **41**, 435–464.

Davies, D.G. and McFeters, G.A. (1988). Growth and comparative physiology of *Klebsiella oxytoca* attached to granular activated carbon particles in liquid media. *Microb. Ecol.* **15**, 165–175.

Davies, D.G., Chakrabarty, A.M. and Geesey, G.G. (1993). Exopolysaccharide production in biofilms: substratum activation of alginate gene expression by *Pseudomonas aeruginosa*. *Appl. Environl Microbiol.* **59**, 1181–1186.

Dempsey, M. (1990). Ethanol production by *Zymomonas mobilis* in a fluidised bed fermenter. In: J.A.M. de Bont, J. Visser, B. Matiasson and J. Tramper (eds) *Physiology of Immobilized Cells*, pp. 137–148. Elsevier Science Publishers B.V., Amsterdam.

Fan, L.-S., Leyva-Ramos, R., Wisecarver, K.D. and Zehner, B.J. (1990). Diffusion of phenol through a biofilm grown on activated carbon particles in a draft-tube three-phase fluidized-bed bioreactor. *Biotechnol. Bioeng.* **35**, 279–286.

Flemming, H.-C. (1997). Reverse osmosis membrane biofouling. *Exp. Therm. Fluid Sci.* **14**, 382–391.

Geesey, G. (1991). What is biocorrosion?. In: H.-C. Flemming and G. Geesey (eds) *Biofouling and Biocorrosion in Industrial Water Systems*, pp. 155–164. Springer-Verlag, Berlin.

Gjaltema, A. (1996). Biofilm development: growth versus detachment. PhD Thesis, Biochemical Engineering Department, Technical University of Delft, Delft, The Netherlands.

Gönenç, I.E. and Harremöes, P. (1985). Nitrification in rotating disk systems. I, Criteria for transition from oxygen to ammonia rate limitation. *Wat. Res.* **19**, 1119–1127.

Gönenç, I.E., Orhon, D. and Baikal, B.B. (1991). Application of biofilm kinetics to anaerobic fixed bed reactors. *Water Sci. Technol.* **23**, 1319–1326.

Grasmick, A., Elmaleh, S. and Ben Aim, R. (1980). Experimental study of submerged biological filtration. *Wat. Res.* **14**, 613–626.

Hamoda, M.F. and Kennedy, K.J. (1987). Biomass retention and performance in anaerobic fixed-film reactors treating acetic acid wastewater. *Biotechnol. Bioeng.* **30**, 272–281.

Hermanowics, S.W. (1999). Two-dimensional simulations of biofilm development: effects of external environmental conditions. *Water Sci. Technol.* **39**, 107–114.

Harremöes, P. (1978). Biofilm kinetics. In: R. Mitchell (ed.) *Water Pollution Microbiology*, Vol. 2, pp. 82–109. John Wiley and Sons, New York.

Harremöes, P. and Henze, M. (1995). Biofilters. In: M. Henze, P. Harremöes, J. de la Cour Jansen and E. Arvin (eds) *Wastewater Treatment: Biological and Chemical Processes*, pp. 143–193. Springer-Verlag, Berlin.

Jansen, J. la Cour (1982). Fixed film kinetics – kinetics of soluble substrates. PhD Thesis, Department of Environmental Engineering, Technical University of Denmark, Lyngby.

Karamanev, D.G. (1991). Model of the biofilm structure of *Thiobacillus ferrooxidans*. *J. Biotechnol.* **20**, 51–64.

Keevil, C.W. and Walker, J.T. (1992). Nomarski DIC microscopy and image analysis of biofilms. *Binary: Comput. Microbiol.* **4**, 93–95.

Kuehn, M., Hausner, M., Bungartz, H.-J. *et al.* (1998). Automated confocal laser scanning microscopy and semiautomated image processing for analysis of biofilms. *Appl. Environ. Microbiol.* **64**, 4115–4127.

Lappin-Scott, H.M. and Costerton, J.W. (1995). *Microbial Films*. Cambridge University Press, Cambridge.

Lawrence, J.R., Delaquis, P.J., Korber, D.R. and Cladwell, D.E. (1987). Behaviour of *Pseudomonas fluorescens* within the hydrodynamic boundary layer of surface microenvironments. *Microb. Ecology*, **14**, 1–14.

Lawrence, J.R. and Korber, D.R. (1993). Aspects of microbial surface colonization behaviour. In: Guerrero, R. and Pedro-Alios (eds), *Trends in Microbial Ecology*. Barcelona, Spanish Society for Microbiology, pp. 113–118.

Lazarova, V. and Manem, J. (1994). Advances in biofilm aerobic reactors ensuring effective biofilm activity control. *Water Sci. Technol.* **29**, 319–327.

Lazarova, V. and Manem, J. (1997). An innovative process for waste water treatment: the circulating floating bed reactor. *Water Sci. Technol.* **34**, 89–99.

Letting, G. and Hulshoff-Pol, L.W. (1992). UASB-process design for various types of wastewaters. In: J. Malina and F. Pohland (eds) *Design of anaerobic processes for the treatment of industrial and municipal wastes*, pp. 119–145. Technomic Pub., Lancaster.

McCarter, L.L., Showalter, R.E. and Silverman, M.R. (1992). Genetic analysis of surface sensing in *Vibrio parahaemolyticus*. *Biofouling* **5**, 163–175.

Melo, L.F. and Bott, T.R. (1997). Biofouling in water systems. *Exp. Therm. Fluid Sci.* **14**, 375–381.

Melo, L.F. and Oliveira, R. (2001). Biofilm reactors In: J.S. Cabral, J. Tramper and M. Mota (eds) *Multiphase Bioreactor Design*, pp. 271–308. Taylor and Francis, London.

Melo, L.F. and Vieira, M.J. (1999). Physical stability and biological activity of biofilms under turbulent flow and low substrate concentration. *Bioproc. Eng.* **20**, 363–368.

Metcalf & Eddy, Inc. (1987). *Wastewater Engineering: Treatment and Disposal*, 2nd edn. Tata McGraw-Hill Publishing Company, New Delhi.

Neu, T. (2000). In: H.-C. Flemming, U. Szewzyck and T. Griebe (eds) Confocal laser scanning microscopy (CLSM) of biofilms. *Biofilms: Investigative Methods and Applications*, pp. 211–224. Technomic Publishing Co., Inc., Lancaster.

Neu, T.R. and Lawrence, J.R. (1997). Development and structure of microbial biofilms in river water studied by confocal laser scanning microscopy. *FEMS Microbiol. Ecol.* **24**, 11–25.

Nogueira, R., Lazarova, V., Manem, J. and Melo, L.F. (1998). Influence of dissolved oxygen on the nitrification kinetics in a circulating bed biofilm reactor. *Bioproc. Eng.* **19**, 441–449.

Nogueira, R., Melo, L.F., Purkhold, U. *et al.* (2002). Microbial population dynamics versus nitrification performance in biofilm reactors: effects of hydraulic residence time and the presence of organic carbon. *Wat. Res.* **36**(2), 469–481.

Noguera, D.R., Pizarro, G., Stahl, D.A. and Ritmann, B.E. (1999). Simulation of multispecies biofilm development in three dimensions. *Water Sci. Techn.* **39**, 123–130.

Pederson, A.R. and Arvin, E. (1996). Toluene removal in a biofilm reactor from waste gas treatment, *3rd International IAWQ Special Conference on Biofilm Systems*, Copenhagen.

Pereira, M.O., Vieira, M.J., Beleza, V.M. and Melo, L.F. (2001). Comparison of two biocides – carbamate and glutaraldehyde – in the control of fouling in pulp and paper industry. *Environ. Technol.* **22**, 781–790.

Pereira, M.O., Kuehn, M., Wuertz, S. *et al.* (2002). Effect of flow regime on the architecture of a *Pseudomonas fluorescens* biofilm. *Biotechnol. Bioeng.* **78**, 164–171.

Picioreanu, C., van Loosdrecht, M.C.M. and Heijnen, J.J. (1999). Discrete-differential modelling of biofilm structure. *Water Sci. Technol.* **39**, 115–122.

Stewart, P.S. (1998). A review of experimental measurements of effective diffusive permeabilities and effective diffusion coefficients in biofilms. *Biotechnol. Bioeng.* **59**, 261–272.

Stoodley, P., de Beer, D. and Lewandowski, Z. (1994). Liquid flow in biofilm systems. *Appl. Environ. Microbiol.* **60**, 2711–2716.

Tijhuis, L., van Loosdrech, M.C.M. and Heijnen, J.J. (1994). Formation and growth of heterotrophic aerobic biofilms on small suspended particles in airlift reactors. *Biotechnol. Bioeng.* **44**, 595–608.

Tijhuis, L., Huisman, J.L., Hekkelman, H.D. *et al.* (1995). Formation of nitrifying biofilms on small suspended particles in airlift reactors. *Biotechnol. Bioeng.* **47**, 585–595.

van Loosdrecht, M.C.M., Eikelboom, D., Gjaltema, A. *et al.* (1995). Biofilm structure. *Water Sci. Technol.* **32**, 35–43.

Vandevivere, P. and Kirchman, D.L. (1993). Attachment stimulates exopolysaccharide synthesis by a bacterium. *Appl. Environ. Microbiol.* **59**, 3280–3286.

Veiga, M.C., Mendez, R. and Lema, J.M. (1992). Development and stability of biofilms in bioreactors. In: L.F. Melo, T.R. Bott, M. Fletcher and B. Capdeville (eds) *Biofilms – Science and Technology*, pp. 421–434. Kluwer Academic Publishers, Dordrecht.

Vieira, M.J. and Melo, L.F. (1995). Effect of clay particles on the behaviour of biofilms formed by *Pseudomonas fluorescens*. *Wat. Sci. Tech.* **32**(8), 45–52.

Vieira, M.J. and Melo, L.F. (1999). Intrinsic kinetics of biofilms formed under turbulent flow and low substrate concentration. *Bioproc. Eng.* **20**, 369–375.

Vieira, M.J., Melo, L.F. and Pinheiro, M.M. (1993). Biofilm formation: hydrodynamic effects on internal diffusion and structure. *Biofouling* **7**, 67–80.

Watanabe, Y. and Ishiguro, M. (1978). Denitrification kinetics in a submerged rotating biological disk unit. *Progr. Water Technol.* **5**, 187–195.

Wilderer, P. (1995). Technology of membrane biofilm reactors operated under periodically changing process conditions. *Water Sci. Technol.* **31**, 173–183.

Wobus, A., Röske, K. and Röske, I. (2000). Investigations of spatial and temporal gradients in fixed-bed biofilm reactors for wastewater treatment. In: H.-C. Flemming, U. Szewzyck and T. Griebe (eds) *Biofilms: Investigative Methods and Applications*, pp. 165–194. Technomic Publishing Co., Inc., Lancaster.

21

Suspended growth processes

Nigel Horan

School of Civil Engineering, University of Leeds, Leeds LS2 9JT, UK

1 INTRODUCTION

Suspended growth systems comprise aggregates of microorganisms generally growing as flocs in intimate contact with the wastewater they are treating. The aggregates or flocs are responsible for the removal of polluting material and comprise a wide range of microbial species. The most prevalent and important of these microorganisms are the bacteria, the protozoa and the metazoa. Fungi and viruses are also found, but probably contribute little to the treatment of the wastewater. Suspended growth treatment systems permit the exploitation of the full range of microbial metabolic capabilities. The full spectrum of redox environments from aerobic, through anoxic to anaerobic can be found within the floc itself, but they can also be created by appropriate process reactor design. These environments allow the growth of both the organoheterotrophs (which oxidize organic carbon and remove BOD) and the lithoautotrophs (which are responsible for ammonia oxidation). Indeed, in waste stabilization pond systems, phototrophic organisms, which utilize a range of electron acceptors, can be exploited to achieve good treatment with negligible energy input.

The food source for all the above organisms comes either in the soluble form dissolved in the wastewater, or as particulate material that is first solubilized by microbial action.

Treatment of wastewaters in suspended growth environments offers many process advantages, which has led to the proliferation of this type of treatment system. Waste stabilization ponds, aerated lagoons and activated sludge are all examples of treatment options that rely on the actions of microorganisms growing in a suspension of the wastewater under treatment.

A suspended growth wastewater treatment processes is a biological reactor which has been engineered to encourage the growth of specific types of microorganisms that are able to undertake the reactions necessary to achieve purification of the influent wastewater. Their successful design requires the provision of:

1. a reactor (or series of reactors) of sufficient capacity to retain the wastewater long enough for the microorganisms to undertake the biological interconversions necessary to achieve the required effluent standard
2. facilities to ensure that the microorganisms are retained in the reactor long enough to grow and divide, thus maintaining a stable population
3. the correct redox environment to achieve the required biological reactions.

The most effective and innovative reactor design thus requires a full understanding of the biochemical basis for the growth of microorganisms using wastewater as the substrate. It is apparent from factors 1 to 3 above that the key design issues for a suspended growth process are the requirements of the treatment system in terms of its effluent quality.

The Handbook of Water and Wastewater Microbiology
ISBN 0-12-470100-0

2 EFFLUENT QUALITY REQUIREMENTS

The sole role of a wastewater treatment process is to protect the receiving watercourse from environmental degradation and ensure that the impact of the effluent from the plant is minimized. This is normally achieved by means of standards, which are set by the appropriate regulatory authority, to describe a minimum quality that the effluent must attain. Ideally these standards should be evaluated based on the characteristics of the receiving watercourse, the likely benefits that will accrue from maintaining a healthy watercourse and the ability of the community to finance treatment. Standards always contain a requirement for the treatment plant to achieve a certain effluent quality in terms of the biochemical oxygen demand (BOD) and nearly always have a standard for suspended solids. Depending on the quality of the receiving watercourse they may then contain requirements for ammonia-nitrogen, total nitrogen and phosphorus. The minimum standards that apply throughout the European Community are summarized in Table 21.1, however, member states may impose stricter standards than this where it is thought appropriate.

TABLE 21.1 The minimum treatment standards for sewage effluents that must be achieved by EC member states

Parameter	*Concentration (mg/l)*	*Minimum % of reduction*[a]
(a) Requirements for discharges from urban waste water treatment plants		
BOD_5 at 20°C without nitrification	25	70–90
COD	125	75
Total suspended solids	35[b] (>10 000 pe)	90
	60 (2000–10 000 pe)	70
(b) Discharges to sensitive areas which are subject to eutrophication		
Total phosphorus	2 mg P/l (10 000–100 000 PE)	80
	1 mg P/l (>100 000 PE)	
Total nitrogen[c]	15 mg N/l (10 000–100 000 PE)	70–80
	10 mg N/l (>100 000 PE)	

[a] Reduction in relation to the load of the influent.
[b] Optional requirement.
[c] Organic and ammonical nitrogen and nitrite-nitrogen.

3 ACHIEVING EFFLUENT QUALITY STANDARDS – MICROBIAL REACTIONS IN SUSPENDED GROWTH SYSTEMS

For each of the parameters listed in Table 21.1(a) and (b) there is one or more species of bacteria which can be exploited to metabolize the parameter using it either as a source of electrons or as a terminal electron acceptor. These reactions are summarized briefly below but are covered in more detail in other chapters in this volume.

3.1 BOD removal

The biochemical oxygen demand measures the amount of oxygen required to oxidize the organic carbon present in a wastewater according to the equation:

$$C_6H_{12}O_6 + 6O_6 \rightarrow 6CO_2 + 6H_2O \quad (1)$$

This oxidation reaction, referred to as respiration or catabolism, provides the energy necessary for bacterial growth and reproduction and there is a vast range of heterotrophic species which will carry it out in suspended growth processes. In addition to energy, the microorganisms need a source of carbon to build new cell material. This is also provided by the organic material measured in the BOD test according to Equation (2), where $C_5H_7O_2$ represents a new bacterial cell.

$$C_6H_{12}O_6 \rightarrow C_5H_7O_2 \quad (2)$$

All biological wastewater treatment systems are designed to remove BOD, both in its particulate and soluble form. However, as much as 40% of the total BOD of a wastewater is particulate, consequently, it is generally most cost effective to remove this fraction by sedimentation. That is the role of the primary sedimentation tank, septic tank or anaerobic pond. The fraction that remains after sedimentation is a mixture of colloidal and soluble BOD that can only be removed biologically. The soluble BOD is usually biodegraded very rapidly, generally in less than one hour. The colloidal fraction is entrapped in the sludge floc and is degraded more slowly. It is apparent from Equation (1) that breakdown of BOD to CO_2 and H_2O, removes BOD without the generation of new bacteria. Thus, if this reaction were encouraged, sewage treatment would take place without the generation of sludge. However, a lot of oxygen would be required and thus power consumption would be high. By contrast, Equation (2) shows that BOD can be removed by conversion to new bacteria without the need for oxygen. Thus, if this reaction were encouraged, treatment would take place without the need for aeration, however, the sludge generated would be very high. In practice there is a balance between the two reactions and treatment plants are operated with some sludge generated and some oxygen required.

It is worth bearing in mind Equations (1) and (2) when evaluating processes that claim a reduced sludge yield or a reduced power requirement. It is easy to reduce sludge yield by increasing the amount of oxygen provided and increasing the sludge age (extended aeration and aerobic digestion are examples of this). Similarly, it is easy to reduce the power requirement by a high rate process which generates more sludge. It is not possible to reduce both the sludge yield and the oxygen requirement!

3.2 Ammonia removal

Ammonia is removed by the action of two groups of bacteria, collectively termed the nitrifying bacteria, which catalyse the reactions of nitrification. Nitrification is the process of ammonia oxidation in which ammonia is oxidized ultimately to nitrate in two reactions carried out by distinct groups of obligately aerobic bacteria. The first intermediate is nitrite and this reaction is catalysed by the genus *Nitrosomonas*:

$$NH_3 + 1.5O_2 \rightarrow NO_2^- + H^+ + H_2O \quad (3)$$

Nitrite is further oxidized to nitrate by *Nitrobacter*:

$$NO_2^- + 0.5O_2 \rightarrow NO_3^- \quad (4)$$

Nitrosomonas and *Nitrobacter* are both autotrophic genera which reduce carbon dioxide (in the form of bicarbonate or carbonate) as a source of cellular carbon. Assuming a gross cell composition for a typical nitrifying bacteria of $C_5H_7NO_2$, then the overall reaction for the oxidation of ammonia, coupled to the synthesis of new nitrifying bacteria, can be represented as:

$$NH_3 + 1.83O_2 + 1.98HCO_3 \rightarrow 0.021C_5H_7NO_2 + 1.041H_2O + 0.98NO_3 + 1.88H_2CO_3 \quad (5)$$

The energy expenditure required to achieve the reduction of bicarbonate is relatively high, yet the nitrifying bacteria can achieve only a low yield of energy from the oxidation of their chosen substrates, ammonia and nitrite. Consequently, these organisms demonstrate very low growth yields and require a long retention time in the aeration basin to ensure they can divide and maintain a stable population. They are also very susceptible to temperature changes and below 20°C their reaction rate slows dramatically. This is illustrated in Fig. 21.1 which shows the sludge age required at a range of temperatures, in order to ensure that nitrification is achieved.

3.3 Nitrogen removal

Nitrate itself is able to act as a terminal electron accceptor. In the absence of a supply of dissolved oxygen, the utilization of oxygen for respiration (Equation (1)) cannot take place. However, certain chemo-organotrophs are capable of replacing O_2 with NO_3 as an oxidizing agent and respiration can proceed with the reduction of nitrate to nitrite, nitric oxide, nitrous oxide or nitrogen. Equation (6) demonstrates the stoichiometric reaction for the reduction of nitrate using methanol as a source of electrons, with the production of

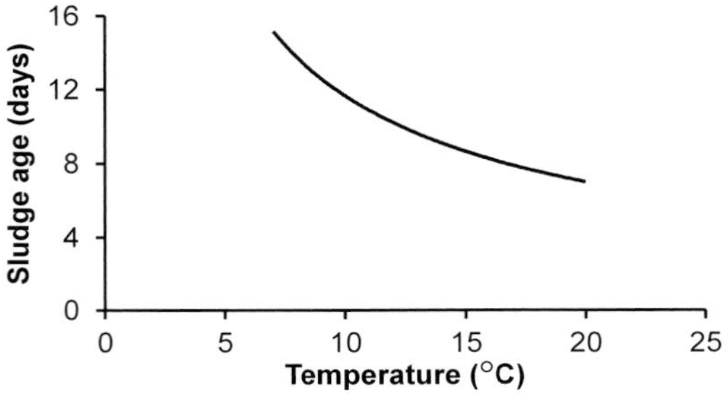

Fig. 21.1 The minimum design sludge age required at a range of sewage temperatures.

nitrogen gas and new cell material. This process is known as anaerobic or nitrate respiration and is carried out by a variety of bacteria such as *Alcaligenes*, *Achromobacter*, *Micrococcus* and *Pseudomonas*. Not all these genera are capable of complete oxidation to nitrogen and a variety of gaseous products can be produced.

$$\begin{aligned} &NO_3 + 1.08CH_3OH + 0.24H_2CO_3 \\ &\quad \rightarrow 0.06C_5H_7NO_2 + 0.47N_2 + 1.68H_2O \\ &\quad + HCO_3 \qquad (6) \end{aligned}$$

As denitrification is an oxidation reaction, it needs an electron donor and an electron acceptor. Nitrate is the electron acceptor and organic carbon generally provides the source of electrons. This is usually from the BOD present in the sewage, or as a supplementary carbon source such as methanol or ethanol. It generally requires 8 mg COD to achieve the removal of 1 mg nitrate and it is advisable to aim for a nitrate concentration <12 mg/l in the final effluent to prevent sludge settlement problems.

3.4 Phosphate removal

Of the nutrients that are capable of supporting luxuriant growths of algae in a receiving water, phosphorus is rate limiting, and a concentration of 10 μg/l is required before algal growth will occur. It has been argued, therefore, that control over phosphorus-containing compounds in aquatic ecosystems presents a means of controlling the deleterious effects of eutrophication. Consequently, if the small concentration of phosphorus that is present in sewage effluents can be removed, then algae will not be able to flourish, regardless of the nitrogen concentration. Phosphorus load control has been demonstrated as one of the most effective ways of dealing with cultural eutrophication. A typical phosphorus standard is 1 mg/l dissolved orthophosphate (as phosphorus). The major sources of phosphorus in domestic wastewaters are from human excreta (50–65%) and synthetic detergents (30–50%) and typical concentrations are in the range 10–30 mg/l as phosphorus.

4 ENHANCED BIOLOGICAL PHOSPHORUS REMOVED IN ACTIVATED SLUDGE SYSTEMS

The uptake and removal of phosphorus from a wastewater by activated sludge followed from the observation that if an activated sludge is allowed to become anaerobic, then the amount of phosphorus (as phosphate, PO_4^{3-}) in the supernatant increases. Upon resumption of aeration, however, there is a rapid uptake of phosphate by the sludge which is in excess of that released during anaerobiosis. This 'luxury uptake' of phosphate results in a phosphate depleted mixed liquor and a phosphate-rich sludge. The ability to release or store phosphorus in a sludge by manipulating the prevailing oxygen concentration, has been exploited in several processes, often referred to as enhanced nutrient removal processes.

The mechanism by which phosphorus is stored and released helps in understanding the operation of phosphorus-removal activated sludge plants. Removal and release of phosphorus within a sludge is thought to be the result of a single genus of bacteria known as *Acinetobacter calcoaceticus.* They are only able to take up and metabolize compounds contributing to the BOD under aerobic conditions. Under anaerobic conditions they cannot metabolize compounds but they can take up volatile fatty acids, in particular acetate, and store them intracellularly in the form of a compound known as phb. Phosphate can also be stored intracellularly as polyphosphate granules known as volutin (Equation (7)):

$$n(PO_4^{3-}) + energy \Leftrightarrow (PO_4^{3-})_n \quad (7)$$

Thus, if a wastewater is allowed to become anaerobic, *Acinetobacter calcoaceticus* will start to take up the volatile fatty acids that are generated under anaerobic conditions. The energy required to achieve this is obtained from the degradation of polyphosphate according to Equation (7), which releases soluble phosphate into the wastewater. The volatile fatty acids are then converted to phb and stored within the bacterial cell.

If the wastewater is now aerated the *Acinetobacter calcoaceticus* reverses this procedure and degrades the phb to volatile fatty acids. However, as oxygen is now present, aerobic metabolism is possible and the bacterium can break down the acetate to generate energy. This energy is used to take up phosphate and generate polyphosphate, but far more phosphate is taken up than is released in the anaerobic phase. The key to biological phosphate removal is thus to achieve anaerobic conditions and generate phosphate release. The more anaerobic the wastewater becomes, the more phosphate released and thus the more taken up in luxury phosphate uptake. In order to achieve the necessary anaerobiosis a lot of BOD is needed in the influent wastewater (around 12 mg BOD for every mg P removed). Often this is not available, particularly during the winter months (or during periods of heavy rainfall), and consequently additional chemical dosing capacity must be provided to ensure consent is met.

5 MANIPULATING REDOX ENVIRONMENTS

It is quite straightforward in a suspended growth wastewater treatment process to vary the electron source and the terminal electron acceptor and in such a way control the redox potential within the reactor. By doing so this allows a wide range of reactions to take place (Table 21.2). Up to three reactor types are exploited; these are defined, based on the source of electrons, as aerobic (or oxic), anoxic and anaerobic.

6 AEROBIC REACTORS

An aerobic reactor exploits oxygen as the terminal electron acceptor and is able to sustain a large number of important reactions that use a number of different electron donors. Principal among these are: oxidation of organic material, which reduces the BOD; oxidation of ammonia to nitrate, which reduces the ammonia concentration; and the luxury uptake of phosphate with the synthesis of polyphosphate, which reduces effluent phosphate concentrations.

A variety of options are available to achieve aerobic conditions in wastewaters. The strength of the incoming wastewater dictates the amount of oxygen that must be introduced into the reactor and guideline values are around 1.2 mg oxygen for every mg of BOD to be removed and 4.8 mg of oxygen for every mg ammonia to be removed. Thus, for a wastewater which has a BOD of 180 mg/l and ammonia of 25 mg/l and which must achieve an effluent quality of 20 mg/l BOD and 1 mg/l ammonia, it is necessary to introduce

TABLE 21.2 The reactions that can be achieved by manipulating the redox environment in a suspended growth reactor

Reaction	*Electron donor*	*Electron acceptor*	*Redox potential*
BOD removal	Organic material	Oxygen	$>+200$ mV
Nitrification	Ammonia	Oxygen	$>+300$ mV
Denitrification	Organic material	Nitrate	-100 mV to $+150$ mV
Phosphorus release	Polyphosphate	phb	<-300 mV
Phosphorus uptake	Acetate	Phosphate	>-150 mV

TABLE 21.3 Key design parameters for suspended growth processes

Reactor type	*Biomass (mg/l)*	*Hydraulic retention time (days)*	*Volumetric loading (kg BOD/m^3)*	*Effluent quality*[a]			
				BOD	*NH_4*	*P*	*SS*
Waste stabilization pond	200	>20	0.002–0.008	30	4	5	[b]
Aerated lagoon	800	2–4	0.04–0.08	30	10	5	30
Activated sludge	3500	0.3–0.5	0.3–0.6	10	<1	<1	10
Membrane bioreactor	>12 000	0.4–0.6	0.6	2	<1	<1	<2

[a] Measured at a 95%ile.
[b] Pond effluents are high in suspended solids, but as this is largely algal biomass it is permitted up to a concentration of 150 mg/l.

307 mg O_2/l. Natural processes such as atmospheric diffusion and photosynthesis are able to introduce oxygen to a wastewater, although very slowly and thus the wastewater must be retained for a long time to achieve the necessary oxygen transfer. As a consequence, natural treatment options, such as waste stabilization ponds, have long hydraulic retention times, typically between 15 and 40 days, with large reactor volumes. By contrast as the rate of oxygen transfer is increased by mechanical means, the retention time of the wastewater can be reduced to as little as a few hours in the activated sludge process. However, as aeration intensity increases, so does the cost of the treatment process and high intensity aeration is both capital and operating cost intensive.

When wastewaters are aerated with a residual dissolved oxygen >1 mg/l, BOD removal occurs at a rate that is proportional to the amount of microorganisms held in the reactor. The more microorganisms that can be retained in the system the more rapid the rate of BOD removal will be and, consequently, the smaller the reactor that is required. However, there are two physical limits to the mass of microorganisms that can be retained, the first of these is the ability to transfer oxygen fast enough to match the oxygen uptake of the microorganisms and the second is the ability to remove the microorganisms by flocculation and sedimentation (section 10). This latter is generally the rate-limiting step in treatment. A new generation of suspended growth systems, known as membrane bioreactors, has eliminated this stage. Instead of a sedimentation step, the microorganisms are retained in the system by a membrane, generally with a pore size of around 0.4 μm, which permits the treated effluent to exit while retaining the microbial biomass. Such reactor systems permit a large increase in the amount of solids that can be retained (Table 21.3).

7 ANOXIC REACTORS

The major anoxic reaction of importance in wastewater treatment is the reduction of nitrate to nitrogen gas. This is important for two reasons. First of all, if the nitrate concentration is high in the final sedimentation tank of the activated sludge process, denitrification can occur in the solids that have settled to the base of the tank. The nitrogen gas generated will buoy the settled sludge carrying it to the surface of the sedimentation tank and over the weir into the final effluent. This problem, termed a rising sludge, can be eliminated by ensuring the nitrate concentration in the final effluent does not rise above 12 mg/l.

A second reason is that some treatment plants have effluent discharge standards for total nitrogen (see Table 21.1b), in which case nitrate removal is mandatory.

Anoxic conditions are achieved by eliminating all the residual dissolved oxygen and ensuring an adequate supply of electrons. The final effluent from a nitrifying process has high concentrations of nitrate but lacks a supply of electrons. Thus, if it is directed to a reactor to undergo denitrification, the electron supply needs augmenting. This is achieved either by adding external carbon (usually as methanol or

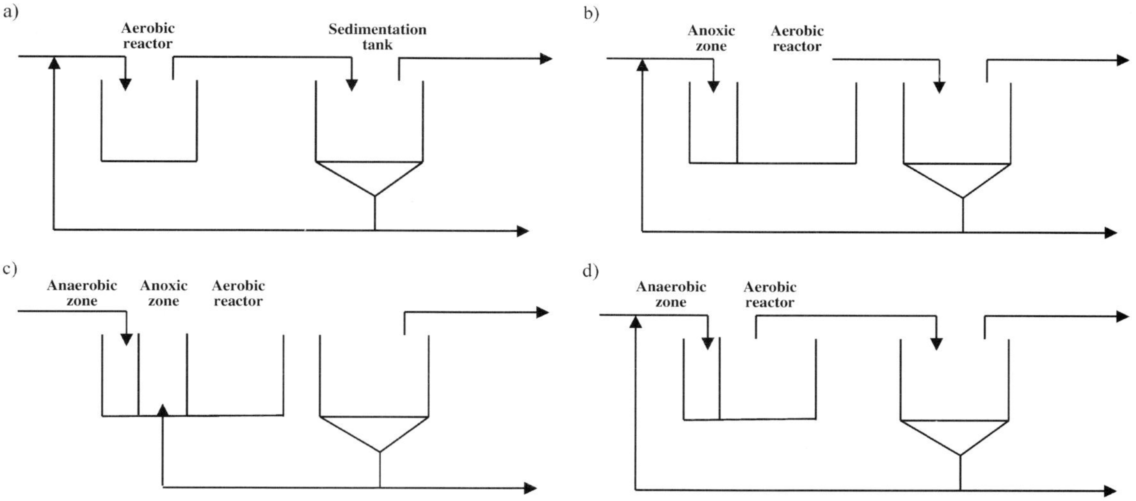

Fig. 21.2 Reactor configurations for manipulating redox conditions. (a) Carbonaceous removal only: an aerobic basin with sludge return direct to the basin; (b) Carbonaceous removal and nitrification: an aerobic basin with sludge return to an anoxic zone which also receives the incoming settled sewage; (c) Carbonaceous removal, nitrification, denitrification and phosphorus removal: an anaerobic zone to permit phosphate release, an anoxic zone which denitrifes the return sludge from the aeration basin and recycles it to the anaerobic zone, and an aerobic basin which achieves carbonaceous removal, nitrification and luxury phosphate uptake; (d) Carbonaceous removal and phosphorus removal. An anaerobic zone which receives return sludge from the aeration basin and permits phosphate release, followed by an aeration basin which achieves carbonaceous removal and luxury phosphate uptake.

a secondary industrial waste such as molasses) or directing a fraction of the settled sewage to the anoxic tanks. The most cost effective way of achieving denitrification is simply to remove the aeration from a small section at the head of the reactor and ensure that both the return sludge and the settled sewage are fed to this section. It requires a retention time of around 1 hour together with gentle mixing to ensure solids do not deposit in the basin (Fig. 21.2).

8 ANAEROBIC ZONES

An anaerobic reactor is necessary to promote the release of phosphate and uptake of acetate by the phosphate-accumulating bacteria. The more anaerobic the conditions then the more acetate is generated. This in turn leads to more phosphate release. Luxury phosphate uptake in the aerobic zone is always proportional to the amount of phosphate released and thus there is less phosphate in the final effluent. In order to ensure anaerobiosis within a reactor, it is essential that there is no aeration device present and that the reactor is highly loaded with BOD such that any dissolved oxygen is removed rapidly. Anaerobic reactors are thus always sited at the head of the reactor treatment train (Fig. 21.2). If the aerobic stage of the treatment plant removes only BOD with no nitrification, then both the return sludge and the settled sewage can be fed to the anaerobic reactor. However, if the aerobic stage achieves nitrification, the nitrate in the return sludge will reduce the extent of anaerobiosis. Under such conditions the return sludge is fed to a separate anoxic reactor to undergo denitrification, and the nitrate free mixed liquor is recycled to the anaerobic reactor (Fig. 21.2).

9 MAINTAINING MICROBIAL GROWTH IN REACTORS

A typical wastewater has a BOD of around 200 mg/l of which 90% is removed during treatment to achieve an effluent quality of

20 mg/l. Using Equation (1), which is based on glucose as the carbon source, each mg of glucose requires 1.06 mg oxygen to oxidize it fully to CO_2. Extending this to a typical wastewater, the microorganisms would have access to only (200–20)/1.06 = 170 mg glucose/l. When growing bacteria in laboratory cultures, it is common to use a glucose concentration of around 10 000 mg/l and thus it is apparent that a wastewater is a very dilute bacterial food source.

In order to determine the value of the three design factors described in the Introduction, it is necessary to know how quickly BOD is metabolized and how fast the bacteria can grow and divide. The rate of BOD removal can be described by the simple equation:

$$-\frac{dS}{dt} = qX \tag{8}$$

which says that the rate of BOD removal (dS/dt) is proportional to the number of microorganisms in the reactor (X in mg/l). The proportionality constant q is the specific BOD uptake rate (/day) and more usually referred to as the Food: Microorganism ratio or F/M.

In a similar way the rate at which bacteria grow can be described by Equation (9):

$$\frac{dX}{dt} = \mu X \tag{9}$$

where the rate of bacterial growth (dX/dt) is proportional to numbers of bacteria present in the reactor. The proportionality constant (μ) is known as the specific growth rate. There is a link between the specific growth rate and the amount of food source available and this is described by the Monod equation, when the food source is the rate limiting substrate (Equation (10)):

$$\mu = \mu_m \frac{S}{k_S + S} \tag{10}$$

where:

μ is the bacterial growth rate (/day)
μ_m is the maximum specific growth rate (/day)
k_S is the saturation coefficient (mg/l), and
S is the amount of rate limiting substrate in the effluent.

For a typical wastewater where the required effluent BOD is 20 mg/l, the value for μ_m is around 0.47/day and k_S is 100 mg/l. This gives a value for μ of 0.42/day (Equation (11)):

$$\mu = 2.5x\frac{20}{100 + 20} = 0.42/d \tag{11}$$

The reciprocal of the specific growth rate will have units of days and it is referred to as the microbial retention time. In other words, it is the amount of time the microorganism must remain in the reactor to ensure that they divide, based on the amount of food available to the microorganisms. Equation (11) can thus be expressed as:

$$\frac{1}{\mu} = \frac{V}{Q} \tag{12}$$

where V is the volume of the reactor (m^3) designed to hold the microorganisms and Q is the flow of sewage (m^3/day) into the reactor.

Substituting the calculated value of μ from Equation (11) into Equation (12) gives:

$$\frac{1}{0.42} = \frac{V}{Q}; \qquad V = 2.4Q \tag{13}$$

Thus, in order to achieve an effluent BOD of 20 mg/l, it is necessary to construct a reactor with a retention time of 2.4 days in order to allow the microorganisms time to divide and grow. This is the principle behind a number of simple suspended growth treatment systems such as aerated lagoons and waste stabilization and they are characterized by large reactors with long residence times.

However, it is possible both to reduce the size of reactors for wastewater treatment, yet at the same time meet the requirements of Equation (10) in terms of bacterial growth. It was noted as long ago as 1912 that:

> *five weeks continuous aeration was required in order to completely nitrify an average sample of Manchester sewage. If, however, the resultant solid matter was allowed to deposit, and the purified sewage removed by*

decantation and replaced by a further sample of crude sewage, complete nitrification of this second dose of sewage ensued within a reduced period of time. It was shown that accumulated deposit resulting from the complete oxidation by prolonged aeration of successive quantities of sewage, which were termed 'activated' sludge, had the property of enormously increasing the purification effected by the simple aeration of sewage.

This observation forms the basis of the activated sludge process and introduced the concept of recycling settled biomass from the sedimentation tank back to the aeration basin. In this way the hydraulic retention time is separated from the solids retention time and the requirements of Equation (10) for a long sludge age (measured in days) can be achieved in an aeration basin that has a short hydraulic retention time (measured in hours).

10 FLOCCULATION AND SEDIMENTATION – THE WEAK LINK IN THE PROCESS

The efficiency of the activated sludge process depends on the fact that, after repeated deposition of solid matter, the microorganisms that comprise these solids begin to change their community structure, and this is most marked for the protozoan and metazoan communities. The new communities which are so formed, acquire flocculating properties such that the sludge rapidly agglomerates and flocculates once aeration has ceased and quiescent conditions are applied. The ability of a floc to settle and thus provide a clear interface layer permits the discharge of a clarified effluent and the recycle of thickened solids back to the aeration basis. It is thus the basis on which the activated sludge process functions.

Floc settlement relies on the formation of large, compact flocs that are able to flocculate into larger aggregates, which have high settling velocities. This is the bottleneck in the treatment process and, although it is essential for ensuring efficient treatment, it is unreliable, unpredictable and uncontrollable. A huge research effort has been expended into understanding the microbiological basis of activated sludge flocculation but negligible progress has been made in our ability to predict the onset of flocculation problems or alleviate them when they do occur.

Our current understanding of flocculation sees extracellular polymeric substances (EPS) as major components of the floc matrix with polymer bridging the mechanism by which flocs are formed. The major components of the EPS are polysaccharide and proteins excreted by bacteria and protozoa. Numerous authors have undertaken correlations that relate the EPS content of the sludge to the settleability as measured by SVI or SSVI, but it has been more difficult to correlate plant operating conditions (sludge age, loading rate, dissolved oxygen concentration etc.) with either EPS or settleability. It would appear that the amount of EPS generated is not as important as the properties of the EPS (namely hydrophobicity, surface charge and composition) in controlling sludge flocculation (Liao *et al.*, 2001).

However, the major settlement problem, which plagues almost all suspended growth processes, is that of filamentous foaming and bulking. This phenomenon is thought to be caused by the proliferation of filamentous bacteria and whenever bulking and foaming do occur these bacteria are almost always observed. Foaming results in the production of voluminous quantities of foam on top of the aeration basin, which can carry over to the secondary sedimentation tank and thus to the final effluent, where it will cause a solids consent failure. It is generally associated with the filament types *Microthrix parvicella* and *Nocardia* spp. Currently, there are no control strategies which will eliminate these filaments while still maintaining a high degree of wastewater treatment. However, it is possible to operate the treatment plant effectively by containing the foam within the aeration basin and preventing its carryover to the sedimentation tank.

Sludge bulking is a very different problem and it is thought that this is caused by the outgrowth of filamentous bacteria from

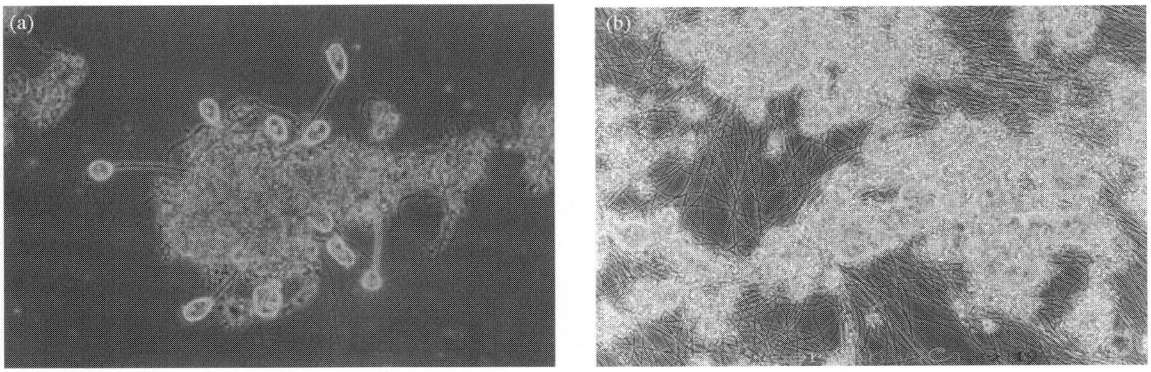

Fig. 21.3 (a) Well-settling sludge with a diverse protozoal community but showing clearly the rigid backbone of filaments supporting the floc biomass; (b) poor settling sludge with a large filament population of Type 021N.

the sludge floc. It is apparent from Fig. 21.3 that these outgrowths will prevent the flocs from approaching close to each other to form compact, well-settling structures. The effect is that the sludge occupies a much larger volume in the sedimentation tank, the sludge blanket is therefore much higher and the recycled sludge has a much lower suspended solids concentration. Where the final sedimentation tank is large with adequate capacity, this is a minor inconvenience which simply requires an increased recycle rate to return adequate solids back to the aeration basin and compensate for the thinner recycled sludge. However, if the sedimentation tank is working close to its capacity (and this is often the case at industrial treatment plants), the increased blanket height may cause solids to be lost over the outlet launder. In extreme cases the whole blanket may overtop the launder with a major loss of solids.

The provision of techniques to identify the causes of filamentous foaming and bulking and strategies to control their proliferation remains one of the challenges in activated sludge research. This problem, which has plagued the process since its inception, is one of the reasons why the membrane bioreactor (MBR) is such an attractive process option as it does not require a sedimentation stage. It is likely that MBRs will become the suspended growth system of choice in the future, as membrane costs reduce and the operating experiences from those plants that have been constructed are disseminated.

REFERENCES

Ardern, E. and Lockett, W.T. (1914). Experiments on oxidation of sewage without the aid of filters. *J. Soc. Chem. Ind.* **33**, 523.

Chudoba, J., Grau, P. and Ottova, V. (1973). Control of activated sludge bulking: II – Selection of microorganisms by means of a selector. *Wat. Res.* **7**, 1389–1406.

Eikelboom, D.H. (1994). The *Microthrix parvicella* puzzle. *Wat. Sci. Tech.* **29**(7), 273–279.

Jenkins, D., Daigger, G.J. and Richard, M.D. (1993). *Manual on the Cause and Control of Activated Sludge Bulking and Foaming*. Lewis Publishers, Chelsea MI.

Liao, B.Q., Allen, D.G., Droppo, I.G. *et al.* (2001). Surface properties of sludge and their role in bioflocculation and settleability. *Wat. Res.* **35**(2), 339–350.

Richard, M. (1989). *The Bench Sheet Monograph on* Activated Sludge Microbiology. WPCF, Alexandria, USA.

WPCF (1990) *Wastewater Biology: The Microlife.* A special publication prepared by the Task Force of Wastewater Biology. Water Pollution Control Federation, Alexandria, USA.

22

Protozoa as indicators of wastewater treatment efficiency

Paolo Madoni

Instituto di Ecologia, Universita di Parma, 43100 Parma, Italy

1 PROTOZOA AS A COMPONENT OF THE ACTIVATED SLUDGE ECOSYSTEM

Today, most wastewater treatment processes make use of the natural self-purification capacity of aquatic ecosystems which results from the presence and action of microbial communities. Thus, biological sewage-treatment plants may be regarded as constructed ecosystems subjected to extreme conditions. As in every other biological system, the community living in the aeration basin of an activated-sludge plant has a precise structure and follows exact dynamics. In activated sludge, abiotic components are represented by the plant and by the sewage, while biotic components are represented by decomposers (bacteria, fungi) which utilize the dissolved organic matter in the wastewater, and by consumers (heterotrophic flagellates, ciliates, rhizopods, and small metazoans) which feed on dispersed bacteria and other organisms. Physical factors that act on biological sewage treatment systems are the climate, temperature and the turbulence to which mixed liquor is subject during the aeration phase, while chemical factors are represented by the composition of the wastewater and by oxygen dissolved by means of the aerators.

Regarding the biotic components, it is well known that activated sludge develops specific communities of protists which are sustained by large populations of bacteria. Ciliated protists are numerous in all types of aerobic biological-treatment systems. They are commonly found at densities of about 10 000 cells/ml of activated sludge mixed liquor and constitute approximately 9% of the dry weight of suspended solids in mixed liquor (Madoni, 1994a). More than 200 species of protists (of which 33 flagellates, 25 rhizopods, 6 actinopods and 160 ciliates) were observed in the various types of aerobic treatment systems. Less than half of them, however, have been observed frequently (Curds and Cockburn, 1970a; Madoni and Ghetti, 1981; Al-Shahwani and Horan, 1991; Amann *et al.*, 1998). This number is evidently only a small part of the thousands of species of freshwater protists that could theoretically be observed in these environments.

The majority of ciliates present in biological wastewater treatment facilities feed upon dispersed populations of bacteria. The bacterivorous ciliates can be subdivided into three functional groups on the basis of behaviour:

- *Free-swimming*: swim in the liquor phase and remain evenly dispersed in the sedimentation tank
- *Crawling*: although they are free forms, inhabit the surface of sludge flocs or biofilm
- *Attached*: firmly fixed to the substrate by means of a stalk. They are strictly associated with the sludge flocs or biofilm and thus settle during sedimentation.

Some ciliates, however, are predators of other ciliates or are omnivorous, feeding upon a variety of organisms including small ciliates, flagellates and dispersed bacteria (Fig. 22.1).

The Handbook of Water and Wastewater Microbiology
ISBN 0-12-470100-0

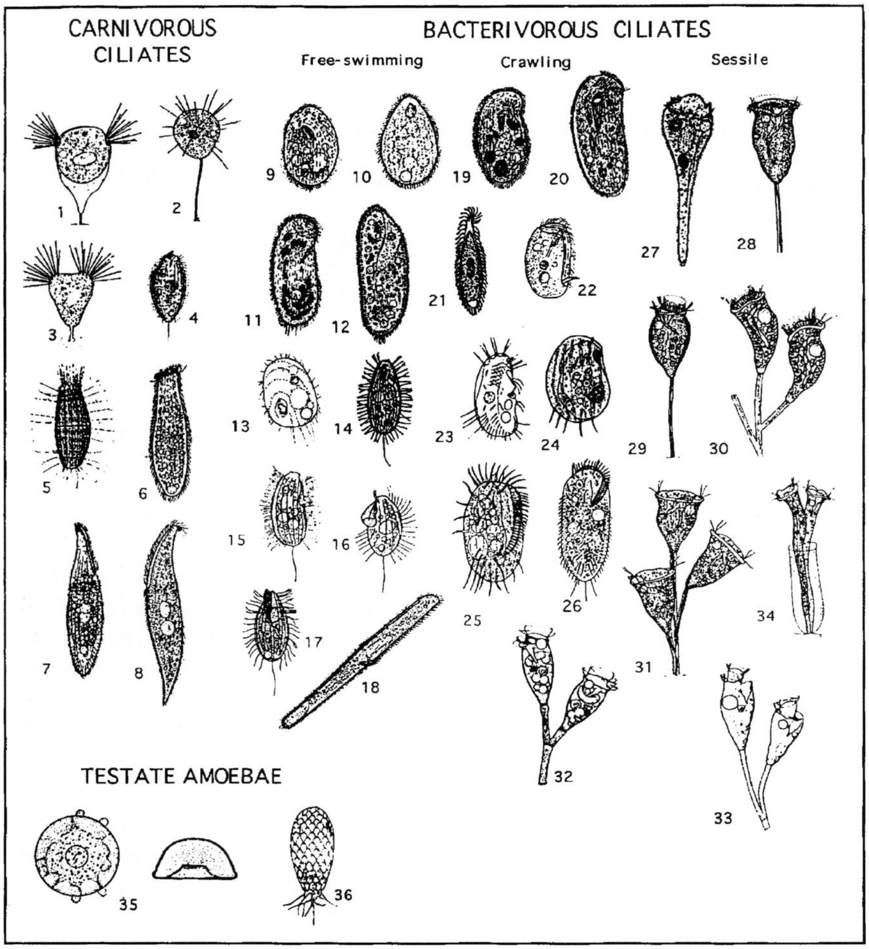

Fig. 22.1 Some common ciliates and testate amoebae of sewage treatment plants and their food habit. 1–8 Carnivorous or omnivorous ciliates: (1) *Acineta* spp.; (2) *Podophrya* spp.; (3) *Tokophrya* spp.; (4) *Plagiocampa rouxi*; (5) *Coleps hirtus*; (6) *Spathidium* spp.; (7) *Litonotus* spp.; (8) *Amphileptus* spp. 9–18 Free-swimming bacterivorous ciliates: (9) *Glaucoma* spp.; (10) *Tetrahymena* spp.; (11) *Colpidum* spp.; (12) *Paramecium* spp.; (13) *Cinetochilum margaritaceum*; (14) *Dexiotricha* spp.; (15) *Uronema* spp.; (16) *Cyclidium* spp.; (17) *Pseudocohnilembus pusillum*; (18) *Spirostomum teres*. 19–26 Crawling bacterivorous ciliates: (19) *Chilodonella* spp.; (20) *Trithigmostoma* spp.; (21) *Acineria uncinata*; (22) *Trochilia minuta*; (23) *Drepanomonas revolute*; (24) *Aspidisca* spp.; (25) *Euplotes* spp.; (26) *Stylonychia* spp. 27–34 Sessile bacterivorous ciliates: (27) *Stentor* spp.; (28) *Vorticella convallaria*; (29) *V. microstoma*; (30) *Carchesium* spp.; (31) *Zoothamnium* spp.; (32) *Epistylis* spp.; (33) *Opercularia* spp.; (34) *Vaginicola* spp. 35–36 Testate amoebae: (35) *Arcella* spp.; (36) *Euglypha* spp.

All bacterivorous ciliates rely upon ciliary currents to force suspended bacteria to the oral region. The subclass Peritrichia is the most important group of ciliates in sewage treatment plants. All peritrichs found in aerobic processes are attached forms. It should be remembered that both activated sludge and attached growth processes rely on the presence of surfaces (sludge or biofilm) upon which the microorganisms can grow. Thus organisms with the ability to attach to, or remain closely associated with, sludge or biofilm have a distinct advantage over organisms that swim freely in the liquid phase and are subject to washout in the effluent. So, while free-swimming and attached ciliates are in competition

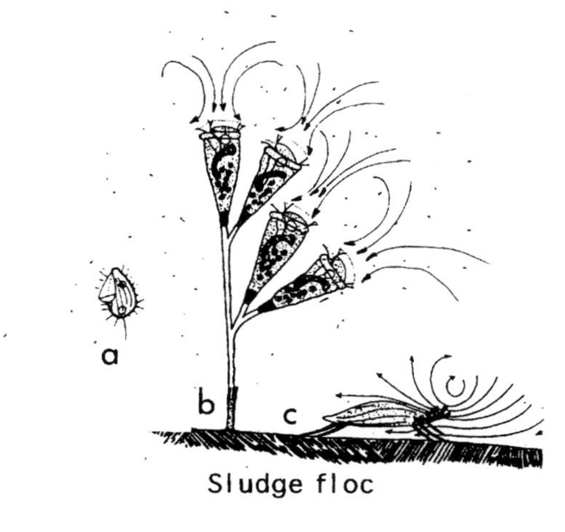

Fig. 22.2 Feeding mechanisms of filter-feeding ciliates in the activated sludge process. Swimming (a) and attached (b) forms filter out dispersed bacteria in the mixed liquor, while crawling forms (c) scrape bacteria from the surface of the flocs.

for bacteria dispersed in the liquid phase, crawling forms that are in close proximity to surface growths feed upon particles that only lightly adhere to the sludge and that are dislodged very easily by the feeding currents (Fig. 22.2).

However, frequent reference to 'scavenger' hypotrichs such as *Aspidisca* and *Euplotes* and cyrtophorids such as *Chilodonella* and *Trochilia* implies that these crawling ciliates can scrape bacteria from surfaces because their mouth is located in the ventral part of the cell.

2 THE ROLE OF PROTISTS IN THE ACTIVATED SLUDGE PROCESS

Large numbers of ciliates are present in all types of aerobic treatment systems and play an important role in the purification process, as well as the overall regulation of the entire community. Ciliated protists improve the quality of the effluent because of their involvement in the regulation of the bacterial biomass by removal, through predation, of the major part of the bacteria dispersed in the mixed liquor (Curds *et al.*, 1968). It is generally assumed that their primary role in the wastewater treatment is the clarification of the effluent. In the absence of ciliated protists, in fact, the effluent from the system has an elevated BOD and is highly turbid due to the presence of many dispersed bacteria.

Ciliates, moreover, feed on pathogenic and faecal bacteria. In the effluent from systems lacking ciliates, the presence of *Escherichia coli* is, on average, equal to 50% of that observed in the sewage entering the aeration tank. This is reduced to 5% when ciliates are present (Curds and Fey, 1969).

3 PROTOZOA DYNAMICS IN ACTIVATED SLUDGE

In the aeration tank of biological processes a true trophic web is established. A simplified diagram of this is illustrated in Fig. 22.3. The biological system of these plants consists of populations in continuous competition with each other for food. The growth of decomposers, prevalently heterotrophic bacteria, depends on the quality and quantity of dissolved organic matter (DOM) in the mixed liquor. For predators, on the other hand, growth depends on the available prey. Dispersed bacteria are thus food for heterotrophic flagellates and bacterivorous ciliates which, in turn, become the prey of carnivorous organisms. The relationships of competition and predation create oscillations and successions of populations until dynamic stability is reached. This is strictly dependent of plant management choices based on design characteristics aimed at guaranteeing optimum efficiency.

Although the ciliates are widely distributed, and many species are able to tolerate precarious environmental conditions, the particular conditions found in sewage treatment processes limit their presence to a restricted number of species (Madoni and Ghetti, 1981). Because of this protists have come to be considered as potential indicators of wastewater treatment plant performance. Curds and Cockburn (1970b) were probably the first to relate the sludge loading to the various forms of ciliated protists. According to these authors,

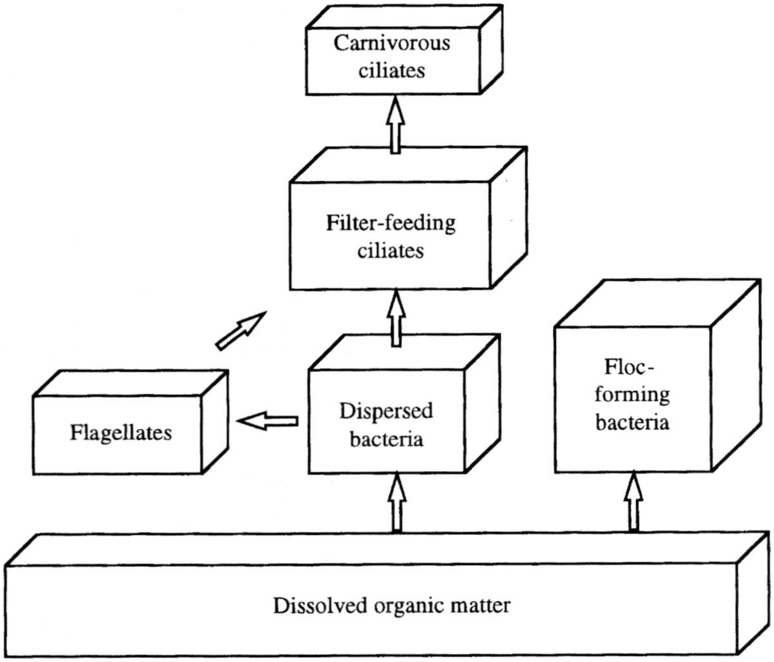

Fig. 22.3 The trophic web in the activated sludge of sewage treatment plants.

crawling ciliates (hypotrichs) decrease with increasing loading (no hypotrichs would be observed in sludge loaded above 0.6 kg BOD/kgMLSS/day), while sessile ciliates (peritrichs) are able to grow throughout a large range of sludge loadings. Nevertheless, at low loadings (0.1–0.3 kg BOD/kgMLSS/day) a wider range of ciliates, in terms of numbers of species with a more uniform distribution between the three functional groups, would be observed. Klimowicz (1970) also found that some ciliated protist species were most abundant on occasions at high sludge loadings (*Paramecium caudatum, Vorticella microstoma, Opercularia coarctata, O. microdiscus*), while some other species occurred in great numbers at medium loadings (*Acineria uncinata, Amphileptus claparedei, Litonotus fasciola*) and some other species were most numerous at low loadings (*Vorticella picta, Zoothamnium, Coleps hirtus*). Such species as *Aspidisca cicada* and *Vorticella convallaria* were equally abundant with various loadings of sludge. Studies on both colonization and population succession in activated sludge emphasized the role of protists as plant performance indicators and demonstrated the effect of environmental conditions in the aeration tank on determining the established ciliate community (Curds, 1971; Madoni, 1982).

Another important result was the identification of three phases in the time span from the beginning to stabilization of the system (Madoni and Antonietti, 1984). The plant starting phase is characterized by the presence of species typical of raw sewage (Fig. 22.4). These 'pioneer' species are represented chiefly by free-swimming bacteriophagous ciliates and flagellates and are thus not linked to the presence of sludge so they cannot be considered typical components of these environments. With the formation of activated sludge, they compete with species better adapted to an aeration tank environment and rapidly decline in numbers. The second phase is one of transition, and is characterized by the strong growth of ciliates typical of the aeration tank habitat. In this phase, the community has a wealth of species, but rapid substitution of the species takes place with the progressive formation of activated sludge. The free-swimming forms, above all the hymenostomes, are involved only during the first days of

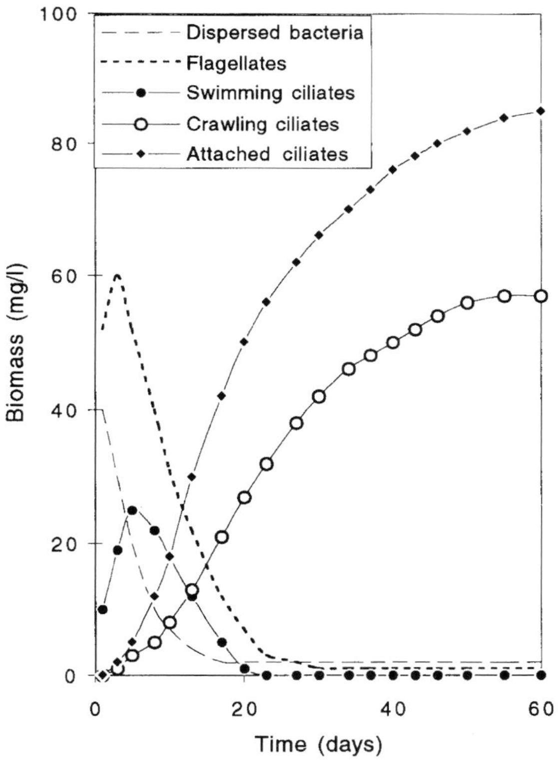

Fig. 22.4 Microorganism succession during colonization of activated sludge.

colonization and are gradually substituted by the attached and crawling forms. The steady-state phase is characterized by a ciliate community whose structure reflects the stable condition of the aeration tank environment with a balance between the organic loading and the sludge that is produced, removed and recycled. Each of the three phases is characterized by a typical species structure. Free-swimming bacterivorous ciliates such as *Colpidium*, *Cyclidium* and *Paramecium* are linked to the first phases of colonization of the plant, while attached peritrichs such as *Vorticella*, *Epistylis*, *Opercularia* and crawling ciliates such as *Aspidisca* are typical of the third phase.

A fully functioning plant need not host species characteristic of one of the colonization phases unless dysfunctions cause regression in the environmental conditions, such as the amount of sludge, the degree of aeration, sewage retention time, and organic loading at input.

The species structure of the microfauna is thus a diagnostic instrument which serves to integrate the parameters on which the evaluation of plant performance is based. By following the general criterion of the biological indicators, the presence of particular species, as well as the overall composition of the microfauna, can be taken as indicators of the performance of an activated sludge plant. With regard to ciliated protists, the groups of species take on particular importance both in systematic and functional terms (free-swimming, crawling, attached).

4 MICROFAUNA AS INDICATOR OF ACTIVATED-SLUDGE SYSTEMS

The numerous studies undertaken have been able to ascertain that the number of ciliated protists living in a normally functioning plant is about 10^6 individuals per litre. When the number falls below 10^4 per litre it indicates insufficient purification. In this case, there is a proliferation of dispersed bacteria which render the effluent turbid and consequently cause a greatly increased BOD in the output water. A high number of ciliates ($>10^7$ per litre) instead, almost always indicates good purification and optimum plant performance. The microfauna of a normally functioning system is almost always highly diversified. That is, it is composed of different groups of organisms, and each group is made up of several species. No group or species is every numerically dominant over the other components, even though the ratios between various groups or species can differ. However, a microfauna that is dominated by one species or group is almost always an index of trophic imbalances due to the existence of limiting factors impeding development of most of the other species and favouring the growth of forms more tolerant to these factors. The chief limiting conditions are generally the presence of a shock load of toxic discharge, under- or overloading, excessive sludge wastage or lack of aeration.

The structure of the microfauna is indeed a valid indicator of plant purification performance and an efficient activated sludge plant has the following characteristics:

- High numbers of microfauna cells ($\geq 10^6$ organisms/l)
- Microfauna composed chiefly of crawling and attached ciliates, with almost no flagellates or swimming ciliates
- The species and ciliates groups are highly diversified and none dominates numerically by a factor greater than 10.

When this is not the case, the identification of the dominant group of the microfauna allows diagnosis of the particular state of functionality of the plant.

Just as observation of the microfauna, identification of flagellates, free-swimming, crawling and attached ciliates and the analysis of their abundance ratios is a useful tool in the diagnosis of plant performance, so more complete information can be obtained by a knowledge of the different species of the protistan community. Thus, if it is true that a microfauna rich in attached and crawling ciliates indicates improved performance with respect to that found when free-swimming ciliates dominate, it is also true that different species of crawling and attached ciliates can be associated with various conditions. Thus, each species is able to add more detailed information than that obtained through simple identification of functional groups. For example, among attached ciliates, *Vorticella convallaria* and *V. microstoma* characterize the first phase of colonization but the latter are then substituted by *Vorticella convallaria* which may reach high numbers during the growth phase of the sludge. In the case of a drastic reduction in the oxygen content in the mixed liquor, the alternation of these two species is observed, due to their different degree of tolerance to the lack of oxygen (Fig. 22.5). Large quantities of *Vorticella microstoma* thus indicate a poorly aerated sludge.

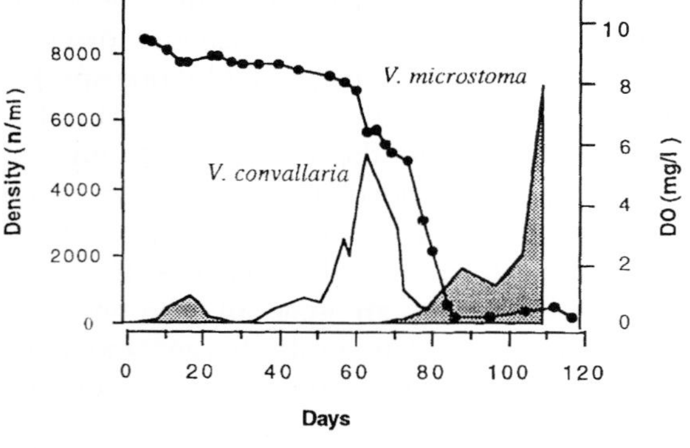

Fig. 22.5 Dynamics of two species of *Vorticella* related to dissolved oxygen (DO) in the aeration tank.

5 THE KEYGROUPS

Quantifying the indicator value of the microfauna is a difficult task because there are groups that are more or less tolerant of a wide range of environmental factors, while the plant performance results from the simultaneous action of many of these external and operational conditions. Nevertheless, investigating the relationship between the various groups of the microfauna and the main operational conditions of the plant in a wide survey, it was possible to select and group microfauna organisms into positive and negative keygroups (Madoni, 1994b). The positive keygroups – positively correlated with high plant performance – are crawling and attached ciliates and testate amoebae. The negative keygroups – negatively correlated with high plant performance – are small flagellates, swimming bacterivorous ciliates, and the sessile ciliates *Vorticella microstoma* and *Opercularia* spp. Density and species richness of the microfauna, moreover, appear to be highly correlated to plant performance (Table 22.1).

5.1 Small flagellates

Small heterotrophic flagellates continuously enter the plant in the sewage influent where they are very numerous. Flagellate species such as *Bodo, Polytoma* and *Tetramitus*, normally dominate the microfauna during the starting phase of the plant when floc-forming bacteria are scarcely present. They feed on dispersed bacteria and, in time, are substituted by bacterivorous ciliates. In a normally-functioning

TABLE 22.1 Positive and negative correlation between protozoans and plant operational conditions ($^{*}P < 0.01$, $^{**}P < 0.001$)

	DO	*Nitrifying ability*	*BOD removed*	*Effluent colour*
Small flagellates	$-^{*}$	$-^{**}$	$-^{**}$	$+^{**}$
Swimming ciliates	$-^{**}$	$-^{**}$	$-^{**}$	$+^{**}$
Crawling ciliates	$+^{**}$	$+^{**}$	$+^{**}$	$-^{**}$
Sessile ciliates	$+$	$-$	$+^{*}$	$-^{**}$
V. microstoma	$-^{**}$	$-^{**}$	$-^{**}$	$+^{**}$
Opercularia spp	$-^{**}$	$-^{**}$	$-^{**}$	$+^{**}$
Testate amoebae	$+^{**}$	$+^{**}$	$+^{**}$	$-^{**}$
Microfauna density	$+^{**}$	$+^{*}$	$+^{**}$	$-^{**}$
Number of species	$+^{**}$	$+^{**}$	$+^{**}$	$-^{**}$

activated sludge, in fact, these protists are outstripped by the bacterivorous ciliates and, in addition, they are strongly subjected to predatory activity of other protists; so their presence in the activated sludge is limited to few individuals (>10 individuals counted along the diagonal in a Fuchs-Rosenthal chamber). By contrast, the massive presence of these protists in a mature activated sludge is associated with a poor performance in the biological purification, due to the following causes: poorly aerated sludge; overloading; fermenting substances involved (Drakides, 1978; Madoni, 1986). The dominance of flagellates becomes apparent when they reach a density of more than 100 individuals along the diagonal in a Fuchs-Rosenthal chamber. In this case the effluent leaving the system has an elevated BOD and is highly turbid due to the presence of these microorganisms. Flagellates become the only protistan form present in highly-loaded sludge (>0.9 kg BOD/kgMLSS/day) (Curds and Cockburn, 1970b).

5.2 Swimming bacterivorous ciliates

Free-swimming bacterivorous ciliates are more abundant in the early phases of a developing plant when sludge flocs are still scarce and consequently attached ciliates are absent. Nevertheless, they are soon substituted by sessile ciliate species owing to competition for bacteria dispersed in the mixed liquor. Sessile forms in fact are filter-feeders and are more efficient than free-swimming ciliates in forcing suspended bacteria into the cell by means of ciliary currents. Small free-swimming ciliates (such as *Uronema, Dexiostoma* and *Tetrahymena*) sometimes dominate the microfauna of plants operating at sludge age which is too low or at both high sludge loadings and a lack of oxygen (Martín-Cereceda *et al.*, 1996). These bacterivorous ciliates require high concentrations of dispersed bacteria but survive better than other protists when there is toxicity in the influent or a lack of oxygen. Swimming ciliates dominate the aeration basin of plants characterized by high sludge loadings (0.6–0.9 BOD/kgMLSS/day).

5.3 Crawling and attached ciliates

These two functional groups normally codominate the protistan community in activated sludge plants. This is due to their different food habits preventing their competition. Nevertheless, the ratio between the two groups tends to change with sludge loading. Crawling ciliates reduce their presence as sludge loading increases, so that above 0.6 kg BOD/kgMLSS/day, most species of this group disappear (Curds and Cockburn, 1970b; Klimowicz, 1970).

Sessile ciliates are normally codominant in the activated sludge. Nevertheless, a massive increase in their number (>80% of the whole microfauna) occurs on occasions in transient situations that reduce the plant performance (Drakides, 1978; Madoni, 1981). Such transient conditions are a rapid increase in the sludge load due to a loss of sludge and a discontinuous input of organic load from the influent. Sessile ciliates are able to grow throughout a large range of sludge loadings; nevertheless, at values ranging from 0.3 to 0.6 kg BOD/kgMLSS/day, these protists dominate and, for sludge loadings of 0.6–0.9, sessile ciliates some forms, such as *Vorticella microstoma* and *Opercularia* spp., can survive and grow in activated sludge subjected to severe conditions (lack of oxygen, presence of toxins). When these species are present in high numbers, they must be considered as separate keygroups.

5.3.1 *Vorticella microstoma*

The peritrich ciliate *V. microstoma* is quoted in the saprobic system as a polysaprobic species (Foissner, 1988). This ciliate is frequently present in the plant during the first phase of colonization but is substituted by other sessile ciliates (*V. convallaria*) during stable conditions when there is a drastic and prolonged reduction in the dissolved oxygen concentration in the mixed liquor. An alternation of the two species can be observed owing to their different degrees of tolerance to the lack of oxygen (Madoni and Antonietti, 1984). *V. microstoma* thus indicates a lack of dissolved oxygen in the aeration tank. Massive growth of this sessile ciliate was also observed at times of high wastewater flow to the sewage plant and low values of mass loading (Esteban *et al.*, 1990).

5.3.2 *Opercularia* spp.

Low numbers of *Opercularia* often occur in activated sludge, where three species are commonly observed: *O. articulata, O. coarctata*, and *O. microdiscus*. These ciliates are quite useful as bioindicators, because their numbers increase when the activated sludge is of poor quality. *Opercularia* spp. are associated with high final effluent BOD concentrations, and are among the most abundant forms at high loadings (Curds and Cockburn, 1970b; Klimowicz, 1970; Esteban *et al.*, 1991; Salvadó *et al.*, 1995). These ciliates moreover can survive in stressed environments better than other protists; in fact, large numbers of *Opercularia* were found in sludge of plants receiving industrial waste containing toxic substances (Antonietti *et al.*, 1982; Becares, 1991; Becares *et al.*, 1994). Moreover, *Opercularia* spp. were associated with high effluent BOD and ammoniacal N concentrations (Poole, 1984; Madoni *et al.*, 1993). *O. coarctata* may be the only component of the microfauna in sludges of plants treating industrial waste containing metal salts. *O. microdiscus* is able to survive the severe lack of oxygen found in sedimentation tanks with a low return-sludge ratio. *Opercularia* spp. are often associated with *V. microstoma*.

5.3.3 *Testate amoebae*

Two genera of testate amoebae, *Arcella* and *Euglypha*, are commonly present in activated sludge. These protists are found normally in the aeration basins of N-removal plants (Poole, 1984). Testate amoebae are more abundant or dominant in sludges characterized by low loading, a long retention time and high DO in aeration tanks that enable complete nitrification (Madoni *et al.*, 1993). Under these conditions, the quality of the effluent is excellent and a high biological performance of the plant is reached. Testate amoebae only colonize plants with a long sludge age since these protists have low growth rates. They are often seasonal, being more common in summer when temperature and growth rates increase. They may be more abundant in activated sludge plants at breweries with a low sludge load and a good quality effluent. Nevertheless, when the sludge load reaches high values (>1 kg BOD/kgMLSS/day) and the COD of the effluent is elevated, these protists are replaced by the peritrich ciliate *Opercularia* and by free-swimming ciliates (Sasahara and Ogawa, 1983).

5.3.4 *Density and diversity*

The number of ciliates living in a normally functioning plant is about 10^3 individuals/ml, and when the number falls below 10^2/ml it indicates insufficient purification (Drakides, 1980; De Marco *et al.*, 1991). In this case, there is a proliferation of dispersed bacteria which make the effluent turbid and consequently causes a greatly increased BOD in the output water. On the other hand, a high number of ciliates ($\geq 10^4$/ml) indicates good purification and optimum biological performance of the plant. The microfauna of a normally functioning plant is almost always highly diversified, namely composed by a high number of species (>10). In this case no species is ever numerically dominant over the other organisms, even if the ratios between various species differ. A microfauna that is dominated by one species is an index of trophic imbalances due to the existence of limiting factors impeding the development of most of the other species and favouring the growth of forms more tolerant

TABLE 22.2 Two-way table to determine the Sludge Biotic Index (modified from Madoni, 1994b). Horizontal entrance in the table on the basis of both keygroup and density. Vertical entrance in the table. Total number of taxa of the microfauna and number of small flagellates F counted along the Fuchs-Rosenthal chamber diagonal

Dominant keygroup	*Density (ind/l)*	*>10*		*8–10*		*5–7*		*<5*	
		F > 10	*10 < F < 100*	*F > 10*	*10 < F < 100*	*F > 10*	*10 < F < 100*	*F > 10*	*10 < F < 100*
Crawling + sessile ciliates[a]	$\geq 10^6$	10	8	9	7	8	6	7	5
and/or testate amoebae	$<10^6$	9	7	8	6	7	5	6	4
Sessile ciliates[a]	$\geq 10^6$	9	7	8	6	7	5	6	4
>80%	$<10^6$	8	6	7	5	6	4	5	3
Opercularia spp.	$\geq 10^6$	7	5	6	4	5	3	4	2
	$<10^6$	6	4	5	3	4	2	3	1
Vorticella microstoma	$\geq 10^6$	6	4	5	3	4	2	3	1
	$<10^6$	5	3	4	2	3	1	2	0
Swimming ciliates	$\geq 10^6$	5	3	4	2	3	1	2	0
	$<10^6$	4	2	3	1	2	0	1	0
Small flagellates (>100)[b]	$\geq 10^6$	4		3		2		1	
	$<10^6$	3		2		1		0	

Conversion of SBI values into four quality classes

SBI Value	*Class*	*Judgement*
8–10	I	Very well colonized and stable sludge; excellent biological activity; very good performance.
6–7	II	Well colonized and stable sludge; biological activity on decrease; good performance.
4–5	III	Insufficient biological purification in the aeration tank; mediocre performance.
0–3	IV	Poor biological purification in the aeration tank; low performance.

[a] *Opercularia* and *V. microstoma* not dominant.
[b] along the Fuchs-Rosenthal chamber diagonal.

to these factors. The number and diversity of ciliate communities change according to the quality of the settled sewage and operating conditions of the plant (Esteban *et al.*, 1991; Esteban and Tellez, 1992). The most common limiting conditions are generally the presence of a shock load of toxic material, the lack of aeration, and the excess sludge wastage.

6 THE SLUDGE BIOTIC INDEX (SBI)

The performance of the plant has to be constantly monitored and is subjected to strict regulation. Nevertheless, malfunctions resulting in decreased purification efficacy are frequent. The SBI, an objective index based on the protistan community, has been devised to monitor activated-sludge plant performance (Madoni, 1994b). This method is based on two principles. First, the dominance of protistan keygroups changes in relation to environmental and operational conditions of the plant. Second, cell density and number of taxa diminish as the efficiency of the plant drops. The SBI enables the operator to define the biological quality of the sludge by means of conventional numerical values (from 0 to 10) that are grouped into four quality classes (Table 22.2). The identification of the various species of protista is important, in order to obtain an accurate SBI value; some keys written specifically about the protista found in sewage treatment processes and polluted waters are available (Bick, 1972; Foissner *et al.*, 1991–95).

Since the SBI was set up specifically for the evaluation of the biological reactor performance, this index is unable to reveal any dysfunction in the final sedimentation tank such as sludge bulking or rising sludge.

REFERENCES

Al-Shahwani, S.M. and Horan, N.J. (1991). The use of protozoa to indicate changes in the performance of activated sludge plants. *Water Research* **25**, 633–638.

Amann, R., Lemmer, H. and Wagner, M. (1998). Monitoring the community structure of wastewater treatment plants: a comparison of old and new techniques. *FEMS Microbiology Ecology* **25**, 205–215.

Antonietti, R., Broglio, P. and Madoni, P. (1982). The evaluation of biological parameters as indicators of purification efficiency in activated sludge plants. *Ingegneria Ambinetale* **11**, 472–477.

Becares, E. (1991). Microfauna of an activated sludge pilot plant treating effluents from a pharmaceutic industry. In: P. Madoni (ed.) *Biological Approach to Sewage Treatment Process: Current Status and Perspectives*, pp. 105–108. Centro Bazzucchi, Perugia.

Becares, E., Romo, S. and Vega, A. (1994). Organic pollutants and microfauna in an industrial wastewater treatment system. *Verh. Internat. Verein. Limnol.* **25**, 2051–2054.

Bick, H. (1972). Ciliated Protozoa. *An illustrated guide to the species used as biological indicators in freshwater biology.* World Health Organization, Geneva.

Curds, C.R. (1971). Computer simulations of microbial population dynamics in the activated sludge process. *Water Research* **5**, 1049–1066.

Curds, C.R. and Cockburn, A. (1970a). Protozoa in biological sewage treatment processes. I. A survey of the protozoan fauna of British percolating filters and activated sludge of sewage and waste treatment plants. *Water Research* **4**, 225–236.

Curds, C.R. and Cockburn, A. (1970b). Protozoa in biological sewage treatment processes. II. Protozoa as indicators in the activated-sludge process. *Water Research* **4**, 237–249.

Curds, C.R. and Fey, G.J. (1969). The effect of ciliated protozoa on the fate of *Escherichia coli* in the activated sludge process. *Water Research* **3**, 853–867.

Curds, C.R., Cockburn, A. and Vandyke, J.M. (1968). An experimental study of the role of the ciliated protozoa in the activated sludge process. *Water Pollution Control* **67**, 312–329.

De Marco, N., Gabelli, A., Cattaruzza, C. and Petronio, L. (1991). Performance of biological sewage treatment plants: some experiences on municipal plants in the province of Pordenone (Italy). In: P. Madoni (ed.) *Biological Approach to Sewage Treatment Process: Current Status and Perspectives*, pp. 247–251. Centro Bazzucchi, Perugia.

Drakides, C. (1978). L'observation microscopique des boues actives appliquée à la surveillance des installations d'épuration: technique d'étude et interpretation. *T.S.M.-L'Eau* **73**, 85–98.

Drakides, C. (1980). La microfaune des boues actives. Etude d'une méthode d'observation et application au suivi d'un pilote en phase de démarrage. *Water Research* **14**, 1199–1207.

Esteban, G. and Tellez, C. (1992). The influence of detergents on the development of ciliate communities in activated sludge. *Water, Air and Soil Pollution* **61**, 185–190.

Esteban, G., Tellez, C. and Bautista, L.M. (1990). Effect of habitat quality on ciliated protozoa communities in sewage treatment plants. *Environmental Technology* **12**, 381–386.

Esteban, G., Tellez, C. and Bautista, L.M. (1991). Dynamics of ciliated protozoa communities in activated-sludge process. *Water Research* **25**, 967–972.

Foissner, W. (1988). Taxonomic and nomenclatural revision of Sladecek's list of ciliates (Protozoa: Ciliophora) as indicators of water quality. *Hydrobiologia* **166**, 1–64.

Foissner, W., Blatterer, H., Berger, H. and Kohmann, F. (1991–95). *Taxonomische und ökologische Revision der Ciliaten des Saprobiensystem*. Band I–IV. Bayerische und Landesamt für Wasserwirtschaft, München.

Klimowicz, H. (1970). Microfauna of activated sludge. Part I. Assemblage of microfauna in laboratory modes of activated sludge. *Acta Hydrobologica Krakow* **12**, 357–376.

Madoni, P. (1981) O Protozoi ciliate degli impianti biologici di depurazione. *C.N.R.* AQ/1/167, Rome, pp. 1–134.

Madoni, P. (1982). Growth and succession of ciliate populations during the establishment of a mature activated sludge. *Acta Hydrobiologica Kraków* **24**, 223–232.

Madoni, P. (1986). Protozoa in waste treatment systems. In: F. Megusar and M. Gantar (eds) *Perspectives in Microbial Ecology*, pp. 86–90. Slovene Society for Microbiology, Ljubljana.

Madoni, P. (1994a). Quantitative importance of ciliated protozoa in activated sludge and biofilm. *Bioresource Technology* **48**, 245–249.

Madoni, P. (1994b). A sludge biotic index (SBI) for the evaluation of the biological performance of activated sludge plants based on the microfauna analysis. *Water Research* **28**, 67–75.

Madoni, P. and Antonietti, R. (1984). Colonization dynamics of ciliated protozoa populations in an activated sludge plant. *Proceeding IV Ital. Symp. Popul. Dynam.* Parma pp. 105–112.

Madoni, P. and Ghetti, P.F. (1981). The structure of ciliated protozoa communities in biological sewage treatment plant. *Hydrobiologia* **83**, 207–215.

Madoni, P., Davoli, D. and Chierici, E. (1993). Comparative analysis of the activated sludge microfauna in several sewage treatment works. *Water Research* **27**, 1485–1491.

Martín-Cereceda, M., Serrano, S. and Guinea, A. (1996). A comparative study of ciliated protozoa communities in activated-sludge plants. *FEMS Microbiology Ecology* **21**, 267–276.

Poole, J.E.P. (1984). A study of the relationship between the mixed liquor fauna and plant performance for a variety of effluent quality in activated sludge sewage treatment works. *Water Research* **18**, 281–287.

Salvadó, H., Gracia, M.P. and Amigó, J.M. (1995). Capability of ciliated protozoa as indicators of effluent quality in activated sludge plants. *Water Research* **29**, 1041–1050.

Sasahara, T. and Ogawa, T. (1983). Treatment of brewery effluent. Part VIII: protozoa and metazoa found in the activated sludge process for brewery effluent. *Monatsschrift für Brauwissenschaft* **11**, 443–448.

23

The microbiology of phosphorus removal in activated sludge

Thomas E. Cloete[*], M.M. Ehlers[*], J. van Heerden[*] and B. Atkinson[†]

[*]*Department of Microbiology and Plant Pathology, University of Pretoria, Transvaal 0001, South Africa;*
[†]*Technikon of Natal, Durban, South Africa*

1 INTRODUCTION

Eutrophication is a natural process which usually occurs in lakes and other quiescent bodies of water through the introduction of the plant nutrients, phosphorus (P) and nitrogen (N), to the impoundment. Without human intervention, the process takes place over hundreds of years, but is greatly accelerated by various human activities in sensitive areas.

Eutrophication results in many undesirable effects, the primary effect being profuse algal blooms and excessive growth of nuisance aquatic plants. These include several species of blue-green (*Cyanobacteria*) and green (*Chlorophyta*) algae, the diatoms and flagellates, water grasses, rooted broad-leaved plants and floating water plants (hyacinth) (Rudd, 1979). Secondary effects, which are a direct result of weed and algal growth, include rapid oxygen uptake from the water causing the lower water to become anaerobic; production of methane and sulphides, which may result in fish kills (especially in winter) or impart a strong rotten egg smell to the water; and the negative aesthetic appeal of the dam due to overturning of the water and the appearance that the water is black; and the water can no longer be used for potable purposes as certain species of the algae are toxic (Rudd, 1979). Costs of purification of such eutrophied waters escalate dramatically.

Many limnological studies have been conducted concerning eutrophication, primarily with its causes and effects and the results have conclusively indicated that eutrophication is promoted if phosphate (P) and nitrogen (N) are released into a reservoir or catchment area (Walmsley and Thornton, 1984; Chutter, 1990; Dillon and Molot, 1996). These studies have also shown that eutrophication can be effectively controlled if the P load to receiving waters is controlled. Gross eutrophication becomes marked when the inorganic soluble N and P concentrations in waters are in excess of 0.3 mgN/l and 0.015 mgP/l, respectively (Lilley *et al.*, 1997) (Table 23.1). It is virtually impossible to control eutrophication by limiting nitrogen due to the ability of the causative agents (algae) to fix and assimilate atmospheric N. Assimilated N is then made available to other aquatic life forms when these cells die and assimilated N is released. The increased awareness that P is the limiting nutrient has led to the introduction of more stringent legislation governing the discharge of P to receiving waters.

To prevent eutrophication, phosphate removal from effluents is necessary, whether it is by chemical and/or biological means (Toerien *et al.*, 1990). The activated sludge process comprises a complex and enriched culture of a mixture of generalists and specialist organisms. The lack of fundamental understanding of the enhanced biological phosphate removal (EBPR) processes, and the requirement to improve processes designed to remove nutrients from the influent have compelled researchers to examine and attempt to optimize the biological

The Handbook of Water and Wastewater Microbiology
ISBN 0-12-470100-0

TABLE 23.1 Relationship of lake trophy status (productivity) to average concentrations of N and P (Wetzel, 1983)

Lake productivity	*Inorganic N (μg/l)*	*Organic N (μg/l)*	*Total P (μgl)*
Ultra-oligotrophic	<200	<200	<5
Oligo-mesotrophic	200–400	200–400	5–10
Meso-eutrophic	300–650	400–700	10–30
Eutrophic	500–1500	700–1200	30–100
Hypereutrophic	>1500	>1200	>100

components of the mixed liquor of activated sludge plants. Many different configurations of activated sludge processes exist. We will discuss only one of these, representing the most common denominators.

1.2 The activated sludge process

Activated sludge systems are based on suspended-growth processes and have become an integral part of municipal wastewater treatment. The process relies on the dense growth of microorganisms in a reactor where air is continuously supplied to allow for carbonaceous oxidation. The term 'activated sludge' refers to an aerobic slurry of microorganisms which can be removed from the process through sedimentation and returned, in quantifiable amounts, to the wastewater stream (Grady and Lim, 1980a, b). All activated sludge systems operate with the following characteristics in common: utilization of a flocculent slurry of microorganisms to remove organic matter from the surrounding wastewater; prior to effluent discharge from the plant, microorganisms are removed through sedimentation thereby reducing outgoing solids loads; settled microorganisms are recycled to the biological reactor via a clarifier underflow; and dependency of plant performance on the mean cell residence time (MCRT or sludge age) in the system (Grady and Lim, 1980a, b). Microbial metabolism of the organic matter present results in the production of oxidized end-products such as carbon dioxide, nitrates, sulphates and phosphates, as well as the biosynthesis of new microbial biomass (Gray, 1989; Horan, 1990; Bitton, 1994; Muyima *et al.*, 1997).

The basic ecological unit of activated sludge is the floc. Microbial floc formation is essential to the success of activated sludge processes as it allows for rapid and efficient separation of sludge from treated wastewater in the sedimentation tank. Although the exact mechanism of floc formation is not well understood, it seems to be almost entirely bacterially mediated (Muyima *et al.*, 1997). As extracellular polysaccharide production gradually continues, other microorganisms and colloidal material become entrapped in the matrix and the floc diameter increases. Surface charges on the microbial cells and bridge formation by polyvalent cations also contribute to flocculation (Gray, 1989). This very rigid floc structure has impeded the quantitative analysis of the activated sludge microbial community structure, as complete dispersion of the floc is extremely difficult. Clumping of cells in the floc leads to an underestimation of the number of active cells present in the mixed liquor when using viable plate count techniques. Problems attributable to poor floc formation can be one of two causes:

1. non-filamentous bulking which describes flocs that do not settle well due to excessive production of extracellular polysaccharides and the formation of loose flocs, i.e. excess of zoogloeal organisms;
2. filamentous bulking, caused by excessive growth of filamentous bacteria (Bitton, 1994). These undesirable characteristics will have negative economic consequences to the plant in question.

1.2.1 Anaerobic zone

The principal function of the anaerobic zone is to establish a facultatively anaerobic microbial community as indicated by the fermentation pattern. During anaerobiosis this bacterial community produces compounds such as ethanol, acetate and succinate, which serves as carbon sources for phosphate-accumulating bacteria (Cloete and Muyima, 1997).

In the anaerobic zone dissolved oxygen and oxidized nitrogen (nitrate or nitrite) are absent (Barnard, 1976; Buchan, 1984). Sludge from the clarifier flows together with the influent wastewater into this zone. The anaerobic zone is essential for the removal of phosphate, because the bacteria in the activated sludge passing through this zone are preconditioned to take up excess phosphate under aerobic conditions (Cloete and Muyima, 1997). The release of a certain quantity of phosphate from the biomass into the solution indicates that the bacteria have been suitably conditioned (Pitman, 1984). The retention time (about 1 h) of the influent wastewater is of extreme importance. Nitrates and dissolved oxygen discharge into the zone must be zero or as near to zero as possible at all times (Cloete and Muyima, 1997). The presence of nitrate in an anaerobic zone interferes with the phosphate removing capability of the activated sludge during aerobiosis (Barnard, 1976; Nicholls and Osborn, 1979; Marais *et al.*, 1983). The reason may be due to competition for substrate between phosphate-accumulating and denitrifying organisms. In the presence of nitrate, the redox potential is too high to produce fatty acids for the release of phosphate (Cloete and Muyima, 1997). However, the use of unsettled influent and the presence of sludge from the sludge treatment in the primary clarifiers, probably producing lower fatty acids, have a positive effect on the phosphate removal (Mulder and Rensink, 1987). The degree of nitrate feedback that can be tolerated depends on the strength of the sewage feed to the anaerobic zone and its readily biodegradable chemical oxygen demand (COD) concentration (Pitman, 1984).

The microorganisms found in the anaerobic zone are normally living in soil and water and are capable of fermentation (species of *Aeromonas*, *Citrobacter*, *Klebsiella*, *Pasteurella*, *Proteus* and *Serratia*). They accumulate and produce organic compounds such as lactic acid, succinic acid, propionic acid, butyric acid, acetic acid and ethanol during fermentation. These organic compounds serve as electron donor and acceptor, but cannot be utilized under anaerobic conditions. These organic compounds will only be consumed in the anoxic and aerobic zones. Therefore it seems as though the anaerobic zone provides substances for the proliferation of aerobic phosphate-accumulating bacteria (Fuhs and Chen, 1975; Buchan, 1984).

Phosphate release or uptake can be induced by certain carbon sources. The addition of acetate to the medium and the lowering of the pH, as well as phosphate starvation under anaerobic conditions, resulted in phosphate release (Fuhs and Chen, 1975; Barnard, 1976; Buchan, 1983). Nicholls and Osborn (1979) confirmed that bacteria relieved from stress conditions rapidly take up phosphate in an overplus reaction immediately on entering the aerobic zone, where stress conditions are relieved and phosphate together with an abundant source of energy is available. Phosphate seemed to be released from the acid-soluble fraction of cells and to a minimal degree from the RNA and DNA (Cloete and Muyima, 1997).

1.2.2 Primary anoxic zone

Anoxic refers to the presence of nitrates and the absence of dissolved oxygen (Buchan, 1984; Pitman, 1984; Streichan *et al.*, 1990). The primary anoxic zone is the main denitrification reactor in the activated sludge process. It is fed by effluent from the anaerobic zone and by mixed liquor recycled from the aerobic zone (Cloete and Muyima, 1997). The absence of oxygen and the presence of nitrate and nitrite lead to the enrichment of denitrifying bacteria, which reduce nitrate or nitrite to molecular nitrogen. Soluble and colloidal biodegradable matter is removed in the primary anoxic zone. It was found that phosphate release was induced in the anoxic zone in the presence of low fatty acids or their salts such as acetate, formate and propionate (Gerber *et al.*, 1986). Phosphate release was not only affected by the substrate but also by the dosage of the substrate concentrations of soluble, readily biodegradable carbon substrates (Cloete and Muyima, 1997).

1.2.3 Primary aerobic zone

The main function of the primary aerobic zone is to oxidize organic material in the sewage, to oxidize ammonia to nitrite and then to nitrate (by chemoautotrophs), and to provide an environment in which the biomass can take up all the phosphate released in the anaerobic zone, plus all the phosphate that enters the process in the feed sewage (Cloete and Muyima, 1997). Ammonia is oxidized to nitrite by *Nitrosomonas*, *Nitrosospira* and *Nitrosolobus* spp. Nitrite is oxidized to nitrate by *Nitrobacter*, *Nitrospira* and *Nitrococcus* spp. (Buchan, 1984). The principal operational determinant of the efficiency of phosphate removal seems to be the aeration rate, which should be sufficient to promote the rapid uptake of released and feed phosphate. The aeration rate should ensure the oxidation of the carbon compounds and ammonia and suppress the growth of filamentous microorganisms that produce poorly settling sludge (Pitman, 1984). High rate aeration resulted in maximum uptake of phosphate within 2 h (Levin and Shapiro, 1965; Shapiro, 1967, Carberry and Tenny, 1973). and colleagues Wentzel *et al.*, (1985) indicated that the excessive phosphate uptake in the aerobic stage was associated directly with the degree of phosphate release during the previous anaerobic phase, therefore, more phosphate released leads to more phosphate uptake. There is a linear relation between phosphate release and uptake (Wentzel and colleagues *et al.*, 1985).

Other parameters also play a role in enhanced phosphate removal such as: the presence of readily biodegradable compounds, especially volatile fatty acids (VFA), produced by fermentative bacteria from organic compounds in the influent. These volatile fatty acids are removed during the anaerobic phase and polymerized at the expense of energy obtained from the breakdown of polyphosphate (Cloete and Muyima, 1997). Poly-β-hydroxybutyric acid (PHB) plays an important role in the mechanism of phosphate uptake and release (Nicholls and Osborn, 1979; Deinema *et al.*, 1980; Lawson and Tonhazy, 1980; Comeau *et al.*, 1986). The presence of PHB in bio-polyphosphate bacteria helps them to grow and rebuild their own polyphosphate by taking up soluble phosphate from the solution (Comeau *et al.*, 1986). PHB and polyphosphate will therefore play a mutually interdependent role to assist aerobic bacteria to survive through an anaerobic period (Nicholls and Osborn, 1979).

1.2.4 Secondary anoxic zone

The function of this zone is the removal of excess nitrates that were not removed in the primary anoxic zone. Denitrification in this zone is very slow and therefore the quantity of nitrate removed is small. Due to the low COD, the retention time in this anoxic zone is relatively long (Cloete and Muyima, 1997).

1.2.5 Secondary aerobic zone and clarifier

The main function of the secondary aerobic zone is to increase the dissolved oxygen to a level between 2 and 4 mg/l in the mixed liquor before it enters the clarifier (Barnard, 1976), and to refine the final effluent by the removal of additional phosphate and the oxidation of residual ammonia (Cloete and Muyima, 1997). Mixed liquor must be aerated for at least 1 h before it passes into the clarifier to promote phosphate uptake and maintain good aeration conditions (Cloete and Muyima, 1997). Excess aeration should be prevented as it would result in the conversion of organically bound nitrogen to nitrate and cause the slow aerobic release of phosphate from the solids (Keay, 1984; Pitman, 1984). Ammonia nitrification as well as phosphate removal which has not been completed, will continue in the secondary aerobic zone (Buchan, 1984).

The function of the clarifier is to produce a clear effluent free of suspended solids, and a thickened sludge for recycling to the inlet of the process. The quality of the underflow sludge should be such that nitrate is not recycled to the anaerobic zone (Ekama *et al.*, 1984).

The biomass of activated sludge is the active agent of biological wastewater treatment, responsible for carbonaceous material oxidation and nutrient removal. To date, process engineering has received the greatest attention and, in the interim, has practically become optimized. Yet systems based on EBPR principles still regularly fail to achieve the desired end result. This is due to limitations in our current understanding of the complexities of microbiological interactions occurring within the sludge, as well as our inadequate knowledge of microbial community structure–function correlations (Wagner *et al.*, 1993).

To describe and control these microbial processes and mechanisms, Wanner (1997) suggests that activated sludge should be characterized from the following viewpoints:

- characterization and quantification of microbial constituents according to metabolic activities
- identification and classification of microorganisms
- activated sludge quality, i.e. settleability, de-waterability.

Microbial–ecological studies of activated sludge are integral to creating a complete and more definitive understanding of the process, diversity and various functions performed by the constituent microflora. A review of our knowledge about the microbiology of biological phosphorus removal will now follow.

2 MICROBIAL POPULATION DYNAMICS OF BIOLOGICAL PHOSPHORUS REMOVAL

The microbial community of activated sludge consists of bacteria, protozoa, fungi, algae and filamentous organisms (Bux *et al.*, 1994; Muyima *et al.*, 1997). Since organic carbon is the most important energy source entering these systems, it can be expected that heterotrophic bacteria will dominate the community structure (Kämpfer *et al.*, 1996). A variety of methods have been used to study the microbial community in activated sludge. The results obtained using these methods today form the basis of our understanding of the microbiology of biological phosphorus removal and will hence be discussed in more detail.

2.1 Culture based techniques

Since enhanced phosphate removal was first postulated to be mediated by a biological mechanism (Srinath *et al.*, 1959; Fuhs and Chen, 1975), much research has centred around identifying and elucidating those organisms responsible for the process (Barnard, 1976; Hart and Melmed, 1982; Buchan, 1983; Cloete and Steyn, 1988a, b; Bosch and Cloete, 1993; Kavanaugh and Randall, 1994; Nakamura *et al.*, 1998).

Some time ago, *Acinetobacter* came to be regarded as the model organism in biochemical models describing the mechanism of EBPR (Wentzel *et al.*, 1986) due mainly to its presence in high numbers and its favourable physiological characteristics when isolated from biological phosphorus removal plants using conventional plating techniques. Bacteriological studies of biological nutrient removal (BNR) systems have emphasized the functional role of *Acinetobacter* in enhanced P removal (Lötter, 1985). Development of enhanced cultures of polyphosphate accumulating organisms (PAOs) has shown tremendous specificity in the population structure when one considers that more than 90% of the organisms cultured aerobically from laboratory-scale UCT and three-stage Bardenpho systems were identified as *Acinetobacter* (Wentzel *et al.*, 1988, 1989).

Since Fuhs and Chen (1975) first implicated *Acinetobacter* spp. as having an important role in EBPR, most subsequent studies have focused on this bacterial genus. The reasons for this attention have not been entirely unjustified. Culture-dependent methods consistently indicated that *Acinetobacter* spp. were the numerically dominant members of EBPR systems. Furthermore, some, but not all, *Acinetobacter* strains isolated from activated sludge accumulated excessive amounts of polyphosphate in pure culture, suggesting their importance in the EBPR process (Bond *et al.*, 1995). However, researchers have difficulties reconciling the carbon and phosphorus transformations in pure cultures of *Acinetobacter* strains with the biochemical model for EBPR. In recent years, serious doubts have been raised as to the significance of *Acinetobacter* spp. in the EBPR process. The most compelling evidence for this change of view has been the recent non-culture-dependent studies of phosphate-removing communities. In all cases, *Acinetobacter* spp. was found to represent only a small portion of the total EBPR microbial population. Instead, other bacterial groups such as the Gram-positive bacteria and the beta-subclass groups of the Proteobacteria were numerically dominant (Bond *et al.*, 1995). While the use of these non-culture-dependent methods reduces the significance of *Acinetobacter* spp. in EBPR processes, the resolution of the methods has not been sufficient to propose alternative EBPR candidate genera (Bond *et al.*, 1995).

A number of heterotrophic organisms have been isolated from EBPR systems using culture dependent methods. These organisms include *Acinetobacter, Moraxella, Pseudomonas, Microlunatus, Achromobacter, Aeromonas, Vibrio, Citrobacter, Pasteurella, Enterobacter, Proteobacter, Klebsiella, Bacillus* and coliforms (*Escherichia coli* and *E. intermedium*) (Hart and Melmed, 1982; Brodisch and Joyner, 1983; Cloete *et al.*, 1985a, b; Lötter, 1985; Lötter and Murphy, 1985; Lötter *et al.*, 1986; Cloete and Steyn, 1988a, b; Kavanaugh and Randall, 1994; Wagner *et al.*, 1994a, b; Momba and Cloete, 1996a, b; Ubukata and Takii, 1998).

We know that there is a large discrepancy between the total direct microscopic counts and viable plate counts (usually less than 1% of the former) for many ecosystems (Cloete and Steyn, 1987; Wagner *et al.*, 1993). Recoveries from activated sludge systems, even with optimized media, are only between 5 and 15% (Wagner *et al.*, 1993). The exact ecological significance of these organisms in EBPR is not understood. This is due to the fact that these organisms make up part of the culturable activated sludge microbial community. Consequently, researchers have been investigating the use of non-culture dependent techniques to obtain a better understanding of EBPR microbiology.

2.2 Fluorescent *in situ* hybridization (FISH)

All life on earth is comprised of three domains, i.e. the *Bacteria* (e.g. *Acinetobacter, Pseudomonas*), the *Archaea* (microorganisms thought to live in extreme environments, i.e. anaerobic, heat and salt) and the *Eukarya* (animals, fungi, plants, protists) (Woese *et al.*, 1990). This phylogenetic system is based on comparative analyses of sequences of the small subunit (16S) rRNA gene.

The domain (super-kingdom) *Bacteria* presently consists of eleven characterized bacterial phyla (divisions) (Woese, 1987; Wagner and Amann, 1997). It is evident that when using the 16S rRNA molecule phylogenetically to structure the domain *Bacteria*, the arrangement within the various phyla becomes somewhat confusing, i.e. photosynthetic species grouped with non-photosynthetic species; anaerobes are paired with aerobes and heterotrophs with chemolithotrophs. Most of the taxa are defined by unique, conserved oligonucleotide sequences or signatures, i.e. rRNA sequence positions that are not found in the rRNA of other groups (Manz *et al.*, 1992). Many of the traditional Gram-negative bacteria are contained in the *Proteobacteria* class (Woese, 1987; Manz *et al.*, 1992). At least four distinct subdivisions exist in the purple bacteria, designated alpha (α), beta (β), gamma (γ) and delta (δ) (Woese, 1987).

Each rRNA-targeted oligonucleotide is chemically linked to a fluorochrome molecule which allows cells hybridized with the fluorescently-labelled oligonucleotide to be directly visualized using epifluorescent microscopy or scanning confocal laser microscopy (Amann, 1995; Amann *et al.*, 1995). The process usually incorporates a total cell count using the DNA intercalating dye 4,6-diamidino-2-phenylindole (DAPI) followed by hybridization with the universal bacterial probe, EUB (the fluorescing stain 3,6-tetramethyl diaminoacridine or acridine orange can also be used to visualize microscopically DNA molecules, but there are a number of difficulties inherent to its application; refer to Porter and Feig (1980) for a brief review). DNA-DAPI complexes fluoresce bright blue when visualized at wavelengths of $>$ 390 nm, while unbound DAPI and non-DNA-DAPI complexes fluoresce a weak yellow which enhances the visibility of DAPI fluorescence (Porter and Feig, 1980).

The EUB/DAPI ratio gives an indication as to the bacterial composition of the sample. Dual EUB/DAPI staining of activated sludge samples (Wagner and Amann, 1997) has revealed that approximately 80% of the microbial cells present in the samples were metabolically active bacteria, of which only 3–19% could be cultivated on optimized media. The specificity of oligoprobes is freely adjustable, dependent upon the requirements of the user. Different phylogenetic levels ranging from kingdom, e.g. *Bacteria* (EUB), to subclass, e.g. gamma (GAM) subclass of *Proteobacteria*, to genus, e.g. *Acinetobacter* (ACA), species and subspecies can be probed, the degree of organization required being the only limitation (Wagner *et al.*, 1994a) (Table. 23.2).

In situ identification of the organisms in a batch-type EBPR sludge has revealed that the four major bacterial groups present were the alpha and beta subclass of *Proteobacteria*, Gram-positive bacteria with a high G+C content (GPBHGC) and bacteria belonging to the *Cytophaga-Flavobacterium* cluster of the *Cytophaga-Flavobacterium-Bacteroides* phylum (Kawaharasaki *et al.*, 1999). However, Wagner *et al.*, (1994a), when characterizing the Proteobacterial microbial consortia in municipal mixed liquor from Hirblingen, Germany, found that the beta subclass dominated over the alpha and gamma subclasses. Simultaneous plating of the sludge on nutrient-rich medium showed dominance of the gamma subclass of *Proteobacteria* emphasizing the bias introduced with cultivation techniques.

In situ hybridization of mixed liquor samples at the family level of organization from the anaerobic and aerobic zones of the same plant in Germany have revealed the following trends (Snaidr *et al.*, 1997):

anaerobic basin	α-*Proteobacteria*	= 11%
	β-*Proteobacteria*	= 24%
	γ-*Proteobacteria*	= 5%
	GPBHGC	= 24%
	Cytophaga-Flavobacterium	= 9%
aerobic basin	α-*Proteobacteria*	= 9%
	β-*Proteobacteria*	= 26%
	γ-*Proteobacteria*	= 10%
	GPBHGC	= 19%
	Cytophaga-Flavobacterium	= 8%

These results show conclusively that there is no markable population shift between zones in an EBPR process when characterized at family level. Whether a shift occurs at species level requires verification, but regardless of any shifts which may occur, the significance of the *Proteobacteria* to efficient EBPR operations is evident.

TABLE 23.2 Phyla and subdivisions (including representative examples) of the domain *Bacteria* (shaded rows indicate those subdivions known to be implicated in EBPR)

***Proteobacteria* (Purple bacteria)**
α subdivision (*Nitrobacter*)
β subdivision (*Alcaligenes, Pseudomonas testoteroni, P. cepacia*, autotrophic nitrifiers)
γ subdivision (*Acinetobacter, Aeromonas*, fluorescent pseudomonads, Enterics, *Vibrio*)
δ subdivision (sulphate-reducing bacteria and myxobacteria)
***Firmicutes*(Gram-positive bacteria)**
A. *Actinobacteria* (high G + C species) (*Arthrobacter, Micrococcus, Microlunatus*)
B. *Bacillus/Clostridium* group (low G + C species) (*Clostridium, Bacillus*)
C. Photosynthetic species (*Heliobacterium*)
D. Species with Gram-negative walls (*Megasphaera*)
Cyanobacteria and chloroplasts
Nostoc, Oscillatoria
Spirochetes and relatives
A. Spirochetes (*Spirochaeta*)
B. Leptospiras (*Leptospira*)
Green sulphur bacteria
Chlorobium
Bacteroides, flavobacteria and relatives
A. Bacteroides (*Bacteroides*)
B. Flavobacterium group (*Flavobacterium, Cytophaga*)
Planctomyces and relatives
A. Planctomyces (*Planctomyces*)
B. Thermophiles (*Isocystis*)
Chlamydiae
Chlamydia
Radioresistant micrococci and relatives
A. Deinococcus group (*Deinococcus radiodurans*)
B. Thermophiles (*Thermus aquaticus*)
Green non-sulphur bacteria and relatives
A. Chloroflexus group (*Chloroflexus*)
B. Thermomicrobium group (*Thermomicrobium roseum*)
Thermotogoa

Probing activated sludge with fluorescently-labelled oligonucleotide probes specific for the alpha, beta and gamma subclasses of the *Proteobacteria* (see section 2.5) has revealed that the microbial consortia are dominated by the *Proteobacteria* (approximately 80%), a phylum containing the majority of the traditional Gram-negative bacteria (the majority of which are heterotrophic) (Wagner *et al.*, 1993).

Staining activated sludge samples with DAPI at elevated concentrations to those used for DNA staining – referred to as the polyphosphate-probing concentration – results in the fluorescence of intracellular volutin and lipid inclusions (Nakamura *et al.*, 1998; Streichan *et al.*, 1990, as cited by Kawaharasaki *et al.*, 1999). Bacteria that accumulate large quantities of poly-P are easily distinguished by colour and intensity of fluorescence due to the following DAPI stain characteristics:

1. DNA-DAPI fluorescence is blue-white
2. polyphosphate-DAPI fluorescence is bright yellow
3. lipid-DAPI fluorescence is weak yellow and fades rapidly (in the space of seconds) (Kawaharasaki *et al.*, 1999).

Dual staining of samples with DAPI at elevated concentrations and EUB will result in the determination of the PAO population in EBPR sludges, i.e. PAO/EUB ratio. *In situ* identification of those bacteria exhibiting strong poly-P accumulation (identified through DAPI staining) can also be achieved through dual staining with class, i.e. *Proteobacteria*, and subclass, i.e. alpha, beta or gamma, oligonucleotide probes which totally negates culture-dependent methods of isolation and/or Neisser (poly-P) staining procedures.

For all its promise and potential, however, there are still technical problems inherent to FISH when applying the technology to microbial systems such as activated sludge. Qualitatively, the protocol of probe hybridization and detection has been optimized at all levels of organization, but due to the nature of activated sludge, quantitative results are often difficult and limited, i.e. complete dispersion of sludge flocs remains a technical problem which limits the application. Other problems include deoxyribonucleic acid (DNA) retrieval for sequence determination, polymerase chain reaction (PCR) biases when amplifying the sequence of interest and an imposed selection of the retrieved or target sequences (Hiraishi *et al.*, 1998). One of the possible solutions directed towards these problems is the combination of molecular and biomarker methods.

2.3 Immunological techniques

The immunofluorescence approach was introduced as an *in situ* identification technique, prior to FISH, in an attempt to avoid culture-dependent techniques and has been used effectively to identify *Acinetobacter* in activated sludge samples (Cloete *et al.*, 1985a, b; Lötter and Murphy, 1985; Cloete and Steyn, 1987, 1988b). Cloete and Steyn (1987) found that less than 10% of microscopic cell counts of activated sludges could be accounted for in agar plate enumerations. Although the technique is highly specific for the bacterium in question, there are a number of limitations associated with it (Wagner and Amann, 1997). First, the presence of extracellular polymeric substances in activated sludge flocs can inhibit the penetration of antibodies to the target cells; the method of raising antibodies in host animals requires initially culturing the bacterium of interest; and cross-reaction of antibody with contaminants does occur, resulting in high levels of background fluorescence (Wagner and Amann, 1997).

Monoclonal antibody production is the method of choice to obtain immunoglobulins against proteins that cannot be purified or that are available in low amounts (Drenckhahn *et al.*, 1993). Antibodies have been used in investigations of samples from activated sludge systems, although such studies are not numerous. Cloete and Steyn (1987) used a fluorescent antibody technique for the identification and enumeration of *Acinetobacter* in activated sludge (Cloete and Steyn, 1988b). Monoclonal antibodies specific for *Nitrobacter* and *Nitrosomonas* respectively were produced and used for estimation of cell numbers in activated sludge (Sandén *et al.*, 1994).

Conventional as well as focused immunization strategies were also used to maximize the possibility of producing antibodies with the required ability to discern between the antigen preparations. A significant finding emerging from this study was based on the observation that antigenic differences were clearly detected between the aerobic zones of two activated sludge systems with differing phosphate removal ability. Characterization of the antigen recognized suggested a protein nature. The antibody showed five compact bands on the blot, with M_r appearing to be multiples of 18 kDa. The epitope recognized probably occurred on an 18 kDa proteinaceous monomer, represented on the blot in incremental steps of subunit association. The *raison d'être* of the unique antigen in the phosphate removing system may be speculated upon: first, did the specific conditions prevailing in the phosphate removing system as opposed to the non-phosphate removing system induce its expression, or did the anaerobic zone select for a unique bacterial population in the phosphate removing system which is recognized by the antibody, but is not necessarily correlated with phosphate removal? Could the state of oligomerization of the 18 kDa protein correlate to the phosphate removing ability of the sludge? All these and other questions still need to be answered, before any valid conclusions can be made using this technique.

2.4 Quinone profiles

Respiratory or isoprenoid quinones are a class of lipids that are constituents of bacterial plasma membranes. They play important roles in electron transport, oxidative phosphorylation and active transport across the membrane (Collins and Jones, 1981). The numerical analysis of lipoquinone profiles has offered an effective method for monitoring population shifts and for classifying bacterial communities in wastewater sludges (Hiraishi *et al.*, 1991). Quinones are usually extracted from an environmental sample using an organic solvent. After evaporation and re-extraction, the concentrated quinone is applied to column chromatography to separate menaquinone and ubiquinone (Hiraishi *et al.*, 1998). Municipal sludges are usually characterized according to their menaquinone and ubiquinone components (Hiraishi *et al.*, 1989, 1998). Quinone components are then identified and quantified using spectrochromatography and mass spectrometry. Numerical analyses of quinone profiles can enhance the information regarding bacterial community dynamics in wastewater ecosystems. The strength of the technique lies not only in its ability to assess taxonomic structure of bacterial communities, but also in that variations in bacterial population structure over space and time can be quantified (Hiraishi *et al.*, 1991). The use of respiratory quinone profiles to characterize the bacterial population structure of the anaerobic–aerobic activated sludge system showed that *Acinetobacter* species were not important in the system (Hiraishi *et al.*, 1989). However, it was shown that *Acinetobacter*, as detected by the biomarker diaminopropane, was the dominant organism only in wastewater treatment plants with low organic loading (Auling *et al.*, 1991).

2.5 Microautoradiography

Autoradiography has classically been used in the medical field but has recently been introduced to environmental sample analyses to determine microbial community structures. Typically, a radio-labelled compound appears in the cell or biological structure of interest through adsorption of a tracer or labelled substrate uptake. The radio-labelled sample is then placed in contact with a radiosensitive emulsion and the emissions from the radioactive sample interact with silver bromide crystals in the emulsion. The emulsion is then developed using standard photographic procedures and the silver grains appear on top of the radioactive structure, which can then be viewed microscopically (Nielsen *et al.*, 1998, 1999). Although autoradiography can successfully be applied to study the *in situ* physiology of various microorganisms, it is limited by its lack of proper identification of the organisms in question. However, Nielsen *et al.*, (1998), through simultaneous use of autoradiography and FISH, were able to correlate function/

activity with identification, which is a tremendous breakthrough for activated sludge identification-diversity-functional studies.

2.6 SDS-PAGE of proteins used for fingerprinting microbial communities

The extraction of proteins directly from environmental samples is desirable for multiple reasons:

1. Analysis of proteins extracted from environmental samples may help characterize the response of microbial communities to stressful conditions such as contamination with toxic chemicals, starvation, heat or oxygen levels
2. Analysis of total proteins extracted from an environmental sample can be employed as a 'fingerprint' to type the diversity in the sample, in a way similar to grouping of bacteria according to enzyme polymorphisms and immunological reactions. Such fingerprints may eventually be used to monitor the deterioration or enrichment of species diversity in microbial communities
3. The abundance of proteins to which specific antibodies are available can be directly measured in total proteins extracted from complex ecosystems and used as an index for monitoring the progress of a biocatalytic reaction *in situ* (Ogunseitan, 1993).

The main advantages of numerical analysis of electrophoretic patterns of large numbers of bacteria are:

1. rapid grouping
2. allocation of unknown microorganisms to a group and its possible identification
3. storage of large numbers of patterns in data banks for reference
4. a quick decision on whether two colony types in a culture are due to variation or contamination
5. information on epidemiology, spreading of animal and plant pathogens
6. DNA:DNA homology determinations of large numbers of strains can be reduced to hybridization of DNAs from the typical representative of each group, previously established by gel electrophoresis (Kersters and De Ley, 1975).

Similarities are calculated between each sample using a suitable similarity coefficient, the Pearson product moment correlation coefficient (r). The resultant matrix is clustered using the average linkage algorithm (also known as unweighted pair group method of arithmetic averages, UPGMA) to provide a sorted similarity matrix or dendrogram.

Valuable information concerning the bacterial population structure of activated sludge was obtained when sodium dodecyl sulphate polyacrylamide gel electrophoresis (SDS-PAGE) was used. The results confirmed previous studies performed by Cloete and Steyn (1987), which indicated that the bacterial population of activated sludge stayed the same throughout the system (Ehlers and Cloete, 1999a). Furthermore, no specific protein pattern due to seasonal changes or between different zones or between N- and P-removing and N-removing systems were observed using SDS-PAGE (Ehlers and Cloete, 1999a, b, c). The main drawback of this technique was that it was not sensitive enough to determine the difference in protein profiles of P-removing and non-P-removing bacterial populations.

2.7 Community-level carbon source utilization

Garland and Mills (1991) introduced the use of community-level carbon source utilization patterns for comparison of microbial communities from different habitats. The Biolog system (Biolog Inc., Hayward, USA) is based on the different utilization of a large number of organic compounds by the test organisms.

The Biolog system has already been used for characterization of naturally occurring bacteria and for classification of bacterial communities of different environments (Frederickson *et al.*, 1991; Garland and Mills, 1991; Verniere *et al.*, 1993; Winding, 1994; Zak *et al.*, 1994). The comparisons of Biolog results with other test systems, i.e. API and Biotype-100, have also been performed (Frederickson *et al.*, 1991; Amy *et al.*, 1992; Klinger *et al.*, 1992; Verniere *et al.*, 1993). Community-level physiological profiles (CLPP) have been used to characterize microbial communities in freshwater, coastal and lagoon areas, as well as soil, rhizosphere, bioreactors and gnotobiotic mixtures (Garland and Mills, 1991; Gorlenko and Kozhevin, 1994; Winding, 1994; Zak *et al.*, 1994; Bossio and Scow, 1995; Ellis *et al.*, 1995; Haack *et al.*, 1994; Lehman *et al.*, 1995; Vahjen *et al.*, 1995; Garland, 1996a, b; Glimm *et al.*, 1997; Engelen *et al.*, 1998). Environ ments such as groundwater, phyllosphere, activated sludge and compost were also studied (Ellis *et al.*, 1995; Heuer *et al.*, 1995; Lehman *et al.*, 1995; Insam *et al.*, 1996; Victorio *et al.*, 1996). For rapidly assessing the dynamics of autochthonous microbial communities, the CLPP techniques have been used (Wünsche and Babel, 1995. Diversity can be calculated using univariate indices which do not capture all potential differences in the CLPP (Garland, 1997).

The purpose of using Biolog was therefore not to try to detect each and every metabolic reaction of all the species in the community, but the collective utilization pattern for a specific community. A high species diversity should lead to a higher relative number of substrates utilized, because there are more possibilities and upon dilution, some organisms will be lost (causing a decrease in species diversity) from the community, depending on their abundance and the relative contribution (perhaps only one metabolic reaction in the system), reducing the number of possibilities. The extent of the reduction of the possibilities upon dilution, should theoretically reflect something about the community structure. The key, therefore, lies in the interpretation of the results.

Differences in microbial community structure in activated sludge systems may exist, but this had no bearing on the effectivity with which these systems removed phosphate. If PO_4^{3-} removal was related to the microbial community composition, a high correlation of the Biolog patterns would be expected among PO_4^{3-} removing systems. However, this was not the case. All the different zones of systems tested indicated a high initial diversity (10^{-1} to 10^{-2} dilutions), due to the high number of substrates utilized. No specific patterns could, however, be identified for PO_4^{3-} removing systems, indicating that PO_4^{3-} removal was not community structure specific. This agrees with previous studies (Dold *et al.*, 1980; Kersters *et al.*, 1997).

2.8 Biomass

Current research in wastewater treatment has been directed towards mathematical modelling of basic design and operational procedures. One important parameter in such models is the amount of viable biomass. For this reason attempts have been made to find simple and reliable methods to determine the biomass in wastewater and activated sludge. The simplest and most often used method is to measure suspended solids or volatile suspended solids. Such methods, however, do not distinguish between living cells and debris of either organic or inorganic origin. Using traditional plate techniques, the problem is normally an underestimation of the biomass due to selectivity of the media (Jørgensen *et al.*, 1992).

One key component of mixed liquor suspended solids (MLSS) is the heterotrophic active biomass. This component mediates the biodegradation processes of COD removal and denitrification. Thus the rates for these processes are directly related to the heterotrophic active biomass present, and the specific rates should be expressed in terms of this parameter to allow a meaningful comparison of the rates measured in different systems. However, the heterotrophic active biomass parameter has been only hypothetical within the structure of these models; it has not been measured directly, primarily due to the lack of suitable simple measurement techniques. Ubisi *et al.*, (1997) have, however, reported close correlation between measured heterotrophic active biomass concentration with those calculated theoretically, thus promoting confidence in the application of the models for design, operation and control of activated sludge systems.

It was indicated that biomass was related to phosphorus removal (Bosch, 1992; Momba, 1995; Muyima, 1995). The higher the biomass, the better the P removal. This suggested that the main difference between P-removing and non-P-removing systems was biomass related and not due to the microbial community structure. An increase in biomass led to an increase in P removal (Momba and Cloete, 1996a). When calculating the quantity of P removal per g of sludge no significant difference was observed, indicating that there was a direct relationship between P removal and MLSS for a specific system. The effects of growth phase and initial cell concentration of *Acinetobacter* spp. on P release and uptake have also been investigated (Momba and Cloete, 1996a, b; Rustrian *et al.*, 1997). As can be expected, Momba and Cloete (1996b) found a linear relationship existed between initial biomass and phosphate uptake and that at low cell densities (10^2–10^5 per ml) there was a net release of P rather than uptake. The physiological state of the cells also determines the response of the biomass to P accumulation. Bosch and Cloete (1993) have observed that P is accumulated at the end of log and during stationary phase once active growth has ceased and concluded that a maximum number of cells in the stationary phase of growth should be accumulated in the aerobic zone to optimize bio-P removal. Rustrian *et al.* (1997) found that *Acinetobacter* released more P under anaerobic conditions when in the stationary phase of growth yet P uptake was equally efficient with cells in the log and stationary phase. These results can have a direct implication on EBPR operations as growth conditions for the polyphosphate accumulating organisms (PAO) community can be better optimized to improve and promote the mechanism.

TABLE 23.3 Numbers of *Acinetobacter* spp. in activated sludge using different numeration techniques

Number of Acinetobacter cells (% of total community)	*Enumeration/ identification technique*	*Reference*
54–66%	API 20E	Hart and Melmed, 1982
90%	total plate counts and API 20E	Lötter *et al.*, 1986a
5–11%	MPN and API 20E	Kavanaugh and Randall, 1994
<10%	fluorescent antibodies	Cloete and Steyn, 1987, 1988b
3–6%	quinone profiles	Hiraishi *et al.*, 1989
<1%	16S rRNA oligonucleotide probes	Bond *et al.*, 1997
6%	16S rRNA oligonucleotide probes	Wagner *et al.*, 1994b

2.8.1 *Microlunatus phosphovorus* – the new model organism?

Due to the controversy surrounding the role of *Acinetobacter* in P removal operations (Table 23.3), microbiologists have, for the past decade, attempted to isolate other bacteria in activated sludge upon which the mechanism can be modelled.

Recently, a new P-removing bacterium from a laboratory-scale activated sludge system in Japan was isolated and identified as *Microlunatus phosphovorus* (*M. phosphovorus*) (Nakamura *et al.*, 1995, as cited by Ubukata and Takii, 1998). The bacterium shows all the physiological traits characteristic of PAOs, i.e. P accumulation mechanism is only induced in sequential anaerobic/aerobic systems, and has the propensity to accumulate P to a maximum of 23% dry weight (luxury uptake). Its carbon and phosphorus transformation patterns coincide with those of EBPR sludge and, as such, the bacterium has been considered as a candidate for the model PAO in EBPR processes. *M. phosphovorus* is a Gram-positive coccus (diameter = 1.7–2.1 μm) and sequencing shows a high G + C genomic DNA content (65.6 mol%; phylogenetically belongs to the Gram-positive high GC (HGC) bacteria) (Ubukata and Takii, 1998).

The excess P accumulation mechanism is inducible in the bacterium, a useful feature when examining and attempting to define bio-P removal processes. However, the bacterium's dominance in EBPR processes has yet to be demonstrated and, because it cannot readily be isolated from activated sludge (as opposed to *Acinetobacter*), its application to BPR studies may be limited.

2.8.2 The 'G' bacteria

When assessing EBPR efficiency, cognisance must be taken of the 'G' bacteria, a group of normal inhabitants found in activated sludge mixed liquor. There is a microbial community in activated sludge which is capable of organic substrate uptake and assimilation in the anaerobic zone with subsequent metabolism of these storage granules in the aerobic zone. This community is composed of two distinct groups of organisms with distinctly different modes of substrate uptake and synthesis of storage granules. The first group are the PAOs, the metabolism of which has been previously discussed. The second group are the 'G' or glycogen accumulating organisms (GAOs). In conventional systems GAOs are involved in normal organic oxidation processes but, in selector systems such as BPR, their presence and impact to system efficiency can become more prominent. These Gram-negative cocci grow as tetrads and are able to out-compete PAOs in anaerobic/aerobic systems by accumulating polysaccharide and not polyphosphate in the aerobic zone (Cech and Hartman, 1993; Cech *et al.*, 1994; Maszenan *et al.*, 1998). The 'G' bacteria are thought to be able to compete effectively with fermentative organisms for RBCOD and PAOs for VFA in the anaerobic zone as they are able to obtain the necessary reducing power and energy required for uptake through glycolysis, i.e. the Embden-Meyerhof pathway. Their proliferation in an EBPR system will eventually lead to a decline so far as phosphate removal is concerned. The 'G' bacteria were able to dominate in an anaerobic-anoxic system when the influent consisted of an acetate-glucose mixture yet, when acetate was used as substrate alone, the PAOs were able to dominate their competitors (Cech and Hartman, 1993). Influent P/COD (P^{ti}/S^{ti}) ratios also affects the microbial community structure of EBPR activated sludge (Lui *et al.*, 1997). A low P^{ti}/S^{ti} ratio enriches for GAOs, while high P^{ti}/S^{ti} ratios promote the growth of PAOs and suppress the proliferation of GAOs in the process (Lui *et al.*, 1997). Competing with the non-PAOs, i.e. fermentative bacteria, for RBCOD in the anaerobic zone reduces VFA production which ultimately influences P release by the PAOs. With the advent of molecular techniques for elucidating bacterial identification and relatedness, it has been shown that all the Gram-negative tetrad cocci in activated sludge are closely related taxonomically, belonging to the alpha subdivision of the *Proteobacteria* (see section 2.5) (Maszenan *et al.*, 1998). According to Blackall *et al.*, (1997), independently isolated 'G' bacterial strains from the Cech and Hartman (1993) study and from full-scale plants in Italy showed near identical homogeneity to one another and 93% homogeneity to *Rhodobacter*(alpha subdivision of *Proteobacteria*), based on 16S rDNA sequences. They proposed the generic and species name, *Tetracococcus cechii*, for the two strains. Maszenan *et al.*, (1998) proposed a new genus (*Amaricoccus*) to house four 'G' bacterial isolates from Australia, Italy and Macau. Phylogenetic classification of these isolates in the *Proteobacteria* phylum of the domain *Bacteria* shows their relatedness to many of the PAOs, indicating that some relationship between the two competing bacteria does exist. This implies that an identical or similar mode of metabolism between the two bacterial types must have existed at some stage of their evolution.

3 CONCLUDING REMARKS

The microbiology of nitrification and denitrification in the activated sludge process is well understood. Nitrification is the process where ammonia, the reduced form of nitrogen is oxidized by autotrophic nitrifying bacteria to nitrite and nitrate. Ammonia is oxidized to nitrite by *Nitrosomonas*, *Nitrosospira* and *Nitrosolobus* spp. (Buchan, 1984). Nitrification is a chemolithotrophic process where microorganisms utilize energy generated from the oxidation of inorganic compounds (Cloete and Muyima, 1997). When reduced nitrogen is incorporated into newly synthesized biomass the process is termed assimilative nitrate reduction (Cloete and Muyima, 1997). When nitrate nitrogen is reduced to elementary nitrogen and serves as an electron acceptor, the process is known as denitrification (Cloete and Muyima, 1997). A wide range of heterotrophic bacteria can accomplish denitrification under anoxic conditions. Nitrite is oxidized to nitrate by *Nitrobacter*, *Nitrospira* and *Nitrococcus* spp. (Buchan, 1984).

Culture dependent methods have yielded a variety of microbial isolates. However, due to the limitations of culture-based techniques, the ecological role of these microorganisms is not well understood.

The immunofluorescence approach was introduced as an *in situ* identification technique, prior to FISH, in an attempt to avoid culture-dependent techniques and has been used effectively to identify *Acinetobacter* in activated sludge samples (Cloete *et al.*, 1985a, b; Lötter and Murphy, 1985; Cloete and Steyn, 1987, 1988b). Cloete and Steyn (1987) found that less than 10% of microscopic cell counts of activated sludges could be accounted for in agar plate enumerations. Although the technique is highly specific for the bacterium in question, there are a number of limitations associated with it (Wagner and Amann, 1997). *In situ* identification of the organisms using FISH in a batch-type EBPR sludge has revealed that the four major bacterial groups present were the alpha and beta subclass of *Proteobacteria*, Gram-positive bacteria with a high G + C content and bacteria belonging to the *Cytophaga-Flavobacterium* cluster of the *Cytophaga-Flavobacterium-Bacteroides* phylum (Kawaharasaki *et al.*, 1999). However, Wagner *et al.*, (1994a), when characterizing the Proteobacterial microbial consortia in municipal mixed liquor from Hirblingen, Germany, found that the beta subclass dominated over the alpha and gamma subclasses. Simultaneous plating of the sludge on nutrient rich medium showed dominance of the gamma subclass of *Proteobacteria* emphasizing the bias introduced with cultivation techniques.

The use of respiratory quinone profiles to characterize the bacterial population structure of the anaerobic–aerobic activated sludge system showed that *Acinetobacter* species were not dominant in the system (Hiraishi *et al.*, 1989). However, it was shown that *Acinetobacter*, as detected by the biomarker diaminopropane, was the dominant organism in wastewater treatment plants with a low organic loading (Auling *et al.*, 1991).

Due to the controversy surrounding the role of *Acinetobacter* in P removal operations microbiologists have, for the past decade, attempted to isolate other bacteria in activated sludge upon which the mechanism can be modelled. Recently, a new P-removing bacterium from a laboratory-scale activated sludge system in Japan was isolated and identified as *Microlunatus phosphovorus* (*M. phosphovorus*) (Nakamura *et al.*, 1995, as cited by Ubukata and Takii, 1998). The bacterium shows all the physiological traits characteristic of PAOs, i.e. P accumulation mechanism is only induced in sequential anaerobic/aerobic systems, and has the propensity to accumulate P to a maximum of 23% dry weight (luxury uptake). However, the bacterium's dominance in EBPR processes has yet to be demonstrated and, because it cannot readily be isolated from activated sludge (as opposed to *Acinetobacter*), its application to BPR studies may be limited.

The 'G' bacteria are thought to be able to compete effectively with fermentative organisms for RBCOD and PAOs for VFA in the anaerobic zone as they are able to obtain the necessary reducing power and energy required for uptake through glycolysis, i.e. the Embden-Meyerhof pathway. Their proliferation in an EBPR system will eventually lead to a decline as far as phosphate removal is concerned.

Bacterial community composition studies using SDS-PAGE, Biolog and biomass indicated that there were no differences in the bacterial community structure between different zones in the same systems, as well as between different activated sludge systems. These results indicated that EBPR was, therefore, not due to a specific bacterial community structure but rather due to the amount of biomass in the system. The higher the biomass, the higher the phosphorus removal capacity of an activated sludge system.

In conclusion the ultimate method to determine the bacterial community structure and function of environmental samples still has to be developed. Therefore, each possible method should be investigated, until one or a combination of methods is found, that can assist in the better understanding of microbial community structure and function in activated sludge and specifically biological phosphorus removal.

REFERENCES

Amann, R. (1995). *In situ* identification of microorganisms by whole cell hybridization with rRNA-targeted nucleic acid probes. *Mol. Micro. Ecol. Man* **3.3.6**, 1–15.

Amann, R., Ludwig, W. and Schleifer, K.-H. (1995). Phylogenetic identification and *in situ* detection of individual microbial cells without cultivation. *Microbiol. Rev.* **59**, 143–169.

Amy, P.S., Haldeman, D.L., Ringelberg, D. *et al.* (1992). Comparison of identification systems for classification of bacteria isolated from water and endolithic habitats within the deep subsurface. *Appl. Environ. Microbiol.* **58**, 3367–3373.

Auling, G., Pilz, F., Busse, H.J. *et al.* (1991). Analysis of polyphosphate accumulation microflora in phosphorus-eliminating, anaerobic-aerobic activated sludge systems by using diaminopropane as a biomarker for rapid estimation of *Acinetobacter* spp. *Appl. Environ. Microbiol.* **57**, 3585–3592.

Barnard, J.L. (1976). A review of biological phosphorus removal in activated sludge in the activated sludge process. *Water SA.* **2**, 136–144.

Bitton, G. (1994). Wastewater Microbiology, pp. 59–303. John Wiley & Sons, Inc, New York.

Blackall, L.L., Rossetti, S., Christensson, C. *et al.* (1997). The characterization and description of representatives of 'G' bacteria for the activated sludge plants. *Lett. Appl. Microbiol.* **25**, 63–69.

Bond, P.L., Hugenholtz, J.K., Keller, J. and Blackall, L.L. (1995). Bacterial community structure of phosphate-removing and non-phosphate-removing activated sludge from sequencing batch reactors. *Appl. Environ. Microbiol.* **61**, 1910–1916.

Bosch, M. (1992) Phosphate removal in activated sludge and its relationship to biomass. MSc. Thesis. University of Pretoria. Pretoria, South Africa.

Bosch, M. and Cloete, T.E. (1993). Research on biological phosphate removal in activated sludge. *Water Research Commission Report No. 314/1/93.*

Bossio, D.A. and Scow, K.M. (1995). Impact of carbon and flooding on the metabolic diversity of microbial communities in soils. *Appl. Environ. Microbiol.* **11**, 4043–4050.

Brodisch, K.E.U. and Joyner, S.J. (1983). The role of microorganisms other than *Acinetobacter* in biological phosphate removal in activated sludge processes. *Water Sci. Technol.* **15**, 117–122.

Buchan, L. (1983). Possible biological mechanism of phosphorus removal. *Wat. Sci. Tech.* **15**, 87–103.

Buchan, L. (1984). Microbiological aspects. In: H.N.S. Wiechers, G.A. Ekama and A. Gerber (eds) *Theory, design and operation of nutrient removal activated sludge processes*, pp. 9.1–9.6. Water Research Commission, Pretoria, South Africa.

Bux, F., Swalaha, F.M. and Kasan, H.C. (1994). Microbiological transformation of metal contaminated effluents. *Water Research Commission Report No. 357/1/94.* .

Carberry, J.B. and Tenny, M.W. (1973). Luxury uptake of phosphate by activated sludge. *J. Wat. Control Fed.* **45**, 2444–2462.

Chutter, F.M. (1990). Evaluation of the impact of the 1 $mgP.l^{-1}$ phosphate P standard on the water quality and trophic status of Hartbeespoort Dam. *Wat. Sew. Eff.* **10**, 29–33.

Cech, J.S. and Hartman, P. (1993). Competition between polyphosphate and polysaccharide accumulating bacteria in enhanced biological phosphate removal systems. *Wat. Res.* **27**, 1219–1225.

Cech, J.S., Hartman, P. and Macek, M. (1994). Bacteria and protozoa populations dynamics in biological phosphate removal systems. *Wat. Sci. Tech.* **29**, 109–117.

Cloete, T.E. and Muyima, N.Y.U. (1997). *Microbial Community Analysis: The key to the design of biological wastewater treatment systems.* International Association on Water Quality, Cambridge.

Cloete, T.E., Steyn, P.L. and Buchan, L. (1985). An autecological study of *Acinetobacter* in activated sludge. *Wat. Sci. Tech.* **17**, 139–146.

Cloete, T.E., Steyn, P.L. and Buchan, L. (1985). An autecological study of *Acinetobacter* in activated sludge. *Environ. Technol. Lett.* **5**, 457–463.

Cloete, T.E. and Steyn, P.L. (1987). A combined fluorescent antibody-membrane filter technique for enumerating *Acinetobacter* in activated sludge. In: R. Ramadori (ed.) *Advances in Water Pollution Control, Biological Phosphate Removal from Wastewaters*, pp. 335–338. Pergamon Press, Oxford.

Cloete, T.E. and Steyn, P.L. (1988a). The role of *Acinetobacter* as a phosphorus removing agent in activated sludge. *Wat. Res.* **22**, 971–976.

Cloete, T.E. and Steyn, P.L. (1988b). A combined membrane filter immunofluorescent technique for the *in situ* identification and enumeration of *Acinetobacter. Wat. Res.* **22**, 961–969.

Collins, M.D. and Jones, D. (1981). Distribution of isoprenoid quinone structural types in bacteria and their taxonomic implications. *Microbiol. Rev.* **45**, 316–354.

Comeau, Y., Hall, K.L., Hancock, R.E.W. and Oldham, W.K. (1986). Biochemical model for enhanced biological phosphorus removal. *Wat. Res.* **20**, 1511–1521.

Deinema, M.H., Habets, L.H.A., Scholten, J. *et al.* (1980). The accumulation of polyphosphate in *Acinetobacter* spp. *FEMS Microbiol. Lett.* **9**, 275–279.

Dillon, P.J. and Molot, L.A. (1996). Long-term phosphorus budgets and an examination of the steady-state mass balance model for central Ontario lakes. *Wat. Res.* **20**, 2273–2280.

Dold, P.L., Ekama, G.A. and Marais, GvR. (1980). A general model for the activated sludge process. *Prog. Water Technol.* **12**, 47–77.

Drenckhahn, D., Jöns, T. and Schmitz, F. (1993). Production of polyclonal antibodies against proteins and peptides. *Meth. Cell Biol.* **37**, 7–56.

Ehlers, M.M. and Cloete, T.E. (1999a). Comparing the protein profiles of 21 different activated sludge systems after SDS-PAGE. *Wat. Res.* **33**, 1181–1186.

Ehlers, M.M. and Cloete, T.E. (1999b). Direct extraction of proteins to monitor an activated sludge system on a weekly basis for 34 weeks using SDS-PAGE. *Water SA* **25**, 57–62.

Ehlers, M.M. and Cloete, T.E. (1999c). Protein profiles of phosphorus-and nitrate-removing activated sludge system. *Water SA* **25**, 351–356.

Ekama, G.A., Marais, G.V.R. and Siebritz, I.P. (1984). Biological excess phosphate removal. In: H.N.S. Wiechers, G.A. Ekama and A. Gerber (eds) *Theory, design and operation of nutrient removal activated sludge processes*, pp. 7.1–7.32. Water Research Commission, Pretoria, South Africa.

Ellis, R.J., Thompson, I.P. and Bailey, M.J. (1995). Metabolic profiling as a means of characterizing plant-associated microbial communities. *FEMS Microbiol. Ecol.* **16**, 9–18.

Engelen, B., Meinken, K., Von Wintzingerode, F. *et al.* (1998). Monitoring impact of a pesticide treatment on bacterial soil communities by metabolic and genetic fingerprinting in addition to conventional testing procedures. *Appl. Environ. Microbiol.* **64**, 2814–2821.

Frederickson, J.K., Balkwill, D.L., Zachara, J.M. *et al.* (1991). Physiological diversity and distributions of heterotrophic bacteria in deep cretaceous sediments of the Atlantic coastal plain. *Appl. Environ. Microbiol.* **57**, 402–411.

Fuhs, G.W. and Chen, M. (1975). Microbiological basis of phosphate removal in the activated sludge process for the treatment of wastewater. *Microb. Ecol.* **2**, 119–138.

Garland, J.L. (1996a). Analytical approaches to the characterization of samples of microbial communities using patterns of potential C source utilization. *Soil. Biol. Biochem.* **28**, 213–221.

Garland, J.L. (1996b). Patterns of potential C source utilization by rhizosphere communities. *Soil. Boil. Biochem.* **28**, 223–230.

Garland, J.L. (1997). Analysis and interpretation of community-level physiological profiles in microbial ecology. *FEMS Microbiol. Ecol.* **24**, 289–300.

Garland, J.L. and Mills, A.L. (1991). Classification and characterization of heterotrophic microbial communities on the basis of patterns of community-level sole-carbon-source utilization. *Appl. Environ. Microbiol.* **57**, 2351–2359.

Gerber, A., Mostert, E.S., Winter, C.T. and de Villiers, R.H. (1986). The effect of acetate and other short-chain carbon compounds on the kinetics of biological nutrient removal. *Water SA.* **12**, 7–12.

Glimm, E., Heuer, H., Engelen, B. *et al.* (1997). Statistical comparisons of community catabolic profiles. *J. Microbiol. Methods.* **30**, 71–80.

Gorlenko, M.V. and Kozhevin, P.A. (1994). Differentiation of soil microbial communities by multisubstrate testing. *Microbiology* **63**, 158–161.

Grady, C.P.L. and Lim, H.C. (1980). Activated sludge. In: P.N. Cheremisinoft (ed.) *Biological wastewater treatment theory and applications*. Marcel Dekker (Inc.), New York.

Grady, C.P.L. and Lim, H.C. (1980). *Biological wastewater treatment–theory and application*. Marcel Dekker, New York and Basel.

Gray, N.F. (1989). *Biology of Wastewater Treatment*. Oxford University Press, Oxford.

Haack, S.K., Garchow, H., Klug, M.J. and Forney, L.J. (1994). Analysis of factors affecting the accuracy, reproducibility, and interpretation of microbial community carbon source utilization patterns. *Appl. Environ. Microbiol.* **61**, 1458–1468.

Hart, M.A. and Melmed, L.N. (1982). Microbiology of nutrient removing activated sludge. *Wat. Sci. Tech.* **14**, 1501–1502.

Heuer, H., Hartung, K., Engelen, B. and Smalla, K. (1995). Studies on microbial communities associated with potato plants by Biolog and TGGE patterns. *Med. Fac. Landbouww. Univ. Gent.* **60**, 2639–2645.

Hiraishi, A., Masamune, K. and Kitamura, H. (1989). Characterization of the bacterial population structure in an anaerobic-aerobic activated sludge system on the basis of respiratory quinone profiles. *Appl. Environ. Microbiol.* **55**, 897–901.

Hiraishi, A., Morishima, Y. and Takeuchi, J.-I. (1991). Numerical analysis of lipoquinone patterns in monitoring bacterial community dynamics in wastewater treatment systems. *J. Gen. Appl. Microbiol.* **37**, 57–70.

Hiraishi, A., Ueda, Y. and Ishihara, J. (1998) Biomarker and molecular approaches to the microbial community analysis of activated sludge. *Proceedings of Microbial Community and Functions in Wastewater Treatment Processes*. The International Symposium of The Centre of Excellence, Department of Urban Engineering, School of Engineering, The University of Tokyo. 10–11 March, 1998.

Horan, N.J. (1990). *Biological Wastewater Treatment Systems: Theory and Operations*. John Wiley and Sons, Chichester.

Insam, H., Amor, K., Renner, M. and Crepaz, C. (1996). Changes in functional abilities of the microbial community during composting of manure. *Microb. Ecol.* **31**, 77–87.

Jørgensen, P.E., Eriksen, T. and Jensen, B.K. (1992). Estimation of viable biomass in activated sludge by determination of ATP, oxygen utilization rate and FDA hydrolysis. *Wat. Res.* **26**, 1495–1501.

Kämpfer, P., Erhart, R., Beimfohr, C. *et al.* (1996). Characterization of bacterial communities from activated sludge: culture-dependent numerical identification versus *in situ* identification using group- and genus-specific rRNA-targeted oligonucleotide probes. *Microb. Ecol.* **32**, 101–121.

Kavanaugh, R.G. and Randall, C.W. (1994). Bacterial populations in a biological nutrient removal plant. *Wat. Sci. Technol.* **29**, 25–34.

Kawaharasaki, M., Tanaka, H., Kanagawa, T. and Nakamura, K. (1999). *In situ* identification of polyphosphate accumulating bacteria in activated sludge by dual staining with rRNA-targeted oligonucleotide probes and 4′,6-diamidino-2-phenylindol (DAPI) at a polyphosphate-probing concentration. *Wat. Res.* **33**, 257–265.

Keay, G.F.P. (1984). Practical design consideration. In: H.N.S. Wiechers, G.A. Ekama and A. Gerber (eds)

Theory, design and operation of nutrient removal activated sludge processes, pp. 10.1–10.12. Water Research Commission, Pretoria, South Africa.

Kersters, K. and De Ley, J. (1975). Identification and grouping of bacteria by numerical analysis of their electrophoretic protein patterns. *J. Gen. Microbiol.* **87**, 333–342.

Kersters, I., Van Vooren, L., Verschuere, L. *et al.* (1997). Utility of the Biolog system for the characterization of heterotrophic microbial communities. *System. Appl. Microbiol.* **20**, 439–447.

Klinger, J.M., Stowe, R.P., Obenhuber, D.C. *et al.* (1992). Evaluation of the Biolog automated microbial identification system. *Appl. Environ. Microbiol.* **58**, 2089–2092..

Lawson, E.N. and Tonhazy, N.E. (1980). Changes in morphology and phosphate uptake patterns of *Acinetobacter calcoaceticus* strains. *Water SA.* **6**, 105–112.

Lehman, R.M., Colwell, F.S., Ringelberg, D.B. and White, D.C. (1995). Combined microbial community-level analysis for quality assurance of terrestrial subsurface cores. *J. Microbiol. Methods.* **22**, 263–281.

Levin, G.V. and Shapiro, J. (1965). Metabolic uptake of phosphorus by wastewater organisms. *J. Water Pollut. Control Fed.* **37**, 800–821.

Lilley, I.D., Pybus, P.J. and Power, S.P.B. (1997) Operating manual for biological nutrient removal wastewater treatment works. *Water Research Commission Report No. TT 83/97.*

Lötter, L.H. (1985). The role of bacterial phosphate metabolism in enhanced phosphorus removal from the activated sludge process. *Wat. Sci. Tech.* **17**, 127–138.

Lötter, L.H. and Murphy, M. (1985). The identification of heterotrophic bacteria in an activated sludge plant with particular reference to polyphosphate accumulation. *Water SA.* **11**, 179–184.

Lötter, L.H., Wentzel, M.C., Loewenthal, R.E. *et al.* (1986). A study of selected characteristics of *Acinetobacter* spp. isolated from activated sludge in anaerobic/anoxic/aerobic and aerobic systems. *Water SA* **12**, 203–208.

Lui, W.-T., Marsh, T.L. and Forney, L.J. (1997) Determination of the microbial diversity of anaerobic-aerobic activated sludge by a novel molecular biological technique. *Proceedings Second International Conference on Microorganisms in Activated Sludge and Biofilm Processes.* IAWQ, 21–23 July, 1997, Berkeley, California.

Manz, W., Amann, R., Ludwig, W. *et al.* (1992). Phylogenetic oligonucleotide probes for the major subclasses of Proteobacteria: Problems and solutions. *System. Appl. Microbiol.* **15**, 593–600.

Maszenan, A.M., Seviour, R.J., Patel, B.K.C. *et al.* (1998). The hunt for the G-bacteria in activated sludge biomass. *Wat. Sci. Tech.* **37**, 65–69.

Marais, GvR., Loewenthal, R.E. and Siebritz, I.P. (1983). Observations supporting phosphate removal by biological excess uptake–a review. *Wat. Sci. Tech* **15**, 15–41.

Momba M.N.B. (1995) Phosphate removal in activated sludge and its relationship to biomass. M Sc. Thesis. University of Pretoria, Pretoria, South Africa.

Momba, M.N.B. and Cloete, T.E. (1996). The relationship of biomass to phosphate uptake by *Acinetobacter junii* in activated sludge mixed liquor. *Wat. Res.* **30**, 364–370.

Momba, M.N.B. and Cloete, T.E. (1996). Biomass relationship to growth and phosphate uptake of *Pseudomonas fluorescens*, *Escherichia coli* and *Acinetobacter radioresistens* in mixed liquor medium. *J. Ind. Microbiol.* **16**, 364–369.

Mulder, J.W. and Rensink, J.H. (1987). Introduction of biological phosphorus removal to an activated sludge plant with practical limitations. In: R. Ramadori (ed.) *Biological phosphate removal from wastewaters. Advances in water pollution control*, pp. 213–223. Pergamon Press, Oxford.

Muyima N.Y.O. (1995) Enhanced biological phosphate removal by immobilized *Acinetobacter* and activated sludge microbial populations. PhD Thesis. University of Pretoria, Pretoria, South Africa.

Muyima, N.Y.O., Momba, M.N.B. and Cloete, T.E. (1997). Biological methods for the treatment of wastewaters. In: T.E. Cloete and N.Y.O. Muyima (eds) *Microbial Community Analysis: The key to the design of biological wastewater treatment systems.* International Association on Water Quality, Cambridge.

Nakamura, K., Kawaharasaki, M., Hanada, S. *et al.* (1998) Characterization of bacterial community constructing anaerobic/aerobic activated sludge by *in situ* identification and cloning of 16S rDNA's. *Proceedings of Microbial Community and Functions in Wastewater treatment Processes.* The International Symposium of the Centre of Excellence, Department of Urban Engineering, School of Engineering, University of Tokyo. 10–11 March, 1998.

Nicholls, H.A. and Osborn, D.W. (1979). Bacterial stress: prerequisite for biological removal of phosphorus. *J. Wat. Pollut. Control Fed.* **51**, 557–569.

Nielsen, P.H., Andreasen, K., Lee, N. *et al.* (1998) Autoradiography for the *in situ* analysis of microbial community structure in wastewater processes. *Proceedings of Microbial Community and Functions in Wastewater treatment Processes.* The International Symposium of the Centre of Excellence, Department of Urban Engineering, School of Engineering, University of Tokyo. 10–11 March, 1998.

Nielsen, P.H., Andreasen, K., Lee, N. and Wagner, M. (1999). Use of microautoradiography and fluorescent *in situ* hybridization for characterization of microbial activity in activated sludge. *Wat. Sci. Tech.* **39**, 1–9.

Ogunseitan, O.A. (1993). Direct extraction of proteins from environmental samples. *J. Microbiol. Methods.* **17**, 273–281.

Pitman, A.R. (1984). Operation of biological nutrient removal plants. In: H.N.S. Wiechers, G.A. Ekama and A. Gerber *et al.* (eds) *Theory, design of nutrient removal activated sludge processes*, pp. 11.1–11.16. Water Research Commission, Pretoria, South Africa.

Porter, K.G. and Feig, Y. (1980). The use of DAPI for identifying and counting aquatic microflora. *Limnol. Oceanogr.* **25**, 943–948.

Rustrian, E., Delgenes, J.P. and Moletta, R. (1997). Phosphorus release by pure cultures of *Acinetobacter* spp.:

effect of the growth stage with cells cultivated on various carbon sources. *Lett. Appl. Microbiol.* **24**, 144–148.

Rudd, R.T. (1979). The necessity for the promulgation of standards for the limitation of nutrients in effluents in sensitive areas in South Africa. In *Nutrient Removal from Municipal Effleunts*, Technology Transfer Seminar, Pretoria, 17th May, 1979.

Sandén, B., Grunditz, C., Hansson, Y. and Dalhammar, G. (1994). Quantification and characterization of *Nitrosomonas* and *Nitrobacter* using monoclonal antibodies. *Wat. Sci. Tech.* **29**, 1–6.

Shapiro, J. (1967). Induced rapid release and uptake of phosphate by microorganisms. *Science* **155**, 1269–1271.

Snaidr, E., Amann, R., Huber, I. *et al.* (1997). Phylogenetic analysis and *in situ* identification of bacteria in activated sludge. *Appl. Environ. Microbiol.* **63**, 2884–2896.

Srinath, E.G., Sastry, C.A. and Pillai, S.C. (1959). Rapid removal of phosphorus from sewage by activated sludge. *Experientia* **15**, 339–340.

Streichan, M., Golecki, J.R. and Schon, G. (1990). Polyphosphate accumulating bacteria from sewage plants with different processes for biological phosphorus removal. *FEMS Microbiol. Ecol.* **73**, 113–124.

Toerien, D.F., Gerber, A., Lötter, L.H. and Cloete, T.E. (1990). Enhanced biological phosphorus removal in activated sludge systems. *Adv. Microb. Ecol.* **11**, 173–230.

Ubisi, M.F., Jood, T.W., Wentzel, M.C. and Ekama, G.A. (1997). Activated sludge mixed liquor heterotrophic active biomass. *Water SA* **23**, 239–248.

Ubukata, Y. and Takii, S. (1998) Some physiological characteristics of phosphate removing bacterium, *Microlunatus phosphovorus*, and a simplified isolation and identification method for phosphate removing bacteria. *Proceedings (Book 1) Water Quality International 1998*. IAWQ 19th Biennial International Conference, 21–26 June, 1998, Vancouver, Canada.

Vahjen, W., Munch, J.C. and Tebbe, C.C. (1995). Carbon source utilization of soil extracted microorganisms supplemented with genetically engineered and non-engineered *Corynebacterium glutamicum* and a recombinant peptide at the community level. *FEMS Microbiol. Ecol.* **18**, 317–328.

Verniere, C., Pruvost, O., Civerolo, E.L. *et al.* (1993). Evaluation of the Biolog substrate utilization system to identify and assess metabolic variation among strains of *Xanthomonas campestris* pv. *citri*. *Appl. Environ. Microbiol.* **59**, 243–249.

Victorio, L., Gilbride, K.A., Allen, D.G. and Liss, S.N. (1996). Phenotypic fingerprinting of microbial communities in wastewater treatment systems. *Wat. Res.* **30**, 1077–1086.

Wagner, M. and Amann, R. 1997. Molecular techniques for determining microbial community structure in activated sludge. In: T.E. Cloete and N.Y.O. Muyima (eds) *Microbial Community Analysis: The Key to the Design of Biological Wastewater Treatment Systems*, pp. 61–72. IAWQ Scientific and Technical Report No. 5. IAWQ, London.

Wagner, M., Amann, R., Lemmer, H. and Schleifer, K. (1993). Probing activated sludge with oligonucleotides specific for proteobacteria: Inadequacy of culture-dependent methods for describing microbial community structure. *Appl. Environ. Microbiol.* **59**, 1520–1525.

Wagner, M., Amann, R., Lemmer, H. *et al.* (1994a). Probing activated sludge with fluorescently labelled rRNA targeted oligonucleotides. *Wat. Sci. Technol.* **29**, 15–23.

Wagner, M., Erhart, R., Manz, W. *et al.* (1994b). Development of an rRNA-targeted oligonucleotide probe specific for the genus *Acinetobacter* and its application for *in situ* monitoring in activated sludge. *Appl. Environ. Microbiol.* **60**, 792–800.

Walmsley, R.D. and Thornton, J.A. (1984). Evaluation of OECD-type phosphorus eutrophication models for predicting the trophic status of southern African man-made lakes. *SAJS* **80**, 257–259.

Wanner, J. 1997. Microbial population dynamics in biological wastewater treatment plants. In: T.E. Cloete and N.Y.O. Muyima (eds) *Microbial Community Analysis: The Key to the Design of Biological Wastewater Treatment Systems*, pp. 35–59. IAWQ Scientific and Technical Report No. 5. IAWQ, London.

Wentzel, M.C., Dold, P.L., Ekama, G.A. and Marais, GvR. (1985). Kinetics of biological phosphorus release. *Wat. Sci. Technol.* **17**, 57–71.

Wentzel, M.C., Lötter, L.H., Loewenthal, R.E. and Marais, GvR. (1986). Metabolic behaviour of *Acinetobacter* spp. in enhanced biological phosphorus removal–A biochemical model. *Water SA* **12**, 209–224.

Wentzel, M.C., Loewenthal, R.E., Ekama, G.A. and Marais, GvR. (1988). Enhanced polyphosphate organism cultures in activated sludge systems–Part I: Enhanced culture development. *Water SA* **14**, 81–92.

Wentzel, M.C., Ekama, G.A., Loewenthal, R.E. *et al.* (1989). Enhanced polyphosphate organism cultures in activated sludge systems–Part II: Experimental behaviour. *Water SA* **15**, 71–88.

Wetzel, R.G. (1983). *Limnology*, 2nd edn. CBS College Publishing, New York.

Winding, A. (1994). Fingerprinting bacterial soil communities using Biolog microtiter plates. In: K. Ritz, K.E. Dighton and K.E. Giller (eds) *Beyond the biomass: compositional and functional analysis of soil microbial communities*. Wiley, Chichester.

Woese, C.R. (1987). Bacterial evolution. *Microbiol. Rev.* **51**, 221–271.

Woese, C.R., Kandler, O. and Wheelis, M.L. (1990). Towards a natural system of organisms: proposal for the domains Archaea, Bacteria and Eukarya. *Proc. Natl. Acad. Sci. USA* **87**, 4576–4579.

Wünsche, L. and Babel, W. (1995). The suitability of the Biolog automated microbial identification system for assessing the taxonomical composition of terrestrial bacterial communities. *Microbiol. Res.* **151**, 133–143.

Zak, J.C., Willig, M.R., Moorhead, D.L. and Wildman, H.G. (1994). Functional diversity of microbial communities: a quantitative approach. *Soil. Biol. Biochem.* **26**, 1101–1108.

24

Anaerobic treatment processes

Ken Anderson*, Paul Sallis* and Sinan Uyanik†

*Department of Civil Engineering, University of Newcastle-upon-Tyne, UK; †Environmental Engineering Department, Harran University, 63300 Yenisehir, Sanlýurfa, Turkey

1 INTRODUCTION

The term anaerobic digestion is used to describe many different anaerobic waste treatment processes. The basic principle common to them all is the fact that the biological degradation of the organic components in the waste is achieved with no requirement for molecular oxygen (air). In most cases, anaerobic digestion processes are also methanogenic, i.e. most of the carbon atoms originating in the waste material are reduced to methane (CH_4), the ultimate product of biological metabolism in anaerobic environments. The economic value of methane produced by anaerobic digestion can be a major consideration in the selection of this treatment technology, allowing significant levels of energy recovery for space heating, lighting and power generation. Although the anaerobic degradation of organic pollutants may be achieved without methane formation, e.g. when denitrifying and sulphate-reducing bacteria utilize nitrate and sulphate as terminal electron acceptors, the term anaerobic digestion generally refers to the methanogenic process in which carbon dioxide acts as the terminal electron acceptor.

The feasibility of anaerobic digestion has been demonstrated with many different types of waste-stream, however, only a relatively small proportion of these have shown consistent advantages over aerobic treatment technologies when implemented at the commercial scale. As a result, the range of proven applications is presently small compared to the number of potential applications and anaerobic digestion is generally restricted to the treatment of high strength wastes having a relatively consistent composition. These include food and paper processing wastewaters, sewage sludge and farm manures, crop residues, and several other specific industrial wastewaters. The maximum organic loading rate employed in these cases is often in excess of an order of magnitude higher than that in an equivalent aerobic process. Furthermore, many organic pollutants, such as pesticides, halogenated compounds and the complex chemical cocktails present in landfill leachates, are not particularly amenable to aerobic degradation pathways but may be readily treated anaerobically.

As an emerging technology, the success of anaerobic digestion was initially hampered by unsatisfactory reactor designs that had been inherited from conventional (aerobic) processes. These failed to optimize treatment efficiency and were often the cause of process failure. Recent developments in anaerobic reactor design have been instrumental in allowing the full potential of anaerobic digestion to be achieved and these are discussed later in more detail.

Until recently, the complex microbial ecology of the sludge (biomass) in anaerobic digestion systems was poorly understood, and this has undoubtedly played its part in the slow move towards the exploitation of this technology in effluent treatment. These issues have now been fully elucidated over the last two decades which has allowed modern full-scale treatment plants to be designed, monitored and controlled with a high degree of confidence in their performance characteristics.

This chapter begins with an overview of the microbiology and biochemistry unique

The Handbook of Water and Wastewater Microbiology
ISBN 0-12-470100-0

to the anaerobic digestion process before describing the key physiological and environmental factors that affect microbial activity and system performance. The final section outlines the range of reactor designs currently employed in the anaerobic digestion of wastewaters.

2 MICROBIOLOGY OF ANAEROBIC DIGESTION

Methane production is the final step in a cascade of biochemical reactions taking place within anaerobic digesters treating the organic components of sludges and wastewaters. However, no individual microorganism is capable of carrying out all of these reactions independently. Consequently, anaerobic treatment processes are complex ecosystems comprising several diverse microbial guilds (organisms that have the same specialist role in their community) that work together in a coordinated manner to convert the organic components to methane and carbon dioxide. Early concepts of anaerobic digestion recognized that the transformation process would involve at least two stages, initial acidification of the complex organic matter followed by gas formation from simpler intermediates (Fig. 24.1).

As a greater understanding of the process developed, alternative new schemes were proposed that detailed both substrate conversion and flux in the conversion of organic materials to methane and carbon dioxide. A well-accepted scheme is presented in Fig. 24.2 (Gujer and Zehnder, 1983) and according to this, six distinct processes may be identified in an anaerobic digester:

1. Hydrolysis of biopolymers:
   (i) hydrolysis of proteins
   (ii) hydrolysis of carbohydrates
   (iii) hydrolysis of lipids
2. Fermentation of amino acids and sugars
3. Anaerobic oxidation of long chain fatty acids and alcohols
4. Anaerobic oxidation of intermediary products such as volatile acids, except acetate
5. Conversion of acetate to methane
6. Conversion of hydrogen to methane.

Since none of the above steps occurs through spontaneous chemical reactions to any significant degree, the individual steps are catalyzed by separate groups of microorganisms, each having its own specific role within the overall process. Such microbial consortia effect complete breakdown of complex organic substrates though series metabolism, a process involving the sequential flow of substrates and products from one microbial guild to another. The following classification illustrates the bacterial consortium of anaerobic digestion and highlights the relationships between each of the bacterial groups concerned.

The first microbial group is composed of hydrolytic bacteria. These break down lipids, complex polymeric molecules (e.g. protein and carbohydrate) and particulate organic matter into simpler soluble components such as short-chain fatty acids, glycerol, peptides, amino acids, oligosaccharides and sugars.

The second group is designated the acid-forming bacteria and consists of both acidogenic (organic acid forming) and acetogenic (acetate forming) bacteria. These acid-forming bacteria convert the end-products from the first group into both the key substrates of methanogenesis, namely acetate, hydrogen, carbon dioxide, and a number of minor intermediary products such as formate, propionate, butyrate, valerate etc.

The sequence is completed by the third group of bacteria, the methanogens, which consume the end-products from the second group of bacteria and convert them to the final

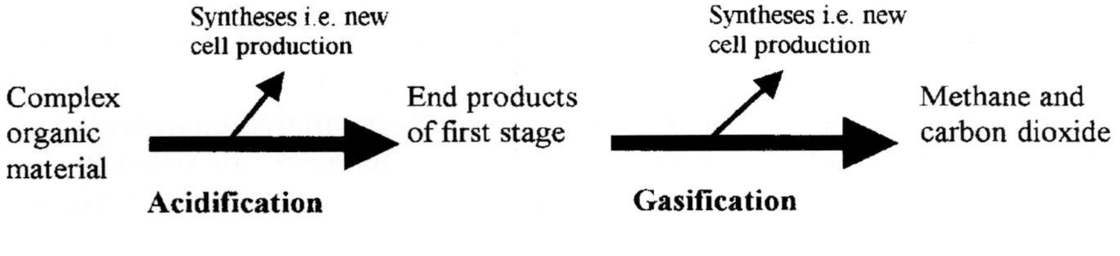

Fig. 24.1 Simple schematic representation of the anaerobic digestion process.

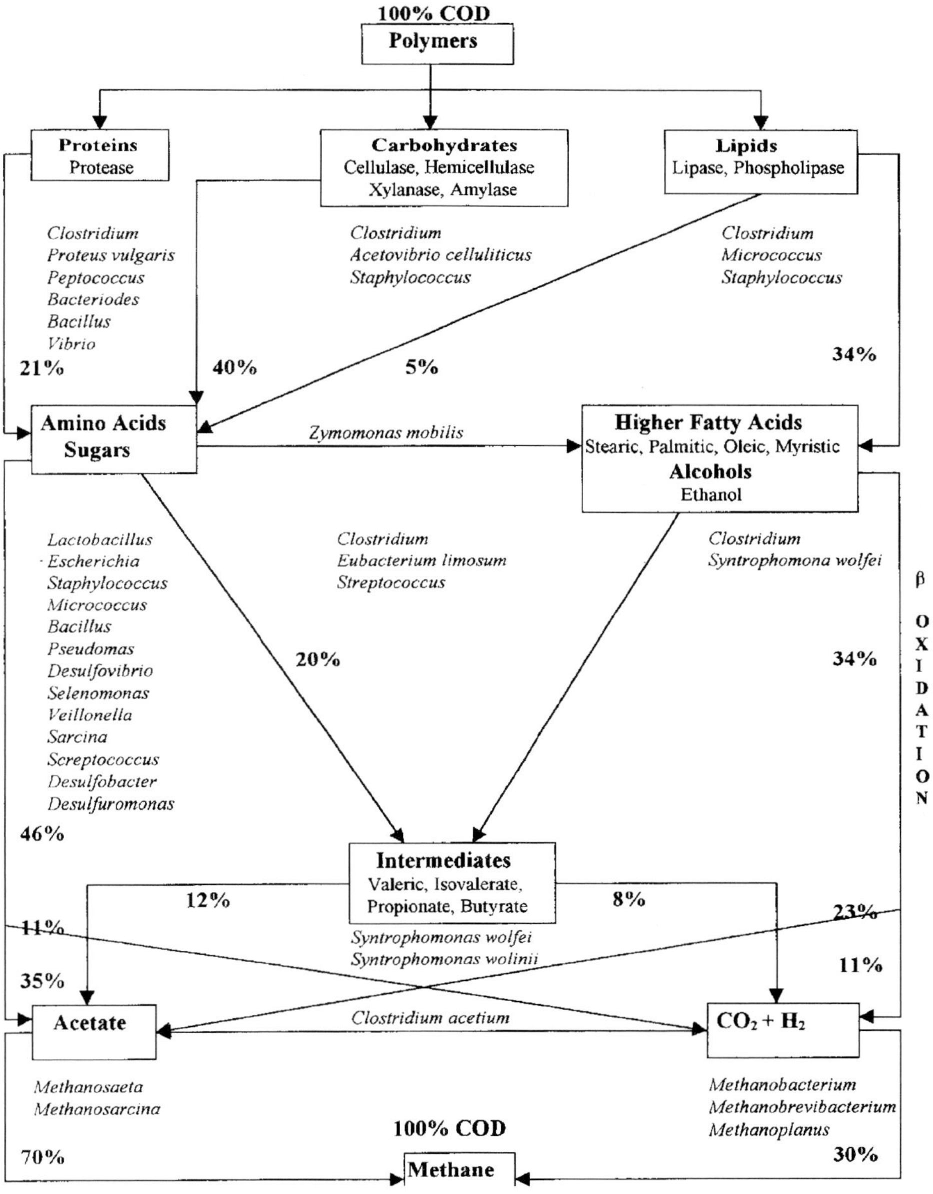

Fig. 24.2 Carbon flow to methane in anaerobic digesters with the microorganisms responsible for each step. Adapted from Gujer and Zehnder, 1983.

end-products (methane and carbon dioxide). The methanogens comprise two physiologically distinct groups of methane-forming bacteria, namely acetoclastic methanogenic bacteria and hydrogen-utilizing methanogenic bacteria. Each of the stages outlined above will now be described in more detail.

2.1 Hydrolytic bacteria

Since methanogenic and acetogenic bacteria are generally incapable of utilizing complex polymeric substrates directly, these materials must first be broken down into soluble monomers before the process of methanogenesis can proceed. Consequently, hydrolysis is the first essential step in the anaerobic degradation of complex biopolymers. A number of extracellular hydrolytic enzymes that are capable of initiating the attack on these complex substrates may be produced within anaerobic digesters by hydrolytic genera such as *Clostridium, Peptococcus, Vibrio, Micrococcus* and *Bacillus*. These include protease, lipase, cellulase, pectinase, amylase, chitinase etc. and their relative composition and activity will reflect the prevalence of their respective substrates in the digester feed. Being extracellular, these enzymes are able to access large substrate molecules that are incapable of crossing the bacterial cell wall due to their size. Anaerobic digesters contain between 10^8–10^9 hydrolytic bacteria per ml comprising both facultative and obligate anaerobes.

In addition to bacteria, protozoa and fungi have been observed in anaerobic digesters; however, on the basis of observed numbers, protozoa are not thought to have an important role in the anaerobic digestion process. The protozoa that have been observed include flagellates belonging to the genera *Trepomonas, Tetramitus* and *Trichomonas*, amoebae belonging to genera *Vahlkampfia* and *Hartmanella*, and ciliates belonging to the genera *Metopus, Trimyema* and *Saprodinium*.

On account of similarly low cell densities, fungi are also believed to play a relatively minor role within anaerobic digesters and much of the recorded fungal biomass probably enters in the digester feed. However, fungi have been shown to be capable of reproduction within operational digesters and to this extent are taking part in the digestion process by consuming nutrients for growth. Genera of fungi observed include *Phycomycetes, Ascomycetes* and fungi imperfecti.

2.2 Acid-forming bacteria

The monomers produced by the hydrolytic bacteria during the first stage of the digestion process are fermented during the second acid-forming stage to produce several intermediate products, namely acetate, propionate, butyrate and hydrogen. Each acidic end-product contains a carboxylic acid group that was introduced during the fermentation, and hence the microorganisms responsible for these fermentations are collectively called the acid-forming bacteria. There are two groups of acid-forming bacteria, namely the acidogenic bacteria and the acetogenic bacteria.

2.2.1 *Acidogenic bacteria*

Acidogens provide important substrates for acetogens and methanogens. They metabolize amino acids and sugars to the intermediary products, acetate, hydrogen and carbon dioxide. The acidogenic stage includes many different fermentative genera and species; among them are *Clostridium, Bacteroides, Ruminococcus, Butyribacterium, Propionibacterium, Eubacterium, Lactobacillus, Streptococcus, Pseudomonas, Desulfobacter, Micrococcus, Bacillus* and *Escherichia.* The facultative members of this group also help protect the oxygen-sensitive methanogens by consuming traces of oxygen that may enter in the feed. Typical cell counts of acidogens in anaerobic digesters range from about 10^6–10^8 per ml (Archer and Kirsop, 1990).

2.2.2 *Acetogenic bacteria*

The main function of acetogenic bacteria in anaerobic digestion is the production of acetate, carbon dioxide and hydrogen, as these are the only substrates that can be metabolized efficiently by the methanogens in the final stage of anaerobic digestion. Two distinct groups of acetogenic bacteria can be distinguished on the basis of their metabolism.

The first group, the obligate hydrogen-producing acetogens (OHPA), also called proton-reducing acetogens, produce acetic acid, carbon dioxide and hydrogen from the major fatty acid intermediates (propionate and butyrate), alcohols and other higher fatty acids (valerate, isovalerate stearate, palmitate and myristate via β-oxidation). OPHA species are particularly important in the β-oxidation of longer-chain fatty acids arising from lipid hydrolysis and are also involved in the anaerobic degradation of aromatic compounds.

Thermodynamic considerations of the free energy (ΔG) available from fatty acid oxidation and hydrogen production, predict that the OHPA will be capable of growth only in environments that maintain a low concentration of the metabolic product hydrogen. Such environments will exist where hydrogen-consuming species like methanogens are able to thrive. This mutualistic interaction between OHPA and hydrogen-removing species (e.g. methanogens and sulphate-reducing bacteria) has been termed *syntrophy*, literally 'eating together', and is a critical requirement for efficient methanogenic anaerobic digestion. The syntrophic relationships in a working anaerobic digester are held in a fairly fragile state of equilibrium because even small perturbations may lead to positive feedback of the inhibitory effects. Specifically, methanogens are inhibited by fatty acids (substrates of the OHPA) and the OHPA are inhibited by hydrogen (a substrate of the methanogens); any significant increase in the level of either of these substrates will eventually lead to the inhibition of both groups of bacteria (Fig. 24.3).

Most methanogenic environments maintain a hydrogen concentration below 10^{-4} atm, which is sufficiently low to stimulate the OHPA and prevent the accumulation of fatty acids. So far, only a limited number of OPHA species have been isolated and identified, namely *Syntrophomonas wolfei* and *Syntrophobacter wolinii*, which oxidize butyrate and propionate, respectively. Mesophilic sludges contain approximately 4.5×10^6 *S. wolfei* per gram of digester sludge (Toerien and Hattingh, 1969).

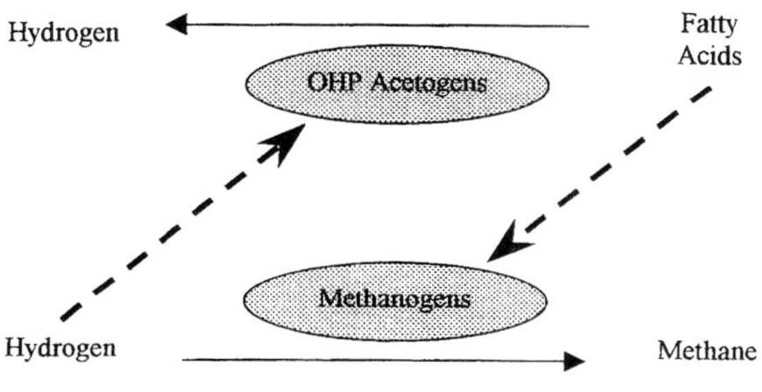

Fig. 24.3 Schematic representation of feedback inhibition (broken lines) among syntrophic bacteria (methanogens) and obligate hydrogen-producing (OHP) acetogens. Substrates and products are shown incompletely.

The second group of acetogenic bacteria are the homoacetogens, which are strictly anaerobic microorganisms catalyzing the formation of acetate from hydrogen and carbon dioxide. Homoacetogens are known in the genera *Acetobacterium, Acetoanaerobium, Acetogenium, Butribacterium, Clostridium* and *Pelobacter.* Homoacetogenic bacteria are also syntrophs because they participate in the interspecies hydrogen transfer process which maintains the low hydrogen concentrations required by the OHPA. However, their importance in this respect, relative to that of the methanogens, is still not clear although the number of homoacetogenic bacteria in anaerobic digesters is considerably lower, at around 1×10^5 per ml (Toerien and Hattingh, 1969), than that of the methanogens, suggesting a relatively minor role.

2.3 Methanogenic bacteria

The methanogens are strict anaerobes and form methane gas as the end-product of their metabolism. They are known to be truly distinct from the typical bacteria (*Eubacteria*) and are classified in a separate kingdom, the *Archaea*. A recent taxonomic scheme of methanogens is given in Table 24.2. They are the key organisms in the production of methane from acetate, hydrogen and carbon dioxide. Without methanogens, the ultimate breakdown of an organic material would not take place due to the accumulation of the end-products of the acid-producing bacteria. Importantly, methane production is considered to be the slowest (rate limiting) step in the anaerobic digestion process. Furthermore, since the methanogens are most active in the pH range of 6.5–8.0, they will be sensitive in environments poorly buffered against acidification caused by the products of the acidogenic and acetogenic bacteria. A limited range of substrates can be utilized by methanogens with acetate, hydrogen and carbon dioxide being the most important. According to their substrate specificity, methanogens are divided into two groups: acetoclastic methanogens and hydrogen-utilizing methanogens. The methanogenic population in anaerobic digesters is typically present at a level of 1×10^6 to 1×10^8 per ml (Toerien and Hattingh, 1969).

2.3.1 *Acetoclastic methanogens*

Of the various end-products produced by acid-forming bacteria, acetate is regarded as the most important precursor of methane production and the source of up to 70% of methane evolved in digesters. In spite of this fact, only two methanogenic genera contain species that are able to utilize acetate (acetoclastic), and these are *Methanosaeta* (formerly known as *Methanothrix*) and *Methanosarcina*. In addition to this acetoclastic activity, *Methanosarcia* spp. are also capable of using methanol, methylamines and sometimes H_2 and CO_2 as growth substrates, while *Methanosaeta* spp. are restricted to growth only on acetate.

Methanosarcina spp. exhibit a higher maximum specific growth rate (shorter doubling times) on acetate than *Methanosaeta* spp., however, the latter has the higher substrate affinity (lower K_s) for acetate. Consequently, *Methanosaeta* will be the dominant acetoclastic species at acetate concentrations below 1 mM, whereas higher acetate concentrations favour *Methanosarcina* due to its faster growth. With acetate as the growth substrate, doubling times of *Methanosarcina* spp. and *Methanosaeta* spp. are 24 hours and 3.5–9.0 days, respectively. The two genera also exhibit different growth physiology; *Methanosaeta* are filamentous organisms, whereas *Methanosarcina* usually grow in aggregates consisting of large numbers of individual cells, each surrounded by a thick cell wall.

2.3.2 *Hydrogen-utilizing methanogenic bacteria*

A significant quantity of the methane production within anaerobic digesters, up to 30% of the total, is produced by hydrogen-utilizing methanogenic bacteria. These methanogens reduce carbon dioxide, formate, methanol and methylamines, using the hydrogen produced fermentatively by the hydrolytic and acid-forming bacteria earlier in the digestion process. In 1906, Soehngen was the first to record the reduction of CO_2 by hydrogen and the subsequent yield of methane.

When utilizing only hydrogen and carbon dioxide, the methanogens grow as chemolithotrophic autotrophs because they derive both their energy and cellular carbon from inorganic chemicals. Interestingly, such energy metabolism does not involve conventional cytochromes for electron transport. Instead, a complex seven-step process has evolved with specific cofactors, such as coenzyme M (CoM), which are unique to the methanogens. CoM is the smallest coenzyme known and is exceptional in its high sulphur content and acidity. The other implication of using only carbon dioxide or other one-carbon (C_2) substrates for growth is the need to generate two-carbon (C_2) building blocks for anabolic processes. Methanogens achieve this in a manner similar to that of the homoacetogenic bacteria (Zeikus *et al.*, 1985). These issues are discussed further in Section 4.1.

3 LABORATORY METHODS FOR ANAEROBIC BACTERIA

3.1 Isolation and cultivation methods

Most anaerobic bacteria are extremely sensitive to oxygen and therefore require stringent anaerobic techniques. In order to maintain an anaerobic environment during handling and incubation periods, oxygen exclusion techniques should be employed. Holland *et al.* (1987) stated that two types of oxygen exclusion techniques can be used in anaerobic microbiology. The first method employs all possible precautions to ensure that oxygen is excluded from every step of the handling procedure, including sampling, media preparation, transfer and incubation. The second technique is a less systematic approach in which only the incubation step is maintained oxygen-free. These workers also noted the advantages and disadvantages of each method. For example, the former is slow, time consuming and demanding on the operator, however, when performed correctly it is the only method that allows successful handling of the strict anaerobes. The latter is relatively quick and easy to carry out, but has an unacceptably high failure rate with sensitive organisms.

Removal of oxygen by chemical means can be achieved by the addition of oxygen-reactive compounds directly into culture media (e.g. cysteine, sodium sulphides) or into traps within the culture vessel in order to absorb molecular oxygen (e.g. the mixture of pyrogallic acid and sodium hydroxide, the use of iron wool dipped in acidified copper sulphate, the combination of chromium and sulphuric acid, and the burning of yellow phosphorus). However, the use of the latter chemicals can be associated with toxicity and inhibition of anaerobic bacteria and may also represent an unacceptable risk to the worker.

Biological methods for establishing low oxygen concentrations in culture media involve the inoculation of aerobic organisms together with anaerobic bacteria in order to reduce any oxygen to carbon dioxide. However, the aerobic organism should be a non-fermentative strain so as not to interfere with the fermentation reactions of the anaerobic bacteria. Organisms such as *Acinetobacter, Pseudomonas aeruginosa* and *Bacillus subtilis* may be suitable for this purpose.

When more sophisticated physical methods of oxygen exclusion are desired, the need for specialized equipment will be inevitable. Typical equipment includes anaerobic jars and anaerobic cabinets that are filled with an inert gas such as nitrogen or helium to maintain an oxygen-free environment. Such equipment has been widely used in the isolation and detailed study of anaerobic microorganisms.

Exclusion of oxygen from culture media by chemical, biological or physical methods has been detailed by several authors (Willis, 1977, 1990; Holland 1987 *et al.*, Levett, 1990). The methodology for culturing strict anaerobic bacteria in anaerobic jars was first developed at the beginning of this century. The standard anaerobic jars can hold 12 petri dishes and contain a sachet holding 1 g deoxo cold catalyst for oxygen removal in the presence of excess hydrogen. Anaerobic conditions develop in the jar when the air is vacuumed out and replaced with a mixture of hydrogen (90%) and carbon dioxide (10%). It is advisable to put an indicator into the jar in order to verify the establishment of anaerobic conditions; resazurin, methylene blue or biological indicators (either a strict aerobe or a strict anaerobe) are frequently used.

Anaerobic cabinets were first employed in isolation work during the 1960s to eliminate the problems encountered with other devices. An anaerobic cabinet, equipped with glove ports and a rigid air lock for transfer of materials into the working area, provides an oxygen-free environment in which conventional aseptic bacteriological techniques can be applied to the isolation and manipulation of obligate anaerobes under conditions of strict and continuous anaerobiosis (Willis, 1990). They offer the microbiologist a large working area with control over temperature, pressure and gas composition. The advent of the anaerobic cabinet greatly facilitated use of petri dish plate cultures, subculturing of anaerobic bacteria, and culture inspection without the continual risk of oxygen exposure. Prolonged culture incubations, handling of multiple sample replicates and the storage of prereduced media are all feasible with this technique. However, microscopic examinations within an anaerobic cabinet remain problematic.

There are several alternative methods to achieve anaerobic culture conditions in the laboratory. These include the Hungate or roll-tube technique (Holland *et al.*, 1987; Levett, 1990), the serum tube method (Balch *et al.*, 1979) and shake-flask cultures and fluid cultures (Willis, 1977, 1990).

3.2 Examination and identification of anaerobic bacteria

The anaerobic bacteria are a diverse group of microorganisms and examples of the major generic groups are given in Table 24.1. However, in the context of anaerobic digestion processes, our interest is focused on those bacteria associated with the mechanism of methane production. Therefore, this section will be limited to the enrichment, identification, enumeration and taxonomy of only

TABLE 24.1 Types of anaerobic bacteria and their generic names (Holland *et al.*, 1987)

1. **Phototrophic bacteria**
   A. Purple sulphur bacteria (Family Chromatiaceae): *Amoebobacter, Chromatium, Lamprocystis, Thiocystis, Thiodictyon, Thiopedia, Thiosarcina, Thiospirillum, Ectothiorhodospira*
   B. Green sulphur bacteria (Family Chlorobiaceae): *Ancalochloris, Chlorobium, Pelodictyon, Prosthecochloris* (Family Chloroflexaceae): *Chloroflexus, Chloronema, Oscillochloris*
2. **Spirochaetes:** *Spirochaeta, Cristispira, Treponema, Borrelia*
3. **Anaerobic Gram-negative bacteria:** *Bacteroides, Fusobacterium, Butyrivibrio, Leptotrichia, Succinivibrio, Succinomonas, Anaerobiospirillum, Wolinella, Anaerovibrio, Pectinatus, Acetovibrio, Selenomonas, Lachnospira, Desulfovibrio* and other sulphate-reducers
4. **Methanogenic bacteria:** *Methanobacterium, Methanobrevibacter, Methanothermus, Methanococcus, Methanomicrobium, Methanogenium, Methanospirillum, Methanosarcina, Methanosaeta, Methanoplanus*
5. **Anaerobic cocci**
   A. Gram-positive: *Streptococcus, Peptococcus, Peptostreptococcus, Gaffkya, Ruminococcus, Sarcina, Coprococcus*
   B. Gram-negative: *Veillonella, Megasphaera, Gemmiger, Acidaminococcus*
6. **Anaerobic Gram-positive non-sporing bacteria:** *Bifidobacterium, Eubacterium,* some *Lactobacillus,* some *Actinomyces, Corynebacterium* (*Bacterionema*), *Propionibacterium*
7. **Anaerobic endospore-forming bacilli:** *Clostridium, Desulfotomaculum*
8. **Cell-wall deficient anaerobes:** *Anaeroplasma* and L-forms
9. **Microaerophilic bacteria:** *Beggiatoa, Campylobacter, Rhodopseudomonas, Spirillum*

the methanogenic bacteria. For a comprehensive treatise on the other anaerobic bacteria, the reader is referred to Holland *et al.* (1987).

3.3 Taxonomy and identification of methanogens

With increasing knowledge of the physiology and structure of methanogens and with the advent of newer techniques for establishing the interrelationship between bacteria, it has been confirmed that the methanogens are a coherent group. Indeed, they are now known to be truly distinct from the typical bacteria (*Eubacteria*) and are classified in a separate kingdom, the *Archaea* (Kirsop, 1984). The phylogenetic evidence for the taxonomic distinction between *Archaea* and *Eubacteria* was derived from 16S ribosomal (rRNA) sequence characterization two decades ago. In addition to this phylogenetic difference, there are also a number of phenotypic differences between them (Kirsop, 1984; Levett, 1990):

1. Eubacterial cell walls contain muramic acid, whereas those of Archaebacteria do not
2. the occurrence of unique transfer RNA and 5S ribosomal RNA in the translation mechanisms of protein synthesis
3. membrane lipids in Eubacteria are glycerol esters of fatty acids, whereas Archaebacterial lipids are diethers of glycerol and isoprenoids
4. Archaebacteria possess tRNA devoid of ribothymidine in the TψC loop
5. Archaebacterial RNA polymerases have distinct subunit structures.

As the basis for any taxonomic work, an identification method should be employed to distinguish the species of methanogenic bacteria. In the early studies, cell shape was used as a primary determinant for taxonomic assignment of methanogenic genera, while physiological and nutritional properties were used to form the basis of species designation. A recent attempt by Kasapgil (1994) to organize the taxonomy of methanogens using cell morphology and nutritional capability is tabulated in Table 24.2. However, an adequate taxonomic classification of methanogens could not be established using only these parameters. A more comprehensive elucidation of the taxonomy of the methanogens was provided using comparative cataloging of 16S rRNA (Balch *et al.*, 1979). In this technique RNA is extracted from the cell and the 16S rRNA is purified. This material is digested by ribonuclease T_2 into oligonucleotides that can then be separated

TABLE 24.2 Methanogenic classification (Kasapgil, 1994)

Order	*Family*	*Genus*	*Species*	*Gram reaction*	*Morphology*	*Substrate*
Methanobacteriales	Methanobacteriaceae	*Methanobacterium*	*M. formcicum*	+	Long rods, filaments	H_2, CO_2, formate
			M. bryanti	+	Short long rods	H_2, CO_2, formate
			M. thermoautotrophicum	+	Long rods, filaments	H_2, CO_2, formate
			M. wolfei	+	Rods	H_2, CO_2
			M. alcaliphilum	+	Rods	H_2, CO_2
			M. uliginosum	–	Rods	H_2, CO_2
			M. thermoformicicum	+	Rods	H_2, CO_2, formate
		Methanobrevibacter	*M. arbophilius*	+	Short rods	H_2, CO_2, formate
			M. ruminantium	+	and short	H_2, CO_2, formate
			M. smithii	+	chains	H_2, CO_2, formate
	Methanothermaceae	*Methanothermus*	*M. fervidus*		Short rods	H_2, CO_2
			M. sociabilis		Rods	H_2, CO_2
Methanococcales	Methanococcaceae	*Methanococcus*	*M. vannielli*	–	Irregular	H_2, CO_2, formate
			M. voltae	–	cocci	H_2, CO_2, formate
			M. maripaludis	–	single or pairs	H_2, CO_2, formate
			M. thermolithotrophicus	–		H_2, CO_2, formate
			M. halophilus	–		Methanol, methylamines
			M. jannaschi	–	Irregular cocci	H_2, CO_2, formate
			M. deltae	–		H_2, CO_2, formate
			M. frisisus	–	Irregular cocci	H_2, CO_2
Methanomicrobiales	Methanomicrobiaceae	*Methanomicrobium*	*M. mobile*	–	Short rods single	H_2, CO_2, formate
			M. paynter	–	Short rods single	H_2, CO_2

(*continued on next page*)

TABLE 24.2 (*continued*)

Order	*Family*	*Genus*	*Species*	*Gram reaction*	*Morphology*	*Substrate*
		Methanogenium	*M. cariaci*	–	Irregular	H_2, CO_2, formate
			M. marisnigri	–	cocci, single or	H_2, CO_2, formate
			M. olentangyi		pairs	H_2, CO_2, formate
			M. thermophilicum	+	Irregular cocci	H_2, CO_2
			M. aggregands	+	Irregular cocci	H_2, CO_2, formate
			M. bourgense	+	Irregular cocci	H_2, CO_2, formate
			M. tationis		Irregular cocci	H_2, CO_2, formate
		Methanospirillum	*M. hungatei*	–	Spirillum, regular rods and filaments	H_2, CO_2, formate
	Methanoplanaceae	*Methanoplanus*			Plated shape	H_2, CO_2, formate
	Methanosarcinaceae	*Methanosarcina*	*M. limicola*	–	Pseudosarcina	H_2, CO_2, formate
			M. barkeri	+	Irregular cocci in	
			M. mazei	+	large aggregates	
			M. thermophila		Pseudosarcina	Acetate
			M. acetivorans	+	Pseudosarcina, coccoid	Methylamines
			M. vacuolate		Pseudosarcina	
		Methanococcoides	*M. methylutents*	+	Irregular cocci	Metahnol, methylamines
		Methanothrix	*M. soehngenii*	+	Irregular cocci sheat forming long filament	Acetate
			M. concilli		Sheated rod	Acetate
		Methanolobus	*M. tindarius*	+	Irregular cocci single or loose	Metahnol, methylamines

TABLE 24.3 Taxonomic treatment of methanogens by Balch *et al.* (1979) based on 16S rRNA comparative cataloguing

Order	*Family*	*Genus*	*Species*
Methanobacteriales	Methanobacteriaceae	*Methanobacterium*	*M. formicum*
			M. bryantii
			M. thermoautotrophicum
		Methanobrevibacter	*M. ruminantium*
			M. orbariphilus
			M. smithii
Methanococcales	Methanococcaceae	*Methanococcus*	*M. vannielii*
			M. voltae
Methanomicrobiales	Methanomicrobiaceae	*Methanomicrobium*	*M. mobile*
		Methanogenium	*M. cariaci*
			M. marisnigri
		Methanospirillum	*M. hungatei*
	Methanosarcinaceae	*Methanosarcina*	*M. barkeri*

and sequenced. Sequences of oligonucleotides with six or more units are compared between pairs of bacteria to show their identity (Taylor, 1982). Using this approach, a new taxonomic treatment for the methanogenic bacteria (Table 24.3) based on 16S rRNA comparative cataloguing was proposed (Balch *et al.*, 1979). Other techniques that have been used to identify and classify the taxonomy of methanogens include:

1. morphology and staining
2. cell wall structure and composition
3. colonial appearance
4. growth inhibition and stimulation
5. gas liquid chromatography
6. distribution and molecular weight of polar lipids
7. antigenic relationships
8. polyamine content.

For more information on the above techniques, the reader is referred to Willis (1977); Balch *et al.* (1979); Whitman *et al.*(1992).

3.4 Enrichment

Enrichment culture is the technique that is used to enhance the population density of a particular group of microorganisms within the total microbial population of a sample. This is achieved by preferentially stimulating the growth of the target group of microorganisms by judicious manipulation of the physiological conditions during the enrichment phase. Methanogenic enrichments are usually performed in media that has a nutrient composition, environmental pH value, temperature and oxygen-free conditions, similar to those of natural methanogenic environments.

Confirmation of methanogens in the enrichment can be achieved by microscopy or gas chromatographic analysis of the headspace gas for methane. There are a number of specialized media used in anaerobic enrichment. Some allow greater initial selectivity for the methanogens against the background populations and are best suited to isolation work, while others have been developed to optimize growth of the isolated strains and are used for routine culture maintenance. Balch *et al.* (1979) described a standard medium suitable for the enrichment of methanogenic bacteria.

Early work on the isolation of pure cultures and the selection of anaerobic and methanogenic bacteria was reported by Toerien and Hattingh (1969). Further details on the isolation and selection of specific anaerobic bacteria are given by Holland *et al.* (1987).

As an example of the nutritional requirements of methanogens, Table 24.4 lists an anaerobic growth medium based on that reported by Balch *et al.* (1979). For more information on other media preparations, the reader is referred to the reports of Willis (1977); Kirsop (1984); Whitman *et al.* (1992); Levett (1990).

TABLE 24.4 Composition of an enrichment media (Balch *et al.*, 1979)

Sodium formate	1200 mg
Sodium acetate	1200 mg
Yeast extract	3000 mg
Sodium carbonate (8% w/v)	15 ml
Resazurin (0.01% w/v)	0.6 ml
Ferrous sulphate (1% w/v)	0.6 ml
Mineral solution 1	15 ml
Mineral solution 2	15 ml
Trace minerals	6 ml
Vitamins	6 ml
Reducing solution	9 ml

Mineral solution 1: Dipotassium hydrogen orthophosphate (0.6 (w/v)); *Mineral solution 2:* Potassium dihydrogen orthophosphate (600 mg), Ammonium sulphate (12000 mg), Sodium chloride (12000 mg), Calcium chloride (2400 mg), Magnesium sulphate (2500 mg); *Trace mineral solution:* Nitrilotricetic acid (1500 mg), Magnesium sulphate (3000 mg), Manganese sulphate (500 mg), Sodium chloride (1000 mg), Ferrous sulphate (100 mg), Cobalt sulphate (100 mg), Calcium chloride (100 mg), Zinc sulphate (100 mg), Copper sulphate (10 mg), Aluminium potassium sulphate (10 mg), Boric acid (10 mg), Sodium molybdate (10 mg), Nickel chloride (10 mg), Sodium selenate (10 mg); *Vitamins:* D-Biotin (2 mg), Folic acid (2 mg), Pyridoxine hydrochlorie (10 mg), Thiamine hyrochloride (5 mg), Riboflavin (5 mg), Nicotinic acid (5 mg), D-Pantothenic (5 mg), Vitamin B_{12} (0.1 mg), p-Aminobenzoic acid (5 mg), DL-6 8 Thiotic acid (5 mg); *Reducing solution:* Cysteine hydrochloride/sodium sulphide (2.5%)*.

* Dissolve 2.5 g cysteine hydrochloride in 40 ml of distilled water. Adjust pH of the solution to 10 using NaOH pellets. Dissolve 2.5 g sodium sulphide in 40 ml of distilled water. Mix the solutions and make up to 100 ml. Heat and cool under O_2 free nitrogen gas. Dispense into Hungate tubes. Store at 4°C until required.

3.5 Bacterial enumeration

The treatment capacity in any anaerobic digester is primarily determined by the concentration of active biomass retained within the system. Maintaining a high biomass concentration in full-scale anaerobic digesters makes anaerobic treatment more stable and the improved performance also makes the process more attractive economically (Morgan *et al.*, 1991). Therefore, it is important to monitor anaerobic reactors from a microbiological point of view in order to understand the operational conditions that promote high levels of active biomass within the system.

Supervision of full-scale and laboratory-scale anaerobic reactors can be greatly assisted by bacterial enumeration of the digester sludge, as this provides an indication as to whether or not:

1. the reactor is overloaded by an excessive feed rate, and/or
2. there are any inhibitory materials in the system (and the significance and magnitude of their effect).

Since the enumeration of anaerobic bacteria requires strict anaerobic conditions, it will be necessary to use some of the techniques mentioned above to exclude oxygen while handling anaerobic bacteria.

Enumeration of anaerobic bacteria can be achieved by the agar plate technique, most probable number (MPN) methods and direct counts by microscopy (utilizing the fluorescent properties of oxidized factor F_{420} and specific staining techniques to show particular groups of bacteria). The basic procedures for the direct count and MPN methods are described in more detail below.

3.5.1 *Microscopic method (direct count)*

Enumeration of the total bacteria and the total fluorescent methanogenic bacteria can be made by using an Epifluorescent microscope (e.g. Zeiss D-7082) with illumination by either a halogen/ultraviolet lamp at 376 nm or with a 50 W high pressure mercury lamp (Kasapgil, 1994). Methanogenic and non-methanogenic bacteria can be differentiated on the basis that only methanogenic bacteria will be seen to emit blue-green fluorescence when the ultraviolet lamp is used.

The following protocol may be used to enumerate methanogens and the acidogenic bacteria. In a sample (Ince *et al.* 1997):

1. Dilute and homogenize the sample to give a count of between 50 and 400 bacteria per field of view
2. Place the sample onto a Neubauer Counting Chamber (graduated and calibrated glass slide)
3. Microscopically count several chambers using (i) visible light illumination to record the total bacteria, and (ii) visible plus

ultraviolet light illumination to record the number of fluorescent cells (methanogenic bacteria)

4. calculate the number of bacteria per ml of sample with the following formula:

Count of bacteria (ml^{-1}) = Y*D/V

where:

Y = mean count per chamber,
D = dilution factor giving counts between 50 and 400 per field of view and
V = volume of a chamber (ml).

Estimate the number of acidogenic bacteria by difference:

Acidogenic bacteria = Total bacteria − Methanogenic bacteria.

3.5.2 *Multiple tube method*

The most probable number (MPN) technique can be used to estimate the number of viable methanogenic bacteria by serial dilution of the sample and inoculation into culture media containing a specific carbon source. This method was described by Siebert and Hattingh (1967, 1987) and used by Morgan *et al.* (1991) and Kasapgil (1994). The most probable number of the methanogens can be calculated by using the probability tables described by Greenberg *et al.*(1985). For the MPN enumeration, the media described by Balch *et al.* (1979) can be used. The media and samples are placed into Hungate tubes for incubation. After inoculation, the head space of each tube should be replaced with an oxygen-free mixture of hydrogen and carbon dioxide (4:1) and tubes should be incubated statically for 4–6 weeks at 37°C. Positive tubes can be identified by the presence of methane gas in the head-space using gas chromatography and each should be confirmed by epifluoresence microscopy.

3.5.3 *Nucleic acid hybridization techniques*

Recently, the molecular methods of rRNA analysis, widely used for identifying phylogenetic relationships between microorganisms, have been developed into a number of emerging techniques. These are capable of both quantifying and visualizing individual groups of microorganisms within complex communities, such as anaerobic granular sludge (Harmsen *et al.*, 1996). The two main techniques both use oligonucleotide probes (or primers) that are specific for conserved regions of rRNA in their target organisms. The probes can be tailored to detect individual species, genera or wider groups of phylogenetically-related microorganisms. Quantification involves extracted rRNA from the sample material being hybridized on a membrane with a series of radio-labelled probes for each target organism (blot technique). The intensity of the radio-label signal is proportional to the concentration of the target group and results can therefore be used to quantify the dynamics of the microbial composition in a sample.

A powerful variation of this technique is fluorescent *in situ* hybridization (FISH) in which fluorescently-labelled probes, similar to those above, are made to react directly with intact microbial cells in the sludge sample (Raskin *et al.*, 1994). This allows fluorescently-stained cells to be visualized microscopically showing the numbers and location of the target cells and their proximity to other cells in the sample. The latter is proving to be a powerful tool in determining the syntrophic relationships between community members in anaerobic sludge granules (Rocheleau *et al.*, 1999).

4 BIOCHEMISTRY OF ANAEROBIC DIGESTION

4.1 Energy conservation

All microorganisms are endowed with a metabolic system that transforms the chemical energy of their feed substrate into biologically useful energy, which is then utilized to perform work for the cell (Thauer *et al.*, 1977). The energy taken up by cells is first used to drive the endergonic synthesis of ATP (adenosine triphosphate) from ADP (adenosine diphosphate) and P_i (inorganic orthophosphate). This ATP then drives anabolic metabolism and other cellular processes (ion transport, motility, etc.).

$$ATP + H_2O \rightarrow ADP + P_i + \textit{useful energy}$$

$$ATP + H_2O \rightarrow AMP + PP_i + \textit{useful energy}$$

In anaerobic cells, ATP is produced through the same two metabolic routes that occur in aerobic metabolism. The first of these is substrate-level phosphorylation, in which the chemical bond energy is transferred directly to ADP from key intermediates in the fermentation pathway, with concomitant ATP generation. Anaerobic fermentation is equivalent to the process of glycolysis in aerobic metabolism, with the exception that fermentation fails to bring about any net oxidation of the substrate during the formation of fermentation products. The second metabolic route is electron transport phosphorylation (ETP), in which electrons flow down an electrochemical gradient between redox carriers (respiratory chain) and the energy in this redox potential drives the translocation of protons (or sodium) across the cell membrane, creating a chemiosmotic proton imbalance. The resulting proton motive force (PMF) is coupled to the phosphorylation of ADP as protons channel back into the cell through a membrane-integrated proton-translocating ATP-ase (Thauer *et al.*, 1977; Hawkes, 1980; Madigan *et al.*, 1997).

Despite the enormous range of electron donors (e.g. H_2, NADH, NH_3, Fe^{2+}, $S_2O_3^{2-}$) and electron acceptors (CO_2, NO_3^-, SO_4^{2-}, Fe^{3+}, O_2) that may donate and accept reducing equivalents (electrons) to the respiratory chains of both aerobic and anaerobic metabolism, many of the membrane redox carriers appear to be ubiquitous (e.g. cytochromes). In contrast, major differences exist in the enzymes and cofactors that are involved in the first and last steps of electron transfer both to and from the respiratory chain. Some examples of these are discussed below for the organisms involved in the final steps of methanogenic anaerobic digestion.

The maximum free energy that is available to the chemolithotrophic autotrophs of anaerobic digestion (i.e. sulphate-reducing bacteria (SRB), methanogens and acetogens) can be determined from thermodynamic consideration of the electron donors and acceptors according to the Gibbs free energy relationship:

$$\Delta G^{0'} = -nF\Delta E^{0'}$$

where:

$\Delta G^{0'}$ is the free energy change under standard conditions, i.e. 25°C and a pressure of 1 atm, at pH 7;
n is the number of electrons transferred;
F is the Faraday constant (96.48 kJ/V);
$E^{0'}$ is the electrochemical oxidation potential;
$\Delta E^{0'}$ is $E^{0'}$ (electron accepting couple) minus $E^{0'}$ (electron donating couple).

Negative values of $\Delta G^{0'}$ indicate an exergonic reaction, i.e. energy is available for the synthesis of ATP. Free energy values for the stoichiometric equations that represent some of the common substrates and intermediates in anaerobic digestion are given in Table 24.5.

The energy available from such reactions is further influenced by the actual concentrations of reactants and products, reactions becoming more energetically favourable when product concentrations are maintained at low levels (see Madigan *et al.*,1997). Consequently, during anaerobic digestion, the energy yield from methane production and acetogenesis will be considerably lower than the quoted values (−135.6 and −104.6 kJ/mol respectively) because the concentration of hydrogen is usually extremely low in most anaerobic digesters.

4.2 Growth substrates

In practice, most wastes that are treated by anaerobic digestion seldom contain the compounds shown in Table 24.5 in their raw state. More typically, they comprise complex or polymeric organic materials such as lipids, cellulose and protein. In order for these to be converted to the immediate substrates of the methanogens (hydrogen, acetate and CO_2), an

TABLE 24.5 Estimated free energy changes of some biological reactions under standard conditions, i.e. 25°C and a pressure of 1 atm

Dehydrogenations i.e. electron donating, oxidation, reactions:	$\Delta G^{0'}$ *(J/mol)*
Butyrate → Acetate	
$CH_3CH_2CH_2COO^- + 2H_2O \rightarrow 2CH_3COO^- + 2H_2 + H^+$	+48.1
Lactate → Acetate	
$CH_3CHOHCOO^- + 2H_2O \rightarrow CH_3COO^- + HCO_3^- + 2H_2 + H^+$	−4.2
Propionate → Acetate	
$CH_3CH_2COO^- + 3H_2O \rightarrow CH_3COO^- + HCO_3^- + 3H_2 + H^+$	+76.1
Ethanol → Acetate	
$CH_3CH_2OH + H_2O \rightarrow CH_3COO^- + 2H_2 + H^+$	+9.6
Formate → HCO_3^-	
$HCOOH + H_2O \rightarrow HCO_3^- + H_2$	+ 1.3
Acetate→Methane	
$CH_3COO^- + H_2O \rightarrow HCO_3^- + CH_4$	−31.0
Hydrogenations, i.e. electron accepting reactions:	
Methanol → Methane	
$CH_3OH + H_2 \rightarrow CH_4 + H_2O$	−112.5
Formate → Methane	
$HCOOH + 3H_2 + H^+ \rightarrow CH_4 + 3H_2O$	−134.3
HCO_3^- → Methane	
$HCO_3^- + 4H_2 + H^+ \rightarrow CH_4 + 3H_2O$	−135.6
HCO_3^- → Acetate	
$2HCO_3^- + 4H_2 + H^+ \rightarrow CH_3COO^- + 4H_2O$	−104.6

initial partial degradation is carried out by non-methanogenic populations (see Table 24.1) and involves the following major steps.

4.2.1 Hydrolytic degradation

Hydrolases cleave their substrate by insertion of a water molecule. They do not require stoichiometric quantities of cofactors or an energy supply and must be able to function extracellularly as their substrates are too large to cross the cell membrane. Examples include protease, which hydrolyses proteins to peptides and amino acids, amylase, cellulase, pectinase and xylanase, which hydrolyse their respective carbohydrate substrates to dextrins and hexose sugars and lipase which hydrolytically releases the glycerol and fatty acids from the parent lipid.

4.2.2 Fermentation pathways

The products of hydrolysis, along with other low molecular weight compounds present in the effluent, are further degraded intracellularly by a number of different fermentation pathways. These involve enzymic steps that may require cofactors (e.g. dehydrogenases) or energy (kinases) to be supplied, and unless reducing equivalents are consumed by a respiratory chain oxidation process, or converted to molecular hydrogen (H_2) by a hydrogenase, the fermentation product will have the same oxidation state as the substrate. Typical fermentation products from sugars and amino acids are acetate, propionate, butyrate and formate. Long chain fatty acids are degraded by β-oxidation, a process that is coupled to hydrogen formation as means of recycling the adenine nucleotide cofactors (NAD and FAD).

4.2.3 Hydrogen production

It is clear from the above considerations that hydrogen is an important intermediate in anaerobic digestion, being produced during the degradation of all major classes of complex substrate. During anaerobic digestion, approximately 30% of all methane production is generated from the reduction of carbon dioxide by hydrogen. In addition, the concentration of hydrogen also affects the pH value in anaerobic reactors. A rise in the level of H_2 leads to a lowering of pH in the digester. Low pH can selectively inhibit the methanogenic bacteria and this in turn may result in a further and catastrophic acidification of the digester as volatile acids accumulate (Archer and Kirsop, 1990).

Hydrogen may also affect the substrate conversion potential of many major anaerobic bacteria. A build-up of hydrogen can inhibit the growth of hydrogen-producing organisms and alter electron flow, resulting in the formation of more reduced products such as lactate, butyrate or ethanol instead of acetate. Inhibition of hydrogen-producing bacteria occurs because hydrogen inhibits the hydrogenase of these bacteria (Oremland, 1988), preventing them from recycling their reduced cofactors.

Briefly, hydrogen imparts two major effects on anaerobic degradation. The first involves the selection process that determines the chemical pathway that is used to degrade organics, and the second is associated with inhibition of syntrophic bacteria responsible for C_3 and C_4 acid oxidation (Harper and Pohland, 1986).

Partial pressure of hydrogen (1 ppm (v/v) H_2 produces 10^{-6} atm pressure at pH 7) has a vital role in methane formation and substrate flow. In other words, thermodynamically, the conversion of VFA to methane depends on the partial pressure of H_2. For example, if the reactions for ethanol, butyrate, propionate are to be thermodynamically favourable methanogenic substrates, the partial pressure of H_2 must be maintained below 10^{-1}, 10^{-3} and 10^{-4} atm, respectively. On the other hand, for H_2 to be converted to methane, a concentration of H_2 greater than 10^{-6} atm is required. Therefore, there is a limited range of hydrogen concentrations that satisfies both these requirements and these are shown in Fig. 24.4.

The partial pressure of H_2 is regulated not only by hydrogen-producing bacteria but also by hydrogen-consuming bacteria, e.g. homoacetogens and methanogens (see Interspecies hydrogen transfer).

Hydrogen level in anaerobic digesters can be used for process control and in the diagnosis of process problems. It can be measured by several methods, such as gas chromatography, spectrometry (Archer and Kirsop, 1990) and polarographic methods, such as the exhaled hydrogen monitor (Collins and Paskins, 1987).

4.3 Interspecies hydrogen transfer

The discovery that *Methanobacterium omelianskii* really consisted of a syntrophic association of two different bacteria helped to reveal the subtle role of H_2 as an intermediate. Originally *M. omelianskii* was believed to be a methanogen that catabolized ethanol according to:

$$2CH_3CH_2OH + CO_2 \rightarrow 2CH_3CO_2H + CH_4$$

Later this was shown not to be the case and resulted from a syntrophic association between an acetogen and a methanogen. The acetogen converts ethanol to acetate and hydrogen while the methanogen uses the hydrogen to reduce CO_2:

$$2CH_3CH_2OH + H_2O \rightarrow 2CH_3CO_2H + 4H_2 \qquad \textit{acetogen}$$

$$CO_2 + 4H_2 \rightarrow CH_4 + 2H_2O \qquad \textit{methanogen}$$

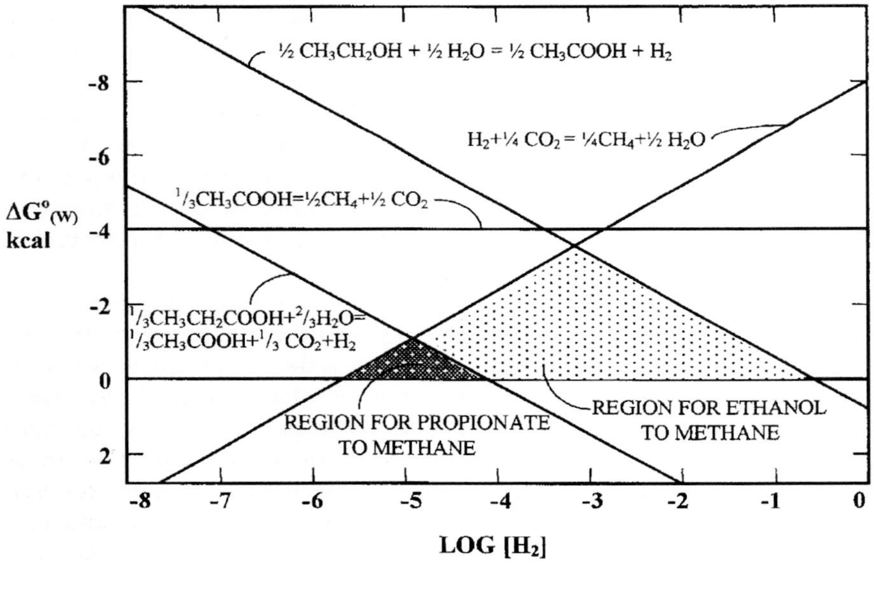

Fig. 24.4 Effect of hydrogen partial pressure on the free energy changes ($\Delta G^{0\prime}{}_{(w)}$) (McCarty, 1981, with permission).

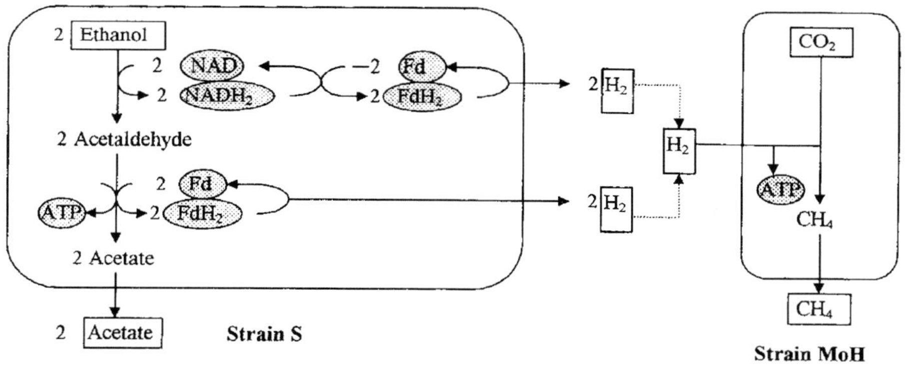

Fig. 24.5 Interspecies hydrogen transfer between two bacteria, Strain S and Strain MoH (Schlegel, 1993, with permission).

It has now been established that other long chain fatty acids are also catabolized by similar associations (Klass, 1984).

In anaerobic digesters, the above mechanism of 'interspecies hydrogen transfer' is usually an obligatory syntrophic association between the anaerobic bacteria forming hydrogen and those converting it to methane. Each member benefits from, or is dependent on, the activity of the other. This is shown schematically in Fig. 24.5.

4.4 Methanogenesis

The final stage in anaerobic digestion is methanogenesis. The production of methane, the most fully reduced form of carbon, is the terminal product since it can undergo no further degradation anaerobically. Methane production brings two major economic benefits to the anaerobic digestion process. First, it is a valuable resource in its own right having a high calorific content, and second, its dissipation from the treatment system prevents a major part of carbon load in the wastewater from being incorporated anabolically into biomass, ensuring low sludge production rates.

The methanogens have evolved as a specialized group of *Archaea* that utilize the CO_2, hydrogen and acetate generated as end-products by the fermentative bacteria. This capability depends on several enzymes and coenzymes found in no other living organisms. By definition coenzymes exist in cells in catalytically small amounts (Hawkes, 1980), however, because of their importance in the energy-yielding pathway and relatively low energy yield, many of the coenzymes are present at high levels in methanogens.

4.4.1 Carrier coenzymes of methanogenesis

The pathway of methane production from carbon dioxide and hydrogen is an intriguing process. First, synthesis of methane appears to offer little energetic reward and second it is intimately linked to the generation of acetate within the cell, the starting intermediate for anabolic pathways in methanogens, SRBs and homoacetogens. The general scheme of these two processes is described below with reference to the key biochemical carriers.

C_1-carriers. During methanogenesis, carbon dioxide is subjected to a series of reductive steps in which hydrogen is the ultimate supplier of reducing electrons, however, the exact nature of many of the immediate electron carriers involved in these steps is still unknown. Several specific coenzymes (carriers) have been elucidated which function to stabilize the carbon atom of carbon dioxide as it is successively reduced to formyl-carbon, methylene-carbon, and methyl-carbon before being finally released as methane.

Methanofuran. Initially, carbon dioxide binds to methanofuran (Fig. 24.6), is reduced to the formyl-level and is then transferred to methanopterin, the second carrier in the pathway. Methanofuran is a low-molecular weight coen-

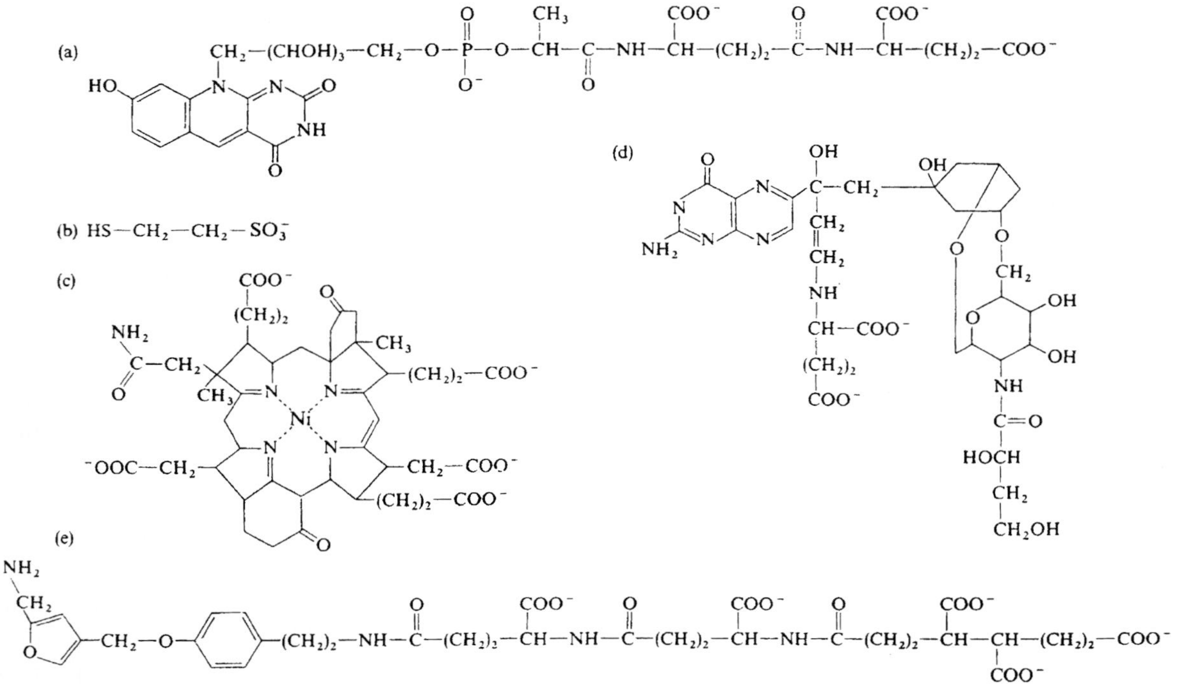

Fig. 24.6 Unique coenzymes of the methanogenic pathway. (a) Factor 420; (b) coenzyme M; (c) Factor 430; (d) methanopterin; (e) methanofuran (Stanier *et al.*, 1986).

zyme which consists of phenol, two glutamic acid molecules, an unusual long chain dicarboxylic fatty acid, and a furan ring.

Methanopterin. Methanopterin then carries the C_1-atom through the next two steps. First, during reduction to the methylene-level by reducing electrons supplied from coenzyme F_{420} and second as it is further reduced to the methyl-level. At this point the methanopterin-bound methyl group may enter the acetyl-CoA biosynthetic pathway (see below) or continue along the methanogenic pathway. The latter involves its transfer to the final carrier, coenzyme M (CoM).

Coenzyme M (CoM). Coenzyme M (CoM) is the simple molecule, 2-mercaptoethanesulphonic acid ($HS\text{-}CH_2.CH_2.SO_3^-$). During the final step, the CoM-bound methyl group is reduced to free methane in a complex reaction involving the enzyme methyl reductase. Apart from releasing the C_1-unit as a free form (CH_4) once again, this final step is also of note because it is the position of energy conservation in the methanogenic pathway, creating a PMF through the proton translocation that occurs concomitantly with reduction of the methyl group.

CoM has been detected in all methanogenic bacteria at an internal concentration of 0.3–50 nmol/mg protein. It is also used as a growth factor (Stanier *et al.*, 1986). The overall scheme in which the C_2 of carbon dioxide is successively reduced to methane is shown schematically in Fig. 24.7.

4.4.2 Other cofactors of methanogenesis

Methyl-CoM and factor$_{430}$ (F_{430}). Methyl-CoM ($CH_3\text{-}S\text{-}CH_2.CH_2.SO_3^-$) is probably the central intermediate in methane formation from CO_2, formate, methanol, methylamines, and acetate. In cell extracts of methanogenic bacteria

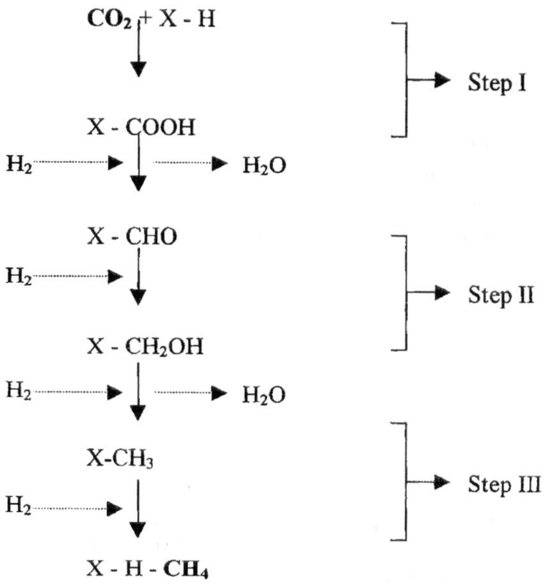

Fig. 24.7 The pathway of methane production from CO_2. X represents the various cofactor carriers that bind the C_1 unit during each successive reduction step. (See text for details.)

methyl-CoM is reduced to methane and CoM. The methyl reductase catalysing this reaction has been purified and shown to contain the cofactor F_{430}, which is a yellow, non-fluorescent compound with an absorption maximum at 430 nm.

Coenzyme F_{420}. Coenzyme F_{420} is an electron carrier and has a blue-green fluorescence with an absorption peak at 420 nm. The fluorescence ability of F_{420} can be used to identify and enumerate methanogenic bacteria. F_{420} interacts with a number of different enzymes in methanogens, including hydrogenase, formate dehydrogenase, NADP reductase, carbon monoxide dehydrogenase, and pyruvate dehydrogenase.

$$H_2 + F_{420\,ox} \rightarrow 2H^+ + F_{420red}$$
(hydrogenase)
$$2H^+ + F_{420red} + NADP \rightarrow NADPH + H^+$$
(NADP reductase)
$$F_{420\,ox} + HCOOH \rightarrow F_{420red} + CO_2 + H_2$$
(formate dehydrogenase)

Coenzyme HS-HTP. Coenzyme HS-HTP is a phosphorylated derivative of the amino acid threonine containing a fatty acid side chain with a terminal SH group. It is the electron donor involved in the last step of CO_2 conversion to CH_4:

$$HS{\cdot}HTP + CoM{\cdot}S{\cdot}CH_3 \rightarrow CH_4 + CoM{\cdot}S{\cdot}S{\cdot}HTP$$

CoM and HS-FTP are regenerated by reduction with H_2:

$$CoM{\cdot}S{\cdot}S{\cdot}HTP + 2H \rightarrow HS{\cdot}HTP + CoM{\cdot}SH$$

4.5 Biosynthesis of acetate

The pathway of methane formation in methanogens is intimately linked to their biosynthetic pathway for acetate synthesis, as all chemolithotrophic autotrophs must produce this as the starting intermediate for their anabolic pathways. Specifically, the methanopterin-bound methyl group in the methanogenic pathway (see above) may be transferred to a C_1-carrier (a corrinoid protein) which is part of the acetyl-CoA (Ljungdahl-Wood) pathway. This methyl group is then attached to a second C_1-unit, also originating from carbon dioxide, in a complex series of reactions involving the enzyme carbon monoxide dehydrogenase (Madigan *et al.*, 1997). The C_2-acetyl group formed by the fusion of these two C_1-fragments is the precursor of all subsequent cellular biosynthesis.

4.6 Alternative electron acceptors

4.6.1 Sulphate and nitrate

Although methanogenesis is the major route for hydrogen oxidation in anaerobic digesters, the terminal electron acceptor of this process, carbon dioxide, is not the only compound likely to exist in such environments with the capacity to support chemolithotrophic metabolism. The most common alternative electron acceptors are sulphate and nitrate, terminal electron acceptors used by sulphate reducing bacteria (SRBs) and denitrifying bacteria respectively. The former are often present at significant concentrations in sewage sludge and industrial effluents, leading to competition for hydrogen, lower methane productivity and the formation of hydrogen sulphide, a potentially toxic compound (Colleran *et al.*, 1995).

4.6.2 *Xenobiotics as electron acceptors*

One of the main advantages of anaerobic digestion over aerobic treatment technologies is the ability of anaerobic bacteria to reductively transform a number of oxidized carbon compounds of environmental significance. These remain largely unaffected as they pass through activated sludge plants because their chemically oxidized structure is difficult to degrade further through oxidative metabolism. Typical examples usually have highly oxidized substituents, like chlorine, on an aromatic nucleus and include pesticides, dyes, solvents, explosives and numerous speciality chemicals. As highly reduced environments are able to promote both the chemical and biological reduction of these xenobiotic compounds, anaerobic digestion will often be the most practical option for their destruction in a treatment process.

It appears that in certain groups of anaerobic bacteria, such compounds may act directly as terminal electron acceptors during growth on electron donors such as VFA. Furthermore, this type of metabolism can yield similar amounts of energy as metabolism with typical electron acceptors like nitrate (Mackiewicz and Wiegel, 1998), making their destruction ecologically feasible in the competitive environment of an anaerobic digester.

5 ENVIRONMENTAL FACTORS AFFECTING ANAEROBIC DIGESTION

As discussed above, effective anaerobic degradation of organic matter requires healthy populations of the relevant bacterial groups to be working in synergy. In all biological wastewater treatment processes, the effective removal of pollutants and contaminants depends not only on the metabolic potential of the microorganisms but also on the existence of suitable environmental conditions to support these activities. In anaerobic treatment processes, mainly as a result of the critical nature of the syntrophic relationships, environmental conditions require stringent monitoring and control if process failure is to be avoided. Consequently, factors such as nutrient composition, temperature, pH, mixing, toxicity and inhibition have all been thoroughly investigated, and the salient points are summarized below.

5.1 Nutrients

The nutritional requirements of anaerobic bacteria are of paramount importance because nutrients supply the basic cellular building blocks for growth and ensure the cell is able to synthesize the enzymes and cofactors that drive the biochemical and metabolic reactions. Nutrients can be divided into two groups, the macronutrients and micronutrients, according to the relative quantities required by the cell. It is essential for both types of nutrients to be present in an available form in the growth environment to allow effective uptake. Ideally, nutrient levels should be in excess of the optimum concentration required as anaerobic bacteria can be severely inhibited by even slight nutrient deficiencies. However, many essential nutrients can become toxic when present in high concentrations (Gunnerson and Stuckey, 1986), a fact that precludes the use of excess amendments. A rough estimate of the theoretical amount of macronutrients, nitrogen (N), phosphorus (P) and sulphur (S), that are required in digesters can be derived from the elemental composition of bacterial cells within the anaerobic sludge (Lettinga, 1995).

Alternatively, Speece (1996) has suggested that since the empirical formula for biomass is $C_5H_7O_2N$, then the nitrogen requirement would be 3–6 kg N/1000 kg of COD consumed or 0.5–10 kg N/60 m^3 of methane produced. However, some workers have cited an optimum C:N ratio of 30:1 (Gunnerson and Stuckey, 1986). Others have preferred to differentiate further the N requirement of a digester according to whether the system is highly loaded or lightly loaded with COD. Using this approach, Henze and Harremoes (1983) recommended that the COD:N ratios of 350:7 and 1000:7 are required for highly loaded and lightly loaded systems, respectively. Nevertheless, the COD:N ratio most commonly recommended and practised in anaerobic digestion is 100:2.5. The above ratios are all consistent in predicting lower nitrogen requirements for anaerobic digestion than those typically used in aerobic processes (COD:N of 20–30:1). It should be noted that true nutrient balances in both aerobic and anaerobic treatment systems are not fixed relationships and will vary according to operating conditions (organic load, temperature, solid retention time (SRT) etc.).

Nitrogen can occur in a great variety of inorganic forms, the most common being ammonia (NH_3), nitrate (NO_3^-), nitrite (NO_2^-), and nitrogen gas (N_2). Ammonia is the most readily utilized of the inorganic forms of nitrogen, existing in the reduced state that is required for anabolic metabolism and an uncharged state that facilitates cellular uptake.

Several values for the phosphorus requirement have been reported on the basis of batch experiments with pure cultures and by estimating the value from the empirical formula for biomass (Archer and Kirsop, 1990; Lettinga, 1995; Speece, 1996). Although most of these estimates predict an N:P ratio of 7:1, some propose a COD:P ratio varying from 80:1 to 200:1. The usual forms taken by phosphorus in aqueous solution include orthophosphate, polyphosphate, and organic phosphate. The orthophosphates, e.g. PO_4^-, HPO_4^{2-}, $H_2PO_4^-$, H_3PO_4, are immediately available for biological metabolism without further modification, organic phosphates must generally be hydrolysed by the cell to release inorganic phosphate before use.

In addition to the two main macronutrients (N and P), the sulphur (S) requirements of anaerobic bacteria should also be satisfied and this can be supplied as sulphur, sulphide, sulphite, thiosulphate, sulphate or amino acids (cysteine and methionine). Optimum digester concentrations of S have been reported between 0.001 and 1.0 mg/l (Speece, 1996).

TABLE 24.6 Typical elemental composition (mg/l) of methanogenic bacteria

C	370 000	–	440 000
H	55 000	–	65 000
N	95 000	–	128 000
Na	3000	–	40 000
K	1300	–	50 000
S	5600	–	12 000
P	5000	–	28 000
Ca	85	–	4500
Mg	900	–	5300
Fe	700	–	2800
Ni	65	–	180
Co	10	–	120
Mo	10	–	70
Zn	50	–	630
Cu	<10	–	160
Mn	<5	–	25

Adapted from Scherer *et al.* (1983).

As for micronutrients, anaerobic bacteria require a range of trace elements for metabolism and growth. These requirements can be inferred from the elemental composition of anaerobic bacteria (Table 24.6). In addition, the physiological effects and working concentrations of individual micronutrients that have been reported to improve process performance are listed in Table 24.7.

5.2 Temperature

Temperature is one of the most influential environmental factors as it controls the activity of all microorganisms through two contrasting effects. Generally, a rise in temperature leads to an increase in the rate of biochemical and enzymatic reactions within cells, causing increased growth rates. However, above a specific temperature, which is characteristic of each species, this phenomenon gives way to one of inhibition, and then mortality, as the proteins and structural components of the cell become irreversibly denatured.

Although methane formation is biologically feasible at all temperatures between 0°C and 100°C, there are two distinct temperature

TABLE 24.7 Effects of micronutrients on some physiological and operational processes of anaerobic digestion (after Ince, 1993; Kasapgil, 1994)

Nutrient	*Concentration required (mg/l)*	*Effects on digestion*
Ca	100–200	Granulation and increase in activity
Mg	75–150	Granulation and increase in activity
Na	100–200	Increase in activity
Fe	20–100	Increase in activity and precipitation of sulphide
K	200–400	Increase in activity
Ba	0.01–0.1	Divalent cation effect hence good granulation
Co	20	Vitamin B_{12} dependent
W	–	Formate dehydrogenase
Se	0.8	Formate dehydrogenase, glycine reductase, hydroxylase, and dehydrogenase dependent
SO_4	0.1–10	Sulphur source of cell synthesis

ranges associated with anaerobic digestion. These are defined as mesophilic, with a temperature optimum at 30–37°C and thermophilic, with an optimum at 55–60°C, the former being most commonly employed in engineered processes of anaerobic treatment. A third range, favoured by psychrophilic organisms has a temperature optimum at 15–20°C (ambient). Although it is not as efficient as high rate mesophilic and thermophilic digestion, it may still have desirable economic trade-offs for the anaerobic treatment of wastewaters in temperate climates.

It is generally believed that thermophilic reactors are more efficient than mesophilic reactors and bench-scale experiments reveal methane production rates in thermophilic reactors can be double that of mesophilic reactors. Thermophilic reactors can also accept higher organic loading rates and produce lower quantities of sludge. However, a number of disadvantages have been observed for full-scale thermophilic digesters which include:

1. they are often less stable than mesophilic reactors
2. they require more energy to heat the reactor
3. they produce high concentrations of VFA in their effluent.

Nevertheless, thermophilic anaerobic digestion is an attractive option for treating warm industrial effluents and slurries of relatively constant composition (Lettinga, 1995).

Compared to many aerobic processes which are relatively robust to temperature variations, anaerobic digestion is sensitive to sudden temperature fluctuations; changes as small as 1–2°C having significant adverse effects on process performance particularly when changes occur rapidly (<2 hours). Should the bacteria become adversely affected by digester temperature variations, several days or even weeks may be required to restore a healthy population once again.

5.3 pH

Anaerobic bacteria, especially methanogens, exhibit a characteristic sensitivity to extremes of pH. Therefore, maintaining a suitable and stable pH within the digester should be a major priority for ensuring efficient methanogenic digestion. This is because the hydrogen ion concentration has a critical influence on the microorganisms responsible for anaerobic digestion, the biochemistry of digestion, alkalinity buffering and several other chemical reactions affecting the solubility and availability of dissolved ions. The best pH range appears to be around neutrality, while the range between 6.5 and 7.8 is generally believed to be optimal. However, there are exceptions for which no satisfactory explanation has been given, such as anaerobic digestion occurring under conditions as low as pH 3 (Zehnder *et al.*, 1982) and as high as pH 9.7 (Oremland, 1988).

There are four major types of chemical and biochemical reaction that influence the pH of a digester. These reactions are:

1. ammonia consumption and release
2. volatile fatty acid production and consumption
3. sulphide release by dissimilatory reduction of sulphate or sulphite

4. conversion of neutral carbonaceous organic carbon to methane and carbon dioxide (Anderson and Yang, 1992).

In an effective working digester, pH reduction can be countered by natural processes such as bicarbonate alkalinity and the consumption of volatile fatty acids by methanogens. However, the latter is dependent on the equilibrium between acidogens and methanogens and this can be easily upset by changes in the operational or environmental conditions (Anderson and Yang, 1992). Should this occur, there are two options to rectify the situation. The first approach is to stop feeding the reactor, giving the methanogens sufficient time to consume excess fatty acids and raise the pH value to an acceptable level. The second option is to dose the reactor with alkali, i.e. NaOH or Na_2CO_3, in order to raise the pH or provide additional buffering capacity. In some cases both options may be used simultaneously (Gunnerson and Stuckey, 1986).

An appreciation of bicarbonate alkalinity and its relationship with the biochemistry of anaerobic digestion must be gained if the pH balance is to be maintained. The alkalinity in water is due principally to salts of weak acids and strong bases, (e.g. bicarbonate ions). Such substances act as buffers when the pH of the environment is approximately equal to their pK_a value, resisting changes in the hydrogen ion concentration that result from acid production or consumption (Sawyer *et al.*,1994). Although the amount of alkalinity required to accommodate VFA increases in a reactor depends on many factors (Speece, 1996), well-established anaerobic reactors treating typical organic loads are likely to contain alkalinity in the range 2000 to 3000 mg/l as $CaCO_3$. This level of alkalinity will impart an improved resistance to acidification caused by short-term fluctuations in feed composition.

5.4 Mixing

Methanogenic anaerobic digestion comprises an inherent degree of mixing from the continuous rise of methane bubbles within the reactor, however, this natural mixing is usually considered to be rate limiting for efficient mass transfer. Consequently, contact between the organic matter and the microorganisms can be improved by enhanced mixing, leading to higher reactor performance. The level and type of mixing also affects the growth rate and distribution of microorganisms within the sludge, substrate availability and utilization rates, granule formation, and gas production. Mixing can be enhanced using:

1. mechanical devices (paddles, turbines and propellers)
2. hydraulic shear force (feed recycle)
3. gas recirculation.

Stafford (1981) reported that so long as adequate mixing was achieved, the method of mixing had little bearing on the digestion rate. However, it was also reported that excessive mixing could actually lead to a reduction in reactor performance. Smith *et al.*(1996) stated that although plug-flow conditions enable feed components to remain in the reactor for one complete hydraulic retention time (HRT), giving maximum theoretical contact time, the potentially high concentration of substrates and their fermentation products that can occur at the reactor inlet may inhibit the biomass. On the other hand, although excessive mixing alleviates this problem, it may result in short-circuiting of the reactor, leading to unconverted substrate appearing in the reactor effluent. In practice, an intermediate degree of mixing appears to give the best substrate conversion by striking a balance between these adverse effects. The relationship between performance and mixing can be quantified using spikes of lithium tracer in the reactor feed to indicate mixing efficiency (Uyanik *et al.*, 2002a).

5.5 Toxicity and inhibition

The definitions of toxicity and inhibition within the context of anaerobic digestion have been defined by Speece (1996) and will be adopted here. According to these, toxicity is an adverse effect (not necessarily lethal) on bacterial metabolism, while inhibition is an impairment of bacterial function. There are many potential substances that may be present, either as components in a reactor feed or as by-products of anaerobic metabolism, which can slow down the rate of digestion (toxicity) or cause process failure (inhibition). Common examples include, heavy metals, alkali and alkaline earth metals, volatile fatty acids, oxygen, ammonia and sulphide.

5.6 Metals

As discussed above, a trace level of many metal ions is required for the function of certain enzymes and coenzymes, however, excessive amounts may result in toxicity or inhibition. Heavy metal toxicity is believed to occur through the structural disruption of enzymes and protein molecules within the cell (Hickey *et al.*,1989). The toxic effects of a number of metal ions, shown as the concentration required for a 50% reduction in gas production rate, are shown in Table 24.8 (Mosey, 1976).

Several workers have studied the effects of metal ion toxicity, McCarty (1964) claiming that iron and aluminium are not particularly toxic because of their low solubility at operational pH values. Mehrotra *et al.* (1987) revealed that the relative toxicity of zinc (Zn), lead (Pb) and chromium (Cr) appeared to decrease in the order Zn > Pb > Cr, while Hickey *et al.*(1989) reported the relative toxicity of copper (Cu), cadmium (Cd) and zinc (Zn) to be Cu > Cd > Zn. In addition to acute toxicity effects, there is also a possibility that heavy metals may accumulate to significant levels within the digester sludge through precipitation even though feed concentrations may be relatively low. The precipitates are largely inert (unavailable) while in the reducing environment of the digester but their potential effect must be considered when sludge is finally disposed to an oxidizing environment (e.g. land).

TABLE 24.8 Heavy metal concentrations (mg/l) that elicit a 50% reduction in the gas production rate of laboratory digesters (Mosey, 1976)

Zinc	163.0
Cadmium	180.0
Copper	170.0
Nickel	0.6
Lead	2.0

TABLE 24.9 Inhibitory concentrations of alkali and alkaline–earth cations (McCarty, 1964)

Cation	*Concentrations in mg/l*	
	Moderately inhibitory	*Strongly inhibitory*
Sodium	3500–5500	8000
Potassium	2500–4500	12 000
Calcium	2500–4500	8000
Magnesium	1000–1500	3000

Alkali and alkaline earth metals, i.e. sodium, potassium, magnesium, and calcium, are stimulatory to anaerobic bacteria unless present at excessive concentrations. The toxicity of salts of these metals is associated with the cation rather than anion (McCarty, 1964), and acclimatization of digester with cations can often increase the toxicity threshold. For an unacclimatized reactor, McCarty (1964) listed the moderately and strongly inhibitory concentration of these cations (Table 24.9).

5.7 Volatile fatty acids

High concentrations of VFA are often associated with the effects of toxicity and inhibition. Volatile fatty acids normally found in anaerobic digestion processes are listed in Table 24.10. It is generally believed that VFA inhibition is due to their accumulation and a consequent reduction in pH value. However, several experiments have shown that the VFA are

TABLE 24.10 VFA generally present in anaerobic digestion processes (after Stafford, 1981)

Formic acid	$HCOOH$
Acetic acid	CH_3COOH
Propionic acid	$CH_3\,CH_2COOH$
Butyric acid	$CH_3\,CH_2\,CH_2COOH$
Valeric acid	$CH_3\,CH_2\,CH_2\,CH_2COOH$
Hexanoic acid	$CH_3\,CH_2\,CH_2\,CH_2\,CH_2COOH$
Heptanoic acid	$CH_3\,CH_2\,CH_2\,CH_2\,CH_2\,CH_2COOH$
Octanoic acid	$CH_3\,CH_2\,CH_2\,CH_2\,CH_2CH_2\,CH_2COOH$

themselves toxic. For example, depending on pH, propionic acid concentrations in the order of grams per litre can be tolerated with a minimal degree of toxicity. However, at low pH values much more of the propionic acid exists in the undissociated HPr form which is much more toxic than propionate ion, Pr^-, due to its greater membrane permeability. In a well-operating digester running with lightly loaded feed, VFA concentration is typically less than 100 mg/l (Hajarnis and Ranade, 1994).

5.8 Oxygen

Strict anaerobes are very sensitive to oxygen exposure. Oxygen can cause an irreversible dissociation of some enzymes and cofactors, such as F_{420}. Another adverse effect of oxygen stems from its capacity to increase the standard redox potential (E_h). Optimum methane production occurs with a redox potential of between −520 and −530 mV with a limiting value of around −350 mV. Therefore, highly reduced environments (absence of oxygen) should be maintained to promote obligate anaerobic bacteria (Pfeffer, 1979). However, methane production is possible even in the presence of oxygen because many of the fermentative bacteria involved in the initial stages are facultative, and will consume the oxygen that may be present in a reactor feed. Furthermore, anaerobic bacteria often exist in structured communities (e.g. granules, biofilms and flocs) where the outer layers of cells are responsible for creating anoxic or anaerobic core microenvironment suitable for the sensitive methanogens.

5.9 Ammonia

Ammonia is released by the fermentation of amino acids and proteins and the breakdown of methylamine and other nitrogenous compounds (Anderson and Yang, 1992). Although ammonia acts to buffer some of the acidity generated by anaerobic digestion, and will be beneficial to anaerobic bacteria at low concentrations, high concentrations can lead to process failure. Depending on the pH of anaerobic reactors, ammonia can be present either in the form of ammonium ion (NH_4^+) or as ammonia gas in solution (NH_3). The equilibrium between them can be expressed as:

$$NH_4^+ \Leftrightarrow NH_3(aq) + H^+ \qquad pK_b \sim 9$$

At pH values around neutrality, most (>99%) of the NH_3-N will be present as NH_4, which is much less toxic than dissolved ammonia. However, at the higher pH values that are compatible with anaerobic digestion (pH 8), the equilibrium shifts towards the more toxic free ammonia, with concentrations of the latter almost an order of magnitude higher than at pH 7. The maximum concentration of free (dissolved) ammonia should not exceed the inhibitory threshold of 150 mg/l, quoted by Kasapgil (1994), although depending on the operating pH, reactors may be acclimatized reliably to NH_3-N concentrations of several thousand mg/l.

5.10 Sulphide

Inorganic forms of sulphur present in reactor feeds, mainly sulphate, are rapidly converted by sulphate-reducing bacteria (SRB) to the reduced forms, sulphide (S^{2-}), and hydrogen sulphide (H_2S), which are ranked as important inhibitors of anaerobic digestion (Anderson *et al.*, 1982). Sulphides in anaerobic treatment can result from other sulphur-containing compounds in the feed, and will be prevalent during anaerobic protein degradation (McCarty, 1964). It is thought that sulphide (S^{2-}) inhibition of anaerobic digestion arises from:

1. competitive consumption of the methanogenic substrates, acetate and hydrogen, as SRB reduce sulphate to sulphide, lowering methane production
2. its toxic, even fatal, effect on anaerobic microorganisms (it reacts chemically with their cytochromes and other iron-containing compounds)
3. Precipitation of essential trace metals (micronutrients) as insoluble metal sulphides.

The degree of sulphide inhibition is largely dependent on pH, temperature, metal ion concentration and operation type. However, inhibition of acetate-utilizing methanogens has been observed at concentrations of 250 mg S/l (Koster *et al.* 1986). Maillacheruvu and Parkin (1996) showed that the H_2S toxicity of the hydrogen-utilizing methanogens ($K_i = 625$ mg S/l) is relatively weaker than for other microbial groups, explaining why methanogenesis from complex substrates can occur even at high concentrations of sulphide.

Hydrogen sulphide inhibition may be mitigated by increasing the pH (H_2S converted to the less toxic HS^- form), chemically scrubbing and recycling the reactor gas, or the addition of metals (e.g. Fe precipitates the sulphide as iron sulphide and Mo inhibits the SRB that generates the H_2S).

In addition to those mentioned above, Oremland (1988) listed several other chemical and organic compounds which can cause inhibition in anaerobic reactors, e.g. chlorinated methanes, 2-bromoethanesulphonic acid, iodopropane, monensin, unsaturated carbon–carbon bonds, halogenated aromatic compounds, and some aromatic chemicals.

In toxicity control, it is important to diagnose the inhibition first, and then take the necessary actions. This requires close process monitoring. Monitoring should be implemented in either the liquid/slurry phase or the gas phase. The former should include measurements of pH, total and individual VFA, alkalinity, COD, BOD and solids. The latter usually involves measurements of gas production rate and gas composition (methane and carbon dioxide). In addition to these chemical analyses, microbiological analyses (enumeration of anaerobic bacteria by microscopy, agar plate count and MPN), and biochemical analyses (ATP, coenzyme F_{420}, specific methanogenic activity (SMA) and dehydrogenic activity) have been undertaken to give an early indication of metabolic inhibition (Kasapgil, 1994).

Gunnerson and Stuckey (1986) have proposed the following measures to control reactor toxicity:

1. remove toxic substances from the feed
2. dilute the feed to bring levels below the toxic threshold value
3. add chemicals to form a non-toxic complex or insoluble precipitate
4. add an antagonistic substance.

Application of one or more of these measures may attenuate the apparent toxicity of an industrial effluent, making it more amenable to anaerobic treatment.

6 REACTOR CONFIGURATIONS

Anaerobic treatment of wastewaters has existed as a practical technology for over 100 years. It gradually evolved from a simple uncontrolled septic tank system to the high rate, completely controlled reactors now used for treating complex industrial effluents. In anaerobic systems, the key microbial populations generally have a lower reproductive growth rate than their counterparts in aerobic reactors. This requires that a longer sludge retention time (SRT) be provided in order to allow a stable equilibrium to be achieved between the diverse microbial community members in the anaerobic sludge. The loading rates permissible in an anaerobic waste treatment process are, therefore, primarily dictated by the concentration of active biomass within the digester. Consequently, the maintenance of a high SRT has been, at least until recently, the major point of interest in the practical application of anaerobic processes (Lettinga *et al.*, 1980; Anderson and Saw, 1986). As the anaerobic technologies advanced, it became evident that high rate anaerobic treatment could be achieved by employing efficient biomass retention methods. These allowed the solids retention time (SRT) to be maintained well in excess of the hydraulic retention time (HRT) resulting in higher biomass densities within the system. Consequently, improved conservation rates per unit volume of reactor were possible and the high biomass densities also conferred greater resistance to any inhibitory substances in the feed. To accomplish the higher treatment efficiency and reliability associated with a long SRT, a number of novel anaerobic reactor configurations have been developed. During the last two decades, a succession of small modifications to existing reactor designs has enabled engineers to develop completely new reactor configurations. Most of these, however, have only been evaluated at bench-scale and pilot-scale. An illustration of anaerobic reactor configurations covered in this section is given in Fig. 24.8.

6.1 Conventional or completely mixed anaerobic digester

The concept of the conventional anaerobic digester goes back to 1881 when the first anaerobic digestion process was developed by the French inventor, M. Louis Mouras. The reactor, called the 'Mouras automatic scavenger', was an air-tight chamber and was used to liquidify the solid components of sewage. In 1891, the first septic tank to retain solids was constructed in England by W.D. Scott Moncrieff. Later, William Travis developed the 'Travis' tank at Hampton in 1904 in order to separate and ferment septic sewage sludge in a separate reaction chamber. This was followed by the development of the 'Imhoff' tank in Germany by Karl Imhoff in 1905. These simple

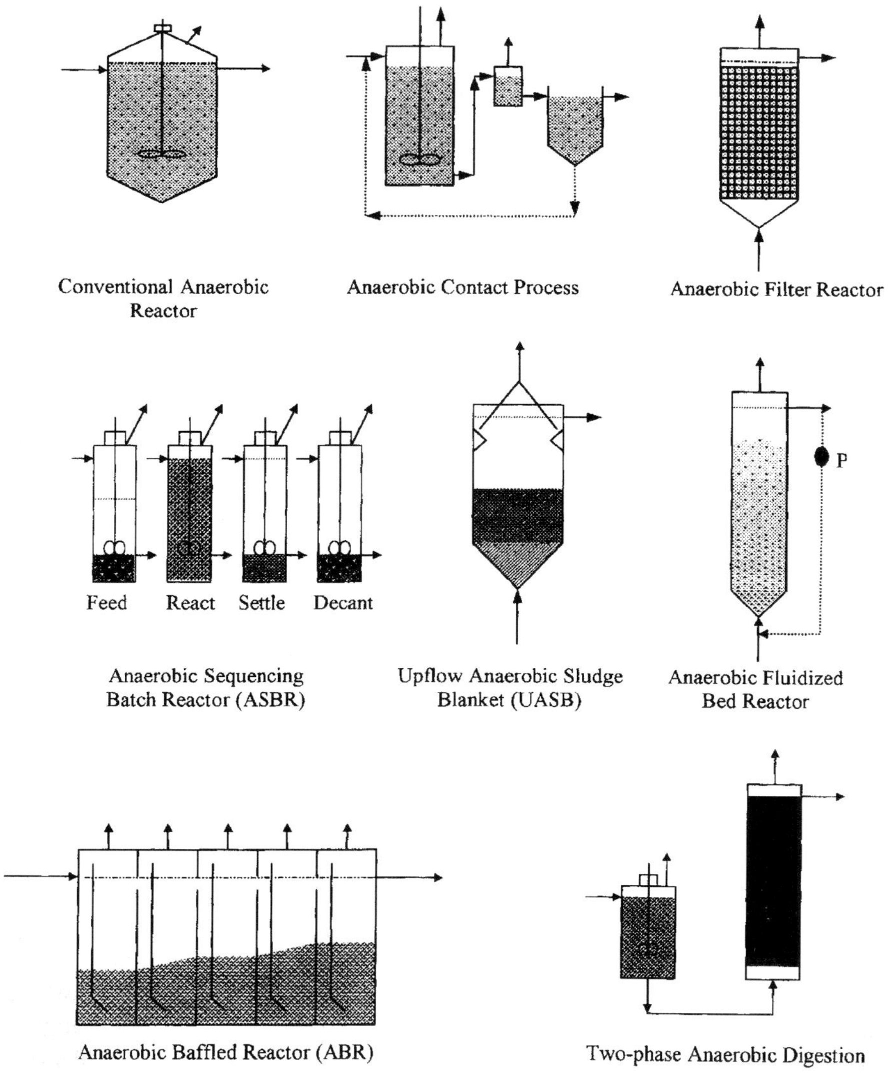

Fig. 24.8 Anaerobic reactor configurations used in wastewater treatment.

unmixed chambers provided a means of enhanced sludge retention within a reactor chamber. Later on, the digesters were also mixed and heated to improve digestion rate. At this time, only the suspended solids fraction of the sewage was subjected to anaerobic treatment and it was not until the development of the anaerobic contact tank by Schroepher in 1955 that anaerobic digestion was used for anaerobic treatment of dilute wastewaters. Thereafter, conventional anaerobic digesters have been used for the degradation of organic matter in both sewage sludges and industrial wastewaters (McCarty, 1985).

The conventional anaerobic digester is a completely mixed reactor with no solids recycle in which the solids retention time equals the hydraulic retention time. In these reactors, wastewater and anaerobic bacteria are mixed together and allowed to react. When the anaerobic bacteria have reduced the organic pollutant load to the desired level, the treated wastewater is then removed from the reactor for disposal. This system can be operated in either batch or continuous mode and depends on the continuous growth of new biomass to replace that lost in the effluent. We know that some key methanogens are slow-growing organisms with doubling times of 5–10 days; hence at least 10 days of solid and hydraulic retention time will be required. This will necessitate a reactor with a very large volume. The time requirement can be reduced by applying phase separation/solid recycle process, i.e. two tanks, one for mixing and reaction, and the other for solids settlement and storage (see Anaerobic Contact Process).

The large volume requirements and wash-out of microorganisms in the effluent pose serious problems downstream and make conventional anaerobic digesters unsuitable for use with most industrial wastewaters. However, they can be used successfully for sludge treatment and for wastewaters that contain high solids and organic matter content.

6.2 Anaerobic contact process

The link between high biomass concentration, greater efficiency and smaller reactor size seeded the idea of the contact tank process. The principle of the anaerobic contact reactor has parallels with the activated sludge process, i.e. settling of microbial sludge in a settling tank and its return back to the reactor allows further contact between biomass and raw waste (see Fig. 24.8). In anaerobic contact reactors, due to the arrangement for solids recycle, the SRT is no longer coupled to the HRT as in conventional digesters. As a result, considerable improvements in treatment efficiency can be achieved. Good mixing is essential within the anaerobic contact digester so that adequate mass transfer occurs between the feed and active biomass.

The major shortcoming for contact reactors is their reliance on favourable settling characteristics of the anaerobic sludge, a requirement which may not be fulfilled at all times. Poor sludge settlement arises from the growth of filamentous anaerobic bacteria and gas formation by anaerobic bacteria in the settling tank. Anderson and Saw (1986) suggested that filamentous growth may be a result of nutrient deficiency and this can be rectified by careful maintenance of the N/P ratio. The problems of gas formation can be minimized by employing vacuum degasification or thermal shock prior to sedimentation, by using flocculating agents in the settling tank or by incorporating inclined plates into the settler design.

Although simple in concept, the individual component units of the anaerobic contact process make it more complex than other high rate anaerobic digester systems, however, the absence of any internal fittings offers some advantages for the treatment of wastes having a high solids content (Wheatley, 1990). Typical performance of the anaerobic contact process is reported in Table 24.11.

6.3 Anaerobic sequencing batch reactor (ASBR)

The anaerobic sequencing batch reactor (ASBR) process was developed by Dague's group at Iowa State University. It is a batch-fed, batch-decanted, suspended growth system and is operated in a cyclic sequence of four stages: feed, react, settle and decant (Wirtz and Dague,

TABLE 24.11 Typical operating conditions of various anaerobic digester configurations

Reactor type	*Load (kg COD/m³day)*	*HRT (hour)*	*COD removal (%)*
Conventional anaerobic reactor	1–5	240–360	60–80
Anaerobic contact reactor	1–6	24–120	70–95
Anaerobic sequencing batch reactor	1–10	6–24	75–90
Anaerobic filter	2–15	10–85	80–95
Fluidized bed	2–50	1–4	80–90
UASB	2–30	2–72	80–95
Anaerobic baffled reactor	3–35	9–32	75–95
Two phase anaerobic digestion	5–30	20–150	70–85

1996). It comprises a single tank in which all events take place, and since a significant part of the cycle-time is spent settling the biomass from the treated wastewater, the reactor volume requirement is higher than for continuous flow processes. However, this disadvantage is largely offset by its simplicity (it requires no additional biomass settling stage or solids recycle) and the absence of feed short-circuiting which often occurs in continuous flow systems. It has also been reported that biomass granulation (see the next section) can occur in an ASBR after long periods (300 days) of operation. Wirtz and Dague (1996) claimed that the time required for granulation could be shortened by approximately 2 months when granulation enhancements such as granular active carbon (GAC), silica, polymers and ferric chloride were utilized. Operational cycle-times for the ASBR can be as short as 6 hours (Wirtz and Dague, 1996; Banik *et al.*, 1997).

6.4 Anaerobic packed bed or anaerobic filter

Introduced by Coulter *et al.* in 1957 and developed by Young and McCarty in 1967, the anaerobic filter is a fixed-film biological wastewater treatment process in which a fixed matrix (support medium) provides an attachment surface that supports the anaerobic microorganisms in the form of a biofilm. Treatment occurs as the wastewater flows upwards through this bed and the dissolved pollutants are absorbed by the biofilm (Young, 1983). Anaerobic filters were the first anaerobic systems that eliminated the need for solids separation and recycle while providing a high SRT: HRT ratio. Various types of support material can be used, such as plastics, granular activated carbon (GAC), sand, reticulated foam polymers, granite, quartz and stone. These are usually particulate, moulded or sheet materials that have been produced specifically for the task, often having exceptionally high surface area to volume ratios ($400\,m^2/m^3$) and low void volumes. The simplicity and robustness of the anaerobic filter, i.e. its resilience to shock loads and operational perturbations (e.g. pH, flow rate inhibitors etc.), make it ideally suited to treatment of both dilute soluble wastewaters and higher-strength soluble wastewaters that can be diluted by recycling (Wheatley, 1990).

The limitations of the anaerobic filter are mostly physical ones related to the deterioration of the bed structure through a gradual accumulation of non-biodegradable solids. This leads eventually to channelling and short-circuiting of flow, and anaerobic filters are therefore unsuitable for wastewaters with a high solids content. Additionally, there is a relatively high cost associated with the packing materials. To overcome some of these problems, packed bed reactors can be operated in a downflow mode which forces the non-biodegradable solids out of the system.

6.5 Anaerobic fluidized and expanded bed reactors

The concerted research efforts of the mid-1970s greatly improved the performance of existing anaerobic technologies and resolved many of the early operational problems, e.g. the difficulties of biomass separation experienced with conventional and contact anaerobic digesters and the blocking problems encountered in anaerobic filters. One such development was the fluidized bed reactor, originally proposed by Jeris in 1974 during denitrification studies (van Haandel and Lettinga, 1994). Adaptation of this system to anaerobic digestion was made by Jewell and his colleagues in the 1970s in an attempt to develop a biological reactor that would accumulate a maximum active attached biomass yet still handle fine suspended solids without blocking (Jewell, 1985). They suggested that if the surface area available for microbial attachment was maximized and the volume occupied by the media minimized, then this should achieve a maximum specific activity of attached biomass for a given reactor volume. In practice, a filter containing extremely small particles (0.5 mm) provides adequate surface area to achieve these benefits. In order to achieve fluidization of the biomass particles, units must be operated in an upflow mode. This simplistic design led to the development of a unit similar to fluidized bed reactors called the 'anaerobic attached film expanded bed reactor' (Jewell, 1980). In fact, it is only the rate of liquid flow and the resulting degree of bed expansion that determines whether the reactor is termed a fluidized bed or expanded bed system (van den Berg and Kennedy, 1983). Expanded bed reactors have a bed expansion of 10% to 20% compared to 30% to 90% in fluidized beds.

In the fluidized bed reactor, the biomass is attached to the surface of small, low specific gravity particles (such as anthracite, high density plastic beads, sand etc.) which are kept in suspension by the upward velocity of the liquid flow (Anderson and Saw, 1986). Effluent is recycled to dilute the incoming waste and to provide sufficient flow-rate to keep particles in suspension. The large surface area of support particles and high degree of mixing that results from the high vertical flows enable a high concentration of biomass to develop and efficient substrate uptake kinetics, respectively. Biomass concentrations between 15 000 and 40 000 mg/l have been reported in these systems (Metcalf and Eddy, 1991).

The greatest risk with the fluidized bed and expanded bed reactors is the loss of biomass particles from the reactor following sudden changes in particle density, flow rate or gas production. If flow is interrupted and the bed allowed to settle, there is a tendency once flow is restarted for the entire bed to move upward in plug-flow rather than fluidizing. Furthermore, in practical applications, considerable difficulties have been experienced in controlling the particle size and density of the flocs due to variable amounts of biomass growth on the particles. Fluidized beds have therefore acquired the reputation for being difficult to operate (Wheatley, 1990; van Haandel and Lettinga, 1994). Such operational drawbacks are often considered to outweigh the inherent performance benefits of these reactors.

6.6 Upflow anaerobic sludge blanket (UASB) reactor

The problem associated with packed bed anaerobic filters and fluidized or expanded bed reactors has led to the development of unpacked reactors that still incorporate an immobilized form of particulate biomass. One of the most successful high-rate unpacked anaerobic digester configurations was developed during the 1970s by Lettinga *et al.* (1980) in the Netherlands, and was descriptively termed the 'upflow anaerobic sludge blanket (UASB) reactor'. This reactor was very similar to the first upflow sludge blanket reactor, the 'biolytic tank', used by Wilson and Phelps in 1910. Lettinga *et al.* (1980) have modified this design and applied it to wastewater treatment in a number of studies; however, recent emphasis has been on its application for warm and more highly concentrated industrial wastewaters and it has been successfully commercialized in this respect (Jewell, 1985). In fact, the UASB reactor is by far the most widely used high-rate anaerobic system for domestic and industrial wastewater treatment (van Haandel and Lettinga, 1994).

The UASB reactor is based on the premise that anaerobic sludge exhibits inherently good settling properties, provided the sludge is not exposed to heavy mechanical agitation. Adequate mixing between sludge and wastewater is provided by an even flow-distribution combined with a sufficiently high upflow velocity, and by the agitation that results from gas production (Lettinga, 1995). The biomass in this type of reactor is retained as a blanket or granular matrix, and is kept in suspension by controlling the upflow velocity. The wastewater flows upwards through an expanded bed of active sludge located in the lower part of the reactor, while the upper part contains a three-phase (solid, liquid, gas) separation system. The latter facilitates the collection of biogas and also provides internal recycling of the sludge by disengaging adherent biogas bubbles from rising sludge particles (Anderson and Saw, 1986). van Haandel and Lettinga (1994) considered the three-phase separation device to be the most characteristic feature of the UASB. This device, which is located at the top of the reactor, creates three separate internal zones: the digestion zone, the settling zone and the gas and liquid separation zone. Typical performance of the UASB is presented in Table 24.11. Although UASB reactors have been built and operated since 1971, it was not until 1974–1976 that a particularly desirable variant of the anaerobic sludge, granular sludge, was observed inside a pilot plant treating sugar beet factory wastewater. The superior settling characteristics of these sludge granules allowed higher sludge concentrations to be retained and consequently permitted the system to achieve much higher organic loading rates. Granular sludge development has now been observed in UASB reactors treating many different types of wastewater (de Zeeuw, 1987).

The phenomenon of granulation is a process in which a non-discrete flocculent biomass begins to form discrete well-defined pellets, or granules. These vary in dimension and appearance depending on the wastewater and reactor conditions, but generally have a flattened spherical geometry with a diameter of 1–3 mm. The mechanism of biomass granulation has been widely studied the objective being that the rate and extent of granule formation could be manipulated, particularly in wastewaters that show little intrinsic propensity to granulate (e.g. fat and oil containing effluents).

The consequences of granulation and its influence on the physiology of anaerobic bacteria and UASB reactor performance have been reviewed by Calleja *et al.* (1984). These are summarized as follows:

1. leads to internal physicochemical gradients within the aggregates
2. leads to heterogeneous structured populations of syntrophic microorganisms
3. affects overall stoichiometry, rates of growth and metabolism
4. allows the manipulation of growth rate independent of the dilution rate
5. allows the manipulation of biomass as a single phase
6. generates a reactor effluent with low suspended solids
7. allows high biomass concentrations in continuous reactors
8. allows reactors to be operated continuously beyond normal washout flow rates.

There are several hypotheses to explain the development of granulation, although it is likely that a complex combination of physical, chemical and biological interactions between the dissolved and suspended solids and the microorganisms is required, rather than a single condition.

The phenomenon of granulation has many aspects in common with biofilm formation, a process that has been studied extensively, and it has been postulated that granule formation is probably dependent on the microorganisms participating in the same critical events (Schmidt and Ahring, 1996). These are:

1. transport of microbial cells to the surface of an uncolonized inert material or other microbial cells (termed 'substratum' in the following points)
2. initial reversible adsorption to the substratum by physicochemical forces
3. irreversible adhesion of the cells to the substratum by microbial appendages and/or polymers attaching the cell to the substratum

4. replication of the cells and the development of the granules.

This conceptual chain of events does not address the more complex and interesting issue of how the microbiological and biochemical processes of substrate (pollutant) utilization are linked spatially within the granule. It is probable that the microbial distribution within the internal architecture of the granules is a highly ordered arrangement, with interdependent syntrophic cells existing in juxtaposition to maximize substrate transport efficiency. Such a spatial organization of microorganisms would in part explain the high specific activity of granular sludge relative to flocculent biomass.

6.7 Anaerobic baffled reactor (ABR)

The anaerobic baffled reactor was developed by Bachmann *et al.* (1983), and evolved from initial studies with an anaerobic rotating biological disc reactor. They described the concept of the reactor as essentially a number of upflow sludge blanket reactors connected in series. The design consists of a series of staggered vertical baffles which the wastewater passes over and under as it flows from inlet to outlet (Grobicki and Stuckey, 1991). In the last 20 years much research has been carried out with the ABR, mostly at laboratory-scale. This work has identified some of the advantages and disadvantages of using anaerobic baffled reactors. The main advantages are:

1. it combines the advantages of the anaerobic filter, which has a high stability and reliability due to attachment of the biological solids onto and between the filter media, and the upflow anaerobic sludge process, in which the microbial mass itself functions as the support medium for microorganism attachment, leading to a high void volume (Bachmann *et al.*, 1983)
2. its unique baffled design enables the system to reduce biomass washout, hence retain a high active biological solids content (Polprasert *et al.*, 1992)
3. the reactor can recover remarkably quickly from hydraulic and organic shock loads e.g. 9 hours (Grobicki and Stuckey, 1991; Nachaiyasit and Stuckey, 1997)
4. it has a simple design and requires no special gas or sludge separation equipment (Bachmann *et al.*, 1983; Polprasert *et al.*, 1992; Yu *et al.*, 1997)
5. owing to its compartmentalized configuration, it may function as a two-phase anaerobic treatment system with some spatial separation of acidogenic and methanogenic biomass (Nachaiyasit and Stuckey, 1997; Uyanik *et al.*, 2002b)
6. the fraction of dead space in the ABR is low compared to other designs of anaerobic digesters (Grobicki and Stuckey, 1992)
7. the ABR can be used for almost all soluble organic wastewater from low to high strength (Polprasert *et al.*, 1992)
8. considering its simple structure and operation, it could be considered a potential reactor system for treating municipal and domestic wastewater in tropical and subtropical areas of developing countries (Yu and Anderson, 1996).

There are few significant drawbacks to the ABR mentioned in the literature. The only real disadvantages that can be levelled against this reactor is that it has not been widely used at full-scale. The ABR has been reviewed extensively by Barber and Stuckey (1999).

6.8 Two-phase anaerobic digestion

Two-phase anaerobic digestion implies a process configuration employing separate reactors for acidification and methanogenesis. These are connected in series, allowing each phase of the digestion process to be optimized independently since the microorganisms concerned have different nutritional requirements, physiological characteristics, pH optima, growth and nutrient uptake kinetics, and tolerances to environmental stress factors (Cohen, 1983). The idea of a two-phase anaerobic digestion process was originally proposed by Pohland and Ghosh (1971). It aims to enhance anaerobic biodegradation through a controlled separation of the major reactions, with the hydrolysis, fermentation and acidification reactions contained within the first phase and the acetogenic and methanogenic reactions predominating in the second phase (Fox and Pohland, 1994). Owing to the fact that bacteria associated with the former reactions have the highest growth and activity rates, the acidogenic reactor will always be smaller than the methanogenic phase reactor.

Several different methods of achieving two-phase separation have been suggested. These include selective inhibition of certain groups of microorganisms, kinetic control, pH control and membrane separation (Anderson and Saw, 1986).

A number of advantages and disadvantages of the two-phase digestion process have been noted by several authors (Cohen, 1983; van den Berg, 1984; Anderson and Saw, 1986; van Haandel and Lettinga, 1994; Kasapgil, 1994; Fox and Pohland, 1994). These are:

Advantages:

1. improvement in process control
2. disposal of excess fast growing acidogenic sludge without any loss of slow growing methanogenic bacteria
3. degradation and attenuation of toxic materials in the first phase (protects the sensitive methanogens)
4. precise pH-control in each reactor
5. higher methane content in the biogas from the methanogenic phase
6. increased loading rate possible for the methanogenic stage
7. balancing tanks in existing treatment plants might be readily converted to acidification tanks for two-phase operation.

Disadvantages:

1. possible disruption of syntrophic relationships
2. high sludge accumulation in the first phase
3. lack of process experience and so more difficult to operate
4. difficulty maintaining a balanced segregation of the phases.

REFERENCES

Anderson, G.K., Donnelly, T. and McKeown, K.J. (1982). Identification and control of inhibition in the anaerobic treatment of industrial wastewater. *Process Biochemistry* **17**, 28–32.

Anderson, G.K. and Saw, C.B. (1986). State of the art of anaerobic digestion for industrial applications in the UK. In: *Proceedings of the 39th Industrial Waste Conference*, pp. 783–793. Purdue University, Indiana.

Anderson, G.K. and Yang, G. (1992). pH control in anaerobic treatment of industrial wastewater. *Journal of Environmental Engineering ASCE* **118**(4), 551–567.

Archer, D.B. and Kirsop, B.H. (1990). The microbiology and control of anaerobic digestion. In: A. Wheatley (ed.) *Anaerobic Digestion: a waste treatment technology*, pp. 43–91. Elsevier Applied Science, London.

Bachmann, A., Beard, V.L. and McCarty, P.L. (1983). Comparison of fixed-film reactors with a modified sludge blanket reactor. In: *Proceedings of the 1st International Conference on Fixed Film Biological Processes*, pp. 1192–1211. Noyes Date Corporation, New Jersey.

Balch, W.E., Fox, G.E. and Magrum, L.J. *et al.* (1979). Methanogens: re-evaluation of a unique biological group. *Microbiological Reviews* **43**, 260–296.

Banik, G.C., Ellis, T.G. and Dague, R.R. (1997). Structure and methanogenic activity of granules from an ASBR treating dilute wastewater at low temperatures. *Water Science and Technology* **36**(6–7), 149–156.

Barber, W.P. and Stuckey, D.C. (1999). The use of the anaerobic baffled reactor (ABR) for wastewater treatment: A review. *Water Research* **33**(7), 1559–1578.

Calleja, G.B., Atkinson, D.R. and Garrod, D.R. *et al.* (1984). Aggregation group report. In: K.C. Marshall (ed.) *Microbial Adhesion and Aggregation*, pp. 303–321. Springer Verlag, Berlin.

Cohen, A. (1983). Two-phase digestion of liquid and solid wastes. In: R.L. Wentworth *et al.* (eds) *Third International Symposium on Anaerobic Digestion*, pp. 123–138. Cambridge University Press, Massachusetts.

Colleran, E., Finnegan, S. and Lens, P. (1995). Anaerobic treatment of sulphate-containing waste streams. *Antonie van Leeuwenhoek* **67**, 29–46.

Collins, L.J. and Paskins, A.R. (1987). Measurement of trace concentrations of hydrogen in biogas from anaerobic digesters using an exhaled hydrogen monitor. *Water Research* **21**, 1567–1572.

de Zeeuw, W. (1987). Granular sludge in UASB reactors. In: *Proceedings of the GASMAT Workshop*, pp. 132–145. Lunteren, The Netherlands.

Fox, P. and Pohland, F.G. (1994). Anaerobic treatment applications and fundamentals: substrate specificity during phase separation. *Water Environment Research* **66**, 716–724.

Greenberg, A.E., Trussell, R.R. and Clisceri, L.S. (eds) (1985). *Standard Methods for Examination of Water and Wastewater*, 16. American Public Health Association, Washington, DC.

Grobicki, A. and Stuckey, D.C. (1991). Performance of the anaerobic baffled reactor under steady-state and shock loading conditions. *Biotechnology and Bioengineering* **37**, 344–355.

Grobicki, A. and Stuckey, D.C. (1992). Hydrodynamic characteristics of the anaerobic baffled reactor. *Water Research* **26**, 371–378.

Gujer, W. and Zehnder, A.J.B. (1983). Conversion processes in anaerobic digestion. *Water Science and Technology* **15**, 127–167.

Gunnerson, C.G. and Stuckey, D.C. (1986). *Anaerobic digestion: principles and practices for biogas system, The World Bank Technical Paper Number 49*, Washington DC.

Hajarnis, S.R. and Ranade, D. R. (1994). Effect of propionate toxicity on some methanogens at different pH values and in combination with butyrate. Paper presented at the 7th International Symposium on Anaerobic Digestion, Cape Town, South Africa.

Harper, S.R. and Pohland, F.G. (1986). Recent developments in hydrogen management during anaerobic biological wastewater treatment. *Biotechnology and Bioengineering* **28**, 585–602.

Harmsen, H.J.M., Akkermans, A.D.L., Stams, A.J.M. and de Vos, W.M. (1996). Population dynamics of propionate-oxidizing bacteria under methanogenic and sulfidogenic conditions in anaerobic granular sludge. *Applied and Environmental Microbiology* **62**(6), 2163–2168.

Hawkes, F.R. (1980). The biochemistry of anaerobic digestion. In: R. Buvet, M.F. Fox, D.J. Picken (eds) *Biomethane, Production and Uses*. Sec.2.2. Roger Bowskill Ltd. Exeter.

Henze, M. and Harremoes, P. (1983). Anaerobic treatment of wastewater in fixed film reactors – a literature review. *Water Science and Technology* **15**(8–9), 1–101.

Hickey, R.F., Vanderwiellen, J. and Switzenbaum, M.S. (1989). The effect of heavy metals on methane production and hydrogen and carbon monoxide levels during batch anaerobic sludge digestion. *Water Research* **23**(2), 207–218.

Holland, K.T., Knapp, J.S. and Shoesmith, J.G. (1987). *Anaerobic bacteria; tertiary level biology.* Blackie and Son Ltd, London.

Ince, O. (1993). *Control of biomass in anaerobic reactors using ultrafiltration membranes*. PhD. Thesis. Newcastle University, Newcastle.

Ince, O., Anderson, G.K. and Kasapgil, B. (1997). Composition of the microbial population in a membrane anaerobic reactor system during start-up. *Water Research* **31**(1), 1–10.

Jewell, W.J. (1980). Future trends in digester design. In: D.A. Stafford, B.I. Wheatley, D.E. Hughes (eds) *Anaerobic Digestion*. Applied Science, London.

Jewell, W.J. (1985). The development of anaerobic wastewater treatment. In: M.S. Switzenbaum (ed.) *Proceedings of the Seminar/Workshop Anaerobic Treatment of Sewage*, pp. 17–54. National Science Foundation, Amherst.

Kasapgil, B. (1994). *Two-phase anaerobic digestion of dairy wastewater*. PhD Thesis. Newcastle University, Newcastle.

Kirsop, B.H. (1984). Methanogenesis. *CRC Critical Reviews in Biotechnology* **1**, 109–160.

Klass, D.L. (1984). Methane from anaerobic science. *Science* **223**(4640), 1021–1028.

Koster, I.W., Rinzema, A., de Vegt, A.L. and Lettinga, G. (1986). Sulfide inhibition of the methanogenic activity of granular sludge at various pH levels. *Water Research* **20**(12), 1561–1567.

Lettinga, G. (1995). Anaerobic digestion and wastewater treatment systems. *Antonie van Leeuwenhoek* **67**, 3–28.

Lettinga, G., van Velsen, A.F.M., Hobma, S.W. *et al.* (1980). Use of upflow sludge blanket (USB) reactor concept for biological wastewater treatment, especially for anaerobic treatment. *Biotechnology and Bioengineering* **22**, 699–734.

Levett, P.N. (1990). *Anaerobic bacteria; a functional biology.* Open University Press, Milton Keynes and Philadelphia.

Mackiewicz, N. and Wiegel, J. (1998). Comparison of energy and growth yields for *Desulfitobacterium dehalogenansi* during utilization of chlorophenol and various traditional electron acceptors. *Applied Environmental Microbiology* **64**(1), 352–355.

Madigan, M.T., Martinko, J.M. and Parker, J. (1997). *Brock's Biology of Microorganisms*. Prentice-Hall Int. Inc., New Jersey.

Maillacheruvu, K.Y. and Parkin, G.F. (1996). Kinetics of growth, substrate utilization and sulfide toxicity for propionate, acetate and hydrogen utilizers in anaerobic systems. *Water Environmental Resesearch* **68**, 1099–1106.

McCarty P.L. (1964). Anaerobic waste treatment fundamentals: part one, part three and part four. *Public Works* **95** (Part 1 September, 107–112; Part 3 November, 91–94; Part 4 December, 95–099.).

McCarty, P.L. (1981). One hundred years of anaerobic treatment. In: D.E. Hughes *et al.* (eds) *Anaerobic Digestion 1981*, pp. 3–23. Elsevier Biomedical Press, Amsterdam.

McCarty, P.L. (1985). Historical trends in the treatment of dilute wastewaters. In: M.S. Switzenbaum (ed.) *Proceedings of the Seminar/Workshop Anaerobic Treatment of Sewage*, pp. 3–16. National Science Foundation, Amherst.

Mehrotra, I., Alibhai, K.R.K. and Forster, C.F. (1987). The removal of heavy metals in anaerobic upflow sludge blanket reactors. *Journal of Chemical Technology and Biotechnology* **37**, 195–202.

Metcalf and Eddy, Inc. (1991). *Wastewater engineering; treatment, disposal and reuse*. McGraw-Hill Book Co., Singapore.

Morgan, J.W., Evison, L.M. and Forster, C.F. (1991). Changes to the microbial ecology in anaerobic digesters treating ice-cream wastewater during start-up. *Water Research* **25**, 639–653.

Mosey, F.E. (1976). Assessment of the maximum concentration of heavy metals in crude sewage which will not inhibit the anaerobic digestion of sludge. *Journal of Water Pollution Control* **75**, 10–20.

Nachaiyasit, S. and Stuckey, D.C. (1997). The effect of shock loads on the performance of an anaerobic baffled reactor (ABR): 1. Step changes in feed concentration at constant time. *Water Research* **31**(11), 2737–2746.

Oremland, R.S. (1988). Biochemistry of methanogenic bacteria. In: A.J.B. Zehnder (ed.) *Biology of Anaerobic Microorganisms*, pp. 641–705. John Wiley and Sons Inc., New York.

Pfeffer, J.T. (1979). Anaerobic digestion processes. In: D.A. Stafford, B.I. Wheatley, D.E. Hughes (eds) *Anaerobic Digestion 1979 (1. International Symposium)*, pp. 15–33. Applied Science Publishers, Cardiff.

Pohland, F.G. and Ghosh, S. (1971). Developments in anaerobic stabilization of organic wastes: the two-phase concept. *Environmental Letters* **1**, 255–266.

Polprasert, C., Kemmadamrong, P. and Tran, F.T. (1992). Anaerobic baffle reactor (ABR) process treating a slaughterhouse wastewater. *Environmental Technology* **13**, 857–865.

Raskin, L., Poulsen, L.K., Noguera, D.L. *et al.* (1994). Quantification of methanogenic groups in anaerobic biological reactors by oligonucleotide probe hybridization. *Applied and Environmental Microbiology* **60**(4), 1241–1248.

Rocheleau, S., Greer, C.W., Lawrence, J.R. *et al.* (1999). Differentiation of Methanosaeta concilii and Methanosarcina barkeri in anaerobic mesophilic granular sludge by fluorescent in situ hybridization and confocal scanning laser microscopy. *Applied and Environmental Microbiology* **65**(5), 2222–2229.

Sawyer, C.N., McCarty, P.L. and Parkin, G.F. (1994). *Chemistry for environmental engineering*, 4th edn. McGraw-Hill Inc., Singapore.

Scherer, P., Lippert, H. and Wolff, G. (1983). Composition of the major elements and trace elements of 10 methanogenic bacteria determined by inductively coupled plasma emission spectrometry. *Biological Trace Element Research* **5**, 149–163.

Schlegel, H.G. (1993). *General Microbiology* (translated by Kogut, M.) 2nd edn. Cambridge University Press, Cambridge.

Schmidt, J.E. and Ahring, B.K. (1996). Granular sludge formation in upflow anaerobic sludge blanket (UASB) reactors. *Biotechnology and Bioengineering* **49**, 229–246.

Siebert, M.L. and Hattingh, W.H.J. (1967). Estimation of methane-producing bacterial numbers by the most probable number (MPN) technique. *Water Research* **1**, 13–19.

Siebert, M.L. and Hattingh, W.H.J. (1987). Estimation of methane-producing bacterial numbers by the most probable number (MPN) technique. *Water Research* **27**, 13–19.

Smith, L.C., Elliot, D.J. and James, A. (1996). Mixing in upflow anaerobic filters and its influence on performance and scale-up. *Water Research* **30**(12), 3061–3073.

Speece, R.E. (1996). *Anaerobic Biotechnology for Industrial Wastewater.* Archea Press, Tennessee.

Stafford, D.A. (1981). The effects of mixing and volatile fatty acid concentration on anaerobic digester performance. *Trib. Cebedeau* **34**(456), 493–500.

Stanier, R.Y., Ingraham, J.L., Wheelis, M.L. and Painter, P.R. (1986). *General Microbiology*, 5th edn. Prentice Hall, New Jersey.

Taylor, G.T. (1982). The methanogenic bacteria. *Progress in Industrial Microbiology* **16**, 233–329.

Thauer, R.K., Jungermann, K. and Decker, K. (1977). Energy conservation in chemotrophic anaerobic bacteria. *Bacteriological Reviews* **41**, 100–180.

Toerien, D.F. and Hattingh, W.H.J. (1969). Anaerobic digestion 1. The microbiology of anaerobic digestion. *Water Research* **3**, 385–416.

Uyanik, S., Sallis, P.J. and Anderson, G.K. (2002a). Improved split feed anaerobic baffled reactor (SFABR) for shorter start-up period and higher process performance. *Water Science & Technology* **46**: 223–230.

Uyanik, S., Sallis, P.J. and Anderson, G.K. (2002b). The effect of polymer addition on granulation in an anaerobic baffled reactor (ABR). Part II: Compartmentalization of bacterial populations. *Water Research* **36**: 944–955.

van den Berg, L. (1984). Developments in methanogenesis from industrial wastewater. *Canadian Journal of Microbiology* **30**, 975–989.

van den Berg, L. and Kennedy, K.J. (1983). Comparison of advanced anaerobic reactors. In: R.L. Wentworth (ed.) *Third International Symposium on Anaerobic Digestion*, pp. 71–79. Cambridge University Press, Massachusetts.

van Haandel, A.C. and Lettinga, G. (1994). *Anaerobic sewage treatment: a practical guide for regions with a hot climate.* John Wiley and Sons, Chichester.

Wheatley, A.D. (ed.) (1990). Anaerobic digestion: industrial waste treatment. *Anaerobic Digestion: a Waste Treatment Technology*, pp. 171–224. Elsevier Science, London.

Whitman, W.B., Bowen, T.L. and Boone, D.R. (1992). The methanogenic bacteria. In: A. Balows (ed.) *The Prokaryotes: a handbook of bacteria ecophysiology, isolation, identification, application*, 2nd edn. Springer-Verlag, New York.

Willis, A.T. (1977). *Anaerobic bacteriology; clinical and laboratory practice.* Butterworths, London.

Willis, A.T. (1990). Anaerobic culture methods. In: P.N. Levett (ed.) *Anaerobic bacteria; a functional biology*, pp. 1–12. Open University Press, Milton Keynes.

Wirtz, R.A. and Dague, R.R. (1996). Enhancement of granulation and start-up in the anaerobic sequencing batch reactor. *Water Environment Research* **68**, 883–892.

Young, J.C. (1983). The anaerobic filter-past, present and future. In: R.L. Wentworth *et al.* (eds) *Third International Symposium on Anaerobic Digestion*, pp. 91–106. Cambridge University Press, Massachusetts.

Yu, H. and Anderson, G.K. (1996). Performance of a combined anaerobic reactor for municipal wastewater treatment at ambient temperature. *Resources Conservation and Recycling* **17**, 259–271.

Yu, H., Tay, J. and Wilson, F. (1997). A sustainable municipal wastewater treatment process for tropical and subtropical regions in developing countries. *Water Science and Technology* **35**(9), 191–198.

Zehnder, A.J.B., Ingvorsen, K. and Marti, T. (1982). Microbiology of methane bacteria. In: D.E. Hughes *et al.* (eds) *Anaerobic Digestion 1981*, pp. 45–68. Elsevier Biomedical Press B.V.

Zeikus, J.G., Kerby, R. and Krzycki, J.A. (1985). Single-carbon chemistry of acetogenic and methanogenic bacteria. *Science* **227**, 1167–1173.

25

The nitrogen cycle and its application in wastewater treatment

Chien Hiet Wong*, Geoff W. Barton† and John P. Barford‡

*Environmental Engineering Division, CPG Consultants Pte Ltd, Novena Square, Singapore 307685; †Department of Chemical Engineering, University of Sydney, Australia; ‡Department of Chemical Engineering, Hong Kong University, Kowloon, Hong Kong

1 THE BIOLOGICAL NITROGEN CYCLE

Nitrogen is a necessary element for all living organisms. Living cells contain up to 14% nitrogen, which forms an essential part of several key cell components, such as protein and DNA.

The biological nitrogen cycle can be summarized as in Fig. 25.1. Various forms of nitrogenous compounds exist in nature with oxidation states ranging from −3 to +5. Even though (di)nitrogen gas makes up around 79% of the atmosphere, it is chemically extremely inert and generally unavailable to living organisms. Only a select group of microorganisms (*Rhizobium, Azotobacter, Clostridium*) are capable of fixing nitrogen gas directly into a biological usable form (Postgate, 1998). The product of such nitrogen fixation is ammonia, and this is the essential 'gateway' for the formation of organic nitrogenous compounds, such as plant proteins. In the various assimilatory pathways, nitrogenous compounds of higher oxidation states (such as nitrate and nitrite) can also be reduced to organic nitrogen via ammonia.

Degradation and decay of plant and animal material returns nitrogen to an inorganic form, the principal component of which again is ammonia (via ammonification). In nitrification, another select group of microorganisms is capable of oxidizing free ammonia to nitrogenous compounds of higher oxidation states. The end product of nitrification is nitrate, while in denitrification, nitrate is reduced by a group of respiratory microorganisms to nitrogen gas.

1.1 Imbalance within the nitrogen cycle

As can be seen in Fig. 25.1, the balance of the nitrogen cycle depends on the combined activities of nitrogen fixation, nitrification, denitrification, nitrogen assimilation and dissimilation. Nitrogen fixation and assimilatory pathways form that part of the cycle (moving from left to right in Fig. 25.1) that is most useful for agriculture, given that the success of commercial plant growth is usually limited by the source of nitrogen. Industrial processes (such as the Haber process) have, thus, been used for the manufacture of nitrogen-based fertilizers, which can be used to improve agricultural performance.

Increased urbanization and industrialization, however, lead to intense cultivation practices that can locally distort the natural balance of the nitrogen cycle and may result in the accumulation of nitrogenous compounds (such as ammonia and nitrate) in rivers and lakes. Typical nitrogen-rich wastewater includes municipal sewage, as well as landfill leachate and wastewater from abattoirs and livestock farms. Table 25.1 summarizes typical concentrations of nitrogenous compounds in these forms of wastewater.

Ammonia, nitrite and nitrate are toxic to aquatic life if present at sufficiently high

The Handbook of Water and Wastewater Microbiology
ISBN 0-12-470100-0

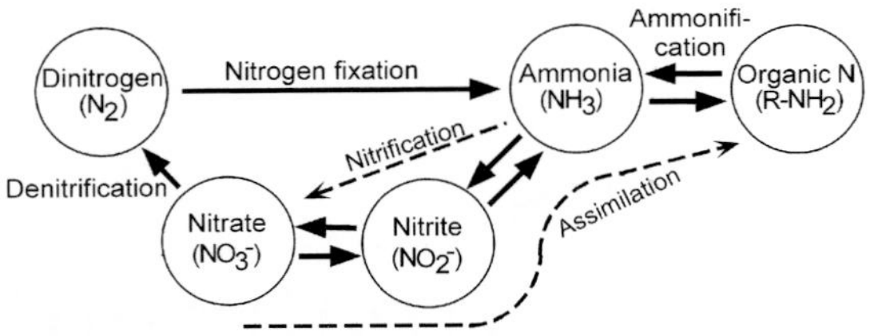

Fig. 25.1 The biological nitrogen cycle.

concentrations. In addition, such nitrogenous material is also known to stimulate growth of algae and other photosynthetic aquatic life, which can lead to eutrophication via excessive loss of available oxygen, and other undesirable changes in the aquatic ecological system. Hence, it is necessary to prevent the accumulation of such nitrogenous compounds in aquatic systems.

To restore the balance of an ecological system, it is natural to exploit the biological mechanisms within the nitrogen cycle to eliminate any accumulation of ammonia and nitrate. More specifically, nitrogen removal from wastewater is achieved via nitrification and denitrification, converting harmful ammonia and nitrate to (essentially) harmless nitrogen gas (i.e. by moving from right to left in Fig. 25.1).

2 BIOLOGICAL TRANSFORMATIONS OF NITROGEN IN WASTEWATER TREATMENT

Nitrogen in municipal and agricultural sewage is usually present in the form of organic nitrogen or ammonia. Fig. 25.2 depicts the fate of such nitrogen within a wastewater treatment plant. In this figure, bold solid lines indicate that the reaction is coupled to the generation of biomass (anabolism). The rectangular boundary signifies the physical limits of the wastewater treatment plant, while the horizontal dotted line divides the wastewater treatment plant into solid and liquid phases.

Solid hydrolysis converts organic nitrogen (in both soluble and insoluble forms) to ammonia. Biological removal of ammonia is then achieved by first oxidizing ammonia to nitrite/nitrate (via nitrification) which, in turn, is reduced to nitrogen gas (via denitrification). In summary, then, the role of a wastewater treatment plant is to convert nitrogen-rich influent to nitrogen-lean 'clean' effluent, with solid sludge as a by-product of the overall process.

As shown in Fig. 25.2, several nitrification and denitrification routes are possible, depending on the environmental conditions existing within the treatment plant. These will now be discussed separately.

TABLE 25.1 Typical nitrogen concentrations in municipal and agricultural wastewater

Source	*Concentration N (mg/l)*	*Reference*
Municipal sewage	20–85	Metcalf and Eddy (1991)
Landfill leachate	1825–2985	Inanc *et al.* (2000)
Livestock (swine)	390–690	Edgerton *et al.* (2000)
Abattoir run-off	160–280	Pochana and Keller (1999)

2.1 Nitrification

Nitrification is the aerobic conversion of reduced nitrogenous compounds such as ammonia, hydroxylamine (NH_2OH) and nitrite to more oxidized products. Instead of the ammonium ion, ammonia is the real substrate for the nitrifiers as it is easier to transport the neutral ammonia molecule across the cell membrane (Wiesmann, 1994). Although heterotrophic nitrification has been reported (Robertson *et al.*, 1988), the dominant mechanism for nitrification in wastewater treatment plants is autotrophic in nature.

Chemo-litho-autotrophic bacteria oxidize ammonia to nitrate in two separate stages. The oxidation of ammonia to nitrite is carried out by *Nitrosomonas, Nitrosospira, Nitrosovibrio* and *Nitrosococcus*, while nitrite oxidizers include *Nitrobacter, Nitrospira* and *Nitrococcus* species (Bock *et al.*, 1986; Burrell *et al.*, 1999).

In this oxidation process, ammonia is first oxidized to hydroxylamine with ammonium monooxygenase as the catalyst. The second, energy generating, step is catalysed by hydroxylamine oxidoreductase, which also delivers

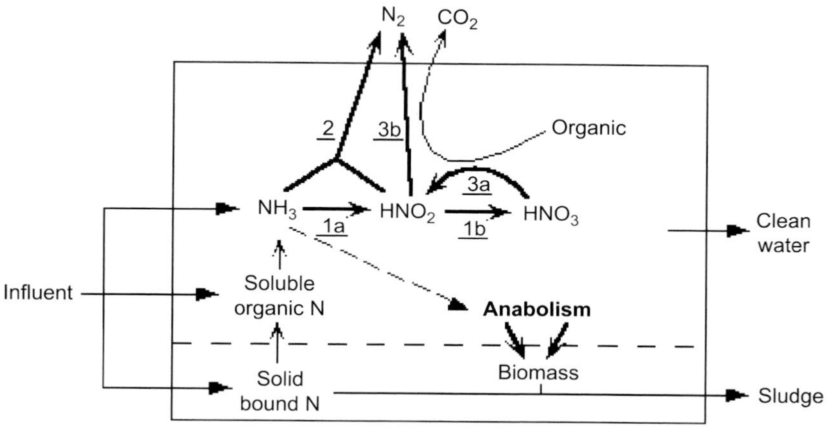

Fig. 25.2 Fate of nitrogen in a sewage treatment plant. (1) Autotrophic nitrification (oxygen consumption is not shown); (2) autotrophic anaerobic oxidation of ammonia; (3) heterotrophic denitrification.

the reducing agents required for the first step. The source of the second oxygen atom in the nitrite is water.

$$NH_3 + O_2 + 2H \rightarrow NH_2OH + H_2O$$
$$NH_2OH + H_2O \rightarrow HNO_2 + 4H$$
$$0.5O_2 + 2H \rightarrow H_2O$$

$$NH_3 + 1.5O_2 \rightarrow HNO_2 + H_2O + 240 - 350KJ \quad (1)$$

Nitrite is then oxidized to nitrate (by the nitrite oxidizers) with nitrite oxidoreductase catalysing this reaction:

$$HNO_2 + H_2O \rightarrow HNO_3 + 2H$$
$$0.5O_2 + 2H \rightarrow H_2O$$
$$HNO_2 + 0.5O_2 \rightarrow HNO_3 + 65\text{–}90\ KJ \quad (2)$$

Reducing agents [H] generated in both the oxidation of ammonia and nitrite are oxidized in the bacteria's respiratory chain in order to generate the energy necessary for growth. The terminal electron acceptor for the respiratory chain of both the ammonia and nitrite oxidizers is oxygen, a fact that emphasizes the importance of oxygen to autotrophic nitrification.

In the case of anabolism, inorganic carbon (such as bicarbonate) is the primary carbon source for the autotrophic nitrifiers. Such carbon fixation is achieved via the Calvin cycle and takes up about 80% of the energy generated from catabolism (Wood, 1986). This energy requirement contributes to the slow growth rate of both the ammonia oxidizers (about 21 hours doubling time) and the nitrite oxidizers (some 15 hours doubling time).

The overall stoichiometry for nitrification can be derived assuming an average bacteria composition of $CH_{1.4}O_{0.4}N_{0.2}$ with ammonia as the only source of cell nitrogen:

$$11NH_3 + H_2CO_3 + 15.2O_2 \rightarrow CH_{1.4}O_{0.4}N_{0.2} + 11.4H_2O + 10.8HNO_2 \quad (3)$$

$$80HNO_2 + 0.2NH_3 + H_2CO_3 + 39O_2 \rightarrow CH_{1.4}O_{0.4}N_{0.2} + 0.6H_2O + 80HNO_3 \quad (4)$$

In the absence of substrate, bacteria survive via the process of endogenous respiration:

$$CH_{1.4}O_{0.4}N_{0.2} + O_2 \rightarrow 0.2NH_3 + 0.4H_2O + CO_2 + \text{Energy} \quad (5)$$

From Equations (3) and (4), the following yields of biomass from substrate can be derived: 0.15 g-biomass/g-NH_3—N for ammonia oxidizers and 0.02 g-biomass/g-HNO_2—N for nitrite oxidizers. In terms of the yields of biomass on an oxygen basis, the following can also be derived: 0.05 g-biomass/g-O_2 for ammonia oxidizers and 0.02 g-biomass/g-O_2 for nitrite oxidizers.

From the above stoichiometry, it is clear that a significant amount of oxygen is needed for nitrification while the wastewater pH would also decrease as a result of this process.

It should also be noted that the sludge yield from nitrification (0.17 g-biomass/g-NH_3—N) is also low compared to heterotrophic growth (0.40–0.50 g-biomass/g-COD).

2.2 Denitrification

Denitrification is the reduction of oxidized nitrogenous compounds to nitrogen gas (Payne, 1981). This is generally a facultative trait and is carried out by a variety of respiratory bacteria that can utilize oxidized nitrogen compounds (instead of oxygen) as the electron acceptor. Typical denitrifiers include the species *Pseudomonas*, *Alcaligenes* and *Paracoccus*.

Reduction of nitrate is carried out in two stages. First, nitrate reductase catalyses reduction to nitrite while nitrite reductase completes the reduction to nitrogen gas.

The possible electron donors for denitrification include organic (both extracellular and intracellular) and inorganic ions such as ammonia, sulphide and hydrogen. Each class is discussed below.

2.2.1 Organic denitrification

Denitrification with an organic material relies on facultative heterotrophs that are capable of using nitrite and/or nitrate as the terminal electron acceptors in their respiratory chain. In the coupled redox reactions, nitrate/nitrite is reduced to nitrogen gas while the organic material is oxidized to carbon dioxide and water. Virtually all bacteria that are able to reduce nitrate are also able to reduce nitrite. Thus, for simplicity, sometimes a single group of denitrifiers capable of both nitrite and nitrate reduction is assumed.

Organic carbon is the source of cell carbon. Thus, for example, with acetate as the energy source and ammonia as the source of cell nitrogen, the overall stoichiometry for catabolism and anabolism is:

$$1.77CH_2O + 0.62HNO_3 + 0.2NH_3 \rightarrow CH_{1.4}O_{0.4}N_{0.2} + 0.77CO_2 + 1.68H_2O + 0.305N_2 \quad (6)$$

The stoichiometry of endogenous metabolism can be derived as for nitrification, except that nitrate is the electron acceptor in this case:

$$CH_{1.4}O_{0.4}N_{0.2} + 0.8HNO_3 \rightarrow 0.2NH_3 + 0.8H_2O + CO_2 + 0.4N_2 \quad (7)$$

The actual growth yield can be derived from Equation (7), considering only one group of denitrifiers, as 0.43-g biomass/g-CH_2O, or 0.40-g biomass/g-COD. The sludge yield from such heterotrophic denitrification is higher than that for nitrification, while denitrification generates alkalinity and, thus, the wastewater pH will increase as a result of this process.

Apart from extracellular substrates, denitrification can also occur using intracellular organic polymer. Heinemann and Müller (1991) co-immobilized denitrifying bacteria with poly-β-hydroxylbutyric (PHB) acid in a biopolymeric matrix to denitrify water. The possibility of using PHB for denitrification allows for the combined removal of nitrogen and phosphorus from sewage (Kuba *et al.*, 1996).

2.2.2 Inorganic denitrification

Autotrophic denitrification can be achieved with sulphide, hydrogen and ammonia as the electron donor. In general, however, the concentrations of hydrogen and sulphide in wastewater are too low for them to be regarded as significant electron donors.

Van de Graaf (1997) showed that ammonia could be an electron donor for the reduction of nitrite to nitrogen gas. The ammonia is oxidized under anaerobic (or strictly speaking, anoxic or denitrifying) conditions by nitrite, with both species contributing equally to the nitrogen content in the resultant nitrogen gas.

$$HNO_2 + NH_3 \rightarrow N_2 + 2H_2O \quad (8)$$

This process is termed anaerobic ammonia oxidation (Anammox) and is classified as denitrification (rather than nitrification) given that nitrogen gas is generated in the reaction. The bacteria responsible for Anammox are autotrophic, with inorganic carbon as the source of their cell carbon.

2.3 Summary of biological nitrogen cycle in wastewater treatment plants

As shown in Fig. 25.2, removal of ammonia-nitrogen in wastewater treatment plants is dominated by two mechanisms:

1. A conversion pathway to nitrogen gas via the related processes of nitrification and denitrification.
2. An assimilatory pathway for the formation of bacterial biomass.

Sludge production and its subsequent removal is usually a major operating cost in any wastewater treatment plant. To reduce sludge production, endogenous respiratory conditions are often encouraged in a downstream sludge digester. However, sludge digestion means that any assimilated nitrogen is returned as soluble components to the wastewater again (see Equation (5)). Hence, rather than sludge production, the major objective of biological nitrogen removal should be to maximize the conversion of ammonia to nitrogen gas via nitrification and denitrification.

For combined nitrification-denitrification, Fig. 25.2 also shows that there are at least two routes:

- Anaerobic ammonia oxidation (i.e. the Anammox process)
- Autotrophic nitrification followed by heterotrophic denitrification.

In theory, Anammox promises to deliver combined removal of both ammonia and nitrite. In the case of ammonia removal from wastewater, if half of the influent ammonia is first oxidized to nitrite, Anammox could then be used to convert ammonia to nitrogen gas without any aeration and without any organic carbon source. The autotrophic nature of the reaction also promises a low sludge yield. The growth rate of the bacteria responsible for Anammox is, however, extremely low with a doubling time of around 29 days (Van de Graaf, 1997), a fact that translates to a very lengthy start-up time for a full-scale Anammox plant. In addition, for organic-rich wastewater, the slow growth rate of Anammox microorganisms implies that they would be easily outgrown by organic-consuming heterotrophs.

The application of Anammox has, however, been shown to be feasible given the right operating conditions. For example, for digester effluent at moderately high temperatures (25–35°C, to promote a higher growth rate) and with a low organic concentration, Anammox has been shown to be excellent option for the removal of ammonia (Strous *et al.*, 1997).

Under typical wastewater treatment conditions (10–20°C, and with the presence of organic material), autotrophic nitrification followed by heterotrophic denitrification is the dominant mechanism for biological nitrogen removal. As such, these are the subjects of further discussion for the remainder of this chapter – with 'nitrification' referring to autotrophic nitrification while 'denitrification' refers to heterotrophic denitrification, unless stated otherwise.

From a pH control perspective, nitrification consumes alkalinity while denitrification generates alkalinity. In theory, combining nitrification with denitrification minimizes the need for pH control. There are, however, several other considerations relating to combined nitrification and denitrification, and these are discussed in the following section.

3 KINETICS OF NITRIFICATION AND DENITRIFICATION

3.1 General considerations

Real bacterial yield is defined as,

$$Y^{o}_{X/S} = \frac{r^{g}_{X}}{r_{S}} \tag{9}$$

with the specific value given explicitly by the stoichiometry of the reaction. In Equation (9), subscripts '*X*' and '*S*' refer to bacteria and substrate, respectively. The specific growth rate is usually assumed to be first-order with respect to the bacterial concentration:

$$r^{g}_{X} = \mu^{g} \cdot X \tag{10}$$

Considering the loss of biomass by endogenous respiration to also be first-order with respect to the bacterial concentration,

$$r_X^e = \mu^e \cdot X \qquad (11)$$

the observed yield is, thus, given by:

$$Y_{X/S} = \frac{r_X^g - r_X^e}{r_S} = Y_{X/S}^o \cdot \left(1 - \frac{\mu^e}{\mu^g}\right) \qquad (12)$$

Experimentally, the real yield ($Y_{X/S}^0$) cannot be directly observed. It is a 'theoretical upper limit' that can only be estimated for cases where μ^g is significantly larger than μ^e.

The specific growth rate (μ^g) is, however, not a constant. It is a function of the substrate concentration, inhibitor concentration (if any) and other environmental factors such as pH and temperature.

3.2 Autotrophic nitrification

Substrate limitation and inhibition have both been reported for nitrification. Haldane-kinetics are commonly used to describe the inhibition process (Wiesmann, 1994):

$$\mu^g = \mu_{max}^g \cdot \frac{c(NH_3-N)}{K_S + c(NH_3-N) + \frac{c(NH_3-N)^2}{K_I}} \times \frac{c(O_2)}{K_{O2} + c(O_2)} \qquad (13)$$

The maximum specific growth rate is given by μ_{max}^g, while the second factor in Equation (13) describes any oxygen limitation. K_S and K_{O2} are the half-saturation constants describing substrate and oxygen limitation, respectively, while K_I is a constant related to substrate inhibition.

Fig. 25.3 illustrates the effect of substrate concentration on the relative growth rate (note that the x-axis here is a logarithm scale). The specific growth rate of ammonia oxidizers is, thus, substrate limited at low concentrations and substrate inhibited at high concentrations.

Substrate inhibition is caused by the toxicity of ammonia, which is the reason that the concentration of free ammonia is used in Equation (13). In practice, however, the total ammonium and ammonia concentration,

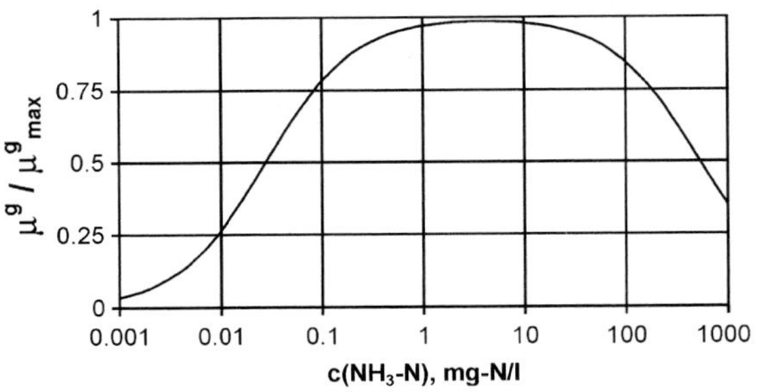

Fig. 25.3 Effect of substrate based on Haldane-kinetics.

c(NH), is usually what is experimentally measured. Given that the ammonia-ammonium equilibrium is pH-dependent,

$$NH_3 + H^+ \leftrightarrow NH_4^+ \qquad (14)$$

with an equilibrium constant K_a, we get the following:

$$c(NH) = c(NH_3) + c(NH_4^+) = c(NH_3) \cdot (1 + K_a \cdot 10^{-pH}) \qquad (15)$$

Replacing $c(NH_3 - N)$ with $c(NH - N)$, as given by Equation (15), the substrate term in Equation (13) becomes:

$$\frac{c(NH-N)}{K_S \cdot (1+K_a \cdot 10^{-pH}) + c(NH-N) + \frac{c(NH-N)^2}{K_I \cdot (1+K_a \cdot 10^{-pH})}} \qquad (16)$$

In this case, the new half-saturation constant, $K_S \cdot (1 + K_a 10^{-pH})$ is pH-dependent. For the ammonia-ammonium equilibrium, pK_a equals -9.26 at 25°C. For pH < 9 and $c(NH - N) < 65$ mg- N./l, the concentration of free ammonia is negligible and the substrate inhibition term can be ignored (Wiesmann, 1994). The effect of the substrate is thus reduced to a limitation at low concentrations and can be modelled using Monod kinetics:

$$\mu^g = \mu_{max}^g \frac{c(NH_3-N)}{K_S + c(NH_3-N)} \frac{c(O_2)}{K_{O2} + c(O_2)} \qquad (17)$$

A similar analysis can be carried out for nitrite oxidation and the following expression can be derived:

$$\mu^g = \mu^g_{max} \frac{c(HNO_2 - N)}{K_S + c(HNO_2 - N) + \frac{c(HNO_2 - N)^2}{K_I}} \times \frac{c(O_2)}{K_{O2} + c(O_2)} \quad (18)$$

3.2.1 *Effect of other environmental factors*

Both ammonia monooxygenase and nitrite oxidoreductase are photosensitive. Thus, under strong light conditions, the nitrifiers will be out-competed for the inorganic carbon source by photosynthetic autotrophs or algae (Wood, 1986). Thiourea or allylthiourea have been used to inhibit nitrification in BOD tests, while the toxicity of other chemicals to nitrification has been summarized by Henze *et al.* (1995).

The effects of temperature and pH could be analysed in terms of an increase or decrease in the value of the maximum specific growth rate (μ^g_{max}). This rate increases with temperature before the enzymes responsible for nitrification are deactivated over the range 32–35°C. Meanwhile, the optimum pH for nitrification is in the 8–9 range. As already mentioned, the effect of pH on the growth rate is also linked to substrate inhibition, especially at pH values above 9. It has also been reported that pH values lower than about 6.0 are unfavourable for nitrification (Henze *et al.*, 1995).

3.3 Heterotrophic denitrification

Here, the specific growth rate is limited by both the concentration of the substrate and the electron acceptor (HNO_2 or HNO_3), and double Monod-kinetics is usually used. Thus, for the denitrification of nitrate, the specific growth rate is given by,

$$\mu^g = \mu^g_{max} \frac{c(COD)}{K_S + c(COD)} \times \frac{c(HNO_3 - N)}{K_{HNO3-N} + c(HNO_3 - N)} \quad (19)$$

while a similar expression holds for the denitrification of nitrite:

$$\mu^g = \mu^g_{max} \frac{c(COD)}{K_{SH} + c(COD)} \times \frac{c(HNO_2 - N)}{K_{HNO2-N} + c(HNO_2 - N)} \quad (20)$$

3.3.1 *Effect of other environmental factors*

Molecular oxygen has been shown to repress the enzymes responsible for denitrification (Gottschalk, 1979). In the presence of oxygen, the denitrification reaction is inhibited and the denitrifying bacteria switch to using oxygen as their terminal electron acceptor. The inhibition mechanism is, however, reversible and the response of the enzymatic system to the switch of electron acceptor has been shown to only be in the order of minutes (Kornaros and Lyberatos, 1998).

Including the effect of non-competitive inhibition by oxygen, Equations (19) and (20) become, respectively:

$$\mu^g = \mu^g_{max} \frac{c(COD)}{K_S + c(COD)} \times \frac{c(HNO_3 - N)}{K_{HNO3-N} + c(HNO_3 - N)} \times \frac{K_{O2,I}}{K_{O2,I} + c(O_2)} \quad (21)$$

$$\mu^g = \mu^g_{max} \frac{c(COD)}{K_{SH} + c(COD)} \times \frac{c(HNO_2 - N)}{K_{HNO2-N} + c(HNO_2 - N)} \times \frac{K_{O2,I}}{K_{O2,I} + c(O_2)} \quad (22)$$

with $K_{O2,I}$ as a constant related to oxygen inhibition.

The effects of temperature and pH on denitrification are similar to those for nitrification. The optimum pH for denitrification is, however, somewhat broader than for nitrification, falling in the range 7–9 (Henze *et al.*, 1995).

3.4 Summary of kinetic data

Average kinetic coefficients for nitrification and denitrification (obtained from the literature) are summarized in Table 25.2. Values of μ^g_{max}, K_I, K_{O2}, K_{HNOx} and $Y^O_{X/S}$ are adapted from Wiesmann (1994), μ^e and K_S are from Wiesmann (1994); and Henze *et al.* (1995), $Y^O_{O2/S}$ and $Y^O_{NOx/S}$ are derived from stoichiometry, while $K_{O2,I}$ are from Gujer *et al.* (1999).

Note that the coefficients related to organic degradation are strongly dependent on the nature of the organic substrate, i.e. μ^g_{max} and K_S for aerobic and denitrifying organic degradation. Note that the values given here are only representative. For more specific values, these parameters should be experimentally measured for the wastewater of interest.

The data given in Table 25.2 highlight several significant differences between nitrification, aerobic organic degradation and denitrification:

1. Nitrification requires oxygen while denitrification is inhibited by oxygen.
2. Both sludge yield and maximum specific growth rate are significantly lower for nitrification than for organic degradation.

The different effects of oxygen link directly to the aeration pattern in nitrogen-removing plants. Moreover, the differences in sludge yield and maximum specific growth rate imply that the nitrifiers are usually outgrown by organic-consuming heterotrophs, unless a two-sludge system is used. Design consideration of nitrogen-removing plants will be discussed in the next section.

4 DESIGN OF BIOLOGICAL NITROGEN REMOVAL PLANTS

4.1 Overview

The primary design objective in the design of a nitrogen-removing plant is to devise a system that is capable of reducing the concentrations of nitrogenous compounds to the desired levels, although clearly the capital and operating costs are also important considerations in the final decision. The capital cost is usually dominated by the size of the reactor used and its associated solid-liquid separation unit, while the dominant factors in the operating cost include:

- Aeration
- Sludge removal
- Pumping cost (primarily for any recycle stream required)
- Chemical dosage requirement.

Over the past few decades, a variety of process flowsheets for nitrogen-removing plants have been proposed, with the basic structure shown in Fig. 25.4. For such a flowsheet structure,

TABLE 25.2 Typical kinetic coefficients (20°C and a pH of 8)

	NH_3 oxidation	*HNO_2 oxidation*	*Aerobic degradation of organic*	*HNO_3 reduction*	*HNO_2 reduction*
μ^g_{max}	0.77/d	1.08/d	7.2/d	2.6/d	1.5/d
μ^e	0.03/d	0.03/d	0.2/d	0.1/d	0.1/d
K_S	0.028 mg-NH_3-N/l	3.2 10^{-5}mg-HNO_2-N/l	10–100 mg-COD/l	5–25 mg-COD/l	–
K_I	540 mg-NH_3-N/l	0.26 mg-HNO_2-N/l	–	–	–
$K_{O2,I}$	–	–	–	0.20 mg-O_2/l	0.20 mg-O_2/l
K_{O2}	0.3 mg O_2/l	1.1 mg O_2/l	0.08 mg O_2/l	–	–
K_{HNOx}	–	–	–	$\leq$0.14 mg-HNO_3-N/l	$\leq$0.12 mg-HNO_2-N/l
$Y^0_{X/S}$	0.147 g-cell/ g-NH_3-N	0.02 g-cell/ g HNO_2-N	0.43 g-cell/g-COD	0.40 g-cell/g-COD	0.40 g-cell/g-COD
$Y^0_{O2/S}$	3.16 g-O_2/ g-NH_3-N	1.11-g O_2/ g-HNO_2-N	0.40 g-O_2/g-COD	–	–
$Y^0_{NOx/S}$	–	–	–	0.378 g-HNO_3-N/ g-COD	0.255 g-HNO_2-N/ g-COD

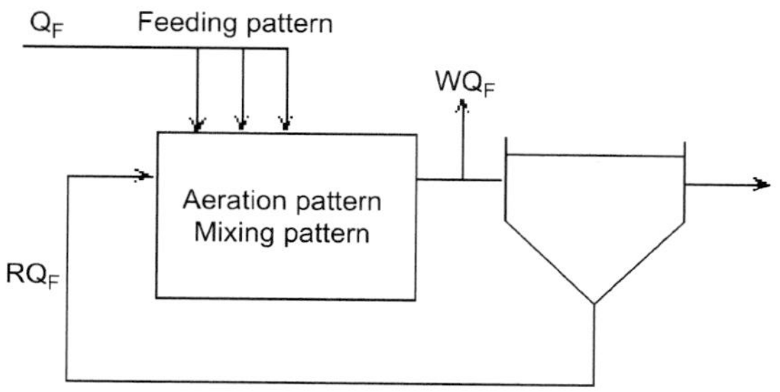

Fig. 25.4 Basic flow sheet for biological nitrogen removal.

the operating decisions available to a design engineer might be summarized as follows:

- Solids retention time
- Reactor mixing pattern
- Recycle ratio
- Feeding pattern to the reactor
- Aeration pattern in the reactor.

The effect of each of these design variables on the overall performance of a nitrogen-removing plant can be analysed from a kinetic point of view.

4.2 Impact of individual design decision

If the effect of endogenous respiration on substrate removal is ignored, then the combination of Equations (9) and (10) shows that the rate of substrate removal is proportional to both the specific growth rate (μ^g) and the bacteria concentration (X):

$$r_S^g = \frac{1}{Y_{X/S}^o} \cdot \mu^g \cdot X \qquad (23)$$

X is controlled by the solids retention time. μ^g is a function of the substrate, electron acceptor and inhibitor concentrations, as given by Equations (13) and (18) for nitrification, and Equations (21) and (22) for denitrification. As a result, the average value of μ^g in the reactor is influenced by the recycle ratio, mixing pattern, feeding pattern and aeration pattern. The impact of pH and temperature changes on μ_{max}^g is often ignored in the first instance, assuming constant values for both.

4.2.1 Solids retention time

The solids retention time or SRT controls the concentrations of bacteria throughout the treatment system. A higher SRT contributes to a higher bacterial concentration in the reactor, which gives rise to:

- Smaller reactor size
- Larger separator size
- Reduced sludge production
- Higher aeration requirements due to the extra oxygen required for endogenous respiration.

Clearly, an optimum SRT exists, resulting from a trade-off between the gains and losses in the various cost terms. For municipal sewage treatment plants performing combined nitrification–denitrification, typical wasting ratios generally fall in the range 0.025–0.10 for a hydraulic retention time of 12–24 hours.

4.2.2 Mixing pattern

Two types of idealized mixing pattern are possible, i.e. plug-flow and mixed-flow. For zero-order kinetics, both mixing patterns give exactly the same reactor size. For first-order kinetics, however, a plug flow reactor is always smaller in volume than a mixed flow reactor carrying out the same amount of reaction (Levenspiel, 1972).

Monod-kinetics are mixed-order, being close to zero-order at high substrate (or electron acceptor) concentration and essentially first-order at low substrate (or electron acceptor) concentration. The defining concentrations are given by the half-saturation constants K_S, K_{O2} and K_{HNOx} (refer to Table 25.2 for typical values).

For nitrification, K_S is a function of pH – at a pH value of 7 (and a temperature of 25°C), this parameter equals 5.1 mg NH – N/l. This suggests that if an effluent ammonia–ammonium concentration of less than 1 mg NH – N/l is required (as in the case of most municipal sewage treatment plants), a plug-flow reactor is the preferred option. A similar argument holds for organics removal if an effluent COD concentration of less than 10 mg COD/l is required.

For denitrification, values of K_{HNOx} generally fall in the range 0.10–0.15 mg HNOx – N/l, which are low in comparison to typical effluent requirements of 1–10 mg HNOx-N/l. This suggests that the kinetics will be primarily zero-order with respect to the nitrate concentration, and both plug-flow and mixed-flow reactors give similar performance if organic material is adequate for growth.

4.2.3 *Recycle ratio, feeding pattern, aeration pattern*

Any recycle stream affects the overall treatment system in two ways – it dilutes the influent concentration, and increases the hydraulic loading on both the reactor and the separator. The first effect reduces the required reactor size, while the second effect reduces the effective hydraulic retention time of the reactor and separator. In addition, pumping costs increase with the recycle ratio.

For modestly loaded municipal sewage treatment plants, the optimum recycle ratio is usually in the range of 0.5–1.0. However, for wastewater with a high ammonia concentration, such as many industrial effluent streams, a higher recycle ratio is required to dilute the influent ammonia concentration to avoid substrate inhibition, which will reduce the effective rate of removal. As shown in Fig. 25.3, the desired influent ammonia concentration in the reactor is in the range 5–30 mg NH_3–N/l.

As an alternative, step feeding can be used to maximize exposure of the reactor volume to the optimal substrate concentration ranges, and thereby reduce the overall reactor size.

The aeration pattern to be used in the reactor is also strongly related to the wastewater characteristics, as will be discussed in the next section.

4.3 Process flowsheet selection based on wastewater composition

The flowsheet requirements of a treatment system differ quite markedly depending on the characteristics of the nitrogenous compounds in the wastewater, which can be broadly categorized as follows:

- Nitrate-rich wastewater
- Ammonia-rich wastewater
- Ammonia and organic-rich wastewater.

4.3.1 *Nitrate-rich wastewater*

Nitrate-rich wastewater includes wastewater from fertilizer industry and metal finishing plants where nitric acid is used as a cleaning agent. For this class of wastewater, an anoxic-aerobic system is commonly used. The source of cell nitrogen in this case is nitrate, which can be achieved through an assimilatory pathway as depicted in Fig. 25.1.

The exact flowsheet of the treatment plant depends on the organic to nitrate ratio of the wastewater. From stoichiometry, 3–5 g of COD is needed for every gram of NO_3–N reduced (Henze *et al.*, 1995). Based on this ratio, if nitrate is in excess, aeration is not needed, while an organic supplement (methanol is a typical organic supplement for denitrification) may be needed to reduce the nitrate concentration to the desired level (see Fig. 25.5a).

If organic material is in excess, then an anoxic-aerobic system is used first to remove nitrate in the anoxic reactor before any excess organic is oxidized aerobically in the second reactor to meet the desired organic effluent concentration (see Fig. 25.5b).

4.3.2 *Ammonia-rich wastewater*

For ammonia-rich wastewater that is depleted of organic carbon (such as digester effluent and wastewater from fertilizer industry), Anammox is potentially useful if faster growth rate can be promoted, e.g. if the wastewater temperature is high.

For dilute wastewater, heating up volume of wastewater is cost-prohibitive. In this case, an aerobic-anoxic reactor can be used. In the aerobic reactor, ammonia is oxidized to nitrate while nitrate is reduced in the second anoxic reactor with supplement of organic carbon source (see Fig. 25.5c).

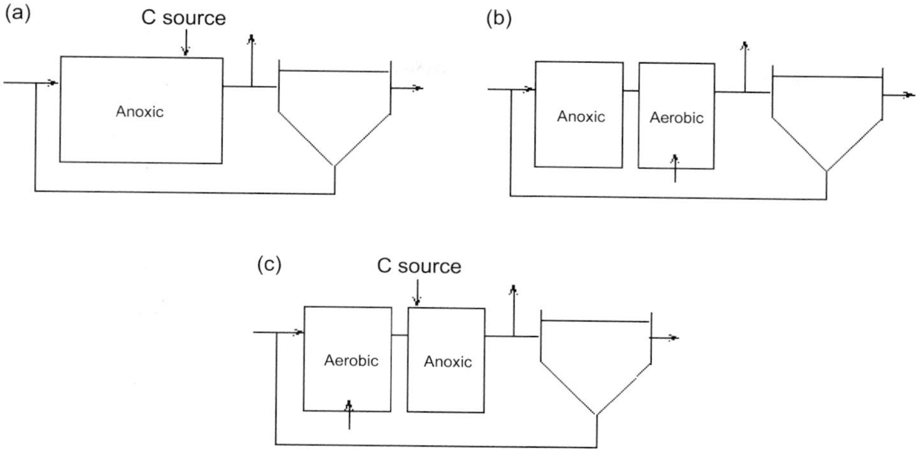

Fig. 25.5 Treatment systems for nitrate-rich and ammonia-rich wastewater.

4.3.3 Ammonia and organic-rich wastewater

Most wastewater (such as municipal sewage, leachate and agricultural run-off) is rich in both ammonia and organic material. For combined removal of organic and nitrogenous material, the conflicting oxygen requirements for nitrification and denitrification usually mean that complete nitrogen removal cannot be obtained. Nitrification produces nitrate that is reduced in the denitrification stage, however, the aerobic nitrifying conditions also promote aerobic degradation of organic material. As a result, denitrification is usually limited by the organic carbon source, unless an organic supplement is added.

If the organic material in wastewater is in excess of the requirements for denitrification, a single mixed aerobic-anoxic system could be used (see Fig. 25.6a). In such systems, the dissolved oxygen concentration in the reactor is controlled at a low level (typically 0.5–2.0 mg-O_2/l) to allow simultaneous nitrification, denitrification and aerobic organic degradation. Examples of such systems include the 'oxidation ditch' and various attached

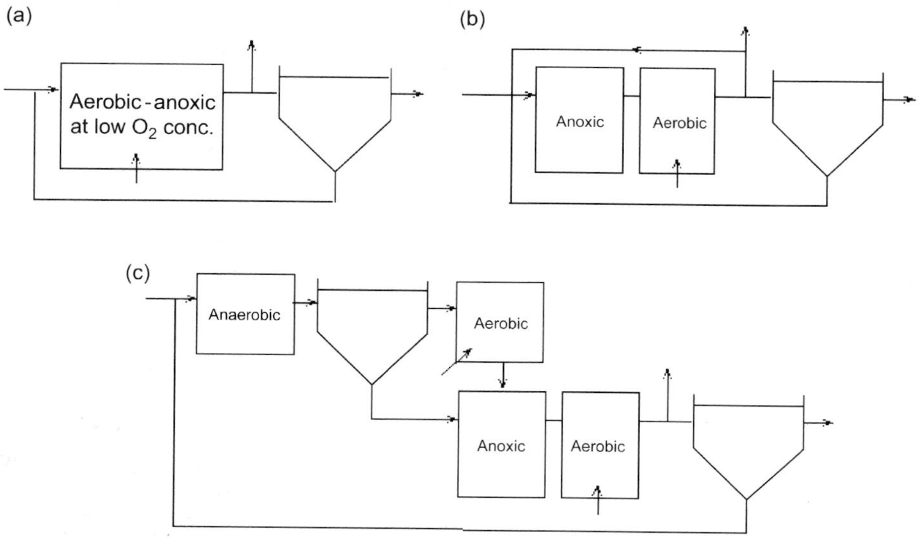

Fig. 25.6 Process flowsheets for ammonia/organic-rich wastewater.

growth systems – the latter often being the preferable option, given that a thicker biofilm layer can be formed which encourages the formation of an anoxic zone for denitrification.

If the ratio of organic to nitrate is not in excess of the stoichiometry required for denitrification, it is necessary to avoid aerobic degradation of the organic material. In this case, an anoxic–aerobic-recycle system can be used (see Fig. 25.6b; note that this is often referred to as the modified Ludzack-Ettinger configuration). Nitrification occurs in the second (aerobic) reactor. Internal recycling of mixed liquor returns nitrate to the front of the reactor system where it is reduced with the influent organic under anoxic conditions. Clearly, more nitrate can be reduced if a higher internal recycle is used, but this can lead to higher pumping costs and an increased hydraulic loading on the reactor (which may in turn require a larger reactor). However, it should be noted that internal recycle ratios in excess of 4–6 have not been reported in the literature for municipal sewage treatment plants.

Internal recycling can be avoided if the organic material can be separated from the ammonia. This may be achieved via sorption of such material onto bacterial flocs (Wanner *et al.*, 1992; Kuba *et al.*, 1996). Subsequent solid–liquid separation allows ammonia-rich supernatant to be nitrified separately, before it is returned as a nitrate-rich stream and reduced with the organic-rich sludge. An aerobic zone is then used to degrade any residual organic material (see Fig. 25.6c). This form of flowsheet is useful for wastewaters with a low organic to nitrogen ratio. Given that there is no internal recycling used, the reactor size is minimal despite the fact that there are several anaerobic/anoxic/aerobic zones. The only (minor) drawback with this configuration is the presence of an additional settler.

5 CONCLUSIONS

Nitrification and denitrification both play important roles in the balance of the nitrogen cycle. In wastewater treatment plants, nitrification followed by denitrification is commonly used to convert potentially harmful nitrogenous compounds into nitrogen gas. In this chapter the characteristics of the dominant pathways for nitrification and denitrification were reviewed, key kinetic data were summarized (in Table 25.2), before moving on to consider the evaluation of the various design variables for typical nitrogen-removing treatment plants.

REFERENCES

Bock, E., Koops, H.-P. and Harms, H. (1986). Cell biology of nitrifying bacteria. In: J.A. Prosser (ed.) *Nitrification*, pp. 17–38. IRL Press, Oxford.

Burrell, P., Keller, J. and Blackall, L.L. (1999). Characterisation of the bacterial consortium involved in nitrite oxidation in activated sludge. *Wat. Sci. Tech.* **39**(6), 45–52.

Edgerton, B.D., McNevin, D., Wong, C.H., Menoud, P., Barford, J.P. and Mitchell, C.A. (2000). Strategies for dealing with piggery effluent in Australia: The sequencing batch reactor as a solution. *Wat. Sci. Tech.* **41**(1), 123–126.

Gottschalk, G. (1979). *Bacterial metabolism.*. Springer-Verlag, New York.

Gujer, W., Henze, M., Mino, T. and van Loodrecht, M. (1999). Activated sludge model no.3. *Wat. Sci. Tech.* **39**(1), 183–193.

Heinemann, A. and Müller, W.-R. (1991). Co-immobilization of bacteria and biodegradable solid substrates in biopolymer-matrices for denitrification of water. In: *Proceedings International Symposium on Environmental Biotechnology*, Vol. 1, pp. 37–39, Oostende, Belgium, 22–25 April 1991. Royal Flemish Society of Engineers, Belgium.

Henze, M., Harremoes, P., Jansen, J.L.C. and Arvin, E. (1995). *Wastewater treatment: Biological and chemical processes*. Springer-Verlag, Berlin.

Inanc, B., Calli, B. and Saatci, A. (2000). Characterization and anaerobic treatment of the sanitary landfill leachate in Istanbul. *Wat. Sci. Tech.* **41**(3), 223–230.

Kornaros, M. and Lyberatos, G. (1998). Kinetic modelling of *Pseudomonas denitrificans* growth and denitrification under aerobic, anoxic and transient operating conditions. *Wat. Res.* **32**(6), 1912–1922.

Kuba, T., van Loosdrecht, M.C.M. and Heijnen, J.J. (1996). Phosphorus and nitrogen removal with minimal COD requirement by integration of denitrifying dephosphatation and nitrification in a two-sludge system. *Wat. Res.* **30**(7), 1702–1710.

Levenspiel, O. (1972). *Chemical reaction engineering*, 2nd edn. John Wiley and Sons, Singapore.

Metcalf & Eddy Inc. (1991). *Wastewater engineering: Treatment, disposal, reuse*, 3rd edn. McGraw-Hill, Singapore.

Payne, W.J. (1981). *Denitrification*. John Wiley and Sons.

Pochana, K. and Keller, J. (1999). Study of factors affecting simultaneous nitrification and denitrification (SND). *Wat. Sci. Tech.* **39**(6), 61–68.

Postgate, J. (1998). *Nitrogen fixation*, 3rd edn. Cambridge University Press, Cambridge.

Robertson, L.A., van Niel, E.W.J., Torremans, R.A.M. and Kuenen, J.G. (1988). Simultaneous nitrification and denitrification in aerobic chemostat cultures of *Thiosphaera pantotropha*. *Appl. Env. Microb.* **54**(11), 2812–2818.

Strous, M., van Gerven, E., Zheng, P., Kuenen, J.G. and Jetten, M.S.M. (1997). Ammonia removal from concentrated waste streams with the anaerobic ammonium oxidation (ANAMMOX) process in different reactor configurations. *Wat. Res.* **31**(8), 1955–1962.

Van de Graaf, A.A. (1997). Biological anaerobic ammonium oxidation. PhD thesis, Delft University of Technology, Delft, The Netherlands.

Wanner, J., Cech, J.S. and Kos, M. (1992). New process design for biological nutrient removal. *Wat. Sci. Tech.* **25**(4–5), 445–448.

Wiesmann, U. (1994). Biological nitrogen removal from wastewater. In: A. Fiechter (ed.) *Advances in Biochemical Engineering Biotechnology*, Vol. 51, pp. 113–154. Springer-Verlag, Berlin and Heidelberg.

Wood, P.M. (1986). Nitrification as a bacterial energy source. In: J.A. Prosser (ed.) *Nitrification*, pp. 39–62. IRL Press, Oxford.

26

Low-cost treatment systems

Duncan Mara

School of Civil Engineering, University of Leeds, Leeds LS2 9JT, UK

1 INTRODUCTION

Low-cost wastewater treatment systems comprise:

1. Waste stabilization ponds (WSP)
2. Wastewater treatment and storage reservoirs (WSTR)
3. Constructed wetlands.

They are often called 'natural' treatment systems as they do not rely on electromechanical plant, such as aerators, for their operation. The oxygen needed by the heterotrophic bacteria for the oxidation of organic compounds comes mainly from the photosynthetic activity of the algae in WSP and WSTR and of the plants in constructed wetlands. Since this activity depends on sunlight (which arrives at the earth's surface), these processes require a greater surface area of land than electromechanical treatment systems. In developing countries this is often not a problem; in industrialized countries, on the other hand, this generally means that these low-cost treatment processes are used for small rural communities, under around 5000 population.

2 WASTE STABILIZATION PONDS

WSP systems comprise a series of anaerobic, facultative and maturation ponds. The primary function of anaerobic and facultative ponds is BOD removal. Maturation ponds are primarily for the removal of excreted pathogens, although some pathogen removal occurs in anaerobic and facultative ponds (particularly the removal of *Vibrio cholerae*, helminth eggs and protozoan cysts), and some BOD removal occurs in maturation ponds.

WSP design is detailed in EPA (1983), Mara *et al.* (1991), Mara (1997) and Mara and Pearson (1998). Arthur (1983) presents a methodology for comparative costing of wastewater treatment options, and in his case study (based on the city of Sana'a in the Yemen Arab Republic), WSP were found to be the least-cost technology at land prices of US$ 50 000–150 000 per ha, depending on the discount rate used (5–15%).

WSP are widely used in France and Germany, where there are around 2500 and 1100 systems, respectively (Bucksteeg, 1987; CEMAGREF *et al.*, 1997), and also in the USA, which has around 7000 systems (EPA, 1983). The UK has, in contrast, only 19 WSP systems (Mara *et al.*, 1998).

2.1 Anaerobic ponds

Anaerobic ponds (2–4 m deep) function much like open septic tanks: they receive such a high organic loading (100–350 g BOD/m^3/day, depending on temperature) that they are devoid of both dissolved oxygen and algae (although occasionally a thin surface film of *Chlamydomonas* may be present). In warm climates they are extremely efficient in removing BOD – up to 75% at 25°C and a retention

The Handbook of Water and Wastewater Microbiology
ISBN 0-12-470100-0

time of 1 day for wastewaters with a BOD up to 350 mg/l (Silva, 1982; Pearson *et al.*, 1996a). Performance is lower at lower temperatures, but even in European winters BOD removal is 40–50% (Bucksteeg, 1987). Odour (due to H_2S release) is not a problem provided the recommended BOD loadings are not exceeded (see Mara and Pearson, 1998) and the wastewater contains less than 500 mg SO_4^{2-}/l (Gloyna and Espino, 1969); the latter should not be a problem with purely domestic wastewater as drinking waters should contain no more than 250 mg SO_4^{2-}/l (WHO, 1993), although in coastal areas salt water intrusion into sewers can increase the sulphate concentration of the wastewater.

Vibrio cholerae is very sensitive to sulphide concentrations as low as 3 mg/l, and its removal in anaerobic ponds is consequently very good; Oragui *et al.* (1993) report a reduction from 485 to 28/l.

Helminth eggs and protozoan cysts are removed in WSP by sedimentation. Compliance with the WHO (1989) recommendation for crop irrigation of less than 1 egg per litre is often achieved in a series of anaerobic and facultative ponds, although this depends on the number of eggs in the raw wastewater (which in turn depends on the endemicity of helminthic infections in the community producing the wastewater) and the WSP retention times (see Ayres *et al.*, 1992).

2.2 Facultative ponds

Facultative ponds (1.5–1.8 m deep) (Fig. 26.1) receive organic loads in the range 100–400 kg BOD/ha day and, due to the profuse growth of algae, they appear dark green, although when slightly overloaded they may appear red or purple due to the predominance of anaerobic sulphide-oxidizing photosynthetic bacteria (see Chapter 27). The pond algae and the pond heterotrophic bacteria depend on each other: the algae act as oxygen generators and the bacteria use this algal oxygen for BOD removal, and the bacterially-produced carbon dioxide is fixed by the algae (Fig. 26.2). Some oxygen and some carbon dioxide enter the pond through its surface from the atmosphere, but most of the oxygen comes from algal photosynthesis and most of the carbon dioxide from bacterial metabolism.

BOD removal in facultative ponds is around 80–90% on a *filtered* basis, i.e. (and as permitted by the European Directive on Urban Wastewater Treatment – Council of the European Communities (CEC), 1991) effluent samples are filtered prior to BOD analysis to remove algal solids (and hence algal BOD). Filtration is through glass fibre filters of the type normally used for suspended solids (SS) determination (e.g. Whatman GF/C). In Europe (CEC, 1991) WSP effluents must contain $\not>$25 mg filtered BOD/l and $\not>$150 mg SS/l, and this is often achieved by facultative pond effluents (see Pearson *et al.*, 1995; CEMAGREF *et al.*, 1997).

2.3 Maturation ponds

The function of maturation ponds (1–1.5 m deep) is the destruction of faecal bacteria, particularly pathogenic members of the Enterobacteriaceae, and faecal viruses. As shown in Table 26.1, this is extremely efficient in warm climates. Faecal coliform (FC) bacteria are commonly used as indicators of faecal bacterial pathogens and, as shown by Marais (1974), FC removal in WSP can be adequately modelled by first order kinetics in a series of completely mixed reactors.

Faecal viruses are removed by adsorption on to settleable solids and subsequent sedimentation, including the adsorption on to algae in facultative and maturation ponds and the subsequent sedimentation of the algae when they die.

Faecal bacterial removal (except, as noted above, that of *V. cholerae*) is largely due to the activities of the pond algae. When the algae are photosynthesizing rapidly, their demand for CO_2 outstrips its supply from bacterial metabolism (see Fig. 26.1) and this leads to carbonate and bicarbonate dissociation:

$$2HCO_3^- \rightarrow CO_3^{2-} + H_2O \rightarrow + CO_2$$

$$CO_3^{2-} + H_2O \rightarrow 2OH^- + CO_2$$

The resulting CO_2 is fixed by the algae and the hydroxyl ions accumulate to raise the pH to

Fig. 26.1 A facultative pond at Botton Village in North Yorkshire, England.

often above 9, which is rapidly lethal to faecal bacteria (Parhad and Rao, 1974; Pearson *et al.*, 1987). Curtis *et al.* (1992) showed that high levels of dissolved oxygen (due to algal photosynthesis) and high visible light intensities, in the presence of dissolved humic substances, which act as very efficient exogenous photosensitizers, are rapidly fatal to faecal bacteria, a process that is exacerbated by high pH (Fig. 26.3).

A properly designed series of maturation ponds, that is also well operated and maintained (good O & M is principally the removal of vegetation and any floating materials to permit unfettered sunlight penetration for algal photosynthesis) can easily reduce FC levels to below 1000 per 100 ml, which is the

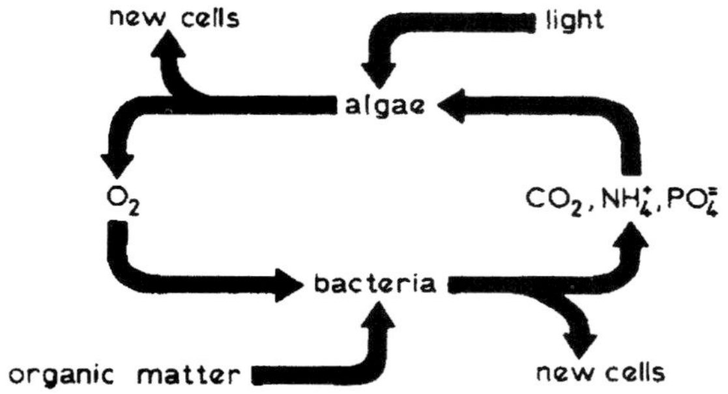

Fig. 26.2 Algal-bacterial mutualism in facultative ponds.

TABLE 26.1 Management strategy for three WSTR in parallel for an irrigation season of 6 months

Month[a]	*WSTR 1*	*WSTR 2*	*WSTR 3*
January	Rest	Fill (1)	Empty
February	Rest	Fill (1)	Empty
March	Rest	Rest (or Fill (1))	Fill (1) (or Empty)
April	Rest	Rest	Fill (1)
May	**Use**	Rest	Fill (1)
June	**Use**	Rest	Fill (1)
July	Fill (1)[b]	**Use**	Rest
August	Fill (1)	**Use**	Rest
September	Fill ($\frac{1}{2}$)	Fill ($\frac{1}{2}$)	**Use**
October	Fill ($\frac{1}{2}$)	Fill ($\frac{1}{2}$)	**Use**
November	Fill ($\frac{1}{2}$)	Fill ($\frac{1}{2}$)	Empty
December	Fill ($\frac{1}{2}$)	Fill ($\frac{1}{2}$)	Empty
Volume[c]	4	4 (or 5)	4 (or 3)

[a] July and August are the hottest months, so WSTR No. 3 has the minimum rest period of two months at this time. The other two WSTR have rest periods of 4 months to ensure FC die-off to <1000 per 100 ml during the cooler months.

[b] Proportion of monthly flow discharged into each WSTR.

[c] WSTR volume expressed as multiple of monthly wastewater flow. The WSTR need not have equal volumes as the volume of treated wastewater required for irrigation may vary throughout the irrigation season.

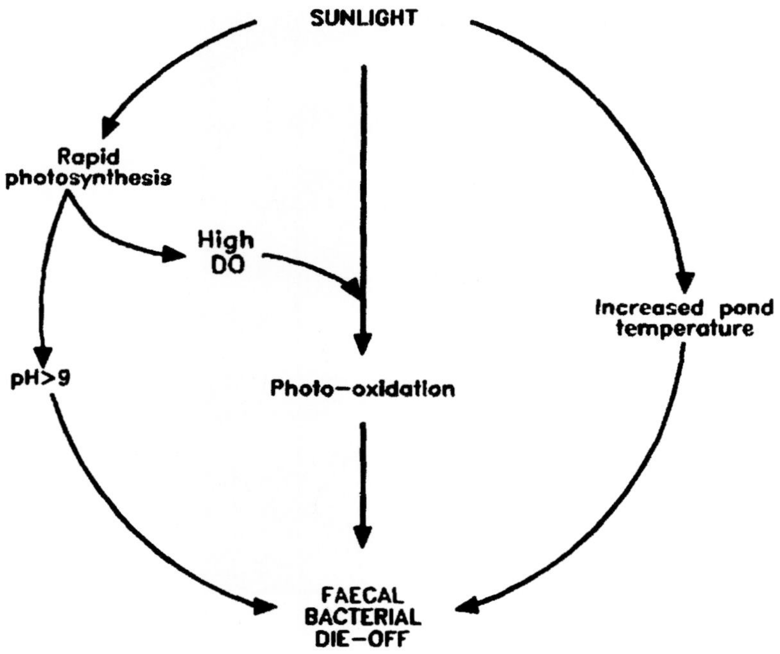

Fig. 26.3 Conceptual mechanisms for faecal bacterial die-off in facultative and maturation ponds.

WHO (1989) recommendation for unrestricted crop irrigation (see Chapter 15). However, WSP systems designed to produce effluents suitable for unrestricted irrigation generally require about twice the area of land required for those designed to produce effluents suitable for restricted irrigation. Thus decisions regarding unrestricted versus restricted irrigation should not be taken lightly and are best discussed with the local farmers.

3 WASTEWATER STORAGE AND TREATMENT RESERVOIRS

In arid or semi-arid areas where agricultural production is limited by the amount of water, including treated wastewater, available for irrigation, it is often sensible to use WSTR as these permit the whole year's wastewater to be used for irrigation, rather than just that produced during the irrigation season. WSTR are 5–15 m deep and function much like facultative ponds (Mara *et al.*,1996). Single WSTR are used if their contents are used for restricted irrigation as, for example in Israel where WSTR were developed. Israeli practice has been to treat the wastewater in an anaerobic pond and to discharge its effluent into a single WSTR which is 5–15 m deep (Fig. 26.4a). The irrigation season in Israel is 4–6 months long, and so the single WSTR has a storage capacity equivalent to 6–8 months' wastewater flow. It is full at the start of the irrigation season, and empty at the end of it. In this way two to three times as much land can be irrigated, and two to three times as many crops produced. Further details are given in Juanico and Shelef (1991, 1994). The long retention time in the WSTR ensures the removal of all helminth eggs, so the WSTR contents can be safely used for restricted irrigation.

For unrestricted irrigation, sequential batch-fed WSTR can be used (Mara and Pearson, 1992). This system comprises an anaerobic pond and, depending on the length of the irrigation season, three or four WSTR in parallel (but usually only three) (Mara and Pearson, 1992; Mara *et al.*, 1997) (Fig. 26.4b). Each WSTR is operated sequentially on a cycle of fill-rest-use (see Table 26.1), and faecal coliform reduction to <1000 per 100 ml occurs

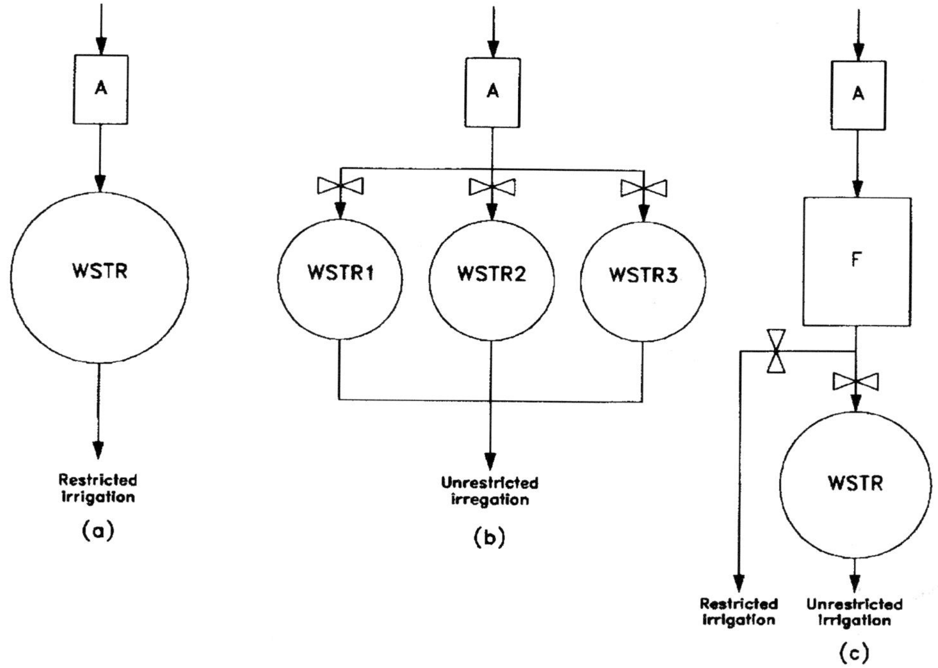

Fig. 26.4 (a) Single WSTR system for restricted irrigation, (b) sequential batch-fed WSTRs in parallel for unrestricted irrigation, and (c) hybrid WSP-WSTR system for both restricted and unrestricted irrigation. A, anaerobic pond; F, facultative pond.

rapidly during the fill and rest phases (Pearson *et al.*, 1996b). Thus the whole year's wastewater is available for unrestricted irrigation during the irrigation season.

WSTR design is detailed in Mara (1997) and Mara and Pearson (1998). Mara and Pearson (1999) describe a hybrid WSP–WSTR system to produce effluents for both restricted and unrestricted irrigation (Fig. 26.4c).

4 CONSTRUCTED WETLANDS

Constructed wetlands generally referred to as rooted macrophyte ponds, e.g. reedbeds, although the term can be used to describe an ecologically more diverse pond system with both rooted and floating macrophytes. The latter is often more to provide a pleasant recreational aquatic habitat (see EPA, 1993; Gearheart, 1996), rather than being specifically for wastewater treatment. This section will therefore address only reedbeds.

4.1 Reedbeds

Usually following septic tanks, anaerobic ponds or rotating biological contactors, reedbeds are long thin beds of reeds (*Phragmites australis*) planted either in the soil or in a gravel bed (alternatively, or in addition, other plants such as bulrushes (*Typha latifolia*) or rushes (*Scirpus lacustris*) can be used) (Fig. 26.5). If planted in the soil, the bed usually has an open water surface and this promotes the breeding of mosquitoes, both culicines (the larvae of which obtain their oxygen by breathing through the water surface) and *Mansonia* spp. (the larvae of which bore into the stem of the plant to obtain air from the air lacunae which transport air from the leaves to the roots). For this reason soil reedbeds are generally not much favoured, especially in developing countries where culicine mosquitoes are the vector of Bancroftian filariasis.

Gravel-bed reedbeds are commonly horizontal flow reactors with the water surface

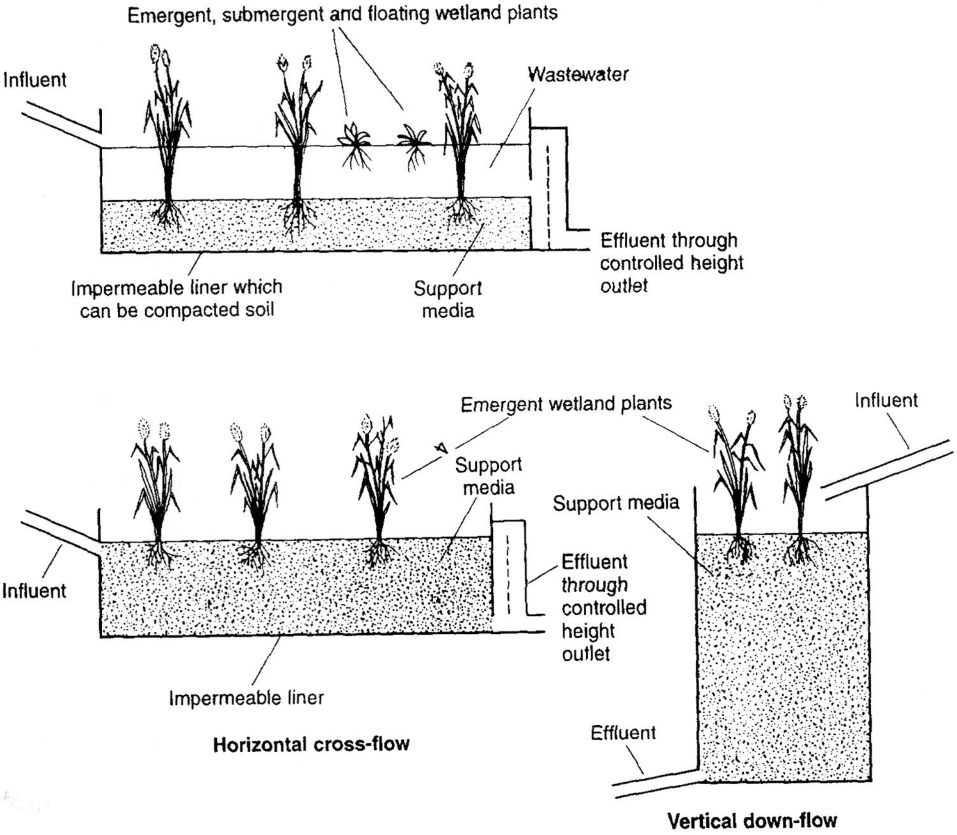

Fig. 26.5 Surface-flow soil-based reedbed with free water surface (top) and horizontal- and vertical-flow gravel-bed reedbeds (bottom) (Nuttal *et al.*, 1997).

below the top of the gravel bed to avoid mosquito breeding. Bed depth is 0.6 m, varying from 0.3 m at the inlet to 0.9 m at the outlet. Gravel size is 5–15 mm. The reeds (or other rooted macrophytes) (4–8 plants/m^2) help to oxygenate the bed, and heterotrophic bacterial activity within it reduces the BOD by around 90% (Williams *et al.*, 1995; Cooper *et al.*, 1996; Nuttal *et al.*, 1997). Bed size varies according to application: 3.7–5 m^2/person for secondary treatment, 1–2 m^2/person for tertiary treatment and 0.3–0.5 m^2/person for stormwater treatment (Nuttal *et al.*, 1997).

Vertical-flow reedbeds are used at higher organic loadings than horizontal flow beds, often where space is at a premium. Bed depth is 0.5–1.5 m. An area of around 1 m^2/person is required for 90% BOD removal and 2 m^2/person for 70% ammonia removal (Nuttal *et al.*, 1997).

Cooper (2001) should be consulted for an up-to-date review of reedbed design.

REFERENCES

Arthur, J.P. (1983). Notes on the Design and Operation of Waste Stabilization Ponds in Warm Climates of Developing Countries. *Technical Paper No. 7*. The World Bank, Washington, DC.

Ayres, R.M., Alabaster, G.P., Mara, D.D. and Lee, D.L. (1992). A design equation for human intestinal nematode egg removal in waste stabilization ponds. *Water Research* **26**(6), 863–865.

Bucksteeg, K. (1987). German experiences with sewage treatment ponds. *Water Science and Technology* **19**(12), 17–23.

CEMAGREF, SATESE, Ecole National de la Santé Publique and Agences de l'Eau (1997). *Le Lagunage Naturel: Les Leçons Tirées de 15 Ans de Pratique en France*. Centre National de Machinisme Agricole, du Génie Rural, des Eaux et des Fôrets, Lyon.

Cooper, P. (2001). Constructed wetlands and reed-beds: mature technology for the treatment of wastewater from small populations. *Water and Environmental Management* **15**(2), 79–85.

Cooper, P.F., Job, G.D., Green, M.B. and Shutes, R.B.E. (1996). *Reed Beds and Constructed Wetlands for Wastewater Treatment*. Water Research Centre, Swindon.

Council of the European Communities (1991). Council directive of 21 May 1991 concerning urban waste water treatment (91/271/EEC). *Official Journal of the European Communities*, No. L 135/40-52 (30 May).

Curtis, T.P., Mara, D.D. and Silva, S.A. (1992). Influence of pH, oxygen, and humic substances on ability of sunlight to damage fecal coliforms in waste stabilization pond water. *Applied and Environmental Microbiology* **58**(4), 1335–1343.

EPA (1983). *Design Manual: Municipal Wastewater Stabilization Ponds*, Report No. EPA-625/1-83-015. Environmental Protection Agency, Center for Environmental Research Information, Cincinnati.

EPA (1993). *Constructed Wetlands for Wastewater Treatment and Wildlife Habitat: 17 Case Studies*. Report No. EPA-832-R-93-005. Environmental Protection Agency. Washington, DC.

Gearheart, R.A. (1996). Watersheds, wetlands, wastewater management. In: R. Ramadori, L. Cingolani and L. Cameroni (eds) *Natural and Constructed Wetlands for Wastewater Treatment and Reuse: Experiences, Goals and Limits*, pp. 19–37. Centro Studi, Perugia.

Gloyna, E.F. and Espino, E. (1969). Sulphide production in waste stabilization ponds. *Journal of the Sanitary Engineering Division, American Society of Civil Engineers* **95**(SA3), 607–628.

Juanico, M. and Shelef, G. (1991). The performance of stablization reservoirs as a function of design and operation parameters. *Water Science and Technology* **23**(7/9), 1509–1516.

Juanico, M. and Shelef, G. (1994). Design operation and performance of stabilization reservoirs for wastewater irrigation in Israel. *Water Research* **28**, 175–186.

Liran, A., Juanico, M. and Shelef, G. (1994). Coliform removal in a stablization reservoir for wastewater irrigation in Israel. *Water Research* **28**, 1305–1314.

Mara, D.D., Pearson, H.W., Alabaster, G.P. and Mills, S.W. (1991). *Waste Stabilization Ponds: A Design Manual for Eastern Africa*. Lagoon Technology International, Leeds.

Mara, D.D. (1997). *Design Manual for Waste Stabilization Ponds in India*. Lagoon Technology International Ltd, Leeds.

Mara, D.D. and Pearson, H.W. (1992). Sequential batch-fed effluent storage reservoirs: a new concept of wastewater treatment prior to unrestricted crop irrigation. *Water Science and Technology* **26**(7/8), 1459–1464.

Mara, D.D. and Pearson, H.W. (1998). *Design Manual for Waste Stabilization Ponds in Mediterranean Countries*. Lagoon Technology International Ltd, Leeds.

Mara, D.D., Pearson, H.W., Oragui, J.I. and Cawley, L.R. (1997). *Wastewater Storage and Treatment Reservoirs in Northeast Brazil*. TPHE Research Monograph No. 12. University of Leeds, Leeds.

Mara, D.D. and Pearson, H.W. (1999). A hybrid waste stabilization pond and wastewater storage and treatment reservoir system for wastewater reuse for both restricted and unrestricted crop irrigation. *Water Research* **33**(2), 591–594.

Mara, D.D., Alabaster, G.P., Pearson, H.W. and Mills, S.W. (1992). *Waste Stabilization Ponds: A Design Manual for Eastern Africa*. Lagoon Technology International Ltd, Leeds.

Mara, D.D., Pearson, H.W., Oragui, J.I. *et al.* (1996). *Wastewater Storage and Treatment Reservoirs in Northeast Brazil*. TPHE Research Monograph No. 12. Department of Civil Engineering, University of Leeds, Leeds.

Mara, D.D., Cogman, A., Simkins, P. and Schembri, M.C.A.C. (1998). Performance of the Burwarton Estate waste stabilization ponds. *Water and Environmental Management* **12**(4), 260–264.

Marais, G.v.R. (1974). Faecal bacterial kinetics in waste stabilization ponds. *Journal of the Environmental Engineering Division, American Society of Civil Engineers* **100**(EE1), 119–139.

Nuttal, P.M., Boon, A.G. and Rowell, M.R. (1997). Review of the Design and Management of Constructed Wetlands. *Report No. 180*. Construction Industry Research and Information, London.

Oragui, J.I., Curtis, T.P., Silva, S.A. and Mara, D.D. (1987). The removal of excreted bacteria and viruses in deep waste stabilization ponds in northeast Brazil. *Water Science and Technology* **19**(Rio), 569–573.

Oragui, J.I., Arridge, H., Mara, D.D., Pearson, H.W. and Silva, S.A. (1993). *Vibrio cholerae* O1 (El Tor) removal in waste stabilization ponds in northeast Brazil. *Water Research* **27**, 727–728.

Parhad, N. and Rao, N.V. (1974). Effect of pH on survival of *E.coli*. *Journal of the Water Pollution Control Federation* **46**, 980–986.

Pearson, H.W., Mara, D.D. and Arridge, H.M. (1995). The influence of pond geometry and configuration on facultative and maturation waste stablization pond performance and efficiency. *Water Science and Technology* **31**(12), 129–139.

Pearson, H.W., Mara, D.D., Mills, S.W. and Smallman, D.J. (1987). Factors determining algal populations in waste stabilization ponds and the influence of algae on pond performance. *Water Science and Technology* **19**(12), 131–140.

Pearson, H.W., Avery, S.T., Mills, S.W. *et al.* (1996a). Performance of the Phase II Dandora waste stablization ponds, the largest in Africa: the case for anaerobic ponds. *Water Science and Technology* **33**(7), 91–98.

Pearson, H.W., Mara, D.D., Arridge, H. and Cawley, L.R. (1996b). Pathogen removal in experimental deep effluent storage reservoirs. *Water Science and Technology* **33**(7), 251–260.

Silva, S.A. (1982). *On the Treatment of Domestic Sewage in Waste Stabilization Ponds in Northeast Brazil*. PhD Thesis. University of Dundee, Dundee.

Williams, J., Ashworth, R., Ford, M. *et al.* (1995). *Physical and Chemical Aspects of Sewage Treatment in Gravel Bed Hydroponic (GBH) Systems.* Research Monograph in Wastewater Treatment and Reuse in Developing Countries No. 2. Department of Civil Engineering, University of Portsmouth, Portsmouth.

WHO (1989). *Health Guidelines for the Use of Wastewater in Agriculture and Aquaculture.* Technical Report Series No. 778. Geneva: World Health Organization.

WHO (1993). *Guidelines for Drinking Water Quality – I. Recommendations*, 2nd edn. World Health Organization, Geneva.

27

Microbial interactions in facultative and maturation ponds

Howard Pearson

Department of Civil Engineering, Federal University of Rio Grande do Norte, Natal, Brazil

1 INTRODUCTION

The key component that sets facultative and maturation ponds apart from other wastewater treatment systems in terms of process microbiology is the presence of microalgae. These microorganisms basically control treatment efficiency and effluent quality and ponds must be designed to optimize both the concentration and species diversity of algae present. This factor is frequently overlooked by sanitary engineers accustomed to designing other types of treatment facility and it is the main cause of poor pond design. Thus an understanding of the dynamics of algal–bacterial interactions in ponds is fundamental both to the design and efficient operation of such systems. Even so, correctly designed pond systems are highly robust and easy to maintain.

2 ALGAL BIOMASS AND ALGAL DIVERSITY

It has been estimated that approximately 80% of the dissolved oxygen in waste stabilization ponds (WSP) results from the photosynthetic activity of the microalgae forming the phytoplankton population and thus the aeration of ponds depends heavily on algal activity rather than on surface reaeration. The maintenance of a healthy algal population is therefore fundamental to the efficient oxidation of organic material by the bacterial population.

This relationship between the phototrophic microalgae and the aerobic chemo-organotrophic bacteria is often illustrated as a mutualistic relationship (see Chapter 26, Fig. 26.2) in which the bacteria benefit from the oxygen produced by algal photosynthesis to metabolize aerobically organic material for growth and energy production. The algae benefit by utilizing the carbon dioxide produced by bacterial respiration along with the released nutrients (N and P) to derive energy and fix carbon for growth via photosynthesis. This view of the role of the microalgae is, however, somewhat simplistic since the algae are now known to be strongly implicated in the natural disinfection processes in maturation ponds that are responsible for destroying viral and bacterial pathogens as will be discussed later in this chapter.

The algal genera and species that predominate in a pond appear to be a function of the surface organic loading the pond receives and, in general, at higher organic loadings species diversity decreases (Konig, 1984; Konig *et al.*, 1987). Thus facultative ponds have fewer algal genera than maturation ponds and flagellate genera tend to predominate, whereas in maturation ponds diversity is greater and non-flagellates frequently predominate (Table 27.1). This fact serves as a quick on-site means of determining whether the ponds in a series are underloaded or overloaded as the dominant algal species can be quickly identified with a simple field microscope. Algal biomass decreases with increased BOD surface loadings

The Handbook of Water and Wastewater Microbiology
ISBN 0-12-470100-0

TABLE 27.1 Examples of algal genera present in facultative and maturation ponds

Algal genus	*Facultative ponds*	*Maturation ponds*
Euglenophyta		
Euglena	+	+
Phacus	+	+
Chlorophyta		
Chlamydomonas	+	+
Chlorogonium	+	+
Eudorina	+	+
Pandorina	+	+
Pyrobotrys	+	+
Ankistrodesmus	−	+
Chlorella	+	+
Micractinium	−	+
Scenedesmus	−	+
Selenastrum	−	+
Carteria	+	+
Coelastrum	−	+
Dictyosphaerium	−	+
Oocystis	−	+
Cryptophyta		
Rhodomonas	−	+
Volvox	+	−
Chrysophyta		
Navicula	+	+
Cyclotella	−	+
Cyanobacteria[a]		
Oscillatoria	+	+
Arthrospira	+	+

+ = present; − = absent.
[a] photosynthetic prokaryotes.

in facultative ponds (Fig. 27.1) and the impact appears greater in shallow ponds than in deeper ones. At high organic loads the algal population tends towards monoculture of flagellate genera with *Chlamydomonas* proving to be the most resistant.

The recent work of Athayde (2001), studying algal dynamics in various series of waste stabilization ponds operating at different surface organic loadings at Extrabes in Northeast Brazil, where light intensities remain relatively constant and water temperatures are normally around 25°C, identified 28 different algal genera with the Chlorophyta dominant irrespective of organic loading. The genera *Chlamydomonas*, *Pyrobotrys*, *Phacus* and *Euglena* were among the most resistant algae to high organic loadings, being present at the highest loadings encountered of 770 kg BOD_5/ha day. In contrast, *Ankistrodesmus* and *Scenedesmus* were sensitive to high organic loadings and were most abundant at surface organic loadings below 75 and 50 kg BOD_5/ha day, respectively. High concentrations of ammonia and sulphide are associated with high organic loadings and it is these rather than the organic load (BOD) which control algal biomass and algal species dominance (Abelovich and Azov 1976; Konig 1984; Konig *et al.*, 1987; Mills 1987; Pearson *et al.*, 1987a; Athayde 2001). Since it is the non-ionic species of both ammonia (NH_3) (which is the increasingly dominant form at pH above 7) and sulphide (H_2S) (which predominates at neutral to acid conditions) that rapidly enter algal cells leading to toxicity, the prevailing pH of the pond water (which is influenced

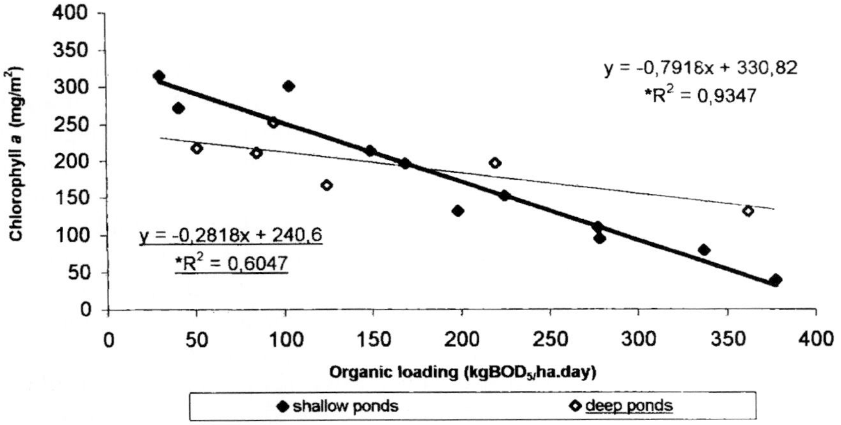

Fig. 27.1 The effect of surface organic loading on chlorophyll *a* values expressed on an area basis in shallow (1.0 m) and deep (2.20 m) pond series in Northeast Brazil (Athayde, 2001).

by algal photosynthetic activity), as well as the overall concentrations of these substances, is important.

The genus *Chlorella* is an enigma in that it is probably the most persistent genus, occurring across a very wide range of organic loadings. However, in terms of maximum biomass concentration it exhibits a preference for low loadings in the region of 28 kg BOD_5/ha day, normally found in maturation ponds. Athayde (2001) concluded that changes in the biomass (cell numbers) of *Chlamydomonas*, as the most pollution tolerant genus, and *Scenedesmus*, which favours low organic loadings, were the best indicator algae to use in monitoring changes in loading conditions in pond systems (Figs 27.2 and 27.3).

There is much still to do in terms of identifying pond algae as most studies have limited algal identification only down to the genus level. This is not surprising since, for example, there are allegedly over 200 species of *Euglena* and, according to Lund and Lund (1995), more than 400 species of *Chlamydomonas*. Only by the use of modern molecular techniques will it be possible to determine whether a species is, for example, miniaturized or modified in form due to prevailing conditions or represents a different species of the same genus which is proliferating under the changing conditions. Such studies will help considerably in understanding the complex microbiology of the waste stabilization pond environment.

Total algal biomass as determined by chlorophyll *a* concentration is also higher in facultative ponds than in the subsequent maturation ponds of a series. This probably reflects the reduction in available nutrients and also the increased grazing pressures by the zooplankton population which is larger in the more aerobic conditions prevailing in maturation ponds.

The degree of grazing of algae by the invertebrate fauna in ponds can affect effluent quality in two ways, one being beneficial and the other detrimental. A reduction in the algal standing crop in the final maturation pond of a series by grazing will improve effluent quality in terms of BOD since the algal concentration leaving the pond will be diminished. In fact, zooplankton grazing can theoretically reduce algal BOD in the effluent by 80–90% since the energy transfer from algae to the next trophic level, the grazers, will be between 10 and 20%, with 80% energy loss via heat and detritus sedimentation.

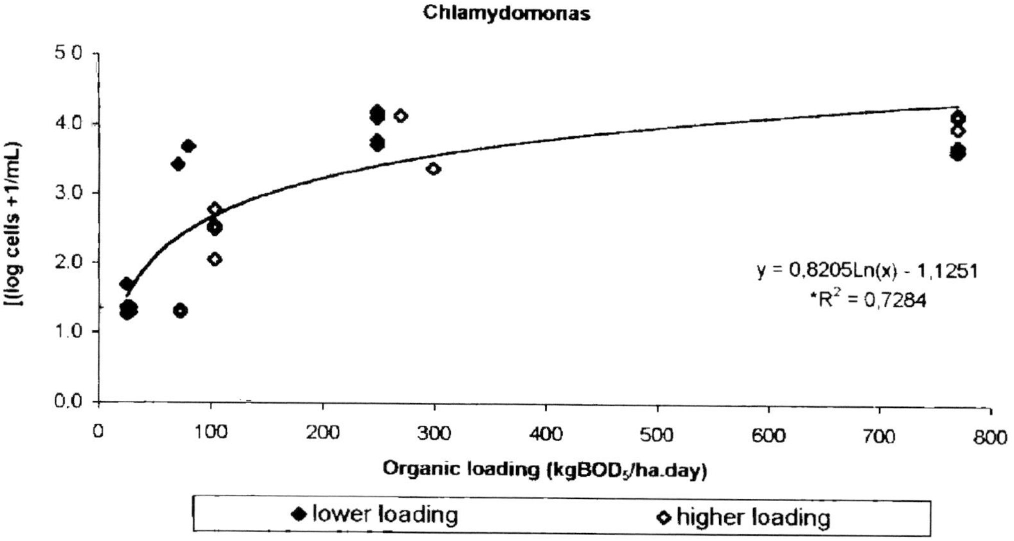

Fig. 27.2 *Chlamydomonas* biomass concentration plotted against surface organic loadings under both lower (250 kg BOD/ha day) and higher (770 kg BOD/ha day) loadings applied to the secondary facultative pond of the 5 pond series which included an anaerobic pretreatment pond (Athayde, 2001).

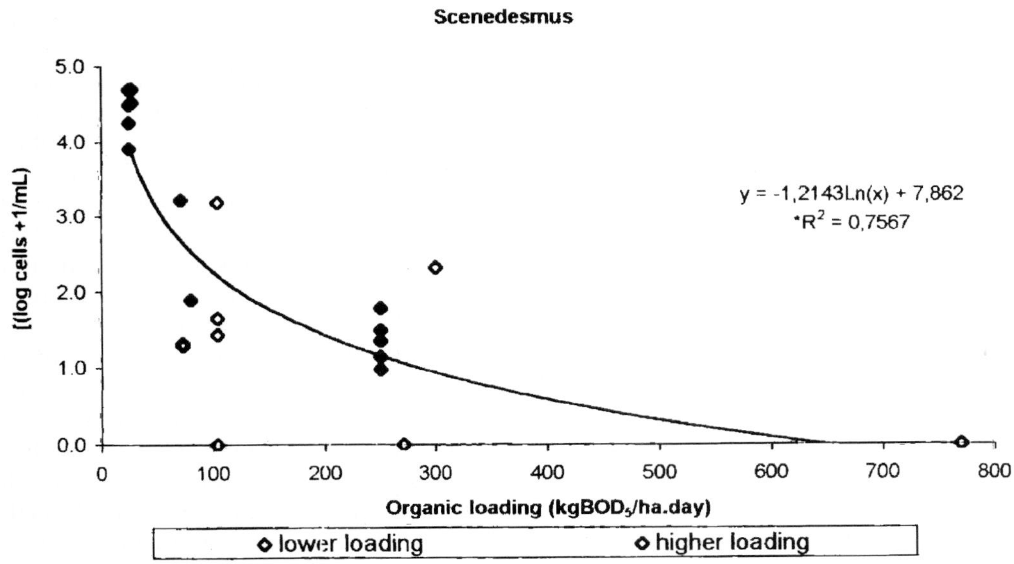

Fig. 27.3 *Scenedesmus* biomass concentration plotted against surface organic loadings under both lower (250 kg BOD/ha day) and higher (770 kg BOD/ha day) applied to the secondary facultative pond of the 5 pond series which included an anaerobic pretreatment pond (Athayde, 2001).

In contrast, excessive zooplankton grazing can reduce the effectiveness of the natural disinfection process which is reliant upon an active algal population to produce the high oxygen concentrations and elevated pH which are fundamental to the overall process (see *Pathogen removal* below).

Several algal genera common in facultative ponds, including *Euglena*, are capable of heterotrophic growth on organic substrates such as acetate released as a result of anaerobic degradation of organic material in the anaerobic sediments (Mills 1987; Pearson *et al.*, 1987a). *Euglena* species have also been observed migrating to the lower anaerobic organic-rich sediments in facultative ponds at night (Konig, 1984). However, the significance of this in terms of overall organic carbon removal in ponds is not clear.

3 SULPHATE-REDUCING BACTERIA AND PHOTOSYNTHETIC SULPHUR BACTERIA

The production of H_2S and thus the risk of bad odour in ponds is the result of the activity of sulphate-reducing bacteria (SRB) of, for example, the genera *Desulfovibrio* and *Desulfobacter*. These are obligately anaerobic bacteria and are present in the anaerobic layer and sediments of the facultative ponds (as well as in anaerobic ponds). They require organic material (e.g. organic acids) or hydrogen as a source of reductant, and sulphate (or sulphur or sulphite) as the terminal electron acceptor to reoxidize their electron transport chains under anoxic conditions during the production of energy (ATP) required for growth. Examples of equations for H_2S production by SRB are as follows:

$$CH_3COO^- + SO_4^2 + 3H^+ = 2CO_2 + \mathbf{H_2S}$$
$$+ 2H_2O4H_2 + SO_4^{2-} + H^+$$
$$= \mathbf{HS}^- + 4H_2O$$

Thus both excess sulphate and excess organic material (organic overloading) will stimulate the growth and activity of SRB. Acidic conditions ($pH < 6$) or alkaline conditions ($pH > 8$) will favour SRB proliferation over pH-sensitive methane-producing bacteria, which are also found in the anaerobic sediments since they compete for similar organic substrates. This results in more H_2S production and greater (or complete) inhibition of methane production.

The photosynthetic purple and green sulphur bacteria utilize hydrogen sulphide (H_2S), generated by SRB activity, as an electron donor for CO_2 reduction in photosynthesis. The sulphide is oxidized to elemental sulphur (S^0) and the sulphur is stored within the cells or, more accurately, in the periplasm (the space between the outer membrane and cytoplasmic membrane) of the cell in the case of the purple forms, but is deposited outside the cells in the case of the greens. If no H_2S is available, the purple forms utilize the stored elemental sulphur oxidizing it to sulphate. They can also utilize other reduced sulphur compounds such as thiosulphate and sulphite. Unlike the cyanobacteria and microalgae they do not utilize water as an electron donor and thus do not liberate oxygen as a by-product of photosynthesis. This non-oxygen-evolving photosynthetic process is termed anoxygenic photosynthesis, as opposed to oxygen-evolving or oxygenic photosynthesis.

The purple sulphur bacteria (e.g. the genus *Thiopedia*) are common members of the microbial flora of properly functioning facultative lagoons and are normally located in the water column at a depth of approximately 50 cm in the anoxic but illuminated zone below the surface microalgal layer. They do not compete for light with the algae as they utilize wavelengths longer than 800 nm, which pass unabsorbed through the algal zone. They normally obtain H_2S by diffusion of the gas through the water column from the lagoon sediments where it has been produced by the activity of sulphate-reducing bacteria as part of the normal anaerobic activity occurring in the base of facultative lagoons. In fact, under normal conditions in ponds, purple and green sulphur bacteria are important components of the natural odour filtration system as they oxidize a proportion of the H_2S before it reaches the aerobic surface waters of the pond where the photosynthetically produced oxygen completes the process. They also protect the algae from photosynthetic inhibition by sulphide (Houghton and Mara, 1992).

On occasions facultative (and indeed maturation lagoons) turn purple as a result of an increase in the purple bacterial population to the exclusion of the algal population. This usually results from the presence of high H_2S concentrations generated by SRB activity in the sediments sufficient to support the photosynthetic requirements of a large population of photosynthetic bacteria and to remove all the dissolved oxygen from the pond water column so providing a larger than normal anoxic illuminated (photic) zone, in which for example, purple photosynthetic bacteria such as *Thiopedia* can develop. Concentrations of H_2S greater than ≈ 8 mg/l are known to inhibit oxygenic photosynthesis, leading to the death of many pond microalgae (Mills, 1987; Pearson *et al.*, 1987a), so further exacerbating the presence of anaerobic conditions. However, it is interesting to note that in Morocco, WSP systems close to the sea with purple ponds continue to function well in terms of BOD removal. These ponds were dominated by *Thiopedia* as this genus is favoured over other photosynthetic purple sulphur bacteria in illuminated anoxic saline environments. Thus purple lagoons are often an indicator not only of organic overloading or high sulphate concentrations, but also of saline intrusion into coastal sewers.

4 HETEROTROPHIC BACTERIAL POPULATION IN WSP

A wide range of aerobic heterotrophic bacterial genera was found in ponds by Gann *et al.* (1968), including *Pseudomonas*, *Achromobacter*, *Flavobacterium* and *Bacillus*, although much less is known of their activities than those of the photosynthetic organisms. The bacteria present in the aerobic conditions of facultative ponds are deemed to be essentially those saprophytes present in the incoming wastewater and also include *Beggiatoa*, *Sphaerotilus*, and *Alcaligenes*, in addition to those already mentioned (Environmental Protection Agency, 1983). In other words, most aquatic bacterial groups are present and it is assumed that the microbial degradation of organic matter in ponds is similar to that in other biological wastewater treatment systems, although the biomass concentrations are presumably lower and thus account for the longer hydraulic retention times required for effective treatment.

5 PATHOGEN REMOVAL AND THE NATURAL BIOLOGICAL DISINFECTION PROCESS IN WSP

In comparison with other secondary wastewater treatment processes, ponds have been known to be very effective at removing most types of helminths, protozoa, bacteria and viruses for some years. Indeed, Caldwell (1946) reported on the excellent removal of coliforms in WSP, and Marais produced an engineering design equation for coliform removal in WSP based on time and temperature. It assumes first order kinetics for bacterial removal with the first order removal coefficient based on air temperature and that the ponds are completely mixed (Marais and Shaw, 1961; Marais, 1966, 1974). Several workers have criticized the concept that WSP are completely mixed reactors as demanded by the Marais design approach.

Dissanayake 1980 (Polprasert *et al.*, 1983), assuming first order removal kinetics for bacterial die-off, developed a dispersed flow model which corresponds more closely with the hydraulic regime found in ponds (Mara, 1976; Marecos do Monte and Mara, 1987). Yanez (1984) and Saenz (1992) also favoured the inclusion of a dispersion number in their equations to explain better bacterial removal in WSP and suggested dispersion values based on basic pond geometry. While such an approach may well better explain the kinetics of bacterial removal in ponds, and in this context is of value, it does little to aid pond design directly since a range of variables affects the value of the hydraulic dispersion number of a pond. These include pond shape, the number and positioning of inlets and outlets, shear stresses at the sides and the bottom, wind speed and direction, and thus basically it can only be determined with any accuracy by the use of tracers once the pond has been constructed (Polprasert *et al.*, 1983; Marecos do Monte and Mara, 1987; Vorkas and Lloyd, 2000). Indeed Shilton and Harrison (2002) suggest that inlet design (including baffles) is a key factor in determining the hydraulic regime in pond systems, and in comparison the influence of outlet design and the relative significance of the wind can be considered to be only of secondary importance.

Despite the criticisms made of the Marais design approach in terms of pond hydraulics, it nevertheless seems to work well in practice in terms of predicting the effluent quality of a WSP series. Pearson *et al.* (1995), studying various series of experimental ponds in Northeast Brazil where mean air and water temperatures vary little throughout the year, calculated first order bacterial removal constants (K_b) for faecal coliforms using the Marais equations and found that in the five facultative ponds studied the K_b values ranged between 2.5 and 4.0/day and were thus lower than the theoretical value of 6.2/day for a temperature of 25°C. In contrast, the tertiary maturation ponds had elevated K_b values in the range 10.0–13.8/day. Despite these discrepancies, when actual log numbers of faecal coliforms in individual pond effluents in a series were plotted against the theoretical effluent values that could be predicted from raw wastewater values using the Marais design approach, linear regressions with highly significant positive correlations ($r^2 = 0.99$) were obtained (Fig. 27.4). Thus, while the Marais approach may have its limitations, it nevertheless works well when there are three or more ponds in series since over- and under-estimations of K_b balance out, giving effluent values very close to predicted values using the Marais pond design equations. Furthermore, there was a good negative correlation between total hydraulic retention time (HRT) and final effluent concentrations of faecal coliforms in the five experimental pond series at depths of 1.0 m and in the deeper 2.2 m studied by Athayde (2001) in Northeast Brazil (Fig. 27.5).

Brissaud *et al.* (2000) found, in tracer tests using sodium iodide on a maturation lagoon in the South of France to determine residence time distribution and the simultaneous determination of the die-off constant in pilot lagoons

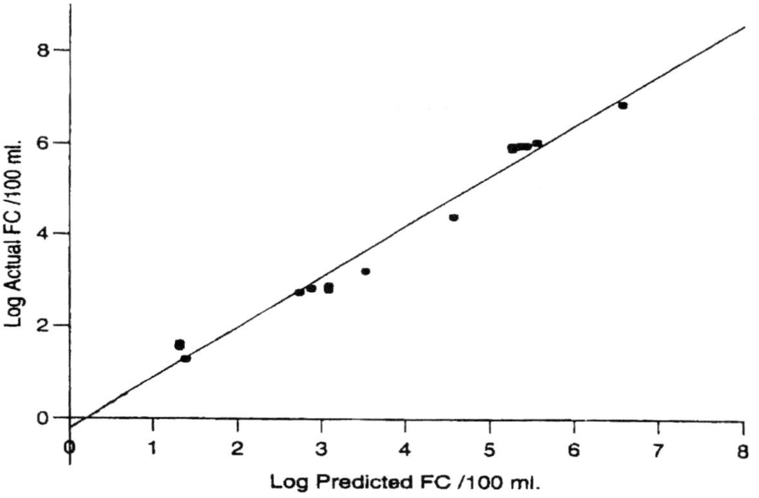

Fig. 27.4 Comparison of log actual and log predicted faecal coliform (FC) numbers in individual pond effluents. Log predicted values were derived using the Marais equation for determining effluent concentrations in a series of ponds and the FC concentration of the raw wastewater.

assuming a first order faecal coliform decay in a completely mixed reactor, that the Marais equations gave a good prediction of the actual faecal coliform numbers in the effluent. The pond in general mixed quickly throughout most of its depth. They thus verified that faecal coliform removal could be predicted from observed residence time assuming first-order faecal coliform die-off constant in a completely mixed reactor.

Despite these useful engineering studies on pond kinetics, the underlying mechanisms involved in the natural disinfection process have only become apparent relatively recently. It was Parhad and Rao (1974) who, on observing high pH values in ponds in India, investigated the role of microalgae on *E. coli* die-off. When algal photosynthesis raised the pH above 9.4, *E. coli* were removed very rapidly. Buffering the flasks containing *E. coli* and the algal cultures to keep the pH near neutral prevented die-off, and conversely raising the pH above 9.4 with alkali in the absence of algae also caused rapid bacterial die-off. They therefore concluded that high pH conditions brought about by rapid algal photosynthesis were the key mechanism for killing bacterial pathogens in WSP. Pearson *et al.* (1987b) also reported the rapid die-off of various faecal coliform populations isolated from WSP in Portugal at pH values above 8.6. The rate of die-off under laboratory conditions was independent of both light and oxygen concentration (up to 100% saturation), but die-off accelerated with increasing temperatures in the range 10–30°C (however, it should be

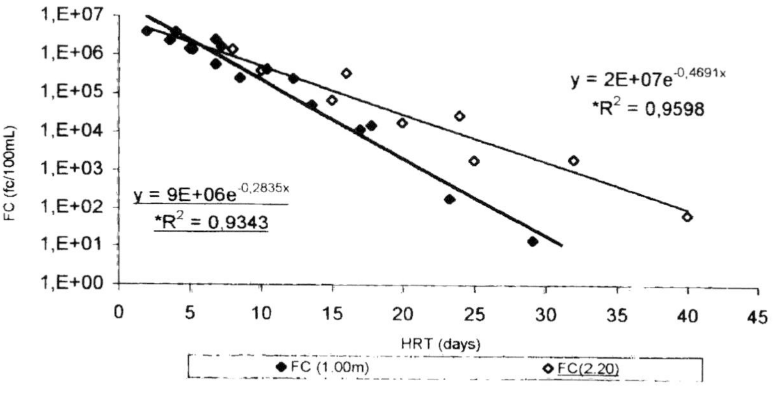

Fig. 27.5 Faecal coliform concentration plotted against the hydraulic retention time (HRT) in shallow (1.0 m deep) and deep pond series (2.2 m deep) (Athayde, 2001).

mentioned that only relatively low light intensities were used from 'Growlux' fluorescent tubes). The effects were independent of the presence of algae and they also concluded that pH was a key factor in the die-off of bacterial pathogens. They also noted that pH values and die-off were generally higher in maturation ponds than in facultative ponds.

The rise in pH is a direct result of rapid photosynthetic carbon fixation by the algae, which removes dissolved CO_2 from the pond water more rapidly than it can be replaced by either bacterial respiration or from the air across the pond–air interface. This results in a shift in the carbonate–bicarbonate equilibrium to produce CO_2 and hydroxyl ions. This CO_2 is fixed by the microalgae and hydroxyl ions accumulate with a consequent rise in pH:

$$2HCO_3^- \leftrightarrow 2CO_3^{2-} + H_2O + CO_2$$

$$CO_3^{2-} + H_2O \leftrightarrow 2OH^- + CO_2$$

This process can lead to pH values of above 10, especially in maturation ponds.

Smallman (1986) and Mills (1987) working together placed samples of pond water in dialysis bags and suspended them in various maturation and facultative ponds at different depths in the water column and found that when the daytime dissolved oxygen concentration (DO) was less than 20 mg/l and the pH was 8.5 or less the faecal coliform removal rate was roughly uniform at all depths. However, when the pH exceeded 9.0 and DO exceeded 20 mg/l the removal rate was highest at the surface. In further studies on the effluent quality of two maturation ponds in series, one, in which Daphnia had consumed the bulk of the algal population and thus had good light penetration but low pH and DO, had an effluent faecal coliform concentration that was higher than that of the other maturation pond which did not have a grazing Daphnia population and thus had a high algal concentration, lower light penetration and high pH and oxygen concentrations.

Curtis (1990) concluded that, while light did not seem to be directly responsible for faecal coliform die-off in ponds, high pH was also not an explanation for all situations in which good faecal coliform die-off rates occurred. Troussellier *et al.* (1986), working on ponds in the south of France, correlated faecal coliform die-off with three parameters, namely light, pH and chlorophyll *a* and provided the first clear evidence that more than a single parameter might be responsible for faecal coliform removal in ponds. Curtis *et al.* (1992a,b), in a series of laboratory field experiments, established that light between 425 and 700 nm (i.e. photosynthetically active radiation) killed faecal coliforms in ponds principally through an oxygen-mediated exogenous photosensitization process. The light is absorbed by dissolved humic substances in the pond water and these then enter an excited state for long enough to damage the bacterial cells. Light-mediated faecal coliform die-off was completely dependent on the presence of oxygen and is greatly enhanced by high pH levels (both of which are products of algal photosynthesis). Curtis *et al.* (1992b) have summarized conceptual mechanisms for faecal coliform die-off in WSP (See Chapter 26, Fig. 26.3) and concluded that any models explaining the bactericidal activity of light in ponds should include dissolved oxygen concentration and pH.

Recently Davies-Colley *et al.* (2000), in a series of experiments using WSP effluent in small, stirred reactors showed that different faecal indicators were inactivated by different components of the solar spectrum, and the rates of sunlight inactivation had differing dependencies on physicochemical conditions. For example, F-specific DNA phage was inactivated only by solar UVB (300–320 nm) at a rate unaffected by other factors, whereas enterococci and F-specific RNA phage were inactivated by a wide range of wavelengths (300–550 nm) by DO-dependent photo-oxidation. Sunlight inactivation of faecal coliforms was particularly complicated: at pH <8.5 only solar UVB (300–320 nm) caused (slow) inactivation, but at higher pH, the inactivation rate increased and a wider range of wavelengths (300–550 nm) contributed, suggesting photo-oxidation damage to membranes, which sensitizes faecal coliforms to high external pH.

In general therefore it seems likely that the biological disinfection process implicating light, pH and oxygen is a key mechanism controlling most faecal bacterial and at least some viral removal in ponds, rather than the mechanisms of starvation and predation. A notable bacterial exception appears to be *Vibrio cholerae* O1 which is tolerant of high pH but seems more vulnerable to sulphide inhibition and was more rapidly removed in anaerobic ponds (Arridge *et al.*, 1995). There is therefore a good reason for including anaerobic ponds in the treatment system where cholera is common. Also it would seem that RNA viruses may be less vulnerable to the photo-oxidative process, which accounts for why Oragui *et al.* (1995) found rotavirus removal to be slow in ponds.

REFERENCES

Abelovich, A. and Azov, Y. (1976). Toxicity of ammonia to algae in wastewater oxidation ponds. *Applied and Environmental Microbiology* **31**, 801–806.

Arridge, H., Oragui, J.I., Pearson, H.W. *et al.* (1995). *Vibrio cholerae* O1 and *Salmonella* spp. removal compared with the die-off of faecal indicator organisms in waste stabilization ponds in northeast Brazil. *Water Science and Technology* **31**(12), 249–256.

Athayde, S.T.S. (2001). *Algal and Bacterial Dynamics in Waste Stabilization Ponds and Wastewater Storage and Treatment Reservoirs*. PhD Thesis. University of Liverpool, Liverpool.

Brissaud, F., Lazarova, V., Ducoup, C. *et al.* (2000). Hydrodynamic behaviour and faecal coliform removal in a maturation pond. *Water Science and Technology* **42**(10–11), 119–126.

Curtis, T.P. (1990). *Mechanisms of Removal of Faecal Coliforms from Waste Stabilization Ponds*. PhD Thesis. University of Leeds, Leeds.

Curtis, T.P., Mara, D.D. and Silva, S.A. (1992a). Influence of pH, oxygen and humic substances on the ability of sunlight to damage fecal coliforms in waste stabilization pond water. *Applied and Environmental Microbiology* **58**, 1335–1343.

Curtis, T.P., Mara, D.D. and Silva, S.A. (1992b). The effect of sunlight on fecal coliforms in ponds: implications for research and design. *Water Science and Technology* **26**(7–8), 1729–1738.

Curtis, T.P., Mara, D.D., Dixo, N.G.H. and Silva, S.A. (1994). Light penetration in waste stabilization ponds. *Water Research* **28**, 1031–1038.

Davies-Colley, R.J., Donnison, A.M. and Speed, D.J. (2000). Towards a mechanistic understanding of pond disinfection. *Water Science and Technology* **42**(10–11), 149–158.

Dissanayake, M.G. (1980). *Kinetics of Bacterial Die-off in Waste Stabilization ponds*. PhD Dissertation No. Ev-80-2. Asian Institute of Technology, Bangkok.

Environmental Protection Agency (1983) *Design Manual for Municipal Wastewater Stabilization Ponds*. Report No. EPA 625/1-83-015. Office of Wastewater Management, EPA, Washington, DC.

Gann, J.D., Collier, R.E. and Lawrence, C.H. (1968). Aerobic bacteriology of waste stabilization ponds. *Journal of the Water Pollution Control Federation* **40**, 185–191.

Houghton, S.R. and Mara, D.D. (1992). The effects of sulphide generation in waste stabilization ponds on photosynthetic populations and effluent quality. *Water Science and Technology* **26**(7–8), 1759–1768.

Konig, A. (1984). *Ecophysiological Studies on Some Algae and Bacteria of Waste Stabilization Ponds*. PhD Thesis. University of Liverpool, Liverpool.

Konig, A., Pearson, H.W. and Silva, S.A. (1987). Ammonia toxicity to algal growth in waste stabilization ponds. *Water Science and Technology* **19**(12), 115–122.

Mara, D.D. (1976). *Wastewater treatment in Hot Climates*. John Wiley and Sons, Chichester.

Marais, G.v.R. (1966). New factors in the design, operation and performance of waste stabilization ponds. *Bulletin of the World Health Organization* **34**, 737–763.

Marais, G.v.R. (1970). Dynamic behaviour of oxidation ponds. In: R.E. McKinney (ed.) *Proceedings of Second International Symposium for Waste Treatment Lagoons*. University of Kansas, Laurence.

Marais, G.v.R. (1974). Faecal bacterial kinetics in waste stabilization ponds. *Journal of Environmental Engineering Division, American Society of Civil Engineers* **100**, 119–140.

Marais, G.v.R. and Shaw, V.R. (1961). A rational theory for the design of wastewater stabilization ponds in central and South Africa. *Transactions of the South African Institution of Civil Engineers* **3**, 205–227.

Marecos do Monte, M.H. and Mara, D.D. (1987). The hydraulic performance of waste stabilization ponds in Portugal. *Water Science and Technology* **19**(12), 219–227.

Mills, S.W. (1987). *Wastewater Treatment in Waste Stabilization Ponds: Physiological Studies on the Microalgal and Faecal Coliform Populations*. PhD Thesis. University of Liverpool, Liverpool.

Oragui, J.I., Arridge, H., Mara, D.D. *et al.* (1995). Rotavirus removal in experimental waste stabilization pond systems with different geometries and configurations. *Water Science and Technology* **31**(12), 285–290.

Parhad, N.M. and Rao, N.U. (1974). Effect of pH on survival of *Escherichia coli*. *Journal of the Water Pollution Control Federation* **46**, 980–986.

Pearson, H.W., Mara, D.D. and Arridge, H.A. (1995). The influence of pond geometry and configuration on facultative and maturation pond performance and efficiency. *Water Science and Technology* **31**(12), 129–139.

Pearson, H.W., Mara, D.D., Mills, S.W. and Smallman, D.J. (1987a). Factors determining algal populations in waste stabilization ponds and the influence of algae on pond performance. *Water Science and Technology* **19**(12), 131–140.

Pearson, H.W., Mara, D.D., Mills, S.W. and Smallman, D.J. (1987b). Physiochemical parameters influencing faecal bacterial survival in waste stabilization ponds. *Water Science and Technology* **19**(12), 131–140.

Polprasert, G., Dissanayake, M.G. and Thanh, N.C. (1983). Bacterial die-off kinetics in waste stabilization ponds. *Journal of the Water Pollution Control Federation* **55**, 285–296.

Saenz, R. (1992). *Predicción de la Calidad del Efluente en Lagunas de Estabilización*. Pan American Health Organization, Washington DC.

Shilton, A. and Harrison, J. (2002). Development of guidelines for improved hydraulic design of waste stabilization ponds. In: *Proceedings of the Fifth International IWA Specialist Group Conference on Waste Stabilization Ponds: Pond Technology for the New Millenium, Conference*, pp. 469–475. New Zealand Water and Wastes Association, Auckland.

Troussellier, M., Legendre, P. and Baleux, B. (1986). Modelling the evolution of bacterial densities in a eutrophic ecosystem (wastewater lagoons). *Microbial Ecology* **17**, 227–235.

Vorkas, C.A. and Lloyd, B.J. (2000). The application of a diagnostic methodology for the identification of hydraulic design deficiencies affecting pathogen removal. *Water Science and Technology* **42**(10–11), 99–109.

Yanez, F.A. (1984). Reduccíon de organismos patogenos y diseño de lagunas de estabilización en paises en desarrollo. *XIX Congresso Interamericano de Ingeneria Sanitaria y Ambiental*, Santiago, Chile.

28

Sulphate-reducing bacteria

Oliver J. Hao

Department of Civil Engineering, University of Maryland, MD 20742, USA

1 INTRODUCTION

The corrosion of sewers and the control of odour are the major operational and maintenance problems in wastewater collection and treatment systems. Significant capital investment is required for extensive sewer rehabilitation as reflected by the fact that the corrosion rate is about 5 mm/year in certain sections of the sewers in Japan (Mori *et al.*, 1992) and Los Angeles County (EPA, 1991). The public demand for an odour-free environment presents one of the major problems in the wastewater field.

The odour problems result from the generation of hydrogen sulphide (H_2S) by a specialized group of anaerobes, collectively called sulphate-reducing bacteria (SRB). Different sulphide species (H_2S, HS^- and S^{2-}) coexist in equilibrium in wastewater and the distribution of each species is a function of pH. For instance, approximately 45% of the total sulphide exists as H_2S_{aq} at pH 7. There is a further equilibrium between aqueous and gaseous H_2S governed by Henry's law; and the escape of H_2S_{gas} depends on the degree of mixing, and surrounding ambient H_2S_{gas} concentration. H_2S_{gas} exhibits a typical rotten-egg smell.

The corrosion problem results from the subsequent biological process in which sulphide is converted to sulphuric acid by aerobic autotrophic bacteria (sulphide oxidizing bacteria, SOB). The H_2SO_4 produced reacts with the various materials in sewer systems, resulting in eventual sewer corrosion. Additionally, H_2S itself, under anaerobic conditions, also causes the so-called depolarization of iron, hence the corrosion of iron (e.g. Bryant *et al.*, 1991). The formation of different copper sulphides (e.g. CuS and Cu_2S) is further indicative of SRB-induced corrosion of copper and copper alloys (McNeil *et al.*, 1991).

The practical control of corrosion and odour problems essentially involves the prevention of sulphide generation induced by SRB. Obviously, an adequate flow velocity in sewers may prevent sediment deposits which provide the ideal habitat for SRB. Also, adequate dissolved oxygen (DO) in wastewater can inhibit growth of the strictly anaerobic SRB; a few SRB species, however, can survive under aerobic conditions as discussed later. These practices for controlling odour problems, however, may be limited by the sewer system and treatment plant related constraints.

Normally, heavy metals present in the wastewater above certain threshold levels can be toxic to microorganisms, including SRB, thus restricting SRB growth and sulphide generation. Conversely, heavy metals readily react with sulphide to form metal sulphide precipitates, resulting in a lower sulphide concentration and reduced metal toxicity. For example, Dezham *et al.* (1988) used iron salts to form FeS in anaerobic digesters, thereby reducing the available sulphide, to control the H_2S present in the digester gas for both corrosion and air pollution control purposes.

Although the SRB and SOB cause some environmental problems of odour and corrosion, the activities of these bacteria may also alleviate some environmental problems in nature and in man-made engineered systems. Specifically, SRB can minimize metal mobility and toxicity because of insoluble metal

The Handbook of Water and Wastewater Microbiology
ISBN 0-12-470100-0

sulphide precipitates. A potential widespread utilization of SRB in removing heavy metals is envisioned for the future. Furthermore, as laboratory experiments are transformed into pilot and full scale works, the microbial transformation of oxyanions such as As(V)/As(III), Se(VI)/Se(IV) and Cr(VI) by SRB may provide a cost-effective means in bioremediation of these hazardous oxyanions.

This chapter presents pertinent information about the microbiology of SRB in general and a brief review of utilizing SRB for removing heavy metals and reducing metal oxyanions in particular.

2 SULPHUR CYCLE

The biological sulphur cycle, similar to the biological nitrogen cycle described in chapter 25, is driven by microbial activities. The sulphate ion (SO_4^{2-}), the most common sulphur species present in wastewater, is biologically reduced to sulphide. H_2S is the most common and prevalent odourous inorganic substance associated with wastewater collection and treatment systems. The threshold odour concentration is extremely low (0.0005 ppm, Leonardos *et al.*, 1969). Adverse physiological effects to humans can occur at concentrations greater than 10 ppm. Finally, H_2S may reach life-threatening levels at concentrations greater than 300 ppm (Mori *et al.*, 1992). The odour of H_2S is so strong that it is readily perceived in the environment long before the toxic level is reached. Unfortunately, human deaths due to sulphide toxicity are not uncommon, especially for sewage workers and others working in enclosed systems where H_2S might build up (Brock and Madigan, 1991). In sewer systems, H_2S_{gas} concentration ranges from 3 to an extremely high 400 ppm near a manhole (Mori *et al.*, 1992). Ambient H_2S levels at a wastewater treatment plant were found to be higher (>1 ppm) around the inlet structure and primary clarifiers (Koe, 1985). The conditions leading to H_2S formation generally favour the production of other malodourous organic compounds. Thus, solving H_2S odour problems can often solve other odour problems as well.

Microbial corrosion of materials is the result of oxidative reactions initiated by the metabolic activity of SOB, of which *Thiobacillus thiooxidans* is the predominant species. The corrosive action of H_2SO_4 results in damage to the concrete walls of reactors and sewer systems as well as steel pipelines. (Refer to Section 4 for biologically mediated corrosion problems in water distribution and treatment systems.)

Although SOB are aerobic and SRB are anaerobic microbes, the presence of SRB in aerobic environments as well as that of SOB within the anaerobic microbial community is well documented.

3 SULPHATE-REDUCING BACTERIA

A relatively wide range of genera of SRB has been identified. The major characteristics of the presently classified SRB are summarized by Widdel (1988). All SRB are Gram-negative except *Desulfonema* species. The two most prevalent genera of SRB are *Desulfotomaculum*, comprised of spore-forming, straight or curved rods, and the non-sporing genus *Desulfovibrio* with curved motile vibrios or rods (Gibson, 1990).

3.1 Metabolism

SRB are obligate anaerobes which obtain energy for growth by oxidation of organic substrates and use sulphate as the terminal electron acceptor as:

$$SO_4^{2-} + 2C + 2H_2O \xrightarrow{SRB} H_2S + 2HCO_3^- \quad (1)$$

Equation 1 is similar to nitrate reduction in that alkalinity is produced and there is a stoichiometric relationship between mg carbon required per mg SO_4^{2-} reduced. However, the sulphate reducers, unlike the denitrifiers, which are facultative organisms and prefer an aerobic environment, can only use sulphate in the absence of oxygen or nitrate.

3.1.1 Carbon source

The preferred carbon sources for SRB are low-molecular-weight compounds such as organic acids (e.g. lactate, pyruvate, formate and malate), fatty acid (e.g. acetate), and alcohols (e.g. ethanol, propanol, methanol and butanol). Nearly all of these compounds are fermentation products from anaerobic degradation of carbohydrates, proteins and lipids. Thus, SRB are terminal degraders and their role is analogous to that of methanogenic bacteria that produce methane and CO_2 as final products. The sulphide production rate in sewers has been correlated with dissolved carbohydrate and volatile fatty acids (Nielsen *et al.*, 1998). The well known *Desulfovibrio* genus does not utilize acetate, i.e. only degrading lactate to acetate (commonly referred to as incomplete SRB), while different *Desulfotomaculum* species are acetate-utilizing bacteria (Postgate, 1984; Widdel, 1988; Brock and Madigan, 1991).

A few species can even grow autotrophically using CO_2 as a sole source of carbon. Some strains of *Desulfotomaculum* can utilize glucose, but this is rather rare among sulphate reducers in general. Degradation of aromatic and saturated cyclic organic compounds, as well as long-chain n-alkanes, has also been reported with some sulphate reducers (Widdel, 1988; Caldwell *et al.*, 1998; Häggblom, 1998). SRB metabolic reactions and free energy changes with different carbon substrates are shown in Table 28.1.

TABLE 28.1 Metabolic reactions and free energy changes for SRB (Hao *et al.*, 1996)

Reactions	*Free energy (kJ/mol)*
$3LA^- \rightarrow 2PA^- + HA^- + HCO_4^- + H^+$	−165
$2LA^- + SO_4^{2-} + H^+ \rightarrow 2HA^- + 2CO_2 + HS^- + 2H_2O$	−189
$4PA^- + 3SO_4^{2-} \rightarrow 4HA^- + 4HCO_4^- + 3HS^- + H^+$	−151
$HA^- + SO_4^{2-} \rightarrow 2HCO_4^- + HS^-$	−60
$HA^- + 4S + 3H_2O \rightarrow 4H^+ + HCO_4^- + 4HS^- + CO_2$	−24
$4H_2 + SO_4^{2-} + CO_2 \rightarrow 3H_2O + HS^- + HCO_4^-$	−152

Note: LA^-, $CH_4COOHCOO^-$ (lactate); PA^-, $CH_4CH_2COO^-$ (propionate); HA^-, CH_4COO^- (acetate).

3.1.2 *Sulphur source*

In the presence of organic electron donors and thiosulphate or sulphite, SRB may first break down the thiosulphate or sulphite compounds to sulphate (equations 2 and 3) and then oxidize the organic substrates with the newly formed sulphate (Widdel, 1988):

$$S_2O_3^{2-} + H_2O \rightarrow SO_4^{2-} + HS^- + H^+ \quad \Delta G^{0'} = -22\ \text{kJ} \tag{2}$$

$$4SO_3^{2-} + H^+ \rightarrow 3SO_4^{2-} + HS^- \quad \Delta G^{0'} = -236\ \text{kJ} \tag{3}$$

These disproportionation reactions can provide energy for cell synthesis (Bak and Pfennig, 1987; Brock and Madigan, 1991). In fact, some SRB prefer thiosulphate to sulphate. The sulphide production rates from both thiosulphate and sulphite were found to be considerably higher than those from sulphate (Nielsen, 1991). It is assumed that sulphate would not be a limiting factor for sulphide production in most wastewaters at concentrations greater than 20 mg/l (EPA, 1974).

3.2 Environmental factors

A knowledge of the various factors affecting SRB growth is essential to control their microbial reactions in wastewater environments.

3.2.1 *Dissolved oxygen*

Although sulphate-reducers are obligate anaerobes, they may survive a temporary exposure to oxygen and again become active under anaerobic conditions. Because of the O_2 mass transfer, the critical DO concentration in the wastewater below which sulphate reduction can occur is 0.1–1.0 mg/l (EPA, 1985). If aerobic sediments or waters are rich in organic particles, SRB may be active in anaerobic microniches despite the aerobic surroundings (Widdel, 1988). Several SRB strains, including *Desulfovibrio* and *Desulfobacter* species, could be enriched from aerobic activated sludge. A few strains of SRB have been reported to be able to utilize oxygen without growth (Dilling and Cypionka, 1990). Sulphate reduction occurs consistently within the oxygenated photosynthetic microbial mats (Canfield and Des Marais, 1991; Frund and Cohen, 1992). In general, DO concentrations above certain levels inhibit SRB; thus, maintenance of relatively high DO levels is recommended in the design and operation of sewer and treatment systems to prevent odour problems.

3.2.2 *Temperature*

Sulphate reduction rates depend strongly on temperature; sulphate reduction typically increases 2- to 3.9-fold with a temperature increase of 10°C (Nielsen, 1987; Widdel, 1988). The optimum temperature for most pure cultures of SRB ranges between 28 and 32°C; the lower optima among SRB are observed with some *Desulfobacterium* strains and a curved *Desulfobacter* strain at 24–28°C and the highest is around 70°C for *Thermodesulfobacterium commune*. Most species of SRB die rapidly at temperatures above 45°C (Postgate, 1984; Widdel, 1988).

Okabe and Characklis (1992) reported that the maximum specific growth rate (μ_{max}) of *Desulfovibrio desulphuricans* varied between 0.38 and 0.55/h at 25°C and 43°C, respectively, and the odours dramatically decreased outside this temperature range. A temperature shock of 45°C imposed in a mesophilic (30°C) upflow sludge bed system treating sulphate-containing wastewater did not show any detrimental effect; temperature shocks to 55 and 65°C, however, significantly reduced the treatment efficiency (Visser *et al.*, 1993).

3.2.3 *pH*

SRB prefer an environment around pH 7 and are usually inhibited at pH values below 5.5 or above 9 (EPA, 1974). Nevertheless, sulphate reduction has been observed in a peat bog and acid mine water that exhibit pH values of about 3 to 4. SRB isolated from these habitats, however, were inhibited below pH 6. Therefore, it was hypothesized that the SRB in acidic environments were present in microniches, where higher and more favourable pH conditions could exist. Recent studies further demonstrated that two isolated strains of SRB could even grow at pH < 4.5 (Hard *et al.*, 1997), and SRB are capable of sulphate reduction and production of alkalinity at pH values as low as 3.3 (Elliott *et al.*, 1998). The effect of low pH could also be minimized in sulphate-reducing systems, since the sulphate-reduction process generates additional bicarbonate alkalinity.

A *Desulfovibrio* culture was inoculated with lactate and sulphate in the pH range of 5.8–7.0 and both the growth yield coefficient (Y) and μ_{max} increased with pH; maximum values of 0.15 g cell/g SO_4^{2-} and 0.14/h were obtained at pH 6.6 (Reis *et al.*, 1992). Changing waste pH with acid or alkali has been suggested as a method for diminishing sulphate reduction in industrial plants.

3.2.4 *Sulphide*

Sulphide is known to inhibit SRB, a classic example of the product inhibition often encountered in chemical and biological processes. Sulphide inhibits SRB probably when anyone of the sulphide species (H_2S, HS^- or S^{2-}) combines with the iron of ferredoxin, cytochromes or other essential iron-containing compounds in the cell, causing the electron transport systems to cease activity (Okabe *et al.*, 1992). However, literature data with respect to inhibitory concentrations of H_2S or total sulphide vary due to different environmental conditions (e.g. pH, temperature), wastewater characteristics (e.g. presence of iron salts and type of carbon source) and reactor systems (e.g. fixed film).

3.3 Ecology

SRB are widespread in sewers and wastewater treatment plants. In sewer systems, SRB are normally present in slime layers (EPA, 1985; Holder, 1986) and sediments (Mori *et al.*, 1991). Sulphate reduction rates increase proportionally with the increase in the SS (suspended solids) concentration; e.g. an increase from 250 to 1500 mg/l SS results in a sixfold increased rate (Attal *et al.*, 1992). Because of their slow

TABLE 28.2 SRB count in sewers, wastewater and sludge samples

Sample type	*MPN count (per 100 ml unless otherwise noted)*	*References*
Raw wastewater	$>2\times10^8$	Lish, 1993
Primary effluent	10^7	Morton *et al.*, 1991
Aerobic fixed film (trickling filter and rotating biological contactor)	6×10^7–10^8 (per g VS)	Lens *et al.*, 1995
Activated sludge (aerobic)	2×10^5–10^6 (per g VS)	Lish, 1993; Yamamoto-Ikemoto *et al.*, 1994; Lens *et al.*, 1995
Activated sludge (anoxic)	7×10^5	Lish, 1993
UASB (upflow anaerobic sludge blanket) sludge	3×10^7– 5×10^{10} (per g VSS)	Harada *et al.*, 1994; Lens *et al.*, 1995; Fang *et al.*, 1997
Anaerobic digester sludge	6×10^5– 5×10^7	Postgate, 1963; Lish, 1993; Harada *et al.*, 1994
Acidogenic anaerobic sludge	up to 10^{11}	Mizuno *et al.*, 1998
Enriched sludge for sulphate reduction	10^8	Lish, 1993
Pure SRB culture	up to 8×10^{10}	Postgate, 1963
Stream	6×10^2	Postgate, 1963
Sediment in sewer	3×10^4–10^8 (per g VSS)	Mori *et al.*, 1991; Schmitt and Seyfried, 1992
Landfill	8×10^6 (per g)	Fairweather and Barlaz, 1998

growth characteristics and carbon preference, SRB are usually outnumbered by other types of microbes, except in special environments. In anaerobic sludge digesters, there exists a competition between SRB and MPB (methane-producing bacteria), since these two groups of bacteria exhibit a similar affinity for organic substrates. In general, factors such as concentration of sulphide, type of fermented organics, COD/SO_4^{2-} ratios, and type and concentration of other inhibitory substances affect the relative prevalence of the bacterial types.

Despite their strict anaerobic requirement, the presence of SRB has been detected in many aerobic regions (Gibson, 1990), including activated sludge and aerobic fixed film. Table 28.2 presents the results of SRB counts in various environmental samples. The differences in SRB number are obvious, even with the same type of samples. Many factors can contribute to the differences, e.g. depth of sediment samples (a three-order of magnitude difference (7×10^4 to 3×10^7/g) in sewer trunk sediments (Mori *et al.*, 1991), type of substrate (e.g. benzoate or H_2) used, and system objective (e.g. maximum of MPB) in various SRB studies.

4 METAL REMOVAL

The role of sulphide in preventing heavy metal toxicity under anaerobic conditions has long been recognized (Lawrence and McCarty, 1965). The insoluble metal sulphide precipitates (Table 28.3), due to sulphide generated *in situ*, are responsible for immobilization of heavy metals in natural environments of soils and sediments and in engineering systems as:

$$Me^{2+} + S^{2-} \rightarrow MeS \quad (4)$$

TABLE 28.3 Solubility product of metal sulphide and metal hydroxide precipitates (Zumdahl, 1989)

Metal	*Ksp*	
	Metal sulphide	*Metal hydroxide*
Cu	8.5×10^{-45}	1.6×10^{-19}
Zn	1.2×10^{-23}	8.5×10^{-17}
Pb	3.4×10^{-28}	1.2×10^{-15}
Cd	3.5×10^{-29}	5.9×10^{-15}
Fe	3.7×10^{-19}	1.8×10^{-15}
Ni	1.6×10^{-16}	1.6×10^{-16}
Cr	–	6.7×10^{-31}

As mentioned earlier, the external addition of iron salts may be used to precipitate sulphide for odour control (Sercombe, 1995), reduce metal/sulphide toxicity (Edwards *et al.*, 1997), reduce H_2S gas in anaerobic digesters (Dezham *et al.*, 1988), and control dissolved sulphide in sewers (Padival *et al.*, 1995). In fact, Jin *et al.* (1998) even added sulphide into a methanogenic system to reduce the effect of copper toxicity. In wastewater, if iron salts are present in large quantities, FeS precipitation normally occurs. The black colour of sediments as well as primary and digested sludge, where sulphate reduction occurs, is due to the accumulation of FeS.

The use of SRB for metal removal has been applied to different industrial wastes, including contaminated groundwater (Barnes *et al.*, 1991), landfill leachate (DeWalle *et al.*, 1979), plating wastes (Bewtra *et al.*, 1995), leachate from contaminated soils (White *et al.*, 1998) and mining effluent (Maree *et al.*, 1987), in a variety of processes including anaerobic filters, sludge blanket reactors, and contact systems. The different metal removal schemes using SRB are shown in Fig. 28.1. Essentially, metal-laden wastes are supplemented with organics and other nutrients to provide adequate sulphate/COD ratios, and other conditions are used, e.g. lower hydraulic detention times and neutral pH, to favour the growth of SRB. Once sulphide is generated *in situ*, it readily reacts with metals to form MeS precipitates. In general, high efficiencies of metal removal can be achieved. For example, an upflow sludge blanket reactor (9 m^3) was installed near a zinc refining site in the Netherlands to treat metal-contaminated ground water (Barnes *et al.*, 1991). With external addition of nutrients (N and P) and carbon/energy substrate (ethanol), the system reduced Zn concentrations of 107–1070 mg/l to below 1 mg/l.

Haas and Polprasert (1993) used the recycling stream containing bio-generated sulphide to remove Cu, similar to the scheme shown in Fig. 28.1c. The system produced an effluent with consistently low COD and metal content. Hammack *et al.* (1994) used a similar system (shown in Fig. 28.1d) to treat mine waste, with a limestone neutralization reactor between the metal precipitator and sulphate reduction reactor. The countercurrent metal precipitator removed metal and the limestone reactor increased pH to 6. Sodium lactate was injected directly into the sulphate reduction reactor. The process removed more than 99% of the initial concentrations of Fe (620 mg/l), Cu (178 mg/l), Zn (530 mg/l), and Al (278 mg/l).

The potential use of SRB for pollution abatement of acid mine drainage (AMD) has long been recognized (Tuttle *et al.*, 1969; Wakao *et al.*, 1979). Since the 1980s, laboratory, pilot-scale, and full-scale evaluations have demonstrated the capability of using SRB to remove sulphate and metals from AMD (Table 28.4). In general, the amount and type of organics added, as well as pH adjustment, are critical for metal removal. Most metals are precipitated as MeS, although some are retained as hydroxides or carbonates.

Because sulphate reduction may be significant in artificial or natural wetlands, there has been some keen interest in using wetlands for the treatment of the AMD (e.g. Webb *et al.*, 1998), due to low cost and low-maintenance requirements. A survey conducted in 1989 indicated that a total of 142 wetlands in the eastern USA were constructed specifically for AMD treatment (Wieder, 1989). However, the effectiveness of these wetland treatments of AMD has been variable and unpredictable (Wieder, 1989).

4.1 Metal toxicity

Toxicity studies for SRB are often reported as a function of the initial metal concentrations. Because of complicated biological processes, some metals may precipitate and/or physically entrap within the biomass. Thus, the residual soluble metal concentrations should be determined to account for the observed toxicity phenomenon. Furthermore, metals present in free form, and not complexed forms due to naturally and biologically generated surfactants, exhibit a toxic effect; and different forms of metals, e.g. As(V) versus As(III), may exhibit

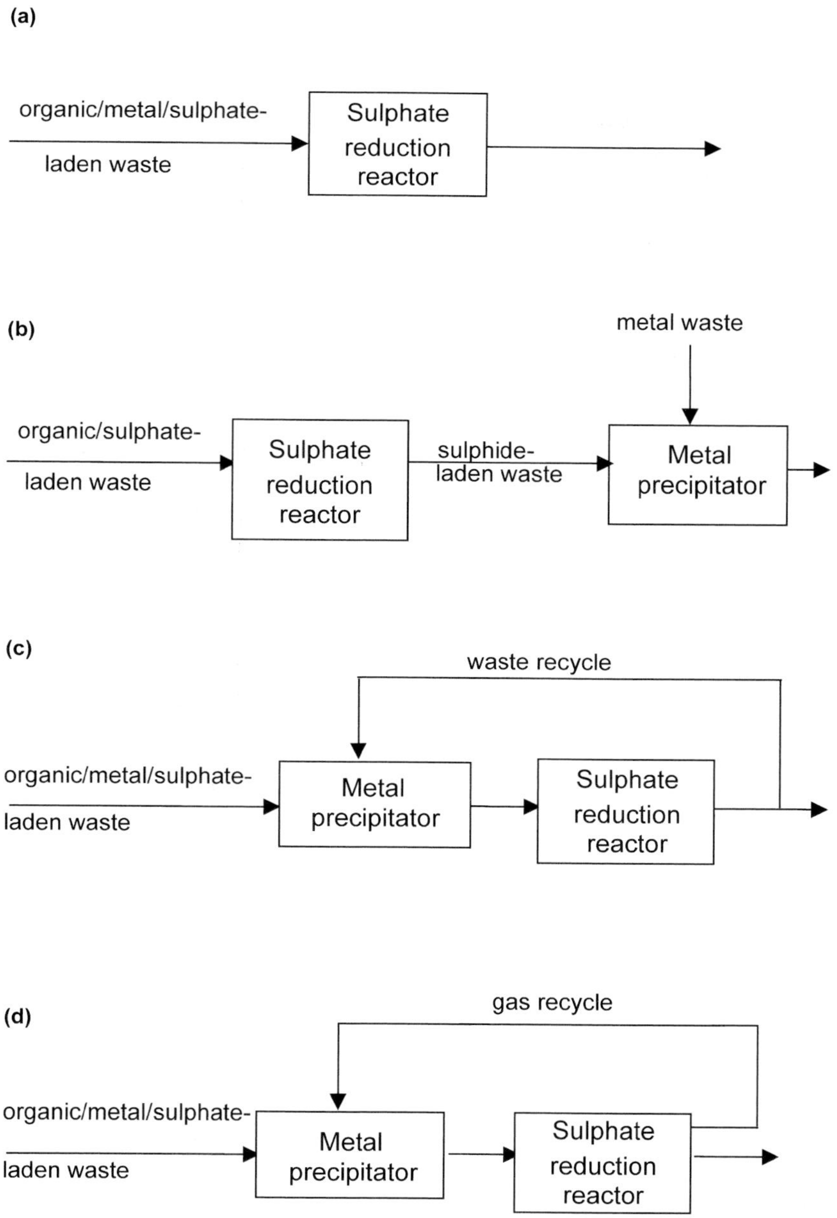

Fig. 28.1 Potential schemes for removal of heavy metals by sulphate-reducing bacteria

different effects. Consequently, highly soluble metal concentrations may not exert inhibitory effects in anaerobic sulphate-reducing systems which generate biosurfactants rendering metals less toxic.

The correlation between metal toxicity and sulphide content in sediments has been extensively studied (e.g. Di Toro *et al.*, 1992) in terms of the ratio of simultaneously extracted metal (SEM) to the acid volatile sulphide (AVS). Typically, sediment/sludge samples with SEM/AVS ratios of metals (e.g. Cd) greater than one may be toxic to the test organism, whereas samples with ratios less than one are not toxic, due to insoluble MeS.

TABLE 28.4 Treatment of acid mine drainage by SRB

Type of mine wastes	*Systems*	*Efficiency*	*References*
Ni simulated waste supplemented with lactate and sulphate	Columns with mushroom compost	540 mg Ni/kg compost-day	Hammack and Edenborn, 1992
Coal mine dump and smelting residues	3–200 l reactors in series, filled with spent mushroom compost and a single 4500 l reactor	95% removal efficiencies for Al, Cd, Fe, Mn, and Zn	Dvorak *et al.*, 1992
Copper-zinc mine waste supplemented with whey	Batch treatment in cylinders for 200 days	Significant removal of Cu, Fe, Al and Zn; pH increased and redox decreased	Christensen *et al.*, 1996
Mine contaminated groundwater	Reactive wall (40% municipal compost, 50% leaf compost and 20% wood chips) installed in the path of waste	Fe decreased from 250–1300 to 10–40 mg/l; pH increased from 5.8 to 7; alkalinity from 50 up to 2000 mg/l	Benner *et al.*, 1997

5 METALLOID OXYANION REDUCTION

Many strains of SRB are capable of reducing the soluble oxidized oxyanions (e.g. MoO_4^{2-}) to insoluble phases (e.g. MoS_2 or MoO_2). Growth of these microbes may occur with these oxyanions as the sole electron acceptors (Lovley, 1993; Newman *et al.*, 1997; Tebo and Obraztsova, 1998), although some SRB strains cannot grow with oxyanions as sole electron acceptors (Lovley and Phillips, 1994). It is generally established that the enzymatically mediated reactions are responsible for oxyanion reduction, e.g. both hydrogenase and cytochrome c were responsible for Cr(VI) reduction (Lovley and Phillips, 1994) and c_3 cytochrome for the U(VI) reduction (Lovley

TABLE 28.5 Reduction of metalloid oxyanions by SRB strains

Soluble oxyanions	*Insoluble species*	*Comments*	*SRB species*	*References*
U(VI)	U(IV), $(UCa)O_2$	From 235 to less than 2 mg U/l	*Shewanella putrefaciens*	Abdelouas *et al.*, 1998
	Mo(IV) MoS_2(s)	Either lactate or H_2 as electron donor; MoS_2(s) formed in the presence of sulphide	*Desulfovibrio desulphuricans*, *Desulfovibrio vulgaris*	Tucker *et al.*, 1997
MoO_4^{2-}, SeO_4^{2-}, $UO_2(CO_4)_2^{2-}$, $UO_2(CH_4COO)_2$, Cr(VI)	MoS_2, UO_2, Se^0	High removal efficiency in an immobilized system; formate or lactate as electron donor	*Desulfovibrio desulphuricans*	Tucker *et al.*, 1998a, b
SeO_4^2 or SeO_4^2	Se^0	Formate as electron donor, sulphate or fumarate as electron acceptor	*Desulfovibrio desulphuricans*	Tomei *et al.*, 1995
U(VI)	U(IV)	Half saturation coefficient = 127 mg/l, yield coefficient = 0.02 g/mol pyruvate	*Desulfovibrio desulphuricans*	Tucker *et al.*, 1996
U(VI), Cr(VI), Mo(VI)	U(IV), Cr(III), Mo(IV),	First-order rate constants for U(VI), Cr(VI), and Mo(VI) = 0.8, 0.15, and 0.02/h, respectively	*Desulfovibrio desulphuricans*	Lovley and Phillips, 1992; Fude *et al.*, 1994; Tucker *et al.*, 1997

et al., 1993). The role of sulphide (Fude et al., 1994) and, to a lesser extent, SRB adsoprtive capability (Mohagheghi et al., 1985) in oxyanion reduction has, however, has been attributed in removing these oxidized oxyanions.

A summary of the reduction of metalloid oxyanions by several SRB strains is shown in Table 28.5.

6 CLOSING REMARKS

The SRB directly and indirectly cause many environmental problems, including odour and sewer corrosion. However, the activities of SRB may also alleviate some environmental problems in nature and in man-made engineered systems, specifically for the minimization of metal mobility and removal of heavy metals because of insoluble metal sulphide precipitates. Furthermore, future usage of SRB for microbial transformation of hazardous oxyanions such as As(V)/ As(III), Se(VI)/-Se(IV), Cr(VI), Mo(VI), and U(VI) may establish SRB as one of the most important microbial species in bioremediation.

REFERENCES

Abdelouas, A., Lu, Y., Lutze, W. and Nuttall, H.E. (1998). Reduction of U(VI) to U(IV) by indigenous bacteria in contaminated ground water. *J. Contam. Hydrol.* **35**, 217–233.

Attal, A., Brigodiot, M., Camacho, P. and Manem, J. (1992). Biological mechanisms of H_2S formation in sewer pipes. *Wat. Sci. Technol.* **26**, 907–914.

Bak, F. and Pfennig, N. (1987). Chemolithotrophic growth of *Desulfovibrio sulfodismutans* sp. nov. by disproportionation of inorganic sulphur compounds. *Arch. Microbiol.* **147**, 184–189.

Barnes, L.J., Janssen, F.J., Sherren, J. *et al.* (1991). A new process for the microbial removal of sulphate and heavy metals from contaminated waters extracted by a geohydrological control system. *Trans. Inst. Chem. Eng.* **69**, 184–186.

Benner, S.G., Blowes, D.W. and Ptacek, C.J. (1997). A full-scale porous reactive wall for prevention of acid mine drainage. *Ground Wat. Monit. Remed.* **17**, 99–107.

Bewtra, J.K., Biswas, N., Henderson, W.D. and Nicell, J.A. (1995). Recent advances in treatment of selected hazardous wastes. *Wat. Pollut. Res. J. Can.* **30**, 115–125.

Brock, T.D. and Madigan, M.T. (1991). *Biology of Microorganisms*, 6th edn. Prentice-Hall, Englewood Cliffs, NJ.

Bryant, R.D., Jansen, W., Boivin, J. *et al.* (1991). Effect of hydrogenase and mixed SRB populations on the corrosion of steel. *Appl. Environ. Microbiol.* **57**, 2804–2809.

Caldwell, M.E., Garrett, R.M., Prince, R.C. and Suflita, J.M. (1998). Anaerobic biodegradation of long-chain n-alkanes under sulphate-reducing conditions. *Environ. Sci. Technol.* **32**, 2191–2195.

Canfield, D.E. and Des Marais, D.J. (1991). Aerobic sulphate reduction in microbial mats. *Science* **251**, 1471–1473.

Christensen, B., Laake, M. and Lien, T. (1996). Treatment of acid mine water by sulphate-reducing bacteria: results from a bench scale experiment. *Wat. Res.* **30**, 1617–1624.

DeWalle, F.B., Chian, E.S.K. and Brush, J. (1979). Heavy metal removal with completely mixed anaerobic filter. *J. Wat. Pollut. Control Fed.* **51**, 22–36.

Dezham, P., Rosenblum, F. and Jenkins, D. (1988). Digester gas H_2S control using iron salts. *J. Wat. Pollut. Control Fed.* **60**, 514–517.

Di Toro, D.M., Mahony, J.D., Hansen, D.J. *et al.* (1992). Acid volatile sulphide predicts the acute toxicity of cadmium and nickel in sediments. *Environ. Sci. Technol.* **26**, 96–101.

Dilling, W. and Cypionka, H. (1990). Aerobic respiration in sulphate-reducing bacteria. *FEMS Microbiol. Lett.* **71**, 123–128.

Dvorak, D.H., Hedin, R.S., Edenborn, H.M. and McIntire, P.E. (1992). Treatment of metal-contaminated water using bacterial sulphate reduction: results from pilot scale reactors. *Biotechnol. Bioeng.* **40**, 609–616.

Edwards, M., Courtney, B., Heppler, P.S. and Hernandez, M. (1997). Beneficial discharge of iron coagulation sludge to sewers. *J. Environ. Eng.* **123**, 1027–1032.

Elliott, P., Ragusa, S. and Catcheside, D. (1998). Growth of sulphate-reducing bacteria under acidic conditions in an upflow anaerobic bioreactor as a treatment system for acid mine drainage. *Wat. Res.* **32**, 3724–3730.

EPA (1974). *Process Design Manual for Sulfide Control in Sanitary Sewerage Systems.* Center for Environmental Research Information, US Environmental Protection Agency, Cincinnati.

EPA (1985). *Odor and Corrosion Control in Sanitary Sewerage Systems and Treatment Plant.* Office of Research and Development, US Environmental Protection Agency, Washington, DC.

EPA (1991). *Hydrogen Sulphide Corrosion in Wastewater Collection and Treatment Systems.* Office of Water, US Environmental Protection Agency, Washington, DC.

Fairweather, R.J. and Barlaz, M.A. (1998). Hydrogen sulphide production during decomposition of landfill inputs. *J. Environ. Eng.* **124**, 353–361.

Fang, H.H.P., Liu, Y. and Chen, T. (1997). Effects of sulphate on anaerobic degradation of benzoate in UASB reactors. *J. Environ. Eng.* **123**, 320–328.

Frund, C. and Cohen, Y. (1992). Diurnal cycles of sulphate reduction under oxic conditions in cyanobacterial mats. *Appl. Environ. Microbiol.* **58**, 70–77.

Fude, L., Harris, B., Urrutia, M.M. and Beveridge, T.T. (1994). Reduction of Cr(VI) by a consortium of sulphate reducing bacteria (SRB III). *Appl. Environ. Microbiol.* **60**, 1525–1531.

Gibson, G.R. (1990). A review: physiology and ecology of the sulphate-reducing bacteria. *J. Appl. Bact.* **69**, 769–797.

Haas, C.N. and Polprasert, C. (1993). Biological sulphide prestripping for metal and COD removal. *Wat. Environ. Res.* **65**, 645–649.

Häggblom, M.M. (1998). Reductive dechlorination of halogenated phenols by a sulphate-reducing consortium. *FEMS Microbiol. Ecol.* **26**, 35–41.

Hammack, R.W. and Edenborn, H.M. (1992). The removal of nickel from mine waters using bacterial sulphate reduction. *Appl. Microbiol. Biotechnol.* **37**, 674–678.

Hammack, R.W., Edenborn, H.M. and Dvorak, D.H. (1994). Treatment of water from an open-pit copper mine using biogenic sulphide and limestone: a feasibility study. *Wat. Res.* **28**, 2321–2329.

Hao, O.J., Chen, J.M., Huang, L. and Buglass, R.L. (1996). Sulphate-reducing bacteria. *Crit. Rev. Environ. Sci. Technol.* **26**, 155–187.

Harada, H., Uemura, S. and Momonoi, K. (1994). Interaction between sulphate-reducing bacteria and methane-producing bacteria in UASB reactors fed with low strength wastes containing different levels of sulphate. *Wat. Res.* **28**, 355–367.

Hard, B.C., Friedrich, S. and Babel, W. (1997). Bioremediation of acid mine water using facultatively methylotrophic metal-tolerant sulphate-reducing bacteria. *Microbiol. Res.* **152**, 65–73.

Holder, G.A. (1986). Prediction of sulphide build-up in filled sanitary sewers. *J. Environ. Eng.* **112**, 199–209.

Jin, P., Bhattacharya, K., Williams, C.J. and Zhang, H. (1998). Effects of sulphide addition on copper inhibition in methanogenic systems. *Wat. Res.* **32**, 977–988.

Koe, L.C.C. (1985). Ambient hydrogen sulphide levels at a wastewater treatment plant. *Environ. Monit. Assess.* **5**, 101–108.

Lawrence, A.W. and McCarty, P.L. (1965). The role of sulphide in preventing heavy metal toxicity in anaerobic treatment. *J. Wat. Pollut. Control Fed.* **37**, 392–405.

Lens, P.N., De Poorter, M.-P., Cronenberg, C.C. and Verstraete, W.H. (1995). Sulphate reducing and methane producing bacteria in aerobic wastewater treatment systems. *Wat. Res.* **29**, 871–880.

Leonardos, G., Kendall, D. and Bardard, N. (1969). Odour threshold determinations of 53 odourant chemicals. *J. Air Pollut. Control Assoc.* **19**, 91–95.

Lish, I.J. (1993). *Environmental Survey of the Sulphate Reducing Bacteria.* Scholarly Paper, Dept. of Civil Eng., University of Maryland, College Park, MD.

Lovley, D.R. (1993). Anaerobes into heavy metals: dissimilatory metal reduction in anoxic environments. *Trends Ecol. Evo.* **8**, 213–217.

Lovley, D.R. and Phillips, E.J.P. (1992). Reduction of uranium by *Desulfovibrio desulphuricans*. *Appl. Environ. Microbiol.* **58**, 850–856.

Lovley, D.R., Widman, P.K., Woodward, J.C. and Phillips, E.J.P. (1993). Reduction of uranium by cytochrome c_4 of *Desulfovibrio vulgaris*. *Appl. Environ. Microbiol.* **59**, 3572–3576.

Lovley, D.R. and Phillips, E.J.P. (1994). Reduction of chromate by *Desulfovibrio vulgaris*. *Appl. Environ. Microbiol.* **60**, 726–728.

Maree, J.P., Gerber, A. and Hill, E. (1987). An integrated process for biological treatment of sulphate-containing industrial effluents. *J. Wat. Pollut. Control Fed.* **59**, 1069–1074.

McNeil, M.B., Jones, J.M. and Little, B.J. (1991). Production of sulphide minerals by sulphate-reducing bacteria during microbiologically influenced corrosion of copper. *Corrosion* **47**, 674–677.

Mizuno, O., Li, Y.Y. and Noike, T. (1998). The behavior of sulphate-reducing bacteria in acidogenic phase of anaerobic digestion. *Wat. Res.* **32**, 1626–1634.

Mohagheghi, A., Updegraff, D.M. and Goldhaber, M.B. (1985). The role of sulphate-reducing bacteria in the deposition of sedimentary uranium ores. *Geomicrobiol. J.* **4**, 153–173.

Mori, T., Koga, M., Hikosaka, Y. *et al.* (1991). Microbial corrosion of concrete sewer pipes, H_2S production from sediments and determination of corrosion rate. *Wat. Sci. Technol.* **23**, 1275–1282.

Mori, T., Nonaka, T., Tazaki, K. *et al.* (1992). Interaction of nutrients, moisture and pH on microbial corrosion of concrete sewer pipes. *Wat. Res.* **26**, 29–37.

Morton, R.L., Yanko, W.A., Graham, D.W. and Arnold, R.G. (1991). Relationships between metal concentration and crown corrosion in LA county sewers. *Res. J. Wat. Pollut. Control Fed.* **63**, 789–798.

Newman, D.K., Kennedy, E.K., Coates, J.D. *et al.* (1997). Dissimilatory arsenate and sulphate reduction in *Desulfotomaculum auripigmentum* sp. nov. *Arch. Microbiol.* **168**, 380–388.

Nielsen, P.H. (1987). Biofilm dynamics and kinetics during high-rate sulphate reduction under anaerobic conditions. *Appl. Environ. Microbiol.* **53**, 27–32.

Nielsen, P.H. (1991). Sulphur sources for hydrogen sulphide production in biofilms from sewer systems. *Wat. Sci. Technol.* **23**, 1265–1274.

Nielsen, P.H., Raunkjer, K. and Hvitved-Jacobsen, T. (1998). Sulphide production and wastewater quality in pressure mains. *Wat. Sci. Technol.* **37**, 97–104.

Okabe, S. and Characklis, W.G. (1992). Effects of temperature and phosphorus concentration on microbial sulphate reduction by *Desulfovibrio desulphuricans*. *Biotechnol. Bioeng.* **39**, 1031–1042.

Okabe, S., Nielsen, P.H. and Characklis, W.G. (1992). Factors affecting microbial sulphate reduction by *Desulfovibrio desulphuricans* in continuous culture: limiting nutrients and sulphide concentration. *Biotechnol. Bioeng.* **40**, 725–734.

Padival, N.A., Kimbell, W.A. and Redner, J.A. (1995). Use of iron salts to control dissolved sulphide in truck sewers. *J. Environ. Eng.* **121**, 824–829.

Postgate, J.R. (1963). Versatile medium for the enumeration of sulphate-reducing bacteria. *J. Appl. Microbiol.* **11**, 265–267.

Postgate, J.R. (1984). *The Sulphate Reducing Bacteria*, 2nd edn. Cambridge University Press, Cambridge.

Reis, M.A.M., Almeida, J.S., Lemos, P.C. and Carrondo, M.J.T. (1992). Effect of hydrogen sulphide on growth of sulphate reducing bacteria. *Biotechnol. Bioeng.* **40**, 593–600.

Schmitt, F. and Seyfried, C.F. (1992). Sulphate reduction in sewer sediments. *Wat. Sci. Technol.* **25**, 83–90.

Sercombe, D.C.W. (1995). The control of septicity and odours in sewerage systems and at sewage treatment works operated by Anglian Water Services Ltd. *Wat. Sci. Technol.* **31**, 283–292.

Tebo, B.M. and Obraztsova, A.Y. (1998). Sulphate-reducing bacterium grows with Cr(VI), U(VI), Mn(IV), and Fe(III) as electron acceptors. *FEMS Microbiol. Lett.* **162**, 193–198.

Tomei, F.A., Barton, L.L., Lemanski, C.L. *et al.* (1995). Transformation of selenate and selenite to elemental selenium by *Desulfovibrio desulphuricans*. *J. Ind. Microbiol.* **14**, 329–336.

Tucker, M.D., Barton, L.L. and Thomson, B.M. (1996). Kinetic coefficients for simultaneous reduction of sulphate and uranium by *Desulfovibrio desulphuricans*. *Appl. Microbiol. Biotechnol.* **46**, 74–77.

Tucker, M.D., Barton, L.L. and Thomson, B.M. (1997). Reduction and immobilization of molybdenum by *Desulfovibrio desulphuricans*. *J. Environ. Qual.* **26**, 1146–1152.

Tucker, M.D., Barton, L.L. and Thomson, B.M. (1998a). Reduction of Cr, Mo, Se and U by *Desulfovibrio desulphuricans* immobilized in polyacrylamide gels. *J. Ind. Microbiol.* **20**, 13–19.

Tucker, M.D., Barton, L.L. and Thomson, B.M. (1998). Removal of U and Mo from water by immobilized *Desulfovibrio desulphuricans* in column reactors. *Biotechnol. Bioeng.* **60**, 88–96.

Tuttle, J.H., Dugan, P.R. and Randles, C.I. (1969). Microbial sulphate reduction and its utility as an acid mine water pollution abatement procedure. *Appl. Microbiol.* **17**, 297–302.

Visser, A., Gao, Y. and Lettinga, G. (1993). Effects of short-term temperature increases on the mesophilic anaerobic breakdown of sulphate containing synthetic waste-water. *Wat. Res.* **27**, 541–550.

Wakao, N., Takahashi, T., Sakurai, Y. and Shiota, H. (1979). A treatment of acid mine waster using sulphate-reducing bacteria. *J. Ferment. Technol.* **57**, 445–452.

Webb, J.S., McGinness, S. and Lappin-Scott, H.M. (1998). Metal removal by sulphate-reducing bacteria from natural and constructed wetlands. *J. Appl. Microbiol.* **84**, 240–248.

White, C., Sharman, A.K. and Gadd, G.M. (1998). An integrated microbial process for the bioremediation of soil contaminated with toxic metals. *Nat. Biotechnol.* **16**, 572–575.

Widdel, F. (1988). Microbiology and Ecology of Sulphate- and Sulphur-Reducing Bacteria. In: A.J.B. Zehnder (ed.) *Biology of Anaerobic Microorganisms*. Wiley, New York.

Wieder, R.K. (1989). A survey of constructed wetlands for acid mine drainage treatment in the Eastern US. *Wetlands* **9**, 299–315.

Yamamoto-Ikemoto, R., Matsui, S. and Komori, T. (1994). Ecological interactions among denitrification, poly-P-accumulation, sulphate reduction, and filamentous sulphur bacteria in activated sludge. *Wat. Sci. Technol.* **30**, 201–210.

Zumdahl, S.S. (1989). *Chemistry*. D.C. Heath and Company, Lexington, MA.

Behaviour of Pathogens in Wastewater Treatment Processes

29

Viruses in faeces

John Oragui

Department of Microbiology, Harefield Hospital, Harefield, Middlesex UB9 6JH, UK

1 INTRODUCTION

Many viruses are excreted by humans and animals. Some of these are present in very large numbers but cannot be grown, others can only be grown with difficulty in cell cultures. Enteroviruses, hepatitis A virus, rotaviruses, parvovirus-like viruses, astroviruses, caliciviruses, adenoviruses and coronaviruses may be present in human excreta. Rotaviruses are the commonest cause of acute non-bacterial gastroenteritis in infancy and childhood. In children aged 6 months to 10 years, infection by rotaviruses occurs more frequently during winter months in temperate countries. In tropical countries, there is considerable variation in the incidence of rotavirus infections among children.

Enteroviruses include the polio, coxsackie A and B and echoviruses. These belong to the family of Picornaviridae. They contain single-stranded RNA. Poliomyelitis is rare in developed countries, as a result of the widespread use of oral polio vaccines, but continues to circulate in developing countries. Similarly, hepatitis A is an RNA virus and has the physicochemical characteristics of a typical enterovirus. It is transmitted from person to person by the faeco-oral route. Outbreaks may originate from viral contamination of food, water, milk and shellfish. Rotaviruses and parvovirus-like viruses cause gastroenteritis. The role of adenoviruses, astroviruses, caliciviruses and coronaviruses is not well-established. However, it is now recognized that these viruses can cause outbreaks of acute diarrhoeal disease. Enteric adenoviruses represent a separate serotype distinct from adenoviruses associated with acute respiratory infection and have been associated with acute diarrhoeal disease.

Besides parvovirus-like viruses, other small rounded viruses have been detected in faecal samples from patients with acute diarrhoeal disease. These include:

1. astroviruses, encountered in outbreaks in children's wards (Kurtz *et al.*, 1977), stools of lamb with diarrhoea and calf faeces
2. caliciviruses, detected in stools in children in winter and staff with diarrhoea and also asymptomatic children (McSwiggan *et al.*, 1978)
3. coronaviruses, which cause severe diarrhoea in pigs and calves; there is no conclusive evidence that these viruses cause diarrhoea in humans.

2 VIRUSES IN WASTEWATER

The types and numbers of human enteric viruses in raw wastewater depend on the origin and nature of the wastewater. Enteroviruses, including hepatitis A virus, adenovirus and rotaviruses, are frequently present in domestic wastewater. These viruses reflect those circulating in the community and depend on a number of factors, which include population, climate, seasonal and diurnal fluctuations and the presence of chemical effluents. Raw wastewater contains viral numbers which vary widely from country to country. For example,

The Handbook of Water and Wastewater Microbiology
ISBN 0-12-470100-0

in raw wastewater in Israel, higher numbers are generally present than in the USA, although the numbers reported depend on the detection and concentration methods used.

Animal viruses may also be present in raw wastewater from the faeces of household pets and effluents from farms, abattoirs and stormwater. The isolation of animal viruses from wastewater, especially abattoir wastes, has been reported by several workers. However, more emphasis has been on human viruses due to their potential for causing human diseases.

Viruses may be categorized or divided into three main groups:

1. viruses associated with humans
2. viruses associated with higher plants and animals
3. viruses associated with the microbial flora (bacteriophages)

Over a hundred different types of viruses are known (WHO, 1979). Excreta is the commonest source, but by no means the only source. Nose and throat secretions also contribute some viruses to wastewater, and viruses from skin lesions and blood (which may find its way through baths and toilets during acute stages of infection giving rise to viraemia) can also be found in wastewater. These sources, although they do not contribute much, may in certain situations be important.

Viruses associated with plants and animals may be important for economic reasons, especially in countries where the reuse of wastewater for crop irrigation is practised. Some of these viruses come from the wastewater itself, whereas others may enter the sewer via wastewater from farms, slaughterhouses, food-processing factories and stormwater. Faecal materials from birds, dogs and other animals will be washed from roofs and roads by rain and find their way ultimately into the wastewater.

The third group comprises the bacteriophages, which are viruses that parasitize bacteria and, because of the huge variety of bacteria in wastewater, which may all have their own phages, phage numbers in wastewater can be very high.

3 VIRUS RECOVERY FROM WASTEWATERS

One of the major problems encountered in the enumeration of viruses in water and wastewater before and, particularly, after treatment is the large quantity of sample that has to be examined in order to detect and count them. This is made even more difficult by the fact that the enumeration technique usually employed for rotavirus assays utilizes only 25–200 μl of sample. For enterovirus assays, 0.1–1 ml can usually be assayed conveniently. The consequent dilution of viruses in faeces when discharged into receiving waters or sewage means the sample must be concentrated in order to analyse them. Various techniques are available for concentrating viruses (see Oragui and Mara, 1996).

One of the difficulties encountered with concentrated wastewater samples is the toxic effects that they have on the cell lines used for virus propagation and enumeration. Concentrates of raw wastewater and, in particular those obtained from anaerobic waste stabilization ponds, are very toxic to the cells used for rotavirus assay. Toxic effects are indicated by the destruction or detachment of cells during rotavirus assay by the indirect immunofluorescence technique. A simple method for the detoxification of concentrate has been developed and is described by Oragui and Mara (1989).

4 VIRUS REMOVAL AND SURVIVAL IN WASTEWATER TREATMENT

Conventional wastewater treatment plants are not specifically designed to reduce the number of excreted pathogens, including excreted viruses. However, there is usually some incidental removal of viruses – primary sedimentation ranges from 30 to 65% (Berg, 1973; Rao *et al.*, 1977); in activated sludge from 80 to >90% (Clarke *et al.*, 1961; Malina, 1976); and in trickling filters from 10 to 20% (Sherman *et al.*, 1975), although Berg (1973) reported removals of 15–100%. Disinfection processes, principally chlorination and ozonation, for the removal of faecal bacteria, have not been

TABLE 29.1 Ranges and geometric means of faecal coliforms and excreted rotaviruses in raw wastewater and pond effluents, September–December 1988

Sample source	*No. of samples*	*Faecal coliforms (per 100 ml)*	*Rotaviruses (per litre)*	*Percentage removal*	
				Faecal coliforms	*Rotavirus*
Raw wastewater	12	3.5×10^7–7.8×10^4 (6.12×10^4)	1.06×10^4–2.66×10^5 (1.13×10^4)	–	–
Anaerobic pond	12	2.1×10^6–4.2×10^6 (3.13×10^6)	1.02×10^3–7.27×10^4 (5.87×10^3)	94.9	94.8
Facultative pond	12	1.1×10^5–1.4×10^6 (8.91×10^5)	8.5×10^2–3.2×10^4 (1.41×10^3)	71.5	76.0
1st Maturation pond	12	8.5×10^4–2.75×10^5 (1.50×10^6)	1.06×10^2–6.6×10^3 (1.87×10^2)	83.2	86.7
2nd Maturation pond	12	2.5×10^4–5.85×10^3 (4.37×10^4)	0.9×10^2–6.6×10^2 (0.234×10^2)	70.8	87.5
3rd Maturation pond	11	5.3×10^2–2.75×10^3 (1.41×10^3)	0.09×10^2 (3.9)	96.8	83.3

Retention time: 5 days in each pond.
Geometric means are given in parentheses.
Source: Oragui and Mara (1996)

widely adopted other than in a few countries, notably the USA, Israel and South Africa, because of the detrimental effects on the natural fauna of receiving waters and the production and discharge of toxic and carcinogenic chlorination by-products.

In a study on rotavirus removal in a series of five 3 m deep waste stabilization ponds in northeast Brazil, Oragui and Mara (1996) found four log unit removals of both rotaviruses and faecal coliform bacteria (Table 29.1). The highest rotavirus removal was found in the anaerobic pond (95%), with lower removal in the facultative and maturation ponds (76–88%).

4.1 Factors affecting rotavirus removal in ponds

Several factors may explain viral inactivation in waste stabilization ponds, including solar radiation, temperature, pH, adsorption onto solids, heavy metals, algal and bacterial activity, and the action of certain chemicals, notably ammonia and sulphide.

The unionized ammonia (dissolved NH_3 gas) is toxic to bacteria (Pearson *et al.*, 1987), with toxicity increasing with increasing pH. The effect of ammonia toxicity on rotavirus inactivation has also been shown to be pH dependent (Oragui and Mara, 1996): at pH 9.0 there was a reduction of just over half a log unit at all ammonia concentrations (20–80 mg N/l), whereas at pH of 6.9, the log reduction ranged from 1.3 to 2.2. In contrast to the effect of ammonia on bacteria, ammonia is more toxic to rotaviruses at pH 6.9, when it is essentially all present as the ammonium ion, than at pH 9.

Sulphide is one of the main toxicants present in raw wastewater and anaerobic pond effluents. Sulphide was found to exert the most effect on simian rotaviruses at pH 6.9, with log reductions of 0.1–3.9; at pH 8 reduction was 0.04–0.09, but at pH 9 it was 0.03–0.46. In contrast to ammonia, increasing sulphide concentrations in the range 4–16 mg S/l increased the degree of virus inactivation.

In 2 m deep facultative ponds in-pond rotavirus numbers decreased principally in the top 1 m of the pond, with little difference found in rotavirus numbers in the 1–2 m lower layer. This indicates that, as the surface layers experience algae-induced pH changes, the toxic effects of ammonia and sulphide are likely to be the main mechanism of removal of rotaviruses, and possibly of all excreted virus, although adsorption on to the algae and subsequent settlement when the algae die may be an important secondary mechanism.

REFERENCES

Berg, G. (1973). Removal of viruses from sewage effluents and water: a review. *Bull. World Hlth Org.* **49**, 451–460.
Clarke, N.A., Stevensen, R.E., Chang, S.L. and Kabler, P.N. (1961). Removal of enteric viruses from sewage by

activated sludge treatment. *Am. J. Public Hlth* **51**, 1118–1129.

Kurtz, J.B., Lee, T.W. and Pickering, D. (1977). Astrovirus associated gastro-enteritis in a children's ward. *J. Clin. Pathol.* **30**, 948–952.

Malina, J.F. Jr (1976). Viral pathogen inactivation during treatment of municipal wastewater. In: L.B. Baldwin, J.M. Davidson and J.F. Gerber (eds) *Virus Aspects of Applying Municipal Wastes to Land*, pp. 9–23. Gainesville, Center for Environmental Programs, University of Florida.

McSwiggan, D.A., Cabitt, D. and Moore, W. (1976). Caliciviruses associated with winter vomiting disease. *Lancet* **1**, 1215–1216.

Oragui, J.I. and Mara, D.D. (1989). Simple method for the detoxification of wastewater ultrafiltration concentrates for rotavirus assay by indirect immunofluorescence. *Appl. Environ. Microbiol.* **55**, 401–405.

Oragui, J.I. and Mara, D.D. (1996). *Enumeration of Rotaviruses in Tropical Wastewaters*. Research Monographs in Tropical Public Health Engineering No. 10. University of Leeds, Leeds.

Pearson, H.W., Mara, D.D., Mills, S.W. and Smallman, D.J. (1987). Physico-chemical parameters influencing faecal bacterial survival in waste stabilisation ponds. *Wat. Sci. Technol.* **19**, 145–152.

Rao, V.C., Lakhe, S.B., Waghmane, S.V. and Dube, P. (1977). Virus removal in activated sludge treatment. *Prog. Wat. Technol.* **9**, 113–127.

Sherman, V.R., Kawata, K., Oliveris, V.P. and Naparstek, J.D. (1975). Virus removals in trickling filter plants. *Wat. Sewage Works* **122**, R36–R44.

WHO (1979). *Human Viruses in Water, Wastewater and Soil*. Technical Report Series No. 639. World Health Organization, Geneva.

30

Bacterial pathogen removal in wastewater treatment plants

Tom Curtis

Department of Civil Engineering, University of Newcastle-upon-Tyne, NE1 7RU, UK

1 INTRODUCTION

No widespread outbreak of enteric bacterial disease would be complete without a call for improved wastewater treatment (McCabe, 1970; Nichols 1991, Tauxe *et al.*, 1994). These statements were presumably made in the sincere belief that wastewater treatment systems are generally good at removing pathogenic bacteria and that this makes a significant contribution to public health in the richer countries. Unfortunately, pathogen removal is poor in most wastewater treatment plants. Thankfully, this usually has little impact on public health in most countries as other measures (notably water treatment and supply) protect us from the pathogenic bacteria we excrete.

However, where pathogen removal is important it can be very important indeed. For example, wastewater reuse in agriculture and aquaculture will be vital to balancing the water demand and supply in the 21st century. This will mean removing bacterial pathogens before the water is used for food production. We already know that reuse without treatment can lead to significant excess disease (Shuval *et al.*, 1986; Swerdlow *et al.*, 1992) and (as in Peru in the early 1990s) to significant economic losses. Economic issues may also be an important factor in compliance with recreational water treatment standards. The costs of recreational water quality standards may be borne by relatively small coastal communities who may be ill-equipped to pay for large centralized disinfection schemes.

2 PATHOGEN RANGE AND LOAD DICTATE THE HAZARD AND RISK IN UNTREATED WASTEWATER

The numbers and range of pathogens found in wastewater are significant considerations. The range of bacterial pathogens is important because it dictates the nature of the hazard posed by the wastewater. On the other hand, the numbers of pathogens have a direct bearing on the risk those pathogens pose and the magnitude of the task of reducing that risk.

A knowledge of the range of pathogens also tells us something of the ecology of the organism we are trying to remove (Table 30.1). This table will probably be incomplete within a few years as, at present, it appears that significant new enteric bacterial pathogens are being discovered every decade or so. Some, such as *Helicobacter pylori*, which can cause ulcers, pose unexpected hazards. Others such as the highly infectious *Escherichia coli* 0157 are extraordinarily virulent agents with familiar symptoms (Collier *et al.*, 1998).

The wastewater of any significant community can be assumed to have virtually all the pathogens normally found in excreta at any one time. Even in the wealthiest countries the annual incidence of infectious intestinal disease may exceed 20% (Wheeler *et al.*, 1999). Organisms such as *Vibrio cholerae* O1 or O139. which cause cholera, may be present in non-endemic areas in carriers and the mildly ill can make a substantial contribution to the contamination of wastewater without coming to the attention of the medical authorities (Curtis,

The Handbook of Water and Wastewater Microbiology
ISBN 0-12-470100-0

TABLE 30.1 The reported concentrations of bacterial pathogens in untreated domestic wastewater

Organism	*Disease*	*Symptoms*	*Reported concentrations/ 100 ml wastewater*	*References*
Genus *Campylobacter*	Campylobacter enteritis	Diarrhoea	70–1630	Oragui *et al.*, 1987; Stampi *et al.*, 1992
Enterhaemorragic *E. coli*	Haemorrhagic colitis	Diarrhoea, cramp, loss of kidney function	Ubiquitous but no quantitative reports	Muniesa and Jofre, 2000
Other enterovirulent *E. coli*	Diarrhoea and bacillary dysentery	Diarrhoea and mild dysentery	Detected in wastewater but no quantitative reports	Ohno *et al.*, 1997
Helicobacter pylori		Achlorhydria stomach ulcers	No quantitative reports	Morton and Bardhan, 2000
Salmonella typhi	Typhoid	Septicaemia and fever	No quantitative reports	Shuval *et al.*, 1986
Salmonella enteriditis	Salmonellosis	Diarrhoea	20– >1800	Langeland, 1982; Oragui *et al.*, 1987; Yaziz and Lloyd, 1979
Genus *Shigella*	Shigellosis	Diarrhoea and dysentery	No quantitative reports	
Vibrio cholerae O1 and O130	Cholera	Diarrhoea	$1–10^7$	Curtis, 1996

1996). In England and Wales there are typically over 50 cases of imported cholera reported to the Public Health Authority each year (Communicable Disease Surveillance Centre, 2000). Given the degree of under-reporting of enteric disease, the true number of cases might be an order of magnitude higher.

Our knowledge of the ecology of these bacteria is poor, even though they represent some of the most intensively studied organisms on the planet and the importance of them in environmental transmission is beyond doubt. This gap reflects both cultural and scientific problems. On the technical side, although the entire genome of *V. cholerae* has been sequenced (Heidelberg *et al.*, 2000), there is probably not a scientist alive who could enumerate 2000 such organisms in 100 ml of raw wastewater. The pathogenic organism would be hidden among a much larger number of virtually identical non-pathogenic types of *V. cholerae* that are themselves a numerically insignificant proportion of the bacterial diversity of the wastewater.

Thus the enumeration of pathogenic enteric bacteria is often like to trying to find a needle in a needlestack that has been thrown into a haystack! These severe technical limitations underlie the paucity of quantitative information on many pathogens in raw sewage. Many ecological studies rely on pure cultures seeded into beakers and buckets (microcosms) from at best, the environment and at worst the laboratory. These studies tend to fall short of the highest standards in microbial ecology (Brock, 1987). Furthermore, such technical shortcomings are compounded by the general perception (with honourable exceptions) that the study of the microbial ecology of bacterial pathogens is not a prestigious area of medicine or ecology (though, in truth, infectious disease epidemiology and biological waste treatment are both branches of microbial ecology).

3 COMPARATIVE ECOLOGY OF BACTERIAL PATHOGENS

Engineers and scientists in the field tend to fall back on a small range of indicator organisms (Table 30.2) because of the difficulties of measuring bacterial pathogens in wastewater.

TABLE 30.2 Typical indicator values in untreated raw sewage

Indicator	*Concentration in typical wastewater (/100 ml)*	*Comments*	*References*
Faecal coliforms	10^6–10^7	Gram-negative	Oragui *et al.*, 1987
Faecal enterococci	10^5–10^6	Gram-positive	Feachem *et al.*, 1983
Clostridium perfringens spores	10^4–10^5	Gram-positive	
Bifidobacteria	10^7	Oxygen sensitive	

They typically employ faecal coliforms, *E. coli* or Gram-positive enterococci to monitor bacterial pathogen removal in wastewater treatment. These organisms have been borrowed from the water supply industry where they were used to indicate the presence of faecal contamination. Sewage is, of course, faecally contaminated so nothing can be inferred from the mere presence of the indicator organisms in wastewater. However, these indicator organisms occur in large numbers and are relatively easy to enumerate and are (usually) non-pathogenic. The organisms are therefore monitored on the assumption that they will act as analogues of these numerous and more difficult to count pathogenic bacteria. The term pathogen indicator has been used to describe the use of these organisms to monitor pathogens (Feachem and Mara, 1979).

Ideally, pathogen indicators should be used in situations where there is evidence that the indicator is a valid analogue of the pathogens of concern. However, because the pathogens of concern are often difficult, or impossible, to enumerate, this condition is rarely fulfilled. In the absence of such evidence some indication of the validity of the indicator may be sought by determining how the pathogen and the indicator are related or by comparing the known physiology of the pathogen and the indicator. Faecal coliforms, the most widely used indicators, are usually >95% *E. coli* in temperate and tropical wastewaters which facilitates such comparisons.

E. coli are mesophilic Gram-negative bacteria in the gamma subdivision of the proteobacteria (Woese *et al.*, 1985). They are capable of using oxygen to oxidize carbon and of growing fermentatively, can withstand some oxidative damage and find pH values of about 7.75 most agreeable. pH values more than 1 unit above or below this value may be considered stressful.

Shigella and (obviously) enterovirulent *E. coli* are essentially subspecies to *E. coli* (Collier *et al.*, 1998) and so are most likely to find an adequate analogue in faecal coliforms. However, we cannot discount the possibility that the extra genetic material associated with pathogenicity may affect the survival of the organism in the environment. For example *E. coli* O157 retains the ability to make toxin on a lysogenic (is capable of lysing the cell) bacteriophage encoded gene (Muniesa and Jofre. 2000). The phage may be induced, and the cell lysed, by exposure to oxidative stress (e.g. sunlight). This may confer a disadvantage on this organism outside the host. *Salmonella* spp. are Gram-negative rods and members of the family Enterobacteriaceae. They are thought to have similar sensitivities to pH, oxygen and temperature to *E. coli*, to which they are relatively closely related (Feachem *et al.*, 1983).

Campylobacter and *Helicobacter pylori* are only very distant relatives of *E. coli*, being from the epsilon subgroup of the proteobacteria. Nevertheless, they are known to grow well at moderate temperatures and salinities favouring *E. coli*. However, they are both microaerophilic (only able to tolerate low oxygen concentrations) and susceptible to oxidative conditions and sunlight. In addition, *Helicobacter* are tolerant of high pH values (Bergey and Holt, 1994).

E. coli and *V. cholerae* are fundamentally different because the latter is not really a gut organism at all. Although the two organisms have a common evolutionary ancestry, as they both fall within the gamma subdivision of the proteobacteria (Woese *et al.* 1985), they have diverged considerably from their common ancestor. Consequently, *E. coli* and *V. cholerae* are members of distinct and varied bacterial families, the Enterobacteriaceae and the Vibrionaceae (MacDonell *et al.*, 1986). Only two serotypes of *V. cholerae* (O1 and O139) are capable of causing the eponymous disease. There are well over 100 other serotypes, which are relatively harmless and ubiquitous members of the natural bacterial flora of brackish water. Indeed, the disease-causing serotypes of the organism are now considered to be aquatic organisms, which infected humans, rather than a human pathogen that occasionally entered bodies of water (Feachem, 1982; West, 1986). Not surprisingly, the organism survives better in slightly brackish water at elevated pH values and has an absolute requirement for sodium ions (Singleton *et al.*, 1982a; Miller *et al.*, 1984). There is evidence that nutrient levels typical of wastewater may permit growth in fresh water (Singleton *et al.*, 1982b; Nair *et al.*, 1988). The optimum pH for survival was found to increase with decreasing salinity. At salinities equivalent to about 0.1% NaCI the survival is best at pH values between 7.5 and 9.0 (Miller *et al.*, 1984). In addition, *V. cholerae* is thought to have specific relationships with higher organisms, some of which may be found in wastewater treatment plants, including phytoplankton (Tamplin *et al.*, 1990; Islam *et al.*, 1994), water hyacinth (Spira *et al.*, 1981) and freshwater amoebae (Thom *et al.*, 1992). *V. cholerae* are not thought to be particularly sensitive to sunlight (Lema *et al.*, 1979; MacKenzie *et al.*, 1992), a form of oxidative stress. However, it is now realized that the cholera toxin gene is encoded on a filamentous bacteriophage (Faruque *et al.*, 2000) and that the phage may be induced by sunlight. Though the phage does not lyse the cells (they bud off), it seems plausible that the budding phage will impose a metabolic burden on the cell that may place it at a disadvantage in relation to its non-pathogenic competitors in the environment.

In principle, other pathogen indicators could be compared with pathogens in the same way. However, faecal enterococci are more heterogeneous than faecal coliforms and not particularly closely related to any of the known pathogenic organisms present in wastewater. Thus, the most we can say with certainty is that faecal enterococci are different from the organisms that we hope they will mimic and that different species may be different in slightly different ways. Similarly *Clostridium perfringens* is very different from the important excreta-transmitted pathogens. Consequently, there is no *a priori* reason to suppose that either of these indicators will be good analogues of the bacterial pathogens known to be present in faeces.

Although these comparisons represent mere thought experiments in ecology, they do seem broadly to tally with existing data on the comparative survival rates of different indicators, at least in waste stabilization ponds (a high pH, oxidizing environment). The faecal coliforms appear to be adequate indicators of salmonella. Campylobacters appear to die quickly. Faecal enterococci and *Cl. perfringens* spores are indeed different and appear to be more conservative indicators (Fig. 30.1).

The range of pathogen concentration (see Table 30.1) represents snap shots of wastewaters at different times and places. In reality we know that pathogen numbers vary over the day (Yaziz and Lloyd, 1979) because of diurnal patterns in defaecation. In addition, the point prevalence for a particular disease will vary within and between communities because of a myriad of other epidemiological factors. Seasonal changes of prevalence are extremely common in bacterial diseases (Feachem *et al.*, 1983), e.g. in temperate climates faecal oral bacterial infections tend to peak in the summer. In addition, widespread outbreaks of disease can lead to significant (two to three orders of magnitude) and rapid changes in pathogen concentration.

The concentration of pathogens in wastewater may be related to point prevalence (the number of people with a disease on any given day) by equation 1:

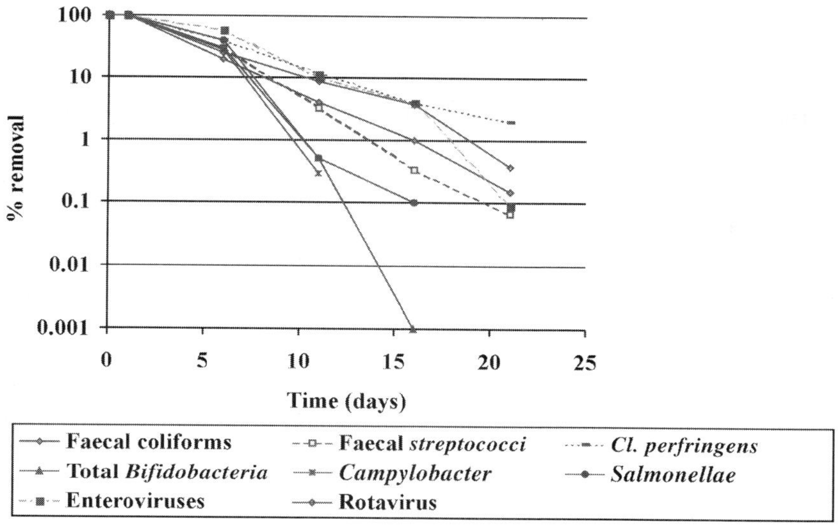

Fig. 30.1 A comparison of the removal of some enteric pathogens and typical indicator organisms in a series of deep waste stabilization ponds in the north-east of Brazil (Oragui *et al.*, 1987).

Pathogen/100 ml of sewage

$$= ((m \times f)/w \times 10) \times p \qquad (1)$$

where:

m is a weighted mean pathogen/g faeces
f = mean weight of faeces/person/day
w = litres mean water used/person/day
p = the proportion of the population excreting the organism and 10 is a correction factor.

The value of m will depend on the severity of the illness while p reflects the extent of disease in the community.

We can use this formula to consider the likely load of *V. cholerae* on a wastewater treatment plant. The values of the variables will depend on whether the excretors being considered are carriers or patients. For the carriers: $m = 7 \times 10^4$ organisms/g (Dizon *et al.*, 1967); $f = 100$ g/day (Feachem *et al.*, 1983) and p is unlikely to exceed 0.01 (a point prevalence of 1%), $w = 50$ litres/day. Experimental infections have suggested that most cholera patients have mild symptoms (Levine *et al.*, 1979), for such patients: $w = 10^8$ organisms/g (Dizon *et al.*, 1967), $f = 4000$ g/day (that is 4 litres of diarrhoea, assumed to have a density of 1: Levine *et al.*, 1979), $w = 200$ litres/day and p is unlikely to exceed 0.002 (a point prevalence of 0.2%). Cholera patients with severe symptoms are unlikely to use the toilet and may be discounted in the calculations.

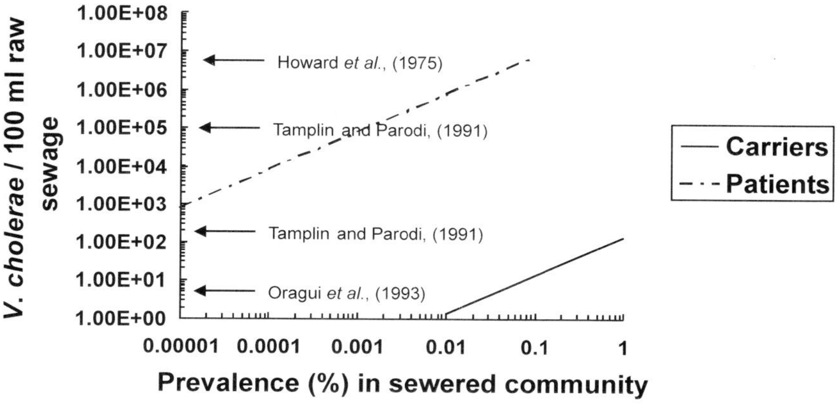

Fig. 30.2 The relationship between prevalence and estimated levels of *V. cholerae* O1 (lines) in raw sewage and reported levels of *V. cholerae* O1 in raw sewage.

These estimates are consistent with published levels or *V. cholerae* in raw sewage (Fig. 30.2). One mildly ill individual in a city of a million could contaminate the wastewater for a whole town without ever coming to the attention of the medical authorities. At the other extreme, a hospital treating cholera victims may have more toxigenic *Vibrio* than *E. coli* in the wastewater.

In certain exceptional circumstances, extremely high pathogen concentration may assume particular significance. Recreational and wastewater reuse standards assume that the pathogens occur at much lower densities than the indicator and that the reduction of the indicator by 3–4 logs results in the elimination of the bacterial pathogens. While this is nearly always true, it is possible that some breakthrough could occur in a severe outbreak.

4 THE FATE OF PATHOGENIC BACTERIA IN WASTEWATER TREATMENT PLANTS

Wastewater treatment can be divided into anaerobic systems (septic tanks, anaerobic ponds and anaerobic digesters), aerobic systems with attached growth (typically trickling filters) and aerobic suspended growth systems (typically activated sludge). A primary sedimentation tank usually precedes these systems. An increasingly popular alternative is the reed bed or wetland. Another well-established alternative technology is the waste stabilization pond (WSP) – a sort of very wetland. Both ponds and wetlands are relatively easy to construct, biologically sophisticated and contain both aerobic and anaerobic zones. Waste stabilization ponds and wetlands can be designed to optimize pathogen removal. In this sense they are different from the other wastewater treatment systems we will consider. In most wastewater treatment systems any pathogen removal that occurs is a fortuitous by-product of the principal design objective (usually organic carbon removal). Pathogen removal cannot usually be refined in these systems without compromising this objective or at least the costs of meeting this objective.

For any given system, there are essentially two factors in pathogen removal: how long the pathogen stays in the system and how quickly it dies. The former is governed by the hydraulic flow regime and the latter depends on the ecology of the reactor.

The simplest conception of hydraulic flow in a reactor is as a completely stirred reactor in which any material entering into a system is immediately uniformly dispersed within the reaction vessel. The effluent quality of a completely stirred reactor is the same as the reactor quality and therefore some of the pathogens pass almost directly from the inlet to the outlet. However, if a number of completely mixed reactors are placed in series then they can begin to approximate plug flow reactors. In plug flow reactors the material entering into the reactor proceeds through the wastewater treatment plant without mixing with the material that has entered before or after. In reality some mixing does occur, this is known as dispersed flow.

Low rate systems such as ponds and wetlands have long retention times (typically days) and thus more opportunity to remove pathogens. However, they may have inefficient hydraulic flow regimes. Processes such as activated sludge have much lower hydraulic retention times, though some plants may be hydraulically very efficient indeed.

5 PRIMARY SEDIMENTATION TANKS

Most conventional wastewater treatment plants have some form of primary sedimentation tank to remove solids. Although the retention time of such tanks is short (typically 2–6 hours), 30–50% faecal coliform removal and 29–99% (mean 79%) salmonellae removal have been reported (Yaziz and Lloyd, 1979; Wheater *et al.*, 1980). The salmonellae study found removal to be positively correlated with suspended solids removal which plausibly implies that the organisms are associated with solids. Since other pathogens are also likely to be associated with solids, we may assume that faecal coliforms will be adequate indicators of bacterial pathogen removal in primary sedimentation tanks. However, indicator removal is quite variable (Feachem *et al.*, 1983) and so the lower figure should be used for design purposes.

6 ACTIVATED SLUDGE AND OTHER SUSPENDED GROWTH SYSTEMS

In suspended growth systems the biomass is formed into flocs that are typically separated from the water by sedimentation (or, more recently, membranes) allowing the retention time of the biomass to be varied independently of the hydraulic retention time. Hydraulic retention times are therefore short, typically 6–12 hours in a conventional plant, rising to 30 hours in certain designs. The opportunities for bacterial pathogen removal are therefore necessarily limited. Between 90 and 99% removal of faecal coliforms (Feachem *et al.*, 1983) and salmonellae (Yaziz and Lloyd, 1979) has been reported. Stampi *et al.* (1993) reported 99% removal of the microaerophilic *Campylobacter*. This is not very different from the removal of aerobic organisms. Clearly, there is not enough oxygen in the activated sludge significantly to prejudice the survival of this microaerophilic organism. There are no reports of the removal of *V. cholerae* in working activated sludge plants. Streeter (1930) reported that Courmont and Rochaix (1922) had seeded a batch of activated sludge and sewage with *V. cholerae* and found that they could not recover the organisms after 5–8 hours aeration. The same review stated that Bruns and Sierp (1927) had conducted analogous experiments and found that 98% of the seeded *V. cholerae* were removed. It is doubtful if even this level of removal could be achieved in practice since most suspended growth systems operate on a continuous flow basis and consequently a proportion of the influent will leave the aeration basin before the mean hydraulic retention time.

The underlying mechanisms for the removal of bacterial pathogens in activated sludge have not been rigorously investigated. However, they are likely to be related to the adsorption to solids and predation. These mechanisms should not confer a particular advantage or disadvantage on any bacterial pathogen or indicator organism. However, it is unlikely that the system can be engineered to enhance the removal mechanisms and predation-based removal is likely to be inherently variable due to the variation inherent in predator–prey systems (Case, 2000). Moreover, if solids retention is poor, pathogen removal will dip below the 90% removal suggested above. Increasing the hydraulic retention time of the system may enhance removal of the pathogens. The impact is likely to be marginal, especially in completely mixed reactors.

Membrane reactors may be very different. A side effect of the use of membranes (rather than secondary sedimentation tanks) to retain biomass is the efficient retention of virtually all bacterial pathogens. The use of membranes will impose significant capital and running costs and, at present, the technology is still relatively new. However, where the costs can be borne they may well prove to be an excellent technology for the removal of bacterial pathogens (Ueda *et al.*, 1996; Ueda and Hata, 1999). In principle, membrane bioreactors should not be effective against viruses which cause the low-grade diarrhoea associated with recreational waters and the hepatitis A occasionally spread by shellfish. A technology that just removes bacteria without removing viruses could allow an effluent to meet a health-based standard expressed in terms of an indicator while not actually protecting health. However, in practice, it appears that model scale membrane bioreactors can remove viruses, once a biofilm has built up on the membrane (Winnen *et al.*, 1996; Ueda and Horan, 2000).

7 FIXED FILM REACTORS

Fixed film reactors are also designs that allow short hydraulic retention times and long biomass retention times. Fixed film secondary treatment processes work by passing the wastewater over a biofilm, which is capable of removing the oxygen demand. Units in which the water is distributed over a stationary media include tricking filters and biological aerated filters (though no filtration occurs), while units in which the biofilm is passed through the wastewater are called rotating biological contactors.

In trickling filters, the removal of indicator bacteria (Wheater *et al.*, 1980) and salmonellae (Yaziz and Lloyd, 1979) is poor. By contrast, Daniel and Lloyd (1980b) reported a three-log increase in the numbers of *V. cholerae* non-O1 (the non-pathogenic kind) when they passed the effluent from a dual chambered septic tank over an improvised trickling filter. The authors showed that *V. cholerae* non-O1 would reproduce in batch samples of the trickling filter effluent at 20–25°C and suggested that ponding in the filter was permitting the growth of the organism. However, when the survival of *V. cholerae* El Tor was studied in identical batches of effluent, the number of culturable organisms was observed to decline by a factor 10 over a period of 7 hours and then to maintain itself at the reduced level for a further 41 hours.The apparent discrepancy between the two forms of *V. cholerae* could not be readily explained.

Factors likely to affect the fate of bacterial pathogens in fixed film systems include binding to the biofilm, predation by protozoa and competition for nutrients. It would be difficult to engineer such a system to optimize the apparently unimpressive removal mechanisms. The short retention times characteristic of fixed film reactors could not be easily extended. It seems, therefore, that attached growth systems will never be any good at removing bacterial pathogens.

8 ANAEROBIC REACTORS

Four principal kinds of anaerobic system are used to treat domestic wastewaters anaerobically: the septic tank, the anaerobic pond, anaerobic filters and the upward flow anaerobic sludge blanket (UASB). Retention times vary with the design and with the level of maintenance (which may be very low). However, for septic tanks and anaerobic ponds, retention times are typically a matter of 1–5 days, while for anaerobic filters and UASBs retention time may be only a matter of hours.

A well-designed and well-maintained septic tank (the minority) with a 3-day retention time might remove 50–95% of the indicator bacteria (Feachem *et al.*, 1983). Because septic tanks serve small populations, they cannot be expected reliably to contain pathogens. However, the removal of *V. cholerae* O1 and *Salmonella* (Howard *et al.*, 1975) and *V. cholerae* non-O1 (Daniel and Lloyd, 1980a) have been studied in double chambered septic tanks with a total mean hydraulic retention time of 12–15 days. The performance was relatively poor with an overall removal of 99.8% for *V. cholerae* O1 and 98.8% for salmonellae. The conditions in the tanks were inhospitable (anaerobic, with pH values of 6.3–6.7 and 40–90 mg/I of ammonia), so the poor removal may have been attributable to short-circuiting.

Anaerobic filters have been used as post-treatment systems for septic tanks. Cullimore and Viraraghavan (1994) report that indicator removal in such systems is poor. However, UASB and anaerobic ponds are being used for the domestic treatment of wastewater in a number of countries. Indeed, where the year-round ambient temperatures are high and the wastewater is reasonably strong, anaerobic treatment must be the pre-treatment method of choice. The pre-treatment contributes to pathogen removal in two ways: by removing pathogens directly; and by promoting the aerobic conditions required for removal in the subsequent post-treatment stages.

Dixo (Dixo, 1994; Dixo *et al.*,1995) reported that a UASB reactor in northeast Brazil removed 67% of the faecal coliforms in the influent. The retention time was short (about 8 hours) and it is likely that the performance probably represents the removal of the solids-associated organisms. This mechanism is not likely to be specific to a particular organism and therefore the UASB can be expected to achieve this level of performance for bacterial pathogens. Although, superficially disappointing, the apparently low bacterial indicator removal of the UASB must be set against the excellent BOD removal these systems can achieve. As WSP are frequently used as a post-treatment and pathogen removal in ponds is related to the organic load, a UASB can make a substantial indirect contribution to bacterial pathogen removal.

Anaerobic ponds should not have retention times of much less than a day, even in the warmest climates, to avoid the washout of the treatment bacteria. Faecal coliform removals of 80 and 95% have been reported for anaerobic ponds in northeast Brazil with retention times of 1 and 5 days respectively (Oragui *et al.*, 1987; Silva, 1982). Oragui *et al.* (1987) reported similar levels of removal of salmonellae, campylobacter, bifidobacteria and faecal enterococci. The removal observed in anaerobic ponds must, in part, reflect a one-off elimination of solids-associated organisms that would be independent of the organism or the climate and would not occur in a second identical pond in series.

9 WASTE STABILIZATION PONDS

Waste stabilization ponds are a series of one or more hypereutrophic basins, typically 1–2 m deep. Owing to the algal activity, WSP will usually have high dissolved oxygen levels and high pH values. They have very low running costs and are simple to construct, the primary expense usually being the cost of the land. The physical simplicity of WSP belies the complexity of the pathogen removal mechanisms that take place within them. For, unlike the systems we have considered above, WSP can be designed to optimize pathogen removal and the performance of two systems at the same site can vary by 3–4 orders of magnitude. Our increasing appreciation of the mechanism by which this is achieved has yet to be fully incorporated into the design equations and the design of such systems is still comparatively crude. Nevertheless, a proper understanding of these mechanisms is invaluable in the design of WSP.

High rate algal ponds (HRAP) are specially designed waste stabilization ponds in which the algal population is gently stirred mechanically, usually in some form of race-track configuration to optimize algal production. Though more difficult and expensive to operate than algal ponds, HRAP are, from a bacterial pathogens' point of view, not dissimilar to a WSP. HRAP also achieve high (sometimes higher) pH and dissolved oxygen concentrations and have correspondingly high indicator removal rates (Bahlaoui *et al.*, 1997).

Faecal coliforms are the organisms most commonly used to monitor the removal of pathogens from wastewater treatment plants. This indicator has been compared with other indicators and salmonellae, campylobacter (see Fig. 30.1) and with *V. cholerae* 01 (Curtis, 1996) and found to be adequate. The oxygen-sensitive campylobacter presumably die due to the high oxygen concentrations. The removal of *V. cholerae* Ol is slightly more mysterious, because the closely related non-pathogenic *V. cholerae* have been reported to grow in WSP (see Curtis, 1996 for a comprehensive review). It has recently been discovered that the toxin gene that distinguishes the pathogenic organisms from the non-pathogenic form is encoded by a filamentous bacteriophage that is incorporated into the organism chromosome. This phage is induced by sunlight (Faruque *et al.*, 2000). This decreases the growth rate of the organisms and might leave the toxigenic organism at a sufficient disadvantage to inhibit its survival in a WSP.[1]

The mechanisms of faecal coliform removal are not surprisingly, multifactorial (Troussellier *et al.*, 1986). However, the factors can be divided into fast and slow. The fast (3–0.1 log/h) factors are photo-oxidation and pH and are associated with light and thus the upper parts of the pond during the hours of daylight.

[1] It does not appear that the engineer has much scope for increasing the inherent removal rates in wetland systems, beyond ensuring a low organic load to encourage predation. However, as with WSP, considerable improvements in the performance may be achieved by improving the hydraulic flow regime in the reed bed with consequent reduction in short-circuiting.

The slow (<0.1 log/h) factors are less well defined, and presumably include starvation, predation and adsorption to solids. The slow processes are independent of light and are thought to function at all times at all depths. Although most of research has focused on the fast factors, their relative importance depends on the depth and location of the WSP. WSP are heterogeneous in space and time. Thus the fate of a pathogenic bacterium in a WSP depends not only how long it remains in that pond, but which part of the pond it passes through and at what time of day it passes through it.

In photo-oxidation light kills an organism when the energy in the light is absorbed by a chemical or sensitizer. The sensitizer enters a short-lived excited state in which it can pass the light energy on to oxygen and form toxic forms of oxygen, including singlet oxygen, hydroxyl radicals, super-oxide radicals or hydrogen peroxide. The toxic forms of oxygen kill the cell by oxidizing vital cellular components (Curtis *et al.*,1992).

The rate of photo-oxidation increases linearly with oxygen concentration, presumably because the increasing oxygen concentration increases the probability of the excited sensitizer meeting an oxygen molecule. In addition, the process is synergistic with pH, at least for some organisms (Davies-Colley *et al.*, 1999). The basis for the synergism is not clear; it might be because the oxygen radicals damage the membrane, thus rendering the cells sensitive to high pH values. It could also be because high pH values prolong the life of hydroxyl radicals. Since even long wavelength red light has enough energy to create singlet oxygen, any wavelength of visible or ultraviolet light can initiate photo-oxidative damage. Thus the relative importance of a given wavelength is governed by the nature of the sensitizer, the wavelengths of light involved and the relative ability of these wavelengths to penetrate into a pond. In WSP the principal sensitizer for faecal coliforms is the humic substances that occur naturally in wastewater. Other organisms may have other sensitizers, including sensitizers within the cell (Davies-Colley *et al.*, 1999).

High pH values are also thought to be responsible for some rapid removal processes. I believe that death occurs when the bacterium is no longer able to acidify its own cytoplasm. This would explain why the relationship between removal rate and pH is non-linear, with pH values below a certain threshold value having little or no effect on bacterial indicator removal. Working with real pond water, Parhad and Rao (1974) and Curtis (1991) reported a threshold pH value of about the 9.3, while Pearson *et al.* (1987), working in buffer, report a threshold of pH 8.9. A considerable amount of energy is required for a cell to acidify its cytoplasm (Booth, 1985), thus the threshold probably represents the point at which this homeostatic mechanism is overwhelmed. By using low nutrient conditions, Pearson *et al.* (1987) artificially lowered the threshold pH value. Thus, in real ponds, very rapid pH related removal will only occur above the higher threshold.

More and better work (especially quantitative work and modelling) remains to be done on the fast processes. However, it is clear that the excellent removal seen in WSP in warm climates is intimately linked with the hypereutrophic conditions observed in WSP. It follows that algal concentrations really should feature in design for bacterial pathogen removal and that proposals for very clear ponds (James, 1987) would probably not work.

The light independent processes could be driven by a variety of factors, the most obvious of which are starvation, predation and adsorption to algae. Whatever the dark mechanisms are they will certainly be very important wherever pH and photo-oxidation effects are retarded by the depth or location of the pond, e.g. in deep VJSP or WSP in winter in temperate climates. Unfortunately, there has been essentially no systematic study of these factors in ponds. In an excellent paper, Trousellier *et al.* (1986) sought linear relationships between a number of interrelated factors in WSP and faecal coliform removal in the South of France. A weak negative relationship was found between algal concentrations and faecal coliforms but not between indicator organism and the BOD_5 or the rotifer concentration. However, if the effects were non-linear or relatively unimportant at this study site, compared to the light and pH effects (which were significant), then the study would not find them to be significant.

Of course the fastest removal processes will be ineffective if the bacterial pathogens do not spend very long in the system. Conversely, even a slow rate of removal may be adequate if endured for long enough. There has always been uncertainty about the nature of the hydraulic regime in WSP. Recently, a number of authors have pointed to the importance of short-circuiting in the removal of pathogens in WSP (James, 1987; Fredrick and Lloyd; 1996; Salter *et al.*, 1999). However, it should be noted that short-circuiting is implicit in Marais' (1974) original assumption of completed mixing (some of the influent will instantaneously find itself in the effluent). Moreover, recent survey work has shown that ponds crudely approximate to complete mix reactors (Nameche and Vasel, 1998) and, for the time being, this assumption may be retained and perhaps used in conjunction with a small safety factor.

At present, in practice, the best way to reduce the impact of short-circuiting is to use more than one pond in series (Mara, 1976). Other ways of improving WSP hydrodynamics, e.g. baffles, are conceivable. However, to be usable the improvements must not prejudice other features of the WSP. For example, by promoting the build-up of solids and sulphide generation baffles in facultative WSP, baffles may cause odour and prejudice photosynthesis. In addition to being used in design, the improvements must be predictable. I hope that the current interest in the use of computational fluid dynamics will yield verified models that are sufficiently cheap to employ to justify their widespread use in the optimization of WSP design for pathogen removal.

10 WETLANDS AND REED BEDS

Wetland is a term used to describe a variety of formats of wastewater treatment systems in which wastewater is fed from, through, or over a substrate planted with reeds. Though in many ways analogous to the nineteenth century land treatment of wastewater, wetlands have become more and more popular. Like ponds, they may be specifically designed to remove pathogenic organisms. Indeed, wetlands have been advocated for use by small communities in temperate climates trying to meet pathogen or pathogen indicator standards (Cooper *et al.*, 1996).

There have been a number of reports of pathogen removal in wetlands in various formats (Coombes and Colett, 1995; Cooper *et al.*, 1996; Green *et al.*, 1997). The general capabilities of wetlands may be inferred from work on an interesting design of intermittently fed shallow (~0.3 m), narrow (~1.3 m) gravel-filled hydroponics beds fed settled (secondary treatment) or treated sewage (tertiary treatment) in the UK and Egypt. The researchers expressed the performance as a decimal reduction distance or DRD, i.e. the distance required to effect a 90% reduction in pathogen or indicator concentration. The treatment appears to be partially related to organic load with better faecal coliform removal in tertiary treatment than in secondary treatment (DRD values of 51 m and 95 m respectively) and climate, removal was better in Egypt than in the UK (DRD for secondary treatment 30 m). The removal of salmonella (in the UK) was slightly better than the removal of faecal coliforms, however, the removal of *Vibrio cholerae* (presumably non-O1) was twice as slow as the removal of the indicator (Stott *et al.*, 1996).

Cooper *et al.* (1996) have usefully reinterpreted these and other reports and concluded that the rate of removal of faecal coliforms for secondary beds is about 0.02 log/h and about 0.08–0.04 log/h tertiary beds. These removal rates are subject to modest climatic (Stott *et al.*, 1996) and seasonal effects (Rivera *et al.*, 1995) and are comparable to those seen in the dark in WSP.

Speculation that protozoal predation (Green *et al.*, 1997) is important in bacterial pathogen removal has proven to be well founded. Decamp and Oliver (1998) found that observed levels of ciliate bacterivores could account for all the observed removal of faecal coliforms. The effect of organic loading may be accounted for by the reduced level of predation observed at low oxygen tensions (Decamp *et al.*, 1999) and the climatic and seasonal variation by documented temperature effects (Iriberri *et al.*, 1995). However, Decamp and Oliver (1998) did not discount the possibility of other biotic and abiotic factors (bacteriophage attack, adsorption to solids, starvation and sedimentation) being significant in the removal of bacterial pathogens.

REFERENCES

Bahlaoui, M.A., Baleux, B. and Troussellier, M. (1997). Dynamics of pollution-indicator and pathogenic bacteria in high-rate oxidation wastewater treatment ponds. *Water Research* **31**, 630–683.

Bergey, D.H. and Holt, J.G. (1994). *Bergey's Manual of Determinative Bacteriology*, 9th edn. Williams &Wilkins, Baltimore.

Booth, I.R. (1985). Regulation of cytoplasmic pH in bacteria. *Microbiological Reviews*, 359–378.

Brock, T.D. (1987). The study of microorganisms *in situ*: progress and problems. In: M. Fletcher, T.R.G. Gray and J.G. Jones (eds) *Ecology of Microbial Communities*. Cambridge University Press, Cambridge, UK.

Bruns, H. and Sierp, F. (1927). Einflub der schlammbelebung des abwassers auf pathogene keime. *Zeitschrift fur Hygiene*. Leipzig (later Berlin) **107**, 571–584.

Case, T.J. (2000). *An Illustrated Guide to Theoretical Ecology*. Oxford University Press.

Collier, L., Balows, A. and Sussman, M. (eds) (1998). *Topley and Wilson's Microbiology and Microbial Infections*, 9th edn. Arnold, London.

Communicable Disease Surveillance Centre (2000). *Notification of Infectious Disease*. Communicable Disease Report, 10, 119.

Courmont, P. and Rochaix, A. (1922). Sur l'epuration bacterienne des eaux d'egouts. par le procede des 'boues activiees'. Revue d'Hygiene **44**, 907–919.

Coombes, C. and Collett, P.J. (1995). Use of constructed wetland to protect bathing water quality. *Water Science and Technology* **32**, 149–158.

Cooper, P.F., Job, G.D., Green, M.B. and Shutes, R.B.E. (1996). *Reed Beds and Constructed Wetlands for Wastewater Treatment*. WRC Swindon, UK.

Cullimore, D.R. and Viraraghavan, T. (1994). Microbiological aspects of anaerobic filter treatment of septic tank effluent at low temperatures. *Environmental Technology* **15**, 165–173.

Curtis, T.P. (1991). *Mechanisms of removal of faecal coliforms from waste stabilisation ponds*. Phd Thesis, University of Leeds, UK.

Curtis, T.P., Mara, D.D. and Silva, S.A. (1992). Influence of pH, oxygen and humic substances on ability of sunlight to damage fecal coliforms in waste stabilization pond water. *Applied and Environmental Microbiology* **58**, 1335–1343.

Curtis, T.P. (1996). The fate of *Vibrio cholerae* in wastewater treatment plants. In: B. Drasar and B.D. Forrest (eds) *Ecology and the Ecology of Vibrio cholerae*. Chapman and Hall, London.

Daniel, R.R. and Lloyd, B.J. (1980a). Microbiological studies on two Oxfam sanitation units operating in Bengali refugee camps. *Water Research* **14**, 1567–1571.

Daniel, R.R. and Lloyd, B.J. (1980b). A note on the fate of El Tor cholera and other vibrios in percolating filters. *Journal of Applied Bacteriology* **48**, 207–209.

Davies-Colley, R.J., Donnison, A.M., Speed, D.J. *et al.* (1999). Inactivation of faecal indicator microorganisms in waste stabilisation ponds: Interactions of environmental factors with sunlight. *Water Research* **33**, 1220–1230.

Decamp, O. and Warren, A. (1998). Bacterivory in ciliates isolated from constructed wetlands (reed beds) used for wastewater treatment. *Water Research* **32**, 1989–1996.

Decamp, O., Warren, A. and Sanchez, R. (1999). The role of ciliated protozoa in subsurface flow wetlands and their potential as bioindicators. *Water Science and Technology* **40**, 91–98.

Dixo, N.G.H. (1994). The removal of *V. cholerae* by water treatment systems in developing countries. MSc Thesis. University of London, UK.

Dixo, N.G.H., Gambrill, M.P., Catunda, P.F.C. and van Haandel (1995). Removal of pathogenic organisms from the effluent of an upflow anaerobic digester using waste stabilisation ponds. *Water Science and Technology* **31**, 275–284.

Dizon, J.J., Fukimi, H., Barua, D. *et al.* (1967). Studies on cholera carriers. *Bulletin of the World Health Organisation* **37**, 737–743.

Faruque, S.M., Asadulghani, Rahman, M.M. *et al.* (2000). Sunlight-induced propagation of the lysogenic phage encoding cholera toxin. *Infection and Immunity* **68**, 4795–4801.

Feachem, R.G. (1982). Environmental aspects of cholera epidemiology, III. Transmission and control. *Tropical Diseases Bulletin* **79**, 1–47.

Feachem, R.G., Bradley, D.J., Garelick, H.D. and Mara, D.D. (1983). *Sanitation and Disease, Health Aspects of Excreta and Waste Management and Disease*. World Bank Studies in Water Supply and Sanitation 3. John Wiley and Sons Ltd, Chichester, UK.

Feachem, R.G. and Mara, D.D. (1979). A reappraisal of the role of faecal indicator organisms in tropical waste treatment processes. *The Public Health Engineer* **7**(1), 31–33.

Frederick, G.L. and Lloyd, B.J. (1996). An evaluation of retention time and short-circuiting in waste stabilisation ponds using *Serratia marcescens* bacteriophage as a tracer. *Water Science and Technology* **33**, 49–56.

Green, M.B., Griffin, P., Seabridge, J.K. and Dhobie, D. (1997). Removal of bacteria in subsurface flow wetlands. *Water Science and Technology* **35**, 109–116.

James, A. (1987). An alternative approach to the design of waste stabilisation ponds. *Water Science and Technology* **19**, 213–218.

Heidelberg, J.F., Eisen, J.A., Nelson, W.C. *et al.* (2000). DNA sequence of both chromosomes of the cholera pathogen *Vibrio cholerae*. *Nature* **406**, 477–483.

Howard, J., Lloyd, B. and Webber, D. (1975). Oxfams Sanitation Unit, *The design and testing of a sanitation and sewage treatment unit for disasters and long term use.* 2nd edn, Oxfam Technical Paper, Oxfam, Oxfam House, 274 Banbury road, OX2, 7DZ, UK.

Iriberri, J., Ayo, B., Santamaria, E. *et al.* (1995). Influence of bacterial density and water temperature on the grazing activity of 2 freshwater ciliates. *Freshwater Biology* **33**, 223–231.

Islam, M.S., Drasar, B.S. and Sack, R.B. (1994). The aquatic flora and fauna as resevoirs of Vibrio cholerae: a review. *Journal of Diarrhoeal Disease* **12**, 87–96.

Langeland, G. (1982). *Salmonella* spp. in the working environment of sewage treatment plants in Oslo, Norway. *Applied & Environmental Microbiology* **43**, 1111–1115.

Lema, O., Ogwa, M. and Mhalu, F.S. (1979). Survival of El Tor cholera vibrio in local water sources and beverages in Tanzania. *East African Medical Journal* **56**, 504–508.

Levine, M.M., Nalin, D.R., Rennels, M.B. *et al.* (1979). Genetic susceptibility to cholera. *Annals of Human Biology* **6**, 369–374.

Mara, D.D. (1976). *Sewage Treatment in Hot Climates.* John Wiley and Sons, Chichester.

Marais, G.V.R. (1974). Faecal bacterial kinetics in stabilisation ponds. *Journal of Environmental Engineering Division, American Society of Civil Engineers* **100**, 119–140.

MacDonell, M.T., Swartz, D.G., Ortiz-Conde *et al.* (1986). Ribosomal RNA phylogenies for the vibrio-enteric group of eubacteria. *Microbiological Sciences* **3**, 172–178.

Mackenzie, T.D., Ellison, R.T. and Mostow, S.R. (1992). Sunlight and cholera. *Lancet* **340**, 367.

Miller, C.J., Drasar, B.S. and Feachem, R.G. (1984). Response of toxigenic Vibrio cholerae O1 to physico-chemical stresses in aquatic environments. *Journal of Hygiene (Cambridge)* **93**, 475–495.

Mcabe, D.B. (1970). Water and waste-water systems to combat cholera in East Pakistan. *Journal of Water Pollution Control Federation* **42**, 1968–1981.

Morton, D. and Bardhan, K.D. (2000). The presence of bacteriophages active against Helicobacter pylori in UK sewage – Natures eradicator? *Gut* **46**, A69–A69 (Suppl.).

Muniesa, M. and Jofre, J. (2000). Occurrence of phages infecting *Escherichia coli* O157:H7 carrying the Stx 2 gene in sewage from different countries. *FEMS Microbiology Letters* **183**, 197–200.

Nair, G.B., Sarkar, B.L., De, S.P. *et al.* (1988). Ecology of *Vibrio cholerae* in the freshwater environs of Calcutta, India. *Microbial Ecology* **15**, 203–215.

Nameche, T. and Vasel, J.L. (1998). Hydrodynamic studies and modelization for aerated lagoons and waste stabilization ponds. *Water Research* **32**, 3039–3045.

Nichols, A.B. (1991). Sanitation woes contribute to Peruvian epidemic. *Water Environment and Technology* **7**, 13–14.

Ohno, A., Marui, A., Castro, E.S. *et al.* (1997). Enteropathogenic bacteria in the La Paz River of Bolivia. *American Journal of Tropical Medicine and Hygiene* **57**, 438–444.

Oragui, J.I., Curtis, T.P., Silva, S.A. and Mara, D.D. (1987). The removal of excreted bacteria and viruses in deep waste stabilization ponds. *Water Science and Technology* **19**, 569–573.

Oragui, J.I., Arridge, H., Mara, D.D. *et al.* (1993). *Vibrio cholerae* 01 (El Tor) removal in waste stabilization ponds in Northeast Brazil. *Water Research* **27**, 727–728.

Parhad, N.M. and Rao, N.U. (1974). Effect of pH on survival of *Escherichia coli*. *Journal of the Water Pollution Control* **46**, 980–986.

Pearson, H.W., Mara, D.D., Mills, S.W. and Smallman, D.J. (1987). Physicochemical factors influencing bacterial survival in waste stabilisation ponds. *Water Science and Technology* **19**, 145–152.

Rivera, F., Warren, A., Ramirez, E. *et al.* (1995). Removal of pathogens from wastewaters by the root zone method (RZM). *Water Science and Technology* **32**, 211–218.

Salter, H.E., Boyle, L., Ouki, S.K. *et al.* (1999). Tracer study and profiling of a tertiary lagoon in the United Kingdom: II. *Water Research* **33**, 3782–3788.

Shuval, H.I., Adin, A., Fattal, B. *et al.* (1986). *Wastewater irrigation in developing countries. Health effects and Sanitation Solutions.* World Bank Technical Paper 51. World Bank, Washington DC, USA.

Silva, S.A. (1982). *On the treatment of domestic sewage in waste stabilization ponds in NE Brazil.* PhD Thesis, University of Dundee.

Singleton, F.L., Attwell, R., Jangi, S. and Colwell, R.R. (1982a). Effects of temperature and salinity on Vibrio cholerae growth. *Applied and Environmental Microbiology* **44**, 1047–1058.

Singleton, F.L., Attwell, R.W., Jangi, M.S. and Colwell, R.R. (1982b). Influence of salinity and organic nutrient concentration on survival and growth of Vibrio cholerae in aquatic microcosms. *Applied and Environmental Microbiology* **43**, 1080–1085.

Spira, W.M., Huq, A., Ahmed, Q.S. and Saeed, Y.A. (1981). Uptake of Vibrio cholerae biotype eltor from contaminated water by water hyacinth (Eichornia crassipes). *Applied and Environmental Microbiology* **42**, 550–553.

Stampi, S., Varoli, O., de Luca, G. and Zanetti, F. (1992). Occurrence, removal and seasonal variation of 'thermophilic' campylobacters in a sewage treatment plant in Italy. *Zentralblatt fur Hygiene und Umweltmedizin* **193**(3), 199–210.

Stampi, S., Varoli, O., Zanetti, F. and Deluca, G. (1993). Arcobacter-cryaerophilus and thermophillic campylobacters in a sewage treatment plant in Italy – two secondary treatments compared. *Epidemiology and Infection* **110**, 633–639.

Stott, R., Jenkins, T., Williams, J. *et al.* (1996). *Pathogen removal and microbial ecology in Gravel Bed Hydroponic (GBH) treatment of wastewater.* Research Monographs in Wastewater Reuse in Developing Countries, Monograph Number 4. University of Portsmouth, UK.

Streeter, H.W. (1930). The effects of activated sludge on pathogenic organisms. *Sewage Works Journal* **2**, 292–295.

Swerdlow, D.L., Mintz, E.D., Rodriguez, M. *et al.* (1992). Waterborne transmission of epidemic cholera in Trujillo, Peru: lessons for a continent at risk. *Lancet* **340**, 28–32.

Tamplin, M.L., Gauzens, A.L., Huq, A. *et al.* (1990). Attachment of Vibrio cholerae serogroup O1 to zooplankton and phytoplankton of Bangladesh waters. *Applied and Environmental Microbiology* **56**, 1977–1980.

Tamplin, M.L. and Parodi, C.C. (1991). Environmental spread of Vibrio cholerae in Peru. *Lancet* **338**, 1216–1217.

Tauxe, R., Blake, P., Olsvik, O. and Wachsmuth, I.K. (1994). The future of cholera: persistance, change, and an expanding research agenda. In: I.K. Wachsmuth, P.A. Blake and O. Olsvik (eds) *Vibrio cholerae and cholera*. American Society for Microbiology, Washington, USA.

Thom, S., Warhurst, D. and Drasar, B.S. (1992). Association of *Vibrio cholerae* from fresh water amoebae. *Journal of Medical Microbiology* **36**, 303–306.

Troussellier, M., Legendre, P. and Baleux, B. (1986). Modelling the evolution of bacterial densities in an eutrophic ecosystem (sewage lagoons). *Microbial Ecology* **12**, 355–379.

Ueda, T. and Hata, K. (1999). Domestic wastewater treatment by a submerged membrane bioreactor with gravitational filtration. *Water Research* **33**, 2888–2892.

Ueda, T., Hata, K. and Kikuoka, Y. (1996). Treatment of domestic sewage from rural settlements by a membrane bioreactor. *Water Science and Technology* **34**, 189–196.

Ueda, T. and Horan, N.J. (2000). Fate of indigenous bacteriophage in a membrane bioreactor. *Water Research* **34**, 2151–2159.

Wheater, D.W.F., Mara, D.D., Jawad, L. and Oragui, J. (1980). *Pseudomonas aeruginosa* and *Escherichia coli* in sewage and fresh water. *Water Research* **14**, 713–721.

Wheeler, J.G., Sethi, D., Cowden, J.M. *et al.* (1999). Study of infectious intestinal disease in England: rates in the community, presenting to general practice, and reported to national surveillance. *British Medical Journal* **318**, 1046–1050.

Winnen, H., Suidan, M.T., Scarpino, P.V. *et al.* (1996). Effectiveness of the membrane bioreactor in the biodegradation of high molecular-weight compounds. *Water Science and Technology* **34**, 197–203.

West, P.A., Brayton, P.R., Bryant, T.N. and Colwell, R.R. (1986). Numerical taxonomy of vibrios isolated from the aquatic environments. *International Journal of Systematic Bacteriology* **36**, 531–543.

Woese, C.R., Weisburg, W.G., Hahn, C.M. *et al.* (1985). The phylogeny of purple bacteria: the gamma subdivision. *Systematic and Applied Microbiology* **6**, 25–33.

Yaziz, M. and Lloyd, B.J. (1979). The removal of salmonellas in conventional sewage treatment processes. *Journal of Applied Bacteriology* **46**, 131–142.

31

Fate and behaviour of parasites in wastewater treatment systems

Rebecca Stott

Department of Civil Engineering, University of Portsmouth, Portsmouth PO1 3HF, UK

1 INTRODUCTION

Diseases caused by intestinal parasites (helminths[1] and protozoa) are often a principal cause of human morbidity. Recent estimates suggest that at least 50% of the world's population may be infected with one or more helminth species (Chan, 1997), while prevalence rates of protozoan diseases ranging between 15 and 30% are not uncommon in tropical regions and may range between <1% and 10% in temperate areas (Mata, 1986; Casemore *et al.*, 1997). Transmission of many of the parasitic diseases, particularly helminthic infections, is often as a result of inadequate water supplies, lack of sanitation or disposal of insufficiently treated wastewater (Shuval *et al.*, 1986; Cairncross and Feachem, 1993). However, the association of faecally-contaminated drinking water supplies with recent waterborne outbreaks of emerging protozoan diseases is becoming of increasing concern (Smith and Rose, 1998; Huston and Petri, 2001).

Parasites are considered wastewater-associated pathogens of great public health importance due principally to their environmentally persistent transmissive stages, a low infective dose, limited or transient acquired immunity and morbidity, particularly in immunocompromised hosts (Anderson and May, 1979; Chan *et al.*, 1994; Guerrant *et al.*, 1999; Stephenson *et al.*, 2000b; Stanley, 2001). It is apparent that while intestinal parasites continue to have a significant worldwide impact on human health, wastewater can be a major contributor to the transmission of parasitic diseases to susceptible hosts, either directly through occupational exposure and consumption of wastewater irrigated crops (Schlosser *et al.*, 1999; Blumenthal *et al.*, 2000), or indirectly through ingestion of contaminated water and exposure to polluted recreational water (Furtado *et al.*, 1998). However, the performance of wastewater treatment systems in removing parasites (thereby reducing the potential risk of disease transmission resulting from wastewater disposal or reuse) is rarely considered.

This chapter reports on current information about the occurrence of parasites in wastewater and their fate and behaviour during wastewater treatment. The removal and destruction of parasites by treatment processes is considered for 'conventional' (mechanical) and 'natural' wastewater treatment systems.

2 PARASITES IN RAW WASTEWATER

Wastewaters can contain a wide variety of excreted parasites depending on the source of wastewater and diseases present in the sewage contributing community (Table 31.1). The principal groups of helminth parasites include nematodes, cestodes and trematodes, while that of protozoa includes coccidia, flagellates,

[1] Helminth is the collective term for 'worm' which can be classified further into three main groups: roundworms, tapeworms and flatworms.

The Handbook of Water and Wastewater Microbiology
ISBN 0-12-470100-0

TABLE 31.1 Principal intestinal parasites found in raw wastewaters

Parasite		*Species*	*Common name*
HELMINTH	Nematode (roundworms)	*Ascaris lumbricoides*	human roundworm
		Trichuris trichiura	human whipworm
		Ancylostoma duodenale	hookworm
		Necator americanus	hookworm
		Enterobius vermicularis	pin worm/threadworm
		Strongyloides stercoralis	small roundworm[a]
		Toxocara spp.	
	Cestode (tapeworms)	*Taenia solium*	pork tapeworm
		Taenia saginata	beef tapeworm
		Hymenolepis nana	dwarf tapeworm
		Hymenolepis diminuta	rodent tapeworm
	Trematode (flatworms/flukes)	*Schistosoma* spp.	blood fluke
		Clonorchis sinensis	oriental liver fluke
		Fasciola hepatica	liver fluke
		Paragonimus westermani	lung fluke
PROTOZOA	Coccidia	*Cryptosporidium* spp.	cryptosporidiosis
		Cyclospora spp.	cyclosporiosis
		Isosopora belli	isosporiasis
	Flagellate	*Giardia* spp.	giardiasis
	Amoebae	*Entamoeba histolytica*	amoebic dysentery
	Ciliate	*Balantidium coli*	balantidiasis
		Microsporidia spp.	Microsporidiosis

[a] may also be called threadworm

amoebae and ciliates. General information about parasites can be found in Smith (1994) and Bogitsh and Cheng (1998).

Intestinal helminth parasites are usually found in wastewaters as eggs, although some species of helminths (notably *Strongyloides*) may be present as the larval form. Protozoan parasites are found in wastewater as cysts or oocysts depending on the species. Disease transmission to humans for the parasites shown in Table 31.1 can be through a variety of routes. In helminth infections, the infective stage may be an embryonated egg containing a larva which is then transmitted directly to other human hosts by ingestion (e.g. *Ascaris* spp.) or the larva hatches, developing to an infective filariform larva in the environment, which then penetrates the skin of unsuspecting hosts (e.g. hookworm). Some helminth infections require transmission to humans via one or more intermediate hosts (e.g. trematodes such as *Schistosoma* spp. or cestodes such as *Taenia* spp.). Protozoan (oo)cysts excreted in faeces are usually infective immediately or within a few hours of being shed in faeces; a few species may require a longer incubation period of several days or weeks at ambient temperature. Disease transmission occurs through ingestion of (oo)cysts usually from faecally-contaminated water or food.

Humans can also become infected accidentally by some intestinal parasitic diseases normally transmitted between wild or domestic animals or agricultural livestock (Faust *et al.*, 1976; McLauchlin *et al.*, 2000; Robertson *et al.*, 2000a). These 'zoonotic' diseases include helminths such as *Fasciola, Paragonimus* and *Toxocara* (eggs of which are also sometimes found in raw municipal wastewaters) and also the protozoa *Giardia duodenalis* and *Cryptosporidium parvum*. Thus effective treatment of animal wastes with removal of excreted parasites can also be an important human health-related treatment objective.

2.1 Parasite occurrence in raw wastewaters

The occurrence and concentration of parasites in raw and influent wastewaters depends on a variety of factors, including the source of wastewater (e.g. domestic/agricultural), prevalence, intensity and duration of infection in the contributing population, population size, per capita water consumption and proportion of population sewered, fecundity or multiplication rate of the parasite species and survival of the excreted stage in wastewaters. Point prevalence surveys indicate that parasites eggs and (oo)cysts can commonly occur in wastewaters worldwide with species diversity often reflected in sludge surveys (Reimers *et al.*, 1982; Schwartzbrod *et al.*, 1987; Paulino *et al.*, 2001). Eggs of nematodes (roundworms) and cestodes (tapeworms) are typically found in wastewaters (Fig. 31.1) with eggs of *Ascaris* usually predominating in raw wastewaters followed by *Trichuris*, hookworm and *Hymenolepis* spp. (Bouhoum *et al.*, 1997; Stott *et al.*, 1997). *Taenia* eggs are sometimes found as well as eggs of *Toxocara* and *Enterobius*. However, eggs of trematodes (flukes) are seldom reported as they tend to hatch on contact with water.

Protozoan parasites are also reported in wastewaters, with *Giardia* cysts and *Cryptosporidium* oocysts being the most frequently identified (Fig. 31.2). The relative frequency of protozoan species may vary depending on catchment characteristics. In England, the majority of protozoa found in sewage influent to treatment works were that of *Giardia*, while a greater prevalence of *Cryptosporidium* was found in agricultural impacted wastewaters (Bukhari *et al.*, 1997). Generally, although both species often coexist in wastewaters, *Giardia* usually exceeds *Cryptosporidium* in the number of positive samples detected and concentration. Point prevalence surveys carried out in developed countries, and reviewed by Robertson *et al.* (1999), indicated that over 66% of influent samples collected from wastewater treatment sites contained *Cryptosporidium* oocysts in comparison to more than 98% of influent samples containing *Giardia*. Cysts of *Entamoeba histolytica* have been identified in

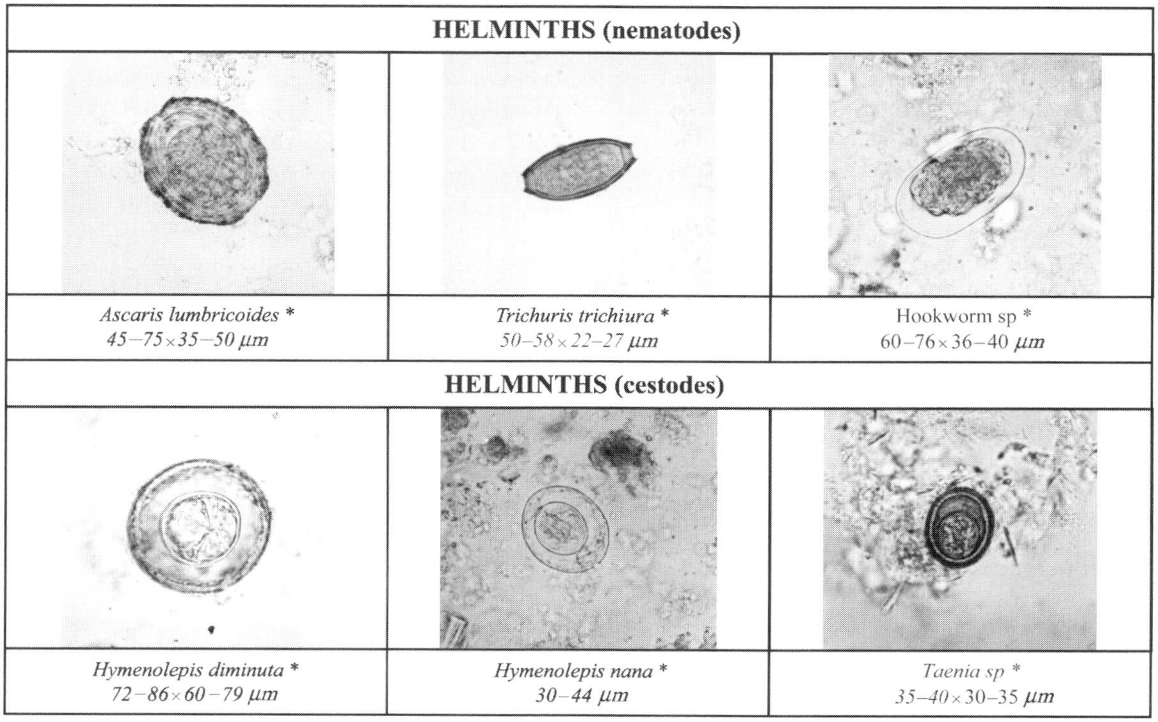

Fig. 1 Helminth parasites frequently found in raw and influent wastewaters.
* Printed with permission by R Ayres. Not to scale.

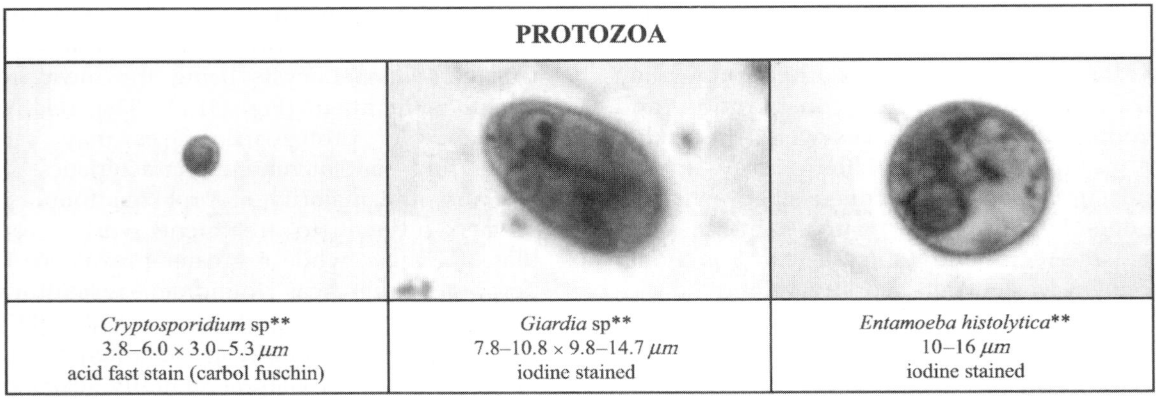

Fig. 2 Protozoan parasites frequently found in raw and influent wastewaters.
** Image source R Stott. Not to scale
NB. Parasite images can also be found on the web pages of the Parasitology Department at Chiang Mai University, Thailand. (www.medicine.cmu.au.th/dept/parasite/official/image.htm

raw wastewaters from developing countries (Panicker and Krishnamoorthi, 1978a) but have not been reported in developed countries probably due to the low level of amoebiasis in the community. There is limited information on the occurrence of other 'emerging' enteric protozoan parasites such as *Isospora, Cyclospora* and microsporidia in wastewater. Sturbaum *et al.* (1998) reported unsporulated (i.e. undeveloped non-infective) oocysts of *Cyclospora* in 64% of wastewater samples and sporulated (i.e. infective) oocysts in 18% of samples collected from a shantytown in Peru where cyclosporiasis was widespread. Spores of microsporidia have been reported in raw wastewater in one study in which the human-specific *Encephalitozoon intestinalis* was identified (Dowd *et al.*, 1998).

2.2 Parasite concentration in raw wastewaters

The concentration of parasites commonly found in raw wastewater and influents to treatment systems can vary greatly. High prevalences and levels of infection for diseases exacerbated by partial sanitation, poverty and low water usage per capita (Cairncross and Feachem 1993; Chan, 1997) and the potentially large number of parasite eggs or (oo)cysts excreted daily by infected individuals and zoonotic hosts (10^2–10^4 eggs/g and 10^5–10^7 (oo)cysts/g (Feachem *et al.*, 1983; Smith and Rose, 1998) can all contribute to high parasite loading rates in wastewaters. Dixo *et al.* (1995) reported up to 35 000 helminth eggs/l in wastewaters from an urban slum area of NE Brazil, while Lloyd and Frederick (2000) report concentrations of *Ascaris* of $>$100 000/l in a high density slum area in Bangladesh. However, maximum concentrations of around 1000 eggs/l are perhaps more typical in urban wastewaters. In India and Africa, concentrations ranging from 120 to around 900 eggs/l have been reported in raw wastewaters (Panicker and Krishnamoorthi, 1981; Ayres *et al.*, 1993). Ayres (1991) also found comparable numbers of eggs in Brazil with relatively high numbers of *Ascaris* eggs (up to 700 eggs/l) and up to 19 *Trichuris* eggs/l, 8 hookworm eggs/l and 20 *Hymenolepis* spp eggs/l. Of all the helminth eggs likely to be present in wastewaters, *Ascaris, Trichuris* and hookworm are of particular public health concern because of severe socioeconomic consequences of an estimated 39 million DALYs (disability-adjusted life year) lost to these infections (Chan, 1997; Crompton, 2000; O'Lorcain and Holland, 2000; Stephenson *et al.*, 2000a). There is limited information on protozoan (oo)cysts in raw wastewaters in developing countries. Concentrations of 25 000 cysts per litre have been found in Africa (Grimason *et al.*, 1996b), although, in general, lower concentrations rarely exceeding 1000 (oo)cysts/l have been reported (Panicker and Krishnamoorthi, 1981; Jimenez *et al.*, 2001). However, it is likely that higher concentrations may be present due to high reported incidences of protozoan infection (exceeding 50% in some communities) in developing countries (Zu and Guerrant, 1993).

Relatively few studies have assessed parasite egg numbers in raw sewage in developed countries. Where reported, helminth eggs occur in lower concentrations, rarely exceeding 100 eggs/l, probably reflecting a generally low incidence of helminthiasis in communities. In contrast, protozoan (oo)cysts have been found in very high numbers in several developed countries with concentrations frequently exceeding 10 000 (oo)cysts/l; up to 44 000 cysts/l have been reported in UK influent wastewaters (Robertson *et al.*, 1995). Significantly higher numbers of *Giardia* than *Cryptosporidium* have been detected in influent domestic wastewaters in the USA, UK and Europe (Mayer and Palmer, 1996; Robertson *et al.*, 2000b; Medema and Schijven, 2001). However, higher numbers of *Cryptosporidium*, more than 10^4–10^5 oocysts/l, have been reported in US livestock slaughterhouse and UK holding pen effluents (Robertson *et al.*, 1999).

The magnitude of parasite concentrations in raw wastewaters may also fluctuate temporally. Diurnal and seasonal variations in parasite egg and (oo)cyst numbers have been observed (Casson *et al.*, 1990; Gassman and Schwartzbrod, 1991; Sykora *et al.*, 1991; Capizzi and Schwartzbrod, 1998) possibly reflecting human activity factors and periods of increased incidence of infection in communities (Casemore *et al.*, 1997; Chai *et al.*, 2001). Parasite concentrations in wastewaters can also be considerably affected by rainfall and storm events (Panicker and Krishnamoorthi, 1978a; Atherholt *et al.*, 1998; Capizzi and Schwartzbrod 1998). Parasite loading rates into wastewater treatment systems can thus be highly variable with short- and longer-term trends in excreted parasite concentrations possible.

In general, raw wastewaters in developed countries may contain helminth egg numbers of up to 10^2 eggs/l and protozoa ranging from 10 to 10^5(oo)cysts/l. In comparison, concentrations of helminths of 10^2– $< 10^4$ eggs/l and protozoa of 10^2–$10^{3\,(\text{or }4)}$ (oo)cysts/l have been reported for developing countries. Since parasites are highly infectious with infective doses as few as 1–30 eggs/(oo)cysts capable of causing infections (Feachem *et al.*, 1983; Okhuysen *et al.*, 1999; Kothary and Babu, 2001), a high degree of parasite removal (at least 2–4 log units, i.e. >99%) may be required by sewage treatment processes for public health protection and compliance with current recommended quality criteria of 0.1–1 egg/l for wastewater irrigation (WHO, 1989; Blumenthal *et al.*, 2000).

2.3 Parasite removal in wastewater treatment systems

The effect of treatment processes on the fate and behaviour of parasites has been evaluated for high rate mechanical processes (conventional) and low rate (natural) treatment systems. Parasite removal is commonly assessed by comparing the microbiological quality of influent and effluent wastewaters. Removal rates reported for treatment systems may thus only refer to the removal of parasite eggs or (oo)cysts from suspension in wastewaters into sediments and sludge, and not to their destruction or inactivation as a result of the treatment process. In addition, non-detection of parasites in wastewaters does not guarantee that effluents are completely free of parasites since none of the enumeration and detection methods for parasitological analysis of wastewaters can guarantee a 100% recovery of parasites in processed samples. In relation to inactivation, reports usually refer to an assessment of viability rather than infectivity of eggs or (oo)cysts. Viability of helminth eggs is commonly determined by a morphological assessment of egg development to an embryonated stage containing a larva (e.g. *Ascaris)* or the hatchability of eggs (e.g. hookworm). The viability of protozoan (oo)cysts is typically by dye inclusion/exclusion or excystation methods (e.g. *Cryptosporidium*).

The contribution of each treatment process on removal or destruction of parasite is discussed below for the various levels of wastewater treatment, which usually consist of a sequence of physical, biological and sometimes chemical processes generally classified as preliminary, primary, secondary and tertiary or advanced. The fate of parasites in sludge treatment processes is covered comprehensively in Lewis-Jones and Winkler (1991).

2.4 Preliminary treatment – screening and grit removal

Preliminary treatment processes are used to improve downstream operations and typically consist of physical processes for screening and grit separation to remove large solids and rubbish. It is unlikely that parasites will be removed at this stage of treatment. The use of coarse screens with bar spacing usually at >10 mm and even fine screens (0.2–10 mm spacing) will not impede the passage of parasite eggs and cysts which range in size from 0.001 to 0.15 mm. Short detention times and high velocity flow of wastewater through grit chambers are also unlikely to facilitate removal of parasites.

2.5 Primary treatment

Parasite removal is highly variable between primary treatment processes. Generally, helminth eggs are removed more effectively than protozoan (oo)cysts. Performance for primary treatment processes for helminth removal range from 60 to 99% and where reported, primary effluents can contain between 0 and 1500 eggs/l. Removal of protozoa is more variable ranging from 4 to 93% with effluents containing $10–10^5$ (oo)cysts/l.

2.5.1 *Primary settlement tanks*

The aim of primary treatment is to remove settleable solids of 0.05–10 mm by gravity in sedimentation tanks. The principal mechanism for parasite removal during primary treatment is also assumed to be by sedimentation, and possibly by adsorption onto solids that are settling out of suspension.

2.5.1.1 Plain/simple (non-assisted) sedimentation. Parasite sedimentation rates (estimated using Stoke's law for discrete particle settling) will vary between parasite species depending on the specific gravity and dimensions of the parasite and liquid density (and temperature) suggesting that eggs of *Ascaris* and *Trichuris* will be removed more effectively than eggs with slower rates of settling velocity, such as hookworm and (oo)cysts of protozoa (Table 31.2). Panicker and Krishnamoorthi (1978b) reported *Ascaris* and *Trichuris* removal rates of 96% and 90% respectively in comparison to removal rates of 80% for hookworm eggs during primary sedimentation. Stokes' law, however, does not hold for *Giardia, Entamoeba histolytica* and *Cryptosporidium* (oo)cysts as their Reynolds' numbers are below 10^{-4} indicating that protozoan (oo)cysts are unlikely to be effectively removed by sedimentation. Nevertheless, protozoan removal has been reported during primary sedimentation in operational works, although removal performance is poor. Removal rates of between 4 and 47% have been reported by Robertson *et al.* (2000b) for protozoa in primary sedimentation tanks with estimated retention times of up to 25 hours. *Cryptosporidium* were less efficiently removed (19% average removal) than *Giardia* (38% removal) probably as a result of their smaller size and slower settlement rates.

A variety of factors can hinder the removal of parasites during primary sedimentation. Free-falling settling velocities of helminth eggs (<0.8 m/h excluding *Schistosoma* spp.) and protozoan (oo)cysts (<0.01 m/h at 20°C) are not much greater than the upward flow velocity of wastewater within tanks (conventionally between 0.5 and 1.5 m/h) which can prevent parasite eggs and (oo)cysts from settling out of suspension unless assisted. Adsorption,

TABLE 31.2 Estimated settling velocities of helminth eggs and protozoan (oo)cysts (in water 5–20°C)

Parasite	*Dimensions (μm) average*	*Specific gravity (g/cm³)*	*Settling velocity (m/h)*		
			5°C	15°C	20°C
Helminths					
Ascaris lumbricoides – fertile	60 × 45	1.11	0.51	0.48	0.77
Ascaris lumbricoides – infertile	90 × 40	1.2	2.09		3.15
Trichuris trichiura	50 × 22	1.15	0.48		0.73
Hookworm	60 × 40	1.055	0.26		0.39
Taenia saginata	40 × 30	1.23	0.48	0.83[a]	0.72
Schistosoma mansoni	150 × 55	1.18	5.24		7.87
Protozoa					
Cryptosporidium parvum	5.0 – 4.5	1.045	0.0014	0.0035[b]	0.002
Giardia lamblia	12.2 × 9.3	1.036	0.007		0.01
Entamoeba histolytica	12 – 10	1.07		0.018	

References: Thienpoint *et al.*, 1986; Fayer *et al.*, 1997; Medema *et al.*, 1998; David and Lindquist, 1982; Shuval *et al.*, 1986; Pike, 1990

[a] based on Taenia Sp gravity 1.3

[b] based on Crypto Sp gravity 1.08

aggregation or attachment to larger or denser particles may increase settling velocities and enhance parasite settling during clarification treatment processes. Whitmore and Robertson (1995) and Medema *et al.* (1998) present evidence that this could be an important removal mechanism for protozoan (oo)cysts. Settling velocities of *Cryptosporidium* oocysts during primary sedimentation in laboratory trials (0.022–0.028 m/h) exceeded rates in water suggesting attachment of oocysts to particles with greater settling. Protozoan (oo)cysts were also found to attach readily to wastewater particles in secondary sedimentation effluents, with maximum velocities of 0.25 m/h observed when (oo)cysts attached to the largest particles (>200 μm; Medema *et al.*, 1998). In contrast, no significant association has been found between suspended solids and parasite eggs numbers in wastewaters suggesting a lack of solid-associated removal for helminth eggs (Ayres, 1991; Stott *et al.*, 1999). Interestingly, Casson *et al.* (1990) also reported a lack of correlation between the removal of *Giardia* cysts and suspended solids during primary treatment.

Detention time can also affect the parasite removal performance of primary sedimentation. Sedimentation tanks are usually 1.5–4 m deep with detention times typically 1.5–2.5 h, although primary settling tanks preceding biological treatment may be designed for shorter detention periods of 0.5–1 h (Metcalf and Eddy, 1991; Gray, 1999). Parasite removal increases with retention time, with settlement times of around 2 h observed for effective removal of parasite eggs. Parasite removal rates after 1.5 h sedimentation of, on average, 52% for helminth eggs and 27% for protozoan cysts, were improved after 2 h detention to 74% and 67% for parasite eggs and cysts respectively (Bhaskaran, *et al.*, 1956; Panicker and Krishnamoorthi, 1978b). Longer retention times thus facilitate the removal of free-falling parasite eggs and also allow a greater degree of flocculation, which may assist in the settlement of protozoan (oo)cysts.

Sedimentation tanks are generally designed to remove all particles which have a terminal velocity equal or greater than the surface overflow rate ($m^3 \cdot m^{-2}$/day, i.e. m/day). Factors such as detention time, water velocity, turbulence (from sewage inflow), wind, and tank design can greatly influence sedimentation rates. Consequently, primary removal of parasites is highly variable. Protozoa removal rarely exceeds 50%, while removal rates for helminth eggs average 70%. Numbers of parasites in primary treated effluents can also be considerable, particularly for protozoa due to limited removal. In some studies protozoan (oo)cysts concentrations have exceeded 10^4 *Giardia* cysts/l and 10^5 *Cryptopsporidium* oocysts/l (Robertson *et al.*, 2000b; Medema and Schijven, 2001). Generally, protozoan (oo)cysts can average 2400/l (general range 17–6370) in comparison to around 10 helminth eggs/l (range 0–15) reported in primary effluents (Wang and Dunlop, 1954; Robertson *et al.*, 1995; Stott *et al.*, 1996; Chauret *et al.*, 1999; Medema and Schijven, 2001).

2.5.1.2 Chemically assisted sedimentation. Chemically assisted sedimentation, using coagulants (ferric chloride, alum or lime) to assist flocculation and enhance solids removal, can also facilitate concomitant enhanced removal of parasites from suspension, particularly protozoa. Pilot studies in Brazil, using lime treatment, reported removal rates of 99.99% for helminth eggs (4.01 log units; Taylor *et al.*, 1994). The viability of eggs was not assessed but the effect of lime treatment on inactivation of helminth eggs is time, temperature and pH dependent. Elevated pH (average of 11.5) may not alone have had a significant detrimental effect, as helminth eggs such as *Ascaris* can survive more than 120 days at pH > 12 (Storm *et al.*, 1981 cited Taylor *et al.*, 1994). In Mexico, pilot studies of advanced primary treatment (APT), using coagulants (typically aluminium sulphate), anionic polymer flocculants and a high rate sedimentation basin with a hydraulic retention time ranging from 15 to 50 min, demonstrated removal rates of 95% for helminth eggs (effluent contained around 1.2 helminth eggs/l) at influent wastewater concentrations of 23 eggs/l (Jimenez *et al.*, 2000). A removal efficiency of 71% was also reported for protozoan cysts with APT effluent containing almost 400 cysts/l when influent wastewater contained around 1300 cysts/l (Jimenez *et al.*, 2001). Comparable protozoan removal rates were reported by Payment *et al.* (2001) for a large operational works in Canada in which physicochemical treatment removed, on average, 76% *Giardia* cysts with effluent containing up to 350 cysts/l, when influent contained around 1500 cysts/l. Oocysts of *Cryptosporidium* were less effectively removed with removal rates averaging 27% and effluent containing around 10 oocysts/l. Higher removal rates for *Giardia* of 80–93% have been reported for a plant in the USA treating wastewater containing up to 13000 cysts/l (Mayer and Palmer, 1996).

Rates of helminth egg removal during chemical assisted sedimentation are comparable to non-assisted/plain sedimentation but are achieved with shorter detention periods. However, protozoan removal rates of, on average, 40% during non-assisted sedimentation (due mainly to removal of solids) may be improved to around 70% removal efficiency using chemical assisted sedimentation.

2.5.2 Septic tanks and anaerobic digesters

There is limited information available on the removal of parasites from septic tanks. Sedimentation is the most likely removal mechanism and thus, depending on tank conditions and extended retention times, septic tanks are likely to be relatively effective for the removal of parasites. Feachem *et al.* (1983) suggested that septic tanks with a normal retention time of 1–3 days would reduce protozoa and helminths by 0–2 $\log_{10}$ units. These rates are supported by Bhaskaran *et al.* (1956), who reported helminth removal rates of 99.4% in an experimental septic tank system with a 3-day retention and by Lloyd and Frederick (2000) of 99.95–100% helminth egg removal for an OXFAM emergency system of two tanks in series providing anaerobic settling conditions and a hydraulic retention time of 2–3 days. However, Bhaskaran *et al.* (1956) also reported concentrations of *Ascaris* and hookworm eggs of 30 eggs/l in effluent from an operational system. Further treatment of the septic tank effluent through a rock filter removed more than 70% of parasite eggs and reduced the concentration of eggs to 8 and 13 eggs/l for *Ascaris* and hookworm respectively. Chlorination of the rock filter effluent further reduced the egg concentration to around 4 eggs/l and improved overall parasite egg removal to 95%. Lloyd and Frederick (2000) also report disappointing rates of parasite egg removal of 84.5–89.5% in a double vaulted septic tank indicating poor sedimentation/retention time. However, the system received high numbers of eggs at 4×10^4–2×10^5 eggs/l, which were reduced to 2×10^4 after septic tank treatment and reduced further to 5×10^3 after treatment in two ponds (overall reduction of 97.5%).

The anaerobic digestion process is generally less effective in parasite removal. Paulino *et al.* (2001) found removal rates of between 59.7% and 93% for helminth eggs and protozoan cysts during anaerobic digestion. Black *et al.* (1982) reported that less than 25% of *Ascaris* eggs were destroyed in anaerobic digesters, but no effect was found on *Trichuris* or *Toxocara* eggs. Furthermore, the proportion of viable eggs was higher after treatment indicating that non-viable eggs were preferentially removed/destroyed in the process. Other studies have reported that mesophilic digestion does not significantly reduce numbers of protozoan (oo)cysts, although high densities of *Giardia* and *Cryptosporidium* of up to 11 800 cysts/100 g and 3810 oocysts/100 g can be present in sludge (Sykora *et al.*, 1991; Chauret *et al.*, 1999). However, several studies have shown that anaerobic digestion can significantly reduce parasite viability with the effect of anaerobic digestion on parasite inactivation related to temperature, exposure time and parasite species. Cram (1943) reported that the viability of *Ascaris* eggs was unaffected after 3 months mesophilic anaerobic digestion at 20–30°C, but was significantly reduced after 6 months when only 10% of *Ascaris* eggs were still viable. Hookworm eggs were more susceptible surviving only 40 days at 30°C. In contrast, thermophilic digestion (>50°C) can significantly reduce egg viability with 15 minutes at 55°C reportedly required for 99% inactivation of *Ascaris* eggs (Pike *et al.*, 1988 cited by Lewis-Jones and Winkler, 1991). Although mesophilic digestion appears to be ineffective in removing viable helminth eggs, the process has a greater effect on protozoan cyst inactivation. Cram (1943) reported that cysts of *Entamoeba histolytica* did not survive the digestion process. *Giardia* and *Cryptosporidium* (oo)cysts are also readily inactivated. Gavaghan *et al.* (1993) and Stadterman *et al.* (1995) reported that between 40 and 50% of *Giardia* and *Cryptospodium* (oo)cysts were inactivated (non-viable) after 4 hours exposure to mesophilic digestion at 37°C and more than 99% of (oo)cysts were non-viable after 24 h exposure in laboratory studies. However, Whitmore and Robertson (1995) reported that 10% of *Cryptosporidium* oocysts were still viable after 18 days exposure to mesophilic anaerobic digestion and suggested that it was likely that viable oocysts would still be present after 3 weeks.

Clearly, mesophilic anaerobic digestion may not completely remove parasite eggs and (oo)cysts from wastewaters. However, with detention times varying from 30 to 60 days for the standard rate anaerobic digestion process and up to 15 days for high rate digestion (Metcalf and Eddy, 1991), significant inactivation of parasites may occur, especially for protozoa and reduction in (oo)cyst viability.

2.5.3 *Upflow anaerobic sludge blanket (UASB)*

High rate anaerobic processes, such as UASB reactors, have been used as a pretreatment stage for a variety of processes, including activated sludge and waste stabilization ponds. In one study, Dixo *et al.* (1995) reported that a UASB digester reduced parasite egg concentrations in raw wastewater from 17 000 eggs/l to 1740 eggs/l (89.6% removal) after a theoretical hydraulic retention time of 7 hours. With a volumetric egg loading of 5.8×10^7/m^3/day, the UASB performance was better than that of an anaerobic pond loaded at 4.5×10^5/m^3/day (with a retention time of 10 h), which demonstrated a rate of removal of only 26.6%. The authors suggested that the majority of parasite eggs were removed in the UASB by filtration and aggregation as the influent flowed up through the sludge blanket. Sedimentation is unlikely to play an important role in removal as upflow velocities (0.6–0.9 m/h) used to keep the sludge blanket in suspension (Metcalf and Eddy, 1991) are higher than most parasite settling velocities. A high removal efficiency in the UASB reactor was attributed by Dixo *et al.* (1995) to a hydraulic flow regime approximating plug flow ensuring all influent was strained through the sludge blanket limiting short-circuiting. The eggs in the UASB effluent were sequentially removed in experimental waste stabilization ponds incorporating a secondary facultative pond (3.5 eggs/l) and completely removed after the first tertiary maturation pond. In another study, a UASB demonstrated removal rates of 70–82% for helminth eggs with effluent containing on average 18 eggs/l (range 14–21 eggs/l) after treatment of raw wastewater containing much lower parasite concentrations of between 47 and 121 eggs/l (Chernicharo *et al.*, 2001).

2.6 Secondary treatment

Primary sedimentation (assisted or non-assisted) is not an effective process for complete removal of parasite eggs or (oo)cysts. However, primary treatment combined with secondary treatment can significantly improve parasite removal efficiency especially for protozoa. Robertson *et al.* (2000b) reported that removal of *Giardia* and *Cryptosporidium* occurred predominantly during secondary treatment (>60% removal) rather than during primary treatment (<40% removal).

2.6.1 *Conventional wastewater treatment*

Secondary treatment generally consists of an aerobic biological treatment process for the removal of organic material and nutrients. Typical conventional secondary treatment systems utilize mechanical high rate processes such as activated sludge and trickling filters. Generally, conventional treatment systems are not particularly effective in parasite removal, especially for protozoa. In general, secondary treatment plants (STP) achieve 80–100% removal of helminth eggs (<1–60 eggs/l) and 5–99% removal of protozoan cysts (10–< 8000 (oo)cysts/l).

Comparisons between activated sludge and trickling filter plants suggest parasite removal is highly variable. Generally, activated sludge plants (including secondary sedimentation) demonstrate slightly higher removal rates than trickling filters. *Giardia* removal in activated sludge plants varied from 97 to 99.8% in comparison to 92–< 98.3% removal in trickling filter plants (Casson *et al.*, 2000; Wiandt *et al.*, 2000). Robertson *et al.* (2000b) reported more variable protozoan removal of 15–99% in activated sludge compared to 5–85% removal in trickling filters but no significant difference in efficacy of protozoan removal was found between the two processes. Helminth egg removal rates are comparable for activated sludge and trickling filter plants with overall removal performance typically 75–100% (Bhaskaran *et al.*, 1956; Panicker and Krishnamoorthi, 1978b; Rose *et al.*, 1996).

2.6.1.1 Activated sludge. The activated sludge process is one of the most commonly used for secondary wastewater treatment. As a suspended-growth biological treatment process, activated sludge utilizes a dense microbial culture in suspension to biodegrade organic material under aerobic conditions and form a biological floc for solid separation in the settling units. Diffused or mechanical aeration maintains the aerobic environment in the reactor. Typical retention times are 5–14 hours in conventional units rising to 24–72 in low rate systems (Gray, 1999).

Performance of activated sludge systems in removing parasites varies with parasite species. Bhaskaran *et al.* (1956) reported greater removal of *Ascaris* and *Trichuris* (96–97%) than for hookworm (88%). Similarly, higher rates of removal have been reported for *Giardia* cysts (>97%) than for cysts of *Entamoeba histolytica* (<85%) or *Cryptosporidium* oocysts (around 80%) (Panicker and Krishnamoorthi, 1978b; Feachem *et al.*, 1983; Madore *et al.*, 1987; Casson *et al.*, 1990; Mayer and Palmer, 1996).

Efficacy of removal reported may depend on whether effluents were sampled after aeration and sludge separation or after secondary sedimentation following activated sludge treatment. Parasite removal during activated sludge treatment occurs predominantly during secondary sedimentation with free and assisted sedimentation likely to be occurring in a similar manner to that during primary sedimentation.

Medema *et al.* (1998) reported that 70% of protozoan (oo)cysts attached to particles 0–200 μm in secondary settled sewage after 24 h of activated sludge treatment suggesting enhanced settling of protozoa during secondary sedimentation. Higher removal efficiencies for protozoan (oo)cysts have been reported after activated sludge treatment (including secondary sedimentation) than during primary sedimentation. At two activated sludge plants, primary sedimentation removed 4–42% protozoan (oo)cysts in comparison to 53–98% removal in activated sludge units (including secondary sedimentation) (Robertson *et al.*, 2000b). Overall the activated sludge plants removed <99% of protozoan (oo)cysts (primary and secondary sedimentation included). Similarly, Mayer and Palmer (1996) found that most protozoan removal occurred after activated sludge treatment (84.5–99.5%; lowest rates were that of *Cryptosporidium*); removal performance improved to 98.9–99.9% for the activated sludge plant (complete works including primary sedimentation). In contrast, both primary and secondary sedimentation may be important for helminth egg removal. In an activated sludge plant not preceded by primary sedimentation, removal rates of 75% helminth eggs and 93% protozoan cysts were reported. Where primary and secondary sedimentation was considered, activated sludge plants (complete works) removed 97–100% *Ascaris* eggs and 99.7% *Schistosoma mansoni* eggs (Rowan, 1964). Effective removal of *Ascaris* is most likely due to sedimentation. The high removal rates for *Schistosoma* may have been facilitated by egg hatching since *Schistosoma* eggs have been found to hatch readily in experimental wastewater tanks (Jones *et al.*, 1947).

Relatively high rates of protozoan removal during activated sludge treatment may be due to mixing enhancing attachment of (oo)cysts to flocs for subsequent solid assisted sedimentation in secondary clarifiers and/or flocculated material acting as a settling blanket during secondary sedimentation. In laboratory studies, Stadterman *et al.* (1995) found that *Cryptosporidium* removal was higher for activated sludge effluents compared to trickling filter after secondary sedimentation for 2.8 h (92% removal activated sludge) and 3.5 h (50% removal trickling filter). The authors suggested that protozoan removal during activated sludge occurred in the hindered (zone) settling region in which flocculated material forms a sludge blanket and filters suspended particles.

Despite relatively high removal rates, parasites can survive the activated sludge process and are usually not completely removed. Helminth eggs of *Taenia saginata*, *Ascaris*, *Trichuris*, hookworm and *Hymenolepis* spp. and protozoan (oo)cysts of *Giardia* spp., *Entamoeba* spp., *Cryptosporidium* spp. as well as microsporidia spores have all been reported in activated sludge effluents (Bhaskaran *et al.*, 1956; Panicker and Krishnamoorthi, 1978b; Mayer and Palmer, 1996; Dowd *et al.*, 1998). The concentration of parasites in activated sludge treated effluents is usually low at less than 5 eggs/l, though higher numbers of around 60 *Ascaris* eggs/l have been reported (Bhaskaran *et al.*, 1956). Numbers of protozoan (oo)cysts in activated sludge effluents can be high despite relatively high removal rates. Concentrations have ranged from 11 to 7600 (oo)cysts/l, while 0.05–< 20 (oo)cysts/l have been specifically reported in activated sludge effluents following secondary sedimentation (Madore *et al.*, 1987; Sykora *et al.*, 1991; Mayer and Palmer, 1996; Chauret *et al.*, 1999; Robertson *et al.*, 2000b).

The activated sludge process has little effect on the viability and infectivity of parasite eggs that pass through the system (Cram, 1943; Newton *et al.*, 1949), but may affect (oo)cysts. Robertson *et al.* (2000b) reported no reduction in viability for *Cryptosporidium* oocysts during activated sludge treatment. Oocyst viability after 7 days exposure was 68% in raw wastewaters and was comparable in primary sedimentation effluent (77% viability), activated sludge effluent (73% viability) and secondary sedimentation effluent (77% viability). In contrast, Carraro *et al.* (2000) reported a presumed reduction in protozoan (oo)cysts' viability from 66 to 70% in raw wastewater to 46–50% in filtered activated sludge effluent. They suggested that activated sludge treatment favours the excystation of (oo)cysts. Infectivity of *Cryptosporidium* oocysts was also reduced after laboratory simulation of activated sludge, although some oocysts retained infectivity after treatment to initiate infection (Villacorta-Martinez de Maturana *et al.*, 1992). Incidence of infection in mice was variable (5–6% and 74–95% in two trials) but intensity of infection was reduced to 82–99%.

In general, activated sludge systems demonstrate average removal rates of 92% and 87% for parasite eggs and cysts respectively (Panicker and Krishnamoorthi, 1978b; Madore *et al.*, 1987; Schwartzbrod *et al.*, 1989; Casson *et al.*, 1990; Rose *et al.*, 1996; Robertson *et al.*, 2000b). For the activated sludge process itself, 0–2 $\log_{10}$ units of removal may be expected for protozoa and helminths: activated sludge including primary and secondary sedimentation improves the removal performance to 0–3 $\log_{10}$ units of removal for parasites.

2.6.1.2 Trickling filter. Trickling filtration is an attached growth process which utilizes filter beds, consisting of highly permeable media to which microorganisms are attached, to treat wastewater which is percolated or trickled through the bed. Following filtration, effluent is treated in settling tanks where remaining solids and detached biofilm from the filter bed are separated from the treated wastewater. In the trickling filter process, biodegradation occurs due to microbial activities associated with the biofilm attached to the filter media.

Biological filter beds as a unit process generally contribute little to the removal of parasites from wastewaters and parasite eggs and cysts pass through into effluents. Panicker and Krishnamoorthi (1978b) reported low rates of helminth egg removal by trickling filter units ranging from 18 to 35% and trickling filter bed effluent concentrations ranging from 22 to 41 eggs/l. In contrast, higher removal rates of 98–100% for parasite eggs have been found in India after treatment of filter bed effluents in humus tanks (i.e. after secondary sedimentation): all hookworm eggs were removed and final effluents contained less than 1 *Ascaris* egg/l when raw wastewater contained 466 eggs/l (Bhaskaran *et al.*, 1956). Treatment of wastewater through a trickling filter plant in the Ukraine reduced *Ascaris* egg concentrations by 97%. Raw wastewater containing 60 eggs/l was reduced to 20 egg/l after primary sedimentation (67% removal) to 13 eggs/l after trickling filtration (35% removal) and finally to 2 eggs/l after secondary sedimentation (85% removal) (Vishneuskaya, 1938 cited Feachem *et al.*, 1983).

Protozoan (oo)cysts may not be so efficiently removed. In several trickling filter plants, rates of protozoan (oo)cyst removal averaged 84% after secondary sedimentation in comparison to higher removal rates of 91% for helminth eggs (Kott and Kott, 1967; Panicker and Krishnamoorthi, 1978b): *Ascaris* and *Trichuris* eggs were predominantly removed (95%) followed by hookworm (87%), *Giardia* (85%), *Hymenolepis nana* (84%) and *Entamoeba histolytica* (83%) (Panicker and Krishnamoorthi, 1978b).

Bukhari *et al.* (1997) reported highest removal rates for *Giardia* when wastewater treatment works included biological filtration (26–94%) in comparison to activated sludge treatment systems (26–71.7%). Trickling filter effluents contained <10–50 cysts/l, whereas activated sludge effluents contained <10–210 cysts/l. Feachem *et al.* (1983) reported similar removal rates of 71–94% for *Entamoeba histolytica* cysts in trickling filters. However, oocysts of *Cryptosporidium* may not be so favourably removed. Poor removal rates of 5% have been shown for trickling filters with effluents containing up to 1000 oocysts/l (and also up to 3800 *Giardia* cysts/l; Robertson *et al.*, 2000b).

Parasite removal mechanisms may be similar to those that contribute to the removal of suspended particulate matter in trickling filter beds, such as mechanical straining, sedimentation or inertial impaction, interception, adhesion or flocculation (Metcalf and Eddy, 1991). Biofilms on filter media may also provide attenuation sites for parasites as wastewater percolates through the trickling filter with grazing by predatory fauna associated with biofilms possibly contributing to removal and inactivation (Stott *et al.*, 2001). Panicker and Krishnamoorthi (1978b) proposed that predation by oligochaetes (which colonize the filter beds) may reduce pathogen survival.

Although the majority of parasites are removed throughout the treatment works up to secondary sedimentation, parasites are rarely completely removed, and parasite cysts and eggs can pass through trickling filters, with filter effluents potentially containing significant parasite concentrations (Nupen and Villiers, 1975; Bukhari *et al.*, 1997). Treated effluents may contain, on average, <5 helminth eggs /l (range 0–13) and up to 4000 protozoan (oo)cyts/l. Trickling filters are generally less effective in removing parasites than activated sludge processes. From the literature, trickling filters (including primary and secondary sedimentation) demonstrate average removal efficiencies of 92% for parasite eggs and 77% for protozoan (oo)cysts. In isolation, trickling filters may demonstrate 0–1 $\log_{10}$ units of reduction for helminth and protozoan (oo)cysts (Feachem *et al.*, 1983). Higher parasite removal rates are reported after secondary sedimentation of trickling filter effluent of around 0–2 $\log_{10}$ units.

2.6.1.3 Aerated lagoons. Fully aerated lagoons or ponds may be considered as a simple modification of the activated sludge process with screened rather than settled sewage aerated in ponds with no sludge return. Retention times are typically 2–6 days in these lagoons followed by further treatment in a settling pond (5–10 days retention) or secondary sedimentation tank (usually >2 hours retention) (Feachem *et al.*, 1983).

Parasites are not completely removed during treatment despite longer retention times in aerated lagoons than in conventional activated sludge plants. Parasite removal rates ranging from 50 to 100% have been reported for lagooning without secondary sedimentation (Panicker and Krishnamoorthi, 1978b, 1981). Average removal rates for helminth eggs were 88% and 85% for protozoan cysts. Smaller eggs with lower specific gravity and hence slower settling velocities (such as hookworm) are usually not removed as effectively as the larger and heavier eggs (such as *Ascaris*). Panicker and Krishnamoorthi (1981) reported 0–40 hookworm eggs/l, 0–20 *Ascaris* eggs/l and up to 96 *Entamoeba histolytica* cysts/l in lagoon effluents. Settling ponds following aerated lagoons are likely to enhance parasite removal due to extended retention times. Feachem *et al.* (1983) suggested aerated lagoons (including settling ponds) could achieve 1–3 log and 0–1 log removal for helminth and protozoa respectively.

2.6.1.4 Oxidation ditches. Oxidation ditches are also another modification of the activated sludge process. Screened sewage is aerated and circulated around an oval ditch (0.25–0.35 m/s). Oxidation ditches usually operate with long detention and solids retention times (ditch retention time of 1–3 days). Effluent is usually then treated in secondary sedimentation tanks with settled sludge returned to the oxidation ditch (Feachem *et al.*, 1983).

Panicker and Krishnamoorthi (1978b, 1981) reported that under optimum operation, oxidation ditches removed, on average, 92% (80–100%) of parasite ova and cysts; minimum rates of removal were 72% helminths and 60% protozoa, with effluents containing 4–14 helminth eggs/l and up to 120 cysts/l of *Entamoeba histolytica* (Panicker and Krishnamoorthi, 1981). Sedimentation was considered to be the principal removal process despite water turbulence with removal rates enhanced by floc settlement with parasites attached. Rates of removal may vary with helminth species and velocity of water flow.

2.7 Tertiary treatment

Wastewater treatment plants may use tertiary treatment measures for improving the quality of effluents from conventional treatment works. Processes utilized include sand filtration and disinfection. Effects of tertiary treatment on parasite removal are variable. Where tertiary treatment has been employed, the majority of parasite removal occurs in the preceding primary and secondary treatment processes. Overall, tertiary treatment plants (TTP) remove >99% helminth eggs (10 eggs/l) and 95–99% protozoa (0.1–< 20 (oo)cysts/l).

2.7.1 *Sand filtration*

Sand filtration enhances a reduction of parasites, although removal rates may be highly variable depending on grain size and hydraulic loading rates (Logan *et al.*, 2001). In general, sand filtration is effective in removing helminths with almost 100% removal reported for helminth eggs (Rose *et al.*, 1996) and specifically eggs of *Ascaris* and hookworm (Cram, 1943), *Schistosoma japonicum* (Jones *et al.*, 1947) and *Taenia saginata* (Newton *et al.*, 1949). Sand filters have demonstrated a great capacity for helminth egg removal. Schwartzbrod *et al.* (1989) reported 99% removal of helminth eggs (for secondary filtration) with effluent containing 10 eggs/l when high numbers of eggs (<900 eggs/l) were present in influent wastewaters. Sand filtration may also be effective in improving protozoan removal. Rose *et al.* (1996) reported that rapid sand filtration improved upon activated sludge removal by a further 7–10% with removal rates of >98% observed for the unit process. However, protozoa were still present in tertiary filtered effluents of around 4 (oo)cysts/l. Filtration is slightly more effective for *Giardia* removal than for *Cryptosporidium* (Rose *et al.*, 1996; Robertson *et al.*, 2000b).

Rates of removal for tertiary filtration units range from 97.9 to 99% for rapid sand filters (including coagulation/flocculation) and multimedia filtration (gravel, sand and carbon) with filtered effluents containing <1–18(oo)cysts/l and 0.1–2.4 (oo)cysts/l respectively (Madore *et al.*, 1987; Rose *et al.*, 1996; Carraro *et al.*, 2000). However, higher numbers of protozoa were reported in effluents from rapid sand filters treating activated sludge effluents in the UK (<10–435 (oo)cysts/l) and also from sandfilters treating trickling filter effluents (<10–60 (oo)cysts/l) (Bukhari *et al.*, 1997).

2.7.2 *Disinfection*

Parasitic eggs and (oo)cysts are extremely resistant to various forms of disinfection and few disinfectants are effective in inactivating parasites (Korich *et al.*, 1990). Eggs of *Ascaris lumbricoides* are very resistant to the effects of chemical disinfection (Krishnaswami and Post, 1968), which may be due to its relatively impermeable egg shell membrane (Wharton, 1980). However *Schistosoma* eggs are susceptible to chlorination at concentrations of 3.9–11 mg/l residual chlorine for 30 min (Jones and Hummel, 1947). Chlorination is particularly ineffective for removing protozoan (oo)cysts as free chlorine levels of up to 16 000 mg/l are required to inactivate completely *Cryptosporidium* oocysts (Smith, 1990), levels far in excess of dosages routinely used for wastewater disinfection of primary effluents (5–20 mg/l), secondary effluents (2–15 mg/l) and filtered secondary effluents (1–5 mg/l; Metcalf and Eddy, 1991). Chlorination is also less effective for *Cryptosporidium* removal than for *Giardia*. Rose *et al.* (1996) reported removal rates of 61% and 78% for *Cryptosporidium* and *Giardia*, respectively, after chlorination of filtered tertiary effluents. Disinfection of effluents was not effective since removal was only improved by 0.05–0.1% after chlorination and protozoan (oo)cysts were still present in 17–25% of samples of chlorinated effluents at levels of 1.8 *Crytposporidium* oocysts/l and 1.0 *Giardia* cysts/l. McHarry (1984) reported similar concentrations of 0.8 *Giardia* cysts/l (0.3–1.2 cysts/l) after tertiary chlorination of trickling filter effluents. Chlorination of filtered effluent following advanced primary treatment demonstrated similar rates of removal of 67% for protozoan cysts after 1 h exposure to 10–14 mg/l (Jimenez *et al.*, 2001). Extended exposure improved removal to 89% after 4 h, but chlorination was not capable of complete cyst removal and removal rates of 94% were observed after 48 h (17 cysts/l). Protozoan (oo)cysts have also been detected in final effluents following UV and chlorination (range <0.26–7 oocysts/l and 0.3–104 cysts/l) and ozone disinfection of tertiary effluents; the viability of cysts was not determined (Rose *et al.*, 1996; Oswald *et al.*, 2000; Liberti *et al.*, 2000; Wiandt *et al.*, 2000). Spores of human microsporidia species also present in tertiary treated filtered and chlorinated effluent indicate that microsporidia can survive wastewater treatment processes including disinfection (Dowd *et al.*, 1998). Microsporidia spores may also be resistant to disinfection inactivation since spore structure (Weber *et al.*, 1994 cited in Dowd *et al.*, 1998) is reportedly similar to that of *Giardia* spp. cysts and *Cryptosporidium* spp. oocysts, which are already known to be highly resistant to disinfection processes.

Ozone may offer greater potential for parasite inactivation than chlorination. Ozone, at initial levels of 1.1 mg/l, resulted in a 90% loss of viability for Cryptosporidium oocysts (Somiya *et al.*, 2000) and at levels of 4 mg/l is able to inactivate *Schistosoma mansoni* eggs (Mercado-Burgos *et al.*, 1975). However, ozone appears to have no effect on *Ascaris* eggs (Burleson and Pollard, 1976 cited by Reimers, 1989). UV irradiation has also shown potential for inactivating protozoan cysts by 90–99% depending on intensity and duration of exposure (Campbell *et al.*, 1995).

2.8 Natural wastewater treatment

Waste stabilization ponds (WSP) and constructed wetlands (CW) utilize physical and ecological processes to remove nutrients and pathogens. These natural wastewater treatment systems have been used as primary, secondary and tertiary systems to treat a variety of wastewaters worldwide.

2.8.1 Anaerobic lagoons

Anaerobic ponds are commonly used for the treatment of high strength wastewaters (e.g. agricultural) and can have high detention times of 20–50 days (Metcalf and Eddy, 1991). They are typically deep earthen ponds and are anaerobic throughout their depth. Partially clarified effluents are passed into other unit processes for further treatment. For domestic wastewater treatment, anaerobic ponds are usually used in series with a facultative pond followed by at least one maturation pond (see waste stabilization ponds below).

Information relating to parasite removal in agricultural anaerobic lagoons is limited. In one study, effective removal of protozoan (oo)cysts was reported for an anaerobic lagoon (60 days detention) treating dairy wastewaters (Karpiscak *et al.*, 2001). Removal rates of 99.99% for *Cryptosporidium* oocysts were found with almost complete removal of 2.22×10^4 oocysts/l in influent wastewaters reduced to <10 oocysts/l in lagoon effluents. Ayres *et al.* (1992) reported that helminth egg numbers can be substantially reduced in anaerobic ponds (up to 99% removal) with detention times of <7 days. Anaerobic lagoons with retention times typically in excess of 50 days thus offer enhanced opportunities for sedimentation removal of parasite eggs.

2.8.2 Waste stabilization ponds

A variety of studies has shown that waste stabilization ponds (WSP) can effectively remove parasites from wastewaters, although wide variability in removal performance has been reported in the literature. All types of WSP (anaerobic, facultative and maturation) have demonstrated high rates of parasite removal, with the principal removal mechanism considered to be sedimentation facilitated by the long hydraulic retention times of 1–5 + days (anaerobic ponds), 5–40 days (facultative ponds) and 3–10 days (maturation ponds) (Feachem *et al.*, 1983; Ayres *et al.*, 1992; Mara *et al.*, 1992).

In a study of single ponds in India, Veerannan (1977) reported parasite egg removal rates of 62.9%, 88.5% and 93.3% for three systems (unknown retention times) with effluent concentrations of 56, 61 and 181 eggs/l. Protozoan removal rates of 84–100% and 94–100% for cysts of *Entamoeba* spp. and *Giardia* spp. respectively have also been reported for single ponds in India (Panicker and Krishnamoorthi, 1978b). However, generally, a higher efficiency for parasite removal is found for a series of ponds rather than single ponds with comparable overall retention times, although removal rates may still vary with parasite species.

In India, complete removal of parasite eggs and cysts was reported for three ponds in series with a total retention time of 6–7 days, although hookworm larvae (non-infective) were found in final pond effluents (Lakshminarayana and Abdulappa, 1969; Arceivela, 1970 cited Bouhoum *et al.*, 2000). Similarly, parasite eggs were completely removed after 9 days retention in two ponds in Colombia (Madera *et al.*, 2002) and after 16 days retention in an experimental two pond system in Morocco; protozoan cysts were also completely removed (Bouhoum *et al.*, 2000).

Feachem *et al.* (1983) suggested that a well designed series of three or more ponds with a total retention time greater than 20 days would completely remove all nematode eggs and protozoan cysts. Grimason *et al.* (1993) suggested that for complete removal of protozoan cysts, a minimum retention period of 37 days was required. Mara and Silva (1986) reported that an 11-day, three-pond series comprising an anaerobic pond (1 day retention), and a facultative and maturation pond (each of 5 days retention) would remove most parasite eggs and achieve WHO (1989) reuse irrigation guidelines of ≤ 1 nematode egg/l, while Saqqar and Pescod (1992) suggested that 14 days were required to achieve the WHO quality criteria. Complete removal of parasites or attaining a final effluent quality of less than 1 parasite egg/l has been reported in several studies. Stott *et al.* (2002a) reported pond effluent compliance with WHO criteria after 10 days retention with reductions of parasite egg concentrations from around 1000 eggs/l in raw wastewater to 54 eggs/l after 5 days retention in an anaerobic pond (94.6% removal) and to 0.2 eggs/l after 5 days retention in a facultative pond (99.6% removal); eggs were not completely removed until after the second tertiary maturation pond (total of 20 days retention). Eggs of *Ascaris*, *Trichuris* and hookworm were found in anaerobic pond effluents, but only *Ascaris* eggs were present in effluent from facultative and the first maturation pond. Madera *et al.* (2002) reported that parasite eggs in raw wastewater (440 eggs/l) were completely removed after 12 days in a three-pond series incorporating an anaerobic pond (2.5 days), facultative pond (6.5 days) and maturation pond (3 days).

However, complete removal of parasites or achieving the nematode criterion is not guaranteed in multicelled systems. In a two-pond system in Sudan, only 90% of hookworm eggs were removed after 16.4 days total retention, although no *Ascaris* eggs were detected in the final pond effluent (Klutse and Baleux, 1995). Hookworm eggs have also been detected in final effluent from a 17-day retention five-pond series (Mara and Silva, 1986), and from a 23-day, three/five-pond series (33–690 eggs/l), although *Ascaris*, *Trichuris* and *Giardia* were effectively removed (Ellis *et al.*, 1993). Nematode eggs were also found (6 eggs/l) in final effluent from a six-pond series with a total retention time of 25 days (Saqqar and Pescod, 1991), while protozoan cysts have been found in final maturation pond effluent at concentrations ranging from 0.1 to 90 cysts/l from a series of three and four ponds with total retention times of 18–40 days (Grimason *et al.*, 1996b).

A number of studies have shown that parasite eggs and cysts are detected with relative frequency in primary pond effluents and to a lesser extent in secondary and tertiary effluents. Grimason *et al.* (1996a) studied parasite egg removal in 10 WSP systems in Kenya where raw wastewater contained between 18 and 133 *Ascaris* eggs/l. Eggs were detected in 30% of primary anaerobic pond effluents which contained 3.6–88.9 eggs/l after 1–6 days estimated hydraulic retention (85.2–93.3% removal), while effluent from an 18.4-day primary facultative pond contained 0.7 eggs/l (98.9% removal). Effluent from a secondary facultative pond contained 2 eggs/l after 9.3 days retention (97.7% removal) but no helminth eggs were detected in any maturation pond effluents from any of the systems. In contrast, primary anaerobic, primary and secondary facultative and first tertiary maturation pond effluents contained protozoan cysts from pond series in France and Kenya (Grimason *et al.*, 1996b). Cysts were reduced from around 4×10^3/l in raw wastewater to 225 cysts/l (primary anaerobic: 92.5% removal), 13 cysts/l (primary facultative: 99.7% removal), 2–120 cysts/l (secondary facultative: 46.7–84.6% removal), 70 cysts/l (first tertiary maturation: 41.7% removal) and 1–34 cysts/l (final tertiary maturation: 50–51.4% removal). Helminth eggs were completely removed in the Kenya pond system after the primary facultative pond unlike protozoan cysts, indicating that the removal of helminth eggs is not a reliable indicator for protozoan cyst removal. Pond performance data also show that primary ponds (anaerobic and facultative) remove the greatest numbers of protozoan cysts, smaller oocysts of *Cryptosporidium* are not reliably removed in comparison to *Giardia* cysts and low numbers of (oo)cysts in influents are difficult to remove completely especially in secondary and tertiary ponds. Smith *et al.* (1992, cited Grimason *et al.*, 1993) detected *Cryptosporidium* spp. oocysts and *Giardia* spp. cysts in raw wastewaters in Brazil (60 oocysts/l, 1.1×10^4 cysts/l) in effluents from primary facultative ponds (15 oocysts/l, 75% removal; 490 cysts/l, 95% removal) and in effluents from secondary maturation ponds (25 oocysts/l; 0 cysts/l, 100%). Similar observations have been found for parasite egg loading and performance. Stott *et al.* (2002a) reported that effluent from a 5-day maturation pond still contained 0.1 egg/l despite influent containing only 0.2 eggs/l.

Parasite numbers in pond effluents vary widely in the literature. Helminth eggs in effluents can typically range from 1.5 to 89 eggs/l in effluents from anaerobic ponds, 0–88 eggs/l in facultative ponds, 0–0.1 eggs/l in first and second maturation ponds and 0 eggs/l in final maturation pond effluents (Mara and Silva, 1986; Ouazzani *et al.*, 1995; Grimason *et al.*, 1996a; Stott *et al.*, 2002a). Grimason *et al.* (1993) analysed raw and treated wastewaters from 11 pond systems in Kenya and detected protozoan parasites in raw wastewaters (13–73 *Cryptosporidium* oocysts/l; 213–6213 *Giardia* cysts/l), anaerobic pond effluents (2.3–50 oocyst/l; 133–231 cysts/l), facultative pond effluents (3.3–17 oocysts/l; 1.4–193 cysts/l), first and second maturation pond effluents (2.5–4.5 oocysts/l; 3.1–177 cysts/l) and final maturation pond effluents (0 oocysts; 40–50 cysts/l from one site).

The sequential reduction in parasite numbers with subsequent pond effluents indicates that removal (presumed by sedimentation) and/or destruction of parasite eggs and cysts is related to the cumulative hydraulic retention time. The majority of eggs and cysts are removed in anaerobic ponds, although lower percentage rates of removal are usually seen probably due to shorter hydraulic retention times in comparison to facultative/maturation ponds. Ayres *et al.* (1992) reported egg removal in anaerobic ponds with 1–4 days retention of 77–98%. In comparison, facultative/maturation ponds with 3–18 days retention demonstrated parasite egg removal rates of 83–99.99%. The primary function of maturation ponds is usually pathogen removal. However, removal rates for parasite pathogens may vary considerably between tertiary maturation pond systems with removal rates for different parasite species ranging between 36 and 100% (Maynard *et al.*, 1999).

Ayres *et al.* (1992) developed a simple empirical model for predicting nematode egg removal as a function of hydraulic retention time (HRT). The lower 95% confidence limits of the model can be used to determine the number and retention times of ponds needed to reduce the parasite egg level in effluents to ≤1 nematode egg/l for restricted irrigation based on initial concentrations of nematode eggs in raw wastewaters. There are presently no design guidelines for protozoan removal.

$$\text{Nematode Egg Removal \%} = 100\,[1 - 0.41\exp(-0.49\,\text{HRT} + 0.0085\,\text{HRT}^2)]$$ (Ayres *et al.*, 1992)

Using the model above, Ayres *et al.* (1992) predicted 1–2 day retention in ponds would achieve 75–84% egg removal, 3–5 days retention 89.8–95.6% removal, ≤10 days up to 99.3% removal and ≤20 days retention, up to 99.9% egg removal. Overall, 1–4 log units of reduction for protozoan and helminth parasites can be achieved in a well-designed series of WSP.

Multiple pond systems with sequential environments demonstrate higher parasite removal performance than conventional treatment systems (see Section 2.6.1). Parasite removal in ponds is attributed primarily to sedimentation facilitated by long HRT. Gravitational settling of free-falling helminth eggs with relatively fast settling velocities favours their partition from liquid to sludge (see

Section 2.5.1). The predominant removal of protozoa in primary ponds suggests that the principal removal mechanism for protozoan (oo)cysts is the adsorption/attachment of (oo)cysts onto settleable solids (Grimason *et al.*, 1993). Relatively high rates of removal can be achieved for eggs of *Ascaris* and *Trichuris* and sometimes for *Giardia* cysts, but WSP are less successful in removing hookworm eggs and *Cryptosporidium* oocysts. The removal efficiency of ponds can be affected by various factors retarding the sedimentation process and settling velocity of parasites. Ellis *et al.* (1993) reported that resuspension of parasite eggs due to gas production and floating scum passing through the ponds was a likely factor affecting the removal of hookworm eggs in their pond system. However, Lloyd and Frederick (2000) also implicate the use of aerators in the facultative ponds and pronounced short-circuiting (4.5 days hydraulic retention time compared to nominal 23 days) as a factor for the under-performance of the ponds. Yanez *et al.* (1980) also suggested that temperature inversions within ponds could affect protozoan cyst removal. Thermal stratification and accumulation of sludge are also thought to increase the likelihood of short-circuiting, thereby reducing the hydraulic retention time and facilitating the transport of parasites through the ponds (Bartone, 1985). Even if high sedimentation rates are achieved, parasite eggs can survive for at least 2–7 years in sludge (O'Donnell *et al.*, 1984; Schwartzbrod *et al.*, 1987; Nelson and Jimenez, 2000) and thus disturbance of sediments such as high inflow from storm events or wildlife, can potentially release high concentrations of parasites back into suspension.

Optimizing parasite reduction in pond effluents thus requires strategies to maximize sedimentation and minimize parasite resuspension and transport. WSP can achieve very high rates of parasite removal providing that hydraulic retention times are attained and that measures are taken to prevent hydraulic short-circuiting which can cause WSP to fail microbiological quality criteria for wastewater irrigation. Ayres *et al.* (1992) recommend that a larger number of smaller ponds in series is used as a design approach to removing parasite eggs in WSP in order to minimize hydraulic short-circuiting effects and maximize removal. Other simple approaches suggested include the careful use of baffles to prevent breakthrough of solids and retain floatable scum and to increase hydraulic retention time (Shilton and Harrison, 2002) and non-shading pond windbreaks to reduce wind effects and short-circuiting in maturation ponds (Lloyd *et al.*, 2002).

2.8.3 Hyacinth and duckweed ponds

Macrophyte ponds have been utilized for wastewater treatment and show relatively effective removal of parasite eggs and cysts though removal performance is likely related to retention times of up to 1 week.

Mandi (1994) and Ouazanni *et al.* (1995) reported complete removal of parasite eggs in water hyacinth ponds with 7 days retention in Morocco when influent contained between 0 and 120 eggs/l. Performance compared favourably with that of facultative ponds with 50 days retention (also 100% removal) and was better than that for a 0.4 day retention anaerobic pond achieving 79–100% egg removal.

The removal of protozoa was size related in duckweed ponds treating secondary effluents (Falabi *et al.*, 2002). Cysts of *Giardia* were removed more efficiently than *Cryptosporidium* oocysts from wastewater with removal rates of 98% and 89% reported for a 9-day detention *Lemna gibba* covered pond. Influent concentrations were reduced from 16 *Giardia* cysts/l and 1.6 *Cryptosporidium* oocysts/l to 0.4 cysts/l and 0.2 oocysts/l respectively.

2.8.4 Advanced pond systems and high rate algal ponds

Advanced pond systems (APS) are an adaptation of waste stabilization ponds and optimize conditions for algal photosynthetic oxygen production to enhance treatment efficiency and reduce pond area. APS usually incorporate an advanced facultative pond followed by a high rate algal pond (HRAP) consisting of a shallow raceway around which pond contents are circulated by a paddle wheel to maintain aerobic conditions (1–3 day retention), an algal settling pond (2–3 day retention) and two to three maturation ponds (3–5 day retention) for further disinfection. Variations of APS have used anaerobic digesters or anaerobic ponds instead of facultative ponds for smaller footprint systems and constructed wetlands instead of maturation ponds.

HRAP systems can demonstrate high rates of parasite removal especially when preceded by settling ponds/basins. El Hamouri *et al.* (1994) reported a removal of 99.3% for helminth eggs in a 36 m long HRAP system (including 24 settling) in Morroco. However, the majority of helminth eggs (96.4%) were removed from raw wastewater (113 eggs/l) in the preceding settling basin and, in isolation, the HRAP achieved a removal of 80% with effluent still containing 0.8 eggs/l despite low influent numbers from the settling basin of 4 eggs/l. Thus the HRAP improved egg removal by around 3%. A variety of eggs detected in the raw wastewater, including *Ascaris* and *Trichuris* (15% and 6% occurrence), were completely removed by sedimentation, but eggs of *Toxocara* spp. and *Enterobius* spp. passed through the system and were also present in the HRAP effluent. In a later study, a 100% removal rate was reported for a HRAP with an increased raceway of 788 m and treating effluent from a 4-day anaerobic lagoon that contained only 2 eggs/l (El-Hamouri *et al.*, 1995). Information relating to helminth egg removal in HRAP receiving high parasite loading rates and protozoan removal in these systems is still required in order to assess the capability of HRAP systems for parasite removal.

Egg removal processes were not investigated by the authors, though sedimentation and die-off were suggested as egg removal mechanisms. Some sedimentation of helminth eggs may occur in the HRAP despite continuous mixing of the pond contents and a surface water flow rate of 0.15 m/s. Settlement of helminth eggs (and protozoan cysts) has been found in wastewater flowing at 0.7 m/s along an 2 km open channel in Morocco (Bouhoum *et al.*, 1997). Parasites were reduced from 145 eggs/l and 1.6×10^5 cysts/l to 33.4 egg/l and 2.7×10^4 cysts/l after 2 km with greatest reductions found for *Ascaris* eggs (85%). Sediment samples showed that *Ascaris* eggs settled first at 80 m, but *Enterobius* eggs remained longer in suspension and were not detected until 560 m. Sediment samples at 80 m contained 9.6 eggs and 1.7×10^4 cysts/g yet, at 2 km, densities in sediments increased to 78 eggs and 1.4×10^5 cysts/g.

Parasite viability in HRAP ponds was investigated by Araki *et al.* (2000, 2001). Viability of helminth eggs (*Parascaris equorum*: horse roundworm) was reduced by 60% and 90% after 4 and 10 days exposure within HRAP ponds respectively. However, the authors also reported a 25% loss of egg viability in sterile water control suspensions after 10 days at 4°C (presumably refrigerated and in the dark) which suggests that the pond inactivation results for helminth eggs may be overestimated. Infectivity of *Cryptosporidium* oocysts retained in HRAP ponds was also significantly reduced by 97–99.9% after 3 and 10 days exposure, but some oocysts remained viable and infective since prevalence rates of infection in mice were 40% and 20% respectively. The authors implicated the effect of pH, ammonia and/or light as factors affecting inactivation (predation factors were excluded from the experimental studies). Ammonia levels in HRAP wastewaters (1.2–1.7N-NH_4 mg/l reported) are unlikely to have a deleterious effect on parasite survival. Biological wastewater treatment had no effect on *Cryptosporidium* oocyst viability (Robertson *et al.*, 2000b), although ammonia concentrations in effluents were likely to range up to 20 mg/l. Furthermore, very high levels of >1000 mg/l (levels far in excess of those reported in HRAP wastewater) are reportedly required for >99.9% inactivation of *Cryptosporidium* oocysts after a similar 8 days exposure (Jenkins *et al.*, 1998).

2.8.5 Constructed wetlands (CW)

Artificially constructed wetlands (or reedbeds) are of two main types – vertical and horizontal. Horizontal systems may be further classified depending on the pathway of water flow as surface and subsurface flow systems. Vertical wetlands can be used as primary treatment processes, whereas horizontal systems are commonly used for secondary or tertiary treatment. Detention times in wetland systems are comparatively shorter than in WSP and are typically hours to several days.

Constructed wetlands have also demonstrated effective removal of parasites with horizontal subsurface flow (HSSF) gravel based systems usually more effective in parasite removal than equivalent soil based wetland systems (Rivera *et al.*, 1995). Greater rates of parasite removal are typically found for subsurface flow than surface flow wetlands. Gerba *et al.* (1999) reported protozoan removal rates of 69–88% in subsurface flow (SSF) wetlands compared to 58–73% removal in surface flow (SF) systems each with a retention of less than 4 days; SSF wetland effluents contained, on average, 0.01–0.07 (oo)cysts/l and SF effluents 4.9–16 (oo)cysts/l. Higher removal performance of subsurface systems was attributed to greater surface area for adsorption and filtration/straining. Parasite removal in wetland systems may also be related to size since cysts of *Giardia* are more effectively removed than smaller *Cryptosporidium* oocysts. Thurston *et al.* (2001) reported protozoan removal in HSSF receiving tertiary treated effluents of 87.8% for *Giardia* and lower rates of 64.2% removal for *Cryptosporidium*. Helminth removal in HSSF reedbeds in Morocco (1–4 h retention) was related to bed length (Mandi *et al.*, 1996) and season (Mandi *et al.*, 1998). Removal performance was greater with longer bed lengths of 50 m (95%) than 20 m (71%) and higher for all beds 20–50 m in the hot season (89.5%) than in the cold season (77.5%). In Egypt, HSSF reedbeds demonstrated helminth egg removal rates of 90% (0.02 eggs/l) in 50 m beds and 100% in 100 m beds also indicating that parasite egg removal is improved with increasing bed length and concomitant reduction in hydraulic loading (Stott *et al.*, 2002a). When challenged with artificially high numbers of eggs, equivalent to 100–500 eggs/l and a daily loading rate of $0.5–4 \times 10^4$ eggs/m^2/day, 100 m HSSF beds demonstrated a substantial capacity (up to 100%) for helminth egg removal (Stott *et al.*, 1999). All eggs recovered from sediments were considered to be non-viable. Parasite cysts were less effectively removed and a few cysts of *Entamoeba* spp. were detected in 100 m bed effluents (Stott *et al.*, 1997). Generally, subsurface wetlands can achieve between 0.4 and 3 log units of removal in horizontal flow systems. There is currently no information available on parasite removal in vertical flow constructed wetland systems.

Sedimentation is considered to be one of the principal egg removal mechanisms in pond systems with long retention times of days, yet the role of this or other physical processes, such as filtration, in the removal of parasite eggs in CW with a short retention time of <1 day and horizontal water flow is unclear. Parasite eggs and (oo)cysts might be removed in wetland systems by entrapment and sedimentation within the bed and root matrix as wastewater percolates through the bed substrate. Wastewater and sediment profile analysis along 100 m reedbeds in Egypt has shown that helminth egg removal increases with distance along the bed with the majority of eggs removed within the first 10 to 25 m of the bed (Stott *et al.*, 1999). Particulate matter has also been found to accumulate predominantly within the first 10 to 20 m of the 100 m reed beds (Williams *et al.*, 1995) suggesting that parasite eggs (and cysts) in subsurface flow systems might be removed by mechanical straining and sedimentation as well as by similar processes described previously for particulate matter removal in trickling filters (see Section 2.6.1.2). Attenuation and entrapment within wetland biofilms with subsequent grazing by predatory fauna may further improve removal and inactivation, especially of protozoan (oo)cysts (Stott *et al.*, 2001).

The influence of vegetation in constructed wetlands on parasite removal is unclear. In surface flow systems, removal of protozoan parasites was improved in planted beds (95% removal) compared to unplanted beds (88–92% removal) for both *Giardia* and *Cryptosporidium* (Quiñonez-Diaz *et al.*, 2001). In soil based subsurface systems in Mexico, planted soil mesocosms showed a 50% improvement in helminth egg removal compared to unplanted mesocosms (Rivera *et al.*, 1995). However, in a field-scale gravel based wetland trial in Egypt, vegetation had no apparent effect on helminth egg removal (Stott *et al.*, 1996). Vegetation in surface flow systems probably provides quiescent zones where sedimentation of parasites can occur. While the presence of hydrophytes in subsurface flow gravel beds may not be a significant factor for the removal of parasites, planted beds generally perform better than unplanted beds in removing other pathogens (Gersberg *et al.*, 1987).

Rates of removal in CW compare favourably with that in ponds and CW can provide complementary technology to WSP. CW usually have relative short hydraulic retention times (typically hours to a few days) compared to 20–30 days in ponds. Furthermore, the efficiency of sediment retention in CW is higher and thus there is less chance of resuspension. Slow water velocities and effective entrapment within short distances facilitate high rates of removal. Multicomponent hybrid systems may offer further opportunities for parasite removal and inactivation. Hybrid systems incorporating aquatic ponds and subsurface flow wetlands gave the greatest removal rates of *Cryptosporidium* (95.3%) and *Giardia* (99.94%; Gerba *et al.*, 1999) and hybrid systems can exceed rates of removal for ponds with similar retention times (Bavor *et al.*, 1987).

2.8.6 *Overland flow/land application*

Overland flow systems have been used as a polishing process for conventional and natural treatment effluents. Information on parasite removal is limited. An overland flow system consisting of three slopes (each 25 m long by 3 m wide and gradient of 4%) and receiving UASB effluent, demonstrated complete removal of parasite eggs when receiving a constant flow of influent containing 14 eggs/l (Chernicharo *et al.*, 2001). In another trial, slope removal efficiency was 99% under intermittent influent flow conditions with all species of helminths removed except hookworm. Slope effluent contained, on average, 0.2 eggs/l (maximum 2 eggs/l) with hookworm eggs detected in 18.5% of effluent samples. The authors conclude that the overland flow process with application rates of 0.4–0.5 $m^3/m/h$ is a satisfactory post-treatment system capable of producing effluents with low concentrations of helminth eggs, although the slope length should be extended to at least 35 m for intermittent inflow operations.

3 PARASITE SURVIVAL IN WASTEWATER TREATMENT SYSTEMS

The survival of parasites in wastewaters depends on exposure to prevailing abiotic and biotic variables. Factors that affect survival include parasite species (structure of egg/(oo)cyst cell wall confers protection), stage of development (developed infective stage may be more resilient than undeveloped stages), temperature (longer survival at lower temperatures), humidity (longer survival at high humidity), pH (extremes reduce survival), sunlight (solar radiation can be detrimental to survival depending on exposure duration and intensity), protection by vegetation (from sunlight mediated disinfection), and presence of predatory fauna. Although survival of parasites can vary widely, typical survival times for protozoa in sewage are less than 30 days, while eggs of helminths (especially *Ascaris*) can reportedly survive for many months in manures, raw wastewaters and sludges (Feachem *et al.*, 1983; Schwartzbrod *et al.*, 1987). Protozoan (oo)cysts and helminth eggs are thus both likely to survive beyond typical retention times for treatment processes.

Sedimentation processes within conventional and natural wastewater treatment plants are unlikely to have much effect on parasite survival. Ambient temperatures in wastewaters (typically 10–25°C) and pH are unlikely to have a detrimental effect on survival of parasite eggs or cysts. *Cryptosporidium* oocysts can retain viability and infectivity after freezing (−15°C) and at temperatures up to 68°C (Fayer, 1994; Fayer and Nerad, 1996). Helminth eggs can also tolerate similar temperature ranges of −27°C to <65°C (Cram, 1943; Feachem *et al.*, 1983). Since helminth eggs and protozoan cysts can survive in wastewaters for at least 1–2 weeks without loss of viability (Feachem *et al.*, 1983; Robertson *et al.*, 2000b), a retention time of several hours to days within sedimentation tanks will thus have a very limited effect on inactivation for parasites still in suspension. Anaerobic environments may also prolong survival of parasites. Parasites that are removed from suspension can remain in sludges for at least 1 year with very little loss of viability (Gaspard *et al.*, 1995; Gibbs *et al.*, 1995). The development of *Ascaris* eggs is suppressed at low oxygen concentrations (Brown, 1928) and at low temperatures of 4°C at least 50% of *Ascaris* eggs can remain viable after 33 months in sludge (O'Donnell *et al.*, 1984). Therefore, resuspension of sludge from sedimentation reactors and ponds can also release viable parasites back into effluents.

Conventional secondary biological treatment processes do not inactivate parasites. Field studies have shown that trickling filters, intermittent sand filters and activated sludge processes can promote embryonation of helminth eggs and the hatching of hookworm and *Schistosoma* eggs (Cram, 1943; Jones *et al.*, 1947). Protozoan (oo)cysts are also unlikely to be affected during treatment. Robertson *et al.* (2000b) reported no reduction in *Cryptosporidium* oocyst viability during aerobic biological treatment (activated sludge and trickling filters); between 72 and 77% of oocysts remained viable in raw, primary settled and secondary effluents. Robertson *et al.* (1999) also reported higher proportions of viable *Giardia* cysts in treated effluents (>20%) than in influents (<10%), indicating that wastewater treatment processes may selectively remove non-viable cysts.

Treatment in natural systems such as waste stabilization ponds (WSP) may have a limited effect on parasite inactivation, depending on the type and species of parasite and type of pond. Viable eggs of *Ascaris* have been reported in WSP effluents and may embryonate in aerobic conditions (Ayres, 1991). Detection of viable eggs in pond sludges by Nelson and Jimenez (2000) shows that eggs removed from influent wastewaters by WSP are not eliminated and that pond sludges may serve as parasite reservoirs. Hookworm and *Schistosoma* eggs may also survive anaerobic WSP conditions, although cellular development and hatching may be significantly reduced or inhibited. Hookworm eggs can survive anaerobic conditions for 2 weeks without loss of hatchability (Ayres, 1991) and can pass into aerobic ponds where they may hatch into non-infective free-living forms, although development to infective stage larvae in either anaerobic or aerobic environments was not found in experimental studies. Similarly, *Schistosoma* eggs can hatch in facultative or maturation ponds (Kawata and Kluse, 1966), although ecological factors within facultative ponds can reportedly affect hatchability of *Schistosoma* eggs and infectivity of miracidia (Bunnag *et al.*, 1978). The maximum survival time of hatched *Schistosoma* miracidia is less than 10 hours, which is less than typical retention times in ponds (Kawata and Kluse, 1966). Ponds may therefore provide an effective barrier for schistosomiasis transmission. Information relating to protozoan viability in WSP wastewaters is extremely limited. Sporulated (i.e. potentially infective) oocyts of *Cyclospora cayetanensis* detected in 18% of samples collected from primary ponds indicate that WSP environments do not have a significant detrimental effect on protozoan survival either on development or on the infective stage (Sturbaum *et al.*, 1998). However, information relating to the viability of more ubiquitous protozoa in wastewaters such as *Giardia* and *Cryptosporidium* is needed.

Algae-based shallow pond systems such as HRAP, however, seem to have a deleterious effect on the survival of protozoan oocysts and perhaps to a lesser degree on nematode eggs. Experimental studies suggest that parasite inactivation in these systems is higher than in conventional treatment systems and may be rapid with exposure of only a few days required for significant reductions in protozoan oocyst infectivity. Further work is clearly needed to elucidate the contribution of physicochemical conditions within HRAP/APS systems and the consequences of stirring pond contents on parasite removal and inactivation. It is interesting to note that activated sludge which also utilizes mixing processes, can remove protozoan (oo)cysts but may not appear to have any effect on parasite viability. All protozoan oocysts in activated sludge effluent from a model system were viable despite removal rates of 80% from influent wastewaters (Villacorta-Martinez de Maturana *et al.*, 1992). However, it has been found that dead oocysts show greater adhesion to each other and to debris (Bukhari and Smith, 1995), which thus may facilitate the removal of non-viable oocysts during secondary sedimentation, while the passage of viable oocysts through the treatment process is unaffected. In HRAP, mixing of the water column and shallow pond depths may facilitate sunlight mediated disinfection processes on parasite inactivation and die-off so that environmentally 'stressed' cells may be more susceptible to removal and destruction. Curtis *et al.* (1992) reported that light mediated destruction of faecal coliforms in WSP was dependent on oxygen and sensitive to elevated pH, physicochemical conditions that are found within APS and WSP systems. Ayres (1991) reported depth dependent effects on hookworm development and larvae die off in WSP with faster embryonation of eggs and significantly shorter survival of larvae at 10–20 cm below water surface of a maturation pond compared to depths of 0.5–1.1 m. pH, temperature and dissolved oxygen varied near the surface (pH 8–9; 6–20 mg/l DO; 23.5–26°C) but were constant and lower at deeper depths (pH 8; 2–6 mg/l DO; 23.5°C).

The effect of UV radiation and sunlight exposure on helminth inactivation can depend on the stage of egg development exposed. Undeveloped eggs and larvae are more susceptible to the effects of UV radiation and sunlight than fully developed eggs (Spindler, 1940; Tromba, 1978). Elevated pH within HRAP (and facultative ponds) is unlikely to have much effect on helminth egg inactivation and removal. *Ascaris* egg development is unaffected in alkaline pH <12 (Lewis-Jones and Winkler, 1991) and eggs of hookworm can hatch and develop to infective stages over the pH range of 4.6–9.4 (Udonsi and Atata, 1987). The detrimental effects associated with photo-oxidation are likely to be limited for most helminth eggs.

Protozoa survive differently under similar conditions and appear more susceptible to environmental factors such as pH and solar radiation. Sunlight can significantly inactivate *Cryptosporidium* oocysts in comparison to dark environments (Huffman *et al.*, 2000), and UV irradiation and high pH can have a significant impact on viability of *Cryptosporidium* oocysts (Robertson *et al.*, 1992; Campbell *et al.*, 1995). Jenkins *et al.* (1998) suggested that pH indirectly affects oocyst survival by affecting oocyst wall permeability, the effect of which is time dependent. Elevated pH may therefore facilitate synergistic inactivation by other factors such as sunlight and UV exposure. Wiandt *et al.* (1995) reported seasonal variation in protozoan removal in WSP with highest removals of *Giardia* in spring and summer compared to relatively poor removal rates in winter, possibly due to sunlight exposure and inactivation. Further work is needed on elucidating the contribution of light (and dark) removal and die-off processes for parasites, especially protozoa and the influence of environmental factors, wastewater characteristics and process operation in mediating these effects.

Other processes likely to contribute to the removal and inactivation of parasites during wastewater treatment are predation and microbial attack. Ovicidal fungi have been shown to penetrate and destroy helminth eggs (Lysek and Bacovsky, 1979; Sobenina, 1978), while ingestion by insects (Miller *et al.*, 1961) and gastropods (Asitinskaya, 1979) can contribute to egg removal and inactivation. Predation by free-living protozoa and metazoa (surface associated and suspension feeders) may also play a role in the removal of protozoan (oo)cysts (Stott *et al.*, 2002b) where sedimentation and sunlight exposure are similar to that found for virus removal in biological wastewater treatment systems (Kim *et al.*, 1996). The ecological significance of invertebrates and microbial communities and their grazing activities requires further investigation to determine the role of predation in removal and inactivation of parasites.

4 CONCLUSIONS

Parasite removal can vary considerably in wastewater treatment plants (WTP). Eggs and (oo)cysts of helminth and protozoan parasites have been detected in final effluents from a variety of WTP despite advanced levels of treatment. Factors affecting removal include:

1. the *type* and *level of treatment* (conventional and natural systems; primary, secondary, tertiary treatment)
2. *unit processes* (e.g. activated sludge, trickling filters, WSP and CW)
3. *parasite type and species* (helminth or protozoa; *Ascaris*, hookworm, *Giardia*, *Cryptosporidium* etc.)
4. *parasite loading* in influent wastewaters (Table 31.3).

Natural wastewater treatment systems with longer retention times are generally more effective at parasite removal than 'conventional' (mechanical) systems. In conventional WTP, removal of helminth eggs occurs predominantly during primary treatment. In comparison, primary sedimentation is relatively ineffective in removing protozoan (oo)cysts and (oo)cysts are preferentially removed during secondary treatment (including secondary clarification). Tertiary treatment plants, which usually incorporate filtration and chlorination, may improve removal but, unless influent concentrations are low, they will not completely remove parasites or produce effluents satisfying reuse criteria (0.1–1 egg/l; 1 cyst/40 l; Rose and Gerba, 1990; Blumenthal *et al.*, 2000). Generally, eggs of *Ascaris* and *Giardia* cysts are more readily removed in WTP than other helminth and protozoan species such as hookworm and *Cryptosporidium*, respectively.

The use of multiple stages with long retention times (days) and exposure to different environmental conditions (pH, DO, sunlight exposure) in natural treatment systems, such as ponds and wetlands, appears to be more effective in removing and inactivating parasites. In waste stabilization ponds, reduction of helminth and protozoan parasites numbers occurs principally in primary ponds (typically anaerobic). Removal can be improved for helminths and, to a lesser degree, for protozoa, in sequential ponds. Parasite removal in constructed wetlands is also size related and wetland systems are especially effective for helminth egg removal. Increasing bed length and reducing hydraulic loading improves removal performance.

Although a variety of processes has been identified in the removal of bacterial and viral pathogens during wastewater treatment, there is a lack of detailed studies of parasite removal and inactivation mechanisms, and qualitative assessment of the relative importance of each. Parasite removal during wastewater treatment is likely due to a combination of physical, chemical and biological factors. Key removal mechanisms include:

1. direct sedimentation
2. solid assisted sedimentation and entrapment in flocs
3. 'filtration' through reactor substrates and attached biofilm or biological flocs
4. sunlight inactivation
5. grazing by predatory protozoa, metazoa and invertebrates.

Short-circuiting and buoyant sludge particles (e.g. in anaerobic or overloaded facultative ponds) will reduce removal rates and lead to passage of parasites into effluents. Enhancing opportunities for sunlight-mediated disinfection and natural 'die-off', e.g. by mixing and/or shallow reactors may offer improved inactivation, especially for protozoa.

Surveys indicate that 25–100% of wastewater effluents may contain protozoan (oo)cysts with densities ranging from <1 to 10^3 (oo)cysts/l. Protozoan concentrations in treated effluents indicate that further wastewater treatment may be necessary for effluent discharge into receiving waters sourced for drinking water, especially during low flow in

TABLE 31.3 Summary of reported parasite removal and effects of wastewater treatment[a]

Unit process	*Parasite removal (%) mean and (range)*		*Effect*
	Helminth eggs	*Protozoan (oo)cysts*	
Sedimentation:			Removal depends on operating conditions;
Plain	73 (35–96)	38 (4–68)	H & P: No parasite inactivation; H & P: Inactivation may
Chemical assisted	97 (95– >99.99)	70 (27–93)	depend on elevated temperatures, pH, and time
Anaerobic digestion Mesophilic	60 (5–90)	31 (0–50)	Removal and inactivation depend on time, temperature and parasite species; H: limited effect: 52% inactivation after 20 days; > 90% inactivation after 6 month; P: Some inactivation: 40–50% inactivation after 4 h, 99.9% inactivation after 18–24 h; < 10% still viable after 18 days
Septic tanks	94.7 (85–100)		Anaerobic conditions may prolong survival; H: no effect
Imhoff tanks	97		
UASB	80 (70–90)		
Trickling filter (Complete works)	60 (5–90)	77 (5–93)	No effect on inactivation; H: promotes egg development, embryonation and hatching; P: promotes development
Activated sludge (Complete works)	92 (75–100)	87 (15–>99.9)	No effect on parasite inactivation; H: promotes egg development and hatching; P may promote excystation
Aerated Lagoon	83 (48–100)	85 (84–87)	
Oxidation ditch	93 (72–100)	81 (60–91.3)	
Filtration (Sand/ multimedia)	92 (78–99.6)	72 (40–99)	H: promotes egg development and hatching
Disinfection:	81	69 (61–78)	Removal and inactivation depend on contact time, dose/intensity and parasite species; H & P: Very little effect on parasite inactivation; P: >50% non-viable after 2 min at > 0.3 mg ozone/ l; H limited effect: P: >90% non-viable at > 80 mW s/cm^2
Chlorination			
Ozonation			
UV irradiation			
Anaerobic lagoons	<99	99.99	Anaerobic conditions may prolong survival
Macrophyte ponds: Hyacinth Duckweed	100	86 (69–98)	
Waste Stabilization Ponds			
WSP (complete works)	99 (88–100)	98 (87–100)	Anaerobic conditions may prolong survival
Anaerobic pond	89 (58–100)	71 (32–98)	Aerobic conditions may promote egg development and hatching
Facultative pond	92 (50–100)	76 (16–99.9)	
(1–2) Maturation ponds	92 (50–100)	62.4 (8–100)	WSP: H: No effect
Final Maturation pond	99.8 (98.9–100)	83 (72–100)	WSP: P limited effect. Solar disinfection?
HRAP (complete works)	99.7 (99.3–100)		H: limited effect: 60% inactivated after 4 days
HRAP unit only	90 (80–100)		P: >97% inactivation after 3 days
Constructed Wetlands:			
Surface flow (SF)	89 (71–100)	85 (58–99.9)	
Subsurface flow (SSF)		80 (32–99.8)	

[a] relates to information in text: H: Helminth; P: Protozoa
In most cases, removal rates may range from 0 to 100%. Parasite removal rates in the table refer to minimum removals where reported.

dry seasons. Furthermore, where wastewater is intended for reuse, it is advised that effluents should contain less than 1 parasite egg/l for public health protection. Restrictions regarding protozoan concentration in reclaimed wastewaters may also apply.

To meet these parasitological treatment objectives for effluent quality, tertiary treated conventional WTP or natural systems that are properly hydraulically loaded and designed should be sufficient. Conventional secondary WTP are unlikely consistently to meet these criteria when treating high influent loadings of $<10^3$ helminth eggs/l and 10^2–10^4 protozoan (oo)cysts/l in raw wastewaters and thus additional wastewater treatment may be required. Multicomponent wastewater treatment systems are more effective for parasite removal. Natural wastewater treatment systems have potential for offering flexible wastewater treatment systems in situations unsuitable for conventional plants and/or can be used as a component of hybrid, and modular systems in combination with other natural or conventional treatment processes where parasite removal is required.

Although information on parasite removal in wastewater treatment is available, further work is still needed. Understanding processes that cause parasite inactivation, the effect of mixing wastewaters on parasite removal and inactivation, multifactorial sunlight-mediated disinfection on parasite viability in illuminated surface waters and the role of grazing by predatory fauna would be of value. By fully understanding parasite fate and behaviour during wastewater treatment, the design of treatment systems can be improved and strategies undertaken to minimize the release of excreta-related parasitic diseases and safeguard public health.

REFERENCES

Anderson, R.M. and May, R.M. (1979). Population biology of infectious diseases: part I. *Nature* **280**, 361.

Araki, S., González, J.M., Luis, E. and Bécares, E. (2000). Viability of nematode eggs in high rate algae ponds. The effect of physico-chemical conditions. *Water Science and Technology* **42**(10–11), 371–374.

Araki, S., Martin-Gomez, S., Bécares, E. *et al.* (2001). Effect of high-rate algal ponds on viability of *Cryptosporidium parvum* oocysts. *Applied and Environmental Microbiology* **67**(7), 3322–3324.

Arthur, R.G., Fitzgerald, P.R. and Fox, J.C. (1981). Parasite ova in anaerobically digested sludge. *Journal of the Water Pollution Control Federaton* **53**, 1334–1338.

Asitinskaya, S.F. (1979). The role of freshwater gastropods in removing Ascaris eggs from water. *Abstract in Helminthological Abstract Series A* **48**(2), 89.

Atherholt, T.B., LeChevallier, M.W., Norton, W. and Rosen, J.S. (1998). Effect of rainfall on Giardia and Crypto. *Journal of the American Water Works Association* **90**(9), 66–80.

Ayres, R.M. (1991). *On the removal of nematode eggs in waste stabilization ponds and consequent potential health risks from effluent reuse*. Ph.D Thesis, University of Leeds, UK.

Ayres, R.M., Alabaster, G.P., Mara, D.D. and Lee, D.L. (1992). A design equation for human intestinal nematode egg removal in waste stabilisation ponds. *Water Research* **26**(6), 863–865.

Ayres, R.M., Lee, D.L., Mara, D.D. and Silva, S. (1993). The accumulation, distribution and viability of human parasitic nematode eggs in the sludge of a primary facultative waste stabilisation pond. *Transactions of the Royal Society of Tropical Medicine and Hygiene* **87**(3), 256–258.

Bartone, C.R. (1985). Reuse of wastewater at the San Juan de Miraflores stabilisation ponds. *Bulletin of the Pan American Health Organisation* **19**, 147–164.

Black, M.I., Scarpino, P.V., O'Donnell, C.J. *et al.* (1982). Survival of parasite eggs in sludge during aerobic and anaerobic digestion. *Applied and Environmental Microbiology* **44**(5), 1138–1143.

Bhaskaran, T.R., Sampathkumaran, M.A., Sur, T.C. and Radhakrishnan, I. (1956). Studies on the effect of sewage treatment processes on the survival of intestinal parasites. *Indian Journal of Medical Research* **44**, 163–180.

Blumenthal, U.J., Mara, D.D., Peasey, A. *et al.* (2000). Approaches to establishing microbiological quality guidelines for treated wastewater use in agriculture: recommendations for the revision of the current WHO guidelines. *WHO Bulletin* **78**(9), 1104–1116.

Bogitsh, B.J. and Cheng, T. (1998). *Human Parasitology*, 2nd edn. Academic Press.

Bouhoum, K., Amahmid, O., Habbari, Kh. and Schwartzbrod, J. (1997). Fate of helminth eggs and protozoan cysts in an open channel receiving raw wastewater from Marrakech. *Revue des sciences de l'eau* **2**, 217–232.

Bouhoum, K., Amahmid, O. and Asmama, S. (2000). Occurrence and removal of protozoan cysts and helminth eggs in waste stabilisation ponds in Marrakech. *Water Science and Technology* **42**(10–11), 159–164.

Brown, H.W. (1928). A quantitative study of the influence of oxygen and temperature on the embryonic development of the pig Ascaris (*Ascaris suum*, Goetze). *Journal of Parasitology* **14**(3), 141–160.

Bukhari, Z. and Smith, H.V. (1995). Effect of three concentration techniques on viability of *Cryptosporidium parvum* oocysts recovered from bovine feces. *Journal of Clinical Microbiology* **33**, 2592–2595.

Bukhari, Z., Smith, H.V., Sykes, N. *et al.* (1997). Occurrence of *Cryptosporidium* spp oocysts and *Giardia* spp cysts in sewage influents and effluents from treatment plants in England. *Water Science and Technology* **35**(11–12), 385–390.

Bunnag, T., Rabello de Freitas, J. and Scott, H.G. (1978). Sewage stabilization pond: The effects on *Schistosoma mansoni* transmission. *The Southeast Asian Journal of Tropical Medicine and Public Health* **9**(1), 41–47.

Cairncross, S. and Feachem, R.G. (1993). *Environmental Health Engineering in the Tropics*, 2nd edn. John Wiley and Sons, Chichester.

Campbell, A.T., Robertson, L.J., Snowball, M.R. and Smith, H.V. (1995). Inactivation of oocysts of *Cryptosprodium parvum* by ultraviolet irradiation. *Water Research* **29**(11), 2583–2586.

Capizzi, S. and Schwartzbrod, J. (1998). Helminth egg concentration in wastewaters: influence of rainwater. *Water Science and Technology* **38**(12), 77–82.

Carraro, E., Fea, E., Salva, S. and Gilli, G. (2000). Impact of a wastwater treatment plant on Cryptosporidium oocyst and Giardia cysts occurring in a surface water. *Water Science and Technology* **41**(7), 31–37.

Casemore, D.P., Wright, S.E. and Coop, R.L. (1997). Cryptosporidiosis – human and animal epidemiology. In: R. Fayer (ed.) *Cryptosporidium and cryptosporidiosis*, pp. 65–92. CRC Press, Boca Raton.

Casson, L.W., Sorber, C.A., Sykora, J.L. *et al.* (1990). Giardia in wastewater-effect of treatment. *Water Pollution Control Federation Research Journal* **62**, 670–675.

Chai, J.Y., Kim, N.Y., Guk, S.M. *et al.* (2001). High prevalence and seasonality of cryptosporidiosis in a small rural village occupied predominantly by aged people in the Republic of Korea. *American Journal of Tropical Medicine and Hygiene* **65**(5), 518–522.

Chan, M.S., Medley, G.F., Jamison, D. and Bundy, D.A. (1994). The evaluation of potential global morbidity attributable to intestinal helminth infections. *Parasitology* **109**, 373–387.

Chan, M.S. (1997). The global burden of intestinal nematode infections – Fifty years on. *Parasitology Today* **13**(11), 438–443.

Chauret, C., Springthorpe, S. and Sattar, S. (1999). Fate of *Cryptosporidium* oocysts, *Giardia* cysts, and microbial indicators during wastewater treatment and anaerobic sludge digestion. *Canadian Journal of Microbiology* **45**, 257–262.

Chernicharo, C.A.L., da Silveira Cota, R., Zerbini, A.M., von Sperling, M. and Novy de Castro Brito, L.H. (2001). Post-treatment of anaerobic effluents in an overland flow system. *Water Science and Technology* **44**(4), 229–236.

Cram, E.B. (1943). The effect of various treatment processes on the survival of helminth ova and protozoan cysts in sewage. *Sewage Works Journal* **15**, 1119–1138.

Crompton, D.W.T. (2000). The public health importance of hookworm disease. *Parasitology* **121**, S39–S50.

Curtis, T.P., Mara, D.D. and Silva, S.A. (1992). Influence of pH, oxygen and humic substances on ability of sunlight to damage faecal coliforms in waste stabilisation pond water. *Applied and Environmental Microbiology* **58**(4), 1335–1343.

David, E.D. and Lindquist, W.D. (1982). Determination of the specific gravity of certain helminth eggs using sucrose gradient centrifugation. *Journal of Parasitology* **68**(5), 916–919.

Dixo, N.G.H., Gambrill, M.P., Catunda, P.F.C. and van Haandal, A.C. (1995). Removal of pathogenic organisms from the effluent of an upflow anaerobic digester using waste stablisation ponds. *Water Science and Technology* **31**(12), 275–284.

Dowd, S.E., Gerba, C.P. and Pepper, I.L. (1998). Confirmation of the Human-Pathogenic Microsporidia *Enterocytozoon bieneusi, Encephalitozoon intestinalis* and *Vittaforma corneae* in Water. *Applied and Environmental Microbiology* **64**(9), 3332–3335.

El-Hamouri, B., Khallayoune, K., Bouzoubaa, N., Rhallabi, N. and Chalabi, M. (1994). High rate algal pond performances in faecal coliforms and helminth egg removals. *Water Research* **28**(1), 171–174.

El-Hamouri, B., Jellal, J., Outabiht, H. *et al.* (1995). The performance of a high-rate algal pond in the Moroccan climate. *Water Science and Technology* **31**(12), 67–74.

Ellis, K.V., Rodrigues, P.C.C. and Gomez, C.L. (1993). Parasite ova and cysts in Waste Stabilisation Ponds. *Water Research* **27**(9), 1455–1460.

Falabi, J.A., Gerba, C.P. and Karpiscak, M.M. (2002). *Giardia* and *Cryptosporidium* removal from waste-water by a duckweed (Lemna gibba L.) covered pond. *Letters in Applied Microbiology* **34**(5), 384–387.

Faust, E.C., Beaver, P.C. and Jung, R.C. (1976). *Animal Agents and Vectors of Human Diseases*. Lea and Febiger, Philadelphia.

Fayer, R. (1994). Effect of high temperature on infectivity of *Cryptosporidium parvum* oocysts in water. *Applied and Environmental Microbiology* **60**(8), 2732–2735.

Fayer, R. and Nerad, T. (1996). Effects of low temperature on viability of *Cryptosporidium parvum* oocysts. *Applied and Environmental Microbiology* **62**(4), 1431–1433.

Fayer, R., Speer, C.A. and Dubey, J.P. (1997). The General Biology of *Cryptosporidium*. In: R. Fayer (ed.) *Cryptosporidium and cryptosporidiosis*, pp. 1–41. CRC Press, Boca Raton.

Feachem, R., Bradley, D.J., Garelick, H. and Mara, D. (1983). *Sanitation and Disease: Health Aspects of Excreta and Wastewater Management*. John Wiley and Sons, Chichester.

Furtado, C., Adak, G.J., Stuart, J.M. *et al.* (1998). Outbreaks of waterborne infectious intestinal disease in England and Wales 1992–1995. *Epidmiology Infection* **121**, 109–119.

Gassman, L. and Schwartzbrod, J. (1991). Wastewater and Giardia Cysts. *Water Science and Technology* **24**, 183–186.

Gaspard, P.G., Wiart, J. and Schwartzbrod, J. (1995). Urban sludge reuse in agriculture – waste treatment and parasitological risk. *Bioresource Technology* **51**(1), 37–40.

Gavaghan, P.D., Sykora, J.L., Jakubowski, W. *et al.* (1993). Inactivation of *Giardia* by anaerobic digestion of sludge. *Water Science and Technology* **27**(3–4), 111–114.

Gerba, C.P., Thurston, J.A., Falabi, J.A., Watt, P.M. and Karpiscak, M.M. (1999). Optimisation of artificial wetland design for removal of indicator microorganisms and pathogenic protozoa. *Water Science and Technology* **40**(4–5), 363–368.

Gersberg, R.M., Lyon, S.R., Brenner, R. and Elkins, B.V. (1987). Fate of viruses in artificial wetlands. *Applied and Environmental Mcirobiology* **53**(4), 731–736.

Gibbs, R.A., Hu, C.J., Ho, G.E., Phillips, P.A. and Unkovich, I. (1995). Pathogen die-off in stored wastewater sludge. *Water Science and Technology* **31**(5–6), 91–95.

Gray, N.F. (1999). *Water Technology: An introduction for environmental scientist and engineers*, Arnold, London.

Grimason, A.M., Smith, H.V., Thitai, W.N. *et al.* (1993). Occurrence and removal of *Cryptosporidium* spp oocysts and *Giardia* spp oocysts in Kenya waste stabilisation ponds. *Water Science and Technology* **27**(3–4), 97–104.

Grimason, A.M., Smith, H.V., Young, G. and Thitai, W.N. (1996a). Occurrence and removal of *Ascaris* sp.ova by waste stabilisation ponds in Kenya. *Water Science and Technology* **33**(7), 75–82.

Grimason, A.M., Wiandt, S., Baleux, B. *et al.* (1996b). Occurrence and removal of *Giardia* spp cysts by Kenyan and French waste stabilisation pond systems. *Water Science and Technology* **33**(7), 83–89.

Guerrant, D.I., Moore, S.R., Lima, A.A.M. *et al.* (1999). Association of early childhood diarrhea and cryptosporidiosis with impaired physical fitness and cognitive function four-seven years later in a poor urban community in northeast Brazil. *American Journal of Tropical Medicine and Hygiene* **61**(5), 707–713.

Huffman, D.E., Slifco, T.R., Salisbury, K. and Rose, J.B. (2000). Inactivation of bacteria, virus and Cryptopsoridium by a point-of-use device using pulsed broad spectrum white light. *Water Research* **34**, 2491–2498.

Huston, C.D. and Petri, W.A. (2001). Emerging and reemerging intestinal protozoa. *Current Opinion in Gastroenterology* **17**(1), 17–23.

Jenkins, M.B., Bowman, D.D. and Ghiorse, W.C. (1998). Inactivation of Cryptosporidium parvum oocysts by ammonia. *Applied and Environmental Microbiology* **64**(2), 784–788.

Jimenez, B., Chavez, A., Leyva, A. and Tchobanoglous, G. (2000). Sand and synthetic medium filtration of advanced primary treatment effluent from Mexico City. *Water Research* **34**(2), 473–480.

Jimenez, B., Chavez, A., Maya, C. and Jardines, L. (2001). Removal of microorganisms in different stages of wastewater treatment for Mexico City. *Water Science and Technology* **43**(10), 155–162.

Jones, M.F., Newton, W.W., Weibel, S.R. *et al.* (1947). Studies on schistosomiasis. The effects of sewage treatment processes on the ova and miracidia of *Schistosoma japonicum*. *National Institute of Health Bulletin* **189**, 137–172.

Jones, M.F. and Hummel, M.S. (1947). The effect of chlorine and chloramine on *Schistosoma* ova and miracidia. *National Institute of Health Bulletin* **189**, 173.

Karpiscak, M.M., Sanchez, L.R., Freitas, R.J. and Gerba, C.P. (2001). Removal of bacterial indicators and pathogens from dairy wastewaters by a multi-component treatment system. *Water Science and Technology* **44**(11–12), 183–190.

Kawata, K. and Kluse, C.W. (1966). The effect of sewage stabilisation ponds on the eggs and miracidia of *Schistosoma mansoni*. *American Journal of Tropical Medicine and Hygiene* **15**, 896–901.

Kim, T.D. and Unno, H. (1996). The roles of microbes in the removal and inactivation of viruses in a biological wastewater treatment system. *Water Science and Technology* **33**(10–11), 243–250.

Klutse, A. and Baleux, B. (1995). Nematode egg and protozoan cyst removal in microphytic waste stabilisation ponds in Sudan-Sahel area. *Revue des sciences de l'eau* **8**(4), 563–577.

Korich, D.G., Mead, J.R., Madore, M.S., Sinclair, N.A. and Sterling, C.R. (1990). Effects of ozone, chlorine dioxide, chlorine, and monochloramine on *Cryptosporodium parvum* oocyst viability. *Journal of Applied and Environmental Microbiology* **56**, 1423–1428.

Kothary, M.H. and Babu, U.S. (2001). Infective dose of foodborne pathogens in volunteers: a review. *Journal of Food Safety* **21**(1), 49–73.

Kott, H. and Kott, Y. (1967). Detection and variability of *Entamoeba histolytica* cysts in sewage effluents. *Water Sewage Works* **114**, 177–180.

Krishnaswami, S.K. and Post, F.J. (1968). Effects of chlorine on Ascaris (Nematoda) eggs. *Health Laboratory Science* **5**, 225–232.

Lakshminarayana, J.S.S. and Abdulappa, M.K. (1969). The effect of the sewage stabilization ponds on Helminths. *Proceedings of Symposium on Central Public Health Engineering Resarch Institute, Nagpur India* Oct 27–29, pp. 270–276.

Lewis-Jones, R. and Winkler, M. (1991). *Sludge parasites and other pathogens*, Ellis Horwood Ltd, Chichester.

Liberti, L., Notarnicola, M. and Lopez, A. (2000). Advanced treatment for municipal wastewater reuse in agriculture. III – Ozone disinfection. *Ozone – Science and Engineering* **22**(2), 151–166.

Lloyd, B.J. and Frederick, G.L. (2000). Parasite removal by waste stabilisation pond systems and the relationship between concentrations in sewage and prevalence in the community. *Water Science and Technology* **42**(10–11), 375–386.

Lloyd, B.J., Vorkas, C.A. and Guganesharajah, R.K. (2002). Reducing hydraulic short-circuiting in maturation ponds to maximise pathogen removal using channels and wind breaks, *Proceedings of 5th International IWA Conference on Waste Stabilisation Ponds*, 2–5 April, 2002, Auckland, New Zealand.

Logan, A.J., Stevik, T.K., Siegrist, R.L. and Rønn, R.M. (2001). Transport and fate of *Cryptosporidium parvum* oocysts in intermittant sand filters. *Water Research* **35**(18), 4359–4369.

Lysek, H. and Bacovsky, J. (1979). Penetration of ovicidal fungi into altered eggs of *Ascaris lumbricoides*. *Folia Parasitologica (PRAHA)* **26**, 139–142.

McHarry, M.J. (1984). Detection of *Giardia* in sewage effluent. *Jourrnal of Protozology* **31**(2), 362–364.

McLauchlin, J., Amar, C., Pedraza-Diaz, S. and Nichols, G.L. (2000). Molecular epidemiological analysis of *Cryptosporidium* spp. in the United Kingdom: Results of genotyping cryptosporidium spp. in 1705 fecal samples from humans and 105 fecal samples from livestock animals. *Journal of Clinical Microbiology* **38**(11), 3984–3990.

Madera, C.A., Peña, M.R. and Mara, D.D. (2002). Microbiological quality of a waste stabilisation pond effluent used for restricted irrigation in Valle Del Cauca, Colombia. *Water Science and Technology* **45**(1), 139–143.

Madore, M.S., Rose, J.B., Gerba, C.P., Arrowood, M.J. and Sterling, C.R. (1987). Occurrence of *Cryptosporidium* oocysts in sewage effluents and selected surface waters. *Journal of Parasitology* **73**, 702–705.

Mandi, L. (1994). Marrakesh waste-water purification experiment using vascular aquatic plants *Eichhornia-crassipes* and *Lemna gibba*. *Water Science and Technology* **29**(4), 283–287.

Mandi, L., Bouhoum, B., Asmama, S. and Schwartzbrod, J. (1996). Wastewater treatment by reedbeds. An experimental approach. *Water Research* **30**(9), 2009–2016.

Mandi, L., Bouhoum, K. and Ouazzani, N. (1998). Application of constructed wetlands for domestic wastewater treatment in an arid climate. *Water Science and Technology* **38**(1), 379–387.

Mara, D.D. and Silva, S.S. (1986). Removal of intestinal nematode eggs in tropical waste stabilisation ponds. *Journal of Tropical Medicine and Hygiene*. **89**, 71–74.

Mara, D.D., Mills, S.W., Pearson, H.W. and Alabaster, G.P. (1992). Waste Stabilisation Ponds. A viable alternative for small community treatment systems. *Journal of International Water and Environmental Management (JIWEM)* **6**, 72–78.

Mata, L. (1986). *Cryptosporidium* and other protozoa in diarrheal disease in less developed countries. *Pediatric Infectious Disease* **5**(1), S117–S130.

Mayer, C.L. and Palmer, C.J. (1996). Evaluation of PCR, Nested PCR and Fluorescent Antibodies for Detection of *Giardia* and *Cryptosporidium* speices in wastewater. *Applied and Environmental Microbiology* **62**(6), 2081–2085.

Maynard, H.E., Ouki, S.K. and Williams, S.C. (1999). Tertiary lagoons: A review of removal mechanisms and performance. *Water Research* **33**(1), 1–13.

Medema, G.J., Schets, F.M., Teunis, P.F.M. and Havelaar, A.H. (1998). Sedimentation of Free and Attached *Cryptosporidium* Oocysts and *Giardia* Cysts in Water. *Applied and Environmental Microbiology* **64**, 4460–4466.

Medema, G.J. and Schijven, J.F. (2001). Modelling the sewage discharge and dispersion of *Cryptopsoridium* and *Giardia* in surface water. *Water Research* **35**(18), 4307–4316.

Mercado-Burgos, N., Hoehn, R.C. and Holliman, R.B. (1975). Effects of halogens and ozone on *Schistosoma* ova. *Water Pollution control Federation* **47**, 2411–2419.

Metcalf and Eddy (1991). *Wastewater Engineering: Treatment, Disposal and Reuse*. McGraw-Hill Book Co.

Miller, A., Chi-Rodriquez, E. and Nichols, R.L. (1961). The fate of helminth eggs and protozoan cysts in human faeces ingested by dung beetles (*Coleoptera: Scarabaeidae*). *American Journal of Tropical Medicine and Hygiene* **10**(4), 748–754.

Nelson, K.L. and Jimenez, B.C. (2000). Sludge accumulation, properties and degradation in a waste stabilisation pond in Mexico. *Water Science and Technology* **42**(10–11), 231–236.

Newton, W.L., Bennett, H.J. and Figgat, W.B. (1949). Observations of the effects of various sewage treatment processes upon eggs of *Taenia saginata*. *American Journal of Hygiene* **49**(2), 166–175.

Nupen, E.M. and Villiers, R.H. (1975). *The Evaluation of pathogenic parasites in water environments*. Project Report No 10, South Africa National Committee for Water Research.

O'Donnell, C.J., Meyer, K.B., Jones, J.V. *et al.* (1984). Survival of parasite eggs upon storage in sludge. *Applied and Environmental Microbiology* **48**(3), 618–625.

O'Lorcain, P. and Holland, C.V. (2000). The public health importance of *Ascaris lumbricoides*. *Parasitology* **121**, S51–S71.

Okhuysen, P.C., Chappell, C.L., Crabb, J.H., Sterling, C.R. and DuPont, H.L. (1999). Virulence of three distinct *Cryptosporidum parvum* Isolates for Healthy Adults. *The Journal of Infectious Diseases* **180**(4), 1275–1281.

Oswald, A.M., Gerba, C.P. and Karpiscak, M.M. (2000). Removal of enteric microorganisms from secondary effluent and backwash filter water by artificial wetlands. *Proceedings of 1st IWA World Water Congress, Wastewater Reclamation, Recycling and Reuse*, Paris, 3–7 July.

Ouazzani, N., Bouhoum, K., Mandi, L. *et al.* (1995). Waste-Water treatment by stabilization pond – Marrakesh Experiment. *Water Science and Technology* **31**(12), 75–80.

Panicker, P.V.R.C. and Krishnamoorthi, K.P. (1978a). Elimination of enteric parasites during sewage treatment processes. *International Association of Water Pollution Control (IAWPC) Technology Annual*, **5**, 130–138.

Panicker, P.V.R.C. and Krishnamoorthi, K.P. (1978). Studies on intestinal helminthic eggs and protozoan cysts in sewage. *Indian Journal of Environmental Health* **20**, 75–78.

Panicker, P.V.R.C. and Krishnamoorthi, K.P. (1981). Parasitic egg and cyst reduction in oxidation ditchs and aerated lagoon. *Journal of Water Pollution Control Federation* **53**, 1413–1419.

Paulino, R.C., Castro, E.A. and Thomaz-Soccol, V. (2001). Helminth eggs and protozoan cysts in sludge obtained by the anaerobic digestion process. *Revista da Sociedad Brasileira de Medicina Tropcial* **34**(5), 421–428.

Payment, P., Plante, R. and Cejka, P. (2001). Removal of indicator bacteria, human enteric viruses, giardia cysts, and Cryptosporidum oocysts at a large wastewater primary treatment facility. *Canadian Journal of Microbiology* **47**, 188–193.

Pike, E.B. (1990). The Removal of Cryptosporidial Oocysts During Sewage Treatment. In: *Cryptosporidium in Water Supplies*. Dept of Env and Dept of Health Report of the Group of Experts. Chair Sir John Badenoch. Part II Paper X, pp. 205–208. HMSO, London.

Quiñonez-Diaz, M., Karpiscak, M.M., Ellman, E.D. and Gerba, C.P. (2001). Removal of pathogenic and indicator micoorganisms by a constructed wetland receiving untreated domestic wastewater. *Journal of Enviornmental Science and Health. Part A – Toxic/Hazardous Substances and Environmental Engineering* **36**(7), 1311–1320.

Reimers, R.S., Little, N.D., Englander, A.J., Leftwich, D.B., Bowman, D.D. and Wilkinson, R.F. (1982). Parasites in southern sludges and disinfection by standard sludge treatment. *Water Science and Technology* **20**(11–12), 101–107.

Rivera, F., Warren, A., Ramirez, E. *et al.* (1995). Removal of pathogens from wastewaters by the root zone method (RZM). *Water Science and Technology* **32**(3), 211–218.

Robertson, I.D., Irwin, P.J., Lymbery, A.J. and Thompson, R.C.A. (2000a). The role of companion animals in the emergence of parasitic zoonoses. *International Journal for Parasitology* **30**(12–13), 1369–1377.

Robertson, L.J., Campbell, A.T. and Smith, H.V. (1992). Survival of Cryptosporidium parvum oocysts under various environmental pressures. *Applied and Environmental Microbiology* **58**(11), 3494–3500.

Robertson, L.J., Smith, H.V. and Paton, C.A. (1995). Occurrence of *Giardia* cysts and Cryptosproidium oocysts in sewage influent in six sewage treatment plants in Scotland and prevalence of cryptosporidiosis and giardiasis in the communities served by those plants. In: W.B. Betts, D. Casemore, C. Fricker, H. Smith and J. Watkins (eds) *Protozoan Parasites and Water*, pp. 47–49. The Royal Society of Chemistry, Cambridge.

Robertson, L.J., Smith, P.G., Grimason, A.T. and Smith, H.V. (1999). Removal and destruction of intestinal parasitic protozoans by sewage treatment processes. *International Journal of Environmental Health Research* **9**, 85–96.

Robertson, L.J., Paton, C.A., Campbell, A.T. *et al.* (2000b). Giardia cysts and Cryptosporidium oocysts at sewage treatment works in Scotland, UK. *Water Research* **34**(8), 2310–2322.

Rose, J.B. and Gerba, C.P. (1990). Assessing potential health risks from viruses and parasites in reclaimed water in Arizona and Florida. *Water Science and Technology* **23**, 2091–2098.

Rose, J.B., Dickson, L.J., Farrah, S.R. and Carnahan, R.P. (1996). Removal of pathogenic and indicator microorganisms by a full-scale water reclamation facility. *Water Research* **30**(11), 2785–2797.

Rowan, W.B. (1964). Sewage treatment and schistosome eggs. *American Journal of Tropical Medicine and Hygiene* **13**, 572–576.

Saqqar, M. and Pescod, M.B. (1991). Microbiological performance of multi-stage stabilisation ponds from effluent use in agriculture. *Water Science and Technology* **23**(7–9), 1517–1524.

Saqqar, M. and Pescod, M.B. (1992). Modelling nematode egg elimination in wastewater stabilisation ponds. *Water Science and Technology* **26**(7–8), 1659–1665.

Schlosser, O., Grall, D. and Laurenceau, M.N. (1999). Intestinal parasite carriage in workers exposed to sewage. *European Journal of Epidemiology* **15**(3), 261–265.

Schwartzbrod, J., Mathieu, C., Thévenot, M.T., Baradel, J. and Schwartzbrod, L. (1987). Wastewater sludge: parasitological and virological contamination. *Water, Science and Technology* **19**(8), 33–40.

Schwartzbrod, J., Stien, J.L., Bouhoum, K. and Baleux, B. (1989). Impact of wastewater treatment on helminth eggs. *Water Science and Technology* **21**, 295–297.

Schwartzbrod, J. (1993) Wastewater treatment by stabilisation ponds: Marrakech experience. *Proc 2nd IAWQ Special conference on Waste Stabilisation Ponds and the Reuse of Effluents, Brazil..*

Shepard, M.R.N. (1971). The role of sewage treatment in the control of human helminthiasis. *Helminthological Abstracts* **40**, 1–16.

Shilton, A. and Harrison, J. (2002). Development of guidelines for improved hydraulic design of waste stabilisation ponds. *Proceedings of 5th International IWA conference on Waste Stabilisation Ponds*, pp. 2–5 April. Auckland, New Zealand.

Shuval, H.I., Adin, A., Fattal, B., Rawitz, E. and Yekutiel, P. (1986). Evaluation of epidemiological evidence of human health effects associated with wastewater irrigation. In: *Wastewater Irrigation in Developing Countries: Health effects and Technical solutions. Technical Paper Number 51*, World Bank, Washington DC, 324 p.

Smith, H.V. (1990). Cryptosporidium and Water: A Review. *Journal of International Water and Environmnental Management (JIWEM)* **6**, 443–451.

Smith, H. and Rose, J.B. (1998). Waterborne Cryptosporidiosis: Current Status. *Parasitology Today* **14**(1), 14–22.

Smith, J.D. (1994). *Introduction to Animal Parasitology*, 3rd edn. Cambridge University Press, Cambridge.

Sobenina, G.G. (1978). Study of the effect of some fungi on the embryogenesis and survival of Ascaris ova. *Helminthological Abstract Series A* **47**(10), 440.

Somiya, I., Fujii, S., Kishimoto, N. and Kim, R.H. (2000). Development of a mathematical model of Cryptopsoridum inactivation by ozonation. *Water Science and Technology* **41**(7), 173–180.

Spindler, L.A. (1940). Effects of tropical sunlight on eggs of Ascaris suis, (nematode) the large intestinal roundworm of swine. *Journal of Parasitology* **26**, 323–331.

Stanley, S.L. (2001). Protective immunity to amebiasis: new insights and new challenges. *Journal of Infectious Diseases* **184**(4), 504–506.

Stadterman, K.L., Sninsky, A.M., Sykora, J.L. and Jakubowski, W. (1995). Removal and inactivation of *Cryptosporidium* oocysts by activated sludge treatment and anaerobic digestion. *Water Science and Technology* **31**(5–6), 97–104.

Stephenson, L.S., Holland, C.V. and Cooper, E.S. (2000). The public health significance of *Trichuris trichiura*. *Parasitology* **121**, S73–S95.

Stephenson, L.S., Latham, M.C. and Ottesen, E.A. (2000b). Malnutrition and parasitic helminth infections. *Parasitology* **121**, S23–S38.

Stott, R. Jenkins, T. Williams, J. *et al.* (1996). *Pathogen removal and microbial ecology in Gravel Bed Hydroponic (GBH) treatment of wastewater.* Research Monograph 4, Dept Civil Engineering, University of Portsmouth.

Stott, R., Jenkins, T., Shabana, M., May, E. and Butler, J. (1997). A survey of the microbial quality of wastewaters in Ismailia, Egypt and the implications for wastewater reuse. *Water Science and Technology* **35**(11–12), 211–217.

Stott, R., Jenkins, T., Bahgat, M. and Shalaby, I. (1999). Capacity of Constructed Wetlands to remove Parasite eggs from Wastewaters in Egypt. *Water Science and Technology* **40**(3), 117–123.

Stott, R., May, E., Matsushita, E. and Warren, A. (2001). Protozoan predation as a mechanism for the removal of *Cryptosporidium* oocysts from wastewaters in constructed wetlands. *Water Science and Technology* **44**(11–12), 194–198.

Stott, R., May, E. and Mara, D.D. (2002a). Parasite removal by natural wastewater treatment systems: Performance of waste stabilisation ponds and constructed wetlands. *Proceedings of the 5th IWA Waste Stabilisation Ponds*, 2–5 April. Auckland, New Zealand.

Stott, R. May, E. Ramirez, E. and Warren, A. (2002b). Predation of *Cryptosporidium* oocysts by protozoa and rotifers: implications for water quality and public health. *Proceedings of IWA World Water Congress (Health Related Water Microbiology Symposium)* 7–12 April, Melbourne, Australia.

Sturbaum, G.D., Ortega, Y.R., Gilman, R.H. *et al.* (1998). Detection of *Cyclospora cayetanensis* in Wastewater. *Applied and Environmental Microbiology* **64**(6), 2284–2286.

Sykora, J.L., Sorber, C.A., Jakubowski, W. *et al.* (1991). Distribution of giardia cysts in wastewater. *Water Science and Technology* **24**(2), 187–192.

Taylor, H.D. Gambrill, M.P. and Mara, D.D. (1994) *Lime Treatment of Municipal Wastewater.* Research Monograph No. 3, School of Civil Engineering, University of Leeds, Leeds.

Thienpoint, D., Rochette, F. and Vanparij, O.F.J. (1986). Diagnosing helminthais by coprological examination. Janssen Research Foundation, Belgium.

Thurston, J.A., Gerba, C.P., Foster, K.E. and Karpiscak, M.M. (2001). Fate of indicator microorganisms Giardia and Cryptosporidium in sub surface flow constructed wetlands. *Water Research* **35**(6), 1547–1551.

Tromba, F.G. (1978). Effects of UV radiation on the infective stages of *Ascaris suum* and *Stephanarus dentatus* with a comparison of the relative susceptibilities of some parasitic nematodes to UV. *Journal of Parasitology* **64**(2), 245–252.

Udonsi, J.K. and Atata, G. (1987). *Necator americanus*: Temperature, pH, light and larval development, longevity and desiccation tolerance. *Experimental Parasitology* **63**, 136–142.

Veerannan, K.M. (1977). Effect of sewage treatment by stabilisation pond method on the survival of intestinal parasites. *Indian Journal of Environmental Health* **19**, 100–106.

Villacorta-Martinez de Maturana, I., Ares-Mazas, M.E., Duran-Oreiro, D. and Lorenzo-Lorenzo, M.J. (1992). Efficacy of activated sludge in removing *Cryptosporidium parvum* oocysts from Sewage. *Applied and Environmental Microbiology* **58**(11), 3514–3516.

Wang, W.L. and Dunlop, S.C. (1954). Animal parasites in sewage and irrigation waters. *Sewage and Industrial Wastes* **26**, 1020–1032.

Wharton, D.A. (1980). Nematode egg-shells. *Parasitology* **81**, 447–463.

WHO (World Health Organisation) (1989) Health guide-lines for the use of wastewater in agriculture and aquaculture. *WHO Technical Report Series No 778*, Geneva.

Whitmore, T.N. and Robertson, L.J. (1995). The effect of sewage sludge treatment processes on oocysts of Cryptosporidium parvum. *Journal of Applied Bacteriology* **78**, 34–38.

Wiandt, S., Baleux, B., Casellas, C. and Bontoux, J. (1995). Occurrence of *Giardia* sp cysts during a waste-water treatment by a stabilization pond in the south of France. *Water Science and Technology* **31**(12), 257–265.

Wiandt, S., Grimason, A.M., Baleux, B. and Bontoux, J. (2000). Efficiency of wastewater treatment plants at removing Giardia sp cysts in southern France. *Schriftenreihe Des Vereins Fur Wasser-, Boden-Und Lufthygiene* **105**, 35–42.

Williams J., Ashworth, R., Ford, N. *et al.* (1995). Physical and chemical aspects of sewage treatment in Gravel Bed Hydroponic (GBH) systems. *Research Monograph No 2*, Dept. Civil Engineering, University of Portsmouth.

Yanez, F.A. Rojas, R. Castro, M.L. and Mayo, C. (1980). Evaluation of the San Juan stabilisation ponds. Final research report on the first phase, PATHO/CEPIS, Lima, Peru.

Zu, S.X. and Guerrant, R.L. (1993). Cryptosporidiosis. *Journal of Tropical Pediatrics* **39**(3), 132–136.

Problems in Wastewater Treatment Processes

32

Activated sludge bulking and foaming: microbes and myths

R.J. Foot and M.S. Robinson

Wessex Water, UK

1 HISTORY OF BULKING AND FOAMING

1.1 Activated sludge bulking

When the activated sludge process was first developed in 1919 the sludge was described by Arden as 'flocculent in character', indicating its natural ability to settle well. This property of good settlement following a period of aeration allowed development of batch experiments into a full-scale process involving a continuous flow of wastewater and a continuous recycle of settled sludge from a subsequent settlement process. Poor settleability of activated sludge does not appear to be a widespread or serious problem during the early years of developing activated sludge to treat domestic wastewater.

However, in the next decade there are early signs of future problems. 'Undesireable filamentous organisms' were identified by Morgan and Beck (1928) as being responsible for poorly settling activated sludge. These filamentous growths were identified as *Sphraerotilus natans* and were stimulated by the presence of excessive carbohydrates in the wastewater.

A few years later, Donaldson (1932) described filamentous growths as 'the weeds of activated sludge' and firmly laid the blame for the cause of poor settleability at their presence. By this time two types of filament were identified, *S. natans* and *Leucothrix*. While this is now known to be an over-simplification, the mechanisms by which filamentous growths affect activated sludge settleability was becoming known. The filaments physically keep the floc forming bacteria apart leading to a sludge of high water content.

The number of activated sludge plants in the UK increased rapidly after the Second World War, but sludge bulking did not present a serious problem except for plants with a high proportion of industrial wastewater.

The picture had changed dramatically by the 1970s. A survey by Eikelboom (1975) and others of several hundred activated sludge plants showed that 50% of them had problems at one time or another with bulking sludge. By this time identification of the causative filamentous organisms had also improved by Eikelboom's simple microscopic technique. By the 1980s at least 17 different filamentous bacteria had been separately identified in bulking sludges on three continents.

In the 1980s attempts were made to characterize these filaments into groups showing common morphological, physiological and metabolic similarity. This approach, by Wanner and Grau (1989), significantly incorporated operational conditions and problems:

- S *Sphaerotilus*-like oxic zone growers
- C Cyanophyte-like oxic zone growers
- A *Microthrix parvicella* types – all zone growers
- F Foam-forming filaments.

More recently, attempts have been made to characterize further specific filaments in

The Handbook of Water and Wastewater Microbiology
ISBN 0-12-470100-0

terms of their preferred conditions for growth and triggers for proliferation. This has finally led to a more scientific approach to reduction or elimination.

1.2 Activated sludge foaming

One of the earliest reports of foam formation is the so-called Milwaukee mystery (Anon., 1969). Surveys in the USA showed that 66% of activated sludge plants expressed some types of foaming as long ago as 1979 (Pitt and Jenkins, 1990). In a French study of some 6000 plants, over half had experienced foaming (Pujol *et al.*, 1991).

Foaming became more widespread in the UK in the early 1980s as reported in the South-East Dorset area of England (Foot, 1992). In this study foaming was not confined to any particular type, size or configuration of activated sludge plant and was liable to foam at any time. Biological foaming is mostly reported in nutrient removal and extended aeration plants (Pujol *et al.*, 1991).

Foaming appears to be connected to the presence of three types of filamentous micoorganisms, as identified by Wanner in 1994.

- *Microthrix parvicella*, the most common (Fig. 32.1)

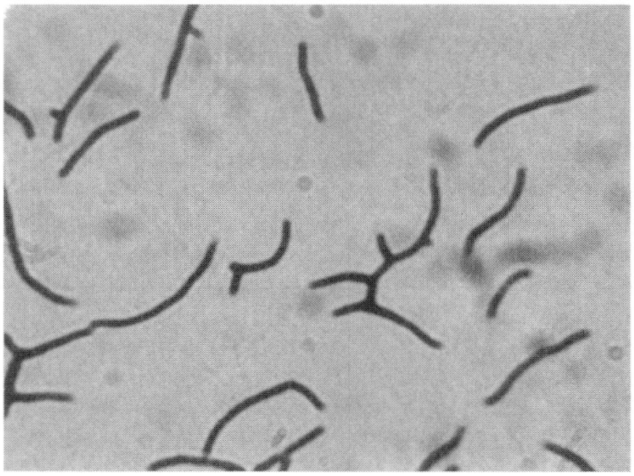

Fig. 32.2 Branched nocardiaforms grown in pure culture. × 2000 magnification. Courtesy of Dr E. Kocianova, Environmental Consultancy Services, Cornwall.

- *Nocardia* (now called *Gordona*) *amarae*-like organisms, including *Rhodococcus* spp. (Fig. 32.2)
- *Nostocoidia limilocola*, Type 0041 and Type 0092 (Fig. 32.3).

Studies have shown that the distribution of filamentous organisms in biological foams at wastewater treatment plants varies from country to country, season to season and is changing with time (Wanner, 1998).

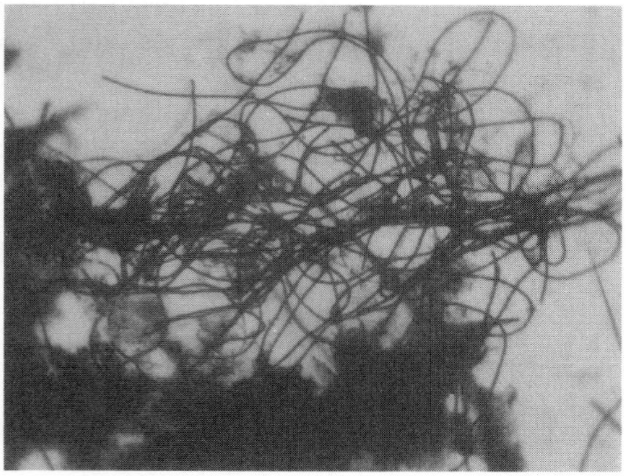

Fig. 32.1 Gram-positive *Microthrix parvicella* filaments coiled in typical 'bootlace' fashion. ×1000 magnification. Courtesy of Dr E. Kocianova, Environmental Consultancy Services, Cornwall.

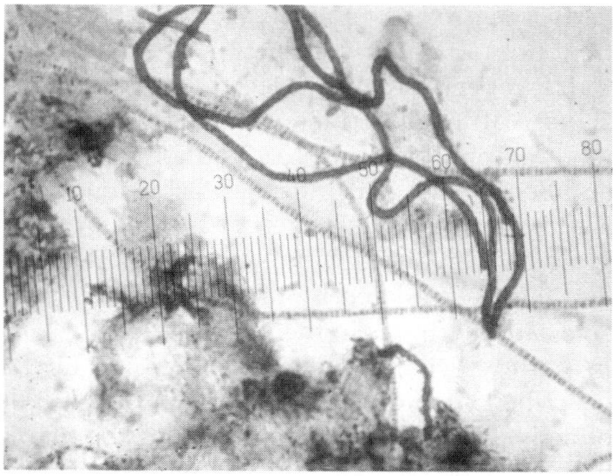

Fig. 32.3 Coiled Neisser positive *Nostocoidia limilocola* and extended filaments of Type O21N (or *Thiothrix* spp.). Courtesy of Dr E. Kocianova, Environmental Consultancy Services, Cornwall.

1.3 Anaerobic digester foaming

Foaming within anaerobic digesters is not uncommon, but is usually associated with physical causes, such as gas recirculation or simple overloading. In the latter, an excess of volatile fatty acids causes foaming. Such foaming is usually short-lived and dependent upon the continued presence of the volatile fatty acids.

However, in recent years, the presence of stable foams at times of acceptable digester loading rate has been observed. When the sludge has been investigated, filamentous bacteria have been found. In Stockholm (Dillner Westlund *et al.*, 1998) identified *Microthrix parvicella* as the causative agent of a stable anaerobic digester foam. Interestingly, the ability to create such a foam was eliminated when the feed sludge was heated for 5 minutes at 70°C. In this study, the source of the *Microthrix parvicella* was waste activated sludge. The same study also found that adding aluminium salts to the feed sludge caused the foam to collapse. An American study (Pagilla *et al.*, 1997) identified *Nocardia* as being responsible for anaerobic digester foaming in Sacramento. Again, the source of the *Nocardia* was waste activated sludge. A later study by the same authors (Pagilla *et al.*, 1998) concluded that chlorination of the waste activated sludge stream was ineffective for controlling such foams, even when dose rates were as high as 200 mg/l. Anaerobic digester foaming caused by filamentous bacteria appears to be a consequence of the presence of the same bacteria in waste activated sludge fed to the digester.

2 DEFINITIONS OF BULKING AND FOAMING

2.1 Activated sludge bulking

'Bulking' is a term referring to the phenomenon of a poorly settling sludge. Typically, suspended solids become visible in the bulk of a final settlement tank and, in more acute cases, will be lost over outlet weirs to pollute the effluent discharge. Ironically, chronic bulking can often be associated with high quality effluent as the sludge blanket captures and retains small floc particles.

Bulking should not be confused with the loss of suspended solids from physical causes. There are four main possible causes as follows:

- Failure of recycle RAS (return activated sludge) pump. This will allow solids to accumulate in the final settlement tank which will eventually be lost *en-masse* in a gross pollution of the final effluent.
- Excessive hydraulic loading, e.g. from failure of inflow control. This will usually be a problem in wet weather conditions only. However, an under-sized final settlement tank will exhibit similar characteristics.
- A high concentration of MLSS (mixed liquor suspended solids) caused by an insufficient wastage rate of SAS (surplus activated sludge). Gross pollution of the final effluent is likely if this is not resolved and is often accompanied by reduced nitrification due to the excess oxygen demand exerted by the excess mixed liquor solids.
- Denitrification within the final tanks, caused, for example by too long a solids retention time in the final settlement tank, e.g. failure of a final tank scraper.

Bulking sludges can be defined by reference to the measurement of sludge settleability or SSVI (stirred specific volume index) as shown in Table 32.1.

Normally settling activated sludge separates quickly form the supernatant liquor which will have an excellent clarity. When the SSVI approaches 120 ml/g then pin-floc solids can be observed in the bulk of the final settlement tank, often migrating to the peripheral weir. Unfortunately, pin-floc can also be caused by shock loads from excess sludge dewatering, trade effluents, or even over-aeration in

TABLE 32.1 Definition of sludge bulking by measurement of SSVI

SSVI (ml/g)	*Condition*
80–100	Normal
>120	'Solids' appear in final settlement tank
>150	On-set of bulking
>180	On-set of chronic bulking

the aeration tank. The division between blanket level and supernatant liquor becomes ill-defined because the sludge does not flocculate properly and hence does not compact. Often, this process is accompanied by denitrification, due to the prolonged solids retention time, and a surface film forms on the surface of the final settlement tank. Unfortunately, surface films can be caused by other processes, particularly foam-forming microorganisms.

There are two known biological causes of bulking, or poorly settling sludge:

- By far the most common is the presence of excessive growths of filamentous organisms.
- The remaining type is often referred to as non-filamentous bulking or zoogleal bulking. This occurs when common floc-forming bacteria produce extracellular polysaccharides which attract water, resulting in a sludge that has a high degree of hydration.

The incidence of non-filamentous bulking is comparatively rare. Activated sludge bulking has been observed world-wide and is a phenomenon affecting both municipal and industrial wastewater treatment plants. A survey of activated sludge plants in the UK in 1976 by Tomlinson revealed that 63% had experienced bulking.

2.2 Definition of activated sludge foaming

The presence of foam in an activated sludge plant is obvious; frequent points of generation or accumulation of microbiological foams are:

- On the aeration tank surfaces
- On the surface of final settlement tanks
- On the interconnection channels and chambers
- From pipeline chambers.

Biological foaming is often referred to as stable foam, scum or mousse. Biological foaming should not be confused with chemical foaming caused by the presence of excess surfactants from trade effluent discharges or during the start-up of new plant.

Minor foaming is usually present in all activated sludge plants at one time or another.

Filamentous foaming is caused by the presence of foam-forming filamentous microorganisms in the activated sludge. When these filaments become dominant, foam is produced. The critical factors involved in the formation of foam are cell-surface hydrophobicity coupled with biosurfactant production and the presence of gas bubbles from denitrification or aeration.

These three factors combine to produce a stable three-phase system of air, water and cell material. In extreme cases, foaming can escape from its point of generation and spill into unwanted areas such as effluent channels, electrical equipment and on roads and paths. This is illustrated in Fig. 32.4. It is also a health hazard and risk to safety and effluent quality.

Fig. 32.4 A *Microthrix parvicella* foam illustrating the risk placed upon effluent quality and safety.

Fig. 32.5 A biological (nocardioform) foam similar in appearance to a detergent foam.

Fig. 32.6 Foam can be managed to an acceptable level and retained on the surface of the aeration tank.

One way to differentiate a detergent (surfactant) foam from a biological foam is from an estimate of the half-life (the time for 50% of the foam to degrade). A surfactant foam will have a typically short half-life of a few hours or less, whereas a microbiological foam will have a half-life of between 15 and 55 hours depending on the species. The effects of detergent foams are often spectacular and similar in appearance to the foam illustrated in Fig. 32.5. On this occasion the foam was biological in origin (*Rhodococcus* spp.).

It should be noted that the presence of biological foams is not necessarily related to settlement characteristics of the activated sludge. It is quite possible to have chronic foaming that can be kept in check by good management, without noticing any ill effect on sludge settleability or effluent quality (Fig. 32.6).

Sludge floc particles appear to be necessary to stabilize foam structure and hence give it a long half-life. As foam stabilizes apart from the main body of activated sludge biomass, further foam formation is limited only by the growth of foam-forming microorganisms. These provide a key to their control. De-selection of foam-forming microorganisms and removal or suppression of the foam will eventually prevent its proliferation.

2.3 Definition of anaerobic digester foaming

The presence of stable foam within an anaerobic digester is characterized by a foam layer up to 2.4 m deep on its surface. The foam will usually cause a spill of digested sludge onto walkways and the surrounding ground area if it occurs on a floating dome gas-holder design. The foam will enter gas draw-off pipework and compromise gas mixing systems and heating systems that rely upon biogas combustion.

Over-feeding or reducing the operating temperature can induce a similar, but less stable, foaming. In this case, the presence of volatile fatty acids will cause an unpleasant smell. If foaming is noticed with no unpleasant smell then the presence of filamentous bacteria should be investigated.

3 CAUSES OF BULKING AND FOAMING

Given the nature of the activated sludge process we are unlikely ever to be in the position to define the precise cause of bulking and foaming. The process itself forms the largest biotechnology industry in the world, providing economic treatment to an infinitely variable range of effluents and environmental conditions. It is not surprising, therefore, that for every report identifying a single cause there exists another, suggesting an alternative. The challenge for the microbiologist is first to dispel the myths surrounding activated sludge bulking and foaming and then to present a reasoned case to both operators and engineers for adequate controls and treatment capacity.

3.1 Common causes of activated sludge bulking and foaming

Sludge bulking and foaming is invariably associated with:

- Industrial effluents or domestic waste with a large proportion of industrial waste
- Plant loading
- Plant configuration
- Process control.

Throughout the world the filamentous organisms commonly associated with bulking and foaming are the nocardiaforms and *Microthrix parvicella* (Pitman, 1996).

3.1.1 Industrial and high strength waste

Little needs to be said about industrial waste other than to be aware that filamentous bacteria are capable of exploiting selective substrates. The science is well known and can be applied to new works design or nutrient dosing programme. Non-biological foams are generated by the introduction of surfactants and unused or unsuitable polyelectrolytes in recycle flows.

Even the best designed and operated treatment plant will occasionally suffer the effects of filamentous bacteria, often from external causes such as a sudden discharge of high strength waste (e.g. recycle liquors or a sludge spill, use of glycol, sugar). When this occurs, filamentous growth is rapid and its effects long lasting.

Process air requirements also need to allow for sewages with variable treatability (alpha factor) caused by salinity, septicity and industrial waste.

3.1.2 Salinity

Saline intrusion into coastal towns can also result in poor sludge settleability. While the organisms in waste treatment are able to adapt to increased levels of chloride, the rate of change in chloride concentration through a spring tide can be more than activated sludge can accommodate. Acceptable limits would be a chloride concentration no greater than 6 g/l and an hourly change no higher than 50% of the average chloride concentration of the previous 24 hours. Large salinity swings cause dispersed floc and poor treatability. Chronic infiltration can cause chloride concentration to go from 0.1 g/l to 17 g/l within a matter of hours. Investing in the sewerage infrastructure gives additional cost benefits by reducing pumping, pump maintenance and overall treatment costs.

3.1.3 Septicity

Increased sulphate concentrations from saline intrusion also contribute to sewage septicity. Stormwater attenuation systems and long pumping mains cause sulphide to develop in the sewage, increasing malodours, concrete corrosion and sludge production. Oxygen consumption can increase by 30% (Dormoy *et al.*, 1999). Septicity bulking follows the growth of sulphur-oxidizing filamentous bacteria.

3.1.4 Plant loading

Previous work on foaming and non-foaming plants shows that plants with a low aeration capacity (relative to the organic loading rate) are more likely to foam (Foot *et al.*, 1993). In Fig. 32.7, installed aeration capacity (as kg $O_2/m^3/day$) is compared with the organic loading rate of the plant (as kg BOD/kg MLSS/day). Those plants that fall below the regression line are low loaded plants with an ample supply of oxygen, whereas those above represent plants with a high organic load but without a corresponding increase in the amount of oxygen available. Stable foam production occurs on most of the plants which are positioned above the regression line. Therefore, plants with a high F/M (food to mass) to O_2/m^3 ratio are more likely to foam than those with a low ratio.

3.1.5 Plant configuration

The relationship between the settling properties of the activated sludge and the configuration of the aeration tanks is well established. As early as the 1960s plug-flow systems with a high substrate gradient were shown to have had better settling characteristics than completely mixed systems with a low substrate gradient (Water Pollution Research Laboratory,

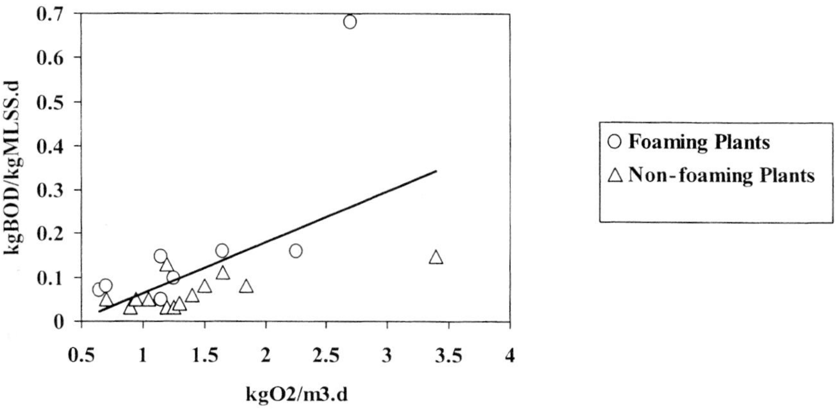

Fig. 32.7 Foam production as a function of F:M ratio and available oxygen. Adapted from Foot *et al.* (1993).

1967). Absence of a substrate gradient, initial contact zone (selector) or longitudinal mixing is invariably linked to poor settlement and foam formation. A substrate gradient can be achieved in both conventional plant and oxidation ditches where:

- The aeration tank is separated into compartments, or
- Flow is channelled.

Compartmentalization is key to control, with the initial compartments exerting the all important species selection pressures. These have been defined as kinetic (based on microbial competition) and metabolic (based on cultivation conditions), the principles of which are combined in the design of modern activated sludge plants to achieve good settling properties (Wanner, 1998).

3.1.6 *Process control*

Lack of process control leads not only to poor settlement and foam formation but also to inefficient treatment. In turn, inefficient treatment results in poor effluent quality and high energy costs. For example, a plant designed to treat peak flow over a 5-year design horizon could incur an increase in power consumption of 50% if operated without process control. SCADA (supervision control and data acquisition), PLC (process logic controller) or simple timer controls usually control plants' dissolved oxygen levels and sludge recycle rates. Insufficient knowledge of the treatment process will ultimately lead to treatment problems.

3.2 No effect without cause

As discussed previously, solids loss from a final settlement tank, caused by plant failure, can easily be mistaken for a bulking or foaming incident. The effects of a pump failure or blockage can look very similar to that caused by a bulking or foaming sludge (i.e. rising sludge blanket or floating solids on the final tank). Filamentous microorganisms are present in most activated sludges in varying amounts and their presence does not necessarily mean that they are the cause of the solid loss. Correctly identifying the dominant filaments and getting a measure of their abundance is the first step in establishing this. Furthermore, an understanding of the mechanisms and drivers behind the predominance of filamentous microorganisms in the biomass will lead to control strategies for their eventual elimination.

4 ORGANISMS RESPONSIBLE: MECHANISMS AND DRIVERS

4.1 Taxonomic history

4.1.1 *Diagnostics keys*

The ecology, or balance of microbial species in activated sludge, has long been held as a significant factor in providing explanations for the attributes of the activated sludge biomass. With attention being drawn to filamentous species and the absence of diagnostic keys to identify these organisms, came an oversimplification of the nature of the problem and the organisms involved. *Sphaerotilus natans* (commonly known as sewage fungus) was aptly described as the 'weeds' of the activated sludge process and was alone blamed for activated sludge bulking. It was not until Eikelboom and van Buijsen (1983) devised an analytical technique for filament identification, that species other than *S. natans* became associated with bulking.

A similar pattern emerged with the arrival of the 'foaming' phenomenon, which also had a microbial origin. Like bulking, this problem is also associated with a dominance of filamentous species and originally it was also considered that a single species, *Nocardia amarae*, was responsible. Armed with additional diagnostic keys (Jenkins *et al.*, 1986), microbiologists set about identifying (and enumerating) a range of filamentous microbes associated with both bulking and foam formation.

Being in a position to identify individual types of filamentous microbe, patterns soon emerged, linking filaments with plant operation conditions (Table 32.2). While many had not yet been taxonomically defined (being

TABLE 32.2 Filamentous microbes in activated sludge grouped (Forster, 2000)

Conditions	*Likely filaments*
Low loading, extended aeration plants	*M. parvicella* (winter) *0092* (summer)
Conventional loading; full/partial nitrification	*021N, M. parvicella, 1701, 0803, H. hydrossis*
Conventional loading, no nitrification	*021N, S. natans, H. hydrossis*
Low DO	*1701, S. natans, H. hydrossis*
Septicity	*Thiothrix, Beggiatoa, 021N*
Nutrient imbalance	*Thiothrix, S. natans, 021N*
Low F:M	*M. parvicella, H. hydrossis, 021N, 0031, Nocardia, 0675. 0092, 0961, 0803*

know only by type rather than species or genus), this work provided vital clues to cause and, therefore, control mechanisms.

Comparisons drawn between filament populations associated with bulking and foaming events soon identified a hard-core of filaments. Rankings based on these reports have shown similar patterns as shown in Fig. 32.8(a) and (b). However, rather than clarifying the situation these data show that (although some species have a unique function), there is a small group of filaments that can induce either bulking or foaming, i.e. they appear in both bulking and foaming sludges. One of these is *Microthrix parvicella*, which is seen to be just as common in bulking sludges as Type 021N, in addition to being dominant in almost half the foaming sludges. While not complete, the data are indicative and show how some species are capable of exploiting more than one set of operating conditions.

4.1.2 *Genetic classification*

Mirothrix parvicella was one of the first to be defined using genetic techniques (Blackall *et al.*, 1994). Eikelboom type 021N became linked to *Thiothrix* (Howarth *et al.*, 1999) and type 1701 to *Sphaerotilus natans* (Howarth *et al.*, 1998) using 16S rRNA sequencing. Nocardioforms were

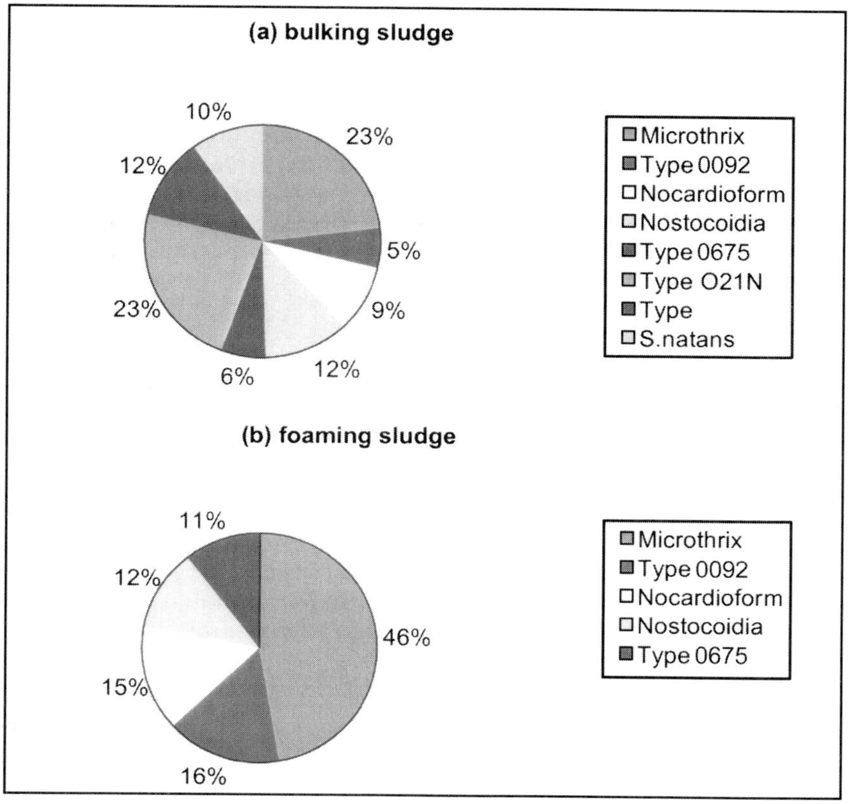

Fig. 32.8 Ranking of filamentous microorganisms in (a) bulking sludge and (b) foaming sludge.

placed under similar scrutiny using cluster analysis of their morphological and physiological characteristics (Soddell and Seviour, 1998).

With this approach comes the use of acronyms (e.g. NALO for *Nocardia amarae*-like organismns, PTLO for pine tree-like organisms and GALO for *Gordona amarae*-like organisms) underlining a desire to become technically correct when describing a species.

While essential for research into bulking and foaming, the diagnostic tools available to water industry practitioners (microscopes and stains) are much less sophisticated, but still fit for purpose. In this application the skill lies in knowing what first to look out for, appreciating the significance of subtle changes and then being able to act on that information.

4.2 Mechanisms and drivers

4.2.1 *Selection pressures*

The principles set out in Tomlinson and Chambers' (1984) benchmark paper 'Control strategies for bulking sludge' demonstrated how the substrate model developed by Chudoba and his co-workers (1985) worked in practice. Understanding the physiometabolic grouping of filamentous organisms responsible for bulking and foaming (Wanner and Grau, 1989) and the definition of the physicochemical properties of foam-forming sludges (Kocianova *et al.*, 1992) has gone some way to explaining how scum forms and control strategies work. Then, by quantifying design criteria for initial contact zones (Albertson, 1991) came the means to design selectors specific to the organisms involved (Foot and Forster, 1997; Wanner, 1998).

4.2.2 *Bubbles and buoyancy*

Biological foam on activated sludge plant consists of gas bubbles held in a dense matrix of fibrous material. Filamentous microbes form the basis of this structure which, when compared to sludge of a similar density, have a greater viscosity (Foot *et al.*, 1993). Therefore, the network of intermeshed microbial filaments within the foam layer provides structure, stability and buoyancy to flocs intended for use as a suspension, in a mixed reactor.

The relationship between the sludge ecology and foam formation is described thus:

- The affinity foam-forming nocardioforms have for hydrophobic substrates (Soddell *et al.*, 1998) explains why, once filamentous microbes exploit the foam environment, their domination becomes absolute.
- Subtle changes in the composition of the mixed liquor may provide the initial trigger to stimulate foam formation (Stratton *et al.*, 1998).
- Foams also have the ability to denitrify depending upon the dominating filamentous microbes and the type of substrate (Wanner *et al.*, 1998) and associated nitrogen gas would further increase the separation process.

4.2.3 *Bridging*

Bridging describes the physical separation of floc by filamentous microbes, which extend beyond the floc into the aqueous phase of the liquor. While surrounded by floc-forming bacteria the filamentous microbes provide structure to the floc but excessive filament growth within the floc leads to a dispersed floc. Bridging also places filamentous microbes in a position of advantage, where they are able to absorb substrate (sewage) in preference with little competition from the floc-forming bacteria.

4.2.4 *Exploitation*

The selection of filamentous microbes is to a large extent dependent upon the nutrient and metabolic conditions placed upon the biomass by the process itself. Eliminating the controllable variables such as oxygen, mixing patterns and solids retention time leaves us to consider temperature and substrate.

Species such as *M. parvicella* are known to predominate at certain times of the year (Wanner, 1998) but, unfortunately, little can be done about this other than have a strategy to deal with it.

Substrate appears to offer the greatest potential for control as selective substrates have long been associated with microbial selection, in particular surfactants. Recent investigations into micronutrient supplements

(Burgess *et al.*, 1999) returns research to an area better understood by biotechnologists with the potential to explain trigger mechanisms and enhanced treatment performance.

5 TREATMENT STRATEGIES

5.1 Background

While the types of filamentous organisms responsible for bulking and foaming appear not to have changed in recent years, treatment standards and waste treatment processes have. Anticipated problems associated with biological nutrient removal have not yet materialized in the UK as they have done elsewhere in the world because of the general adoption of chemical phosphorus removal. However, achieving compliance with UWWTD (Urban Waste Water Treatment Directive) standards has presented a new challenge when providing full treatment on restricted sites in coastal areas. The further requirement to disinfect final effluents discharging close to bathing waters extends the need for consistent treatment to a high standard. Given the robustness of the process, it is with little surprise that activated sludge (and its variants) remains a popular treatment option.

Treatment plant with suspended biomass, SBRs (sequencing batch-reactors), fixed film biomass BAF (biological aerated filter) plant, SBCs (submerged biological contactors) and more recently MBRs (membrane bioreactors) with a combined fixed and suspended biomass have been used in this application. Table 32.3 explores the relative merits of these plants.

Just as increased urbanization led to the development of conventional activated sludge processes in the middle of the last century so has the demand for high quality effluent treatment today resulted in the development of more intensive processes. Sewage treatability has also changed. For example, in coastal areas there is a likelihood of increased chloride levels, increased septicity and increased levels of industrial waste. Unless controlled, these factors will increase the tendency to sludge bulking and foaming.

Some plants remain unaffected by the symptoms of bulking and foaming, whether by design or by accident. The industry would suggest that control features now being incorporated into new plant designs avoid features that would have previously made plants vulnerable to bulking and foaming. Our challenge is to ensure that current experience influences future plant design.

5.2 Conventional sludge bulking and foaming control

5.2.1 *Managing expectation*

Tipping (1995) concluded, in his operator's overview of the activated sludge process, that 'The operational and ecological factors which are responsible for inducing microbial growth, sufficient for foam formation, are largely unresolved and likely to be site specific'. This resulted in the frequent application of physical control measures along with the expectation that the problem would likely recur. While the temptation exists to 'bend the biology' of activated sludge by expecting more from the process than it can deliver, what we now have is a greater awareness of the requirements (and hence limitations) of the organisms involved in waste treatment.

TABLE 32.3 Design features of treatment plant

	Conventional	*SBR*	*BAF*	*SBC*	*MBR*
Primary treatment required	✓	✗	✓	✓	✗
Final tanks required	✓	✗	✗	✓	✗
Disinfected effluent	✗	✗	✗	✗	✓
Relatively easy to cover	✗	✓	✓	✓	✓
Simple controls	✓	✗	✗	✓	✓

5.2.2 *Solution pathway*

Conventional wisdom applied to resolve a sludge bulking or foaming event has been to:

- Review operating and control procedures and make required corrections
- Employ short-term remedial action if necessary.

And if the problem persists:

- Review plant design parameters
- Implement design changes.

The first of these two steps requires an initial understanding of the process and the key performance measures. A greater understanding of the process is obtained when both cause and effect are measured, thus leading to an overall improvement in performance. Only when this initial cycle fails does it become necessary to take the second step of reviewing plant design. Improved awareness of the requirements and limitations of activated sludge organisms gives structure to the control measures employed and direction to the recommendations made for plant design.

This approach has formed the basis for improvements to plant design and operation in resolving persistent bulking and foaming problems. However, if significant changes occur in the incoming sewage or the process control, then bulking and foaming will still happen, irrespective of how well the plant is designed or operated. It is therefore important to know the relative efficiency and the cost of available solutions.

5.2.3 *Common cures*

Given that bulking and foaming can seriously compromise effluent quality, it is essential to reduce the risk of discharge consent compliance failure as soon as possible. Table 32.4 illustrates the most commonly applied bulking and foaming control measures for conventional activated sludge plant along with an estimated response time and cost. Not all solutions apply to every type of plant (e.g. importing MLSS to large plants), nor are some of them sustainable for prolonged periods of time (e.g. increased RAS pumping and chlorine dosing). Response times also vary depending upon the nature of the activity. While control changes can have an immediate effect on solids loss, measures exerting selective pressures on the biomass (selector or MLSS reduction) can take at least one cycle of the solids retention time (sludge age) to take effect.

5.2.4 *Best working practice*

Building your own knowledge database is relatively easy to do and essential to assure that best practice is employed when first assessing the nature of the problem. Making sure that this approach is transferred through time and to new processes is more difficult to achieve. However, problems can be resolved by different means, sometimes novel and sometimes with little reference to perceived wisdom.

5.2.5 *Better by design*

Good aeration plant design is only the first step in the elimination and prevention of activated sludge bulking or foaming. Basic principles include:

TABLE 32.4 Summary of commonly used cures for sludge bulking and foaming

	Response period	*Cost*	*Application*
Reduce MLSS	1–3 weeks	Nil	all plant
Import MLSS	Immediate	£0.50 per capita	small plant
Increase RAS rate	Immediate	Nil	all plant
Nutrient dose	Immediate	£1.00 per trade p.e./annum	all plant
Chlorine dose	1–10 days	£0.01 per capita	large plant
Sprays	Immediate	£3k per tank	all plant
Selector	1 month	Site specific	all plant

- Attention to the layout of new designs (e.g. avoiding complete-mix systems in favour of plug-flow) to promote the growth of good settling bacteria.
- It has also been observed that plants with insufficient aeration capacity to cope with diurnal load variations or breakdown of aeration equipment are more prone to outbreaks of filamentous bacteria.
- Ability to return a maximum of 150% dry weather flow as RAS (return activated sludge) is also important in counteracting the effects of poorly settling sludge.
- Spray bars fitted to final tank scrapers are also useful for dispersing and avoiding the proliferation of foams.
- Final tanks also need to be large enough to cope with a peak SSVI of 120 ml/g. However, if selector zones are incorporated into the plant, then final tanks can be sized to a lower value of 100 ml/g.
- Selectors designed purely to discourage filamentous growth would have an average retention time of between 10 and 20 minutes (Wanner, 1998), while tanks designed to remove nitrate would have a much longer retention time.

Success rates for selectors vary depending on the application and ability to document accurately the effect due to plant modifications that are invariably linked to construction work. The design favoured by many has a high F:M (food to mass) loading rate of 12 kg BOD/kgMLSS/day in the initial contact zone and 3 kg BOD/kgMLSS/day overall.

5.2.6 *Better by operation*

While good design will increase the tendency towards a good, settleable activated sludge plant, operators need to be given flexibility to change operating parameters as conditions demand. Adjustable dissolved oxygen targets, avoidance of prolonged over-aeration, variable RAS and SAS rates, control of final tank blanket levels, etc. are all useful tools for operators to employ to reduce the effects of foaming and bulking. Standard operational process changes can be very effective in combating filamentous growths if they occur. These would normally be to:

- Reduce mixed liquors concentration (to avoid solids loss)
- Increasing/decreasing dissolved oxygen levels (to correct under/over-aeration)
- Increase RAS rate (to reduce final tank blanket levels)
- Taking final tanks out of service (to reduce the effects of denitrification).

There is often the temptation (and pressure) to carry out all these changes at once, but this makes it difficult to evaluate any effects. As these changes act on the biomass slowly, they are often (on their own) insufficient to prevent a series of effluent discharge consent violations.

One fallacy is that chemicals will provide a permanent quick fix. Chemicals employed at wastewater treatment plants to combat the effects of filamentous bacteria include settlement aids such as polymers and coagulants. Examples of the latter include, iron and aluminium compounds and even talcum powder! Bio-augmentation with enzymes and other cultured bacterial mixes have also been on trial, but the results are often inconclusive. Inorganic chemicals, such as bleach and hydrogen peroxide, are also proven in disrupting bacterial filaments. A timely application of chlorine or bleach (Jenkins *et al.*, 1986) encourages a rapid return to the biomass equilibrium associated with a well-designed and operated plant.

Control strategies for foaming in anaerobic digesters follow similar patterns:

- Dosing with poly-aluminium salts and chlorine
- Mechanical disruption
- Decrease feed rate
- Lower reactor level.

5.3 Solution review

Reviewing sludge bulking and foaming control for conventional activated sludge processes, it is apparent that new control measures come from a better understanding of the sludge

TABLE 32.5 Bulking and foaming control for treatment plant

	Conventional	*SBR*	*BAF*	*SBC*	*MBR*
Reduce MLSS	✓	✓	✓	**n/a**	✓
Import MLSS	✓	✓	✗	**n/a**	✓
Increase RAS rate	✓	✗	✗	**n/a**	✗
Nutrient dose	✓	✓	✓	✓	✓
Chlorine dose	✓	✓	✗	**n/a**	✓
Sprays	✓	✓	✓	**n/a**	✓
Selector	✓	✓	✗	**n/a**	✓

Note: **n/a** = control not applicable.

microbiology. When all filaments in activated sludge were generically termed 'sewage fungus', little could be done to exploit the metabolic preferences of the organisms comprising that group. Thankfully that is not now the case and the operator has a range of options to consider.

Looking at the new processes now employed in treating sewage from coastal areas, it can be seen that not all of the old solutions employed to combat bulking and foaming are appropriate or even necessary (Table 32.5). While combined fixed film and suspended biomass systems can experience forms of sludge bulking and foaming the absence of some process flows (e.g. sludge recycle) reduces the options available for control.

The activated sludge process has been suitably challenged following tighter effluent quality standards such as N and P (nitrogen and phosphorus) removal and disinfection. Irrespective of how well the plant is designed, problems will occur. Conventional methods for the prevention or elimination of sludge bulking and foaming have proved adequate provided that:

- Decisions are made in a logical order
- Contingency plans are in place to offer a quick response
- The treatability of sewages are correctly assessed before designing new plant
- Salinity and septicity of sewages from coastal sites are brought within reasonable limits
- The treatment process operating procedures are flexible
- More use is made of existing data to identify problems before they impact on plant performance.

6 CASE STUDIES

Examples of bulking and foaming in the South East Dorset area of the UK demonstrate how these problems can be resolved. Table 32.6

TABLE 32.6 Summary of bulking and foaming case histories in the UK

STW design	*Problem*	*Cause*	*Solution*
Surface aeration	Filamentous bulking	Excessive aeration in anoxic zone	Chlorine dose
Fbda	Microthrix foaming	Excessive aeration in anoxic zone	Reduce aeration
Fbda	Biopolymer production	Low DO levels	Spray bars, intensive aeration
Ditch	Nocardia foaming	Digested brewery waste	Chlorine dosing
Surface aeration	Microthrix bulking	Sludge spillage	Chlorine dosing
Ditch	Filamentous foaming	Overloaded, low DO	Additional aerators
Surface aeration	Non-settling sludge	Metal dust discharge	Re-seed
Package plant	Flow bulking	Hydraulic flushing	Re-seed and tankering
Surface aeration	Filamentous bulking	No known cause	Selector zone and spray bars
Ditch	Filamentous bulking	No known cause	Selector zone and spray bars
SBR	Flow bulking	Poor hydraulic design	Reduce mlss and process review

Fig. 32.9 Swanage MBR plant showing, (a) site excavation, (b) empty aeration tank with diffusers but no membranes, (c) membranes and connecting pipework.

summarizes how recent bulking and foaming events have been tackled to ensure compliance with effluent quality standards.

More recent experience with secondary treatment on UWWTD (urban waste water treatment directive) sites has led to an awareness of several important issues not yet fully understood.

Space restrictions for constructing wastewater treatment plant on the coast are illustrated in Fig. 32.9(a). The area under excavation defined the limit of the land available to install secondary treatment to a resort town on the south coast of the UK. Proximity to bathing beaches required the effluent to be disinfected. The MBR plant (19 000 connected population) is now complete and enclosed. Aeration tank detail includes diffusers positioned along the length of the channel (Fig. 32.9(b). Fig. 32.9(c) illustrates the type of membrane used and the associated pipework through which treated effluent would pass.

Compact plant designs make good process control an essential prerequisite. For example, taking tanks out of service for re-seeding an SBR plant with activated sludge is not a viable option as both bioreactor and settlement capacity are lost. Similarly, an MBR and BAF plant require lead-in time to build up sufficient biomass or seed the media. Not that these processes are inflexible rather, remedial controls should be incorporated into the design as a precautionary measure. Installing spray bars and chlorine dosing lines is more easily done during construction, chlorine is often available for odour treatment or chemical cleaning.

7 FURTHER DEVELOPMENT

7.1 Potential options for the future

Understanding the interactions between cause and effect in the activated sludge process is not easy, especially when there may be a dozen or so reasons for a bulking or foaming incident to occur. The process represents the largest biotechnology industry in the world and the combinations of inputs, control variables and configurations are daunting. Mapping the processes involved is seen as one way of putting structure to the apparent chaos of the operating process. Operating parameters are often documented within procedures (held in site manuals) and the implementation of changes to the plant is controlled by standard work instructions. The discipline of managing these activities is fundamental to the efficient running of treatment plants and control of quality and costs.

7.1.1 Process mapping

By way of an overview, the process map presented in Fig. 32.10 illustrates how the attributes of a plant, such as influent, is linked to operating variables such as flow strength and composition. Bulking and foaming symptoms are then placed against each of these variables along with the suggested solution to resolve the problem. Controls, system configuration, and treatment capacity attributes have also been applied to the same matrix of conditions, symptoms and solutions. While not exhaustive in terms of detail the process map offers a mechanism by which both:

- The myths behind the science can be dispelled and
- The gaps in our knowledge can be filled.

The next logical step would then be to review improvements that have been made, measure them in some way and incorporate them into new working practices or designs.

7.1.2 Process modelling

Process models have been used for many years and models (such as STOAT and GPSX) are readily available and often used for new works designs. Modelling is particularly relevant when an existing plant is to be extended and its performance assessed. However, to minimize the risk of bulking and foaming, process models need to be capable of defining the risk or probability for this occurring.

7.1.3 Performance management

Two areas where advances have been made are in that of process monitoring and information

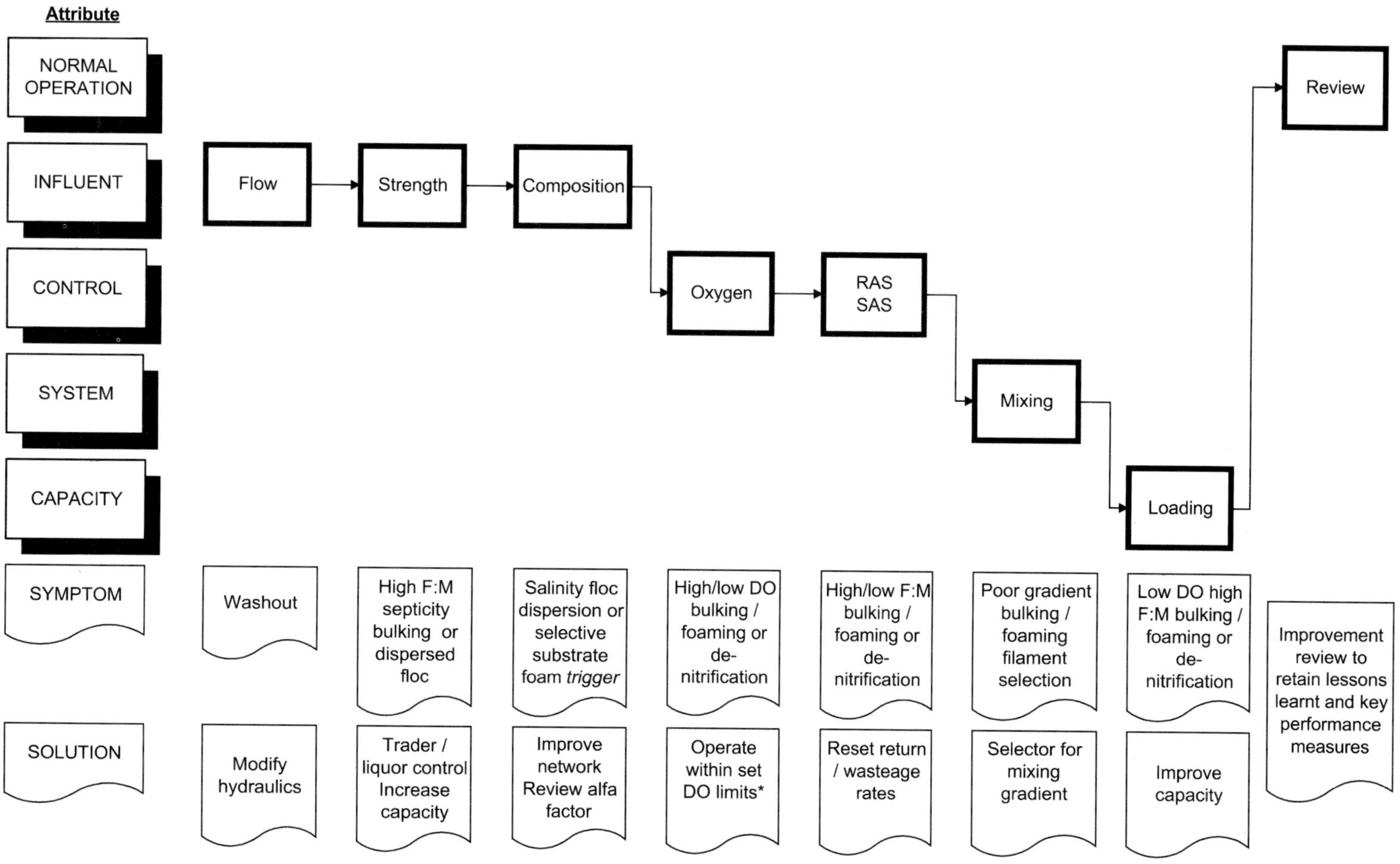

Fig. 32.10 Process flow for the control of bulking and foaming.

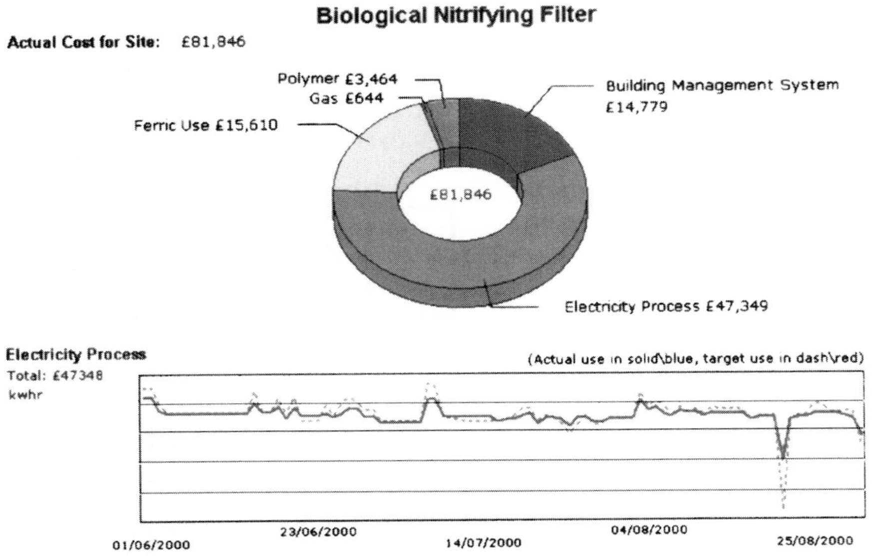

Fig. 32.11 The relative costs of operating a modern waste treatment process. Courtesy of Meniscus, UK.

networks. It is now possible to measure plant performance against benchmark values and treatment quality to provide unit process cost (*http://www.meniscus.co.uk*). This information can then be relayed to plant operators and managers for appropriate action. Activated sludge treatment is basically the same regardless of the type of plant discussed, which is probably why many of the remedies for bulking and foaming in conventional plant now apply to new plant with suspended biomass. Similar pressures are exerted upon the biomass irrespective of the type of process involved and the same filamentous microbes that cause separation problems in conventional plant are present also in the new treatment processes.

Variables include sewage (such as flow, load and substrate type), consumables (power and oxygen requirement), while outputs include waste solids settlement (and disposal) and effluent quality. Broadly speaking it is a case of balancing inputs with outputs and, although the configuration might change the values to be measured remain the same. Fig. 32.11 compares the relative costs of these components in a modern sewage treatment process.

Benchmarking has already indicated that the typical aeration cost for an activated sludge plant is £0.009 per m^3 treated (based on an electricity price of £0.035 per kWh), which is a valuable starting point in evaluating the cost performance of a works. One would expect an optimum unit treatment cost for any given process and that an excessively high or low unit cost would reflect an imbalance in the inputs or outputs. This will help identify conditions that may give rise to sludge bulking or foaming.

By combining expert systems, advanced data management and summary data from telemetry, it is now feasible to identify known operating problems at a site and to suggest appropriate remedial action within minutes of inputting the original raw data.

With advances in internet and mobile phone technology the day is not far off when operators will enter site readings into their mobile phone and, using the Internet, view process and cost exceptions, operating suggestions, trends and telemetry data on their phone while they are still on site.

REFERENCES

Albertson, O. (1991). Bulking sludge control – progress, practice and problems. *Wat. Sci. Tech.* **23**, 835–846.

Anon. (1969) Milwaukee mystery: unusual operating problems develop. *Water and Sewage Works*, 116–213.

Blackall, L., Seviour, E., Bradford, D., Cunningham, M., Seviour, R. and Hugenholtz, P. (1994). *Microthrix parvicella* is a novel, deep branching member of the Actinomycete subphyllum. *Sys. Appl. Microbiol.* **17**, 513–518.

Burgess, J., Quarmby, J. and Stevenson, T. (1999). Micronutrient supplements to enhance the biological wastewater treatment of phosphorus-limited industrial effluents. *Trans. I. Chem. E.* **77**, 199–204.

Chudoba, J., Cech, J. and Grau, P. (1985). Control of activated sludge filamentous bulking – experimental verification of a kinetic selection theory. *Wat. Res.* **19**, 191–196.

Chartered Institution of Water and Environmental Management (1997). *Handbooks of UK Wastewater Practice: Activated Sludge Treatment.* 15 John Street, London.

Dillner Westlund, A., Hagland, E. and Rothman, M. (1998). Foaming in anaerobic digesters caused by *Microthrix parvicella*. *Wat. Sci. Tech.* **37**(4/5), 51–55.

Dormoy, T., Tisserand, B. and Herremans, L. (1999). Impact of the volume of rain water on the operating constraints for a treament plant. *Wat. Sci. Tech.* **39**(2), 145–150.

Donaldson, W. (1932). Some notes on the operation of sewage treatment works. *Sew. Works J.* **4**, 48–59.

Eikelboom, D. (1975). Filamentous organisms observed in bulking activated sludge. *Wat. Res.* **6**, 345–388.

Eikelboom, D. and van Buijsen, H. (1983). *Microscopic Sludge Investigation Manual.* TNO Institutre, Delft.

Foot, R. (1992). The effects of process control parameters on the composition and stability of activated sludge. *J. IWEM* **6**(4), 215–278.

Foot, R. and Forster, C. (1997). The operation of a selector for the control of foam forming bacteria in activated sludge. *Environ. Technol.* **18**, 237–241.

Foot, R., Robinson, M. and Forster, C. (1993). Operational aspects of three 'selectors' in relation to aeration tank ecology and stable foam formation. *J. IWEM* **7**(6), 304–309.

Foot, R., Kocianova, E., Forster, C. and Wilson, A. (1993). An examination into the structure of stable foams formed on activated sludge plants. *J. Chem. Tech. Biotech.* **56**, 21–24.

Forster, C. (2000). Understanding the microbiology of the activated sludge process. In *Proceedings Lessons of the Last Decade.* CIWEM Seminar, Nottingham.

Howarth, R., Head, I. and Unz, R. (1998). Phylogenetic assessment of five filamentous bacteria isolated from bulking activated sludges. *Wat. Sci. Tech.* **37**(4/5), 303–306.

Howarth, R., Unz, R., Seviour, E. *et al.* (1999). Phylogenetic relationships of filamentous sulfur bacteria (*Thiothrix* spp and Eikelboom type 021N bacteria) isolated from wastewater treatment plants and description of *Thiothrix eikelboomii* sp nov., *Thiothrix unzii* sp nov., *Thiothrix fructosivorans* sp nov. and *Thiothrix defluvii* sp nov. *Int. J. Sys. Bact.* **49**, 1817–1827.

Jenkins, D., Richard, M. and Daigger, G. (1986). *Manual on the Causes and Control of Activated Sludge Bulking and Foaming.* Water Research Commission, Pretoria.

Kocianova, E., Foot, R. and Forster, C. (1992). Physicochemical aspects of activated sludge in relation to stable foam formation. *J. Instn. Wat. Envir. Mangt.* **6**(3), 342–350.

Morgan, E. and Beck, A. (1928). Carbohydrate waste stimulates growth of undesirable filamentous organisms in activated sludge. *Sew. Wks. J.* **1**, 46–51.

Pagilla, K., Craney, K. and Kido, W. (1997). Causes and effects of foaming in anaerobic sludge digesters. *Wat. Sci. Tech.* **36**(6/7), 463–470.

Pagilla, K., Jenkins, D. and Kido, W. (1998). Nocardia effects in waste activated sludge. *Wat. Sci. Tech.* **38**(2), 49–54.

Pitman, A. (1996). Bulking and foaming in BNR plants in Johannesburg, problems and solutions. *Wat. Sci. Tech.* **34**(3/4), 291–298.

Pitt, P. and Jenkins, D. (1990). Causes and control of *Nocardia* in activated sludge. *J. Wat. Pollut. Control Fed.* **62**, 143–150.

Pujol, R., Duchene, Ph., Schetrite, S. and Canler, J. (1991). Biological foams in activated sludge plants: characterisation and situation. *Wat. Res.* **25**(11), 1399–1404.

Soddell, J. and Seviour, R. (1998). Numerical taxonomy of Skermania piniformis and related isolates from activated sludge. *J. Appl. Microbiol.* **84**(2), 272–284.

Soddell, J., Seviour, R., Blackall, L. and Hugenholtz, P. (1998). New foam-forming nocardioforms found in activated sludge. *Wat. Sci. Tech.* **37**(4/5), 495–502.

Stratton, H., Seviour, B. and Brooks, P. (1998). Activated sludge foaming: what causes hydrophobicity and can it be manipulated to control foaming? *Wat. Sci. Tech.* **37**(4/5), 503–509.

Tipping, P. (1995). Foaming in activated-sludge processes: an operator's overview. *J. Instn. Wat. Envir. Mangt.* **9**(6), 281–289.

Tomlinson, E. (1976). *Bulking – A survey of activated sludge plants.* Water Research Centre, Medmenham. Tech. Rpt., TR35.

Tomlinson, E. and Chambers, B. (1984). Control strategies for bulking sludge. *Wat. Sci. Tech.* **16**(10/12), 15–34.

Wanner, J. (1994). *Activated Sludge Bulking and Foaming Control.* Technomatic Publishing Co. Inc, Lancaster.

Wanner, J. (1998). Stable foams and sludge bulking: the largest remaining problems. *J. CIWEM* **12**(10), 368–374.

Wanner, J. and Grau, P. (1989). Identification of filamentous micro-organisms from activated sludge: a compromise between wishes needs and possibilities. *Wat Res.* **23**, 883–891.

Wanner, J., Ruzickova, I., Jetmarova, P., Krhutkova, O. and Paraniakova, J. (1998). A national survey of activated sludge separation problems in the Czech Republic: filaments, floc characteristics and activated sludge metabolic properties. *Wat. Sci.Tech.* **37**(4/5), 271–279.

Water Pollution Research Laboratory (1967). Effect of hydraulic mixing characteristics of aeration units on the physical properties of the activated sludge. In *Water Pollution Research 1966*, pp. 69–71. HMSO, London.

33

Odour generation and control

Arthur G. Boon* and Alison J. Vincent†

*Halcrow Water, Cranley; †Hyder Consulting, Cardiff, UK

1 INTRODUCTION

Fresh sewage has a characteristic musty smell resulting from its components; a mixture of discharges from toilets, baths, sinks, dishwashers and washing machines, and industrial wastes. The mixture of malodours typically contains compounds utilized in cleaning agents used in the home (such as limonene), solvents (such as tetrachloroethylene), petrol derivatives (such as benzene) as well as the odours associated with human waste, such as urea and ammonia from urine and skatole and indole (breakdown products of tryptophan) from faeces.

Although the smell of fresh sewage is not pleasant, the odorous compounds given above rarely give rise to odour problems. It is the strong objectionable malodours, in particular hydrogen sulphide, that develop when sewage and sewage sludges are stored, which give rise to odour problems. These strong objectionable malodours are nearly always associated with the development of anaerobic conditions and the activity of anaerobic bacteria as soon as nitrate (and nitrite) and dissolved oxygen are exhausted. This anaerobic condition of sewage or sludge is often termed septicity. The chemicals commonly responsible for sewage odour are given in Table 33.1. All the compounds given in Table 33.1 are produced by the action of bacteria on components of wastewater. Table 33.1 also gives the concentrations in the air at which they can be detected by the human nose (the threshold odour concentration).

Septicity can occur whenever sewage or sludge is retained for more than a few hours without replenishment of dissolved oxygen and in the absence of nitrate (or nitrite) salts. Typically sewage becomes septic in pumped (rising-main) sewers, primary settlement tanks, storm storage tanks and flow-balancing tanks. Primary sludges and mixed primary and biological sludges become septic in primary settlement tanks, storage tanks, and during anaerobic digestion. Sludges containing primary sludge are likely to become septic very rapidly because the numbers of microorganisms and the available substrates per unit volume are greater than in sewage.

The consequences of sewage or sludge becoming septic are important for a number of reasons, in addition to the potential for odour nuisance:

1. Hydrogen sulphide in the atmosphere is an extremely toxic gas at concentrations above 300 ml/m^3 (ppm) and has been the cause of deaths in enclosed areas. It is a particularly dangerous gas as, although it can be smelt at very low levels (0.5 μl/m^3 (ppb)), it cannot be smelt at concentrations above 150–250 ppm, and at the concentrations at which it is toxic (World Health Organization, 1987).
2. Hydrogen sulphide that has been released into the atmosphere can cause corrosion of metals (such as copper (and many of its alloys), silver as well as iron or steel) and is a significant cause of corrosion of concrete, as it may be biochemically oxidized on moist surfaces to sulphuric acid by various autotrophic *Thiobacillus* spp. such as *Thiobacillus concretivorus* (Parker, 1947).

The Handbook of Water and Wastewater Microbiology
ISBN 0-12-470100-0

TABLE 33.1 Threshold odour concentrations of most common sewage odour compounds (Woodfield and Hall, 1994; Bonnin *et al.*, 1990)

Class	*Compound*	*Chemical formula*	*Description*	*Threshold odour concentration*[a] *($\mu l/m^3$)*
Sulphides	Hydrogen sulphide	H_2S	Rotten egg	0.5
	Methylmecaptan (methanethiol)	CH_3SH	Cabbage, garlic	0.0014–18
	Ethylmercaptan (ethanethiol)	C_2H_5SH	Rotten cabbage	0.02
	(Di)methylsulphide	$(CH_3)_2$-S	Rotten vegetable	0.12–0.4
	(Di)methyl disulphide	$(CH_3)_2$-S_2	Putrefaction	0.3–11
Nitrogen containing	Ammonia	NH_3	Very biting, irritating	130–15300
	Methylamine	CH_3NH_2	Rotten fish	0.9–53
	Dimethylamine	$(CH_3)_2$-NH	Fish	23–80
	Indole	C_8H_6-NH	Faecal, nauseating	1.4
	Skatole (3-methyl indole)	C_9H_8-NH	Faecal, nauseating	0.002–0.06
Acid	Acetic	CH_3-COOH	Vinegar	16
	Butyric (butanoic)	C_3H_7-COOH	Rancid	0.09–20
	Valeric	C_4H_9-COOH	Sweat, transpiration	1.8–2630
Aldehydes and	Formaldehyde	H-CHO	Acrid, suffocating	370
Ketones	Butyraldehyde	C_3H_7-CHO	Rancid	4.6

[a] The threshold odour concentration of a specific compound is the concentration at which it can only just be detected by a panel of at least 8 people. A similar technique, olfactometry, can be used to measure the strength of an unknown odorous gas containing a mixture of compounds. In this case the odour strength (in units of ou/m^3) is defined as the number of dilutions of an odorous gas required before it can be detected by only 50% of a panel of at least 8 people (CEN (1999) European Committee for Standardization). A range of values is given by Woodfield where quality of data is uncertain.

However great the concentration of odorous compounds that develop in sewage or sludge, they have impact only when they are transferred into the atmosphere, particularly beyond the boundary of the works and to the vicinity of a potential complainant. Therefore, in addition to the factors affecting the production of malodours, the factors leading to transport to locations where they may cause nuisance, and the factors affecting their release also need to be considered when dealing with a potential odour problem. Hydrogen sulphide and organic sulphides that remain dissolved and are subsequently oxidized chemically or biochemically while in solution will not create odour problems.

An important factor in the amount released is the chemical state of the odorous compound within the liquid phase, which is affected by the pH value of the sewage or sludge. Factors affecting the pH value of the sewage or sludge are, therefore, also important in the consideration of the potential for odour nuisance. Low pH values favour the emission of H_2S, mercaptans and volatile fatty acids, while high pH values favour the emission of ammonia and amines. Although odours at high pH values can be unpleasant, they are generally less intense, and increasing the pH value of sludges by chemical addition has been used as an odour control technique.

The solubility of H_2S in water varies with temperature from 4 ml/ml at 4°C to 2.55 ml/ml at 20°C, although the amount dissolved will also be directly proportional to the total pressure and to the partial pressure of H_2S in the gaseous phase, according to Henry's law. H_2S is a weak dibasic acid which dissociates as follows:

$$H_2S \Leftrightarrow H^+ + HS^- \Leftrightarrow 2H^+ + S^{2-}$$

The proportions of H_2S, HS^- and S^{2-} are significantly affected by pH value and will also be slightly affected by temperature and ionic strength (EPA Process Design Manual for Sulphide Control and Sanitary Sewerage Systems, 1974). The proportion of dissolved sulphide, in an associated (non-ionic) form as H_2S, can vary from 75% at pH 6.5 to 50% at pH 7 and only 10% at pH 8. Between pH 9 and 10, the dissolved sulphide would be almost entirely present as HS^- ions. Some of the sulphide may also be present as insoluble metal sulphides, having combined with metals in the sewage. In the UK, the pH value of sewage is normally within the range 6.5 to 8 (depending on the alkalinity of the water) and the formation of volatile fatty acids that will occur under anaerobic conditions can cause the pH value to decrease below 6.

2 DEVELOPMENT OF MALODOURS

Sewage and sludge malodours are predominantly the result of the action of anaerobic and facultative anaerobic bacteria. These bacteria are derived from the original wastewater. Anaerobic bacteria, which grow more slowly and have a lower cell yield than their aerobic counterparts, are selectively encouraged to grow by the conditions in the various stages and processes. The steps of increasing levels of anaerobicity leading to the formation of malodours include:

1. Depletion of dissolved oxygen and nitrate by microorganisms present in sewage and slimes
2. Microbial reduction of organic sulphides and nitrogen compounds by anaerobic and facultative anaerobic proteolytic bacteria under anaerobic conditions, giving rise to dimethyl sulphide, ammonia, amines, mercaptans and other malodorous compounds
3. Production of volatile fatty acids (e.g. butyric, propionic, lactic, acetic) by microbial lipolysis by anaerobic and facultative anaerobic bacteria, also leading to a reduction in pH value and increased release of any sulphur containing malodours
4. Microbial respiration of sulphate by strict anaerobes (the sulphate-reducing bacteria) giving rise to H_2S. This is the most important of the odour-generating processes
5. Microbial breakdown of volatile fatty acids giving rise to methane (anaerobic digestion) in digester gas. In the digester, methanogenesis occurs concurrently with hydrolysis, acidogenesis, proteolysis and sulphate reduction and biogas will contain hydrogen sulphide and other malodorous sulphur compounds. However, the residual odour of well-digested sludge, after cooling, is less than that of primary sludge.

These steps are accompanied by a decrease in the redox potential of the sewage, as illustrated in Fig. 33.1 (Boon, 1995). Steps 2, 3, 4 and 5 are of particular importance with respect to sludge odours. Step 4 is of particular importance with respect to sewage malodours.

2.1 Depletion of dissolved oxygen and nitrate

Electron receptors for microbial respiration are (in order of decreasing preference) oxygen, nitrate, sulphate and carbonate. Their utilization under increasingly anaerobic conditions produces respectively water, nitrogen, hydrogen sulphide and methane (Fig. 33.1). Anaerobicity, or 'septicity', is considered to occur in sewage or sludge when the microorganisms that are present have utilized all the dissolved oxygen (DO) and any nitrates (or nitrites) that may be present.

The rate at which the DO will be used by the microorganisms (aerobic respiration) depends on the 'age' and temperature of the sewage and the area of the submerged surfaces. The 'age' of sewage affects the respiration rate of the sewage because the numbers of microorganisms will increase while sewage remains in a sewer under aerobic conditions. Under anaerobic conditions, complex soluble, suspended and colloidal organic compounds, such as carbohydrates and protein will be broken down into readily biodegradeable and soluble volatile fatty and amino acids. The respiration rate will eventually reach a maximum as readily biodegradeable substrates become limiting and it will then start to decline. In practice, the rate of uptake of DO by aerobic sewage has been found to vary from 2 to 3 mg/l/h for 'young' domestic sewage and up to 14 mg/l/h at 15°C, as the sewage 'ages' during its passage through the sewerage system

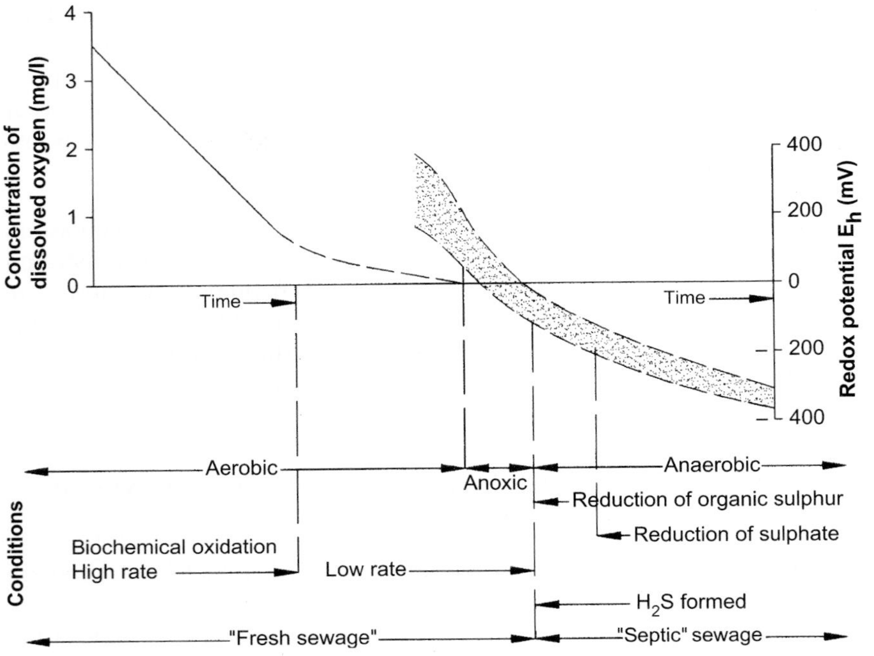

Fig. 33.1 Variations in condition of sewage in relation to concentration of dissolved oxygen and redox potential.

(Boon, 1995). A typical average rate in the UK is about 6 mg/l/h at 15°C. Bacterial slimes that adhere to submerged surfaces in sewers are generally not limited by available substrate; they have been found to consume DO at a rate of about 700 mg/m^2/h at 15°C (Boon and Lister, 1975). The consequence of respiration in both the sewage and slimes occurring simultaneously is that the sewage will rapidly become devoid of DO, particularly in a rising-main sewer of small diameter as the surface area per unit volume will be large compared with a sewer of large diameter.

In gravity sewers, the slope of the sewer is designed so that grit and other debris do not accumulate, and under average flow-rate conditions, the velocity of sewage generated by the slope will normally create sufficient turbulence to ensure that the rate at which oxygen dissolves from the atmosphere in the sewer exceeds the respiration rate of the microorganisms growing in the sewage and on the submerged surfaces. Under such conditions, septicity will not occur, although volatile and odorous compounds already present in sewage will be stripped from solution and into the atmosphere. However, such compounds are unlikely to cause a nuisance, unless the sewage had become septic upstream of the turbulent conditions.

If aerobic sewage (containing DO) enters a rising-main sewer or other process stage, where there is no opportunity for the sewage to be re-aerated, the concentration of dissolved oxygen will decrease due to the respiration of sewage and slimes. The rate that microorganisms consume DO will remain fairly constant (see Fig. 33.1) until the concentration reaches about 0.2 to 0.4 mg/l. At about 0.2 mg/l, the rate will decline and asymptote to zero at zero DO. However, as the DO becomes rate limiting, any oxidized nitrogen present in the sewage will provide an alternative electron acceptor for the dissimilation of organic matter (in the same process as denitrification in the anoxic zone of an activate-sludge process). Under such conditions, the microorganisms will continue to 'respire' and oxidize substrate but at a slower rate of about 40% of the aerobic rate. Under these 'anoxic' conditions, the redox potential will decline from about +50 mV to about −100 mV (E_h). Nitrate levels are rarely high in domestic sewage, but addition of nitrate salts (normally of calcium or magnesium) has been used as an effective means of septicity control at a large number of sites. When all DO and nitrate is utilized the sewage then becomes septic. Time taken for sewage to cease being aerobic and to become anaerobic will vary according to the concentrations of microorganisms and the substrates present; it is likely to be in the range from 30 minutes to 2–3 hours (Boon and Lister, 1975).

Septicity is most likely to occur during warm conditions, when an increase in temperature of the sewage leads to increased respiration and growth of the microorganisms and rapid depletion of a reduced concentration of dissolved oxygen. As the activity of anaerobic bacteria also increases with temperature, odour problems are most frequently associated with warm summer months.

The causes of septicity developing in primary sludges are similar to those given above for sewage in sewers and storage tanks, although the effect of microorganisms growing on submerged surfaces is likely to be less than for sewage in pipes. In primary sludges, depletion of dissolved oxygen and nitrate occurs very rapidly because the numbers of microorganisms in the sludge are several orders of magnitude higher than in sewage and the availability of substrate per unit volume is also much greater. It is unlikely that primary sludge will contain dissolved oxygen or nitrate.

Surplus activated or humus sludges, derived from aerobic biological treatment processes, may contain dissolved oxygen and nitrate (particularly if the aerobic biological processes achieved complete nitrification). Such sludges become septic less rapidly than primary sludges, unless co-settled with primary sludge. The depletion of nitrate, by biological sludges in settlement tanks, can be a cause of problems of rising sludge, due to the formation of micro-bubbles of evolved nitrogen derived from nitrate. Humus sludge, from a nitrifying biological filter, is unlikely to contain nitrate as the retention period in the humus settling tank is likely to be long and any nitrate in the filter effluent will have been 'respired' by heterotrophic bacteria in the sludge.

2.2 Microbial reduction of organic sulphides and nitrogen compounds under anaerobic conditions

In the absence of DO, proteolytic bacteria (which may be facultative anaerobes or anaerobes) break down proteins in the sewage into amino acids, and subsequently degrade the sulphur-containing amino acids, cysteine, cystine and methionine to produce thioles (mercaptans), and other organic sulphides and disulphides. These compounds smell very obnoxious when released to the atmosphere and have very low odour thresholds (see Table 33.1). The proteolytic bacteria are active at a higher redox potential than those which subsequently reduce sulphate to form H_2S.

Domestic sewage normally contains about 3–6 mg/l of organic sulphur, which is present mainly as proteinaceous matter, and additional organic sulphur in the form of sulphonates (about 4 mg/l), derived from household detergents. The ability to produce sulphide from protein is quite common among bacteria, for instance *Proteus* spp., *Bacteroides* spp. and some *Clostridium* spp., all of which can grow anaerobically (Crowther and Harkness, 1975).

Volatile sulphides produced in this way are a small proportion of the malodours associated with sewage but can be very significant in the odours from sludges where the concentration of the bacteria and substrates are higher. Recent measurements on-site have shown that concentrations of dimethyl sulphide and/or thiols in air and/or biogas in contact with sludge may be occasionally greater than the concentration of hydrogen sulphide in the air and/or biogas.

2.3 Production of volatile acids

In addition to the action of proteolytic bacteria under anaerobic conditions in the breakdown of protein, there are also the bacterial actions of hydrolysis, fermentation and lipolysis on carbohydrates, fats and oils leading to the formation of short-chain volatile fatty acids.

Under aerobic conditions in sewers and primary settlement tanks, the concentration of the volatile fatty acids (VFAs) developing in sewage is relatively low, and is rarely the main cause of odour nuisance. Under anaerobic conditions, the formation of VFAs may lead to a lowering of the pH value, resulting in the enhancement of release of any hydrogen sulphide formed under such conditions. Significant concentrations of VFAs may develop in process stages which deliberately introduce anaerobic conditions, e.g. the use of a primary sludge fermentation stage to increase the concentration of readily biodegradeable BOD to enhance biological phosphorus removal (see Chapter 23), or the use of anaerobic treatment processes for the partial treatment of high BOD wastewaters or sludges.

In sludges, the reduction of carbohydrates (including polysaccharides), fats, oils and greases is the 'acid forming' stage of the anaerobic digestion process as described in Chapter 24. When the process is contained within a heated anaerobic digester, the volatile acids are rapidly converted to methane. However, in primary sludges the presence of volatile fatty acids contributes to the characteristic and obnoxious smell of septicity. The reduction in volatile acids during the anaerobic digestion process can be accompanied by a reduction in the odour level of the sludge of about 90% (Hobson, 1995).

The concentrations of volatile fatty acids that can develop in sludges are much higher than in sewage and the decrease in pH value is more marked with values of about 5.5 having been observed. At such low pH value, hydrogen sulphide and other malodorous sulphur-containing compounds, which will also be present in the sludge, are almost completely associated (non-ionic) and available for release into the atmosphere.

2.4 Reduction of sulphate

Hydrogen sulphide, produced from sewage or sludge by the reduction of sulphates by anaerobic bacteria, is the most significant cause of odour problems related to sewage and sludge treatment. Even when hydrogen sulphide is not the main cause of malodour, it is nearly always present where there are sewage or sludge odours. For this reason, hydrogen sulphide is often used as a 'marker' for the detection of sewage-related malodours. Factors that affect the generation and control of hydrogen sulphide in sewers have been studied by numerous research workers (Thistlethwayte, 1972; Boon, 1995). In addition, the use of sulphate-reducing bacteria in the treatment of strong wastes has also been the subject of numerous papers, as discussed in Chapter 28.

The bacteria responsible for sulphide formation are in a class which 'respires' sulphate to provide energy for the dissimilation of organic matter. This is dissimilatory sulphate reduction, as opposed to assimilatory reduction of inorganic sulphate which occurs naturally in an analogous process to nitrogen fixation. The process of dissimilatory sulphate reduction is analogous to 'nitrate respiration', associated with denitrification in an anoxic environment. A very small amount of reduced sulphur is assimilated by the bacteria, but most is released into solution as sulphide ion, usually substantially as molecular (non-ionic) H_2S. These bacteria are termed 'sulphate-reducing' (Postgate, 1959, 1984).

The sulphate-reducing bacteria are strict anaerobes and cannot metabolize while dissolved oxygen or nitrate (or nitrite) remain in the sewage. However, they can survive adverse conditions of temperature, aerobicity, salinity and pressure, and are widely distributed in the environment, ready to become active whenever local conditions become anaerobic, such as within sewage or sludges and also in bacterial slimes on the submerged surfaces of sewers and holding-tanks and in the undisturbed sediments of sewers, rivers and estuaries. They are heterotrophic bacteria and grow slowly at a rate which is about 7% of the growth rate of aerobic microorganisms. The relative rates of growth can be calculated from the rate of sulphide formation in a rising-main sewer in comparison with the rate of demand for dissolved oxygen in the same sewer that would be required to prevent septicity (Boon, 1995).

A number of genera and species of sulphate-reducing bacteria have been isolated, as listed in Table 33.2 (after Postgate, 1984).

TABLE 33.2 Sulphate-reducing bacteria (after Postgate, 1984)

Genus	*Species*	*Strains*
Desulfomaculum	*nigrificans*	7
	orientis	2
	ruminis	2
	acetoxidans	1
Desulfovibrio	*africanus*	3
	desulfuricans subsp. aestuarii	1
	desulfuricans subsp. desulfuricans	18
	gigas	2
	salexigens	7
	vulgaris subsp. oxamicus	1
	vulgaris subsp. vulgaris	13
	Desulfovibrio spp.	7
	baarsii	1
	sapovorans	1
Desulfobacter	*postgatei*	1
Desulfobulbus	*propionicus*	1
Desulfococcus	*multivorans*	1
Desulfonema	*limicola*	1
	magnam	1
Desulfosarcina	*variabilis*	1

The best studied is the non-spore forming genus *Desulfovibrio*, of which the main type associated with the production of sulphide from sewage and sludge is *Desulfovibrio desulfuricans* (Crowther and Harkness, 1975). *Desulfovibrio* species are Gram-negative, polarly-flagellated, curved rods (Stanier *et al.*, 1976).

The number of sulphate-reducing bacteria present in sewage or sludges is dependent on the length of time that the sewage or sludge has been retained under anaerobic conditions. Heukelekian (1948) found typical numbers of sulphate-reducing bacteria of 25 000/ml in sludges and 60 to 600/ml in crude sewage, with numbers increasing, under anaerobic conditions, from 100/ml to nearly 100 000/ml over 14 days. Toerian *et al.* (1968) found counts of 27 500 to 53 000/ml in raw sewage sludge. Numbers increased when sulphate rich water was added to the sludge. Schmitt and Seyfied (1992) found concentrations in sewer sediment samples of 2.4×10^7–11.4×10^7 per g volatile solids (volatile solid content of the sediment was less than 5% in a cleaned sewer, and between 3 and 27% in two sewers that were not cleaned).

The biochemistry of sulphate-reducing bacteria has been reviewed by Postgate (1984). Ammonia is the principal source of nitrogen for the bacteria and pre-formed organic matter (acetate in particular) is required for growth of *Desulfovibrio* species. Postgate also indicated an exceptionally high requirement for inorganic iron. The sulphide ion generated by sulphate reduction will also precipitate iron and other heavy metals as insoluble sulphides. This accounts for the darker colour of septic sewage and sludges.

The reduction of sulphate to sulphide by *Desulfovibrio* involves the transfer of four pairs of electrons (Stanier *et al.*, 1976). The process, described in Postgate's review (1984), is a four step process, with sulphite, trithionate and thiosulphate as intermediates. The reduction to sulphide is mediated by five enzymes and the catalyst desulfovibrin with a key intermediate of adenylphosphosulphate (APS). The process is inhibited by structural analogues of sulphate, including selenate (SeO_4^{2-}), and monofluorophosphate (FPO_3^{2-}). Chromate and molybdate ions are also inhibitors, probably acting by forming an unstable analogue of APS.

The sulphate-reducing bacteria require a redox potential of about -200 mV for the initiation of growth. Sulphate reduction occurs at a redox potential within the range −200 to −300 mV (E_h), depending on the pH value of the sewage being within the range 6.5–8, respectively. The growth of sulphate-reducing bacteria is accompanied by a further drop in redox potential (Postgate, 1984).

The concentration of sulphate in sewage and its resultant sludges can vary greatly from area to area, depending, for example, on the hardness of the ground water (which may infiltrate the sewerage system) or the potable water supply, the method of treatment of potable water and on the input of industrial wastewaters. In coastal areas, where there may be infiltration of the sewerage system with seawater, and in certain volcanic parts of the world, concentrations may be very much higher and extreme problems with malodours, corrosion and inhibition of anaerobic digestion may be experienced. Sulphate concentrations in sewage in inland areas of the UK are generally over 10 mgS/l, with typically about 20 mgS/l present.

In wastewater, the concentration of sulphate (at about 20 mgS/l) rarely limits the sulphide production and, in the range 10–60 mgS/l, does not affect the rate of sulphide formation (Boon, 1995). The amount of sulphide formed in a rising-main sewer has been found to be directly proportional to the retention time of the sewage in the main. The longer the time, the greater the amount formed. As the pipe remains full of sewage, the amount formed daily is directly related to the volume of the sewage in the pipe, the internal surface area of the pipe, the chemical oxygen demand (COD) and temperature of the sewage and the availability of inorganic and organic sulphur-containing compounds (Boon, 1995; Hvitved-Jacobson *et al.*, 1988; Nielson *et al.*, 1998). Changing the rate or frequency of pumping will have little or no effect on the total amount formed daily, although flow variations and changes in COD will account for variations in concentration of sulphide during the day. Increases in temperature will increase the rate of formation; the rate is likely to double for a 10°C increase in temperature within the range of about 5–25°C. At temperatures above 25°C, increases in temperature will start to have an adverse effect on growth rate of microorganisms, with the rate declining to zero as the pasteurization temperature of 45°C is reached.

A number of empirical relations have been published to estimate the production of sulphide in rising-main sewers and gravity sewers, and to derive the conditions necessary to prevent sulphide formation. These have recently been reviewed (Boon, 1995). Equations based on these relations, together with others which are traditionally used for the hydraulic design of a sewerage system, have been linked together to form a mathematical model (Boon *et al.*, 1998).

In sediments and thick slimes, sulphate concentration may become limiting across the depth (Schmitt and Seyfied, 1992). The rate of sulphate reduction in sediments was found in experiments on sewer sediments in Hannover (Schmitt and Seyfied, 1992) to be up to 0.64 g $S/m^2/h$. Maximum rates of sulphate reduction were found at sediment depths of between 5 and 7.5 cm.

In sludges, much higher concentrations of dissolved H_2S may develop compared with sewage, particularly where there are high sulphate concentrations. The rate at which septicity develops depends on the solids content and temperature of the sludge and the availability of sulphur-containing organic and inorganic compounds. Total sulphur in sludge can account for between 1 and 2% of the dry matter (Sommers *et al.*, 1977), although the form of the sulphur as inorganic and organic compounds is variable. Storage of primary sludge for more than 1–2 hours will generally result in hydrogen sulphide being produced by sulphate reduction. In addition, other malodorous compounds will be formed by the action of the proteolytic and acid-forming bacteria as described above, producing compounds such as methyl sulphide, dimethyl sulphide, methyl disulphide, thiols and volatile organic fatty acids, such as acetic, propionic and butyric. Lowering of the pH value increases the potential for release of volatile sulphides to the atmosphere.

In the anaerobic digestion process, the sulphate-reducing bacteria operate in parallel, and in competition, with methanogenic bacteria. Both sulphate-reducing bacteria and methanogenic bacteria are strict anaerobes. The anaerobic digestion process breaks down a wide range of odorous compounds with a reduction in the volatile fatty acid content and an increase in pH value to approximately neutral. However, sulphate reduction during the anaerobic digestion process will lead to H_2S being released in the biogas and being present in the digested sludge. In addition, other malodorous sulphides, formed by the action of proteolytic bacteria, will be present in both digesting and digested sludge.

The release of malodours during the digestion process is controlled because the process of anaerobic digestion is carried out within totally enclosed vessels. Biogas produced by the process is combusted in a boiler, combined heat and power (CHP) plant or flare-stack, with the consequent oxidation of odorous compounds. H_2S concentrations in biogas can be within the range 100 to 3000 ppm (v/v). Severe odour nuisance can occur if biogas is released directly into the atmosphere, through pressure-relief valves or if there are problems with the gas-handling or storage systems. Malodours may also be released when the digested sludge is transferred to a secondary digester, if care is not taken to avoid turbulence of the sludge after digestion. Hydrogen sulphide in the biogas can cause corrosion problems with CHP or boiler equipment.

In addition to exacerbating odour and corrosion problems, high levels of sulphate in digesting sludge and the resultant development of sulphate-reducing bacteria, have been found to have an adverse effect on the methane-producing bacteria. The apparent incompatibility of sulphate-reducing and methane-producing bacteria (Postgate, 1984) is thought to be partly due to the following:

- the toxic effects of sulphide produced (concentrations of soluble sulphide are toxic to methanogenic bacteria in concentrations above 200 mg/l (Lawrence *et al.*, 1964)
- competition by some species of sulphate-reducing bacteria for acetate (e.g. *Desulfotomaculum acetoxidans* and *Desulfobacter postgateii*). However, *Desulfovibrio* species do not oxidize acetate, and the end-products of carbon oxidation, coupled to sulphate reduction by most *Desulfovibrio* species, are acetate, water and carbon dioxide
- competition by the sulphate-reducing bacteria for molecular hydrogen (Pfennig and Widdel, 1982; Postgate, 1984).

Various methods of control of sulphate-reducing bacteria during digestion have been tried, including the use of specific inhibitors of the sulphate-reducing bacteria and the addition of iron salts to precipitate sulphide as it is formed.

2.5 Summary of common causes of odour nuisance

The main causes of odour problems associated with different stages of wastewater and wastewater sludge handling and treatment are summarized in Table 33.3.

3 EMISSION OF MALODOURS AND THEIR CONTROL

Control of malodours is achieved by a combination of measures ranging from prevention to reduction of septicity, together with minimization of the release and potential impact of malodours. In most circumstances, total prevention of septicity at all times and under all conditions would not be practical or economical.

3.1 Prevention of septicity

Prevention, or a reduction in malodour formation may be achieved by the following:

- Maintaining conditions that prevent the development or action of anaerobic bacteria. This is most effectively achieved by maintaining aerobic or anoxic conditions and by minimizing retention in systems where there is no aeration, such as in holding tanks or rising-main sewers.

TABLE 33.3 Summary of the main causes of odour production at different process stages

Process stage	*Main odorous compounds*	*Bacteria responsible*	*Possible locations of odour release*
SEWAGE			
Rising main sewer	H_2S	Sulphate-reducing bacteria in sewage and in slimes on sewer walls	At and downstream of the discharge point
Sediments	H_2S	Sulphate-reducing bacteria in sediments	From sewage when sediments are disturbed
Primary, storm storage or balancing tanks	H_2S	Sulphate-reducing bacteria in sewage and sludge	Tank weirs and outlet channels
Biological filters	H_2S	Sulphate-reducing bacteria in slimes if excessive slime growth (particularly high rate filters)	Ventilation air from the biofilter
SLUDGES AND IMPORTED WASTES			
Primary sedimentation and primary sludge storage tanks	H_2S Organic sulphides	Sulphate-reducing bacteria Proteolytic bacteria in sludge	De-sludging wells, discharges to storage tanks, dewatering area, return liquor sumps and discharge points, tanker collection/delivery points
	Volatile fatty acids which also lower pH value and increase potential release of sulphides	Lipolytic bacteria in sludges	At overflow weirs and other turbulent areas downstream
	Ammonia in the form of ammonium salts	Formed by the hydrolysis of urea and proteolysis and released if lime dosing employed	Downstream of alkaline additions
Anaerobic digesters	H_2S Organic sulphides	Sulphate-reducing bacteria Proteolytic bacteria	Discharge wells, sludge-digestion tanks, pressure relief valves for biogas
	Ammonia	Proteolytic bacteria in well-digested sludge	Ammonia from discharge of liquors after sludge pressing

Aerobic conditions can be maintained in some cases by the addition of air, oxygen, or hydrogen peroxide or anoxic conditions can be achieved by the addition of nitrate salts

- Reduction in the input of sulphate, or modifications to the treatment processes, in order to reduce the formation of sulphide in the treatment and handling of sludges, particularly if there were saline intrusion in the sewerage system or an industrial discharge with a high sulphate content. A reduction in sulphate concentration is unlikely to have a significant impact on sulphide formation in sewage or the malodours associated with septic sewage
- Addition of bactericides (such as chlorine or hypochlorite) has been used to reduce the growth of sewer wall slimes and to inhibit the respiration rate of bacteria in sewage. These have had variable success and may have potentially harmful effects on subsequent biological treatment stages or the quality of effluent discharge to receiving waters
- Addition of specific inhibitors of sulphate-reducing bacteria have also been used (particularly for anaerobic digestion stages), and proprietary products are available, although their use is not widespread in the water industry. However, these tend to be successful in the short term only, as species may be selected which are resistant to the inhibitor.

3.2 Treatment of septicity

Treatment of sewage or sludges that have already become septic can be achieved either by bacterial oxidation of the odorous compounds or by addition of chemicals to oxidize or precipitate sulphides, or to increase the pH value so that they are not emitted to the atmosphere.

- Hydrogen sulphide and volatile organic sulphides in sewage or sludge will be oxidized when conditions become aerobic (such as in a gravity sewer following a rising-main sewer, or in an aerated activated-sludge plant following a primary sedimentation tank) and this may be used as a control method, if release of sulphide prior to the aerobic step can be avoided, e.g. by reducing turbulence or provision of covers for treatment stages upstream of the aerobic stage.
- Hydrogen sulphide may also be controlled by the addition of precipitants, such as iron salts, or oxidizing agents, such as peroxide.

3.3 Prevention of release of malodorous compounds

The amount of hydrogen sulphide and other malodorous compounds released into the atmosphere will depend on the interfacial area between liquid and air and the solubility of the odorous gases, which in turn is affected by temperature, total pressure and partial pressure. Under static enclosed conditions, e.g. under the covers of enclosed storage tanks, gases will be released from solution at a slow rate to maintain chemical equilibrium with the gas phase according to Henry's law. However, significant release of odour is likely to occur where the interfacial area is large, such as settling tanks or sludge lagoons and at points of turbulence, where the flow breaks up into droplets or air is entrained into the liquid (e.g. at the discharge end of a rising-main sewer, at discharge weirs of primary sedimentation tanks or during sludge handling). Malodours may also be released during disposal of sludges to agricultural land.

Odour release can be minimized, for example, by avoiding cascades and turbulence, retaining sludges in pipes (rather than in open channels) and covering distribution chambers, channels and tanks.

In the case of ionic species of volatile organic chemicals, the pH value will also have an important impact on the amount and type of odorous compounds available to be released. Many of the odorous compounds in sewage (including sulphides and some nitrogenous compounds) are present in sewage in unionized (dissolved gas) and ionized forms (see above). Only the unionized form is available for transfer to the atmosphere and release can be prevented by maintaining the compounds in the ionized form. Neutral or acidic conditions (pH 7 or below) favour the release of 'acidic' malodours including organic sulphides and hydrogen sulphide and volatile, organic, fatty acids, such as acetic, propionic and butyric (see Table 33.1). Under alkaline conditions there will be an increase in the release of ammonia and amines, together with other volatile organic compounds such as skatole and indole. These 'alkaline' odours have much higher odour threshold concentrations than acidic sulphides but are more persistent.

Measures to control septicity by addition of alkali should avoid increasing the pH value above about 8.5 or the problem will not be resolved, as malodorous alkaline odours would prevail.

3.4 Treatment of malodorous gases

Where covers are provided for tanks or other odour-emitting stages, odorous gases may need to be vented to an odour treatment unit. A wide range of odour treatment processes are available, including chemical and biochemical scrubbing and absorption processes. One form of control for treated gases employs biological processes in the form of a biofilter or a bioscrubber, in which sulphide-oxidizing bacteria (*Thiobacilli*) reoxidize sulphide, mercaptans, and other malodorous compounds. The *Thiobacilli* are chemoautotrophic and utilize the oxidation of reduced sulphur compounds to support growth (Sublette, 1992). The end product is sulphate. In this process the autotrophic bacteria, as well as heterotrophic bacteria, are allowed to develop on the surface of a media (e.g. a natural medium such as peat, shells or coir) or in the recirculation liquor and on the exposed surfaces of a plastic medium bioscrubber. In both cases, sufficient moisture must be provided to maintain the biological film and ensure that the sulphide and other soluble odorous compounds transfer from the gas phase. Control is required to ensure that the biofilter does not become too acidic. Recently, two-stage bioscrubbers have been developed and successfully used (Joyce and Leach, 1998) to remove carbon- (and nitrogen-) based odours in the first stage (by the growth of heterotrophic bacteria) and to remove sulphides in the second stage (by maintaining the growth of the slowly-growing autotrophic bacteria). In a development of these methods, the nitrifying activated-sludge process has been used to treat odorous air, by ducting malodorous air to the intake of the blowers that are normally provided for aerating the process (Stillwell *et al.*, 1994; Ostojic and O'Brien, 1994).

An effective and low-cost catalytic hydrated iron oxide chemical filter has been recently developed (Boon and Boon, 1998), which has been successfully used to remove up to 95% of H_2S from air, together with lower percentages (about 30 to 80%) of organic sulphides. The design of this filter is based on the fact that H_2S will react chemically with hydrated iron oxide to form water and sulphur. The conditions that this reaction requires have been investigated and an application for a patent has been submitted to provide commercial benefit for the future design and operation of such filters.

3.5 Summary of control methods

Odour problems can be minimized by:

1. minimizing retention under anaerobic conditions, including avoiding excessive accumulation of debris and grit in pipes, channels and tanks
2. avoiding unnecessary contact of sewage and sludge with the atmosphere and minimizing turbulence
3. retaining malodours dissolved in solution until they can be biochemically oxidized naturally, e.g. in an aerobic stretch of sewer, or in an aerated activated-sludge plant
4. covering units and venting of contained air to an odour treatment unit
5. addition of chemicals such as:
   - oxygen or nitrate to maintain aerobic or anoxic conditions
   - oxidant chemicals, such as hypochlorite, hydrogen peroxide or potassium permanganate to reduce microbial activity and oxidize previously formed sulphides
   - iron salts to precipitate sulphides. Iron salts are also used as a catalyst for the chemical and biochemical oxidation of sulphide
   - alkali to convert H_2S to HS^- and S^{2-}, the pH value should not exceed 8.5, to avoid release of alkaline odours (e.g. ammonia, amines, skatole and indole)
   - specific inhibitors for sulphate reduction, these may be effective in the short term only, as adaptation of microbial species may occur.

REFERENCES

Bonnin, C., Loborie, A. and Paillard, H. (1990). Odor nuisances created by sludge treatment: problems and solutions. *Water Sci. Tech.* **22**(12), 65–74.

Boon, A.G. (1995). Septicity in the sewers: causes, consequences and containment. *Wat. Sci. Tech.* **31**(7), 237–253.

Boon, A.G., Boon, K. (1998). Catalytic-iron filters for effective and low-cost treatment of odorous air. Paper presented at the WEF/EWPCA/CIWEM Conference – Innovation 2000, Churchill College, Cambridge. 7–10 July, 1998.

Boon, A.G. and Lister, A.R. (1975). Formation of sulphide in a rising-main sewer and its prevention by injection of oxygen. *Prog. Wat. Technol.* **7**(2), 289–300.

Boon, A.G., Vincent, A.J. and Boon, K.G. (1998). Avoiding the problems of septic sewage. *Wat. Sci. Tech.* **37**(1), 223–231.

CEN (1999) European Committee for Standardisation, CEN/TC264/WG2 Odours (draft standard), Central Secretariat, rue de Stassart 36, B-1050, Brussels.

Crowther, R.F. and Harkness, N. (1975). Anaerobic bacteria. In: C.R. Curds and H.A. Hawkes (eds) *Ecological Aspects of Used Water Treatment Volume 1, The Organisms and their Ecology.* Academic Press, London.

EPA Process Design Manual for Sulphide Control and Sanitary Sewerage Systems. (1974). United States EPA 625/1-64-005 Technology Transfer, Washington DC.

Heukelekian, H. (1948). Some bacteriological aspects of hydrogen sulphide production from sewage. *Sewage Wks J.* **20**, 490–498.

Hobson, J. (1995). The odour potential, a new tool for odour management. *J Ints. Wat. Envir. Mangt.* **9**(5), 458.

Hvitved-Jacobson, T., Jutte, B., Neilson, P.H. and Jensen, N.A. (1988). Hydrogen sulfide control in municipal sewers. In: H.H Hahn and R. Klute (eds) *Pretreatment in Chemical Water and Wastewater Treatment. Proceedings of the 3rd International Gothenburg Symposium*, Gothenburg, Sweden, 1–3 June 1988, pp. 239–247. Springer-Verlag, Berlin.

Joyce, J. and Leach, K. (1998). Desgin of bioscrubbers (not biofilters) for improved odor control flexibility and operational effectiveness. Paper presented to 71st Water Environment Federation Conference, Orlando, Florida. 3–7 October 1998.

Lawrence, A.W., McCarty, P.L. and Guerin, F.J.A. (1964). The effects of sulphide on anaerobic treatment. *Proceedings of the 19th Industrial Waste Conference*, Purdue University, Engineering Extension Series, No. 117, pp. 343–357.

Nielson, P.H., Raunkjaer, K. and Hvitved-Jacobson, T. (1998). Sulphide production and wastewater quality in pressure mains. *Wat. Sci. Tech.* **37**(1), 97–104.

Ostojic, N. and O'Brien, M. (1994) Control of odors from sludge composting using wet scrubbing, biofiltration and activated sludge treatment. *Proceedings of Water Environment Federation Conference: Odor and volatile compound emission control for municipal and industrial treatment facilities.* Jacksonville, Florida, April 1994, pp. 5-9–5-19.

Parker, C.D. (1947). Species of bacteria associated with the corrosion of concrete. *Nature* **159**, 439.

Pfennig, P. and Widdel, F. (1982). The bacteria of the sulphur cycle. *Phil. Trans. R. Soc. Lond.* **B298**, 433–441.

Postgate, J.R. (1959). Sulphate reduction by bacteria. *Ann. Rev. Microbiol.* **12**, 505–520.

Postgate, J.R. (1984). *The Sulphate-Reducing Bacteria.* Cambridge University Press, Cambridge.

Schmitt, F. and Seyfied, C.F. (1992). Sulfate reduction in sewer sediments. *Wat. Sci. Tech.* **25**(8), 83–90.

Sommers, L.E., Tabatabai, M.A. and Nelson, D.W. (1977). Forms of sulfur in sewage sludge. *J. Environ. Qual.* **6**(1), 42–46.

Stanier, R.Y., Adelberg, E.A. and Ingraham, J.L. (1976). *General Microbiology.* Prentice-Hall Inc.

Stillwell, S.A., Hans, D.E. and Katen, P.C. (1994) Biological scrubbing of foul air in activated sludge treatement reduces odors and ROGs from headworks and primary clarifiers. *Proceedings of Water Environment Federation Conference: Odor and volatile compound emission control for municipal and industrial treatment facilities.* Jacksonville, Florida, April 1994, pp. 5-1–5-8.

Sublette, K.L. (1992) A review of the oxidation of hydrogen sulphide by *Thiobacillus denitrificans* with a case study: microbial removal of hydrogen sulfide from biogas. Second International Symposium on Waste Management Problems in Agro-Industries, Istanbul, Turkey.

Thistlethwayte, D.K.B. (ed.) (1972). *The Control of Sulphides in Sewerage Systems.* Butterworths, London.

Toerian, D.F., Thiel, P.G. and Hattingh, M.M. (1968). Enumeration, isolation and identification of sulphate-reducing bacteria of anaerobic digestion. *Wat. Res.* **2**, 505–513.

Woodfield, M. and Hall, D. (ed.) (1994). Odour Measurement and Control – An Update, A.E.A. Technology, National Environmental Technology Centre AEA/CS/REMA 038, ISBN 0856248258, August 1994.

World Health Organisation (1987) *Air Quality guidelines for Europe.* WHO Regional Publications Series No. 23, Regional Office for Europe, Copenhagen.

34

Recalcitrant organic compounds

J.S. Knapp and K.C.A. Bromley-Challoner

Department of Microbiology, University of Leeds, Leeds LS2 9JT, UK

1 INTRODUCTION

The question of recalcitrant organic compounds first surfaced as a real public and scientific issue with the publication of *Silent Spring* (Carson, 1962). This book raised awareness of the actual and potential problems caused by the accumulation in the environment of organic compounds which were not readily destroyed by biological activity and so built up to concentrations which were toxic to man or wildlife or had some other unacceptable effects. Perhaps the most infamous of these compounds are the polychlorinated insecticides such as DDT, aldrin and dieldrin, which are subject to bioaccumulation in the food chain and which caused devastating effects on the populations of predatory birds in the 1950s and 1960s. A more visible example was the problem of water pollution by so-called 'hard' detergents during the same period. The realization of the consequences of the release of these chemicals in the environment led to much research, which is still on-going, into the fate of these materials in the wider environment.

In the 1950s there was a strongly held view among microbiologists, often called the 'principle of microbial infallibility', that all chemicals were susceptible to microbial degradation if the right organism and conditions could be identified. The metabolic capabilities of microbes were 'all powerful'. The growing realization that many chemicals that found their way into the environment were not being degraded and were persisting for many years put an end to such thinking. Half-lives of polychlorinated organic insecticides in the order of tens of years were reported and there was considerable fear that such compounds could be with us forever. With these rather frightening statistics in view, Martin Alexander reviewed biodegradation in 1965 and raised the possibility that microbes were, in fact, fallible and that, through a combination of a chemical's structure and the environment it found itself in, a chemical might survive in the environment for prolonged periods. Since then there has been intensive research into this problem world-wide in both academia and industry. Although there are still many causes for serious concern, there is now hope that some of the mistakes of the past will not be repeated and, indeed, that some old pollution problems may be cleared up. Particularly important developments have included:

1. An understanding of how different types of recalcitrant chemicals are degraded and what structural factors make them recalcitrant. This has allowed the design of important chemicals with degradability in mind.
2. The discovery of co-metabolism.
3. The realization of the role of microbial communities rather than pure cultures in degradative processes.
4. The realization that degradation of many chemicals can occur in anaerobic environments.
5. The discovery that white rot fungi have the ability to degrade a huge range of recalcitrant chemicals.

The Handbook of Water and Wastewater Microbiology
ISBN 0-12-470100-0

2 DEFINITIONS

It is useful in starting this chapter to define and discuss a few important terms that will be used repeatedly.

2.1.1 *Recalcitrant*

This term means difficult or obstinate, not easy to control. In the current context, recalcitrant means *difficult, but not necessarily impossible, to degrade*. Recalcitrant chemicals may resist biodegradation for a whole range of reasons and some are more resistant than others. Many recalcitrant chemicals are xenobiotic in nature, but by no means all xenobiotics are recalcitrant. Furthermore, many natural compounds and materials (e.g. lignin) are degraded only with difficulty.

2.1.2 *Xenobiotic*

Xenobiotic means 'foreign to life'. It is a term applied to many recalcitrant organic chemicals. These are synthetic chemicals and are not found in nature. They will contain structural elements that are unknown or rare in nature and/or are assembled in a structure which is not of natural occurrence. Examples of such structural elements are the aromatic sulphonic acids (found for example in alkyl benzene sulphonate surfactants and many dyes); polychlorination of an alkane or aromatic compound; and the diazo bond. Examples are given in Fig. 34.1.

Some researchers also use the term *'xenobiotic concentration'*, meaning a concentration that is not found in nature – so we may have a natural compound present at a xenobiotic concentration. Such a concentration might occur as the result of, for example, an oil spill or discharge of an industrial effluent.

2.1.3 *'Hard' and 'soft'*

The terms *'hard'* and *'soft'* are used quite widely to describe, respectively, recalcitrant chemicals and ones that can easily be degraded.

2.1.4 *Biodegradation*

Biodegradation is the term given to the breakdown of organic chemicals by the biological action of a living organism. In the environmental context, generally microorganisms are the most important agents of biodegradation. Although extensive degradation of some xenobiotic chemicals can occur in mammals (usually in the liver), they are not particularly important in degradation of environmental pollutants. The heterotrophic bacteria are often considered to be of prime importance, however, the role of fungi is being increasingly recognized. Algae and cyanobacteria can catalyse some biodegradative processes but are probably of only limited importance. Varying degrees of biodegradation can occur according to how much the compound has been modified.

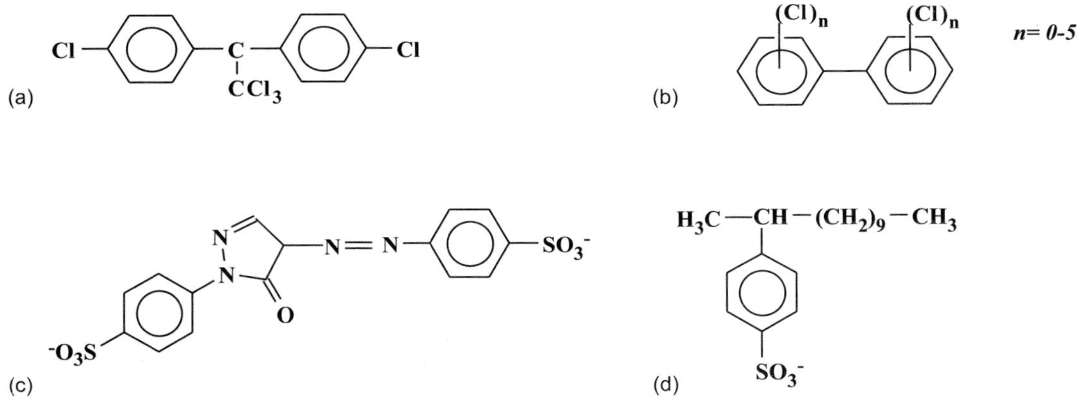

Fig. 34.1 Examples of recalcitrant xenobiotic compounds. (a) DDT (1,1,1-trichloro-2,2-di(chlorophenyl)ethane) an insecticide; b) PCB (polychlorinatedbiphenyl) used in electrical transformers; (c) Tartrazine, an azo dye used as a food colourant; (d) Linear alkyl benzene sulphonate, a surfactant.

2.1.5 *Primary biodegradation*

This term is used to describe processes in which a compound loses its characteristic properties but may be little altered in terms of its size or complexity. The compound will have lost some characteristics and may no longer respond in a particular assay. Good examples of primary biodegradation are to be found in the degradation of surfactants (Fig. 34.2). Anionic surfactants are often assayed by the methylene blue active substance (MBAS) test. Positive results are given by compounds that contain a large hydrophobic group linked to a hydrophilic group. If the hydrophobic and hydrophilic groups in a compound are separated then it will no longer give a response in the assay and will no longer have surfactant properties.

Another example (Fig. 34.2) might be the reductive decolorization of an azo dye. This primary biodegradation leads to a change in colour or its removal, but two aromatic amines are released. Depending on which amines they are, these could be much more harmful to the biota than the original dye. Several aromatic amines are, for example, actual or suspect carcinogens.

Thus, a large portion of a molecule may remain intact – this portion may have a BOD and COD and may be toxic or may have other environmental effects, it may also be recalcitrant to further degradation. In conclusion, primary biodegradation may herald the complete destruction of a pollutant or may cause even more problems.

2.1.6 *Mineralization*

Mineralization is synonymous with *ultimate biodegradation* or *complete biodegradation*. It describes the degradation of a compound to its mineral components, i.e. carbon dioxide and water. Depending on the compound's composition, other minerals may be released; these might include sulphide, sulphate or sulphite; ammonia, nitrite or nitrate; phosphate or phosphite; chloride; fluoride etc. If the mineralization is anaerobic, methane may be a product. In addition to the carbon dioxide etc. released, some of the mass of the chemical may be converted to biomass (cellular components). In fact this is usually, but by no means always, the case. Thus the complete destruction of a chemical may result in only 50–80% of the carbon being converted to CO_2.

2.1.7 *Co-metabolism*

This term is used (and often mis-used) to apply to a range of different types of degradative process. Perhaps the best definition is *the degradation of a compound in the obligate presence of another compound(s)*. It is used to describe a situation in which a microbial culture is only able to degrade a substance in the presence

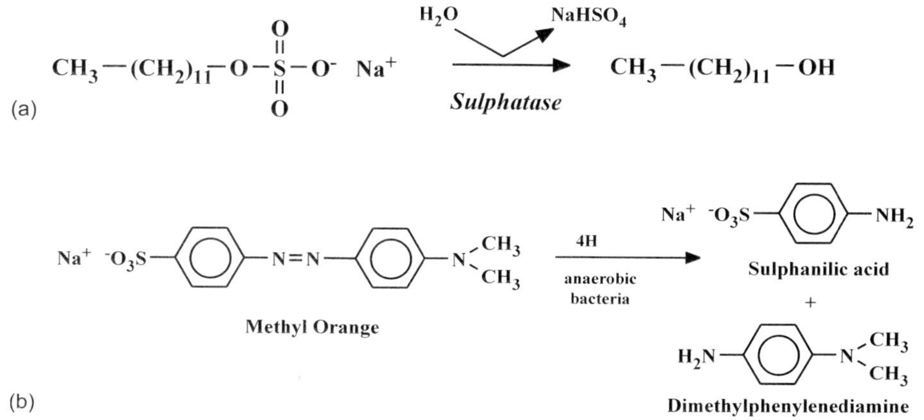

Fig. 34.2 Examples of primary biodegradation. (a) The surfactant sodium dodecyl sulphate (sodium lauryl sulphate) is hydrolysed to dodecanol and sulphate. Removal of the sulphate ester group prevents the molecule acting as a surfactant and giving a positive MBAS test. (b) Reduction of methyl orange by anaerobic intestinal bacteria results in decolorization and production of sulphanilic acid and N,N-dimethyl-4-phenylenediamine.

of other compounds. These compounds may be specific in some cases, while in others a wide variety of *co-substrates* (additional compounds) may be suitable.

Specific co-substrates are required in the case of a process often called *analogue enrichment*. This is where the recalcitrant compound and the co-substrate are structurally related and the co-substrate has a role in inducing the production of some, or all, of the enzymes required to degrade the recalcitrant compound which cannot itself act as an inducer. This process can only work if the degradative enzymes are sufficiently non-specific to degrade compounds other than their natural substrate. The target compound may be mineralized or only suffer partial biodegradation. For example, the addition of biphenyl has been reported to encourage biodegradation of the analogous polychlorinated biphenyls (PCBs).

Non-specific co-metabolism generally occurs in cases in which the degrading organisms derive no obvious benefit from the degradation of the recalcitrant compound. The degrading organisms cannot use energy or carbon resulting from the degradation process and therefore need to be provided with a source of carbon and energy on which they can grow. Such processes are sometimes referred to as *gratuitous degradation* (i.e. it is for free) or alternatively as *fortuitous* or *accidental degradation* (i.e. it occurs by chance). There are many examples of this type of co-metabolism. Occasionally degradative processes are reported to involve co-metabolism when the degradative organism simply requires small amounts of a growth factor (e.g. a vitamin or amino acid) – this is not really an appropriate use of the term.

2.1.8 Microcosms

The term *microcosms* has been used to describe mixed microbial populations which carry out degradative reactions. It really means little more than the terms 'mixed cultures' or consortia. It would be more helpful perhaps to use the term *community degradation*. Some compounds can be degraded completely by single microbial strains in pure cultures, but this is not always the case. It is quite common to find that a compound can only be degraded by a mixed culture; none of the individual strains present being able to use the recalcitrant compound as a sole source of carbon and energy. In some cases, this requirement for a mixed culture is apparently obligate while in others researchers can sometimes find pure culture degradation and in other circumstances find a need for a mixed culture. There are several possible reasons for the requirement for a mixed culture:

1. *Cross-feeding of growth factors*. In this process one organism may produce a growth factor, such as a vitamin or amino acid, which is required for the primary degradative organism to grow. The process may be mutual, i.e. two organisms will produce growth factors that each promote the other's growth. Cross-feeding between bacterial strains often occurs in nature and also in activated sludge populations.
2. *Removal of accumulated toxic intermediates*. Often in degradation processes intermediates can accumulate in the organism or the growth medium. Such intermediates may be toxic to the degradative organism and so the growth in the mixed culture of another organism that can remove these intermediates, thus relieving toxicity, will be of value. The requirement for such a detoxifying organism will clearly depend on the concentration of the initial substrate and the accumulated toxin. Thus, at high initial substrate concentrations, this interaction may be obligatory while at low concentrations it may not.
3. *Sequential degradation*. It is not uncommon to find that an organism can only catalyse certain initial steps in a degradation or only degrade certain parts of a recalcitrant molecule. In these circumstances, complete degradation of the compound can only be achieved by the action of two or more microbial strains working together and carrying out different parts of the degradative process.

Some consortia/mixed cultures may involve two or more of these interactions. In nature, as opposed to the research laboratory, it is very likely that most biodegradative processes are

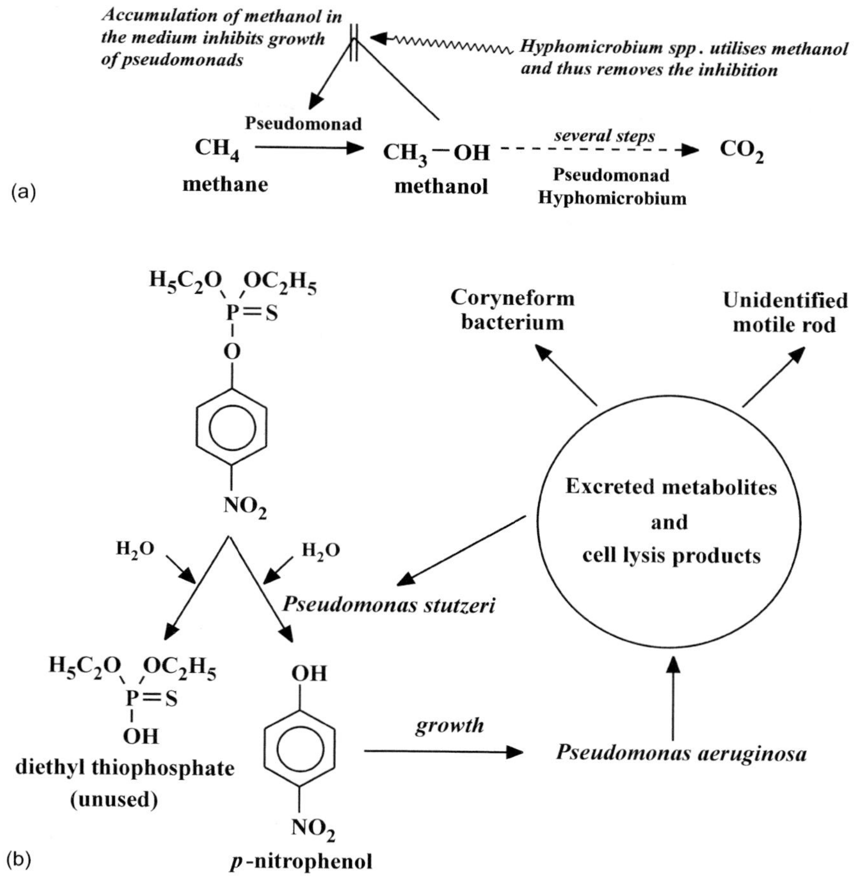

Fig. 34.3 Examples of community degradation processes. (a) Relief of inhibition during growth of a mixed culture on methane. (b) Degradation of parathion by a mixed microbial culture. Initial degradation was catalysed by a strain of *Pseudomonas stutzeri* which could not utilize the hydrolysis products. This organism and the other two were dependent on metabolites of the *p*-nitrophenol-utilizing *P. aeruginosa*.

dependent on mixed cultures. Pure cultures are rarely found outside the laboratory, however, they are of great scientific importance in allowing us to identify the biochemical and genetic mechanisms that are of importance in degradation processes. Some examples of the involvement of mixed cultures in biodegradation are given in Fig. 34.3.

3 FACTORS INFLUENCING THE BIODEGRADATION OF RECALCITRANT COMPOUNDS

In order to understand why some recalcitrant organic compounds can persist for a prolonged time in the environment, or survive effluent treatment processes, it is essential to study the factors that influence their degradation. These can be crudely separated into factors associated with the compound and factors associated with the environment. They are summarized in Table 34.1.

The failure of a compound to be degraded, or its slow biodegradation, may be due to environment-related factors or to chemical-related factors or to a combination of both. Sometimes several of the factors listed from both categories may be involved to varying degrees. It is perhaps useful to give more details and some examples of the effects of these various factors on degradative processes.

TABLE 34.1 Factors which influence the biodegradation of recalcitrant compounds

1) Chemical-specific factors	
State	Gas, liquid or solid Surface area
Solubility	Aqueous
Hydrophobicity	Ability to dissolve in hydrophobic (lipophilic) solvents, hydrophobic compounds will have relatively low water solubility
Adsorbability	Ability to adsorb to and complex with organic and inorganic fractions in soil, sediments and water
Size and shape	Can affect ability to permeate cell membrane and to interact with enzymes
Charge	Can affect ability to permeate cell membrane and to interact with enzymes
Toxicity	May be a specific effect or a general one
Detailed molecular structure	Important factors include: i) Presence of an 'easily metabolizable' structural unit (s) ii) Presence of a 'difficult to metabolize' structural unit (s) iii) Presence of an 'unnatural' (xenobiotic) structural unit (s) iv) Degree of branching v) Nature of substituents vi) Number of substituents vii) Position of substituents
Concentration	Too high or too low
2) Environment-specific factors	
Biotic factors	Presence of suitable or 'potentially suitable' organisms
Abiotic factors	
Physical	Temperature or pressure
Chemical	Nutrients – minerals, growth factors Presence of oxygen as: i) terminal electron acceptor ii) substrate iii) inhibitor Presence of alternative electron acceptors: i) nitrate ii) sulphate iii) carbon dioxide PH Inhibitory materials Soil type Moisture level Type of water – fresh, brackish, saline

3.1 Chemical-specific factors

3.1.1 State/solubility/hydrophobicity

These factors are grouped together since they all influence the availability of a compound. For biodegradation to occur a chemical must either be in intimate contact with an organism or its enzymes (the enzymes may be intracellular or extracellular). Biological, enzyme-catalysed reactions normally occur in an aqueous milieu with the enzyme and its substrate being dissolved in water. The availability of a compound may be influenced by its physical state. Gases permeate well but may only be present at very low concentrations (see below). Liquids and solids are often present at apparently very high local concentration but, if they are not readily soluble, then the concentration available to microbes may in reality be very low and this may retard degradation. In such cases the rate at which a compound is metabolized may depend on the surface area available, as this will dictate the rate at which it can dissolve into the aqueous phase and become available to the enzymes which degrade it.

Sometimes biodegradative reactions involve the adhesion of the degrading organisms to their insoluble substrate – this is often the case for lipids and for hydrocarbons, such as mineral oils. When this is so, surface area is again important and if the material is finely divided into small particles, a greater surface area will be available for the attachment of degrading microbes. This will apply in the cases of environmental contamination during 'oil spills' – if the oil is dispersed it will degrade faster than if it accumulates in large masses. For solid hydrocarbons, such as polyaromatics, surface area will be important. Materials that aid solubilization (e.g. surfactants) may promote biodegradation. In fact, some microbes can themselves produce surfactants which improve degradation reactions. Evidence on the effects of adding synthetic surfactants on rates of biodegradation is equivocal.

There is a tendency for materials that are of limited water solubility to be relatively resistant to biodegradation, but it must be emphasized that this is not always the case. For example, toluene is only slightly soluble in water but is fairly degradable while some sulphonated aromatic compounds are very soluble but highly resistant to degradation. Hydrophobicity is inversely related to water solubility and, as such, affects the availability of compounds. Also hydrophobic materials tend to coalesce into droplets of low surface area, again restricting their availability.

A further complication is that hydrophobic organic compounds have a tendency to *bioaccumulate* in the fatty tissues of higher organisms. Thus the concentration in the higher organism will be greater than that in the surrounding environment. Once present in these tissues, *biomagnification* or *bioconcentration* may occur, a process by which the concentration of a material increases in organisms higher up the food chain. Sometimes a compound may bioaccumulate in an animal to levels thousands or even millions of times greater than those found in its environment. In one example, PCBs were present in water at a concentration of 2×10^{-6} ppm, in phytoplankton at 2.5×10^{-3} ppm, in zooplankton at 0.123 ppm, in rainbow trout smolt at 1.0 ppm, in lake trout at 4.83 ppm and in herring gulls at 124 ppm. This represents a biomagnification in the order 10^8.

A compound may cause little damage at lower trophic levels, but as it increases in concentration the damage increases. The biomagnification of organochlorine pesticides had disastrous effects on the populations of birds of prey and fish-eating birds in the 1950 and 1960s. Populations plummeted and it took many years for these populations to recover after the withdrawal of the pesticides. Another problem with bioaccumulation is that once accumulated in fatty tissues of animals or birds, a recalcitrant compound may be sequestered from microbial degradative processes until the animal dies and thus it may survive for a more prolonged period in the environment. Thus bioaccumulation can increase the persistence of a compound in the environment. The hydrophobicity of a compound and the likelihood of bioaccumulation can be assessed by determination of its octanol:water partition coefficient (often referred to as K_{ow}) – the higher the coefficient the more likely it is that bioaccumulation will occur. Often values of K_{ow} are so high that they are given as logarithm of K_{ow}. The types of compounds that are likely to bioaccumulate are hydrocarbons and organochlorines (e.g. DDT, lindane, aldrin and polychlorinated biphenyls – PCBs).

3.1.2 Adsorbability

Some chemicals have a propensity to adsorb to other materials, e.g. particular fractions of soils. If the adsorption is 'tight' then it may reduce the availability of a compound to microbes and enzymes thereby restricting or preventing its biodegradation. A classic example of this are the herbicides paraquat and diquat, which adsorb very tightly to certain clay fractions (notably montmorillonite) in soils. In these circumstances virtually no biodegradation occurs even in the presence of organisms which can degrade the herbicide. The amount of degradation will depend on the amount of the herbicide and the adsorbent material present in the soil. Some chemicals may adsorb to the organic (humus) fraction rather than to clay minerals.

3.1.3 *Size and shape*

For a chemical to be degraded it will usually have come into contact with an enzyme. Often, but not always, the degradative enzymes are intracellular, i.e. they are located inside a microbial cell and therefore within the cytoplasmic membrane – a selective permeability barrier that controls access to the cell. Some chemicals, by dint of their size, cannot pass across the membrane and therefore cannot come into contact with the enzymes therein. Thus, if they cannot be attacked by extracellular enzymes they will remain undegraded.

For an enzymic reaction to occur the substrate must enter into intimate contact with the enzyme and form an enzyme/substrate complex. Some recalcitrant compounds may be unable to complex with the active site of the enzyme due to their size, and probably more so, to their shape.

3.1.4 *Charge*

Highly charged molecules, particularly those with large negative charges, have great difficulty in penetrating cell membranes. Compounds with many sulphonic acid groups, e.g. some azo dyes and surfactants may come into this category.

3.1.5 *Toxicity*

It is fairly obvious that there may be problems in the biodegradation of chemicals that are toxic to microbes. Their biodegradation will clearly depend on their concentration and the susceptibility of the microbes. There will be similar problems with chemicals that are not generally toxic but inhibit a specific key enzyme in the degradative pathway. Many recalcitrant chemicals are toxic to some extent to some or most microbes; examples include most phenols (especially chloro- and nitro- derivatives) and fungicides. Toxic compounds may kill susceptible microbes (-cidal effects) or merely inhibit their growth (-static effects). The mechanisms of toxicity will vary considerably and can include inhibition of vital processes (like respiration), protein denaturation and membrane disruption. It may in some cases be possible for cells to evolve to become resistant, or tolerant, to a toxic compound. So, for example, most bacteria are inhibited or killed by toluene, which damages their membrane structure, but some can develop resistance and some bacteria have been reported to survive in the presence of 50% v/v toluene!

3.1.6 *Concentration*

Concentration is a major factor in the fate of recalcitrant compounds in the environment. If concentration is too high, problems of toxicity may arise (see above). However, concentrations may be too low. For example, in one study on degradation of the 2,4-D in stream water, about 75% mineralization occurred in 8 days when the herbicide was present at 2.2×10^{-2} or 2.2×10^{-4} g/l. However, at 2.2×10^{-6} or 2.2×10^{-8} g/l there was less than 10% mineralization. The lack of biodegradation of compound at very low concentrations can be accounted for in two ways. First, at low concentrations the compound (or its metabolites) may not be present at a sufficient level to induce the formation of the enzymes required to degrade it. Second, if evolution of new metabolic capacities is required, low concentrations of recalcitrant compounds may not provide enough carbon and energy to give newly evolved organisms a selective advantage. It is therefore highly desirable that compounds are degraded when present in relatively high concentrations (e.g. in industrial effluents) rather than escaping undegraded into the wider environment in which they will be greatly diluted and their degradation will be less assured.

3.1.7 *Detailed molecular structure*

Certain features of a chemical's structure are likely to make it either easier or more difficult for microorganisms to degrade. Ultimately, it is the properties of an entire molecule, not its components, that decides its fate. However, it is important to know how different subunits may influence the degradation process.

1. Easily metabolizable units: certain structures are easily metabolized by microbial enzymes. These include the ester (carboxylic,

sulphate and phosphate) and amide bonds, which are common and widely distributed in nature. Furthermore, all are susceptible to hydrolysis and thus do not need special coenzymes. Fig. 34.4 shows some of these easily metabolizable structures and their degradation.

Another relatively easily degradable bond is the amine bond. However, amines are much less degradable than amides and, in general, are cleaved oxidatively rather than hydrolytically. Amine-degrading enzymes are usually oxygenases, oxidases or dehydrogenases. As amine cleavage involves a redox reaction, the enzymes often need co-factors and they are therefore less likely to operate extracellularly.

(a) carboxylic ester $\xrightarrow{H_2O, \text{ esterase}}$ carboxylic acid + alcohol

$$R-C(=O)-O-R' \xrightarrow{H_2O, \text{ esterase}} R-C(=O)-OH + R'-OH$$

(b) sulphate ester $\xrightarrow{H_2O, \text{ sulphatase}}$ alcohol + HSO_4^-

$$R-O-S(=O)_2-O^- \xrightarrow{H_2O, \text{ sulphatase}} R-OH + HSO_4^-$$

(c) phosphate ester $\xrightarrow{H_2O, \text{ phosphatase}}$ alcohol + HPO_4^{2-}

$$R-O-P(=O)(O^-)-O^- \xrightarrow{H_2O, \text{ phosphatase}} R-OH + HPO_4^{2-}$$

(d) amide $\xrightarrow{H_2O, \text{ amidase}}$ carboxylic acid + amine or ammonia

$$R-C(=O)-NR'R'' \xrightarrow{H_2O, \text{ amidase}} R-C(=O)-OH + H-NR'R''$$

example

N,N-Dimethylformamide (DMF) $\xrightarrow{H_2O}$ Formic acid + Dimethylamine

$$H-C(=O)-N(CH_3)_2 \xrightarrow{H_2O} H-C(=O)-OH + H-N(CH_3)_2$$

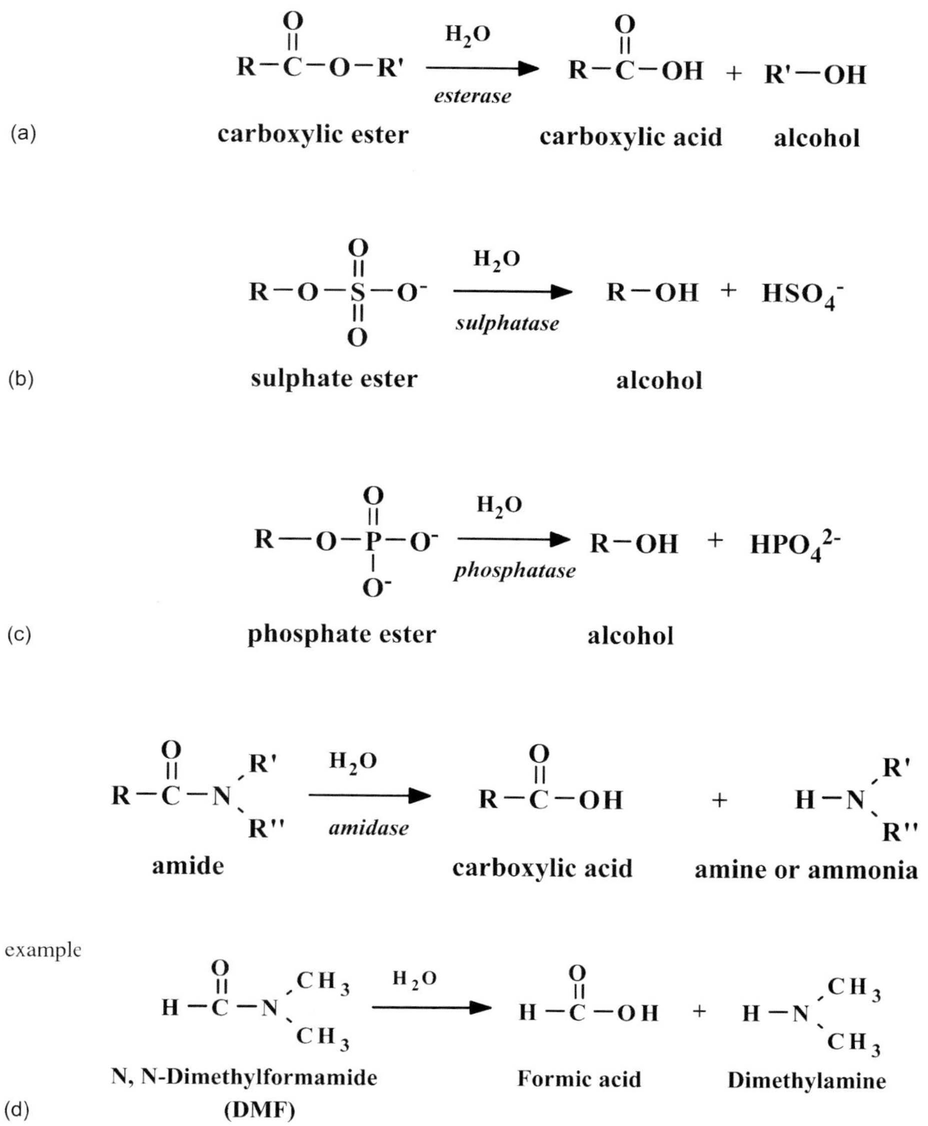

Fig. 34.4 Examples of easily degradable structural units. (a) Carboxylic ester linkage. (b) Sulphate ester linkage (example given in Figure 34.2). (c) Phosphate ester linkage. (d) Amide linkage (where R, R' and R" can be an alkyl group or hydrogen).

2. Units which are difficult to metabolize: some structural units are difficult to metabolize because they are not amenable to the normal processes of metabolism. A good example is the quaternary carbon atom, i.e. a carbon atom that has attached to it four other carbon atoms. Such structures can occur in alkyl chains and are resistant to degradation because they cannot participate in the process of β-oxidation (they prevent the formation of a carbon–carbon double bond) which is used to degrade alkanes and fatty acid or alcohols. Quaternary carbon structures can occur in, and restrict the biodegradability of, alkyl benzene sulphonate surfactants or hydrocarbons.
3. Presence of xenobiotic structural units: by definition, xenobiotic structural units have not been present in our biosphere for a long period. Such units only became common with the rise of synthetic organic chemistry in the 19th century and so microbes will have only had a relatively short interval in which to adapt to their presence compared with the geological time-scale that has been available for evolution to degrade 'natural chemicals'. Nevertheless, microbes, and particularly bacteria, can evolve very rapidly and some enzymes have evolved to degrade xenobiotic compounds. In other cases organisms may happen to possess enzymes of broad specificity, which can catalyse fortuitous metabolism of xenobiotics (with no certain advantage to the microbe).

   It is difficult to be certain whether a particular structure is, or is not, truly xenobiotic. So, for example, carbon-halogen bonds are not common in nature and natural compounds containing them only make up a tiny proportion of the organic matter on earth. Chloramphenicol, trifluoroacetic acid and thyroxine are examples of natural compounds with C-Cl, C-F and C-I bonds. Most organohalogens are truly xenobiotic and polychlorination of organic molecules is not natural. Nitro groups do occur naturally but are uncommon. Probably the best examples of xenobiotic structural units are the diazo linkage, aromatic (but not aliphatic) sulphonic acids and polychlorinated aromatic or aliphatic compounds; all these features retard biodegradation. Methyl, methoxy, amino and hydroxyl groups are common in nature but may, of course, be part of a xenobiotic compound.
4. Degree of branching: in alkanes or compounds with alkyl chains, the degree of branching is an important factor influencing biodegradability. Simply stated, the more branched a compound the less degradable it will be. Furthermore, branching in certain positions tends be particularly problematic. So, for example, two branches on the same carbon atom (see quaternary carbon above) or on adjacent carbon atoms are particularly problematical. The effects of branching on biodegradability have been noted particularly for alkanes and for alkyl benzene sulphonate surfactants (see Section 4).
5. Nature of substituents: some substituents of a basic chemical structure have the effect of increasing its biodegradability and others of decreasing degradability. The effects of substituents can be direct or indirect. So, for example, a substituent may indirectly decrease degradability by increasing toxicity, by decreasing solubility, by increasing hydrophobicity, by increasing ability to adsorb to soil or by altering charge. Direct effects tend to be related to changes in electron distribution, steric effects or hydrophobicity, which alter the compound's ability to participate in enzyme reactions.

   Electron distribution is always an important consideration; enzyme reactions tend to be either electrophilic or nucleophilic and the possibility of a reaction occurring (and its rate) depends, *inter alia*, on the density of negative or positive charges at certain positions in the molecule. Chemical substituents usually either donate or withdraw electrons, thus altering charge density. The magnitude and position of these changes in charge depend on the chemical and its substituent. There is no simple rule that electron donation or withdrawal is beneficial or adverse, it depends entirely on the nature of the enzyme reaction under consideration. Some will be accelerated by substituents that withdraw electrons and others by substituents that donate electrons.

Steric effects are related to the bulk of substituents. If a substituent is large it may cause *steric hindrance*. This is a phenomenon whereby a substituent may interfere with a reaction simply by dint of its size and position. Large substituents may prevent a potential substrate entering the active site of the enzyme or prevent it taking up the correct orientation at the enzyme's active site. For a reaction to occur, certain parts of the substrate must be aligned with certain regions of the enzyme, if a large structural unit is present this alignment may be impossible and thus the reaction may not occur.

The correct alignment of a substrate at the active site of an enzyme may also depend on hydrophobic or charge interactions or on hydrogen bonding. Changes in hydrophobicity or charge may prevent this alignment and thus prevent formation of the enzyme–substrate complex prior to catalysis.

A good example of the effects of substituents on biodegradation is provided by a study of the aerobic degradation of substituted benzene derivatives by soil microflora. The time taken for decomposition of benzoate, phenol, aniline, anisole, benzenesulphonate and nitrobenzene was 1, 1, 4, 8, 16, and >64 days respectively. Thus in terms of their ability to improve degradation:

$$-CO_2^- = -OH > -NH_2 > -OCH_3 > -SO_3^- > -NO_2$$

Clearly nitro- and sulphonic acid groups tend to retard degradation (the same is also true of chloro- and other halogen groups). For aerobic attack on aromatic rings, structural groups that have an electron withdrawing effect (e.g. chloro-, nitro- and sulphonate) often decrease the rate of degradation because in withdrawing electrons they inactivate the aromatic ring to electrophilic attack.

It must be emphasized that electron withdrawing groups do not always have an adverse effect on degradation, so, for example, when the initial attack is nucleophilic, electron withdrawing groups may increase reaction rates. This is the case, for example, for the anaerobic reduction of the diazo bond in azo dyes.

Very good quantitative correlations have been obtained in some instances between the coefficients that describe electronic effects of substituents and biodegradation rates. However, this is not always the case because other factors, like steric effects and hydrophobicity, can complicate the picture. The effects of particular substituents may well be different in aerobic and anaerobic conditions.

6. Number of substituents: as a general rule, although not an absolute one, biodegradability decreases with increasing number of substituents. Aerobic degradation often involves the introduction of hydroxyl groups onto the carbon skeleton. The more substituents there are, the less free carbons there will be for hydroxylation. If an unsubstituted carbon cannot be hydroxylated then a substituent will need to be removed, which may be more difficult.

   A good example of the effect of number of substituents is to be found in the aerobic degradation of the chlorinated phenoxyacetate herbicides (Fig. 34.5). Many pure microbial cultures have been described which can readily degrade 2,4-D and these can be easily obtained by enrichment culture. 2,4,5-T differs only in the presence of one extra chlorine atom, but it is much more resistant to biodegradation. It takes much longer to isolate 2,4,5-T-degrading bacteria by enrichment culture and only a few pure cultures have been isolated that can degrade it as a sole carbon source.

   PCBs (polychlorinated biphenyls) (see Fig. 34.1) were widely used industrial chemicals, but are now banned for most applications. The commercial preparations were mixtures of many congeners (isomers). Several studies have investigated the relationship between the number and pattern of substitution and biodegradability. Aerobic biodegradation is favoured by:

   i) the presence of fewer chlorine atoms
   ii) the presence of chlorine substituents on only one aromatic ring

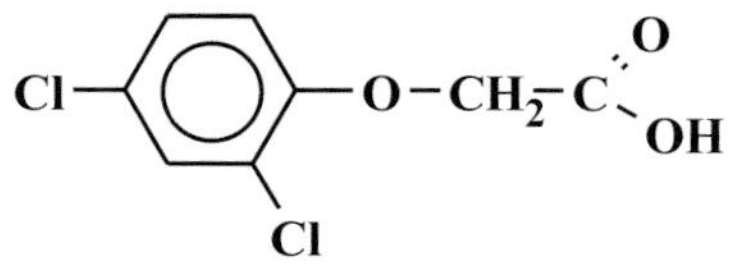

2,4-D - 2,4-dichlorophenoxyacetic acid (readily biodegradable)

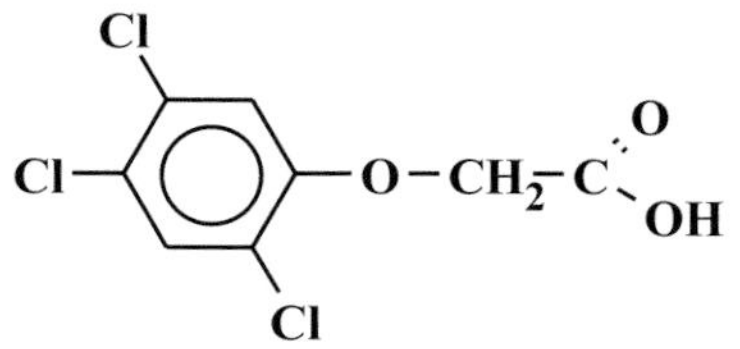

2,4,5-T - 2,4,5-trichlorophenoxyacetic acid (highly recalcitrant)

Fig. 34.5 The structure of 2,4-D and 2,4,5-T.

iii) the absence of chlorine atoms in certain positions (the 2, 6, 2′ or 6′).

Conversely in anaerobic degradation of chlorinated compounds, particularly aromatics, it often appears that reductive dechlorination is easier with the more highly chlorinated compounds.

7. Position of substituents: as has already been mentioned for branching, the position of substituents can have a major bearing on degradability of a compound. This effect can be due to either steric or electronic factors. Perhaps the best examples of positional effects are found with positional isomers of disubstituted aromatic compounds and of the insecticide lindane.

Disubstituted aromatic compounds have three isomers usually known as *ortho, meta* and *para* or 1, 2, and 3, respectively (Fig. 34.6).

A study on the effect of position of substitution on the aerobic degradation of a range of disubstituted benzenes by soil microflora showed large differences in the degradability of *o*, *m* and *p* isomers. In general, *o* and *p* isomers of a compound were of similar degradability to each other, while the *m* isomer differed in its

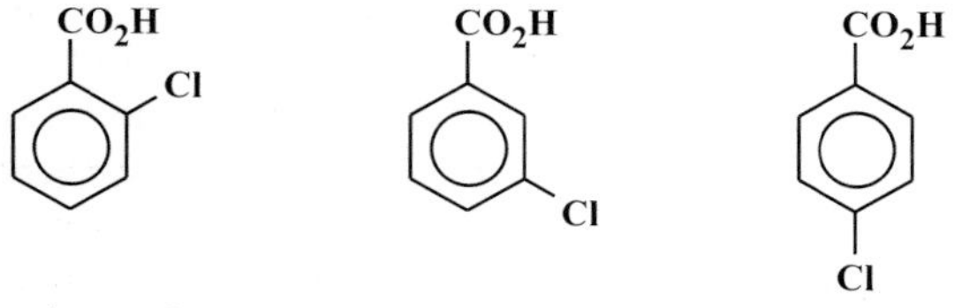

ortho-, o-, 2- *meta-, m-,* 3-, *para-, p-,* 4-,

Fig. 34.6 Structural isomers of chlorobenzoic acids. Typical disubstituted derivatives of benzene.

degradability. This is because the patterns of electron distribution are similar for *o* and *p* isomers but differ in the *m* isomer. Often *o* and *p* isomers are more degradable than the *meta* isomer, but this is not always so and it clearly depends very much on the effects of particular combinations of substituents on the electron density in particular parts of the molecule. A typical example is given in Fig. 34.7. In some cases, while *o* and *p* isomers of substituted benzoic acids are degraded in similar times, the *ortho* isomer is somewhat more recalcitrant. This is probably due to steric effects.

Lindane is the trade name of the γ-isomer of hexachlorocyclohexane (HCCH), formerly, and inaccurately, known as benzenehexachloride (BHC). This recalcitrant compound consists of a cyclohexane ring (in the chair conformation) substituted at each carbon atom with a chlorine atom and a hydrogen atom. At first sight it may appear that all carbon atoms are equivalent to each other and therefore that no isomers would exist. This is not so. HCCH is a planar molecule and the chlorines and hydrogens can be disposed either roughly in line with the plane of the ring or at an angle to it or both and several positional isomers are possible (Fig. 34.8).

It has been shown that there are considerable differences between isomers in terms of their degradability (Fig. 34.9). These differences cannot be accounted for by differences in electron distribution and must be related to steric effects – the distribution of the bulky chlorine atoms presumably interfering with the degradative process in some isomers.

In multisubstituted chloroaromatic compounds, e.g. chlorinated phenols or benzoic

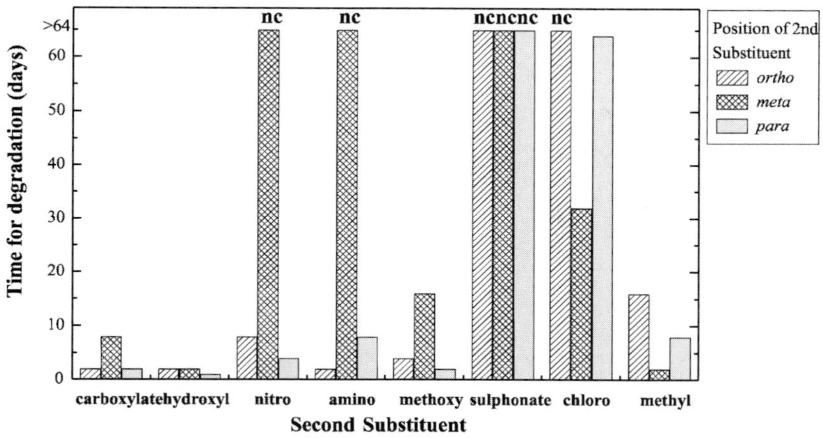

Fig. 34.7 The effect of the position of substitution in benzoic acid derivatives on the rate of biodegradation by soil microorganisms. The time for degradation is the time taken for the total loss of ultraviolet light absorption which equates to cleavage of the aromatic ring. NC indicates that a compound was not completely degraded in 64 days.

acids, it may be possible to predict which bonds will be broken and in which order. Bond strength of carbon-chlorine bonds can be predicted using computer models and, on this basis, the likelihood of particular bonds being cleaved reductively by unacclimated anaerobic microbial consortia has been determined. Very good correlations between predictions and experimental results have been obtained, with the most negatively charged carbon-chlorine bond being the first to be reduced.

In conclusion, it should be emphasized that while the susceptibility of a compound to biodegradation can sometimes be accounted for by one structural factor, more commonly a combination of several factors governs its fate. In some cases a compound may be less degradable than might have been expected on theoretical grounds. Such deviations from expected relationships may be due to factors like steric hindrance and toxicity.

3.2 Environment-specific factors

A whole range of environment-specific factors can influence a compound's fate in the environment. These can act alone or in concert and are discussed in detail below.

3.2.1 Presence of suitable organisms

For a compound to be degraded the presence of suitable organisms is essential. By 'suitable' we mean organisms that already have some ability to degrade the target compound. 'Potentially suitable organisms' (see Section 6 on evolution) are ones which initially cannot degrade a compound but have the potential to evolve to carry out the required biodegradation. Clearly, degradation of a target chemical will occur more rapidly after its exposure to 'suitable' rather than 'potentially suitable' organisms, because the time period during which evolution occurs will not be required. Nevertheless, these 'suitable' organisms are often present in very small numbers and, until a large population of suitable organisms has developed, little degradation will be observed. The 'lag' period before degradation is observed is known as the *acclimation* period and it may be due to the time taken for evolution to occur or the time taken for a large population to develop, or both. The time taken for the microbial population to increase to a 'critical mass' will depend on both the original number of suitable organisms present and on their growth rate. It will also depend on the number of organisms needed to make an observable effect on substrate concentration and clearly

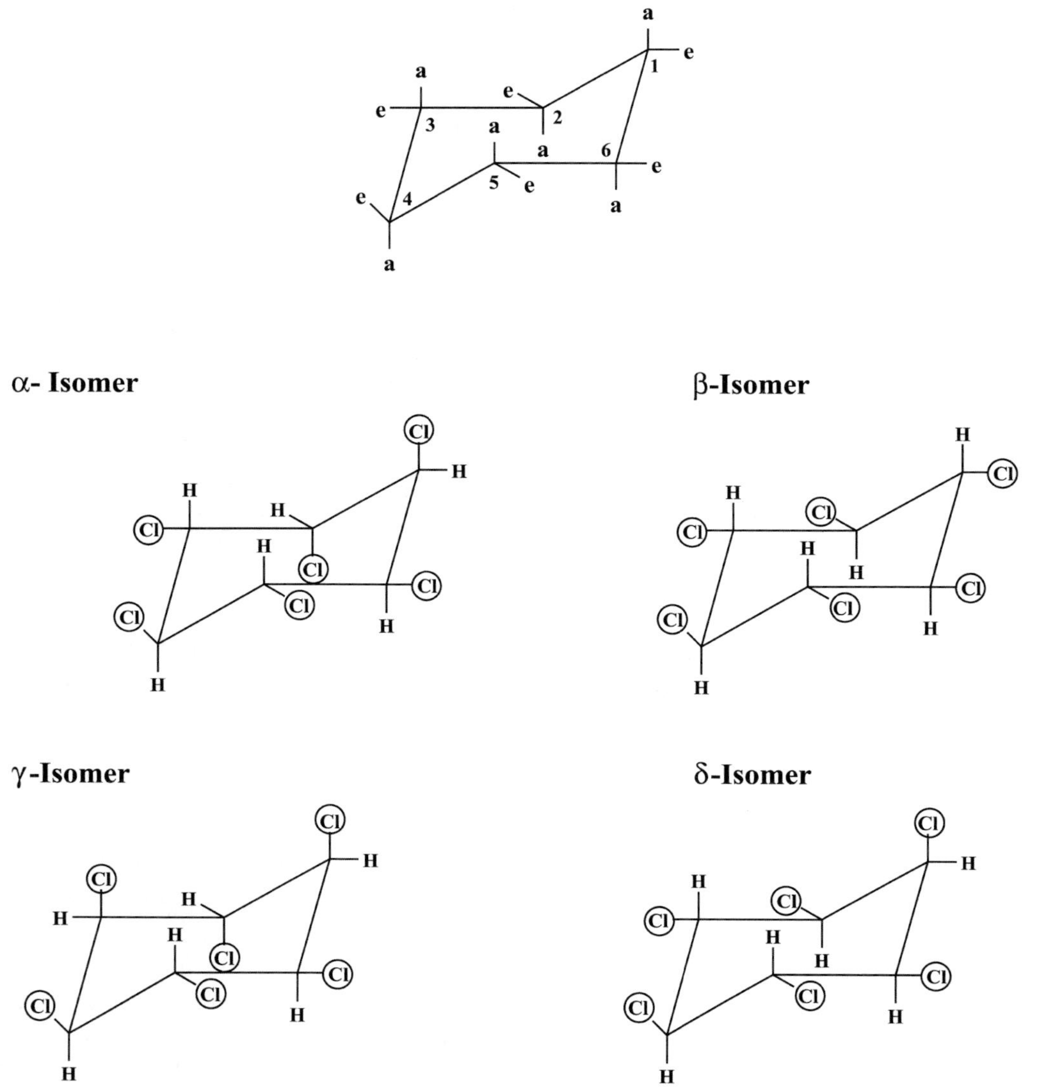

Fig. 34.8 Structural isomers of hexachlorocyclohexane (Lindane). Each carbon atom carries one chlorine atom; the position of the chlorine atom is either approximately in the plane of the ring (equatorial – e) or at an angle to the plane (axial – a). There are several possible isomers. The four most abundant are α, β, γ and δ.

this will be a function of the sensitivity of the analytical methods used. Obviously it is very difficult to generalize about this, but the number needed to make an impact may be in the order of 10^4–10^6 bacteria/ml.

Thus, if a chemical enters an environment with very few organisms in it then degradation will either be slow or will not occur. Pristine water in the upper reaches of a stream, ground water (in aquifers) and subsoil all have only very small microbial communities, thus recalcitrant compounds entering these environments will be degraded slowly, if at all. The presence of acclimated organisms, which have previously been exposed to a target compound, can greatly accelerate biodegradation. This has been shown repeatedly with degradation of herbicides, surfactants, industrial chemicals and mineral oils, in both soil and in river systems. Microbes in river water that has

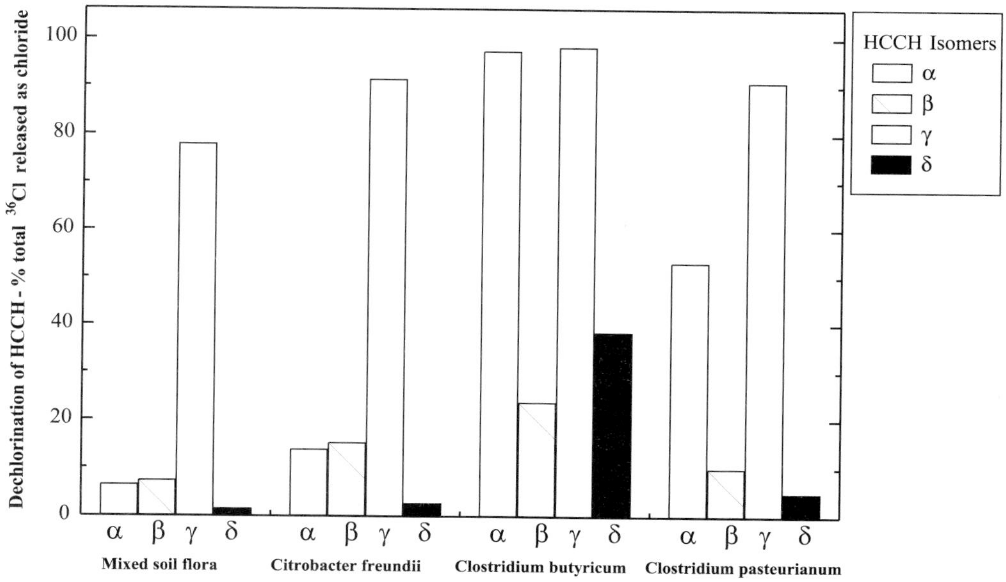

Fig. 34.9 The effect of isomer structure on the anaerobic degradation of hexachlorocyclohexane (HCCH) by microbial cultures. Biodegradation was co-metabolic with glucose as co-substrate. Cultures were incubated for 6 days at 25°C with 10 mg/1 HCCH. Degradation was assessed by measuring the release of radio-labelled ^{36}Cl from HCCH as chloride ions.

suffered industrial pollution will degrade target chemicals faster than those from unpolluted streams.

3.2.2 *Physical factors*

In general, temperatures that are too high or too low will retard, or even prevent, biodegradation. All organisms have their own optimum, maximum and minimum temperature for growth and degradative reactions. It should be noted that in regions of very high (e.g. hot springs) or low (e.g. polar regions) temperatures, it is likely there will only be a restricted and specialized microflora and thus the chances of organisms with the requisite degradative capabilities being present is small. However, there has been little research specifically on the biodegradation of recalcitrant organics by thermophiles and psychrophiles.

In addition to influencing the composition of the microflora and rates of reaction, changes in temperature may alter the physical state of a compound or its solubility – such changes may be beneficial or adverse to the chances of degradation.

Another physical factor is barometric pressure; high pressures, such as those found in the abyssal depths of the oceans, will restrict the diversity of the microflora with similar consequences to extremes of temperature.

3.2.3 *Nutrient availability*

In order to grow, microbes need a range of nutrients. The simplest requirements are for nothing but mineral salts and a source for carbon and energy. Most microbes that degrade organic pollutants are heterotrophs and the carbon and energy source can be a single compound (often a target recalcitrant compound). The major elements required are carbon, nitrogen, phosphorus, sulphur, oxygen and hydrogen, with smaller amounts of iron, magnesium, calcium and potassium. Sometimes specific metals are needed for particular enzymic reactions. For example, cobalt will be required if the degradative enzyme contains cyanocobalamin (Vitamin B_{12}) as a prosthetic group (this cofactor is important in some dechlorination reactions in certain organisms). Manganese will be required by white rot fungi if manganese peroxidase is important. If important elements are missing or in short supply growth will be impossible or unbalanced and degradative reactions may be slow or impossible. Some microbes also need the presence of growth factors; these are organic chemicals which the organism cannot make itself and must have if it is to grow. Growth factors may include specific amino acids, nucleotides or vitamins/cofactors. Some cofactors may not be essential but may, nonetheless, increase the rate of specific degradative reactions. For example, in some cases, the addition of flavine nucleotides (e.g. FAD) can increase the rate of anaerobic reduction of azo dyes.

3.2.4 *Presence of oxygen*

Oxygen has several roles in microbial growth and in each of them its presence can profoundly influence the occurrence and rate of degradative reactions.

1. Quantitatively the main use of oxygen is as a terminal electron acceptor during respiration in aerobic growth. Biological oxidations often involve the removal of hydrogen and electrons (reducing equivalents), which must be coupled to an oxidizing agent – the terminal electron acceptor. In this process energy is produced and intermediate electron acceptors are regenerated so that more oxidation can occur. Strict aerobes cannot grow without oxygen and if it is in short supply, their growth will be slower than normal and metabolism may be atypical. Thus, in the absence of oxygen, biodegradative reactions that depend on strict aerobes will not occur. At one time it was thought that oxygen was essential for degradation of most recalcitrant xenobiotic compounds, it is now known that this is not the case. Indeed, many compounds are degraded preferentially, if not uniquely, in anaerobic conditions. There are a few examples of compounds whose degradation requires oxygen, e.g. the degradation of lignin and the complex lignosulphonates produced in wood pulping appears to have an obligate requirement for oxygen.

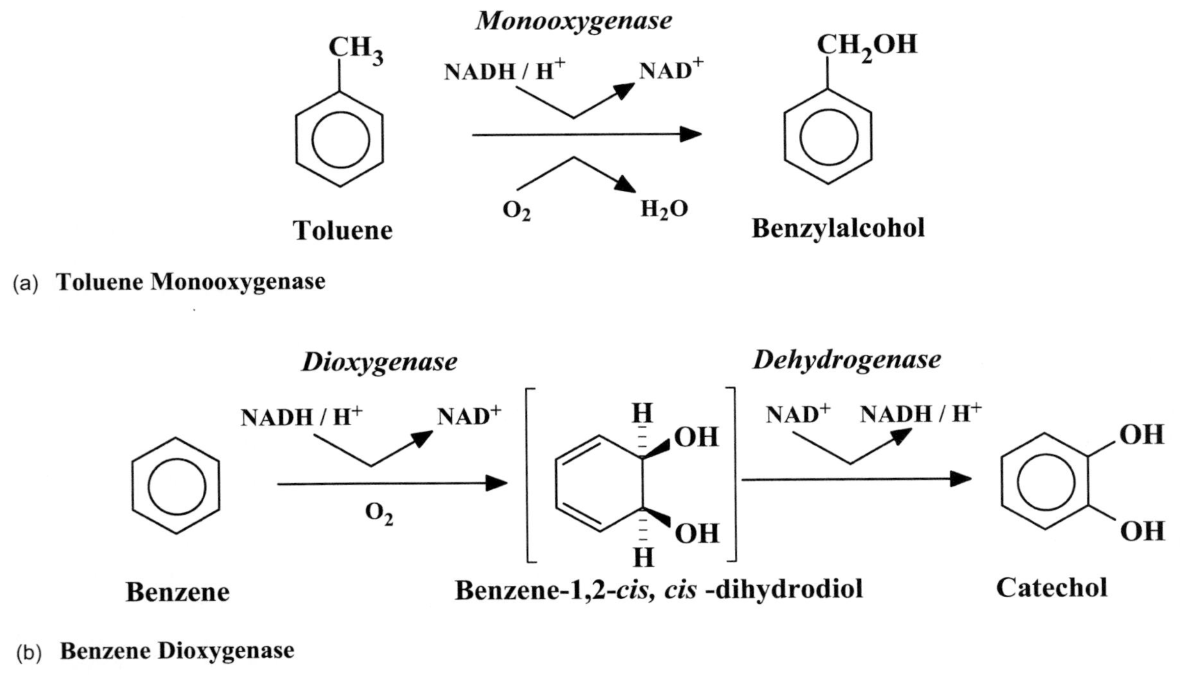

Fig. 34.10 Examples of monooxygenase and dioxygenase reactions.

2. Oxygen may also react directly as a substrate in certain enzymic oxidations. There are many examples of this type of involvement in which molecular oxygen is incorporated into an organic molecule – these are called oxygenations. Such reactions are often used to activate a relatively inert molecule and lead to further degradation, they are important in many processes for biodegradation of recalcitrant organics. Monooxygenations incorporate one atom of oxygen and dioxygenations two. Some examples are given in Fig. 34.10.
3. Finally, oxygen can act as an inhibitor and thus have a deleterious effect on certain biodegradative reactions. It has been observed in recent years that not only can many xenobiotic compounds be degraded very well in the absence of oxygen, but that some biodegradative reactions are inhibited by oxygen or occur only in its absence. Although some such reactions are catalysed by strict anaerobes, this is by no means always the case and degradative processes catalysed by facultative anaerobes or even obligate respiratory organisms (like pseudomonads) can occur preferentially in the absence of oxygen. For example, many polychlorinated organic compounds like DDT or lindane are degraded better anaerobically than aerobically. This is also the case for cleavage of the diazo bond. While some simple azo dyes can be decolourized in the presence of oxygen, much better results are obtained anaerobically. It seems that what most of these reactions, which are preferentially anaerobic, have in common is that they are reductive reactions. As such they will only be successful if there is an excess of reducing power. In the presence of oxygen, electrons may be diverted into reduction of oxygen rather than the target compound.

3.2.5 Presence of alternative electron acceptors

In the absence of molecular oxygen some organisms can carry out respiratory metabolism using so-called 'alternative electron acceptors'. These are inorganic compounds that can be reduced, thus allowing energy generation by oxidative phosphorylation. The materials that can be used in this way are listed in Table 34.2. In some cases the organisms using these alternate electron acceptors are normally aerobic, only resorting to this type of metabolism in the absence of molecular oxygen. In other cases the organisms are strictly anaerobic and are often very intolerant of oxygen – dying in its presence. There are now many examples of the degradation of

TABLE 34.2 Use of 'alternative' electron acceptors in biodegradation

Electron acceptor and/or process	*Reduced to*	*Organisms' usual relation to oxygen*	*Degradation of recalcitrant organics in pure culture*	*Involvement in centre mixed cultures*
Nitrate and oxides of nitrogen–*denitrification, nitrate reduction*	Nitrogen, N_2O	normally aerobes	√ e.g. toluene, phenol	possible
Sulphate, sulphur, and some oxides of sulphur–*sulphate reduction, sulphidogenesis*	Hydrogen sulphide	Always strict anaerobes	√ e.g. aniline, phenol	may act as electron sink; *Desulfomonile tiedjei* can dechlorinate 3-chlorobenzoate
Ferric ion Fe (III)–*iron reduction*	Ferrous ion Fe (II)	both normal aerobes and strict anaerobes	√e.g. toluene	possible
Manganic ion Mn (IV)–*manganese reduction*	Manganous ion Mn (II)	both normal aerobes and strict anaerobes	√ e.g. toluene	Possible
Carbon dioxide–*methanogenesis*	Methane	Always strict anaerobes	no	methanogens act as an electron sink

recalcitrant organics under strictly anaerobic conditions. Sometimes a pure culture can be isolated while, in others, consortia are involved. The role of sulphate reducers and methanogens in consortia may be to act as an electron sink. They are able to remove hydrogen, which is inhibitory to the fermentative organisms that produced it. These fermentative organisms may degrade recalcitrant xenobiotics with production of hydrogen, but often these reactions appear to be thermodynamically unfavourable without the removal of hydrogen. Some sulphidogens can also dechlorinate or even cleave aromatic rings.

For many years it was considered that some compounds, like aromatic hydrocarbons, were only degraded aerobically. It has now been shown that a range of the monocyclic aromatics can be degraded under a range of different types of anaerobic conditions. This gives hope that eventually such materials which have contaminated oxygen-deficient or anoxic environmental niches like ground water may ultimately be removed, especially if suitable electron acceptors are present. When considering bioremediation of hydrocarbon-contaminated aquifers it may be easy to encourage degradation processes by the supply of nitrate (which is highly soluble) rather than sparingly soluble oxygen.

3.2.6 pH

Clearly all organisms have an optimum pH for growth and for any degradative processes that they catalyse. Changes from optimal pH will clearly decrease rates of biodegradation for particular reactions. It has to be said that most research has concentrated on metabolism at near neutral pH (6–8) and there is very little literature on biodegradative reactions at very high or very low pH. The role of acidity in preservation is well known, common examples being preservation of pickled foods with acetic or lactic acids, silage production and yoghurt manufacture. The acid conditions in peat bogs are also a contributory factor to preservation of materials in that environment. Many fungi have acidic pH optima for growth and some can survive and metabolize at very low pHs – cellulolytic and saccharolytic fungi, for example, will often reduce the pH of their own growth media to between 2 and 3. However, as a group, the fungi are less able to utilize recalcitrant organics as growth substrates than are the bacteria. Having said this, the white rot fungi are known to be able to degrade a huge range of recalcitrant xenobiotics but, in general, these processes are fortuitous degradations from which the fungi derive little obvious benefit from the degradative process. White rots often operate best in the laboratory as agents of biodegradation at around pH 5. Although these organisms are widespread in the environment, they only grow in a restricted ecological niche (dead wood or dying trees).

Although quite a range of fungi can utilize aromatic compounds at subneutral pH, there has been little systematic study of their abilities. Similarly, despite recent research that has established the ability of many organisms, notably bacteria, to grow at very alkaline pHs, the degradative capacities of these alkalophiles has rarely been determined. As with any other extreme environment we can only expect to find a restricted microflora at extremes of pH and so it is likely that the range of biodegradative capacities of these specially adapted organisms will be less than that found with neutralophiles.

3.2.7 Inhibitory materials

Natural environments and man-made ones (like effluent treatment plants) may contain a wide range of toxic, inhibitory materials, which may either kill microorganisms or inhibit their degradative processes. Some toxic chemicals (e.g. phenol, toluene or cyanide) may inhibit a wide range of microbes, others may be much more specific, only killing particular taxa (e.g. fungi), groups of microbes (e.g. nitrifying bacteria) or even species. Some inhibitors are naturally occurring (e.g. plant phenolics derived from decomposing plant tissues, or heavy metals derived from mineral deposits), others are present as the result of environmental pollution. The concentration of these inhibitors is crucial in determining their effects. Phenol and toluene are degradable at low concentrations but generally extremely toxic at high concentrations. As always with microbes, there is the possibility that given constant exposure some will evolve to become more resistant to the inhibitory materials. This is certainly true for toluene and for many heavy metals, e.g. mercury.

Salinity is of course an important factor, as increasing salinity generally inhibits microbial growth. There are some obligate halophiles, which not only survive in, but actually need, very high salt concentrations, but this is a small group of specialized organisms, with relatively low genetic diversity. The metabolic capabilities of halophiles have been studied to some extent and appear, as expected, to be much more limited than those of organisms living in less extreme conditions.

3.2.8 Soil type

The composition of a soil may profoundly affect the degradation of materials in the soil. Some soil fractions can adsorb and sequester recalcitrant compounds (see Section 3.1.2 on adsorbability). It is also possible for soil fractions (notably some clays) to bind extracellular enzymes, thus inactivating them and allowing their substrates to survive for a longer time in the soil than would otherwise be possible. This has been shown to occur with materials like proteins, which are normally perishable and so greater problems might be expected with recalcitrant organics that require extracellular enzymes for their degradation. Similarly, some soils contain tannins and similar plant phenolics (derived from decomposing plant remains) which can complex with and inactivate enzymes in a non-specific fashion.

4 'DESIGNER DETERGENTS'

Artificial surfactants, the active agents in detergent preparations, first came into widespread use after the Second World War when they replaced soaps. There are four categories of surfactants: anionic, cationic, non-ionic and zwitterionic and within each category there are many variations on the theme. All consist of a hydrophobic (lipophilic) group coupled to a hydrophilic group; this latter may have a negative or positive charge, no charge or a variable charge according to pH. By far the most important group in terms of amount used is the anionics of which there are two main types, linear alkyl sulphates and alkyl benzene sulphonates (ABS). Alkyl benzene sulphonates are the principal type of laundry surfactants and these gave rise to enormous problems when they were introduced. The type of formulation used was a material called tetrapropylene alkyl benzene sulphonates (TPS), which were made by polymerizing four molecules of propylene. This gave rise to an indeterminate mixture of many different isomers (all branched). A high proportion of these isomers were recalcitrant to degradation and so they passed through effluent treatment plants unscathed by the microbial populations. On reaching the rivers they did what surfactants are best at – they foamed, especially in regions of great turbulence like weirs and waterfalls. Huge layers of foam developed and were carried down the rivers like sheets of meringue! The foam could be a few feet deep, and it would blow off the rivers and into surrounding countryside or riverside towns. Not only was this an eyesore, but there were also problems of the toxic effects of the surfactants to aquatic life. Furthermore, the foam picked up and transported pathogenic microbes from the sewage polluted rivers. Action was called for and an intense research programme was established to determine how a biodegradable surfactant could be produced.

The following observations were made:

1. the presence of a quaternary carbon near the end of the alkyl chain prevented degradation
2. the degree of branching was important; the more branches present, the worse the degradation – a preparation of TPS will typically be very resistant to biodegradation with only a small part of the mixture succumbing easily to degradation
3. in ABS surfactants with straight alkyl chains, the length of the chain affects degradability. Chains with about 12 carbon

atoms being optimal, longer or shorter chain lengths delayed but did not prevent biodegradation

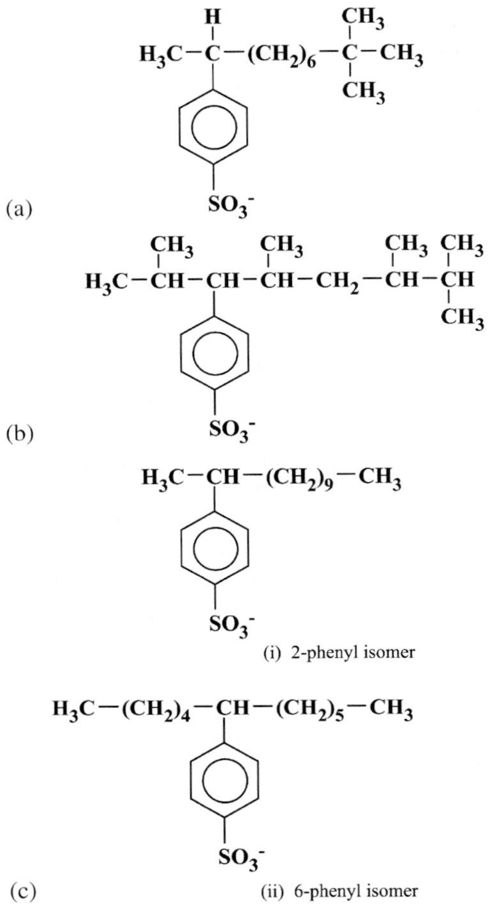

Fig. 34.11 The effect of structure on degradability of alkylbenzene sulphonate detergents. (a) An alkyl benzene sulphonate detergent with a terminal quaternary carbon atom. This is highly resistant to biodegradation since the inital attack is at the end of the alkyl chain and the quaternary carbon prevents β-oxidation. (b) This structure is a typical tetrapropylene alkylbenzene sulphonate (TPS). TPS comprises of a mixture of isomers which differ in the position of the methyl branches and the phenyl group. Most isomers are very resistant to biodegradation. (c) These structures represent typical linear alkylbenzene sulphonates (LABS) the phenyl group can be on any alkyl carbon except carbon-1. The 2-phenyl derivative (i) is more degradable than the 6-phenyl isomer (ii).

4. in straight chain ABS molecules the benzene sulphonic acid group can be attached in several positions. It is generally the case that degradability is enhanced by attachment close to the end of the alkyl chain and decreased by attachment near the middle of the chain.

Thus the ideal ABS surfactant will have a straight alkyl chain about 12 carbon atoms long with the benzene ring attached near the end of the alkyl chain and there will be no branching. Typical structures of these surfactants are shown in Fig. 34.11.

By investigating the effect of the various structural elements on degradability of a surfactant it was possible to work out the optimum structure for a readily degradable alkyl benzene sulphonate surfactant. So-called 'soft' surfactants were produced and subjected to trials in certain parts of the UK with excellent results. 'Hard' ABS surfactants were largely replaced following a voluntary agreement in 1964 between manufacturers and the UK government.

Surfactant pollution is now largely a thing of the past on most rivers in the UK. The only real exception being found on a few rivers where the local textile industry still uses certain types of hard non-ionic surfactants. The problem compounds are commonly known as 'APES' (alkylphenol polyethoxylates). They are still required for certain functions, although for many applications they have been replaced by the 'softer' alkyl ethoxylates (Fig. 34.12).

Among the ethoxylate non-ionic surfactants, degradation is favoured by:

1. a linear hydrophobe rather than a branched one
2. an alkyl rather than an alkyl phenol hydrophobe
3. a relatively short ethoxylate chain.

Non-ionic surfactants can include propoxy or butoxy groups as well as or instead of ethoxylate groups. These larger alkoxy groups tend to decrease degradability.

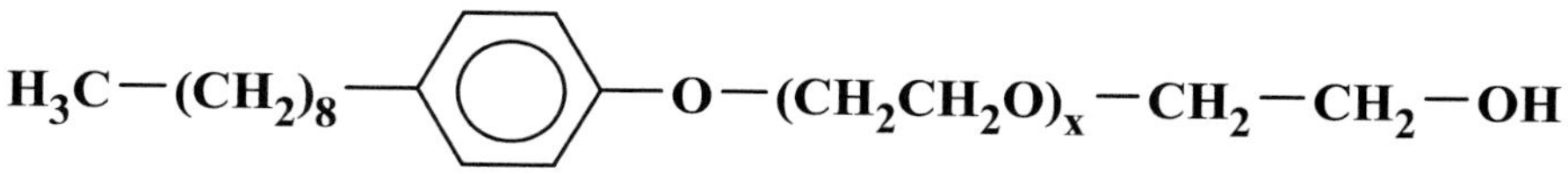

(a) Nonyl phenol ethoxylate – a typical alkyl phenol ethoxylate surfactant (APE). The number of ethoxyl units (x) can be in the range 2 to 50.

$$H_3C-(CH_2)_y-CH_2-O-(CH_2CH_2O)_z-CH_2-CH_2-OH$$

(b) An alkyl ethoxylate (fatty alcohol ethoxylate surfactant - FAE). The number of methylene units (y) can be in the range 7 to 16. The number of ethoxyl units (z) can be in the range 2 to 30.

Fig. 34.12 Structures of ethoxylate non-ionic detergents.

5 ACCLIMATION OF MICROBIAL POPULATIONS TO DEGRADE RECALCITRANT ORGANICS

Acclimation is the process by which a microbial population adapts to degrade a compound to which it is exposed. Given a population which, on initial exposure, appears to have no significant degradative ability against a target compound, acclimation may involve three different processes. The acclimation or 'lag' period can vary greatly in duration. The different processes and possible timescales are summarized in Table 34.3.

Acclimation to degrade a target compound can be lost as well as gained. Loss of acclimation can occur not only for xenobiotics, but even for central metabolites like glucose.

TABLE 34.3 The mechanisms of acclimation and the processes involved

Mechanism	*Processes involved*	*Relative speed*	*Possible timescale*	*Factors influencing rate*
Induction of enzymes or transport systems	The organisms already have potential to make the required enzymes etc. but need presence of the target compound to stimulate production	Fastest	Minutes to hours	Rate of metabolism
Growth of population of suitable microbes	A few 'suitable organisms' are present but have to grow until the population is large enough to make an observable change in the concentration of the target compound	Intermediate	Hours to weeks	Growth rate; initial number of organisms
Evolution (see Section 6)	Mutations required to change an organism so that it can now degrade the target compound. One or more mutations may be needed in one or more genes. Once mutations have occurred the size of the mutant population has to increase – see above	Slowest	Weeks to months	Number of organisms with 'potential'; mutation rate; number of mutations needed; growth rate

It can occur for three reasons: absence of degradative enzymes in the absence of inducer (see above); genetic instability (see Section 11); and loss of a specific group of degradative organisms (see Section 14.3).

6 EVOLUTION OF DEGRADATIVE ABILITIES

Assuming that it is not toxic, the ability of an organism to degrade a compound depends on the ability of the compound to come into contact with an enzyme or a series of enzymes which can degrade it.

This is affected by three things:

1. access of the compound to the enzymes
2. ability of the enzyme to catalyse a degradative reaction
3. production of the enzyme in suitable quantities.

All three of these may be subject to evolutionary change.

Evolution may involve the mutation of pre-existing genes and thus the production of new, altered proteins. It can also involve the acquisition of new genetic information from other organisms. This new information may come via plasmid transfer, transposons (jumping genes) or by uptake of DNA from the environment.

Sometimes extracellular enzymes are involved in a biodegradation, in which case the ability of compounds to enter the cell is not a problem. If enzymes are intracellular, then the compound needs to cross the cell membrane. This can involve free diffusion or use of a permease or other transport system. If the compound cannot easily enter the cell, there is the possibility for improvement in access to the cytoplasm, due either to alterations in the structure of the cell membrane (this could involve changes in either proteins or lipids) or to changes in the specificity of the permease proteins which catalyse translocation of compounds across membranes.

Enzymes may already exist that catalyse metabolism of similar compounds. Small changes in the structure of the protein may alter its substrate specificity such that a new substrate can be metabolized. Such changes can occur either in the regions which catalyse the reaction or in regions responsible for binding of the substrate into the enzyme/substrate complex. It should be noted that some enzymes are naturally of broad substrate specificity (see Section 9).

Where an organism already possesses genes for the production of an enzyme which can metabolize a target compound, degradation may not occur due to a failure of the organism to produce the requisite enzymes. This can be overcome by mutations that lead to constitutive production of the enzyme (i.e. the enzyme is produced all the time) or to a change in the inducer specificity for the enzyme such that the target compound is now an inducer.

It should be noted here that there are often large differences in the specificity of transport systems and enzymes, and also of the control systems which regulate their production. Thus, when a new metabolic activity arises in an organism, it may be due to mutation (followed, of course, by selection) in one or several of the functions mentioned above. Studies on the evolution of new metabolic activities in organisms have shown that often the early stages of evolution involve selection of constitutive mutants or mutants that produce very large amounts of an enzyme and this is then followed by the selection of mutants with altered enzyme activity. Alterations in enzyme activity are more likely to involve changes in the regions of the protein that involve substrate binding than changes in catalytic sites.

One stage in the evolution of an organism to degrade a toxic compound may involve the organism becoming tolerant to the toxic effects. This tolerance may be due to a range of different adaptations depending on the mechanism by which the compound causes inhibition.

7 METABOLIC VERSATILITY

It is interesting to observe that, although some microbes are very specialized in the type of compounds they can degrade or utilize as growth substrates, others may not be. This is particularly so for bacteria in a few genera which show extreme nutritional versatility, i.e. the ability to degrade many organic compounds. It is not uncommon to find individual bacterial *strains* (not species) that can utilize over 100 defined organic compounds as sole source of carbon and energy. This trait is most famous in the aerobic pseudomonads. The old genus *Pseudomonas* has now been dismantled by taxonomists and split into a range of new genera. Some of the most versatile species include *Pseudomonas aeruginosa, P. putida, P. fluorescens, P. stutzeri, Burkholderia cepacia* and *Comamonas testosteroni*. However, the pseudomonads are not the only group possessed of nutritional versatility. Other noted Gram-negative bacteria include *Acinetobacter, Alcaligenes, Moraxella, Achromobacter* and *Flavobacterium* spp. For many years the qualities of the Gram-positives were rather overlooked – possibly because the nutritionally versatile species are often rather slow-growing in comparison to the pseudomonads. The Gram-positives most noted for their ability to degrade recalcitrant chemicals are all in the actinomycete line and include *Mycobacterium, Nocardia, Rhodococcus* and *Arthrobacter* spp. The organisms most frequently identified as agents of biodegradation tend to belong to the genera named above, although other genera, e.g. *Streptomyces*, are occasionally encountered.

All of the organisms listed above are primarily aerobes, however, with the increasing appreciation of biodegradation in anaerobic conditions, it is likely that there will soon be a considerable number of strict or facultative anaerobic bacteria to add to the list. For example, degradation of aromatic compounds under denitrifying conditions has been demonstrated for members of the genera *Azoarcus* and *Thauera*, organisms which were little known until recently but are likely to prove important. Similarly, the significance of sulphate reducers in biodegradation has only just become apparent with the identification of new organisms like the aromatic-degrading *Desulfobacula toluolica*.

Fungi are to some extent the 'Cinderellas' of biodegradation, their role having been largely ignored. In the last 10 years their abilities have been more recognized with the realization of the potential of white rots (discussed in more detail in Section 12). However, outside the white rots they are still largely ignored.

8 METABOLIC PATHWAYS

Clearly with the plethora of different recalcitrant chemicals, a very wide range of metabolic pathways are involved in biodegradation. If a compound is to serve as a carbon and energy source, then it has to be converted into a form that can enter central metabolism. Normally this involves converting it into one, or more, low molecular weight intermediates of the tricarboxylic acid (TCA) cycle (otherwise known as Kreb's or the citric acid cycle) or compounds that feed into it. The means by which this is done obviously varies considerably – typical intermediates include acetate (or acetyl CoA), acetaldehyde, pyruvate, succinate or fumarate. The length of metabolic pathways varies enormously according to the complexity of the target compound; the number of steps ranges from one or two to over a dozen. In some cases target compounds require very little modification before they can enter a pre-existing pathway.

It should be stressed that the metabolic pathways used for a degradation may vary not only according to the environmental conditions (e.g. aerobic or anaerobic) but also according to the type of organism. There may be a wide range of pathways available, some of which differ from each other subtly and others markedly. It is quite likely that different strains of the same species will employ different metabolic pathways for the same degradative process. For this reason it is of little value to reproduce here details of lots of metabolic pathways, as no assumptions can be made as to which pathway will be used for a particular degradative process without proper testing being done. Generally, there is no obvious reason why one strain uses one pathway and another strain a different pathway. Indeed, in effluent treatment plants there does not appear to be any evidence as to which pathways are in practice used by the microflora.

One of the best examples of multiple pathways is found with toluene degradation (Fig. 34.13). Two pathways have been proposed for anaerobic degradation and at least five for aerobic degradation. Two of the five are from different strains of *Pseudomonas putida*.

Many recalcitrant chemicals are aromatic and it is perhaps useful to look briefly at how aromatics are degraded aerobically. There are basically three different sets of pathways. To begin with the aromatic compound is converted to a ring cleavage intermediate which normally has two hydroxyl substituents. These may be next to, or opposite, each other. This intermediate is then cleaved by a dioxygenase to give a straight chain intermediate. If the hydroxyl groups are next to each other then the ring is either cleaved between them (*ortho* cleavage) or to one side of them (*meta* cleavage). Where the hydroxyls are opposite each other the ring is cleaved immediately to one side of one of the hydroxyl groups. The straight chain ring cleavage product is degraded to give smaller units, which will enter the tricarboxylic acid cycle. The initial pathway which produces the ring cleavage substrate may be simple, involving only one step (e.g. hydroxylation of phenol to produce catechol) or complex involving many steps. Fig. 34.14 shows the early stages of the three main types of pathway. It should be noted that a range of analogues may act as ring cleavage substrates, but often (not always) distinct, specific enzymes are involved. So, for example, in *Pseudomonas putida*, catechol 1,2-oxygenase employed to cleave catechol is distinct from protocatechuate 3,4-oxygenase which cleaves its carboxylated analogue in the *ortho* pathway.

Anaerobic degradation of aromatics proceeds in a very different way since it is not possible to activate the benzene ring by the addition of oxygen. Instead the ring is activated by converting alkyl side chains, or adding carbon dioxide, to give a carboxylated ring, which is then converted to a coenzyme A ester. Benzoyl CoA or its *para* hydroxy derivative appears to be a common intermediate and this is progressively reduced to a cyclohexyl derivative which is primed for ring cleavage by the addition of water across a double bond next to the carboxyl group. After ring cleavage the straight chain intermediate is degraded to acetyl CoA by β-oxidation. Thus, in this type of attack the oxygen required is introduced from water and oxidation is accomplished by removal of hydrogen/electrons. These are generally disposed of by the reduction of nitrate or sulphate or the generation of hydrogen, which is utilized in consortia by methano- or sulphido-gens.

9 NON-SPECIFIC ENZYMES

Some types of co-metabolism depend on the fact that certain enzymes are sufficiently non-specific to catalyse degradation of compounds other than their 'normal' substrate. Several enzymes are well known for the wide range of substrates that they can modify, some of which have little resemblance to their 'normal' substrate. These enzymes include methane monooxygenase, ammonia monooxygenase and toluene dioxygenase. Examples of some of the substrates they modify and reactions they catalyse are given in Table 34.4.

10 REDUCTION BY CO-ENZYMES

Co-enzymes are relatively small organic molecules often derived from vitamins. Some are reactants in enzyme-catalysed reactions while others are essential parts of many enzymes (sometimes called prosthetic groups). Co-enzymes are often important in enzymes carrying out oxidation or reduction reactions and they may participate as electron carriers.

It has been observed that in some cases the co-enzyme on its own can catalyse important biodegradative reactions with recalcitrant compounds. For example, cyanocobalamin and other similar derivatives of vitamin B_{12} are able to catalyse reductive dechlorination reactions. Compounds dechlorinated in this way include PCBs, hexachlorobenzene, carbon tetrachloride, chloroform, dichloromethane and chlorinated ethanes. Some other co-enzymes involving transition metals, e.g. nickel and

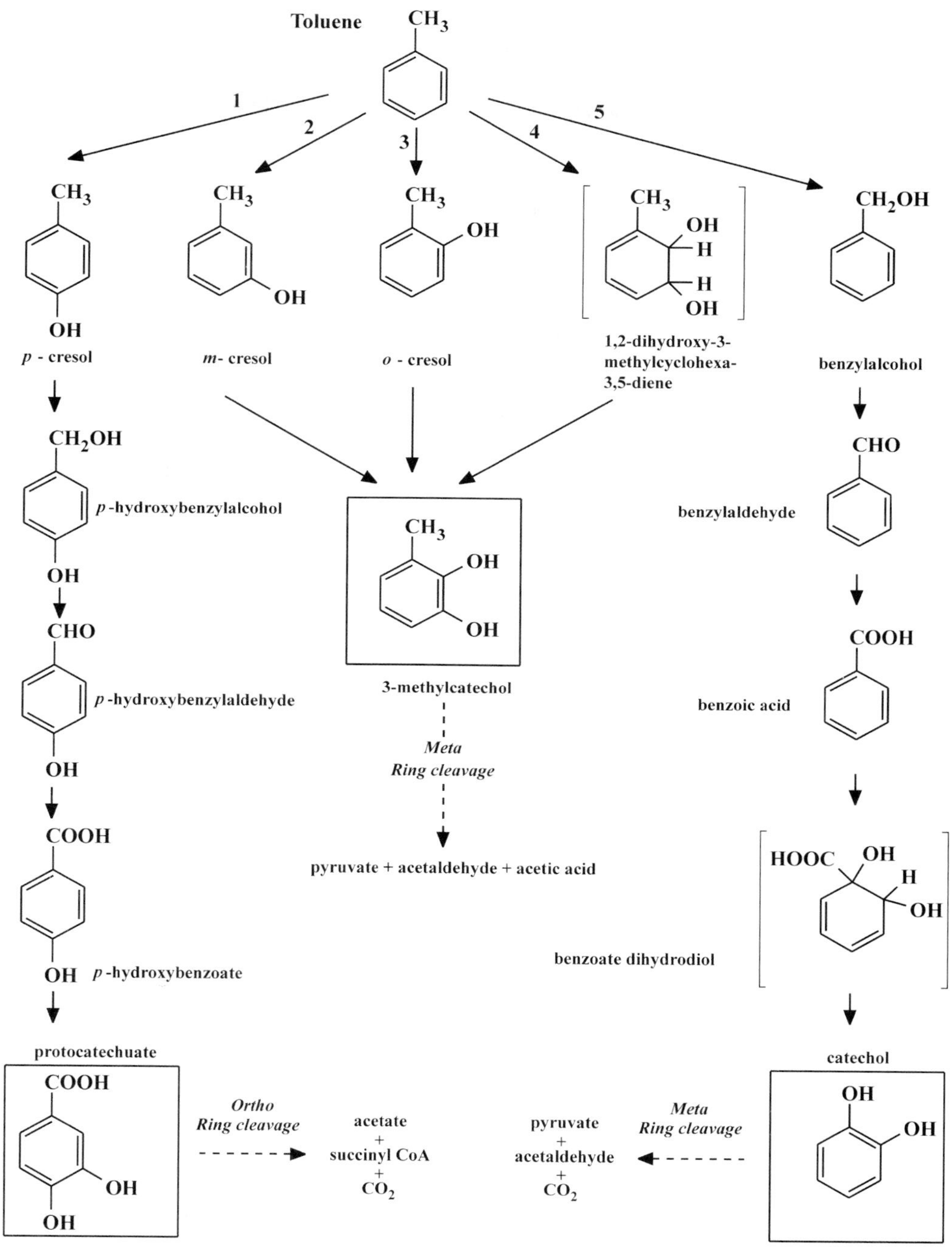

Toluene is degraded via 5 different routes by the following bacteria; *Pseudomonas mendocina* KR1 (1), *P. pickettii* PKO1 (2), *Burkholderia* (formerly *Pseudomonas*) *cepacia* G4 (3), *P. putida* F1 (4), *P. putida* mt-2 (5). Postulated intermediates are shown in parentheses. Intermediates immediately prior to ring cleavage are shown in boxes. ------▶ indicates several steps.

Fig. 34.13 Different pathways for the aerobic biodegradation of toluene.

Name of pathway	ortho-cleavage intra-diol cleavage; β-ketoadipate pathway	meta-cleavage; extra-diol cleavage	gentisate/ homogentisate pathway
Typical ring cleavage substrate	catechol; protocatechuate; 3 or 4 methyl catechol	catechol; protocatechuate; 3 or 4 methyl catechol	gentisate; homogentisate
Typical ring cleavage product	*cis,cis* -muconate; *β*- carboxy-*cis,cis* -muconate	2-hydroxymuconic semialdehyde; 2-hydroxy-4-carboxymuconic semialdehyde	maleylpyruvate; maleylacetoacetate
Typical degradation products	Acetyl CoA, succinate		fumarate; pyruvate; acetoacetate; acetate

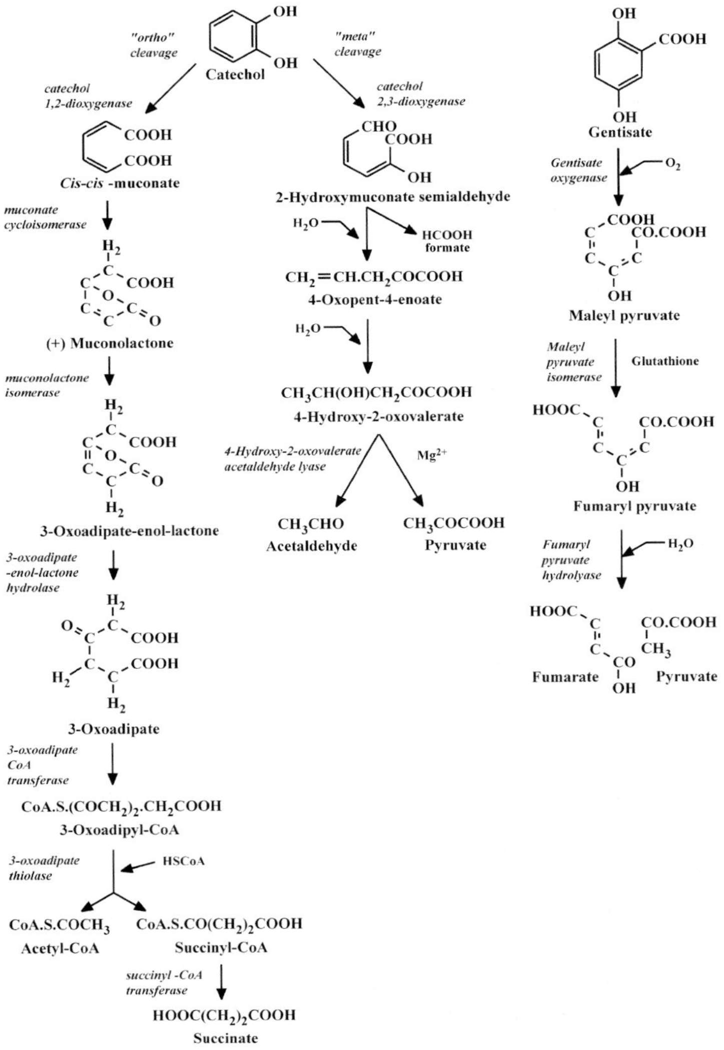

Fig. 34.14 Shows the major metabolites involved in aerobic aromatic degradation pathways.

TABLE 34.4 Examples of 'non-specific' enzymes and the reactions they can catalyse

Enzyme	*Organism*	*Normal reaction*	*Examples of substrates metabolized or reactions catalysed*
Ammonia monooxygenase	*Nitrosomonas europaea*	$NH_3 \rightarrow NH_2OH$	dichloroethylene; trichloroethylene; vinyl chloride
Methane monooxygenase	Methanotrophs e.g. *Methylosinus trichosporium; Methylococcus capsulatus*	$CH_4 \rightarrow CH_3OH$	chloromethane to formaldehyde; chloroform to CO_2; dichloromethane to CO; pyridine to pyridine N-oxide; toluene to benzyl alcohol
$P450_{cam}$	*Pseudomonas putida*	camphor to hydroxy camphor	reductive dechlorination of chlorinated methanes and ethanes
Propane monooxygenase	*Mycobacterium vaccae*	propane to propanol	cyclohexane to cyclohexanol; dichloroethylene; trichloroethylene; vinyl chloride
Toluene dioxygenase	*Pseudomonas putida* F1	Toluene → 3 methylcatechol	trichloroethylene to glyoxylate, formate and chloride

iron, can have similar effects. Reduced flavin nucleotides (FMN and FAD) are thought in some cases to mediate azo dye reduction.

It is possible that use may be made of some of these processes at least to facilitate primary biodegradation of some pollutants. The co-enzymes may have certain advantages over cell-free enzymes as biocatalysts. Notably they may be more stable. It is possible to immobilize some co-enzymes for use in bioreactors. One factor that has to be borne in mind is that most of the reactions catalysed are redox reactions and that for continuous use it will be necessary to return the co-enzyme to the correct oxidation state. In some cases this can be accomplished using chemical reagents, in others use of microbes is a possibility.

11 GENETIC ASPECTS

The genes that encode for biodegradation of many recalcitrant organic chemicals are often encoded on plasmids. Sometimes only one or two enzymes are encoded on a plasmid with the rest on the bacterial chromosome, while in others, an entire pathway is plasmid encoded.

Plasmids are small genetic elements separate from the chromosome and are often unstable. There is a range of causes for their instability but often a plasmid will be lost from an organism if it is cultivated in the absence of the substrate whose degradation it encodes. Plasmids tend to remain stable in a population when their presence confers a selective advantage (e.g. ability to utilize a substrate which is present) on individual cells which carry them. When no advantage accrues to the organism, plasmids can be lost because their maintenance in the population represents a drain of resources which can impose (at least in the short term) a selective disadvantage.

Plasmids can often be transferred easily between organisms, even ones that are not closely related, and this transfer offers a means by which genetic recombination can occur. This happens in nature but it can also be made to happen in the laboratory giving scientists the potential to develop new bacterial strains with altered and enhanced degradative ability. Various researchers have genetically engineered so-called 'super bugs', which can degrade a wider range of, for example, hydrocarbons. The potential application of these strains in clearing up oil spills or other pollution incidents has been patented and much 'hyped'. In practice, it seems unlikely that most of these genetically engineered organisms will really offer much advantage over mixed cultures developed by conventional enrichment culture, or indeed over the native microflora. Such strains are more likely to be of use in the more carefully controlled conditions of special treatment plants.

Genetic engineering may have a role in creating bacterial strains with entirely new metabolic activities by combining genes from different organisms to produce an entirely new metabolic pathway. An example of this is the creation of a new strain of *Pseudomonas* to degrade chlorobenzoates by using part of the *ortho* pathway from one organism with part of the *meta* cleavage pathway from another. It is likely that such strains will also arise as a result of 'natural' genetic recombination.

Table 34.5 gives a list of some of the recalcitrant organics which can, at least in some organisms, be degraded by plasmid-encoded enzymes.

It is important to remember that enzymes that are plasmid encoded in one organism are not necessarily plasmid encoded in other organisms, or even other strains of the same species.

TABLE 34.5 Examples of substrates whose degradation is, at least in part, plasmid-encoded

Substrate degraded	*Organism*
toluate, toluene, xylenes	*Pseudomonas putida*
p-toluidine	*Pseudomonas putida*
salicylate	*Pseudomonas putida*
octane, hexane, decane	*Pseudomonas oleovorans*
camphor	*Pseudomonas putida*
nicotine, nicotinate	*Pseudomonas convexa*
2,4-D/methylchloro phenoxyacetate	*Alcaligenes eutrophus*, *A. paradoxa*
aniline	*Pseudomonas* spp.
naphthalene	*Pseudomonas putida*
fluoroacetate	*Moraxella* spp.
chlorinated biphenyls	*Alcaligenes*, *Arthrobacter*
S-ethyl N,N–di-isopropyl thiocarbamate	*Arthrobacter*
parathion	*Pseudomonas diminuta*, *Flavobacterium* spp.
geraniol	*Pseudomonas putida*
phenanthrene	*Mycobacterium* spp.
morpholine	*Mycobacterium chelonei*
chloropropionic/ chloroacetic acid	*Alcaligenes*
chlorobenzoate	*Alcaligenes*

12 WHITE ROT FUNGI

One of the most interesting areas of study in the field of biodegradation in recent years has been that of degradation by white rot fungi. These basidiomycete fungi are commonly found growing on dead or dying trees. Their main growth substrates are cellulose and hemicellulose, but to gain access to these substrates they have first to remove the complex aromatic polymer lignin, which encases it. The organisms cannot utilize lignin as a growth substrate, and indeed to remove it they need the presence of an easily degraded substrate. White rots have evolved a very interesting mechanism for the removal of lignin, which has been likened to 'enzymic combustion'. Lignin is not a typical polymer, it has several kinds of subunits and these are assembled randomly in a 3D-structure with several different types of bonds. It is not amenable to simple hydrolysis.

White rots produce a variety of extracellular enzymes which, between them, generate, with the aid of hydrogen peroxide (made by the fungus), chemical oxidants that destabilize the lignin and render it liable to oxidation or hydrolysis. The enzymes vary from fungus to fungus but generally include manganese peroxidase and/or lignin peroxidase and sometimes laccase. The chemical oxidants include Mn^{3+} ion chelates and free radicals or radical cations of veratryl alcohol but others are almost certainly involved. This system has evolved to degrade one of nature's most complex and recalcitrant materials and it has to be non-specific to deal with the wide range of structural units and bonds in lignin. It comes as no real surprise that this system is sufficiently non-specific to allow it to degrade a huge range of recalcitrant xenobiotics – many, but not all, of which are aromatic in nature. The list of chemicals degraded by white rots is long and some examples are given in Table 34.6.

TABLE 34.6 A list of some of the recalcitrant compounds and materials reported to be degraded by white rot fungi

o-cresol	PCBs
Catechol	DDT
Creosote	Lindane
Chlorophenols, pentachlorophenol	Polycyclic aromatics – e.g. pyrene, benzopyrene, anthracene
2,4,5-T	Trinitrotoluene
Azo dyes – e.g. Orange II	Dioxins
Heterocyclic dyes	Chlorolignins
Triphenyl methane dyes – e.g. crystal violet	Lignosulphonates
	Kraft pulp effluent
Chloro anilines	Cotton black liquor

The type of information available on these biodegradations varies very much. In some cases it is clear that target compounds are mineralized extensively while in others only primary degradation may be occurring. Some chemicals, notably dyes, can be metabolized at relatively high concentrations (grams per litre) while others have only been studied at microgram per litre concentrations. What is clear is that the range of chemical structures degraded is large and, in terms of their degradative abilities, white rots are among the most versatile of microbes. The type of compounds most studied in this context are the chlorinated hydrocarbons, polyaromatics and dyes and coloured effluents. With coloured effluents, the main aim is to decolorize but, in some cases, COD removal is also possible. There is evidence to suggest that with some materials part of the COD may be metabolized but not removed. However, it may be converted to a form that is susceptible to biodegradation by other organisms such as those in activated sludge. Thus, although the white rots may not always remove COD, they may be able to 'soften' hard COD, thus enabling its ultimate removal.

One drawback to the use of these organisms is their obligate requirement for the provision of an easily degradable co-substrate like glucose or starch. The co-substrate is required to maintain viability and to provide reducing equivalents for hydrogen peroxide production. Some white rots can only work effectively in nitrogen-limited conditions (which promote synthesis of their ligninolytic enzyme systems), however, this is not the universal requirement it was once thought to be and there is considerable variation between fungal strains. The fungal mycelia once produced can be re-used many times and can be employed without loss of effectiveness in continuous or semi-continuous reactors for several months. Another advantage is that once produced, the mycelium can be kept in cold storage for several months without loss of activity.

There has been much interest in the use of white rot fungi to clean-up land polluted by recalcitrant organics. Studies involve inoculating the soil with a suitable white rot fungus (usually *Phanerochaete chrysosporium*) and providing a readily degradable carbon source to encourage its growth. Trials involving both *in situ* and *ex situ* bioremediation have been performed. The success of these trials has been rather patchy and it seems likely that white rots are not well adapted to the soil environment. They are, after all, normally inhabitants of wood, a more specialized ecological niche. In the soil there will be much more competition for the 'readily degraded nutrients' and it is likely that the relatively slow growing white rotters will be unable to compete effectively. Without their easily degraded carbon source they will be unable to degrade lignin or the recalcitrant target chemicals in the soil.They also work best under pH conditions (e.g. pH 4–5) which are considerably more acidic than those pertaining in normal soils.

One problem in this research field is that there has been too much emphasis on one organism, *Phanerochaete chrysosporium*, which is not a typical white rot. More recently a wider range of organisms, e.g. *Coriolus (Trametes) versicolor, Pleurotus ostreatus* and *Bjerkandera adjusta* have been studied and been found to be as good as or better than *P. chrysosporium*. Notably some of the other white rots appear to be able to degrade a wider range of target compounds and to be more resistant to inhibition than is *Phanerochaete*. It may be possible to isolate other lignolytic fungi (perhaps leaf litter degraders) that are better adapted than the white rots to the soil environment.

White rot fungi are perhaps better adapted to growth in aseptic conditions, without competition. This factor together with their relatively slow growth suggested that they will not be a useful option for treatment of large volumes of dilute effluents. However, they clearly have considerable potential for the pre-treatment of particularly difficult waste streams prior to a further treatment by conventional processes. Some industrial waste streams arise at such high temperature and extreme pH that they are essentially sterile and so the costs of separate sterilization may be avoided.

13 ANAEROBIC-AEROBIC TREATMENT

Many compounds are resistant or highly resistant to biodegradation under aerobic conditions, often because their substituents (e.g. nitro, chloro or diazo units) have a strong electron withdrawing effect making them less susceptible to electrophilic attack under oxidative conditions. Such compounds are often relatively easily metabolized by reductive attack. In some cases this anaerobic attack may only result in primary biodegradation but may produce a product that is readily degradable under aerobic conditions. There are plenty of examples of anaerobic treatment followed by aerobic treatment that can either lead to complete mineralization or to substantial degradation of a molecule which would otherwise be difficult or impossible to degrade. In some of the reported studies the microbial populations are very specific and highly selected while in others no selection is required. In most examples of this phenomenon the anaerobic and aerobic stages are carried out in different reactors or in the same reactor but sequentially. However, in one study a mixed population of bacteria was immobilized in gel beads. Reduction occurred in the anaerobic centre of the beads and oxidative metabolism in the aerobic outer zone. It is thus possible that anaerobic and aerobic reactions can occur at the same time in different regions.

In one promising process, nitrobenzene, which is very recalcitrant to aerobic degradation, is anaerobically reduced to aniline. This gratuitous reduction is encouraged by the addition of glucose or a mixture of industrial solvents. The resultant aniline is easily degraded in a conventional activated sludge process.

With azo dyes, the ability to cleave the diazo bond reductively is very widely distributed and can probably be found in most activated sludges – the problem is that some, though by no means all, of the resulting aromatic amines are still difficult to degrade. In one study, the azo dye Mordant Yellow 3 is reduced anaerobically by a mixed culture of two pseudomonads to yield 6-aminonaphthalene-2-sulphonate and 5-aminosalicylate, with glucose added to promote reduction. When the culture is made aerobic, *Pseudomonas* BN6 partly degrades the naphthalene derivative, converting it to another molecule of 5-aminosalicylate, this is then completely degraded by *Pseudomonas* BN9. Thus the entire dye is mineralized. This is an encouraging observation. However, Mordant Yellow 3 is a simple dye, and most of the commercially used azo dyes are more complex and highly sulphonated giving rise to aryl amines that are more complex and difficult for bacteria to degrade. This technology may not be universally applicable and its utility has to be established on a case-by-case basis. However, for azo dyes that have at least one easily metabolizable moiety (e.g. aniline) anaerobic reduction may be useful and may lead eventually to a net reduction in effluent COD.

Sequential anaerobic/aerobic treatment has also been demonstrated to be of potential value in the treatment of chlorinated aliphatics, trichlorophenol and polychlorinated benzenes. It appears that for monocyclic chlorinated aromatics reductive dechlorination occurs more readily with more highly chlorinated compounds. These are reduced in a stepwise fashion but, generally, some mono- or dichlorinated compounds are left which are much more readily degraded by aerobic cultures than their polychlorinated precursors.

It has been suggested that for polychlorinated pesticides, like DDT, alternate anaerobic/aerobic degradation may be a means of achieving mineralization. This may be so, but DDT is rarely a problem in effluent treatment as it occurs mainly as a diffuse pollutant, rather than at point sources. Flooding of soils which are heavily contaminated with polychlorinated compounds may produce anaerobic conditions which could allow reductive primary biodegradation yielding metabolites that can be aerobically degraded when the soil becomes aerobic again. It has been shown that DDT can be converted anaerobically by the common coliform bacterium *Aerobacter aerogenes* to a range of less chlorinated products including 4, 4′-dichloro benzophenone (DBP). DBP is much more amenable to aerobic attack than is DDT.

14 DEGRADATION OF RECALCITRANT ORGANICS IN EFFLUENT TREATMENT PLANTS

Many recalcitrant organic chemicals find their way into effluent treatment plants either from domestic use or from industry, principally from the chemical, petrochemical and textile industries. There is no doubt that it is much better for such chemicals to be degraded in a treatment plant than to enter the wider environment in which degradation is even less certain. In the UK it is still the case that most industrial liquid effluents are treated at plants which treat mixtures of domestic and industrial wastes – special treatment plants dedicated to particular industrial enterprises are a distinct rarity. Often consent conditions for discharge of effluent are very general (COD, BOD etc.) and usually do not include lists of particular chemicals. If specific chemicals are effectively to be removed from effluents then routine monitoring is essential to establish how efficient the treatment plant is and what increases or decreases its efficiency. Limits on COD are becoming more common and will probably soon be imposed on all effluents – if these are to be met it will be necessary to ensure that all potentially degradable materials are in fact degraded. This can be difficult as in many situations the composition of effluents can vary enormously and fluctuations can be irregular or on any time basis (hourly, weekly etc.). The factors that control the possibility of degradation have been spelt out in general terms in Section 3, but some specific examples in the context of effluent treatment are now given.

14.1 Mean sludge retention time

Morpholine is a simple heterocyclic compound (Fig. 34.15) that has a reputation for recalcitrance and failure to obtain biodegradation of morpholine has been reported regularly.

It has now been clearly established that it is degradable but the only organisms that have been shown to degrade it in pure culture are all *Mycobacterium* or related species and all are intrinsically slow growing. Their growth rates are generally about a quarter to a tenth of those typical of *Pseudomonas* spp., which are often considered the 'classical' degradative bacteria. The slow growth rate of these morpholine degraders means that they can easily be lost from activated sludge treatment systems if sludge is wasted at too great a rate. Sludge wastage is described by the mean solids retention time (MSRT) or sludge age; this is a measurement of the mean length of time a particle of activated sludge is retained within the treatment system. It has been shown for morpholine that if MSRT is less than 8 days then morpholine degradation will be incomplete and if it less than 3 days there will be no morpholine removal at all. Similar findings have been made for some of the non-ionic surfactants.

It is possible to predict the MSRT needed to ensure degradation of a chemical, and indeed to predict the concentration of the chemical in the effluent of an activated sludge plant, if certain parameters, such as the specific growth rate, are known.

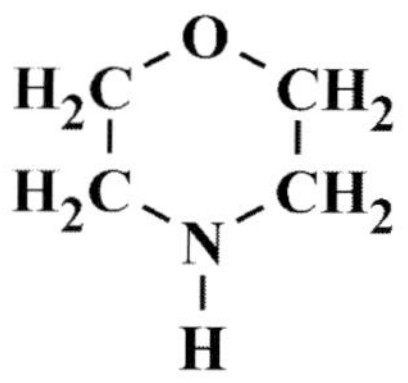

Fig. 34.15 The structure of morpholine.

S_1 is the concentration of the target compound in the treatment plant effluent and it can be obtained from the following equation:

$$S_1 = \frac{K_s(1 - K_d\theta_s)}{\theta_s(\mu_m - K_d) - 1}$$

K_s is the saturation constant for the target compound with the competent organisms; K_d is the decay constant for the microbial population; μ_m is the maximum specific growth rate for the competent microorganisms; θ_s is the mean sludge retention time.

It is notable that the only plant operating parameter in this equation is the MSRT. Both μ_m and K_s are constants only under specified conditions and may (will) change with environmental conditions (see temperature below). Similar relationships between MSRT and activated sludge plant performance have been demonstrated for ammonia oxidation, a similar process in that it relies on growth of a very specific group of slow-growing bacteria (the nitrifying bacteria). There is no doubt at all that the same will be true for the degradation of a great many xenobiotic recalcitrant chemicals. Many of these chemicals are degraded only by a limited range of slow-growing, specialized bacteria. Longer MSRTs may lead to improvement in removal of some of these chemicals, however, it should be noted that MSRTs can be too long and can lead to settling, and other problems.

14.2 Temperature

As detailed above, temperature is an important determinant in biodegradation reactions. In the context of effluent treatment, temperature will control the rates of metabolism and of microbial growth. It has been shown that at higher temperatures shorter MSRTs are required for degradation of ammonia and also for recalcitrant non-ionic surfactants. Low temperature is a real problem in domestic effluent treatment plants in the winter as the effluent is cold after passing through several miles of sewers and heating large volumes of dilute effluent is not economically viable. This can result in large variations in effluent quality over the course of a year, especially with regards to materials degraded by slow-growing microbes. Dedicated industrial effluent treatment plants are less susceptible to low temperature problems than domestic plants, partly because the effluent is in relatively smaller volume, is more concentrated and is usually warmer. Furthermore, cheap sources of heat may be available to maintain higher temperatures in the treatment plant. Thus seasonal fluctuations in temperature are smaller and less important for industrial activated sludge plants.

14.3 Specialist degraders

Some chemicals can be degraded by a wide variety of different microbes and this is true of compounds like toluene, phenol and benzoate. Quite often, however, a compound is only degraded by a small group of specialist organisms. When this is the case the ability of an activated sludge to degrade the compound will be more precarious and more likely to be lost. The specialists, for example, may be susceptible to a toxic effluent component that may eliminate the whole population or repress its activity. A related example in which specialist populations are involved is that of nitrifying bacteria which, as a group, are susceptible to inhibition by a wide range of organic chemicals.

In the treatment of chemical industry effluents, one potential problem is that products are often made on a campaign basis. Consequently, a recalcitrant chemical may be present in the factory effluent for several months or weeks (or even less) and then absent for a prolonged period. When this happens the microbial ecology of the treatment plant may be affected. If a 'specialist' group of degradative microbes are the agents of degradation, these may suffer if their substrate is no longer present. This has been demonstrated for morpholine degradation. In a laboratory treatment plant the population of potential morpholine degraders declined markedly with the absence of morpholine in the influent. When the activated sludge was re-exposed to morpholine the lag period before its degradation greatly increased – from nothing to about a day after 7 weeks, and about 2 days after 14 weeks, without morpholine in the feed.

This finding agrees with observations from an industrial activated sludge plant serving a chemical factory where it was noted that if morpholine had been absent from the factory effluent for a time, when it was reintroduced the activated sludge plant had lost the ability to degrade it. To overcome this, when morpholine was absent from the effluent it was added artificially to help maintain the sludge's acclimation to degrade it. The consequences of failure to degrade being that morpholine would pass through into the final effluent undegraded, giving an avoidable increase in COD. We know of another chemical factory that occasionally uses morpholine as a solvent and has developed a special microbiological pre-treatment process for the morpholine-containing waste stream. This has proved essential as, with only occasional presence of morpholine, is was not possible to develop and maintain a morpholine-degrading capability in the main treatment plant.

Specific pre-treatment of minor effluents generated during campaigns may be a useful approach, but is likely to be an expensive option. Another possibility is the addition, as and when required, of specially adapted microbial cultures or activated sludges to the main treatment plant. For this to be successful, storage of large volumes of cultures or sludge would be required; success cannot be guaranteed as it may take a considerable time for the added cultures to become established in the treatment plant. The artificial addition to the effluent of likely problem compounds to help maintain the biodegradative capabilities of the activated sludge between campaigns may be a more reliable and cheaper option, although it may go against the grain actually to add chemicals to an effluent!

14.4 Analysis and monitoring

It is clearly desirable for the recalcitrant organic compounds found in industrial effluents to be degraded in effluent treatment plants. Indeed, if companies are to meet the ever-tightening restrictions it is essential. One factor that mitigates against the success of such treatment plants is the lack of detailed knowledge about specific aspects of their performance. This problem is particularly bad in treatment plants that principally treat domestic sewage with a variable portion of different industrial effluents. It is important that plant operators have available to them information not only on gross characteristics, like BOD, COD or TOC, but also on the concentrations of specific chemicals in the effluent. It should be pointed out that BOD is not a good indicator of the degree of removal of recalcitrant organics from effluents. It may give an idea of relatively short-term effects of an effluent on a water course but not of the extent to which any recalcitrant chemicals have been removed (often recalcitrant chemicals will not exhibit a BOD using the standard test).

At the end of the treatment process effluents will contain a proportion of 'hard COD' which is difficult to remove. If progress is to be made operators must have detailed knowledge of what this residual material comprises and the extent to which the microflora is able to degrade individual components of it. Such analytical data need to be available from regular time intervals (or composite samples) spread over a considerable period of time. Operators will then be able to look for trends in the degree of degradation of specific problem chemicals. These trends can then be correlated with overall plant performance, variations in operating practice (e.g. sludge wastage) and parameters (e.g. concentration of mixed liquor suspended solids and MSRT), changes in pH or oxygen levels, effluent composition (e.g. concentration of potentially toxic components) and environmental factors (e.g. temperature). Such correlations may allow operators to optimize removal of potential problem compounds. If problem compounds or waste streams are identified then details of their nature, and effects, can be determined and research undertaken to develop strategies for their more effective treatment. In extreme cases this may involve special pre-treatments (biological or chemical) for particular waste streams.

Detailed chemical analysis has always been expensive but is becoming more and more important. Techniques such as GC/MS and even HPLC/MS are becoming cheaper and more readily available and offer the prospect of relatively easy identification and quantification of a wide range of chemicals without the employment of an army of analysts. Part of the price for the successful treatment of industrial effluents is constant analysis and monitoring of data.

15 MEASUREMENT OF BIODEGRADABILITY

There are many different methods of assessing the biodegradability of a chemical, but there have been efforts to standardize them by transnational bodies such as the Organization for Economic Cooperation and Development (OECD) or the European Community. Pollution, after all, is no respecter of international boundaries and it is important that official tests are consistent between different countries. There is no room to describe all the different tests and readers are directed to the early chapters of Karsa and Porter (1995) which give an excellent analysis of the different methods and their 'pros and cons'. When choosing a test for biodegradability it is important to ensure that the incubation time is sufficiently long. Many biodegradable compounds have in the past been described as completely recalcitrant because incubations for biodegradability tests were too short. To be effective incubations may need to be continued for anything from a few weeks to several months. This is particularly so when un-acclimated seed organisms are used. In the BOD test, use is often made of sewage effluent and, while this is adequate for many compounds, it will not usually be a good seed for recalcitrant chemicals. For example, it proved impossible to demonstrate a BOD for morpholine until a seed was used which had previously been acclimated for over a hundred days! In other studies the lag period prior to observed degradation of morpholine was often between 20 and 60 days, shorter lag periods were only encountered when the inoculum came from a site likely to have been exposed previously.

It can be seen that in order to assess the inherent, or potential, biodegradability of a compound, considerable persistence may be required on behalf of the scientist! Preferably a range of tests should be performed which can give evidence of the type of biodegradation, as well as its rate; e.g. is the biodegradation merely primary or has mineralization occurred? Is the compound readily degradable or is it degraded only slowly and after a prolonged lag period? All this information will be of value in assessing the likely fate of a compound in an effluent treatment plant or in the wider environment.

16 BIOREMEDIATION

This term means making things better by biological means and is applied to processes used to solve problems of environmental pollution. The term bioremediation is used in a variety of contexts. It is applied, for example, to the 'cleaning-up' of land polluted by industry, attempts to remove pollutants from contaminated aquifers and the biological treatment of contaminated effluents. In our view this term is best reserved for processes that are used to solve past incidents of pollution rather than those used to treat new effluents. There are many problems of 'historic' pollution including old chemical factory sites that have been grossly polluted over decades due to ignorance of the effects of materials which were spilled onto them accidentally or deliberately. The process of coal gasification has also left a legacy of land polluted, for example, with phenols and pyridine derivatives and cyanides. Pollution of land and aquifers by spilling or disposal of solvents (such as the chlorinated aliphatics used for de-greasing and dry-cleaning) or jet fuel is also widespread and threatens water supplies in some areas. The military have also been responsible for contamination of land with explosives like TNT. Acute problems of recent pollution are also regrettably common, with the spillage of crude oil at sea or on coastlines, due to the sinking of oil tankers, being the best-known example.

Bioremediation processes are generally split into two broad categories – *in situ* and *ex situ* processes. In the former, the pollution problem is tackled where it occurs and in the latter contaminated materials, be they soil or ground water, or the contaminants extracted from them, are removed and treated in a separate location.

Ex situ treatment has the advantage that it is possible to exercise much more control over the process with regard to environmental conditions such as pH, temperature and moisture. It is also possible to ensure more easily that aeration (if necessary) is adequate and that nutrients, if required, are available in the correct amounts. The disadvantage of *ex situ* processes is the expense involved in removing and transporting the contaminated materials to the treatment site. Once at the treatment site, a variety of approaches can be used including the composting of polluted soil, the treatment of the soil as a slurry (in lagoons or reactors) and the extraction of pollutants into water with treatment of the resulting solution by activated sludge. This list is not exhaustive. Contaminated ground water or water from aquifers can be pumped up and treated by a conventional biological treatment processes. Volatile compounds can also be removed from below ground by air stripping and subsequently removed from the effluent gases. Aerobic treatment of soil or water may lead to evaporation of volatiles. It is undesirable simply to volatilize the pollutants and transfer them to the atmosphere and so arrangements may need to be made to collect vapours and to remove volatile organics from them – possibly by biological treatment. Compounds like toluene and trichloroethylene can be treated in the vapour phase by a trickling filter-type apparatus. Care has to be taken not to pollute the surrounding soil and so the treatments have to be conducted on lined, impervious surfaces.

The *ex situ* approach may be impossible, undesirable or unnecessary. *In situ* processes are usually less intensive. They may involve the addition of suitable microorganisms where none exist, the addition of nutrients that are in short supply or the addition of air or oxygen (or an alternate electron acceptor, like nitrate). The simplest of such treatments are processes such as nutrient addition and ploughing of oil-contaminated soils (to improve aeration). More difficult processes involve the injection of oxygen or mineral nutrients into aquifers.

The addition of competent microbial cultures in bioremediation processes ('bioaugmentation') is somewhat controversial. In some situations it has been shown that such additions are unnecessary, for example, this appears to be so in the case of oil pollution. Several studies have shown that the addition of microbial cultures gives little or no benefit. The most effective approach for this type of pollution is the addition of mineral nutrients, notably nitrogen and phosphorus. The metabolism of such a concentrated carbon source as crude oil will rapidly deplete nutrient supplies and, in the case of soil pollution, supplies of oxygen as well. Nutrients have to be provided in an appropriate form and it has been found effective in treating oil spills to use oleophilic fertilizer preparations that will bind to beached oil and are not washed away by the large volumes of water. In bioremediation it is clearly important to identify the rate-limiting factor and address that in whatever way is feasible. Alterations in other factors may be of little benefit. Processes in which conditions for pre-existing organisms are improved by addition of nutrients/O_2 etc. are often referred to as 'biostimulation'.

Often the native microflora will already have the desired capabilities or will be capable of evolving to degrade the added pollutants. The addition of laboratory strains of microorganisms may be of little use. Such strains are often habituated to laboratory life and may have a very limited ability to survive in the more challenging and competitive environments of the soil or seawater. This is particularly true for genetically-engineered organisms which may be rendered uncompetitive in many situations by the carriage of additional genetic information (which will impose a metabolic cost on the organism). In some situations in which the native microflora is very small or non-existent, e.g. in deep subsoil or aquifers, the addition of competent microbes may be of use. However, it must be established that the organisms can cope with, and carry out the required metabolism under the prevailing conditions (which may include a shortage, or absence, of oxygen). It has been proposed that white rot fungi may be suitable for the bioremediation of contaminated land – their use in this context is discussed in Section 12.

For a more in-depth treatment of bioremediation readers are referred to the August 1993 edition (Volume 11) of the journal *Trends in Biotechnology*, most of which is devoted to this topic, or to the series of books *Remedial Treatment of Contaminated Land* (see reference list).

17 CONCLUSIONS

After the gloom of impending disaster and environmental calamity that was present in the 1960s, the threat from recalcitrant organic chemicals has diminished to some extent. Extensive microbiological and biochemical research has explained the reasons for recalcitrance and suggested routes by which problem compounds may be degraded. New processes of biodegradation (e.g. by anaerobes and white rot fungi) have come to light and methods of utilizing microbial activity have been developed or improved. An understanding of the effects of chemical structure on biodegradability means that we now have the knowledge to design chemicals that are more amenable to biodegradation in the environment, *if we wish to*. If progress is to be made it will be essential to gain an improved knowledge of the microbiology and biochemistry of the treatment of industrial effluents, and the biodegradation of their components, so that it can become more effective. If tighter environmental protection legislation is to be applied effectively and fairly then research will be required to ensure that industry can comply with more stringent limits on the COD of effluents and direct toxicity assessments.

Environmental pollution is a transnational phenomenon and Western countries will not be able to escape its consequences by simply manufacturing products in developing countries where legislation may be less strict or less strictly enforced. Although DDT is now less of a problem in developed countries (in which its use has generally been banned) it has not entirely disappeared and is still in use in many poorer tropical countries. It is well known that DDT has been found in regions of the world in which it has never been used. This compound is still effective in malaria control, it is cheaper than, and has some advantages over, alternatives. Although it can be slowly degraded there is no *easy* biological treatment solution for it, especially as it arises as a diffuse pollutant and spreads. Clearly, if the threat of DDT is to be lifted completely then developing countries will need aid, either to buy more expensive but less persistent alternatives or to develop new insecticides. Currently there is a thriving illegal international trade in serious pollutant chemicals (like CFCs) which are banned in the West but are manufactured elsewhere. It is vital that new chemicals being developed are given more extensive testing to ensure their biodegradability before they are put into commercial use. Export of pollution is not morally acceptable.

As we enter the 21st century, further scientific investigation is essential, but its beneficial effects will be nullified without a transnational legal framework to outlaw serious pollutant chemicals and generous financial aid to make it effective across the globe.

FURTHER READING

Alexander, M. (1965). Biodegradation: problems of molecular recalcitrance and microbial fallibility. *Advances in Applied Microbiology* **7**, 35–80.

Alexander, M. (1999). *Biodegradation and Bioremediation*, 2nd edn. Academic Press, San Diego.

Beek, B. (ed.) (2000). *Bioaccumulation: New aspects and developments*. (Volume 2 part J of the Handbook of Environmental Chemistry). Springer-Verlag, Berlin. The University of Minnesota Biocatalysis/Biodegradation (Database at http://umbbd.ahc.umn.edu/).

Beek, B. (ed.) (2001). *Biodegradation and persistence*. (Volume 2 part K of the Handbook of Environmental Chemistry). Springer-Verlag, Berlin.

Carson, R. (1962). *Silent Spring*. Houghton Mifflin, Boston.

Chaudhry, G.R. (ed.) (1994). *Biological Degradation and Bioremediation of Toxic Chemicals*. Chapman & Hall, London.

Gadd, G.M. (ed.) (2001). *Fungi in Bioremediation*. Cambridge University Press, Cambridge.

Karsa, D.R. and Porter, M.R. (eds) (1995). *Biodegradation of Surfactants*. Blackie Academic & Professional, Glasgow.

Pitter, P. and Chudoba, J. (1990). *Biodegradability of Organic Substances in the Aquatic Environment*. CRC Press, Boca Raton.

Remedial Treatment for Contaminated Land (1995). Construction Industry Research Information Association, London. This is a series of twelve books, volumes VII, VIIII and IX are particularly useful.

Swisher, R.D. (1987). *Surfactant Biodegradation*, 2nd edn. Marcel Dekker, New York.

Trends in Biotechnology (1993). Volume 11 – most of this volume is devoted to bioremediation.

Wackett, L.P. and Hershberger, C.D. (2001). *Biocatalysis and Biodegradation – microbial transformations of organic compounds*. ASM Press, Washington, DC.

Young, L.Y. and Cerniglia, C.E. (eds) (1995). *Microbial Transformation and Degradation of Toxic organic Chemicals*. Wiley-Liss, New York.

35

Heavy metals in wastewater treatment processes

John Binkley and J.A. Simpson

Product Design and Development, Bolton Institute, Bolton BL3 5AB, UK

1 INTRODUCTION

The term heavy metals is used to denote a large group of elements. Widely accepted definitions are those of Phipps (1981) and Weast (1984) defining the heavy metals as those elements with a density greater than 6 g/cm^3 and 5 g/cm^3 respectively. From an environmental standpoint, the definition of a heavy metal is fairly broad and includes metalloid substances which exhibit similar properties in the environment; typically high toxicity at low concentrations and with a long residence time in soil. From this standpoint the definition of Weast is the closest.

Some of the heavy metals are micronutrients and are required in trace amounts by living organisms for their normal metabolic function. At elevated concentrations in the environment they exhibit a toxic effect. Elements such as cadmium and mercury have no known metabolic function and are considered to exhibit some toxicity at all concentrations. The toxicity of the heavy metals depends upon the chemical form in which they are present (Rochow, 1964). Organometallic complexes may well show the phenomenon of bioconcentration due to the greater solubility, and hence preferential uptake, of the substance in lipid rather than water.

Besides directly causing non-infectious diseases in man, heavy metals may also have a significant impact when released into the environment. They are a serious and persistent pollutant of terrestrial and aquatic ecosystems (Bourg, 1995). Unlike various refractory organic compounds there is no possibility of destruction of these elements in the biosphere. Contamination of soil by heavy metals is a long-term problem. Estimation of the half-lives of some of the elements in soil range from 15 to 1100 years for cadmium, 310 to 1500 years for copper and 740 to 5900 years for lead, the wide range of values being due to differing soil conditions (Alloway, 1995).

Heavy metals are present as trace contaminants in most materials; more concentrated sources of these elements are found in fossil fuels and mineral ores. Anthropogenic sources have greatly accelerated biogeochemical cycles of heavy metals (Lombi *et al.*, 1998). Atmospheric release may generally disperse contaminants, such as lead from the, now banned, use of tetraethyl lead as an 'anti-knocking' agent in petrol or pollutants may be more concentrated on a local level, e.g. cadmium from pre-industrial mining and smelting operations. Pollution from heavy metals can generally be traced back to a point source.

Many industrial processes are reliant on the use of heavy metals as process reagents, or they are present as contaminants in their raw materials. Elimination of compounds from many sources of waste is often either impossible or needlessly expensive.

For the operator of a wastewater treatment plant the problems of heavy metals in the wastewater are twofold; their presence may significantly affect the efficiency of the plant, and discharges containing metals may have

The Handbook of Water and Wastewater Microbiology
ISBN 0-12-470100-0

a serious effect on the environment or man. This is reflected in the legislation concerning such emissions.

2 LEGISLATION

Global warming and the increasing world population will place an increased stress on the agricultural and water supply systems. Industrial activities pollute these resources and reduce the future capacity. Concern over the impact of the activities of man on the natural environment has led to international agreements limiting resource usage and environmental pollution. With these concerns in mind legislation is now being implemented which not only requires an installation to meet specific discharge criteria, but also to consider to Best Environmental Practice (BEP), Best Available Technology (BAT) and Integrated Pollution Prevention and Control (IPPC). In order to enforce these regulations enforcing bodies require some form of criteria and indicators to measure environmental damage and, to conform to such regulations, the operators of polluting or potentially polluting installations will most probably have to implement some form of environmental management programme. The heavy metals of concern, as have been identified in European Union and US legislation, are those which are present in sufficient quantity in the earth's crust and are of sufficient solubility in water to be determined an environmental hazard. European Union Directives concerning these dangerous substances list the pollutants as being of concern in the aquatic environment.

The use of heavy metals in industry is governed by health and safety, transportation and storage legislation. These will also cover operators of wastewater treatment facilities. They may impose exposure limits to gaseous concentrations, requirements for personal protection equipment and storage, handling and disposal procedures.

The legislation covering discharges of heavy metals into the environment can be split into three broad categories:

1. Discharges to water
2. Discharges to land
3. Release to the atmosphere.

Note that heavy metals in ground water may be considered analogous to contaminated land where there is not a threat to a potable water supply.

2.1 Discharges to water

Limitations on the quantity of heavy metals discharged into water are required to protect potable water supplies, fisheries and, more recently, the aquatic environment. Environmental concerns over the marine environment, once seen as a suitable dumping ground for radionuclides and raw sewerage have seen this route of disposal curtailed.

World Health Organization recommendations on the maximum acceptable concentration of various priority pollutants have been accepted internationally and provide the basis of European Union and US legislation on the protection of potable water supplies. These are designed to ensure that drinking water is fit for human consumption (see Table 35.1 for metal concentration limits on discharge and land loading rates).

Municipal sewerage works have to protect the receiving medium from heavy metals in the aqueous discharge. This has traditionally been achieved by the dilution of metal-bearing industrial waste with less contaminated wastewater from domestic sources. Biological treatment systems as well as their primary role of reducing the BOD of the effluent, also have a great affinity for heavy metals and provide a convenient method of removal. The solid wastes generated from these treatment processes must also be disposed of. Recycling of these wastes is seen as advantageous from an environmental standpoint, but the metal content may preclude this option. Disposing of metals to landfill and incineration are the other two major options available, but are set to become increasingly more expensive. When constructing a wastewater treatment facility, consideration should be given to the disposal of the solid waste generated. The European Union

TABLE 35.1 Limitations on discharges into the environment of heavy metals

Heavy metal	*British EQS μgl^{-1}*	*EU limit values for the concentration of metals in sludge $mg\,kg^{-1}dm$*	*EU annual loading rates for agricultural land receiving MSW sludge[a] $gha^{-1}y^{-1}$*	*US part 501 Ceiling concentration of metals in sludge $mgkg^{-1}$*	*Annual pollutant loading rates $kgha^{-1}y^{-1}$*
Arsenic	50			75	2.0
Cadmium	5	20–40	150	85	1.9
Chromium	20				
Copper	10	1000–1750	12 000	4300	75
Iron	1000				
Lead	10	750–1200	15 000	840	15
Manganese					
Mercury (total)	1	16–25	30	57	0.85
Molybdenum				75	
Nickel	150	300–400	3000	420	21
Selenium				100	5.0
Vanadium	20				
Zinc	75	2500–4000	30 000	7500	140

Based on a 10 year average.
[a] MSW Municipal Sewage Works.

Directives give an indication of future legislative requirements on national governments.

2.2 Discharges to land

The European Union Landfill Directive (EU (1999), implemented in the UK in July 2001, will further increase the pressure to recycle biological waste to land with a commitment to reduce, in stages, the landfill of biological waste to 25% of its current level by 2010.

The application of waste and industrial activated sludge to agricultural land as a soil conditioner and to contaminated land, is regulated under the Waste Management Licensing Regulations 1994 (HMG, 1994). Under these rules the spreading of various wastes, including those from industrial and biological treatment plant origins, is exempt from waste management licensing, provided that agricultural or ecological improvement can be demonstrated. This waste application is limited to a maximum annual application of 250 tonnes/hectare.

The Environment Agency of England and Wales (EA) and the Scottish Environmental Protection Agency (SEPA) have both issued reports expressing concern over the spreading of industrial waste to land (Anon, 1998a,b). In these reports they have recommended the tightening of regulation to ensure that industrial waste which is spread to land is of a similar quality to that of municipal sewage waste (MSW). The application of potentially toxic elements from MSW to agricultural land is well regulated. The Department of Environment, Transport and Regions (DETR) has recently reviewed the scientific evidence relating to the agricultural use of sewerage sludge, including the scientific basis for the limits of the potentially toxic elements (PTEs) Cd, Cu, Zn, Pb, Hg, Cr, As, Se, Mo, and Fe (DETR, 1998). Application of excess activated sludge from industrial waste treatment that contains PTEs above these limits is unlikely to be acceptable.

The textile industry, despite recent reports in the press, is still a major manufacturing industry in the UK and will continue to be so for the foreseeable future (Barker, 2000). In October 1999 the European Commissions Environment Directorate produced a draft working paper outlining plans to revise the 1986 EC Sludge use in agriculture directive (EC, 2000; EEC, 1986). Included in the plans is a proposal to reduce the limit values for metals in sludge in 2005 and 2010, which may cause 21% and 72% failure of excess municipal sewerage works (MSW) return activated sludge (RAS) to meet the limits (Anon, 1999). RAS is the suspended solid fraction found in the bottom of a secondary clarifier, composed mostly of microorganisms, which is returned to the aeration treatment pond and mixed with influent waste.

Certain sections of the textile industry, wet processing in particular, engage in practices that may have a significant environmental impact. The wet processing sector of the textile industry is among the top four or five major polluters, together with pharmaceuticals, the paper industry and smokeless fuel production in the UK (Pierce, 1994). Of these, approximately 10% have their own effluent treatment plants. The majority of the remaining effluent is combined with other domestic and industrial sources for treatment at a municipal sewerage works.

Solid waste from water treatment plants must be disposed of or recycled. Three practical routes exist: incineration, applying to land as a fertilizer or soil conditioner, and dumping in landfill. Incineration leaves ash, which requires disposal either by one of the two above methods or its incorporation into building materials.

Biosolids are rich in phosphorus and, as such, are suitable for application to land. Authorities in Australia, for example, limit use to land especially when applied to acidic agricultural land. Bioleaching occurs, transporting the heavy metals to other areas. The metals end up bioleaching in this manner as a result of bioacidification. Heavy metals can therefore, be removed from aqueous effluent wastewaters by bioacidification prior to disposal of biosolids to land. Unfortunately, the nutrient concentration is also reduced significantly in the process (Shanableh and Ginige, 1999).

2.3 Discharges to air

Within the workplace limit values are applied to various hazardous chemicals establishing short- and long-term exposure concentrations, which provide no or an acceptable health risk. Discharges to atmosphere are also regulated, particularly important in industrialized urban areas.

2.4 Concerns for the future

In the EU, discharges of dangerous substances to water, as listed in EC directives 76/-464EEC (Dangerous Substances Directive) and 86/440/EEC, will be banned.

2.4.1 Legislative concerns for the future

Climate change due to global warming is now widely accepted internationally by most industrialized nations. The goal of sustainable development arising and the further transcription of the precautionary and polluter pays principles into law will necessitate more binding environmental legislation than is currently in force. In the European Union, future requirements upon industry can be found in directives which have not yet been transcribed into national law. When considering the disposal of metal-containing sludge an EU policy document advocates the introduction of much more restrictive practices.

2.4.2 The EU water framework directive

The Directive 2000/60/EC will integrate existing water quality legislation (EU, 2000a). An obligation is placed on member states to ensure that surface and ground waters reach a good condition by 2010. Polluters will be required to pay the full cost of the damage they cause, rather than leave it to society to bear the cost. This may have important cost implications for industries discharging heavy-metal-bearing industrial wastes to sewerage treatment works.

Member states will be required to establish river basin management plans, overseen by a competent authority. Previous water management regulations will be integrated into the water framework directive. The discharge limits of heavy metals will then be set by environmental quality standards (EQSs) specific to the river basin management plan. The EQSs will be set at a level intended to protect the biota, both terrestrial and aquatic, of the river basin. The EQSs will also have to ensure that naturally occurring dangerous ubstances will not be present in the marine nvironment at levels significantly above the ackground concentration.

.4.3 *Proposed sludge application to land evisions*

'he third EU draft working document on ludge contains proposals that will affect ıost treatment works that recycle excess ludge to land as a fertilizer. A significant roposal is that all operations producing ludge are to be brought under similar reguıtions to those currently regulating sewerage eatment works. Current limit values on the cceptable sludge metal content, soil metal ɔading rates and the total acceptable metal ɔading on the soil will be lowered. Reductions re proposed in the short, medium (≈2015) nd long term (≈2025). If these medium-term alues are applied, as suggested, the new limit alues would preclude the use of many MSW ludges currently recycled to land. The new ghter threshold limit values for the amount of llowable heavy metal in receiving soils will educe the available arable land.

Under the proposal, it will not be possible to educe the concentration of heavy metals in ludge for application to land by combining it ith another sludge source. The producers of ludge will also be under obligations to ensure ıat the sludge they generate is suitable for ınd application, even if the distribution and pplication of the sludge is performed by a ıird party. The producer will also be required ɔ implement a quality system over the whole rocess, from reduction of pollutants into the ⁄aste treatment, to the supply of relevant ocumentation to the delivery site. The produer of sludge for arable land application, will be esponsible for testing for the presence of toxic lements.

There are several outstanding features ⁄hich may be observed from Table 35.1, the ıost significant being the relatively high ɔading rates of these metals. Generally this ⁄ould be worthy of considerable discussion. Iowever, it is not considered to be approriate in this text. It is apparent that the elative toxicity of metals common to the UK, EU and the USA are of the following order:

$$\xrightarrow[\text{Decreasing toxicity}]{\text{Hg} > \text{Cd} > \text{Ni/Pb} > \text{Cu} > \text{Zn}}$$

From Table 35.1 there is general agreement that Hg and Cd are the most highly unacceptable pollutants of the metals considered.

3 CHEMISTRY OF HEAVY METALS IN WATER

Heavy metals consist of a large group of elements. They may be present in water in a variety of states including free metal ions, chelated compounds, organometallic complexes and elemental metal. These substances may be dissolved in the water, associated with colloidal compounds or present in association with particulate matter. The oxidation state of the metal ion may also be of importance; chromium exists in water in two oxidation states. The Cr^{3+} is relatively non-toxic and is required as a trace element by animals. The Cr^{6+} ion, however, exhibits carcinogenic properties. The majority of the heavy metals are cations in solution. At acidic pH they tend to exist as free metal ions. Around neutral pH 6–9 some precipitate as hydroxides or other insoluble species if the appropriate co-ion is available.

4 DETECTION OF HEAVY METALS

The current method of choice for quantitative heavy metal detection is inductively coupled plasma mass spectroscopy (ICPMS), which can simultaneously detect multiple elements, in some cases down to low parts per trillion concentrations for many metals. The equipment required is expensive to purchase and operate. Electrochemical methods of detection, including polography and anodic stripping voltimetry, are capable of detecting multiple elements. They can also yield valuable information about the availability of metal ions, and are most suitable for research purposes. Where the element of interest is known, then the use of traditional chemical reagents for metal detection provides a useful tool. These reagents are available as proprietary kits, which may be used onsite.

5 SOURCES OF HEAVY METALS IN WASTEWATER

Industrial waste contains pollutants that may preclude the use of excess activated sludge, derived from industrial waste treatment, as an agricultural fertilizer. Recent legislative changes have considerably reduced the options available for the disposal of solid waste from biological treatment plants (McCann, 1998). Consequently, it is now more economically favourable to recycle organic wastes to land. Heavy metals may gain entry into waste streams from a variety of sources. The presence of metals in some dyestuffs is well known, but these contaminants may also be present on the raw materials and as a contaminant of other process auxiliary chemicals, where it has been involved in the compound's synthesis (Charlton, 1999). The effluent from cotton wet processing typically has a highly alkaline pH. This must be lowered to a pH value generally between 6 and 8 before the effluent is processed by aerobic biological treatment.

Heavy metals are present in wastewater from industrial and domestic sources. Concentrated highly contaminated wastes can be found in wastes from metal processing industries such as electroplating. Chemical works, textile wet processing, tanneries, photographic industries and mining may also produce wastes with a significant metal content.

Sources of heavy metals in domestic waste may come from metal piping (lead and copper), galvanic corrosion (zinc), cosmetics and household cleaning agents. Surface water runoff from roads and fire water may also contribute to the metal load on a wastewater treatment plant when water from these sources enters the sewer.

6 THE EFFECT OF HEAVY METALS ON BIOLOGICAL TREATMENT PROCESSES

Cu^{2+}, Pb^{2+}, Cd^{2+}, Ni^{2+}, Zn^{2+} and Cr^{6+} are known to inhibit anaerobic digestion (Lin, 1992; Mueller and Steiner, 1992). The relative degree of anaerobic inhibition in municipal sludge has been found to be:

$$\xrightarrow{\text{Ni} > \text{Cu} > \text{Cd} > \text{Cr} > \text{Pb}}$$

Decreasing anaerobic inhibition

The toxicity of the metal is directly related t its solubility in the presence of the sludge. If th affinity of the metal for the sludge is high the the toxicity is reduced (Mueller and Steine 1992). Metal toxicity in such systems may b reduced by precipitating the metal out of sol ution as the insoluble sulphide or hydroxide etc. Metals such as Ni, Co and Mo, howeve are known to promote anaerobic digestio (Shonheit *et al.*, 1979; Whitman and Wolfe 1980; Murray and van den Berg, 1981; Bitton 1994).

Toxicity is expressed as 'inhibition of respir atory activity of microorganisms' present in th activated sludge'. Cr^{3+} and Cr^{6+} were used in study in order to determine their effect on th activated sludge. Oxygen concentration an biological oxygen demand were measured a an indication of inhibition in the presence c the metal. Exposure times and activated sludg biosolids concentration was examined and thei effect on EC_{50} recorded. The 1-hour EC_{50} fo Cr^{6+} was found to be 40–90 mg/l. This effec was not observed for Cr^{3+} alone, only th combined effect of the two ions (Vanková *et al* 1999).

In another study, the three metal ions Cu^{2+} Zn^{2+} and Ni^{2+} were found to inhibit β-gluco sidase enzyme activity. The pH and buffer typ were investigated and were found to hav considerable effect on the enzyme activity Under optimal conditions, a metal concen tration of 0.6 mM and a pH of 5, the presence c Zn and Ni decreased the enzyme activity by 2 to 30%, whereas Cu reduced it by more tha 90%. The citrate buffer was not inhibited at al even at the higher Cu concentrations. Th inhibition by the Zn and Ni was found to b less pH dependent – range 4 to 5.5. Chemica speciation models were used to describ inhibition by buffer and pH for all three metals Here it was assumed that the enzyme activit depended on protonation of the amino acid a the reactive site complexation by the heav metal cation (Geiger *et al.*, 1999).

In natural unpolluted environments, heavy metals are present in trace concentrations. The most abundant may have a metabolic function. The inability of most living organisms to cope with the toxic effects of heavy metals is due to lack of evolutionary selection pressures to develop these mechanisms.

Some fungi have shown toxic effects to heavy metals. An example here is the thermochemical study of some heavy metals. Thermogenetic growth curves of *Rhizopus nigricans* have shown how a variety of metals have inhibited growth rate constants and inhibitory ratios, although low Cu concentrations promote these properties. The method of analysis, microcalorimetric analysis bioassay, for acute cellular toxicity is based on metabolic heat evolution from cultured cells (Cheng Nong *et al.*, 1999).

Here the effect of heavy metals in activated sludge was examined when used as a food supplement in Wister rats. The toxic effect of heavy metals as contaminants from sludge-supplemented diets on male Wistar rats has been studied (Bag *et al.*, 1999). Poultry and cattle feed was supplemented with activated sludge as it is well known that it is high in nitrogenous matter and is inexpensive. The toxic effect of using domestic sewage sludge, which is contaminated with heavy metals, pesticides etc. on male Wistar rats was undertaken, and the metal content of the dried sludge before mixing with the cattle feed was determined using atomic absorption. The toxicity was determined by analysis of enzyme activity in serum, liver, muscle and the brain of the Wistar rats. It was concluded that the mixing with animal feeds should be implemented with caution because of the growing awareness that biomagnification of heavy metals may result in the food chain. This begs the question that, if awareness of the biomagnification effect was not growing, would we implement without caution?

5.1 Effect of heavy metals on treatment efficiency

Heavy metals have a deleterious effect on living organisms if present in concentrations above that which naturally occurs in the organisms' environment. In general, heavy metals tend to have a bacteriostatic effect. Increasing concentrations lead to mortality. Some microorganisms have developed mechanisms to cope with elevated concentrations of heavy metals in their environment.

A recent study has shown that, at concentrations well below that where toxic effects are observed, and not uncommon in most combined industrial and domestic wastewater treatment works, the presence of heavy metals can considerably affect the efficiency of the plant, reducing the chemical oxygen demand adsorption capacity (CAC). The presence of heavy metals may also affect the settling characteristic of sludge. This may be due to a direct effect upon the extracellular biopolymer or to the relative promotion of filamentous bacteria. Subsequent anaerobic digestion of excess metal contaminated sludge flocs from activated sludge treatment may be inhibited or prevented by the presence of heavy metals. Anaerobic treatment processes are, generally, most sensitive to contamination by heavy metals and the material entering a sludge digester may contain many times the concentrations of heavy metals compared to the influent plant waste stream.

Trace levels of Cu, Cr and Zn have been investigated on the performance of activated sludge. Adsorption onto batch flocs followed both Langmuir and Freundlich isotherms. The presence of the three metals affected organic matter adsorption and the CAC. The metal ions reduced CAC by competition for adsorption sites on bioflocs in a sequence batch reactor (SBR) depending on hydraulic retention time (HRT). Heavy metals were adsorbed quicker in shorter HRTs (2 days) as opposed to longer HRTs (5 days). Again, indicative of the greater mobility of the single metal ion despite its high atomic mass, although this being quite small compared with the organic competitors (Sin *et al.*, 2000).

There is an excellent review on microbial heavy metal resistance, which discusses metabolism of heavy metals to a degree of detail (Nies, 1999). Even though heavy metals form a major part of the elements in the periodic table, as mentioned above, a lot of them are essential as trace elemental nutrients, although, at high enough concentrations most are toxic. Transport of the most important heavy metals is compared after discussion of the principles of homeostasis for all heavy metal ions.

The uptake and control of heavy metals across cell membranes has been investigated and studies have shown that heavy metals are transported across the outer microbial cell membrane via two types of transport system:

1. Fast unspecific uptake driven by a chemostatic gradient across a cell membrane. These are constitutionally expressed mechanisms of which two general types have been recognized. Divalent metal cations enter through one type of channel that is present to allow the ingress of micronutrient metal cations, e.g. Zn^{2+}. Oxyanions, e.g. chromate, gain entry through unspecific uptake systems allowing the passage of phosphate and sulphate ions.
2. Systems showing high substrate specificity. This may be utilized against a concentration gradient, sometimes use an ATP source as an energy source; they are inducible.

The toxic mode of action of heavy metals is generally within the cell itself. This may be by the substitution of the metal ion for a structural homologue micronutrient or binding to material within the cell, especially heavy metals with a high atomic mass showing a great affinity for thiol constituent groups on amino acids, such as cysteine. Both mechanisms can cause considerable oxidative stress on the cell.

7 NATURAL MICROBIAL RESISTANCE OR DETOXIFICATION MECHANISMS

Three general mechanisms are available to the microbial cell:

1. Reduction to a less toxic form
2. Effusion from the cell generally requiring a specific transport mechanism working against a concentration gradient coupled to an energy releasing reduction
3. Complexing the heavy metal to form an inert relatively non-reactive species.

Microbial cells generally adopt two or all three of the above mechanisms.

The rate of oxygen uptake inhibition with PACT sludge has been studied in the presence of a variety of inhibitors (Sher, 2000). However, in this particular study, Zn was the only metal used and its inhibitory effect was compared with those of phenol and one of its derivatives. The sludges were unacclimated and their ages were 4 to 12 days old. For the poorly adsorbed Zn, the IC_{50} defined values and sorption did not show any change. However, the 12-day-old sludge showed much more resistance to Zn exposure. Changes in biomass were the only apparent reason for this behaviour. PACT was mixed with the activated sludge and is an activated carbon adsorbent.

7.1 Metabolic activity and metal biotransformations

The metabolic activity of microorganisms can result in solubilization, precipitation, chelation, biomethylation or volatilization of heavy metals (Iverson and Brinckman, 1978). Microbial activity may result in the following:

- Strong acid production, such as sulphuric acid by chemoautotrophic bacteria, e.g. *Thiobacillus*, which dissolves minerals
- Weak organic acid production, such as citric acid, which dissolves and chelates metals forming organometallic compounds
- Ammonia production or organic bases which precipitate heavy metals as the insoluble hydroxides
- Hydrogen sulphide production by sulphate-reducing bacteria, which precipitates heavy metals as insoluble sulphides
- Extracellular polysaccharide production which can chelate heavy metals and reduce toxicity (Bitton and Freihoffer, 1978)
- Certain bacteria (sheathed filamentous) fi[x] Fe and Mn on their surface in the form o[f] hydroxides or other insoluble metal salts

- Biotransformation of certain bacteria that have the ability biomethylate or volatilize heavy metals.

7.1.1 Metal biotransformation

Heavy metal sources in wastewater treatment plants include mainly industrial discharges and urban storm water runoff. Biological treatment processes such as activated sludge, biological filtration and oxidation ponds remove 24% (e.g. Cd) to 82% (e.g. Cu, Cr) of metals (Hannah *et al.*, 1986). Toxic metals may adversely affect biological treatment processes as well as the quality of receiving waters. They are inhibitory to both aerobic and anaerobic processes in wastewater treatment (Barth *et al.*, 1965).

Biomethylation of Hg was first reported in 1969. Since then, many bacterial, e.g. *Pseudomonas fluorescens, E. coli, Clostridium* spp., and fungal, e.g. *Aspergillus niger, Saccharomyces cerevisiae*, isolates have been shown to methylate Hg^{2+} to methyl mercury. Vitamin B_{12} stimulates the production of dimethyl mercury $(CH_3)_2Hg$ by microorganisms:

$$Hg^{2+} \rightarrow CH_3Hg^{+} \rightarrow (CH_3)_2Hg$$

Methyl mercury accumulates under anaerobic conditions and is degraded under aerobic conditions. Sulphate-reducing bacteria are also capable of biomethylating mercury (Compeau and Bartha, 1985). Molybdate, an inhibitor of sulphate reducers, decreases mercury methylation in anoxic environments by 95%. Microorganisms can also transform Hg^{2+} or organic mercury compounds, such as $(CH_3)_2Hg$, phenylmercuric acetate and ethylmercuric phosphate to the volatile metal Hg^0. Mercury volatilization is a detoxification mechanism, the genetic control of which has been elucidated. Plasmid genes code resistance to metals such as Hg or Cd. Resistance to Hg is carried out by the *mer* operon, which consists of a series of genes (merA, merC, merD, merR, merT). The merA gene is responsible for the production of mercuric reductase enzyme, which transforms Hg^{2+} to Hg^0 (Silver and Misra, 1988).

Cadmium, Cd^{2+}, can be accumulated by bacteria, e.g. *E. coli, B. cereus* and fungi, e.g. *Aspergillus niger*. This metal can also be volatilized in the presence of vitamin B_{12}. Like Hg, Pb can be methylated by bacteria such as *Pseudomonas, Alcaligenes, Flavobacterium*, to $(CH_3)_4Pb$. Fungi, e.g. *Aspergillus, Fusarium*, are capable of transforming As to $(CH_3)_3As$, a volatile form with a garlic-like odour. Selenium is another metal which is methylated through the metabolic activity of bacteria, e.g. *Aeromonas, Flavobacterium* and fungi such as *Penicillium, Aspergillus*.

Cr-rich wastewaters are generated by industrial processes such as leather tanning, metal plating and cleaning. An *Enterobacter cloacae* strain, isolated from municipal wastewater, was able to reduce Cr^{6+} to Cr^{3+}, which precipitates as metal hydroxide at pH 7, thus reducing the bioavailability and toxicity of this metal (Ohtake and Hardoyo, 1992).

8 HEAVY METALS REMOVAL

The methods of metal removal are numerous and selection of the appropriate techniques depend upon factors such as the particular metal(s) involved, the chemical composition of the wastestream, the volume requiring treatment, variability in composition and flow of the wastestream and level of treatment required.

Both physical and chemical processes are employed for the removal of heavy metals. Such processes include ion exchange, oxidation/reduction, precipitation, ultrafiltration and many others. During primary treatment much of the particulate-associated metal is removed by sedimentation.

Microorganisms offer an alternative to physical/chemical methods for metal removal and recovery (Forster and Wase, 1987). Metal removal or immobilization is carried out by the biomass and depends on the biomass solids concentration and sludge age. The older the sludge the higher the biomass solids concentration and therefore the greater the metal removal.

Cation exchanger textiles (CET) have been used for the removal of heavy metals from industrial wastewaters in batch reactors.

The CETs carry carboxylate, sulphonate and phosphate functional groups. The influence of counter-ion and functional group, metal concentration and the presence of different competing ions in solution on the exchange capacity and selectivity coefficients was studied for Cd^{2+} and Cu^{2+}. Comparisons of optimum operating conditions for the CETs in this particular study with conventional ion exchange resins were carried out. Regeneration of the CETs used enables them to be an attractive alternative for wastewater treatment (Lacour *et al.*, 2001).

9 METAL REMOVAL BY WASTEWATER MICROORGANISMS

The extracellular polymer produced by microorganisms such as *Zooglea ramigera* and *Bacillus licheniformis*, for example, is known to have great affinity for a range of metals, e.g. Fe, Cu, Cd, Ni and U. The metals form complexes with the extracellular polymer and can readily accumulate. The bioaccumulation behaviour and toxicity of Cd in a protozoan community has also been studied (Fernandes-Leborans and Herrero, 1999). They found that Cd was accumulated by protozoa and bacteria at different Cd concentrations. The protozoa accumulating approximately 15 to 70 times more than the bacteria. The presence of Cd caused the abundance of protozoa to decrease down to a level at which recovery of the community was then suspected to be unrealistic. The abundance of the bacteria was not affected. The metals above, however, can easily be removed by treating with hydrochloric acid. Citric acid was used as a more sustainable acid, because of its inherent biodegradability to remove Cu and Zn and competing metals Ca and Fe from activated sludge prior to composting and use as fertilizer (Veeken and Hamelers, 1999). The metals were increasingly removed at higher temperatures and citric acid concentrations. At pH 3–4 the Cu removal was up to 60–70%, whereas with the Zn as much as 90–100% removal was achieved. The microorganism, *Zooglea ramigera*, was capable of removing Cu by 0.17 g/g biomass (Norberg and Persson, 1984; Norberg and Rydin, 1984).

The same bacterium when immobilized in alginate beads can accumulate as much as 250 mg/l Cd from solution. The alginate beads themselves actually remove some of the Cd (Kuhn and Pfister, 1990). Heavy metals have been removed from anaerobically digested sewage sludge by isolated indigenous iron-oxidizing bacteria (Xiang *et al.*, 2000). The bacteria were used in an anaerobic batch reactor for the removal of Cr, Cu, Zn, Ni and Pb. Solubilization of the metal ions was promoted by inoculation with the iron-oxidizing bacteria and $FeSO_4$. It was confirmed that the technique was effective, although the effectiveness varied between the different metals, e.g. 92% removal for Cu and only 16% for Pb.

RAS has a high metal affinity and may contain hundreds of times the concentration of heavy metals of the initial waste from which it was derived (Tyagi *et al.*, 1993). The secreted products of microbial anabolism (bacterial extracellular polymers) play the major significant role in the uptake of heavy metals from the aqueous medium (Nelson *et al.*, 1981). Binding sites in the secreted anionic and neutral biopolymers may combine with the cationic metal ions in solution (Brown and Lester, 1982). Studies into the removal of heavy metals from activated sludge solids have concentrated on the release of these metal ions from the sludge matrix by lowering the pH of the sludge with mineral and organic acids, typically below pH 4, the metal cations being in competition with H^+ ions (Sreekrishnan *et al.*, 1993; Du *et al.*, 1994). Hence lowering the pH lowers the affinity of the sludge for the metal ions and these are released into solution. Above neutral pH the dominant insoluble form of many heavy metals becomes the metal hydroxide. These are only sparingly soluble in alkaline conditions and lead to the precipitation of the heavy metal hydroxide for aqueous concentrations above their limit of solubility (Apak *et al.*, 1999). The pH of the medium has a significant effect on the solubilities of the metal precipitates. Most metals may also be precipitated as sulphides as shown in the proprietary Sulfex process (Feigenbaum, 1978):

$$Cr_2O_7^{2-} + 2FeS + 7H_2O \xrightarrow{@pH = 8} 2Fe(OH)_3 + 2Cr(OH)_3 + 2S^0 + 2OH^-$$

some as carbonates, phosphates or elemental (S^0) metal precipitation with sodium borohydrate as the reducing agent. The use of sodium borohydrate as a reagent is limited by its expense. Sulphide precipitation has been used as an effective alternative to hydroxide precipitation. It has the advantage of producing a precipitate even in conditions where the presence of a chelating agent would prevent the use of metal hydroxide precipitation. A serious disadvantage to the precipitation of metal sulphides is the potential for production of toxic hydrogen sulphide as a by-product. In activated sludge the potential number of ligand sites is often in excess of the metal ions present (Nelson *et al.*, 1981). The presence of free soluble heavy metal ions is therefore low except in acidic conditions. However, metals may still be present in the soluble phase attached to soluble ligands. By raising the pH of the activated sludge to favour the formation of metal hydroxides, or other insoluble precipitates, over metal ligand complexes, crystals of metal hydroxide should form within the sludge matrix which proves more amenable to physical separation techniques.

Studies have shown that inactivated microbial biomass in water treatment works takes up heavy metals from solution in a similar manner and concentration to live organisms. The main method of concentration of the heavy metals on to biosolids is by cation exchange with acidic functional groups on the excreted extracellular biopolymer surrounding the bacteria within a sludge floc.

Siderophores are low molecular mass molecules that are able to chelate with heavy metals and allow transport into the microorganism cells. Some microorganisms produce these especially to promote Fe^{3+} uptake from solution, the ferric ion being only sparingly soluble (Lundgren and Dean, 1979). Immobilized microorganisms (bacteria, fungi and algae), e.g. Algasorb – algae cells embedded in silica gel polymer – is able to remove heavy metals including uranium (Anon, 1991). Fungal mycelia such as *Aspergillus* and *Penicillium* have also been used for metal removal from wastewater (Galun *et al.*, 1982) and are therefore a suitable alternative for detoxifying effluents. Metals are adsorbed from solution by fungi rather than by intracellular uptake, the latter process being much slower and energy dependent. Biosorption column studies have shown that immobilized *Aspergillus oryzae* remove Cd efficiently from solution. It has also been shown that detergent treatment of fungal biomass considerably improves metal removal (Ross and Townsley, 1986). Cd seems to adsorb onto the fungal biomass but active uptake of the metal by the fungus does not occur to any significant extent and is therefore probably not a metabolic process (Kiff and Little, 1986). Artificial wastewater has been used for the removal of the metals Cd^{2+}, Pb^{2+} and Cu^{2+} by biosorption onto dry fungal biomass. The metals were in the 5–500 mg/l concentration range. Biosorption was established after a period of 6 hours. Good agreement between experimental data for the three metals and the Langmuir model was found (Say *et al*, 2001).

Table 35.2 below gives a list of some microorganisms used for removal and/or recovery of metals from industrial wastewaters (Eccles and Hunt, 1986).

9.1 Intracellular accumulation of metals

Microbial cells can accumulate metals, which gain entry into the cell by specific transport systems.

9.2 Adsorption onto cell surface

Interactions between metal ions and the negatively charged microbe surfaces cause microorganisms to bind metals to the surface. Gram-positive bacteria are particularly suited to metal binding. Fungal and algal cells also have a good affinity for heavy metals (Darnall *et al.*, 1986; Ross and Townsley, 1986).

TABLE 35.2 Some microorganisms involved in metal removal/recovery from industrial wastewaters

Microorganisms	*Metal*
Zooglea remigera	Cu
Saccharamyces cerevisieae	U and others
Rhizopus arrhizus	U
Chlorella vulgaris	Au, Zn, Cu, Hg
Aspergillus oryzae	Cd
Aspergillus niger	Cu, Cd, Zn
Penicillium spinulosum	Cu, Cd, Zn
Trichiderma viride	Cu
AMT Bioclaim™	Biotechnology-based use of granulated product derived from biomass.

Removal of metals from solution may take place by various mechanisms (Brierley *et al.*, 1989; Sterrit and Lester, 1986; Trevors, 1989; Trevors *et al.*, 1985).

Adsorption of metals by activated sludge biomass has been found to follow Langmuir and Freundlich isotherms (Mullen *et al.*, 1989).

The sublethal effects of the heavy metals Cu, Cr, Pb and Zn have been studied and they were shown to adsorb onto biomass much faster than organic competitors. This is probably due to the increased mobility of the metal ion over the organic molecules due to their relative size and greater charge. The metal ions, being much more mobile, are able to seek out adsorption sites quicker than their larger organic competitors on the bioflocs.

'Biosorption being a process that utilizes inexpensive dead biomass to sequester toxic heavy metals and is particularly useful for the removal of contaminants from industrial effluents'(Kratochvil and Volesky, 1998).

9.3 Complexation

Microorganisms are known to produce weak acids such as citric acid, which may form metal chelates. Metals may also be complexed by carbonyl groups found in microbial polysaccharides and other polymers. This is of importance in wastewater treatment plants, particularly those using the activated sludge process, where industrial wastes are treated (Bitton and Freihoffer, 1978; Brown and Lester, 1979, 1982; Rudd *et al.*, 1984; Sterrit and Lester, 1986; McLean *et al.*, 1990). *Pseudomonas putida* has a cystein-rich protein that readily binds with Cd (Higham *et al.*, 1984).

9.4 Precipitation

Some bacteria promote metal precipitation producing ammonia, organic bases, or hydrogen sulphide, which precipitate metals as hydroxides or sulphides. Sulphate-reducing bacteria transform SO_4 to H_2S, which promotes the extracellular precipitation of metals from solution. *Klebsiella aerogenes* is able to detoxify Cd to CdS, which precipitates as electron-dense granules at the cell surface. This process is induced by Cd (Aiking *et al.*, 1982).

It has also been shown, using ICP, that by using the pyridine-thiol ligand Cu and Cd can be removed from solution at pH values of 4.5 and 6.0 respectively by as much as 99.99%; the starting concentrations being 50 ppm. The multiple bonding sites on the pyridine-thiol ligands are responsible for the metal precipitation (Matlock *et al.*, 2001).

9.5 Volatilization

Some metals are transformed to volatile species as a result of microbial action. Bacterially mediated methylation converts Hg^{2+} to $(CH_3)_2Hg$, a volatile compound. Some bacteria have the ability to detoxify Hg by transforming Hg^{2+} to Hg^0, which again is volatile. This detoxification process is plasmid-encoded and is regulated by an operon consisting of several genes. The most important gene is the merA gene, which is responsible for the production of mercuric reductase, the enzyme that catalyses the transformation Hg^{2+} to Hg^0.

REFERENCES

Aiking, H. *et al.* (1982). Adaption to cadmium by *Klebsiella aerogenes* growing in continuos culture proceeds mainly via the formation of cadmium sulphate. *Applied Environmental Microbiology* **44**, 938–944.

Alloway, B.J. (1995). The origins of heavy metals in soils. In: B.J. Alloway (ed.) *Heavy Metals in Soils*, 2nd edn, pp. 39–57. Blackie Academic & Professional, London.

American Dye Manufacturers Institute (1972). The contribution of dyes to the metal content of textile mill effluents. *Textile Chemist and Colorist*, **14**(12), 29–31.

Anon (1998a). Clampdown on the horizon for the land spreading of industrial wastes. *ENDS Report* **281**, 27–29.

Anon (1998b). SEPA pushes for tighter controls on land spreading of wastes. *ENDS Report* **286**, 36.

Anon (1999). Anon Revision of EC sludge directive challenges landspreading. *ENDS Report* **299**, 29–31.

Apak, R., Hizal, J. and Ustaer, C. (1999). Correlation between the limiting pH of metal ion solubility and total metal concentration. *Journal of Colloid and Interface Science* **211**, 185–192.

Bag, S. *et al.* (1999). Ecotoxicol Environ Saf. 1999 Feb **42**(2), 163–170.

Barker, T. (2000). Textiles bid by US company highlights lack of appetite for sector after years of decline and the loss of 500,000: Thorold Barker explains why a once proud name is likely to disappear. *Financial Times* 15 February. CD-ROM. Financial Times on CD ROM 01 January-30 April. Serfert Software Engineering.

Barth, E.F. *et al.* (1965). Summary report on the effects of heavy metals on biological treatment processes. *Journal of the Water Pollution Control Federation* **37**, 86–96.

Bitton, G. (1994). *Wastewater Microbiology.* Wiley-Liss Inc, New York.

Bitton, G. and Freihoffer, V. (1978). Influence of extracellular polysaccharides on the toxicity of copper and cadmium towards *Klebsiella aerogenes*. *Microbial Ecology* **4**, 119–125.

Bourg, A.C.M. (1995). Speciation of heavy metals in soils and groundwater and implications for their natural and provoked mobility. In: Salomons, Förstner and Mader (eds) *Heavy Metals Problems and Solutions*, pp. 19–25. Whittaker.

Brierley, C.L. *et al.* (1989). Applied microbial processes for metal recovery and removal from wastewater. In: T.J. Beverage and R.J. Doyle (eds) *Metal Ions and Bacteria.* Wiley, New York.

Brown, M.J. and Lester, J.N. (1979). Metal removal in activated sludge: The role of bacterial extracellular polymers. *Water Research* **13**, 817–837.

Brown, M.J. and Lester, J.N. (1982). Role of bacterial extracellular polymers in metal uptake in pure bacterial culture and activated sludge. I. Effect of metal concentration. *Water Research* **16**, 1539–1548.

Charlton, S. (1999). Heavy metal threat to cotton dyers. *International Dyer* **184**(10), 27–30.

Cheng Nong, Y. *et al.* (1999). Thermochemical studies of the toxic actions of heavy metal ions on Rhizopus nigricans. *Chemosphere* **38**(4), 891–898.

Darnall, D.W. *et al.* (1986). Selective recovery of gold and other metal ions from algal biomass. *Environmental Science and Technology* **20**, 206–210.

Compeau, G. and Bartha, R. (1985). Sulfate reducing bacteria: principal methylators of Hg in anoxic estuarine sediments. *Appl. Environ, Microbiol.* **50**, 498–502.

DETR (1998) Part 1 – Evidence Underlying the 1989 Department of the Environment Code of Practice for Agricultural use of Sludge and the Sludge (Use in Agriculture Regulations). *Review of the Scientific Evidence Relating to the Controls on the Agricultural use of Sewage Sludge*, DETR 4415/3.HMSO, London.

Du, Y.G., Sreekrishnan, R.D., Tyagi, R.D. and Campbell, P.G.C. (1994). Effect of pH on metal solubilisation from sewage sludge: a neural-net-based approach. *Canadian Journal of Civil Engineering* **21**, 728–735.

Eccles, H. and Hunt, S. (1986). *Immobilisation of Ions by Bio-sorption.* Ellis Horwood, Chichester.

EC (2000) COM(1999) 752 final Report from the Commission to the Council and the European Parliament on the Implementation of Community Waste Legislation Directive 75/442/EEC on waste, Directive 91/689/EEC on hazardous waste, Directive 75/439/EEC on waste oils and Directive 86/278/EEC on sewage sludge for the Period 1995–1997. Online. Available: http://europa.eu.int/eur-lex/en/com/pdf/1999/com1999_0752en01.pdf(27 February 2000).

EEC (1986) Council Directive 86/278/EEC of 12 June 1986 on the protection of the environment, and in particular of the soil, when sewage sludge is used in agriculture. *Official Journal L* 181, 04/07/1986, pp. 0006–0012.

EU (1999) Waste management – The landfill of waste. Online. Available: http://europa.eu.int/scadplus/leg/lvb/121208.htm (30 July 1999).

EU (2000a) Directive 2000/60/EC of the European Parliament and of the Council of 23 October 2000 establishing a framework for Community action in the field of water policy. *Official Journal of the European Communities* OJ: L327.

EU (2000b) ENV.E.3/LM 3rd Draft Working Document on Sludge. Online. Available: europa.eu.int/comm/environment/waste/sludge_en.pdf (20 January 2001).

Feigenbaum, H.N. (1978). Process for removal of heavy metals from textile waste streams. *American Dyestuff Reporter* **46**, 43–44.

Fernandes Leborans, G. and Olalla Herrero, Y. (1999). *Ecotoxicol Environ Saf* **43**(3), 292–300.

Forster, C.F. and Wase, D.A.J. (1987). *Environmental Biotechnology.* Ellis Horwood, Chichester.

Galun, M. *et al.* (1982). Removal of uranium (VI) from solution by fungal biomass and fungal wall-related biopolymers. *Science* **219**, 285–286.

Geiger, G. *et al.* (1999). *Enzyme Inhib* **14**(5), 365–379.

Hannah, S.A. *et al.* (1986). Comparative removal of toxic pollutants by six wastewater treatment processes. *Journal of the Water Pollution and Control Federation* **58**, 27–34.

Higham, D.P. *et al.* (1984). Cadmium resistant *Pseudomonas putida* synthesizes novel cadmium binding proteins. *Science* **225**, 1043–1046.

HMG (1994). *Statutory instruments: 1994: 1056 The Waste Management LicensingRegulations 1994 Environmental protection.* HMSO, London.

Iverson, W.P. and Brinckman, F.E. (1978). Microbial metabolism of heavy metals. In: R. Mitchell (ed.) *Water Pollution Microbiology*, Vol. 2, pp. 201–232. Wiley, New York.

Kratochvil, D. and Volesky, B. (1998). Advances in the biosorption of heavy metals. *Trends in Biotechnology* **16**(7), 291–300.

Kuhn, S.P. and Pfister, R.M. (1990). Accumulation of cadmium by immobilized *Zoogloea ramigera* 115. *J. Ind. Microbiol.* Amsterdam, Vol. 6, pp. 123–138.

Lacour, S., Bollinger, J.C., Serpaud, B. *et al.* (2001). Removal of heavy metals in industrial wastewaters by ion-exchanger grafted textiles. **428**(1), 121–132.

Lin, C.-Y. (1992). The effect of heavy metals on volatile fatty acid degredation in anerobic digestion. *Water Research* **26**, 177–183.

Lombi, E., Wenzel, W.W. and Adriano, D.C. (1998). Soil contamination, risk reduction and remediation. *Land Contamination & Reclamation* **6**(4), 183–197.

Lundgren, D.G. and Dean, W. (1979). Biochemistry of iron. In: P.A. Trudinger and D.J. Swaine (eds) *Biogeochemical Cycling of Mineral-Forming Elements*, pp. 211–233. Elsevier, Amsterdam.

McCann, B. (1998). Disposable outcome. *Water* **19**, 11.

McLean, R.J.C. *et al.* (1990). Metal-binding characteristics of the gamma-glutamyl capsular polymer of *Bacillus licheniformis* ATTC 9945. *Applied Environmental Microbiology* **56**, 3671–3677.

Matlock, M.M., Howerton, B.S., Henke, K.R. and Atwood, D.A. (2001). A pyridine-thiol; ligand with multiple bonding sites for heavy metal precipitation. **82**(1), 55–63.

Mueller, R.F. and Steiner, A. (1992). Inhibition of anaerobic digestion by heavy metals. *Water Science and Technology* **26**, 835–846.

Mullen, M.D. *et al.* (1989). Bacterial sorption of heavy metals. *Applied Environmental Microbiology* **55**, 3143–3149.

Murray, W.D. and van den Berg, L. (1981). Effect of nickel, cobalt and molybdenum on performance of methanogenic fixed film reactors. *Applied Environmental Microbiology* **32**, 502–505.

Nelson, P.O., Chung, A.K. and Hudson, M.C. (1981). Factors affecting the fate of heavy metals in the activated sludge process. *Journal of the Water Pollution Control Federation* **53**(8), 1323–1333.

Nies, D.H. (1999). Microbial heavy metal resistance. *Applied Microbial Biotechnology* **51**(6), 730–750.

Norberg, A.B. and Persson, H. (1984). Accumulation of heavy metal ions by *Zooglea ramigera*. *Biotechnology and Bioengineering* **26**, 239–246.

Norberg, A.B. and Rydin, S. (1984). Development of a continuous process for metal accumulation by *Zooglea ramigera*. *Biotechnology and Bioengineering* **26**, 265–268.

Ohtake, H. and Hardoyo (1992). New biological method for the detoxification and removal of hexavalent chromium. *Water Science and Technology* **25**, 395–402.

Phipps, D.A. (1981). Chemistry and biochemistry of trace metals in biological systems, in *Effect of Heavy Metal Pollutions on Plants*, N.W. Lepp (ed), Applied Science Publishers, Barking.

Pierce, J. (1994). Colour in textile effluent – the origin of the problem. *Journal of the Society of Dyers and Colourists* **110**(4), 131.

Rochow, E.G. (1964). *Organometallic Chemistry*, pp. 79–80. Chapman and Hall, London.

Ross, I.S. and Townsley, C.C. (1986). The uptake of heavy metals by filamentous fungi. In: H. Eccles and S. Hunt (eds) *Immobilisation of Ions by Bio-Sorption*, pp. 49–58. Ellis Horwood, Chichester.

Rudd, T. *et al.* (1984). Complexation of heavy metals by extracellular polymers in the activated sludge process. *Journal of the Water Pollution Control Federation* **56**, 1260–1268.

Say, R., Denizli, A. and Yakup Arica, M. (2001). Biosorption of cadmium(II), lead(II) and copper(II) with the filamentous fungus Phanerochaete chrysosporium. *Bioresource Technology* **76**(1), 67–70.

Shanableh, A. and Ginige, P. (1999). Impact of metals bioleaching on the nutrient value of biological nutrient removal biosolids. *Water Science and Technology* **39**(6), 175–181.

Shonheit, P. *et al.* (1979). Kinetic mechanism for the ability of sulphate reducers to outcompete methanogens for acetate. *Archives of Microbiology* **132**, 285–288.

Silver, S. and Misra, T.K. (1988). Plasmid-mediated heavy metal resistance. *Annual Review of Microbiology* **42**, 717–743.

Sin, S.N. *et al.* (2000). *Appl Biochem Biotechnol* **84–86**, 487–500.

Sreekrishnan, T.R., Tyagi, R.D., Blais, J.F. and Campbell, P.G.C. (1993). Kinetics of heavy metal bioleaching from sewage sludge I, effects of process parameters. *Water Research* **27**(11), 1641–1651.

Sterrit, R.M. and Lester, J.N. (1986). Heavy metals immobilisation by bacterial extracellular polymers. In: H. Eccles and S. Hunt (eds) *Immobilisation of Ions by Bio-Sorption*, pp. 121–134. Ellis Horwood, Chichester.

Trevors, J.T. (1989). The role of microbial metal resistance and detoxification mechanisms in environmental bioassay research. *Hydrobiology* **188**, 143–147.

Trevors, J.T. *et al.* (1985). Metal resistance in bacteria. *FEMS Microbiological Review* **32**, 39–54.

Tyagi, R.D., Shreekrishnan, T.R., Campbell, P.G.C. and Blais, J.F. (1993). Kinetics of heavy metal bioleaching from sewage sludge – II Mathematical model. *Water Research* **27**(11), 1653–1661.

Vanková, S. *et al.* (1999). *Ecotoxicology and Enviromental Safety* **42**(1), 16–21.

Veeken, A.H.M. and Hamelers, H.V.M. (1999). *Water Science and Technology* **40**(1), 129–136.

Weast, R.C. (1984). CRC Handbook of Chemistry and Physics, 64th Edn., CRC Press Inc., Boca Raton, Florida USA.

Whitman, W.B. and Wolfe, R.S. (1980). Presence of nickel in factor F_{430} from *Methanobacterium bryantii*. *Biochemical and Biophysical Research Communication* **92**, 1196–1201.

Xiang, L., Chan, L.C. and Wong, J.W. (2000). Removal of heavy metals from anaerobically digested sewage sludge by isolated indigenous iron-oxidizing bacteria. *Chemosphere* **41**(1–2), 283–287.

Part IV Drinking Water Microbiology

36

Surface waters

Huw Taylor

EPHRLL, School of the Environment, University of Brighton, BN2 4GJ, UK

1 INTRODUCTION

Surface water is a general term describing any water body that is found flowing or standing on the earth's surface, such as streams, rivers, ponds, lakes and reservoirs. In this chapter, the microbial pollution of surface waters other than reservoirs is reviewed.

River flow is affected by surface run-off, direct precipitation, inter-flow (excessive soil moisture draining into the river) and water-table discharge. The relationship between flow and precipitation is therefore not always immediately obvious. In temperate climates flow is normally higher in winter months than n summer months and in tropical and sub-ropical climates flow is higher during the ainy season. Geology, however, is also a factor n the flow rate of a river at any particular time. The 'run-off ratio' describes the percentage of ainfall yielding river flow. In the UK, this value has been estimated to reach 80% in parts of Scotland and Wales compared with 30% in owland England (Gray, 1999). Since ground-water inputs also influence river flow, summer roughts are often modified by storage build-p in the aquifer during a previously wet winter. The chemical, physical and biological omposition of these various flow inputs obviously influences the quality of water in vers and other surface waters.

River water represents a readily available ource of water for human activity and storically many civilizations have relied on e ample supplies of freshwater found in ajor river catchments for irrigation of arid d semi-arid environments. Throughout the world, developing human settlements have made use of streams, rivers and lakes to provide water for human activity and to dispose of water-borne wastes.

Surface water quality is, however, often unreliable and is more likely to be heavily contaminated by faecal microorganisms than groundwater that has undergone a natural process of physical, chemical and biological filtration through the soil and substrata. Surface run-off is a major component of the soft water of rivers in areas of impervious geology and may be highly turbid. In chalk lowland rivers, water-table discharge is more significant and water quality is generally less turbid and of a more consistently high quality.

Turbidity is the result of suspended particles that are too small to settle out readily under quiescent conditions. These particles include both organic and inorganic material and originate in surface waters, predominantly from recent surface run-off to rivers and resuspension of settled sediments. Planktonic microorganisms contribute to this turbidity. Larger assemblages of microorganisms in 'flocs' or those adhered to the surface of soil particles may also remain in suspension for a considerable time, especially in fast flowing rivers.

Surface water quality may be divided into three categories, or trophic levels, according to nutrient levels and microbial populations:

(i) oligotrophic (low nutrients, minimal microbiological activity)
(ii) mesotrophic (moderate nutrients, moderate microbiological activity)

e Handbook of Water and Wastewater Microbiology
N 0-12-470100-0

(iii) eutrophic (high nutrients, high microbiological activity (Margolin, 1997).

Eutrophic conditions reduce the suitability of river water as a source of raw drinking water as high microbial numbers reduce the efficacy of filtration and high organic levels can reduce the efficacy of chlorination and increase the survival times of pathogenic organisms. Specific health issues associated with eutrophic waters include the formation of trihalomethanes (THM) during the water-treatment process and the development of blooms of cyanobacteria (Reinert and Hroncich, 1990).

Water-borne pathogenic organisms of enteric origin include viruses, bacteria and protozoan and helminthic parasites. The host specificity of many of these organisms suggests that the presence of human faeces in river water poses a greater potential risk to human health than the presence of the faeces of other animals. However, this is not always the case as some pathogens (the zoonoses) may cross the host species divide. The presence of faeces from feral animals and those from agriculture in surface waters therefore also present a potential risk to human health.

Established methods for assessing the sanitary quality of water rely on the enumeration of bacterial indicator species, but these methods do not readily identify the source of contamination. Section 6 of this chapter therefore summarizes recent techniques used to track sources of faecal pollution in surface waters.

The relative contribution of the various sources of faecal pollution to a surface water catchment varies according to the characteristics of the specific catchment. Differences are also temporal with the run-off following periods of rainfall leading to increases in the levels of both microbial pollutants and nutrients. In tropical environments, the hygienic quality of surface waters is often crucial to the protection of human health as water-related diseases are often the major cause of morbidity and premature mortality. Run-off carries faecal material from humans and non-human animals into the water body and can be a significant source of pathogens in many catchments.

2 SOURCES OF ENTERIC ORGANISMS IN SURFACE WATERS

Temporal changes to the hygienic quality of surface waters are often significant. Kistemann *et al.* (2002) noted that hygienic and microbial examinations of watercourses are usually not carried out during heavy rainfall and runoff events. In their study they found that a substantial component of total microbial load in watercourses and in drinking water reservoirs results from rainfall and extreme run-off events.

The contributions of point and non-point sources of microbial contaminants in freshwater have previously been reviewed by White and Godfree (1985) and by Geldreich (1990).

2.1 Human wastes and municipal wastewaters

The composition of municipal wastewaters discharging to surface waters varies enormously. Often these are composed of both domestic and industrial effluents. Wastewater flow and the concentration of human faecal material in these wastewaters shows both diurnal and seasonal variations, the latter being particularly pronounced in areas that have fluctuating seasonal human populations such as holiday resorts and agricultural areas dependent on migratory labour. 'Combined sewerage' systems are common in many countries. These carry both wastewaters and the urban stormwater run-off. The composition of combined wastewater may therefore be extremely variable.

In rural areas and throughout much of the developing world, human wastes are less likely to enter a sewerage system. Where it exists sanitation provision is more likely to consist of a more simple 'on-site' form of technology. Where reliable running water is unavailable sanitation may consist of simple pit latrines or a suitable alternative. The aim of such provision may be to treat safely human excreta, preferably with the production of a usable 'end-product' such as a fertilizer for agricultural use. No wastewater effluent results from many 'on-site' forms of sanitation and, if the facility is designed and located with care, surface waters should not become contaminated.

In the developed world (and the urbanized developing world), flush toilets are more common. Where small rural communities are not connected to a mains sewerage system, domestic wastewater is often partially treated in a septic tank or similar. The stored and partially treated sludge is normally collected at intervals by tanker and taken to a wastewater treatment facility where it is added to municipal wastewater from sewered communities. Liquid effluents from septic tanks are allowed to leach through the soil, again with the potential risk of river pollution. If well located and maintained however, septic tanks and similar on-site facilities should not significantly contaminate nearby watercourses.

The total load of pathogenic microorganisms of human origin that enter a surface water catchment depends on a number of factors. These factors include the total human population of the catchment, the proportion of the population using sewerage systems that eventually discharge into the river, the level of wastewater treatment and the operating efficacy of these systems. The flow level of the recipient river water will not be constant and during periods of dry weather flow, municipal wastewater discharges may represent a significant proportion of the total flow and contribute an equally significant proportion of faecal microorganism levels to the river water.

During periods of heavy rainfall, the wastewater of combined sewers entering treatment facilities often does not undergo full treatment and may be discharged to the river with little more than partial sedimentation in stormwater tanks. However, combined raw wastewater after storm events is normally more dilute and enters surface waters under high flow conditions so that the impact is less marked than might be expected. In many countries, discharge consents permit the occasional discharge of partially treated wastewaters to surface waters from wastewater treatment plants during periods of very high rainfall.

2.2 Other wastewater discharges

Not all industrial wastewaters enter a municipal sewerage system and many such wastewaters, such as those of food, beverage, meat packing, wood pulp, and paper wastes, often contain faecal material that may include pathogens (Geldreich, 1990). Abattoir wastes are obviously a potential source of animal faeces and may be highly polluting both in terms of pathogens and organic matter.

2.3 Agricultural effluents and run-off

Agricultural communities have traditionally developed in close proximity to rivers, streams or other readily accessible sources of freshwater. However, in the late twentieth century, intensive animal husbandry in the developed world led to increasingly concentrated loads of animal faeces from agriculture and consequently incidents of significant organic pollution to river water from agricultural practice have increased. Table 36.1 shows the *E. coli* concentrations found in the faeces of various animals.

Figures for 1989/90 (Anon., 1991) suggest that over 10% of all recorded incidents of river water pollution in England and Wales (UK) during this period were attributed to organic farm waste. Such figures generally only record specific incidents of 'gross' point source pollution and do not attempt to identify the contribution of agriculture to river pollution as a result of farmland run-off following heavy rainfall. In a faecal indicator budget study of two UK lowland pastoral catchments, Crowther *et al.* (2002) recorded greater than tenfold elevations in geometric mean faecal indicator concentrations following high-flow conditions. Although inputs from diffuse and point sources of pollution were not quantified, point sources (such as channelled run-off from farmyards) seemed likely to be significant.

TABLE 36.1 Levels of *E. coli* excreted by birds and other animals

	Faecal production g/day	*Average number* E. coli/g	*Daily load* E. coli
Man	150	13×10^6	1.9×10^9
Cow	23 600	0.23×10^6	5.4×10^9
Hog	2700	3.3×10^6	8.9×10^9
Sheep	1130	16×10^6	18.1×10^9
Duck	336	33×10^6	11.1×10^9
Turkey	448	0.3×10^6	0.13×10^9
Chicken	182	1.3×10^6	0.24×10^9
Gull	15.3	131.2×10^6	2×10^9
		E. coli concentrations/100 ml	
Sewage		3.4×10^5–2.8×10^7	
Sewage effluent		1×10^3–1×10^7	

Source: Jones and White (1984).

In an earlier study in the USA, Geldreich (1972) reviewed the faecal pollution from stormwater run-off from animal feedlots where thousands of cattle were confined and found that it equated to the discharge of raw wastewater from cities that approached 10 000 people.

During recent years attempts have been made in many countries to control the impact of agricultural effluents and run-off on surface waters. In the UK, government guidelines outline practical ways of disposing of, or recycling farm wastes that lead to the effective destruction of pathogenic organisms before they are able to contaminate surface waters (Anon., 1991).

Most pathogens express a degree of host specificity so the pathogens egested in human faeces normally pose the greatest risk to human health. However, some pathogens of animal origin may infect humans. A number of these diseases (known as zoonoses) are often endemic in farm animals, and agricultural wastes may therefore pose a specific risk of human infection. Examples include *Cryptosporidium* spp. and various toxigenic strains of *Escherichia coli*.

2.4 Stormwater and urban surface run-off

Where urban run-off is not directed to combined sewerage systems, separate systems may deliver this water directly to surface waters with no prior treatment. Discharges from such systems may contain high levels of faecal indicator bacteria (Ellis, 1993). In South Africa, faecal coliform levels as high as 10^4/100 ml have been measured in stormwater run-off entering streams in low-income urban residential developments (Jagals, 1997). In tropical and subtropical environments, river water quality deteriorates markedly during the dry season often making the production of potable water economically unfeasible (Bordalo *et al.*, 2001).

2.5 Avian sources

The faeces of birds often contain higher concentrations of the faecal indicator bacteria *E. coli* and intestinal enterococci and specific pathogens such as *Salmonella* spp. and *Campylobacter* spp. than the faeces of humans. Therefore, wild bird populations may represent a significant source of enteric microorganisms in many surface waters, although the areas affected tend to be localized and specific (Godfree, 1993). Obiri-Danso and Jones (1999) discuss the possible contribution of avian sources in a UK freshwater catchment.

2.6 Feral mammals

In urbanized catchments the contribution of feral mammals to the faecal pollution of surface waters is often of low significance compared with the faecal contribution from agriculture and municipal wastewaters. However, in rural catchments throughout the world indigenous animals may pose a potential source of pathogens to surface waters used as a source of drinking water.

In North America, infection by the protozoan parasite *Giardia* is relatively common in upland catchments. Although the parasite is found in the faeces of many animals, including wild beaver populations, no conclusive link has been made with human infections (Rusin *et al.*, 2000).

2.7 Recreational use of surface waters

Faecal contamination of recreational waters may derive from their use by bathers, especially infants. In crowded bathing waters this source may pose a significant risk to human health. Although boating and other recreational use of reservoirs is common in the developed world, bathing is generally discouraged in order to protect source waters. In developing countries the distinction between recreational waters and source waters is clearly less distinct.

In the European Union as a whole many freshwater sites have been designated as bathing waters and are monitored for compliance with the EU Directive on the quality of bathing water. In the UK, however, few freshwater sites have been designated as bathing waters under the EU directive. The impact of 'bather shedding' is therefore unlikely to have any significant impact on the quality of source waters in the UK.

2.8 Recirculation of sediments

Surface water sediments may act as a reservoir of enteric microorganisms deriving from faecal pollution. In some instances, regrowth of some species may be possible in these relatively protected and nutrient rich environments. High numbers of organisms are consequently released during periods of high flow and as a result of mechanical dredging of sediments. Recirculation of sediment organisms in lakes may occur as a result of thermal stratification followed by seasonal inversion of the water body.

2.9 Subsurface sources

Poorly designed and managed landfill of municipal and industrial solid waste poses a risk of pollution to surface and groundwaters. The leachate or liquid waste that is a natural product of the landfill treatment process normally has a very high organic load (although this reduces with time) and often contains high levels of toxic substances and enteric organisms. In older waste landfill sites, where no lining system is employed, leachate passes directly into the local subsurface environment where further biological, physical and chemical stabilization occurs (Williams, 1998). However, the risk of contamination of ground and surface waters from such systems is potentially great.

Modern landfill sites are designed to minimize the risk of water pollution. Leachates from modern landfill sites in the European Union states are commonly designed to be channelled for discharge to a suitable sewerage system, usually without prior treatment. On-site treatment of leachate is not common but is an increasing practice (Hjelmar *et al.*, 1995). Where on-site pretreatment occurs, it may consist of physicochemical processes, attached growth biological process, land treatment or leachate recirculation (Williams, 1998).

3 PATHOGENIC AND INDICATOR ORGANISMS IN SURFACE WATERS

Enteric organisms entering surface waters encounter a highly antagonistic environment. Factors that may directly or indirectly influence cell death or sublethal damage include solar radiation, pH, water temperature, concentration of humic substances, predation, adsorption/sedimentation, and salinity. Time of exposure is also a factor and so the higher residence time of slow-flowing rivers leads to more effective microbial decay. The processes that lead to the mortality of particular pathogens and faecal indicators vary and are not fully understood, but recent studies have suggested the possible mechanisms that might predominate in surface waters.

3.1 Bacteria

Studies of bacterial decay generally establish a numerical value for mortality rate (T_{90}; the time required for 90% mortality). Usually the population dynamics of allochtonous bacteria are modelled by first order kinetics $dN(t)/dt = -kN\ (t)$ with $k > 0$, i.e. an exponential decay of N with time t. This model is both simple and efficient in practical situations. Its single parameter, i.e. the decay coefficient k, is often replaced by $K = k/2.3$, which corresponds to the use of decimal logarithms for bacterial counts: $\log N(t) - \log N(t_0) = -K(t - t_0)$. K, usually expressed in h^{-1}, is thus the inverse of the period T_{90} (h) necessary for reducing the bacterial population by 90%.

In a study of coliform bacteria in the River Seine in France, it was estimated (George *et al.*, 2001) that grazing by protozoa was responsible for 47–99% of the mortality of coliforms in the river. Attachment of coliforms to suspended matter was another factor that could have been important in controlling the dynamics of coliforms in the river, particularly in its estuary.

Beaudeau *et al.* (2001) measured and modelled the decay of *Escherichia coli in situ* in small streams in the north of France. They showed that in these environments the set of significant variables did not include light indicators. *E. coli* decay was inversely related to river flow and the relationship became highly significant below 0.3 m^3/s. The positive effect of small flows on die-off was increased by water temperature over 15°C, whereas it could be reduced by the presence of suspended particulate matter. The major covariable of the model was an empiric composite variable integrating the effect of flow and temperature. The authors interpreted this as an expression of predation by benthic micrograzers that, they suggest, could be the main cause of *E. coli* die-off in small streams in temperate climates.

The observation that river waters generally contain higher levels of enteric bacteria than lakes and reservoirs is often not only the consequence of greater point and non-point sources of faecal pollution but is also linked to the shorter residence time of water in rivers. Pathogens and faecal indicators tend to spend less time in the water body between pollution source and abstraction point and the self-purification processes summarized above have less time to take effect.

Obiri-Danso and Jones (1999) monitored two freshwater bathing sites on the River Lune in the north-west of England over a 2-year period. They found that neither site showed a seasonal variation in faecal indicator levels (faecal coliforms and faecal streptococci) and neither site satisfied the guidelines or imperative standards of the EU Bathing Water Quality Directive (76/160/EEC). Indicator levels in the sediments were of an order of magnitude higher. Campylobacter counts were higher during the winter although numbers were low.

Polo *et al.* (1998) investigated the presence of salmonellae and its relationship with indicators of faecal pollution in aquatic habitats in Spain. They observed the highest frequency of *Salmonella* spp. in rivers (58.7% of samples), followed by freshwater reservoirs (14.8%) and seawater (5.9%).

Niemi *et al.* (1997) investigated long-term trends in the bacteriological quality of Finnish rivers and lakes. Rivers were consistently more polluted than lakes and those in the more highly populated south of the country were more polluted than those in the sparsely populated areas of the north. Levels of thermotolerant coliform bacteria and faecal streptococci showed a decreasing trend during the 1960s and 1970s during a period of major pollution control measures but, by 1990, this trend had ceased.

Estimating bacterial fluxes can demonstrate the impact of rainfall on river water pollution. Baudart *et al.* (2000) demonstrated high loads of *Salmonella* spp. during storm events and the annual loads were higher than those estimated from a coastal wastewater outfall. Bacterial loads from the river were associated with smal clay particles that trap bacteria in the sediment during the lowest water levels and were resuspended during storm events.

3.1.1 Recovery of injured bacteria in aquatic environments

Enteric indicator organisms and pathogens may become sublethally damaged in surface waters as a result of the factors described above. Such bacteria are unable to multiply to form visible colonies on selective agars in the various standard methodologies that are used to enumerate pathogen and indicator organisms. Since many of these organisms may retain their pathogenicity, many methods of enumeration may underestimate the risk to public health of a water sample. The effect is most noticeable in waters treated with chlorine, although the impact of UV radiation on surface waters may produce similar sublethal damage (McFeters, 1990; McFeters and Singh, 1991).

3.1.2 Multiplication of 'enteric' bacteria in aquatic environments

One criterion of an 'ideal' faecal indicator organism is that it should not grow and multiply outside the intestinal tract of its host (Bonde, 1977). For many years this was thought to hold true for the two classic bacterial indicators of faecal pollution, the faecal coliforms and the faecal streptococci. Recent studies, however, have demonstrated that regrowth of these organisms may occur under some conditions in the natural environment (Pepper *et al.*, 1993; Gibbs *et al.*, 1997; Momba and Kaleni, 2002). Growth in surface waters of the bacterium *Vibrio cholerae*, the causative agent of cholera, is discussed in Section 4.

3.2 Viruses

It has become increasingly evident in recent years that viruses are a leading cause of waterborne gastroenteritis and five major groups of human gastroenteritis virus have been identified: rotavirus, enteric adenovirus, Norwalk virus, calicivirus and astrovirus (Rusin *et al.*, 2000). In general, enteric viruses survive longer than faecal bacteria in natural freshwaters (Rao and Melnick, 1986) and longer in cooler climates than in hot, sunny regions (Ward *et al.*, 1986; Hurst *et al.*, 1989). West (1991) presents a review of pathogenic viruses in the water cycle.

Other studies have investigated the survival of bacteriophages of interest as potential indicators of faecal pollution in surface waters. Studies comparing the levels of somatic coliphages, F-specific phages and phages of *Bacteroides fragilis* in freshwater environments with recent and persistent faecal pollution have indicated that phages of *B. fragilis* are the most resistant to natural inactivation processes (Araujo *et al.*, 1997a,b).

3.3 Parasitology

Parasites in freshwaters include protozoan microorganisms such as species of the genera *Cryptosporidium*, *Cyclospora*, *Toxoplasma*, *Entamoeba* and *Giardia* and the multicellular helminths. The genus *Cryptosporidium* is a protozoan parasite that can infect cattle and other farm animals. Oocysts of *Cryptosporidium* may enter surface water via point sources such as cattle slaughterhouses or as a result of diffuse run-off from farmland. Many animals, especially beavers, are reservoirs of *Giardia*, although no direct evidence has been provided that animal sources of *Giardia* lead to human infection (Rusin *et al.*, 2000). Contamination of freshwaters from human wastewaters may therefore be the predominant cause for concern. *Giardia* cysts may survive for long periods in temperate freshwaters (DeRegnier *et al.*, 1989). West (1991) presents a review of parasites in the water cycle.

3.4 Cyanobacteria

The cyanobacteria are prokaryotic organisms that, like the eukaryotic green algae, are able to carry out photosynthesis with the production of oxygen. Until their prokaryotic cell structure was recognized they were termed 'blue-green algae'. They are not enteric in origin and cannot multiply in the gut but their presence in large numbers in surface waters may reduce the potability of the water and may pose a risk to the health of those ingesting or having skin-contact with the water. Cyanobacteria occur in a wide variety of habitats and partly because of their ability to fix atmospheric nitrogen, act as 'primary colonizers' of terrestrial habitats in which few other organisms can multiply. In eutrophic waters, especially those with levels of phosphorus greater than 0.01 mg/l and levels of ammonia or nitrate-nitrogen greater than 0.1 mg/l, they may multiply rapidly to form blooms (Rusin *et al.*, 2000). Although they are most noticeable when they form scums on the surface of waters, they normally live a planktonic existence and circulate at a depth that gives them their preferred level of light (Hunter, 1997).

The most common complaints related to the presence of cyanobacterial blooms in surface waters used as a raw source of drinking water are those of odour and taste. Cyanobacterial compounds causing off-tastes and odour include geosmin and 2-methylisoborneol (MIB) that can produce odours at levels as low as 1.3–10 and 6.3–29 ng/l, respectively (Young *et al.*, 1996). Over 20 species of cyanobacteria have been associated with adverse health effects. Acute health effects relate to the ability of several species to produce toxins that may cause liver damage, neural damage, and gastrointestinal (GI) disturbances (Rusin *et al.*, 2000). Cyanobacterial toxins are of three main types: lipopolysaccharide endotoxins, hepatotoxins and neurotoxins. It has been suggested that liver cancer may be a long-term chronic effect of the ingestion of cyanobacterial hepatotoxins in drinking water. Cyanobacterial toxins are reviewed by Carmichael (1992, 1994).

4 SURFACE WATER QUALITY AND DISEASE: THE CASE OF CHOLERA

Cholera is caused by the ingestion of the bacterium *Vibrio cholerae*. It is a disease of poverty and 'the crowded populations of the developing world's suburban slums ... still bear the burden of cholera disease' (Mintz *et al.*, 1998, cited in Gatrell, 2001). Over 100 serotypes of *V. cholerae* exist but only two (O1 and the recently identified O139) have been responsible for the major recorded epidemics. Cholera was the first disease to be demonstrably linked to the ingestion of contaminated water when, in 1854, Dr Jon Snow mapped the cases of cholera in the Soho district of London during that year's epidemic and implicated a particular groundwater pump in Broad Street. The water from this source was subsequently found to have been contaminated by faecal material from a leaking cesspool. The incidence of cholera in districts of London varied markedly depending on their levels of poverty and also on their source of drinking water. Cholera was particularly high in communities supplied with water by two companies that used the then grossly polluted River Thames as their source of raw water. It was not until the 1880s that Robert Koch identified the bacterium *Vibrio cholerae* as the causative organism of the disease.

The infectious dose of cholera is normally high (10^6–10^8 organisms) but may be as low as 10^3 organisms where gastric acidity is neutralized. Its pathogenicity results from its production of a potent enterotoxin as it colonizes the lining of the gut (Hunter, 1997). The effect of this enterotoxin on the cells of the gut lining is to cause a massive loss of water and electrolytes into the gut, causing watery diarrhoea. The effect can be life threatening within hours as a result of complications such as renal and cardiac failure resulting from rapid loss of body fluid.

Man is the only known host of *Vibrio cholerae* and rapid person-to-person transmission may occur in the unhygienic conditions that are often found in low-income urban communities. However, the key to understanding its spread throughout the world in pandemics may lie in its survival characteristics in water. West (1989) reviewed the distribution, survival and ecology of these organisms in the aquatic environment.

Hughes *et al.* (1982) studied the role of water in the transmission of *V. cholerae* in a rural area of Bangladesh. They demonstrated that surface waters in close proximity to communities infected with cholera were frequently contaminated with *V. cholerae*. In neighbourhoods where there were people infected with cholera, 44% of surface water sources were contaminated with *V. cholerae*, whereas only 2% of surface sources were positive in areas where there were no clinical infections. Samples from canals, rivers and storage tanks were more frequently positive for *V. cholerae* than were samples from water wells.

An investigation carried out in Indonesia during a cholera outbreak reported the presence of *V. cholerae* in all samples of water taken from a river located near the place where the outbreak occurred. In contrast no samples of water taken from other rivers adjacent to areas that were free of cholera were contaminated by *V. cholerae* (Glass *et al.*, 1984).

The water-borne transmission of cholera as a result of surface water contamination has also been documented in South Africa (Sinclair, 1982), Bolivia (Gonçalves *et al.*, 1992, cited by Tauxe *et al.*, 1994), El Salvador (Tauxe *et al.*, 1994), Argentina (Mazzafero *et al.*, 1995), Nepal (Pokharel *et al.*, 1996) and Burundi (Birmingham *et al.*, 1997).

Until fairly recently it was thought that humans were the only reservoir of *V. cholerae*. Although these organisms were isolated from many different water environments their presence was always related to pollution by contaminated faeces. This implied that contaminated water required continual recontamination with faeces from infected individuals. This assumption fitted with the results of studies that suggested that survival of *V. cholerae* in water was relatively short. However, the standard methods used in these studies were based on clinical practice and did not take into consideration the adaptation that the bacterium may undergo in conditions of low nutrients and variations in parameters such as pH and temperature.

During the last three decades, the use of novel techniques, such as fluorescent antibodies and DNA probes, have demonstrated that *V. cholerae* may be present in water in cholera-free areas (Table 36.2), thus demonstrating that water may be a natural reservoir in which the organism may multiply. The problem is compounded by the ability of *V. cholerae* to survive in a non-culturable state, in which they nevertheless remain viable and pathogenic (Dixo, 1994).

Miller *et al.* (1985) established a model for the transmission of cholera that recognized the hypothesis of an aquatic reservoir for *V. cholerae* and which incorporated two forms of transmission: primary and secondary. If a natural aquatic reservoir of *V. cholerae* is used as a source of drinking water, then the water may be a route for the primary transmission of cholera. The secondary transmission of cholera would be related to the use of faecally-contaminated water.

5 PROTECTION OF SURFACE WATERS FROM FAECAL POLLUTION

Prevention of water-borne infection is dependent on a 'multiple barrier' approach that involves preventing or minimizing the passage of pathogenic organisms at several stages in the environmental transport of water. Actions

TABLE 36.2 Isolation of *V. cholerae* from surface waters in locations with no registered cholera cases

Country	*Year*	*Water source implicated*
UK	1976/79	Ponds, ditches and canals
USA	1978	Canals and lakes
USA	1977/80	Bay, sewers
Italy	1979	Lakes
Australia	1977	Rivers
Australia	1977/88	Rivers
USA	1985	Rivers, creeks and canals
Malaysia	1990	River
Bangladesh	1988	Ponds

Source: Dixo, 1994.

include water source protection, treatment of drinking water, and treatment and safe disposal of wastewaters. In the long term, water resource management should focus on pollution prevention from point sources of waste discharges and the spread of pathogens from diffuse sources.

Where a surface water is considered to be a suitable source of raw drinking water, a full assessment should be carried out of pollution sources and how these might be minimized. In upland streams faecal pollution is likely to be limited under most conditions and the major sources of any pollution easy to ascertain. However, in many rivers, faecal pollution may be high and the sources less easy to assess.

Rivers often act as a conduit for the indirect reuse of treated and untreated wastewater. Abstraction of water for potable purposes downstream of a wastewater discharge presents a potential risk to human health. The risk is minimized by application of the multiple barrier approach to break the 'faeco-oral route'. Hazard Analysis Critical Control Points (HACCP) is a preventive management system used in the food industry. Important steps in every HACCP procedure are:

(i) setting up and verification of the process flow
(ii) executing the hazard analysis (i.e. hazard identification and risk analysis) and defining the control measures
(iii) identification of the critical control points (CCPs), (defining standards and critical limits and establishing a monitoring system
(iv) establishing corrective actions (Dewettinck *et al.*, 2001).

6 IDENTIFICATION OF FAECAL POLLUTION SOURCES IN SURFACE WATERS

Given the importance of surface water as a source of raw drinking water, tracking the source of faecal pollutants in surface water catchments is clearly important to human health. As improvements are made to the treatment of wastewaters discharging to surface waters, especially in the developed world, non-point sources are often the major source of faecal contamination of surface waters. Successful faecal source tracking within a catchment may lead to a sustainable means of protecting drinking water sources and recreational water bodies. The most significant sources of enteric organisms in river water used as a source of drinking water may be assessed by studying the impact of known point sources of wastewaters and studying land use and rainfall patterns in the watershed. Fisher *et al.* (2000) related data sets representing surface water quality at selected sites in a watershed to the predominant land use in that portion of the watershed. The study identified agricultural impacts and areas of focus for future mitigation measures.

Catchment studies may indicate impact of diffuse sources of faecal material. One study of the Buffalo River watershed in the USA (Pettibone and Irvine, 1996) suggested that fluctuations and relative magnitude of indicator bacteria levels were similar in the Buffalo River and at upstream sites in the absence of combined sewer overflow events. Bacteria densities were greatest during storm events, suggesting run-off as an important source pathway (Irvine and Pettibone, 1996). Total suspended solids were strongly correlated with faecal coliform levels in the upper watershed during the summer months when flow velocities were greatest and solids were thought to play an important role in transporting bacteria into the River Buffalo. Similar results were obtained in another American study of the Rouge River in Michigan (Murray *et al.*, 2001). The mean levels of faecal coliforms and faecal streptococci showed no correlation with combined sewer overflow locations and levels were often greater at sites upstream. The highest faecal coliform levels coincided with rainfall events.

Venter *et al.* (1997) determined the water and land use in a South African peri-urban catchment experiencing microbial water quality problems. The main sources of pollution were identified and the effects of dilution and bacterial die-off on water quality were evaluated by modelling the level of faecal coliforms along the length of the river using the QUAL2E model. Assessment of indicator organism and pathogen analyses indicated that the main factors affecting the microbial quality were discharges from wastewater treatment plans and run-off from informal settlement areas. The industrial activities in the catchment did not have a major effect. Geographic information systems (GIS) based transport models have been applied to watershed management studies and have shown potential to predict the relative contribution of diverse livestock operations (Fraser *et al.*, 1998).

Monitoring levels of traditional faecal indicator bacteria gives only limited information on the potential source of the pollution and research during the past 30 years has sought to find effective methods of distinguishing human from non-human faecal pollution in water. Investigations into the use of microorganisms for faecal source identification have tended to involve four basic approaches:

(i) isolation and identification of species that may be host specific
(ii) ratios between various faecal bacterial groups or species
(iii) phenotypic analysis of mixed microbial communities and comparison with those found in the faeces of humans and various animals
(iv) genotypic identification of faecal microorganisms.

(i) Ideally, we would wish to enumerate groups or species of faecal bacteria that have high host specificity. Suggested microbial groups for this purpose have included:

Streptococcus bovis (Geldreich, 1976; Oragui, 1982)
Rhodococcus coprophilus (Mara and Oragui, 1981)
Sorbitol-fermenting bifidobacteria (Mara and Oragui, 1983)
Enterococcus faecalis (Mead, 1964)
Clostridium perfringens (Sorensen *et al.*, 1989)
Pseudomonas aeruginosa (Oragui, 1982)
The *Bacteroides fragilis* group (Kreader, 1995)
Male-specific RNA coliphages (FRNA phage) (Havelaar and Hogeboom, 1984)
Bacteroides fragilis phage (Tartera and Jofre, 1987)
Bovine enteroviruses (Ley *et al.*, 2002).

Each group has its particular limitations as a source identifier such as short survival times in aquatic environments and significant geographical differences in host specificity. However, enumerating a selection of microbial groups may provide more accurate information on possible sources.

(ii) One early proposal for distinguishing animal from human faecal contamination of water was to calculate the ratio of the two major faecal indicator groups, namely, the faecal coliforms and the faecal streptococci (FC:FS). Geldreich and Kenner (1969) enumerated faecal coliforms and faecal streptococci in water and faecal samples from a wide range of sources including many animals and birds. They suggested that an FC:FS ratios of four or higher was indicative of human faecal pollution. A ratio of 0.7 or less was considered indicative of animal pollution (Geldreich, 1976). Although this approach is simple to perform, its limitations soon became clear. McFeters *et al.* (1974) pointed out that the marked difference in survival rates of the two bacterial groups invalidated the use of the ratio as a source tracking technique for faecal wastes that had discharged more than 24 hours before samples were taken. The survival of faecal streptococci is also variable with *Streptococcus bovis* and *S. equinus* dying off more rapidly than *S. faecalis* and *S. faecium*. The latter two organisms are now identified as members of the *Enterococcus* genus and 'intestinal enterococci' are now considered a better faecal indicator than the broader faecal streptococcus group.

Other ratios that have been proposed as tools in faecal source tracking have included *Streptoccus bovis* to faecal streptococci and *Rhodococcus coprophilus* to total actinomycetes, and sorbitol-fermenting bifidobacteria to total bifidobacteria (Oragui, 1982).

(iii) Microbial source tracking techniques aim to study mixed microbial communities and identify those that are most likely to have derived from a particular host. However, source tracking is made more difficult because, at species level at least, host specificity among many faecal microorganisms is limited. For example, isolates of intestinal enterococci from water samples may be identified to species level. Table 36.3 outlines the findings of work by Devriese *et al.* (1987) suggesting the predominant sources of each species of the genus *Streptococcus*. However, as host specificity is low this approach may give misleading results.

Although the host specificity of individual species may be very limited, certain subspecies or biotypes often predominate in the faeces of particular animals. Comparison of the highly mixed microbial populations found in faecally polluted surface waters with those found in the faeces of different animals is potentially time consuming as many isolates must be analysed to achieve a representative sample size. Kühn *et al.* (1997) used a phenotypic analysis approach using a method based on the measurement of the kinetics of biochemical tests performed in microplates to type coliform bacteria isolated from a polluted river in Sweden and from the wastewater outlets of three paper mills that were suspected contamination sources. A population similarity coefficient was used to interpret the data from over 1000 isolates. The phenotypes isolated from the factory effluents were rarely recovered from river water, which was also reflected in low similarities between bacterial populations in river water and factory outlets. In contrast, bacterial populations from sampling points close to each other were more similar to each other indicating the presence of several, diffuse contamination sources, possibly from animal or human faecal material.

TABLE 36.3 Species of the genera *Enterococcus* and *Streptococcus* of faecal origin

Species	*Intestinal origin*
Enterococcus	
faecium	humans, cattle, pigs, birds
faecalis	humans, cattle, pigs, birds
durans	humans, pigs, birds
hirae	humans, pig, birds
avium	humans, cattle, pigs, birds
gallinarum	humans, birds
cecorum	cattle, pigs, birds
columbae	cattle, pigs, birds
Streptococcus	
bovis	humans, cattle pigs
equines	humans, cattle, pigs
alactolyticus	pigs, birds
hyointestinalis	pigs
intestinalis	pigs
suis	pigs

After Devriese *et al.* (1987); Anon., (1995).

In the more grossly polluted river waters of catchments receiving many potential sources of faecal pollution under highly variable meteorological conditions, the phenotypic analysis of indicator bacteria in river water and human and animal faeces gives less clear results (Taylor and Wallis, 2001). However, Kühn's simplified method of phenotyping large numbers of enterococcal isolates proved useful in identifying specific biotypes that might be potential source identifiers (Wallis and Taylor 2002).

Another phenotypic approach is the use of discriminant analysis of patterns of antibiotic resistance in faecal bacteria. Wiggins (1996) analysed 1435 faecal streptococcus isolates from 17 samples of cattle, poultry, human and wild-animal wastes for their ability to grow on four concentrations of five antibiotics. An average of 74% of isolates were correctly classified into one of six possible sources and 92% of human isolates were correctly identified. Human and non-human isolates were correctly classified at an average rate of 95%.

Despite its successes, the multiple antibiotic resistance (MAR) methodology is currently time and labour intensive, requiring the isolation and analysis of a large library of sour isolates. Parveen *et al.* (1999) also note that antibiotic resistance patterns of bacteria are influenced by selective pressure and may thus show significant temporal and spatial variation.

Recent developments in genotyping methodologies potentially offer a high degree of discrimination between organisms of faecal origin in natural waters. One recent approach is to determine the distribution of subgroups (genotypes) of F-specific RNA bacteriophages in surface waters (Schaper and Jofre, 2000). DNA methods for the differentiation of *Giardia*, adenoviruses and enteroviruses have also been proposed. Genotypic characterization of *E. coli* populations in surface waters is also a potential tool for tracing faecal pollution (Farnleitner *et al.*, 2000; Parveen *et al.*, 2001).

An alternative to the analysis of microbial communities is the study of chemicals of faecal origin. Such studies have mainly concentrated on the distribution patterns of faecal sterol isomers (Leeming *et al.*, 1996), long-chain alkylbenzenes used in commercial detergents (Eganhouse, 1986) and caffeine (Standley *et al.*, 2000).

It seems that no single approach to source tracking of faecal pollution is useful in all situations, but multivariate statistical analysis of appropriate 'baskets' of microbial and chemical determinants may offer the possibility of identifying and apportioning human and animal faecal inputs to natural waters. However, accurate assessments of a catchment require very many samples to be taken under varying meteorological conditions. One promising approach to the analysis of such complex data sets from source tracking studies is the application of an artificial neural-network (ANN) classification scheme (Brion *et al.*, 2002).

DISCUSSION

In many environments river water is the most logical source of a community's drinking water. However, the hygienic quality of river water may be poor and temporally unpredictable. Low residence time reduces the impact of self-purification processes that lead to the die-off of pathogens and indicators in surface waters. It is important therefore that abstraction at the drinking water treatment plant is followed by a period of storage and sedimentation prior to further treatment. This allows time for the continued reduction in levels of enteric organisms by natural processes. Storage also allows for contamination peaks following heavy rainfall or accidental point discharges to be diluted prior to use.

Catchment protection plays an important role in the multiple barrier approach to breaking the faeco-oral route. Successful improvements to surface water quality management in the future may benefit from the application of novel management systems such as Hazard Analysis Critical Control Points (HACCP) and the use of recently developed microbiological and chemical methods to identify the sources of faecal contamination within particular surface water catchments.

REFERENCES

Anon. (1991). *Code of Good Agricultural Practice for the Protection of Water.* Ministry of Agriculture Fisheries and Food and the Welsh Office Agriculture Department, London.

Anon. (1995). *Performances of Methods for the Microbiological Examination of Bathing Water.* Part 1 EUR 16601 EN, DG XII, European Commission, Brussels.

Araujo, R., Lasobras, J., Puig, A. *et al.* (1997). Abundance of bacteriophages of enteric bacteria in different freshwater environments. *Water Science and Technology* **35**(11–12), 125–128.

Araujo, R., Puig, A., Lasobras, J. *et al.* (1997b). Phages of enteric bacteria in fresh water with different levels of faecal pollution. *Journal of Applied Microbiology* **82**(3), 281–286.

Baudart, J., Grabulos, J., Barusseau, J.P. and Lebaron, P. (2000). *Salmonella* spp. and fecal coliform loads in coastal waters from a point vs. nonpoint source of pollution. *Journal of Environmental Quality* **29**(1), 241–250.

Beaudeau, P., Tousset, N., Bruchon, F. *et al.* (2001). *In situ* measurement and statistical modelling of *Escherichia coli* decay in small rivers. *Water Research* **35**(13), 3168–3178.

Birmingham, M.E., Lee, L.A., Ndayimirije, N. *et al.* (1997). Epidemic cholera in Burundi: patterns of transmission in the Great Rift Valley Lake region. *Lancet* **349**(9057), 981–985.

Bonde, G.J. (1977). Bacterial indication of water pollution. *Advances in Aquatic Microbiology* **1**, 273–364.

Bordalo, A.A., Nilsumranchit, W. and Chalermwat, K. (2001). Water quality and uses of the Bagpakong River (Eastern Thailand). *Water Research* **35**(15), 3635–3642.

Brion, G.M., Neelakantan, T.R. and Lingireddy, S. (2002). A neural-network-based classification scheme for sorting sources and ages of fecal contamination in water. *Water Research* **36**, 3765–3774.

Carmichael, W.W. (1992). Cyanobacteria secondary metabolites – the cyanotoxins. *Journal of Applied Bacteriology* **72**, 445–459.

Carmichael, W.W. (1994). The toxins of cyanobacteria. *Scientific American* **270**, 78–86.

Crowther, J., Kay, D. and Wyer, M.D. (2002). Faecal-indicator concentrations in waters draining lowland pastoral catchments in the UK: relationship with land use and farming practices. *Water Research* **36**(7), 1725–1734.

DeRegnier, D., Cole, L., Shupp, D. and Erlandsen, S. (1989). Viability of *Giardia* cysts suspended in lake, river and tap water. *Applied and Environmental Microbiology* **55**, 1223–1229.

Devriese, L.A., Van de Kerckhove, A., Kilpper-Balz, R. and Schliefer, K.H. (1987). Characterisation and identification of *Enterococcus* species isolated from the intestines of animals. *International Journal of Systematic Bacteriology* **37**, 257–259.

Dewettinck, T., Van Houtte, E., Geenens, D. *et al.* (2001). HACCP (hazard analysis and critical control points) to guarantee safe water reuse and drinking water protection – a case study. *Water Science and Technology* **43**(12), 31–38.

Dixo, N.G.H. (1994). *The Removal of* V. cholerae *by Water Treatment Systems in Developing Countries*. MSc Thesis. Imperial College of Science, Technology and Medicine. London.

Eganhouse, R.P. (1986). Long-chain benzenes: their analytical chemistry, environmental occurrence and fate. *International Journal of Environmental Chemistry* **26**, 241–263.

Ellis, J.B. (1993). Achieving standards for recreational use of urban receiving waters. In: D. Kay and R. Hanbury (eds) *Recreational Water Quality Management. Volume 2 Fresh Waters*, Ellis Horwood, Chichester.

Farnleiter, A.H., Kreuzinger, N., Kavka, G.G. *et al.* (2000). Simultaneous detection and differentiation of *Escherichia coli* populations from environmental freshwaters by means of sequence variations in a fragment of the beta-D-glucuronidase gene. *Applied and Environmental Microbiology* **66**(4), 1340–1346.

Fisher, D.S., Steiner, J.L., Endale, D.M. *et al.* (2000). The relationship of land use practices to surface water quality in the Upper Oconee watershed of Georgia. *Forest Ecology and Management* **128**(1–2), 39–48.

Fraser, R.H., Barten, P.K. and Pinney, D.A.K. (1998). Predicting stream pathogen loading from livestock using a geographical information system-based delivery model. *Journal of Environmental Quality* **27**(4), 935–945.

Gatrell, A.C. (2001). Geographies of Health. In: *Water Quality and Health*. Chapter 8. Blackwell, Oxford.

Geldreich, E.E. (1972). Buffalo Lake recreational water quality: a study of bacteriological data interpretation. *Water Research* **6**, 913–924.

Geldreich, E.E. (1976). Faecal coliform and faecal streptococcus density relationships in waste discharge and receiving water. *CRC Critical Reviews in Environmental Control* **6**, 349–369.

Geldreich, E.E. (1990). Microbiological quality of source waters for water supply. In: G.A. McFeters (ed.) *Drinking Water Microbiology Progress and Recent Developments*, pp. 3–31. Springer Verlag, New York.

Geldreich, E.E. and Kenner, B.A. (1969). Concepts of faecal streptococci in stream pollution. *Journal of Water Pollution Control Federation* **41**, R336–R352.

George, I., Petit, M., Theate, C. and Servais, P. (2001). Distribution of coliforms in the Seine River and estuary (France) studies by enzymatic methods and plate counts. *Estuaries* **24**(6B), 994–1002.

Gibbs, R.A., Hu, C.J., Ho, G.E. and Unkovich, I. (1997). Regrowth of faecal coliforms and salmonellae in stored biosolids and soil amended with biosolids. *Water Science and Technology* **35**(11–12), 269–275.

Glass, R.I., Alim, A., Eusof, A. *et al.* (1984). Cholera in Indonesia: epidemiologic studies of transmission in Aceh Province. *American Journal of Tropical Medicine and Hygiene* **33**(5), 933–939.

Godfree, A. (1993). Sources and fates of microbial contaminants. In: D. Kay and R. Hanbury (eds) *Recreational Water Quality Management Volume 2 Fresh Waters*, Ellis Horwood, Chichester.

Gray, N.F. (1999). *Water Technology. An Introduction for Environmental Scientists and Engineers*. Arnold, London.

Havelaar, A.H. and Hogeboom, W.M. (1984). A method for the enumeration of male-specific bacteriophages in sewage. *Journal of Applied Bacteriology* **69**, 30–37.

Hjelmar, O., Johannesson, L.M., Knox, K. *et al.* (1995) Composition and management of leachate from landfills within the EU. In: T.H. Christensen, R. Cossu, R Stegmann *et al.* (eds.) *Sardinia 95, Fifth International Landfill Symposium (Cagliari, Sardinia, October 1995)*, CISA-Environmental Sanitary Engineering Centre Cagliari.

Hughes, J.M., Boyce, J.M., Levine, R.J. *et al.* (1982 Epidemiology of El Tor cholera in rural Bangladesh importance of surface water in transmission. *Bulletin of the World Health Organization* **60**(3), 395–404.

Hunter, P.R. (1997). *Waterborne Disease: Epidemiology and Ecology*. John Wiley, Chichester.

Hurst, C.J., Benton, W.H. and McClellan, K.A. (1989 Thermal and water source effects upon the stability of enteroviruses in surface freshwaters. *Canadian Journal of Microbiology* **35**, 474–480.

Irvine, K.N. and Pettibone, G.W. (1996). Planning level evaluation of densities and sources of indicator bacteria in a mixed land use watershed. *Environmental Technology* **17**(1), 1–12.

Jagals, P. (1997). Stormwater run-off from typical developed and developing South African urban developments: definitely not for swimming. *Water Science and Technology* **35**(11–12), 133–140.

Jones, F. and White, W.R. (1984). Health and amenity aspects of surface waters. *Water Pollution Control* **83**, 215–225.

Kistemann, T., Classen, T., Koch, C. *et al.* (2002). Microbial load of drinking water reservoir tributaries during extreme rainfall and runoff. *Applied and Environmental Microbiology* **68**(5), 2188–2197.

Kreader, C. (1995). Design and evaluation of *Bacteroides* DNA probes for the scientific detection of human faecal pollution. *Applied and Environmental Microbiology* **61**(4), 1171–1179.

Kühn, I., Allestram, G., Engdahl, M. and Stenstrom, T.A. (1997). Biochemical fingerprinting of coliform bacterial populations – comparisons between polluted river water and factory effluents. *Water Science and Technology* **35**(11–12), 343–350.

Leeming, R., Ball, A., Ashbolt, N. and Nichols, P. (1996). Using fecal sterols from humans and animals to distinguish faecal pollution in surface waters. *Water Research* **30**(12), 2893–2900.

Ley, V., Higgins, J. and Fayer, R. (2002). Bovine enteroviruses of fecal contamination. *Applied and Environmental Microbiology* **68**(7), 3455–3461.

Mara, D.D. and Oragui, J.I. (1983). Sorbitol-fermenting bifidobacteria as specific indicators of human faecal pollution. *Journal of Applied Bacteriology* **55**, 349–357.

Margolin, A.B. (1997). Control of microorganisms in source water and drinking water. In: C.J. Hurst, G.R. Knudsen and M.J. McInerney *et al.* (eds) *Manual of Environmental Microbiology*. American Society for Microbiology, ASM Press, Washington DC.

Mazzafero, V.E., Wyszynski, D.F., Marconi, E. and Giacomini, H. (1995). Epidemic of cholera among the aborigines of northern Argentina. *Journal of Diarrhoeal Diseases Research* **13**(2), 95–98.

McFeters, G., Bissonnette, G., Jezeski, J. *et al.* (1974). Comparative survival of indicator bacteria and enteric pathogens in well-water. *Applied Microbiology* **27**, 823–829.

McFeters, G.A. (1990). Enumeration, occurrence and significance of injured bacteria in drinking water. In: G.A. McFeters (ed.) *Drinking Water Microbiology*, pp. 478–492. Springer-Verlag, New York.

McFeters, G.A. and Singh, A. (1991). Effect of aquatic environmental stress on enteric bacterial pathogens. *Journal of Applied Bacteriology, Symposium Supplement* **70**, 115S–120S.

Mead, G.C. (1964). Isolation and significance of *Streptococcus faecalis, sensu strictu*. *Nature* **204**, 1224–1225.

Miller, C.J., Feachem, R.G. and Drasar, B.S. (1985). Cholera epidemiology in developed and developing countries: new thoughts on transmission, seasonality and control. *Lancet* **1**(8423), 261–263.

Momba, M.N.B. and Kaleni, P. (2002). Regrowth and survival of indicator microorganisms on the surfaces of household containers used for the storage of drinking water in rural communities of South Africa. *Water Research* **36**(12), 3023–3028.

Murray, K.S., Fisher, L.E., Therrien, J. *et al.* (2001). Assessment and use of indicator bacteria to determine sources of pollution to an urban river. *Journal of Great Lakes Research* **27**(2), 220–229.

Niemi, J.S., Niemi, R.M., Malin, V. and Poikolainen, M.L. (1997). Bacteriological quality of Finnish rivers and lakes. *Environmental Toxicology and Water Quality* **12**(1), 15–21.

Obiri-Danso, K. and Jones, K. (1999). Distribution and seasonality of microbial indicators and thermophilic campylobacters in two freshwater bathing sites on the River Lune in northwest England. *Journal of Applied Microbiology* **87**(6), 822–832.

Oragui, J.I. (1982). *Bacteriological methods for the distinction between human and animal faecal pollution*. PhD Thesis, University of Leeds, UK.

Parveen, S., Portier, K.M., Robinson, K. *et al.* (1999). Discriminant analysis of ribotype profiles of *Escherichia coli* for differentiating human and nonhuman sources of fecal pollution. *Applied and Environmental Microbiology* **65**(7), 3142–3147.

Parveen, S., Hodge, N.C., Stall, R.E. *et al.* (2001). Phenotypic and genotypic characterization of human and nonhuman *Escherichia coli*. *Water Research* **35**(2), 379–386.

Pepper, I.L., Josephson, K.L., Bailey, R.L. *et al.* (1993). Survival of indicator organisms in Sonoran Desert soil amended with sewage-sludge. *Journal of Environmental Science and Health Part A: Environmental Science and Engineering & Toxic and Hazardous Substance Control* **28**(6), 1287–1302.

Pettibone, G.W. and Irvine, K.N. (1996). Levels and sources of indicator bacteria associated with the Buffalo River 'Area of Concern', Buffalo, New York. *Journal of Great Lakes Research* **22**(4), 896–905.

Pokharel, B.M., Rawal, S., Shrestha, R.S. and Dhakwa, J.R. (1996). Outbreaks of cholera in Kathmandu valley in Nepal. *Journal of Tropical Pediatrics* **42**(5), 305–307.

Polo, F., Figueras, M.J., Inza, I. *et al.* (1998). Relationship between presence of salmonella and indicators of faecal pollution in aquatic habitats. *FEMS Microbiology Letters* **160**(2), 253–256.

Rao, V.C. and Melnick, J.L. (1986). *Environmental Virology*. Van Nostrand Reinhold, Wokingham.

Reinert, P.E. and Hroncich, J.A. (1990). Source water quality management. In: F.W. Pontius (ed.) *Water Quality and Treatment: a Handbook of Community Water Supplies*, pp. 189–268. McGraw-Hill, New York.

Rusin, P., Enriquez, C.E., Johnson, D. and Gerba, C.P. (2000). Environmentally transmitted pathogens. In: R.M. Maier, I.L. Pepper and C.P. Gerba (eds) *Environmental Microbiology*. Academic Press, San Diego.

Schaper, M. and Jofre, J. (2000). Comparison of methods for detecting genotypes of F-specific RNA bacteriophages and fingerprinting the origin of faecal pollution in surface water. *Journal of Virological Methods* **89**, 1–10.

Sinclair, G.S., Mphalele, M., Duvenhage, H. *et al.* (1982). Determination of the modes of transmission of cholera in Lebowa – an epidemiological investigation. *South African Medical Journal* **62**(21), 753–755.

Sorensen, D., Eberl, S. and Dicksa, R. (1989). *Clostridium perfringens* as a point source indicator in non-polluted streams. *Water Resources* **23**(2), 191–197.

Standley, L.J., Kaplan, L.A. and Smith, D. (2000). Molecular tracer of organic matter sources to surface water resources. *Environmental Science and Technology* **34**(15), 3124–3130.

Tartera, C. and Jofre, J. (1987). Bacteriophages active against *Bacteroides fragilis* in sewage-polluted waters. *Applied and Environmental Microbiology* **53**(7), 1632–1637.

Tauxe, R., Seminario, L., Tapia, R. and Libel, M. (1994). The Latin American epidemic. In: I. Kay Wachsmuth, P.A. Blake and O. Olsvik (eds) *Vibrio cholerae and Cholera: Molecular to Global Perspectives*. American Society for Microbiology, Washington DC.

Taylor, H.D. and Wallis, J.L. (2001). *Sources of Faecal Pollution on the Fylde Coast*. Report prepared for the UK Department for the Environment, Transport and the Regions.

Venter, S.N., Steynberg, M.C., deWet, C.M.E. *et al.* (1997). A situational analysis of the microbial water quality in a peri-urban catchment in South Africa. *Water Science and Technology* **35**(11–12), 119–124.

Wallis, J.L. and Taylor, H.D. (2002) Phenotypic population characteristics of the enterococci in human and animal faeces: implications for the new European directive on the quality of bathing waters. Paper presented at the Xth IWA Symposium on Health-related Water Microbiology, Melbourne, Australia, April.

Ward, R.L., Knowlton, D.R. and Winston, P.E. (1986). Mechanism of inactivation of enteric viruses in fresh water. *Applied and Environmental Microbiology* **52**, 450–459.

West, P.A. (1989). The human pathogenic vibrios – a public health update with environmental perspectives. *Epidemiology and Infection* **103**, 1–34.

West, P.A. (1991). Human pathogenic viruses and parasites: emerging pathogens in the water cycle. *Journal of applied Bacteriology Symposium Supplement* **70**, 107S–114S.

White, W.R. and Godfree, A.F. (1985). Pollution of freshwater and estuaries. *Journal of Applied Microbiology Supplement*, 67S–79S.

Wiggins, B.A. (1996). Discriminant analysis of antibiotic resistance patterns in fecal streptococci, a method to differentiate human and animal sources of fecal pollution in natural waters. *Applied and Environmental Microbiology* **62**(11), 3997–4002.

Williams, P.T. (1998). *Waste Treatment and Disposal*. John Wiley & Sons, Chichester.

Young, W.F., Horth, H., Crane, R. *et al.* (1996). Taste and odor threshold concentrations of potential potable water contaminants. *Water Research* **30**, 331–340.

37

Stored water (rainjars and raintanks)

John Pinfold

Brighton Marina Village, Brighton BN2 5XJ, UK

Storing water for domestic usage is a common feature in developing countries. There is generally a need to store water when people have to carry water from the source to their home. Even where the level of water supply service involves piped supplies with house connections, water storage may be necessary because supplies are often intermittent. Containers used to store water vary widely in size, shape and material and may include clay/cement jars, concrete tanks, metal drums, or even the plastic jerrycans actually used for carrying the water.

Although rain tends to be seasonal, rainwater can provide a convenient alternative water source when rains are abundant. For domestic water consumption, the most common method of rainwater harvesting is by channelling water from the roof into a large vessel. This generally entails a zinc sheet or tiled roof with some form of guttering (however primitive), but people will collect rainwater by any means available (Fig. 37.1). Grass roofs tend to discolour the water making it less attractive for drinking purposes but acceptable

. **37.1** Rainwater collection from roof to metal drum in rural Uganda.

Handbook of Water and Wastewater Microbiology
N 0-12-470100-0

Fig. 37.2 A typical rainjar with a tap in northeast Thailand.

for certain washing activities. Rainwater is especially valued as a drinking water source when alternative drinking water sources are of poor physical (e.g. groundwater salinity) or bacteriological quality (particularly surface waters).

By far the most common route of contamination of stored water is through water handling. Usually contamination occurs when water is regularly obtained by using a 'dipper' (often a plastic/metal bowl or gourd). Hands are in regular contact with the local surroundings and act as a potential conduit for transferring microorganisms from contaminated sites within the home (and compound) to the stored water, either via the 'dipper' or through direct contact with the water (Pinfold, 1990). Consequently, the protective features of the water container become a major factor in influencing the microbiological quality of stored water. Table 37.1 illustrates this point by comparing water from large rainjars where water is drawn through taps located near the base (Figure 37.2), with water containers where water is obtained by a dipper.

Water may be stored in just one vessel usec for all domestic activities, but in many case drinking water will be stored separately fron that used for washing/cleaning activities. Eve water used for washing/cleaning activitie may be stored in different vessels convenientl located close to where the activity takes plac Stored water can act as a reservoir for micrc organisms, as cross-contamination occur during the water-related activity. Althoug some studies have suggested that increase faecal contamination of stored water may b the result of bacterial growth (Black *et al.*, 198

TABLE 37.1 Faecal contamination of stored ra water by type of container

Type of container	*Number of samples*	*Geometric mean of* E. Coli *(150 ml)*	*Geometric mean of* F. Streptoco *(150 ml)*
Rainjars with tap	196	1.8	12.1
Other containers	1,120	7.5	26.1

Source: Pinfold (1989).

Kirchoff *et al.*, 1985), the majority of evidence suggests that the main cause of contamination to stored water is through mechanical means (Feachem *et al.*, 1978; Khairy *et al.*, 1982; Attar *et al.*, 1982). Furthermore, as Table 37.2 illustrates, certain water-related activities are likely to lead to greater levels of contamination than others. Here, water used for washing dishes and food preparation activities was nearly as highly contaminated as that used for anal cleansing, while activities such as drinking and washing clothes resulted in much lower levels of contamination. As cross-contamination occurs then the level of contamination that reaches stored water through water handling only reflects a small proportion of the amount of contamination occurring in the local environment. Thus the high level of contamination associated with the food-related activities and the consequent danger of consuming contaminated foods is of particular concern when considering faeco-oral disease transmission. Studies in Zimbabwe (Simango *et al.*, 1992), in Myanmar (Oo *et al.*, 1991), in Liberia (Molbak *et al.*, 1989) and in Bangladesh (Henry, 1990) also suggest a greater potential health risk from food handling than from handling stored drinking water.

Rain itself is free from bacteria. Contamination levels of stored rainwater are generally low where containers are well protected from water handling, such as with the tap on a large rainjar. The presence of a lid may offer some additional protection but this was not found to be a significant factor in the Thailand study, although householders felt a lid was important in protecting water from dust. Further improvements to water quality were apparent in rainjars fitted with fine mosquito netting (Table 37.3). Protecting the rainjar with nets is primarily to prevent mosquito breeding in the stored water (particularly important in areas where dengue and malaria are endemic) rather than preventing bacterial contamination. However, it is likely that a side effect of the netting prevents access and subsequent faecal pollution from small lizards. These geckoes were frequently found in rainjars without nets as they were able to squeeze underneath the rainjar lid and appeared to like the moist warm air and safety offered. Although cold-blooded reptiles, their faecal pellets were found to contain *E. coli* and faecal streptococci, albeit in much smaller numbers than from warm-blooded mammals (Pinfold *et al.*, 1993).

Other more obvious sources of contamination to collected runoff from roofs and gutters are birds and possibly small rodents. Koplan *et al.* (1978) postulated roof-collected rainwater as a possible cause of an outbreak of salmonellosis in Trinidad, West Indies. Appan (1997) concluded that water contamination of roof-collected rainwater was largely animal in origin given the high faecal coliform/faecal streptococci ratio from a number of separate studies in Southeast Asian countries. Some of these studies have also reported the presence of *Salmonella* (Wirojanagud *et al.*, 1989; Fujioka *et al.*, 1991) and there are many other references to pathogens, including *Clostridium perfringens, Campylobacter, Cryptosporidium* and *Giardia*, having been isolated from rainwater samples

TABLE 37.2 Stored water contamination by water use

Water use	*Number of samples*	*Geometric means of bacteria (150 ml)*	
		E. Coli	F. Streptococci
Drinking only	106	4.2	15.0
Drinking and cooking	128	6.7	18.5
Washing dishes/ cleaning food	70	37.2	68.2
Washing clothes	17	4.6	31.7
Bathing	58	22.4	43.4
Toilet	75	44.2	42.7

Source: Pinfold (1989). This table also includes water sources other than rain, but overall contamination of these sources was less than with rainwater.

TABLE 37.3 *E. coli* contamination of rainjars with and without nets

Rainjars	*Number of samples*	*Geometric mean count of E. Coli (per 100 ml)*
Without nets	86	1.6
With nets	40	0.5*

*$P < 0.05$ (Mann-Whitney Z value = −2.2).
Source: Pinfold *et al.* (1993).

(Fujioka *et al.*, 1991; Brodribb *et al.*, 1995; Crabtree *et al.*, 1996). However, apart from the Koplan study, evidence of actual disease outbreak resulting from drinking roof-collected rainwater polluted by pathogens is extremely rare.

Under normal circumstances bacteria would be expected to die off naturally in stored water. Factors that affect the die-off rate include temperature, availability of nutrients and solar radiation. Direct sunlight increases the die-off rate and solar water disinfection (SODIS) has been proposed as an appropriate water treatment method for stored drinking water (Appan, 1997; Wegelin and Sommer, 1998). However, in order for the process to achieve disinfection without encouraging algal growth, it does involve a certain amount of additional effort and expense for the users. As the temperature rises, so the bacterial die-off rate increases (so long as nutrients are absent), but generally inhabitants from warm/hot climates prefer to drink cool water and tend to decant drinking water into vessels located away from the sun (often prefering a slightly porous pot where evaporation induces cooling). Despite this, Tompkins and Drasar (1978), in rural Nigeria, found that overnight storage in earthenware containers led to a marked fall in coliforms; presumably there was no water taken at night thus allowing the natural process of bacterial die-off. Others in Bangladesh (Deb *et al.*, 1986), Bolivia (Quick *et al.*, 1996) and Guatemala (Sobel *et al.*, 1998) developed narrowed necked vessels for drinking water in order to eliminate recontamination of the drinking water through water handling.

Bacterial growth may occur when water is physically 'dirty' and bacteria have sufficient nutrients to multiply. This has particular significance to warm/hot climates as many pathogenic bacteria also require a high ambient temperature for regrowth. Regrowth of *E. coli* in water has been reported when associated with rotting vegetation at elevated temperatures (Taylor, 1972). Water used for soaking dishes in Northeast Thailand has also been shown to provide a favourable environment for growth for both *E. coli* and faecal streptococci resulting in very high levels of contamination (Pinfold, 1990). However, these studies refer to water that is visibly dirty and there is no clear evidence found to suggest bacterial growth in visibly clean drinking water, although Ahmed *et al.* (1998), in Bangladesh, did find evidence of bacterial growth on the internal surface at the base of storage containers. Presumably sedimentation of small amounts of food particles entering the water during handling would lead to a build-up of nutrients in the base of the water container.

The individual nature of rainwater collection makes public health control measures extremely difficult. Gould (1999) describes a variety of methods for improving rainwater quality but suggests that good system design as well as operation and maintenance are the simplest and most effective means of ensuring good water quality, while water treatment (including on site chlorination, boiling water, filtration and SODIS) is mainly appropriate as a remedial action if contamination is suspected. The type of methods for roof-collected rainwater that are adopted will depend on circumstances within different countries. In Thailand, gradual improvements to the design of rainjars (see Fig. 37.2) over a number of years, affecting shape, size, material, presence of tap and drainage plug, has helped increase their popularity and improve water quality (Tunyavanich and Hewison, 1989). As seen from the above tables, the introduction of a tap together with an increase in the height of the rainjars has led to a dramatic impact on water quality. In northeast Thailand, it is also common for householders to protect stored water by allowing the first rains to clean away the dust/dirt that accumulates during the dry season before water is channelled into the rainjar. Attempts to introduce water treatment methods, such as on-site chlorination, have met with strong resistance here as they are very particular about the taste of the drinking water. Even in Australia many residents who have to rely on collecting rainwater for drinking prefer the taste of this to that of the local town (piped) water (Noosa Shire Environmental Health Officers, 1994; Thurman, 1995).

Much is made of the fact that roof-collected rainwater cannot always meet the WHO guidelines. When considering appropriate measures to improve drinking water quality it is crucial to understand the relative risk of disease transmission from contaminated drinking water in relation to other faeco-oral transmission routes. In countries where diarrhoeal disease is common, but low levels of contamination exist in drinking water, then further efforts to improve drinking water quality may not be the most cost-effective way to reduce disease transmission. Moe *et al.* (1991), in the Philippines, provide evidence to support this by showing the incidence of diarrhoea in young children to be significantly related to drinking water containing high levels (>1000 *E. coli*/100 ml of contamination, but little difference was observed between the illness rates of children using either good quality drinking water (<1 *E. coli*/100 ml) or moderately contaminated drinking water (2–100 *E. coli*/100 ml). Evidence suggests that improving hygiene behaviour and sanitation are often neglected, but are nevertheless just as important as improving drinking water quality in the fight against disease (Esrey, 1991).

Rainwater collection is practised by individuals in almost every country in the world (Lye, 1992). In developed countries, where diarrhoeal disease is much less common and public health standards higher, then more sophisticated methods for protecting rainwater, including water treatment, may be appropriate. However, although more sophisticated systems are available such as the first-flush device (which automatically does the same as is done manually by users in northeast Thailand), there is little evidence of routine water treatment being necessary except as a remedial action. Gould (1999) quotes the official line from the Australian government (National Environmental Health Forum) as summarized by Cunliffe (1998): 'Providing the rainwater is clear, has little taste or smell and is from a well maintained system, it is probably safe and unlikely to cause any illness for most users'.

If roof-collected rainwater is found to be a serious threat to health, then any proposed public health control measure should first assess whether it is likely to be effective. Changing the behaviour of water users is not an easy process, and it is important that any expected behaviour change requires very little extra effort or cost and that health messages are simple and few in number (Pinfold and Horan, 1996; Pinfold, 1999). It is advisable first to conduct 'behavioural trials' in order to assess their feasibility and subsequently assist in the development of a large-scale campaign.

REFERENCES

Ahmed, S.A., Hoque, B.A. and Mahmud, A. (1998). Water management practices in rural and urban homes: a case study from Bangladesh on ingestion of polluted water. *Public Health* **112**, 317–321.

Appan, A. (1997). Roof water collection systems in some Southeast Asian countries: status and water quality levels. *Journal of the Royal Society of Health* **117**, 319–323.

Attar, L.E., Gawad, A.A., Khairy, A.E.M. and Sebaie, O.E. (1982). The sanitary condition of rural drinking water in a Nile Delta village: II Bacterial contamination of drinking water in a Nile Delta village. *Journal of Hygiene* **88**, 63–67.

Black, R.E., Brown, K.H., Becker, S. *et al.* (1982). Contamination of weaning foods and transmission of enterotoxigenic *Escherichia coli* diarrhoea in children in rural Bangladesh. *Transactions of the Royal Society of Tropical Medicine and Hygiene* **76**, 259–264.

Brodribb, R., Webster, P. and Farrell, D. (1995). Recurrent *Campylobacter fetus* subspecies bacteraemia in a febrile neutropaenic patient linked to tank water. *Communicable Disease Intelligence* **19**, 312–313.

Crabtree, K., Ruskin, R., Shaw, S. and Rose, J. (1996). The detection of Cryptosporidium oocysts and Giardia cysts in cistern water in the U.S. Virgin Islands. *Water Research* **30**, 208–216.

Cunliffe, D. (1998). *Guidance on the Use of Rainwater Tanks.* National Environmental Health Forum Monographs, Water Series 3. Public and Environmental Health Service, Dept. of Human Services, PO Box 6, Rundle Mall SA 5000, Australia.

Deb, B.C., Sircar, B.K., Sengupta, P.G. *et al.* (1986). Studies on interventions to prevent eltor cholera transmission in urban slums. *Bulletin of the World Health Organization* **64**, 127–131.

Esrey, S.A., Potash, J.B., Roberts, L. and Shiff, C. (1991). Effects of improved water supply and sanitation on ascariasis, diarrhoea, dracunculiasis, hookworm infection, schistosomiasis, and trachoma. *Bulletin of World Health Organization* **69**, 609–621.

Feachem, R.G., Burns, E. and Cairncross, S. (1978). Water, health and development: An interdisciplinary evaluation. Tri-med, London.

Fujioka, R.S., Inserra, S.G. and Chinn, R.D. (1991). The bacterial content of cistern water in Tantalus area of Honolulu, Hawaii. *Proceedings of the 5th International Conference on RWCS, Keelung, Taiwan, 10 August*.

Gould, J. (1999). Is rainwater safe to drink? A review of recent findings. *Proceedings of the 9th International Conference on Rain Water Cistern Systems*, Brazil.

Henry, F.J., Patwary, Y., Huttly, S.R. and Aziz, K.M. (1990). Bacterial contamination of weaning foods and drinking water in rural Bangladesh. *Epidemiology and Infection* **104**, 79–85.

Khairy, A.E.M., Sebaie, O.E., Gawad, A.A. and Attar, L.E. (1982). The sanitary condition of rural drinking water in a Nile Delta village: I. Parasitological assessment of 'sir' stored and direct tap water. *Journal of Hygiene* **88**, 57–61.

Kirchoff, L.V., McClelland, K.E., Pinho, M.D.C. *et al.* (1985). Feasibility and efficacy of in-home water chlorination in rural North-eastern Brazil. *Journal of Hygiene* **94**, 173–180.

Koplan, J., Deen, R., Swanston, W. and Tora, B. (1978). Contaminated roof-conducted rainwater as a possible case of an outbreak of Salmonell osis. *Journal of Hygiene*, 303–309.

Lye, D.J. (1992). Microbiology of rainwater cistern systems. *Journal of Environmental Science and Health* **A27**, 2123–2166.

Moe, C.L., Sobsey, M.D., Samsa, G.P. and Mesolo, V. (1991). Bacterial indicators of risk of diarrhoeal disease from drinking-water in the Philippines. *Bulletin of the World Health Organization* **69**, 305–317.

Molbak, K., Hojlyng, N., Jepsen, S. and Gaarslev, K. (1989). Bacterial contamination of stored water and stored food: a potential source of diarrhoeal disease in West Africa. *Epidemiology and Infection* **102**, 309–316.

NoosaShire Environmental Health Officers (1994). A microbiological investigation into the degree of contamination of water in domestic rain water storage tanks with residents of Noosa Shire, solely supported by tank water. *Proceedings of the 54th Annual State Conference of the Australian Environmental Health.*

Oo, K.N., Han, A.M., Hlaing, T. and Aye, T. (1991). Bacteriologic studies of food and water consumed by children in Myanar: I. The nature of contamination. *Journal of Diarrhoeal Disease Research* **9**, 87–90.

Pinfold, J.V. (1989). *Assessment of the effects of low-cost water supply and sanitation initiatives on faeco-oral disease transmission*. PhD Thesis, University of Leeds, UK.

Pinfold, J.V. (1990). Faecal contamination of water and fingertip-rinses as a method for evaluating the effect of low-cost water supply and sanitation activities on faeco-oral disease transmission. I. A case study in rural northeast Thailand. *Epidemiology and Infection* **105**, 363–375.

Pinfold, J.V. (1999). Analysis of different communication channels for promoting hygiene behaviour. *Health Education Research: Theory and Practice* **14**, 629–639.

Pinfold, J.V. and Horan, N.J. (1996). Measuring the effect of a hygiene behaviour intervention by indicators of behaviour and diarrhoeal disease. *Transactions of the Royal Society of Tropical Medicine and Hygiene* **90**, 366–371.

Pinfold, J.V., Horan, N.J., Wiroganagud, W. and Mara, D.D. (1993). The bacteriological quality of rainjar water in rural northeast Thailand. *Water Research* **27**, 297–302.

Quick, R.E., Venczel, L.V. and Gonzalez, O. (1996). Narrow-mouthed water storage vessels and *in situ* chlorination in a Bolivian community: A simple method to improve drinking water quality. *American Journal of Tropical Medicine and Hygiene* **54**, 511–516.

Simango, C., Dindiwe, J. and Rukure, G. (1992). Bacterial contamination of food and household stored drinking water in a farmworker community in Zimbabwe. *Central African Journal of Medicine* **38**, 143–149.

Sobel, J., Mahon, B. and Mendoza, C.E. (1998). Reduction of fecal contamination of street-vended beverages in Guatemala by a simple system for water purification and storage, handwashing and beverage storage. *American Journal of Tropical Medicine and Hygiene* **59**, 380–387.

Taylor, E.W. (1972). Report on the results of the bacteriological chemical and biological examination of London waters, 1969–1970. *Report of the Metropolitan Water Board* **44**, 22.

Thurman, R. (1995). Evaluation of rainwater stored in collection tanks. *Australian Microbiologist, March*, 20–22.

Tompkins, A.M. and Drasar, B.S. (1978). Water supply and nutritional status in rural northern Nigeria. *Transactions of the Royal Society of Tropical Medicine and Hygiene* **72**, 239–243.

Tunyavanich, N. and Hewison, K. (1989). Water supply and water use behaviour: the use of cement rainwater jars in N.E. Thailand. *Proceedings of the 4th International Rainwater Cisten systems Conference*. Manila, **C2**, 1–23.

Wegelin, M. and Sommer, B. (1998). Solar water disinfection (SODIS) – destined for worldwide use? *Waterlines* **16**, 30–32.

Wirojanagud, W., Hovichitr, P., Mungkarndee, P. *et al.* (1989). *Evaluation of Rainwater Quality: Heavy Metals and Pathogens.* IDRC, Ottawa.

38

Coagulation and filtration

Caroline S. Fitzpatrick and John Gregory

Department of Civil Engineering, University College London, WC1E 6BT, UK

1 INTRODUCTION

Physical separation of particles, including microbes, from water can be achieved by a few basic processes: sedimentation (including centrifugation), flotation and various forms of filtration. All particle separation processes depend greatly on particle size, and different techniques are appropriate for different size ranges.

The diagram in Fig. 38.1 shows examples of various particles, including microbes, over a wide range of size, together with appropriate separation techniques. The conventional boundary between *colloidal* and *suspended* particles is shown at 1 μm, but this is a rather arbitrary distinction. The most common processes used in water and wastewater treatment are gravity sedimentation, flotation (usually *dissolved air flotation, DAF*) and filtration through beds of granular media, such as sand (*depth filtration*). All of these become more effective for larger particles. Removal of particles of a few micrometres or less is quite difficult by these processes.

By using membranes with suitable pore dimensions it is possible, in principle, to remove particles and dissolved impurities of any size. Depending on the type of membrane used, and the applied pressure, the processes of *microfiltration, ultrafiltration, nanofiltration* and *hyperfiltration* (or *reverse osmosis*) are possible. These are *absolute* filtration techniques, in that all particles and molecules above a certain cut-off size are removed, unless failure occurs.

Depth filtration, through beds of granular filter media, is a very widely-used technique in practice. In very many cases the particles to be removed are considerably smaller than the effective pore size, so that simple straining is not the operative mechanism. Instead, removal occurs by *particle capture* on the surfaces of filter grains, or of existing deposits of particles. Depth filtration is not an absolute separation method since some particles may avoid capture and penetrate right through the bed.

Since conventional separation techniques are not especially effective for particles of colloidal dimensions (including bacteria and viruses), it is necessary to increase their size in some way. The only practical method is to cause particles to *aggregate*, forming larger units. The aggregation process may be known variously as *agglomeration, coagulation* or *flocculation*, depending on the field of application and on the supposed mechanism. In water and wastewater treatment, the terms 'coagulation' and 'flocculation' are in common use and the distinction will be explained later.

In general, aggregates can be separated much more readily than the original particles. Sedimentation rates, for instance, can be increased many fold. In a typical water treatment sequence, a coagulation/flocculation process is followed by some form of sedimentation, which removes most of the particulate impurities. The water then passes through a granular filter, where the remaining particles are nearly all removed. Effective coagulation is vital to the success of such a procedure.

In membrane filtration, the separated particles can build up on the membrane, forming a filter cake, which may play a large part in the removal of further particles. Small particles, especially those in the colloidal size range, give

The Handbook of Water and Wastewater Microbiology
ISBN 0-12-470100-0

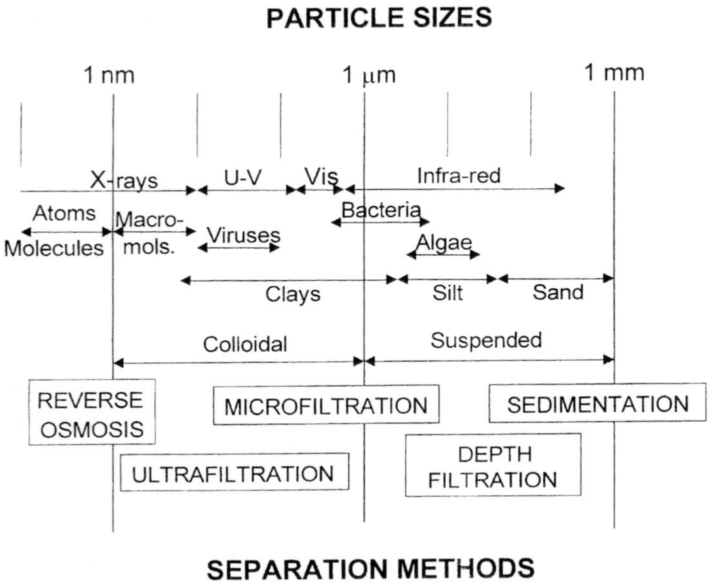

Fig. 38.1 Particle size ranges and appropriate separation methods.

filter cakes of very low permeability and hence low filtration rates. Again, aggregation can give a marked improvement.

In this chapter we shall review the more important aspects of coagulation/flocculation and depth filtration processes. No further discussion of membrane filtration will be given. There is a large and growing body of literature on this subject (see Mallevialle *et al.*, 1996).

The subject of *colloid interactions* is very important in both coagulation and filtration, since attachment of particles to each other or to filter media is essential.

2 COLLOIDAL INTERACTIONS AND COLLOID STABILITY

2.1 General

For effective coagulation and filtration, there must be attractive forces between particles or between particles and filter media, so that attachment can occur.

Colloidal particles in water interact with each other in various ways to give either repulsion or attraction. If there is significant repulsion between particles, then they will not be able to come into close contact and aggregation will be prevented. Such particles are said to be *colloidally stable*, in the sense that they remain dispersed over long times. The most common reason for colloid stability is the fact that most particles in water are charged and hence repel each other electrically. Similar considerations apply to the attachment of particles to filter media.

If there is only limited repulsion between particles, or if there is an attraction, then aggregation can occur when particles collide and the suspension is said to be unstable.

The important types of interaction between particles in water are:

1. van der Waals attraction
2. Electrical interaction, usually giving a repulsion
3. Steric repulsion, associated with adsorbed layers on particles
4. Bridging by adsorbed polymers giving an attraction
5. Hydrophobic attraction.

All of these interactions (except possibly hydrophobic attraction) are of rather short range. Typically, they act over a range considerably less than the particle size. This means that they only come into play when particles are nearly in contact and they have little influence over the transport of particles in coagulation or filtration processes.

Furthermore, the magnitude of these interactions is roughly proportional to the particle size, which has very important implications in practice. Other forces, such as gravity and hydrodynamic drag, have different dependence on particle size. Thus the gravitational attraction on a particle depends on the mass of the particle and is hence proportional to the *cube* of the size. The drag force on a particle in a flowing fluid is dependent on the *square* of the particle size. It follows that these 'external' forces become very important for larger particles and can be much larger than the 'colloidal' interactions listed above. Conversely, colloidal interactions usually predominate for smaller particles, typically of the order of a few micrometres or less. This is the reason for the term 'colloidal interactions'.

Although all of the interactions 1–5 above can be important in practice, we shall here concentrate on van der Waals attraction and electrical repulsion, since these form the basis of the *DLVO theory* of colloid stability, which will be reviewed briefly below. Polymer bridging will be discussed separately, under the heading of Polymeric flocculants.

2.2 van der Waals attraction

The universal attractive forces between atoms and molecules, originally postulated by van der Waals, have long been known to act between all materials (e.g. Israelachvili, 1991). Between equal spherical particles of radius a, separated by a distance d, the van der Waals interaction energy, V_A, is given approximately by:

$$V_A = -\frac{Aa}{12d} \qquad (1)$$

where A is the *Hamaker constant*, which depends on the properties of the particles and water. For aqueous suspensions, this ranges between about 5 and 100 $\times\ 10^{-21}$ J. The lower values are typical of biological particles such as bacteria and algae. Dense mineral particles have much higher values. An expression of similar form to equation (1), but with a factor 1/6, applies to the interaction between a spherical particle and a flat surface, which is a reasonable model for a colloidal particle approaching the surface of a much larger filter grain.

Equation (1) applies only to very close approach, where $d << a$, but the interaction is often negligible at larger distances. The negative sign implies, by convention, an attraction. The energy is proportional to particle size, as mentioned above, and depends inversely on separation distance, becoming very large as particles come into contact.

van der Waals attraction is essential in coagulation and filtration, and we have to be aware of its presence. However, there is no realistic method by which the attraction can be significantly altered in practical systems and we shall not go any further into theoretical aspects.

2.3 Electrical interaction

Particles in aqueous suspensions usually carry a surface charge. This may arise for several reasons, including ionization of surface groups and adsorption of certain ions. Biological surfaces are usually charged through acid–base interactions, such as the ionization of carboxylic acid groups and the protonation of amine groups. Oxide surfaces are charged by virtue of protonation/deprotonation of surface hydroxyl groups. In these cases the surface charge depends greatly on the pH of the solution, with positive values at low pH and negative charge at high pH. There is a characteristic pH value at which the surface charge is zero, with equal numbers of positive and negative surface groups. This is the *point of zero charge* (*pzc*), which is a very important surface characteristic. The charging of a hypothetical biological surface is shown schematically in Fig. 38.2.

Whatever the origin of surface charge, it must be associated with an appropriate number of oppositely charged ions (*counterions*) in solution, so that overall the system has no net charge. A surface charge, together with counterions in solution gives an *electrical double layer*, the structure of which is very important in

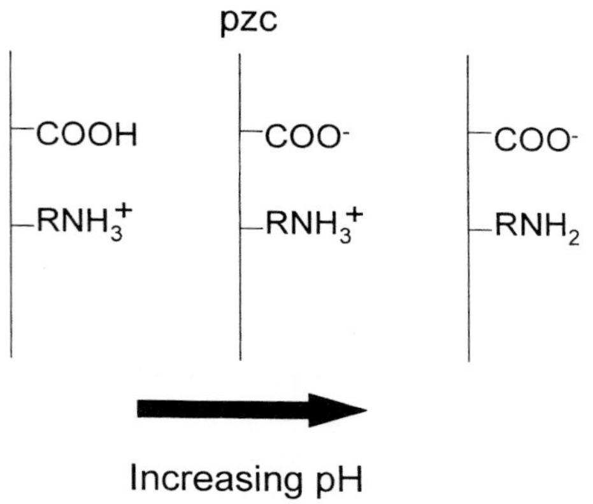

Fig. 38.2 Surface charge development by ionization of surface groups. The point of zero charge (pzc) occurs at a characteristic value.

governing the interaction between charged particles. A simple model of the double layer is shown in Fig. 38.3. This shows the variation of electric potential, ψ, with distance from the surface. At the surface the value is ψ_0 (the surface potential). For a short distance from the surface (of the order of the size of a hydrated ion), the potential drops quite rapidly to a value ψ_δ (the *Stern potential*). (The region close to the surface is known as the *Stern layer*.) Further from the surface, the potential decays in an approximately exponential manner, according to:

$$\psi = \psi_\delta \exp(-\kappa x) \tag{2}$$

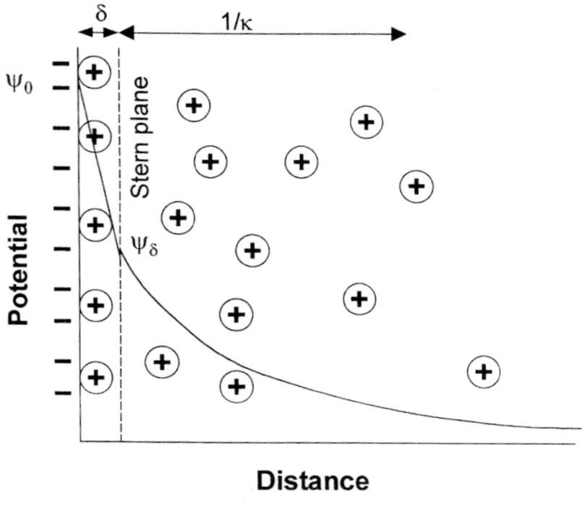

Fig. 38.3 A simple model of the electrical double layer.

where

x is the distance from the Stern plane and

κ is a parameter which determines the rate of decrease of potential with distance.

κ is known as the *Debye-Hückel parameter* and has a great influence on colloid interactions. It has the dimensions of reciprocal length (m^{-1}) and depends on the valence and concentration of ions present in solution. For aqueous solutions at 25°C, it is given by:

$$\kappa = 2.32 \times 10^9 \sqrt{\sum c_i z_i^2} \tag{3}$$

The distance $1/\kappa$, effectively the 'thickness' of the double layer, ranges from around 1 to 100 nm in typical salt solutions. In 10^{-4} M NaCl the value is about 30 nm. In a natural water with a high level of dissolved salts the distance can be much lower, for instance in River Thames water the value is about 4 nm.

Experimentally, charged particles can be investigated by a range of *electrokinetic* techniques, such as microelectrophoresis, in which the velocity of a particle in an electric field is determined. This gives a value of electric potential at the plane of shear, or slipping plane, between the particle and the fluid: the zeta potential, ζ. The slipping plane is generally assumed to be close to the Stern plane, so that the zeta potential can be used instead of the Stern potential (which is not directly measurable).

When two charged particles approach each other their diffuse layers overlap and this leads to a repulsion or attraction, depending on the signs of charge. For similar spherical particles, radius a, there is a repulsion and the interaction energy is given approximately by:

$$V_R = 2\pi e a \zeta^2 \exp(-\kappa d) \tag{4}$$

where

e is the electron charge and

d is the separation distance between the particles.

The magnitude of the repulsion depends greatly on the zeta potential, but the range of interaction is determined by the parameter κ.

For the case of a sphere, radius a, approaching a flat surface (representing a particle and the surface of a filter grain), the corresponding expression is:

$$4\pi\varepsilon a\zeta_1\zeta_2 \exp(-\kappa d) \qquad (5)$$

where ζ_1 and ζ_2 are the zeta potentials of the particle and surface.

With different signs of zeta potentials equation (5) predicts an attraction between particle and surface. Thus particles of opposite charge to filter media should be able to attach without difficulty, which is intuitively reasonable. However, this condition does not necessarily lead to good operation of deep bed filters (see below).

2.4 DLVO theory of colloid stability

The first truly quantitative treatment of colloid stability and coagulation was developed independently by Deryagin and Landau (1941) and Verwey and Overbeek (1948). This is now generally known simply as *DLVO theory.* Essentially, van der Waals attraction and electrical repulsion are assumed to be additive and combine to give the total energy of interaction between particles as a function of separation distance.

From equations (1) and (4), the total interaction energy between similar particles, radius a, separated by a distance d is:

$$V_T = 2\pi\varepsilon a\zeta^2 \exp(-\kappa d) - \frac{Aa}{12d} \qquad (6)$$

When the interaction energy is plotted as a function of separation distance, a characteristic *potential energy diagram* is obtained (Fig. 38.4). Depending on the relative contributions of the repulsion and attraction terms, there may be a significant energy barrier which prevents close approach of the particles. Under these conditions, the particles are said to be *stable* and coagulation would occur only very slowly, if at all. This situation would arise for particles of high zeta potential and in solutions of low ionic strength. Ionic strength affects the parameter κ and hence the range of electrical repulsion. As ionic strength is increased, repulsion occurs over a more limited range and so the attraction term becomes relatively important. Another factor is that, at higher salt concentrations, the zeta potential is usually reduced, which also has the effect of reducing repulsion. Thus increasing ionic strength eventually causes

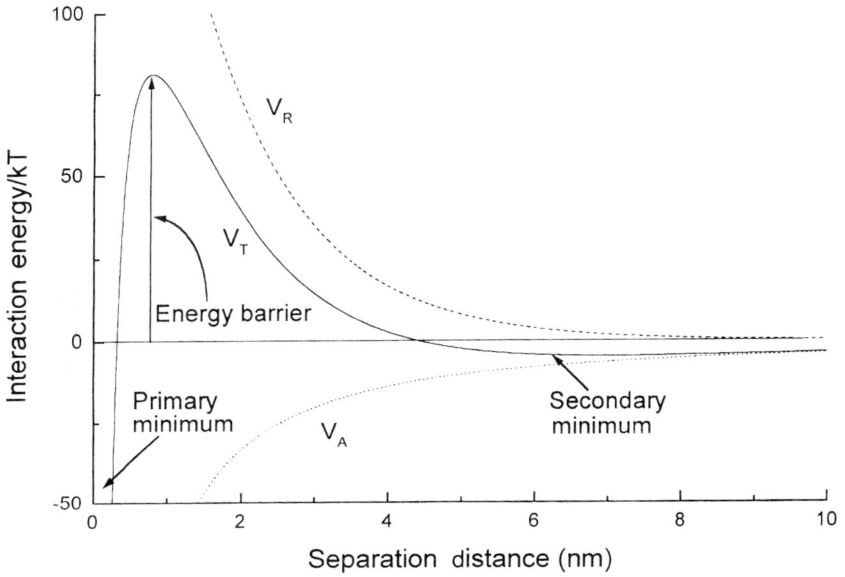

Fig. 38.4 Potential energy diagram for the interaction of spherical colloidal particles, diameter 1 μm and zeta potential 25 mV. The Hamaker constant is assumed to be 2 kT and the medium is an aqueous solution of 1-1 electrolyte at a concentration of 50 mM.

the repulsion to be reduced sufficiently to eliminate the potential energy barrier, allowing particles to coagulate at the most rapid rate.

The energies in Fig. 38.4 are plotted in units of kT, where k is Boltzmann's constant and T is the absolute temperature. For barriers greater than about 10 kT it would be very unlikely that colliding particles would have sufficient thermal energy to surmount the barrier. However, in practical flocculation processes, energy imparted by fluid motion may be important.

Once the energy barrier is removed, particles should be able to come into close contact and be held in a deep *primary energy minimum*. In principle, the van der Waals attraction should become infinite on contact of particles when $d = 0$ (see equation (1)). However, certain types of short-range repulsion (for instance hydration effects) prevent true contact and the attraction remains finite, so that coagulated particles can often be re-dispersed.

Another aspect of the potential energy diagram is the presence of a *secondary minimum* in the interaction energy. This arises because the electrical repulsion decreases exponentially with distance, equation (4), and van der Waals attraction has an inverse distance dependence, equation (1). The attraction term will always be greater at sufficiently large separation. For particles larger than about 1 μm, this can lead to a significant attraction, even when there is a high energy barrier and the particles should be colloidally stable. The secondary minimum can give rather weak aggregates, which are easily re-dispersed, but may be important in some cases.

The DLVO theory allows conditions of stability and coagulation to be predicted, provided that the relevant information is available. In particular it is possible to calculate the salt concentration needed to eliminate the energy barrier and give rapid coagulation. This is the *critical coagulation concentration*, which will be discussed in the next section.

The concepts outlined above apply also to the attachment of particles to filter media. In this case, the particles and filter media may have different signs of charge and there is the possibility of an electrical attraction, which would give strong attachment. However, in building up layers of deposited particles, as is usually the case, particle–particle interaction is relevant and this still needs to be favourable for attachment. Thus negative particles would have no difficulty in attaching to a positively-charged surface, but further particles would not attach to those already deposited if there were significant repulsion between them.

3 COAGULATION AND FLOCCULATION

3.1 Effects of salts

In order to modify colloid stability and promote coagulation, it is usually necessary to reduce electrical repulsion between particles. (In practice, very little can be done to increase van der Waals attraction.) As shown above, increasing ionic strength causes a reduced repulsion. Salts which act simply by reducing the range of electrical repulsion, possibly also with some reduction in zeta potential, are known as *indifferent electrolytes*. Their effect can be examined by considering equation (6). If this is differentiated with respect to separation distance, it is possible to find the condition where the potential energy barrier just disappears, i.e. where:

$$\frac{dV}{dd} = 0 \text{ and } V_T = 0$$

This condition occurs at the critical coagulation concentration, ccc, and it can be shown that:

$$\text{ccc} \propto \frac{\zeta^4}{A^2\, z^2} \qquad (7)$$

This shows a strong dependence on zeta potential, the Hamaker constant A and the valence of the ions, z. The dependence on z is significant, since it shows that salts with highly charged ions (especially counterions) will be more effective in promoting coagulation. Such ions also tend to have a larger effect on zeta potential, which increases their coagulating effect.

Equation (7) is based on an approximation for fairly low values of zeta potential. Another result from DLVO theory suggests that, for high zeta potentials there should be a much stronger dependence on valence (ccc $\propto 1/z^6$) and this has become known as the *Schulze-Hardy rule*. Experimentally, it is well established that the ccc depends strongly on counterion valence, although quantitative agreement with the $1/z^6$ is not often found.

Another important effect occurs with salts containing *specifically adsorbing counterions*. These bind to surfaces not only by electrostatic attraction, but also by some form of 'chemical' interaction. This has the very important consequence that the surface charge of particles can be reversed by specifically-adsorbed counterions. The fact that this 'super-equivalent' adsorption must occur against an electrostatic repulsion means that there must be an additional attractive component. There are many examples, such as the adsorption of calcium ions on surfaces with carboxylate groups.

Salts with specifically-adsorbing counterions are more effective in destabilizing charged particles than indifferent electrolytes and so can cause coagulation at lower concentrations than expected from simple DLVO theory. Furthermore, the ccc should depend on the particle concentration (or, more strictly, the particle surface area), which is not the case for indifferent electrolytes, whose effect depends simply on ionic strength. Charge reversal implies that particles can become *restabilized* when excess counterions are adsorbed and this is a very important effect in practice. For coagulants of this type there is a critical coagulation concentration (ccc) and a critical restabilization concentration (crc) when charge reversal has occurred and the particles repel each other. Between these two concentrations there is an *optimum coagulation concentration* or *optimum dosage* of coagulant. The optimum dosage region is found to correspond with counterion concentrations which give charge neutralization. In the case of surfaces with a pH-dependent surface charge, such as oxides changing pH can bring about coagulation and the optimum pH region is around the pzc.

3.1.1 Action of hydrolysing metal coagulants

In water and wastewater treatment salts of aluminium and iron are widely used as coagulants. In water these give Al^{3+} and Fe^{3+}, which might be thought to be very effective for coagulation of negative particles by virtue of their high positive charge. However, this explanation is greatly oversimplified. Highly-charged cations in water become strongly hydrated, giving, for instance, $Al(H_2O)_6^{3+}$. The high charge on the central cation causes polarization of the hydration layer and a tendency to release protons. This causes a progressive replacement over water molecules by hydroxyl ions in the hydration layer, giving the following sequence (omitting hydration water for simplicity):

$$Al^{3+} \rightarrow Al(OH)^{2+} \rightarrow Al(OH)_2^+ \rightarrow Al(OH)_3 \rightarrow Al(OH)_4^-$$

These hydrolysis reactions are strongly influenced by pH: the higher the pH the greater the tendency for formation of hydroxylated species. The uncharged hydroxide $Al(OH)_3$ has very low solubility in water and hence forms a precipitate. This occurs most readily at around neutral pH. Fig. 38.5 shows the concentration of various hydrolysis species of Al in equilibrium with amorphous $Al(OH)_3$. Similar hydrolysis reactions occur with ferric salts in water, although ferric hydroxide is considerably less soluble than $Al(OH)_3$ and precipitates over a wide pH range.

In addition to monomeric hydrolysis products, it is known that dimers, such as $Al_2(OH)_2^{4+}$ and larger units can form. The tridecamer $Al_{13}O_4(OH)_{24}^{7+}$ or 'Al_{13}' has been reported in many studies and identified by NMR techniques (Bertsch *et al.*, 1986). These are also shown in Fig. 38.5.

Most of the important hydrolysis products are positively charged and they are known to adsorb strongly on negative surfaces, with specific interactions probably playing an important role (Stumm, 1992). They are thus able to neutralize the charge of particles in water and this is one way in which particles can be destabilized. This often occurs at quite

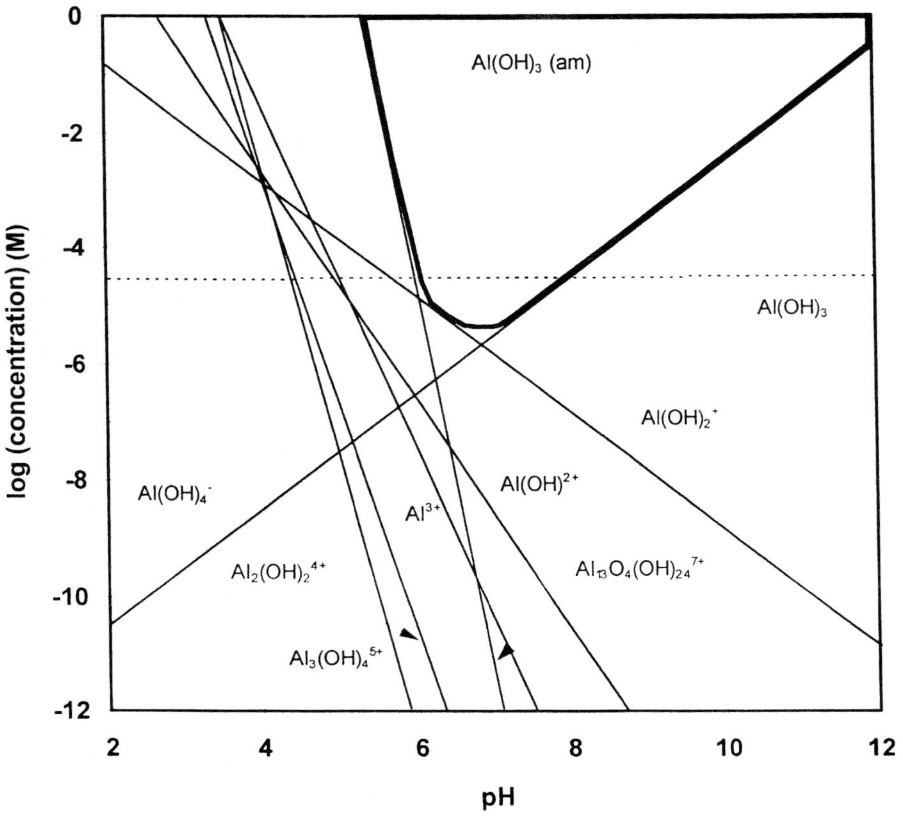

Fig. 38.5 Diagram showing soluble aluminium species, in equilibrium with amorphous aluminium hydroxide, as a function of pH.

low coagulant concentrations and may be followed by restabilization at somewhat higher dosages. For kinetic reasons (see later) coagulation may not be satisfactory in the charge neutralization region and at still higher dosages another effect becomes important, related to hydroxide precipitation. This is commonly known as 'sweep coagulation' and is thought to occur by enmeshment of impurity particles in growing hydroxide precipitates. This can lead to quite large flocs, although they are usually of low density, containing large amounts of water. For this reason, sludges from water treatment plants can present significant dewatering problems.

Another type of coagulant, now widely used in practice, is based on pre-hydrolysed forms of aluminium and iron. Those based on aluminium are most common commercially and typical products are known as *polyaluminium chloride* (PAC). These are made by controlled neutralization of concentrated $AlCl_3$ solutions with base. The degree of neutralization is very important and ranges from about 1 to 2.5 equivalents of base per mole of Al. Three equivalents of base would give $Al(OH)_3$ and hydroxide precipitation. By limiting the degree of neutralization, intermediate, soluble, hydrolysis products are formed, including polymerized forms such as 'Al_{13}', mentioned above. Under the right conditions, the Al_{13} product can be the predominant form. PAC is a very effective coagulant and can give more rapid coagulation and larger flocs than conventional 'alum'. These products are found to be especially useful at low temperatures, where conventional coagulants have reduced efficiency. Part of their effect is due to the high charge of the polycations and their strong adsorption on negatively charged surfaces. This would give effective charge neutralization, but it is highly likely that precipitation also plays an important role. The detailed mode of action of PAC and similar coagulants has yet to be elucidated.

Hydrolysing coagulants are also effective in removing dissolved organic matter, such as humic substances, from water. The mechanism may be charge neutralization or adsorption on precipitated hydroxide (Bose and Reckhow, 1998).

3.2 Polymeric flocculants

Organic polymers, both natural and synthetic, can adsorb on a wide range of surfaces and have a great effect on particle interactions and hence on colloid stability (see Gregory, 1996). Polymers may give enhanced stability, for instance by steric stabilization, or may promote aggregation of particles. We shall only be concerned with the latter aspect here.

Extracellular polymers, such as polysaccharides, are produced and exuded by many bacteria and play a large part in *bioflocculation*, for instance in the activated sludge process (Morgan *et al.*, 1990). Also, natural polymers, extracted from certain plant products (especially nuts and seeds) have been found to act as flocculants (e.g. Muyibi and Evison, 1995). However, most commercial applications use synthetic polymers.

3.2.1 Polymer bridging

Polymers of high molecular weight may adsorb on surfaces from solutions in such a way that parts of the polymer chain extend into the solution. This gives the possibility that parts of the same polymer chain can adsorb on different particles, leading to *polymer bridging* as a means of binding particles together. This is illustrated schematically in Fig. 38.6a. Polymer bridging is now recognized as a very important mechanism in practice. It is most effective with linear, high molecular weight (several millions) polymers. Aggregation brought about in this way is often known as *flocculation*, in order to distinguish it from *coagulation* caused by salts. However, this distinction is not universally used and the terms are often used interchangeably.

Many types of polymer are effective as flocculants, but only relatively few are used commercially. Polymers may be non-ionic, anionic or cationic, and if there are ionic groups, the products are known as polyelectrolytes. By far the most important are those based on polyacrylamide, which can be produced with very high molecular weights (20 million or more). Nominally, polyacrylamide is non-ionic, but controlled hydrolysis of amide groups gives carboxylic acid groups, which can ionize at around neutral pH and higher to give anionic sites. Cationic polyelectrolytes can be synthesized by copolymerization of acrylamide with suitable cationic monomers. This is a common route to commercial cationic polyelectrolytes of high molecular weight. The most important characteristics of polymeric flocculants are the molecular weight and, in the case of polyelectrolytes, the charge density.

For effective bridging flocculation polymers must adsorb on particles to give only partial surface coverage, with segments extending

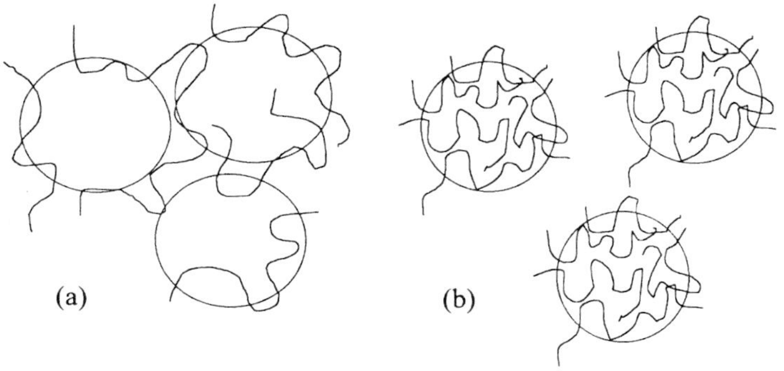

Fig. 38.6 Schematic illustration of (a) bridging flocculation and (b) restabilization by adsorbed polymers.

some distance from the surface. If too much polymer is adsorbed then there are not enough surface sites remaining to form links with other particles. This is another form of restabilization (Fig. 38.6b) and means that there is usually an *optimum dosage* for polymeric flocculants. It is often assumed that optimum bridging flocculation corresponds to conditions where about half of the available surface area is occupied by adsorbed polymer, although this concept is quite difficult to verify.

A major advantage of polymeric flocculants is that they can produce aggregates (flocs) which are much stronger (and hence larger) than those produced simply by the effect of salts. For this reason they are used when flocs are subject to high shear rates such as in certain types of centrifuge. In conjunction with hydrolysing salts, polymeric flocculants at very low dosages can give greatly enhanced floc strength.

3.2.2 Electrical effects

Very often it is found that the most effective flocculants are polyelectrolytes of opposite sign to the particle charge. For instance, negatively charged particles can be readily flocculated by cationic polyelectrolytes. The opposite charges provide an electrostatic attraction, which promotes adsorption, but there is also the possibility of charge neutralization reducing the colloid stability, as in the case of specifically-adsorbing counterions. The optimum dosage region would then correspond to amounts of polymer needed to neutralize the particle charge. In many cases polyelectrolytes of low molecular weight and high charge density are effective. In this case, bridging is unlikely because of the short chain length and the tendency for adsorbed polyelectrolytes to adopt a rather flat configuration at the surface. These products are sometimes called 'coagulants', to distinguish them from flocculants, which act by a bridging mechanism.

A variant of the simple charge neutralization picture is the *Electrostatic patch* mechanism (Gregory, 1973). This is illustrated schematically in Fig. 38.7, for the case of a cationic polyelectrolyte adsorbing on a negatively-charged particle surface. This model applies to a system

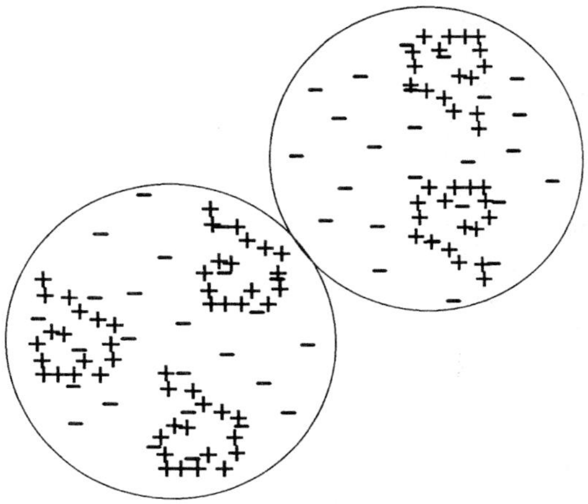

Fig. 38.7 'Electrostatic patch' interaction with cationic polyelectrolyte adsorbed on negatively charged particles.

where the surface charge density of the particles is quite low and the polyelectrolyte has a high charge density. In such a case the separation between charges along the polymer chain is less than the distance between charged sites on the particle surface. This means that, when the particle charge is balanced overall by adsorbed polyelectrolyte, the surface sites cannot all be neutralized individually. The surface then has adsorbed 'islands' or 'patches' of excess positive charge, surrounded by uncoated, negative regions. When particles with polyelectrolyte adsorbed in a 'patchwise' manner approach each other there can be a significant electrical attraction between the oppositely-charged surface regions. This gives stronger flocs than would be expected on the basis of simple charge neutralization, although not generally as strong as flocs produced as a result of bridging by long-chain polymers.

The electrostatic patch model is now widely accepted and appears to be relevant to many practical applications.

3.3 Coagulation kinetics

Particles must collide with each other in order to coagulate, so that the rate of coagulation depends on the particle collision frequency. Depending on the interactions between colliding particles, collisions may or may not result in the formation of aggregates. The fraction of successful collisions is known as the *collision efficiency*, α. When there is significant repulsion between particles the collision efficiency will be very low, but when repulsion has been eliminated, by suitable chemical treatment, α can approach unity, so that nearly every collision results in attachment.

Detailed accounts of this subject can be found elsewhere (e.g. Elimelech *et al.*, 1995). Only a brief and selective treatment can be given here.

The frequency of collisions between particles of type i and j depends on the product of their concentrations, the collision efficiency and a *rate coefficient*, k_{ij}, which depends on the mode of transport of particles (see below):

$$J_{ij} = \alpha k_{ij} n_i n_j \qquad (8)$$

where

J_{ij} is the number of collisions occurring between i and j particles and

n_i and n_j are their number concentrations.

The rate coefficient depends on the particle sizes and on certain physical factors. Relative motion between particles, leading to collisions, may caused by:

1. Brownian motion (*perikinetic coagulation*)
2. Fluid motion (*orthokinetic coagulation*)
3. Differential settling.

Of these, Brownian motion is important only for quite small particles and is not usually able to produce the large flocs that are required in most practical applications. Usually, some form of agitation, such as in stirred tanks or flow-through reactors is applied, causing particle collisions. With a range of particle sizes, different settling rates cause relative motion and hence collisions. This can become significant as flocs grow quite large.

In the most important case of orthokinetic collisions, the rate of coagulation can be written in terms of the rate of reduction of particle number concentration, since each successful collision results in the net loss of one particle:

$$-\frac{dn}{dt} = \frac{16}{3}\alpha n^2 \; G a^3 \qquad (9)$$

This expression is based on the assumption of spherical particles, radius a, in a laminar shear field with a uniform shear rate G. The particle number concentration is n. These are highly restrictive conditions, which would not apply in practical coagulation processes. Nevertheless, this approach leads to a simple result that gives an idea of the important parameters.

No practical coagulation processes are carried out under laminar conditions; turbulent flow is almost always involved. In this case an effective shear rate can be defined, in terms of the power input to the vessel (such as a stirred tank), P, the volume of suspension, V and the viscosity, μ. The result due to Camp and Stein (1943) is:

$$\bar{G} = \sqrt{\frac{P}{\mu V}} \qquad (10)$$

The effective shear rate can then be inserted in the appropriate rate expression instead of the laminar shear rate. Although this approach is undoubtedly oversimplified, it gives a result which is quite similar to that from a more rigorous approach.

Equation (9) shows that the rate of coagulation depends on the square of the particle concentration, as expected for a second-order rate process. There is also a very strong dependence on particle size, which explains why orthokinetic coagulation becomes much more important for larger particles.

The form of equation (9) is such that a simple transformation gives the result in terms of the volume fraction of particles, which is simply the total volume of particles per unit volume of suspension. For equal spherical particles this is simply given by:

$$\phi = \frac{4}{3}\pi a^3 n \qquad (11)$$

If it is assumed that the volume of particles remains constant during the coagulation process, and the collision efficiency, α = 1 (i.e. for a fully-destabilized suspension), then equations (9) and (11) lead to:

$$\frac{n}{n_0} = \exp\left(\frac{-4G\phi t}{\pi}\right) \qquad (12)$$

Although this expression is based on a number of simplifying assumptions, it is very important in showing the main parameters influencing the rate of orthokinetic coagulation. The term n/n_0 represents the loss of particles by coagulation and the reciprocal, n_0/n can be regarded as the average floc size at time t. (Effectively, it is the average number of primary particles per floc.)

Equation (12) shows that the extent of coagulation increases with the shear rate G, the volume fraction of particles and the time t. The product Gt is a dimensionless number, sometimes called the *Camp number*. In water treatment plants, flocculation units are often designed to have a Camp number in the region of 30 000–50 000. In principle, the same degree of coagulation should be attained for any combination of G and t, provided that the product remains constant. A high shear rate for a short time or a low shear rate for a correspondingly longer time should give the same result. However, this ignores the break-up of flocs, which becomes very important at high shear rates. Also, for hydrodynamic reasons, the effective collision efficiency at high shear rates can become quite low (van de Ven and Mason, 1977).

The growth of flocs, for different shear rates is shown schematically in Fig. 38.8. At low shear rate, the coagulation rate is low, but the flocs can eventually reach quite large sizes. By contrast, at high shear rates, the initial floc growth will be faster, but the limiting floc size will be less. There is, as yet, no quantitative theory of floc strength, but a common empirical approach is to relate the limiting floc size to the effective shear rate (Mühle, 1993):

$$d_{\max} = CG^{-\gamma} \tag{13}$$

where $d_{\max}$ is the maximum floc size for a shear rate G and C and γ are empirical constants. Reported values of γ are in the range 0.2–0.5.

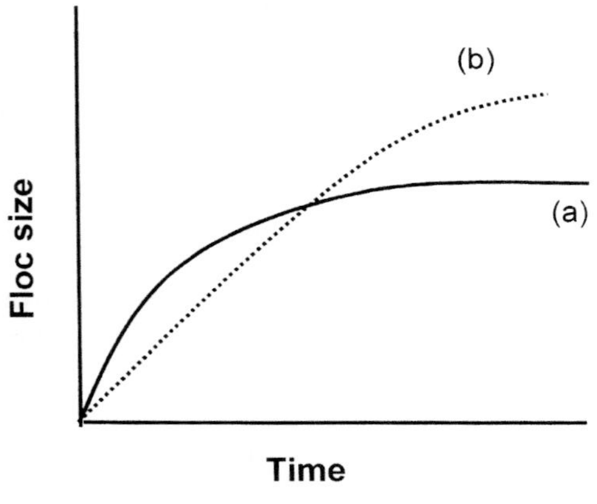

Fig. 38.8 Growth of flocs under (a) high shear and (b) low shear conditions.

Another factor which greatly affects coagulation rate is the solids concentration, or the volume fraction, ϕ. Equation (12) shows that this parameter is of equal importance to G and t. In practical units it is of great advantage to maintain a zone of high solids concentration to exploit this effect. Recycling of flocs (as in activated sludge units) or establishing a 'floc blanket' (as in upflow clarifiers) are effective means of achieving high solids concentrations. The use of hydrolysing coagulants under 'sweep floc' conditions also increases the effective volume fraction by the formation of a voluminous hydroxide precipitate.

3.4 Properties of flocs

The main purpose of carrying out a coagulation process is to achieve more efficient solid–liquid separation. Certain floc properties, notably *floc density*, can greatly influence separation processes.

3.4.1 *Fractal nature of flocs*

Aggregates or flocs are now recognized as *self-similar, fractal* objects (Meakin, 1988). Self-similarity simply means that an aggregate appears to have a similar structure, independent of the scale of observation. Fractal character implies that the mass of an aggregate (or the number of primary particles within it) scales as the size raised to some power, d_F (or the *fractal dimension*).

$$M \propto L^{d_F} \tag{14}$$

Of course, a solid, three-dimensional object has a mass which depends on the cube of its size. Aggregates effectively have fractional dimensions; hence the term 'fractal'.

Early models of aggregate structure were based on the addition of single particles to growing clusters (Fig. 38.9a). This tends to give quite compact aggregates with a fractal dimension of around 2.5. Later simulations (e.g. Lin *et al.*, 1989) allowed for cluster–cluster collisions, which is a more realistic model in many practical cases. This gives much more open aggregate structures (Fig. 38.9b), with d_F about 1.8. The reason for the different structures should be clear from Fig. 38.9. In the particle–cluster case, the approaching particle is able to penetrate some way into the cluster before making contact. When two clusters approach each other it is likely that contact will occur before the clusters have interpenetrated significantly, giving a much more open structure.

These models are based on two important assumptions:

1. There is no repulsion between particles, so that every collision leads to permanent contact
2. Aggregation occurs by diffusion, as in perikinetic coagulation (*diffusion-limited aggregation*).

In cases where there is repulsion between particles the process is known as *reaction-limited aggregation*, which leads to rather more compact structures (d_F about 2.1).

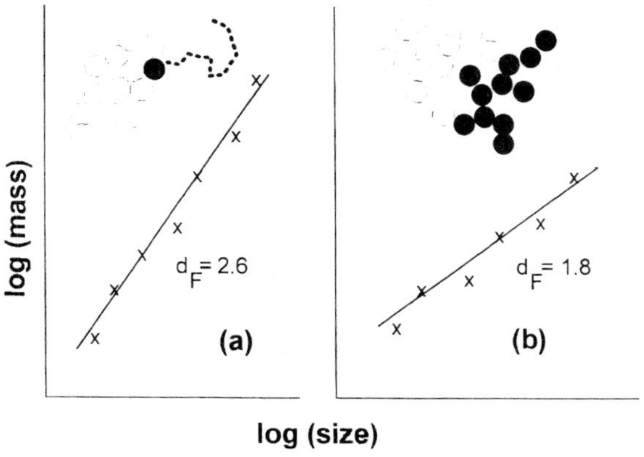

Fig. 38.9 Formation of fractal aggregates by (a) particle–cluster and (b) cluster–cluster aggregation.

In agitated suspensions (orthokinetic aggregation), there is a tendency for more compact aggregates to be formed, as a result of 'restructuring'. This may be a result of floc breakage and re-formation, or deformation and rearrangement of floc structures under shear conditions.

3.4.2 Floc density

An inevitable consequence of the fractal nature of flocs is that floc density decreases as floc size increases. It has been found experimentally (Tambo and Watanabe, 1979) that the effective (buoyant) floc density, ρ_e, decreases with floc size, a, and the data are usually plotted in log–log form (Fig. 38.10). Although there is usually a large degree of scatter, such plots show a linear trend, which implies a relationship of the form:

$$\rho_E = Ba^{-y} \tag{15}$$

where B and y are constants.

It is easy to show that there is a simple connection between the slope of plots such as Fig. 38.10 and the fractal dimension of the flocs:

$$d_F = 3 - y \tag{16}$$

Experimental values of the slope, y, for various systems of practical interest, are in the range 1–1.4, corresponding to fractal dimensions of 2–1.6, which are in line with determinations by other methods.

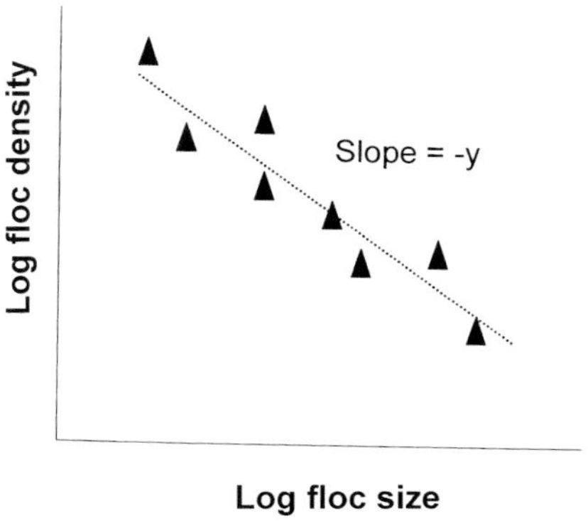

Fig. 38.10 Change of floc density with floc size.

3.5 Practical aspects

Coagulation/flocculation processes are usually carried out in some form of flow-through unit, such as stirred tanks or upflow clarifiers. The purpose is to encourage particle collision by induced fluid motion (i.e. orthokinetic coagulation). It is necessary to ensure that the particles are adequately destabilized by the addition of suitable coagulants, such as metal salts or polymers.

Additives need to be thoroughly and rapidly mixed throughout the suspension and this is achieved by a rapid mix unit at the point of coagulant dosing. Only a short time of rapid mix (a few seconds) is needed and the dosed suspension then flows to the flocculation unit, where the agitation is gentler, allowing flocs to form. The energy input to the flocculator, or the effective shear (see equation **12**) determines the rate of flocculation, but too high a value will result in floc breakage. A compromise between rapid floc growth and the avoidance of high shear rates is the process known as taper flocculation. This involves a series of stirred tanks or compartments with decreasing power input. In the early stages, high shear causes rapid growth of small flocs, which are then allowed to grow larger in the subsequent stages, as the shear rate is reduced.

The most important factors influencing the rate of floc growth are the shear rate and the solids content. A high solids content can be achieved by recirculating flocculated material or in a form of clarifier, where incoming particles pass through a layer of previously-formed flocs (a *floc blanket*).

The density of flocs is an extremely important property (Gregory, 1997) which determines the rate of floc growth. The larger the floc, for a given mass, the greater the rate of collision with other flocs. For similar reasons, the capture of flocs by filter grains and by bubbles in flotation processes is more effective with flocs of lower density. However, there are other separation processes, notably sedimentation and cake filtration (such as in sludge dewatering), where flocs of high density are preferred. For a given mass, a higher density floc will have a smaller size and hence will experience less fluid drag. This gives more rapid sedimentation and more permeable filter cakes.

The addition of small particles to existing flocs promotes higher floc density and this explains why high floc recirculation or floc blanket clarifiers tend to give quite dense flocs. An extreme case is the so-called *pellet flocculation* process (e.g. Yusa and Igarishi, C, 1984) where compact flocs of very high density can be produced, without the usual decrease of density with increasing floc size. (This implies a fractal dimension close to 3.) The fundamentals of this process are not yet fully understood, but it depends on a zone of high solids content with some mechanical agitation (to promote restructuring of flocs) and usually involves the use of high molecular weight polymers to give strong bonding.

4 FILTRATION

4.1 Introduction

Filtration through granular media has been practised since Roman times. It is a physical process to remove suspended particles from water. The suspended particles may be inorganic or organic, including microbes. In the light of problems with *Cryptosporidium* and the pressure to improve the bacterial quality of treated water, a great deal of attention has been paid to filtration processes, especially optimization through the correct choice of operational procedures.

There are three main types of granular media filter used in water treatment:

- slow sand filters
- rapid gravity filters
- pressure filters.

4.2 Slow sand filters

As the name implies, slow sand filters are operated at low filtration rates, typically 0.1–1 m/h. (Note: filtration rate is usually expressed as *approach velocity*, e.g. m/h, which is equivalent to a volume flow rate per unit area, m^3/m^2h.) Slow sand filters use rather fine sand (less than 0.5 mm) and generally give the best removal performance of all granular media filters. This is only partly due to the low filtration rate and small filter grains, since in many cases there is significant biological action, restricted to a fairly thin layer at the surface – the *Schmutzdecke*. Aerobic bacteria in the top few centimetres of sand produce extracellular polymers which form an adhesive network, giving enhanced removal of fine particles. It is likely that microbes captured in the schmutzdecke are subject to predation by other organisms which colonize this layer. In most cases slow sand filters are operated without prior chemical treatment or coagulation. They are cleaned periodically by scraping off the top 2–3 cm of sand, which is washed and returned to the filter.

The disadvantages of slow sand filtration include the large land area required to give high volumetric production and the inconvenient cleaning method.

We shall now restrict attention to rapid filtration methods, which rely predominantly on physical removal mechanisms. There is no fundamental difference in principle between the action of rapid gravity and pressure filters. The former is operated by gravity flow and the latter under applied pressure in closed vessels.

4.3 Rapid filtration

Rapid gravity filtration developed from slow sand filtration in response to requirements for greater volumetric production from less land area. By using coarser media than for slow sand filters, depth penetration of deposits is possible, thereby using the whole bed depth. Filtration rates are usually in the range 5–30 m/h. In conventional operation rapid filtration is preceded by coagulation/flocculation and sedimentation, which greatly reduces the load on the filters. However, in cases where the suspended particle concentration is low, the sedimentation step can be omitted and the process is then known as *direct filtration*.

Granular media in rapid filters typically have grain sizes in the range 0.5–2 mm and the pores are of the same order of size. Particles to be removed are usually very much smaller than the pores and so straining is not a significant removal mechanism. There are two essential aspects of particle removal in granular filters:

- Particle transport to filter grains
- Attachment of particles to grain surfaces or to existing deposits.

These steps together result in *particle capture* by a filter grain (often called a *collector* in the fundamental filtration literature).

Attachment is governed by colloid interactions, which have been discussed earlier. If there is significant repulsion between particles and filter grains then essentially no removal can occur. In most cases coagulant is added before filtration, so that particles usually have quite a low charge and there would be little electrical repulsion. In that case practically all encounters between particles and filter grains (or 'collectors') would result in attachment and the collision efficiency would be close to 1. If this is not the case, then only a fraction of collisions would be effective and the filtration efficiency would be reduced.

In discussing transport mechanisms it is convenient to assume that there is no repulsion ($\alpha = 1$). Another important point is that colloid interactions are of rather short range (typically below 0.1 μm and hence less than the size of most particles). This means that the transport of a particle from the flowing water in a filter pore to the surface of a filter grain is almost entirely unaffected by colloid forces. Although van der Waals attraction may be vital to ensure particle attachment, it is of much too short a range to influence the transport step. A particle would have to be practically in contact with a grain for the attraction to be felt. This justifies treating transport and attachment separately.

4.3.1 *Transport mechanisms*

The flow rate in most conventional filters is low enough to give laminar conditions and particles can be imagined to follow streamlines unless influenced by some other effect.

There are various ways in which a particle can come into contact with a grain surface, and the most important in water filtration are (see Fig. 38.11):

- **interception** – a particle whose centre is on a streamline passing close enough to a collector is intercepted and captured
- **sedimentation** – as a result of the gravitational force pulling a suspended particle closer to a collector as the water flows round it
- **diffusion** – random Brownian motion of particles causes them to depart from streamlines so that they may collide with collectors.

Without going into theoretical detail (see (Elimelech *et al.*, 1995), we can point out the important influences on these mechanisms. They are:

- Particle size and density
- Grain size
- Flow rate

Particle density is only important for the sedimentation mechanism, where higher density gives greater sedimentation rate and an increased chance of particle capture. Fairly simple theoretical approaches lead to the following conclusions regarding the effect of the other parameters on particle capture rate:

Effect of increasing	Effect on transport mechanism		
	Interception	Sedimentation	Diffusion
Particle size	Increases	Increases	Decreases
Grain size	Decreases	Little effect	Decreases
Flow rate	Little effect	Decreases	Decreases

The main conclusions are that particle capture will be more likely as the filter grain size and the flow rate are decreased. However, increasing particle size should improve particle capture by sedimentation and interception, but have the opposite effect on capture by diffusion, since diffusion rate is greater for smaller particles.

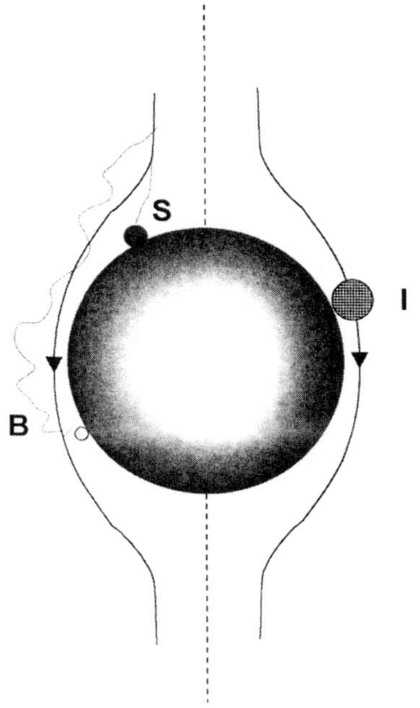

Fig. 38.11 Capture of particles on a single collector by different mechanisms: Interception (I); Sedimentation (S); Brownian diffusion (B).

The different effects of particle size have a very important practical consequence: there is a characteristic size, depending on other conditions, where the capture rate passes through a minimum. Larger and smaller particles are more readily captured in the filter. A schematic plot is given in Fig. 38.12, showing a dimensionless 'collection efficiency' for spherical collectors, against particle size. Although this plot is based on an idealized model, it suggests that filtration through granular media should be least effective for particles of a few micrometres diameter. Smaller particles would be more readily captured by diffusion and larger ones by interception and sedimentation. The particle size around the minimum covers the range of bacteria (0.5–2 μm) and *Cryptosporidium* (4–5 μm), which means that these are more difficult to remove by filtration than other particles. This is one of the reasons why coagulation/flocculation is such an important

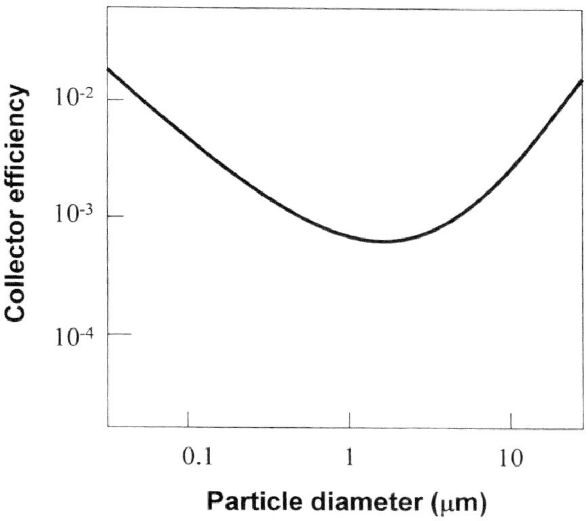

Fig. 38.12 Variation of single collector efficiency with particle size.

preliminary process, since particles in the minimum region can be increased in size to a range where capture is more likely.

4.3.2 Filter performance and operation

During operation of a rapid filter, deposits collect throughout the depth of the bed. If the bed contains uniform-sized media grains then the probability of capture of a particle is the same at any depth. The amount of deposit retained on the grains therefore decreases with depth into the bed.

The filter coefficient, λ is defined as the proportion removed per unit depth of the filter. It is closely related to the collector efficiency mentioned above.

$$\lambda = \frac{-\delta C}{C}\frac{1}{\delta L} \tag{17}$$

where C is the concentration of particles in the water at depth L in the filter bed.

Rearranging gives:

$$\frac{\partial C}{\partial L} = -\lambda C \tag{18}$$

which is known as the Iwasaki equation. The equation has the form of a partial derivative because the deposits, and hence the filter coefficient, vary with time as well as bed depth.

In the early stages of filtration, before significant deposits have built up, the filter coefficient can be assumed constant throughout the bed – the 'clean bed' value, λ_0. Equation (18) can then be integrated to show an exponential decline in suspended particles with depth:

$$C = C_0 \exp(-\lambda_0 L) \tag{19}$$

The implication of this is that the deposit profile with bed depth will also follow an exponential decay pattern, i.e. substantially more deposits near the surface and less at depth. However, as deposits build up on filter grains, the filter coefficient changes. Initially, the coefficient may increase because deposited particles provide extra sites for further deposition. This is responsible for the 'filter ripening' effect (see below). Further build up of deposits causes local constriction of pores and hence an increase in local flow velocity (assuming a constant volume flow rate) and a reduction in λ. As the filter run progresses deposits in the upper bed may reach a limiting level so that no more removal occurs in the top layer. This causes a deposit 'front' to move down through the bed. Eventually this front emerges from the bottom of the bed and there is then a rapid rise in particle concentration in the filtrate. This is known as 'breakthrough'. These concepts are illustrated in Fig. 38.13.

Bacteria and other microbes will also be removed according to this pattern. However, should a biofilm start to develop on the grains, the distribution with depth may vary in a different way, depending on which areas of the bed are more favourable for biofilm formation.

Occasionally, as a result of surface straining, a surface mat develops, which can result in a high head loss (see below) and inefficient use of bed depth.

4.3.3 Head loss

The head loss, or hydraulic, gradient (h/L) is given by the Kozeny-Carman equation, which, for laminar flow in a packed bed of spherical grains, has the form:

$$\frac{h}{L} = 5\frac{v}{g}\frac{\mu}{\rho}\frac{(1-\epsilon)^2}{\epsilon^3}\frac{36}{d^2} \tag{20}$$

where v

is the approach velocity (volumetric flow rate/filter area)

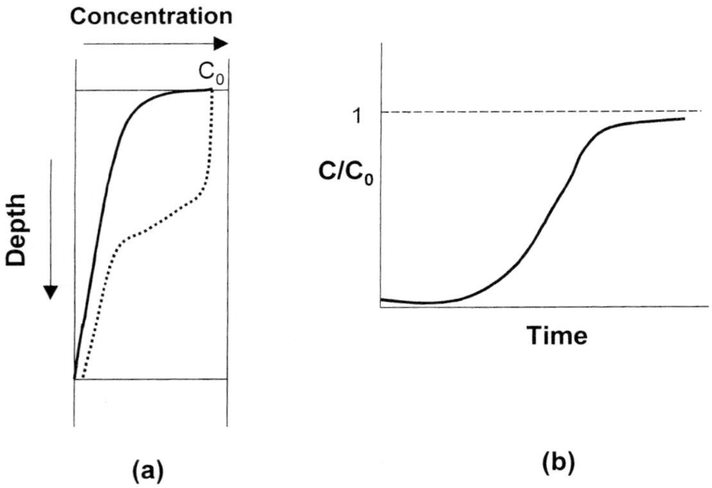

Fig. 38.13 (a) Particle concentration profile through a filter column, initially (full line) and after operating for some time (broken line). (b) Breakthrough curve showing fractional particle concentration in the filtrate as a function of time.

μ is the dynamic viscosity, d the grain diameter

ρ the density of water and

ϵ the porosity of the bed (typically around 0.5).

This expression applies to a clean filter bed. As deposits are formed, the pores become occupied by solids and so the effective porosity decreases, leading to an increase in head loss, if the flow rate is maintained constant. It is not easy to estimate the local change in porosity, since the volume occupied by deposits is uncertain. If particles are deposited singly, fairly compact deposits may form, giving only a slow increase in head loss. However, if deposits have a less dense structure, as expected for fractal flocs then there will be a greater effect on head loss.

Since deposits form preferentially in the upper part of the bed, the clogging can be quite pronounced there and lead to a rapid build-up of head loss. At some point the head loss may become so large that the filter operation has to be terminated and the bed has to be cleaned. Often, particle breakthrough occurs before the head loss limit is reached, but cleaning is necessary in any case. In practice, filters are usually backwashed on a time basis but, should a limiting head loss or filtrate quality be reached, a backwash is initiated automatically.

4.4 Filter backwashing

Granular media filters need regular backwashing to remove clogging deposits and maintain efficient operation. Backwashing of a gravity filter involves flow reversal to dislodge the deposits. This usually gives significant expansion and *fluidization* of the bed. Once expansion occurs the grains are mobile and deposits can be detached and flushed out.

The various techniques include:

- fluidizing water wash
- fluidizing water wash plus surface jets
- air scour followed by a fluidizing water wash
- simultaneous air and water wash followed by a fluidizing water rinse.

All of the above backwash procedures are conducted at varying upflow rates and bed expansions. The water flow required depends on the physical characteristics of the media and the degree of bed expansion required.

4.4.1 *Backwashing with water*

Fluidized beds. A fluidized bed consists of solid particles or grains suspended by a fluidizing medium which can be a gas or a liquid.

Consider a vessel such as a filter shell filled with a granular material, e.g. sand. A fluidizing medium (usually water) enters at the base of the vessel and flows upwards through the bed of sand with a superficial velocity, v. The granular material has a fixed bed porosity of ϵ. When the bed is fluidized the particles or grains are suspended in equilibrium by the fluid drag forces exerted on them. The bed expands to occupy a greater depth and a consequent increased porosity ϵ, i.e. the grains are further apart.

At equilibrium in a fluidized bed, the downward force on the grains (i.e. their weight in water) is just balanced by the force caused by the upward flow of water. We can equate those forces and then substitute for the head loss gradient in the bed at the point of fluidization, using the Kozeny Carman equation above. We then arrive at an expression for the minimum upflow velocity needed to fluidize the bed:

$$\mathrm{v}_{\mathrm{mf}} = \frac{g}{180}\frac{(\rho_s - \rho)}{\mu}\frac{\epsilon^3 d^2}{(1-\epsilon)} \qquad (21)$$

where:

g is the acceleration due to gravity
ρ_s is the density of the grains
ρ the density of water. The porosity, ϵ, grain diameter, d, water viscosity, μ have been defined previously.

From this expression, it is clear that the minimum fluidization velocity depends on:

- media grain size
- media density
- media porosity and packing
- temperature (as it affects the water viscosity).

Changes in any of these parameters affect bed expansion and fluidization conditions, e.g. expansion behaviour during filter backwashing.

If we examine a granular bed subjected to upward flow of water, then we see that as we increase the flow rate, the head loss increases linearly with velocity while the bed remains fixed, i.e. Darcy's law is obeyed. Once we have reached v_{mf}, then the bed starts to expand upwards with the flow and the head loss becomes constant. The head loss is constant as there is no more loss of energy as drag on the grains. All the energy is used in supporting the grains. Typical fluidization curves are shown in Fig. 38.14.

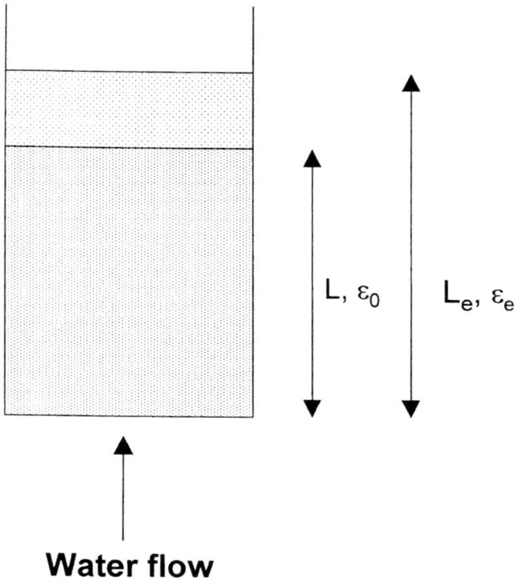

Fig. 38.14 Filter bed expansion by fluidization.

A simple experiment in a small column is a useful way of finding v_{mf} and the expansion characteristics for a particular filter medium under specific conditions.

A minimum expansion is required in order to flush out the deposits and get the whole bed mobile. The expansion required for this varies with filter media size.

A very important aspect of fluidizing a filter bed is that when the upward flow is stopped, grains settle at rates dependent on their size. In all practical cases, filter grains have a range of sizes and the larger ones settle fastest. This means that, after backwashing, smaller grains tend to be nearer the top of the bed, with larger ones towards the bottom. This is unfortunate, since it gives greater filter efficiency at the top, where there is already a greater removal of particles. Thus, the non-uniform distribution of deposited particles throughout the bed (Fig. 38.15) is accentuated. This effect may be partly overcome by the use of *multilayer filters* (see below).

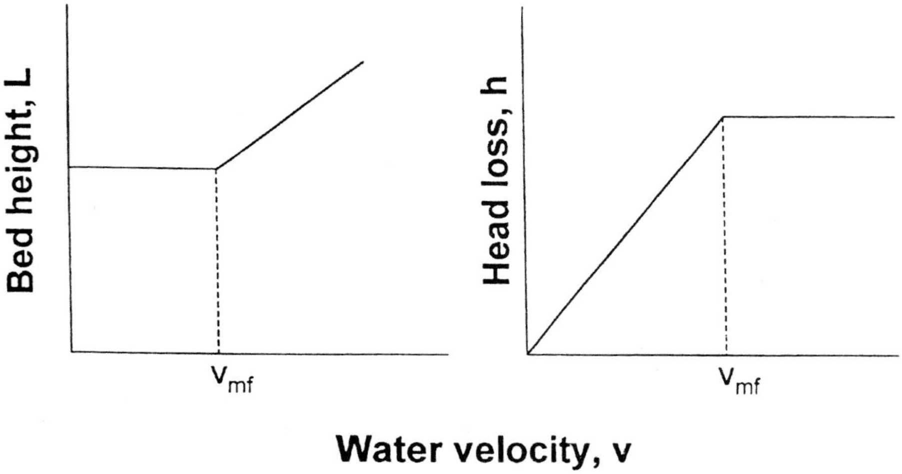

Fig. 38.15 Changes of filter bed height and head loss with increasing backwash flow rate.

Cleaning mechanisms. When a filter is being backwashed, the mechanisms that detach deposits and get the filter grains clean are:

- fluid shear forces
- grain collisions
- abrasion between grains
- forces associated with the air/water interface.

Research has shown that water-only backwash is a weak cleaning process as grain collisions are minimal. It is more effective to include air flow in the cleaning process.

4.4.2 The use of air scour

When air scour precedes the water wash, it is intended to break up and detach deposits ready to be flushed out.

Research has shown that while air scour causes a lot of agitation to grains in the top few centimetres of the bed, there is very little agitation deep down. The agitation is caused by bubbles erupting at the surface of the media. If a filter bed retains most of its deposits at the surface then air scour alone is adequate. If, however, it acts as a deep bed filter, as intended, then deeper deposits may not be dislodged.

4.4.3 Combined air and water wash

A combined wash means that the air and the water flow simultaneously up through the bed. At particular combinations of air and water flow rates for a given medium a phenomenon known as *collapse-pulsing* (Amirtharajah, 1993) is observed. This corresponds to the onset of three-phase fluidization.

What this means in practice is that fluidized bed conditions are achieved for a much lower water rate – typically less than 0.5 v_{mf}. Flow rates when a water only wash is used are greater than v_{mf}. In addition, during collapse-pulsing there is a very high degree of bed agitation, as pockets of air form and collapse within the media. This ensures that deposits are detached from the grain surfaces and can be particularly good for removal of biofilms.

4.4.4 Media attrition and loss during backwash

Media loss may result from inappropriate wash rates for the type of media in use. It is usually due to the lighter grains being carried out over the washwater weir with the backwash water. Lower density materials, e.g. anthracite and activated carbon, are washed out more easily despite the fact they are backwashed at lower rates.

Attrition of more friable filter media may result from a vigorous backwash, particularly if air scour is used.

Fines generated by media attrition can cause problems in drinking water treatment as they potentially shield bacteria from disinfection. An appropriate backwash regime has therefore to be carefully selected; we need to maximize cleaning without causing too much attrition of the media.

4.5 Improving particulate removal

Overall particulate removal can be improved by using finer and/or deeper filters or by improving the coagulation/flocculation stage. During the *ripening* period, particulate removal can be improved by optimizing backwash and filter start-up procedures.

An effective way of achieving more effective filtration is to use two or three layers of progressively finer media, as in multilayer filters.

4.5.1 *Filter ripening and start up*

When a filter is put back on line after backwashing, it usually takes some time for it to reach optimum removal performance. Particulate concentration in the filtrate during this phase is higher than the average level during the filter run. This is known as the *filter ripening period* and has been an area of concern in recent years as it may provide an opportunity for pathogens, such as *Cryptosporidium*, to pass into supply.

Water passing through the filter at start-up initially consists of backwash water remaining within the underdrain system and then the pores of the filter bed. Following this the remaining backwash water above the bed but below the washwater overflow level, passes through the filter. Finally, the new influent for filtration reaches the filter.

Ripening is thought to be due to the gradual build-up of deposits on the filter grain surface. These deposits then provide increased surface area for the accumulation and adhesion of subsequent deposits.

Following publication of the 1990 Badenoch Report on the *Cryptosporidium* problem, there has been much research into filter ripening in the UK. Options to reduce particulate passage during the ripening phase include:

- filter to waste
- slow start
- delayed start
- increased chemical dosing (e.g. during final stage of backwash)
- backwash optimization.

4.5.2 *Filter media*

When designing a filter it is important to select a medium or combination of media with appropriate physical properties for the particular application. The physical properties of filter media that affect performance during filtration and backwashing include:

- size distribution
- bed porosity or voidage
- grain density
- shape, e.g. spherical or angular
- surface characteristics and microporosity
- settling velocity and minimum fluidization velocity
- strength and friability of media grains
- acid solubility
- wettability.

The British Effluent and Water Association recently proposed a standard for filter materials (BEWA, 1993).

It is important to design filters with appropriate media, e.g. if we want to encourage biological growth then perhaps a rough surface is required for adhesion of microorganisms. A rough surface may also reduce the filter ripening period, but then may be harder to get clean during backwashing.

It may be important to minimize water consumption in backwashing so a lighter material might be desirable.

The shape of the grains, and their size distribution affect the bed porosity. Angular grains generally have a higher porosity, e.g. anthracite with about 0.55. Rounded grains, e.g. Leighton Buzzard sand, have a porosity of about 0.43. Porosity also depends on the packing of the grains and the degree of stratification of the media if it has a range of sizes.

4.5.3 *Multilayer filters*

Ideally filtration should take place through progressively decreasing media size, from coarse to fine, in order to maximize usage of the whole bed depth. However, as discussed above, when a filter is backwashed a single graded medium will stratify so that finer grains are at the top and coarser grains at the bottom of the bed, i.e. the opposite of what is required for a gravity or downward flow filter.

The idea behind dual and triple media filters is that we can achieve a progressively finer filter by having finer, denser media below coarser, heavier media. The difference in densities of the media should ensure that the finer material stays at the bottom during and after backwashing.

Dual media filters usually consist of a layer of coarser anthracite above a layer of finer sand. Triple media filters consist of anthracite, sand and then an even finer grained layer of garnet. Other configurations of media have appeared around the world, e.g. a triple media filter consisting of granular activated carbon (GAC) above German anthracite with a layer of fine sand at the bottom.

4.5.4 Backwashing of dual and triple media filters

Dual and triple media filters are usually designed so that they remain stratified after backwashing. It is relatively easy to maintain stratification between anthracite and sand, for the size ranges typically employed, however, mixing between the sand and fine garnet layers in triple media filters is difficult to avoid. This may not actually be detrimental to filter performance but further research is required on this subject.

For a given backwash flow rate it is likely that the different media layers will expand by different amounts.

4.6 Monitoring filter performance

It is very important to check that filters operate correctly, especially with regard to particle removal. Particles in the size range of *Cryptosporidium* oocysts are of special concern and their numbers in filtered water should be as low as possible. Oocysts that pass through a filter are not destroyed by conventional disinfectants at normal dosages, so it is essential that they are removed physically in water treatment processes.

The usual method of monitoring filtered water is by measuring turbidity. Current standards for drinking water turbidity are around 1 NTU (a conventional turbidity unit), but more stringent standards may be applied – perhaps down to 0.1 NTU. Such low levels are approaching the limits of detection of commercial turbidity meters, yet may still correspond to particle concentrations of several hundred per ml. Furthermore, turbidity is a measure of light scattered by the water sample and this is not a sensitive measure of particles much larger than the light wavelength, such as *Cryptosporidium* oocysts.

It has become quite common to apply more sensitive techniques, such as particle counting, to the monitoring of filtered water quality (Gregory, 1998). Typical particle counters can resolve particles down to about 1–2 μm in size. These methods can give an earlier indication of particle breakthrough than conventional turbidity monitoring.

Even a simple visual inspection can be helpful, particularly of the backwash and the filter bed surface where mudballs may be evident.

REFERENCES

Amirtharajah, A. (1993). Optimum backwashing of filters with air scour – a review. *Wat. Sci. Technol.* **27**, 195–211.

Badenoch Report (1990). Cryptosporidium *in water supplies.* Report of the Group of Experts. DoE, DoH, HMSO.

Bertsch, P.M., Thomas, G.W. and Barnhisel, R.I. (1986). Characterization of hydroxyaluminum solutions by Al-27 nuclear- magnetic-resonance spectroscopy. *Soil Sci. Soc. Am. J.* **50**, 825–830.

BEWA (1993). *Standard for the specification, approval and testing of granular filtering materials.* British Effluent and Water Association, BEWA, p. 18.93.

Bose, P. and Reckhow, D.A. (1998). Adsorption of natural organic matter on preformed aluminum hydroxide flocs. *J. Environ. Eng. – ASCE* **124**, 803–811.

Camp, T.R. and Stein, P.C. (1943). Velocity gradients and internal work in fluid motion. *Proc. ASCE* **79**, 1–18.

Deryagin, B.V. and Landau, L.D. (1941). Theory of the stability of strongly charged lyophobic sols and of the adhesion of strongly charged particles in solutions of electrolytes. *Acta Physicochim. URSS* **14**, 733–762.

Elimelech, M., Gregory, J., Jia, X. and Williams, R.A. (1995). *Particle Deposition and Aggregation. Measurement, modelling and simulation*. Butterworth-Heinemann, Oxford.

Gregory, J. (1973). Rates of flocculation of latex particles by cationic polymers. *J.Colloid Interface Sci.* **42**, 448–456.

Gregory, J. (1996). Polymer adsorption and flocculation. In: C.A. Finch (ed.) *Industrial Water Soluble Polymers*, pp. 62–75. Royal Society of Chemistry, Cambridge.

Gregory, J. (1997). The density of particle aggregates. *Wat. Sci. Tech.* **36**, 1–13.

Gregory, J. (1998). Turbidity and beyond. *Filtration Separation* **35**, 63–67.

Israelachvili, J.N. (1991). *Intermolecular and Surface Forces*. Academic Press, London.

Lin, M.Y., Lindsay, H.M., Weitz, D.A. *et al.* (1989). Universality in colloid aggregation. *Nature* **339**, 360–362.

Mallevialle, J. Odendaal, P.E. and Wiesner, M.R. (eds) (1996). *Water Treatment Membrane Processes*. McGraw Hill, New York.

Meakin, P. (1988). Fractal aggregates. *Adv. Colloid Interface Sci.* **28**, 249–331.

Morgan, J.W., Forster, C.F. and Evison, L. (1990). A comparative-study of the nature of biopolymers extracted from anaerobic and activated sludges. *Wat. Res.* **24**, 743–750.

Muyibi, S.A. and Evison, L.M. (1995). Optimizing physical parameters affecting coagulation of turbid water with moringa-oleifera seeds. *Wat. Res.* **29**, 2689–2695.

Mühle, K. (1993). Floc stability in laminar and turbulent flow. In: B. Dobiás (ed.) *Coagulation and Flocculation*, pp. 355–390. Marcel Dekker, New York.

Stumm, W. (1992). *Chemistry of the Solid-Water Interface*. Wiley Interscience, New York.

Tambo, N. and Watanabe, Y. (1979). Physical aspects of flocculation. I. The floc density function and aluminium floc. *Wat. Res.* **13**, 409–419.

van de Ven, T.G.M. and Mason, S.G. (1977). The microrheology of colloidal suspensions. VII. Orthokinetic doublet formation of spheres. *Colloid Polymer Sci.* **255**, 468–479.

Verwey, E.J.W. and Overbeek, J.Th.G. (1948). *Theory of the Stability of Lyophobic Colloids*. Elsevier, Amsterdam.

Yusa, M. and Igarishi, C. (1984). Compaction of flocculated material. *Wat. Res.* **18**, 811–816.

39

Microbial response to disinfectants

Jordi Morató[*], Jaume Mir[*], Francesc Codony[†], Jordi Mas[†] and Ferran Ribas[‡]

[*]*Laboratory of Microbiology, Universitat Politècnica de Catalunya, Terrassa-08222;* [†]*Universitat Autònoma de Barcelona, 08193-Bellaterra;* [‡]*AGBAR, (Sociedad General de Aguas de Barcelona) 08009, Barcelona, Spain*

1 INTRODUCTION

The main objective of disinfection is to guarantee the salubrity of drinking water to avoid its use as a source of disease. In fact, the World Health Organization estimates that nearly one half the population in developing countries is suffering from health problems associated with contaminated water, and in the poorest developing countries morbidity and mortality from waterborne diseases cause an immense burden of disease (Galal-Gorchev, 1993). Many infective diseases are transmitted through contact with poor quality water, especially gastrointestinal diseases. Particularly, rotavirus are the major contributors to death (infant and global) world-wide. Pathologies related to contact or ingestion of contaminated water are responsible for around 16 million deaths per year, and diarrhoea was associated with the death of more than 3 million children under the age of 5 during 1990 (Diamond, 1992).

On the other hand, there is a firm conviction that the elimination of dangerous bacteria in well-chlorinated water does not imply special problems, in spite of the resistance mechanisms that we will describe later in this chapter. Although in the last few years, water chlorination has attracted some criticism due to its secondary effects (chlorination by-products, as trihalomethanes), there is no doubt that it has been one of the greatest advances in water purification. In fact, it permits water authorities to use many supplies which would otherwise not have attained the required standard of purity (Mir *et al.*, 1997). Recently, alternative disinfectants to chlorine have been utilized, due especially to two main reasons:

1. that, in spite of its usefulness for the removal of pathogenic bacteria, it can create many problems concerning the generation of dangerous by-products due to their toxicity or carcinogenic effect. The debate between the need to chlorinate to remove pathogenic bacteria and the need to reduce chlorination to limit the generation of dangerous by-products is well known (Glaze *et al.*, 1993; Miller, 1993; Fielding, 1996; McCann, 2001). Even though this debate has no sense in developing countries, where the eradication of pathogens decimating populations is a priority, the approach is basically correct in industralized countries. This has led some authors to propose that, for waters that are not contaminated originally and that are distributed through a non-vulnerable distribution system, no chlorination, nor any other type of disinfection, is carried out (Levi, 1992; Hambsch, 1999; Van der Kooij *et al.*, 1999).
2. that chlorine is, at doses compatible with water consumption, a disinfectant insufficient for the removal of emerging non-bacterial pathogens, such as viruses and the cysts and oocysts of pathogenic protozoa.

Therefore, in water treatment, it would be a question of increasingly using alternative disinfectants that, at doses compatible with human consumption, are efficient for the removal of viruses and protozoa and minimize the generation of dangerous by-products.

The Handbook of Water and Wastewater Microbiology
ISBN 0-12-470100-0

However, classical studies, which have been included in the reviews of the most important disinfection treatises (Russell *et al.*, 1999; Block, 2001), mainly refer to disinfectants that are not normally used in water treatment, which is the basic object of the present chapter because, for wastewater, disinfection is not such a priority process. However, it is obvious that the comprehension of the mechanisms of action studied for the above-mentioned disinfectants can contribute to clarify the possible mode of action of the ones most commonly used in treatment.

2 TYPES OF DISINFECTANTS AND DISINFECTION MECHANISMS

2.1 Desired characteristics in water disinfectants

Risks associated with contaminated drinking water include bacterial, viral and protozoan diseases whose manifestation may include asymptomatic infection, slight discomfort or severe illness, and which may result in death depending upon the pathogenic agent and host response. On the other hand, water disinfectants must fit some requirements:

- *Disinfecting power: range of application.* Disinfectants must destroy or inactivate, in a specified contact time, the greater amount of classes and number of pathogenic microorganisms that can be present in raw water.
- *Easy to use: availability.* Disinfectants must be reasonably safe and convenient in their handling and application.
- *Easy to control: simple analytical determination.* Analytical methods to determine disinfectant concentration must be precise, simple, fast and economic.
- *Persistence: cost.* The disinfectant has to keep a suitable residual concentration in the water distribution system, to avoid recontamination or reproduction of any microorganisms present. The cost, installation, operation, handling, maintenance and repair of the equipment must be reasonable.
- *Secondary effects: toxicity.* As far as possible, disinfectants must not introduce or produce toxic substances. Otherwise, toxic substances must be maintained below guidance values or limits. Disinfectants must not change water characteristics in such a way that water will not be suitable for human consumption.

2.2 Types of disinfectants and their applications

Application, chemical structure and microbiological range are among the most usual criteria to classify disinfectants used in water treatment (Russell *et al.*, 1999; Block, 2001). In Table 39.1, a non-exhaustive list of the most common disinfectants used for drinking water treatment is presented. Other disinfectants, like those used in water for industrial uses, are not described here, because of their poor application in drinking water due to different reasons like practical, economical or safe handling.

It has to be noted that, as there is not a universal 'general purpose' and 'good for all conditions' disinfectant, disinfectants can be used in combination, to optimize their advantages and decrease their liabilities. The most usual combinations with synergic effect include chlorine-chlorine dioxide and chloramines, ozone and chlorine or chloramines, or UV radiation and chlorine or chloramine.

The choice of disinfectant depends on the initial water quality and its use:

- *Wastewater.* Common treatment of wastewater is usually limited to eliminate or to prevent its enhanced environmental or sanitary risks, and any disinfection step is not usually planned. Originally, ozone was used for wastewater disinfection as an alternative to chlorination. The growing interest for wastewater reuse and reclamation as a means to supply additional water has increased these techniques (Janex *et al.*, 2000).
- *Industrial water.* The main approach is to prevent the damage that water can cause or promote to the installation that contains it.

TABLE 39.1 Main disinfection systems utilized in water treatment (ND, not described)

Type/Name	*Application & advantages*	*Mode or target of action*	*Requirements & disadvantages*	*Concent. & contact time*	*Persist.*	*By-products*
Chlorine	Reduction of pathogens and bacterial counts. Versatility, economic and easy to use and control with residual effect (depends on pH range)	Oxidize cell components (-SH groups, nucleic acid) Oxidative destruction of cell wall, that lends the release of vital cytoplasmic compounds	React with organic impurities, converting them into carcinogens such as THMs. Microbial resistance to chlorination. Generates taste and odour	0.2–2.0 ppm 3–5 min	+++	THMs. Oxygen species such as hydroxyl radical, singlet oxygen and chlorine oxide radical
Chloramines	Chloraminated water presents lower heterotrophic plate counts, less taste and odour. More effective against *Legionella* and other bacteria than chlorine, for its better capacity to penetrate biofilms. It doesn' t originate THMs, and has residual effect	Inactivation of energy-producing enzymes. Inhibition of respiration processes after inactivation of phosphotransferase	The ratio between chlorine and ammonia species at the dosage step must be carefully controlled, to prevent the risk of nitrification, that increases the level of nitrites and promotes microbial regrowth	0.8–3.0 ppm 2–6 min	++++	Can generate nitrification
Chlorine dioxide	Very persistent, used in water disinfection as an alternative to chlorination. It doesn't generate trihalometanes (THMs), and seems to reduce halohydrocarbon compounds	Inhibition of respiration processes after inactivation of phosphotransferase. Reaction with internal RNA in viruses, after adsorption and penetration of the protein capsomeres	Destroys phenols, but not ammonia. Preparation and control operation are complex	0.05–0.75 ppm, 2–5 min	+++	Generates chlorates and toxic chlorites
Ozone	More efficient than free chlorine for viruses, bacterial spores and cysts. Efficiency independent of pH range. Destroys phenols, but not ammonia	Like chlorine, ozone promotes oxidation processes, although its oxidation potential is lower. The reaction with cytoplasmic substances carries the degradation of chromosomal DNA in bacteria and viruses, and the damage of the protein coat in viruses	Very unstable, rapidly decomposes in water. Its lifetime is too short to ensure a residual. High cost of operation, must be generated when used. A final disinfection step with chlorine or chlorine dioxide is also needed	0.2–0.4 ppm	No	Several active oxygen species. Bromate and hydrogen peroxide Biodegradable organic matter, that acts as a carbon source for microbial regrowth

(*continued on next page*)

TABLE 39.1 (*continued*)

Type/Name	*Application & advantages*	*Mode or target of action*	*Requirements & disadvantages*	*Concent. & contact time*	*Persist.*	*By-products*
Peroxides and other oxidant substances	Hydrogen peroxide, peracetic acid and potassium permanganate have been used in combination with some of the major disinfectants, like ozone or chlorine	Produce hydroxyl free radicals which can attack membrane lipids, DNA and other cell components	Permanganate has been described as very active against *Vibrio cholerae*, but not against other pathogens. UV radiation potentiates the activity of hydrogen peroxide	PA, 10 ppm–10 min	No	Active oxygen species
Iodine and derivatives (triiodine, pentaiodine or hypoiodous acid)	Simplier to use than chlorine, but less corrosive and less influenced by pH changes. Has better stable chemical storage characteristics. Better disinfectant in water with high turbidity levels	Iodination of essential molecules in cell (thiol groups). Affects the protein coat of viruses, by attacking aminoacids tyrosine and histidine	A weaker oxidant than chlorine. It has been used only for single and very small on-site systems, like single-family water supplies or water for astronauts	0.5–3.0 ppm 15 min	++	No
Metals	Used specially in water recirculation systems like those in hospitals, hotels, etc., or reservoirs, because of its higher contact times needed	Affect thiol groups and the electron transport chain. Copper interferes at the metal enzyme level in metabolic pathways, while silver adsorbs on surfaces, interfering with microbial metabolism	Its combination is more effective than that seen with either ion alone, especially against *Legionella pneumophila*	Cu: 0.04–0.4 ppm 2 h Ag: 0.02–0.1 ppm 10 h	+++	No
UV radiation	Small supply systems or as a finisher for larger ones	Generation of reactive oxygen species. DNA and RNA absorb UV light. UV radiation alters their nitrogenous heterocyclic basis, which generates dimers and trimers, that rends microorganisms unable to replicate	Application limited when microorganisms are capable of photoreactivation (self-repair). Turbidity level must be very low	35 mW.s/ cm^2	No	Several active oxygen species

Temperature	Very small amounts of water or closed circuits, surgical and pharmaceutical uses	Freezing water near congelation or heating it up from 80°C to boiling point inactivates or destroys microbes	High costs due to the amount of energy required	Up to 80°C	No	No
Filtration	Eliminates all microbial material (in food, electronics or pharmaceutical industries). Doesn't leave chemicals to water, nor modifies its taste and odour. Efficiency is independant of raw water quality	Physical separation, where efficiency depends only on the size of the pores of the membrane. The driving force can be a difference of pressure, electricity or concentration potential	High energy costs	1–100 nm of pore (ultra and nanofilter.) RO even lower	No	No
Electrochemistry	For small close systems or reservoirs. Very useful for disinfection of carbon filtering devices colonized by resistant microbes	Generation of toxic substances and by electrochemical oxidation of intracellular coenzimeA (CoA) at lower potential, using carbon-cloth electrodes, what rends a dimer by disulfide bond formation	High energy costs. Not developed for water distribution systems, used only for disinfection of filtration devices	+1.0 to −1.0 V	No	ND
Photocatalysis	Photocatalysis using TiO_2 and several light sources. Method to treat water without adding chemicals	Generation of oxidizing species like hydroxyl radical or superoxide	Effective against *Cryptosporidium parvum*	380 nm light 30 min	No	ND

THM: trihalomethanes

To maintain water quality, some of these treatments include final disinfection. As this water will not be consumed for drinking, considerations about safety risks include only those related to the environment and safe handling (Brözel and Cloete, 1993). Curiously, shock treatments that are recommended (or obliged in many cases) include chlorine, as the case of *Legionella* elimination from cooling towers.

- *Drinking water.* Drinking water also includes water for bathing and hygiene, and for hydrotherapy. Although chlorine is still the disinfectant most commonly used around the world, its use has been disputed in more developed countries (Bellamy *et al.*, 2000), due to the harmful by-products that it leaves, not only for the taste, but also for health (trihalomethanes, chloramines or chlorophenols) (AWWA, 2000a). In many small or medium size distribution systems, chlorine has been totally substituted by other disinfectants (Parrotta and Bekdash, 1998; AWWA, 2000b) and also in large distribution systems its use is combined with other disinfectants (Schmidt *et al.*, 2000).
- *Water for the chemical industry.* When used as a solvent or reagent for chemical reactions, water must fit specific requirements that depend mainly on the type of process, going from bulk to fine chemicals. The electronics industry could be considered within this group, although in some cases absolute purity is needed. Usually, treatment is required to obtain water with a very low content of salts and organic matter. Nevertheless, as fine chemicals are mainly related to the pharmaceutical industry, the tendency is to move to pharmaceutical requirements.
- *Water for the food industry.* A very high level of disinfection, or sterility in many cases, is required. Microbial growth must be prevented, from the preparation of the product to its consumption. Best extended treatments include heat and filtration.
- *Water for the pharmaceutical industry.* Except for the production of nutraceutical products, sterility is required, at least during the final production steps. The pharmaceutical industry obtains products that must be consumed inside living organisms (drugs and parenteral solutions), so quality requirements are very high (Cross and Phil, 1996; Salazar, 1999) and are regulated by international standards (i.e. GMP, Good Manufacturing Practices). So, not only must the presence of microorganisms in water be prevented, but also their remaining structures after inactivation or destruction, because they can promote allergenic responses.

2.3 Disinfection mechanisms: principles of disinfection activity

Biocides are likely to have multiple target sites within the microbial cell and the overall damage to these target sites results in the bactericidal effect (Denyer and Stewart, 1998; Maillard, 2002). Bacteriostatic effects, usually achieved by a lower concentration of a biocide, might correspond to a reversible activity on the cytoplasmic membrane and/or the impairment of enzymatic activity (Maillard, 2002). Even though the present knowledge on resistance mechanisms by microorganisms is probably higher than that on disinfection mechanisms, it is believed that the three primary mechanisms of pathogen inactivation are:

1. to destroy or impair cellular structural organization by attacking major cell constituents
2. to interfere with energy-yielding metabolism rendering enzymes non-functional
3. to interfere with biosynthesis and growth by preventing synthesis of normal proteins, nucleic acids, coenzymes, or synthesis of the cell wall.

In potable water treatment, it is believed that the primary factors controlling disinfection efficiency are the ability of the disinfectant to oxidize or break the cell wall and its ability to diffuse into the cell and interfere with cellular activity (Montgomery, 1985; Ribas and Matía, 1999). In fact, because most disinfectants used in water treatment are oxidants, it could be reasonable to suppose that their action is primarily due to a redox chemical reaction. Even though this affirmation may be considered as globally true, it also presents some exceptions, e.g. hydrogen peroxide which is of high redox potential but low disinfecting power. On the other hand, the difference in oxidizing power between an ion and the corresponding acid (e.g. hypochlorous acid is more disinfectant than hypochlorite ion) is often bigger than it would be expected from the differences of redox potential, because electric charges surrounding bacteria and viruses constitute a first barrier to the penetration of oxidizing anions (Cramer *et al.*, 1976; Block, 1982; Rumeau, 1982).

In the case of resistant dormant bodies, such as bacterial spores or cysts/oocysts of pathogenic protozoa, it is necessary to consider, additionally, especially resistant protecting structures.

2.3.1 *Inactivation kinetics*

As early as 1897, Kronig and Paul were the first authors to apply the emerging rules of chemical kinetics to the disinfection process, considering that it must be a chemical one (Russell *et al.*, 1999). They plotted for the first time the logarithm of the surviving organisms (ln Nt/No) against time.

Madsen and Nyman (1907) developed the theme of linear log survival/time, and concluded that the different rates of bacterial destruction were essentially determined by the variability of resistance among the cells of the population. Other authors, such as Chick (1908), Watson (1908), Knaysi (1930), Jordan and Jacobs (1944), Rahn (1945) and Berry and Michaels (1947), continued the studies on bacterial survival against disinfectants. Eddy (1953) investigated the death rate of a species of *Klebsiella* under chemical stress induced by several disinfectants and no evidence was found that cells in a given culture possessed variable resistance, and survivors gave rise to a population no more resistant than the originals. The lag phase, observed before the decrease in viability proceeded logarithmically, suggests that a number of events must occur before a significant number of cells begin to die. The number of these events may be the variable factor.

Prokop and Humphrey (1970) revised the most usual previous mathematical models, considering five typical surviving curves (Fig. 39.1). Casolari (1981) described a mathematical model in which some functions can describe the varying shapes of the survival curves, which represent interactions of chemicals having differing cellular targets and modes of action with highly complex microorganisms at different stages of growth, and with different structures and chemical compositions. This subject has been reviewed by Hugo and Denyer (1987).

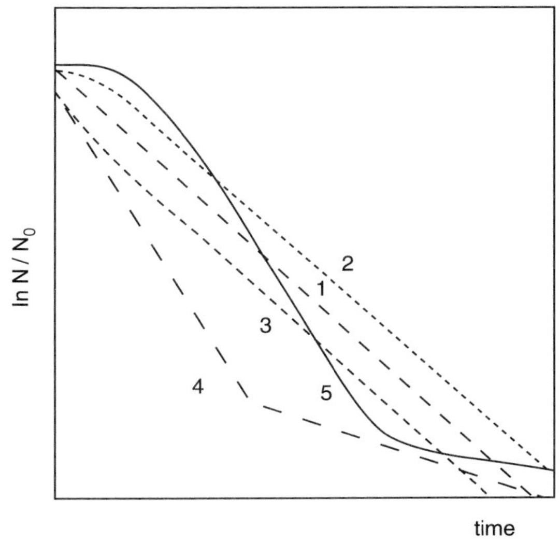

Fig. 39.1 Survival curves, Ln N/N_0 versus time. 1, exponential kinetics; 2, concave upward kinetics; 3, concave downward kinetics; 4, dichotomous (multiple) kinetics; 5, sigmoidal kinetics (adapted from Prokop and Humphrey, 1970).

Mir *et al.* (1997), in studies of inactivation by chlorine (with or without chloramphenicol) of different bacterial strains, Gram-positive and Gram-negative, introduced the kinetic concept of Ki, the inactivation rate constant, by comparison with Ks, in Michaelis-Menten enzyme kinetics (considering enzymic saturation), or with Ks in Monod growth kinetics (considering limiting rates of transport and metabolism of substrates). Ki is the disinfectant concentration at which D (the inactivation decay rate, which can be related to the inactivation speed) is half of the maximum. In their experiments, a maximum decay rate of 1.5 is considered, because this value is the slope calculated for the highest inactivation that can be obtained if all bacteria added to the flask are inactivated at the shortest contact time (2 min) tested.

The Chick-Watson (Chick, 1908; Watson, 1908; Haas and Karra, 1984) pseudo first-order rate law stated:

$$\frac{dN}{dt} = -kNC^n \qquad (1)$$

where:

dN/dt is the rate of inactivation
k is a pseudo-first-order rate constant found experimentally
N is the number of survivors
C is the disinfectant concentration and
n is a dilution factor or empirical constant.

Integration of equation (1) yields:

$$Log(N/N_0) = -kC^nT \qquad (2)$$

If the exponent coefficient $n = 1$, the product CT is constant for a given combination of disinfectant and microorganism. If $n > 1$ the concentration is dependent upon time and if $n < 1$ the time is more important than the concentration. Chlorine and water bacteria seem to adjust enough to this model, assuming that n is very similar to 1. However, the inactivation of *Giardia* cysts by chlorine yields a coefficient $n < 1$. For this reason, the lists given in the literature about CT products for different microorganisms and disinfectants are controversial.

While the Chick-Watson model seems to fit well for the inactivation by chlorine, ozone behaviour seems to be quite different: an initial rapid phase followed by a slower one. Finch *et al.* (1994) have studied the inactivation of *Giardia* cysts and *Cryptosporidium* oocysts by using an alternative model to that of Chick and Watson, proposed by Hom (1972), and expressed by the differential equation:

$$\frac{dN}{dt} = -kmNC^nt^{m-1} \qquad (3)$$

where:

m is an empirical constant, and
t is the contact time.

Assuming that the concentration of the disinfectant remains approximately constant during the course of the contact time, equation (3) is integrated as follows:

$$Log(N/N_0) = -kC^nT^m \qquad (4)$$

According to the experiments of Finch *et al.* (1994), for *Giardia* cysts $n = 0{,}72$ and $m = 0{,}24$, and for *Cryptosporidium* oocysts $n = 0.23$ and $m = 0.64$.

In a study comparing the effect of various disinfection methods on the inactivation of *Cryptosporidium* (Finch *et al.*, 1997), the oxidant was added into a batch reactor in an ideal plug flow manner (Lev and Regli, 1992), i.e. concentration of disinfectant in the flow segment decreases with time as the segment passes along the reactor. Assuming first-order decay, the number of disinfectant molecules destroyed per unit time is proportional to the number of molecules remaining, and the concentration of disinfectant is given by:

$$C = C_0e^{-k't} \qquad (5)$$

where:

C is the final ozone concentration at the end of the contact time
C_0 is the initial concentration at time zero and k^{st} is the first-order rate constant.

Substitution and integration of equation 5 into the Hom rate law gives:

$$Log(N/N_0) = -kmC_0^n \int_0^T e^{-k'tn}t^{m-1}dt \qquad (6)$$

Named the integral Hom model, it is based on three assumptions:

1. the rate of decomposition of the oxidant is not a function of the organisms present
2. the chemical disinfectant decay is independent of the inactivation reaction
3. and rate law parameter estimates are independent of the water characteristics.

This expression cannot be integrated to closed-form, as the limits of integration and integrand are both dependent on the contact time. Nevertheless, equation (6) can be solved numerically.

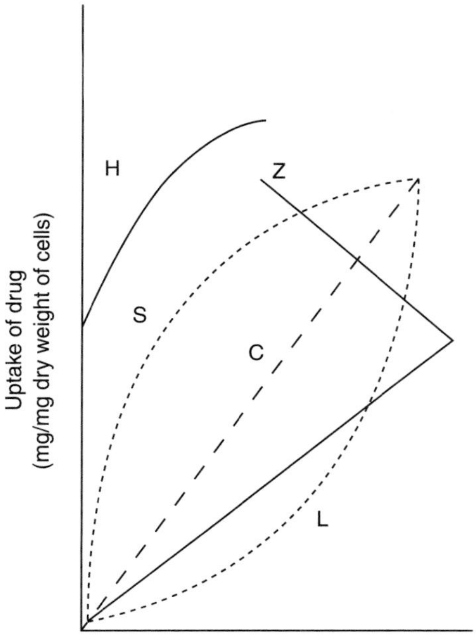

Fig. 39.2 Pattern of adsorption isotherms (adapted from Hugo, 1999).

2.3.2 *Interactions with the whole cell*

A biocide must reach and interact with its microbial target site(s) to be effective. Although different disinfectants can be selective in their action for some cellular structures (organelles) or enzymes, their first apparent interaction is with the whole cell.

Herzog and Betzel (1911) investigated the role of *adsorption* in the disinfection process, using baker's yeast as a target microorganism, and many other authors have continued with these studies. Adsorption isotherms may be plotted and some notion of the adsorptive mechanism can be deduced. Five main patterns of adsorption can be considered: S-shaped (S), Langmuir (L), high-affinity (H), constant partition (C) and Z pattern (Fig. 39.2).

With respect to the changes in *electrophoretic mobility*, we can say that, so far as is known, bacterial cells are normally charged and, if suspended in water on a suitable electrolyte solution containing electrodes to which a potential has been applied, the cells will migrate to the positively charged electrode. The effects of disinfectants on cell mobility can be studied and, from the data obtained, some idea of the disinfectant-cell interaction and the effects of disinfectants on the charged cell surface can be deduced (Lerch, 1953; James, 1972; Richmond and Fisher, 1973).

Although the primary mode of inactivation by potassium permanganate is the direct oxidation of cell material or specific enzyme destruction (Webber and Posselt, 1972), a unique mode of action for permanganate is the *precipitation* of manganese dioxide. This mechanism of disinfectant interaction with the whole cell represents an additional method for the removal of microorganisms from potable water (Cleasby *et al.*, 1964). In colloidal form, the manganese dioxide precipitate has an outer layer of exposed OH^- groups, which are capable of adsorbing charged species and particles in addition to neutral molecules (Posselt *et al.*, 1967). As the precipitate is formed, microorganisms can be adsorbed into the colloids and settled.

2.3.3 *Interactions with external structures (cell wall and cytoplasmic membrane)*

When non-specific and oxidant disinfectants, such as the different forms of chlorine, are utilized the disinfection action will focus first on the outer most structures of the cell (cell wall and cytoplasmic membrane).

All eubacteria (except mycoplasmae) have a cell wall, which confers mechanical rigidity to the cell and protects the underlying cytoplasmic membrane. In Gram-positive bacteria, the cell wall consists of a thick layer of peptidoglycan interspersed with teichoic acid. A typical Gram-negative rod has a much thinner peptidoglycan layer closer to the cytoplasmic membrane. Besides these structures and external to the peptidoglycan, Gram-negative bacteria have a periplasmic space, while the outermost region is termed the outer membrane. This complex, outer part, exclusive of Gram-negative bacteria, contains lipoprotein (covalently linked to the peptidoglycan layer), lypopolysaccharide and phospholipids.

Gram-negative bacteria are generally less sensitive to biocides than Gram-positive bacteria because of their outer membrane (Table 39.2). However, Gram-negative bacteria seem to be more resistant to chlorine (Trueman, 1971; Mir *et al.*, 1997; Norton and LeChevallier, 2000). Membrane permeability loss will be the first lethal effect of chlorine on *E. coli*, due to damage to the external membrane, producing an imbalance in the transmembrane ionic gradient (Berg *et al.*, 1986). When sublethal chlorine levels are present, respiration inhibition will be the first effect (McFeters and Camper, 1978, Haas and Englebrecht, 1980).

Among the disinfectants used in water treatment, chlorine dioxide has been mentioned as disrupting the permeability of the outer bacterial membrane (Aieta and Berg, 1986). These results were supported by previous findings (Olivieri *et al.*, 1985), which found that the outer membrane proteins and lipids were sufficiently altered by chlorine dioxide to increase permeability, allowing it to reach the internal structures.

At the level of the outer membrane, there is also the effect of chlorine on adhesions, which are responsible for bacterial attachment to a host, where bacteria have pathogenic characteristics (Walsh and Bissonette, 1983).

EDTA is not a powerful bactericide, but it has some effect on Gram-negative bacteria, such as *Pseudomonas aeruginosa*, and it is now known to be inhibitory to Gram-positive bacteria as well (Russell, 1971; Wilkinson, 1975; Kraniak and Shelef, 1988; Vaara, 1991; Russell and Chopra, 1996). However, it can potentiate the activity of chemically unrelated antibacterial compounds against Gram-negative bacteria (Russell, 1971; Wilkinson, 1975). Although EDTA has an effect on Gram-positive bacteria, it does not increase their sensitivity to antibacterial agents (Russell, 1971; Russell and Furr, 1977).

TABLE 39.2 Relative microbial susceptibility to biocides (−, least resistant, +, most resistant)

Enveloped virus	−	
Gram-positive bacteria		Free cells
Non-enveloped virus		
Fungi		Aggregates
Gram-negative bacteria		
Prokaryotic-eukaryotic interactions		Attached cells
Mycobacteria		Biofilms
Spores		
Cysts		
Oocysts		
Prions	+	

The first sites attacked by ozone are the cell wall and membrane, either through glycoproteins, glycolipids or certain amino acids as tryptophane (Scott and Lesher, 1963; Goldstein and McDonagh, 1975). Some studies suggest that ozone alters proteins and unsaturated bonds of fatty acids in the membrane structures, leading to cell lysis (Scott and Lesher, 1963; Pryor *et al.*, 1983). Inactivation of bacteria by ozone is attributed to an oxidation reaction (Bringmann, 1954; Chang, 1971). However, some researchers believe that ozone disinfection is a result of direct ozone reaction, while others believe that the hydroxyl radical mechanism is the most important one for disinfection (Hoigné and Bader, 1977).

Once the cell wall is overcome, the cytoplasmic membrane is the preferred target of disinfectants at low concentrations (Haas and Englebrecht, 1980; Duguet, 1981). Several substances are known to disrupt the cytoplasmic membrane. These agents may cause a leakage of intracellular materials and modification of cell permeability, although other effects can also occur (Lambert, 1978; Denyer and Stewart, 1998; Hugo, 1999). Bacterial lysis, by disruption of the wall, is quite a strange phenomenon that only takes place with massive doses of disinfectant, mainly ozone, among those normally used in water treatment. However, low concentrations of phenol, formaldehyde, hypochlorite and mercuric chloride have been reported to produce lysis of *E. coli* (Hugo, 1999). Chlorhexidine induces leakage at a low concentration and, at bacteriostatic concentrations, inhibits membrane-bound ATPase (Harold *et al.*, 1969). Chlorhexidine affects the outer membrane of Gram-negative bacteria with the release of periplasmic enzymes and disturbs the functionality of the inner membrane (Kuyyakanond and Quesnal, 1992; Barret, 1994).

Once the cell membrane is damaged, leakage of cell material is promoted and can affect inner cell constituents. However, leakage is best considered as a measure of the disruption of the cell permeability barrier, and it might reflect a bacteriostatic effect rather than cell death (Maillard, 2002). Once the permeabilization of the membrane has been carried out, the respiratory enzymes are oxidized via their sulphhydryl groups: this action has been observed both for chlorine (Rumeau, 1982) and ozone (Block, 1982; Langlais *et al.*, 1991), but is only accepted as an *in vitro* event by some authors (McFeters and Camper, 1983). On the other hand, some disinfectants can provoke interference with membrane enzymes and hexachlorophene inhibits part of the membrane-bound electron transport chain in *Bacillus megaterium* (Frederick *et al.*, 1974). Mercuric salts act at low concentrations (10^{-6} M), by combining with membrane enzymes containing thiol groups (Hugo, 1999). Bronopol (2-bromo-2-nitro-propanol-3-diol) oxidizes thiol groups to disulphides in bacteria. These compounds also cause membrane damage as indicated by leakage of material, which absorbs at 260 nm.

It has been known for many years that several agents can uncouple oxidative phosphorylation. These uncoupling agents inhibit ATP synthesis in a different way from ATPase inhibitors (Hammond, 1979). Uncoupling agents such as 2,4-dinitrophenol, tetrachlorsalicylanilide (TCS) and carbonylcyanide-m-chlorophenyl-hydrazone are lipid-soluble; they dissolve in biological membranes, dissociating oxidation from phosphorylation, effectively short-circuiting the proton-motive force. It has been known for nearly a century that nitrophenols interfere with oxidative phosphorylation without inhibiting other metabolic processes. Nitrophenols short-circuit the membrane, causing a rapid backflow of protons (so they are also called proton ionophores) into the cell and are active at concentrations of 10^{-4} M. Some other phenolic compounds cause both backflow of protons and leakage of materials. Hammond (1979) has divided the true inhibitors of ATPase into two groups: those acting on both the soluble and membrane-bound ATPase, and those inhibiting only the membrane-bound enzyme.

2.3.4 Interactions with the inner cell constituents

There are four main targets for antibacterial drugs: the cytoplasm itself, cytoplasmic enzymes, nucleic acids and ribosomes. The irreversible coagulation of cytoplasmic constituents is a drastic lesion that is usually seen at drug concentrations far higher than those causing general lysis or leakage. Bancroft and Richter (1931) observed coagulation of bacterial cell protein by means of microscopic observations.

As for the effects of disinfectants on metabolism and cytoplasmic enzymes, despite the large volume of work over many years, and still being prosecuted, the general conclusion is that enzyme inactivation is only one of many events caused by chemical stress and is not likely to be a prime mechanism of bacterial death. However, in some circumstances, it may be a cause of bacteriostasis. This subject has been reviewed by Rahn and Shroeder (1941), Roberts and Rahn (1946) and Hugo (1957).

Silver salts, sometimes used for disinfection of potable water and especially swimming pools, combine with thiol (-SH) groups in proteins but also interact with DNA (Russell and Hugo, 1994).

Although there are a number of antibiotics which affect the biosynthesis and functioning of nucleic acids, only a few other drugs affecting these targets have been identified: acridine dyes, formaldehyde, phenylethanol and ethylene oxide (Hugo, 1999; Russell, 2001a).

Ribosomes are associated with the formation of peptides from amino acids ordered by messenger RNA and assembled by transfer RNA. This process, that is a target for many antibiotics, is also sensitive to some non-antibiotic antibacterial agents such as toluene (Jackson and De Moss, 1965) and possibly EDTA (Russell, 1971). Hydrogen peroxide dissociates the 30S and 50S subunits of the 70S ribosomes in *E. coli* (Nakamura and Tamaoki, 1968) and p-chloromercuribenzoate (0.5 mM) dissociates the 100S ribosomes of *E. coli* into 70S monomers (Wang and Matheson, 1967). However, the ribosome cannot be considered as a prime target for the specific selective action of any known antimicrobial agent, although it may be destroyed by some disinfectants.

The inhibition of the protein synthesis has been pointed out by Benarde *et al.* (1967), by using chlorine dioxide, and by Haas and Englebrecht (1980), who used free chlorine. At higher doses, ozone modifies the structure of nucleic acids.

In bacteria, chlorine was found adversely to affect cell respiration, transport and DNA activity (Haas and Englebrecht, 1980). Water chlorination caused an immediate decrease in oxygen utilization in both bacteria (*Escherichia coli, Mycobacterium fortuitum*) and fungi (*Candida parapsilopsis*).

The mechanisms by which chloramines inactivate microorganisms have been studied to a lesser degree than chlorine. Once it has penetrated the bacterial cell, monochloramine readily reacts with four aminoacids (cysteine, cystine, methionine and tryptophan (Jacangelo *et al.*, 1987). The inactivation mechanism for chloramine is therefore thought to involve inhibition of proteins or protein-mediated processes such as respiration.

Studies on the inactivation mechanisms of microorganisms by chlorine dioxide have focused on two more subtle mechanisms that lead to this inactivation: determining specific chemical reactions between chlorine dioxide and biomolecules, and observing the effect that chlorine dioxide has on physiological functions. The first disinfection mechanism has been studied more in viruses (see later). The second type of disinfection mechanism focuses on the effect of chlorine dioxide on physiological functions. Although it has been suggested that the primary mechanism was the disruption of protein synthesis (Benarde *et al.*, 1967), later studies reported that this might not be the primary inactivation mechanism (US EPA, 1999). As mentioned before, a more recent study reported that chlorine dioxide disrupted the permeability of the outer membrane (Aieta and Berg, 1986).

Beyond the bacterial cell membrane and cell wall, ozone may affect both puric and pyrimidinic bases in nucleic acids (Scott and Lesher, 1963). In addition, ozone disrupts enzymatic activity by acting on the sulphhydryl groups of certain enzymes.

Besides chemical agents, the physical agent most used for disinfection in water treatment is UV radiation, which is efficient at inactivating vegetative and resistant forms of bacteria, viruses and vegetative and resistant forms of bacteria, viruses and pathogenic protoza (Gerba *et al.*, 2002; Morita *et al.*, 2002; Linden *et al.*, 2002). For ultraviolet radiation, the target is always nucleic acid (Harris *et al.*, 1987). Electromagnetic radiation in the wavelengths ranging from 240 to 280 nm effectively inactivates microorganisms by damaging their nucleic acid. The most potent wavelength for damaging DNA is 253.7 nm (Wolfe, 1990).

The germicidal effects of UV involve photochemical damage to RNA and DNA within the microorganisms. Nucleic acids are the most important adsorbers of light energy in the wavelength 240–280 nm interval (Jagger, 1967). Damage often results from the dimerization of pyrimidine molecules. Cytosine (in DNA and RNA), thymine (only in DNA) and uracil (only in RNA) are the three sensitive types of pyrimidinic bases. Replication of the nucleic acid becomes very difficult once the pyrimidinic molecules are bonded together, due to the distortion of the DNA helical structure by UV radiation (Snider *et al.*, 1991). Moreover, if replication does occur, mutant cells that are unable to replicate will be produced.

Under certain conditions, some organisms are capable of repairing damaged DNA and reverting back to an active state in which reproduction is again possible. Typical photoreactivation occurs as a consequence of the catalysing effects of sunlight at visible wavelengths outside the effective disinfecting range. The extent of reactivation depends on different organisms. Coliforms and some bacterial pathogens, such as *Shigella*, have exhibited the photoreactivation mechanism; however, viruses and other bacteria cannot photoreactivate (Hazen and Sawyer, 1992). Medium-pressure mercury UV lamps have recently been considered an effective alternative to low-pressure lamps. Under laboratory conditions, *E. coli* underwent photorepair following exposure to the low-pressure UV source, but no repair was detectable following exposure to the medium-pressure UV source at the initial doses examined (Zimmer and Slawson, 2002).

2.3.5 *Disinfection of eukaryotic microorganisms*

Several biocides possess significant antifungal activity, but their mechanisms of action against yeasts and moulds are not well understood. Although it is assumed that the mechanisms are similar to those responsible for bacteria, it is not necessarily so (Russell and Furr, 1996; Russell, 1999b). Glutaraldehyde combines strongly with amino groups at the cell surface and the presence of polymers, such as chitin, indicates a potentially reactive site (Gorman *et al.*, 1980). The cytoplasmic membrane is probably the major target for many agents, such as chlorhexidine, organic acids, and esters and alcohols (Hiom *et al.*, 1995; Russell and Furr, 1996). Ethanol also induces leakage of intracellular material from *Saccharomyces cerevisiae* (Salueiro *et al.*, 1988). Organic acids are rapidly taken up by yeasts and act as lipophilic acids that damage the cytoplasmic membrane (Macris, 1974; Krebs *et al.*, 1983). It is very probable that organic acids and esters produce inhibition of the proton-motive force across the cytoplasmic membrane, either in bacteria or in fungi (Hunter and Segal, 1973).

Inactivation mechanisms of pathogenic protozoa have been less studied than those of viruses and bacteria, and until recently, ozone had been studied exclusively. Experiments using trophozoites of *Naegleria* and *Acanthamoeba* showed that they were rapidly destroyed and the cell membrane was ruptured by ozone (Perrine *et al.*, 1984). In the case of amoeba cysts, the targets of ozone seem to be mucopolysaccharide plugs distributed over their surface (Perrine and Langlais, 1989). It seems reasonable that ozone initially affects the *Giardia* cyst wall and makes it more permeable (Wickramanayake *et al.*, 1984). When the wall barrier is overcome, aqueous ozone penetrates into the cyst and damages the plasma membranes and, eventually, affects the nucleus, ribosomes and other ultrastructural components. Little information can be found on the mode of action of ozone on protozoan oocysts. However, it has been suggested that ozone causes the oocyst density to decrease and alters the oocyst structure (Wickramanayake *et al.*, 1984; Wallis *et al.*, 1990).

Recent studies performed by using different disinfectants and combinations of disinfectants against *Cryptosporidium* oocysts only refer to disinfection kinetics and disinfection efficiencies. However, no reference to disinfection mechanisms can be found (Finch *et al.*, 1997). Ozone is a very good disinfectant, but ozone followed by monochloramine or free chlorine has an enhanced disinfection effect greater than the sum of the individual disinfectants (synergistic effect). Chlorine dioxide is an effective disinfectant, which also exhibits synergism with monochloramine and free chlorine. The CT approach derived from the work of Chick and Watson in 1908 is not a suitable approach for disinfection process design, but the integral Hom kinetic modelling appears to be more flexible.

2.3.6 *Mechanisms of virucidal action*

The mechanisms of the virucidal action of biocides are poorly understood, although recent progress has been encouraging (Maillard and Russell, 1997; Maillard, 1999). Because viruses have a simpler structure than microorganisms, they present fewer targets for biocide action. In addition, viruses lack metabolic activity. These target sites are the envelope (if present), the capsid and the viral genome.

The viral envelope is usually derived from the host cell cytoplasmic membrane and contains a large amount of lipids. The capsid is responsible for viral shape and protects viral nucleic acid (only RNA or DNA) from harmful external influences, including disinfection. Capsid constituents are primarily protein in nature. Destruction of the viral capsid results in the release of potentially infectious nucleic acid. As this nucleic acid is the infectious component of viruses, viral inactivation is only completed when this is destroyed.

The first studies on virus disinfection were performed by using bacteriophages.

These bacterial viruses offer a rapid means of identifying target sites for biocide action (Maillard *et al.*, 1994, 1996). However, in the last few years significant studies on the inactivation of poliovirus and echovirus have also been undertaken (Chambon *et al.*, 1992, 1994). Most biocides are capable of altering viral capsid structure, but only a few can alter viral nucleic acid. Structural damage to the capsid might not always represent a loss of viral infectivity.

For example, ozone is the only disinfectant used in water treatment which is able to modify, at the doses used, the chemical structure of lipids, by oxidation of the fatty acids, altering the envelope of the viruses belonging to the groups that have a lipid membrane (Riesser *et al.*, 1976).

The first site of action for the inactivation of viruses without a lipid envelope are the proteins of virus capsid, whose modification causes increased permeability, both as for chlorine (Calvet Churn *et al.*, 1983) and ozone (Cronholm *et al.*, 1976; Riesser *et al.*, 1976; Langlais *et al.*, 1991). Ozone seems to modify the capsid sites that the virion uses to fix on cell surfaces. However, high concentrations of ozone dissociate the whole capsid and the nucleic acid (RNA or DNA) is liberated. In fact, the modification of the nucleic acid seems to be the transformation of consequences more important to the inactivation of viruses, due to the action of chlorine (Fauris *et al.*, 1986), chlorine dioxide (Hauchman *et al.*, 1986) and ozone (Roy *et al.*, 1981).

Once the phage coat was broken into many pieces and the absorption to the host pili was disrupted by ozone, the release of RNA from the bacteriophage F2 particles was observed (Kim *et al.*, 1980). Further, the naked RNA may be inactivated by ozone at a lower rate than that for RNA within the intact phage. Quite similar observations have been made for DNA phages, e.g. bacteriophage T4. In contrast, more recent work on the tobacco mosaic virus shows an additional specific effect of ozone on RNA: the damaged RNA cross-links with amino acids of the coat protein subunits, losing its infectivity because of its loss of protein coating (US EPA, 1999). It is generally believed that chlorine inactivates some viruses, such as Hepatitis A virus (HAV), by damaging the nucleic acid of the virus (Dennis *et al.*, 1979; O'Brien and Newman, 1979). However, although complete inactivation of infectivity of HAV could be measured after 30 min of exposure to 10–20 mg of chlorine per litre, antigenicity was not completely destroyed under these conditions (Li *et al.*, 2002).

Few studies have been performed to determine the mechanism for viral inactivation by chloramines. The primary mechanism for poliovirus inactivation involved the protein coat (Fujioka *et al.*, 1983), but the initial site for destruction of bacteriophage F2 seems to be RNA (US EPA, 1999). Similar to free chlorine, the mechanism of viral inactivation by chloramine depends not only on disinfectant concentration but also on virus type.

Chlorine dioxide reacts readily with the amino acids cysteine, tryptophan and tyrosine, but not with viral RNA (Noss *et al.*, 1983; Olivieri *et al.*, 1985). From this research, it was concluded that chlorine dioxide inactivated viruses by altering the viral capsid proteins. However, chlorine dioxide impairs RNA synthesis in poliovirus (Alvarez and O'Brien, 1982). It has also been shown that chlorine dioxide reacts with free fatty acids. Although it is still unclear whether the primary mode of inactivation for chlorine dioxide lies in the peripheral structures or nucleic acids, it is very probable that reactions at both levels contribute to pathogen inactivation.

2.4 Factors influencing the efficiency of disinfectants

Although efficiency of a disinfectant is strongly affected by organism-dependent characteristics as discussed elsewhere in this chapter, it depends to a large extent on environmental factors such as pH, temperature or the availability of the active disinfecting species modulated by the presence of interfering compounds.

As a rule, the efficiency of most disinfectants decreases with time due to inactivation of the product through reaction with environmental chemicals or organic matter. When the amount of microbial contaminants is very high, this decrease in efficiency can lead to the survival of a significant amount of contaminants (Gardner and Peel, 1991; Russell and Hugo, 1987; Scott and Gorman, 1987). Physical removal of the contaminants or maintenance of a constant concentration of active contaminant are required under these conditions. Efficiency of disinfection also depends on the concentration of disinfectant. In most disinfectants used in water and wastewater treatment, doubling the concentration decreases by half the time required to reduce the microbial population by a certain factor (Gardner and Peel, 1991).

pH can play a significant role in determining the activity of disinfectants. When using chlorine and chlorine-releasing compounds (hypochlorites or chlorinated isocianurates), the active disinfecting species is hypochlorous acid (HOCl) while hypochlorite ions (OCl^-) have virtually no activity as disinfectants (Brazis *et al.*, 1958). Partition of these forms is a function of pH with 96.5% of chlorine as HOCl at pH 6 and 78.5% of chlorine as OCl^- at pH 8 (Gardner and Peel, 1991). In general, active forms of chlorine are predominant at pH below neutrality. For other disinfectants, the effect of pH is the opposite, and quaternary ammonium compounds are active only when present as cations. This requires a pH above neutrality and, as a rule, these disinfectants have poor activity under acidic conditions (Lawrence, 1950). Chlorine dioxide is a strong oxidizer which does not require conversion into hypochlorous acid to be effective and thus remains active between pH 6 and pH 10 (Ridenour and Ingols, 1947).

For most disinfectants, increasing temperature results in a decrease of the time required to reduce contaminant levels by a certain factor. In chlorinated disinfectants, a 10°C increase in temperature results in a 50% reduction of the contact time required to kill the organisms (Gardner and Peel, 1991).

Most disinfectants are inactivated in the presence of organic matter. Quaternary ammonium compounds adsorb to the surface of particles and colloids (Gélinas and Goulet, 1983), while chlorinated compounds and oxidizers react both with organic compounds and with reduced inorganic compounds such as ferrous ions, nitrite, ammonia or sulphide (Dychdala, 1991). As a general rule, disinfection in the presence of large amounts of reactive materials usually requires either pretreatment or a more energetic addition of the disinfecting agent.

Efficiency in water disinfection has usually been measured in terms of ratio elimination of faecal indicators, as total coliforms, faecal coliforms or faecal streptococci. Nevertheless, there is no fixed relationship between those groups and real pathogens, although a good safety level can be presumed because real pathogens are usually less resistant than indicators (Lazarova *et al.*, 1998; Le Gudayer *et al.*, 1994). Another important factor to determine pathogens is the ability of detection methods. A lot of stressed bacteria are viable but nonculturable (Roszack and Colwell, 1987). The main stress types, like chemical disinfection and starvation, can promote a lack of culturability, in short or long periods (Oliver, 1992). Nonculturable organisms can retain their pathogenicity and can be still dangerous for the human being (McKay, 1992; Oliver and Bockian, 1995).

3 RESISTANCE (REDUCED SUSCEPTIBILITY) TO DISINFECTANTS

The occurrence of coliform bacteria in otherwise high quality drinking water with a high chlorine content has recently shown that the maintenance of a chlorine residual cannot be relied on to prevent public health problems (Olivieri *et al.*, 1985; LeChevallier *et al.*, 1988b). Also, it has been known for several years that not all bacterial strains show the same sensitivity to chlorine. Selective pressures of water treatment can produce microorganisms with resistance mechanisms favouring survival in an otherwise restrictive environment (LeChevallier *et al.*, 1988b). Bacteria from the chlorinated systems were more resistant than those from the unchlorinated systems, suggesting that there may be selection for more disinfectant-tolerant microorganisms in chlorinated waters. Conceivably, some mechanisms may favourably influence the survival of various opportunistic pathogenic microorganisms in waters containing relatively high concentrations of disinfectant, such as chlorine (LeChevallier *et al.*, 1980).

A diverse range of responses to disinfection can be distinguished, and a single resistance mechanism does not exist. The presence of undesirable microorganisms in water in spite of disinfection has to be related to the interplay of several resistance mechanisms. In some cases several of these mechanisms were synergistic and they increased the microbial resistance to disinfectants (LeChevallier *et al.*, 1988a).

Biocides interact with microorganisms initially at the cell surface. Resistance is thus significantly influenced by cell wall composition and components of the outer surface, which determine this interaction and subsequent uptake by the cell. As summarized in Table 39.2, bacteria may differ considerably in their responses to biocides because the Gram-negative cell wall presents a more significant barrier to entry. The most resistant are undoubtedly bacterial spores, followed, in order, by mycobacteria, prokaryotic and eukaryotic interactions, Gram-negative bacteria and then Gram-positive bacteria. This general response classification, however, masks widespread differences (Russell, 2001b). For example, within Gram-negative bacteria as a group, organisms such as pseudomonads, *Proteus spp.* and *Providencia spp.* may possess above-average resistance to biocides (Stickler and King, 1999).

Why do different types of bacteria show such a wide diversity in response to biocide action? Over the last few years, several authors have suggested a number of mechanisms by which microorganisms may become resistant to or protected from disinfection. On the basis of the information available today, it is possible to distinguish two general types of mechanisms: passive and active resistance. The boundary between both mechanisms may not always be clear, because its separation has been done by microbiologists according to their partial vision of the true microbial world.

3.1 Passive resistance

Passive resistance mechanisms are a consequence of the previous (innate) existence of structural or functional components, and unlike active mechanisms, they do not originate as a response to disinfection (Russell and Chopra, 1996; McDonnell and Russell, 1999).

Passive resistance can be divided into two categories. First, the individual resistance of each microorganism as a consequence of their structural components, and this is shown, for example, by bacterial spores, mycobacteria and several other Gram-negative bacilli. Second, resistance as a result of arrangement produced by aggregation of microorganisms or by attachment to surfaces.

3.1.1 Individual resistance

Individual passive resistance is frequently associated with cellular impermeability imparted by the outer layers of the cell that limit the uptake of biocides (Poole, 2002).

The cell envelope. Peptidoglycan is the basic unit of the cell wall in bacteria, which confers mechanical rigidity to the cell, protects the cytoplasmic membrane and determines the cell form. In Gram-positive bacteria, a thick coat of peptidoglycan combined with teichoic acid constitutes the basic structure of the cell wall. On the other hand, Gram-negative bacteria possess a more complex structure, with a much thinner peptidoglycan layer closer to the cytoplasmic membrane. Moreover, the periplasmic space is situated outermost and delimited externally by an asymmetric outer membrane with the outer surface constituted in essence by lipopolysaccharide (LPS). The resultant layer provides a formidable barrier that can restrict the uptake of biocides (Russell, 2001b). This is particularly true for organisms such as *Pseudomonas aeruginosa*, where the high Mg^{2+} content of the outer membrane aids in producing strong LPS-LPS links. On the other hand, in *Proteus spp.*, the presence of a less acidic type of lipopolysaccharide is a contributory factor to its resistance to chlorhexidine and other cationic biocides (Russell and Chopra, 1996). Moreover, the outer membrane of Gram-negative bacteria acts as a permeability barrier because the narrow porin channels limit the penetration of hydrophilic molecules and the low fluidity of the LPS leaflet slows down the inward diffusion of lipophilic compounds (Beumer *et al.*, 2000).

Therefore, generally speaking, Gram-negative bacteria are more resistant to disinfection than Gram-positive bacteria (Russell, 1998, 1999b). Nevertheless, Gram-positive bacteria are more resistant to chlorine (Trueman, 1971; Mir *et al.*, 1997). In a pilot study investigating bacteriological population changes through potable water treatment, Norton and LeChevallier (2000) demonstrate the effect of disinfection with chlorine on the reduction of the total viable counts and selection of Gram-positive bacteria. In fact, chlorination becomes a limiting factor for microbial diversity in drinking water systems (Maki *et al.*, 1986). In these cases, Gram-positive bacteria levels were three-fold higher than those found in raw water, and the compact and thick cell wall has been attributed as one of the possible mechanisms conferring major resistance to disinfection by chlorine (LeChevallier *et al.*, 1980).

Apart from the usual cell wall structure, other cell wall configurations must be taken into account, such as the one present in mycobacteria. Mycobacterial cell walls are lipid-rich and complex structures that consist of peptidoglycan, a mycolate of arabinogalactan, various lipids and peptides. In this group, the peptidoglycan layer differs from the classical form in its linked polysaccharide side chains, esterified at their distal ends with mycolic acids. The presence of mycolic acids on the cell wall is not an exclusive characteristic of mycobacteria, because they could be present (although in a slightly different manner) in other genera such as *Corynebacterium*, *Nocardia* and *Rhodococcus* (Brennan and Nikaido, 1995).

Its special composition configures an excellent barrier with an extremely low permeability coefficient (Jarlier and Nikaido, 1994) and a very hydrophobic structure, determining a significant restriction on diffusion of hydrophilic biocides. The presence and persistence of *Mycobacteria* in drinking water systems suggests that this treatment could have little effect on them (Collins *et al.*, 1984). In addition to their resistance to chemical disinfection, some mycobacteria show some degree of reduced heat susceptibility. Schulze-Röbbecke and Buchholtz (1992) demonstrate that thermal measures to control *Legionella pneumophila* may not be sufficient to control several mycobacterial species in contaminated water systems. Considering their susceptibility to disinfection, mycobacteria show an intermediate position between bacteria and resistance forms such as bacterial spores or protozoan cysts and oocysts.

Furthermore, bacteria possess efflux pumps of low specificity that can extrude by pumping out mostly lipophilic or amphipathic molecules. These chromosomally-encoded multidrug resistance (MDRs) pumps are widely distributed (Lewis, 1994) and are increasingly implicated as a resistance mechanism (Levy, 2002). They are very important in defining the susceptibility to biocides and antibiotics, especially of Gram-negative bacteria, but also play a role in the development of multiresistance. Their presence in pathogens such as *S. aureus* and *Pseudomonas aeruginosa* poses a serious threat to public health (Littlejohn *et al.*, 1991). For example, the Smr (staphylococcal multidrug resistance) that extrudes membrane-permeable cations such as ethidium bromide and tetraphenilphosponium, the QacA pump of *S. aureus* involved in the extrusion of quaternary ammonium compounds and the MexEF in *P. aeruginosa* and the AcrAB in *E. coli* (Lewis, 1994; Nikaido, 1994).

In fact, current data suggest that efflux pumps are part of the natural defence mechanisms against toxic compounds that exist in the environment. Although Gram-negative bacteria defend against large hydrophilic molecules by utilizing the narrow porin channels in the outer membrane, the lipopolysaccharide-containing bilayer still allows slow diffusion of lipophilic agents. Thus, it has been hypothesized that efflux pumps originally evolved to allow bacterial populations to respond to changes in their environment but nowadays, faced with increasing threat from biocides, mutant strains in which efflux systems are constitutively expressed are the stable state (Beumer *et al.*, 2000).

The glycocalyx. Some bacteria have additional outmost extracellular structures, basically constituted by polysaccharide materials in the form of glycocalyx and capsules. These structures may limit disinfectant action by reaction with disinfectant and/or limit their diffusion. Nevertheless, it has been pointed out that encapsulation, by itself, did not increase disinfection resistance, but the form of the capsule (which appeared to be related to growth condition) may affect the bacterial susceptibility to chlorine. In fact, under nutrient-limiting conditions, the capsule appeared to be threefold more resistant to chlorine (LeChevallier *et al.*, 1988a).

This extracellular polymeric compound has another important role because it is one of the significant constituents of biofilm matrix (see 3.1.2).

Resistance and dispersion forms. In bacterial spores, the outer and inner spore coats limit the uptake of biocidal agents by preventing penetration of the agent to its site of action, although the cortex also has a role to play (Russell, 1992, 2001b; Bloomfield and Arthur, 1994). In fact, spores may be as much as 100 000 times more resistant to disinfectants than vegetative bacterial cells (Phillips, 1952).

Protozoa can present complex life cycles with resistance and dispersion forms such as cysts or oocysts. In drinking water systems disinfection resistance of *Giardia* cysts and *Cryptosporidium* oocysts constitutes a public health concern, and is responsible for outbreaks around the world. Oocysts of *Cryptosporidium parvum* are more resistant to chlorination than bacterial spores (Venczel *et al.*, 1997; Chauret *et al.*, 2001) and, for this reason, using a single disinfectant may not be sufficient in potable waters. In fact, Korich and co-workers (1990) concluded that with the possible exception of ozone, the use of disinfectants alone should not be expected to inactivate *C. parvum* in drinking water. However, the use of ozone as the last disinfectant step before the inlet of water to the distribution system is not recommended, because it may increase the biodegradable organic matter content, with the result of undesirable bacterial regrowth.

The viruses. As a simplification, the virus particle can be described as a protein capsid containing the genetic and enzymatic material. Form and integrity of virus are determined by proteins from the capsid. Viruses are inactive forms when they are free in the environment, without the capacity to repair the sublethal damage induced by disinfectant action or by chemical ageing (this reason may also be valid for spores, cysts and oocysts).

Virus inactivation is a function of several parameters, such as pH, temperature, disinfectant type and strain (Maillard *et al.*, 1994; Meng and Gerba, 1996), and does not require the destruction of the entire virus particle because it can be the consequence of local effects (Fauris *et al.*, 1986). In fact, virus resistance to disinfection is determined first by the capacity of mechanisms involved in the infection to resist disinfectant action. If the external proteins related to infection are damaged, the virus particle becomes inactivated although the rest of virus particle rests intact.

Shaffer *et al.* (1980) demonstrated that poliovirus isolates obtained from fully treated, chlorinated drinking water were several orders of magnitude more resistant to free chlorine than two stock laboratory strains which were used for comparison. On the other hand, human enteroviruses are more resistant to disinfection than water enteropathogenic bacteria (Gerba and Rose, 1990).

Biocide degradation. The most diffuse boundary between active and passive mechanisms is present in the resistance mechanisms generated as a consequence of enzymatic activity. Some bacteria can degrade biocides, such as chlorhexidine (Ogase *et al.*, 1992), QAC (Nishihara *et al.*, 2000) and H_2O_2 (Ichise *et al.*, 1999). Although this capacity to degrade some disinfectants may also be regarded as an active mechanism, different examples of the presence of basal levels of enzymes previous to biocide apparition have been demonstrated. Catalase is an appropriate enzyme to illustrate this fact. Constitutive levels of this enzyme can be used to resist external and sublethal levels of H_2O_2. Nevertheless, as a consequence to phenotypic adaptation to this oxidative stress,

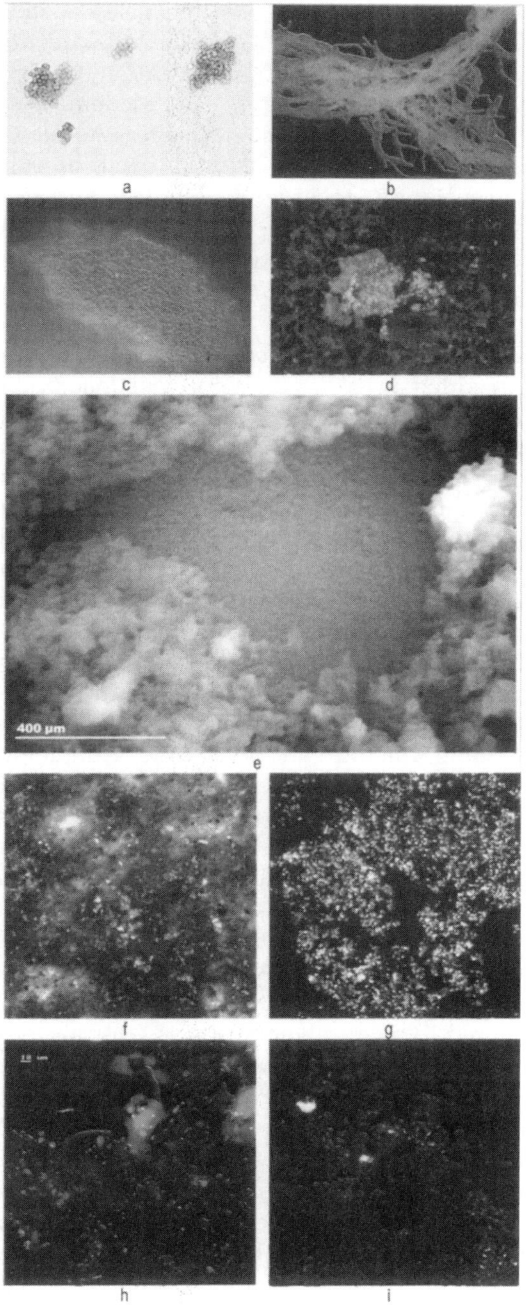

Fig. 39.3 Resistance mechanisms. (a), (b) Bacterial aggregates; (c), (d), (e) bacterial attachment to surfaces; (f), (g), (h) biofilms and (i) biofilm from water distribution system.

catalase levels can be increased, developing active resistance to disinfection. For example, *Pseudomonas aeruginosa* possess two kinds of catalase, KatA is constitutively expressed and KatB is induced in response to H_2O_2 insult (Elkins *et al.*, 1999).

Cooperative interactions. Resistance to disinfection may also be related to interaction among different microorganisms, as is the case of the interaction between bacteria and protozoa. In this sense, several bacteria have the facility to resist protozoal ingestion and, in the protozoa, they increase their resistance to disinfection (King *et al.*, 1988). Moreover, some protozoa such as *Acanthamoeba* spp. produce vesicles prior to cyst formation, which may contain bacteria. Inclusion of *Legionella pneumophila* into these vesicles considerably increases their resistance to disinfection, and this mechanism could help *Legionella* to survive an exposure of 24 h, commonly used in cooling tower treatments (Berk *et al.*, 1998).

3.1.2 Collective resistance

Collective passive resistance to disinfection is a result of direct interactions among microorganisms and/or surfaces, as bacterial aggregates (Fig. 39.3a, b) or more complex conformations on surfaces, microorganisms, organic and/or inorganic matter deposits on static surfaces such as pipe surfaces (Fig. 39.3d, e) or suspended particles such as carbon fines (Fig. 39.3c).

Attachment to surfaces. Most bacterial surfaces are charged at neutral pH because of the ionization of surrounding reactive chemical groups, and for this reason they are highly interactive with their environment. This fact facilitates biogeochemical interactions with metallic ions and silicates (Beveridge *et al.*, 1997). Frequently, this high interactivity of bacteria with their environment may also include other microorganisms and surfaces.

Microbial adhesion to inert substratum surfaces as well as to other microbial cell surfaces is governed by the interplay of two interaction kinds (reviewed in Bos *et al.*, 1999). First, specific interactions mediated by lecithin-like adhesins or similar structures, which are highly directional, spatially confined among molecular groups and operative only over small distances (less than 5 nm). Second, non-specific interactions as a result of the interplay among all molecules of the whole cell and its substratum, operative over longer distances than specific interactions. Although these mechanisms may be described and classified in detail, both always originate from the same fundamental forces: Lifshitz–van der Waals forces, electrostatic forces and acid–base interactions.

Microbial adhesion to solid surfaces is frequently mediated by extracellular mucopolysaccharides (Costerton *et al.*, 1987), but thick biofilms are not required to provide disinfection resistance. Even sparsely distributed attached cells were hundreds of times more resistant to free chlorine than were monodispersed, suspended bacteria (LeChevallier *et al.*, 1988b; see Fig. 39.3i). On the other hand, the growth rate also appears to modulate the hydrophobicity of the cell surface and thereby to influence the colonization of surfaces (Allison *et al.*, 1990).

Numerous researchers have shown that increased resistance to biocides may result from attachment to various surfaces, including macroinvertebrates, algae, organic and inorganic particles (LeChevallier *et al.*, 1981, 1984; Camper *et al.*, 1986). Ridgway and Olson (1982) have shown that the majority of viable bacteria in chlorinated drinking water are attached to surfaces.

In the aqueous phase, two alternatives to increase resistance to disinfection could be distinguished: aggregation and attachment to suspended particles. Aggregation with other microorganisms is one of the major protection factors of bacteria against antimicrobial agents (Matin and Harakeh, 1990) and without doubt, it is clear that bacteria able to aggregate have an increased resistance to disinfection, attributed to restricted penetration of disinfectant to the cells (Stewart and Olson, 1992). In this respect, researchers found that aggregation increased resistance of the bacteria to hypochlorous acid over 100-fold (LeChevallier *et al.*, 1988c).

Attachment to suspended particles (see Fig. 39.3c) usually produces better disinfection resistance than microbial aggregation (Camper *et al.*, 1986; Hoff and Akin, 1986; Medema *et al.*, 1998). In fact, a surface will alter the way a disinfectant interacts with a bacterium by concentration of organic solutes causing a transport limitation of the disinfection to the cell due to a disinfectant demand at the interface (LeChevallier *et al.*, 1988b, c).

Biofilms. A biofilm is a community of microbes embedded in an organic polymer matrix, the glycocalyx, adhering to the surface (Carpentier and Cerf, 1993). The films of microbial growth that develop from these adhered cells are commonly mixed communities containing several different species (see Fig. 39.3h). Biofilm grows by internal replication and by recruitment of other microorganisms from the aqueous phase (Costerton *et al.*, 1987). On the other hand, biofilms are heterogeneous structures changing in time and space with horizontal and vertical discontinuities affecting microorganisms' biotic and abiotic components (Hamilton, 1988), and with an important presence of networks of water channels, responsible for the exchange of nutrients and metabolites with the bulk fluid (see Fig. 39.3f, g).

Biofilm formation is a dynamic process involving association, adherence, microcolony development and maturation and it is considered to be a universal strategy for bacterial survival (Costerton *et al.*, 1987). In many natural habitats, bacteria are able to adhere to and colonize available surfaces. Undoubtedly, the contribution of biofilms to water contamination can be significant. Recurrent episodes of coliforms in some networks cannot all be explained by accidental contamination, broken water mains or obvious mistakes in the treatment processes (LeChevallier, 1990; Camper, 1994). Their greater prevalence during warm weather (higher temperatures and higher levels of biodegradable organic matter) suggests that low levels of coliforms have entered the networks and that these bacteria have been able to grow in the distribution networks (Fass *et al.*, 1996). Although direct proof of such growth is still rare, it seems clear that *E. coli* strains are able to grow at 20°C in the absence of residual chlorine in a distribution network system largely colonized by an autochthonous population. Therefore, enteric bacteria can grow in a rather dilute and cold environment such as distribution networks, compared to the digestive tract of mammals, although its colonization seems to be just partial and transient (Fass *et al.*, 1996).

In this mode of growth, bacteria exhibit some characteristics that can be quite different from those of the same cells grown in suspension (Costerton *et al.*, 1987). Sessile populations are metabolically quiescent, relative to their planktonic counterparts, as reflected by their slow growth rates (Gilbert *et al.*, 1987).

Upon adhesion, synthesis of exopolysaccharide (EPS) is coupled to cell division, which leads to the formation of microcolonies. For example, in *P. aeruginosa*, the adhesion of cells triggers secretion of alginate, allowing concentration of nutrients and adhesion of additional bacterial cells (Hoyle *et al.*, 1993). This material is very sticky, so that when the cell reproduces and divides the two new cells remain in close contact with each other. As this continues, a considerable film of organisms, polysaccharides, and precipitated process materials accumulates. *P. aeruginosa* biofilm and free-floating (planktonic) cells produce similar amounts of EPS. However, the distribution of the glycocalyx is different, with biofilm cells cemented to one another by the EPS matrix and planktonic cells having a compressed, incomplete glycocalyx (Brown *et al.*, 1995). Therefore, glycocalyx plays an important role in the persistence of bacteria within biofilms (Hoyle *et al.*, 1990), because it excludes and/or influences the access of biocides to the underlying organisms. The polyanionic nature of the glycocalyx matrix may contribute significantly to this role of the biofilm in resistance (Sutherland, 1977), because it can function as an ion-exchange column and exclude large, highly charged molecules, imposing a barrier to the inward diffusion of cationic molecules. Furthermore, within the biofilm glycocalyx the sessile bacteria are shielded from the effects of surfactants (Govan, 1975).

The biofilm matrix may also limit diffusion of disinfectants by simple molecular sieving which is a function of the molecular size of the disinfectant and biofilm matrix density. As a consequence of glycocalyx viscosity, which produces a disinfectant saturation in the biofilm, it prevents attainment of lethal concentrations (Brown *et al.*, 1995). In a similar vein, activities of chemically highly reactive biocides such as chlorine or iodine are substantially reduced by the presence of protective exopolymers. In such instances, not only will the polymers act as adsorption sites but they will also react chemically with, and neutralize, biocides (Brown and Gilbert, 1993). In this respect, the limitation of diffusion has been indicated as the principal limiting factor in disinfection (LeChevallier *et al.*, 1988c).

As the biofilm progresses from early colonization events to a thick mature film, physiological and metabolic changes, for example, in the type and rate of polymer production and in specific growth rate, are to be expected (Costerton *et al.*, 1987; LeChevallier *et al.*, 1988a; Srinivasan *et al.*, 1995). In fact, cells in different parts of the biofilm are likely to experience different nutrient and physicochemical environments, which will in turn affect their physiology. Diffusion limitation by the glycocalyx together with localized high densities of cells creates gradients across the biofilm (Brown and Gilbert, 1993). Biofilms will therefore not only consist of mixed species, but also of mixed phenotypes of a given species. Indeed, physiological diversity may contribute to the establishment of 'guard cells' in the upper regions of the biofilm, which enzymically neutralize or exclude biocides by virtue of their altered surfaces (Nichols *et al.*, 1989). In fact, it can be argued that development of the adherent population within the glycocalyx blanket triggers the microheterogeneities of the sessile population and growth rate variations (Hoyle *et al.*, 1990), which are likely to be slow in the depths of biofilms and increase with the nutrient gradient. Disinfectant penetration and diffusion, such as in the case of chlorine, may be restricted by reaction between disinfectant and biofilm matrix components (Huang *et al.*, 1995), varying at a local level as a result of the structure heterogeneity and hydrodynamic conditions variability (de Beer *et al.*, 1994).

It is well known that all such changes can alter susceptibility to antimicrobial agents (Brown and Gilbert, 1993). Therefore, when microorganisms grow attached to a surface in the form of a biofilm, they exhibit remarkable resistance to all types of antimicrobial challenge when compared with the same microorganisms grown as freely suspended cells (LeChevallier *et al.*, 1988b,c). Field and laboratory experiments confirm this nearly universally observed reduced susceptibility of biofilms to disinfection. Cells in biofilms are protected from biocide action and are killed only at biocide concentrations several orders of magnitude higher than those necessary to kill suspended cells. The observed resistance increased with increasing biofilm age (Cochran *et al.*, 2000).

Detachment from biofilms is an important process in the contamination of water distribution systems. Detachment can alter the efficiency of disinfection. Although single cells and small clusters detach more frequently, larger aggregates with a reduced susceptibility to disinfectants contain a disproportionately high fraction of total detached biomass (Stoodley *et al.*, 2001). On the other hand, biofilms also capture suspended particles, which become incorporated into the biofilm and are retained (Bouwer, 1987). The decreased efficacy against particulate-containing biofilms could be due to destructive reactions between the particles and the biocide, alteration of local pH or altered biofilm structure. The effect of inorganic particles has important practical implications because biofilms in field systems, such as pipelines and cooling towers, can be dominated by inorganic matter (Mortensen and Conley, 1994). Additionally, nutrient sequestration would reduce nutrient availability, and hence influence the phenotypic response of the bacteria (Hoyle *et al.*, 1990).

Finally, biofilms also show a protective property by physical shielding against UVC, UVB and UVA. Absorption of UV light by the glycocalyx matrix (alginate) translated into a higher survival rate than observed on planktonic cells at similar input fluences (Elasri and Miller, 1999).

Although biocides are widely used to control the detrimental formation of microbial biofilms, their efficacy is always disappointing when compared with disinfection assays performed with planktonic microorganisms (Costerton *et al.*, 1987; Brown and Gilbert, 1993). In fact, different treatment plants have periodically experienced erratic disinfection and persistence of faecal coliform bacteria in the presence of apparently adequate levels of disinfectant in the effluent (Scully *et al.*, 1999). Much effort is being directed towards elucidating those features that determine the resistance of bacterial biofilms towards biocides. Nevertheless, the fundamental physical, chemical and biological mechanisms by which biofilm microorganisms escape killing by biocides and antibiotics are still incompletely understood because the underlying basis for this resistance is not completely established (Brown and Gilbert, 1993; Cochran *et al.*, 2000).

Most studies have examined individual mechanisms of disinfection resistance. It was of interest to know if combined resistance mechanisms were additive or multiplicative. In drinking water distribution systems, encapsulated aggregated bacteria attached to pipe surfaces grow under low nutrient conditions for long periods of time. Bacteria grown under these conditions were approximately 600-fold more resistant than unattached bacteria grown in rich media, demonstrating that combined resistance mechanisms were multiplicative (LeChevallier *et al.*, 1988a).

3.2 Active resistance

As a general rule, active mechanisms are the result of enzymatic activity, active synthesis of cellular components or a certain metabolic status produced in response to stimulation, injury or stress caused by disinfection, based on genetic and/or phenotypic changes. Both mechanisms, genotype (better described by the term resistance) and phenotypic (adaptation), are involved in the bacterial survival after different disinfection processes (Abu-Shkara *et al.*, 1998). Active resistance can also be the result of an individual cell response to the attack of disinfectants or as a collective response due to the adhesion to surfaces.

3.2.1 Individual resistance

Reduced susceptibility to biocides may be acquired through a phenotypic adaptative response to environmental changes or to genotypic changes through mutation or by the acquisition of a plasmid or transposon.

Physiological (phenotypic) adaptative responses to environmental changes. Many microbial processes are concerned with their constantly changing external environment to which the organism must respond quickly in order to compete successfully. Microorganisms respond to various environmental stimuli at the level of genes that are eventually transcribed and translated into functional polypeptides, which in turn catalyse necessary physiological responses. Thus, as a response to harmful environmental conditions, the cell may produce additional proteins, often referred to as stress proteins. This resistance is not permanent, but a temporary response and can be reversed by withdrawal of the biocide (Beumer *et al.*, 2000).

Since most aquatic environments are characterized by low bioavailability of nutrients – oligotrophic environments – it is understandable that aquatic organisms have developed a starvation response mechanism. For instance, the dissolved organic carbon in the oceans ranges between 0.3 and 1.2 mg/l, or even lower (Morita, 1982). Consequently, bacteria in nature often experience severe nutrient limitations and non-growth or growth only at submaximal rates owing to nutrient scarcity is the norm for natural populations (Matin and Harakeh, 1990).

Starvation is one of the most important factors that influence the sensitivity of bacteria to disinfectants (Berg *et al.*, 1982; Kuchta *et al.*, 1984; Matin and Harakeh, 1990; Blom *et al.*, 1992; Stewart and Olson, 1992; Lisle *et al.*, 1998). In fact, adaptation to conditions of low nutrient availability has been shown to allow long-term survival, as it has been best characterized for *Escherichia coli* (Matin, 1990), *Salmonella typhimurium* (Spector, 1990), Enterococcus faecalis (Hartke *et al.*, 1998) and the marine *Vibrio* strain S14 (Albertson *et al.*, 1990). However, only a few investigators have addressed the question of bacterial starvation under environmental conditions, where microorganisms are frequently confronted with multiple starvations. The survival of microorganisms in their oligotrophic aquatic ecosystem is affected by complex environmental stresses and killing agents, reducing the culturability of allochthonous bacteria (Morita, 1992). Nevertheless, it has been shown that non-culturable cells may exhibit various degrees of metabolic activity (Roszack and Colwell, 1987), because of the mobilization of endogenous reserves (Hartke *et al.*, 1998).

A characteristic response of a population of replicating microorganisms when an adverse environmental change occurs is to reduce the growth rate (Brown *et al.*, 1990). Evidence has been accumulating over the years that bacteria grown at submaximal rates under nutrient limitation or those subjected to nutrient starvation (complete absence of a nutrient) are more resistant to various deleterious agents. Thus, slowly growing organisms isolated from natural environments are markedly more resistant to disinfection agents and may repair sublethal damage more readily than the same strains replicating quickly after growth in laboratory batch cultures (Hurst, 1977; Berg *et al.*, 1982). For example, *Legionella pneumophila* grown in a low-nutrient natural environment has been reported to be six to nine times more resistant than cells grown on agar (Kuchta *et al.*, 1984). However, the way in which bacteria develop resistance is poorly understood.

Recent advances in studying the physiological response of non-differentiating bacteria following starvation of different individual nutrients has led to an understanding that such bacteria undergo a concerted rapid change in the pattern of gene expression (Kjelleberg, 1993), suggesting a global response of gene regulation. Although the induction of these genes has been documented primarily in the context of starvation, it appears that they are not confined to situations of complete nutrient exhaustion and they are probably also expressed during nutrient limitation (Matin and Harakeh, 1990). The synthesis of these unique proteins is an orderly programmed sequence of events, which equips a cell to survive a stressful change in its environment and may reflect a more global adaptation to aquatic environments (Morton and Oliver, 1994; Hartke *et al.*, 1998). Furthermore, under these conditions of nutrient starvation, bacteria have been found to alter cellular nucleic acid concentrations as well as protein concentration and composition (Caldwell *et al.*, 1989).

The synthesis of new proteins (including membrane as well as cytoplasmic proteins) protects the cell against deleterious influences, and it alters the cells in two fundamental ways. First, it makes them more efficient scavengers of the scarce nutrient, thereby enabling them to escape starvation. Second, it confers on them a more stress-resistant phenotype increasing their resistance to a number of stresses (Groat and Matin, 1986; Groat *et al.*, 1986; Matin *et al.*, 1989). As an example, Greenaway and England (1999) have demonstrated the intracellular accumulation of guanosine tetraphosphate (ppGpp) in pseudomonads exposed to a nutrient limitation or to a thiazolinone. Therefore, imposition of different nutrient limitations gives rise not only to populations which divide slowly, but also to cells with physiologies and cell envelopes which are radically different from those of cells grown under nutrient-rich conditions (Brown and Gilbert, 1993).

The nature of the physiological response is influenced by the particular nature of the adversity, be it the lack of an essential nutrient(s) or the presence of an inhibitor. In the last resort, the precise cellular response to a specific chemical stress could be characteristic for that chemical by production of particular polypeptides. The fact that individual chemicals induce unique proteins can conceivably provide a means of identifying pollutants in an environment (Blom *et al.*, 1992; Ogunseitan, 1996).

On the other hand, as a consequence of the enhanced resistance phenotype, cells under a starvation state develop cross-protection against different stresses (Hartke *et al.*, 1998). Strikingly, starved cells exhibited greater resistance to heat or oxidation than cells specifically preadapted to these stresses (Jenkins *et al.*, 1988). For example, oxidative stress (sublethal doses of H_2O_2) made *E. coli* more resistant to chlorine (Dukan and Touati, 1996).

The degree of cross-protection conferred by starvation increased with the time of starvation prior to exposure to the stress (Matin *et al.*, 1989). Even though a majority of the proteins required for starvation survival are produced early, continued protein synthesis is required (Morton and Oliver, 1994). Indeed, the temporal distribution of synthesis of individual proteins probably accounts for the increasing degree of resistance to deleterious agents with increasing period of starvation (Matin and Harakeh, 1990).

Major biochemical and morphological changes have been shown to occur during this starvation response, including changes in cell envelope components such as fatty acids and phospholipids, metal cations, envelope proteins and extracellular enzymes and polysaccharides (Holme, 1972; Elwood and Tempest, 1972). Protein changes during long-term starvation survival involve synthesis and maintenance both of specific proteins whose functions can include detoxification of cell membrane components to exclude the agent or modified biosynthetic pathways (Foster, 1983).

Starved cells can also exhibit different alterations in their usual structure, reducing their cell size or modifying their shapes (Lange and Hengge-Aronis, 1991; Van Overbeek *et al.*, 1995). Cell size is altered widely as a function of the specific growth rate, and causes changes in the cell surface area/volume ratio (Brown *et al.*, 1990).

Other phenotypic changes linked to cell membrane components, such as the ratio of saturated to unsaturated fatty acid, suggest possible differences in membrane permeability (Stewart and Olson, 1992), decreasing in stressed cells. In fact, lipids and particularly fatty acids play a major role in membrane fluidity and consequently its permeability to a large variety of chemicals such as disinfectants. The consensus is that the individual resistance of starved bacteria to chlorine is primarily due to limited chlorine transfer to intracellular target sites because of these changes in the cell membrane permeability (Lisle *et al.*, 1998).

In response to deprivation of a specific nutrient, bacteria generally increase the level of enzymes involved in the capture of the nutrient concerned, synthesize alternative proteins possessing a higher affinity for the nutrient and/or acquire new capacities to obtain the scarce nutrient from a range of different substrates. A variety of unusual nucleotides, such as cyclic AMP (cAMP) have been suggested as possible factors in cellular responses to nutrient deprivation (Matin *et al.*, 1989).

Comparative analysis of the proteomes from biofilm and planktonic cells indicated that there were distinct differences between the protein profiles (Oosthuizen *et al.*, 2002). On the other hand, cells are equipped with several defence systems to protect them from the harmful effects of the reactive oxygen species (Storz *et al.*, 1990). For example, high oxygenation of *E. coli* shifts glutathione homeostasis towards high concentration of intracellular reduced glutathione (GSH) and increases resistance to chlorine (Chesney *et al.*, 1996; Saby *et al.*, 1999). In fact, *E. coli* possesses a general response to oxidative stress that is independent of the nature of the oxidant (Saby *et al.*, 1999). The GSH homeostasis plays a key role in protecting bacteria against chlorination by acting as an oxidant scavenger (Chesney *et al.*, 1996; Komanapalli *et al.*, 1997; Saby *et al.*, 1999) and perhaps by activation of the sigma factor (σ^S), the major regulator of the general starvation response in *E. coli*, produced by the rpoS gene (Loewen and Hengge-Aronis, 1994).

As an outcome of all these mechanisms and because potable water treatment includes the inactivation of bacteria by ozonation and chlorination and the reduction of organic matter, the resulting low nutrient level and the exposition of bacteria to oxidants could make the final chlorination less effective. Consequently, the ability of bacteria such as *E. coli* to remain physiologically active and develop resistance to chlorine after sublethal stresses has potential public health implications and may require further changes in water treatment practices, particularly when the disinfectant residual in the water is less than 0.1 mg of chlorine/litre (Saby *et al.*, 1999).

Mutational resistance to disinfectants. Although there are relatively few studies of the role of mutation in conferring resistance to biocides, exposure of bacteria to a disinfectant may select the presence of pre-existing biocide-resistant mutants. Chaplin (1951,1952) was probably the first to show exposure of *E. coli* to gradually increasing concentrations of a disinfectant led to the development of resistant cells, although the level of resistance expressed in most cases is low and might probably result from phenotypic adaptation.

Nevertheless, a large number of potential mutants altered in outer membrane permeability have been isolated, such as porin-deficient mutants (Hancock, 1984). Moreover, Dagostino *et al.* (1991) isolated a number of mutations which did not express a gene in association with liquid culture, but which did so when cells were attached to a polystyrene surface. As discussed earlier (see Section 3.1.2), attachment of bacteria can significantly increase resistance to disinfection.

On the other hand, recent investigations show that mutations which affect the AcrAB and MexAB multidrug efflux pumps in *E. coli* and *P. aeruginosa* respectively, are associated with reduced susceptibility to some biocides (McMurry *et al.*, 1998; Schweizer, 1998).

Finally, several studies have shown potential links between the development of biocide resistance and the development of reduced susceptibility to antibiotics (Russell, 2001b). Different studies show that increased percentages of the population carry antibiotic-resistant coliforms in their faeces, and other studies have found these antibiotic-resistant coliforms in both drinking water and wastewater. Nevertheless, there is currently no evidence that biocide exposure is a significant factor in the development of antibiotic resistance in clinical practice (Beumer *et al.*, 2000). However, multidrug resistance is a well-known phenomenon in Gram-negative bacteria, being a term used to describe a resistance mechanism by induction of genes that comprise part of the normal cell genome, or by their mutation caused by some types of stress (George, 1996). For this reason it is important to ensure that biocides are used responsibly in order to avoid the possibility of any impact on antimicrobial resistance in the future. Especially, they should be used in a way that, so far as possible avoids the build-up of residues of biocide which might encourage the selection of resistant strains. However, it is important to ensure that biocide use is not discouraged in situations where there is real benefit in terms of preventing infection transmission.

Plasmid-mediated resistance. Many resistant microorganisms carry the genes for resistance on plasmids. These are small and stable genetic elements that are easily transferred to other organisms, thereby spreading the problem. Gene transfer in aquatic environments has been a scientific concern since resistance(s) to antibiotics was discovered to be mediated by plasmid transfer among microorganisms (Mach and Grimes, 1982; Trevors *et al.*, 1987). Nowadays there are numerous examples linking the presence of plasmids to reduced susceptibility to disinfectant. For example, genetic determinants of resistance to heavy metals, chlorhexidine, QACs, triclosan and others are often found on plasmids and transposons (Russell, *et al.*, 2001b). These plasmids can be transferred among species and even genera of bacteria, although some transfer to dissimilar populations has been demonstrated (Yamamoto *et al.*, 1988). However, the more similar an organism is, the higher the potential for transfer.

The genetic transfer of resistance depends on organism survival, plasmid stability, available nutrients and cell numbers, all under the influence of biological, chemical and physical factors (Trevors *et al.*, 1987). In a study, the survival of *Pseudomonas* and *Klebsiella* species and the stability of their plasmids was reported one year after introduction into agricultural drainage water, suggesting that long-term plasmid maintenance occurs naturally (Caldwell *et al.*, 1989).

Resistance to biocides in *S. aureus* is encoded by at least five multidrug resistance determinants, namely *qac* A–E genes, carried on broad-host range plasmids (Littlejohn *et al.*, 1991; Behr *et al.*, 1994) and involved in the extrusion of membrane permeable cations. The *qac* A/B family of genes encodes export proteins that show significant homology to other energy-dependent transporters, such as tetracycline exporters found in tetracycline-resistant bacteria (Rouch, 1990). For example, methicillin-resistant *S. aureus* (MRSA) strains with plasmids carrying genes encoding resistance to gentamicin show increased resistance to cationic biocides such as chlorhexidine, QACs, acridines, diamidines and ethidium bromide (Lyon and Skurray, 1987). Efflux appears to be the predominant biocide resistance mechanism (Paulsen *et al.*, 1996; Moken *et al.*, 1997).

Plasmid-mediated resistance to biocides has also been observed in Gram-negative bacteria, generally associated with envelope changes. For example, decreased susceptibility of *E. coli* and *Serratia marcescens* to formaldehyde is associated with outer membrane changes encoded by plasmids, although it may also arise because of increased synthesis of the enzyme formaldehyde dehydrogenase, which metabolizes formaldehyde to a non-toxic metabolite (Kummerle *et al.*, 1996).

3.2.2 *Collective resistance*

Aggregation and attachment to surfaces. Both processes can have a significant role in the occurrence of coliform bacteria in otherwise high quality drinking water. Under low-nutrient growth conditions, bacterial cells were smaller, but also extensively aggregated compared with cells grown under high-nutrient conditions (Abu-Shkara *et al.*, 1998). Increased adhesion rates, polymer production and the accumulation of the exopolymers on the bacterial cell surface, formation of appendages mediating adhesion and aggregation have been demonstrated to be among the different responses to starvation (Dawson *et al.*, 1981).

Sloughing of cell aggregates from treatment filters or pipe walls has been suggested as a possible mechanism by which coliform bacteria occur in drinking water supplies (LeChevallier *et al.*, 1988b, c; Holt, 1993).

Biofilms. The tendency to ascribe biofilm resistance to either the physiological status of the sessile bacteria or the barrier effect of the glycocalyx presumes a static nature for the biofilm in diverse environments. In fact, the tacit assumption that glycocalyx exists as a static, soluble and homogeneous matrix has been re-examined. Clearly, sessile bacteria do not remain static, but exhibit an adaptable, environmentally responsive phenotype. This response has been extensively documented for *Pseudomonas*, where it seems reasonable to speculate that *Pseudomonas* alginate will react with environmental factors and that the nature of this response influences the potential of glycocalyx to serve as a penetration barrier (Govan, 1975; Deretic *et al.*, 1989). Therefore, recognition that glycocalyx is not static, but rather a plastic, environmentally-responsive matrix unites glycocalyx, the sessile bacteria and the environment in a functional triad (Hoyle *et al.*, 1990). Consideration of the barrier properties of the biofilm must take this interrelationship into account.

On the other hand, bacteria living in an apparently loosely organized microbial community may cooperatively interact with each other and consequently, the processes of that community level are not restricted to highly organized microbial communities (Caldwell and Costerton, 1996; Moller *et al.*, 1998). With the maturation of a biofilm, recruitment of other organisms from the bulk phase can produce a consortium where physiological cooperation involving substrate transfer is facilitated (Costerton *et al.*, 1987).

This mode of growth may help bacteria forming biofilms to increase their resistance following long-term exposure to biocides (Brözel and Cloete, 1993). Therefore, certain bacteria could adapt and grow in the presence of bactericides, which could lead to these bacteria becoming the dominant species in that particular community under those conditions (McDonald *et al.*, 2000). Furthermore, since the killing of biofilm bacteria does not necessarily bring about removal of a biofilm from the surface, the largely dead biofilm could still be detrimental to the system by promoting regrowth and by fouling the surface (Flemming, 1991).

Resistance of microbial biofilms to a wide variety of biocide agents is clearly associated with the organization of cells within an extensive exoplymer matrix. Such organization is able to moderate the concentrations of antimicrobial agents to which the more deeply lying members of the biofilm community are exposed (Gilbert *et al.*, 2002). Such cells are coincidentally slow-growing, starved and express stressed phenotypes.

However, there are three main recognized mechanisms of reduced biofilm susceptibility to antimicrobial agents:

1. penetration failure
2. differences in the physiological state associated with lower growth rate due to slow-growing or starved state
3. the adoption of a biofilm phenotype.

Retarded delivery of biocide or the simple failure to penetrate into the biofilm is unquestionably one of the mechanisms to explain biofilm resistance to biocides, rendering biofilms less susceptible to disinfection by chlorine (de Beer *et al.*, 1994; Xu *et al.*, 1996). Decreased action of biocides such as chlorine against biofilms is due to limited penetration stemming from a reaction–diffusion interaction when the biocide is neutralized by reaction with biofilm constituents faster than it diffuses into the biofilm. Because of the rapid reaction of chlorine with biomass, particularly its nitrogenous components, the biocide may easily be reduced to ineffectual concentrations in the interior of biofilm (de Beer *et al.*, 1994; Xu *et al.*, 1996; Stewart *et al.*, 1998). Eventually, the biocide can penetrate the biofilm fully, but only after depleting the neutralizing ability of the biofilm.

The degree of transport limitation depends most significantly on the biofilm thickness, density of neutralizing sites in the biofilm, bulk fluid biocide concentration and the reaction rate between biocide and neutralizing biomass (Stewart *et al.*, 1998). Thus, penetration failure is most viable as a resistance mechanism when dealing with thick biofilms and highly reactive antimicrobials (Cochran *et al.*, 2000). As the biofilm thickness increases, the time required for a solute to penetrate into the biofilm increases. As the cell density increases, the biocide neutralizing capacity increases, which in turn leads to poorer penetration. Such a transport-based explanation could account for the decreased efficacy of monochloramine against biofilms with the highest areal cell densities (Srinivasan *et al.*, 1995). The determined values of such transport limitation can reach significant values. For example, the time required to attain 1 mg/l of chlorine at the substratum beneath a 500 μm thick biofilm exposed to a bulk fluid concentration of 2 mg/l chlorine would be 9300 times longer (114 h) than the calculated penetration time if there was no reaction between chlorine and biomass (Chen and Stewart, 1996).

By coupling a biofilm cryosectioning technique with a fluorogenic probe that is specific for bacterial respiratory activity, Huang *et al.* (1995) were able to visualize spatial patterns of respiratory activity within a heterogeneous biofilm community during disinfection. While the spatial distribution of respiratory activity within untreated biofilms was relatively uniform, a spatially non-uniform loss of microbial respiratory activity was observed after treatment with the disinfectant. Such activity gradients after disinfection could be explained, as we did earlier, by the depletion of the disinfectant in the biofilm interior through a reaction–diffusion interaction, and also by the induction of adaptative stress responses in the entrapped bacterial system.

Nevertheless, the presence of local differences within biofilms with respect to resistance to disinfectants could be caused by local differences in hydrodynamics resulting in different chlorine transfer rates from the bulk liquid, higher cell density, subpopulations with higher reducing capacity per cell, higher density of extracellular polymeric substances or extracellular polymeric substances with higher reducing potential (de Beer *et al.*, 1994). The phenomenon of rapid regrowth after biocide treatment may originate from such highly resistant spots (de Beer *et al.*, 1994).

However, the recalcitrance of biofilms of very low areal cell density could be indicative of a resistant subpopulation. A second mechanism of biofilm reduced susceptibility is the induction of adaptative stress responses in the entrapped bacterial system. This mechanism requires that at least some of the cells within a biofilm experience a nutrient limitation that causes them to enter a slow-growing or starved state (Brown *et al.*, 1988). As stated earlier, slow or non-growing cells have been shown to be less susceptible to a variety of antimicrobial agents when compared with cells grown in rich media at high specific growth rates.

Such resistant subpopulations could be genotypically or phenotypically distinct, or could be a manifestation of biofilm structural heterogeneity. There is considerable support for striking spatial heterogeneity in the physiological status of bacteria within relatively thick (100 μm) biofilms (Wentland *et al.*, 1996; Xu *et al.*, 1998), but even in apparently thin biofilm, there could be occasional thick microcolonies that are protected by transport limitations of biocide (Srinivasan *et al.*, 1995).

The third mechanism of reduced biofilm susceptibility is that at least some of the cells in a biofilm adopt a distinct and relatively protected biofilm phenotype. This inherent phenotypic change is not the result of a nutrient limitation, and it implies that reduced susceptibility of biofilms is genetically programmed (Cochran *et al.*, 2000). Recent reports are now providing the first glimpse of the genetic basis for biofilm formation (Davies *et al.*, 1993, 1998; Stickler *et al.*, 1998).

Certain bacteria communicate with each other to form structured macroscopic groups. Recently, it has become apparent that, in appropriate environments, common bacteria exhibit similar social behaviour (Davies *et al.*, 1998). These sessile microbial biofilm populations have a complicated structural architecture, with thick layers consisting of differentiated mushroom and pillar-like structures separated by water-filled spaces (Costerton *et al.*, 1995). The structures consist primarily of an EPS matrix or glycocalyx in which the bacterial cells are embedded.

The finding that *P. aeruginosa* produces at least two extracellular signals involved in cell-to-cell communication and cell density-dependent expression of many secreted virulence factors suggests cell-to-cell signalling could be involved in the differentiation of *P. aeruginosa* biofilms (Davies *et al.*, 1998). Moreover, similar signals have been detected in naturally occurring biofilms (McLean *et al.*, 1997). At sufficient population densities, these self-produced signals reach the concentrations required for gene activation. This type of gene regulation has been termed *quorum sensing* and *response* (Fuqua *et al.*, 1994). Quorum sensing systems are required for the differentiation of individual cells into complex multicellular structures. The differentiation is triggered when the cell mass produces a sufficient amount of the quorum-sensing signal (Davies *et al.*, 1998).

Therefore, it seems that new genes are expressed when bacteria attach to a surface and begin to form a biofilm, and some of the resulting gene products reduce the susceptibility of the cell to antimicrobial agents, including oxidative biocides such as monochloramine and hydrogen peroxide (Cochran *et al.*, 2000). The possibility that the biofilm phenotype should be genetically determined may open new perspectives to control biofilm fouling and infections. However, some authors reported that the structure of laboratory-grown biofilms is often highly variable (Heydorn *et al.*, 2000). A complicating factor in the role of quorum-sensing in biofilm formation is the possible effect that an overlying flowing fluid will have on the concentration of signal molecules within a biofilm. Recent investigations suggest that quorum-sensing alone is not necessarily required for biofilm formation and that other factors of the growth environment, such as nutrients and hydrodynamic conditions, can play a role of equal if not greater significance in determining the biofilm structure (Purevdorj *et al.*, 2002; Heydorn *et al.*, 2002; Redfield, 2002).

An understanding of the effects of quorum-sensing on biofilm structure and behaviour under different flow conditions has important applications in industry and medicine. However, interference in cell signalling pathways may not be the final solution for biofilm control as it was initially thought to be.

REFERENCES

Abu-Shkara, F., Neeman, I., Sheinman, R. and Armon, R. (1998). The effect of fatty acid alteration in coliform bacteria on disinfection resistance and/or adaptation. *Wat. Sci. Tech.* **38**(12), 133–139.

Aieta, E. and Berg, J.D. (1986). A review of chlorine dioxide in drinking water treatment. *JAWWA* **78**(6), 62–72.

Albertson, N.H., Nyström, T. and Kjelleberg, S. (1990). Starvation-induced modulations in binding protein-dependent glucose transport by the marine *Vibrio sp.* S14. *FEMS Microbiol. Lett.* **70**, 205–210.

Allison, D.G., Evans, D.J., Brown, M.R.W. and Gilbert, P. (1990). Possible involvement of the division cycle in dispersal of *Escherichia coli* from biofilms. *J. Bacteriol.* **172**, 1667–1669.

Alvarez, M.E. and O'Brien, R.T. (1982). Mechanism of inactivation of poliovirus by chlorine dioxide and iodine. *Appl. Environ. Microbiol.* **44**, 1064–1071.

AWWA Water Quality Division Disinfection Systems Comittee (2000a). Committee report: disinfection at large and medium-size systems. *JAWWA.* **92** (5), 32–43.

AWWA Water Quality Division Disinfection Systems Comittee (2000b). Committee report: disinfection at small systems. *JAWWA.* **92** (5), 24–31.

Bancroft, W.D. and Richter, G.H. (1931). The chemistry of disinfection. *J. Phys.Chem.* **35**, 511–530.

Barret, B.K., Newboult, L. and Edwards, S. (1994). The membrane destabilising action of the antibacterial agent chlorhexidine. *FEMS Microbiol. Lett.* **119**, 249–254.

Behr, H., Reverdy, M.E. and Mabilat, C. (1994). Relation entre le niveau des concentrations minimales inhibitices de cinq antiseptiques et la presence du gene *qac*A chez *Staphylococcus aureus*. *Pathol. Biol.* **42**, 438–444.

Bellamy, W., Carlson, K., Pier, D. *et al.* (2000). Determinig disinfection needs. *JAWWA* **92**(5), 44–52.

Benarde, M.A., Snow, W.B., Olivieri, V.P. and Davidson, B. (1967). Kinetics and mechanism of bacterial disinfection by chlorine dioxide. *Appl. Microbiol.* **15**(2), 257–265.

Berg, J.D., Roberts, P.V. and Matin, A. (1986). Effect of chlorine on selected membrane functions of Escherichia coli. *J. Appl. Microbiol.* **60**, 213–220.

Berg, J.D., Matin, A. and Roberts, P.V. (1982). Effects of the antecedent growth conditions on sensitivity of *Escherichia coli* to chlorine dioxide. *Appl. Environ. Microbiol.* **144**, 814–819.

Berk, S.G., Ting, R.S., Turner, G.W. and Ashburn, R.J. (1998). Production of respirable vesicles containing live *Legionella pneumophila* cells by two *Acanthamoeba* spp. *Appl. Environ. Microbiol.* **64**, 279–286.

Berry, H. and Michaels, I. (1947). The evaluation of the bactericidal activity of ethylene glycol and some of its monoalkylethers against Bacterium coli. *Q. J. Pharm. Pharmacol.* **20**, 331–347.

Beumer, R. Bloomfield, S.F. Exner, M. *et al.* (2000). Microbial resistance and biocides. International Scientific Forum on home hygiene (IFH).

Beveridge, T.J., Makin, S.A., Kadurugamuwa, J.L. and Li, Z. (1997). Interactions between biofilms and environment. *FEMS Microb. Rev.* **20**, 291–303.

Block, J.C. (1982). Mécanismes d'inactivation des micro-organismes par les oxydants. *TSM- l'Eau* **77**, 521–524.

Block, S.S. (2001). *Disinfection, Sterilization and Preservation*, 5th edn. Lippincott Williams & Wilkins, Philadelphia.

Blom, A., Harder, W. and Matin, A. (1992). Unique and overlapping stress proteins of *Escherichia coli*. *Appl. Environ. Microbiol.* **58**, 331–334.

Bloomfield, S.F. and Arthur, M. (1994). Mechanisms of inactivation and resistance of spores to chemical biocides. *J. Appl. Bacteriol* **76**(Suppl.), 91S–104S.

Bos, R., Van der Mei, H.C. and Busscher, H.J. (1999). Physico-chemistry of initial microbial adhesive interactions – its mechanisms and methods for study. *FEMS Microb. Rev.* **23**, 179–230.

Bouwer, E.J. (1987). Theoretical investigation of particle deposition in biofilm systems. *Water Res.* **21**, 1489–1498.

Brazis, A.R.L., Kabler, P.W. and Woodward, B.L. (1958). The inactivation of spores of *Bacillus globigii* and *Bacillus anthracis* by free available chlorine. *Appl. Microbiol.* **6**, 338–342.

Brennan, P.J. and Nikaido, H. (1995). The envelope of Mycobacteria. *Annu. Rev. Biochem.* **64**, 29–63.

Bringmann, G. (1954). Determination of the lethal activity of chlorine and ozone on *E. coli*. *Z.F.Hydiene* **139**, 130–139.

Brown, M.L., Aldrich, H.C. and Gauthier, J.J. (1995). Relationship between glycocalyx and povidone-iodine resistance in *Pseudomoas aeruginosa* (ATCC 27853) biofilms. *Appl. Environ. Microbiol.* **61**, 187–193.

Brown, M.R.W. and Gilbert, P. (1993). Sensitivity of biofilms to antimicrobial agents. *J. Appl. Bacteriol* **74**(Suppl.), 87S–97S.

Brown, M.R.W., Allison, D.G. and Gilbert, P. (1988). Resistance of bacterial biofilms to antibiotics: a growth-rate related effect? *J. Antimicrob. Chem.* **22**, 777–780.

Brown, M.R.W., Collier, P.J. and Gilbert, P. (1990). Influence of growth rate on susceptibility to antimicrobial agents: modification of the cell envelope and batch and continuous culture studies. *Antim. Ag. Chem.* **34**(9), 1623–1628.

Brözel, V.S. and Cloete, T.E. (1993). Bacterial resistance to conventional water treatment biocides. *Biodeterioration Abstracts* **7**, 387–395.

Caldwell, D.E. and Costerton, J.W. (1996). Are bacterial biofilms constrained to Darwin's concept of evolution through natural selection? *Microbiol. SEM.* **12**, 347–358.

Caldwell, B.A., Ye, C., Griffiths, R.P. *et al.* (1989). Plasmid expression and maintenance during long-term starvation-survival of bacteria in well water. *Appl. Environ. Microbiol.* **55**, 1860–1864.

Calvet Churn, C., Bates, R.C. and Broadman, G.D. (1983). Mechanism of chlorine inactivation of DNA-containing Parvovirus H-1. *Appl. Environ. Microbiol* **50**, 1378–1382.

Camper, A.K. (1994). Coliform regrowth and biofilm accumulation in drinking water systems: a review. In: G.G. Geesey, M. Lewandowski and H.C. Flemming (eds) *Biofouling and Biocorrosion in Industrial Water Systems*, pp. 91–105. Lewis Publishers, New York.

Camper, A.K., LeChevallier, M.W., Broadaway, S.C. and McFeters, G.A. (1986). Bacteria associated with granular activated carbon particles in drinking water. *Appl. Environ. Microbiol.* **52**, 434–438.

Carpentier, B. and Cerf, O. (1993). Biofilms and their consequences, with particular reference to hygiene in the food industry. *J. Appl. Bacteriol.* **75**, 499–511.

Casolari, A. (1981). A model describing microbial inactivation and growth kinetics. *J. Theoret. Biol.* **88**, 1–34.

Chambon, M., Bailly, J.L. and Peigue-Lafeuille, H. (1992). Activity of glutaraldehyde at low concentrations against capsid proteins of poliovirus type 1 and echovirus type 25. *Appl. Environ. Microbiol.* **59**, 3517–3521.

Chambon, M., Bailly, J.L. and Peigue-Lafeuille, H. (1994). Comparative sensitivity of the echovirus 25 JV-4 prototype strain and two recent isolates to glutaraldehyde at low concentrations. *Appl. Environ. Microbiol.* **60**, 387–392.

Chang, S.L. (1971). Modern concept of disinfection. *J. Sanit. Engin. Div.* **97**, 689–707.

Chaplin, C.E. (1951). Observations on quaternary ammonium disinfectants. *Can. J. Bot.* **29**, 373–382.

Chaplin, C.E. (1952). Bacterial resistance to quaternary ammonium disinfectants. *J. Bacteriol.* **63**, 453–458.

Chauret, C.P., Radziminski, C.Z., Lepuil, M. *et al.* (2001). Chlorine dioxide inactivation of *Cryptosporidiu parvum* oocysts and bacterial spore indicators. *Appl. Environ. Microbiol.* **67**, 2993–3001.

Chen, X. and Stewart, P.S. (1996). Chlorine penetration into artificial biofilm is limited by a reaction-diffusion interaction. *Environ. Sci. Tech.* **30**, 2078–2083.

Chesney, J.A., Eaton, J.W. and Mahoney, J.R. (1996). Bacterial glutathione: a sacrificial defense against chlorine compounds. *J. Bacteriol.* **178**, 2131–2135.

Chick, H. (1908). An investigation of the laws of disinfection. *J. Hyg. Cambridge* **8**, 92–99.

Cleasby, J.L., Baumann, E.R. and Black, C.D. (1964). Effectiveness of potassium permanganate for disinfection. *JAWWA* **56**, 466–474.

Cochran, W.L., McFeters, G.A. and Stewart, P.S. (2000). Reduced susceptibility of thin *Pseudomonas aeruginosa* biofilms to hydrogen peroxide and monochloramine. *J. Appl. Microbiol.* **88**, 22–30.

Collins, C.H., Grange, J.M. and Yates, M.D. (1984). Mycobacteria in water. *J. Appl. Bacteriol.* **57**, 193–211.

Costerton, J.W., Cheng, K.-J., Geesey, G.G. *et al.* (1987). Bacterial biofilms in nature and disease. *Annu. Rev. Microbiol.* **41**, 435–464.

Costerton, J.W., Lewandowski, Z., Caldwell, D.E., Korber, D.R. and Lappin-Scott, H.M. (1995). Microbial biofilms. *Annu. Rev. Microbiol.* **49**, 711–745.

Cramer, W.N., Kawata, K. and Kruse, C.W. (1976). Chlorination and iodination of poliovirus and F-2. *J. Water Poll. Contr. Fed.* **48**, 61–76.

Cronholm, L.S. McCammon, J.R. and Kruse, C.W. (1976). Enteric virus survival in package plants and the upgrading of the small treatment plants using ozone. Res. Rep. No. 98, Wat. Resources Res. Instit. Univ. Kentucky, Lexington.

Cross, J. and Phil, D. (1996). Upgrading a pharmaceutical water purification system to meet FDA requirements. *Microbiol. Eur.* **4**(3), 16–17.

Dagostino, L., Goodman, A.E. and Marshall, K.C. (1991). Physiological responses induced in bacteria adhering to surfaces. *Biofouling* **4**, 113–119.

Davies, D.G., Chakrabarty, A.M. and Geesey, G.G. (1993). Exopolysaccharide production in biofilms: substratum activation of alginate gene expression by *Pseudomonas aeruginosa*. *Appl. Environ. Microbiol.* **59**, 1181–1186.

Davies, D.G., Parsek, M.R., Pearson, J.P. *et al.* (1998). The involvement of cell-to-cell signals in the development of a bacterial biofilm. *Science* **280**, 295–298.

Dawson, M.P., Humphrey, B.A. and Marshall, K.C. (1981). Adhesion: a tactic in the survival strategy of a marine vibrio during starvation. *Curr. Microbiol.* **6**, 195–199.

de Beer, D., Srinivasan, R. and Stewart, P.S. (1994). Direct measurement of chlorine penetration into biofilms during disinfection. *Appl. Environ. Microbiol.* **60**, 4339–4344.

Dennis, W.H., Olivieri, V.P. and Krause, C.W. (1979). The reaction of nucleotides with aqueous hypochlorous acid. *Water Res.* **13**, 357–362.

Denyer, S.P. and Stewart, G.S.A.B. (1998). Mechanisms of action of disinfectants. *Int. Biodet. Biodeg.* **41**, 261–268.

Deretic, V., Dikshit, R., Konyecsni, W.M. *et al.* (1989). The algR gene, which regulates mucoidy in *Pseudomonas aeruginosa*, belongs to a class of environmentally responsive genes. *J. Bacteriol.* **171**, 1278–1283.

Diamond, J. (1992). The return of cholera. *Discover*, 60–66.

Duguet, J.P. (1981). Contribution a l'étude du traitement par l'ozone des eaux résiduaires. Thesis no. 18, INSA, Toulouse.

Dukan, S. and Touati, D. (1996). Hypochlorous acid stress in *Escherichia coli:* resistance. *DNA damage and comparison with hydrogen peroxide stress. J. Bacteriol.* **178**, 6145–6150.

Dychdala, G.R. (1991). Chlorine and chlorine compounds. In: S.S. Block (ed.) *Disinfection, Sterilization and preservation*, 4th edn. Lea and Febiger, Philadelphia.

Eddy, A.A. (1953). Death rate of populations of *Bact. lactis aerogenes*. III. Interpretation of survival curves. *Proc. Roy. Soc. Lond. Series B* **141**, 137–145.

Elasri, M.O. and Miller, R.V. (1999). Study of the response of a biofilm bacterial community to UV radiation. *Appl. Environ. Microbiol.* **65**, 2025–2031.

Elkins, J.G., Hassett, D.J., Stewart, P.S. *et al.* (1999). Protective role of catalase in *Pseudomonas aeruginosa* biofilm resistance to hydrogen peroxide. *Appl. Environ. Microbiol.* **65**, 4594–4600.

Elwood, D.C. and Tempest, D.W. (1972). Effects of environment on bacterial cell wall content and composition. *Adv. Microb. Ecol.* **7**, 83–117.

Fass, S., Dincher, M.L., Reasoner, D.J. *et al.* (1996). Fate of *Escherichia coli* experimentally injected in a drinking water distribution pilot system. *Water Res.* **30**, 2215–2221.

Fauris, G., Danglot, C. and Vilagines, R. (1986). Inactivation of viruses by disinfection methods. *Wat. Supply* **4**, 19–41.

Fielding, M. (1996). La désinfection, ses sous-produits et leur contrôle. *TSM-l'Eau* **91**, 524–529.

Finch, G.R., Black, E.K., Gyürék, L. and Belosevic, M. (1994). Ozone disinfection of *Giardia and Cryptosporidium*. AWWA Research Foundation and AWWA, Denver.

Finch, G.R., Gyürék, L., Liyanage, R.J. and Belosevic, M. (1997). *Effect of various disinfection methods on the inactivation of Cryptosporidium*. AWWA Research Foundation and AWWA, Denver.

Flemming, H.C. (1991). Biofouling in water treatment. In: H.C. Flemming and G. Geesey (eds) *Biofouling and Biocorrosion in Industrial Cooling Water Systems*, pp. 47–80. Springer-Verlag, Berlin.

Foster, T.J. (1983). Plasmid-determined resistance to antimicrobial drugs and toxic metal ions in bacteria. *Microbial. Rev.* **47**, 361–409.

Frederick, J.J., Corner, T.R. and Russell, A.D. (1974). Antimicrobial actions of hexachlorophene: inhibition of respiratrion in Bacillus megaterium. *Antimicrob. Agents Chemother.* **6**, 712–721.

Fujioka, R.S., Tenno, K.M. and Loh, P.C. (1983). Mechanism of chloramine inactivation of Poliovirus. A concern for regulation. In: R.L. Jolley (ed.) *Water Chlorination: Environmental Impacts and Health Effects*, Vol. 4. Ann Arbor Science Publishers, Ann Arbor.

Fuqua, W.C., Winans, S.C. and Greenberg, E.P. (1994). Quorum sensing in bacteria: the LuxR/Luxl family of cell density responsive transcriptional regulators. *J. Bacteriol.* **176**, 269–275.

Galal-Gorchev, H. (1993). WHO guidelines for drinking water quality. In: G.F. Craun (ed.) *Safety of Water Disinfection*. ILSI Press, Washington.

Gardner, J.F. and Peel, M.M. (1991). *Introduction to Sterilization, Disinfection and Infection Control*, 2nd edn. Churchill Livingstone, Melbourne.

Gélinas, P. and Goulet, J. (1983). Neutralization of the activity of eight disinfectants by organic matter. *J. Appl. Bacteriol.* **54**, 243–247.

George, A.M. (1996). Multidrug resistance in enteric and other gram-negative bacteria. *FEMS Microbiol. Lett.* **139**, 1–10.

Gerba, C.P. and Rose, J.B. (1990). Virus in source and drinking water. In: G.A. McFetters (ed.) *Drinking Water Microbiology: Progress and Recent Developments*, pp. 380–396. Springer-Verlag, New York.

Gerba, C.P., Gramos, D.M. and Nwachuku, N. (2002). Comparative inactivation of enteroviruses and adenovirus 2 by UV light. *Appl. Environ. Microbiol.* **68**, 5167–5169.

Gilbert, P., Brown, M.R.W. and Costerton, J.M. (1987). Inocula for antimicrobial sensitivity testing: a critical review. *J. Antimicrob. Chem.* **20**, 147–154.

Gilbert, P., Allison, D.G. and McBain, A.J. (2002). Biofilms *in vitro* and *in vivo*: do singular mechanisms imply cross-resistance? *J. Appl. Microbiol. Symp. Suppl.* **92**, 98S–110S.

Glaze, W.H., Andelman, J.B., Bull, R.J. *et al.* (1993). Determining health risks associated with disinfectants and disinfection by-products: research needs. *JAWWA* **85**, 53–56.

Goldstein, B.D. and McDonagh, E.M. (1975). Effect of ozone on cell membrane protein fluorescence I. In vitro studies utilizing the red cell membrane. *Environ. Res.* **9**, 179–186.

Gorman, S.P., Scott, E.M. and Russell, A.D. (1980). Antimicrobial activity, uses, and mechanisms of action of glutaraldehyde. *J. Appl. Bacteriol.* **48**, 161–190.

Govan, J.R.W. (1975). Mucoid strains of *Pseudomonas aeruginosa*: the influence of culture medium on the stability of mucus production. *J. Med. Microbiol.* **8**, 513–522.

Greenaway, D.L.A. and England, R.R. (1999). ppGpp accumulation in *Pseudomonas aeruginosa* and *Pseudomonas fluorescens* subjected to nutrient limitation and biocide exposure. *Lett. Appl. Microbiol.* **29**, 298–302.

Groat, R.G. and Matin, A. (1986). Synthesis of unique proteins at the onset of carbon starvation in *Escherichia coli*. *J. Ind. Microbiol.* **1**, 69–73.

Groat, R.G., Schultz, J.E., Zychlinsky, E. *et al.* (1986). Starvation proteins in *Escherichia coli*: kinetics of synthesis and role in starvation survival. *J. Bacteriol.* **168**, 486–493.

Haas, C.N. and Englebrecht, R.S. (1980). Physiological alterations of vegetative microorganisms resulting from chlorination. *J. Water Poll. Cont. Fed.* **52**, 1976–1989.

Haas, C.N. and Karra, S.B. (1984). Kinetics of microbial inactivation by chlorine: I. Review of results in demand-free systems. *Wat. Res.* **18**, 1443–1449.

Hambsch, B. (1999). Distributing groundwater without a disinfectant residual. *JAWWA* **91**, 81–85.

Hamilton, A. (1988). Biofilms at the interface between microbiology and engineering. *Trends Biotechnol.* **7**, 19–20.

Hammond, S.M. (1979). Inhibitors of enzymes of microbial membranes: agents affecting Mg^{2+}-activated adenosine triphosphatase. *Prog. Med. Chem.* **16**, 223–256.

Hancock, R.E.W. (1984). Alterations in outer membrane permeability. *Ann. Rev. Microbiol.* **38**, 237–264.

Harold, F.M., Baarda, J.R. and Baron, C. (1969). Dio 9 and chlorhexidine: inhibitors of membrane bound ATPase and of cation transport in *Streptococcus faecalis*. *Biochim. Biophys. Acta* **183**, 129–136.

Harris, G.D., Adams, V.D., Sorensen, D.L. and Curtis, M.S. (1987). Ultraviolet inactivation of selected bacteria and viruses with photoreactivation of the bacteria. *Water Res.* **21**, 687–692.

Hartke, A., Giard, J-C., Laplace, J-M. and Auffray, Y. (1998). Survival of *Enterococcus faecalis* in an oligotrophic microcosm: changes in morphology, development of general stress resistance and analysis of protein synthesis. *Appl. Environ. Microbiol.* **64**, 4238–4245.

Hauchman, F.S., Noss, C.I. and Olivieri, V.P. (1986). Chlorine dioxide reactivity with nucleic acids. *Water Res.* **20**, 357–361.

Hazen, E.L. and Sawyer, C.N. (1992). *Disinfection Alternatives for Safe Drinking Water.* Van Nostrand Reinhold, New York.

Herzog, R.A. and Betzel, R. (1911). Zür Theorie der Dissinfektion. *Physiol. Chem.* **74**, 221–226.

Heydorn, A., Nielsen, A.T., Hentzer, M., Sternberg, C., Givskov, M., Ersboll, B.K. and Molin, S. (2000). Quantification of biofilm structures by the novel computer program COMSTAT. *Microbiology* **146**, 2395–2407.

Heydorn, A., Ersboll, B.K., Kato, J., Hentzer, M., Parsek, M.R., Nielsen, A.T., Givskov, M. and Molin, S. (2002). Statistical analysis of *Pseudomonas aeruginosa* biofilm development: impact of mutations in genes involved in twitching motility, cell-to-cell signaling, and stationary phase sigma factor expression. *Appl. Environ. Microbiol.* **68**, 2008–2017.

Hiom, S.J., Furr, J.R. and Russell, A.D. (1995). Uptake of 14C-chlorhexidine gluconate by *Saccharomyces cerevisiae, Candida albicans and Candida glabrata. Lett. Appl. Microbiol.* **21**, 20–22.

Hoff, J.C. and Akin, E.W. (1986). Microbial resistance to disinfectants: mechanisms and significance. *Environ. Health. Perspect.* **69**, 7–13.

Hoigné, J. and Bader, H. (1977). The role of hydroxyl radicals reactions in ozonating processes in aqueous solutions. *Water Res.* **10**, 377–386.

Holme, T. (1972). Influence of the environment on the content and composition of bacterial envelopes. *J. Appl. Chem. Biotech.* **22**, 391–399.

Holt, D.M. (1993). *Removal of microorganisms by filtration processes in water treatment works.* Thames Water Plc Group Research and Development Technical Report.

Hom, L.W. (1972). Kinetics of chlorine disinfection in an ecosystem. J. of the Sanitary Engineering Division,. *Proc. Am. Soc. Civil Eng.* **98**(SA1), 183–193.

Hoyle, B.D., Jass, J. and Costerton, J.W. (1990). The biofilm glycocalyx as a resistance factor. *J. Antimicrob. Chemoth.* **26**, 1–6.

Hoyle, B.D., Williams, L.J. and Costerton, J.W. (1993). Production of mucoid exopolysaccharide during development of *Pseudomonas aeruginosa* biofilms. *Infect. Immun.* **61**, 777–780.

Huang, C-T., Yu, F.P., McFeters, G.A. and Stewart, P.S. (1995). Nonuniform spatial patterns of respiratory activity within biofilms during disinfection. *Appl. Environ. Microbiol.* **61**, 2252–2256.

Hugo, W.B. (1957). The mode of action of antiseptics. *J. Pharm. Pharmacol.* **9**, 145–161.

Hugo, W.B. (1999). Disinfection mechanisms. In: A.D.S. Russell, W.B. Hugo and G.A.J. Ayliffe (eds) *Principles and Pratice of Disinfection, Preservation, and Sterilization,* 3rd edn, pp. 258–283. Blackwell Scientific Publications, Oxford.

Hugo, W.B. and Denyer, S.P. (1987). The concentration exponent of disinfectants and preservatives (Biocides). In: R.G. Board, M.C. Allwood and J.G. Banks (eds) *Preservatives in the Food, Pharmaceutical and Environmental Industries*, pp. 281–291. Society for Applied Bacteriology Technical Series No. 22. Blackwell Scientific Publications, Oxford.

Hunter, D.R. and Segal, I.H. (1973). Effect of weak acids on amino acids by *Penicillium chrysogenum*: evidence for a proton or charge gradient as the driving force. *J. Bacteriol.* **113**, 1184–1192.

Hurst, A. (1977). Bacterial injury: a review. *Can. J. Microbiol.* **23**, 935–944.

Ichise, N., Morita, N., Hoshino, T. *et al.* (1999). A mechanism of resitance to hydrogen peroxide in *Vibrio rumoiensis* S-1. *Appl. Environ. Microbiol.* **65**, 73–79.

Jacangelo, J.G., Olivieri, V.P. and Kawata, K. (1987). Mechanism of inactivation of microorganisms by combined chlorine. AWWA Research Foundation, Denver.

Jackson, R.W. and De Moss, J.A. (1965). Effect of toluene on *Escherichia coli*. *J. Bacteriol.* **90**, 1420–1425.

Jagger, J. (1967). *Introduction to Research in Ultraviolet Photobiology*. Prentice-Hall, Englewood Cliffs.

James, A.M. (1972). *The Electrochemistry of Bacterial Surfaces*. Inaugural Lecture. Bedford College, University of London.

Janex, M.L., Savoye, P., Roustan, M. *et al.* (2000). Wastewater disinfection by ozone: influence of water quality and kinetics modelling. *Ozone Sci Eng.* **22**, 113–121.

Jarlier, V. and Nikaido, H. (1994). Mycobacterial cell wall: structure and role in natural resistance to antibiotics. *FEMS Microbiol. Lett.* **123**, 11–18.

Jenkins, D.E., Schultz, J.E. and Matin, A. (1988). Starvation-induced cross protection against heat or H_2O_2 challenge in *Escherichia coli*. *J. Bacteriol.* **170**, 3910–3914.

Jordan, R.C. and Jacobs, S.E. (1944). Studies on the dynamics of disinfection. I. New data on the reaction between phenol and *Bact. coli* using an improved technique, together with an analysis of the distribution of resistance amongst the cells of the bacterial population studied. *J. Hyg. Cambridge* **43**, 275–289.

Kim, C.K., Gentile, D.M. and Sproul, O.J. (1980). Mechanism of ozone inactivatioPn of bacteriophage f2. *Appl. Environ. Microbiol.* **39**, 210–218.

King, C.H., Shots, E.B. Jr, Wooley, R.E. and Porter, K.G. (1988). Survival of coliforms and bacterial pathogens within protozoa during chlorination. *Appl. Environ. Microbiol.* **54**, 3023–3033.

Kjelleberg, S. (1993). *Starvation in Bacteria*. Plenum Press. Inc, New York.

Knaysi, G. (1930). Disinfection I. The development of our knowledge of disinfection. *J. Infect. Dis.* **47**, 303–317.

Komanapalli, I.R., Mudd, J.B. and Benjamin, H.S.L. (1997). Effect of ozone on metabolic activities of *Escherichia coli* K-12. *Toxicol. Lett.* **90**, 61–66.

Korich, D.G., Mead, J.R., Madore, M.S.C.R. *et al.* (1990). Effects of ozone, chlorine dioxide, chlorine, and monochloramine on *Cryptosporidium parvum* oocyst viability. *Appl. Environ. Microbiol.* **56**, 1423–1428.

Kraniak, J.M. and Shelef, L.A. (1988). Effect of ethylenediaminetetraacetic acid (EDTA) and metal ions on growth of *Staphylococcus aureus* 196E in culture media. *J. Food Sci.* **53**, 910–913.

Krebs, H.A., Wiggins, D. and Stubbs, M. (1983). Studies of the antifungal action of benzoate. *Biochem. J.* **214**, 657–663.

Kuchta, J.M., States, S.J., McGlaughlin, J.E. *et al.* (1984). Enhanced chlorine resistance of tap water adapted *Legionella pneumophila* as compared with agar medium passed strains. *Appl. Environ. Microbiol.* **50**, 21–26.

Kummerle, N., Feucht, H.H. and Kaulfers, P.M. (1996). Plasmid-mediated formaldehyde resistance in *Escherichia coli*: characterization of the resistance gene. *Antimicrob. Agents Chemother.* **40**, 2276–2279.

Kuyyakanond, T. and Quesnal, L.B. (1992). The mechanism of action of chlorhexidine. *FEMS Microbiol. Lett.* **100**, 211–216.

Lambert, P.A. (1978). Membrane-active antimicrobial agents. *Prog. Med. Chem.* **15**, 87–124.

Lange, R. and Hengge-Aronis, R. (1991). Growth phase regulated expression of *bol*A and morphology of stationary phase *Escherichia coli* cells is controlled by the novel sigma factor σs (rpoS). *J. Bacteriol.* **173**, 4474–4481.

Langlais, B., Reckhow, D.A. and Brink, D.R. (1991). *Ozone in Water Treatment*. Lewis, Chelsea.

Lawrence, C.A. (1950). *Surface Active Quaternary Ammonium Germicides*. Academic Press, New York.

Lazarova, V., Janex, M.L., Fiksdahl, L. *et al.* (1998). Advanced wastewater disinfection technologies: short and long term efficiency. *Wat. Sci. Tech.* **38**(12), 109–117.

LeChevallier, M.W. (1990). Coliform regrowth in drinking water: a review. *JAWWA* **82**, 74–86.

LeChevallier, M.W., Seidler, R.J. and Evans, T.M. (1980). Enumeration and characterization of standard plate count bacteria in chlorinated and raw water supplies. *Appl. Environ. Microbiol.* **40**, 922–930.

LeChevallier, M.W., Evans, T.M. and Seidler, R.J. (1981). Effect of turbidity on chlorination efficiency and bacterial persistence in drinking water. *Appl. Environ. Microbiol.* **42**, 159–167.

LeChevallier, M.W., Hassenauer, T.S., Camper, A.K. and McFeters, G.A. (1984). Disinfection of bacteria attached to granular activated carbon. *Appl. Environ. Microbiol.* **48**, 918–923.

LeChevallier, M.W., Cawthon, C.D. and Lee, R.G. (1988a). Mechanisms of bacterial survival in drinking water. *Wat. Sci. Tech.* **20**(11/12), 145–151.

LeChevallier, M.W., Cawthon, C.D. and Lee, R.G. (1988b). Factors promoting survival of bacteria in chlorinated water supplies. *Appl. Environ. Microbiol.* **54**, 649–654.

LeChevallier, M.W., Cawthon, C.D. and Lee, R.G. (1988c). Inactivation of biofilm bacteria. *Appl. Environ. Microbiol.* **54**, 2492–2499.

Le Gudayer, F., Dubois, E., Menard, D. and Pommepuy, M. (1994). Detection of hepatits A virus. Rotavirus and enterovirus in naturally contaminated shellfish and sediment by reverse transcription seminested PCR. *Appl. Environ. Microbiol.* **60**, 3665–3671.

Lerch, C. (1953). Electrophoresis of *Micrococcus pyogenes* var. *aureus*. *Acta Pathol. Microbiol. Scand.* **98**(Suppl.), 1–94.

Lev, O. and Regli, S. (1992). Evaluation of ozone disinfection systems: characteristic concentration. *C. J. Environ. Eng.* **118**, 477–494.

Levi, Y. (1992). Distribuer une eau non chlorée: le défi de la qualité. *Sci. tech. de l'eau* **26**, 251–255.

Levy, S.B. (2002). Active efflux, a common mechanism for biocide and antibiotic resistance. *J. Appl. Microbiol. Symp. Suppl.* **92**, 65S–71S.

Lewis, K. (1994). Multidrug resistance pumps in bacteria: variations on a theme. *Trends Biochem. Sci.* **19**, 119–123.

Li, J.W., Xin, Z.T., Wang, X.W., Zheng, J.L. and Chao, F.H. (2002). Mechanisms of inactivation of Hepatitis A virus by chlorine. *Appl. Environ. Microbiol.*, **68**, 4951–4955.

Linden, K.G., Shin, G.-A., Faubert, G., Cairns, W. and Sobsey, M.D. (2002). UV disinfection of *Giardia lamblia* cysts in water. *Environ. Sci. Technol.*, **36**, 2519–2522.

Lisle, J.T., Broadaway, S.C., Prescott, A.M. *et al.* (1998). Effects of starvation on physiological activity and chlorine disinfection resistance in *Escherichia coli* O157: H7. *Apppl. Environ. Microbiol.* **64**, 4658–4662.

Littlejohn, T.G., DiBernardino, D., Messerotti, J. *et al.* (1991). Structure and evolution of a family of genes encoding antiseptic and disinfectant resistance in *Staphylococcus aureus*. *Gene* **101**, 59–66.

Loewen, P.C. and Hengge-Aronis, R. (1994). The role of the sigma factor σ^s (KatF) in bacterial global regulation. *Annu. Rev. Microbiol.* **48**, 53–80.

Lyon, B.R. and Skurray, R.A. (1987). Antimicrobial resistance of *Staphylococcus aureus*: genetic basis. *Microbial Rev.* **51**, 88–134.

Mach, P.A. and Grimes, D.J. (1982). R-plasmid transfer in a wastewater treatment plant. *Appl. Environ. Microbiol.* **44**, 1395–1403.

Macris, B.J. (1974). Mechanism of benzoic acid uptake by *Saccharomyces cerevisiae*. *Appl. Microbiol.* **30**, 503–506.

Madsen, T. and Nyman, M. (1907). Zur Theorie der Desinfektion. *I. Z. Hyg. Infektionskrank.* **57**, 388–395.

Maillard, J.Y. (1999). Viricidal activity of biocides: D. Mechanisms of viricidal action. In: A.D. Russell, W.B. Hugo and G.A.J. Ayliffe (eds) *Principles and Practice of Disinfection, Preservation and Sterilization*, 3rd edn. pp. 227–231. Blackwell Scientific Publications, Oxford.

Maillard, J.Y. (2002). Bacterial target sites for biocide action. *J. Appl. Microbiol,* **92** (Suppl.), 16S–27S.

Maillard, J.Y. and Russell, A.D. (1997). Viricidal activity and mechanisms of action of biocides. *Sci. Ptrog.* **80**, 287–315.

Maillard, J.-Y., Beggs, T.S., Day, M.J. *et al.* (1994). Effect of biocides on MS2 and K Coliphages. *Appl. Environ. Microbiol.* **60**, 2205–2206.

Maillard, J.Y., Beggs, T.S., Day, M.J. *et al.* (1996). Damage to *Pseudomonas aeruginosa* PA01 bacteriophage F116 DNA by biocides. *J. Appl. Bacteriol.* **80**, 540–544.

Maki, J.S., LaCroix, S.J., Hopkins, B.S. and Staley, J.T. (1986). Recovery and diversity of heterotrophic bacteria from chlorinated drinking waters. *Appl. Environ. Microbiol.* **51**, 1047–1055.

Matin, A. (1990). Molecular analysis of the starvation stress in *Escherichia coli*. *FEMS Microbiol. Ecol.* **74**, 184–196.

Matin, A. and Harakeh, S. (1990). Effect of starvation on bacterial resistance to disinfectants. In: G.A. McFetters (ed.) *Drinking Water Microbiology: Progress andRecent Developments*, pp. 88–103. Springer-Verlag, New York.

Matin, A., Auger, E.A., Blum, P.H. and Schultz, J.E. (1989). Genetic basis of starvation survival in nondifferentiating bacteria. *Annu. Rev. Microbiol.* **43**, 293–316.

McCann, B. (2001). Disinfection debate. *Water*, **21**, 29–30. June 2001.

McDonald, R., Santa, M. and Brözel, V.S. (2000). The response of a bacterial biofilm community in a simulated industrial cooling water system to treatment with an anionic dispersant. *J. Appl. Microbiol.* **89**, 225–235.

McDonnell, G. and Russell, A.D. (1999). Antiseptics and disinfectants: activity, action and resistance. *Clin. Microbial. Rev.* **12**, 147–179.

McFeters, G.A. and Camper, A.K. (1978). Chlorine injury and the enumeration of waterborne coliform bacteria. *App. Environ. Microbiol.* **37**, 633–654.

McFeters, G.A. and Camper, A.K. (1983). Enumeration of bacteria exposed to chlorine. *Adv. Appl. Microbiol.* **29**, 177–193.

McKay, A.M. (1992). Viable but non-culturable forms of potentially pathogenic in water. *J Appl. Microbiol.* **14**, 129–135.

McLean, R.J.C., Whiteley, M., Stickler, D.J. and Fuqua, W.C. (1997). Evidence of autoinducer activity in naturally occurring biofilms. *FEMS Microbiol. Lett.* **154**, 259–263.

McMurry, L.M., Oethinger, M. and Levy, S.B. (1998). Overexpression of marA, soxS or acrAB produces resistance to triclosan in laboratory and clinical strains of *Escherichia coli*. *FEMS Microbiol. Lett.* **166**, 305–309.

Medema, G.J., Schets, F.M., Yeunis, P.F.M. and Havelaar, A.H. (1998). Sedimentation of free and attached *Cryptosporidium* oocysts and *Giardia* cysts in water. *Appl. Environ. Microbiol.* **64**, 4460–4466.

Meng, Q.S. and Gerba, C.P. (1996). Comparative inactivation of enteric adenoviruses, poliovirus and coliphages by ultraviolet irradiation. *Wat. Res.* **30**, 2665–2668.

Miller, S. (1993). Disinfection products in water treatment. *Environ. Sci. Technol.* **27**, 2292–2294.

Mir, J., Morató, J. and Ribas, F. (1997). Resistance to chlorine of freshwater bacterial strains. *J. Appl. Microbiol.* **82**, 7–18.

Moken, M.C., McMurry, L.M. and Levy, S.B. (1997). Selection of multiple-antibiotic-resistant (*Mar*) mutants of *Escherichia coli* by using the disinfectant pine oil: roles of the *mar* and *acr*AB loci. *Antimicrob. Agents Chemother.* **41**, 2770–2772.

Moller, S., Sternberg, C., Anderson, J.B. *et al.* (1998). In situ gene expression in mixed-culture biofilms: evidence of metabolic interactions between community members. *Appl. Environ. Microbiol.* **64**, 721–732.

Montgomery, J.M. (1985). *Water treatment Principles and Design*. John Wiley & Sons, New York.

Morita, R.Y. (1982). Starvation-survival of heterotrophs in the marine environment. *Adv. Microb. Ecol.* **6**, 171–197.

Morita, R.Y. (1992). Low nutrient environments In: *Encyclopedia of Microbiology*, Vol. 2, pp. 617–624. Academic Press Inc, New York.

Morita, S., Namikoshi, A., Hirata, T., Oguma, K., Katayama, H., Ohgaki, S., Motoyama, N. and Fujiwara, M. (2002). Efficacy of UV irradiation in inactivating *Cryptosporidium parvum* oocysts. *Appl. Environ. Microbiol.* **68**, 5387–5393.

Mortensen, K.P. and Conley, S.N. (1994). Film fill fouling in counter-flow cooling towers: mechanism and design. *CTIJ* **15**, 10–25.

Morton, D.S. and Oliver, J.D. (1994). Induction of carbon starvation-induced proteins in *Vibrio vulnificus*. *Appl. Environ. Microbiol.* **60**, 3653–3659.

Nakamura, K. and Tamaoki, T. (1968). Reversible dissociation of *Escherichia coli* ribosomes by hydrogen peroxide. *Biochim. Biophys. Acta* **161**, 368–376.

Nichols, W.W., Evans, M.J., Slack, M.P.E. and Walmsley, H.L. (1989). The penetration of antibiotics into aggregates of mucoid and nonmucoid *Pseudomonas aeruginosa*. *J. Gen. Microbiol.* **135**, 1291–1303.

Nikaido, H. (1994). Prevention of drug access to bacterial targets: permeability barriers and active efflux. *Science* **264**, 382–388.

Nishihara, T., Okamoto, T. and Nashiyana, N. (2000). Biodegradation of didecyl-dimethylammonium chloride by *Pseudomonas fluorescens* TN4 isolated from activated sludge. *J. Appl. Microbiol.* **88**, 641–647.

Norton, C.D. and LeChevallier, M.W. (2000). A pilot study of bacteriological population changes through potable water treatment. *Appl. Environ. Microbiol.* **66**, 268–276.

Noss, C.I., Dennis, W.H. and Olivieri, V.P. (1983). Reactivity of chlorine dioxide with nucleic acids and proteins. In: R.L. Jolley (ed.) *Water Chlorination: Environmental Impact and Health Effects*. Lewis Publishers, Chelsea.

O'Brien, R.T. and Newman, J. (1979). Structural and compositional changes associated with chlorine inactivation of polioviruses. *Appl. Environ. Microbiol.* **38**, 1034–1039.

Ogase, H., Nagae, I., Kameda, K. *et al.* (1992). Identification and quantitative analysis of degradation products of chlorhexidine with chlorhexidine-resistant bacteria with three dimentional high performance liquid chromatography. *J. Appl. Bacteriol.* **73**, 71–78.

Ogunseitan, O.A. (1996). Protein profile variation in cultivated and native freshwater microorganisms exposed to chemical environmental pollutants. *Microb. Ecol.* **31**, 291–304.

Oliver, J.D. (1992). Formation of viable but non-culturable cells. In: S. Kjelleberg (ed.) *Starvation in Bacteria*, pp. 239–272. Plenum Press, New York.

Oliver, J.D. and Bockian, R. (1995). In vivo resuscitation and virulence towards mice of viable but non-culturable cells of *Vibrio vulnificans*. *Appl. Env. Microbiol.* **61**, 2620–2623.

Olivieri, V.P., Hauchman, F.S. and Noss, C.I. (1985). Mode of action of chlorine dioxide on selected viruses. In: R.L. Jolley (ed.) *Water chlorination: environmental impact and health effects*. Lewis Publishers, Chelsea.

Oosthuizen, M.C., Steyn, B., Theron, J., Cosette, P., Lindsay, D., von Holy, and A., Brözel, V. S. (2002). Proteomic analysis reveals differential protein expression by Bacillus cereus during biofilm formation. *Appl. Environ. Microbiol.* **68**, 2770–2780.

Parrotta, M.J. and Bekdash, F. (1998). UV disinfection of small groundwater supplies. *JAWWA* **90**(2), 71–81.

Paulsen, I.T., Brown, M.H. and Littlejohn, T.G. (1996). The SMR family: a novel family of multidrug efflux proteins involved with the efflux of lipophilic drugs. *Mol. Microbiol.* **19**, 1167–1175.

Perrine, D. and Langlais, B. (1989). Etude du mécanisme de l'action kysticide de l'ozone sur les amibes libres. *TSM-l'Eau* **84**, 214–218.

Perrine, D., Barbier, D. and Georges, P. (1984). Action de l'ozone sur les trophozoÏtes d'amibes libres pathogènes ou non. *Bull. Soc. Franç. Parasitol.* **3**, 81–84.

Phillips, C.R. (1952). Relative resistance of bacterial spores and vegetative bacteria to disinfectants. *Bacteriol. Rev.* **16**, 135–138.

Poole, K. (2002). Mechanisms of bacterial biocide and antibiotic resistance. *J. Appl. Microbiol. Symp.* **92** (Suppl.), 55S–64S.

Posselt, H.S. Anderson, F.J. and Webber, W.J. (1967) The surface chemistry of hydrous manganese dioxide. Presented at meeting of Water, Air, and Waste Chemistry Division, American Chemical Society, Bar Harbor.

Prokop, A. and Humphrey, Q.E. (1970). Kinetics of disinfection. In: M.A. Benarde (ed.) *Disinfection*, pp. 61–83. Marcel Dekker, New York.

Pryor, W.A., Dooley, M.M. and Church, D.F. (1983). Mechanisms for the reaction of ozone with biological molecules: the source of the toxic effects of ozone. In: M.G. Mustafa and M.A. Mehlman (eds) *Advances in Modern Environmental Toxicology.* Ann Arbor Science Publishers, Ann Arbor.

Purevdorj, B., Costerton, J.W. and Stoodley, P. (2002). Influence of hydrodynamics and cell signaling on the structure and behavior of *Pseudomonas aeruginosa* biofilms. *Appl. Environ. Microbiol.* **68**, 4457–4464.

Rahn, O. (1945). Factors affecting the rate of disinfection. Bacteriol. Rev. **9**, 1–47.

Rahn, O. and Shroeder, W.R. (1941). Inactivation of enzymes as the cause of death in bacteria. *Byodynamica* **3**, 199–208.

Redfield, R.J. (2002). Is quorum sensing a side effect of diffusion sensing? *Trends Microbiol.* **10**, 365–370.

Ribas, F. and Matía, L. (1999). La desinfección del agua y la eliminación de los microorganismos patógenos. *Ibérica* **422**, 443–447.

Richmond, D.V. and Fisher, D.J. (1973). The electrophoretic mobility of microorganisms. *Adv. Microb. Physiol.* **9**, 1–29.

Ridenour, G.M. and Ingols, R.S. (1947). Bactericidal properties of chlorine diocide. *J. Am. Water Works Assoc.* **41**, 537–550.

Ridgway, H.F. and Olson, B.H. (1982). Chlorine resistance patterns of bacteria from two drinking water distribution systems. *Appl. Environ. Microbiol.* **44**, 972–987.

Riesser, V.W. Perrich, J.R. Silver, B.B. and McCammon, J.R. (1976) Possible mechanisms of poliovirus inactivation by ozone. Internat. Ozone Instit. Forum on Ozone Disinfection, Chicago.

Roberts, M.H. and Rahn, O. (1946). The amount of enzyme inactivation at bacteriostatic and bactericidal concentration of disinfectants. *J. Bacteriol.* **52**, 639–644.

Roszack, D.B. and Colwell, R.R. (1987). Survival strategies of bacteria in the natural environment. *Microbiol. Rev.* **51**, 365–379.

Rouch, D.A. (1990). Efflux mediated antiseptic gene qacA from Staphylococcus aureus: common ancestry with tetracycline- and sugar-transport proteins. *Mol. Microbiol.* **4**, 2051–2062.

Roy, D., Wong, P.K.Y., Engelbrecht, R.S. and Chian, E.S.K. (1981). Mechanism of enteroviral inactivation by ozone. *Appl. Environ. Microbiol.* **41**, 718–723.

Rumeau, M. (1982). Mécanismes d'action de divers bactéricides. In: G. Martin (ed.) *Point sur l'epuration et le traitement des effluents*. Lavoisier, Paris.

Russell, A.D. (1971). In: W.B. Hugo (ed.) *Inhibition and Destruction of the Microbial Cell*. Ethylenediaminetetra-acetic acid. pp. 209–225. Academic Press, London.

Russell, A.D. (1992). Resistance of bacterial spores to chemical agents. In: A.D. Russell, W.B. Hugo and G.A.J. Ayliffe (eds) *Principles and Practice of Disinfection, Preservation and Sterilization*, pp. 230–245. Blackwell Scientific Publications, Oxford.

Russell, A.D. (1998). Resistance to non-antibiotic antimicrobial agents. In: W.B. Hugo and A.D. Russell (eds) *Pharmaceutical Microbiology*, pp. 263–277. Blackwell Scientific Publications, Oxford.

Russell, A.D. (1999a). Bacterial resistance to disinfectants: present knowledge and future problems. *J. Hosp. Infect.* **43**(Suppl.), 57S–68S..

Russell, A.D. (1999b). Antifungal activity of biocides. In: A.D. Russell, W.B. Hugo and G.A.J. Ayliffe (eds) *Principles and Practice of Disinfection, Preservation and Sterilization*, pp. 149–167. Blackwell Scientific Publications, Oxford.

Russell, A.D. (2001a). Principles of antimicrobial activity and resistance. In: S.S. Block (ed.) *Disinfection, Sterilization, and Preservation*, 5th edn. pp. 31–55. Lippincott Williams & Wilkins, Philadelphia.

Russell, A.D. (2001b). Mechanisms of bacterial insusceptibility to biocides. *AJIC* **29**(4), 259–261.

Russell, A.D. and Chopra, I. (1996). *Understanding Antibacterial Action and Resistance*, 2nd edn. Ellis Horwood, Chichester.

Russell, A.D. and Furr, J.R. (1977). The antibacterial activity of a new chloroxylenol preparation containing ethylenediamide tetraacetic acid. *J. Appl. Bacteriol.* **43**, 253–260.

Russell, A.D. and Furr, J.R. (1996). Biocides: mechanisms of antifungal action and fungal resistance. *Sci. Prog.* **79**, 27–48.

Russell, A.D. and Hugo, W.B. (1987). Chemical disinfectants. In: A.H. Linton, W.B. Hugo and A.D. Russell (eds) *Disinfection in Veterinary and Farm Animal Practice*. Blackwell Scientific Publications, Oxford.

Russell, A.D. and Hugo, W.B. (1994). Antibacterial activity and action of silver. *Prog. Med. Chem.* **31**, 351–370.

Russell, A.D., Hugo, W.B. and Ayliffe, G.A.F. (eds) (1999). *Principles and Practice of Disinfection, Preservation and Sterilization*. 3rd edn. Blacwell Scientific Publications, Oxford.

Saby, S., Leroy, P. and Block, J.C. (1999). *Escherichia coli* resistance to chlorine and glutathione synthesis in response to oxygenation and starvation. *Appl. Environ. Microbiol.* **65**, 5600–5603.

Salazar, R. (ed.) (1999). *Validación Industrial*. Glatt Labortechnik.

Salueiro, S.P., Correia, I. and Novias, J.M. (1988). Ethanol-induced leakage in *Saccharomyces cerevisiae*: kinetics and relationship to yeast ethanol tolerance and alcohol fermentation productivity. *Appl. Environ. Microbiol.* **54**, 903–909.

Schmidt, W., Böhme, U., Sacher, F. and Brauch, H.J. (2000). Minimization of disinfection by-products formation in water purification process using chlorine dioxide – case studies. *Ozone Sci. Eng.* **22**, 215–226.

Scott, D.B.M. and Lesher, E.C. (1963). Effect of ozone on survival and permeability of *Escherichia coli*. *J. Bacteriol.* **85**, 567–576.

Scott, E.M. and Gorman, S.P. (1987). Chemical disinfectants, antiseptics and preservatives. In: W.B. Hugo and A.D. Russell (eds) *Pharmaceutical Microbiology*, 4th edn. Blackwell Scientific Publications, Oxford.

Scully, F.E., Hogg, P.A., Kennedy, G. *et al.* (1999). Development of disinfection-resistant bacteria during wastewater treatment. *Wat. Environ. Res.* **71**(3), 277–281.

Schulze-Röbbecke, R. and Buchholtz, K. (1992). Heat susceptibility of aquatic mycobacteria. *Appl. Environ. Microbiol.* **58**, 1869–1873.

Schweizer, H.P. (1998). Intrinsic resistance to inhibitors of fatty acid biosynthesis in *Pseudomonas aeruginosa* is due to efflux: application of a novel technique for generation of unmarked chromosomal mutations for the study of efflux systems. *Antimicrob. Ag. Chem.* **42**, 394–398.

Shaffer, P.T.B., Metcalf, T.G. and Sproul, O.J. (1980). Chlorine resistance of of poliovirus isolants recovered from drinking water. *Appl. Environ. Microbiol.* **4**, 1115–1121.

Snider, K.E., Darby, J.L. and Tchobanoglous, G. (1991). Evaluation of ultraviolet disinfection for wastewater reuse applications in California. Department of Civil Engineering, University of California, Davis.

Spector, M.P. (1990). Gene expression in response to multiple nutrient-starvation conditions in *Salmonella typhimurium*. *FEMS Microbiol. Ecol.* **74**, 175–184.

Srinivasan, R., Stewart, P.S., Griebe, T. *et al.* (1995). Biofilm parameters influencing biocide efficacy. *Biotechnol. Bioeng.* **46**, 553–560.

Stewart, M.H. and Olson, B.H. (1992). Physiological studies of chloramine resistance developed by *Klebsiella pneumoniae* under low-nutrient growth conditions. *Appl. Environ. Microbiol.* **58**, 2918–2927.

Stewart, P.S., Grab, L. and Diemer, J.A. (1998). Analysis of biocide transport limitation in an artificial biofilm system. *J. Appl. Microbiol.* **85**, 495–500.

Stickler, D.J., Morris, N.S., McLean, R.J.C. and Fuqua, W.C. (1998). Biofilm on indwelling urethral catheters produce quorum-sensing signal molecules in situ and in vitro. *Appl. Environ. Microbiol.* **64**, 3486–3490.

Stickler, D.J. and King, J.B. (1999). Bacterial sensitivity and resistance. A. Intrinsic resistance. In: A.D. Russell, W.B. Hugo and G.A.J. Ayliffe (eds) *Principles and Practice of Disinfection, Preservation and Sterilization*, pp. 284–296. Blackwell Scientific, Oxford.

Stoodley, P., Wilson, S., Hall-Stoodley, L. Boyle, J.D. Lappin-Scott, H.M. and Costerton, J.W. (2001). Growth and detachment of cell clusters from mature mixed-species biofilms. *Appl. Environ. Microbiol.* **67**, 5608–5613.

Storz, G., Tartaglia, L.A., Farr, S.B. and Ames, B.N. (1990). Bacterial defense against oxidative stress. *Science* **248**, 189–192.

Sutherland, I.W. (1977). Bacterial exopolysaccharides – their nature and production. In: *Surface Carbohydrates of the Prokaryotic Cell*, pp. 27–96. Academic Press, New York.

Trevors, J.T., Barkay, T. and Bourquin, A.W. (1987). Gene transfer among bacteria in soil and aquatic environments: a review. *Can. J. Microbiol.* **33**, 191–198.

Trueman, J.R. (1971). The halogens. In: W.B. Hugo (ed.) *Inhibition and Destruction of the Microbial Cell*, pp. 137–183. Academic Press, London.

US Envitronmental Protection Agency (1999). Alternative disinfectants and oxidants guidance manual. EPA. 815-R-99-014.

Vaara, M. (1991). Agents that increase the permeability of the outer membrane. *Microbiol. Rev.* **56**, 395–411.

Van der Kooij, D., van Lieverloo, J.H.M., Schellart, J.A. and Hiemstra, P. (1999). Distributing drinking water without disinfectant: highest achievement or height of folly? *J. Water SRT-Aqua* **48**, 31–37.

Van Overbeek, L.S., Eberl, L., Givskov, M. *et al.* (1995). Survival of, and induced stress resistance in, carbon starved *Pseudomonas fluorescens* cells residing in soil. *Appl. Environ. Microbiol.* **61**, 4202–4208.

Venczel, L.V., Arrowood, M., Hurd, M. and Sobsey, M.D. (1997). Inactivation of *Cryptosporidium parvum oocysts* and *Clostridium perfringens* spores by a mixed-oxidant disinfectat and by free chlorine. *Appl. Environ. Microbiol.* **63**, 1598–1601.

Walsh, S.M. and Bissonette, G.K. (1983). Chlorine induced damage to surface adhesines during sublethal injury of enterotoxigenic *Escherichia coli. Appl. Environ. Microbiol.* **45**, 1060–1065.

Walsh, S.M. and Bissonette, G.K. (1987). Effect of chlorine injury on heat-labile enterotoxin production in enterotoxicogenic *E. coli. Can. J. Microbiol.* **33**, 1091–1096.

Wallis, P.M., van Roodselaar, A., Neuwirth, M. *et al.* (1990). *Inactivation of Giardia cysts in a pilot plant using chlorine dioxide and ozone.* AWWA Water Quality Technology Conference, Philadelphia.

Wang, J.H. and Matheson, A.T. (1967). The possible role of sulphydryl groups in the dimerization of 70S ribosomes from *Escherichia coli. Biochem. Biophys. Res. Comm.* **23**, 740–744.

Watson, H.E. (1908). A note on the variation of the rate of disinfection with change in the concentration of the disinfectant. *J. Hyg. Cambridge* **8**, 536–592.

Webber, W.J. Jr and Posselt, H.S. (1972). Disinfection. In: W.J. Webber (ed.) *Physicochemical Processes in Water Quality Control.* John Wiley & Sons, New York.

Wentland, E.J., Stewart, P.S., Huang, C.-T. and McFeters, G.A. (1996). Spatial variations in growth rate within *Klebsiella pneumoniae* colonies and biofilm. *Biotechnol. Prog.* **12**, 316–321.

Wickramanayake, G.B., Rubin, A.J. and Sproul, O.J. (1984). Inactivation of *Naegleria* and *Giardia* cysts in water by ozonation. *J. Water Poll. Contr. Fed.* **56**, 983–988.

Wilkinson, S.G. (1975). Sensitivity to ethylenediamine tetraacetic acid. In: M.R.W. Brown (ed.) *Resistance of Pseudomonas aeruginosa*, pp. 145–188. John Wiley & Sons, London.

Wolfe, R.L. (1990). Ultraviolet disinfection of potable water. *Environ. Sci. Tech.* **24**, 768–773.

Xu, K.X., Stewart, P., Xia, F. *et al.* (1998). Spatial phyisiological heterogeneity in *Pseudomonas aeruginosa* biofilm is determined by oxygen availability. *Appl. Environ. Microbiol.* **64**, 4035–4039.

Xu, X., Stewart, P.S. and Chen, X. (1996). Transport limitation of chlorine disinfection of Pseudomonas aeruginosa entrapped in alginate beads. *Biotech. Bioeng.* **49**, 93–100.

Yamamoto, T.Y., Tamura, Y. and Yokoto, T. (1988). Antiseptic and antibiotic resistance plasmids in Staphylococcus aureus that possess ability to confer chlorhexidine and acrinol resistance. *Antimicrob. Ag. Chemother.* **32**, 932–935.

Zimmer, J.L., and Slawson, R.M. (2002). Potential repair of *Escherichia coli* DNA following exposure to UV Radiation from both medium- and low-pressure UV sources used in drinking water treatment. *Appl. Environ. Microbiol.* **68**, 3293–3299.

40

Giardia and *Cryptosporidium* in water and wastewater

H.V. Smith and A.M. Grimason

Department of Civil Engineering, University of Strathclyde, Glasgow G4 ONG, UK

1 INTRODUCTION

Reducing human parasitic infection by breaking the cycle of transmission of parasites transmitted by the faeco-oral route has been one of the most significant interventions in public health medicine and, embodied at its most practical level, this has been the role of the sanitary engineer. In the last 30 years, there has been increasing epidemiological evidence linking two protozoan parasites of the intestinal tract of humans and other hosts with frequent outbreaks of waterborne disease. The parasites, *Giardia duodenalis* and *Cryptosporidium parvum* are the two most commonly reported parasites of human beings world-wide. Infectious organisms, present in contaminated potable water, have the potential to infect large numbers of people from one contamination event, and these two protozoan parasites have been responsible for over 140 outbreaks of waterborne disease affecting over 450 000 individuals. The transmissive stages (cysts of *Giardia duodenalis* and oocysts of *Cryptosporidium parvum* (oo)cysts) are frequent inhabitants of raw water sources used for the abstraction of potable water and their importance is heightened because, coupled to their low infectious doses (10–1000 organisms), conventional water treatment processes, including chemical disinfection, cannot guarantee their removal or destruction completely. Furthermore, due to their chlorine insensitivity, the coliform standard cannot be relied upon as an indicator of either the presence or viability of *Giardia* cysts and *Cryptosporidium* oocysts. For these reasons, robust, sensitive and specific methods are required for the recovery and identification of (oo)cysts in water concentrates.

The endemnicity of giardiasis and cryptosporidiosis, the large numbers of (oo)cysts excreted in faeces, their small size and disinfection insensitivity also have implications for wastewater treatment. Discharges of (oo)cyst-contaminated wastewater effluent into receiving raw waters used for the abstraction of potable water, for crop irrigation, leading to crop contamination, and into receiving fresh, estuarine and coastal marine waters, used for recreational use, is of public health concern. Furthermore, fresh water and marine filter-feeders, such as shellfish, which are eaten raw or lightly cooked, accumulate viable (oo)cysts in their tissues. The disposal of (oo)cyst-contaminated sewage sludge onto land raises issues of (oo)cyst percolation through soils and substrata and the subsequent contamination of groundwater sources by viable or non-viable organisms.

In order to safeguard public water supplies from (oo)cyst contamination, various recommendations have been made and the sampling and monitoring of water and environmental samples for (oo)cysts has become a concern of increasing importance for the water industry and other interested bodies. Water microbiologists and epidemiologists require knowledge on the source and level of contamination, the viability of the organisms, the relationship to indicator organisms,

The Handbook of Water and Wastewater Microbiology
ISBN 0-12-470100-0

and the reservoirs of infection, while engineers and utility operators require knowledge on (oo)cyst removal and inactivation by treatment processes. Regulators of drinking and wastewater programmes require to know where and when these organisms occur in water, the suitability and availability of monitoring methods, and whether treatment requirements should be standardized.

For the reasons stated above, standardized methods are required for the isolation and identification of (oo)cysts in order to provide useful information for interested professionals, and to enable effective comparison of results from diverse areas of the world. Much emphasis has been placed on developing 'standard' methods for (oo)cyst isolation and identification in raw and potable water, however, the goal of method 'standardization' has yet to be achieved for wastewaters. Thus, the significance of data accrued from various wastewater surveys remains at best dubious, which in turn impedes our understanding of the impact of wastewater discharges on waterborne and foodborne giardiasis and cryptosporidiosis.

2 THE PARASITES, CLASSIFICATION, HOST SPECIFICITY, INFECTIOUS DOSE, TRANSMISSION AND DISEASE

One species of *Giardia* (*G. duodenalis*) and five species of *Cryptosporidium* (*C. parvum*, *C. hominis*, *C. meleagridis*, *C. felis* and *C. muris*) are of particular importance to public health. With the exception of *C. hominis*, these parasites are also known to infect a range of other non-human hosts. The flagellate parasite *Giardia* is a member of the phylum Metamonada, class Trepomonadea, order Diplomonadida, family Hexamitidae. The coccidian parasite *Cryptosporidium* is a member of the phylum Apicomplexa, class Sporozoasida, subclass Coccidiasina, order Eucoccidiorida, suborder Eimeriorina, family Cryptosporidiidae.

2.1 *Giardia*

Three 'type' species of the flagellate, *Giardia*, were described by Filice (1952), based upon trophozoite morphology and morphometry namely, *G. agilis*, in amphibians, *G. muris* in rodents, birds and reptiles, and *G. duodenalis* in humans, other mammals and birds. Two additional species, *G. psittaci* from budgerigars and *G. ardeae* from herons have also been described (Erlandsen and Bemrick, 1987; Erlandsen *et al.*, 1990a; Thompson *et al.*, 1993). Although identified some 300 years ago by van Leeuwenhoek when analysing his stools, species nomenclature within the genus remains confused. Not all *G. duodenalis* isolates cause infection in humans, and some researchers use the species names '*lamblia*' and '*intestinalis*' to identify those parasites which infect human beings. In the USA, '*lamblia*' is the preferred species name for parasites infecting humans. Many researchers deem *G. intestinalis* to be a race of *G. duodenalis*. As *G. duodenalis* genotypes that infect humans can also be detected in various mammalian species, confusion over the use of '*intestinalis*' and '*duodenalis*' can be overcome by adopting Filice's proposal, until more discriminatory systems are developed. Reviews of some of the evidence in this taxonomic and nomenclatural debate are given by Thompson *et al.* (1990a, 1993, 2000) and Smith *et al.* (1995).

Based on genetic criteria, *Giardia duodenalis* recovered from humans and other mammalian species fall into one of two major genetic assemblages, namely Assemblage A which infect humans, livestock, cats, dogs, beavers, guinea pigs, slow loris and Assemblage B which infects humans, slow loris, chinchillas, dogs, beavers, rats and siamang (Thompson *et al.*, 2000). Parasites from both assemblages can cause human disease, but there is strong correlation between isolates which cause mild, intermittent diarrhoea and Assemblage A and similar strong correlation between isolates which cause severe, acute/persistent diarrhoea and Assemblage B (Homan and Mank, 2001).

The *G. duodenalis* life cycle is direct (Fig. 40.1), requiring no intermediate host. *Giardia* exists in two distinct morphological forms: the reproductive, pear-shaped trophozoite, which is non-invasive, and attaches onto the enterocytes of the upper small intestine, and the environmentally resistant cyst, voided in the faeces, which is the infective and disseminating stage. Reproduction is by binary fission. *G. duodenalis*

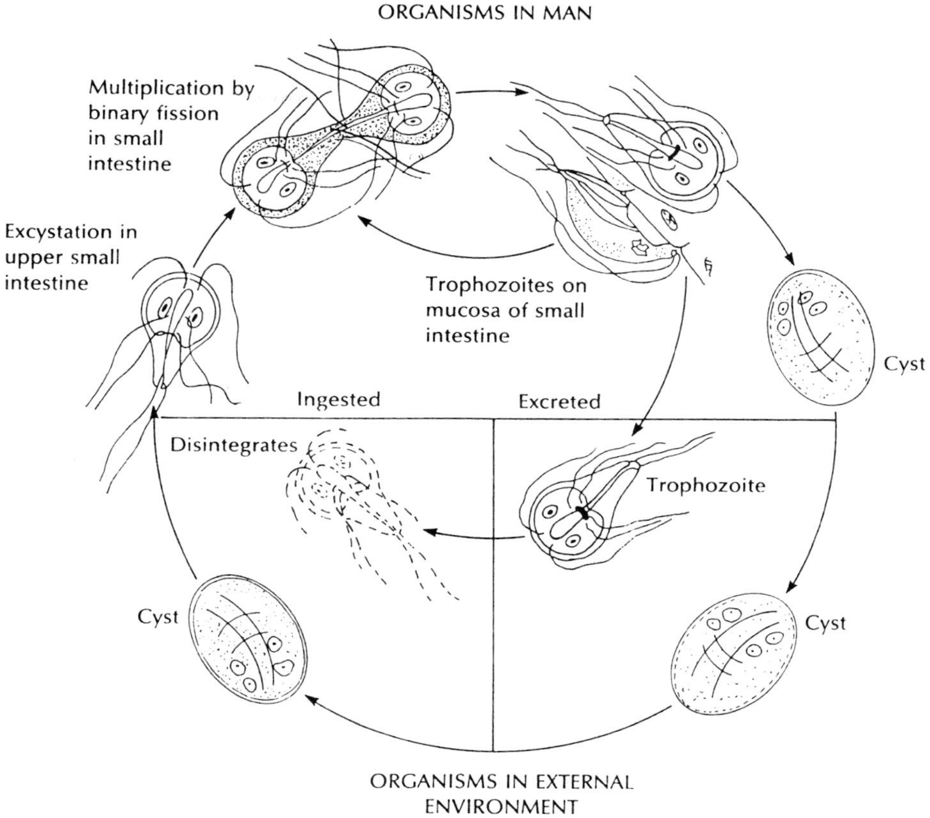

Fig. 40.1 Lifecycle of *Giardia*. The lifecycle is direct, requiring no intermediate host, and the parasite exists in two distinct morphological forms, namely the cyst and trophozoite. Ingested viable cysts excyst in the upper small intestine, following exposure to gastric acid, pepsin and the alkaline environment of the small intestine, releasing the trophozoite which parasitizes the enterocytes of the upper small intestine. Trophozoites either remain attached onto the enterocyte brush border or are motile. Motile trophozoites exhibit forward movement during which the organism tends to rotate around its longitudinal axis, displaying both a tumbling movement resembling that of a falling leaf and an up and down movement referred to as 'skipping'. Reproduction is by binary fission. Exposure to bile salts and alkaline pH as the trophozoite passes down the small intestine, induces the trophozoite stage to encyst, whereby it becomes rounded, forming the binucleate immature cyst. Mature cysts contain four nuclei located at one pole and other structures found within the trophozoite (i.e. axostyles and crescentic bodies (probably remnants of the ventral disc)). Redrawn with permission from Oxford University Press, Meyer and Jarroll (1980) Giardia and cryptosporidium in water, Vol. 111, *American Journal of Epidemiology* **111:** p. 197.

is responsible for 200 million symptomatic cases in Asia, Africa and Latin America, with some 500 000 new cases per annum (World Health Organization, 1996). In the USA, *Giardia* is the most commonly reported human intestinal parasitic infection, being responsible for more than 4000 hospital admissions annually, costing more than US$5 million (Warhurst and Smith, 1992).

2.2 *Cryptosporidium*

Cryptosporidium has a complex life cycle, involving both asexual and sexual reproduction, which is completed within an individual host, and transmission is through an environmentally robust oocyst excreted in the faeces of the infected host. Currently, there is debate concerning the number of species within the genus *Cryptosporidium*. *C. parvum*, *C. muris*, *C. andersoni*, *C. felis*, *C. Canis* and *C. wrairi* infect mammals, while *C. baileyi* and *C. meleagridis* infect birds (Iseki, 1979; Current, 1988; Fayer, 1997; Chrisp, 1990), *C. serpentis* reptiles (Levine, 1980), *C. saurophilum* lizards and *C. nasorum* fish (Hoover *et al.*, 1981; Fayer, 1997; Xiao *et al.*, 2000). *C. parvum* is the major species responsible for clinical disease in man and domestic mammals (Current, 1988; Current and Garcia, 1991). Based on molecular and cross-transmission evidence, within *C. parvum* there are two distinct genotypes (genotypes 1 and 2) with different mammalian host ranges. While both genotypes can infect humans, genotype 1 isolates primarily infect human hosts. Genotype 2 isolates infect both human and non-human hosts. Recently, evidence for the two major genotypes within *C. parvum* (genotypes 1 and 2) being separate species has become significant. Isolates previously referred to as *C. parvum* genotype 1 (which primarily, but not exclusively, infect human hosts) are now renamed *C. hominis*, whereas isolates previously referred to as *C. parvum* genotype 2 (which infect both human and non-human hosts) retain their original species name (*C. parvum*) following their original description in mice. There is increasing molecular evidence suggesting that *Cryptosporidium* species other than *C. parvum* can infect both immunocompetent (*C. meleagridis*) and immunocompromised (*C. felis* and *C. muris*) individuals.

The intracellular reproductive stages of *C. parvum* are extracytoplasmic and reside in a parasitophorous vacuole in the brush borders of enterocytes where they interfere with fluid and nutrient absorption. Autoinfection occurs within the life cycle ensuring that large numbers of infective oocysts are excreted in faeces (Fig. 40.2). *C. parvum* is responsible for between 250 to 500 million infections annually in Asia, Africa and Latin America (Current and Garcia, 1991), and can be life-threatening in immunocompromised hosts (e.g. individuals affected by acquired immune deficiency syndrome, AIDS).

2.3 Infectious dose

The infectious dose for both *Giardia* and *Cryptosporidium* is small. In a human volunteer study, the median infectious dose for *Giardia* was between 25 and 100 cysts, although as few as 10 cysts initiated infection in two out of two volunteers (Rendtorff, 1979). However, one volunteer study demonstrated that a human-source isolate can vary in its ability to colonize other humans (Nash *et al.*, 1987), suggesting that certain isolates may be less infectious to humans, or cause fewer clinical signs and symptoms than others.

To date, all human volunteer infectivity studies have used *C. parvum* genotype 2 oocyst isolates. Of 29 healthy human volunteers, with no evidence of previous *Cryptosporidium* infection, 20% became infected following an oral dose of 30 *C. parvum* (Iowa isolate, bovine, genotype 2) oocysts (DuPont *et al.*, 1995). A dose of 300 oocysts caused infection in 88%, and 1000 oocysts produced infection in 100% of volunteers tested. The median infective dose was calculated to be 132 oocysts. Of the volunteers who excreted oocysts, 39% developed diarrhoea and one other enteric symptom. Those with diarrhoea excreted more oocysts than those without diarrhoea, and were more likely to excrete oocysts on consecutive days (Chappell *et al.*, 1996). Previous exposure confers some protection against reinfection. A 14-fold increase in ID_{50} occurred in volunteers with pre-existing anti-*C. parvum* serum IgG (Chappell *et al.*, 1999).

The infectivity of different *C. parvum* isolates can vary in healthy human adult volunteers. Isolates differed in their ID_{50}, in their attack rate, and in the duration of diarrhoea they induced (Okhuysen *et al.*, 1999). The median infectious dose is nine oocysts for the TAMU (equine, genotype 2) isolate, 132 oocysts for the Iowa isolate and 1042 oocysts for the UCP (bovine, genotype 2) isolate of *C. parvum* (Okhuysen *et al.*, 1999).

Ernest *et al.* (1987) reported that 100 oocysts produced infection in 22% of mice exposed, and Korich *et al.* (2000) calculated that the ID_{50} in outbred CD1 neonatal mice was between 87 and 60 oocysts of the Iowa isolate of *C. parvum*.

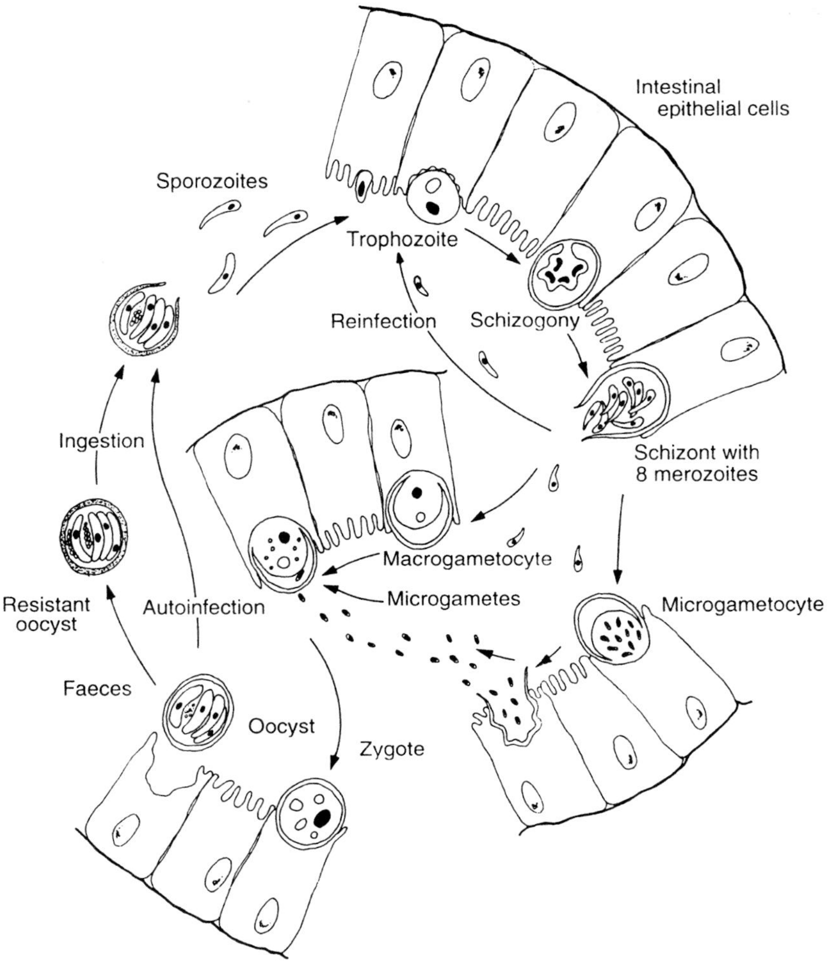

Fig. 40.2 Life cycle of *Cryptosporidium parvum*. The infectious stage is the sporulated oocyst which contains four, naked, motile sporozoites. The banana-shaped sporozoites are released through the suture on the oocyst wall following exposure to body temperature, acid, trypsin and bile salts and attach themselves intimately to the surface of adjacent enterocytes (the epithelial cells which line the gastrointestinal tract). Sporozoites invade enterocytes to initiate the asexual cycle of development. Sporozoites and all subsequent endogenous asexual and sexual stages develop within a parasitophorous vacuole which is intracellular, but extracytoplasmic. Sporozoites differentiate into spherical trophozoites and nuclear division results in the production of the multinucleated schizont stage (schizogony). Type I schizonts contain six to eight nuclei which mature into six to eight merozoites. Merozoites from type I schizonts can either infect neighbouring cells where they recycle and undergo an asexual multiplication cycle (similar to that described for the trophozoite stage) and produce further type I merozoite progeny, or they can develop into a type II schizont. Each maturing type II schizont develops into four type II merozoites, which are thought to initiate the sexual cycle. In sexual multiplication (gametogony), individual merozoites produce either microgamonts or macrogamonts. Nuclear division in the microgamont leads to the production of numerous microgametes which are released from the parasitophorous vacuole and each macrogamont is fertilized by a microgamete. The product of fertilization, the zygote, develops into an oocyst. The zygote differentiates into four sporozoites (sporogony) within the oocyst and fully sporulated oocysts (each containing four sporozoites) are released into the lumen of the intestine and pass out of the body in faeces where they are infectious for other susceptible hosts. Some of the oocysts released in the lumen of the gut have been reported to cause autoinfection in the same parasitized host by liberating their sporozoites in the gut lumen. The released sporozoites undergo the (*continued on next page*)

Miller *et al.* (1990) found that 10 oocysts produced infection in two out of two infant non-human primates tested and Blewett *et al.* (1993) demonstrated that five oocysts (cervine/ovine, MD isolate, genotype 2) produce disease in gnotobiotic lambs.

2.4 Transmission

Transmission occurs via any route by which material contaminated with viable (oo)cysts excreted by infected hosts can reach the intestine of a susceptible host. Person-to-person transmission, via the faecal-oral route, is a major route and has been documented between family/household members, sexual partners, health workers and their patients, and children in day-care centres (probably due to the lower standards of personal hygiene exhibited by pre-school children) and other institutions. The most important route of environmental transmission is through the contamination of water by (oo)cysts (Craun, 1990; Levine *et al.*, 1991; Robertson *et al.*, 1994; Rose and Lisle, 1995; Smith *et al.*, 1995b; Rose *et al.*, 1997; Smith and Rose, 1998). Water borne giardiasis and cryptosporidiosis, associated with community water systems and immersion watersports, has been reported primarily from North America and Europe (Craun, 1988; Wallis, 1994; Rose and Lisle, 1995; Smith *et al.*, 1995b; Smith and Rose, 1998).

Giardia is the most commonly identified agent of waterborne disease in the USA, with over 120 waterborne outbreaks affecting more than 25 000 persons since 1965 (Girdwood and Smith, 1999a). According to Bennett *et al.* (1987), 60% of all *Giardia* infections in the USA are acquired from contaminated water. In the USA, *Giardia* remains the most commonly identified pathogen in waterborne outbreaks, accounting for 18% of outbreaks resulting in 24 124 cases during the period 1971–85 (Craun, 1990). In outbreaks where an aetiological agent was identified, 37% of these outbreaks, accounting for 49% of the cases of illness, were attributable to *Giardia* (Craun, 1990). In the period 1976–80, waterborne outbreaks of giardiasis accounted for 25% of optionally reported cases (Craun, 1986) and since 1965, over 100 outbreaks of waterborne giardiasis have occurred in the USA alone (Craun, 1990; Levine *et al.*, 1991). Jakubowski (1990) estimated that as many as 21 million people in the USA may be at risk of giardiasis since their potable water comes from unfiltered water supplies.

Cryptosporidium is more resistant to disinfectants used in the water industry and has been implicated in >20 waterborne outbreaks, affecting more than an estimated 427 100 individuals (MacKenzie *et al.*, 1994; Smith and Rose, 1998; Girdwood and Smith, 1999b). The Milwaukee outbreak resulted in the death of 104 of 403 000 cases (MacKenzie *et al.*, 1994), and the direct cost was estimated at $53 million and $100 million in loss of life.

2.4.1 *Zoonotic transmission*

The zoonotic potential of *Giardia* remains somewhat controversial. While the widespread distribution of *Giardia* infection in domestic and wild animals indicates the potential for zoonotic transmission (Thompson, 2002), definitive evidence that this route of transmission occurs widely and its significance remains elusive. Evidence, based on circumstance, biochemical and genetic typing, has been advanced both for and against zoonotic transmission (Faubert, 1988; Bemrick and Erlandsen, 1988; Kasprsak and Paulowski, 1989; Healey, 1990; Thompson *et al.*, 1990b; Isaac-Renton *et al.*,

developmental processes of schizogony, gametogony and sporogony in enterocytes of the same infected host. In this monoxenous life cycle, both the recycling of merozoites to produce further type I generations of schizonts and endogenous reinfection from thin-walled oocytes ensure that large numbers of infective (thick-walled) oocysts are excreted in faeces. The intracellular reproductive stages, present in the brush borders of enterocytes, interfere with fluid and nutrient absorption. *Cyptosporidium parvum* oocysts are spherical and their modal size measurement is 4.5 × 5.0 μm (range 4–6 μm). Reproduced with permission from Smith and Rose (1998) Waterborne cryptosporidiosis: current status. *Parasitology Today* **14:** 14–22.

1993; Isaac-Renton, 1994; Erlandsen, 1994), however, in the absence of definitive information identifying its significance, all viable, waterborne cysts are considered to have the potential to infect human beings.

The zoonotic potential of *C. parvum* is better understood. Cryptosporidiosis has been reported in a variety of domesticated animals, livestock and wildlife, including companion animals, which may be reservoirs for human infection (Smith and Rose, 1990; Current and Garcia, 1991). Zoonotic transmission (genotype 2 organisms) has been documented in children on farm visits, and also from laboratory animals and household pets. Animals can be infected, experimentally, with genotype 2 oocysts of human origin, and may acquire infection naturally from man.

2.4.2 *Transport hosts*

A variety of transport hosts has been implicated in the transmission of *Giardia* and *Cryptosporidium* (oo)cysts, including coprophagous animals, birds (seagulls, ducks) and filth flies (Slifko *et al.*, 2000).

2.5 Disease

Whereas effective drug treatments exist for giardiasis, at present, no specific drug treatment exists for cryptosporidiosis.

2.5.1 *Giardiasis*

In most people the disease is self-limiting, exhibiting an acute followed by a chronic phase. An asymptomatic, cyst-passing phase of unknown duration can also occur. The acute phase is usually short-lived, and is characterized by flatulence with sometimes sulphurous belching and abdominal distension with cramps. Diarrhoea is initially frequent and watery but later becomes bulky, sometimes frothy, greasy and offensive and the stools may float on water. Blood and mucus are usually absent and pus cells are not a feature on microscopy. In the chronic stages of disease, malaise, weight loss and other features of malabsorption may become prominent. By this time, stools are usually pale or yellow, being frequent and of small volume. Occasionally episodes of constipation intervene with nausea and diarrhoea precipitated by the ingestion of food. Malabsorption of vitamins A and B_{12} and D-xylose can occur. Disaccharidase deficiencies (most commonly lactase) are frequently detected in chronic cases. In young children, 'failure to thrive' is frequently due to giardiasis, and all infants being investigated for causes of malabsorption should have a diagnosis of giardiasis excluded.

Cyst excretion can approach 10^7 per g faeces (Danciger and Lopez, 1975). The prepatent period (time from infection to the initial detection of parasites in stools) is on average 9.1 days (Rendtorff, 1979). The incubation period (time from ingestion of organisms to the manifestation of symptoms) is usually 1–2 weeks. As the prepatent period can exceed the incubation period, initially a patient can have symptoms in the absence of cysts in the faeces.

2.5.2 *Cryptosporidiosis*

Cryptosporidiosis is associated with profuse watery diarrhoea, rapid weight loss, dehydration and abdominal cramps. Less frequent symptoms also include low-grade fever, nausea, vomiting, anorexia and general fatigue. In immunocompetent individuals, both duration and severity of disease can vary but the diarrhoea is self-limiting and the infection is limited to the small and large intestine. In immunocompromised individuals, especially those with AIDS, infection may lead to dehydration, electrolyte imbalance, and eventually death. Infection in the immunocompromised may spread to the oesophagus, stomach, gall bladder, common bile duct, rectum, appendix and into the respiratory tract (Soave and Armstrong, 1986).

Young children are more susceptible to infection and have more severe clinical signs due to an immature immune system and poor hygiene habits. Children infected with *Cryptosporidium* can suffer retarded growth. Asymptomatic infections appear to be more prevalent in children, young adults and AIDS patients in endemic areas of disease.

Oocyst excretion can be in excess of 10^{10}, with up to 10^7 per g faeces. In immunocompetent individuals, the incubation period (time from ingestion of organisms to the manifestation of symptoms) ranges from 2 to 10 days, but can be as long as 28 days. Experimentally infected neonatal calves can excrete up to 10^9 oocysts daily for up to 14 days (Blewett, 1989).

3 DETECTION IN WATER

Specific methods are required for the isolation and enumeration of (oo)cysts from water for the following reasons:

1. they can pass through physical and chemical barriers in water treatment, are chlorine insensitive and have been detected in potable water supplies in the absence of indicator organisms
2. they can cause epidemic disease in consumers of contaminated potable water
3. as *Giardia* and *Cryptosporidium* are obligate parasites, their numbers cannot be augmented by conventional *in vitro* culture methods, consistently prior to identification
4. the minimum infectious dose for humans is low.

With the exception of some wastewaters and filter backwash waters, (oo)cysts occur at low densities in the aquatic environment and therefore a system which enables their efficient recovery from large volumes of water is required. Diverse techniques employing filtering large volumes of sample followed by sample concentration have been employed to concentrate (oo)cysts and a variety of methods from immunofluorescence to polymerase chain reaction (PCR) have been used to determine their presence. Currently, filtration followed by immunofluorescence detection are regarded as the most effective methods for isolating and enumerating waterborne (oo)cysts. They are also the only methods recognized by the UK government (Anon., 1999a) and the US Environmental Protection Agency (Anon., 1998a, b) for monitoring treated water.

3.1 Raw and treated waters

Many methods have been developed over the last 30 years, each with specific advantages and disadvantages (Jakubowski *et al.*, 1978; Jakubowski and Eriksen, 1979; Ongerth and Stibbs, 1987; Rose *et al.*, 1988; Smith *et al.*, 1989a; Gilmour *et al.*, 1989; Anon., 1990, 1994, 1999a, b; Smith, 1995, 1996; Fricker, 1995; Smith and Hayes, 1996; Smith and Rose, 1998; Smith, 1998). There is no universally accepted procedure, but all methods can be subdivided into the following elements: (a) sampling; (b) elution, clarification and concentration; (c) identification. 'Standardized' methods, which are continually evolving, are available in the UK and the USA (Anon., 1990, 1994, 1999a, b, c).

3.1.1 Sampling

Both large and small volume (grab) sampling methods have been adopted. Large volume sampling entails the collection of a sample over a period of hours at a defined flow rate whereas, in small volume sampling, a volume of 10–20 l, typically, is taken as a grab sample. Where little is known of the occurrence or temporal distribution of (oo)cysts in the matrix tested, large volume sampling is useful as the sample is taken over a long time period. In contrast, grab samples can provide higher recovery efficiencies than large volume sampling and are collected readily. A compromise between both regimes is the collection of numerous grab samples over the large volume sampling period in order to generate one composite sample.

3.1.1.1 Large volume sampling. Here, a large volume (≈100–1000 l) of water is filtered through a depth filter (e.g. yarn wound (CUNO, flow rate 1.5 l/min (Anon., 1990, 1999a)), polypropylene (Filterite, flow rate 4 l/min (Anon, 1994)), compressed foam (Genera Technologies Filta-Max™ Crypto-Dtec, flow rate 1–2 l/min (Sartory *et al.*, 1998), etc.) Pall Envirochek®, HV pleated membrane cartridge filter, flow rate 4 litres/min which entraps (oo)cysts and other particulates of similar and larger size (Smith, 1998). (Oo)cysts and other particulates entrapped in yarn wound and polypropylene cartridges are eluted by immersing the cut, teased filter in large volumes of a mild detergent (0.01% Tween 80 in deionized water containing an antifoaming agent (antifoam A)). The compressed foam cartridge requires expensive and dedicated manufacturer's equipment to elute (oo)cysts which are concentrated onto flatbed membranes and eluted by massaging the membrane in a dilute detergent solution (Sartory *et al.*, 1998; Anon., 1999a).

3.1.1.2 Small volume sampling. (Oo)cysts present in grab samples can be concentrated either by membrane filtration or flocculation. Typically, the sample is filtered through either a flatbed 142 mm, 1.2–2 μm cellulose acetate or polycarbonate membrane (flow rate ≈ 150 ml/min (Ongerth and Stibbs, 1987; Watkins *et al.*, 1995; Shephard and Wynne-Jones, 1996)) or a pleated membrane capsule (e.g. Gelman Envirochek, flow rate 2 l/min (Mattheson *et al.*, 1998)) using a peristaltic pump. The flocculation of grab samples is described below.

Both the large and small volume sampling methods described are included in the current UK Department of the Environment, Standing Committee of Analysts' (UKSCA) provisional recommended methods (Anon., 1990, 1999b). Statutory treated water samples in the UK are collected using the Idexx Genera Technologies Filta-Max™ compressed foam cartridge (Anon., 1999a). Large volume sampling is included in the US Environmental Protection Agency's (USEPA) ICR method (Anon., 1994) and small volume sampling is identified in USEPA Methods 1922 1623 (Anon., 1999c) or the Pall Envirochek® pleated membrane cartridge filter .

3.1.2 (Oo)cyst concentration

All sampling methods generate large volumes of extraneous particles which interfere with (oo)cyst detection and identification. The number of particles that are of a similar size to *Cryptosporidium* oocysts can be in excess of 10^7 per litre of raw water (Smith *et al.*, 1995b), and methods to reduce this interference have been developed. Similarly, the likelihood of masking all or part of an (oo)cyst is high in a sample concentrate where the ratio of particles to (oo)cysts can be $\geq 10^9$:1 (in 100 litres of raw water) (Smith *et al.*, 1995b). Eluted organisms and particulates are concentrated to a small volume (about 10 ml), and subjected to further clarification or, more recently, immunomagnetizable separation (see below) prior to analysis by epifluorescence and DIC microscopy.

3.1.2.1 Membrane filtration. (Oo)cysts trapped on flat-bed membranes are eluted either by scraping or massaging them off the membrane surface with a dilute detergent solution, while (oo)cysts trapped on pleated membrane capsules are released into a mild detergent mixture using a mechanical wrist action shaker. Samples are concentrated to a minimum volume by centrifugation or immunomagnetizable separation (see below) and analysed by epifluorescence and Normarski differential interference contrast (DIC) microscopy.

3.1.2.2 Flocculation. Particles in grab samples are flocculated following the production of a $CaCO_3$ floc generated by the addition of $CaCl_2$, $NaHCO_3$ and NaOH at pH 10. As the floc settles by gravity, it causes (oo)cysts and other particles to settle with it, concentrating them. When settled (≥4 hours, or overnight), the floc is dissolved in sulfamic acid and the particulates in the sample concentrated by centrifugation to a minimum volume (Vesey *et al.*, 1993). The concentrate, or an aliquot thereof, is analysed by epifluorescence and DIC microscopy.

3.1.2.3 Flotation. Non-covalent interactions between (oo)cyst surfaces and other contaminants are reduced by adding detergents and surfactants, and particles with a greater density than (oo)cysts can be clarified by centrifugation through a solution of a predetermined specific gravity (e.g. sucrose (1.18 sp. gr.) or Percoll-sucrose (1.1 sp. gr.)) on which (oo)cysts float. Many believe clarification to be inefficient, leading to (oo)cyst loss, and prefer to omit it if the water concentrate is not too turbid. Non-viable (oo)cysts are more likely to penetrate the sucrose flotation interface than viable (oo)cysts, and viable organisms concentrate on the sucrose density interface (Bukhari and Smith, 1995). The 'inefficiency' of sucrose flotation may be a reflection of the numbers of non-viable (oo)cysts in a water concentrate. Furthermore, (oo)cysts attached onto particles will have a higher combined density which will cause them to pass through the density interface. Clarification by sucrose flotation should be regarded as a method for enriching unattached viable organisms rather than as a method for concentrating and/or purifying all (oo)cysts present in a water concentrate.

Excess flotation fluid can interfere with oocyst attachment onto microscope slides and the immunoreactivity of the detecting antibody and must be removed by washing. Finally, the sample is examined microscopically for the presence of (oo)cysts. Because the infective dose to human beings is small, it is important (a) to have a preparation as free as possible of inorganic and organic debris which might mask the presence of the organisms or interfere with their identification and (b) to be able to identify small numbers of organisms accurately.

3.1.2.4 Immunomagnetizable separation (IMS). The concentration methods described above are not organism specific, relying primarily on biophysical and biochemical parameters to concentrate organisms. Methods based on the specific immunoreactivity between (oo)cyst surface-exposed epitopes and monoclonal antibodies (mAbs) to *Giardia* and *Cryptosporidium* (oo)cysts have also been developed. The availability of these mAbs enabled antibody-based technologies to be developed, which can increase the sensitivity of detection and reduce processing time (Smith, 1998). IMS techniques, where (oo)cysts are bound immunologically to inert, magnetizable beads coated with commercially available mAbs and which are then concentrated by a magnet, concentrate (oo)cysts effectively from contaminating particulates (Bifulco and Schaefer, 1993; Smith *et al.*, 1993a; Parker and Smith, 1994; Campbell and Smith, 1997).

Both paramagnetic colloidal magnetite particles (40 nm) and iron-cored latex beads have been used to concentrate (oo)cysts selectively from water concentrates. Using antibody-coated paramagnetic colloidal magnetite particles, an average of 82% of mAb-coated *Giardia* cysts (seeded at 500 cysts/ml into water concentrates with turbidities from 6 to 6000 NTU), were recovered (Bifulco and Schaefer, 1993). Turbidities of >600 NTU interfered with cyst recoveries, and significantly higher recoveries occurred with water samples of 600 NTU or less. Two benefits of using colloidal magnetite particles were identified by Bifulco and Schaefer (1993). First, the size of the colloidal paramagnetic particle is beyond the resolving power of a light microscope and does not interfere with the microscopic identification of an organism. Second, the surface area to volume ratio of these particles is larger than that for larger magnetizable particles and permits, theoretically, more antibody-binding sites per particle, producing a more reactive particle. Iron-cored latex beads coated with anti-*Giardia* mAb are a recent commercial development, but few comparisons of performance are available at the time of writing.

Approaches to IMS for *Cryptosporidium* have been developed in the Scottish Parasite Diagnostic laboratory over the last 10 years with recoveries ranging from 40 to >99%, dependent upon the matrix tested (Smith *et al.*, 1993a). In one trial, iron-cored latex beads coated with anti-*Cryptosporidium* mAb produced recovery efficiencies ranging from 46 to 87% of oocysts seeded into raw, potable and mineral waters. Each matrix was seeded with 5, 50 or 100 oocysts per 1, 10 or 15 ml sample. A prototype method was developed by Campbell and Smith (1997), but performance was compromised in turbidities ≥40 NTU (Smith and Hayes, 1996). Later commercialization produced better recoveries. Currently, two IMS kits, which use different antibody isotypes for oocyst capture, are available commercially.

The effectiveness of IMS is dependent upon the affinity of the mAb chosen. Two factors which affect paratope-epitope interactions are turbidity and divalent cation concentration. IMS performance is frequently better with less turbid samples but judicious manipulation of magnetizable bead-antibody conditions enhances IMS performance. Where high turbidities are encountered, such as in some raw water and wastewater concentrates, kits containing mAb paratopes with higher affinities for their epitopes can outperform kits with lower affinity mAbs. Most commercially available mAbs for *Cryptosporidium* and *Giardia* are of the IgM isotype, but one commercial kit incorporates an IgG_3 *Cryptosporidium* genus specific mAb which has a higher affinity than commercially available IgM mAbs. Our data indicate that the higher affinity IgG antibody isotype kit outperformed the IgM antibody isotype kit both in low (60 NTU) and high turbidity (60–14 160 NTU) concentrates (Smith, 1998; Paton *et al.*, 2001).

Combination *Giardia* and *Cryptosporidium* IMS kits are recent additions to the current methods available for concentration. IMS is included in the UK regulatory method for treated water (Anon., 1999a), the current UKSCA's provisional recommended methods (Anon, 1999b) and in the USEPA's Method 1623 (Anon., 1999c). Benefits of IMS include separating and concentrating (oo)cysts from contaminating debris and resuspending them in a particulate-reduced medium. This makes identification easier and quicker and enables a larger proportion of a turbid water concentrate to be analysed. These advantages are important where accuracy in identifying small numbers of organisms is a prerequisite. IMS has also been incorporated into methods such as detection by PCR and *in vitro* infectivity (Smith, 1995, 1998; Rochelle *et al.*, 1999).

3.1.3 *Recovery efficiencies*

Recovery efficiency varies from sample to sample and matrix to matrix. A disadvantage of large volume (yarn wound and polypropylene) filtration is that the recovery efficiency can be low (<1–40%). Recovery using the compressed foam cartridge, albeit on the small number of samples tested, was reported as 88–90% (Sartory *et al.*, 1998). Recoveries using flatbed membranes are approximately 5–60%, while recoveries for the pleated membrane capsule range from 58 to 81%, again on a small number of samples (Mattheson *et al.*, 1998). Lower recoveries 2.8% to 63% for *C. parvum* oocysts and 29.1% to 73.2% for *G. duodenalis* cysts (n = 46) were reported in the review by Smith (1998). Recovery using $CaCO_3$ flocculation method can be variable (<20–> 70%) (Vesey *et al.*, 1993; Campbell *et al.*, 1994) and, because of the elevated pH required to generate the floc, (oo)cyst viability is reduced when using this method (Campbell *et al.*, 1994).

3.1.4 *(Oo)cyst identification*

The benefits of fluorescence microscopy for detecting (oo)cysts have been addressed by various researchers (e.g. Rose *et al.*, 1989; Smith *et al.*, 1989b; Erlandsen *et al.*, 1990b) and antibodies which recognize surface-exposed epitopes on (oo)cysts are used (e.g. Riggs, 1983; Sauch, 1985; Rose *et al.*, 1988, 1989; Sterling *et al.*, 1988; Smith *et al.*, 1989b; Smith and Rose, 1990; Stibbs, 1993; Wallis, 1994). MAbs of defined paratope specificity and affinity have consolidated the usefulness of fluorescence antibody detection. Because the antibody paratopes bind surface-exposed (oo)cyst epitopes, the fluorescence visualized defines the maximum dimensions of the organism, enabling morphometric analyses to be undertaken. With the exception of one commercially available species specific mAb (*G. muris*-specific, Waterborne Inc. New Orleans, USA), commercially available mAbs are only specific at the genus level. For *Giardia* and *Cryptosporidium* (oo)cysts, epifluorescence microscopy using fluorescence-labelled mAbs has greatly enhanced our ability to provide more useful data (see below), but antibody specificity and affinity issues require to be addressed.

For raw and treated waters, the only method acceptable to UK and US regulators for determining the presence of (oo)cysts in a sample is microscopy, and is dependent upon defined fluorescence, morphometric (the accurate measurement of size) and morphological criteria. (Oo)cysts are dried onto welled microscope slides, or deposited onto membranes, and commercially available, genus specific, fluorescein isothiocyanate (FITC)-labelled mAbs (FITC-mAbs), reactive with exposed epitopes on the (oo)cyst wall, are used according to the manufacturer's instructions. Application of the fluorogen 4′6-diamidino-2-phenyl indole (DAPI) enhances the visualization of nuclei in (oo)cysts (Grimason *et al.*, 1994; Smith, 1995, 1996; Smith *et al.*, 2002).

Putative (oo)cysts should be examined under DIC microscopy in order to determine whether

TABLE 40.1 Characteristic features of *G. duodenalis* and *C. parvum* cysts by epifluorescence microscopy and Nomarski differential interference contrast (DIC) microscopy.
a) Appearance under the FITC filters of an epifluorescence microscope
G. duodenalis (oo)cysts and *C. parvum* oocysts. The putative organism must conform to the following fluorescent criteria:
Uniform apple green fluorescence, often with an increased intensity of fluorescence on the outer perimeter of an object of the appropriate size and shape (see below).
b) Appearance under Nomarski differential interference contrast (DIC) microscopy

Giardia duodenalis *cysts*	Cryptosporidium parvum *oocysts*
Ellipsoid to oval, smooth walled, colourless and refractile	Spherical or slightly ovoid, smooth, thick walled, colourless and refractile
8–19 × 7–10 μm (length × width)	4.5–5.5 μm
Mature cysts contain one trophozoite with four nuclei displaced to one pole of the organism	Sporulated oocysts contain four sporozoites, each containing one nucleus
Axostyle (flagellar axonemes) lying diagonally across the long axis of the cyst (see Fig. 40.1)	Four elongated, naked (*i.e.* not within a sporocyst(s)) sporozoites and a cytoplasmic residual body within the oocyst (see Fig. 40.2)
Two 'claw-hammer'-shaped bodies lying transversely in the mid-portion of the organism (see Fig. 40.1)	

organelles can be identified. Definitive criteria for epifluorescence and DIC microscopy are adopted for identifying the presence of (oo)cysts. The putative organism must conform to criteria for fluorescence, morphometry and morphology, identified in Table 40.1 (Anon., 1990, 1994, 1999a, b).

A choice of commercially available FITC-mAbs is available for both *Giardia* and *Cryptosporidium* and reagents are available as individual (e.g. *Giardia* only) or combination (e.g. *Giardia* and *Cryptosporidium*) kits, dependent upon requirement.

3.1.5 Limitations

Isolation and enumeration procedures are affected by different matrices, resulting in the underestimation of the occurrence of (oo)cysts in our environment (see section on recovery efficiencies). Water quality, especially the presence of algae, suspended solids, clays and turbidity influence recoveries and decrease accurate identification (Smith and Rose, 1990; Rodgers *et al.*, 1995; Smith and Hayes, 1996). IMS provides concentrated (oo)cyst suspensions, with less contaminating debris, however, the adverse environment of the water concentrate (turbidity, colloids and divalent cations) plays a significant role in determining the effectiveness of antibody binding. Antibody affinity can influence (oo)cyst recovery. High recoveries in low turbidity waters and lower recoveries in high turbidity waters can occur as epitope capture and release are trade-offs. Local water compositions and divalent cation content can have a major influence on performance, and these effects must be determined empirically for each commercial kit (Bukhari *et al.*, 1998; Rochelle *et al.*, 1999; Smith and Girdwood, 1999; Smith, 1998; Paton *et al.*, 2001).

Specificity and sensitivity are of paramount importance when attempting to detect small numbers of (oo)cysts in water concentrates. Waterborne (oo)cysts can be a mixture of both recently voided and aged organisms which may have been subjected to a variety of environmental pressures, including water treatment. Environmental and water treatment stresses as well as sample processing can alter the physical and chemical properties of (oo)cysts, making their appearance atypical. Distortion, contraction, collapse and rupture of the (oo)cyst are consequences of these stresses and lead to difficulty in identifying (oo)cysts, as the typical morphological features seen in Table 40.1 are degraded or lost (e.g. empty (oo)cysts). Inability to fulfil identification criteria results in the under-reporting of 'positives'.

The use of commercially available FITC-mAbs, whose paratopes bind surface-exposed (oo)cyst epitopes and define the maximum dimensions of an organism, enable measurements to be performed readily under near dark field conditions. Similarly, the inclusion of DAPI, to highlight nuclei, is a major advance in assisting analysts to recognize and confirm the presence of (oo)cysts, but is only useful when (oo)cysts contain nuclei. The usefulness of FITC-mAbs for detecting (oo)cysts in water-related samples depends on their ability to react effectively with epitopes on (oo)cyst walls which are resistant to a variety of environmental conditions, including environmental degradation, chlorine disinfection and ageing. Exposure of (oo)cysts to water treatment processes and/or the aquatic environment can affect epitope expression, and hence the intensity of FITC emission of the FITC-mAbs used for detection (Vesey *et al.*, 1993; Smith, 1996, 1998; Moore *et al.*, 1998).

The dearth of published comparative data makes it difficult to draw useful conclusions about the performance of different methods. The situation is further complicated by the use of variants of published methods. Failure to standardize will result in the evolution of hybrid methods that produce occurrence data which are not comparable with 'standardized' methods. The lack of comparable data reduces the significance of occurrence data for medical, environmental, regulatory and other professional bodies involved in the protection of public health. Standardization of isolation and identification procedures is imperative for the correct interpretation of results. Current limitations in our technologies lead to a continuing underestimation of environmental contamination as well as to confusion from the detection of organisms that have no significance to human health.

Alternative concentration and detection methods are available but are beyond the remit of this chapter. The reader is referred to Smith (1996, 1998) and Fricker and Crabb (1998) for a review of the current applicability of these methods.

4 OCCURRENCE IN RAW AND DRINKING WATER

Surveys of occurrence in numerous countries indicate that (oo)cysts can occur commonly in the aquatic environment, frequently in the absence of waterborne disease (Tables 40.2 and 40.3). (Oo)cysts have been detected in surface waters, ground water, springs and drinking water samples including those treated by disinfection alone, filtration, direct filtration, and conventional methods. The presence of (oo)cysts in groundwater, normally considered at lower risk of microbiological contamination than surface water, is of concern. In a UK survey, low-density positive results were reported for *Cryptosporidium* oocysts (Anon, 1992). In a US study (Hancock *et al.*, 1998), *Cryptosporidium* oocysts were detected in 5% (7/149) of vertical wells, 20% (7/35) of springs, 50% (2/4) of infiltration galleries and 36% (4/11) of horizontal wells. In 1997, in the UK, a waterborne outbreak of cryptosporidiosis, strongly associated with an oocyst contaminated deep chalk bore well, occurred with 345 confirmed cases reported. Infiltration though natural interstices of the chalk from a nearby river or through cracks were suggested as contamination routes. Unusually high rainfall and cold weather were also believed to have been contributing factors (Willocks *et al.*, 1998).

4.1 Occurrence in estuarine and coastal waters

Estuarine and coastal marine waters receive (oo)cysts derived from agricultural and livestock practices, run off from land, defaecation by infected, non-human hosts, and wastewater effluent discharges. Naranjo *et al.* (1989) found that marine sewage outfall effluents contained average concentrations of 0.199 cyst/l and 0.066 oocyst/l, while Ho and Tam (1998) noted that cysts were observed with greater frequency in samples of lower grade beach water compared with higher grade beach water, with concentrations generally less than 10 cysts/l. The occurrence of (oo)cysts in marine waters and sediments (Naranjo *et al.*, 1990; Johnson *et al.*, 1995; Ferguson *et al.*, 1996; Ho and Tam, 1998) is

TABLE 40.2 Some examples of the occurrence and density of *Cryptosporidium* oocysts and *Giardia* cysts in surface waters (adapted from Gold and Smith, 2001)

Country	*Number of samples*	*Occurrence of* Cryptosporidium *oocysts (% samples positive)*	*Density of* Cryptosporidium *oocysts (oocyst/l)*	*Occurrence of* Giardia *cysts (% samples positive)*	*Density of* Giardia *cysts (cysts/l)*	*Year*
USA	11	100	2–112	–	–	1987
USA	222	–	–	43	0.5–1	1989
USA	101	24	0.005–252.7	–	–	1990
UK (Scotland)	262	40.5	0.006–2.3	–	–	1990
USA	35	97.1	0.18–63.5	–	–	1991
Germany	9	78	–	–	–	1991
UK	691	52.2	0.04–3	–	–	1992
UK	375	4.4	0.07–2.75	–	–	1992
UK (Scotland)	53	–	–	33	0.01–1.05	1993
Canada	22	0	–	32	–	1993
Spain	8	50	<0.01–0.31	63	<0.01–0.21	1993
Canada	249	–	0.005–0.34	100	0.005–0.34	1996
Canada	1760	6.1	–	21	–	1996
Honduras	–	–	0.58–2.6	–	3.8–21	1998
Taiwan	31	72.2	–	77.8	–	1999
Czech Republic	–	–	0–74	–	0–4.85	2000
Venezuela	12	75	0.15[a]	33	0.008[a]	2000
Japan	156	47	0.07–0.12	–	–	2000

[a] geometric mean.

TABLE 40.3 Some examples of the occurrence and density of *Cryptosporidium* oocysts and *Giardia* cysts in treated waters[a] (adapted from Gold and Smith, 2001)

Country	*Number of samples*	*Occurrence of* Cryptosporidium *oocysts (% samples positive)*	*Density of* Cryptosporidium *oocysts (oocysts/l)*	*Occurrence of* Giardia *cysts (% samples positive)*	*Density of* Giardia *cysts (cysts/l)*	*Year*
USA	36	17	0.005–0.017	0	–	1991
USA	82	26.8	–	16.9	–	1991
UK (Scotland)	15	7	0.006	–	–	1995
UK (Scotland)	106	–	–	19	0.01–1.67	1993
Spain	9	33	<0.01–0.02	22	<0.01–0.03	1993
Brazil	18	22.2	–	–	–	1993
Canada	42	5	–	17	–	1993
Canada	249	–	–	98.5	0.045–1.72	1996
Canada	1760	3.5	–	18.2	–	1996
Germany	12	66.7	0.008–1.09	83.3	0.02–1.03	1996
UK	209	37	0.007–1.36	–	–	1998
Taiwan	31	38.5	–	77	–	1999
Venezuela	11	90	0.004[a]	36	0.013[b]	2000

[a] Waters for potable supply receive different treatments in different areas of the world; whereas some of the waters in this table have received a number of treatments before being considered usable for potable supply, others may have received minimal treatment.

[b] geometric mean.

of public health concern especially to recreational users of these watercourses. In addition, benthic filter-feeders such as shellfish, which are eaten raw or lightly cooked, can accumulate (oo)cysts in their tissues. Recent studies have shown that shellfish can harbour (oo)cysts (Chalmers *et al.*, 1997; Graczyk *et al.*, 1998, 1999; Fayer *et al.*, 1999; Friere-Santos *et al.*, 2000) and may be of value as biological monitors of the presence of (oo)cysts in both fresh and marine waters. *C. parvum* and *G. duodenalis* (oo)cysts have been detected in the faeces of Californian sea lions based upon morphological, immunological and DNA characterization (Deng *et al.*, 2000).

5 REMOVAL IN WATER TREATMENT

The control of microbiological contaminants in potable waters requires an integrated multiple barrier approach to source water protection and water treatment. Water treatment ranges from disinfection only (e.g. groundwater), microstraining and disinfection (e.g. impounding reservoir), to full conventional treatment, i.e. bankside storage, coagulation, flocculation, sedimentation, filtration, and disinfection (e.g. lowland rivers), depending upon the quality of water to be treated. Studies on the removal of *Cryptosporidium* and *Giardia* by chemical and physical treatment processes are comparable, although cysts appear to be removed slightly more efficiently than oocysts. This may, in part, be due to their larger size. Both (oo)cyst structure and organization contribute significantly to the different requirements for control by disinfection, however, their physical characteristics, including size, settling velocities and surface charge, which affect behaviour in physical treatment processes, are similar. Therefore, much of the data on cyst behaviour in treatment processes can be used to predict oocyst behaviour.

While trends can be drawn from the studies on (oo)cyst removal in water treatment processes outlined below, it must be remembered that much of the work performed contains many variables which complicates direct comparison between published data. These include water temperature, uniformity coefficient of sand, (oo)cyst seeding density and procedure, (oo)cyst history (species used, isolate and origin, age and storage conditions, genotype, source – human or non-human, etc.), (oo)cyst viability (e.g. dead (heat killed, formalin preserved), viable) and the recovery and detection method used. Optimized coagulation, the key to effective filtration, depends on water temperature, pH, the constituents of the raw water, dosages of coagulant and coagulant aid, mixing time and mixing intensity.

5.1 Reservoir storage

Bankside storage of abstracted river water provides a retention period within either a closed or open reservoir which facilitates the settlement of suspended particulate matter prior to treatment and distribution. As (oo)-cysts settle slowly (<0.02 m/h) and have a specific gravity ($\approx 1.05-1.08$) similar to that of water, insufficient retention time (RT) is usually available for their complete settlement and removal, unless attached to heavier particulate material (Grimason *et al.*, 1993; Medema *et al.*, 1998). Sedimentation performance data without the aid of coagulants and coagulant aids are scarce at both the pilot and full-scale levels. Existing data are based on ambient levels of oocysts in raw waters before and after sedimentation. A monitoring study of two full-scale conventional plants showed 0.5–0.8 log removal by sedimentation (Kelley *et al.*, 1994). Data from States *et al.* (1997) confirm this with 0.6 $\log_{10}$ removal by sedimentation.

Open reservoirs are subject to potentially greater contamination with (oo)cysts derived from non-point sources of pollution (e.g. agricultural) and liquefied faecal matter from indigenous animal species washed into nearby waters during periods of prolonged rainfall or snow melting events. Studies identified increased oocyst and cyst concentrations in surface and estuarine waters following prolonged rainfall events (Ferguson *et al.*, 1996; Atherholt *et al.*, 1998). In contrast, pristine lakes and protected reservoirs have lower oocyst (and probably cyst) densities than unprotected supplies (Edzwald and Kelley, 1998).

The impact of (oo)cyst contamination of potable water stored in open reservoirs in the USA was assessed by LeChevallier *et al.* (1997) by examining the occurrence and concentration of (oo)cysts in inlet and outlet water samples from six open finished (filtered) water reservoirs. Oocyst and/or cysts were detected in the outlet of all of the reservoirs with theoretical retention times (RT) ranging from 1 to 63 days. Pooled (oo)cyst concentrations revealed a significant increase in the occurrence of these organisms at the outlet. Oocysts, but not cysts, were detected at significantly higher densities at outlets compared with inlets. The majority of cysts (44/45) and oocysts (12/14) detected from all reservoirs were found to be either void of internal organelles or had amorphous structures using phase contrast and DIC microscopy (LeChevallier *et al.*, 1997).

In contrast, a study conducted by van Breemen *et al.* (1998) found that storage reservoirs markedly reduced *Giardia* and *Cryptosporidium* densities in outlet waters, compared with levels detected in abstracted river water. Oocyst removal rates from a reservoir with a RT of 24 weeks ranged from 1.4 to 2.0 $\log_{10}$ units, and cysts from 2.3 to 2.6 $\log_{10}$ units. A large natural storage reservoir or lake (RT = 25 to 52 weeks) followed by an artificial reservoir (RT = 10 weeks), with a maximum cumulative retention period of 62 weeks reduced oocyst and cyst densities by 1.3 and 0.8 $\log_{10}$ units, respectively. Bertolucci *et al.* (1998) also detected (oo)cysts in storage reservoirs with a RT of 18 days. Factors which affect (oo)cyst occurrence, density, settlement and removal in large bodies of water are outlined in the section on waste stabilization ponds.

5.2 Microstraining

Microstrainers are generally designed with pore sizes of 20 μm and greater, and are not expected to remove (oo)cysts, given their considerably smaller size. Despite this, microstraining (aperture pore size of 25 μm) followed by chlorination resulted in approximately 60% removal of oocysts and 80% removal of cysts (Parker, 1993; Grimason *et al.*, 1995). A reduction in pore size as the mesh/screen becomes blocked, or attachment to material filtered on the surface of the screens might account for the observed reductions, but microstraining of water is not an effective method for (oo)cyst removal.

5.3 Coagulation, flocculation and filter media

Full conventional treatment, employing an array of chemical and physical treatment unit processes, in combination and in series, is capable of achieving high, but not complete, (oo)cyst removal. This is borne out both by the studies outlined in Table 40.4 and by waterborne outbreaks, associated with water which underwent full, conventional treatment (Smith and Rose, 1998). Coagulation and flocculation are effective in reducing fine particulate (including (oo)cysts) and colloidal matter challenging the subsequent filtration and disinfection processes. The mechanism of (oo)cyst removal is probably similar to that for other negatively charged particles in water. Cationic coagulants are added to water and mixed rapidly to (i) reduce the large repulsive forces between negatively charged particles and (ii) cause them to collide and form aggregates of greater mass. The process is enhanced by flocculation and sedimentation; slowly mixing the coagulated aggregates to form heavier flocs, which settle out of suspension more readily. Therefore, knowledge of the hydrophobic properties and surface charge of (oo)cysts provides useful information that can be used to optimize (oo)cyst removal during coagulation and flocculation. Measurements of electrophoretic mobility have demonstrated zeta potentials for both *Cryptosporidium* oocysts and *Giardia* cysts (−25 to −35 mV at neutral pH) to be similar to other particles in surface water (Engeset, 1984; Ongerth and Pecoraro, 1995; Drozd and Schwartzbrod, 1996). (Oo)cysts should coagulate and flocculate, as do other naturally occurring particles, in response to conventional coagulation and flocculation practices.

Many investigators report (oo)cyst removal efficiencies as cumulative totals in treatment,

TABLE 40.4 Removal of *Giardia* and *Cryptosporidium* by water treatment processes

Treatment	Giardia (*log removal*)	Cryptosporidium (*log removal*)	*Scale*	*Reference*
Slow sand filtration	2-> 4	>4	Pilot	Schuler *et al.*, 1991
Slow sand filtration	4	–	Pilot	Timms *et al.*, 1995
Sand and DM (S/A) filters	2.7->4.7	2.7–3.9	Pilot	Swertfeger *et al.*, 1999
DEF (pore sizes 23 & 26 μm)	> 3	>3	Pilot	Schuler *et al.*, 1991
C, DEF (pore size 26 μm)	> 3	>3	Pilot	Schuler *et al.*, 1991
DEF (median pore size 5 μm, low flow rate)	–	mean 6.24–6.31	Bench	Ongerth and Hutton, 1997
DEF (median pore size 5 μm, high flow rate)	–	mean 6.33–6.68	Bench	Ongerth and Hutton, 1997
DEF (median pore size 7 μm, low flow rate)	–	mean 5.93–6.00	Bench	Ongerth and Hutton, 1997
DEF (median pore size 7 μm, high flow rate)	–	mean 6.23–6.31	Bench	Ongerth and Hutton, 1997
DEF (median pore size 13 μm, low flow rate)	–	mean 3.84–3.94	Bench	Ongerth and Hutton, 1997
DEF (median pore size 13 μm, high flow rate)	–	mean 5.38–5.64	Bench	Ongerth and Hutton, 1997
Direct filtration (C, DM (S/A) Filt.)	2.9–4	1.31–3.78	Pilot	Nieminski and Ongerth, 1995
C, F, MM Filtration	3.05–3.6	2.7–3.1	Pilot	Ongerth and Pecoraro, 1995
Coagulation and dissolved-air flotation	–	max. 2.6–3.7	Bench	Plummer *et al.*, 1995
Coagulation and sedimentation	–	<1	Bench	Plummer *et al.*, 1995
Conventional (C,F,S,DM (S/A) Filt.)	2.2–3.9	1.94–3.98	Pilot	Nieminski and Ongerth, 1995
Microfiltration (nom. pore size 0.25 μm)	–	>7	Pilot	Hirata and Hashimoto, 1998
Ultrafiltration (nom. MW 13 000 daltons)	–	>7	Pilot	Hirata and Hashimoto, 1998
Microfiltration (pore size 0.1–0.2 μm)	4.6-> 5.2	4.2- >4.9	Bench	Jacangelo *et al.*, 1995
Ultrafiltration (MW 1×10^5–5×10^5 daltons)	>4.7-> 5.2	>4.4->4.9	Bench	Jacangelo *et al.*, 1995
Microfiltration (pore size 0.1–0.2 μm)	>6.4-> 7	>6->6.9	Pilot	Jacangelo *et al.*, 1995
Ultrafiltration (MW 1×10^5–5×10^5daltons)	>6.4-> 6.9	>6.3–7	Pilot	Jacangelo *et al.*, 1995
Crossflow microfiltration (pore size 0.2 μm)	–	>4.3	Pilot	Drozd and Schwartzbrod, 1995
Microstraining and chlorination	<1	–	Full	Grimason *et al.*, 1995
Microstraining and chlorination	–	<1	Full	Parker, 1993
Slow sand filtration	1	0.5	Full	Fogel *et al.*, 1993
Coagulation and dm s/a filtration	mean 3.87	mean 2.79	Full	Nieminski and Ongerth, 1995
Plant A–conventional treatment	5.0	4.7	Full	Payment and Franco, 1993
Plant B–conventional treatment	>5	>1.2	Full	Payment and Franco, 1993
Plant C–conventional treatment	> 5	>1.2	Full	Payment and Franco, 1993
Plant D–conventional treatment	'complete'	'complete'	Full	Chauret *et al.*, 1995

(*continued on next page*)

Table 40.4 (*continued*)

Treatment	Giardia (*log removal*)	Cryptosporidium (*log removal*)	*Scale*	*Reference*
Plant E–conventional treatment	>2.30	>2.38	Full	LeChevallier and Norton, 1992
Plant F–conventional treatment	>2.78	>2.45	Full	LeChevallier and Norton, 1992
Plant G–conventional treatment	>2.24	>2.30	Full	LeChevallier and Norton, 1992
Plant H–conventional treatment	2.82–3.70	1.89–2.78	Full	Nieminski and Ongerth, 1995
Sedimentation	1		Bench / Pilot	Edzwald and Kelley, 1998
Dissolved air flotation (DAF)	3		Bench / Pilot	Edzwald and Kelley, 1998
Sedimentation and DAF	4 to 5		Bench / Pilot	Edzwald and Kelley, 1998
Sedimentation and filtration	3 to 4		Bench / Pilot	Edzwald and Kelley, 1998
Coagulated, flocculated and sedimentation	1	1	Pilot	J.E. Ongerth (personal communication)

Plant A = activated charcoal, aluminium silicate, and activated silica flashed with river water, flocculated, sedimented, raid sand filtered, and chlorinated.
Plant B = flocculated, sedimented, rapid filtered on sand-anthracite, ozonated, refiltered on a biological filter (granular activated charcoal) chlorine dioxide disinfection.
Plant C = alum or polyelectrolyte coagulation, settling, sand-anthracite filtration, ozonation and final disinfection with chlorine dioxide.
Plant D = alum coagulation, activated silica, settling, slow (sand-anthracite) filtration and chloramination.
Plant E = pre-sedimentation, coagulation, flocculation, softening, dual media sand filtration and chloramination.
Plant F = pre-chlorination, coagulation, flocculation, sedimentation, dual media granular activate carbon filtration and chlorination.
Plant G = coagulation, flocculation, sedimentation, polyelectrolyte addition and dual media sand filtration.
Plant H = coagulation, flocculation, sedimentation, dual media (sand/anthracite) filtration. C = coagulation, F = flocculation, S = sedimentation, DM (S/A) Filt = dual media (sand/anthracite) filtration, MM Filt = multimedia (silica sand, anthracite and garnet sand) filtration, DEF = Diatomaceous earth filtration.

and not in relation to specific unit processes (e.g. coagulation and flocculation). However, studies conducted on both pilot and full-scale plants revealed that approximately 1 $\log_{10}$ removal of (oo)cysts can be expected for raw waters which are coagulated and flocculated effectively (Nieminski and Ongerth, 1995; J.E. Ongerth and P.E. Hutton, personal communication). Numerous other studies involving coagulation and flocculation, prior to clarification and filtration, indicate that at least 2 and up to 5 $\log_{10}$ removal rates are attainable (Table 40.4). When monitoring two, small, full-scale filtration plants, during normal operation, at least a 2 $\log_{10}$ removal was achieved for *Giardia* and *Cryptosporidium* (Ongerth, 1990; Karanis *et al.*, 1998). These studies also highlight the importance of effective chemical conditioning and attention to operating details in order to maximize (oo)cyst removal. However, in one of the studies, a low coagulant dose, followed by a sudden increase in the raw water quality resulted in the breakthrough of *Cryptosporidium* into the treated water (Karanis *et al.*, 1998).

Huck *et al.* (2000) examined *Cryptosporidium* removal from low turbidity waters (0.6–2.5 NTU) by granular media filtration, focusing on the impact of high (35 mg/l alum and 2 mg/l activated silica) and low (5 mg/l alum and 1.5 mg/l cationic polymer) coagulation dosage at two pilot plants. Under optimized coagulation conditions (filter effluent turbidity <0.1 NTU), >5 $\log_{10}$ removal was achieved with the high dose and >3 $\log_{10}$ with the lower dose. Suboptimal coagulation (40–60% of optimum coagulant dose) incurred a substantial deterioration (2 $\log_{10}$) in oocyst removal efficiency at both plants. When no coagulant or coagulant aid was used (i.e. direct filtration), removal efficiencies significantly declined at both plants (0.2–0.3 $\log_{10}$ reduction). The authors concluded that while turbidity may be of value for assessing coagulation impacts on *Cryptosporidium*, particle counts may be a more sensitive parameter in this regard.

Dissolved air flotation (DAF) can be a better alternative than sedimentation when a water source contains low density particles which have a tendency to settle very slowly or float (Fukushi *et al.*, 1998). For reasons outlined earlier, oocysts will not settle out of suspension naturally in sedimentation tanks, unless attached to heavier particulates. Thus, as a clarification technique, DAF should be suitable for removing small, low density particles, including protozoan parasites. A bench-scale study conducted in the USA appears to ratify DAF as an effective means of removing *Cryptosporidium* from surface waters, with reported removal rates of 3 $\log_{10}$ (with three different coagulants) compared with 1 $\log_{10}$ removal by sedimentation (Edzwald and Kelley, 1998). Pilot-scale studies on dual media filtration and DAF provided two effective barriers to oocysts with cumulative $\log_{10}$ removals of 4–5 compared to $\log_{10}$ removals of 3–4 by sedimentation and filtration at both high (14.6 m/h) and low (7.3 m/h) flow rates (Edzwald and Kelley, 1998). The authors concluded that particle counts (range 2–8 μm) and turbidities were of little use to predict oocyst occurrence and concentration in raw waters. However, filtered water turbidities of <0.1 NTU and particle counts of ≤ 50/ml were indicators of good treatment for controlling *Cryptosporidium*.

These results confirm those of earlier, bench-scale, studies, undertaken to compare the effectiveness of DAF versus sedimentation for the removal of *Cryptosporidium* oocysts from raw water. In a series of bench scale experiments, individually addressing coagulant dose, pH, flocculation time and recycle ratio, Plummer *et al.* (1995) found that maximum oocyst removal occurred with ferric chloride at 5 mg/l; pH 5; a flocculation time of 10 min; and a recycle ratio of 10%. Water, clarified by coagulation and flotation was shown to be 'superior' (range 2.6 $\log_{10}$–3.7 $\log_{10}$ oocyst removal) compared with coagulation and sedimentation ($<1\log_{10}$ oocyst removal) under all conditions tested. Reasonable correlations ($r \geq 0.7$) were noted between oocyst removal and turbidity, ultraviolet absorbance at 254 nm (UV_{254}) and dissolved organic carbon. The authors concluded that those conditions for coagulation and DAF that minimize residual turbidity, and maximize removal of organic matter, may optimize the removal of *Cryptosporidium* oocysts.

Various filter media, configurations and operating rates have been evaluated for the removal of (oo)cysts under pilot-scale and full-scale conditions (see Table 40.4). A study conducted by LeChevallier and Norton (1992) compared the efficiency of three full-scale rapid sand filters (coagulation and dual media or dual media granular activated carbon) to remove (oo)cysts from differing water quality watersheds, ranging from pristine (<0.3–5 NTU) to highly polluted (<5–240 NTU). While the lowest removal efficiencies were recorded in pristine waters (>2.24 log), higher (oo)cyst removal efficiencies occurred in polluted (>2.45 log) and highly polluted (>−2.3 log) waters. Significant correlations were observed between turbidity removal ($r > 0.77$) and particle counts of at least 5 μm ($r > 0.83$) and *Giardia* and *Cryptosporidium* levels in treated waters. A pilot plant scale study, using anthracite over sand as filtration media, was reported to produce *Cryptosporidium* removals from 2.7 to 3.1 $\log_{10}$ and *Giardia* removals from 3.1 to 3.6 $\log_{10}$ (Ongerth and Pecoraro, 1995). The importance of chemical conditioning in maximizing (oo)cyst removal was illustrated by the reduction in $\log_{10}$ removals for both organisms from about 3 to 1.5 when the filters were operated at one half the optimal coagulant dose (i.e. 5 mg alum/l). Low (oo)cyst removals coincided with poor turbidity reductions. In a similar study, removals were approximately 3–3.5 $\log_{10}$ for *Cryptosporidium* and approximately 3.8 $\log_{10}$ for *Giardia* in both pilot and full-scale filtration plants with sedimentation, operating under optimal conditions (Ongerth and Pecoraro, 1995; Nieminski and Ongerth, 1995). During filter ripening periods and turbidity breakthrough at the end of filter runs, removals of both organisms were 0.5–1.5 $\log_{10}$ less than during optimal periods (Ongerth, 1990; Ongerth and Pecoraro, 1995; Nieminski and Ongerth, 1995; Ongerth and Hutton, 2001).

A detailed investigation of *Cryptosporidium* and *Giardia* removal in an automatic backwash filter reported higher removals than those cited above (Wilczak *et al.*,1994). The study focused on a 57 l/min pilot plant including, independently, alum-poly and ferric-poly coagulation, tapered flocculation, tube-settling and sand filters operated at 4.8 m/h with raw turbidities in two ranges, approximately 1 NTU and approximately 10–12 NTU. The pilot plant was seeded continuously during testing and experiments were conducted over a 12-month period. Overall removals for *Cryptosporidium* averaged approximately 4.5 $\log_{10}$ and for *Giardia* averaged approximately 4.2 $\log_{10}$. Removals of *Cryptosporidium* and *Giardia* in settling averaged from 1 to 2 $\log_{10}$ and approximately 0.5 $\log_{10}$ respectively.

Evaluating the effectiveness of various filter media (sand only, typical dual anthracite-sand and deep dual anthracite-sand configurations) to remove heat-inactivated (oo)cysts seeded into precoagulated and settled water, Swertfeger *et al.* (1999) calculated that between 2.7 and >4.5 $\log_{10}$ removal of cysts and 2.7–3.9 $\log_{10}$ removal of oocysts was achievable. Similar (oo)cyst removal efficiencies have been observed by other investigators (see Table 40.4). In general, most investigators identify that *Giardia* removals are consistently greater than *Cryptosporidium* removals, by at least 0.5 $\log_{10}$.

To compare the effectiveness of conventional treatment and direct filtration to remove (oo)cysts, Nieminski and Ongerth (1995) conducted a series of pilot and full-scale studies. Conventional treatment included coagulation, flocculation, sedimentation and dual media (sand/anthracite) filtration, while direct filtration included coagulation and dual media filtration only. In the pilot-scale studies, little difference was noted between conventional treatment for the removal of *Giardia* (mean 3.4 $\log_{10}$) and direct filtration (mean 3.3 $\log_{10}$). This was also true for *Cryptosporidium*, with average oocyst removals of slightly less than 3 $\log_{10}$ observed for both treatment schemes. Based upon the average $\log_{10}$ removals reported, full-scale tests revealed that direct filtration (cysts 3.87 $\log_{10}$; oocysts 2.79 $\log_{10}$) removed significantly more organisms (at least 0.6 $\log_{10}$) compared with conventional treatment (cysts 3.26 $\log_{10}$; oocysts 2.25 $\log_{10}$). Cysts were removed with greater efficiency than oocysts (0.3 and 1 $\log_{10}$ greater) in both systems tested at pilot and full-scale levels. Unlike previously cited studies, this study utilized chemical tracers (rhodamine and sodium chloride) to establish optimum sampling times, thus enabling a more accurate calculation of removal efficiencies. A reasonable correlation was found between cyst ($r = 0.8$) and oocyst ($r = 0.7$) removal rates and removal of (oo)cyst sized particles in water. However, little or no correlation was found between (oo)cyst removal and the removal of turbidity or heterotrophic bacteria.

(Oo)cyst numbers penetrating such systems can increase during operational practices such as backwashing and ripening of sand filters. Backwashing is responsible for the penetration of coliform bacteria through this treatment barrier (Bucklin *et al.*, 1991). Backwash water from rapid sand filters of a treatment plant using surface water collected from small rivers in Germany was shown to contain cysts (range 0.014–3.74/ l) and oocysts (range 0.008–2.52/ l) in the majority (85%) of samples analysed (Karanis *et al.*, 1996). Overall, *Giardia*, *Cryptosporidium*, or both were detected in 92% of the backwash water samples and the supernatant, returned to the raw water source after sedimentation, still contained cysts and oocysts. Turbidity removal is typically low following backwash, as the filter bed becomes reconditioned by the chemically conditioned water passing through it. Pilot scale trials indicate that cyst-size particles can pass through filters in relatively large numbers during this period (Ongerth, 1990). Logsdon *et al.* (1985) provided results from a pilot scale study indicating that concentrations of *Giardia* cysts in final water increased as much as 20–40-fold during the first portion of the filter run that follows the backwash cycle.

Following clarification and filtration, flocculated particles (including (oo)cysts) are trapped in the sand bed and hence concentrated in the backwash water. Background interference, due to the enmeshment of (oo)cysts in flocculated material, can lead to an underestimation of both occurrence and densities reported in such samples (Hall and Pressdee, 1995). Effective separation and accurate identification of (oo)cysts from liquid and semi-solid material could also explain why Quennell and West (1995) found that 10% of filter backwash water samples contained oocysts (5/50; max. 10 oocysts/l), but found none in 19 samples of waterworks sludge. Clumping of (oo)cysts, prefiltration, assists their removal, while clumping, post-filtration, can lead to a higher risk of contracting infection because of the greater likelihood of ingesting an infectious dose (Smith and Rose, 1990).

Hall and Pressdee (1995) assessed the changes in filtrate quality of water that occurred during the ripening period, and thereafter, in a pilot-scale rapid sand filter operating under a hydraulic loading of 900 l/h and challenged with 1500 oocysts/l (Fig. 40.3). This study also evaluated the impact of stopping the filter after a few hours into the run and restarting after a backwash. Peaks in particle counts (2 μm = 2000/ml; 5 μm = 400/ml) and turbidity (0.4 NTU), observed during the early part of the run (<1hour), were paralleled by a peak in oocyst density (6.3 oocysts/l). This suggests that the ripening phase is the critical part of a filtration cycle, and is the time of highest potential risk of passing oocysts into supply. Similar conclusions were drawn by Hillis and Colton (1995), based upon particle monitoring data. Restarting the filter after backwashing generated peaks in particle counts and turbidity of approximately half those reported by Hall and Pressdee (1995), however, no oocysts were detected in filtered water. For the remainder of the run, turbidity was around 0.1 NTU, rising to 0.5 NTU with particle breakthrough at the end of the run. Oocyst density in a filtrate sample of 1175 litres taken over the remainder of the run, following restart, was 0.6/l. Similar effects were found for the dual media (media not stated) and GAC beds evaluated.

The contribution of oocyst-contaminated reclaimed backwash water to the numbers of oocysts in the influent was assessed in an unseeded 50 Ml/day complete treatment plant in Sydney, Australia (Orchard Hills) which operated daily, for 1 month, under normal operating conditions at flows of 60–70 Ml/day (P.E. Hutton pers. comm.). The plant includes alum (20–30 mg/l) with 0.1 mg/l non-ionic polyelectrolyte coagulation, tapered flocculation, rectangular sedimentation tanks

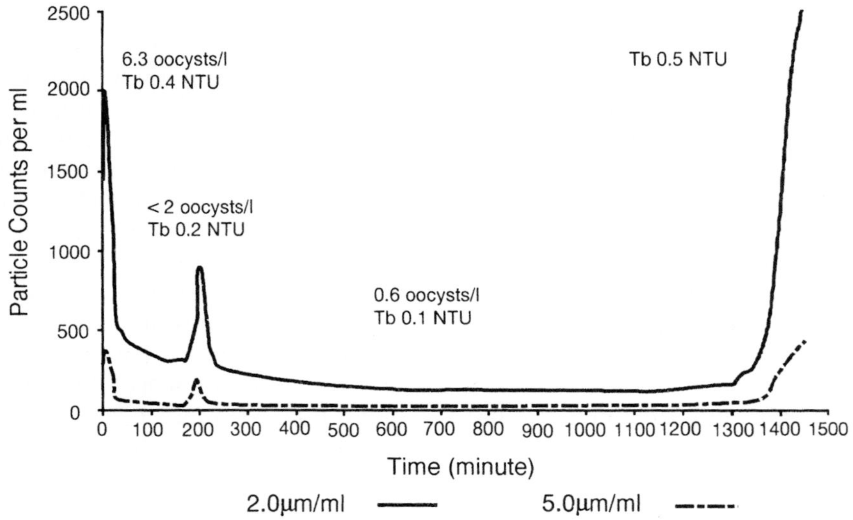

Fig. 40.3 Changes in filtered water quality measured as particle counts, oocyst numbers and turbidity (Tb) during operation of a rapid gravity sand filter on a pilot treatment plant. Reproduced with permission from Hall, T. and Pressdee, M.R. (1995). Removal of *Cryptosporidium* during water treatment. In: Proceedings of workshop on treatment optimization for *Cryptosporidium* removal from water supplies. HMSO.

(RT = 2 h) and anthracite over sand filtration operated at approximately 16 m/h. Samples were taken from the plant influent, plant effluent and backwash water recycle streams three times daily, for 1 month, and oocyst densities were calculated following $CaCO_3$ flocculation and flow cytometry (Vesey *et al.*, 1994). Oocyst densities averaged 0.23/l in raw water and 0.0005/l in filtered water, corresponding to an overall removal of approximately 2.9 $\log_{10}$. The calculated mass balance on the plant, based on flow rates and an average recycle oocyst density of 0.37/l, indicated that the recycle stream did not contribute significantly (approximately 5%) to the numbers of oocysts in the influent.

The results of a survey of *Cryptosporidium* and *Giardia* concentrations found in raw and treated water samples at 66 operating filter plants in the USA and Canada (LeChevallier *et al.*, 1991) complement the studies described, with estimated removal efficiencies ranging from 2 to 2.5 $\log_{10}$ for both organisms. Similar removals were reported by Kelley *et al.* (1994), based on analysis of (oo)cyst densities in samples of raw and filtered water, in a survey of 16 water filtration plants at US Army installations in the eastern USA. In a detailed, 6-month study of two of these 16 treatment plants (a 3.8 Ml/day and a 11.4 Ml/day plant, both with complete treatment) samples were taken monthly and *Cryptosporidium* removals ranged from 2.2 to 2.6 $\log_{10}$ while *Giardia* removals ranged from 1 to 1.8 $\log_{10}$. However, water quality conditions during operating periods with and without settling were not directly comparable. These results emphasize the potential sensitivity of treatment plants to operating conditions, including flow-rates, water temperature, and the nature of particles comprising turbidity, at different times of the year.

The uniformity coefficient (u.c.) of the sand used can affect the efficiency of (oo)cyst removal during slow sand filtration. Fogel *et al.* (1993) evaluated the effectiveness of a full-scale slow sand filter with a u.c. of 3.5–3.8 and noted oocyst and cyst removal efficiencies of 46% and 93%, respectively. They concluded that the large pore size between sand grains may have contributed to the poorer oocyst removal rate. For design purposes, the u.c. of the sand should ideally be <2 to reduce the size of the pore spaces within the filter bed, and to encourage the formation of the schmutzdecke (biological) layer. Here, the poor removal efficiency of the sand filter for *Cryptosporidium* may also have been due to the cold temperature (1°C) under which the plant operated, preventing the build-up of an effective microfauna. Other investigators observed that the removal efficiency of slow sand filters for *Giardia* is affected when temperature and biological activity in the filter are reduced (Schuler *et al.*, 1991; Ongerth and Pecoraro, 1995). Therefore, plant performance should be monitored carefully when water is abstracted from known contaminated water sources under cold conditions. For treatment plants in colder, temperate regions, new slow sand filters should be commissioned during warmer seasons, so that a viable biological community can be established before the onset of winter (Schuler *et al.*, 1991). Otherwise, higher (oo)cyst concentrations might be expected in filter effluents during periods of ripening and end-of-run turbidity breakthrough (Ongerth and Pecoraro, 1995).

The potential for oocysts to be transported through intermittent, unsaturated, laboratory-scale sand filter columns (60 cm, 15°C) used for water and artificial wastewater treatment was investigated by Logan *et al.* (2001). Factorial design analysis indicated that grain size was the variable that most affected oocyst removal efficiencies. Fine-grained sand columns (d_{50} = 0.31 mm) effectively removed (3–4 $\log_{10}$ reduction) oocysts under low (4 cm/day) and high (10 cm/day) hydraulic loading rates (HLRs). At low HLRs the fine-media sand columns demonstrated almost complete removal of the total oocyst load (65 000 oocysts/l). Analysis of the effluent revealed a concentration of 20 oocysts/l in 1 out of 25 samples. Similar removal efficiencies were recorded at high HLRs with 4 of 32 effluent samples containing oocyst concentrations that never exceeded 27 oocysts/l. Increasing the HLR had little effect on effluent concentrations from fine-grained columns but markedly increased effluent oocyst concentrations from the coarse-grained columns. Effluent from columns utilizing the coarse filtration media (d_{50} = 1.40 mm) contained oocysts more frequently and at significantly higher concentrations (max. 400 oocysts/l). An increase in the HLR to 20 cm/day resulted in a reduction in removal efficiency to 2 $\log_{10}$, with oocyst concentrations in coarse-grained effluent (max. 824 oocysts/l) being more than four times that in the influent. Despite the increased HLR, effluent samples from fine-media columns were less likely to contain oocysts (3 of 57) compared with effluent from coarse media columns (25 of 61) suggesting better removal efficiency.

The concentration of oocysts per unit mass of media generally decreased with depth for all columns tested. Analysis of the fine media columns revealed that a larger percentage of the oocysts were retained closest to the surface and oocyst density results below 10 cm were generally negative. Coarse media columns generally demonstrated positive oocyst densities throughout the depth of the columns. Higher loading rates resulted in higher numbers of oocysts per sand mass as compared to an identical column at a lower loading rate. Pore-size calculations indicated that adequate space for oocyst transport existed in the filters. It was concluded that processes (e.g. DVLO mechanisms or sorption processes) other than physical straining mechanisms are mainly responsible for the removal of *C. parvum* oocysts in sand filters. No correlations between turbidity, pH and effluent oocyst concentrations were found.

Hall and Pressdee (1995) determined the changes in water quality (particles of sizes 2 and 5 μm, oocysts and *Clostridium*) and penetration across a pilot-scale slow sand filtration (SSF). Under a hydraulic loading of 400 l/h they found that oocyst density was reduced from 380/l in the inlet feed to 0.7/l at the outlet. At the mid-point (20 cm) of the SSF, 2.5 oocysts/l were recorded, indicating that the majority of oocysts were removed in the upper half of the bed. Similar results were observed for particle and *Clostridium* spore removal, but penetration within the bed occurred. This has operational significance with respect to deciding minimum sand depth before resanding is necessary.

As an alternative to sand, Schuler *et al.* (1991) and Ongerth and Hutton (1997) evaluated the effectiveness of diatomaceous earth (DE) filtration for (oo)cyst removal. Comparing two grades of DE, Schuler *et al.* (1991) noted that, while (oo)cyst removal was excellent (> 99.9%), even with the coarsest DE (pore size 26 μm), as with many studies outlined above, complete removal was not achieved. The use of finer grades of DE (pore size 13 μm) and polymeric flocculant aids (catfloc, magnifloc) yielded similar results, however, complete removal of oocysts was only achieved with alum-treated DE. Enmeshment by, or adsorption onto, the amorphous aluminium hydroxide at pHs above 5.5 was postulated as the most probable reason for the observed increase in removal. In contrast, *Giardia* was completely removed with the coarsest DE (pore size 26 μm) studied: the addition of chemical coagulants offering no added benefit. Studies using three grades of DE indicate that this filtration method may be more effective than conventional or direct granular media filtration for removing oocysts (Ongerth and Hutton, 1997) and, by extrapolation, *Giardia* cysts. At bench scale, the efficiency of oocyst removal was dependent upon flow rate and the grade of DE used (i.e. coarse, moderate, fine). The differences in $\log_{10}$ reductions for the different grades were approximately in inverse proportion to the permeability of the grade. At flow rates of 2.4 m/h, removal efficiencies ranged from 6 to 6.5 $\log_{10}$ for the finest grade (median pore size 5 μm) to 3.60–4.51 $\log_{10}$ for coarser grades (median pore size 13 μm). Doubling the flow rate resulted in an increase in oocyst removal efficiency for all grades of DE tested (see Table 40.4).

In theory, both microfiltration and ultrafiltration should remove (oo)cysts completely, given the small pore size of the membranes and the larger size of (oo)cysts. Seeding studies, conducted at bench and pilot scale (Jacangelo *et al.*, 1995), appear to confirm the theory that size exclusion is probably the major mechanism of (oo)cyst removal. Their data demonstrated that, independent of water quality, (oo)cysts were never detected in the filtrate from a range of micro- (pore size 0.1–0.2 μm) and ultra- (MW 1×10^5–5×10^5 Daltons) filtration systems as long as the membranes remained intact. Removal efficiencies ranged from 6 to 7 $\log_{10}$ (see Table 40.4). Similar removal efficiencies were reported by Hirata and Hashimoto (1998) using microfiltration (nominal pore size 0.25 μm) and ultrafiltration (nominal cut-off 13 kDa) when challenged with very high (10^8) oocyst concentrations. However, in contrast to the findings of Jacangelo *et al.* (1995), oocysts could still be detected in the filtrate from each filter (Hirata and Hashimoto, 1998).

While the above data indicate up to 6 $\log_{10}$ (oo)cyst removal in physical water treatment processes, it remains pertinent to emphasise that it is those oocysts which penetrate such processes that are likely to cause waterborne disease. This is likely to occur when raw water oocyst densities are greater than projected removal rates, or when treatment processes operate sub-optimally. Finally, care should be taken when identifying methods for determining likely oocyst densities in raw water, as methods vary in their sensitivity of detection.

5.4 Disinfection

Practices commonly used to disinfect conventionally treated water prior to distribution include exposure to chlorine, chloride dioxide, monochloramines, ozone and ultraviolet light. Data from numerous studies undertaken to determine the effectiveness of these processes for the inactivation of *Cryptosporidium* and *Giardia* are outlined in Table 40.5. The specific details of experimental procedures and data analyses employed by different investigators have critical effects on both the interpretation and comparability of results, often making comparisons between studies difficult (Smith *et al.*, 1995b). For example, differences in the pH of suspending solution (slightly acidic to alkaline), temperature (0.5–25° C), concentration (low to high) and contact time (minutes to hours) vary between studies. Such differences make it difficult accurately to compare C.t′ (disinfectant concentration × contact time) values and the effects of the dissociated disinfection products. In addition, contact conditions (batch or continuous flow), (oo)cyst concentration used (10^3–10^8), cyst species (*G. duodenalis*, *G. muris*), (oo)cyst viability and the method used to determine viability (*in vitro* excystation, fluorogenic vital dyes, cell culture, animal infectivity) also - vary.

Most studies used recently excreted, purified (oo)cysts to determine disinfection efficacy. Few data are available regarding the effectiveness of disinfectants on environmentally stressed (oo)cysts which have undergone full conventional water treatment, prior to disinfection. Here, further work is required. While most studies have used the species that causes infection in humans (*G. duodenalis, C. parvum*), some have used species (e.g. *G. muris*) not infectious to human beings. Of those studies which use human-infective parasites, some use (oo)cysts derived from laboratory animals (e.g. mice, gerbils, etc.) or experimentally infected calves or lambs. The isolates used to generate experimental infections differ within and between countries, as do the procedures used to purify oocysts from faeces. Given these variables, the possibility exists that (oo)cyst species/subtype/genotype/isolate, age, source and purification procedure can influence the outcome of disinfection studies.

Different parasite species can have different disinfection sensitivities. For example, higher C.t′ values have been reported to achieve similar percentile reductions (99%) for *G. muris* than *G. duodenalis* (Hoff, 1986). This may also be true for different isolates/subtypes of *C. parvum* and *G. duodenalis*. To obtain more meaningful information, future studies should standardize sources and age of (oo)cysts and purification procedures.

Where possible, C.t′ values have been calculated for the various disinfectants based upon at least a 90% (1 $\log_{10}$) reduction in (oo)cyst viability (see Table 40.5), and compared with similar values reported in the literature for other waterborne pathogens (Table 40.6). The disinfectant concentrations and contact times required to achieve complete inactivation of *Cryptosporidium* and *Giardia* indicate that these protozoan parasites are more resistant to inactivation by conventional disinfectants used in the water industry than other waterborne pathogens. Fore example, the C.t′ value of 7200 obtained for 99% inactivation of *C. parvum* by chlorine at 25° C at pH 7 (Korich *et al.*, 1990) is between 480 and 57.6 times greater than the C.t′ value required to achieve the same inactivation of *G. lamblia* at 25° C and 5°C at pH 7.2 (Jarrol *et al.*, 1981) and 180 000 times greater than for *E. coli* at 1°C at pH 7.2 (Scarpino *et al.*, 1974). Given that (oo)cysts are very resistant to disinfectants commonly used in water treatment, it is not surprising to note that waterborne outbreaks have been documented in the absence of faecal indicators.

Based upon the disinfection studies presented in Table 40.5 and documented waterborne outbreaks, the potential exists for infective (oo)cysts to occur in potable water, following conventional disinfection. This emphasizes the need for catchment control and optimization of unit processes such as coagulation, flocculation, DAF and filtration to reduce or remove these organisms prior to disinfection. Recently, investigators have explored the use of sequential inactivation of *C. parvum* oocysts using combined disinfectants (e.g. ozone/free chlorine, ozone/monochloramine, ozone/UV etc.) and electrochemically generated mixed oxidants (Venczel *et al.*, 1997; Casteel *et al.*, 2000; Kanjo *et al.*, 2000; Rennecker *et al.*, 2000a,b; Driedger *et al.*, 2000). Compared with single disinfectants, sequential inactivation, using combined disinfectants, appears to be more effective at inactivating oocysts.

6 VIABILITY AND INFECTIVITY

Little can be inferred about the likely impact of (oo)cysts detected in water concentrates on public health without knowing whether they are alive or dead. In order to determine the efficacy of physical and chemical water treatment processes on the survival of *Giardia* and *Cryptosporidium*, a measure of either their viability or infectivity must be undertaken. Similarly, the exposure of (oo)cysts to various environments and factors during their passage from the infected host into the aquatic environment can also influence their survival. Included in these environments and factors are urine, faeces, slurry, pasture, soil, temperature, pH etc. Methods are also necessary to measure

TABLE 40.5 Comparison of selected studies evaluating the effectiveness of disinfectants for inactivating *Cryptosporidium* and *Giardia* (oo)cysts.

(a) Chlorine	*Ct′ value (concentration × contact time)*	*Result*	*Viability assay*	*Reference*
G. lamblia	15 mg/min/l (1.5 mg/l/10 min) pH6–8 @25°C 30 mg/min/l (3 mg/l/10 min) pH6–8 @15°C 25 mg/min/l (2.5 mg/l/10 min) pH6 @15°C	99.8% reduction 100% reduction 100% reduction	(e)	Jarroll *et al.*, 1981
G. muris	185 mg/min/l (pH6 @ 0.5°C) 289 mg/min/l (pH7 @ 0.5°C) 242 mg/min/l (pH8 @ 0.5°C) 142 mg/min/l (pH6 @ 2.5°C) 252 mg/min/l (pH7 @ 2.5°C) 268 mg/min/l (pH8 @ 2.5°C) 146 mg/min/l (pH6 @ 5°C) 161 mg/min/l (pH7 @ 5°C) 280 mg/min/l (pH8 @ 2.5°C)	99.9 to 99.99% inactivation	(ai)	Hibler *et al.*, 1987
G. duodenalis	65–142 mg/min/l (1–8 mg/6–84 min/l) (pH6@ 5°C) 97–118 mg/min/l (2–8 mg/7–152 min/l) (pH7@ 5°C) 110–142 mg/min/l (2–8 mg/57–164 min/l) (pH8@ 5°C) 20 mg/min/l (2.5–3 mg/7 min/l) (pH6@ 15°C) 32 mg/min/l (2.5–3 mg/6–18 min/l) (pH7@ 15°C) 37 mg/min/l (2.5–3 mg/7–21 min/l) (pH8@ 15°C) <9 mg/min/l (1.5 mg/< 6 min/l) (pH6@ 25°C) <10 mg/min/l (1.5 mg/< 7 min/l) (pH7@ 25°C) <12 mg/min/l (1.5 mg/< 8 min/l) (pH8@ 25°C)	99% inactivation	(e)	Hoff, 1986
G. muris	68 mg/min/l (0.24–1.1 mg/37–297 min/l) (pH6.5@ 3°C) 140 mg/min/l (0.24–1.0 mg/150–770 min/l) (pH7.5@ 3°C) 360 mg/min/l (0.41–2.73 mg/236–467 min/l) (pH7@ 5°C) 66 mg/min/l (4.4–13 mg/4–16 min/l) (pH5@ 25°C) 29 mg/min/l (2.9–7.1 mg/4–16 min/l) (pH7@ 25°C) 206 mg/min/l (11.6–72.6 mg/3–16 min/l) (pH9@ 25°C)	99% inactivation	(e)	Hoff, 1986
G. lamblia	139 mg/min/l (5.8 mg/l/24 min) (pH5@15°C) 153 mg/min/l (3 mg/l/51 min) (pH5@15°C) 176 mg/ min/l (4.5 mg/l/39 min) (pH5@15°C) 182 mg/min/l (0.25 mg/l/729 min) (pH5@15°C) 149 mg/min/l (0.39 mg/l/388 min) (pH7@15°C) 168 mg/min/l (0.86 mg/l/194 min) (pH7@15°C) 231 mg/min/l (2.9 mg/l/231 min) (pH7@15°C) 284 mg/min/l (1.62 mg/l/175 min) (pH7@15°C) 291 mg/min/l (8.2 mg/l/35.5 min) (pH7@15°C) 117 mg/min/l (0.31 mg/l/379 min) (pH9@15°C) 636 mg/min/l (3.55 mg/l/179 min) (pH9@15°C)	2 log reduction	(e)	Rubin *et al.*, 1989

	1495 mg/min/l (16.25 mg/l/92 min) (pH9@15°C)			
G. duodenalis	30 mg/min/l (3 mg/l/10 min) 200 mg/min/l (20 mg/l/10 min)	100% inactivation	(e)	Smith *et al.*, 1989a
C. parvum	11 × 10^6 mg/min/l (8000 mg/l/24 h)	100% inactivation	(e)	Smith *et al.*, 1989b
C. parvum	80 mg/l for 2 h pH7 @25°C 9600 mg/min/l (80 mg/l/2 r) 7200 mg/min/l (80 mg/l/90 min)	 reduced to zero 99% inactivation	(e) & (ai)	Korich *et al.*, 1990
C. parvum	968 mg/l for 24 h @ 10°C (pH 7) 5118 mg/l for 24 h @ 10°C (pH 7)	85.2% reduction 88.1% reduction	(e)	Ransome *et al.*, 1993
C. parvum	5 mg/l/ for 4 hr @ 25°C 5 mg/l for 24 hr @ 25°C	No effect No effect	(e)	Vencel *et al.* 1997
C. parvum	3000 mg/min/l (10 mg/l /5 hr) after ageing in surface water for 14 days (0–0.5°C) after ageing in surface water for 4 days (0.5–2°C) after ageing in surface water for 18 days (0.5–2°C) after ageing in surface water for 5 days (11.2–20.8°C) after ageing in surface water for 26 days (11.2–20.8°C)	 −0.05 log inactivation −0.25 log inactivation −0.14 log inactivation −0.74 log inactivation −0.61 log inactivation	(ie)	Chauret *et al.*, 1998
(b) chlorine dioxide				
G. muris	11.2 mg/min/l	99% inactivation	(e)	Hoff, 1986
C. parvum	4.65 mg/min/l (0.31 mg/min/15 min) 6 mg/min/l (0.4 mg/l/15 min)	mice still became infected	(ai)	Peters *et al.*, 1989
C. parvum	78 mg/min/l (1.3 mg/min/1 h)	as above (90% inactivation)	(e) (ai)	Korich *et al.*, 1990
C. parvum	6.9 mg/min/l (0.46 mg/l/15 min) 75 mg/min/l (4.97 mg/l/15 min) 14.7 mg/min/l (0.49 mg/l/30 min)	28% reduction 96% reduction 64% reduction	(e)	Ransome *et al.*, 1993
C. parvum	60 mg/min/l (2 mg/min/30 min) pH8 @22°C 122 mg/min/l (2 mg/min/61 min) 383 mg/min/l (3.3 mg/min/116 min)	0.99 log inactivation 1.57 log inactivation >3.22 log inactivation	(ai)	Liyanage *et al.*, 1997
C. parvum	120 mg/min/l (2 mg/min/60 min) pH6 @1°C	1.8 log inactivation 2.1 log inactivation	(ai)	Finch and Li, 1999
C. parvum	150 mg/min/l (2 mg/min/60 min) pH8 @20°C	2.0 log inactivation	(e)	Ruffell *et al.*, 2000
C. parvum	1000 mg/min/l pH8 @21°C	2.0 log inactivation	(cc)	Chauret *et al.*, 2001

(*continued on next page*)

TABLE 40.5 (*continued*)

(a) Chlorine	*Ct′ value (concentration × contact time)*	*Result*	*Viability assay*	*Reference*
	1000 mg/min/l (mg/l/time not stated)	0.5 log inactivation	(e)	
	550 mg/min/l (mg/l/time not stated)	2.0 log inactivation	(cc)	
	75 mg/min/l (mg/l/time not stated)	2.0 log inactivation	(cc)	
(c) Monochloramines				
C. parvum	204.6 mg/min/l (3.41 mg/l/1 h)	60% reduction	(e)	Ransome *et al.*, 1993
	342 mg/min/l (2.85 mg/l/2h)	41% reduction		
	636 mg/min/l (2.65 mg/l/14 h)	71% reduction		
	178.6 mg/min/l (0.124 mg/l/24 h)	65% reduction		
	4162 mg/min/l (2.89 mg/l/24 h)	73% reduction		
	6293 mg/min/l (4.37 mg/l/24 h)	81% reduction		
	334.1 mg/min/l (0.116 mg/l/48 h)	52% reduction		
C. parvum	9600 mg/min/l (80 mg/l/2 h)	reduced to zero	(e) (ai)	Korich *et al.*, 1990
C. parvum	3000 mg/min/l (10 mg/l /5 h)		(e)	Chauret *et al.*, 1998
	after ageing in surface water for 14 days (0–0.5°C)	−0.08 log inactivation		
	after ageing in surface water for 4 days (0.5–2°C)	−0.01 log inactivation		
	after ageing in surface water for 18 days (0.5–2°C)	−0.05 log inactivation		
	after ageing in surface water for 5 days (11.2–20.8°C)	−0.74 log inactivation		
	after ageing in surface water for 26 days (11.2–20.8°C)	−0.61 log inactivation		
(d) Ozone				
G. *duodenalis*	0.65 mg/min/l	>2 log inactivation	(ai)	Finch *et al.*, 1993c
	1.23 mg/min/l	3 log inactivation		
	2.57 mg/min/l	4 log inactivation		
G. *muris*	0.24 mg/min/l	2 log inactivation	(ai)	Finch *et al.*, 1993c
	0.45 mg/min/l	3 log inactivation		
	0.86 mg/min/l	4 log inactivation		
C. parvum	6.66 mg/min/l (1.11 mg/l/6 min)	reduced to zero	(ai)	Peters *et al.*, 1989
	18.16 mg/min/l (2.27 mg/l/< 8 min)			
C. parvum	2.64 mg/min/l (0.44 mg/l/6 min)	2 log reduction	(ai)	Perine *et al.*, 1990
	3.2 mg/min/l (0.8 mg/l/4 min)	3 log reduction		
	4.8 mg/min/l (0.6 mg/l/8 min)	4 log reduction		
	4.36 mg/min/l (1.09 mg/l/4 min)	4 log reduction		
C. parvum	3 mg/min/l (1 mg/l/3 min)	1 log reduction	(e) (ai)	Korich *et al.*, 1990
	5 mg/min/l (1 mg/l/5 min)	2 log reduction		
	10 mg/min/l (1 mg/l/10 min)	3 log reduction		

C. parvum	1.5–10 mg/l/min (pH 6 @ 22°C) 3.15–33 mg/l/min (pH 7 @ 22°C) 4.5–15.4 mg/l/min (pH 8 @ 22°C)	1.6–3.3 log eduction 0.5–4.6 log reduction 1.1–2.8 log reduction	(ai)	Gyurek *et al.*, 1999
(e) Ultraviolet	Dose			
G. *duodenalis*	43 mJ/cm² 63 mJ/cm²	<1 log reduction <1 log reduction	(e)	Rice and Hoff., 1981
G. muris	10 mJ/cm² 10 mJ/cm²	2 log reduction 0 reduction	(ai) (e, vd)	Craik *et al*, 2000
C. parvum	42 mJ/cm² 63 mJ/cm² 90 mJ/cm² 120 mJ/cm²	57.6% reduction 80% reduction 95.4% reduction 99.0% reduction	(e)	Ransome *et al.*, 1993
C. parvum	8.75 mJ/cm²	> 2 log reduction	(e, vd)	Campbell *et al.*, 1995
C. parvum	41 mJ/cm² 19 mJ/cm²	3.9 log reduction > 4.5 log reduction	(ai)	Bukhari *et al.*, 1999
C. parvum	10 mJ/ cm² 25 mJ/ cm²	2 log inactivation 3 log inactivation	(ai)	Craik *et al*, 2001

Key: (e) = *in vitro* excystation; (ai) = animal infectivity; (vd) = vital dyes; (cc) = *in vitro* cell culture assay

TABLE 40.6 C.t′ values of chlorine, chlorine dioxide and ozone necessary to achieve 90–99% inactivation of some microorganisms. Comparison with *Cryptosporidium*

Chlorine	*Ct′ value to achieve 99% inactivation*	*Reference*
Cryptosporidium	7200 mg/min/l	Korich *et al.* (1990)
Giardia	60 to 80 times less	Jarrol *et al.* (1981)
Poliovirus	3600 times less	Roy *et al.* (1982)
E. coli	180 000 times less	Scarpino *et al.* (1974)
Chlorine Dioxide	*Ct′ value to achieve 90% inactivation*	
Cryptosporidium	78 mg/min/l	Korich *et al.*(1990)
Giardia	7 times less	Hoff *et al.* (1986)
Ozone	*Ct′ value to achieve 99% inactivation*	
Cryptosporidium	10 mg/min/l	Korich *et al.* (1990)
Giardia	9 to 30 times less	Wickramanayake *et al.* (1984)
Poliovirus	20 to 60 times less	Roy *et al.* (1982)
E. coli	250 times less	Scarpino *et al.* (1974)

the survival of (oo)cysts in these and other environments they encounter, so that a better understanding of the influences on these sources of contamination can be gleaned and useful exposure assessment data can be generated.

The success of treatments to inactivate the (oo)cysts of protozoan parasites is dependent upon observing a reduction in some biological parameter(s) associated with their potential to cause infection. The parasites within (oo)cysts (trophozoites of *Giardia* and sporozoites of *Cryptosporidium*) require a series of well-defined, but poorly understood 'triggers' to excyst, infect and complete their life cycles. Studies for determining parasite survival can be broadly classified into *in vitro* and *in vivo* assays (Table 40.7).

In vitro studies focus upon a subset of parameters required for infection, but which are fundamental to the survival of the (oo)cyst and the ability of living trophozoites or sporozoites to exit, actively, from the cyst or oocyst into the surrounding medium following a series of defined triggers. *In vivo* or infectivity studies describe the ability of living parasites, contained within the (oo)cyst, to cause infection in a susceptible host and to complete their life cycle within that infected host. With respect to infection, *in vitro* studies address some of the series of events which are responsible for the initial release of infective parasites into the surrounding environment but do not address the issues of host infection and replication. Thus, the principles underpinning *in vitro* and *in vivo* studies differ. A comprehensive review of *in vitro* and *in vivo* methods for determining the viability or infectivity of *Giardia* and *Cryptosporidium* can be found in O'Grady and Smith (2001).

6.1 Viability

6.1.1 *In vitro excystation*

Exposure of infectious (oo)cysts to environments in both stomach and intestine (temperature, moisture, acidity, pepsin, intestinal alkalinity) stimulate the release (excystation)

TABLE 40.7 Various *in vitro* and *in vivo* surrogates for *Giardia* and *Cryptosporidium*

Viability	*Infectivity*
In vitro excystation	Experimental human infection
Fluorogenic vital dyes	Experimental murine (neonatal and adult) models
Polymerase chain reaction (PCR); reverse transcription (RT)-PCR	Experimental non-murine models
Fluorescence *in situ* hybridization	*In vitro* infectivity
Electrorotation	

of *Giardia* trophozoites and *Cryptosporidium* sporozoites. Laboratory triggers for excystation include exposure to low pH at 37°C, followed by exposure to a mildly alkaline pH. Specific protocols, reviewed in O'Grady and Smith (2001), are available (e.g. Schaefer III, (1990); Current (1990); Robertson *et al.*, 1993). Although a useful tool, drawbacks in *in vitro* excystation, including asynchronous excystation, aborted excystation, lysis of excysted trophozoites/sporozoites, subjectivity in determining the physiological states of excysted parasites, and the ability of unexcysted *Cryptosporidium* oocysts to infect neonatal mice (Neumann *et al.*, 2000a) have been identified (O'Grady and Smith (2001)).

Neither animal infectivity nor excystation *in vitro* is applicable to the small numbers of organisms found in water concentrates, and much effort has been expended on developing *in vitro* surrogate techniques which assess the viability of individual, or small numbers of (oo)cysts. Presently available surrogates include fluorogenic vital dye uptake and DIC microscopy, *in vitro* infectivity, PCR based methods, fluorescence *in situ* hybridization (FISH), and electrorotation (Tables 40.7 and 40.8).

6.1.2 *Fluorogenic vital dyes*

These methods are based on observing whether specific fluorogenic vital dyes are included in or excluded from *Giardia* or *Cryptosporidium* (oo)cysts as a measure of their viability. A list of vital dyes reported to be useful appears in Table 40.8.

6.1.3 *In vitro infectivity*

The asexual development of *C. parvum* occurs in a variety of *in vitro* cultured cell lines (Upton *et al.*, 1994; Girdwood and Smith, 1999b; O'Grady and Smith, 2001). Infectious oocysts excyst on the 'host' cell monolayers, sporozoites invade the host cells and the endogenous stages, which develop intracellularly, radiate as foci from the point of sporozoite invasion. Slifco *et al.* (1997, 1999) quantified and optimized an *in vitro* infectivity method for *C. parvum*, using a human enterocytic (HCT-8) cell line and, following infection with bleach-treated excysting oocysts, they determined differences in oocyst isolates and ages. Infective foci were detected by labelling endogenous stages with anti-sporozoite/merozoite antibodies followed by secondary detection with a FITC-conjugated anti-species antibody. Foci were enumerated by epifluorescence microscopy and confirmed by DIC microscopy. Dilution studies indicated that levels as low as one infectious oocyst could be detected, allowing a quantitative estimate of oocyst viability to be made.

Caco-2 cell monolayers were used by Rochelle *et al.* (1999) to detect parasite developmental stages by specific reverse

TABLE 40.8 Some vital dyes used to determine the viability of *Giardia* and *Cryptosporidium*

Vital dyes	*Included into living parasites*	*Excluded from living parasites*	*Parasite(s)*	*References*
FDA, PI morphology by DIC	FDA	PI	*G. muris*	Schupp and Erlandsen (1987a,b) Schupp *et al.* (1988a)
FluoroBora 1		FluoroBora 1	*G. muris, G. intestinalis*	Hudson *et al.* (1988)
PI, morphology by DIC		PI	*G. intestinalis*	Smith and Smith (1989)
DAPI, PI	DAPI	DAPI, PI	*C. parvum*	Campbell *et al.* (1992, 1993)
SYTO®9			*G. muris*	Taghi-Kilani *et al.* (1996)
SYTO®9, SYTO®59		SYTO®9, SYTO®59	*C. parvum*	Belosovic *et al.* (1997)
PI		PI	*C. parvum, G. duodenalis*	Dowd and Pillai (1997)
DAPI, PI	DAPI	DAPI, PI	*G. duodenalis* from faeces and wastewater	Thiriat *et al.* (1998)

Key: DAPI = 4′, 6 diamidino-2-phenylindole, DIC = Nomarski differential interference contrast microscopy; FDA = fluorescein diacetate, FluoroBora 1 = 3-[dansylamido-phenyl boronic acid], SYTO®9, (Molecular Probes Live/Dead BacLight kit, Eugene, Oregon, USA), SYTO®59 (Molecular Probes), PI = propidium iodide.

transcriptase PCR (RT-PCR) of extracted messenger RNA (mRNA), targeting the heat shock protein 70 gene (*hsp* 70). A single infectious oocyst could be detected in up to 100 litres of seeded concentrated environmental water following a period of at least 48 h post infection, but the reproducibility of the method is currently unknown.

6.1.4 *Polymerase chain reaction (PCR) methods*

PCR uses molecular probes to target a specific sequence of DNA and a thermostable enzyme which amplifies that sequence so that it can be detected readily. PCR offers great sensitivity and specificity, and is rapid. A positive PCR result, using DNA, gives no indication of viability, as DNA is present whether the parasite is alive or not. To overcome this inherent problem, some researchers include an *in vitro* excystation step as the criterion for viability, followed by PCR amplification of published sequences of DNA. A variant technique, reverse transcription PCR (RT-PCR), targets and amplifies RNA. Messenger RNA (mRNA) is an important target for viability assays: it is produced and functions as an intermediary in protein synthesis, and is thus found only in viable cells. Numerous PCR methods have been used to determine the viability of *Giardia* and *Cryptosporidium*, some of which are identified in Table 40.9.

6.1.5 *Fluorescence in situ hybridization (FISH)*

Vesey *et al.* (1998) developed FISH as a surrogate for *C. parvum* oocyst viability. A fluorescently labelled oligonucleotide probe which targeted a specific sequence in the 18S rRNA of *C. parvum* and caused viable sporozoites, which were capable of *in vitro* excystation, to fluoresce was developed. Dead oocysts and organisms other than *C. parvum* organisms did not fluoresce following *in situ* hybridization.

6.1.6 *Electrorotation*

The biophysical process of electrorotation (ROT) is the result of rotational torque exerted on polarized particles subjected to rotating electrical fields and is dependent upon the relative conductive properties of the particle and the suspending medium. The possibility of characterizing (oo)cysts by measuring relative rotational forces induced by AC electrical fields has been suggested (Smith, 1996; Goater and Pethig, 1998; Dalton *et al.*, 2001). (Oo)cysts which are observed by microscopy in a ROT chamber (which can be manufactured on a reusable glass slide or as a disposable device) must be partially purified and resuspended in a medium of known conductivity. A frequency of around 800 kHz (for oocysts suspended in a 5 μS/cm solution) in the ROT spectrum, where viable and non-viable *C. parvum* oocysts rotate in opposite directions, provided a convenient, single frequency, viability check on individual oocysts (Goater and Pethig, 1998). Oocysts confirmed as viable are unaffected by ROT and are available for further investigations.

6.2 Infectivity

Infectivity reflects a parasite's ability to excyst and cause infection and remains a 'reference standard' for the study of the health risk from ingested (oo)cysts. Animal infectivity is frequently undertaken in laboratory mammals of known genetic constitution, is expensive and requires dedicated facilities and specialist personnel, which bars its use as a routine procedure. Animal infectivity cannot be used for determining the proportion of viable and non-viable organisms in environmental samples containing small numbers of (oo)cysts. Animal infectivity can only be used to determine the number of non-viable organisms in a sample, if samples are compared to a 'standardized' dose-response curve. While its application to surveys and monitoring is limited by economic, practical and ethical considerations, it has proven particularly useful in disinfection studies, especially for assessing logarithmic reductions in parasite killing. Also, animal infectivity can be used to determine the usefulness of preservation methods for (oo)cysts and their developmental stages. Experimental human infections provide

TABLE 40.9 Some PCR approaches to determine *Giardia* and *Cryptosporidium* viability

Approach	*Parasite*	*Reported sensitivity*	*Reference*
Spectrophotometric (A_{260}) differences in giardin mRNA before and after excystation	*G. muris*/ *G. lamblia*	1 cyst	Mahbubani *et al.* (1991)
In vitro excystation followed by PCR of DNA	*C. parvum*	100 sporozoites	Wagner-Wiening and Kimmig (1995)
RT-PCR of heat shock protein (*hsp*70) mRNA	*C. parvum*	1 oocyst in 4 water types	Stinear *et al.* (1996)
RT-PCR of heat shock protein (*hsp*70) mRNA	*G. duodenalis*, *C. parvum*	10 cysts	Abbasedegan *et al.* (1997)
		1 cyst, 1 oocyst	Kaucner and Stinear (1998)
Microbead IMS, *in vitro* excystation followed by nested PCR of DNA	*C. parvum*	10 oocysts in 100 ml apple juice or homogenised milk, 30–100 oocysts inoculated into stool samples.	Deng *et al.* (1997, 2000)
RT-PCR of beta tubulin mRNA and an anonymous mRNA transcript	*C. parvum*	correlated with loss of oocyst infectivity to neonatal mice	Widmer *et al.* (1999)

insights into the doses of (oo)cysts required to cause infection and disease, and the effect of prior infection and/or strain variation on the susceptibility of the human host. Table 40.10 presents some information on the range of animal surrogates used for infectivity, disinfection and environmental studies.

The choice of surrogates for determining the public health significance of waterborne (oo)cysts is broad. That such a variety of *in vitro* and *in vivo* measurements has been devised is evidence that no one measurement is necessarily the best for all studies into the public health significance of (oo)cysts (Table 40.11).

Estimations of (oo)cyst viability and infectivity can generate valuable information for the water industry, regulators and public health officials, particularly in determining risk from water catchments, water treatment processes and potable water supplies. Characterization of (oo)cysts by either viability or infectivity techniques can have its limitations, but a critical definition of the objectives of any proposed study should optimize the choice of surrogate(s) available. Often, the focus of these limitations is (oo)cyst number and, more recently, genotype. In disinfection studies and the validation of disinfection equipment for the water and food industries, where large numbers of purified (oo)cysts are available, many researchers choose animal infectivity as it reflects the ability of treated parasites to establish infection and to complete their life cycles. For these reasons, many regard neonatal mouse infectivity as a 'reference standard' for surrogates, with the other surrogates being seen as 'lesser' standards as only a component of the life cycle is assessed. For *C. parvum*, a significant limitation of *in vivo* surrogates is the inability of readily available and systematically tested murine models to support genotype 1 infections. Here, *in vitro* infectivity studies should supplement animal surrogates when it can be demonstrated that the cell lines used readily support the infection and maintenance of a variety of disparate isolates and genotypes.

While arguments for and against *in vitro* and *in vivo* surrogates abound (see Table 40.11), it is vital to reiterate that the pertinence of surrogates is to reflect, adequately, the data generated from infectivity studies in human beings, particularly volunteer studies, as the issue, in public health terms, is one of infectivity to humans.

7 RISK ASSESSMENT

Risk assessment analysis is a systematic process developed to evaluate the possibility of a given pathogen to reach its host and cause disease. The application of these directives to pathogen risk assessment should consider the complex relationship between the host and the parasite. *Giardia* and *Cryptosporidium* multiply in a susceptible host with various degrees of success, producing infection, disease or, less frequently, fatality. The process of risk assessment consists of four stages. The first stage is the identification of the pathogenic organism; the second stage is the evaluation of its hazard or of its capacity to cause disease to a given population; the third stage is the likelihood of exposure; and the fourth stage is the level of exposure. Information obtained from this analysis leads to a defined assessment of risk or risk characterization. Thereafter, measures that will lessen the risk to a given population can be taken. These last stages are also known as risk management and risk communication. Risk assessment has been applied to *Giardia* and *Cryptosporidium* where susceptible populations include the young, the elderly, the disease impaired, the undernourished, and the immunocompromised.

As large numbers of consumers can be infected by drinking (oo)cyst-contaminated water, the detection of (oo)cysts in water is of paramount importance for risk assessment together with the determination of (oo)cyst viability and infectivity. Based on studies on human volunteers, the minimum dose which causes infection has been estimated. Gibson III *et al.* (1998) and Haas *et al.* (1999) provide a useful insight into risk assessment with respect to *Giardia* and *Cryptosporidium*. Recently, analysis of *C. parvum* dose response data from the human volunteer studies indicated that

TABLE 40.10 Some animal surrogates used for infectivity, disinfection and environmental studies

Animal	*Parasite*	*Aim/Outcome*	*Reference*
Infectivity studies			
Mongolian gerbil (*Meriones unguiculatus*)	*G. lamblia*	ID_{50} 100 cysts	Visvesvara *et al.* (1988)
		ID_{50} 2.45 (probit analysis)	Shaefer *et al.* (1991)
		Parasite strain variation	Abaza *et al.* (1991), Udezulu *et al.* (1992)
		Efficacy of freezing procedures	Dickerson *et al.* (1991)
		Infectivity of *in vitro* derived cysts	Schupp *et al.* (1988b)
C3H/HeN mice	*G. muris*		Labatiuk *et al.* (1991)
C57BL/6 mice	*G. muris*	Period of cyst release and trophozoite burden greater in males	Labatiuk *et al.* (1991); Finch *et al.* 1993a
Outbred neonatal mice	*G. duodenalis*	Pathogenicity of avian isolate	Williamson *et al.* (2000)
Neonatal rat	*G. duodenalis*	Parasite strain variation	Cevallos *et al.* (1995)
Neonatal BALB/c mice	*C. parvum*	Infection dynamics	O'Grady and Smith (2001)
Neonatal CD1/ICR mice	*C. parvum*	Infection dynamics	Finch *et al.* (1993b); O'Grady and Smith (2001)
Neonatal CD Swiss Webster mice	*C. parvum*	Infection dynamics	Ernest *et al.* (1986)
Inbred adult mice	*C. parvum*	Infection dynamics	Enriquez and Sterling (1991); Griffiths *et al.* (1998); Mead and You (1998)
Immunosuppressed adult C57BL/6 mice	*C. parvum*	Commercial oocyst donors	Petry *et al.* (1995)
Neonatal Wistar rats	*C. parvum*	Anti-*Cryptosporidium* drug studies	Armson *et al.* (1999)
New Zealand white rabbits	*C. parvum*	Infection dynamics	Mosier *et al.* (1997)
Disinfection studies			
Mongolian gerbil (*Meriones unguiculatus*)	*G. lamblia*	Ozonation studies	Labatiuk *et al.* (1991); Finch *et al.*, (1993a)
C3H/HeN mice	*G. muris*	Inactivation by heat treatment and chemical disinfection	Taghi-Kilani *et al.* (1996); Labatiuk *et al.* (1991)
C57BL/6 mice	*G. muris*		
Neonatal CD1/ICR mice	*C. parvum*	Disinfection studies	Korich *et al.* (1990); Finch *et al.* (1993b); O'Grady and Smith (2001)
Neonatal CD1/ICR mice	*C. parvum*	SYTO®9 & SYTO®59 comparisons	Neumann *et al.* (2000b)
Neonatal Wistar rats	*C. parvum*	Ozonation studies	Perrine *et al.* (1990)
Environmental studies			
Specific pathogen free dogs	*G. muris*	Developed giardiasis after ingesting raw water concentrates	Shaw *et al.* (1977)
Mongolian gerbil (*Meriones unguiculatus*)	*G. lamblia*, *C. parvum*	Higher densities of infective cysts in raw water concentrates than in chlorinated water. No infections following ingestion of oocysts present in water and sewage.	Isaac-Renton *et al.* (1996); Wallis *et al.* (1996)
Neonatal BALB/c mice	*C. parvum*	Effect of temperature on oocyst storage	Fayer and Nerad (1996), Fayer *et al.* (1998)
Neonatal BALB/c mice	*C. parvum*	Infectivity of oocysts in finished drinking water isolated on cellulose acetate membranes	Graczyk *et al.* (1997)
Neonatal BALB/c mice	*C. parvum*	Effect of storage (−10 to 35°C; 1 – 24 weeks)	Fayer *et al.* (1998)
Neonatal CD1/ICR mice	*C. parvum*	Effect of salinity, temperature and time on oocyst storage	Freire-Santos *et al.* (1999)

TABLE 40.11 Some advantages and disadvantages of various *in vitro* and *in vivo* tests for determining *Giardia* and *Cryptosporidium* viability and infectivity

Technique	*Advantages*	*Disadvantages*
In vitro excystation	• Rapid *in vitro* procedure • Suitable for disinfection and laboratory based survival studies • Suitable for *Giardia* cysts and genotype 1 and 2 *C. parvum* oocysts • Low cost	• Requires large numbers of organisms • Unsuitable for environmental samples • Can only infer infectivity • Requires validation by reference to infectivity studies • Not readily automated procedure
Fluorogenic vital dyes	• Rapid *in vitro* procedure • Suitable for environmental samples • Suitable for disinfection and survival studies • Suitable for individual *Giardia* cysts and genotype 1 & 2 *C. parvum* oocysts • Low cost	• Not suitable for large numbers • Can only infer infectivity • Requires validation by reference to infectivity studies • Not readily automated procedures Automation depends upon assay
Molecular techniques (RT-PCR)	• Relatively rapid *in vitro* procedure • Suitable for small numbers of *Giardia* cysts and genotype 1 and 2 *C. parvum* oocysts • Possibility of differentiating *Giardia* and *Cryptosporidium* genotypes • Readily automated procedure • Medium cost	• Various inhibitors can cause false negative results, especially in environmental samples • Requires further development and comparison to other surrogates, particularly validation by reference to infectivity studies • Can only infer infectivity
In vitro infectivity	• *in vitro* procedure • Determines the ability of *C. parvum* oocysts to infect cell lines • Suitable for disinfection and survival studies • Moderate cost	• Unsuitable for *Giardia* studies • Developed using large numbers of oocysts • Suitability for investigating low densities of environmental oocysts unknown • Requires further validation by reference to infectivity studies • Data generated primarily with genotype 2 *C. parvum* oocysts • Difficult to automate and time consuming
Fluorescence *in situ* hybridization	• Relatively rapid *in vitro* procedure • Suitable for small numbers of *Giardia* cysts and genotype 1 and 2 *C. parvum* oocysts • Possibility of differentiating *Giardia* and *Cryptosporidium* genotypes • Readily automated procedure • Medium cost	• Various inhibitors can cause false negative results, especially in environmental samples • Requires further development and comparison to other surrogates, particularly validation by reference to infectivity studies • Can only infer infectivity

Electro-rotation	• Rapid *in vitro* procedure • Suitable for small numbers of *Giardia* cysts and genotype 1 and 2 *C. parvum* oocysts • Readily automated procedure • Medium cost	• (Oo)cysts require to be purified and suspended in a medium of known conductivity • Requires further development and comparison to other surrogates, particularly validation by reference to infectivity studies • Can only infer infectivity
Neonatal mouse infectivity	• Determines the ability of (oo)cysts to infect mice • *In vivo* demonstration of infectivity and completion of life cycle • Useful for the validation of disinfection procedures and equipment • Suitable for laboratory based survival studies • Can validate the efficacy of viability determinations	• Requires large numbers of (oo)cysts • Not suitable for the routine investigation of low densities of environmental (oo)cysts • Determines the infectivity of genotype 2 *C. parvum* oocysts only • Impossible to automate • Expensive and very time consuming • Variations in mouse strain and parasite isolates
Experimental human infection	• Determines the ability of (oo)cysts to infect humans • *In vivo* demonstration of infectivity and completion of life cycle • Generates significant data on clinical signs, symptoms, host responses, etc • Determines the gold standard for infectivity • Can validate the efficacy of viability determinations	• Requires large numbers of (oo)cysts • Unethical for any investigation of environmental (oo)cysts • Determines the infectivity of *Giardia* and genotype 1 and 2 *C. parvum* oocysts only • Impossible to automate • Very expensive and time consuming • Variations in host status and parasite isolates

the estimated risk of infection from a single oocyst from a mixture of the IOWA, UCP and TAMU isolates tested was 0.018 (Messner *et al.*, 2001).

The identification of (oo)cysts infectious to human beings is also fundamental for risk assessment, as (oo)cysts of various species of *Giardia* and *Cryptosporidium* are found in the aquatic environment. The incorporation of a method, capable of distinguishing between the various species (particularly those infectious to humans) present in the environment, would undoubtedly contribute to a more precise assessment of risk. Currently, (oo)cysts from different *Giardia* (*G. muris, G. duodenalis*) and *Cryptosporidium* (e.g. *C. parvum, C. meleagridis, C. baileyi*) species cannot be differentiated by the standardized methods which employ a genus specific FITC-mAb for detecting oocysts in environmental samples. Although (oo)cysts from different species can differ in size and shape, size overlap can occur and misshapen oocysts are often present in water concentrates.

Molecular techniques including PCR-restriction fragment length polymorphism (PCR-RFLP) and sequencing can offer much needed assistance with species identification and the classification of new genetic types/subtypes of *Giardia* and *Cryptosporidium* associated with different levels of infectivity and virulence. The routine application of PCR for speciation and genotyping will depend on the development of effective, robust and reproducible methods that overcome the inhibitory substances present in environmental samples as well as improved methods for concentrating (oo)cysts from water samples.

Exposure assessment is perhaps the most difficult of the parameters to measure. While dependent upon the detection of (oo)cysts in water and the environment, it also requires an effective understanding of both the transport and survival of (oo)cysts through sewage treatment, agricultural discharges and run-off, soils etc. to reach water courses and to the final elimination or reduction of contamination through water treatment processes. The occurrence of (oo)cysts in various surface and treated waters is well documented (see Tables 40.2 and 40.3) with heavy rain and melting snow contributing to an increase in occurrence of oocysts in the aquatic environment. The presence of (oo)cysts in groundwater has also been reported, however, little is known of the routes by which (oo)cysts migrate through soils and substrata to contaminate groundwaters. The significance of zoonotic versus anthroponotic transmission of *Giardia* and *Cryptosporidium* and whether (oo)cysts from either source are equally infectious are important parameters that could help elucidate transmission routes and contribute significantly towards a better risk assessment.

Currently, optimizing (oo)cyst removal in water treatment is the most effective solution to reducing the risk of waterborne transmission and the use of chemical treatment and filtration processes which can reduce the density of oocysts in treated waters to 0.1 to 72 oocysts per 100 l has been proposed (Rose, 1997; Gibson III *et al.*, 1998a).

8 CONTRIBUTORS OF (OO)CYSTS TO THE ENVIRONMENT

The potential for environmental contamination depends upon a variety of factors including the number of infected human and non-human hosts, seasonal influences and duration of infection, the number of transmissive stages excreted, agricultural practices, host behaviour and activity, socioeconomic and ethnic differences in human behaviour, geographic distribution, sanitation, climate and hydrogeology of the area. The risk of oocysts entering water supply can be determined by reviewing the extent of water treatment available and the integrity of its operation, the latter being the subject of technical audit in order to identify whether any weaknesses in practices exist. Data from numerous countries indicate that *Giardia* and *Cryptosporidium* infections can occur commonly in human, indigenous wildlife, including aquatic and terrestrial mammals, livestock, birds, etc., ensuring a plentiful supply of (oo)cysts in our environment (Meyer, 1985; O'Donoghue, 1995; Smith *et al.*, 1995b; Sturdee *et al.*, 1999; Thompson *et al.*, 2000). (Oo)cysts contaminate both terrestrial and aquatic environments, with a greater likelihood of contamination of the aquatic environment during wet seasons, including snow melt. Both clinically ill and clinically well infected hosts contribute to the (oo)cyst load and the robustness of (oo)cysts to the environment and to water and sewage treatment processes enhance the likelihood of waterborne transmission.

(Oo)cyst contributions into water arise from both point and non-point sources, and knowledge of (oo)cyst sources which contaminate water catchments is helpful when attempting to determine the potential for exposure to (oo)cysts at a water treatment plant. Sewage effluents, livestock farming practices, wastes from livestock markets and abattoirs can release high (oo)cyst densities into the environment from point sources and activities such as slurry spreading and application of farmyard manure and sewage sludge onto land serve to redistribute (oo)cyst populations. Development and regular review of catchment control policies can reduce the potential for the contamination of water courses.

The contribution from infected (symptomatic and asymptomatic) human beings can be assessed by monitoring sewage influent, effluent and sludge (see Tables 40.12 and 40.13). In one study, Smith and Nichols (2003) calculated that, at an average density of 10 oocysts/l of sewage effluent, up to 65×10^6 oocysts per day could be discharged from a sewage treatment works servicing a UK population of 50 000. The potential also exists for introducing oocysts indirectly into water courses, following the disposal of animal waste and sewage sludge to land, therefore, effective pretreatment of such wastes (e.g. thermophilic aerobic digestion, pasteurisation) to inactivate (oo)cysts is necessary. Some oocysts may survive mesophilic digestion and storage and will remain viable in soil.

The contribution from livestock and farming practices is less easy to assess. For example, infections can be clinical in calves, yet subclinical in adult cattle. A clinically ill neonate can excrete $>10^9$ *C. parvum* oocysts daily during the course of infection, whereas a clinically well infected cow can excrete between 760×10^3 and 720×10^6 oocysts daily. The sum total of oocysts contributed into the environment over a 12-month period is similar for both ill and well animals given that immunity prevents the acquisition of further infection in the neonatal host (Smith *et al.*, 1995b).

Contributions from agricultural practices, such as storage and spread of muck and slurry, discharge of oocyst contaminated dirty water to land or to water courses, pasturing of livestock in land adjoining water sources, and from the disposal of faecally-contaminated waste from abattoirs provide data on the potential for livestock to contribute to the (oo)cysts present in water courses. From a dairy farm in the UK with a 12-year history of cryptosporidiosis over 550 oocysts/l were discharged into water courses. Practices which contributed high densities of oocysts into water courses included hosing down calf-rearing pens and sluices (180 oocysts/l) and the contamination of farm drains with slurry and farm yard manure applied onto land ($\approx$370 oocysts/l). Furthermore, practices such as hosing down calf-rearing pens and sluices release recently excreted oocysts into an aquatic environment where survival is prolonged (Smith and Nichols, 2003). Such oocysts are likely to have a higher viability than those excreted onto grazing land which take time to percolate through substrata into water courses.

The contribution from feral animals is also difficult to assess as less is known about the occurrence of *Cryptosporidium* infections in such animals, but contributions from large and small feral mammals as well as seagulls and other transport hosts should also be borne in mind. An estimated overall contamination rate between 0.5 and 32×10^5 oocysts/ha/day from various reaches of a watershed which had uses ranging from recreation only to dairy farming has been calculated (Hansen and Ongerth, 1991).

TABLE 40.12 Occurrence of *Cryptosporidium* and *Giardia* (oo)cysts in untreated sewage using brightfield and immunofluorescence detection methods

Country	*Giardia (cysts/l)*	*Cryptosporidium (oocysts/l)*	*Method*	*Reference*
India	63	not determined	SV & B	Veerannan, 1977
USA	9600–240 000	not determined	SV & B	Jakubowski and Erickson, 1979
India	12–184	not determined	SV & B	Panicker and Krishnamoorthi, 1981
USA	not determined	12 160–197 600	LV & IFA	Rose *et al.*, 1986
USA	not determined	850–13 700	LV & IFA	Madore *et al.*, 1987
USA	3.7 cysts*	521 oocysts*	LV & IFA	DeLeon *et al.*, 1988
USA	200–32 000	not determined	C/LV & IFA	Casson *et al.*, 1990
USA	4–14 000	not determined	C/SV & IFA	Sykora *et al.*, 1991
USA	682–3750	not determined	C/SV & IFA	Jakubowski *et al.*, 1991
France	250–3200	not determined	SV & IFA	Ngo, 1991
France	800–14 000	not determined	SV & B	Gassman and Schwartzbrod, 1991
Canada	26–3 022	0–74		Roach *et al.*, 1993
UK	243–793	2.5–75	LV & IFA	Parker, 1993
UK	–	200–800	LV & IFA	Carrington and Gray, 1993
Cayman Islands	54–> 500 000	not determined	LV & B	Ellis *et al.*, 1993
Kenya	213–6225	12–73	LV & IFA	Grimason *et al.*, 1993
UK	–	1–321	MF & IFA	Dawson *et al.*, 1994
UK	102–43 907	n.d.–800	SV & IFA	Robertson *et al.*, 1995
UK	<10–13 600	<10–170	SV & IFA	Bukhari *et al.*, 1997
France	230–25 000	not determined	LV & B/IFA	Wiandt *et al.*, 1995
Canada	up to 88 000	1–120	SV & IFA	Wallis *et al.*, 1996
USA	mean 13 000	Mean 1500	SVF & IFA	Mayer and Palmer, 1996
France	230–25 000	not determined	LV & IFA	Grimason *et al.*, 1996
Kenya	1000–25 000	not stated	LV & IFA	Grimason *et al.*, 1996
USA	100–13 000/100 l	<6.1–12 000/100 l	LV & IFA	Rose *et al.*, 1996
Israel	5–27.3	8.05–8.3	SV & IFA	Zuckerman *et al.*, 1997
Tunisia	mean 210	not determined	SV & B	Alouini, 1998
Canada	1000–21 000	800–1250	LV & IFA	Chauret *et al.*, 1999
UK	10–52 500**	n.d.–6000**	SV & IFA	Robertson *et al.*, 2000
France		not determined	LV & IFA	Wiandt *et al.*, 2000

Key: LV = large volume sample, SV = small volume sample, C = composite sample, MF = membrane filtration, B = brightfield microscopy and IFA = immunofluorescence antibody test. N.B. American data usually recorded as number of (oo)cysts/100 l (except Casson *et al.*, 1990). * = Geometric mean. n.d. = not detected. ** = (Oo)cyst range dependent upon sample processing method used.

9 WASTEWATER

9.1 Conventional sewage treatment

Although sewage treatment is an important first step in the multibarrier concept for pathogen removal, less is known about the efficiency of sewage treatment processes in removing or inactivating protozoan pathogens than its drinking water counterpart. Further research is required to determine the efficiency of wastewater treatment processes and to optimize the removal and inactivation of these pathogens, prior to effluent discharge into receiving waters. Properly designed and operated wastewater treatment plants should reduce both the risk of water-related disease associated with receiving waters and the number of (oo)cysts challenging water treatment works.

Historically, parasite removal in wastewater has been synonymous with the removal of intestinal nematode ova, particularly those of *Ascaris lumbricoides*, *Trichuris trichiura*, and the hookworms (*Necator americanus* and *Ancylostoma duodenale*). These faecal parasites were chosen because they occur commonly and simple purification and microscopical

TABLE 40.13 Occurrence of *Cryptosporidium* and *Giardia* ((oo)cysts/l) in full-scale sewage treatment plant effluents using the immunofluorescence detection method

Conventional treatment	*Giardia % or log_{10} removal*	*Effluent cyst conc***	*Cryptosporidium % or log_{10} removal*	*Effluent oocyst conc***	*Reference*
Primary					
sedimentation 1	0–89%	300–1,600	not determined	–	Casson *et al.*, 1990
sedimentation 2	63–90%	60–180	not determined	–	Casson *et al.*, 1990
clarification	1 log_{10}	2.6×10^3	1 log_{10}	1.1×10^3	Mayer and Palmer, 1996
sedimentation 1	0–95% (mean 47%)	145–15,100**	0–100% (mean 30%)	n.d.–200**	Robertson *et al.*, 2000
sedimentation 2	0–97% (mean 42%)	–	0–100% (mean 22%)	–	Robertson *et al.*, 2000
sedimentation 3	0–88% (mean 24%)	–	0–100% (mean 4%)	–	Robertson *et al.*, 2000
Secondary					
activated sludge	not determined	–	79%	140–3960	Madore *et al.*, 1987
activated sludge	not determined	–	not stated	5–17	Musial *et al.*,1987
activated sludge	not determined	–	not stated	17–28.4	Rose *et al.*,1988
activated sludge	not determined	–	not stated	0.18	Naranjo *et al.*, 1989
activated sludge	97–99%	< 1–4	not determined	–	Casson *et al.*, 1990
activated sludge	not reported	< 131/40 l	not reported	< 7/40 l	Rose and Gerba, 1991
activated sludge 1	not reported	17 ± 35/40 l*	not reported	1 ± 1.5/40 l*	Enriquez *et al.*, 1995
activated sludge 2	not reported	12 ± 18.3/40 l*	not reported	1.2 ± 1.7/40 l*	Enriquez *et al.*, 1995
activated sludge	2 log_{10}	1.1×10^1	1 log_{10}	1.7×10^1	Mayer and Palmer, 1996
activated sludge	71.7%	50.9 ± 72.5*	0%	6.4 ± 14.4*	Bukhari *et al.*, 1997
activated sludge	93.0%	14–2300	92.8%	25–1100	Rose *et al.*, 1996
activated sludge	1.4 log_{10}	< 5–20.8 /100 l	2.96 log_{10}	18.8–750/100 l	Chauret *et al.*, 1999
activated sludge 1	99.1–99.8%	7–10	not determined	–	Wiandt *et al.*, 2000
activated sludge 2	99.8–99.9%	1.5–5	not determined	–	Wiandt *et al.*, 2000
activated sludge 3	99.8–99.9%	3.5–13	not determined	–	Wiandt *et al.*, 2000
activated sludge 4	98–99.9%	4–12.5	not determined	–	Wiandt *et al.*, 2000
activated sludge 1	30–100%	n.d.–125**	0–100%	n.d.–30**	Robertson *et al.*, 2000
activated sludge 2	0–100%	n.d.–7,600**	0–100%	n.d.–100**	Robertson *et al.*, 2000
activated sludge 3	0–100%	n.d.–240**	0–100%	n.d.–160**	Robertson *et al.*, 2000
trickling filtration	>92%	4–44	not determined	–	Casson *et al.*, 1990
trickling filtration	93%	8.3 ± 14.7*	93%	2.5 ± 6.2*	Bukhari *et al.*, 1997
trickling filtration 1	96–99.1%	34–77	not determined	–	Wiandt *et al.*, 2000
trickling filtration 2	85.6–98.6%	10–19	not determined	–	Wiandt *et al.*, 2000
trickling filtration 3	46.8–69.7%	23–108	not determined	–	Wiandt *et al.*, 2000
trickling filtration	0–99% (mean 85%)	80–20, 400**	0–100% (mean 5%)	n.d.–40**	Robertson *et al.*, 2000
Tertiary					
(a) Sand filtration(AS)	not determined	–	>99%	10	Madore *et al.*, 1987
(b) Sand filtration(AS)	not reported	Mean 0.32/40 l	not reported		Rose and Gerba, 1991

(*continued on next page*)

TABLE 40.13 (*continued*)

Conventional treatment	*Giardia % or log_{10} removal*	*Effluent cyst conc***	*Cryptosporidium % or log_{10} removal*	*Effluent oocyst conc***	*Reference*
(c) Inert media filtration$^{(AS)}$	99.5%	<4	not determined		Casson *et al.*, 1990
(d) Sand/coal filtration$^{(AS\&D)}$	50%	2.7 ± 5.9/40 l*	not reported	0.7 ± 1.4/40 l*	Enriquez *et al.*, 1995
(e) Rapid sand filtration$^{(AS)}$	33%	34.2 ± 120*	54%	2.3 ± 6*	Bukhari *et al.*, 1997
(f) Sand filtration$^{(AS)}$	99%	<1–18/100 l	97.8%	<1.3–13	Rose *et al.*, 1996
(g) Microstraining$^{(AS)}$	26%	14.2 ± 19.8*	46.4%	8.3 ± 17.5*	Bukhari *et al.*, 1997
(h) Sand filtration$^{(TF)}$	93.8%	10 ± 12*	0%	8.7 ± 15*	Bukhari *et al.*, 1997
(i) Surface aeration/filtration	88.5%	21.7 ± 15.2*	0%	2.5 ± 6.2*	Bukhari *et al.*, 1997
(j) Lagoon retention$^{(TF)}$	93.5%	94.4 ± 201.9*	0%	3.1 ± 6.2*	Bukhari *et al.*, 1997
(k) Sand filtration$^{(TF)}$	mean 74 ± 30%	n.d.–5,800	mean 38 ± 38%	n.d.–100	Robertson *et al.*, 2000
(l) Disinfection					
Ultraviolet radiation$^{(AS)}$	97%	6	not determined	–	Wiandt *et al.*, 2000
Ultraviolet radiation$^{(AS)}$	99.1%	19	not determined	–	Wiandt *et al.*, 2000
Chlorination$^{(AS\&S)}$	78%	<0.34–7.5/100 l	61.1%	<0.26–5/100 l	Rose *et al.*, 1996
Chlorination$^{(TF)}$	99.5%	0.7	not determined	–	Wiandt *et al.*, 2000
Chlorination$^{(TF)}$	99.9%	0.3	not determined	–	Wiandt *et al.*, 2000
Non-conventional treatment					
Oxidation pond (USA)	not reported	mean 140/40 l	not reported		Rose and Gerba, 1991
WSP effluent$^{(FIN)}$ (Kenya)	not stated	40–50	100%	–	Grimason *et al.*, 1993
Lagoon effluent (Canada)		2–3,511	not reported	n.d.–333	Roach *et al.*, 1993
WSP effluent$^{(FIN)}$ (France)	< 100	< 1–2.5	not determined	–	Wiandt *et al.*, 2000
WSP effluent$^{(FIN)}$ (Kenya)	99.1%	21–90	not reported	–	Grimason *et al.*,1996
Artificial wetland					
Duckweed pond	98%	–	89%	–	Gerba *et al.*,1999
Multispecies surface flow	73%	–	58%	–	Gerba *et al.*,1999
WSP effluent$^{(FIN)}$ (France)	99.7%	0.1–2.5	not determined	–	Wiandt *et al.*,2000
WSP effluent$^{(FIN)}$ (Malawi)	100%	0	< 100%	120–360	Grimason, unpublished

Key: AS = post activated sludge, TF = post trickle filtration, D = Disinfection and FIN = final effluent, * = Mean (oo)cyst density ± standard deviation, ** = (Oo)cyst range dependent upon sample processing method used.

identification methods, acquired from the clinical parasitology laboratory, were available to determine their presence. Classical methods for isolating and identifying helminth ova and larvae from wastewaters and sludges are well tested and have much to offer. For many parasites in these environments, bright field microscopy remains a valuable tool but can be time consuming and frequently supplies insufficient information on organism viability.

Nematode ova may not be good indicators of protozoan parasite occurrence or removal because of size differences, which affect settling velocities (Grimason *et al.*, 1996). For example, small nematode ova, such as *Trichuris trichiura* measure 52 × 22 μm while the larger, commonly occurring, protozoan (oo)cysts measure 19 × 12 μm or less. Being smaller than ova, low densities of protozoan (oo)cysts are more difficult to detect within the myriad of particles present in wastewater concentrates by bright field microscopy. Although initial estimations of contamination of wastewaters with (oo)cysts were undertaken using bright field microscopy, the benefits of using epifluorescence microscopy, outlined earlier, become even more apparent in the analysis of wastewater.

9.1.1 *Wastewater treatment*

Municipal raw sewage consists of grey and foul wastewater from the community at large and includes domestic, industrial and storm water run-off sources, all of which can harbour *Cryptosporidium* and *Giardia*. Occurrence data, from several developed and developing countries, indicate that *Cryptosporidium* and *Giardia* are frequent contaminants of raw sewage, with (oo)cyst concentrations of up to 500 000/l being reported (see Table 40.6). Although no definitive data exist, it appears that *Giardia* cysts occur in raw sewage at significantly higher densities, and with greater frequency, than *Cryptosporidium* oocysts. Therefore, *Giardia* cysts are ideally suited as a surrogate for the pathogenic protozoa. Factors which impact upon the occurrence and concentrations of (oo)cysts in raw sewage vary greatly between studies, and some of these are identified below:

- the locality and size of the population (e.g. rural/urban/city)
- socioeconomic status of the population (developed versus developing countries)
- the prevalence of infection and disease in the community (symptomatic and asymptomatic)
- the percentage of population sewered (<30% to >90%)
- sewage type (domestic, commercial, industrial)
- shock loadings (e.g. discharge of pit latrine tankers; slaughterhouse effluent)
- the provision of a combined or separate sewage system (impact of combined sewer overflows)
- seasonal factors (e.g. wet versus dry season, dry weather flow)
- the potential for contamination of sewage with non-human faeces (e.g. abattoir effluent, farm wastes)
- time of sampling (e.g. hour/day/month/season)
- sample volume and type (small versus large volume, grab versus composite sample)
- the efficiency of the method used by investigators (often not stated).

Given these variables, the occurrence and concentration of (oo)cysts in raw sewage, reported from various countries, cannot be compared directly due to the inherent differences between studies. The studies presented below are restricted mainly to those in which investigators used FITC-labelled anti-*Cryptosporidium* and anti-*Giardia* mAbs to detect (oo)cysts in raw and treated sewage.

Conventional treatment of sewage usually involves a combination of preliminary (screening and grit removal), primary (sedimentation) and secondary treatment (biological oxidation) processes to remove particulate matter and stabilize the organic strength of the final effluent before discharge to a receiving watercourse. If the receiving watercourse is classified as being 'sensitive' or is used for recreational purposes then sewage undertakers may be obliged to include some form of tertiary treatment (e.g. slow sand filtration and/or disinfection) to ensure compliance with discharge standards. While most conventional sewage treatment works entail at least primary sewage treatment prior to discharge, most investigators report upon the cumulative (oo)cyst removal after secondary treatment.

Primary treatment. Primary treatment usually entails a short period of sedimentation (2–8 h) under relatively quiescent conditions to facilitate the deposition of settleable particulate (faecal) matter to the bottom of the tank, and the removal of hydrophobic compounds (primarily fats and greases) from the surface layer of the tank, by skimming. Few investigators have assessed the efficiency of primary sedimentation for the removal of (oo)cysts (Casson *et al*., 1990; Robertson *et al*., 2000).

Casson *et al*. (1990) compared cyst removal efficiencies over a number of months based upon grab and flow-weighted composites (8 h) taken from two sewage treatment plants in the USA, and noted significant reductions in cyst densities in clarified effluent. From time to time it was noted that higher cyst densities were detected in the effluent compared with the influent resulting in negative efficiencies being recorded. Although negative efficiencies suggest a nett increase in cyst densities after clarification, this clearly was not the case. Their results are more likely to be due to the time of sampling and the frequent fluctuations in cyst concentrations in raw and treated effluent (Gassmann and Schwartzbrod, 1991). These reported increases are no less accurate than observed decreases of equivalent magnitude due to the inherent problems associated with sampling and detection methodologies (Robertson *et al*., 2000). While similar observations have been noted by other investigators, results are usually recorded as 'zero' percent removal (see Table 40.7). Casson *et al*. (1990) reported that primary clarified effluent from these tanks (retention time not stated), challenged with high and low cyst concentrations, contained $\leq$1600 cysts/l (max. 50% removal) and 180 cysts/l (max. 89% removal), respectively. In comparison, analyses of flow-weighted composite samples revealed cyst concentrations (and removal efficiencies) ranging from 736 to 1156 cysts/l (18–67%) and 75 to 148 cysts/l (73–80%) in clarified effluent. An attempt was made to compare the removal efficiency of suspended solids with the removal of cysts by primary settlement, but no correlation was detected between these removals at either plant.

In the most comprehensive study undertaken to date, Robertson *et al*. (2000) evaluated the efficiency of five sewage treatment plants (STPs), over a 3-year period, in removing *Cryptosporidium* and *Giardia*. Increasing retention time within primary sedimentation tanks might be expected to increase the percentage of (oo)cysts removed, however, this was not corroborated in the Robertson *et al*. (2000) study. Three primary sedimentation tanks with design hydraulic retention times of 7 h, 18 h and 25 h, removed on average 94 $\pm$ 11%, 42 $\pm$ 33% and 24 $\pm$ 32% of cysts and 98 $\pm$ 10%, 22 $\pm$ 38% and 4 $\pm$ 22% of oocysts, respectively. The large standard deviations associated with percentage removal provides an insight into the highly fluctuating (oo)cyst densities detected in influent and effluent. One STP that received primary treatment only, was found to contain up to 15 000 cysts/l and 200 oocysts/l in final effluent which was discharged to the receiving watercourse. Both the studies of Casson *et al*. (1990) and Robertson *et al*. (2000) conclude that secondary treatment of sewage is essential to reduce the concentration of (oo)cysts significantly in final effluent.

Factors that can influence removal rates in sedimentation tanks include the terminal settling velocity and zeta potential of the (oo)cyst, and sediment disturbance (e.g. methane bubbling). The time taken for (oo)cysts to settle depends upon a number of factors, including the density of the (oo)cyst and the turbulence within the system. According to Stokes Law, Ives (1990) estimated that *Giardia* cysts, with a theoretical settling velocity of 5.5 cm/h, would settle at 11 times the rate of *Cryptosporidium* oocysts (0.35 cm/h), assuming undisturbed settling. Based upon these theoretical settling rates and the average depth of sedimentation tanks, the theoretical hydraulic retention time required for the complete settlement of (oo)cysts that are not attached to faecal solids would have to be measured in days rather than hours. Sedimentation tanks are usually designed to remove particles with sedimentation velocities of between 0.5 and 1.5 m/h and, as such, would not be expected to remove (oo)cysts completely. This suggests that sedimentation alone, unassisted by coagulation, or attachment to larger or denser particles, present in screened sewage, would fail to remove (oo)cysts completely during primary sedimentation and would explain why many investigators detected (oo)cysts in primary treated effluents.

Laboratory-simulated primary sedimentation of sewage has a low removal efficiency for *Cryptosporidium* oocysts, with a sedimentation velocity of between 2.2 and 2.8 cm/h (Whitmore and Robertson, 1995); an eightfold increase in the rate calculated for oocyst sedimentation in water by Ives (Ives, 1990). Research using a continuous-flow model of an activated sludge STP, indicated oocyst removal efficiencies in primary settlement of approximately 83% and of approximately 91% in secondary sedimentation (Stadterman *et al.*, 1995). These investigators also reported that greater oocyst removal efficiencies were observed in raw (62%) as opposed to treated (activated sludge (55%), trickle filtration (36%) and biodisc (38.5%)) effluent when mixed and allowed to settle over a 2-hour period. Increasing the RT within each system to 2.8, 3.5 and 4.5 hours, respectively improved removal efficiencies (activated sludge (92%), trickle filtration (50%) and biodisc (44%)). Again, this occurs presumably as a result of oocyst attachment to heavier faecal particulate matter present in raw sewage and settleable solids in treated effluent.

Finally, some symptoms of giardiasis and cryptosporidiosis, such as diarrhoea and malabsorption of fats (steatorrhoea) might increase transport of (oo)cysts through sewage treatment processes. Diarrhoeic stools disperse more rapidly in wastewater than formed stools and, as such, may increase (oo)cyst transport through the primary sedimentation process. Excreted fats from patients with steatorrhoea can contain attached (oo)cysts which may aid their transport to the surface of the sedimentation tank. If not removed by surface skimming, (oo)cysts may be ingested by surface scavengers such as seagulls, etc. which may act as transport hosts, depositing their faeces many miles from the point of ingestion. Seagull faeces have been associated with a reduction in the microbial quality of impounding reservoirs (Gould and Fletcher, 1978) and seagulls act as transport hosts for *Cryptosporidium* spp. oocysts (Smith *et al.*, 1993b).

Secondary treatment. Settled sewage from primary sedimentation tanks contains a mixture of suspended colloidal and dissolved organic matter which usually undergoes some form of secondary biological treatment to stabilize the effluent prior to discharge. The two most common methods of biological treatment are activated sludge treatment and trickling filtration. Both systems encourage the growth of a diverse range of heterotrophic microbes including motile bacteria, free-living protozoa and nematodes and rotifers that absorb, engulf and stabilize (biological oxidation) the suspended colloidal and dissolved organic matter, converting them into inert end-products. Biological oxidation of primary effluent occurs over a few hours and the settled sewage is allowed a further period of sedimentation in a secondary sedimentation tank before the effluent is discharged to the receiving watercourse.

Most studies on (oo)cyst removal have focused on the efficiency of secondary treated sewage systems, and are reported as the cumulative removal total for both primary and secondary processes, and not for individual unit processes (see Table 40.7). Some investigators have estimated removal efficiencies, based their results on the mean (arithmetic or geometric) concentration of (oo)cysts detected in the raw and final effluent, while others have simply reported the concentration of organisms detected in final effluent and made no attempt to determine the efficiency of (oo)cyst removal. Nevertheless, it is clear that secondary treatment can reduce the concentration of (oo)cysts in raw sewage by 2–3 $\log_{10}$ (see Table 40.7).

A number of comparative studies have concluded that the activated sludge treatment process is more effective at removing *Giardia* than trickling filtration (Sykora *et al.*, 1988, 1991; Casson *et al.*, 1990; Wiandt *et al.*, 2000). In a survey of 11 US wastewater treatment plants, Sykora *et al.*, (1988, 1991) indicated that the highest cyst removal efficiencies occurred in activated sludge plants (90–100%) compared with trickling filtration plants (40–60%). Based upon the results of a 6-month study in the USA, Casson *et al.* (1990) reported that activated sludge treatment could achieve removal efficiencies of between 97 and 99%; slightly higher than those obtained by trickle filtration (92%). The geometric mean cyst concentrations in secondary treated effluents were 4 and 11 cysts/l, for activated sludge treatment and trickle filtration, respectively. The findings of Wiandt *et al.* (2000) confirmed that higher removal efficiencies occurred in activated sludge treatment plants (n = 4; 98–99.9%) compared with trickling filtration plants (n = 3; 46.8–99.1%). In addition, Wiandt *et al.* (2000) also noted that significantly higher cyst concentrations were detected in trickle filtration effluent (range 10–108 cysts/l) compared with effluent from activated sludge treatment (range 1.5–13 cysts/l).

In contrast, Robertson *et al.* (2000) found no significant difference either in the removal of *Giardia* or *Cryptosporidium*, by either treatment system in a 3-year study which compared (oo)cyst removal efficiencies of three activated sludge plants and two trickle filtration plants in the UK (see Table 40.7). Despite substantial reductions in (oo)cyst concentrations in secondary treated effluent, high (oo)cyst concentrations were detected in activated sludge effluent (maximum 7600 cysts/l; 160 oocysts/l) and trickle filtration effluent (maximum 20 500 cysts/l; 40 oocysts/l) from time to time. In a study of nine activated sludge plants in the USA, Madore *et al.* (1987) found oocyst concentrations of between 140 and 3960 oocysts/l in treated effluent with estimated removals of up to 79% reported (see Table 40.7). The infectivity of *C. parvum* oocysts, suspended in primary clarifier effluent and subjected to a laboratory simulation of activated-sludge treatment was reduced by 91% in naïve, neonatal CD-1 mice, compared with control oocysts (Villacorta-Martinez de Maturana *et al.*, 1992).

Factors which contribute to (oo)cyst removal by activated sludge include the production of bacterial extracellular slimes which induce bacterial flocculation and floc formation. Mechanical aeration and uniform mixing of the mixed liquor cause (oo)cysts to collide with, and become attached to, heavier flocculated material, which cause them to settle out of suspension in the secondary settlement tank more readily. Rotifers also aid floc formation by producing mucus coated faecal pellets which may enhance (oo)cyst attachment and settlement. The potential for similar removal mechanisms exists within the trickling filtration process as settled sewage is slowly filtered through the gelatinous biofilm which is attached to the impervious media. While no data exist for *Cryptosporidium* oocysts, returned activated sludge can contain between 200 and 900 *Giarida* cysts/l (Casson *et al.*, 1990). Predation and toxin production by heterotrophic microbes may reduce both (oo)cyst viability and number. Unlike trickle filtration, during activated sludge treatment a proportion of the settled sludge in the secondary tank is recirculated back to the inlet of the activated sludge tank: this makes it difficult to determine the 'true' removal efficiency of the process. While (oo)cysts appear to be removed by such systems little is known what effect, if any, they have upon (oo)cyst viability, although Bukhari *et al.* (1997) recovered viable oocysts from secondary treated sewage in a UK study.

Tertiary treatment. As a result of the concern expressed about the quality of freshwater, estuarine and coastal waters, additional treatment of secondary treated sewage may be required before permission is granted to discharge into a controlled water zone. Many tertiary (advanced) treatment processes exist, the process of choice being dependent upon the physical (e.g. slow sand filtration), chemical (e.g. phosphorous removal) and microbial (e.g. disinfection) standards that the sewage undertaker is obliged to meet. (Oo)cyst removal by sand filtration and other physical processes and (oo)cyst inactivation by disinfection have been addressed earlier, in relation to water treatment, however, few studies address these issues in wastewater. The studies outlined in Table 40.7 indicate that while (oo)cyst concentrations are reduced significantly after sand filtration, (oo)cysts can still be detected in final effluent discharged to receiving waters. In general, the combination of secondary treatment and sand filtration can reduce the (oo)cyst concentration in the final effluent by 2–3 $\log_{10}$ (see Table 40.7).

In recent years, the major investment in tertiary treatment has been the inclusion of disinfection processes. Disinfection of final effluents by chlorine and chlorine-related compounds has been practised in the USA and mainland Europe for several years, and is now gaining wider acceptance in the UK. Concern over the potential environmental impact and perceived health effects of chlorine and chlorine by-products has encouraged the development of alternative disinfectants. Many of these are still at the experimental stage, but those recognized as practical for wastewater disinfection include chlorination, hypochlorination, chlorine dioxide, bromine chloride, ozone, ultraviolet (UV) irradiation and peracetic acid (Gross and Murphy, 1993). With ever increasing demands for more environmentally sensitive processes, UV disinfection is gaining wider acceptance for the treatment of final effluents discharged into recreational waters. The effectiveness of a number of these disinfectants against (oo)cysts in drinking water has already been identified (see Table 40.5). As yet, no pertinent studies have been conducted upon the efficacy of sewage disinfection and (oo)cyst inactivation.

9.2 Unconventional sewage treatment

Removal of pathogens is considered a major advantage when using waste stabilization pond systems (WSPs) for domestic wastewater treatment, particularly in developing countries where the public health risk from parasitic infections can be high (World Health Organization (WHO), 1989). Unlike the RT of sewage within a conventional treatment system, which is usually measured in hours, the retention time within WSPs is usually measured in days or weeks. The long retention times of primary ponds facilitate the settlement of much of the settleable solids, including pathogens, and similar reductions in the biochemical oxidation demand to those attained by primary sedimentation tanks during conventional treatment can be achieved. Primary ponds can be anaerobic, aerated, and/or facultative, and operate in a similar manner to a suspended growth reactor. Facultative pond effluents are usually 'polished' in one or more maturation ponds before the final effluent is discharged. Maturation ponds are thought to be responsible for pathogen inactivation due to the build-up of algae and the subsequent increase in the pH of the pond water (i.e. pH > 9) (Pearson *et al.*, 1987).

Most studies on parasites in WSPs have concentrated upon the efficiency of removal of nematode ova, and not protozoan (oo)cysts, in compliance with the WHO parasitological guideline ($<$ 1 nematode egg/l) required for unrestricted irrigation (WHO, 1989). As such, existing data are limited to a few studies carried out in India, Kenya, France and the Cayman Islands. Using small volume grab samples and bright field microscopy, early studies conducted in India indicated that WSPs are capable of removing between 95 and 100% of *Giardia* from raw wastewater (Veerannan, 1977; Panicker and Krishnamoorthi, 1978), however, the RTs required to achieve these removals were not stated. In the Cayman Islands, Ellis *et al.* (1993) reported detecting *Giardia* cysts, by brightfield microscopy, from one of two parallel primary facultative ponds (88 cysts/l) after an estimated RT of 14 days. The pond system had an estimated total pond RT of 23 days, and no cysts were detected in any of the subsequent maturation pond effluents examined.

Using large volume sampling techniques and FITC-mAbs, the efficacy of Kenyan and French WSPs in removing (oo)cysts was determined and compared (Grimason *et al.*, 1993, 1996; Wiandt *et al.*, 1994, 2000). In contrast to previous studies, these investigators detected (oo)cysts in a variety of pond effluents examined from anaerobic, facultative and maturation ponds. Whereas data from the Kenyan study indicated that a RT of 38 days may be required to ensure complete removal of (oo)cysts (Grimason *et al.*, 1993), data acquired in France (Wiandt et al., 1994) refuted this, as cysts could still be detected in final effluent after a RT of 40 days. In both studies, *Giardia* cysts were detected in final maturation pond effluents with removal efficiencies estimated to be between 99.1% and 99.7%. In a later study of three French WSPs with cumulative RTs of 44, 103 and 142 days, *Giardia* cysts were detected only in the final effluent from the WSP with the shortest retention period (Wiandt *et al.*, 2000). Recent studies conducted in Malawi have shown that *Cryptosporidium* oocysts can still be detected in final effluent (up to 360 oocysts/l) from a four-pond system with a RT in excess of 56 days (A.M. Grimason, unpublished).

While no protozoan guidelines are recommended by the WHO (Blumenthal *et al.*, 2000) in relation to the use of treated wastewater effluent for unrestricted irrigation, the state of Arizona, USA has laid down a standard of <1 *Giardia* cyst/40 l (Rose and Gerba, 1991). A study comparing effluent compliance with this standard (Rose and Gerba, 1991) reported that while 41% of treated wastewater samples (oxidation pond (mean 140 cysts), activated sludge (mean 48 cysts), activated sludge followed by sand filtration (mean 0.32 cysts)) contained *Giardia* cysts, 35% were in violation of the effluent standard of 1 cyst/40 l required for unrestricted irrigation. Oxidation ponds produced the poorest protozoological quality effluent: even with filtration in place, 33% of samples contained cysts at densities above the Arizona standard. In two of the activated sludge treatment plants examined, 93% of effluent samples contained *Giardia* (max.131 cysts; mean 32 cysts/40 l) and 67% *Cryptosporidium* (max. 7 oocysts; mean 2 oocysts/40 l). The difficulty in determining accurate removal efficiencies by such systems is epitomized by a study of a four-oxidation pond system in Canada (Roach *et al.*, 1993) whereby final effluent from the pond series of unknown retention time was shown to contain significantly higher cyst concentrations (17 cysts/l) compared with the levels detected in raw wastewater (1 cyst/l).

These studies demonstrate that the majority of (oo)cysts are removed by primary ponds in WSP systems, adsorption of (oo)cysts onto settleable solids (faecal) probably being the principal removal mechanism. (Oo)cyst densities detected in subsequent pond effluents decrease sequentially, indicating that the removal (sedimentation) and/or destruction (lysis/predation) of (oo)cysts may also be related to the cumulative effect of hydraulic retention time. (Oo)cyst re-suspension, as a result of sediment disturbance by indigenous aquatic wildlife, or (oo)cyst excretion by infected animals that frequent the pond systems, also contribute to the number of (oo)cysts reported by investigators in various pond effluents. Factors which can aid the transport of protozoan parasites through pond systems, such as short-circuiting and temperature inversion (Yanez *et al.*,1980), thereby reducing the retention time of (oo)cysts within each pond, require further investigation.

The enhancement of wastewater quality by artificial wetland systems is increasingly being employed world-wide. Research on the effectiveness of such systems in removing (oo)cysts is in its infancy. A study of three US wetland systems revealed that cysts (98%) and oocysts (89%) were removed more efficiently by a duckweed pond than a multispecies surface flow wetland system (73% and 58%, respectively) (Gerba *et al.*, 1999). Mechanisms for (oo)cyst removal in these systems include a combination of sedimentation, adsorption and filtration. In order to achieve better performance, it was suggested that a combination of aquatic ponds and subsurface flow wetlands may be required. The potential for these systems to reduce *Giardia* cyst densities effectively from oxidation lagoon effluents has also been demonstrated using water hyacinths. Pertuz *et al.* (1999), found that oxidation lagoon effluent treated with water hyacinths had significantly lower cyst densities compared with the densities detected in lagoon effluent treated without hyacinths.

9.2.1 *Effluent discharge*

Sewage effluents, discharged into freshwater, estuarial and coastal waters present a risk to the user of those waters, should they contain viable (oo)cysts. The densities of (oo)cysts in inland waters receiving sewage effluent can be significantly greater than in those that do not receive such discharges (e.g. Parker, 1993; Medema and Schijven, 2001). During periods of prolonged or high rainfall, the volumetric design capacity in the sewerage network or at the sewage treatment works may be exceeded. In such circumstances, excess sewage is pumped straight to the receiving watercourse having undergone minimal (i.e. preliminary) treatment only, either at the sewage treatment works or within a combined sewer overflow (CSO) system. Gibson III *et al.* (1998b) found up to 40 000 oocysts/100 l and up to 283 000 cysts/100 l in effluent discharged from CSOs.

Effluent irrigation and crop contamination. Although reuse of sewage effluent for crop irrigation is commonplace in arid regions of the world, few studies have assessed levels of protozoan contamination on irrigated crops such as fruit and vegetables, which are consumed raw. Field trials undertaken in Morocco revealed the presence of *Giardia* on coriander (254 cysts/kg), mint (96 cysts/kg), carrots (155 cysts/kg) and radish (59.1 cysts/kg) on crops irrigated with raw wastewater. Potatoes taken from a field irrigated with raw wastewater contained 5.1 cysts/kg (Amahmid *et al.*, 1999). Earlier Mexican studies revealed the presence of *Giardia* cysts and other protozoan parasites on various fruit and vegetables irrigated with raw wastewater (Kowal and Pahren, 1982; Felix *et al.*, 1996). A Costa Rican study (Monge and Chincilla, 1996) found that, of eight different fresh vegetables, commonly consumed raw, coriander roots and leaves, lettuce, radish, tomato, cucumber and carrot were contaminated with *Cryptosporidium* oocysts. *Giardia* was detected on coriander (cilantro) leaves and roots (Monge and Arais, 1996). Samples from vegetables collected at several small markets in a periurban slum in Peru harboured *Cryptosporidium* oocysts (Ortega *et al.*, 1997). Of the vegetables examined, 14.5% contained *C. parvum* oocysts suggesting that washing vegetables does not completely remove *Cryptosporidium* oocysts. Such studies indicate the potential for fruit and vegetables to become contaminated and subsequently act as vehicles for the transmission of foodborne protozoan disease. A survey of lettuce and beetroot sold in a number of Chilean markets (Franjola and Guttierez, 1984) revealed the presence of Iodamoeba (*Pseudolimax*) *butschlii* and *Entamoeba coli*, indicative of faecal contamination (probably untreated wastewater used for irrigation). For vegetables and fruit that are eaten raw, the presence of infectious pathogens presents an unacceptable risk to the consumer.

Based upon available epidemiological evidence, WHO promulgated bacterial and parasitological guidelines for the unrestricted use of effluent for irrigation, i.e. including the irrigation of fruit and vegetables eaten raw (WHO, 1989). These require a faecal coliform count of less than 1000/100 ml and not more than 1 helminth egg/l. Several developing countries have adopted these into their national standards. The state of California, USA requires sewage effluent to be tertiary treated (coagulation, filtration and chlorination) and comply with more stringent effluent standards based upon turbidity (< 2 NTU), total coliform count (median $< 2.2/100$ ml) and virus inactivation (5 $\log_{10}$ reduction) criteria before permission is granted for reuse. Interestingly, neither the WHO guidelines nor the California regulations make reference to protozoan pathogens, in contrast to the state of Arizona (*Giardia* <1 cyst and <1 virus per 40 l for unrestricted irrigation), which also considered a similar standard for *Cryptosporidium* (Rose and Gerba, 1991).

9.2.2 *Sludge treatment and disposal*

Primary and secondary sludge can contain high densities of (oo)cysts. Most studies have concentrated upon (oo)cyst inactivation during sludge treatment rather than (oo)cyst occurrence in raw and treated sludge. Soares *et al.* (1994) provided insight into the densities of *Giardia* cysts in undigested (73 300–3 300 000 cysts/kg) and digested (4360–700 000 cysts/kg) sludge.

In a study of five sludge disposal sites in the UK, Bukhari *et al.* (1997) reported similarly high densities of cysts ($<$10 000–250 000/l) in 100% of sludge samples prepared for disposal to land. Sludge samples from one of the sites also contained high oocyst densities (mean

6700 ± 5800/l). At a Canadian sludge treatment plant, Chauret *et al.* (1999) detected (oo)cysts in thickened, mixed (primary and secondary) sludge, anaerobic, mesophilic, digested sludge and the supernatant. No statistically significant difference was determined between (oo)cyst densities detected in digested (mean 1280 cysts, 265 oocysts/100 g) and undigested (mean 441 cysts, 529 oocysts/100 g) sludge, although a 0.3 $\log_{10}$ reduction was reported for *Cryptosporidium*. In Australia, *Giardia* cyst densities were found to be approximately 900/g wet weight of anaerobically digested sludge (Hu *et al.*, 1996).

The mainstay of sludge stabilization is mesophilic (30–40°C), anaerobic and thermophilic (50–60°C) aerobic digestion but few studies have determined the effectiveness of these processes against *Cryptosporidium* and *Giardia*. Most are laboratory based. Van Praagh *et al.* (1993) indicated that *G. muris* cysts are inactivated at a faster rate when digesters are operated under thermophilic, rather than mesophilic temperatures. Whereas anaerobic digestion at 21°C and 37°C produced 99.9% inactivation in 15.1 and 20.5 days, a similar level of inactivation was achieved at thermophilic temperatures (50°C) in 10.7 minutes. In contrast, Gavaghan *et al.* (1993) reported that 99% inactivation of *G. muris* cysts that had been anaerobically digested at 37°C could be achieved within approximately 18 h. These differences could have been due to the age and viability of the cysts used, or may indicate inherent differences in the sensitivity of different *G. muris* isolates. Further studies should focus on *G. duodenalis* cysts infectious to humans.

Thermophilic, aerobic digestion of sludge has been shown to completely inactivate *C. porvum* oocytes within a 24 h period (Whitmore and Robertson, 1995). However, a small proportion of oocysts that underwent mesophilic anaerobic digestion remained viable after 18 days. This suggests that oocysts may be more robust than cysts at lower (mesophilic) temperatures, but both appear to be equally susceptible at higher (thermophilic) temperatures. In contrast, Stadterman *et al.* (1995) noted that mesophilic anaerobic digestion of oocysts achieved a 1 $\log_{10}$ reduction in viability within 4 h and a 2 $\log_{10}$ reduction after 24 h.

Storage of untreated liquid sludge for at least 3 months prior to reuse is recommended for pathogen removal (Bruce *et al.*, 1990); however, the effect of storage upon (oo)cyst viability is unknown. One study concluded that *Giardia* cyst densities in sludge stored for 12 months were too high for storage to be considered an adequate treatment option (Gibbs *et al.*, 1995). Although these investigators found high cyst densities in stored sludge, their detection method did not determine cyst viability or infectivity. The thermal death point of (oo)cysts is between 55 and 65°C for 10–15 min, and sludge treatment processes which elevate temperature and increase contact time, e.g. pasteurisation (70°C, ≥30 min) and composting (≥40°C for 5 days, ≥55°C for 4 h), should render (oo)cysts non-viable (Fayer, 1994; Gerba *et al.*, 1995; Harp *et al.*, 1996). The use of lime, to increase sludge pH to 12 or greater, is gaining wider acceptance.

As a result of the European Urban Waste Water Treatment Directive the indiscriminate dumping of sludge to sea has been banned from 1999, with the emphasis on costs and beneficial reuse of the end-products without compromising the environment (Bruce *et al.*, 1990; Matthews, 1992; Hall, 1995). The 'best practical environmental option' chosen by many statutory undertakers in the UK is the disposal of sludge by land application and/or landfill disposal. Depending upon the method of sludge treatment, the final product may be disposed of to land by sub surface injection, spraying or mixed through the soil. Untreated sewage sludge can be applied to agricultural land in the UK provided it is injected or rapidly incorporated into the soil. Both the application of sludge to land and the use of sewage effluent for irrigation are risk factors for surface and groundwater contamination. During periods of prolonged rainfall, (oo)cysts can be washed into nearby watercourses or infiltrate further into the substratum and eventually percolate into aquifers. Further laboratory and field trials are required to determine the consequences of applying sludge to land and effluent irrigation on surface and groundwater contamination, including private water supplies and water from wells in close proximity to the application/irrigation area.

Laboratory trials using simulated rainfall onto intact soil cores confirm the potential for *C. parvum* oocysts to leach into groundwater sources and run-off into surface watercourses (Mawdsley *et al.*, 1996a, b). Mawdsley *et al.* (1996a) detected oocysts in leachates from clay loam and silty soils, but not from loamy sand soil. Using a soil-tilting table held at an angle of 7.5°, they found that oocysts could still be detected in surface run-off samples and water that had percolated through the soil after 70 days (Mawdsley *et al.*, 1996b). In both studies, oocyst numbers decreased with increasing soil depth and the majority of oocysts were retained in the top few centimetres of the soil cores. Such oocysts can be eluted over time and remain a pollution threat to surface and groundwater sources. Whether oocysts (and cysts) remain infectious after prolonged wet and dry periods in these natural environments remains to be determined. Recent studies suggest that (oo)cysts can survive in soil for prolonged periods depending upon temperature. Under controlled conditions, *Giardia* cysts survived for up to 7 weeks in soil at 4°C, but <2 weeks at 25°C indicating that survival is greater during colder seasons. At 4°C oocysts were found to survive in soil for >12 weeks but degradation was accelerated at 25°C and in soil containing naturally occurring microorganisms (Olson *et al.*, 1999).

10 CONCLUSIONS

Current studies indicate that (oo)cysts, of both human and non-human origin, occur commonly in both terrestrial and aquatic environments, with both symptomatic and asymptomatic hosts contributing to the burden of environmental contamination. The level of environmental contamination can be further enhanced by zoonotic transmission. (Oo)cyst densities and their frequency of occurrence in waters are highly variable, being a reflection of contamination from either point or non-point sources of pollution or both, seasonal factors, the physicochemical quality of the water/wastewater and the recovery efficiency of the methods used. Importantly, the inefficiency and variability of methods used to determine (oo)cyst densities in raw and treated samples can impinge on the results reported by investigators. For example, for water and wastewater treatment, the efficiencies of unit processes for removing (oo)cysts can either be underestimated or overestimated depending on the method used.

Full conventional water treatment processes are capable of removing up to 5 $\log_{10}$ of (oo)cysts, whereas basic water treatment processes (e.g. microstraining) may remove less than 1 $\log_{10}$. Similar removal efficiencies have been reported during primary (<1 $\log_{10}$) and tertiary (>4 $\log_{10}$) treated sewage treatment processes, with 2–3 $\log_{10}$ removal during secondary biological treatment. While it remains important to determine (oo)cyst removal in water and wastewater treatment systems, we should bear in mind that it is those (oo)cysts which survive through treatment processes that are most likely to pose a threat to public health.

Current disinfection practices for water treatment are ineffective at ensuring complete inactivation of (oo)cysts. However, novel physical and enhanced chemical processes, used in series or sequentially, should enhance inactivation. The numerous and inconsistent variables between many of the studies undertaken make it difficult accurately to compare and contrast findings, removal efficiencies, inactivation rates and their likely impact on public health. Therefore, significant standardization of methods is required to facilitate accurate comparison of the data.

REFERENCES

Abaza, S.M., Sullivan, J.J. and Visvesvara, G.S. (1991). Isoenzyme profiles of four strains of *Giardia lamblia* and their infectivity to birds. *American Journal of Tropical Medicine and Hygiene* **44**, 63–68.

Abbaszadegan, M., Huber, M.S., Gerba, C.P. and Pepper, I.L. (1997). Detection of viable *Giardia* cysts by amplification of heat shock-induced mRNA. *Applied and Environmental Microbiology* **63**, 324–328.

Alouini, Z. (1998). Fate of parasite eggs and cysts in the course of waste water treatment cycle of the Cherguia station in Tunis. *Houille Blanche-Revue Internationale d l'Eau* **53**, 60–64.

Amahmid, O., Asmama, S. and Bouhoum, K. (1999). The effect of wastewater reuse in irrigation on the contamination level of food crops by *Giardia* cysts and *Ascaris* eggs. *International Journal of Food Microbiology* **49**, 19–26.

Anon. (1990). Isolation and identification of *Giardia* cysts, *Cryptosporidium* oocysts and free living pathogenic amoebae in water etc. 1989. *Methods for the examination of waters and associated materials.* HMSO, London.

Anon. (1992). A survey of *Cryptosporidium* oocysts in surface and groundwaters in the UK. *Journal of the Institution of Water and Environmental Management* **6**, 697–703.

Anon. (1994). Proposed ICR protozoan method for detecting *Giardia* cysts and *Cryptosporidium* oocysts in water by a fluorescent antibody procedure. *Federal Register* **59**(28), 6416–6429.

Anon. (1998a). Method 1622, *Cryptosporidium* in water by filtration/IMS/FA. EPA 821-R-98-010. United States Environmental Protection Agency, Office of Water, Washington, DC.

Anon. (1998b). Method 1623, *Cryptosporidium* in water by filtration/IMS/FA. United States Environmental Protection Agency, Office of Water, Washington. Consumer confidence reports final rule. *Federal Register* **63**, 160.

Anon. (1999a). UK Statutory Instruments No. 1524. *The Water Supply (Water Quality) (Amendment) Regulations 1999.* HMSO, London.

Anon. (1999b). Isolation and identification of *Cryptosporidium* oocysts and *Giardia* cysts in waters 1999. *Methods for the examination of waters and associated materials.* HMSO, London.

Anon. (1999c). Method 1623 *Cryptosporidium* in water by filtration/IMS/FA. United States Environmental Protection Agency, Office of Water, Washington. Consumer confidence reports final rule. Federal Register **63**, 160.

Armson, A., Sargent, K., MacDonald, L.M. *et al.* (1999). A comparison of the effects of two dinitroanilines against *Cryptosporidium parvum in vitro* and *in vivo* in neonatal mice and rats. *FEMS Immunology and Medical Microbiology* **26**, 109–113.

Atherholt, T.B., LeChevallier, M.W., Norton, W.D. and Rosen, J.S. (1998). Effect of rainfall on *Giardia* and *Cryptosporidium*. *Journal of the American Water Works Association*, **90**(9), 66–80.

Belosevic, M., Guy, R.A., Taghi-Kilani, R *et al.* (1997). Nucleic acid stains as indicators of *Cryptosporidium parvum* oocyst viability, *International Journal for parasitology* **27**, 787–798.

Bemrick, W.J. and Erlandsen, S.L. (1988). Is giardiasis a true zoonosis? *Parasitology Today* **4**, 66–71.

Bennett, J.V., Holmberg, S.D., Rogers, M.F. and Solomon, S.L. (1987). Infectious and parasitic diseases data selection. *American Journal of Preventative Medicine* **3** Suppl., 102–114.

Bertolucci, G.C., Gilli, G., Carraro, E. *et al.* (1998). Influence of raw water storage on *Giardia, Cryptosporidium* and nematodes. *Water Science and Technology,* **37**(2), 261–267.

Bifulco, J.M. and Schaefer, F.W. III. (1993). Antibody-magnetite method for selective concentration of *Giardia lamblia* cysts from water samples. *Applied and Environmental Microbiology* **59**, 772–776.

Blewett, D.A. (1989). Quantitative techniques in *Cryptosporidium* research. In: K.W. Angus and D.A. Blewett (eds) *Cryptosporidiosis. Proceedings of the First International Workshop.* The Animal Diseases Research Association, Edinburgh.

Blewett, D.A., Wright, S.E., Casemore, D.P. *et al.* (1993). Infective dose size studies on *Cryptosporidium parvum* using gnotobiotic lambs. *Water Science and Technology* **27**, 61–64.

Blumenthal, U.J., Mara, D.D., Peasey, A. *et al.* (2000). Guidelines for the microbiological quality of treated wastewater used in agriculture: recommendations for revising WHO guidelines. *Bulletin of the WHO*, **78**(9), 1104–1116.

Bruce, A.M., Pike, E.B. and Fisher, W.J. (1990). A review of treatment process options to meet the EC Sludge Directive. *Journal of the Institution of Water and Environmental Management* **4**, 1–13.

Bucklin, K.E., McFeters, G.A., Amirtharajah, A. (1991). Penetration of coliforms through municipal drinking water filters. *Wat. Res.,* **25**, 1013–1017.

Bukhari, Z. and Smith, H.V. (1995). Effect of three concentration techniques on viability of *Cryptosporidium parvum* oocysts recovered from bovine faeces. *Journal of Clinical Microbiology* **33**, 2592–2595.

Bukhari, Z., Smith, H.V., Sykes, N., Humphreys, S.W. *et al.* (1997). Occurrence of *Cryptosporidium* spp. oocysts and *Giardia* spp. cysts in sewage influents and effluents from treatment plants in England. *Water Science and Technology* **35**, 385–390.

Bukhari, Z., McCain, R.M., Fricker, C.R. and Clancy, J.L. (1998). Immunomagnetic separation of *Cryptosporidium parvum* from source water samples of various turbidities. *Applied and Environmental Microbiology* **64**, 4495–4499.

Bukhari, Z., Hargy, T.M., Bolton, J.R. *et al.* (1999). Medium-pressure UV for oocyst inactivation. *Journal of the American Water Works Association* **91**, 86–94.

Bukhari, Z., Marshall, M.M., Korich, D.G. *et al.* (2000). Comparison of *Cryptosporidium parvum* viability and infectivity following ozone treatment of oocysts. *Applied and Environmental Microbiology* **66**(70), 2972–2980.

Campbell, A.T. and Smith, H.V. (1997). Immunomagnetisable separation of *Cryptosporidium parvum* oocysts from water samples. *Water Science and Technology* **35**, 397–401.

Campbell, A.T., Robertson, L.J. and Smith, H.V. (1992). Viability of *Cryptosporidium parvum* oocysts: correlation of *in vitro* excystation with inclusion/exclusion of fluorogenic vital dyes. *Applied and Environmental Microbiology* **58**, 3488–3493.

Campbell, A.T., Robertson, L.J. and Smith, H.V. (1993). Effects of preservatives on viability of *Cryptosporidium parvum* oocysts. *Applied and Environmental Microbiology* **58**, 4361–4362.

Campbell, A.T., Robertson, L.J., Smith, H.V. and Girdwood, R.W.A. (1994). Viability of *Cryptosporidium parvum* oocysts concentrated by calcium carbonate flocculation. *Journal of Applied Bacteriology* **76**, 638–639.

Campbell, A.T., Robertson, L.J., Snowball, M.R. and Smith, H.V. (1995). Inactivation of oocysts of *Cryptosporidium parvum* by ultraviolet irradiation. *Water Research* **29**, 2583–2586.

Carrington, E.G. and Gray, P. (1993). The influence of cattle waste and sewage effluent on the levels of *Cryptosporidium* oocysts in surface waters. Report No: FR 0421. Foundation for Water Disease, Marlow.

Casson, L.W., Sorber, C.A., Sykora, J.L. *et al.* (1990). *Giardia* in wastewater – effect of treatment. *Res. J. Water Pollut. Control Fed.* **62**, 670–675.

Casteel, M.J., Sobsey, M.D. and Arrowood, M.J. (2000). Inactivation of *Cryptosporidium parvum* oocysts and other microbes in water and wastewater by electrochemically generated mixed oxidants. *Water Science and Technology* **41**(7), 127–134.

Cevallos, A., Carnaby, S., James, M. and Farthing, J.G. (1995). Small intestinal injury in a neonatal rat model is strain dependent. *Gastroenterology* **109**, 766–773.

Chalmers, R.M., Sturdee, A.P., Mellors, P. *et al.* (1997). *Cryptosporidium parvum* in environmental samples in the Sligo area, Republic of Ireland: a preliminary report. *Letters in Applied Microbiology* **25**(5), 380–384.

Chappell, C.L., Okhuysen, P.C., Sterling, C.R. and DuPont, H.L. (1996). *Cryptosporidium parvum*: intensity of infection and oocyst excretion patterns in healthy volunteers. *Journal of Infectious Diseases* **173**, 232–236.

Chappell, C.L., Okhuysen, P.C., Sterling, C.R. *et al.* (1999). Infectivity of *Cryptosporidium parvum* in healthy adults with pre-existing anti-*C. parvum* serum immunoglobulin G. *American Journal of Tropical Medicine and Hygiene* **60**, 157–164.

Chauret, C., Armstrong, J., Fischer, J. *et al.* (1995). Correlating *Cryptosporidium* and *Giardia* with microbial indicators. *Journal of the American Water Works Association*, 76–84.

Chauret, C., Nolan, K., Chen, P. *et al.* (1998). Ageing of *Cryptosporidium parvum* oocysts in river water and their susceptibility to disinfection by chlorine and monochloramine. *Canadian Journal of Microbiology* **44**, 1154–1160.

Chauret, C., Springthorpe, S. and Satter, S. (1999). Fate of *Cryptosporidium* oocysts, *Giardia* cysts, and microbial indicators during wastewater treatment and anaerobic sludge digestion. *Canadian Journal of Microbiology* **45**, 257–262.

Chauret, C.P., Radziminski, C.Z., Lepuil, M. *et al.* (2001). Chlorine dioxide inactivation of *Cryptosporidium parvum* oocysts and bacterial spore indicators. *Applied and Environmental Microbiology* **67**, 299–3001.

Chrisp, C.E., Reid, W.C., Rush, G.D. *et al.* (1990). Cryptosporidiosis in guinea pigs: an animal model. *Infection and Immunity* **56**, 674–679.

Craik, S.A., Finch, G.R., Bolton, J.R. and Belosevic, M. (2000). Inactivation of *Giardia muris* cysts using medium-pressure ultraviolet radiation in filtered drinking water. *Water Research* **34**, 4325–4332.

Craik, S.A., Weldon, D., Finch, G.R., Bolton, J.R. and Belosevic, M. (2001). Inactivation of *Cryptosporidium* oocysts using medium- and low-pressure ultraviolet radiation. *Water Research* **35**, 1387–1398.

Craun, G.F. (1986). Water not sole source of disease transmission. *Journal of the American Water Works Association* **78**, 4–12.

Craun, G.F. (1988). Surface water supplies and health. *Journal of the American Water Works Association* **80**, 40–52.

Craun, G.F. (1990). Waterborne giardiasis. In: Meyer, E.A. (ed.) Giardiasis, Vol. 3. In: E.J. Ruitenberg and A.J. MacInnes (eds) *Series in Human Parasitic Diseases*, pp. 267–293. Elsevier, New York.

Cryptosporidium oocysts and *Giardia* cysts. In: Ziglio, G. and Palumbo, F. (eds) *Detection methods for algae, protozoa and helminths.* Chapter 12, pp. 193–220. John Wiley and Sons, Chichester, UK.

Current Status and Future Trends. In: H.V. Smith and W.H. Stimson (eds), L.H. Chappel, coordinating (ed.) *Infectious diseases diagnosis: current status and future trends.* Parasitology **117**: S205–212.

Current, W.L. (1988). The biology of *Cryptosporidium*. *American Society of Microbiology News* **54**, 605–611.

Current, W.L. (1990). Techniques and laboratory maintenance of *Cryptosporidium*. In: J.P. Dubey, C.A. Speer and R. Fayer (eds) *Cryptosporidiosis of man and animals*, pp. 31–50. CRC Press, Boca Raton.

Current, W.L. and Garcia, L.S. (1991). Cryptosporidiosis. *Clinical Microbiology Reviews* **4**, 325–358.

Dalton, C.A., Goater, A.D., Pethig, R. and Smith, H.V. (2001). Viability of *Giardia intestinalis* cysts and viability and sporulation state of *Cyclospora cayetanensis* oocysts determined by electrorotation. *Applied and Environmental Microbiology* **67**, 586–590.

Danciger, M. and Lopez, M. (1975). Numbers of *Giardia* in the feces of infected children. *American Journal of Tropical Medicine and Hygiene* **24**, 237–242.

Dawson, D.J., Furness, M.L., Maddocks, M. *et al.* (1994). The impact of catchment events on levels of *Cryptosporidium* and *Giardia* in raw waters. AWWA seminar, Watershed Management and control of infectious organisms. New York, 20 June.

DeLeon, R., Naranjo, J.E., Rose, J.B. and Gerba, C.B. (1988). Enterovirus, *Cryptosporidium* and *Giardia* monitoring of wastewater reuse effluent in Arizona. In: *Implementing Water Reuse*, pp. 833–846. American Water Works Association, Denver.

Deng, M.Q., Cliver, D.O. and Mariam, T.W. (1997). Immunomagnetic capture PCR to detect viable *Cryptosporidium parvum* oocysts from environmental samples. *Applied and Environmental Microbiology* **63**, 3134–3138.

Deng, M.Q., Lam, K.M. and Cliver, D.O. (2000). Immunomagnetic separation of *Cryptosporidium parvum* oocysts using MACS MicroBeads and high gradient separation columns. *Journal of Microbiological Methods* **40**, 11–17.

Deng, M.Q., Peterson, R.P. and Cliver, D.O. (2000). First findings of *Cryptosporidium* and *Giardia* in Californian sea lions (*Zalophus californianus*). *Journal of Parasitology* **86**(3), 490–494.

Dickerson, J.W., Visvesvara, G.S., Walker, E.M. and Feely, D.E. (1991). Infectivity of cryopreserved *Giardia* cysts for Mongolian gerbils (*Meriones unguiculatus*). *Journal of Parasitology* **77**, 688–691.

Dowd, S.E. and Pillai, S.D. (1997). A rapid viability assay for *Cryptosporidium* and *Giardia* cysts for use in conjunction with indirect fluorescent antibody detection. *Canadian Journal of Microbiology* **43**, 658–662.

Driedger, A.M., Rennecker, J.L. and Marinas, B.J. (2000). Sequential inactivation of *Cryptosporidium parvum* oocysts with ozone and free chlorine. *Water Research* **34**, 3591–3597.

Drozd, C. and Schwartzbrod, J. (1996). Hydrophobic and electrostatic cell surface properties of *Cryptosporidium parvum*. *Applied and Environmental Microbiology* **62**, 1227–1232.

Drozd, C. and Schwartzbrod, J. (1997). Removal of *Cryptosporidium* from river water by crossflow microfiltration: A pilot-scale study. *Water Science and Technology* **35**, 392–395.

DuPont, H.L., Chappell, C.L., Sterling, C.R. *et al.* (1995). The infectivity of *Cryptosporidium parvum* in healthy volunteers. *New England Journal of Medicine* **30**, 855–859.

Edzwald, J.K. and Kelley, M.B. (1998). Control of *Cryptosporidium*: From reservoirs to clarifiers to filters. *Water Science and Technology* **37**, 1–8.

Ellis, K.V., Rodrigues, P.C.C. and Gomez, C.L. (1993). Parasite ova and cysts in waste stabilisation ponds. *Water Research* **27**, 1455–1460.

Engeset, J. (1984) Optimisation of drinking water treatment for *G. lamblia* cyst removal. PhD dissertation. University of Washington, Seattle.

Enriquez, V., Rose, J.B., Enriquez, C.E. and Gerba, C. (1995). Occurrence of *Cryptosporidium* and *Giardia* in secondary and tertiary wastewater effluents. In: W.B. Betts, D. Casemore and C. Fricker (eds) *Protozoan Parasites and Water*, pp. 84–86. The Royal Society of Chemistry, Cambridge.

Enriquez, F.J. and Sterling, C.R. (1991). *Cryptosporidium* infections in inbred strains of mice. *Journal of Protozoology* **38**, S100–S102.

Erlandsen, S.L. (1994). Biotic transmission – is giardiasis a zoonosis?. In: R.C.A. Thompson, J.A. Reynoldson and A.J. Lymbery (eds) *Giardia: From Molecules to Disease*, pp. 83–97. CAB International, Oxford.

Erlandsen, S.L. and Bemrick, W.J. (1987). SEM evidence for a new species, *Giardia psittaci*. *Journal of Parasitology* **73**, 623–629.

Erlandsen, S.L., Bemrick, W.J., Wells, C.L. *et al.* (1990a). Axenic culture and characterisation of *Giardia ardeae* from the great blue heron (*Ardea herodias*). *Journal of Parasitology* **76**, 717–724.

Erlandsen, S.L., Sherlock, L.A. and Bemrick, W.J. (1990b). The detection of *Giardia muris* and *Giardia lamblia* by immunofluorescence in animal tissues and fecal samples subjected to cycles of freezing and thawing. *Journal of Parasitology* **76**, 267–271.

Ernest, J.A., Blagburn, B.L., Lindsay, D.S. and Current, W.L. (1986). Infection dynamics of *Cryptosporidium parvum* (*Apicomplexa: Cryptosporiidae*) in neonatal mice (*Mus mucularis*). *Journal of Parasitology* **72**, 796–798.

Ernest, J.A., Blagburn, B.L., Lindsay, D.S. and Current, W.L. (1987). Dynamics of *Cryptosporidium parvum* (Apicomplexa: Cryptosporidiidae) in neonatal mice (*Mus musculus*). *Journal of Parasitology* **75**, 796–798.

Faubert, G.M. (1988). Is giardiasis a true zoonosis? *Parasitology Today* **4**, 66–71.

Fayer, R. (1994). Effect of high-temperature on infectivity of *Cryptosporidium parvum* oocysts in water. *Applied and Environmental Microbiology* **60**, 2732–2735.

Fayer, R. (1997). *Cryptosporidium and cryptosporidiosis*. CRC Press, Boca Raton.

Fayer, R. and Nerad, T. (1996). Effects of low temperatures on viability of *Cryptosporidium parvum* oocysts. *Applied and Environmental Microbiology* **62**, 1431–1433.

Fayer, R., Lewis, E.J., Trout, J.M. *et al.* (1999). *Cryptosporidium parvum* in oysters from commercial harvesting sites in the Chesapeake Bay. *Emerging Infectious Diseases* **5**, 706–710.

Fayer, R., Trout, J.M. and Jenkins, M.C. (1998). Infectivity of *Cryptosporidium parvum* oocysts stored in water at environmental temperatures. *Journal of Parasitology* **84**, 1165–1169.

Felix, N.S., Wastavino, G.R., Artega, I., De *et al.* (1996). Parasite search in strawberries from Irapuato, Guanajuato and Zamora, Michoacan (Mexico). *Archives of Medical Research* **27**, 229–231.

Ferguson, C.M., Coote, B.G., Ashbolt, N.J. and Stevenson, I.M. (1996). Relationships between indicators, pathogens and water quality in an estuarine system. *Water Research* **30**, 2045–2054.

Filice, F.P. (1952). Studies on the cytology and life-history of a Giardia from the laboratory rat. *University of California Publications in Zoology* **57**, 53–145.

Finch, G.R. and Li, H. (1999). Inactivation of *Cryptosporidium* at 1°C using ozone or chlorine dioxide. *Ozone Science and Engineering* **21**, 477–486.

Finch, G.R., Black, E.K., Labatiuk, C.W. *et al.* (1993a). Comparison of *G. lamblia* and *G. muris* cysts inactivation by ozone. *Applied and Environmental Microbiology* **59**, 3674–3680.

Finch, G.R., Daniels, C.W., Black, Schaefer, F.W. III and Belosevic, M. (1993b). Dose response of *Cryptosporidium parvum* in outbred neonatal CD-1 mice. *Applied and Environmental Microbiology* **59**, 3661–3665.

Finch, G.R., Black, E.K., Gyurek, L. and Belosevic, M. (1993c). Ozone inactivation of *Cryptosporidium parvum* in demand-free phosphate buffer determined by in-vitro excystation and animal infectivity. *Applied and Environmental Microbiology* **59**, 4203–4210.

Freire-Santos, F., Oteiza-Lopez, A.M., Vergara-Castiblanco, C.A. and Ares-Mazas, M.E. (1999). Effect of salinity temperature and storage time on mouse experimental infection by *Cryptosporidium parvum*. *Veterinary Parasitology* **87**, 1–7.

Fogel, D., Isaac-Renton, J., Guasparini, R. *et al.* (1993). Removing *Giardia* and *Cryptosporidium* by slow sand filtration. *Journal of the American Water Works Association* **85**, 77–84.

Franjola, R.T. and Guttierez, C.J. (1984). Parasitological study of lettuce and beet roots in the city of Valdivia (Chile). *Rev. Med. Chile* **112**, 57–60.

Fricker, C.F. (1995). Detection of *Cryptosporidium* and *Giardia* in water. In: W.B. Betts, D.P. Casemore and C.R. Fricker (eds) *Protozoan Parasites and Water*, pp. 91–96. Royal Society of Chemistry, Cambridge.

Fricker, C.F. and Crabb, J.H. (1998). Waterborne cryptosporidiosis: detection methods and treatment processes. In: S. Tzipori (ed.) *Opportunistic Protozoa in Humans*, pp. 1–278. 241–280. Advances in Parasitology, Volume 40 (Part 1.). In: J.R. Baker, R. Muller and D. Rollinson (Series eds) *Cryptosporidium parvum and related genera*. Academic Press, London.

Friere-Santos, F., Oteiza-Lopez, A.M., Vergara-Castiblanco, C.A. *et al.* (2000). Detection of *Cryptosporidium* oocysts in bivavlebivalve molluscs destined for human consumption. *Journal of Parasitology* **84**(4), 853–854.

Fukushi, K-I., Matsui, Y. and Tambo, N. (1998). Dissolved air flotation: experiments and kinetic analysis. *Journal of Water Supply Research and Technology – Aqua* **47**, 76–86.

Gassmann, L. and Schwartzbrod, J. (1991). Wastewater and *Giardia* cysts. *Water Science and Technology* **24**, 183–186.

Gavaghan, P.D., Sykora, J.L., Jakubowski, W. *et al.* (1993). Inactivation of *Giardia* by anaerobic digestion of sludge. *Water Science and Technology* **27**, 111–114.

Gerba, C.P., Huber, M.S., Naranjo, J. *et al.* (1995). Occurrence of enteric pathogens in composted domestic solid-waste containing disposable diapers. *Waste and Environmental Management* **13**, 315–324.

Gerba, C.P., Thurston, J.A., Falabi, J.A. *et al.* (1999). Optimization of artificial wetland design for removal of indicator microorganisms and pathogenic protozoa. *Water Science and Technology* **40**, 363–368.

Gibbs, R.A., Hu, C.J., Ho, G.E. *et al.* (1995). Pathogen die-off in stored waste-water sludge. *Water Science and Technology* **31**, 91–95.

Gibson III, C.J. Haas, C.N. and Rose, J.B. (1998a). Risk Assessment of Waterborne Protozoa.

Gibson, C.J., Stadterman, K.L., States, S. and Sykora, J. (1998b). Combined sewer overflows: a source of *Cryptosporidium* and *Giardia*? *Water Science and Technology* **38**, 67–72.

Gilmour, R.A., Smith, H.V., Smith, P.G. *et al.* (1989). A modified method for the detection of *Giardia* spp. cysts in water-related samples. *Communicable Diseases Scotland 89* **33**, 5–11.

Girdwood, R.W. and Smith, H.V. (1999a). *Giardia*. In: R. Robinson, C. Batt and P. Patel (eds) *Encyclopaedia of Food Microbiology*, pp. 946–954. Academic Press, London and New York.

Girdwood, R.W.A. and Smith, H.V. (1999b). *Cryptosporidium*. In: R. Robinson, C. Batt and P. Patel (eds) *Encyclopaedia of Food Microbiology*, pp. 487–497. Academic Press, London and New York.

Goater, A.D. and Pethig, R. (1998). Electrorotation and dielectrophoresis. *Parasitology* **117**, S177–S189.

Gold, D. and Smith, H.V. (2001). Pathogenic protozoa and drinking water. In: G. Ziglio, and F. Palumbo (eds) *Detection methods for algae, protozoa and helminths*, Chapter 12, pp. 143–162. John Wiley and Sons, Chichester, UK.

Gould, D.J. and Fletcher, M.R. (1978). Gull droppings and their effects on water quality. *Water Research* **12**, 665–672.

Graczyk, T.K., Cranfield, M.R., Fayer, R. and Bixler, H. (1999). House flies (*Musca domestica*) as transport hosts of *Cryptosporidium parvum*. *American Journal of Tropical Medicine & Hygiene* **61**(3), 500–504.

Graczyk, T.K., Fayer, R., Cranfield, M.R. and Owens, R. (1997). *Cryptosporidium parvum* oocysts recovered from water by the membrane filter dissolution method retain their infectivity. *Journal of Parasitology* **83**, 111–114.

Graczyk, T.K., Fayer, R., Cranfield, M.R. and Conn, D.B. (1998). Recovery of waterborne *Cryptosporidium parvum* oocysts by freshwater benthic clams (*Corbiula fluminea*). *Applied and Environmental Microbiology* **64**, 427–430.

Graczyk, T.K., Fayer, R., Lewis, E.J. *et al.* (1999). *Cryptosporidium* oocysts in Bent mussels (*Ischadium recurvum*) in the Chesapeake Bay. *Parasitology Research* **85**, 518–521.

Griffiths, J.K., Theodos, C., Paris, M. and Tzipori, S. (1998). The gamma interferon gene knockout mouse: a highly sensitive model for the evaluation of therapeutic agents against *Cryptospordium parvum*. *Journal of Clinical Microbiology* **36**, 2503–2508.

Grimason, A. M. (1992). *The Occurrence and Removal of* Cryptosporidium *sp. Oocysts* and *Giardia* sp. Cysts in Surface, Potable and Wastewater. PhD Thesis, University of Strathclyde, Glasgow.

Grimason, A.M., Parker, J.F.W. and Smith, H.V. (1995). Occurrence of *Giardia* sp. cysts in a Scottish potable water supply and the surface water source used for abstraction. In: W.B. Betts, D.P. Casemore and C.R. Fricker *et al.* (eds) *Protozoan Parasites and Water*, pp. 67–70. Royal Society of Chemistry, Cambridge.

Grimason, A.M., Smith, H.V., Parker, J.F.W. *et al.* (1994). Application of DAPI and immunofluorescence for enhanced identification of *Cryptosporidium* spp. oocysts in water samples. *Water Research* **28**, 733–736.

Grimason, A.M., Smith, H.V., Thitai, W.N. *et al.* (1993). Occurrence and removal of *Cryptosporidium* spp. oocysts and *Giardia* sp. cysts by Kenyan waste stabilisation pond systems. *Water Science and Technology* **27**, 97–104.

Grimason, A.M., Smith, H.V., Young, G. and Thitai, W.N. (1996). Occurrence and removal of *Ascaris* sp. ova by waste stabilisation ponds in Kenya. *Water Science and Technology* **33**, 75–82.

Grimason, A.M., Wiandt, S., Baleux, B. *et al.* (1996). Occurrence and removal of *Giardia* sp. cysts by Kenyan and French waste stabilisation pond systems. *Water Science and Technology* **33**, 83–89.

Gross, T.S.C. and Murphy, R. (1993). Disinfection of sewage effluents: the Jersey experience. *Journal of the Institution of Water and Environmental Management* **7**, 481–491.

Gyurek, L.L., Li, H.B., Belosevic, M. and Finch, G.R. (1999). Ozone inactivation kinetics of *Cryptosporidium* in phosphate buffer. *Journal of Environmental Engineering* **125**, 913–924.

Haas, C.N., Rose, J.B. and Gerba, C.P. (eds) (1999). *Quantitative Microbial Risk Assessment*. John Wiley & Sons, New York.

Hall, J.E. (1995). Sewage sludge production, treatment and disposal in the European Union. *Journal of the Institution of Water and Environmental Management* **9**, 335–343.

Hall, T. and Pressdee, M.R. (1995). Removal of *Cryptosporidium* during water treatment. In *Proceedings of workshop on treatment optimisation for Cryptosporidium removal from water supplies.* (West, P.A., ed.) pp. 25–31. ISBN 0117531790.

Hancock, C.M., Rose, J.B. and Callahan, C.M. (1998). *Crypto* and *Giardia* in US groundwater. *Journal of the American Waterworks Association* **90**, 58–61.

Hansen, J.S. and Ongerth, J.E. (1991). Effects of time and watershed characteristics on the concentration of *Cryptosporidium* oocysts in river water. *Applied and Environmental. Microbiology* **57**, 2790–2795.

Harp, J.A., Fayer, R., Pesch, B.A. and Jackson, G.J. (1996). Effect of pasteurization on infectivity of *Cryptosporidium parvum* oocysts in water and milk. *Applied and Environmental Microbiology* **62**, 2866–2868.

Healey, G.R. (1990). Giardiasis in perspective: the evidence of animals as a source of human *Giardia* infection. In: E.A. Meyer (ed.) *Giardiasis (Meyer, E.A. ed.) pp. 305-313. Vol. 3, Series in Human Parasitic Diseases*. Elsevier, New York.

Hibler, C.P., Hancock, C.M., Perger, L.M. *et al.*(1987). *Inactivation of Giardia Cysts with Chlorine at 0.5°C to 5°C*. American Water Works Association Research Foundation, Denver, Colorado.

Hillis, P. and Colton (1995). Particle monitoring after rapid gravity backwash. In *Proceedings of workshop on treatment optimisation for Cryptosporidium removal from water supplies.* (West, P.A., ed.) pp. 37–41. ISBN 0117531790.

Hirata, T. and Hashimoto, A. (1998). Experimental assessment of the efficacy of microfiltration and ultrafiltration for *Cryptosporidium* removal. *Water Science & Technology* **38**, 103–107.

Hirata, T., Chikuma, D., Shimura, A. *et al.* (2000). Effects of ozonation and chlorination on viability and infectivity of *Cryptosporidium parvum* oocysts. *Water Science and Technology* **41**(7), 39–46.

Ho, B.S.W. and Tam, T.-Y. (1998). Occurrence of *Giardia* cysts in beach water. *Water Science and Technology* **38**, 73–76.

Hoff, J. C. (1986) *Inactivation of Microbial Agents by Chemical Disinfectants*. Report No. EPA-600/2-86-067. US Environmental Protection Agency, Cincinnati.

Hoglund, C.E. and Stenstrom, T.A.B. (1999). Survival of *Cryptosporidium parvum* oocysts in source separated human urine. *Canadian Journal of Microbiology* **45**(9), 740–746.

Homan, W.L. and Mank, T.G. (2001). Human giardiasis: genotype linked differences in clinical symptomatology. *International Journal for Parasitology* **31**, 882–886.

Hoover, D.M., Hoerr, F.J., Carlton, W.W. *et al.* (1981). Enteric cryptosporidiosis in a naso tang, *Naso lituratus* Bloch and Schneider. *Journal of Fish Diseases* **4**, 425.

Hu, C.J., Gibbs, R.A., Mort, N.R. *et al.* (1996). *Giardia* and its implications for sludge disposal. *Water Science and Technology* **34**, 179–186.

Huck, P.M., Coffey, B.M., Emelko, M.B. and O'Melia, C.R. (2000). The importance of coagulation for the removal of *Cryptosporidium* and surrogates by filtration. In: *Chemical Water and Wastewater Treatment VI* (H.H. Hahn, E. Hoffman and H. Odegaard eds), pp. 191–200. ISBN 3540675744.

Hudson, S.J., Sauch, J.F. and Lindmark, D.G. (1988). Fluorescent dye exclusion as a method for determining *Giardia* cyst viability. pp. 255–259. In: *Advances in Giardia Research* (Wallis, P.M. and Hammond, B.R. eds), University of Calgary Press, Calgary, Canada.

Isaac-Renton, J.L. (1994). Giardiasis in British Columbia: studies in an area of endemnicity in Canada. In: R.C.A. Thompson, J.A. Reynoldson and A.J. Lymbery (eds) *Giardia: From Molecules to Disease*, pp. 123–124. CAB International, Oxford.

Isaac-Renton, J.L., Cordeiro, C., Sarafis, K. and Shahriari, H. (1993). Characterisation of *Giardia duodenalis* from a waterborne outbreak. *Journal of Infectious Diseases* **167**, 431–440.

Issac-Renton, J., Moorhead, W. and Ross, A. (1996). Longitudinal studies of Giardia contamination in two community drinking water supplies: cyst levels, parasite viability, and health impact. *Applied and Environmental Microbiology*, **62**, 47–54.

Iseki, M. (1979). *Cryptosporidium felis* sp.n. (Protozoa: Eimeriorina) from the domestic cat. *Japanese Journal of Parasitology* **28**, 285–307.

Ives, K.J. (1990). *Cryptosporidium* and water supplies. Treatment processes and oocyst removal. *Cryptosporidium in Water Supplies*, pp. 154–184. HMSO, London.

Jacangelo, J.G., Adham, S.S. and Laine, J-M. (1995). Mechanism of *Cryptosporidium*, *Giardia*, and MS2 virus removal by MF & UF. *Journal of the American Water Works Association* **87**, 107–121.

Jakubowski, W. (1990). The control of *Giardia* in water supplies. In: E.A. Meyer (ed.) *Giardiasis*, Vol. **3**, pp. 335–353. , E.J. Ruitenberg and A.J. MacInnes (eds) *Series in Human Parasitic Diseases*. Elsevier, New York.

Jakubowski, W. and Eriksen, T.H. (1979). Methods for the detection of *Giardia* cysts in water supplies. In *Waterborne Transmission of Giardiasis* W. Jakubowski, and J.C. Hoff, (eds) pp. 193–210. Report No. EPA 600/9-79-001. US Environmental Protection Agency, Cincinnati.

Jakubowski, W., Chang, S.L., Eriksen, T.H. *et al.* (1978). Large-volume sampling of water supplies for microorganisms. *Journal of the American Water Works Association* **70**, 702–706.

Jakubowski, W., Sykora, J.L., Sorber, C.A. *et al.* (1991). Determining giardiasis prevalence by examination of sewage. *Water Science and Technology* **24**, 173–178.

Jarroll, E.A., Bingham, A.K. and Meyer, E.A. (1981). Effect of chlorine on *Giardia lamblia* cyst viability. *Applied and Environmental Microbiology* **41**, 483–487.

Johnson, D.C., Reynolds, K.A., Gerba, C.P. *et al.* (1995). Detection of *Giardia* and *Cryptosporidium* in marine waters. *Water Science and Technology* **31**, 439–442.

Kanjo, Y., Kimata, I., Iseki, M. *et al.* (2000). Inactivation of *Cryptosporidium* spp. oocysts with ozone and ultraviolet irradiation evaluated by *in vitro* excystation and animal infectivity. *Water Science and Technology* **41**(7), 119–125.

Karanis, P., Schoenen, D. and Seitz, H.M. (1996). *Giardia* and *Cryptosporidium* in backwash water from rapid sand filters used for drinking water production. *Zentrabaltt für Bakteriologie – International Journal of Medical Microbiology Virology, Parasitology and Infectious Diseases* **284**, 107–114.

Karanis, P., Schoenen, D. and Seitz, H.M. (1998). Distribution of *Giardia* and *Cryptosporidium* in water supplies in Germany. *Water Science and Technology* **37**, 9–18.

Kasprsak, W. and Paulowski, Z. (1989). Zoonotic aspects of giardiasis: a review. *Veterinary Parasitology* **32**, 101–108.

Kaucner, C. and Stinear, T. (1998). Sensitive and rapid detection of viable *Giardia* cysts and *Cryptosporidium parvum* oocysts in large-volume water samples with wound fibreglass cartridge filters and reverse transcription-PCR. *Applied and Environmental Microbiology*, **64**, 1743–1749.

Kelley, M.B., Brokaw, J.K., Edzwald, J.K. (1994). *A survey of eastern U.S. Army installation drinking water sources and treatment system for* Giardia *and* Cryptosporidium. American Water Works Association, Denver.

Korich, D.G., Mead, J.R., Madore, M.S. *et al.* (1990). Effects of ozone, chlorine dioxide, chlorine, and monochloramine on *Cryptosporidium parvum* oocyst viability. *Applied and Environmental Microbiology* **56**, 1423–1428.

Korich, D.G., Marshall, M.M., Smith, H.V. *et al.* (2000). Interlaboratory comparison of the CD-1 neonatal mouse logistic dose-response model for *Cryptosporidium parvum* oocysts. *Journal of Eukaryotic Microbiology* **47**, 294–298.

Kowal, N.E. and Pahren, X. (1982). Health effects associated with wastewater treatment and disposal. *Journal of Water Pollution Control Federation* **54**, 677–687.

Labatiuk, C.W., Schaefer, F.W. III, Finch, G.R. and Belosevic, M. (1991). Comparison of animal infectivity, excystation, fluorogenic dye as measures of *Giardia muris* cyst activation by ozone. *Applied and Environmental Microbiology*, **11**, 3187–3192.

Labatiuk, C.W., Belosevic, M. and Finch, G.R. (1992). Factors influencing the infectivity of *Giardia muris* cysts following ozone inactivation in laboratory and natural waters. *Water Research* **26**, 733–743.

LeChevallier, M.W., Norton, W.D. and Lee, R.G. (1991). *Giardia* and *Cryptosporidium* spp. in filtered drinking water supplies. *Applied and Environmental Microbiology* **57**, 2617–2621.

LeChevallier, M.W. and Norton, W.D. (1992). Examining relationships between particle counts and *Giardia*, *Cryptosporidium* and Turbidity. *Journal of the American Water Works Association*, **84**(12), 54–60.

LeChevallier, M.W., Norton, W.D. and Atherholt, T.B. (1997). Protozoa in open reservoirs. *Journal of the American Water Works Association*, **89**(9), 84–96.

Levine, N.D. (1980). Some corrections of coccidian (Apicomplexa: Protozoa) nomenclature. *Journal of Parasitology* **66**, 830.

Levine, W.C., Stephenson, W.T. and Craun, G.F. (1991). Waterborne disease outbreaks, 1986–1988. *Journal of Food Protection* **54**, 71–78.

Lisle, J.T. and Rose, J.B. (1995). *Cryptosporidium* contamination of water in the USA and UK: a mini-review. *Journal of Water Supply Research and Technology – Aqua* **44**, 103–117.

Liyanage, L.R.J., Finch, G.R. and Belosevic, M. (1997a). Sequential disinfection of *Cryptosporidium parvum* by ozone and Chlorine dioxide. *Ozone Science and Engineering* **19**, 409–423.

Liyanage, L.R.J., Finch, G.R. and Belosevic, M. (1997b). Effect of aqueous chlorine and oxychlorine compounds on *Cryptosporidium parvum* oocysts. *Environmental Science and Technology* **31**, 1992–1994.

Logan, A.J., Stevik, T.K., Siegrist, R.L. and Rohn, R.M. (2001). Transport and fate of *Cryptosporidium parvum* oocysts in intermittent sand filters. *Water Research*, **35**(18), 4359–4369.

Logsdon, G.S., Thurman, V.C., Frindt, E.S. and Stoecker, J.G. (1985). Evaluating sedimentation and various filter media for removal of *Giardia* cysts. *Journal of the American Water Works Association*, **77**, 61–66.

MacKenzie, W.R., Hoxie, N.J., Proctor, M.E. *et al.* (1994). Massive waterborne outbreak of *Cryptosporidium* infection associated with a filtered public water supply, Milwaukee, March and April, 1993. *New England Journal of Medicine* **331**, 161–167.

Madore, M.S., Rose, J.B., Gerba, C.P. *et al.* (1987). Occurrence of *Cryptosporidium* oocysts in sewage effluents and selected surface waters. *Journal of Parasitology* **73**, 702–705.

Mahbubani, M.H., Bej, A.K., Perlin, M. *et al.* (1991). Detection of *Giardia* cysts using the polymerase chain reaction and distinguishing live from dead cysts. *Applied and Environmental Microbiology*, **57**, 3456–3461.

Mattheson, Z., Hargy, T.M., McCuin, R.M. *et al.* (1998). An evaluation of the Gelman Envirochek® capsule for the simultaneous concentration of *Cryptosporidium* and *Giardia* from water. *Journal of Applied Microbiology* **85**, 755–761.

Matthews, P.J. (1992). Sewage sludge disposal in the UK: A new challenge for the next twenty years. *Journal of the Institution of Water and Environmental Management* **6**, 551–559.

Mawdsley, J.L., Brooks, A.E. and Merry, R.J. (1996a). Movement of the protozoan pathogen *Cryptosporidium parvum* through three contrasting soil types. *Biology and Fertility of Soils* **21**, 30–36.

Mawdsley, J.L., Brooks, A.E., Merry, R.J. and Pain, B.F. (1996b). Use of a novel soil tilting apparatus to demonstrate the horizontal and vertical movement of the protozoan pathogen *Cryptosporidium parvum* in soil. *Biology and Fertility of Soils* **23**, 215–220.

Mayer, C.L. and Palmer, C.J. (1996). Evaluating of PCR, nested PCR, and fluorescent antibodies for detection of *Giardia* and *Cryptosporidium* species in wastewater. *Appl. Environ. Microbiol.*, **62**, 2081–2085.

Mead, J.R. and You, X. (1998). Susceptibility differences to *Cryptosporidium parvum* infection in two strains of gamma interferon knockout mice. *Journal of Parasitology*, **84**, 1045–1058.

Medema, G.J. and Schijven, J.F. (2001). Modelling the sewage discharge and dispersion of *cryptosporidium* and *giardia* in surface water. *Water Research*, **35**, 4307–4316.

Medema, G.J., Bahar, M. and Schets, F.M. (1997). Survival of *Cryptosporidium parvum*, *Escherichia coli*, faecal enterococci and *Clostridium perfringens* in river water. Influence of temperature and autochthonous microorganisms. *Water Science and Technology* **35**, 249–252.

Medema, G.J., Schets, F.M., Teunis, P.F.M. and Havelaar, H. (1998). Sedimentation of free and attached *Cryptosporidium* and *Giardia* cysts in water. *Applied and Environmental Microbiology* **64**, 4460–4466.

Messner, M.J., Chappell, C.L. and Okhuysen, P.C. (2001). Risk assessment for *Cryptosporidium*: a hierarchical Bayesian analysis of human dose response data. *Water Research* **35**, 3934–3940.

Meyer, E.A. (1985). The epidemiology of giardiasis. *Parasitology Today* **1**, 101–105.

Miller, R.A., Brondson, M.A. and Morton, W.R. (1990). Experimental cryptosporidiosis in a primate model. *Journal of Infectious Diseases* **161**, 312–315.

Monge, R. and Arais, M.L. (1996). Occurrence of some pathogenic microorganisms in fresh vegtables in Costa Rica. *Archivos Latinoamericanos de Nutricion* **46**, 292–294.

Monge, R. and Chincilla, M. (1996). Presence of *Cryptosporidium* in fresh vegetables. *Journal of Food Protection* **59**, 202–203.

Moore, A.G., Vesey, G., Champion, A. *et al.* (1998). Viable *Cryptosporidium parvum* oocysts exposed to chlorine or other oxidising conditions may lack identifying epitopes. *International Journal for Parasitology* **28**, 1205–1212.

Mosier, D.A., Cimon, K.Y., Kulhs, T.L. *et al.* (1997). Experimental cryptosporidiosis in adult and neonatal rabbits. *Veterinary Parasitology*, **69**, 163–169.

Musial, C.E., Arrowood, M.J., Sterling, C.R. and Gerba, C.P. (1987). Detection of *Cryptosporidium* in water by using polypropylene cartridge filters. *Applied and Environmental Microbiology* **53**, 687–692.

Naranjo, J.E., De Leon, R., Gerba, C.P. and Rose, J.B. (1989). Monitoring for viruses and parasites in reclaimed water. *The Bench Sheet* **11**(6), 8–10.

Naranjo, J.E., Toranzos, G.A., Rose, J.B. and Gerba, C.P. (1990). Occurrence of enteric viruses and protozoan parasites in Panama. In: *Proceedings of the Second Biennial Water Quality Symposium*, pp. 15–20.

Nash, T.E., Herrington, D.A., Losonsky, G.A. and Levine, M.M. (1987). Experimental human infections with *Giardia lamblia*. *Journal of Infectious Diseases* **156**, 974–984.

Neumann, N.F., Gyürék, L.L., Gammie, L. *et al.* (2000a). Comparison of animal infectivity and nucleic acid staining for assessment of *Cryptosporidium parvum* viability in water. *Applied and Environmental Microbiology* **66**, 406–412.

Neumann, N.F., Gyürék, L.L., Finch, G.R. and Belosevic, M. (2000b). Intact *Cryptosporidium parvum* oocysts isolated after *in vitro* excystation are infectious to neonatal mice. *FEMS Microbiological Letters* **183**, 331–336.

Ngo, N.H. (1991). *Viabilité des Kystes de Giardia dans les Eaux Usées. Laboratoire de Bactériologie-Parasitologie, Faculté des Sciences Pharmaceutiques et Biologiques.* Université de Nancy, Nancy.

Nieminski, E.C. and Ongerth, J.E. (1995). *Giardia* and *Cryptosporidium* removal by direct filtration and conventional treatment. *Journal of the American Water Works Association* **87**, 96–106.

O'Donoghue, P.J. (1995). *Cryptosporidium* and cryptosporidiosis in man and animals. *International Journal for Parasitology* **25**, 139–195.

O'Grady, J.E. and Smith, H.V. (2001). Methods for determining the viability and infectivity of *Cryptosporidium* oocysts and *Giardia* cysts. In: *Detection methods for algae, protozoa and helminths.* (Ziglio, G. and Palumbo, F. eds.), John Wiley and Sons, Chichester, UK. Chapter 12, pp. 193–220.

Okhuysen, P.C., Chappell, C.L., Crabb, J.H. *et al.* (1999). Virulence of three distinct *Cryptosporidium parvum* isolates for healthy adults. *Journal of Infectious Diseases* **180**, 1275–1281.

Olson, M.E., Goh, J., Phillips, M. *et al.* (1999). *Giardia* cyst and *Cryptosporidium* oocyst survival in water, soil and cattle feces. *Journal of Environmental Quality* **28**, 1991–1996.

Ongerth, J.E. (1990). Evaluation of treatment for removing *Giardia* cysts. *Journal of the American Water Works Association* **82**, 85–96.

Ongerth, J.E. and Hutton, P.E. (1997). DE Filtration to remove *Cryptosporidium*. *Journal of the American Water Works Association* **89**, 39–46.

Ongerth, J.E. and Hutton, P.E. (2001). Removal of *Cryptosporidium* and *Giardia*; filtration with sedimentation. *Journal of Environmental Engineering*. In preparation.

Ongerth, J.E. and Pecoraro, J.P. (1995). Removing *Cryptosporidium* using multi-media filters. Removal of *Cryptosporidium* in rapid sand filtration. *Journal of the American Water Works Association* **87**, 83–89.

Ongerth, J.E. and Stibbs, H.H. (1987). Identification of *Cryptosporidium* oocysts in river water. *Applied and Environmental Microbiology* **53**, 672–676.

Ortega, Y.R., Roxas, C.R., Gilman, R.H. *et al.* (1997). Isolation of *Cryptosporidium parvum* and *Cyclospora cayetanensis* from vegetables collected in markets of an endemic region in Peru. *American Journal of Tropical Medicine & Hygiene* **57**, 683–686.

Panicker, P.V.R.C. and Krishnamoorthi, K.P. (1978). Studies on intestinal helminthic eggs and protozoan cysts in sewage. *Ind. J. Environ. Health*, **20**, 75–78.

Panicker, P.V.R.C. and Krishnamoorthi, K.P. (1981). Parasite egg and cyst reduction in oxidation ditches and aerated lagoons. *Journal of the Water Pollution Control Federation* **53**, 1413–1419.

Parker, J.F.W. (1993). Cryptosporidium *sp. Oocysts in the Aquatic Environment: Occurrence, Removal and Destruction*. PhD thesis. University of Glasgow, Glasgow.

Parker, J.F.W. and Smith, H.V. (1994). The recovery of *Cryptosporidium* sp. oocysts from water samples by immunomagnetic separation. *Transactions of the Royal Society of Tropical Medicine and Hygiene* **88**, 25.

Paton, C.A., Kelsey, D.E., Punter, K. *et al.*(2001). Immunomagnetisable separation for the recovery of *Cryptosporidium* sp. oocysts. In: S. Clark, W. Keevil, C.R. Thompson, M. Smith *et al.* (eds) *Rapid Detection Assays for Food and Water.* Royal Society of Chemistry, Cambridge, UK, pp.38–43.

Payment, P. and Franco, E. (1993). *Clostridium perfringens* and somatic coliphages as indicators of the efficiency of drinking water treatment for viruses and protozoan cysts. *Applied and Environmental Microbiology* **59**, 2418–2424.

Pearson, H.W., Mara, D.D., Smallman, D.J. and Mills, S.W. (1987). Physico-chemical parameters influencing faecal bacterial survival in waste stabilisation ponds. *Water Science and Technology* **19**, 145–152.

Pearson, H.W., Mara, D.D. and Bartone, C.R. (1987). Guidelines for the minimum evaluation of the performance of full-scale waste stabilisation pond systems. *Water Research* **21**, 1067–1075.

Peeters, J.E., Mazas, E.A., Masschelein, W.J. *et al.* (1989). Effect of disinfection of drinking water with ozone or chlorine dioxide on survival of *Cryptosporidium parvum* oocysts. *Applied and Environmental Microbiology* **55**, 1519–1522.

Perrine, D., Georges, P. and Langlais, B. (1990). Water ozonation efficiency of *Cryptosporidium* oocysts inactivation. *Bull. Acadamie Natle. Med.* **174**, 845–850.

Pertuz, S., delaRotta, J., Jimenez, N. *et al.* (1999). Treatment to water from oxidation second lagoon effluents with water hyacinth. *Internacia* **24**, 120–124.

Petry, F., Robinson, H.A. and McDonald, V. (1995). Murine infection model for maintenance and amplification of *Cryptosporidium parvum* oocysts. *Journal of Clinical Microbiology* **33**, 1922–1924.

Plummer, J.D., Edzwald, J.K. and Kelley, M.B. (1995). Removing *Cryptosporidium* by dissolved-air flotation. *Journal of the American Water Works Association* **87**, 85–95.

Quennell, S. and West, P. (1995). *Cryptosporidium* monitoring in the UK. In *Proceedings of Workshop on Treatment Optimisation for Cryptosporidium Removal from Water Supplies.* (Word, P.A., ed.) pp. 1–7. ISBN 0117531790.

Ransome, M.E., Whitmore, T.N. and Carrington, E.G. (1993). Effect of disinfectants on the viability of *Cryptosporidium parvum*. *Water Supply* **11**, 75–89.

Rendtorff, R.C. (1979). The experimental transmission of *Giardia lamblia* among volunteer subjects. In: W. Jakubowski and J.C. Hoff (eds) *Waterborne Transmission of Giardiasis*. Report No. EPA-600/9- 79-001, pp. 64–8. US Environmental Protection Agency, Environmental Research Centre, Cincinnati.

Rennecker, J.L., Corona-Vasquez, B., Driedger, A.M., and Marinas, B.J. (2000a). Synergism in sequential disinfection of *Cryptosporidium parvum*. *Water Science and Technology*, **41**(7), 47–52.

Rennecker, J.L., Driedger, A.M., Rubin, S.A., and Marinas, B.J. (2000b). Synergy in sequential inactivation of *Cryptosporidium parvum* with ozone/free chlorine and ozone/monochloramine. *Water Research*, **34**(17), 4121–4130.

Rennecker, J.L., Marinas, B.J., Owens, J.H. and Rice, E.W. (1999). Inactivation of *Cryptosporidium parvum* oocysts with ozone. *Water Research* **33**, 2481–2488.

Rice, E.W. and Hoff, J.C. (1981). Inactivation of *Giardia lamblia* cysts by ultraviolet irradiation. *Applied and Environmental Microbiology* **42**, 546–547.

Riggs, J.L., Dupuis, K.W., Nakamura, K. and Spath, D.P. (1983). Detection of *Giardia lamblia* by immunofluorescence. *Applied and Environmental Microbiology* **45**, 698–700.

Roach, P.D., Olson, M.E., Whitley, G. and Wallis, P.M. (1993). Waterborne *Giardia* cysts and *Cryptosporidium* oocysts in the Yukon. Canada. *Applied and Environmental Microbiology* **59**, 67–73.

Robertson, L.J., Campbell, A.T. and Smith, H.V. (1993). *In vitro* excystation of *Cryptosporidium parvum*. *Parasitology* **106**, 13–29.

Robertson, L.J., Paton, C.A., Campbell, A.T. *et al.* (2000). *Giardia* cysts and *Cryptosporidium* oocysts at sewage treatment works in Scotland, UK. *Water Research* **34**, 2310–2322.

Robertson, L.J., Smith, H.V. and Ongerth, J.E. (1994). *Cryptosporidium* and cryptosporidiosis. Part 3: Development of water treatment technologies to remove and inactivate oocysts. *Microbiology Europe* **2**, 18–26.

Rochelle, P.A., De Leon, R., Johnson, A. *et al.* (1999). Evaluation of immunomagnetic separation for recovery of infectious *Cryptosporidium parvum* oocysts from environmental samples. *Applied and Environmental Microbiology* **65**, 841–845.

Rogers, M.R., Flanigian, D.J. and Jakubowski, W. (1995). Identification of algae which interfere with the detection of *Giardia* cysts and *Cryptosporidium* oocysts and a method for alleviating this interference. *Applied and Environmental Microbiology* **61**, 3759–3763.

Rose, J.B. (1997). Environmental ecology of *Cryptosporidium* and public health implications. *Annual Reviews of Public Health* **18**, 135–161.

Rose, J. and Gerba, C. (1991). Assessing potential health risks from viruses and parasites in reclaimed water in Arizona and Florida, USA. *Water Science and Technology* **23**, 2091–2098.

Rose, J.B., Clifrino, A., Madore, M.S., Gerba, C.P., Sterling, C.R. and Arrowood, M.J. (1986). Detection of *Cryptosporidium* from wastewater and freshwater environments. *Water Science and Technology* **18**, 233–239.

Rose, J.B., Dickson, L.J., Farrah, S.R. and Carnahan, R.P. (1996). Removal of pathogenic and indicator microorganisms by a full-scale water reclamation facility. *Water Research* **30**, 2785–2797.

Rose, J.B., Farrah, S.R., Friedman, D.M. *et al.* (1999). Public health evaluation of advanced reclaimed water for potable applications. *Water Science and Technology* **40**, 247–252.

Rose, J.B., Kayed, D., Madore, M.S. *et al.*(1988). Methods for the recovery of *Giardia* and *Cryptosporidium* from environmental waters and their comparative occurrence. In: P.M. Wallis, B.R. Hammond *et al.* (eds.) *Advances in Giardia Research*, pp. 205–209. University of Calgary Press, Calgary.

Rose, J.B., Landeen, L.K., Riley, K.R. and Gerba, C.P. (1989). Evaluation of immunofluorescence techniques for detection of *Cryptosporidium* oocysts and *Giardia* cysts from environmental samples. *Applied and Environmental Microbiology* **55**, 3189–3196.

Rose, J.B., Lisle, J.T. and LeChevallier, M. (1997). Waterborne cryptosporidiosis: Incidence, outbreaks and treatment strategies. In: R. Fayer (ed.) *Cryptosporidium* and cryptosporidiosis, pp. 95–111. CRC Press, Boca Raton.

Roy, D., Englebrecht, R.S. and Chian, E.S.K. (1982). Comparative inactivation of six enteroviruses by ozone. *Journal of the American Water Works Association*, **74**, 660–664.

Rubin, A.J., Evers, D.P., Eyman, G.M. and Jarroll, E.J. (1989). Inactivation of gerbil-cultured *Giardia lamblia* cysts by free chlorine. *Applied and Environmental Microbiology* **55**, 2592–2594.

Ruffell, K.M., Rennecker, J.L. and Marinas, B.J. (2000). Inactivation of *Cryptosporidium parvum* oocysts with chlorine dioxide. *Water Research* **34**, 868–876.

Sartory, D., Parton, A., Parton, A.C. *et al.* (1998). Recovery of *Cryptosporidium* oocysts from small and large volume water samples using a compressed foam filter system. *Letters in Applied Microbiology* **27**, 318–322.

Sauch, J.F. (1985). Use of immunofluorescence and phase contrast microscopy for detection and identification of *Giardia* cysts in water samples. *Applied and Environmental Microbiology* **50**, 1434–1438.

Schaefer, F.W. III. (1990). *In vitro* excystation of *Giardia*. In: E.A. Meyer (ed.) *Giardiasis*, pp. 111–136. Elsevier, Amsterdam.

Scarpino, P.V., Lucas, M., Dahling, D.R. *et al.* (1974). Effectiveness of hypochlorous acid and hypochlorite in destruction of viruses and bacteria, pp. 359–368. In: *Chemistry of Water Supply, Treatment, and Distribution*. (A.J. Rubin ed.), Ann. Arbor Science Publishers, Woburn, Mass.

Schuler, P.F., Ghosh, M.M. and Gopalan, P. (1991). Slow sand filtration and diatomaceous filtration of cysts and other particulates. *Water Research* **25**, 995–1005.

Schupp, D.E. and Erlandsen, S.L. (1987a). A new method to determine *Giardia* cyst viability, correlation between fluorescein diacetate/propidium iodide staining and animal infectivity. *Applied and Environmental Microbiology*, **55**, 704–707.

Schupp, D.E. and Erlandsen, S.L. (1987b). Determination of *Giardia muris* cyst viability by differential interference contrast, phase or bright field microscopy. *Journal of Parasitology*, **73**, 723–729.

Schupp, D.E., Januschka, M.M. and Erlandsen, S.L. (1988a). Assessing *Giardia* cyst viability with fluorogenic dyes, comparisons to animal infectivity and cyst morphology by light and electron microscopy. pp. 265–269. In: *Advances in Giardia Research*. (Wallis, P.M. and Hammond, B.R. eds), University of Calgary Press, Calgary, Canada.

Schupp, D.L., Januschka, M.M., Sherlock, L.A. *et al.* (1988b). Production of viable *Giardia* cysts *in vitro*: determination by fluorogenic dye staining, excystation, and animal infectivity in the mouse and Mongolian gerbil. *Gastroenterology* **95**, 1–10.

Shaefer, F.W. III, Johnson, C.H., Hsu, C.H. and Rice, E.W. (1991). Determination of *Giardia lamblia* cyst infective dose for the Mongolian gerbil (*Meriones unguiculatus*). *Applied and Environmental Microbiology*, **57**, 2408–2409.

Shaw, P.K., Bordsky, R.E., Lyman, D.O. *et al.* (1977). A community wide outbreak of giardiasis with evidence of transmission by municipal water supply. *Annals of Internal Medicine*, **87**, 426–432.

Shephard, K.M. and Wyn-Jones, A.P. (1996). An evaluation of methods for the simultaneous detection of *Cryptosporidium* oocysts and *Giardia* cysts from water. *Applied and Environmental Microbiology* **62**, 1317–1322.

Slifko, T.R., Friedman, D., Rose, J.B. and Jakubowski, W. (1997). An *in vitro* method for detecting infectious *Cryptosporidium* oocysts with cell culture. *Applied and Environmental Microbiology* **63**, 3669–3675.

Slifko, T.R., Huffman, D.E. and Rose, J.B. (1999). A most-probable-number assay for enumeration of infectious *Cryptosporidium parvum* oocyst. *Applied and Environmental Microbiology* **65**, 3936–3941.

Slifko, T.R., Smith, H.V. and Rose, J.B. (2000). Emerging parasite zoonoses associated with water and food. *International Journal for Parasitology* **30**, 1379–1393.

Smith, H.V. (1995). Emerging technologies for the detection of protozoan parasites in water. In: W.B. Betts, D. Casemore, C. Fricker *et al.* (eds) *Protozoan Parasites and Water*, pp. 108–114. The Royal Society of Chemistry, Cambridge.

Smith, H.V. (1996). Detection of *Cryptosporidium* and *Giardia* in water. In: R.W. Pickup and J.R. Saunders (eds) *Molecular approaches to environmental microbiology*, pp. 195–225. Ellis-Horwood, Hemel Hempstead.

Smith, H.V. (1998). Detection of parasites in the environment. In: H.V. Smith and W.H. Stimson (eds) *Infectious diseases diagnosis: Current Status and Future Trends, Parasitology*, **117**, pp. S113–S141.

Smith, H.V. and Girdwood, R.W.A. (1999). Waterborne parasites. Detection by conventional and developing techniques. In: R. Robinson, C. Batt and P. Patel (eds) *Encyclopaedia of Food Microbiology*, pp. 2295–2305. Academic Press, London and New York.

Smith, H.V. and Hayes, C.R. (1996). The status of UK methods for the detection of *Cryptosporidium* sp. oocysts and *Giardia* sp. cysts in water concentrates and their relevance to water management. *Water Science and Technology* **35**, 369–376.

Smith, H.V. and Nichols, R.N. (2003). Case study of health effects of *Cryptosporidium* in drinking water. Article 4.12.4.8. UNESCO-EOLSS Encyclopaedia of Life Support Systems. Theme – Environmental Toxicology and Human Health. In press.

Smith, H.V. and Rose, J.B. (1990). Waterborne cryptosporidiosis. *Parasitology Today* **6**, 8–12.

Smith, H.V. and Rose, J.B. (1998). Waterborne cryptosporidiosis, current status. *Parasitology Today* **14**, 14–22.

Smith, H.V., Robertson, L.J., Campbell, A.T. and Girdwood, R.W.A. (1995a). *Giardia* and giardiasis: what's in a name? *Microbiology Europe* **3**, 22–29.

Smith, H.V., Robertson, L.J. and Ongerth, J.E. (1995b). Cryptosporidiosis and giardiasis, the impact of waterborne transmission. *Journal of Water Supply Research and Technology – Aqua* **44**, 258–274.

Smith, H.V., Robertson, L.J. and Campbell, A.T. (1993a). *Cryptosporidium* and cryptosporidiosis Part 2: Future technologies and state of the art research. *European Microbiology* **2**, 22–29.

Smith, H.V., Brown, J. and Coulson, J.C. *et al.* (1993b). Occurrence of *Cryptosporidium* sp. oocysts in *Larus* spp. gulls. *Epidemiology and Infection* **110**, 135–143.

Smith, H.V. Campbell, B.M. Paton, C.A. Nichols, R.A.B. (2002). Significance of enhanced morphological detection of *Cryptosporidium* sp. oocysts in water concentrates using DAPI and immunofluorescence microscopy. *Applied and Environmental Microbiology*, (in press).

Smith, H.V., Parker, J.F.W., Girdwood, R.W.A. *et al.* (1989a). A modified method for the detection of *Cryptosporidium* spp. oocysts in water-related samples. *Communicable Diseases Scotland 89* **15**, 7–11.

Smith, H.V., McDiarmid, A., Smith, A.L. *et al.* (1989b). An analysis of staining methods for the detection of *Cryptosporidium* spp. oocysts in water-related samples. *Parasitology* **99**, 323–327.

Smith, A.L. and Smith, H.V. (1989). A comparison of fluorescein diacetate and propidium iodide staining and *in vitro* excystation for determining *Giardia intestinalis* cyst viability. *Parasitology* **99**; 329–331.

Soares, A.C., Straub, T.M., Pepper, I.L. and Gerba, C.P. (1994). Effect of anaerobic digestion on the occurrence of enteroviruses and *Giardia* cysts in sewage sludge. *Journal of Environmental Science and Health* **A29**, 1887–1897.

Soave, R. and Armstrong, D. (1986). *Cryptosporidium* and cryptosporidiosis. *Reviews of Infectious Diseases* **8**, 1012–1023.

Stadterman, K.L., Sninsky, A.M., Sykora, J.L. and Jakubowski, W. (1995). Removal and inactivation of *Cryptosporidium* oocysts by activated sludge treatment and anaerobic digestion. *Water Science and Technology* **31**, 97–104.

States, S., Stadterman, K., Ammon, L. *et al.* (1997). Protozoa in river water: sources, occurrence, and treatment. *Journal of the American Water Works Association* **89**, 74–83.

Sterling, C.R., Kutob, R.M., Gizinski, M.J. *et al.* (1988). *Giardia* detection using monoclonal antibodies recognising determinants on *in vitro* derived cysts. In: P.M. Wallis, B.R. Hammond *et al.* (eds.) *Advances in Giardia Research*, pp. 219–222. University of Calgary Press, Calgary.

Stibbs, H.H. (1993). Detection and differentiation of *Giardia lamblia* and *Giardia muris* cysts in surface water by immunofluorescence flow cytometry. Abstract no. 663, *American Journal of Tropical Medicine and Hygiene* **49**, 396.

Stinear, T., Matusan, A., Hines, K. and Sandery, M. (1996). Detection of a single viable *Cryptosporidium parvum* oocyst in environmental water concentrates by reverse transcription-PCR. *Applied and Environmental Microbiology* **62**; 3385–3390.

Sturdee, A.P., Chalmers, R.M. and Bull, S.A. (1999). Detection of *Cryptosporidium* oocysts in wild mammals of mainland Britain. *Veterinary Parasitology* **80**, 273–280.

Swertfeger, J., Metz, D.H., De Marco, J. *et al.* (1999). Effect of filter media on cyst and oocyst removal. *Journal of the American Water Works Association* **91**, 90–100.

Sykora, J.L., Bancroft, W.D., States, S.J. *et al.* (1988). *Giardia* cysts in raw and treated sewage. In: G.S. Logsdon *et al.* (eds) *Controlling Waterborne Giardiasis*. Environmental Engineering Division of the American Society for Civil Engineering 22.

Sykora, J.L., Sorber, C.A., Jakubowski, W. *et al.* (1991). Distribution of *Giardia* cysts in wastewater. *Water Science and Technology* **24**, 187–192.

Taghi-Kilani, R., Gyürék, L.L., Millard, P.J. *et al.* (1996). Nucleic acid stains as indicators of *Giardia muris* viability following cyst inactivation. *International Journal for Parasitology* **26**, 637–646.

Thiriat, L., Sidaner, F. and Schwartzbrod, J. (1998). Determination of *Giardia* cyst viability in environmental samples by immunofluorescence, Fluorogenic dye staining and differential interference contrast microscopy. *Letters in Applied Microbiology* **26**, 237–242.

Thompson, R.C.A., Lymberry, A.J. and Meloni, B.P. (1990a). Genetic variation in *Giardia* Kustler, 1882; taxonomic and epidemiological significance. *Protozoological Abstracts* **14**, 1–28.

Thompson, R.C.A., Lymbery, A.J., Meloni, B.P. and Binz, N. (1990b). The zoonotic transmission of *Giardia* species. *Veterinary Record*, 19 May, 513–514.

Thompson, R.C.A., Reynoldson, J.A. and Mendis, A.H.W. (1993). *Giardia* and giardiasis. *Advances in Parasitology* **32**, 71–160.

Thompson, R.C.A., Hopkins, R.M. and Homan, W.L. (2000). Nomenclature and genetic groupings of *Giardia* infecting mammals. *Parasitology Today* **16**, 210–213.

Thompson, R.C.A. (2002). Towards a better understanding of host specificity and the transmission of *Giardia*: the impact of molecular epidemiology. In: Olsen, B.E., Olsen, M.E. and Wallis, P.M., (eds) *Giardia: The Cosmopolitan Parasite*. CAB International, Wallingford, Oxon, UK.

Timms, S., Slade, J.S. and Fricker, C.R. (1995). Removal of *Cryptosporidium* by slow sand filtration. *Water Science and Technology* **31**, 81–84.

Udezulu, I.A., Visvesvara, G.S., Moss, D.M. and Leitch, G.J. (1992). Isolation of two *Giardia lamblia* (WB strain) clones with distinct surface protein profiles and differing infectivity and virulence. *Infection and Immunity* **60**, 2274–2280.

Upton, S.J., Tilley, M. and Brillhart, D.B. (1994). Comparative development of *Cryptosporidium parvum* (Apicomplexa) in 11 continuous host cell lines, *FEMS Microbiology Letters* **118**, 233–236.

van Breemen, L.W.C.A., Ketelaars, H.A.M., Hoogenboezem, W. and Medema, G. (1998). Storage reservoirs – A first barrier for pathogenic micro-organisms in the Netherlands. *Water Science and Technology* **37**, 253–260.

Van Praagh, A.D., Gavaghan, P.D. and Sykora, J.L. (1993). *Giardia muris* cyst inactivation in anaerobic digester sludge. *Water Science and Technology* **27**, 105–109.

Veerannan, K.M. (1977). Effect of sewage treatment by stabilisation pond method on the survival of intestinal parasites. *Indian Journal of Environmental Health* **19**, 100–106.

Venczel, L.V., Arrowood, M., Hurd, M. and Sobsey, M.D. (1997). Inactivation of *Cryptosporidium parvum* oocysts and *Clostridium perfringens* spores by a mixed-oxidant disinfectant and by free chlorine. *Applied and Environmental Microbiology* **63**, 1598.

Vesey, G., Byrne, M., Shepherd, K. *et al.* (1993). Routine monitoring of *Cryptosporidium* oocysts in water using flow cytometry. *Journal of Applied Bacteriology* **75**, 87–90.

Vesey, G., Ashbolt, N., Fricker, E.J. *et al.* (1998). The use of a ribosomal RNA targeted oligonucleotide probe for fluorescent labelling of viable *Cryptosporidium parvum* oocysts. *Journal of Applied Microbiology* **85**, 429–440.

Vesey, G., Hutton, P.E., Champion, A. *et al.* (1994). Application of flow cytometric methods for the routine detection of *Cryptosporidium* and *Giardia* in water. *Cytometry* **16**, 1–6.

Villacorta, I.M.M., Ares-Mazas, M.E., Duran-Oreiro, D. and Lorenzo-Lorenzo, M.J. (1992). Efficacy of activated sludge in removing *Cryptosporidium parvum* oocysts from sewage. *Applied and Environmental Microbiology* **58**, 3514–3516.

Villacorta-Martinez de Maturana, I., Ares-Mazas, M.E., Duran-Oreiro, D. and Lorenzo-Lorenzo, M.J. (1992). Efficacy of activated sludge in removing *Cryptosporidium parvum* oocysts from sewage. *Applied and Environmental Microbiology* **58**, 3514–3516.

Visvesvara, G.S., Dickerson, J.W. and Healy, G.R. (1988). Variable infectivity of human-derived *Giardia lamblia* for Mongolian gerbils (*Meriones unguiculatus*), *International Journal of Microbiology* **26**, 837–841.

Wagner-Wiening, C. and Kimmig, P. (1995). Detection of viable *Cryptosporidium parvum* oocysts by PCR. *Applied and Environmental Microbiology* **61**, 4514–4516.

Wallis, P.M. (1994). Abiotic transmission – is water really significant? In: R.C.A. Thompson, J.A. Reynoldson and A.J. Lymbery (eds) *Giardia: from Molecules to Disease*, pp. 99–122. CAB International, Oxford.

Wallis, P.M., Erlandsen, S.L., Isaac-Renton, J.L. *et al.* (1996). Prevalence of *Giardia* cysts and *Cryptosporidium* oocysts and characterization of *Giardia* spp. isolated from drinking water in Canada. *Applied and Environmental Microbiology* **62**, 2789–2797.

Warhurst, D.C. and Smith, H.V. (1992). Getting to the guts of the problem. *Parasitology Today* **8**, 3292–3293.

Watkins, J., Kemp, P. and Shepherd, K. (1995). Analysis of water for *Cryptosporidium* including the use of flow cytometerycytometry. In: W.B. Betts, D.P. Casemore and C.R. Fricker *et al.* (eds) *Protozoan Parasites and Water*, pp. 115–121. Royal Society of Chemistry, Cambridge.

Whitmore, T.N. and Robertson, L.J. (1995). The effect of sewage sludge treatment processes on oocysts of *Cryptosporidium parvum*. *Journal of Applied Bacteriology* **78**, 34–38.

Wiandt, S., Baleux, B., Casellas, C. and Bontoux, J. (1994). Occurrence of *Giardia* sp. cysts during a waste-water treatment by a stablization pond in the South of France. *Water Science and Technology* **31**, 257–265.

Wiandt, S., Grimason, A.M., Baleux, B. and Bontoux, J. (2000). Efficiency of wastewater treatment plants at removing *Giardia* sp. cysts. In: I. Chorus, U. Ringelband, G. Schlag and O. Schmoll (eds) *Water, Sanitation and Health. Resolving Conflicts between Drinking-water Demands and Pressures from Society's Wastes*, pp. 35–42. IWA Publishing, London.

Widmer, G., Orbacz, E.A. and Tsipori, S. (1999). Beta-tubulin mRNA as a marker of *Cryptosporidium parvum* oocyst viability. *Applied and Environmental Microbiology* **65**, 1584–1588.

Wickramanayake, G.B., Rubin, A.J. and Sproul, O.J. (1984). Inactivation of *Giardia lamblia* cysts with ozone. *Applied and Environmental Microbiology* **48**, 671–672.

Wilczak, A. Gramith, K.M. Oppenheimer, J.A. *et al.* (1994). Effect of treatment conditions on the removal of protozoan cysts and MS2 virus by clarification and automatic backwash filtration. In *Water Quality Technology Conference Proceedings*. American Water Works Association, Denver.

Williamson, A.L., O'Donoghue, P.J., Upcroft, J.A. and Upcroft, P. (2000). Immune and patho-physiological responses to different strains of *Giardia duodenalis* in neonate mice. *International Journal for Parasitology* **30**, 129–136.

Willocks, L., Crampin, A., Milne, L. *et al.* (1998). A large outbreak of cryptosporidiosis associated with a public water supply from a deep chalk borehole. *Communicable Disease and Public Health* **1**, 239–243.

World Health Organisation (1989). *Health guidelines for the Use of Wastewater in Agriculture and Aquaculture*. Technical Report Series No. 778. WHO, Geneva.

World Health Organisation (1996). *The World Health Report, 1996*. WHO, Geneva.

Xiao, L., Morgan, U.M., Fayer, R. *et al.* (2000). *Cryptosporidium* Systematics and Implications for Public Health. *Parasitology Today* **16**, 287–292.

Yanez, F., Rojas, R., Castro, M.L. and Mayo, C. (1980). *Evaluation of the San Juan Stabilisation Ponds: Final Research Report of the First Phase*. Pan American Center for Sanitary Engineering and Environmental Sciences, Lima.

Zuckerman, U., Gold, D., Shelef, G. and Armon, R. (1997). The presence of *Giardia* and *Cryptosporidium* in surface waters and effluents in Israel. *Water Science and Technology* **35**(11–12), 381–384.

41

Biofilms in water distribution systems

Charmain J. Kerr*, Keith S. Osborn*, Alex H. Rickard†, Geoff D. Robson† and Pauline S. Handley†

*United Utilities, Great Sankey, Warrington WA5 3LP; †University of Manchester, Manchester M13 9PL, UK

1 INTRODUCTION

It is essential to deliver high quality potable water from the water treatment works to the customer's tap in order to prevent water-related illness within the population. However, treated water must pass through kilometres of pipes made out of cast iron, copper, lead, medium density polyethylene (MDPE) and unplasticized polyvinylchloride (uPVC) before it reaches the taps. The walls of the pipes in the distribution system provide ideal surfaces for microbial colonization and the biofilms formed cause a number of problems for the water companies. This chapter describes biofilm formation and investigates the relationship between the amount of biofilm and the composition of the pipe material.

Bacteria in the natural environment are either free living or attached to surfaces, growing as biofilms. Biofilm formation at the interface between a solid substratum and a liquid is a common phenomenon in natural, medical and industrial environments. Biofilms have been defined as 'functional consortia of microorganisms organized within extensive exopolymer matrices', (Costerton *et al.*, 1987), a definiton which includes all their essential properties. Bacteria usually predominate in biofilms but protozoa, fungi, algae and viruses will also be present in numbers that depend on the local biofilm environment. Although the processes occurring within biofilms may be beneficial in wastewater treatment works and biodegradation of waste, the presence of biofilms in the distribution pipes causes obstructions, damage to pipe surfaces and can also be a threat to health. Bacterial growth in biofilms is a potential threat to health as it has been linked to increased bacterial counts in treated water (van der Wende and Characklis, 1990) and to gastrointestinal illnesses (Payment *et al.*, 1994, 1997). In spite of the fact that nutrient levels are low in potable water, disinfectants are likely to be present, temperatures are low and shear forces may be high, microorganisms are still able to attach to the pipes and proliferate as biofilms. It is difficult to study pipe wall biofilms *in situ* and most information on biofilm development has been derived from laboratory models where conditions can be easily regulated. The following description of biofilm formation and structure summarizes a fast expanding literature on biofilms, taken from work on a wide range of model biofilm systems.

2 BIOFILM DEVELOPMENT AND BIOFILM PROPERTIES

2.1 Conditioning film formation and attachment

Biofilm formation is preceded by the formation of a conditioning film. The composition of the film is influenced both by the properties of the substratum and by the suspended organic and inorganic molecules in the bulk liquid phase.

The Handbook of Water and Wastewater Microbiology
ISBN 0-12-470100-0

In aquatic ecosystems surfaces become covered in seconds in a conditioning film, the composition of which varies depending on the presence of ions, glycoproteins, polysaccharides and humic and fulvic acids in the liquid phase. The composition and orientation of the molecules in the conditioning film is influenced strongly by the physicochemical properties of the substratum (reviewed by Busscher and van der Mei, 2000). Therefore the composition of the conditioning films on the different pipe materials is likely to vary considerably. It is also possible that the organic components in the conditioning film could be used as substrates for growth. Camper *et al.* (1999) have suggested that the humic substances in water distribution system conditioning films can be utilized when they are in the adsorbed conformation, whereas in the bulk fluid they are less easily degradable. Previously it had been shown that humic substances are biodegradable (Namkung and Rittmann 1987) and that biofilm bacteria can use humic substances (Volk *et al.*, 1997).

The adsorption of microbes to the conditioned surface is a two-stage process and reversible adhesion is followed by irreversible adhesion (Marshall *et al.*, 1971). The organisms make contact with the surface as a result of a number of processes including Brownian movement, sedimentation and convective transport which occurs when the movement of the water brings the bacteria close to the surface. In addition, active transport due to the presence of flagella has recently been shown to be important (Pratt and Kolter, 1998). In addition to flagella, fimbriae and other cell surface proteins are important in bacterial adhesion (reviewed by Dalton and March, 1998). The surface appendages on the bacteria are thought to penetrate through the electrostatic barrier that cells experience when they are held at a separation distance of 10–20 nm from the surface (Busscher *et al.*, 1992). Attachment becomes irreversible if local conditions are suitable, although many initial approaches by cells towards a surface may be abortive and cells will migrate until conditions become more favourable. Attachment is likely to be a trigger for many changes to occur in the cells, and many different genes are known to be transcribed when *Escherichia coli* grows as a biofilm compared to when it is free living (Prigent-Combaret *et al.*, 1999). Although the discovery of attachment-induced changes is a relatively new area in biofilm research, it is likely to be very important in the future.

The relative hydrophobicity and hydrophilicity of bacteria and surfaces are known to influence whether adhesion occurs (Absolom *et al.*, 1983). For example, Fletcher and Loeb (1979) showed a positive linear correlation between the hydrophobicity of the surface and the adhesion of a marine pseudomonad. Other surface properties, such as surface roughness of the substratum, will also influence bacterial adhesion (Quirynen *et al.*, 1989). A rough surface may increase adhesion by providing a larger surface area for attachment, and crevices will shelter bacteria from shear forces and in flowing systems the transport of nutrients to the surface will be increased (Duddridge and Pritchard, 1983).

2.2 The biofilm matrix

After attachment becomes irreversible the cells multiply into microcolonies and begin to produce copious amounts of extracellular polymeric substances (EPS), forming a matrix which surrounds and embeds the cells. These EPS are very complex and are of crucial importance to the biofilm, influencing and controlling many of the properties unique to biofilms. In oligotrophic environments the microcolonies of bacteria forming on the substratum grow into 'stacks' containing cells and EPS (Costerton *et al.*, 1994). In oligotrophic environments open water channels normally develop between the stacks, and the channels act as a circulatory system allowing for the dissemination of nutrients and oxygen and removal of metabolic by-products.

The EPS in the microcolonies have the dual purpose of maintaining adhesion to the substratum and holding the cells together inside the colony. Many different polymers have been detected in EPS. While it is generally agreed that polysaccharides are commonly produced by biofilm cells, proteins, nucleic acids

(DNA and RNA), phospholipids and lipids have also been found in the extracellular matrix (reviewed by Flemming *et al.*, 2000). However, most of the EPS analysis conducted so far has been largely restricted to wastewater biofilms and pure cultures rather than on drinking water biofilms. Most work on EPS has focused on exopolysaccharides (Sutherland, 1997, 2001), although it is still not known whether there are truly unique biofilm polysaccharides.

EPS are often negatively charged, gel-like and highly hydrated and the resulting flexible pillars of cells can therefore adapt and change their structure under different flow conditions (Stoodley *et al.*, 1999b) and 'streamers' or very long flexible stacks form under high shear conditions. The majority of the biofilm carbon may be located in the EPS and in sewer biofilms 70–98% of total organic carbon is extracellular (Jahn and Nielsen, 1998), although equivalent studies are not available for aquatic biofilms. In addition, the EPS can trap biotic and abiotic components from outside the biofilm and sewer biofilms can trap humic substances (Jahn and Nielsen, 1998). There is still much to be learnt about the composition, structure and function of the many different EPS in biofilms in oligotrophic environments as this is an under-researched area.

2.3 Communication between biofilm cells

Many bacteria use chemical signals to monitor their own species population density and control expression of specific genes in response to population density. The EPS are fibrous and porous and will allow low molecular weight, density-dependent cell-signalling molecules to pass between the cells. This type of gene regulation is termed quorum sensing (reviewed by Greenberg, 1999). Several Gram-negative bacteria use *N*- acyl-homoserine lactone (AHL) as signal molecules in quorum sensing. AHLs have been detected in natural river biofilms (McClean *et al.*, 1997) and have also been shown to influence the structural development of *P. aeruginosa PAO1* biofilms (Davies *et al.*, 1998). Evidence is emerging that cell signals from one species may be recognized by other bacterial species (Holden *et al.*,1999). It is highly probable that quorum sensing processes occur frequently in biofilms, where cell densities are high and AHLs may well control the development and maintenance of multispecies biofilms. However, studies on signalling molecules in biofilms are in their infancy and research in this area may well lead to new methods of biofilm control if quorum sensing can be inhibited.

2.4 Coaggregation and coadhesion

During biofilm development (Fig. 41.1) the primary colonizers (pioneer species) attach first and as the conditions in the biofilm change secondary colonizers are able to attach to the already established organisms, finally resulting in a stable, climax community. This colonization sequence has been extensively studied in dental plaque and different genera are continuously being added into an increasingly complex community (reviewed by Marsh and Bowden, 2000). The complexity of plaque is emphasized by the fact that it has been found to contain up to 300 different bacterial taxa. In comparison, very little information is available on the predominant species present in biofilms on different pipe materials in the water distribution system and no succession studies have been carried out to identify the primary and secondary colonizers.

One of the major influences on bacterial succession in plaque is the process of adhesion between cells of different genera and species. When adhesion occurs between cells in suspension it is termed coaggregation (Kolenbrander, 1988), and when it occurs on a surface it is termed coadhesion (Bos *et al.*, 1994). Coaggregation of dental plaque bacteria was first described 30 years ago and in an extensive series of papers Kolenbrander has described an immensely complicated network of intergeneric and intrageneric coaggregation interactions between all the major plaque bacteria (see review by Kolenbrander *et al.*, 2000). Recently, coaggregation has also been detected between strains of aquatic biofilm bacteria isolated from a laboratory model biofilm (Buswell *et al.*, 1997; Rickard *et al.*, 2000; Rickard *et al.*, 2002). Coaggregation between both aquatic biofilm

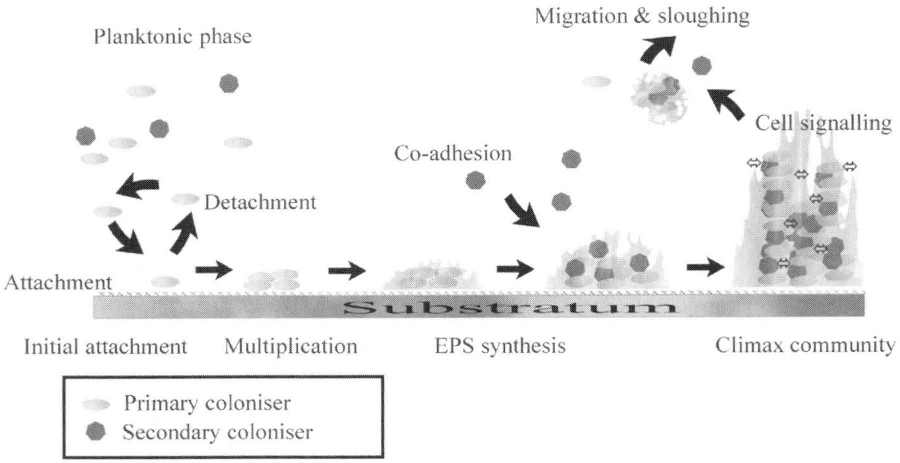

Fig. 41.1 Diagrammatic representation of the stages involved in biofilm development. Primary colonizers attach, multiply and produce EPS around the cells. Secondary colonizers then attach to the established primary colonizers (co-adhesion). Cell signalling is thought to contribute to the process of microcolony (stack) formation. Groups of cells then slough off and migrate from the climax community.

and dental plaque organisms is mediated by specific cell surface polymers (Rickard *et al.*, 2003). These polymers are comprised of a lectin (protein) on one partner and a saccharide containing polymer on the other partner and uniquely coaggregation occurs at the intraspecies (interstrain) level in the aquatic biofilm bacteria (Rickard *et al.*, 2000). Coaggregation is demonstrable when cells are present in high densities *in vitro*, but during biofilm formation cells will be present in much lower numbers and it is probable that coadhesion events occur when the free-living organisms in suspension attach to cells already attached to the substratum (Busscher and van der Mei, 2000). Aquatic biofilm bacteria only coaggregate with each other optimally in stationary phase when cells are in a starved condition (Rickard *et al.*, 2000), which probably reflects the metabolic conditions of the majority of aquatic biofilm cells. The role of coaggregation and coadhesion in aquatic biofilm formation is not yet understood.

2.5 Metabolic interactions and gene transfer

Many different types of interactions will occur between bacteria in any biofilm, in any location. Organisms will compete with each other and some strains will not survive very long in a particular niche. Some strains will show commensalism and others will show syntrophy, where one strain will be dependent on another for growth factors or nutrients (Nielsen *et al.*, 2000). Also it is now known that consortia of different strains cooperate to break down substrates such as mucin in saliva (reviewed by Marsh and Bowden, 2000). It is possible that aquatic biofilm bacteria could cooperate to degrade humic and fulvic acid molecules.

The close proximity of different biofilm strains can also promote genetic exchange between cells (Fry and Day, 1994) and rates of horizontal gene transfer by conjugation are higher between biofilm bacteria than between equivalent planktonic bacteria. This occurs on a range of different substrata including river stones (Fry and Day, 1994) and is thought to occur primarily because of the physical stabilization of cell pairs in the biofilm and the increased length of contact time (reviewed by Ehlers, 2000). The concept of a biofilm that is now emerging is of a community which is complex, organized and interactive (Molin *et al.*, 2000) and there is much more to find out about the complexities of biofilm life.

2.6 Dispersal of biofilm bacteria

Clean surfaces such as new pipes must be colonized by microorganisms flowing downstream and these free-living cells are highly likely to have originated from mature biofilms upstream of the nascent surface. Dispersal of biofilm bacteria is thought to occur as a result of two separate processes. First, through erosion (migration) of individual cells or groups of cells which occurs when they escape from the biofilm as a result of division (Allison *et al.*, 1990) or when enzymes degrade the surrounding EPS matrix. These processes have been termed 'active' detachment (McBain *et al.*,2000). Second, through sloughing, which involves the loss from the biofilm of much larger pieces of mature biofilm, with surrounding EPS matrix, being dislodged by mechanical forces such as cleaning or shear forces due to bulk fluid flow. The influence of shear stress on biofilm sloughing is obviously important in water distribution pipes. Biofilms respond to high shear by developing filamentous stacks known as 'streamers' which develop downstream and can be seen to wave about in the turbulent flow. However, in very high shear stress even the streamers may break off (reviewed by Stoodley *et al.*, 2000). It has been proposed that the frequency of changes in shear stress, as well as the magnitude of the shear, can lead to sloughing (Stoodley *et al.*,1999a).

2.7 Increased resistance of biofilms to antimicrobials

Biofilms have a high level of resistance to a range of antimicrobial substances and they are quoted as being up to 1000 times less susceptible to a wide range of toxic substances compared to their equivalent free-living, planktonic counterparts (reviewed by Allison *et al.*, 2000; Gilbert *et al.*, 2002). Biofilm bacteria may be intrinsically resistant to biocides and antibiotics but they will also acquire resistance genes by transfer of resistance genes (see Section 2.5). However, biofilms have other mechanisms that contribute to this enhanced level of resistance. High resistance is probably multifactorial and contributing factors are decreased diffusion into the biofilm (de Beer *et al.*, 1994), decreased biofilm growth rate (Evans *et al.*, 1990) and recently it has been shown that quorum sensing effects might be important (Hassett *et al.*, 1999).

This enhanced resistance to biocides is very important for effective treatment of biofilms in pipes. Bacteria in biofilms are protected from water disinfectants more than 600-fold, compared to free-living bacteria (LeChevallier *et al.*, 1988b). The synthesis of EPS by *Ps. aeruginosa* is an important factor in decreasing the susceptibility of this microorganism to chlorine disinfection (Geldreich *et al.*, 1972), by acting as an effective barrier preventing the penetration of the biocide (Blenkinsopp and Costerton, 1991). Chlorine is a reactive biocide and the EPS polymers are likely to react chemically with it and neutralize it so that less chlorine reaches the cells (Brown and Gilbert, 1993). The protected cells subsequently serve as a reservoir for the continuous contamination of the water flowing through distribution systems (Vess *et al.*, 1993). Biofilms may also provide an environment where disinfectant-injured cells can repair cellular damage and grow, for example, Waters and McFeters (1990) showed that *Enterobacter cloacae* and *Klebsiella pneumoniae* could recover from chlorine-induced injury within biofilms. The increase in numbers of particles in the water after high rainfall may also provide bacteria with an increased surface area for attachment and protection from disinfection (LeChevallier *et al.*, 1980, 1988a; Ridgway and Olson, 1982; Herson *et al.*, 1987).

3 FACTORS GOVERNING BIOFILM ACCUMULATION IN WATER DISTRIBUTION SYSTEMS

Several factors are known to influence biofilm development including temperature, nutrients, disinfectant residuals, hydraulic regime, and the characteristics of the substratum (Bryers and Characklis, 1981 LeChevallier *et al.*, 1990). The impact that these parameters have on biofilm accumulation in water distribution systems is discussed below.

3.1 Environmental factors

Increased temperature will lead to an increase in the numbers of biofilm bacteria and bacteria in the water. It is known that high temperatures cause an increase in the numbers of bacteria in the water (Donlan and Pipes, 1988), but equivalent studies on the numbers of biofilm bacteria in pipes are lacking. The temperature of the water also influences water treatment plant efficiency (Brazos and O'Connor, 1996), disinfection efficiency, dissipation of disinfectant residuals, corrosion rates, and distribution system hydraulics through changes in customer demand during the day. At lower temperatures bacteria are more likely to be washed out from the distribution system before significant growth has occurred. However, water companies are unable to regulate the water temperature and therefore have little influence over the most important factor controlling bacterial biofilm growth (LeChevallier *et al.*, 1988a).

Nutrient availability also governs the growth rate of the cells and the metabolic activity of the population which will influence the nature of the developed biofilm (LeChevallier *et al.*, 1988b). The principal nutrients for heterotrophic bacteria are carbon, nitrogen and phosphorus which are required in the ratio of approximately 100:10:1 (C:N:P). Therefore, carbon is normally the growth limiting nutrient for bacteria in aquatic environments even though it is present as humic and fulvic acids, carbohydrates, proteins and carboxylic acids in treated water (LeChevallier *et al.*, 1991). By reducing the carbon content biofilm growth should be limited. The coagulation process or membrane filtration at the water treatment works reduces the carbon content of the water entering the distribution system, but not all water treatment works contain these processes. However, van der Kooij (1991) showed that *Aeromonas* spp. can survive at carbon concentrations less than 10 μg/l. Bacterial counts would be reduced by removing carbon, but to lower the carbon content of potable water to less than 10 μg/l would require investment in new treatment technologies and bacteria capable of growing in extreme oligotrophic environments would still persist (LeChevallier *et al.*, 1996).

3.2 Disinfectant residual

On leaving the treatment works, water is treated with a disinfectant, normally chlorine, to inactivate any heterotrophic pathogenic bacteria in the water. Chlorine is added in excess so that if bacteria enter the distribution network, for example through cracks in pipes, they are inactivated. This chlorine excess is known as the residual. As a disinfectant enters a distribution system it reacts immediately with chemical and biological entities present in the network. This results in an inability to maintain a disinfectant residual which leads to bacterial growth in drinking water. However, the maintenance of a disinfection residual alone will not prevent bacterial growth (LeChevallier *et al.*, 1987). Nagy *et al.* (1982) reported that a 1–2 mg/l chlorine residual was necessary to reduce bacterial biofilms by 2 log units, but the biofilm was still present with a viable count of 10^3 colony forming units (cfu)/cm. Even direct contact with chlorine does not stop biofilm formation and Seidler *et al.* (1977) recovered coliforms in a potable water supply one week after scrubbing redwood tanks with a 200 mg/l chlorine solution.

Disinfectant residuals behave differently on different surfaces. Free chlorine is known preferentially to react with ferrous iron to produce the insoluble ferric hydroxide (LeChevallier *et al.*, 1990) and therefore, higher residuals are required for cast iron pipes when compared to the plastic pipes. A greater understanding of the interaction of disinfectants with distribution systems is still required to formulate appropriate strategies for biofilm control.

The disinfectant residual cannot be maintained at high levels because the strong chlorinous taste of the water would result in customer complaints to water companies. High chlorine residuals also promote the formation of trihalomethanes (THMs) which some studies have shown to be carcinogenic at high levels in the mg/l range. Current regulations Anon (2000) require that THM concentrations should not exceed 100 μg/l.

3.3 Hydraulic effects on biofilm build up

The hydraulics of the water system are very much dependent on the consumer and therefore change daily and seasonally. Increasing velocities or flow rates at times of high demand, for example due to fire fighters using water from the hydrants, allows for greater flux of nutrients to the pipe surface, greater transport of disinfectants (Characklis, 1988) and greater shearing of biofilms from the pipe surface (Dumbleton, 1995).

Conversely, stagnation of water in the distribution system can lead to loss of disinfectant residual and sedimentation of particles which can protect bacteria from the disinfectants and provide an increased surface area for attachment. Stagnation of water in service lines supplying customers' taps can result in high bacterial counts in the water when the tap is turned off overnight (LeChevallier *et al.*, 1987). To avoid using water with a high bacterial count the tap should be run for sufficient time to allow any water that has been standing in the pipe overnight to be flushed through.

The reversal of water flow within pipes causes shearing of biofilm into the bulk liquid flow resulting in high bacterial counts at the consumer's tap (LeChevallier *et al.*, 1996). This can occur in drought situations where water is transferred in the opposite direction to normal flow in order to supply areas of high demand with excess water from areas of low demand.

4 PROBLEMS CAUSED BY BIOFILMS IN WATER DISTRIBUTION SYSTEMS

The formation of microbial biofilms on pipe walls can have serious implications in water distribution systems (Donlan and Pipes, 1988; Hauddier *et al.*, 1988; Reasoner *et al.*, 1989; van der Wende and Characklis, 1990). Attached cells represent the major fraction of biomass in a distribution system and contribute to the continuous contamination of the water phase because bacteria are sheared from the surfaces of the pipes (LeChevallier *et al.*, 1988a). Microbial growth on materials gives rise to three categories of problems: those of public health significance, aesthetic problems and those causing detrimental effects on the water distribution network at a major cost to water companies.

4.1 Public health problems associated with biofilm formation

Biofilms are normally seen as a nuisance rather than a threat to public health, however, they are increasingly seen as a prime cause of water quality deterioration in distribution systems due to their ability to harbour opportunistic pathogens. An opportunistic pathogen is an organism that can cause disease in individuals with compromised immune systems, e.g. AIDS patients, burn or transplant patients, the elderly and infants.

Opportunistic bacteria include some species of *Mycobacterium*, *Pseudomonas aeruginosa*, *Klebsiella* spp., *Serratia* spp., *Legionella* spp. and *Flavobacterium* spp. (Ridgway and Olson, 1981; Burke *et al.*, 1984) which can survive free chlorine residuals of 0.5–1.0 mg/l (Ridgway and Olson, 1982). Although these bacteria are ubiquitous in the environment they are normally absent from treated potable water in the distribution system (Colbourne, 1985). However, bacteria and protozoan parasites, e.g. *Cryptosporidium*, can break through the water treatment processes and become incorporated in the biofilm and disease-causing organisms may then be detected sporadically in the distribution system. Rogers *et al.* (1994) have shown that *Legionella* spp. can be incorporated and survive in laboratory biofilms. The presence of *Legionella pneumophila* in domestic water circuits has been implicated in several outbreaks of legionellosis (Walker *et al.*, 1993).

The main organisms associated with gastrointestinal illnesses are the coliforms. The coliforms belong to the Enterobacteriaceae and are found in the intestinal tract of humans and animals. UK Water Quality Regulations state that *E. coli* should not be detectable in a 100 ml sample of drinking water and that for any given distribution system coliform organisms should not be detectable in any more than 5% of routine samples (Anon, 1989a,b, 1991, 2000). All potable water must meet this standard even though less than 4% of the water pumped through the pipe network is consumed (UKWIR, 1998). The presence of coliform bacteria in water may be due to a loss of residual disinfectant, back siphonage, cross connections, pipeline breaks or repairs in the distribution system, survival and recovery of injured organisms within the biofilm or failure of the treatment plant.

The occurrence of pathogenic organisms in the distribution system does not always coincide with epidemics of gastroenteritis in the population, conversely a number of epidemic incidents affecting consumers cannot be traced back to any specific aetiological agent. Payment *et al.* (1997) found that 10–30% of gastrointestinal and respiratory symptoms were associated with consumption of tap water; however, Payment *et al.* (1993) could find no relationship between increases in illness and drinking contaminated water from distribution systems experiencing biofilm growth problems. UKWIR (1998) reviewed the published literature between 1995 and 1997 relating to the health significance of heterotrophic bacteria and came to the conclusion that heterotrophic bacteria did not pose a threat to public health. There is much conflicting evidence and more research is required on the relationship between disease and the consumption of potable water.

4.2 Aesthetic problems

Customer complaints direct to the water industry are mainly due to aesthetic problems, e.g. changes in taste, odour and colour of the water, staining of laundry and appliances and the presence of invertebrates in the drinking water.

Distribution system biofilms have been shown to support growth of *Actinomyces, Nocardia, Streptomyces, Arthrobacter* and certain filamentous fungi which have been linked to taste and odour complaints about potable water (Olson and Nagy, 1984, LeChevallier *et al.*, 1987; Geldreich, 1990). Species within the *Actinomyces* are capable of secreting molecules such as geosmin which give a disagreeable taste to the water even at low concentrations.

The biofilm bacteria can be a starting point for a trophic food web leading to the growth of undesirable higher organisms. Biomass of living and dead bacteria serve as a food source for animals visible with the naked eye, e.g. *Asellus aquaticus*. Amoebae, nematodes, amphipods, copepods and fly larvae (Levy *et al.*, 1986) have been recovered from water systems and have resulted in complaints from customers. Although they pose no health problems themselves (UKWIR, 1998), research has shown that *Legionella* spp. may grow and survive inside certain amoebae (Smith-Somerville *et al.*, 1991).

The build-up of corrosion products (see next section) and the subsequent release into the bulk water can result in discoloured water and staining of laundry resulting in customer complaints to the water company. This corrosion can be microbiologically induced and the build-up of corrosion products can also result in a reduction in the internal diameter of the pipe.

4.3 Problems resulting in additional costs to the water company

Biofilm formation in distribution systems may have a detrimental effect on the water quality and distribution infrastructure resulting in considerable costs to the water company.

4.3.1 Problems of reliable coliform detection

Current UK legislation requires that drinking water should be free from pathogenic and indicator organisms such as coliforms, in general, and *E. coli* specifically (Anon, 1989a, b, 1991, 2000). The total coliform group was selected for assessing microbial quality of drinking water because it is found in high numbers in faeces and can easily be detected in highly diluted contaminated water. It is, therefore, used as a sign that a treatment plant has failed or that the water has become contaminated within the distribution system due to a cross connection or back siphonage. However, there are many environmental strains of coliforms, e.g. *Klebsiella* spp.or *Citrobacter* spp. which do not pose a threat to public health but give a positive result for coliforms in the laboratory (Jones and Bradshaw, 1996). Therefore, the presence of these environmental strains in biofilms on pipe surfaces and their subsequent sloughing into the bulk water may mask the presence of faecal coliforms (*E. coli*) used to show treatment failure, by giving false positive results. The initial stages of detection of coliforms in potable water by the laboratory cannot differentiate between environmental and faecal coliforms. A number of actions must be taken when a positive result is obtained in the laboratory due to the presence of coliforms, e.g. the distribution system can be flushed by increasing the flow, the disinfection level increased and extra samples taken. If failures occur frequently the Drinking Water Inspectorate (DWI) imposes regulatory consequences.

4.3.2 Problems caused by corrosion products

The majority of the distribution network is composed of cast iron pipes which are subject to corrosion by both the water and the microorganisms in the biofilms on the surfaces. Corrosion is the deterioration and leaching of metal from a pipe surface as a result of its reaction with the aquatic environment (LeChevallier *et al.*, 1993) and it may be induced or enhanced by microbial activity (Ford and Mitchell, 1990). Cast iron pipes support the growth of organisms due to the availability of iron as a nutrient source (Camper *et al.*, 1996).

Attachment of bacteria to a metallic pipe surface can also enhance corrosion due to the heterogeneous and patchy distribution of bacteria on the surface. High growth rates in the biofilm can cause oxygen gradients to form within the biofilm, with anaerobic regions present next to the metal surface and oxygen rich areas next to the bulk liquid phase. This produces a differential aeration cell through which oxygen is depleted in an uneven manner near the surface. Pitting occurs under the microcolony as electrons (free Fe^{2+} ions) from the metal surface flow to sites exposed to oxygen. If the pits beneath the colony span the width of the pipe wall, pipe failure and leaking will result. Corrosion can be enhanced further by the presence in anaerobic areas of sulphate-reducing bacteria (SRB) (reviewed by Hamilton, 1985). The pits and nodules provide the organisms with protection from water shear (Allen and Geldreich, 1977) and from disinfection by free chlorine. LeChevallier *et al.* (1993) reported cellular uptake of metal ions within the EPS, and Macrae and Edwards (1972) found that strains of *Pseudomonas, Enterobacter* and *Mycobacteria* formed iron-containing precipitates.

The metal ions can be deposited at a site some distance from where they went into solution. The deposits are known as tubercles and can range in size from a couple of millimetres to several centimetres in diameter and in height (Dumbleton, 1995). They contain 10–15% organic material, a mixed consortium of bacteria and vast amounts of iron bound in EPS by bacteria (Tatnall, 1991). High levels of coliforms were found to be associated with these iron tubercles (LeChevallier *et al.*, 1987).

Build-up of biofilm and scale (calcium, magnesium and silicate scale in areas of hard water) inside a pipeline can reduce the internal diameter or cross-sectional area and increase friction, resulting in reduced velocity and carrying capacity. The roughness effect is magnified by filamentous organisms that become established in the biofilm (McCoy and Costerton, 1982). Constant fluid velocity can be maintained in many systems but pumping costs increase.

Biologically mediated corrosion of cast iron pipes can, therefore, cause substantial damage to the cast iron pipes, service connections, fittings or fixtures due to surface removal, pitting or tuberculation (Chamberlain, 1992). This leads to leaking water pipes and breaks in the water main resulting in major costs for the water company in replacement of damaged pipes.

5 METHODS TO CONTROL BIOFILM GROWTH IN DISTRIBUTION SYSTEMS

There are three main ways to control the proliferation of bacteria within the distribution system. These are to change the disinfectant regime or the disinfectant itself, reduce the carbon content of the water or to change the pipe materials to one that has a lower biofilm-forming potential. The advantages and disadvantages of these three methods are discussed below.

5.1 Changing disinfectant regime or increasing disinfection residuals

Drinking water is routinely disinfected with either chlorine or monochloramine to inactivate bacteria present in drinking water distribution systems. However, there is an instantaneous demand for the disinfectant residual from the biofilm (including bacteria, organic and inorganic material, corrosion by-products), the pipe surface and components in the bulk water (including bacteria, ammonia, humic substances). Maximum distances between consumer and treatment works range from a few kilometres in urbanized areas to more than 50 km in the rural areas (Block, 1992). Payment *et al.* (1988) demonstrated that the number of bacteria increased with the distance from the water filtration plant. To control the number of bacteria in the potable water the disinfection regime or the disinfectant could be changed to provide a more hostile environment for the microorganisms.

5.1.1 Increasing the disinfectant residual

In general, water companies that have experienced coliform problems maintain high free chlorine residuals in an attempt to control coliforms. To increase the disinfectant residual either requires a higher dose of chlorine at the treatment works or rechlorination at booster plants situated near to the customer. However, LeChevallier *et al.* (1987) showed that maintenance of a free chlorine residual did not correlate with reduced bacterial counts in the water column and Nagy *et al.* (1982) found that maintenance of 3–5 mg of chlorine/l of water was necessary to reduce bacterial biofilms by more than 99.9%. The current chlorine residual at the customer's tap is generally in the range 0.3–0.5 mg of chlorine/l, a value far lower than that suggested by Nagy *et al.* (1982) to reduce biofilm formation.

High chlorine residuals can result in new problems for the water industry including trihalomethane (THM) formation due to the combination of chlorine and organic material, customer complaints about chlorinous tastes and odours and increased corrosion rates. THM formation is responsible for tastes and odours of treated water and can be carcinogenic as suggested by epidemiological studies (van der Kooij and Veenendaal, 1994). The presence of a high chlorine residual might also result in selection for chlorine-resistant bacteria in chlorinated systems (Ridgway and Olson, 1982).

It is, therefore, impractical to increase the disinfectant residual without first reducing the organic carbon content of the water to reduce the potential for THM formation.

5.1.2 Changing the disinfectant

At present free chlorine is used as the disinfectant in the majority of water companies in the UK; however, in the USA monochloramine is widely used. Any disinfectant chosen must be capable of penetrating the biofilm and inactivating all of the attached microorganisms. The disinfectant must be stable, potable, persist in the distribution system and must not produce hazardous by-products. Monochloramine may be more effective than free chlorine in water distribution systems (LeChevallier *et al.*, 1988b, 1990). Bacteria may be controlled at levels of 2–4 mg/l, however, the level required will depend on water quality and pipe characteristics. The formation of nitrite in a distribution system containing approximately 1 mg/l chloramine can cause the loss of a disinfectant residual and an increase in heterotrophic plate counts (Woolfe *et al.*, 1988).

The bacteria in the distribution system may be reduced by ozonation of the water at the treatment works. However, ozone does not produce a stable residual and van der Kooij (1987) indicated that, in certain cases, ozonation increased assimilable organic carbon (AOC) and oxygen levels in treated waters which stimulated microbial activity. The efficiency of disinfectant against bacteria in biofilms depends on its penetration ability (LeChevallier, 1990). The inability of a disinfectant to penetrate distribution system biofilms can account for the occurrence of coliform bacteria in highly chlorinated waters. A better understanding of the interaction of disinfectants with cast iron pipe surfaces in distribution systems is necessary before appropriate strategies for biofilm control can be formulated and alternative disinfection reagents need to be studied.

5.2 Reduce the carbon content of the potable water

Controlling the levels of nutrients available for bacterial growth is the most direct means of resolving biofilm problems by limiting the bacterial growth potential. It is the available organic carbon (AOC) which is accessible to microorganisms for growth (LeChevallier *et al.*, 1991). Water companies in the UK currently monitor total organic carbon (TOC) of the water; however, data on AOC are limited. Zacheus and Martikainen (1995) found no correlation between the numbers of microorganisms and the amount of TOC.

The water treatment process is designed to remove nutrients from the source water. Recent changes in the treatment process, such as the introduction of granular activated carbon (GAC) filters, can reduce the amount of carbon (Servais *et al.*, 1991). GAC filters contain porous particles that adsorb and hold organic contaminants and are usually used to remove taste and odour problems (van der Kooij, 1987). However, GAC also has a greater surface area to support biological growth (Camper *et al.*, 1986) and inefficient maintenance of the filters can lead to breakthrough of the bacteria into the network.

LeChevallier *et al.* (1987) showed that AOC levels declined in drinking water in the USA as it flowed through the distribution system. Removal of AOC also occurred very quickly within a short distance from the treatment plant but bacterial growth occurred throughout the distribution network and bacterial numbers increase with distance from the treatment plant. LeChevallier et al. (1987, 1990) recommended total AOC levels less than 100 μg/l for companies trying to control growth of coliform bacteria in distribution system biofilms. The same authors (LeChevallier *et al.*, 1996b) studied AOC levels and coliform occurrences in four distribution systems. There was no relationship between the AOC levels and coliforms as systems with AOC levels between 33 and 93 μg/l had no coliform occurrences, but coliforms were recovered in 11.7% of samples in a system with an AOC level of 74 μg/l. However, the advantage of reducing AOC is twofold. Bacterial growth is limited and less chlorine is consumed in side reactions with organic compounds. To reduce the carbon in the water in the UK to a suitable level to control biofilm growth would require serious investment in new water treatment technologies and, as discussed above, low nutrient levels do not ensure coliform free water.

6 PIPE MATERIALS AND BIOFILM OCCURRENCE

6.1 Changing the pipe materials to substrata with a lower biofilm-forming potential

Since cast iron pipes are deteriorating rapidly and causing so many maintenance problems (Section 4.3.2), the distribution network is currently undergoing an extensive replacement scheme with old, leaking and corroded cast iron pipes being replaced by MDPE and uPVC. These new plastic pipe materials are thought to support fewer bacteria than the old cast iron pipes. Their surface is smoother and therefore the surface area smaller and they are not subject to corrosion or biodeterioration.

In addition, the effectiveness of a disinfectant is greatly influenced by the pipe material.

Biofilms grown on copper or PVC pipe surfaces were inactivated by a 1 mg/l dose of free chlorine or monochloramine. However, on iron pipes 3–4 mg/l of chlorine or monochloramine was ineffective in controlling the biofilm (LeChevallier *et al.*, 1990) because, as discussed before, the chlorine will preferentially react with the iron surface (LeChevallier *et al.*, 1993). It appears that the option of changing pipe materials to ones with lower biofilm-forming potentials would reduce the biofilm problem.

6.2 New pipe materials in use in the UK

Water companies must satisfy themselves that materials used in contact with potable water will not adversely affect water quality. Before any material can be used in contact with potable water it must pass tests to prove that it does not impart a taste or odour to the water, change the appearance of the water, support the growth of microorganisms, release cytotoxic compounds into the water or release metals into the water. New products must be approved by the Secretary of State who is advised by the environment committee on chemicals and materials of construction for use in public water supply and swimming pools.

The water distribution system is required to operate efficiently under pressure, without failure or leakage over a long period of time while simultaneously preserving water quality. Many UK town centres have a distribution system dating from the beginning of the last century, and therefore the majority of the pipes are cast iron and are at the limit of their useful life through corrosion and need to be renewed. A number of pipe materials are now being used to replace the old corroded pipes and their properties are briefly described here.

6.2.1 Unplasticized polyvinyl chloride

uPVC has been used for transporting potable water in Europe for 40 years. The advantages of uPVC include its toughness, corrosion resistance, lightweight, stiffness, high tensile strength and the ability to make pipe joints easily. uPVC is a polymer of vinyl chloride which is derived from oil and salt. The polymer has a major influence on the pipe properties and accounts for 95% of the potable water pipe compound. Compounds are added to the polymer to aid processing and include heat stabilizers, lubricants and pigments (water pipes are coloured blue to allow for identification of water pipes once underground).

6.2.2 Medium density polyethylene

Polyethylene belongs to a family of polymers called polyolefins and is formed by the polymerization of ethylene. The compounds that are added to uPVC are also added to MDPE to aid processing as well as UV stabilizers and antioxidants to prevent deterioration during storage. MDPE is non-corrodible, chemically resistant, lightweight, flexible, tough, strong and resists cracking. The following sections describe trials that have investigated (a) the biofilm forming potential of uPVC and MDPE and (b) whether biofilms can cause any detectable biodegradation of the plastic materials.

6.3 A comparison of biofilm formation on cast iron and plastic pipe materials in a Robbins device

Since uPVC and MDPE pipes are currently replacing the much older cast iron pipes in the system, it is important to be able compare the ability of all the pipe materials to support biofilm growth. Until recently, only limited information on biofilm build-up on the newer plastic materials was available (Pedersen, 1990; Vess *et al.*, 1993; Rogers *et al.*, 1994; Holt 1995; McMath *et al.*, 1997).

To determine the biofilm-forming potential of cast iron, MDPE and uPVC, laboratory trials were carried out using flow-through laboratory model devices (newly modified Robbins devices (nMRD)) (Kerr *et al.*, 2000). Each nMRD contained 25, 7.8 mm diameter discs of one of the pipe materials and three nMRDs were run in parallel, each containing 25 discs of cast iron, MDPE and uPVC in separate nMRDs. Each device was fed simultaneously with potable water, so that each received a very similar inoculum of bacteria at a laminar flow rate of 3 ml/min (Reynolds number 9.05). The total numbers of viable heterotrophs in the biofilms that accumulated on the three pipe materials were recorded for two separate experiments run over 21 days and over a 10-month period respectively (Table 41.1). The biofilm viable counts increased exponentially over the first 11 days on all three materials before reaching a steady state. The mean doubling time of the bacteria on the cast iron was faster than on the plastics (Kerr *et al.*, 1999). The numbers on cast iron continued to increase slowly for the duration of the 10-month trial, but the numbers on MDPE and uPVC did not increase over time.

The rank order of materials according to the numbers of viable heterotrophs in the biofilm was the same for all experiments and the greatest number of biofilm heterotrophs was always found on cast iron, followed in decreasing order by MDPE then uPVC (Table 41.1). There was no significant difference in the viable counts on MDPE and uPVC. After 10 months, MDPE and uPVC supported less than 1% of the population supported by cast iron.

These longitudinal experiments (21 days and 10 months) using the nMRDs were carried out at different times of the year and during each experiment the water temperature, the chlorine residual and the numbers of bacteria in the incoming potable water source varied. Variations in all these parameters influenced the rate at which the biofilm accumulated, but by 21 days the final rank order of populations on the three surfaces was always the same (Kerr *et al.*, 1999).

There was always a positive significant correlation between the numbers in the biofilm and the number of heterotrophs in the water passing through the lumen of the nMRD showing that the biofilm bacteria were shedding into the water (by erosion or sloughing). This would suggest that the lower the biofilm population on the pipe material surface, the lower the number of bacteria in the water and the higher the quality of the water

6.3.1 *Identification of organisms on cast iron, MDPE and uPVC*

Each pipe material had its own distinct population of heterotrophic bacteria (Kerr, 1999). After 21 days when the biofilm populations had reached a stable steady state (Kerr *et al.*, 1999), the dominant colony types were subcultured and identified by partial 16S rRNA gene sequencing using the primers described by Rickard *et al.* (2000). Cast iron supported the greatest diversity of heterotrophs and the 11 most numerous organisms were identified as *Sphingomonas* spp. (3 strains), *Caulobacter* spp., *Erythromicrobium* spp., *Methylobacterium* spp., *Afipia* spp., *Ultramicrobium* spp., *Acinetobacter* spp., *Nevskia* spp. and one unidentifiable strain that had no similar sequences in the EMBL database. In contrast there were only three dominant genera detected on MDPE (*Acinetobacter* spp., *Nevskia* spp. and *Blastobacter* spp.) and four on uPVC (*Sphingomonas* spp.,

TABLE 41.1 Comparison of 21 day and 10-month biofilm heterotroph viable counts (cfu/cm^2) on three pipe materials ranked in order of greatest to lowest numbers of biofilm bacteria recovered (adapted from Kerr *et al.*, 1999)

	Cfu/m^2	*% Relative biofilm viable count (21 days)*[a]	*Cfu/cm^2*	*% Relative biofilm viable count (10 months)*
Cast iron	3.2×10^6	100	4.0×10^7	100
MDPE	2.4×10^5	8	3.9×10^5	1
UPVC	6.6×10^4	3	3.0×10^5	1

[a] The viable count on cast iron was taken to be 100% (i.e. the maximum biofilm population that was developed under those conditions in that period). The relative biofilm viable count supported by the other materials were then calculated in relation to cast iron.

Bradyrhizobium spp., *Nevskia* spp. and a *Micrococcus* spp.). Only two genera, *Sphingomonas* and *Nevskia*, were found on more than one material. The same dominant colony types were found on all 25 discs from the same nMRD. The strains identified on the pipe materials in this study, when a chlorine residual was still present in the water, were predominantly Gram-negative, which is consistent with other reports indicating that the majority of bacteria isolated from water are Gram-negative (Reasoner *et al.*, 1989; Norton and LeChevallier, 2000).

6.4 Effects of biofilm accumulation on cast iron, MDPE and uPVC in a chemostat model

A longitudinal trial was established over 10 months to determine whether biodeterioration of the plastic pipe materials could be caused by the bacterial biofilm population (Kerr, 1999). Biodeterioration occurs as the result of the physical or mechanical breakdown of the material caused by microorganisms. For example, Gu *et al.* (1998) suggested that the process of degradation of polyesters and polyethers involved exodepolymerases secreted by microorganisms that cleave high molecular weight chains. Bacteria can also use the plastic polymer or additives as a source of carbon and energy (Heap and Morrell, 1968; Flemming, 1998; Gu *et al.*, 1998). uPVC contains no plasticizer and should therefore show little or no effect due to the formation of biofilms.

A model system was established to determine if biofilm formation had a detrimental effect on the mechanical properties of the three pipe materials, MDPE and uPVC and cast iron. The model system was based on a chemostat design which allowed small rectangular replicate pieces of pipe material to be exposed to 10 l of a relatively dense mixed bacterial suspension of bacteria from tap water ($\approx 2.7 \times 10^5$ cfu/ml). The chemostat was fed with unfiltered potable water at a dilution rate of 0.048/h, which resulted in a retention time within the chemostat of 20.6 h. Pieces of cast iron, MDPE and uPVC were exposed simultaneously to the suspension for a maximum of 61 weeks.

6.4.1 *Biofilm bacteria isolated from the pipe materials in the chemostat*

During the first 10 weeks cast iron supported 97% more heterotrophs than MDPE and uPVC, but for the remaining 51 weeks, cast iron supported 99% more bacteria than the two plastics (Fig. 41.2), confirming the percentage difference found previously using nMRDs (Kerr *et al.*, 1999). The populations of the dominant biofilm bacteria on each material changed continuously over 15 months as indicated by the changing colony morphology of the predominant isolates. In addition, the greatest diversity of colony types was found on the plastics rather than on the cast iron. After 61 weeks, eight strains were recovered from MDPE, seven from uPVC, but only five different numerically dominant strains were identified on cast iron, when the isolates were identified by partial 16S rDNA sequencing (Rickard *et al.*, 2000) (Table 41.2).

The majority of the biofilm heterotrophs recovered from the biofilms were Gram-negative. Three different strains (one from each material) could not be identified as no similar organisms were found in the database. All the identified genera are common soil and water organisms, which could have originated anywhere in the distribution system. Only *Rhizobium* and *Rhodocyclus* colonized all three materials, but five of the genera were recovered from both the plastics. Since all the materials

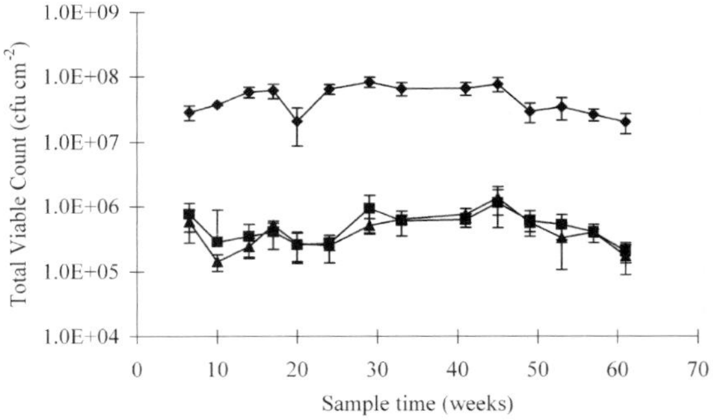

Fig. 41.2 The viable biofilm count (cfu/cm^2) recovered from cast iron (◆), MDPE (■) and uPVC (▲) over 61 weeks in the chemostat model. Each point represents the mean of four replicate pieces of material. The vertical bars represent the standard deviation (data from Kerr 1999).

TABLE 41.2 A comparison of the incidence of the predominant heterotrophic genera recovered from the biofilms on cast iron, MDPE and uPVC after 61 weeks (adapted from Kerr, 1999)

Number of materials[a]	*Genus*	*Cast iron*	*MDPE*	*UPVC*
3 materials	*Rhizobium*	+	+	+
	Rhodocyclus	+	+	+
2 materials	*Bradyrhizobium*		+	+
	Methylobacterium		+	+
	Nevskia		+	+
1 material	*Erythromicrobium*	+		
	Microbacterium			+
	Rhodococcus		+	
	Sphingomonas	+		
	Unidentified[b]	+ (1 isolate)	+ (1 isolate)	+ (1 isolate)

+ indicates that the genus was recovered from the material.
[a] Number of materials the genus was recovered from.
[b] No similar sequence in the EMBL database.

were suspended in the same consortium of planktonic bacteria these data show that both plastics can support very similar biofilm populations.

Taken together the two model systems (nMRD and chemostat) both confirm that biofilms are 99% reduced on the plastics, compared to cast iron. The biofilms in both systems consisted of continuously changing mixed consortia of mainly Gram-negative bacteria from soil and water, and not many of the same genera were found on the same materials in the two experiments. These results emphasize the dynamic nature of biofilm heterotrophic communities on pipe materials.

6.4.2 The effect of biofilm formation on the physical properties of the materials

After 61 weeks of biofilm growth the physical properties of the underlying materials were compared in a series of quantitative tests (Kerr, 1999). Biofilm formation on cast iron caused a significant reduction in weight, strength and stiffness, but had no influence on the same properties for MDPE or uPVC. For example, the cast iron pieces lost 52.6% of their original weight and 64% of their original strength, but there was no significant change in the weight or strength of the MDPE or uPVC pieces when compared to controls. In addition, the growth of biofilm on the materials did not significantly influence surface roughness (R_a), although the surface of the cast iron underneath the corrosion products and the biofilm did show extensive fine cracking in the scanning electron microscope (Kerr, 1999). No visible surface damage was detectable on either MDPE or uPVC.

Overall there were no changes in the physical properties of the two plastic materials over the experimental period, but the biodeterioration of the cast iron was severe and corrosion due to physical and biological processes (see earlier) caused a major loss of physical strength.

These results re-emphasize the importance of the change-over from cast iron pipes to plastic pipes. The distribution system will never be sterile, however, changing pipe materials from cast iron to MDPE or uPVC will result in lower biofilm counts. The quality of the distribution water will therefore be significantly improved by the introduction of new plastic pipe materials.

REFERENCES

Absolom, D.R., Lamberti, F.V., Policova, Z. *et al.* (1983). Surface thermodynamics of bacterial adhesion. *Appl. Environ. Microbiol.* **46**, 90–97.

Allen, M.J. and Geldreich, E.E. (1977). Distribution line sediments and bacterial regrowth. *Proceedings of the AWWA Water Quality Technology Conference*, Kansas City, Kansas, AWWA, Denver.

Allison, D.G., Brown, M.R.W., Evans, D.J. and Gilbert, P. (1990). Surface hydrophobicity and dispersal of *Pseudomonas aeruginosa* from biofilms. *FEMS Microbiol. Lett.* **71**, 101–104.

Allison, D.G., McBain, A.J. and Gilbert, P. (2000). Biofilms: problems of control. In: D.G. Allison, P. Gilbert, H.M. Lappin-Scott and M. Wilson (eds) *Community structure and co-operation in biofilms*, pp. 309–327. Cambridge University Press, Cambridge.

Anon (1989a). The Water Supply (Water Quality) Regulations SI No. 1147.

Anon (1989b). The Water Supply (Water Quality) (Amendment) Regulations SI No. 1384.

Anon (1991). The Water Supply (Water Quality) (Amendment) Regulations SI No. 1837.

Anon (2000). The Water Supply (Water Quality) (England and Wales) Regulations SI No. 3184.

Blenkinsopp, S.A. and Costerton, J.W. (1991). Understanding bacterial biofilms. *Trends Biotechnol.* **9**, 138–143.

Block, J.C. (1992). Biofilms in drinking water distribution systems. In: L.F. Melo, T.R. Bott, M. Fletcher and B. Capdeville (eds) *Biofilms – Science and Technology. Series E. Applied Sciences*, Vol. 223, pp. 469–485. Kluwer Academic Publishers, London.

Bos, R., van der Mei, H.C., Meinders, J.M. and Busscher, H.J. (1994). A quantitative method to study co-adhesion of micro-organisms in a parallel plate flow chamber: basic principles of the analysis. *J. Microbiol. Methods* **20**, 289–305.

Brazos, B.J. and O'Connor, J.T. (1996). Seasonal effects on generation of particle associated bacteria during distribution. *J. Env. Eng.* **122**, 1050–1057.

Brown, M.R.W. and Gilbert, P. (1993). Sensitivity of biofilms to antimicrobial agents. *J. Appl. Bact.* **74**, 87S–97S.

Bryers, J. and Characklis, W.G. (1981). Early fouling biofilm formation in a turbulent flow system: overall kinetics. *Water Res.* **15**, 483–491.

Burke, V., Robinson, J., Gracey, M. *et al.* (1984). Isolation of *Aeromonas hydrophila* from a metropolitan water supply: seasonal correlation with clinical isolates. *Appl. Environ. Microbiol.* **48**, 361–366.

Busscher, H.J. and van der Mei, H.C. (2000). Initial microbial adhesion events: mechanisms and implications. In: D.G. Allison, P. Gilbert, H.M. Lappin-Scott and M. Wilson (eds) *Community, structure and co-operation in biofilms*, pp. 25–36. Cambridge University Press, Cambridge.

Busscher, H.J., Cowan, M.M. and van der Mei, H.C. (1992). On the relative importance of specific and non-specific approaches to oral microbial adhesion. *FEMS Microbiol. Rev.* **88**, 199–209.

Buswell, C.M., Herlihy, Y.M., Marsh, P.D. *et al.* (1997). Co-aggregation amongst aquatic biofilm bacteria. *J. Appl. Microbiol.* **83**, 477.

Camper, A., Burr, M., Ellis, B. *et al.* (1999). Development and structure of drinking water biofilms and techniques for their study. *J. App. Microbiol., Symposium Supplement.* **85**, 1S–12S.

Camper, A.K., Jones, W.L. and Hayes, J.T. (1996). Effect of growth conditions and substratum composition on the persistence of coliforms in mixed population biofilms. *Appl. Environ. Microbiol.* **62**(11), 4014–4018.

Camper, A.K., LeChevallier, M.W., Broadway, S.C. and McFeters, G.A. (1986). Bacteria associated with granular activated carbon particles in drinking water. *Appl. Environ. Microbiol.* **52**(3), 434–438.

Chamberlain, A.H.L. (1992). Biofilms and corrosion. In: L.F. Melo, T.R. Bott, M. Fletcher and B. Capdeville (eds) *Biofilms – Science and Technology. Series E: Applied Sciences*, Vol. 223, pp. 207–217. Kluwer Academic Publishers, London.

Characklis, W.G. (1988). Model biofilm reactors. In: J.W. Wimpenny (ed.) *CRC Handbook of laboratory model systems for microbial ecosystems*, pp. 155–174. CRC Press, Boca Raton.

Colbourne, J.S. (1985). Materials usage and their effects on the microbiological quality of water supplies. *J. Appl. Bacteriol. Symposium Supplement* **57**, 47S–59S.

Costerton, J.W., Cheng, K-J., Geesey, G.G. *et al.* (1987). Bacterial biofilms in nature and disease. *Ann. Rev. Microbiol.* **41**, 435–462.

Costerton, J.W., Lewandowski, Z., de Beer, D. *et al.* (1994). Biofilms, the customised microniche. *J. Bact.* **176**, 2137–2142.

Dalton, H.M. and March, P.E. (1998). Molecular genetics of bacterial attachment and biofouling. *Curr. Opin. Biotechnol.* **9**, 252–255.

Davies, D.G., Parsek, M.R., Pearson, J.P. *et al.* (1998). The involvement of cell-to-cell signals in the development of a bacterial biofilm. *Science* **280**, 295–298.

De Beer, D.R., Srinivasan, R. and Stewart, P.S. (1994). Direct measurement of chlorine penetration into biofilms during disinfection. *App. Envir. Microb.* **60**, 4339–4344.

Donlan, R.M. and Pipes, W.O. (1988). Selected drinking water characteristics and attached microbial population density. *J. Am. Waterworks Assoc.* **80**(11), 70–76.

Duddridge, J.E. and Pritchard, A.M. (1983). Factors affecting the adhesion of bacteria to surfaces. In: *Microbial Corrosion*, pp. 28–35. The Metals Society, London.

Dumbleton, B. (1995). A question of scale and slime. *Water and Wastewater Treatment* **38**, 39, 47.

Ehlers, L.J. (2000). Gene transfer in biofilms. In: D.G. Allison, P. Gilbert, H.M. Lappin-Scott and M. Wilson (eds) *Community, structure and co-operation in biofilms*, pp. 215–256. Cambridge University Press, Cambridge.

Evans, D.J., Brown, M.R.W. and Gilbert, P. (1990). Susceptibility of bacterial biofilms to tobramycin: role of specific growth rate and phase in the division cycle. *J. Antimicrob. Chemother.* **25**, 585–591.

Flemming, H.C. (1998). Relevance of biofilms for the biodeterioration of surfaces of polymeric materials. *Polymer degradation and stability* **59**, 309–315.

Flemming, H.C., Wingender, J., Mayer, C. *et al.* (2000). Cohesiveness in biofilm matrix polymers. In: D.G. Allison, P. Gilbert, H.M. Lappin-Scott, M. Wilson *et al.* (eds) *Community, structure and co-operation in biofilms*, pp. 87–106. Cambridge University Press, Cambridge.

Fletcher, M. and Loeb, G.I. (1979). Influence of substratum characteristics on the attachment of a marine *Pseudomonad* to solid surface. *Appl. Environ. Microbiol.* **37**(1), 67–72.

Ford, T. and Mitchell, R. (1990). The ecology of microbial corrosion. In: K.C. Marshall (ed.) *Advances in Microbiology Ecology,* Vol. 11, pp. 231–262. Plenum Press, New York.

Fry, J.C. and Day, M.J. (1994). Transfer of natural plasmids in a riverine biofilm. In: J. Wimpenny, W. Nichols, D. Stickler and H.M. Lappin-Scott (eds) *Bacterial biofilms and their control in medicine and industry,* pp. 31–36. Bioline, Cardiff.

Geldreich, E.E. (1990). Microbiological quality of source waters for water supply. In: G.A. McFeters (ed.) *Drinking Water Microbiology,* pp. 3–31. Brock-Springer series in Contemporary Bioscience, Springer Verlag, New York.

Geldreich, E.E., Nash, H.D., Reasoner, D.J. and Taylor, R.H. (1972). The necessity of controlling bacterial populations in potable waters: community water supply. *J. Am. Water Works Assoc.* **75**, 568–571.

Gilbert, P., Maira-Litran, T., McBain, A.J. *et al.* (2002). The physiology and collective recalcitrance of microbial biofilm communities. In: Advances in Microbial Physiology, Vol. 46. pp 203–256. Elsevier Science Ltd.

Greenberg, E.P. (1999). Quorum sensing in gram negative bacteria: acylhomo serine lactone signalling and cell-cell communication. In: R. England, G. Hobbs, N. Bainton and Mcl. D. Roberts (eds) *Microbial signalling and communication,* pp. 71–84. Cambridge University Press, Cambridge.

Gu, J.D., Roman, M., Esselman, T. and Mitchell, R. (1998). The role of microbial biofilms in deterioration of space station candidate materials. *International Biodeterioration and Biodegradation* **41**, 25–33.

Hamilton, W.A. (1985). Sulphate reducing bacteria and anaerobic corrosion. *Ann. Rev. Microbiol.* **39**, 195–217.

Hauidier, K., Paquin, J.L., Francais, T. *et al.* (1988). Biofilm growth in a drinking water network: a preliminary industrial pilot plant experiment. *Wat. Sci. Technol.* **20**, 109–115.

Hassett, D.J., Ma, J.F., Elkins, J.G. *et al.* (1999). Quorum sensing in *Pseudomonas aeruginosa* controls expression of catalase and superoxide dismutase genes and mediates biofilm susceptibility to hydrogen peroxide. *Mol. Microbiol.* **34**, 1082–1093.

Heap, W.M. and Morrell, S.H. (1968). Microbiological deterioration of rubbers and plastics. *J. Appl. Chem.* **18**, 189–194.

Herson, D.S., McGoningle, B., Payer, M.A. and Baker, K.H. (1987). Attachment as a factor in the protection of *Enterobacter cloacae* from chlorination. *Appl. Env. Microbiol.* **53**, 1178–1180.

Holden, H.T., Ram Chhabra, S., de Nys, R. *et al.* (1999). Quorum sensing cross-talk isolation and chemical characterisation of cyclic dipeptides from *Pseudomonas aeruginosa* and other gram negative bacteria. *Mol. Microbiol* **33**, 1254–1266.

Holt, D. (1995). Challenge of controlling biofilms in water distribution systems. In:. *The Life and Death of Biofilm* (J. Wimpenny, P. Handley, P. Gilbert and Lappin-Scott, H.M. (eds) pp. 161–166. Contributions made at the Second Meeting of the British Biofilm Club, Gregynog Hall, Powys. 26–28 1995.

Jahn, A. and Nielsen, P.-H. (1998). Cell biomass and exopolymer composition in sewer biofilms. *Wat. Sci. Technol.* **37**, 17–24.

Jones, K. and Bradshaw, S.B. (1996). Biofilm formation by the Enterobacteriaceae: a comparison between *Salmonella enteriditis, Escherichia coli* and a nitrogen fixing strain of *Klebsiella pneumoniae. J. Appl. Bacteriol.* **80**, 458–464.

Kerr, C.J., Jones, C.R., Hillier, V.F. *et al.* (2000). Statistical evaluation of adhesion of a bioluminescent Pseudomonad to Thermanox in a Modified Robbins Device. *Biofouling.* **14**(4), 267–277.

Kerr, C.J. (1999). Biofilm formation on water distribution pipe materials in modal systems. PhD thesis, University of Manchester, Manchester.

Kerr, C.J., Osborn, K.S., Robson, G.D. and Handley, P.S. (1999). The relationship between pipe material and biofilm formation in a laboratory model system. *J. Appl. Microbiol.* **85**, 29S–38S.

Kolenbrander, P.E. (1988). Intergeneric coaggregation among human oral bacteria and ecology of dental plaque. *Ann. Rev. Microbiol.* **42**, 627–656.

Kolenbrander, P.E., Andersen, R.N., Kazmersak, K.M. and Palmer, R.J. Jr (2000). Co-aggregation and co-adhesion in oral biofilms. In: D.G. Allison, P. Gilbert, H.M. Lappin-Scott and M. Wilson (eds) *Community, structure and co-operation in biofilms,* pp. 65–86. Cambridge University Press, Cambridge.

LeChevallier, M.W. (1990). Coliform regrowth in drinking water: a review. *J. Am. Waterworks Assoc.* **82**(11), 74–86.

LeChevallier, M.W., Babcock, T.M. and Lee, R.G. (1987). Examination and characterisation of distribution system biofilms. *Appl. Environ. Microbiol.* **53**(12), 2714–2724.

LeChevallier, M.W., Cawthorn, C.D. and Lee, R.G. (1988). Factors promoting survival of bacteria in chlorinated water supplied. *Appl. Environ. Microbiol.* **54**(3), 649–654.

LeChevallier, M.W., Cawthorn, C.D. and Lee, R.G. (1988). Inactivation of biofilm bacteria. *Appl. Environ. Microbiol.* **54**(10), 2492–2499.

LeChevallier, M.W., Lowry, C.D. and Lee, R.G. (1990). Disinfecting biofilms in a model distribution system. *J. Am. Waterworks Assoc.* **82**(7), 87–99.

LeChevallier, M.W., Lowry, C.D., Lee, R.G. and Gibbon, D.L. (1993). Examining the relationship between iron corrosion and the disinfection of biofilm bacteria. *J. Am. Waterworks Assoc.* **85**(7), 111–123.

LeChevallier, M.W., Schulz, W. and Lee, R.G. (1991). Bacterial nutrients in drinking water. *Appl. Environ. Microbiol.* **57**(3), 857–862.

LeChevallier, M.W., Seidler, R.J. and Evans, T.M. (1980). Enumeration and characterisation of standard plate count bacteria in raw and chlorinated water supplied. *Appl. Environ. Microbiol.* **40**, 922–930.

LeChevallier, M.W., Welch, N.J. and Smith, D.B. (1996). Full-scale studies of factors related to coliform regrowth in drinking water. *Appl. Environ. Microbiol.* **62**(7), 2201–2211.

Levy, R.V., Hart, F.L. and Cheetham, R.D. (1986). Occurrence and public health significance of invertebrates in drinking water distribution systems. *J. Am. WaterWorks Assoc.* **78**, 105–110.

Macrae, I.C. and Edwards, J.F. (1972). Adsorption of colloidal iron to bacteria. *Appl. Microbiol.* **24**, 819–823.

Marsh, P.D. and Bowden, G.H.W. (2000). Microbial community interactions in biofilms. In: D.G. Allison, P. Gilbert, H.M. Lappin-Scott and M. Wilson (eds) *Community structure and co-operation in biofilms*, pp. 167–198. Cambridge University Press, Cambridge.

Marshall, K.C., Stout, R. and Mitchell, R. (1971). Mechanism of the initial events in the sorption of marine bacteria to surfaces. *J. Gen. Microbiol.* **68**, 337–348.

McBain, A.J., Allison, D.G. and Gilbert, P. (2000). Population dynamics in microbial biofilms. In: D.G. Allison, P. Gilbert, H.M. Lappin-Scott and M. Wilson (eds) *Community, structure and co-operation in biofilms*, pp. 257–278. Cambridge University Press, Cambridge.

McClean, R.J., Whiteley, M., Stickler, D.J. and Fugna, W.C. (1997). Evidence of autoinducer activity in naturally occurring biofilms. *FEMS Microbiol. Lett.* **154**, 259–263.

McCoy, W.F. and Costerton, J.W. (1982). Growth of sessile *Sphaerotilus natans* in a tubular recycle system. *Appl. Environ. Microbiol.* **43**(6), 1490–1494.

McMath, S.M., Chamberlain, A.H.L., Lloyd, B.J. and Holt, D.M. (1997). The natural history of biofilms in a model water distribution system. In: J. Wimpenny, P.S. Handley and P. Gilbert (eds) *Biofilms, Community interactions and Control*, pp. 167–174. Cardiff, Bioline.

Molin, S., Haggensen, J.A.J., Barken, K.B. and Sternberg, C. (2000). Microbial communities: aggregates of individuals or co-ordinated systems. In: D.G. Allison, P. Gilbert, H.M. Lappin-Scott and M. Wilson (eds) *Community, structure and co-operation in biofilms*, pp. 199–214. Cambridge University Press, UK.

Nagy, L.A., Kelly, A.J., Thun, M.A. and Olson, B.H. (1982). Biofilm composition, formation and control in the Los Angeles aquaduct system. *Proceedings of the water quality technology conference*, Nashville, Tenessee, AWWA, Denver.

Namkung, E. and Rittmann, B.E. (1987). Removal of taste and odour-causing compounds by biofilms grown on humid substances. *J. Am. Water Works Ass.* **79**, 107–112.

Nielsen, A.T., Tolker-Nielsen, T., Barken, K.B. and Molin, S. (2000). Role of commensal relationships in the spatial structure of a surface attached microbial consortium. *Environ. Microbiol.* **2**, 59–68.

Norton, C.D. and LeChevallier, M.W. (2000). A pilot study of bacteriological population changes through potable water treatment and distribution. *Appl. Env. Microbiol.* **66**, 268–276.

Olson, B.H. and Nagy, L.A. (1984). Microbiology of potable water. *Adv. Appl. Microbiol.* **30**, 73–132.

Payment, P., Coffin, E. and Paquette, G. (1994). Blood agar to detect virulence factors in tap water heterotrophic bacteria. *Appl. Env. Microbiol.* **60**, 1179–1183.

Payment, P., Franco, E. and Siemiatcki, J. (1993). Absence of relationship between health effects due to tap water consumption and drinking water quality parameters. *Wat. Sci. Technol.* **27**(3–4), 137–143.

Payment, P., Gamache, F. and Paquette, G. (1988). Microbiological and virological analysis of water from two water filtration plants and their distribution systems. *Can. J. Microbiol.* **34**, 1304–1309.

Payment, P., Siemiatycki, J., Richardson, L. *et al.* (1997). A prospective epidemiological study of gastrointestinal health effects due to consumption of drinking water. *Int. J. Env. Health Res.* **7**, 5–31.

Pedersen, K. (1990). Biofilm development on stainless steel and PVC surfaces in drinking water. *Wat. Res.* **24**, 239–243.

Pratt, L.A. and Kolter, R. (1998). Genetic analysis of *Escherichia coli* biofilm formation: roles of flagella, motility, chemotaxis and type I pili. *Mol. Microbiol.* **30**, 285–293.

Prigent-Combaret, C., Vidal, O., Dorel, C. and Lejeune, P. (1999). Abiotic surface sensing and biofilm dependent regulation of gene expression in *E. coli*. *J. Bacteriol.* **181**, 5993–6002.

Quirynen, M., Marechal, M., Busscher, H.J. *et al.* (1989). The influence of surface free energy and surface roughness on early plaque formation. *J. Clin. Periodontol.* **16**, 138–144.

Reasoner, D.J., Blannon, J.C., Geldreich, E.E. and Barnick, J. (1989). Nonphotosynthetic pigmented bacteria in a potable water treatment and distribution system. *Appl. Environ. Microbiol.* **55**(4), 912–921.

Rickard, A.H., Leach, S.A., Buswell, C.M., High, N.J. and Handley, P.S. (2000). Co-aggregation between aquatic bacteria is mediated by specific growth phase dependent lectin-saccharide interactions. *Appl. Environ. Microbiol.* **66**, 431–434.

Rickard, A.H., Leach, S.A., Hall, L.S. *et al.* (2002). Phylogenetic relationships and coaggregation ability of freshwater biofilm bacteria. *Appl. Environ. Microbiol.* **68**, 3644–3650.

Rickard, A.H., Gilbert, P., High, N. *et al.* (2003). Bacterial coaggregation: an integral process in the development of multi-species biofilms. *TRENDS Microbiol.* **11**(2), 94–100

Ridgway, H.F. and Olson, B.H. (1981). Scanning electron microscope evidence for bacterial colonisation of a drinking water distribution system. *Appl. Environ. Microbiol.* **41**(1), 274–287.

Ridgway, H.F. and Olson, B.H. (1982). Chlorine resistance patterns of bacteria from two drinking water distribution systems. *Appl. Env. Microbiol.* **44**, 972–987.

Rogers, J., Dowsett, A.B., Dennis, P.J., Lee, J.V. and Keevil, C.W. (1994). Influence of plumbing materials on biofilm formation and growth of *Legionella pneumophila* in potable water systems. *Appl. Environ. Microbiol.* **60**(6), 1842–1851.

Seidler, R.J., Morrow, J.E. and Bagley, S.T. (1977). Klebseillae in drinking water emanating from redwood tanks. *Appl. Environ. Microbiol.* **33**, 893–900.

Servais, P., Billen, G., Ventresque, C. and Bablon, G.P. (1991). Microbial activity in GAC filters at the Choisy-Le-Roi treatment plant. *J. Am. WaterWorks Assoc.* **83**(2), 62–68.

Smith-Somerville, H.E., Huryn, V.B., Walker, C. and Winters, A.L. (1991). Survival of *Legionella pneumophila* in the cold water ciliate *Tetrahymena vorax*. *Appl. Environ. Microbiol.* **57**, 2742–2749.

Stoodley, P., de Beer, D., Boyle, J.D. and Lappin-Scott, H.M. (1999a). Evolving perspectives of biofilm structure. *Biofouling* **14**, 75–94.

Stoodley, P., Dodds, I., Boyle, J.D. and Lappin-Scott, H.M. (1999b). Influence of hydrodynamics and nutrients on biofilm structure. *J. Appl. Microbiol* **85**, 19S–28S.

Stoodley, P., Hall-Stoodley, L. *et al.* (2000). Environmental and genetic factors influencing biofilm structure. In: D.G. Allison, P. Gilbert, H.M. Lappin-Scott, M. Wilson *et al.* (eds) *Community structure and co-operation in biofilms*, pp. 53–64. Cambridge University Press, Cambridge.

Sutherland, I.W. (1997). Microbial biofilm exopolysaccharidges – superglues or velcro?. In: J. Wimpenny, P.S. Handley and P. Gilbert (eds) *Biofilms: community interactions and control*, pp. 33–39. Bioline, Cardiff.

Sutherland, I.W. (2001). Biofilm exopolysaccharides: a strong and sticky framework. *Microbiol.* **147**, 3–9.

Tatnall, R.E. (1991). Case histories: biocorrosion. In: H.C. Flemming and G.G. Geesey (eds) *Biofouling and Biocorrosion in Industrial Water Systems, Proceedings of the International Workshop on Industrial Biofouling and Biocorrosion, September 13–14, 1990*, pp. 167–185. Springer Verlag, Berlin.

UKWIR (1998). Health significance of bacteria in distribution systems – review of literature for 1995–1997. Report No. 98/DW/02/13.

van der Kooij, D. and Veenendaal, H.R. (1994). Biological activity on surfaces in drinking water distribution systems. *Water Supply* **12**(1–2), SS1-1–SS1-7.

van der Kooij, D. (1991). Nutritional requirements of aeromonads and their multiplication in drinking water. *Experimentia* **47**, 444–446.

van der Kooij, D. (1987). The effect of treatment on assimilable organic carbon in drinking water. In: P. Toft (ed.) *Proceedings of the 2nd National conference on drinking water*, Edmonton, Alberta, Canada, 7 and 8 April, 1986. Pergamon Press, London.

van der Wende, E. and Characklis, W.G. (1990). Biofilms in potable water distribution systems. In: G.A. McFeters (ed.) *Drinking Water Microbiology*, pp. 249–268. Springer International, New York.

Vess, R.W., Anderson, R.L., Carr, J.H., Bond, W.W. and Favero, M.S. (1993). The colonisation of solid PVC surfaces and the acquisition of resistance to germicides by water micro-organisms. *J. Appl. Bacteriol.* **74**, 215–221.

Volk, C.J., Volk, C.B. and Kaplan, L.A. (1997). Chemical composition of biodegradable dissolved organic matter in stream water. *Lim. Ocean.* **42**, 39–44.

Walker, J.T., Sonesson, A., Keevil, C.W. and White, D.C. (1993). Detection of *Legionella pneumophila* in biofilms containing a complex microbial consortium by gas chromatography-mass spectrometry analysis of genus specific hydroxy fatty acids. *FEMS Microbiol. Lett.* **113**, 139–144.

Waters, S.K. and McFeters, G.A. (1990). Reactivation of injured bacteria. In: *Assessing and Controlling Bacterial Regrowth in Distribution Systems*. American Waterworks Association and American Water Works Association research foundation, Denver.

Woolfe, R.L., Means, E.G., Davis, M.K. and Barrett, S.E. (1988). Biological nitrification in covered reservoirs containing chloraminated water. *J. Am. Water Works Assoc.* **80**, 109–114.

Zacheus, O.M. and Martikainen, P.J. (1995). Occurrence of heterotrophic bacteria and fungi in cold and hot water distribution systems using water of different quality. *Can. J. Microbiol.* **41**, 1088–1094.

42

Taste and odour problems in potable water

Esther Ortenberg* and Benjamin Telsch†

*Laboratory of Water Chemistry, Israel Ministry of Health, Tel-Aviv, 61082; †Nessin Water Quality Central Laboratory, Mekorot Water Co, Nazareth Illit 17105, Israel

1 INTRODUCTION

Water occupies the greater part of the Earth. It is the source of life for vegetation and animal worlds. However, the quantity of potable water is limited, as only water suitable for the human organism, causing no harm to its health, can be used for drinking purposes. Requirements for potable water quality are given in the corresponding documentation (Federal Register, 1995).

One of the basic requirements to potable water concerns its organoleptic properties.

1.1 Organoleptic properties of potable water

Pure water means a neutral liquid without colour, odour or taste. The presence of odour and taste (or one of them) serves as a dangerous signal for potable water consumers including the beverage industry, foodstuffs, and the aromatic and pharmaceutical industries.

Odour is the result of stimulating substances interacting with corresponding receptors (nasal receptors). This stimulation is chemical and is described by a term, chemical susceptibility, used when determining the organoleptic properties of potable water.

Taste, flavour and bouquet represent a complex of gustatory, flavouring and smelling senses which are also the result of a chemical action on sensory receptors found in the lingual papillae, soft palate and nasal cavity.

There are four kinds of gustatory senses: bitterness, sweetness, salinity and acidity.

When determining organoleptic properties of potable water, the presence of smell is determined first of all. Only pure water can be tasted. In reality, a lot of compounds produce smell before reaching the mouth. Nasal receptors sense concentrations far below those determined by analytical methods. That is why, for qualitative determination of organoleptic properties of potable water, the best of the devices – the human nose – is used.

Odour tests are performed to provide qualitative descriptions and approximate quantitative measurements of odour intensity. The method for odour intensity described here is the threshold odour test, based on a method of limits. This procedure, while not universally preferred (Mallevialle and Suffet, 1987), has definite strength.

Sensory tests are useful for checking on the quality of raw and finished water and for control of odour through the treatment processes. They can assess the effectiveness of different treatments and provide a means for tracing the source of contamination.

2 METHODS FOR DETERMINING POTABLE WATER ODOUR AND TASTE

2.1 Threshold odour test (TOT) (Standard Methods, 1998)

The threshold odour test is used for odour characterization in natural and potable water.

The Handbook of Water and Wastewater Microbiology
ISBN 0-12-470100-0

The principle of the test is the determination of the least concentration of the component possessing odour in the water studied by diluting the water sample with odourless water.

The method is very simple to perform but it is not sufficiently accurate enough when determining concentration. There is no absolute threshold odour concentration because of inherent variations in individual olfactory capability. From 5 to 10 experts take part when conducting the threshold odour test. Experiments on odour determination are carried out in a room free from odour and well ventilated by air filtered through activated carbon. Before determining the threshold odour, samples under study are dechlorinated by means of thiosulphate and heated up to 60°C. Chemical glassware should be odour-free.

Interpretation of the single tester result requires knowledge of the relative activity of that person. Some investigators have used specific odorants (odour reference standard) to calibrate a tester's response (Krasner, 1995).

Measurements of the threshold odour are carried out by using a conventional unit, TON (threshold odour number). The threshold odour number is the greatest dilution of a sample with odour-free water yielding a definitely perceptible odour. The TON is computed in the following way:

$$TON = \frac{A + B}{A}$$

where A is the volume sample (ml) and B is the volume of odour-free water (ml).

Determination is carried out in the following manner: first the approximate TON level is determined by using three to four dilutions, then the TON value is improved by repeating the procedure but with smaller volumes of the sample under study. At this stage two blanks (odour-free water) and two to three odour reference standards are used. Having determined the least concentration still giving the smell, this concentration is improved by studying samples with lower or higher concentrations of the odour-bearing component in water.

A threshold number is not a precise value. In the case of a single observer it represents a judgement at the time of testing. Panel results are more meaningful because individual differences have less influence on the result. One or two observers can develop useful data if comparison with larger panels has been made to check their sensitivity. Comparisons of data from time to time or place to place are not advisable unless all test conditions have been standardized carefully and there is some basis for the comparison of observed intensities. The threshold odour test, when using TON, is one of the numerous methods of arranging and presenting samples for odour determinations. The methods offered here are practical and economical of time and personnel. If extensive tests are planned and statistical analysis of data is required, more accurate methods like FTT are performed.

2.2 Flavour threshold test (FTT) (Standard Methods, 1998)

The FTT has been used extensively and is particularly useful for determining if the overall flavour of a sample of finished water is detectably different from a defined standard.

FTT is used to measure detectable flavour quantitatively. More precisely, the method is used to compare sample flavour objectively with that of specified reference water used as diluent.

The flavour threshold number (FTN) is the greatest dilution of sample with reference water yielding a definitely perceptible difference. The FTN is computed as follows:

$$FTN = \frac{A + B}{A}$$

where A is the sample volume (ml) and B is the reference water (diluent) volume (ml).

Performance of FTT determination is analogous to that of TOT.

2.3 Flavour profile analysis (FPA) (Standard Methods, 1998)

Flavour profile analysis is a technique for identifying sample taste(s) and odour(s). FPA differs from threshold odour number because the sample is not diluted and each taste or odour attribute is individually characterized and assigned its own intensity rating. The single numerical rating obtained in measuring threshold odour is controlled by the most readily perceived odorant or mixture. Sample dilution may change the odour attribute that is measured (Mallevialle and Suffet, 1987). FPA determines the strength or intensity of each perceived taste or odour without dilution or treatment of the sample.

Flavour profile analysis has been applied to drinking water sources, and finished drinking water.

Flavour profile analysis uses a group of four or five trained panellists to examine the sensory characteristics of samples. Flavour attributes are determined by tasting. Odour attributes are determined by sniffing the sample. The method allows more than one flavour, odour attribute, or feeling factor (e.g. drying, burning) to be determined per sample and each attribute's strength to be measured. Flavour profile analysis requires well-trained panellists and data interpreters. Reproducibility of results depends on the training and experience of the panellists.

Initially, panellists record their perceptions without discussion. Once each individual has done an independent assessment of a sample, the panel discusses its findings and reaches a consensus.

Background odours present during analysis affect results. Analyst illness, e.g. cold or allergy, can diminish or otherwise alter perception.

Detailed description of the experiment according to the given method, which is the most important one for determination and identification of potable water taste and odour, is given in Standard Methods (1998).

Identified specific compounds and groups of compounds that produce tastes and odours in drinking water use a new sensory panel flavour profile analysis coupled with gas chromatography (GC)-mass spectrometry (MS) analysis (Giger *et al.*, 1976). Correlations between flavour profile analysis and instrumental analysis of taste-and-odour-producing compounds have been developed. The practical application of the flavour profile analysis technique uses odour reference standards and cross-panel comparisons.

3 TASTE AND ODOUR PROBLEMS IN POTABLE WATER

3.1 Sources of potable water contamination

3.1.1 Biological sources of taste and odour in water

Most of the concerns over taste and odour caused by biological activity are associated with surface water, especially reservoirs, lakes and rivers, rather than groundwater. The organisms most often linked to taste and odour problems are the Actinomycetes and various types of algae, but other aquatic organisms, such as sulphate-reducing bacteria, zooplankton, protozoa and fungi, have been identified from time to time as taste and odour contributors.

Actinomycetes. The Actinomycetes are a group of unicellular filamentous bacteria that form a branching network of filaments and produce spores. They have long been recognized as sources of severe earthy-musty tastes and odours in drinking water (Mallevialle and Suffet, 1987). Gerber and LeChevallier (1965) were the first to isolate and to identify the terpenoide geosmin, *trans*-1,10-dimethyl-*trans*-9-decalol, a potent earthy-smelling compound, from Actinomycetes cultures. Gerber (1969) isolated another terpenoide, a musty-smelling compound named 2-methylisoborneol (MIB), from Actinomycetes. Shortly after, Medsker *et al.* (1969) and Rosen *et al.* (1970) reported independently on this compound. Both geosmin and MIB occur in lakes, reservoirs and rivers (Ridal *et al.*, 1999) where geosmin is more likely to be produced in the water column and MIB is more often produced in sediments (Slater and Block, 1983). They possess musty, earthy odours at very low thresholds, 4 ng/l for geosmin and 15 ng/l for MIB (Gagne *et al.*, 1999).

Studies by Gerber and others have shown that Actinomycetes produce an additional variety of odorous compounds, such as cadin-4-ene-1-ol and 2-isopropyl-3-methoxypyrazine (IPMP) (Gerber, 1971, 1983; Suffet *et al.*, 1999). Pure cultures of Actinomycetes grown by Silvey and Roach (1964) were reported to produce a variety of odours, including earthy, woody, musty, potato-bin-like, hay-like, fishy, and grassy.

The extent to which Actinomycetes actually cause taste and odour problems in water supplies is sometimes difficult to assess. The reasons for this are that they are difficult to culture and that many of the colonies on actinomycete agar plates may originate from spores, which are not involved in odour production (Persson and Sivonen, 1979). Additionally, not all Actinomycetes that can be recovered from lake water or sediments are odour producers. For example, most species of *Micromonospora*, which are readily isolated from lake sediments at the bottom and around shoreline, are not odour producers (Izaguirre and Devall, 1995). The genus primarily involved in odour production is *Streptomyces* (Gerber, 1983).

Actinomycete data for water bodies have to be interpreted with caution and with understanding of organism biology. The causal relationship between Actinomycetes and taste and odour problems is thus presumptive until confirmed (Izaguirre and Devall, 1995). Mallevialle and Suffet (1987) reviewed odour production by *Actinomycetes*.

Algae. Algae are an informal grouping of many simple photosynthetic organisms, most of them within the kingdom Protista, but some of them belong to other taxa (e.g. the prokaryotes cyanobacteria). Algae are given the optimal chance to grow in reservoirs, lakes and rivers and to form algal blooms in the summer months due to excessive sunlight and also when nutrient loading is high. In this case, and under these conditions, taste and odours can often be found in finished water. The importance of algae as contributors of taste and odours in water supplies has long been recognized. Palmer (1959) listed many algal species that caused a variety of tastes and odours in drinking water. More recently, a study of algal problems in the USA and Canada (Casitas Municipal Water District, 1987) reported that of 174 utilities responding to a questionnaire regarding odour problems with algae, 45% had problems with cyanobacteria, 21% with diatoms, 20% with green algae, and 14% with pigmented flagellates.

Cyanobacteria (blue-green algae). The most common cyanobacteria offenders associated with taste and odour episodes are various species of the genera *Oscillatoria*, *Anabaena*, *Aphanizomenon* and *Phormidium*. Numerous *Oscillatoria* and *Anabaena* species produce geosmin and MIB. However, not all species of these genera produce the compounds. Geosmin was first discovered as product of cyanobacteria by Safferman *et al.* (1967) and MIB by Tabachek and Yurkowski (1976). In the last two decades, many new cyanobacteria species have been shown to produce geosmin and MIB (e.g. Izaguirre *et al.*, 1982; Izaguirre, 1992; Hosaka *et al.*, 1995). *Pseudanabaena* was recently linked to off-flavour episodes involving MIB occurring in southern California reservoirs (Izaguirre and Taylor, 1998; Izaguirre *et al.*, 1999).

Certain filamentous cyanobacteria, notably *Oscillatoria* species, form mats on rocks and bottom sediments in reservoirs. Tabachek and Yurkowski (1976) reported the production of strong earthy odours by attached cyanobacteria in Canadian lakes even during the winter. Leventer and Eren (1970) concluded that *Oscillatoria chalybea* is the principal causative agent of the earthy-musty taste and odour episodes in reservoirs of the Israel National Water System, at times when planktonic algal densities were quite low. The major cause of musty off-flavour in fresh water-raised fish is MIB produced by *O. chalybea* (Schrader *et al.*, 1998).

The relationship between the maximum standing crop of musty-odour producing algae and nutrient concentration in the southern basin of Lake Biwa, Japan, was studied by Nakanishi *et al.* (1999) and it was found that the maximum growth number of the algae was limited by soluble phosphorus concentration.

The ability to produce geosmin or MIB seems to be a strain-specific property. Some isolates of *Oscillatoria agardhii* produce geosmin in fresh and brackish water, and some do not (Persson, 1980; Berglind *et al.*, 1983). This strain specificity of odorant production by cyanobacteria partly explains the inconsistency in the literature regarding the odorous products of various species.

Chrysophyceae (Class Bacillariophyceae: diatoms). The genera of diatoms most often mentioned as taste and odour producers include *Asterionella* (geranium and fishy odours), *Cyclotella* (grassy, geranium and fishy odours), *Tabellaria* (grassy, geranium and fishy odours), and *Melosira* (grassy, geranium and musty odours). For the most part, the chemicals responsible for these odours from diatoms have not been identified; however, Kikuchi *et al.* (1974) identified *n*-hexanal and *n*-heptanal in cultures of *Syndera rumpens*.

Chrysophyceae (several classes). Some of the most notorious odour-producing chrysophytes are *Dinobryon*, *Synura*, *Uroglena* and *Uroglenopsis*. *Dinobryon* development in Lake Michigan imparts a prominent fishy odour to the water and produced a TON as high as 13 (Palmer and Poston, 1956). An offensive cod liver oil-like odour from a bloom of *Synura uvella* was attributed to aldehydes (Juttner, 1981; Yano *et al.*, 1988).

Chlorophyceae (green algae). This group of algae grows at its highest levels in the summer, and causes odours reported to be grassy or, sometimes, fishy when the algae are very abundant. The causes of grassy-type odours in natural and drinking water have been identified as *cis*-3-hexan-1-ol and *cis*-3-hexenyl acetate (Khiari *et al.*, 1999).

Dead algae can cause tastes and odours in two ways. First, when algae die and lyse, odorous compounds are liberated. Second, the dead biomass can support bacterial growth, including that of actinomycetes, which produce odorous compounds (MacKenthun and Keup, 1970). Decomposing cyanobacteria can generate a variety of odorous sulphur compounds, including mercaptans and dimethylsulfide and dimethyltrisulfide (Slater and Block, 1983). Obligatory aerobic bacteria feeding on the dinoflagellate algae *Peridinuim gatunense* or their lysis products in Lake Kinneret, Israel, form dimethyloligosulphides that are responsible for malodour episodes in the lake (Ginzburg *et al.*, 1999a).

3.1.2 Other biota

Bacteria. Sulphate-reducing bacteria (e.g. *Desulfovibrio* spp. and *Desulfotomaculum* spp.) create taste and odour problems primarily in deep groundwater, where they release hydrogen sulphide (Lin, 1976). These bacteria also generate sulphide odours in the anaerobic hypolimnia of thermally stratified lakes and reservoirs. 'Swampy' odours produced by other bacteria, notably *Pseudomonas* and *Aeromonas*, have been described in Australian distribution systems (Wajon *et al.*, 1985). This phenomenon is caused by highly odorous organic sulphides, such as dimethylpolysulphide, under anaerobic conditions (Whitfield and Freeman, 1983). Ginzburg *et al.* (1999b), showed formation of inorganic oligosulphides, which affect the odour quality of some drinking and recreational water systems, by bacterial decomposition of sulphur-containing organic matter under aerobic conditions. Ferrobacteria in water distribution systems may produce taste and odours (MacKenthun and Keup, 1970).

Fungi. It has been stated that odour-producing fungi live in all habitats known to be potential sources of earthy taints in potable water (Wood *et al.*, 1983).

Zooplankton. Some zooplankton, such as the crustaceans *Cyclops* and *Daphnia* and the rotifer *Keratella*, can cause odour problems when they are present in large numbers (Lin, 1976).

Nematodes. Nematodes can secrete odorous compounds as well; an oily compound giving an earthy and musty smell was isolated from nematode cultures (Chang *et al.*, 1960).

Protozoa. Some protozoans (amoebae) have been implicated as odour and taste producers (Chang *et al.*, 1960).

3.1.3 Chemical sources of taste and odour in water

Sources of potable water contamination by organic and inorganic compounds, the majority of which offer taste and smell, can be divided into several groups:

(a) Groundwater. Naturally occurring tastes and odours in groundwater, caused either by chemicals contributed to the aquifer by the geological formation or by biological activity that is simulated by these chemicals. These compounds are most often sulphides, oligosulphides and polysulphides, including the simple hydrogen sulphides which are of either an organic or mineral source (Campbell *et al.*, 1994). The formation of these compounds is generally enhanced under anoxic conditions which are quite common in deep groundwater wells.

(b) Industrial and municipal wastewaters. Entrance of industrial and municipal wastewaters into rivers and lakes can serve as one of the sources of potable water contamination. Organoleptic compounds causing organoleptic problems have been found in potable water: alkylbenzenes, chlorobenzenes, alkanes, benzaldehyde and benzothiazole (Pontius, 1998).

(c) Taste and odour-causing compounds obtained during storage and distribution of the water. These compounds may be different or similar to the compounds from the groups mentioned above. New compounds can be added due to leaching of water-pipe linings or storage facilities. In addition, microbial metabolites can be added to treated water during storage in tanks.

3.1.3.4 (d) Disinfectants and the by-products of potable water treatment. When treating raw water for contamination by microbes and for removal taste and odour, disinfectants (oxidants) are used (chlorine, chloramine, chlorine dioxide etc.).

Residual quantities of oxidants, their by-products and other compounds from different sources formed impart taste and odour to water (Table 42.1).

3.2 Organoleptic problems of oxidants and their by-products in water sources treatment

3.2.1 Principal scheme of water sources treatment methods

To obtain potable water, river, lake or groundwater is treated according to a definite technology including a number of standard stages. Depending on the initial composition of water contaminants and presence of taste and odour, the application of the given stages can be varied.

The main stages of water treatment are:

1 Coagulation (flocculation). Preliminary treatment of raw water consists of coagulation. The coagulation method is based on reaction of a soluble salt (alum solution, $Al_2(SO_4)_3$ with a base (NaOH, $Ca(OH)_2$) forming non-soluble base ($Al(OH)_3$). When settling the aluminium base ($Al(OH)_3$) from water, a dispersed phase contaminating water is deposited together with it. The deposit partly adsorbs substances with obnoxious odour, i.e. organic substances with high molecular weight. Thus, coagulation improves water transparency and partly its organoleptic properties and prepares water for further treatment (filtration, adsorption).

2 Filtration. Filtering through sand screens or other media permits the separation of solid particles from water which are then deposited on the screen, thus improving water transparency and its organoleptic properties.

3 Aeration. Aeration is the simplest and cheapest method of treating water from volatile compounds contaminating raw water and imparting taste and odour to it. Aeration has been used to remove hydrogen sulphide, methane, radon, iron, manganese and volatile organic contaminants from drinking water. Aeration also removes carbon dioxide, which directly affects

TABLE 42.1 Odour characteristics of compounds causes in water sources

Compound	*Odour characteristics*	*Concentration (mg/l)*
Dimethyl sulphide	Decaying vegetation, canned corn	0.00005
Dimethyl disulphide	Septic	0.01
Dimethyltri-sulphide	Garlicky, oniony, septic	0.00005
Butyric acid	Putrid, sickening	1.0
Diphenyl ether	Geranium	0.1
D-Limonene	Citrusy	2.0
Hexanal	Lettuce heart, pumpkin, green pistachio	0.2
Benzaldehyde	Sweet almond	1.0
Ethyl-2-methyl-butyrate	Fruity, pineapple	5.0
2-Heptanone		
Hexachloro-1,3-	Banana-like, sweet solventy	0.5
Butadiene	Sweet, minty, Vapo Rub	4.0
Butanol		
Pyridine	Alcohol, solventy	1.0
Chlorine	Sweet, alcohol,	2.0
Dichloramine	Chlorinous	0.5
Monochloramine	Swimming pool chlorine	Ph 7
Toluene	Chlorinous	Ph 7
Indene	Glue, sweet solventy	0.5
Naphthalene	Glue	0.005
Benzofuran	Sweet solventy	0.005
cis-3-Hexene-1-ol	Shoe polish	0.01
	Grassy, green apple	0.5

Adapted from American Water Works Association (1993) Flavour Profile Analysis: Screening and Training of Panelists. AWWA Manual, American Water Works Assoc., Denver, Colo. (*Standard Methods*, 1998).

pH and dissolved inorganic carbon, the parameters that most influence lead and copper solubility. Aeration produces very consistent water quality and may be advantageous (especially to smaller utilities) because of its relatively low costs and simple operational and maintenance needs (Lytle *et al.*, 1998). The main disadvantage of this method lies in the fact that volatile compounds are sent into the atmosphere contaminating the environment and can reappear in raw water.

4 Removal of odour and taste by means of oxidants. Oxidants are more often used for purification of water from taste and odour. The mechanism of taste and odour removal consists of destruction of the organic molecule imparting definite taste and odour to the water. Application of chlorine leads to breakage in organoleptic compound ester bonds which are responsible for odour presence as in the case of raw water rich in algae metabolites – pyrazine derivatives (2-isopropyl-3-methoxy pyrazine, IPMP, and 2-isobutyl-3-methoxy pyrazine, IBMP) (Lalezary *et al.*, 1986). Halogen (chlorine, chloramine, iodine and bromine) application also leads to production of halogen-derivative organic compounds reacting with halogens. When the by-products are odourless, improvement in organoleptic properties occurs. When the by-product possesses odour, water odour is intensified, e.g. formation of chlorophenol intensifies odour (Burttschell *et al.*, 1959).

The mechanism of raw water ozonation consists of destruction of inorganic and large organic compounds, algae metabolites, forming unstable aldegids and organic acids, oxidizing water.

The mechanism of water treatment for taste and odour by chlorine dioxide consists of complete destruction of organic molecules to carbon dioxide and water.

5 Adsorption. To decrease raw water taste and odour, adsorbents are often used: powdered activated carbon (PAC) and granulated activated carbon (GAC). The mechanism of odour and taste removal consists of organic molecule adsorption by a coal porous structure consisting of a system of micro-, meso- and macropores.

Adsorption treatment is rather effective. Odour and taste removal, in relation to adsorption capacity of activated carbon, occurs for 85–100%. However, in the course of operation the adsorptive capacity of carbon decreases and its ability to eliminate odour and taste becomes less effective; it is necessary to change the carbon for a new, fresh portion, thus considerably increasing the cost of adsorption treatment. The choice of oxidant or adsorbent is a difficult task. Effectiveness in potable water treatment is provided by the choice of an oxidant, oxidation mechanism of organic and mineral compounds found in initial water, and the formation of by-products during treatment. By-products are often toxic and harmful for the potable water consumer's health (Blumer and Youngblood, 1975; Pontius, 1998).

6 *Reverse osmosis method*. When using sea and ocean water as sources of potable water, treatment technology includes effective but expensive treatment for mineral salts possessing taste and odour – reverse osmosis method.

3.2.2 *Oxidants and their by-products*

The following are the main oxidants used when treating potable water:

1. Chlorine (Cl_2)
2. Chlorine dioxide (ClO_2)
3. Chloramine (NH_2Cl)
4. Bromine (Br_2)
5. Iodine (I_2)
6. Ozone (O_3)
7. Potassium permanganate ($KMnO_4$)
8. Manganese dioxide (MnO_2).

Chlorine. Chlorine is one of the most widely used water disinfectants. Hydrochlorous acid, HOCl, is an oxidizing agent and represents the product of free chlorine hydrolysis in accordance with equation (1)

$$Cl_2 + H_2O \rightarrow HOCl + HCl \quad (1)$$

Sodium hypochlorite, also serving as an oxidizing agent, is formed in alkaline pH:

$$Cl_2 + NaOH \rightarrow NaOCl + NaCl \quad (2)$$

The OCl^- ion is unstable and is immediately decomposed into atomic oxygen and chlorine radical thus providing for the oxidizing and chlorinating ability of chlorine:

$$OCl^- \rightarrow O + Cl^0 + \bar{e} \quad (3)$$

Good results are attained by applying chlorine for the removal of odour and taste from raw water rich in algae metabolites, such as IBMP and IPMP (Lalezary *et al.*, 1986).

By-products obtained with water chlorination are shown in Table 42.2.

They present the main problems with the treatment. Chlorinated by-products of methane derivatives CH_4 (chloroform, $CHCl_3$; dichloromethane, CH_2Cl_2) are especially known for their carcinogenic properties (for animals).

Rook (1974) and Bellar *et al.* (1974) have shown the presence of chloroform and other trihalogenated derivatives of methane in finished drinking water after chlorination of raw waters during treatment.

Compounds with the general formula CH_3CHOHR or CH_3COR can participate in haloform reactions, as may olefinic substances with the general formula $CH_3CH = CR_1R_2$, which will be oxidized by hypochlorous acid (HOCl), first to secondary alcohols and then to methylketones. The final step involves a hydrolytic cleavage of the trihalogenated carbon to form the THM (Morris, 1975).

The second important problem is the presence of phenol compounds in initial water. Phenolic substance that are present in natural waters are highly varied. The primary concern with phenolic compounds in drinking water supplies has been the offensive tastes that result from their reaction with hypochlorous acid.

Chlorine dioxide. Chlorine dioxide (ClO_2) is an orange-yellow gas with a liquefaction temperature of 9.7°C at atmospheric pressure. It is explosive in either its gaseous or pure liquid state. Dilute aqueous solutions are not explosive.

TABLE 42.2 By-products found in potable water after treatment with chlorine

Water source	*By-products*	*Properties of by-products*	*Reference*
Mississippi River	Chlorophenol, 4-chlororesorcinol $[ClC_6H_3\text{-}1.3\text{-}(OH)_2$	Potentially toxic compounds	Gehrs and Southworth, 1978
Lake Zurich	A-Chloroketones: 2,2-Dichlorobutanone $(CH_3CCl_2COCH_3)$ 2,2-Dichloropentan-3-one $(CH_3CCl_2COC_2H_5)$ 1,1,1-trichloroacetone 3,3-dichlorohexane-4-one $(C_2\ H_5CCl_2COC_2H_5)$	Potentially toxic compounds	Giger *et al.*, 1976
Finished drinking waters	p-Dichlorobenzene $C_6H_4Cl_2$, 1,2-Dichloroethane $(C_2H_4Cl_2)$	Liver, kidney, nervous system effects	Pontius, 1998
Blue Plain pilot plant near Washington	Chlorohexane $C_6H_{11}Cl$ Tetrachloroethane $(C_2H_2Cl_4)$, Pentachloroethane (C_2HCl_5) Hexachlorethane (C_2Cl_6)	Liver, kidney, nervous system effects	Garrison *et al.*, 1976 Pontius, 1998
Finished drinking waters at 20 supplies, USA	Toluene Styrene Xylenes	Liver, kidney, nervous system effects	Sievers *et al.*, 1978 Pontius, 1998
Chlorinated city water supplies	Trichloracetaldehyde (CCl_3CHO) (chloral) Chloral hydrate $(CCl_3CH(OH)_2$	Potentially toxic compounds, highly toxic hypnotic drug	Keith *et al.*, 1976
Source water, USA	Chloroform $(CHCl_3)$	Potential carcinogen	Bellar *et al.*, 1974 Rook, 1974
Ohio River	Trihalomethanes (THM)	Toxic	Bellar *et al.*, 1974 Rook, 1974
Finished potable water	Polynuclear aromatic hydrocarbons (PAH's)	Potent toxins, mutagens, teratogens	Harrison *et al.*, 1975; Blumer and Youngblood, 1975
Lake Baical, Siberia	Haloacetic acids HAAs (monochloroacetic acid, dichloroacetic acid, trichloroacetic acid)	Toxic	Berg *et al.*, 2000
Canadian Lake	Haloacetic acids HAAs	Toxic	Scott *et al.*, 2000 Pomes *et al.*, 2000

Two principal methods are used for the preparation of aqueous solutions of chlorine dioxide for water treatment (Masschelein, 1967; Gall, 1978). The first is the reaction of sodium chlorite ($NaClO_2$) and chlorine (Cl_2) in acidic aqueous solution, primarily by the following equation:

$$HOCl+2ClO_2^-+H^+\rightarrow 2ClO_2+Cl^-+H_2O \quad (4)$$

The other method of preparing chlorine dioxide for use in water treatment is acidification of strong sodium chlorite solution, usually with hydrochloric acid (HCl). The major reaction is:

$$5ClO_2^-+4H^+\rightarrow 4ClO_2+Cl^-+2H_2O \quad (5)$$

Chlorine or hypochlorous acid (HOCl) may be a product of decomposition of chlorous acid ($HClO_2$) under some conditions, however, the method does not assure freedom from contamination by aqueous chlorine (Feuss, 1964). Failure to observe such precautions casts doubts on a number of reported findings of chlorinated products from the reaction of aqueous chlorine dioxide with organic matter.

For industrial use where much greater amounts are used than in water treatment, chlorine dioxide is prepared by reduction of sodium chlorate ($NaClO_3$) with agents such as sulphur dioxide (SO_2), methanol (CH_3OH), or chloride (Cl^-) generally in several molar sulphuric acid (H_2SO_4) solutions (Gall, 1978; Sussman and Rauh, 1978). These processes are more economical than those based on sodium chlorite for the manufacture of large quantities of chlorine dioxide.

The principal side reactions occurring during the preparation of chlorine dioxide solutions are:

$$4HClO_2 \rightarrow 2ClO_2 + ClO_3^- + Cl^- + 2H^+ + H_2O \quad (6)$$

$$HClO_2 + HOCl \rightarrow 2H^+ + ClO_3^- + Cl^- \quad (7)$$

Gordon *et al.* (1972) have shown that as a result of obtaining ClO_2 together with the target product (ClO_2), the intermediate product Cl_2O_2 is present in the water solution and is decomposed according to the following scheme:

$$2Cl_2O_2 \rightarrow Cl_2 + 2ClO_2 \quad (8)$$

Chlorine formed as the result of the reaction gives chlorite.

In neutral and weakly acidic pH, diluted solutions of ClO_2 are stable. With alkaline pH decomposing ClO_2 gives two salts: chlorites and chlorates in accordance with the reaction (Gordon and Feldman, 1964):

$$2ClO_2 + 2OH^- \rightarrow H_2O + ClO_2^- + ClO_3^- \quad (9)$$

In his work, Bray (1906) has shown the second way of obtaining chlorates in water (neutral pH):

$$6ClO_2 + 3H_2O \rightarrow 6H^+ + 5ClO_3^- + Cl^- \quad (10)$$

Chlorine dioxide is also unstable under light (Gordon *et al.*, 1972)

$$ClO_2^{h\nu} \rightarrow ClO^- + O \quad (11)$$

Thus, hydrogen peroxide (H_2O_2), chlorite, chlorate, Cl_2O_2, oxygen (O_2) and chlorine have all been reported as intermediates or products (Gordon *et al.*, 1972).

Chlorine dioxide is a strong oxidizing agent. The normal pathway of action in mildly acidic, neutral and alkaline solutions is:

$$ClO_2(aq) + \bar{e} \rightarrow ClO_2^- \quad (12)$$

Organic compounds and ClO_2 are oxidized to CO_2 and H_2O. When treating potable water by ClO_2 there no problems with secondary organic products.

Inorganic compounds of sulphur (S^{2-}), iodide (I^-), arsenite (AsO_3^-) and bisulphite (HSO_3^-) are some of the inorganic reductants that are oxidized by chlorine dioxide. Manganous ions are also oxidized readily to manganic hydroxides, a reaction that is often used to precipitate manganese as hydroxide in water treatment (Sussman and Rauh, 1978).

Good results are attained with application of chlorine dioxide for removal of taste and odour from raw water, rich in algae metabolites, other biological and chemical compounds.

As demonstrated by Lalezary *et al.* (1986) chlorine dioxide reacts to form a range of algae metabolites such as 2-isopropyl-3-methoxy pyrazine (IPMP), 2-isobutyl-3-methoxy pyrazine (IBMP), 2-methyl isoborneol (MIB), 2,3,6-trichloroanisole (TCA), and trans-1,10-dimethyl-trans-9-decalol (Geosmin).

Thus, the main problem during water treatment for removal of taste and odour by chlorine dioxide consists of obtaining toxic salts of chlorite and chlorate during this treatment.

Chloramine. For taste and smell treatment of water obtained during the process of chlorine treatment chloramine is used (Carlson and Hardy, 1998).

When uniting two agents, chlorine (Cl_2) and ammonia (NH3), a chloramines mixture is formed at a wide range of pH – from 6 to 9 – in accordance with the following reactions:

$$NH_3 + HOCl \rightarrow NH_2Cl + H_2O \quad (13)$$

$$NH_2Cl + HOCl \rightarrow NHCl_2 + H_2O \quad (14)$$

$$NHCl_2 + HOCl \rightarrow NCl_3 + H_2O \quad (15)$$

Formation of different chloramines (monochloramine, NH_2Cl, dichloramine $NHCl_2$ and trichloramine NCl_3) depends on the ratio of the initial quantities of chlorine and ammonia, pH, temperature and speed of the reaction. Monochloramine is the most important of these compounds.

Dichloramine is much less stable than monochloramine. Destruction of dichloramine is simplified in reaction (16):

$$2NHCl_2 + H_2O \rightarrow N_2 + HOCl + 3H^+ + 3Cl^- \quad (16)$$

Monochloramine hydrolysis can occur depending on pH Margerum and Gray, 1978:

$$NH_2Cl + H_2O \rightarrow HOCl + NH_3 \quad (17)$$

$$NH_2Cl + H_2O \rightarrow NH_2OH + Cl^- \quad (18)$$

When treating water with monochloramine, trihalomethanes (THM) and haloacetic acids (HAA) are rarely formed (Stevens *et al.*, 1978; Carlson and Hardy, 1998).

By-products formed when chloramine reacts with organic substances dissolved in the water to be treated are shown in Table 42.3.

Bromine and iodine. Attempts to use bromine and iodine as disinfectant were made in the 1920s–1930s (White, 1972), but by-products formed during the interaction of these halogens with water possess unacceptable odour and taste and, moreover, are rather toxic.

The problem arises in chlorination of natural water containing bromides (Br^-) and iodides (I^-) as well as different organic compounds including metabolites such as methane (CH_4). As the result of oxidation of bromides and iodides by free chlorine bromine and iodine are formed in the water and intensively react with organic compounds forming halogen-organic compounds (Rook *et al.*, 1978; Rav-Acha *et al.*, 1984).

The reaction chain for the halogens interaction with water is given below:

$$Br^{-1} + Cl_2 \rightarrow Br_2 + Cl^- \quad (19)$$

$$Br_2 + H_2O \rightarrow BrO^- \quad (20)$$

$$Cl_2 + CH_4 \rightarrow CHCl_3 \quad (21)$$

$$CHCl_3 + BrO^- \rightarrow CHCl_2Br \quad (22)$$

The trihalomethanes (THM) formed have different compositions (CH_2Cl_2, $CHCl_3$, $CHCl_2$ Br, CCl_3, CBr_3, etc.) but all are carcinogenic substances and possess unacceptable odour and taste.

Bromine and iodine form also bromophenols and bromoingoles. Unpleasant taste and odour can also give a residual quality of halogens.

Some of the halogen-organic by-products found in potable water are given in Table 42.4.

TABLE 42.3 By-products obtained during the process of water treatment with chloramine

Organic substances present in raw water	*By-products*	*Reference*
o-Chlorobenzaldehyde C_7H_5Ocl	o-Chlorobenzalchlorimine, o-Chlorobenzonitrile	Hauser and Hauser, 1930
Formaldehyde H_2CO	N-Trichlorotrimethyl-enetriamine $(CH_2NCl)_3$	Lindsay and Soper, 1946
Phenol C_6H_5OH	p-Aminophenol 4,4-Dihydroxy-diphenylamine Iminoquinone and also quinone chlorimide and indophenol blue	Raschig, 1907; Harwood and Kuhn,1970
Phenol C_6H_5OH	Chlorophenols	Burttschell *et al.*, 1959
Thiol groups R-SH	Sulphides R-S-S-R	Srivastava and Bose, 1975
Thiols R-SH	Sulphenamides R-S-NH_2	Sisler *et al.*, 1970
Cysteine	Cystine	Ingols *et al.*, 1953
Tyrosine	N-Chlorotyrosine	Ingols *et al.*, 1953
Alanine	N-Chloroalanine	
Glycylglycylglycine	Terminal N-Chlorocompound, N-chloroglycylglycylglycine	Ingols *et al.* 1953
Iodide	Iodine and hypoiodous acid	Kinman and Layton, 1976
Activated carbon	N_2 and NO_3^- Organics unidentified	Bauer and Snoeyink, 1973
Natural organic water	Trihalomethans THMs Haloacetic acids HAAs	Carlson and Hardy, 1998

TABLE 42.4 Organobromides and organoiodides found in potable water

By-products	*Reference*
Bromomethane CH_3Br	Shackelford and Keith, 1976
Bromochloromethane CH_2ClBr	
Bromo-chloro-iodomethane $CHClBrI$	
Bromoform $CHBr_3$	
Bromoaceton $BrCH_2COCH_3$	Bean *et al.*, 1978
Dibromo-chloromethane $CHBr_2Cl$	Bunn *et al.*, 1975
Dichloro-iodomethane $CHCl_2I$	Henderson *et al.*, 1976

Ozone. Ozone has been known as a water disinfectant for a long time improving its organoleptic properties and has been used for this purpose for about 100 years (Lawrence and Capelli, 1977). Ozone is a gas and is moderately soluble in water. It is a very strong oxidizer possessing high oxidizing ability relative to inorganic and organic compounds. Ozone's (O_3) oxidizing ability can be explained by the mechanism of its decomposition in water. In water, ozone is decomposed into oxygen (O_2) and hydrogen peroxide (H_2O_2) (Kilpatrick *et al.*, 1956). Other mechanisms of ozone interaction with water are also possible, the result being the formation of the active hydroxyl radical OH^0, which when reacting with dissolved chemical compounds, produces a number of chain reactions and gives active radicals of organic and inorganic compounds (Hoigne and Bader, 1975, 1978). Good results are attained by application of ozone for the removal of odour and taste from water rich with algae metabolites such as geosmin and MIB (Lalezary *et al.*, 1986. After water treatment by ozone a number of by-products are produced in the form of acids (Table 42.5).

Ozone also forms unstable peroxide compounds, gluoxal (HOCCOH), methylgluoxal (CH_3OCCOH) with the presence of aromatic substances in water which also influence health.

Much scientific research has been conducted which has shown the possibility of

TABLE 42.5 By-products of organic compounds obtained with ozonolysis

Organic compound	*By-product*	*Reference*
1,3,5-Trimethyl-benzene Cresols	Formic acid	Cerkinsky and Trahtman, 1972; Bauch *et al.*, 1970
	Acetic acid	
	Maleic acid	
	Acetic acid	
	Oxalic acid	
2,4-Dichlorophenol	Formic acid	Gilbert, 1978
	Oxalic acid	
	Chloride ion	
4-Chloro-o-cresol phenol	Pyruvic acid	Gilbert, 1978
	Fumaric acid	Eisenhauer, 1968
	Oxalic acid	
	Maleic acid	Bauch *et al.*, 1970
	Glyoxylic acid	Mallevialle, 1975
	Catechon (p-Quinone)	
	Hydroquinone	
	Resorcinol	Spanggord and McClurg, 1978
Pyrene	Acetic acid	Kinney and Friedman,1952
	Oxalic acid	
Naphtalene	Salicylic acid	Gilbert, 1978
Naphtalene-2,7-disulphonic acid	Formic acid	
	Oxalic acid	
	Glyoxal	
Natural organic matter with bromide Br^-	Bromoform, bromoacetic acid, bromoaceto-nitrites	Westerhoff *et al.*, 1998

by-product formation with ozonolysis of water where inorganic compounds are dissolved (Table 42.6). Many by-products from ozonolysis possess dangerous toxic properties.

Potassium permanganate and manganese dioxide. Raw water purification from taste and odour by means of potassium permanganate ($KMnO_4$) has been known from the beginning of the nineteenth century. It became popular in the 1960s.

Potassium permanganate is a weak oxidizer when compared with chlorine, chlorine dioxide and ozone. It is low in effectiveness in relation to algae metabolites such as IPMP,

TABLE 42.6 By-products of inorganic compounds obtained with ozonolysis

Compounds found in natural water	*By-products*	*Reference*
Manganese Mn	Manganate MnO_4^{2-}	Kjos *et al.*, 1975
	Manganese dioxide MnO_2	
	Permanganate MnO_4^-	
	Soluble compounds	
Insoluble oxides of metals		Netzer and Bowers, 1975
Organic sulphur	Sulphate	Gunther *et al.*, 1970
Bromine	Bromate	Helz *et al.*, 1978
	Bromoform	
Ammonium ion	Hipobromite	
	Nitrite	Wynn *et al.*, 1973

IBMP, MIB, TCA, Geosmin. (Lalezary *et al.*, 1986. However, potassium permanganate is effective in relation to metabolites of blue-green algae by removing the fishy, septic, grassy, cucumber odour. The mechanism of raw water treatment derives from the fact that when the reducing agent (Na_2SO_3) is added, the Mn^{7+} ion is reduced to ion Mn^{4+}, forming insoluble MnO_2 in water. When depositing MnO_2 from water solution odorants are co-deposited and adsorbed on it. Manganese dioxide is effective in the removal of odour and taste from water with algae metabolites such as TCA (Lalezary *et al.*, 1986.

Inference. Different oxidants are effective in removing odour and taste from biological and chemical sources but, depending on the components dissolved in water, form a number of toxic by-products.

Chlorine forms trihalomethanes (THMs), trichloraceton (CCl_3COCH_3), chloramines, chlorophenols and other chlorinated and oxidized by-products. Chloramine forms aldonic acid from carbohydrates that might be present in water supplies (post-ammoniation). Pre-ammoniation with chloramines (with ammonium addition before chlorination) removes taste and odour from water, thus preventing the appearance of taste and odour from the presence of phenols in water. Organic compounds, ClO_2, are oxidized to CO_2 and H_2O. Chlorine dioxide is formed by chlorinating aromatic compounds as well as inorganic toxic chlorates (ClO_3^-) and chlorites (ClO_2^-). Residual chlordioxide imparts taste and odour to water as well as by-products.

Bromine and iodine form THMs, analogues of chlorinated species bromophenols and bromoingoles. Unpleasant taste and odour can also give a residual quality of halogens. Ozone forms unstable peroxide compounds, gluoxal (HOCCOH), methylgluoxal (CH_3OCCOH) with the presence of aromatic substances in water which also influence health.

The mechanism of the oxidizing process when using potassium permanganate consists of oxidizing Mn^{7+} to Mn^{4+}. Formation of MnO_2 (Mn^{4+}) possessing sorption activity improves the taste and odour of potable water.

4 EXAMPLES OF TASTE AND ODOUR REMOVAL FROM POTABLE WATER

Good results are attained by combining oxidants as purification methods. The Sea of Galilee causes a problem because its waters are enriched by free bromine. When chlorinating water obtained from the Sea of Galilee, THM by-products and bromine analogues are formed (Rav-Acha *et al.*, 1984). When treating the water by means of dichloroxide, THM is not formed but chlorites (ClO_2^-) appear causing haemolytic anaemia (Abdel-Rahman, 1981). Researchers (Rav-Acha *et al.*, 1984) have proposed a solution to the problem, its essence consisting of pre-treatment by ClO_2 at a concentration of 1 mg/l for 2 h, then chlorination at a concentration of 2 mg/l. In this case the quantity of THM is considerably decreased (up to 60%) if compared with chlorinating alone. Chlorine also decreases the quantity of chlorites (up to 90%) when compared with that formed by dichloroxide alone. In the research conducted by Ortenberg *et al.* (2000) the main problems appearing in water source treatment from taste and odour are described as well as methods for solving these problems.

A deep groundwater well was established in a village, in order to supply drinking water for

30 000 inhabitants. Although all water characteristics were within the range permitted by the Israeli water regulations, the operator received complaints about a funny taste and rotten egg odour in water. This was attributed to the presence of 0.38 mg/l hydrogen sulphide (measured as S^{2-}). Treatment with 2 mg/l of chlorine dioxide successfully oxidized hydrogen sulphide and eliminated the odour completely, but produced 1 mg/l of chlorite together with 0.35 mg/l of chlorate. This is above the range permitted by the Israeli water regulations for chlorite (0.5 mg/l). A partial solution to this problem was found by consequential treatment with 2 mg/l of ClO_2 for 10 min, followed by 1.5 mg/l of chlorine (30 min). In this case only 0.5 mg/l chlorite was produced with 0.5 mg/l of chlorate.

5 CONCLUSION

1. Biological sources of taste and odour in raw water are represented by metabolites of actinomycetes and different kinds of algae as well as other aqueous organisms (sulphate-reducing bacteria, zooplankton, protozoa and fungi).
2. Chemical sources of taste and odour in raw water are represented by industrial and municipal wastewaters entering the water source by any means, chemical compounds produced when storing water in reservoirs and with its delivery in pipes, as well as residual quantities of oxidants which are used for water purification and their by-products.
3. Water treatment for taste and odour is carried out by means of adsorbents (mainly activated carbon) and oxidants (chlorine, chlorine dioxide, ozone, etc.).
4. Adsorption treatment is fairly effective. Odour and taste removal in relation to adsorption capacity of activated carbon occurs for 85–100%. Taste and odour removal by means of oxidants is also effective, especially when using chlorine dioxide, ozone and chlorine.
5. However, when treating with oxidants the very important problem of by-products formation, which cause harm to people's health, appears.
6. Complete removal of taste and odour can be attained when combining treatment with activated carbon and oxidants.

With the correct choice of treatment strategy, taking into account the chemical composition of raw water as well as the possibility of toxic by-product production, it is possible to control both taste and odour in water, as well as reducing their quantity of toxic by-product formation to a minimum.

REFERENCES

Abdel-Rahman, H.S., Caurni, D. and Bull, R.J. (1981). Effect of exogenous glutathione ClO_2 and chlorite on osmotic fragility rate of blood *in vitro*. *J. Environ. Path. Tox.* **5**, 864–875.

Bauch, H., Burchard, H. and Arsovic, H.M. (1970). Ozone as an oxidant for phenol degradation in aqueous solutions. *Gesund. Ing.* **91**, 258–262.

Bauer, R.C. and Snoeyink, V.L. (1973). Reaction of chloramines with active carbon. *J. Water Pollut. Control Fed.* **45**, 2290–2301.

Bean, R.M., Riley, R.G. and Ryan, P.W. (1978). Investigation of halogenated components formed from chlorination of marine water. In: R.L. Jolley, H. Gorchev and D.H. Hamilton, Jr (eds) *Water Chlorination: Environmental Impact and Health Effects*, pp. 223–233. Ann Arbour Science Publishers, Inc, Ann Arbor.

Bellar, T.A., Kichtenberg, J.J. and Kroner, R.C. (1974). The occurrence of organohalids in chlorinated drinking water. *J. Am. Water Works Assoc.* **66**, 703–706.

Berg, M., Mueller, S.R., Muelemann, J. *et al.* (2000). Concentrations and mass fluxes of chloracetic acids and trifluoroacetic acid in rain and natural waters in Switzerland. *Environ. Sci. Technol.* **34**, 2675–2683.

Berglind, L., Holtan, H. and Skullberg, O.M. (1983). Case studies on off-flavors in some Norwegian lakes. *Wat. Sci. Tech.* **15**, 199–209.

Blumer, M. and Youngblood, W.W. (1975). Polycylic aromatic hydrocarbons in soils and recent sediments. *Science* **88**, 53–55.

Bray, W. (1906). Beitrage zur Kenntnis der Halogensauerst-off- verbindungen. Abhandlung III. Zur Kenntnis des Chlordioxyds. *Z. Physik. Chem.* **54**, 569–608.

Bunn, W.W., Haas, B.B., Deane, E.R. and Kleopfer, R.D. (1975). Formation of trihalomethanes by chlorination of surface water. *Environ. Lett.* **10**, 205–213.

Burttschell, R.H., Rosen, A.A., Middleton, F.M. and Etinger, M.B. (1959). Chlorine derivatives of phenol causing taste and odor. *J. Am. Water Works Assoc.* **51**, 205–214.

Campbell, A.T., Reade, A.J., Warburton, L. and Wheitman, R.F. (1994). Identification of odour problems in river Dee: A case study. *J. Inst. Water Environ. Manage.* **8**, 52–57.

Carlson, M. and Hardy, D. (1998). Controlling DBPs with monochloramine. *J. Am. Water Works Assoc.* **90**, 95–106.

Casitas Municipal Water District (1987). *Current Methodology for the Control of Algae in Reservoirs*. American Water Works Association Research Foundation, American Water Works Association, Denver.

Chang, S.L., Woodward, R.L. and Kabler, P.W. (1960). Survey of free-living nematodes and amebas in municipal supplies. *J. Am.Water Works Assoc.* **52**, 613–618.

Cerkinsky, S.N. and Trahtman, N. (1972). The present status of research on the disinfection of drinking water in the USSR. *Bull. WHO* **46**, 277–283.

Eisenhauer, H.R. (1968). The ozonation of phenolic wastes. *J. Water Pollut. Control Fed.* **40**, 1887–1899.

Federal Register (1995) *National Primary and Secondary Drinking Water regulations; Analytical methods for regulated drinking water contaminants; Final rule. 40 CFR Part 141. Part X.* Environmental Protection Agency.

Feuss, J.V. (1964). Problems in the determination of chlorine dioxide residuals. *J. Am. Water Works Assoc.* **56**, 607–615.

Gagne, F., Ridal, J.J., Blaise, C. and Brownlee, B. (1999). Toxicological effects of geosmin and 2-methylisoborneol on Rainbow Trout hepatocytes. *Bull. Environ. Contam. Toxicol.* **63**, 174–180.

Gall, R.J. (1978). Chlorine dioxide – an overview of its preparation, properties and uses. In: R.G. Rice and J.A. Cotruvo (eds) *Ozone/Chlorine Dioxide Products of Organic Materials*, pp. 356–382. Ozone Press International, Cleveland.

Garrison, A.W., Pope, J.D. and Allen, F.R. (1976). GC/MS analysis of organic compounds in domestic wastewaters. In: L.H. Keith (ed.) *Identification and Analysis of Organic Pollutants in Water*, pp. 517–556. Ann Arbor Science Publishers, Inc, Ann Arbor.

Gehrs, C.W. and Southworth, G.R. (1978). Investigating the effect of chlorinated organics. In: R.L. Jolley (ed.) *Water Chlorination: Environmental impact and Health Effects*, pp. 329–342. Ann Arbor Science Publishers, Inc, Ann Arbor.

Gerber, N.N. (1969). A volatile metabolite of Actinomycetes, 2-methylisoborneol. *J. Antibiot.* **22**, 508–509.

Gerber, N.N. (1971). Sesquiterpenoids from Actinomycetes: cadin-4-ene-1-ol. *Phytochemistry* **10**, 185–189.

Gerber, N.N. (1983). Volatile substances from Actinomycetes: their role in the odor pollution of water. *Wat. Sci. Tech.* **15**, 115–125.

Gerber, N.N. and LeChevallier, H.A. (1965). Geosmin, an earthy-smelling substance isolated from Actinomycetes. *Appl. Microbiol.* **13**, 935–938.

Giger, W., Reinhard, M., Schaffner, C. and Zurcher, F. (1976). Analyses of organic constituents in water by high-resolution gas chromatography in combination with specific detection and computer-assisted mass spectrometry. In: L.H. Keith (ed.) *Identification and Analysis of Organic Pollutants in Water*, pp. 433–452. Ann Arbor Science Publishers, Ann Arbor.

Gilbert, E. (1978). Reactions of ozone with organic compounds in dilute aqueous solution: identification of their oxidation products. In: R.G. Rice and J.F. Cotruvo (eds) *Ozone/Chlorine Dioxide Oxidation Products of Organic Materials*, pp. 227–242. Ozone Press International, Cleveland.

Ginzburg, B., Chalifa, I., Hadas, O. *et al.* (1999a). Formation of dimethyloligosulfides in Lake Kinneret. *Wat. Res. Tech.* **40**, 73–78.

Ginzburg, B., Dor, I., Chalifa, I. *et al.* (1999b). Formation of dimethyloligosulfides in Lake Kinneret: biogenic formation of inorganic oligosulfide intermediates under oxic conditions. *Environ. Sci. Technol.* **33**, 571–579.

Gordon, G. and Feldman, F. (1964). Stoichiometry of the reaction between uranium (IY) and chlorite. *J. Inorg. Chem.* **3**, 1728–1733.

Gordon, G., Kieffer, R.G. and Rosenblatt, D.H. (1972). The chemistry of chlorine dioxide. In: S.J. Kippard (ed.) *Progress in Inorganic Chemistry*, pp. 201–286. John Wiley & Sons, New York.

Gunther, F.A., Ott, D.E. and Ittig, M. (1970). The oxidation of parathion to paraoxon. II. By use of ozone. *Bull. Environ. Contam. Toxicol.* **5**, 87–94.

Harrison, R.M., Perry, R. and Wellings, R.A. (1975). Review paper: Polynuclear aromatic hydrocarbons in raw, potable and waste waters. *Wat. Res.* **9**, 331–346.

Harwood, J.E. and Kuhn, A.L. (1970). A calorimetric method for ammonia in natural waters. *Wat. Res.* **4**, 805–811.

Hauser, C.R. and Hauser, M.L. (1930). Researches on chloramines. I. Ortho-chlorobenzalchlorimine and anisalchlorimine. *J. Am. Chem. Soc.* **52**, 2050–2054.

Helz, G.R., Hsu, R.Y. and Block, R.M. (1978). Bromoform production by oxidative biocides in marine waters. In: R.G. Rice and J.A. Cotruvo (eds) *Ozone/Chlorine Dioxide Oxidation Products of Organic Materials*, pp. 68–76. Ozone Press International, Cleveland.

Henderson, J.R., Peyton, G.R. and Glaze, W.H. (1976). A convenient liquid-liquid extraction method for the determination of halomethanes in water at the parts-per-billion level. In: L.H. Keith (ed.) *Identification and Analysis of Organic Pollutants in Water*, pp. 105–111. Ann Arbor Science Publishers Inc, Ann Arbor.

Hoigne, J. and Bader, H. (1975). Ozonation of water: role of hydroxyl radicals as oxidizing intermediates. *Science* **190**, 782–784.

Hoigne, J. and Bader, H. (1978). Ozonation of water: kinetics of oxidation of ammonia by ozone and hydroxyl radicals. *Environ. Sci. Technol.* **12**, 79–84.

Hosaka, M., Murata, K., Iikura, Y. *et al.* (1995). Off-flavor problems in drinking water of Tokyo arising from the occurrence of musty odor in a downstream tributary. *Wat. Sci. Tech.* **31**, 29–34.

Ingols, R.S., Wyckoff, H.A., Kethley, T.W. *et al.* (1953). Bactericidal studies of chlorine. *Ind. Eng. Chem.* **45**, 996–1000.

Izaguirre, G. (1992). A copper-tolerant *Phormidium* species from lake Mathews, California, that produces 2-methylisoborneol and geosmin. *Wat. Sci. Tech.* **25**, 217–223.

Izaguirre, G., Hwang, C.J., Krasner, S.W. and McGuire, M.J. (1982). Geosmin and 2-methylisoborneol from cyanobacteria in three water supply systems. *Appl. Envir. Microbiol.* **43**, 708–714.

Izaguirre, G. and Devall, J. (1995). Resources control for management of taste-and-odor problems. In: I.H. Suffet, J. Mallevialle and E. Kawczynski (eds) *Advances in Taste-and-Odor Control*, pp. 23–74. American Water Works Association Research Foundation and Lyonnaise des Eaux. American Water Works Association, Denver.

Izaguirre, G. and Taylor, W.D. (1998). A *Pseudanabaena* species from Castaic Lake, California, that produces 2-methylisoborneol. *Wat. Res.* **32**, 1673–1677.

Izaguirre, G., Taylor, W.D. and Pasek, J. (1999). Off-flavor problems in two resrvoirs, associated with planktonic *Pseudanabaena* species. *Wat. Sci. Tech.* **40**, 85–90.

Juttner, F. (1981). Detection of lipid degradation products in the water of reservoir during a bloom of *Synura uvella*. *Appl. Envir. Microbiol.* **41**, 100–106.

Keith, L.H., Garrison, A.W., Allen, F.R. *et al.* (1976). Identification of organic compounds in drinking water from thirteen US cities. In: L.H. Keith *et al.* (eds) *Identification and Analysis of Organic Pollutants in Water*, pp. 329–373. Ann Arbor Science Publishers, Inc., Ann Arbor.

Khiari, D., Young, C.C., Amah, G. *et al.* (1999). Factors affecting the stability and behavior of grassy odors in drinking water. *Wat. Sci. Tech.* **40**, 287–292.

Kikuchi, T. *et al.* (1974). Metabolites of a diatom, *Syndera rumpens* Kutz, isolated from the water in Lake Biwa. Identification of odorous compounds, *n*-hexanal and *n*-heptanal, and analysis of fatty acids. *Chem. Pharm. Bull. (Tokyo)* **22**, 945.

Kilpatrick, M.L., Herrick, C.C. and Kilpatrick, M. (1956). The decomposition of ozone in aqueous solution. *J. Am. Chem. Soc.* **78**, 1784–1789.

Kinman, R.H. and Layton, R.F. (1976). New method for water disinfection. *J. Am. Water Works Assoc.* **68**, 298–302.

Kinney, C.R. and Friedman, L.D. (1952). Ozonization studies of coal constitution. *J. Am. Chem. Sci.* **74**, 57–61.

Kjos, D.J., Furgason, R.R. and Edwards, L.L. (1975). Ozone treatment of potable water to remove iron and manganese: preliminary pilot plant results and economic evaluation. In: R.G. Rice and M.E. Browning (eds) *First International Symposium on Ozone for Water and Wastewater Treatment*, pp. 194–203. International Ozone Institute, Waterbury.

Krasner, S.W. (1995). The use of reference materials in sensory analysis. *Wat. Sci. Tech.* **31**, 265–272.

Lalezary, Sh., Pirbazari, M. and McGuive, M.J. (1986). Oxidation of five earthy-musty taste and odour compounds. *J. Am. Water Works Assoc* **3**, 62–69.

Lawrence, J. and Capelli, F.P. (1977). Ozone in drinking water treatment: a review. *Sci. Total Environ.* **7**, 99–108.

Leventer, H. and Eren, J. (1970). Taste and odor in the reservoirs of the Israel National Water System. In: H.I. Shuval (ed.) *Developments in Water Quality Research: Proceedings of the Jerusalem International Conference on Water Quality and Pollution Research.* Ann Arbor Science Publishers, Inc., Ann Arbor.

Lin, S.D. (1976). Sources of tastes and odors in water. *Wat. Sewage Works* **1**, 101–104 and **2**, 64–67.

Lindsay, M. and Soper, F.G. (1946). Methylenechloroamine. *J. Chem. Soc.*, 791–792.

Lytle, D.A., Schock, M.R., Clement, J.A. and Spencer, C.M. (1998). Using aeration for corrosion control. *J. Am. Water Works Assoc.* **90**, 74–80.

MacKenthun, K.M. and Keup, L.E. (1970). Biological problems encountered in water supplies. *J. Am. Water Works Assoc.* **62**, 520–526.

Mallevialle, J. (1975). Action de l'ozone dans la degradation des composes phenoliques simples et polymerises: Application aux matieres humique contenues dans les eaux. *Rev. l'Eau Tech. Sci. Munic.* **70**, 49–61.

Mallevialle, J. and Suffet, J.H. (eds) (1981). *The Identification and Treatment of Tastes and Odors in Drinking Water. Appendix A.* American Water Works Association, Denver.

Mallevialle, J. and Suffet, I.H. (eds) (1987). Identification and Treatment of Tastes and Odors in Drinking Water. American Water Works Association Research Foundation and Lyonnaise des Eaux, Denver.

Margerum, W. and Gray, E.T. Jr (1978). Chlorination and the formation of N-chloro-compounds in water treatment. In: *Abstracts of Papers of the 175th American Chemical Society National Meeting*. Anaheim, California. March 13–17. Abstr. No INOR 158.

Masschelein, W. (1967). Development in the chemistry of chlorine dioxide and its applications. *J. Chim. Ind. Genie Chim.* **97**, 49–61.

Medsker, L.L., Jenkins, D., Thomas, J.F. and Koch, C. (1969). Odorous compounds in natural waters: 2-exo-hydroxy-2-methyl-bornane, the major odorous compound produced by several Actinomycetes. *Envir. Sci. Technol.* **3**, 476–477.

Morris, J.C. (1975). Formation of halogenated organics by chlorination of water supplies. (A Review). EPA-600/1-75—002. Office of Research and Development. US Environmental Protection Agency, Washington DC.

Nakanishi, M., Hoson, T., Inoue, Y. and Yagi, M. (1999). Relationship between the maximum standing crop of musty-odor producing algae and nutrient concentrations in the southern basin water of Lake Biwa. *Wat. Sci. Tech.* **40**, 179–184.

Netzer, A. and Bowers, A. (1975). Removal of trace metals from wastewater by lime and ozonation. In: R.G. Rice and M.E. Browning (eds) *First International Symposium on Ozone for Water and Wastewater Treatment*, pp. 731–747. International Ozone Institute, Waterbury.

Ortenberg, E., Groisman, L. and Rav-Acha, C. (2000). Taste and odour removal from an urban groundwater establishment-a case study. *Wat Sci. Tech.* **42**, 123–128.

Palmer C.M. (1959). Taste and odour algae. In: *Algae in Water Supplies*, pp. 18–21.US Public Health Service Pub. No. 657, US Dept. of Health, Education and Welfare, US Public Health Service, Cincinnati.

Palmer, C.M. and Poston, H.W. (1956). Algae and other interference organisms in Indiana ware supplies. *J. Am. Water Works Assoc.* **48**, 1335–1346.

Persson, P.E. (1980). Muddy odour in fish from hypertrophic waters. *Dev. Hydrobiol.* **22**, 203–208.

Persson, P.E. and Sivonen, K. (1979). Notes on muddy odour. V. Actinomycetes as contributors to muddy odour in water. *Aqua Fennica* **9**, 57–61.

Pomes, M.L., Larive, C.K., Green, W.R. *et al.* (2000). Sources and haloacetic acid/ trihalomethane formation potentials of aquatic humic substances in the Wakarusa River and Clinton Lake near Lawrence, Kansas. *Environ. Sci. Technol.* **34**, 4278–4286.

Pontius, F.W. (1998). New horizons in federal regulation. *J. Am. Water Works Assoc.* **90**, 38–49.

Rav-Acha, C., Serri, A., Chosen, E. and Limoni, B. (1984). Disinfection of drinking water rich in bromine with chlorine and chlorine dioxide, while minimizing the formation of undesirable by-products. *Wat. Sci. Tech.* **17**, 611–621.

Raschig, F. (1907). Vorlesungsversuche aus der Chemie der anorganischen Stickstoffverbindungen. *Chem. Ber.* **40**, 4580–4588.

Ridal, J.J., Brownlee, B. and Lean, D.R.S. (1999). Occurrence of odor compounds, 2-methylisoborneol and geosmin in eastern Lake Ontario and upper St. Lawrence River. *J. Great Lakes* **25**, 198–204.

Rook, J.J. (1974). Formation of haloforms during chlorination of natural waters. *Water Treat. Exam.* **23**, 234–243.

Rook, J.J., Gras, A.A., Van der Heijden, B.G. and de Wee, J. (1978). Bromide oxidation and orgnic substitution in water treatment. *J. Environ. Sci. Health* **13**, 91–116.

Rosen, A.A., Mashni, C.I. and Safferman, R.S. (1970). Recent developments in the chemistry of odour in water: the causes of earthy/musty odor. *Wat. Treat. Exam.* **19**, 106–119.

Safferman, R.S., Rosen, A.A., Mashni, C.J. and Morris, M.E. (1967). Earthy-smelling substances from blue-green algae. *Environ. Sci. Technol* **1**, 429–430.

Schrader, K.K., De Regt, M.Q., Tidwell, P.R. *et al.* (1998). Selective growth inhibition of the musty-odor producing cyanobacteria *Oscillatoria cf. Chalybea* by natural compounds. *Bull. Environ. Contam. Toxicol.* **60**, 651–658.

Scott, B.F., Mactavish, D., Spencer, C. *et al.* (2000). Haloacetic acids in Canadian lake waters and precipitation. *Environ. Sci. Technol.* **34**, 4266–4272.

Shackelford, W.M. and Keith, L.H. (1976). Frequency of organic compounds identified in water. EPA-600/4-76-062. US Environmental Protection Agency, Environmental Research Laboratory, Athens, Ga.

Sievers, R.E., Barkley, R.M., Eiceman, G.A. *et al.* (1978). Generation of volatile organic compounds from nonvolatile precursors by treatment with chlorine or ozone. In: R.L. Jolley, H. Gorchev and D.H. Hamilton Jr (eds) *Water Chlorination: Environmental Impact and Health Effects*, pp. 615–624. Ann Arbor Science Publishers, Inc., Ann Arbor.

Silvey, J.K.G. and Roach, A.W. (1964). Studies on microbiotic cycles in surface waters. *J. Am. Water Works Assoc.* **56**, 60–64.

Sisler, H.H., Koua, N.K. and Highsmith, R.E. (1970). The formation of sulfur-sulfur bonds by the chlorination of thiols. *J. Org. Chem.* **15**, 1742–1745.

Slater, G.P. and Block, V.C. (1983). Volatile compounds of cyanophyceae – a review. *Wat. Sci. Tech.* **15**, 181–190.

Spanggord, R.J. and McClurg, V.J. (1978). Ozone methods and ozone chemistry of selected organics in water. 1. Basic chemistry. In: R.G. Rice and J.A. Cotruvo (eds) *Ozone/chlorine dioxide oxidation products of organic materials*. Ozone Press International, Cleveland, Proceedings of a conference held in Cincinnati, Ohio, November 17–19, 1976, pp. 115–125.

Srivastava, A. and Bose, S. (1975). Analytical applications of N-haloamides for the determination of some sulfur-containing functional groups. *J. Indian Chem. Soc.* **LII**, 217–220.

Standard Methods for the Examination of Water and Wastewater (1998). 20th edn. American Public Health Association/American Water Works Association/ Water Environment Federation, Washington DC.

Stevens, A.A., Slocum, C.J., Seeger, D.R. and Pobeck, G.G. (1978). Chlorination of organics in drinking water. In: R.L. Jolley, H. Gorchev and D.H. Hamilton, Jr (eds) *Water Chlorination: Environmental Impact and Health Effects*, pp. 77–104. Ann Arbor Science Publishers, Inc., Ann Arbor.

Suffet, I.H., Khiari, D. and Bruchet, A. (1999). The drinking water Taste and Odor Wheel for the millennium: beyond geosmin and 2-methylisoborneol. *Wat. Sci. Tech.* **40**, 1–13.

Sussman, S. and Rauh, J.S. (1978). Use of chlorine dioxide in water and wastewater treatment. In: R.G. Rice and J.A. Cotruvo (eds) *Ozone/Chlorine Dioxide Oxidation Products of Organic Materials*, pp. 344–355. Ozone Press International, Cleveland.

Tabachek, J.L. and Yurkowski, M. (1976). Isolation and identification of blue-green algae producing muddy odour metabolites, geosmin and 2-methylisoborneol, in salin lakes in Manitoba. *J. Fish Res. Bd. Can.* **33**, 25–35.

Wajon, J.E., Alexander, R. and Kagi, R.I. (1985). Dimethyl trisulfide and objectionable odours in potable water. *Chemosphere* **14**, 85–89.

Westerhoff, P., Song, R., Amy, G. and Minear, R. (1998). NOM's role in bromine and bromate formation during ozonation. *J. Am. Water Works Assoc.* **80**, 82–94.

White, G.C. (1972). *Handbook of Chlorination*. Van Nostrand Reinhold, New York.

Whitfield, F.B. and Freeman, D.J. (1983). Off-flavours in crustaceans caught in Australian coastal waters. *Wat. Sci. Tech.* **15**, 85–95.

Wood, S., Williams, S.T. and White, W.R. (1983). Microbes as a source of earthy flavours in potable water – a review. *Intl. Biodeterioration Bull.* **19**, 83–87.

Wynn, C.S., Kirk, B.S. and McNabney, R. (1973). *Pilot plant for tertiary treatment of wastewater with ozone*. No. EPA-R2-73-146. US Environmental Protection Agency, Washington, DC.

Yano, H., Nakahara, M. and Ito, H. (1988). Water blooms of *Uroglena americana* and the identification of odorous compounds. *Wat. Sci. Tech.* **20**, 75–80.

Useful Websites

GENERAL

American Society for Microbiology (USA)
http://www.asmusa.org

Centres for Disease Control and Prevention (USA)
http://www.cdc.gov

Chartered Institution of Water and Environmental Management (UK)
http://www.ciwem.org.uk

Drinking Water Inspectorate (UK)
http://www.dwi.gov.uk

- *Report No. 71*: www.dwi.gov.uk/regs/pdf/micro.htm

Environmental Protection Agency (USA)
http://www.epa.gov

- Microbiology: www.epa.gov/nerlcwww/index.html
- Office of Water: www.epa.gov/OW/index.html
- Office of Wastewater Management: www.epa.gov/OWM/

European Union
http://europa.eu.int

- Water policy: http://europa.eu.int/comm./environment/water/index.html

European Water Association
http://www.ewpca.de

Federation of European Microbiological Societies
http://www.fems-microbiology.org

International Society for Infectious Diseases (USA)
http://www.isid.org

- at this site you can sign up for ProMed-mail

International Water Association (UK)
http://www.iwahq.org.uk

- Health-related water microbiology: http://www.iwa-microbiology.org

Sanitation Connection (USA)
http://www.sanicon.net

Society for Applied Microbiology (UK)
http://www.sfam.org.uk

Society for General Microbiology (UK)
http://www.sgm.ac.uk

Society for Protozoologists (USA)
http://www.uga.edu/%7eprotozoa

Tropical Public Health Engineering (UK)
http://www.leeds.ac.uk/civil/ceri/water/tphe/tphehome.html

Water Environment Federation (USA)
http://www.wef.org

World Bank
http://www.worldbank.org

- Water and sanitation: www.worldbank.org/watsan/

World Health Organization
http://www.who.int

- Infectious diseases: www.who.int/health-topics/idindex.htm
- Water and Sanitation: www.who.int/water_sanitation_health/index.html

World Water Assessment Programme
http://www.unesco.org/water/wwap

ENVIRONMENTAL LAW

Centre for International Environmental Law (USA and Switzerland)
http://www.ciel.org

Environmental Law Alliance Worldwide (eLaw) (USA)
http://www.elaw.org

Environmental Law Institute (USA)
http://www.eli.org

European Environmental Law (The Netherlands)
http://www.eel.nl

European Union Environmental Legislation
http://europa.eu.int/comm/environment/legis_en.htm

Index

The Handbook of Water and Wastewater Microbiology
ISBN 0-12-470100-0